Christine M. Eckel
Carroll College
Kyla Turpin Ross
Georgia State University
Theresa Stouter Bidle
Hagerstown Community College

Anatomy & Physiology Laboratory Manual

Second Edition

Cat Version

ANATOMY & PHYSIOLOGY LABORATORY MANUAL: CAT VERSION, SECOND EDITION

Published by McGraw-Hill Education, 2 Penn Plaza, New York, NY 10121.

Some ancillaries, including electronic and print components, may not be available to customers outside the United States.

This book is printed on acid-free paper.

1 2 3 4 5 6 7 8 9 0 RMN/RMN 10 9 8 7 6 5

ISBN 978-1-259-13674-0
MHID 1-259-13674-4

Senior Vice President, Products & Markets: *Kurt L. Strand*
Vice President, General Manager, Products & Markets: *Marty Lange*
Vice President, Content Design & Delivery: *Kimberly Meriwether David*
Managing Director: *Michael Hackett*
Director of Digital Content: *Michael G Koot, PhD*
Brand Manager: *Amy Reed*
Director, Product Development: *Rose Koos*
Product Developer: *Donna Nemmers*
Marketing Manager: *Jessica Cannavo*
Digital Product Analyst: *Jake Theobald*
Director, Content Design & Delivery: *Linda Avenarius*
Program Manager: *Angela R. FitzPatrick*
Content Project Managers: *April R. Southwood/Christina Nelson*
Buyer: *Sandy Ludovissy*
Design: *David Hash*
Content Licensing Specialist: *Carrie Burger*
Cover Image: *wheelchair athlete © David Moyer*
Compositor: *MPS North America LLC*
Printer: *R. R. Donnelley*

www.mhhe.com

brief contents

about the authors

With love and thanks to my entire family, including the dogs.

CHRISTINE MARIE ECKEL received her B.A. in integrative biology and M.A. in human biodynamics from the University of California, Berkeley, and her Ph.D. in neurobiology and anatomy at the University of Utah School of Medicine. Christine is associate professor of biology at Carroll College in her hometown of Helena, Montana, where she teaches the two-semester anatomy and physiology course for pre-nursing and pre-health science majors, and an advanced dissection course for premedical students. She also serves as the faculty advisor for pre-physical therapy and pre-physician assistant students. Prior to her position at Carroll College, Christine was associate professor and course director for medical gross anatomy and medical microanatomy at West Virginia School of Osteopathic Medicine (WVSOM). In the 14 years prior to her position at WVSOM, Christine taught undergraduate human anatomy and human physiology courses at Salt Lake Community College and the University of California, Berkeley. She earned outstanding teaching awards at all three of these institutions.

Christine is the author of *Human Anatomy Laboratory Manual,* second edition (McGraw-Hill Education). In addition, her cadaver dissections and photographs are featured in several textbooks, including this laboratory manual.

Christine served as the Western Regional Director for the Human Anatomy & Physiology Society (HAPS) for two terms. She has also served on several committees for both HAPS and the American Association of Anatomists (AAA). Her research is in the field of educational outcomes, and she serves as a peer reviewer for the journals *Anatomical Sciences Education* and *Medical Education*.

With over 25 years of experience engaging with students at all levels, including community college students, medical students, and surgical residents, Christine has a unique appreciation for the learning challenges experienced by students at each level. Christine's passions for anatomy and physiology, teaching, dissection, and photography are evident throughout the pages of this laboratory manual. In her spare time, Christine loves to mountain bike, skate ski, and explore the great Montana outdoors—always with her camera in hand.

To my husband Jim, daughter Ella, and son Cameron: I treasure your constant love and unending support.

KYLA TURPIN ROSS received her undergraduate degree from Louisiana State University in biological and agricultural engineering and her Ph.D. in biomedical engineering from Georgia Institute of Technology and Emory University. Kyla then served as a postdoctoral fellow in the Fellowships in Research and Science Teaching (FIRST) program at Emory University, an NIH-funded program that provides training in both research and teaching. Kyla is now senior academic professional at Georgia State University (GSU), where she teaches and manages the introductory and graduate human anatomy and physiology courses. Kyla has extensive experience developing lecture and laboratory curricula, and incorporates active learning in the classroom as a method to reinforce difficult physiological concepts. In addition, Kyla plays an active role in mentoring GSU faculty and teaching assistants and planning and hosting an annual teaching assistant workshop. She is involved in STEM initiatives at GSU, and serves as the faculty advisor for the Department of Biology Tutorial Center. She has served as a reviewer for numerous publications, and has authored a custom laboratory manual for GSU's human anatomy and physiology course. She is active in several committees within the Human Anatomy & Physiology Society (HAPS). In addition to academic endeavors, Kyla serves on the Decatur Family YMCA board of directors.

With love and thanks to my husband Jay and my daughter Stephanie for their continued support.

TERRI STOUTER BIDLE received her undergraduate degree from Rutgers University and her M.S. degree in biomedical science from Hood College in Maryland, and has completed additional graduate coursework in genetics at the National Institutes of Health. She is a professor at Hagerstown Community College where she teaches anatomy and physiology and genetics to pre-allied health students. Before joining the Hagerstown faculty in 1990, Terri was coordinator of the Science Learning Center, where she developed study materials and a tutoring program for students enrolled in science classes. She has been a developmental reviewer, and has written supplemental materials for both textbooks and laboratory manuals. Terri is a coauthor of *Anatomy & Physiology: An Integrative Approach,* second edition.

contents

preface

Human anatomy and physiology is a complex yet fascinating subject, and is perhaps one of the most personal subjects a student will encounter during his or her education. It is also a subject that can create concern for students because of the sheer volume of material, and the misconception that "it is all about memorization."

The study of human anatomy and physiology really comes to life in the anatomy and physiology laboratory, where students get hands-on experience with human cadavers and bones, classroom models, preserved and fresh animal organs, histology slides of human tissues, and explore the process of scientific discovery through physiology experimentation. Yet, most students are at a loss regarding how to approach the anatomy and physiology laboratory. For example, students are often given numerous lists of structures to identify, histology slides to view, and "wet labs" to conduct, but are given comparatively little direction regarding how to recognize structures, or how to relate what they encounter in the laboratory to the material presented in the lecture. In addition, most laboratory manuals on the market contain little more than material repeated from anatomy and physiology textbooks, which provides no real benefit to a student.

This laboratory manual takes a very focused approach to the laboratory experience, and provides students with tools to make the subject matter more relevant to their own bodies and to the world around them. Rather than providing a recap of material from classroom lectures and the main textbook for the course, this laboratory manual is much more of an interactive workbook for students: a "how-to" guide to learning human anatomy and physiology through touch, dissection, observation, experimentation, and critical thinking exercises. Students are guided to formulate a hypothesis about each experiment before beginning physiology exercises. Diagrams direct students in how to perform experiments, and don't just show the end results. The text is written in a friendly, conversational tone to put students at ease as they discover, organize, and understand the material presented in each chapter.

Organization

Because observation of histology slides, human cadavers or classroom models, and "wet lab" experiments are usually performed in separate physical spaces or at specific times within each laboratory classroom, chapters in this laboratory manual are similarly separated into three sections: Histology, Gross Anatomy, and Physiology. Each exercise within these chapter sections has been designed with the student's actual experience in the anatomy and physiology laboratory in mind. Thus, each exercise covers only a single histology slide, classroom model, region of the human body, or wet lab experiment. At the same time, within-chapter "Concept Connection" and "Clinical View" boxes provide an opportunity to integrate the material from all three sections of each chapter. "Learning Strategies" boxes provide mnemonics, study tips, and other helpful hints to assist students in recall of pertinent information. In addition, "Can You Apply What You've Learned?" and "Can You Synthesize What You've Learned?" questions in Post-Laboratory Worksheets provide further opportunities for students to integrate the information and apply it to clinically relevant and practical situations. Organization of each chapter into a series of discreet exercises makes the laboratory manual easily customizable to any anatomy and physiology classroom, allowing an instructor to assign certain exercises, while telling students to ignore other exercises. Post-Laboratory Worksheets are also organized by exercise and are coded to Learning Objectives within the chapter, which makes it easy for an instructor to assign questions that relate only to the exercises and/or Learning Objectives covered in their classroom.

Changes to the Second Edition

Anatomy & Physiology: An Integrative Approach Laboratory Manual, second edition, continues to serve as a resource for students both in and out of the lab, providing a "how to" guide for learning anatomy and physiology. The interactive pages within serve as a stand-alone manual, while also complementing the textbook, McKinley/O'Loughlin/Bidle: *Anatomy & Physiology: An Integrative Approach*, second edition. Certain changes to the second edition of this lab manual have been applied throughout all chapters.

- Word origins have been added to tables, where relevant.
- Chapter opening pages now include a list of reference tables.
- Ph.I.L.S. exercises throughout the manual have been updated to correlate with Ph.I.L.S. Version 4.0, including new screen captures to illustrate the software.
- Pre-Laboratory Worksheets and Post-Laboratory Worksheets include a broader variety of question types.
- Drawing circles have been enlarged throughout to allow more space for student drawings.
- Tables have been reorganized to include headings and subheadings for ease of learning.
- Chapters 25 and 26 have been reordered so that the urinary system is presented prior to the digestive system, in alignment with the McKinley/O'Loughlin/Bidle textbook.
- Safety icons have been added throughout the manual to alert students to potential hazards in the lab.
- New content has been added in numerous places throughout the manual, including:
  - three new clinical case studies
  - four new BIOPAC exercises
  - seven additional new exercises
  - thirteen new Concept Connection boxes
  - thirty new Clinical View boxes
  - thirty new Learning Strategy boxes

Changes by Chapter

The following is a list of the most significant changes by chapter in the second edition of this lab manual.

Chapter 1

- New Learning Strategy on studying anatomy and physiology
- Safety icons emphasizing safe dissection techniques

Chapter 2

- New Figure 2.1 The Anatomic Position, Body Planes, and Directional Terms
- New Exercise 2.1B Sectioning a Specimen
- New Figure 2.3 Sections Through a Sheep Heart

Chapter 3

- New Figure 3.3 Loading a Microscope Slide
- Revised Figure 3.5 Estimating Specimen Size
- Revised Concept Connection on the structure and function of cells
- New Clinical View: Nail Fungus

Chapter 4

- Revised Exercise 4.2 Observing Mitosis in a Whitefish Embryo to include space for students to sketch the phases of mitosis
- Revised Figure 4.3 Classroom Model of a Generalized Animal Cell
- Revised Exercise 4.3 Observing Classroom Models of Cellular Anatomy to include space for students to sketch a generalized cell with organelles
- Moved Exercise 4.6 Ph.I.L.S. Lesson 1: Osmosis and Diffusion: Varying Extracellular Concentration to immediately follow laboratory exercises on osmosis and diffusion
- New Learning Strategy on active and passive transport mechanisms

Chapter 5

- New Clinical View: Histopathology
- New Table 5.1 Classification of Epithelial Tissue by Number of Cell Layers
- New Learning Strategy on identifying a histological slide of pseudostratified ciliated columnar epithelial tissue

Chapter 6

- New Exercise 6.2 Fingerprinting
- New Clinical View: Fingerprinting
- Revised Figure 6.11 Classroom Model of the Integument
- New Concept Connection on apocrine sweat glands

Chapter 7

- Revised Concept Connection on bone density to include the influence of pregnancy on calcium deposition
- New Learning Strategy on the zones of the epiphyseal plate and bone growth
- New Concept Connection on the influence of hormones on bone formation
- Revised Clinical View: Bones and Mechanical Stress to include quadriplegics and their struggle with bone density loss due to a lack of stress and loading

Chapter 8

- New introductory text on bone markings
- New Table 8.1 Bone Markings
- New Learning Strategy on relating skeletal structure to function
- New Learning Strategy on the word origins of bones and bone markings
- Revised Table 8.2 The Axial Skeleton: Skull Bones and Important Bony Landmarks to include word origins
- Revised Table 8.3 The Axial Skeleton: Anterior View of the Skull to include word origins
- New Learning Strategy on visualizing structures as they travel through the foramina of the skull
- Revised Figure 8.12 The Hyoid Bone
- Revised Table 8.4 The Axial Skeleton: Vertebral Column to include word origins
- New Learning Strategy on learning the number of vertebrae in each region of the vertebral column
- Replaced Clinical View: Spina Bifida with new Clinical View: Spondylolisthesis
- New Concept Connection on the atlas and axis
- New Learning Strategy on identifying vertebrae from each region of the vertebral column
- Revised Table 8.5 The Axial Skeleton: Sternum and Ribs to include word origins
- Revised Figure 8.23 A Typical Rib

Chapter 9

- New Concept Connection on learning the bony features of the appendicular skeleton
- Revised Exercise 9.1 Bones of the Pectoral Girdle
- New Clinical View: Clavicular Fracture
- Revised Exercise 9.2 Bones of the Upper Limb
- Revised Figure 9.8 Surface Anatomy of the Upper Limb
- Revised Exercise 9.4 Bones of the Pelvic Girdle
- Revised Learning Strategy on distinguishing a male versus a female pelvis
- New Learning Strategy on determining distinctive features for each bone
- Revised Exercise 9.5 Bones of the Lower Limb
- New Learning Strategy on remembering the names of the tarsal bones

Chapter 10

- Revised Introduction to more clearly explain joint classification
- Reorganized Exercise 10.1 Fibrous Joints to be consistent with Table 10.2 Classification of Fibrous Joints
- Revised Table 10.4 Components of Synovial Joints to include most relevant terms
- Revised Exercise 10.4 Classification of Synovial Joints to include more detailed description of each type of synovial joint
- New Learning Strategy on distinguishing synchondroses and synovial joints
- New Learning Strategy on the structural classification of synovial joints
- New Clinical View: Bursitis
- Revised Concept Connection on movement of synovial joints to include the relationship between mobility and stability in a synovial joint
- New Clinical View: Low Back Pain

Chapter 11

- Revised Introduction to more concisely summarize the muscular system and chapter organization
- New Concept Connection on comparing the three types of muscle tissue
- New Clinical View: Muscular Dystrophies
- Reorganized the order of chapter topics and exercises: skeletal, cardiac, and smooth muscle
- New Learning Strategy on the movements of synovial joints

- New Learning Strategy on common architecture of skeletal muscles
- Revised introductory text for the physiology of muscle tissue
- Revised Concept Connection on cardiac versus skeletal muscle
- New Exercise 11.15 BIOPAC Electromyography (EMG)

Chapter 12

- Reorganized Exercise 12.1 Muscles of Facial Expression
- New Concept Connection on the facial nerves
- Moved Exercise 12.3 Extrinsic Eye Muscles to Chapter 18
- New Clinical View: Dysphagia
- New Concept Connection on pulmonary ventilation
- New Learning Strategy on learning the external and internal oblique muscles
- New Clinical View: Athletic Pubalgia

Chapter 13

- Revised Gross Anatomy introductory text: Muscles That Act About the Pectoral Girdle/Glenohumeral Joint
- Revised Table 13.1: Muscles That Act About the Pectoral Girdle
- Revised Exercise 13.1: Muscles That Act About the Pectoral Girdle/Glenohumeral Joint to include Exercise 13.1A: Muscles That Act About the Pectoral Girdle and Exercise 13.1B: Muscles That Act About the Glenohumeral Joint
- New Clinical View: Winged Scapula
- Revised Table 13.2: Muscles That Act About the Glenohumeral Joint
- New Learning Strategy for remembering muscles in the forearm
- Reorganized Table 13.6: Posterior (Extensor) Compartment of the Forearm
- Revised Gross Anatomy introductory text: Muscles That Act About the Hip Joint/Thigh
- New Learning Strategy to remember muscles in medial compartment of the thigh
- Revised Table 13.8: Muscles That Act About the Hip Joint/Thigh
- Revised Exercise 13.5: Muscles That Act About the Hip Joint/Thigh
- Revised Exercise 13.6: Compartments of the Thigh
- New Table 13.9: Anterior Compartment of the Thigh
- New Table 13.10: Posterior Compartment of the Thigh
- New Clinical View: Graciloplasty

Chapter 14

- New Concept Connection on somatic motor neurons
- New Concept Connection on the excitability and conductivity of nervous tissue
- Revised Table 14.3 Glial Cells to include headings for the central nervous system and peripheral nervous system
- New Clinical View: Peripheral Nerve Injury
- New Learning Strategy on membrane potential

Chapter 15

- New Clinical View: Meningiomas
- New Exercise 15.3 Circulation of Cerebrospinal Fluid (CSF)
- New Figure 15.5 Cerebrospinal Fluid (CSF) Production and Circulation
- Reorganized Table 15.3 Brain Structures Visible in Superficial Views of Whole or Sagittally Sectioned Brains
- New Clinical View: Vasovagal Syncope
- New Clinical View: Cranial Nerve Assessment
- New Exercise 15.11 BIOPAC Electroencephalography (EEG), and corresponding figures for experiment setup and data collection

Chapter 16

- Chapter renamed The Spinal Cord, Spinal Nerves, and Reflexes
- Revised Table 16.1 Regional Characteristics of the Spinal Cord to include word origins
- Reorganized Table 16.2 Histology of the Spinal Cord in Cross Section
- Reorganized Table 16.5 Major Nerves of the Brachial Plexus
- Revised Exercise 16.5 The Lumbar and Sacral Plexuses
- Reorganized original Table 16.6 into Table 16.6 Major Nerves of the Lumbar Plexus and Table 16.7 Major Nerves of the Sacral Plexus
- New Gross Anatomy section on somatic reflexes
- New Exercise 16.6 Identifying Components of a Reflex on a Classroom Model (moved from chapter 17)
- New section on Reflex Physiology
- New Exercise 16.7 Patellar Reflex (moved from chapter 17)
- New Clinical View: Babinski Reflex (moved from chapter 17)
- New Exercise 16.8 Withdrawal and Crossed-Extensor Reflex
- New Exercise 16.9 Plantar Reflex (moved from chapter 17)

Chapter 17

- Chapter renamed The Autonomic Nervous System
- New Learning Strategy on the two divisions of the autonomic nervous system
- Revised introductory text for the Gross Anatomy of the Autonomic Nervous System
- Revised Figure 17.2 Overview of the Parasympathetic Division of the ANS
- New Clinical View: Pheochromocytoma
- New Exercise 17.4 BIOPAC Galvanic Skin Response, and corresponding figures for experiment setup and data collection

Chapter 18

- Revised Table 18.2 Cells Associated with Taste Buds to include word origins
- New Concept Connection on olfaction and the ethmoid bone
- Revised Figure 18.10 Skin
- Revised Exercise 18.8 Gross Anatomy of the Eye to include Exercise 18.8A Accessory Structures of the Eye and Exercise 18.8B Internal Structures of the Eye
- Revised Figure 18.11 Accessory Structures of the Eye (a) Classroom model
- Revised Figure 18.12 Classroom Model of the Internal Eye
- New Exercise 18.9 Extrinsic Eye Muscles (moved from Chapter 12)
- New Learning Strategy on remembering extrinsic eye muscle innervation
- Revised Figure 18.13 Extrinsic Eye Muscles
- New Clinical View: Pressure Changes in the Middle Ear

- New Exercise 18.17D Color Blindness
- Reorganized Exercise 18.18 Hearing Tests and Exercise 18.19 Equilibrium Tests

Chapter 19

- New Learning Strategy on hormones secreted by the anterior pituitary gland
- New Concept Connection on hormones secreted by the pituitary gland
- Revised Figure 19.6 Adrenal Glands
- New Clinical View: Anabolic Steroids
- New Exercise 19.8 A Clinical Case in Endocrine Physiology

Chapter 20

- Reoriented Table 20.3 Leukocyte Characteristics for better readability
- Revised Blood Diagnostic Tests to provide introductory text with each physiology exercise
- Revised Figure 20.6 Separation of a Whole Blood Sample by Centrifugation
- New Table 20.4 Normal Ranges for Laboratory Blood Tests
- New Learning Strategy for learning the relative abundance of leukocytes in the blood
- New Clinical View: Blood Type Abundance
- New Exercise 20.10 Determination of Blood Glucose
- New Figure 20.12 Blood Glucose Testing
- New Clinical View: Hemoglobin A1c (Glycated Hemoglobin)

Chapter 21

- Revised Exercise 21.3 Location of the Heart and the Pericardium
- New Learning Strategy on remembering the atrioventricular valves on the right versus the left side of the heart
- Reorganized Table 21.3 Arterial Supply to the Heart
- New Clinical View: Myocardial Infarction
- New Exercise 21.9 Electrocardiography Using Standard ECG Apparatus
- New Figure 21.19 Interpreting an ECG Tracing
- New Exercise 21.10 BIOPAC Lesson 5: Electrocardiography I, and corresponding figures for experiment setup and data collection

Chapter 22

- New Concept Connection on endothelium
- New Clinical View: Great Saphenous Vein and Varicose Veins
- New Clinical View: Atherosclerosis in the Internal Carotid Artery
- Revised Figure 22.11 Circulation to the Thoracic and Abdominal Walls
- New Clinical View: Cardiac Catheterization via the Femoral Artery

Chapter 23

- Reorganized the order of chapter topics and exercises: thymus, lymph nodes, and the spleen
- New Clinical View: Appendicitis
- New Table 23.5 Major Lymphatic Vessels of the Body
- Revised Figure 23.13 Lymph Node and Its Components
- New Clinical View: Mononucleosis
- New Physiology section on the immune system
- New Table 23.6 Cells of the Immune System
- New Exercise 23.9 A Clinical Case Study in Immunology

Chapter 24

- New Learning Strategy on structure and function of the trachea
- New Clinical View: Tuberculosis
- New Learning Strategy to remember the lobes of the right versus the left lung
- Revised Exercise 24.12 Pulmonary Function Tests to include Exercise 24.12A Wet Spirometry and Exercise 24.12B BIOPAC Lesson 12: Pulmonary Function Tests

Chapter 25

- Reorganized Table 25.1 Histological Features of the Kidney to include headings and subheadings
- New Clinical View: Glomerulonephritis

Chapter 26

- New Learning Strategy on distinguishing between gastric pits and gastric glands
- New Learning Strategy on distinguishing the three parts of the small intestine
- Revised Figure 26.5 The Small Intestine
- New Learning Strategy on histology of the pancreas
- New Exercise 26.8 Overview of the GI Tract
- New Figure 26.10 Overview of the Digestive System
- Reorganized Table 26.6 Gross Anatomic Regions and Features Associated with the Stomach
- Revised Figure 26.12 Classroom Model of the Stomach
- Reorganized Table 26.7 Gross Anatomic Features of the Liver, Gallbladder, Pancreas, and Their Associated Ducts
- Reorganized Table 26.9 The Cecum, Large Intestine, Rectum, and Anus
- New Figure 26.17 The Cecum, Large Intestine, and Rectum
- Revised Learning Strategy to a Concept Connection on motility in the GI tract
- New Exercise 26.15 A Clinical Case Study in Digestive Physiology

Chapter 27

- New Learning Strategy on recognizing follicles in various developmental stages
- Reorganized Table 27.4 Components of the Uterine Tube
- Reorganized Table 27.5 Phases of the Menstrual Cycle
- New Clinical View: Erectile Dysfunction
- New Concept Connection on lactation
- Revised Table 27.14 Pre-Embryonic Period
- Revised Table 27.15 Stages of Embryonic Development

Chapter 28

- References to structures within thoracic and abdominal cavities changed from "posterior" to "dorsal"
- New Clinical View: Proper Handling and Care of Preserved Cats
- New Learning Strategy to explain color coding for structures in illustrations
- Revised Figure 28.1 Directional Terms
- New Concept Connection comparing structure and function of bones of the cat limbs to those in humans
- Renamed Exercise 28.10 Opening the Thoracic and Abdominal Cavities, and reorganized into two exercises including new Exercise 28.11 The Heart, Lungs, and Mediastinum

the learning system

Features

The Eckel/Ross/Bidle: *Anatomy & Physiology Laboratory Manual* works well as a complement to the McKinley/O'Loughlin/Bidle: *Anatomy & Physiology: An Integrative Approach* textbook, or to accompany any other anatomy and physiology text. Each chapter opener includes an **outline** that lists a set of **learning objectives** for the chapter.

- A chapter **Introduction** opens with a real-life scenario that emphasizes the section of the body covered in the chapter, to connect the anatomy of our bodies with the physiology that helps us to perform day-to-day activities.

CHAPTER 11

The Muscular System: Muscle Structure and Function

OUTLINE AND LEARNING OBJECTIVES

HISTOLOGY 250

Skeletal Muscle Tissue 250

EXERCISE 11.1: HISTOLOGY OF SKELETAL MUSCLE FIBERS 252

1. Identify skeletal muscle tissue through the microscope, and describe the features unique to skeletal muscle tissue
2. Name the visible bands that form the striations in skeletal muscle tissue

EXERCISE 11.2: CONNECTIVE TISSUE COVERINGS OF SKELETAL MUSCLE 253

3. Describe the layers of connective tissue that surround skeletal muscle tissue

The Neuromuscular Junction 253

EXERCISE 11.3: THE NEUROMUSCULAR JUNCTION 255

4. Define motor unit and describe how the concept of a motor unit applies to neuromuscular junctions

Cardiac Muscle Tissue 255

EXERCISE 11.4: CARDIAC MUSCLE TISSUE 255

5. Identify cardiac muscle tissue through the microscope, and describe the features unique to cardiac muscle tissue
6. Compare and contrast the structure of skeletal, smooth, and cardiac muscle tissues

Smooth Muscle Tissue 256

EXERCISE 11.5: SMOOTH MUSCLE TISSUE 256

7. Identify smooth muscle tissue through the microscope, and describe the features unique to smooth muscle tissue
8. Describe how two layers of smooth muscle tissue act as antagonists

GROSS ANATOMY 257

Gross Anatomy of Skeletal Muscles 257

EXERCISE 11.6: NAMING SKELETAL MUSCLES 257

9. Explain some of the logic behind the naming of skeletal muscles

EXERCISE 11.7: ARCHITECTURE OF SKELETAL MUSCLES 260

10. Use anatomic terminology to describe the architecture of skeletal muscles, and describe how the architecture of a skeletal muscle is related to its action

Organization of the Human Musculoskeletal System 261

EXERCISE 11.8: MAJOR MUSCLE GROUPS AND FASCIAL COMPARTMENTS OF THE LIMBS 262

11. Describe the location and major actions of the major muscle groups of the body
12. Describe the fascial compartments of the limbs, and explain the major actions associated with each fascial compartment

PHYSIOLOGY 264

Force Generation of Skeletal Muscle 264

EXERCISE 11.9: MOTOR UNITS AND MUSCLE FATIGUE (HUMAN SUBJECT) 265

13. Compare and contrast the characteristics of Type I, Type IIa, and Type IIb muscle fibers
14. Describe the sequence of motor unit recruitment
15. Explain the relationship between motor unit recruitment, muscle force, and fatigue

EXERCISE 11.10: CONTRACTION OF SKELETAL MUSCLE (WET LAB) 266

16. Describe the effect of adding ATP alone, ATP + salts, or salts only on contraction of glycerinated muscle
17. Explain the relationship between ATP supply to skeletal muscle and rigor mortis

EXERCISE 11.11: Ph.I.L.S. LESSON 5: STIMULUS-DEPENDENT FORCE GENERATION 270

18. Describe the events of a muscle twitch
19. Describe the relationship between stimulus intensity and maximum isometric twitch force
20. Describe the threshold voltage, and explain what happens when suprathreshold stimuli are applied to skeletal muscle

EXERCISE 11.12: Ph.I.L.S. LESSON 7: THE LENGTH-TENSION RELATIONSHIP 271

21. Describe the relationship between muscle length and tension for a maximal isometric contraction
22. Describe what is happening at the sarcomere level when a muscle contracts isometrically at its optimal length

EXERCISE 11.13: Ph.I.L.S. LESSON 8: PRINCIPLES OF SUMMATION AND TETANUS 274

23. Describe the relationship between stimulus frequency and muscle tension
24. Demonstrate the concepts of summation, incomplete tetanus, and complete tetanus

EXERCISE 11.14: Ph.I.L.S. LESSON 9: EMG AND TWITCH AMPLITUDE 275

25. Describe the relationship between muscle tension and EMG amplitude

EXERCISE 11.15: BIOPAC ELECTROMYOGRAPHY (EMG) 276

26. Describe the relationship between motor unit recruitment and EMG amplitude

Anatomy & Physiology REVEALED aprevealed.com MODULE 6: MUSCULAR SYSTEM

247

- The laboratory manual exhibits the highest-quality **photographs and illustrations** of any laboratory manual on the market.

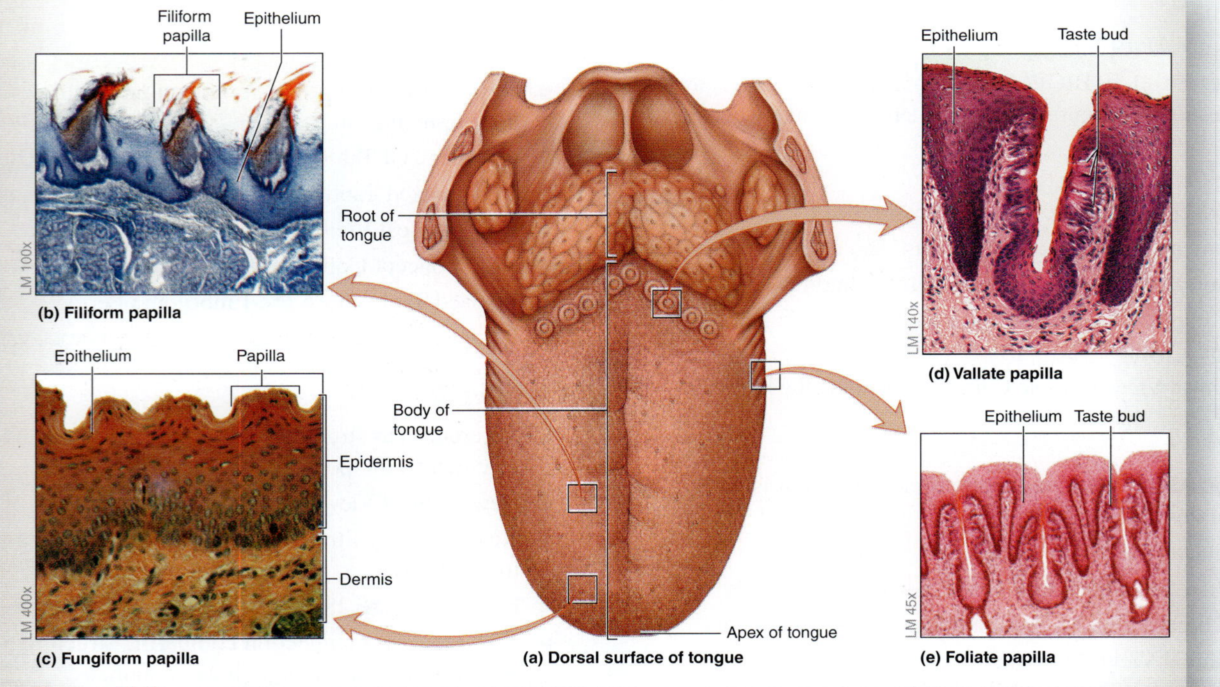

Figure 18.3 **Taste Buds.** Gustation (taste) requires taste buds, which are associated with tongue papillae. (a) Dorsal surface of tongue, (b) Filiform papilla, (c) Fungiform papilla, (d) Vallate papilla, (e) Foliate papilla.

- The content of the laboratory manual is informed by the textbook, and both the textbook and the laboratory manual share similar pedagogic elements: **Concept Connection, Learning Strategy,** and **Clinical View** features from the text are also employed in the laboratory manual.
  - **Integrate: Concept Connection** boxes draw concepts from the classroom into the laboratory for a real-time review of how previously covered concepts relate to body systems.
  - **Integrate: Learning Strategy** boxes offer tried-and-tested learning strategies that consist of everyday analogies, mnemonics, and useful tips to aid understanding and memory.
  - **Integrate: Clinical View** sidebars reinforce facts through a clinical discussion of what happens when the body doesn't perform normally.

INTEGRATE

CONCEPT CONNECTION

Recall that the gray matter of the spinal cord contains predominately neuron cell bodies, dendrites, and unmyelinated axons. In addition, gray matter also contains various types of neurons (e.g., motor neurons, somatic sensory ... which are typically contained within a specified area of the

all of the fibers that it innervates. That *somatic* motor neuron becomes "excited" within the anterior horn of gray matter within the spinal cord, and its axon exits the central nervous system through the anterior root. It then joins a spinal nerve and ultimately travels to individual muscle fibers within a skeletal muscle. An action potential traveling along a somatic ... e in force ... thoracic ... utonomic ... e, smooth ... urons, the ... ystem by

INTEGRATE

LEARNING STRATEGY

To identify superficial muscles and tendons, place your left palm on the medial epicondyle of your right humerus. In this position, the order of the muscles on your right forearm, from lateral to medial, is:

Index finger—pronator teres (PT)

Middle finger—flexor carpi radialis (FCR)

Ring ... us (PL)

Pi ... aris (

W ...

the ...

pa ...

Pronator teres
Flexor carpi radialis
Palmaris longus
Flexor carpi ulnaris

INTEGRATE

CLINICAL VIEW

Piriformis Syndrome

The piriformis muscle is a "pear-shaped" muscle that lies in close proximity to important structures within the gluteal region, such as the sciatic nerve, and the gluteal arteries and nerves. Piriformis syndrome is a painful condition that results from inflammation or overuse of the piriformis muscle. The incidence of piriformis syndrome is relatively common in athletes such as runners and cyclists, who may develop an imbalance in the strength of the piriformis muscle as compared to the gluteal muscles. Specifically, the syndrome occurs when the piriformis muscle (which laterally rotates the thigh) is stronger than the gluteus medius and gluteus minimus muscles (which are responsible for medial rotation of the thigh). As the piriformis muscle becomes inflamed or experiences spasms, it may also compress the underlying sciatic nerve, resulting in sciatica. **Sciatica** is a tingling, painful, or even numbing sensation that travels down the path of the sciatic nerve. Patients complain of shooting pain that runs from the gluteal region down the lateral aspect of the thigh, and toward the leg. Often the pain may be exacerbated when the body is held in certain positions, such as prolonged sitting or standing. The symptoms of piriformis syndrome can be reduced with the administration of anti-inflammatory drugs and through stretching exercises.

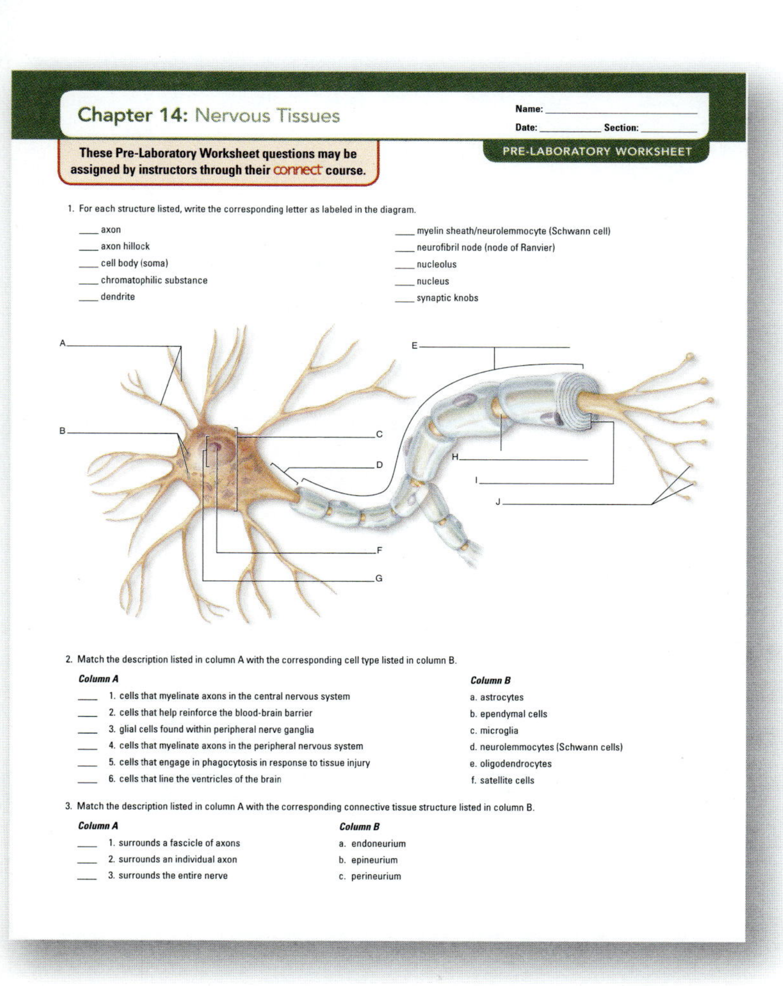

Chapter 14: Nervous Tissues

Name: ______

Date: ______ Section: ______

These Pre-Laboratory Worksheet questions may be assigned by instructors through their connect course.

PRE-LABORATORY WORKSHEET

1. For each structure listed, write the corresponding letter as labeled in the diagram.

____ axon
____ axon hillock
____ cell body (soma)
____ chromatophilic substance
____ dendrite
____ myelin sheath/neurolemmocyte (Schwann cell)
____ neurofibril node (node of Ranvier)
____ nucleolus
____ nucleus
____ synaptic knobs

2. Match the description listed in column A with the corresponding cell type listed in column B.

Column A

____ 1. cells that myelinate axons in the central nervous system
____ 2. cells that help reinforce the blood-brain barrier
____ 3. glial cells found within peripheral nerve ganglia
____ 4. cells that myelinate axons in the peripheral nervous system
____ 5. cells that engage in phagocytosis in response to tissue injury
____ 6. cells that line the ventricles of the brain

Column B

a. astrocytes
b. ependymal cells
c. microglia
d. neurolemmocytes (Schwann cells)
e. oligodendrocytes
f. satellite cells

3. Match the description listed in column A with the corresponding connective tissue structure listed in column B.

Column A

____ 1. surrounds a fascicle of axons
____ 2. surrounds an individual axon
____ 3. surrounds the entire nerve

Column B

a. endoneurium
b. epineurium
c. perineurium

- **Pre-Laboratory Worksheets** at the start of each chapter consist of important refresher points to provide students with a "warm-up" before entering the laboratory classroom. Some questions pertain to previous activities that are relevant to upcoming exercises, while others are basic questions that students should be able to answer if they have read the chapter from their lecture text before coming into the laboratory classroom. The goal of completing these worksheets is to have students arrive at the laboratory prepared to deal with the material they will be covering, so valuable laboratory time isn't lost in reviewing necessary information. All Pre-Laboratory Worksheet questions are assignable within Connect.

- In-chapter **activities** offer a mixture of labeling exercises, sketching activities, table completion exercises, data recording and analysis, palpation of surface anatomy, and other sources of learning. In the gross anatomy exercises of this manual, structures such as cranial bones and muscles of the body are *not* always presented as labeled photos, since students already have labeled photos provided in their anatomy and physiology textbook. Instead, images are presented as labeling activities with a checklist of structures. The checklists serve two purposes: (1) they guide students to items they should be able to identify on classroom models, fresh specimens, or cadavers (if the laboratory uses human cadavers), and (2) they double as a list of terms students can use to complete the labeling activities. Answers to the labeling activities are provided in the Appendix. Thus, if a student does not know what a leader line is pointing to, or cannot remember the correct term, the Appendix serves as a resource for locating the correct answer.
- **Anatomy & Physiology Revealed® (APR)** correlations, indicated by the APR logo, direct students to related content in this cutting-edge software.
- Each chapter contains numerous **tables,** which concisely summarize critical information and key structures and serve as important points of reference while in the laboratory classroom. Most tables contain a column that provides word origins for each structure listed within the table. These word origins are intended to give students continual exposure to the origins of the language of anatomy and physiology, which is critical for learning and retention.
- Numerous **Physiology Interactive Lab Simulations© (Ph.I.L.S.) 4.0** exercises throughout the laboratory manual make otherwise difficult and expensive experiments a breeze, and offer additional opportunities to aid student understanding of physiology.
- **BIOPAC©** exercises are included in chapters 11, 15, 17, 21, 22 and 24.
- **Post-Laboratory Worksheets** at the end of each chapter serve as a review of the materials just covered, and challenge students to apply knowledge gained in the laboratory. The Post-Laboratory Worksheets contain in-depth critical thinking question types, and are perforated so they can be torn out and handed in to the instructor, if so desired. Assessment questions are organized by exercise, and are keyed to the Learning Objectives from the chapter opener outline.
  - **"Do You Know the Basics?"** questions quiz students on the material they have just learned in the chapter, using a variety of question formats including labeling, table completion, matching exercises, and fill-in-the-blank.
  - **"Can You Apply What You've Learned?"** questions are often clinically oriented and expose health-sciences students to problem solving in clinical contexts.
  - **"Can You Synthesize What You've Learned?"** questions combine concepts learned in the chapter to ensure student understanding of each chapter's objectives.

Chapter Ten *Articulations* 235

Synovial Joints

Synovial joints have a complex structure that includes a joint cavity filled with fluid. The term *synovial* literally means "together with egg" (*syn*, together, + *ovum*, egg). This term refers to the fluid inside the joint (the synovial fluid), which has the consistency and appearance of egg white. The joint cavity filled with synovial fluid allows the articulating bones to move easily past one another with very little friction between the bones. The features of synovial joints are covered in detail in the next several exercises. These include exercises covering the general structure of a synovial joint, categories of synovial joints, and the types of movements allowed by synovial joints.

EXERCISE 10.3

GENERAL STRUCTURE OF A SYNOVIAL JOINT

1. Observe a model of a synovial joint, preferably a model of the knee joint. Identify the features of a typical synovial joint listed in **table 10.4.**
2. Label the components of a synovial joint in figure 10.3, using table 10.4 and the textbook as guides.
3. *Optional Activity:* APR 5: **Skeletal System**—Watch the "Synovial Joint" animation for a summary of synovial joint structure and types.

Table 10.4 Components of Synovial Joints

Structure	Description	Word Origin
Articular Capsule	Consists of two layers: an outer fibrous capsule and an inner synovial membrane	*arthron*, a joint, + *capsa*, a box
Articular Cartilage	Hyaline cartilage found on the epiphyses of the articulating bones	*arthron*, a joint
Fibrous Layer of Articular Capsule	Dense irregular connective tissue that anchors the periosteum of the two articulating bones to each other; thickenings of the fibrous capsule form several joint ligaments	*fibra*, fiber
Synovial Cavity	A cavity within the joint that is lined by a synovial membrane and filled with synovial fluid	*syn*, together, + *ovum*, egg, + *cavus*, hollow
Synovial Fluid	A viscous, oily fluid located within the synovial joint; functions as a lubricant, to nourish the articular cartilage, and as a shock absorber	*syn*, together, + *ovum*, egg, + *fluidus*, to flow
Synovial Membrane	Composed primarily of areolar connective tissue that forms the inner lining of the articular capsule and covers internal joint surfaces not covered by cartilage; responsible for the formation of synovial fluid	*syn*, together, + *ovum*, egg, + *membrana*, a skin

1 2 3 4 5 6 7 8

Figure 10.3 Diagram of a Representative Synovial Joint. Use the terms listed to fill in the numbered labels in the figure.

- ☐ articular capsule
- ☐ articular cartilage
- ☐ fibrous layer of articular capsule
- ☐ ligament
- ☐ periosteum
- ☐ synovial (joint) cavity
- ☐ synovial membrane
- ☐ yellow bone marrow

Chapter 9: The Skeletal System: Appendicular Skeleton

Name: ______ Date: ______ Section: ______

POST-LABORATORY WORKSHEET

The ❶ corresponds to the Learning Objective(s) listed in the chapter opener outline.

Do You Know the Basics?

Exercise 9.1: Bones of the Pectoral Girdle

1. The head of the humerus articulates with which bony landmark of the scapula? (Circle one.) ❶
   a. acromion
   b. coracoid process
   c. glenoid fossa
   d. spine
   e. suprascapular notch
2. The only point of articulation between the pectoral girdle and the axial skeleton is the sternoclavicular joint. ______ (True/False) ❷
3. Match the bones listed in column A with the corresponding joint listed in column B. ❶ ❷ ❸

Column A	*Column B*
____ 1. clavicle and scapula	a. acromioclavicular joint
____ 2. humerus and ulna	b. elbow joint
____ 3. scapula and humerus	c. glenohumeral joint
____ 4. ulna, radius, and carpal bones	d. wrist joint

4. Label the following diagram of an articulated shoulder girdle using the terms listed: ❶ ❷

1 2 3 4 5 6 7 Clavicle Scapula 8 9 10 11 12 13 14

Right scapula and clavicle articulation, anterior view

Scapula: acromion, coracoid process, glenoid cavity, inferior angle, infraglenoid tubercle, lateral border, medial border, spine, subscapular fossa, superior angle, supraglenoid tubercle, suprascapular notch

Clavicle: acromial end, sternal end

Chapter Nine *The Skeletal System: Appendicular Skeleton* 225

- **Chapter 28: Cat Dissection Exercises** contains 17 exercises designed to walk students through dissection and identification of cat structures. Students are introduced to the cat skeletal system, and are then provided with dissection instructions that follow a regional approach.

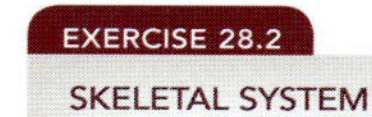
EXERCISE 28.2

SKELETAL SYSTEM

1. Observe a cat skeleton (**figure 28.3**). Be sure to carefully observe all of the bones of the skeleton so that you are able to identify them later (e.g., on a laboratory exam).

2. Record in **table 28.1** the number of vertebrae in each region of the vertebral column in a human. Next, count and record the number of vertebrae in each region of the

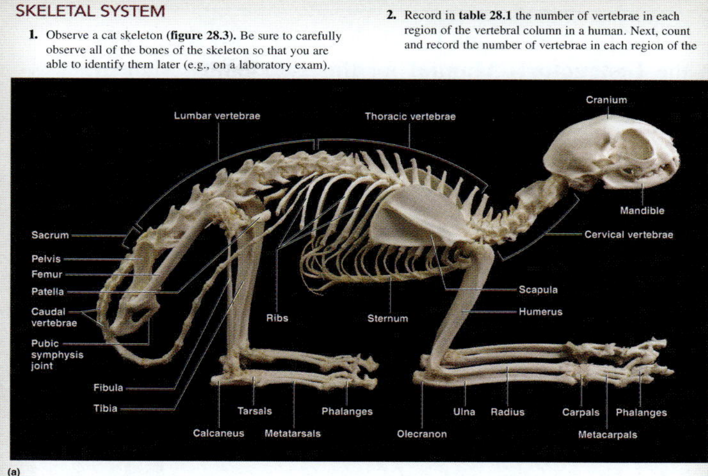

Figure 28.3 The Cat Skeleton. The cat skeleton is similar to the human skeleton in several ways. Several notable differences are the elongation of the pelvis, metacarpals, and metatarsals; the extra vertebrae; and a clavicle that does not articulate with any other bone. (a) Articulated cat skeleton in anatomic position.

846 Chapter Twenty-Eight *Cat Dissection Exercises*

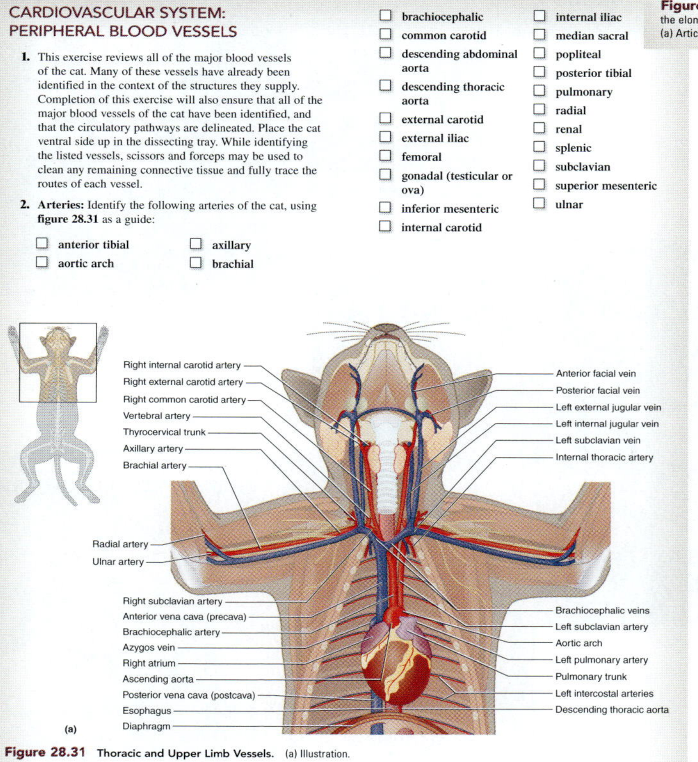
EXERCISE 28.17

CARDIOVASCULAR SYSTEM: PERIPHERAL BLOOD VESSELS

1. This exercise reviews all of the major blood vessels of the cat. Many of these vessels have already been identified in the context of the structures they supply. Completion of this exercise will also ensure that all of the major blood vessels of the cat have been identified, and that the circulatory pathways are delineated. Place the cat ventral side up in the dissecting tray. While identifying the listed vessels, scissors and forceps may be used to clean any remaining connective tissue and fully trace the routes of each vessel.

2. **Arteries:** Identify the following arteries of the cat, using **figure 28.31** as a guide:

- ☐ anterior tibial
- ☐ aortic arch
- ☐ axillary
- ☐ brachial
- ☐ brachiocephalic
- ☐ common carotid
- ☐ descending abdominal aorta
- ☐ descending thoracic aorta
- ☐ external carotid
- ☐ external iliac
- ☐ femoral
- ☐ gonadal (testicular or ova)
- ☐ inferior mesenteric
- ☐ internal carotid
- ☐ internal iliac
- ☐ median sacral
- ☐ popliteal
- ☐ posterior tibial
- ☐ pulmonary
- ☐ radial
- ☐ renal
- ☐ splenic
- ☐ subclavian
- ☐ superior mesenteric
- ☐ ulnar

Figure 28.31 Thoracic and Upper Limb Vessels. (a) Illustration.

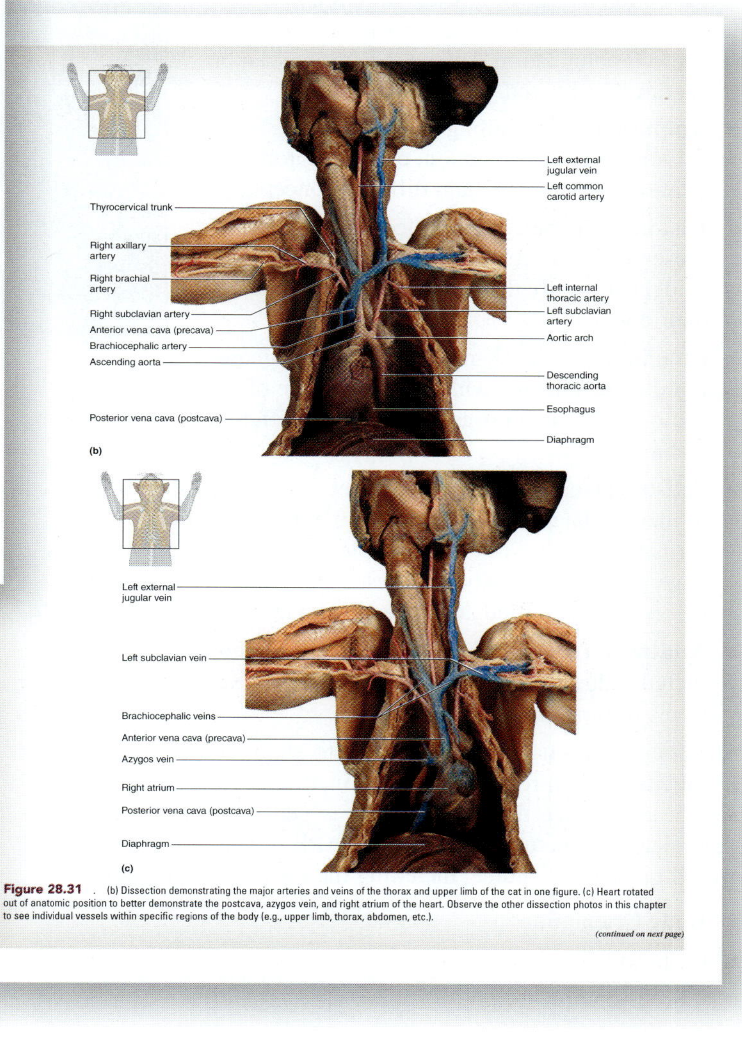

Figure 28.31 . (b) Dissection demonstrating the major arteries and veins of the thorax and upper limb of the cat in one figure. (c) Heart rotated out of anatomic position to better demonstrate the postcava, azygos vein, and right atrium of the heart. Observe the other dissection photos in this chapter to see individual vessels within specific regions of the body (e.g., upper limb, thorax, abdomen, etc.).

(continued on next page)

Teaching Supplements

Answers to the Pre-Laboratory and Post-Laboratory Worksheets can be found in the **Instructor's Manual** for this Laboratory Manual within Connect, by accessing the McKinley/O'Loughlin/Bidle: *Anatomy & Physiology,* 2nd edition Instructor Resources. **Image files** for use in presentations and teaching materials are also provided for instructor use at this location.

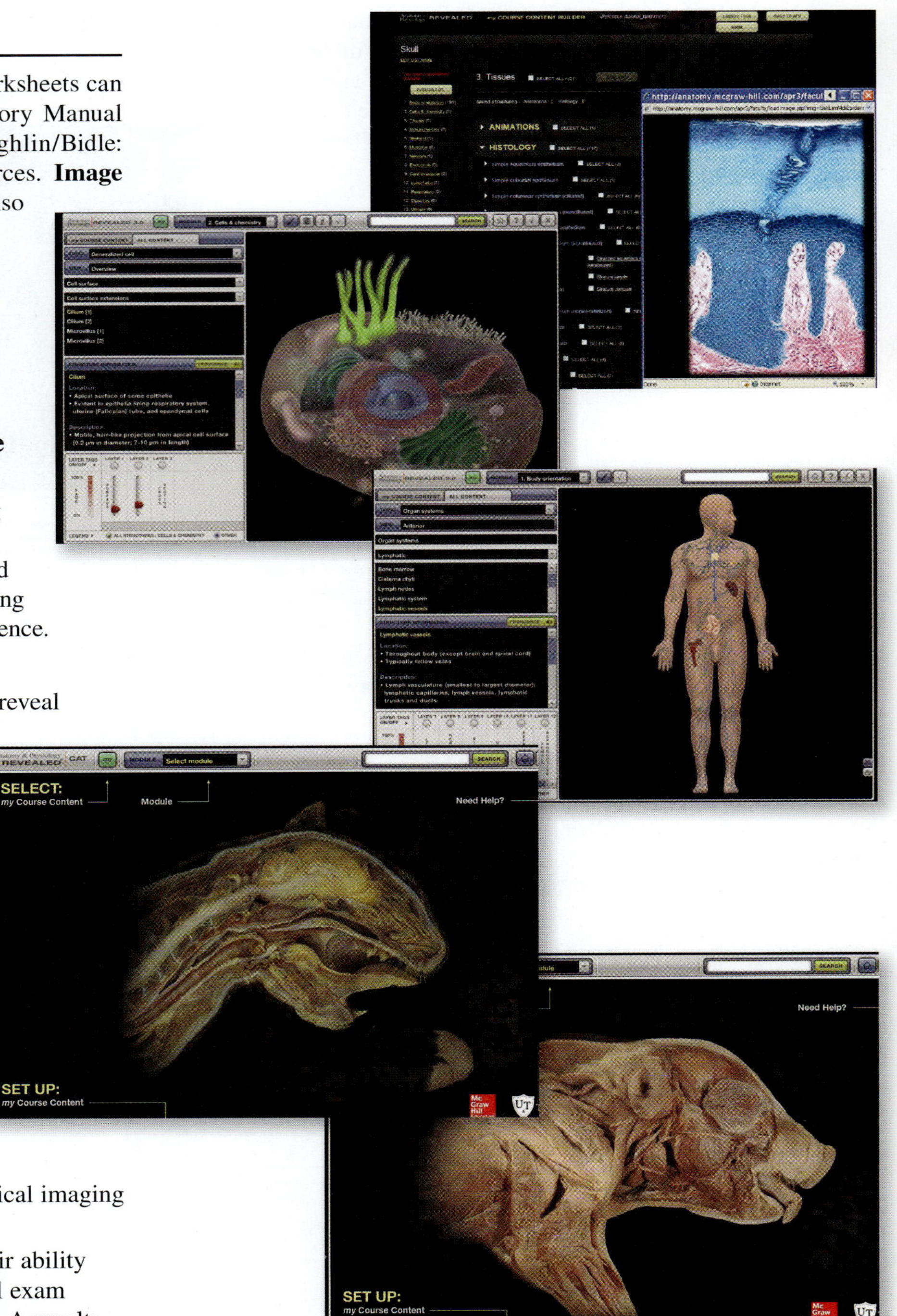

Anatomy & Physiology Revealed®: An Interactive Cadaver Dissection Experience

Available online at www.aprevealed.com, and as an APR application on Apple® and Android™ tablets, this amazing multimedia tool is designed to help students learn and review human anatomy using cadaver specimens. Detailed cadaver photographs blended with a state-of-the-art layering technique provide a uniquely interactive dissection experience. This easy-to-use program features the following sections:

- **Dissection:** Peel away layers of the human body to reveal the structures beneath the surface. Structures can be pinned and labeled, just as in a real dissection lab. Each labeled structure is accompanied by detailed information and an audio pronunciation. Dissection images can be captured and saved.
- **Animation:** Compelling animations demonstrate muscle action, clarify anatomical relationships, and explain difficult concepts.
- **Histology:** Labeled light micrographs presented with each body system allow students to study the cellular detail of tissues at their own pace.
- **Imaging:** Labeled X-ray, MRI, and CT images familiarize students with the appearance of key anatomical structures as seen through different medical imaging techniques.
- **Self-test:** Challenging exercises let students test their ability to identify anatomical structures in a timed practical exam format or with traditional multiple choice questions. A results page provides an analysis of test scores plus links back to all incorrectly identified structures for review.
- **Anatomy Terms:** This visual glossary of general terms includes directional and regional terms, as well as planes and terms of movement.

Instructors may customize APR 3.0 to their course by selecting the specific structures they require in their course, and APR 3.0 does the rest. Once the structure list is generated, APR highlights these selected structures for students. APR contains all the material covered in an A&P course, including these three new modules:

- Body Orientation
- Cells and Chemistry
- Tissues

APR is now available in two new versions!

Anatomy & Physiology Revealed | Cat® and **Anatomy & Physiology Revealed | Fetal Pig®** are online interactive cat dissection and fetal pig dissection experiences that use cat photos or fetal pig photos, combined with a layering technique that allows you to peel away layers to reveal structures beneath the surface. Both **Anatomy & Physiology Revealed | Cat** and **Anatomy & Physiology Revealed | Fetal Pig** offer animations, histologic and radiologic imaging, audio pronunciations, and comprehensive quizzing.

Physiology Interactive Lab Simulations© (Ph.I.L.S.) offers 42 lab simulations that may be used to supplement or substitute for wet labs. Users may adjust variables, view outcomes, make predictions, draw conclusions, and print lab reports.

McGraw-Hill LearnSmart Labs™

THE Virtual Lab Experience

LEARNSMART LABS®

Based on the same world-class super-adaptive technology as LearnSmart, McGraw-Hill's LearnSmart Labs™ are must-see, outcomes-based lab simulations. LearnSmart Labs assess a student's knowledge and adaptively correct deficiencies, allowing the student to learn faster and retain more knowledge with greater success.

First, a student's knowledge is adaptively leveled on core learning outcomes: questioning reveals knowledge deficiencies that are corrected by the delivery of content that is conditional on a student's response. Then, a simulated lab experience requires the student to think and act like a scientist: recording, interpreting, and analyzing data using simulated equipment found in labs and clinics. The student is allowed to make mistakes—a powerful part of the learning experience! A virtual coach provides subtle hints when needed, asks questions about the student's choices, and allows the student to reflect upon and correct those mistakes. Whether your need is to overcome the logistical challenges of a traditional lab, provide better lab prep, improve student performance, or make your online experience one that rivals the real world, LearnSmart Labs accomplishes it all.

ACKNOWLEDGMENTS

This laboratory manual is the product of the excellent work and dedication of a consummate group of talented professionals who have helped lead us through this publishing process. We are forever indebted to all of you for embarking on this journey with us.

We wish to thank McGraw-Hill for providing us the unique opportunity to share our enthusiasm for teaching anatomy and physiology through the pages of this laboratory manual.

From start to finish, this book has been carried through the able hands of product developers Donna Nemmers and Kristine Queck. April Southwood, Content Project Manager, kept everyone on track and managed countless details in the production process.

The beautiful design and line art that make this laboratory manual shine are products of a team of hugely talented designers and artists including David Hash at McGraw-Hill, and the fantastic EPS illustration team. Carrie Burger's direction and guidance were instrumental in bringing out the best of our photography program. Danny Meldung, of Photo Affairs, Inc. researched wonderful photos for this edition.

We thank all of the reviewers of the manual for taking the time to review this manual and provide us with their insight and perspective, gained from years of experience in the classroom. We hope we have honored your suggestions for improvement, and we welcome continued feedback. We also thank the many students we have had the pleasure of interacting with over the years, for teaching *us* what works or does not work in the classroom.

The extent of our gratitude is limitless when it comes to the love, understanding, and support that our families, friends, and colleagues gave to us throughout this process. We are truly honored to live our lives in the presence of such wonderful people.

To the users of this laboratory manual: We sincerely hope we have created a learning resource that not only will excite you about the study of anatomy and physiology, but also will actively engage you in the laboratory as you learn about the wonders of the human body. We welcome your thoughts and suggestions for improvements.

Christine M. Eckel
Department of Life and Environmental Sciences
Carroll College
ceckel@carroll.edu

Kyla Turpin Ross
Department of Biology
Georgia State University
kross@gsu.edu

Terri Stouter Bidle
Science Division
Hagerstown Community College
tsbidle@hagerstowncc.edu

Reviewers

Andrew E. Accardi
Central Carolina Technical College

Kyle Bartow
Indian River State College

Gladys Bolding
Georgia Perimeter College

Pamela K. Elf
University of Minnesota—Crookston

Theresa Felten
Polk State College—Lakeland

Joseph D. Gar
West Kentucky Community and Technical College

Sylvester Hackworth
Bishop State Community College

Steven B. Hammer
Indian River State College

Lesleigh Hastings
Wake Tech Community College

Karen Ramey Hlinka
West Kentucky Community and Technical College

Kathy Jo Ann Jackson
McNeese State University

Scott Johnson
Wake Tech Community College

Karen Dunbar Kareiva
Ivy Tech Community College—Northwest Region

Jessica Burr Lea
Central Carolina Technical College

Tiffany B. McFalls-Smith
Elizabethtown Community and Technical College

Jill Y. O'Malley
Erie Community College

Justicia Opoku
University of Maryland—College Park

Jo Rogers
University of Cincinnati

Connie E. Rye
East Mississippi Community College

Dee Ann Sato
Cypress College

Jackie Thomas
Wake Tech Community College

Joyce E.M. Wall
Housatonic Community College

Charles M. Watson
Midwestern State University

Martha T. Wolfe
Elizabethtown Community and Technical College

Jennie L. Yates
St. Petersburg College—Seminole Campus

The Laboratory Environment

OUTLINE AND LEARNING OBJECTIVES

INTRODUCTION

Welcome to the human anatomy and physiology laboratory! Most students experience both excitement and anxiety about this course. The human body is a fascinating subject, and the study of human anatomy and physiology is an experience that is typically not forgotten.

This laboratory manual is designed for an integrated, systems-based course that combines human histology, gross anatomy, and physiology. **Histology** is the study of tissues, and requires the use of a microscope. **Gross anatomy** is the study of structures that can be seen with the naked eye. This includes any structure that can be seen without the use of a microscope. **Physiology** is the study of body functions. After completing this course, it is our hope that students will have developed an understanding and appreciation for how tissue structure relates to gross structure, and how all levels of structure relate to function. That said, in the laboratory itself, students will often be studying the three somewhat separately. That is, laboratory studies in histology will likely involve observing histology slides with a microscope or using some sort of virtual microscopy system; laboratory studies in gross anatomy will likely involve observing classroom models, dissecting animal specimens, or making observations of human bones and/or human cadavers; and laboratory studies in physiology will involve performing wet lab or virtual (computer software–based) experiments. To assist students in these endeavors, the exercises in this manual are divided into three types of activities: histology, gross anatomy, and physiology activities. Where applicable, each chapter will begin with a section on histology and will end with a section on physiology. Although students will perform the activities somewhat separately, the goal is to integrate what students are learning in each exercise to associate structure with function. "Concept Connection" boxes and questions within exercises in each chapter will assist students with this task.

The purpose of this introductory chapter is to familiarize students with the process of science, systems of measurement, common equipment and dissection techniques encountered in the anatomy and physiology laboratory, proper disposal of laboratory waste materials, and common dissection techniques.

List of Reference Tables

INTEGRATE

CLINICAL VIEW

Use of Human Cadavers in the Anatomy and Physiology Laboratory

Where did that body lying on a table in the human anatomy and physiology laboratory come from? Typically, the body was donated by a person who made special arrangements before the time of death to donate his or her body to a body donor program so it could be used for education or research. Individuals who donate their bodies for these purposes made a conscious decision to do so. Such individuals have given us an incredible gift—the opportunity to learn human anatomy and physiology from an actual human body. It is important to remember that what that person has given is, indeed, a gift. The cadaver deserves the utmost respect at all times. Making jokes about any part of the cadaver or intentionally damaging or "poking" at parts of the cadaver is unacceptable behavior.

The idea that one will be learning anatomy and physiology by observing structures on what was, at one time, a living, breathing human being might make a person feel very uncomfortable at first. It is quite normal to have an emotional response to the cadaver upon first inspection. It takes time and experience to become comfortable around the cadaver. Even if you think you will be just fine around the cadaver when you are to observe it for the first time, it is important to be aware of your initial response and of the responses of fellow classmates. If at any time you feel faint or light-headed, sit down immediately. Fainting, though rare, is a possibility, and can lead to injuries if a fainting person falls down unexpectedly. Be aware of fellow students. If they appear to lose color in their faces or start to look sick—they might need your assistance.

Typically the part of the body that evokes the most emotional response is the face, because it is most indicative of the person that the cadaver once was. Because of this, the face of the cadaver should remain covered most of the time. This does not mean you are not allowed to view it. However, when you wish to do so, make sure that other students in the room know that you will be uncovering the face. If you have a particularly strong emotional response to the cadaver, take a break and come back to it later when you are feeling better.

Individuals with a great deal of experience around cadavers had a similar emotional response during their first time as well. In time one learns to disconnect one's emotions from the experience. Certainly at one time the body that is the cadaver in the laboratory was the home of a living human being. However, now it is just a body. Eventually students do become comfortable using the cadaver and find that it is an invaluable learning tool that is far more useful than any model or picture could ever be. There is nothing quite like the real thing to help students truly understand the structure of the human body. Make the most of this unique opportunity—and give thanks to those who selflessly donated their bodies to provide students with the ultimate learning experience in anatomy and physiology.

Students who are curious about the uses of cadavers in science and research are encouraged to check out the following book from the library: Mary Roach, *Stiff: The Curious Lives of Human Cadavers* (New York: W.W. Norton, 2003).

Chapter 1: The Laboratory Environment

Name: ______________________

Date: __________ Section: __________

These Pre-Laboratory Worksheet questions may be assigned by instructors through their connect course.

PRE-LABORATORY WORKSHEET

1. The study of structures is called ______________________ (anatomy/physiology), whereas the study of functions is called ______________________ (anatomy/physiology).

2. The following metric unit(s) is/are used to report mass. (Check all that apply.)
   _____ a. centimeter
   _____ b. decigram
   _____ c. kiloliter
   _____ d. microgram
   _____ e. millimeter

3. Number the following steps involved in the scientific method in the correct order.
   _____ a. conclusions
   _____ b. data analysis
   _____ c. data collection
   _____ d. experiment
   _____ e. hypothesis

4. When presenting data, the ________________ (dependent/independent) variable is plotted on the X-axis, whereas the ________________ (dependent/independent) variable is plotted on the Y-axis.

5. Converting smaller units to larger units requires dividing by the appropriate power of ten. ________________ (True/False)

6. For each of the following metric prefixes, write the corresponding power of ten in the space provided.
   a. centi-: ________________
   b. deci-: ________________
   c. kilo-: ________________
   d. milli-: ________________
   e. micro-: ________________

7. The following chemical(s) require(s) the use of personal protective equipment. (Check all that apply.)
   _____ a. ethanol
   _____ b. formalin
   _____ c. methylene blue
   _____ d. phenol
   _____ e. potassium permanganate

8. When removing a scalpel blade, be sure to point the blade away from you and others. ______________ (True/False)

9. The following dissecting tool is the most beneficial for attempting to loosen the hold between a specimen's skin and underlying fascia. (Circle one.)
   a. dissecting probe
   b. finger
   c. scalpel
   d. scissors

10. Blunt dissection technique is most useful for separating tissues without damaging delicate structures. ___________ (True/False)

INTEGRATE

LEARNING STRATEGY

The volume of material students cover in an anatomy and physiology course is likely to be greater than that of any other college course. This is the reason it is considered by most to be a very challenging course. Learning how to approach a subject with such a vast amount of material can be quite difficult at first and may require a change of study habits. The list below provides several suggestions that students may want to incorporate into their study habits to make them more effective and efficient. Every individual learns differently, so all tips may not be suitable for all learners.

1. Determine what kind of learner you are by completing a learning inventory such as the VARK® learning styles inventory (http://www.vark-learn.com). This site also provides suggestions for study methods that work best for each learning style, which is one of its major strengths.
2. Study a little bit *every day.* "Cramming" simply does not work in a course such as human anatomy and physiology in which there are a large number of terms to learn. Break down the topics that are to be covered into manageable chunks. There are several tips within this laboratory manual that are designed to assist learners with this task. For example, if a learner sits down to study with the goal of learning the details of every bone in the human body, that learner may feel so overwhelmed that he or she has no idea where to start. On the other hand, if the learner studies with the goal of learning the details of *one* bone (e.g., the humerus), then the task becomes much simpler and the learner is more likely to complete it.
3. Only study with a group when you are *reviewing* material that you have already covered on your own. Because all individuals learn differently, group study when first learning material is very inefficient. On the other hand, having someone ask you questions to review what you know is an excellent way to prepare for exams.
4. Use *active* study methods whenever possible. Active methods include writing, speaking, labeling, and the like. *Passive* methods include reading the book, listening to a lecture, and the like. Many students mistakenly believe they have mastered the material once they feel as if they understand what the instructor lectured about or what they read in the book. Do not make this mistake! Force yourself to recall information by writing it down, drawing it, or telling it to a friend or family member (even the family cat or dog will do as an audience). You will quickly find out what you do or don't know. If it is not in your head to begin with, you won't be able to "bring it out," so to speak.

Be patient and persistent. Have confidence in your ability to learn the material. Think of how AMAZING the human body is, and feel fortunate that you have this opportunity to learn about it. You *can* do this!

GROSS ANATOMY

The Scientific Process of Discovery

What is science? **Science** is a way of knowing about the world. A **scientist** is an individual who engages in research using the scientific method to learn facts about the world. The **scientific method** is a systematic approach to inquiry that assumes that the answer to a question can be explained by phenomena that are *observable* and *measurable*. In the anatomy and physiology laboratory, the scientific method will be used to make observations and conclusions about the structure and function of the human body.

The scientific method is a rigorous and systematic approach to inquiry that requires certain steps be taken. However, the steps need not be followed precisely. That is, there is some flexibility in the order in which the steps take place. Very often several steps take place concurrently as the process of discovery evolves. The steps involved in the scientific method follow the general pattern shown below:

Observation → Hypothesis → Experiment → Data Collection → Data Analysis → Conclusions

Observation

The first step of the scientific method is **observation.** When an unknown phenomenon is observed, the observer often makes a guess as to the cause of that phenomenon. If the guess is based on previous knowledge, it is an *educated* guess as to the cause of the phenomenon. In science this educated guess is called a **hypothesis.** For example, consider the observation that body temperature changes during the day, and an observer of this fact is interested in knowing if body temperature also changes during the course of the night while a person is asleep. An observation has been made (body temperature changes during the day), that was followed up with a question: Does body temperature change during the night? The next step is to formulate a hypothesis.

Hypothesis

The second step of the scientific method is to **formulate a hypothesis.** One of the key features of a good hypothesis is that it must be testable. That is, some aspect of the variable of interest must be measurable. In the current example, body temperature is the variable of interest, and can be measured using a thermometer. Another feature of a good hypothesis is that it must be specific, yet limited in scope. This is not to say that the explanation for a phenomenon is limited, nor that the questions asked are limited. Instead, it means that the testable hypothesis must be limited to something that is measurable while all other conditions are controlled. An example of a simple, testable hypothesis is the following: Body temperature changes over time during the night. Once the hypothesis has been formulated, the next step is to design an experiment to test the hypothesis.

Experiment

One of the most creative and interesting aspects of the scientific method is to **design an experiment** to test a hypothesis. Designing a good experiment to test a hypothesis is a challenging task. The key feature of good experimental design is to attempt to predict any variables that may have an influence on the variables of interest and control for them. In the current example, the variables of interest are body temperature and time during the night. To determine if body temperature changes during the night, an experiment must be designed to measure body temperature at given time intervals during the night.

Variables

A **variable** is a characteristic that may or may not influence the outcome of an experiment. In the current experiment two variables of interest have been identified: body temperature and time. Based on convention, the variables in the experiment must be categorized as independent, dependent, or confounding variables.

The **independent variable** is a variable that is set at the outset of the experiment. It does not change as a result of the experimental procedure. Thus, it is said to be *independent* of the experimental procedure. In this example, time is the independent variable. Note that it is impossible for time to change as a result of any experimental procedure. The **dependent variable** is the unknown variable that is going to be measured. In formulating a hypothesis, the dependent variable is often expected to change as a result of the experimental procedure. Thus, it is said to be *dependent* on the experimental procedure. In the current example, body temperature is the dependent variable. **Confounding variables** are any variables that may affect the variable of interest. In the current example, the only variable of interest that may affect body temperature is time of the night. Thus, all other variables that may affect the measured variable must be controlled. Examples of some confounding variables are the amount and type of clothing the subject is wearing, the type of bedding the subject uses, and how much and what the subject eats or drinks before going to sleep. In setting up an experiment, great efforts are made to control as many confounding variables as possible. Finally, a **control value** with which to compare the measured values of body temperature is required. In this example, a "normal" body temperature of 37 °C is the obvious control value.

Data Collection

Once a controlled experiment has been designed to test the hypothesis, data collection can begin. When a scientist conducts an experiment, he or she begins by performing a statistical test to determine the sample size necessary to get a meaningful result from the experiment. Why? In this example, if body temperature was measured for only one subject, it would be neither reasonable nor appropriate to extrapolate that data to include all individuals, because the person measured may not be typical of most individuals. For experiments conducted in the anatomy and physiology laboratory, the study subjects will likely consist of the students in the class. Thus, the data set will be limited in scope compared to the ideal situation. However, in most cases enough data will have been collected to obtain reasonable results.

Once the sample size has been determined, the next step is to **collect the data.** In this example, body temperature would likely be measured for each of the study subjects at specific time intervals during the night. At the same time, confounding variables would be controlled for by having all study subjects wear the same clothing, use the same bedding, refrain from eating within a certain number of hours before bed, and go to bed at a prescribed time. After data collection comes the fun part: data analysis!

Data Analysis

Once the data has been collected, the experimenter must **analyze the data** in a way that makes sense both to the experimenter(s) and to the rest of the scientific community. There are several ways to present data, and presentation depends somewhat on the variable in question. For any given data point, a mean, median, and standard deviation must be calculated for the value to ensure the value represents the data for the group of study subjects as a whole. The **mean** is the average of all the data points, and is calculated by taking the sum of all the data points and dividing by the number of study subjects. The **median** is the middle value of all the data points. The **standard deviation** is a measure of the variability of the individual data points as compared to the mean. When the standard deviation is small, it means that individual data points are all very close to the mean. When the standard deviation is large, it means there is much variability between individual data points and the mean.

For the purposes of the exercises in this manual, standard deviations will not be reported for experimental data because this generally requires using a computer program to perform the calculations. An easier way of determining variability is to calculate and report the *range* of values. The **range** is simply the difference between the highest and lowest values, and is calculated by subtracting the lowest value from the highest value. If the range is small, it indicates that there is very little variability of individual data points as compared to the mean. If the range is large, it indicates that there was quite a bit of variability of individual data points as compared to the mean. **Table 1.1** is a table showing hypothetical temperature data for five study subjects taken at two different times (12 a.m. and 6 a.m.). The mean, standard deviation, and range of the data are also shown in the table. Notice that even though the standard deviation and range are different for each time, they have the same pattern. That is, for the data taken at 6 a.m. both the standard deviation and the range are higher than for the data taken at 12 a.m. Thus, both of these measures demonstrate there was greater variability in the individual temperature values at 6 a.m. than there was at 12 a.m.

Conclusions

The final step in the scientific method involves **drawing conclusions** based on the results of the experiment. This requires reviewing the hypothesis in light of the data collected. In this example, the hypothesis was that body temperature would change over the course of the night.

Table 1.1 Body Temperature Data for Five Study Subjects

Study Subject	Body Temperature at 12 a.m. (°C)	Body Temperature at 6 a.m. (°C)
Subject A	36.2	35.0
Subject B	36.8	34.0
Subject C	37.2	34.3
Subject D	37.0	35.9
Subject E	36.5	36.0
Mean	**36.7**	**35.0**
Standard Deviation	**0.4**	**0.9**
Range	**1.0**	**2.0**

Figure 1.1 is a sample graph that is based on data from actual studies that looked at the variation in body temperature of a large number of study subjects during a daily cycle. Thus, it can be used as an estimate of data that might have been obtained had the hypothetical experiment been performed. Based on the results, a natural conclusion is that body temperature does change as a result of time of the night. Furthermore, it is possible to be more specific and say *how* body temperature changes during the night. That is, body temperature appears to decrease steadily throughout the night and reaches its lowest point just before waking (5 a.m.).

Often a hypothesis and related experiment will result in a few answers, but even more questions. For example, after concluding that body temperature changes during the night, a follow-up question may be asked: *why* does body temperature change during the night? Using the scientific method, the scientific process can continue through the formulation of a new hypothesis. The new hypothesis will lead to further experimentation, data collection, and analysis. As the process continues more and more details concerning the area of interest emerge.

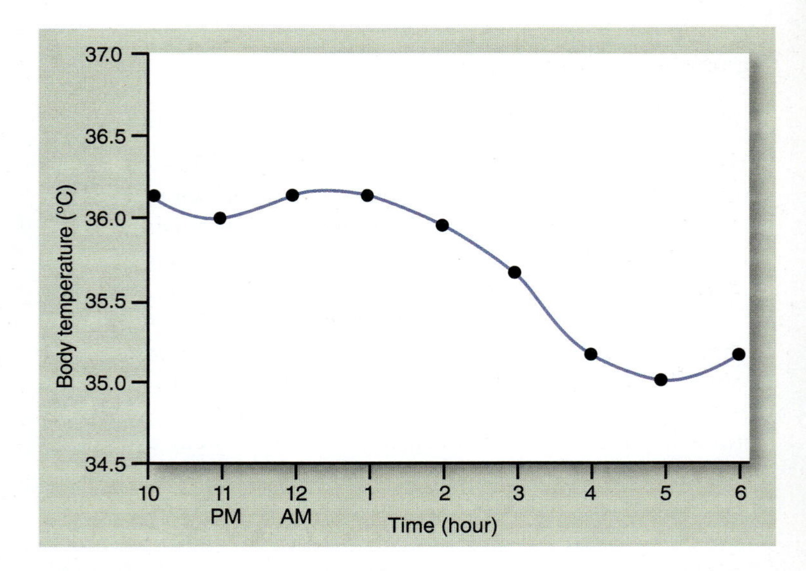

Figure 1.1 Temperature Variance During the Course of the Night. Each data point represents the mean for a number of study subjects.

EXERCISE 1.1

THE SCIENTIFIC METHOD

Let's say that you have observed that heart rate varies with activity level. That is, when an individual exercises heart rate is higher, and when an individual is at rest heart rate is lower. Now, let's say that because you have observed that heart rate is lower when an individual is in a "rested" state, you are now interested in knowing if meditation, an activity that is designed to put the body in a calm/rested state, causes an individual's heart rate to decrease. You decide to perform a scientific experiment to figure out if this is accurate.

1. State your hypothesis: ______________________

2. What is the independent variable? ______________________

3. What is the dependent variable? ______________________

4. Design an experiment to test your hypothesis. In the explanation of your experimental design, describe what variable(s) you will control and what variable(s) you will measure.

5. Finally, describe any confounding variables that may affect the results of the experiment and explain how you might control for such variables.

EXERCISE 1.2

PRESENTING DATA

Once experimental data has been collected, it must be presented in a meaningful way so that others who view the data will be able to draw their own conclusions about the data. In the previous description of the scientific method, data was presented using a data table (table 1.1) and a graph (figure 1.1). These are two of the most common modes of data presentation, and are the modes most often used to present data for the experiments performed in this manual. When presenting data using a **table,** it is common to include mean, number of study subjects, and standard deviation (or range) for each data point (see table 1.1 for an example). When presenting data using a **graph,** there are conventions to follow. Namely, the *independent* variable is plotted on the X-axis (horizontal) and the *dependent* variable is plotted on the Y-axis (vertical). Graphing data this way allows the individual reading the graph to determine what effect, if any, variable X has on variable Y. For the sample experiment looking at the effect of time of the night on body temperature, the data have been plotted in the graph shown in figure 1.1.

1. Why was time plotted on the X-axis? ________________

____

2. Why was temperature plotted on the Y-axis? __________

____

Other items to note when creating a graph are the following:

1. Each axis must be labeled with the appropriate numbers indicating the value of the measurement (in the graph in figure 1.1 these are the measured values for "hour" and "degrees Celsius").
2. Each axis must have a legend and provide the units that are used. In the graph in figure 1.1, the X-axis is labeled "Time" and the unit is "hour"; the Y-axis is labeled "Temperature" and the units are "degrees Celsius."

It is very important to always take care to place labels on the graphs that are created. A common mistake many students make is to assume that their instructor—or whoever is grading their laboratory report—already knows what is supposed to be graphed, therefore the student does not need to put the units on the graph. Do not make this mistake! Always assume that the person who is going to read a graph has no idea what the graph is trying to present. Failing to put units on a graph results in a graph that means nothing. For example, let's say the label "Time" is on the X-axis, but there are no units provided. The reader of the graph is left to wonder if the intervals indicate time in seconds, minutes, hours, or some other unit. Likewise, failing to indicate the units for "Temperature" on the Y-axis leaves the reader wondering if the temperature is given in Celsius (°C) or Fahrenheit (°F).

Sample Graphing Exercise

1. **Table 1.2** represents experimental data obtained by rolling a bowling ball on a track and measuring how far the ball rolls over time.

Table 1.2 Distance Traveled by a Bowling Ball

Time	Distance Traveled
(sec)	(meters)
0	0
1	6
2	8
3	9.5
4	10.5
5	11
6	11.5
7	11.75
8	11.8
9	11.9
10	12

Questions:

a. What is the independent variable? ________________

____

b. What is the dependent variable? ________________

____

2. Using the data in table 1.2, graph the data on the grid provided. Be sure to label the axes and provide the appropriate units.

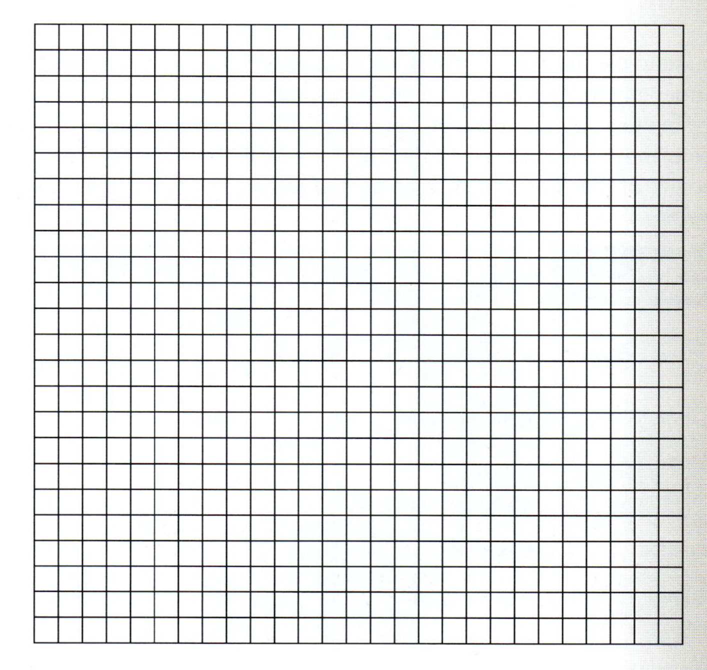

(continued on next page)

(continued from previous page)

3. Now that the data have been graphed, write a brief paragraph in the space provided explaining the results.

Measurement in Science

Systems of Measurement

If a person has ever baked a cake or done another activity that required making measurements, common units of measurement such as cups, gallons, teaspoons, liters, and the like should be familiar. There are two systems of measurement that are most commonly used in the world: English and metric systems. Because of its uniformity and ease of use, the **metric system** is the system most widely used throughout the world. It is also the system of measurement that scientists use when reporting data. Thus, when performing the laboratory activities in this manual, students will be reporting results using metric units.

The metric system is relatively simple to use because all units are given as powers (multiples) of ten. To calculate a power of ten, simply take the superscript (i.e., 10^x, where x is the superscript) and multiply ten by ten that many times. Thus, $10^2 = 10 \times 10 = 100$; $10^3 = 10 \times 10 \times 10 = 1000$. The only unusual numbers are 10^0 and 10^1. When you see these numbers, just remember that ten to the zero power is 1 and ten to the 1st power is ten. In the metric system, specific prefixes denote the power of ten that is used in a measurement. For example, a kilometer is 10^3 meters, and thus represents 1000 meters. **Table 1.3** lists common metric prefixes and the power of ten that each represents.

When converting one metric unit to another, move the decimal point to the right or to the left, depending on if the conversion is from a larger unit to a smaller unit (e.g., kilograms to grams) or vice versa (e.g., grams to kilograms). When converting from a larger unit to a smaller unit, move the decimal to the *right* the same number of units as the power of ten. This is the same as *multiplying* the larger measurement by the appropriate factor of ten. For example, to convert kilometers to meters (1 km = 1000 meters), move the decimal point to the right *three* positions because a kilometer is one meter times ten to the *third* power (10^3). Conversely, when converting from a smaller unit to a larger unit, move the decimal to the *left* the same number of units as the power of ten. This is the same as *dividing* the smaller measurement by the appropriate factor of ten. For example, to convert meters to millimeters (1 mm = 1/1000 meter), move the decimal point to the left *three* positions because a millimeter is one meter times ten to the negative *third* power (10^{-3}). Another way to remember this is that a power of ten that is positive moves the decimal to the right (forward) and a power of ten that is negative moves the decimal to the left (backward).

small unit → larger unit MULTIPLY by appropriate power of ten

larger unit → smaller unit DIVIDE by appropriate power of ten

Practice:

Convert 457 milligrams to grams

Convert 5698 centimeters to decimeters

Convert 4.3 kilometers to meters

Convert 0.5 liter to milliliters

Table 1.3 Metrics

Metric Prefix	Meaning	Symbol	Amount	Power of Ten	Length Measure	Mass Measure	Volume Measure	Word Origin
kilo-	one thousand	k	1000	10^3	kilometer (km)	kilogram (kg)	kiloliter (kL)	*chlioi*, a thousand
No prefix/ base unit	Use the standard unit	–	1	10^0	meter (m)	gram (g)	liter (L)	NA
deci-	one-tenth	d	0.1	10^{-1}	decimeter (dm)	decigram (dg)	deciliter (dL)	*decimus*, tenth
centi-	one-hundredth	c	0.01	10^{-2}	centimeter (cm)	centigram (cg)	centiliter (cL)	*centum*, hundred
milli-	one-thousandth	m	0.001	10^{-3}	millimeter (mm)	milligram (mg)	milliliter (mL)	*mille*, thousand
micro-	one-millionth	µg	0.000001	10^{-6}	micrometer (µm)	microgram (µg)	microliter (µL)	*mikros*, small

Metric Conversions

If measurements are made using the English system, the measurements will need to be converted from English units to metric units for laboratory reports. **Table 1.4** lists some common conversions between English and metric measurements. Some of these conversions are conversions that you should be able to make in your head easily. For example, 1 inch = 2.54 centimeters and 1 pound = 0.45 kilogram.

For example, to convert 16 ounces (English measurement) into milliliters (metric measurement), multiply the number of ounces (16) by 30 (see table 1.4) to get milliliters:

$$16 \text{ ounces} \times 30 = 480 \text{ milliliters}$$

Notice that the conversion in table 1.4 tells how to convert ounces to milliliters. What if we want to convert ounces to liters instead? Recall from table 1.3 that one milliliter is one one-thousandth of a liter, and its power of ten is 10^{-3}. Thus, to convert ounces to liters, first convert to milliliters using the multiplier listed in table 1.4, then convert milliliters to liters by moving the decimal point three places to the left: 480.0 milliliters = 0.48 liter.

Temperature Scales

Just as there are two systems of measurement for volume, length, and mass, there are also two systems of measurement for temperature. In the United States most of our temperatures (such as air temperatures reported by weather stations) are reported as degrees Fahrenheit (°F). However, often temperatures will instead be reported as degrees Celsius (°C, also known as centigrade). In science, it is appropriate to report temperature readings (such as body temperature) in degrees Celsius. To make conversions between these two temperature readings, use the following equations:

To convert degrees Celsius to degrees Fahrenheit:

$$°F = ((°C \times 9) / 5) + 32$$

To convert degrees Fahrenheit to degrees Celsius:

$$°C = ((°F - 32) \times 5) / 9$$

Table 1.4 Common English-Metric Conversions

To Convert From:	To:	Multiply:	By:	Conversion
centimeters	inches	centimeters	0.39	1 cm = 0.39 inch
feet	centimeters	feet	30.48	1 foot = 30.48 cm
feet	meters	feet	0.3	1 foot = 0.3 m
fluid ounces	milliliters	fluid ounces	30	1 oz = 30 mL
gallons	liters	gallons	3.78	1 gallon = 3.78 L
grams	ounces	grams	0.035	1 g = 0.035 oz
inches	centimeters	inches	2.54	1 inch = 2.54 cm
kilograms	pounds	kilograms	2.2	1 kg = 2.2 lb
kilometers	miles	kilometers	0.62	1 km = 0.62 mile
liters	quarts	liters	1.06	1 L = 1.06 qt
liters	gallons	liters	0.26	1 L = 0.26 gallon
meters	yards	meters	0.9	1 m = 1.1 yd
meters	feet	meters	3.3	1 m = 3.3 feet
miles	kilometers	miles	1.61	1 mi = 1.61 km
milliliters	fluid ounces	milliliters	0.03	1 mL = 0.03 oz
millimeters	inches	millimeters	0.039	1 mm = 0.039 inch
fluid ounces	grams	fluid ounces	28.3	1 oz = 28.3 g
fluid ounces	milliliters	fluid ounces	29.6	1 oz = 29.6 mL
pounds	grams	pounds	453.6	1 lb = 453.6 g
pounds	kilograms	pounds	0.45	1 lb = 0.45 kg
quarts	liters	quarts	0.95	1 qt = 0.95 L
yards	meters	yards	0.9	1 yd = 0.9 m

EXERCISE 1.3

UNITS OF MEASUREMENT

1. Make each of the following conversions, using tables 1.3 and 1.4, and the temperature conversion equations given in this chapter as references:

   a. 5 millimeters = ___________ centimeters

   b. 21 grams = ___________ ounces

   c. 500 mL = ___________ liters

   d. 5 inches = ___________ centimeters

   e. 3 meters = ___________ feet

   f. 3 gallons = ___________ liters

   g. 3 liters = ___________ mL

2. The set point for human body temperature is 98.6°F. This means that 98.6°F is the temperature set point that homeostatic mechanisms in the body strive to maintain. It is appropriate in science to report temperature values in degrees Celsius. Convert 98.6°F to degrees Celsius in the space provided. Be sure to show all of your work.

Laboratory Equipment

The typical human anatomy and physiology laboratory classroom consists of laboratory tables or benches that provide ample room for use of microscopes, laboratory equipment, and dissection materials. If human cadavers are used in the classroom, there will also be a space dedicated to the tables where the cadavers are stored. When entering the classroom, look around and familiarize yourself with the environment. Pay particular attention to the location of the sinks, eyewash stations, and safety equipment such as first-aid kits and fire extinguishers. The instructor will provide a detailed introduction specific to the laboratory classroom, safety procedures, and accepted protocol. The main purpose of this chapter is to introduce common safety devices and dissection equipment. *Do not* use the information in this chapter as the sole source of information on laboratory safety, as it is not intended to be a safety manual for the laboratory.

Protective Equipment

The human anatomy and physiology laboratory poses few risks, although it is important to be aware of what these are. The main risks are damage to skin or eyes from exposure to laboratory chemicals (covered in the next section) and cuts from dissection tools. As a general precaution, whenever you are working with fresh or preserved specimens (animal or human), wear protective gloves to keep any potentially infectious or caustic agents from contacting your skin. If there is a risk of squirting fluid, then wear protective eyewear (safety glasses or safety goggles). When wearing gloves, be sure to wear the correct size for your hands. If the gloves are too small, they may tear easily. If they are too big, it may be difficult to handle instruments and tissues. When gloves become excessively dirty, remove them and put on a new pair. When removing gloves, start at the wrist and pull toward your fingers, turning the glove inside-out as it is removed. This will prevent any potentially damaging fluids from contacting your skin during removal of the gloves.

When using dissecting tools there is always a risk of cutting yourself or others. First and foremost—*never* wear open-toed shoes to the laboratory. Dissecting tools are sometimes dropped and will cut your feet if you are not wearing protective footwear. When using sharp tools such as scalpels, always be aware of where the scalpel blades are pointing. They should always be pointed away from you *and* away from others in the laboratory. When dissecting, be aware of where others are standing or sitting, and consider the risk posed to yourself and others if your hand were to slip. *Never* put your hands in the dissecting field when someone else is dissecting. If another person asks for assistance holding tissues while dissecting, use forceps or some other device to hold the tissue so your hands are not within reach of the scalpel blade. Always be aware of the location of any scalpels, particularly when they are not in use. Individuals can be cut accidentally when reaching into a dissecting tray or table to unexpectedly find a scalpel sitting there. Remove dissecting pins from the specimen once the dissection is complete, so that the pins will not poke unsuspecting individuals.

EXERCISE 1.4

IDENTIFICATION OF COMMON DISSECTION INSTRUMENTS

Several dissection instruments are commonly found in the human anatomy and physiology laboratory classroom. **Table 1.5** describes each of these instruments and their uses.

1. Obtain a dissection kit from the laboratory instructor, or use your own dissection kit if you were required to purchase your own materials.
2. Identify the instruments listed in **figure 1.2**, using table 1.5 as a guide. Then label figure 1.2.

Table 1.5 Common Dissection Instruments

Tool	Description and Use	Photo	Word Origin
Blunt Probe	An instrument with a blunt (not sharp) end on it. It is used to pry and poke at tissues without causing damage. Some probes come with a sharper point on the opposite end that can be used for "picking" at tissues.		*proba*, examination
Dissecting Needles	Long, thick needles that have a handle made of wood, plastic, or metal. These needles are used to pick at tissues and to pry small pieces of tissue apart.		*dissectus*, to cut up
Dissecting Pins	"T" shaped pins that are used to pin tissues to a dissecting tray, thus allowing a particular area to be seen more easily		*dissectus*, to cut up
Dissecting Tray	Metal or plastic tray used to hold a specimen. The tray is filled with wax or plastic. The wax and/or plastic is soft enough to pin tissues to.		*dissectus*, to cut up
Forceps	Resemble tweezers, and are used for holding objects. Some are large and have tongs on the ends that assist with grabbing tough tissues. Some are small and fine (needle-nose) for picking up small objects. Forceps may also be straight-tipped or curve-tipped.		*formus*, form + *ceps*, taker
Hemostat	In surgery these are used to compress blood vessels and stop bleeding (hence the name). For dissection they are useful as "grabbing" tools. The handle locks in place, which allows you to pull on tissues without causing fatigue in hand and forearm muscles.		*haimo-*, blood + *statikos*, causing to stop

(continued on next page)

(continued from previous page)

Table 1.5	Common Dissection Instruments (*continued*)		
Scalpel	A sharp cutting tool. Generally the blade and the blade handle will be separate, except when using a disposable scalpel. See specific directions in the text regarding proper use of a scalpel, as they can be dangerous!	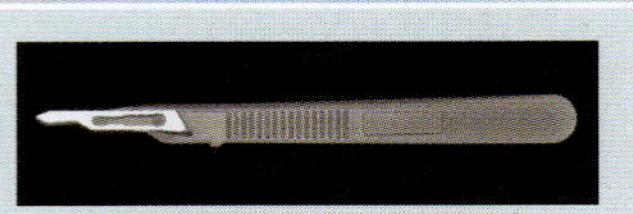	*scalpere,* to scratch
Scalpel Blade	Both the cutting part and the disposable part of a scalpel. The number of the blade indicates the size of the blade, and must be matched with an appropriately numbered blade handle. When a blade becomes dull, it may be removed and replaced with a new blade. Used blades must be disposed of in a sharps container.	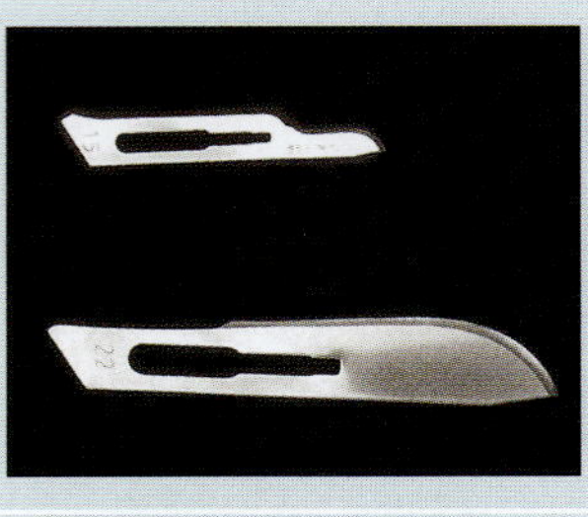	*scalpere,* to scratch
Scalpel Blade Handle	The nondisposable part of a scalpel that is used to hold the blade. The number on the handle indicates the size of the handle and is used to match it with a particular blade size. A scalpel blade handle can be a very useful tool for blunt dissection when used *without* a blade attached.	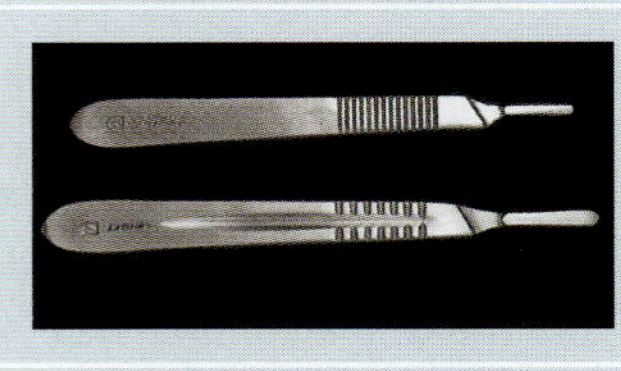	*scalpere,* to scratch
Scissors	Some scissors come with pointed blades and some have one curved (blunt) and one pointed blade. Scissors with the curved/blunt edge are used when you need to be careful not to damage some structures. To use them, direct the curved blade toward the structures you do not want to damage. Pointed-blade scissors are particularly helpful for using "open scissors" technique (see text).	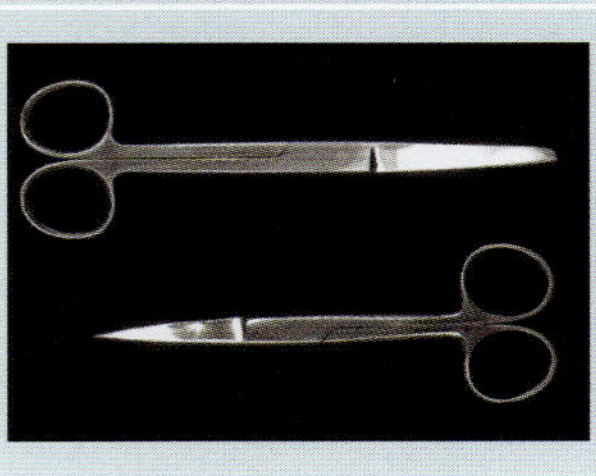	*scindere,* to cut

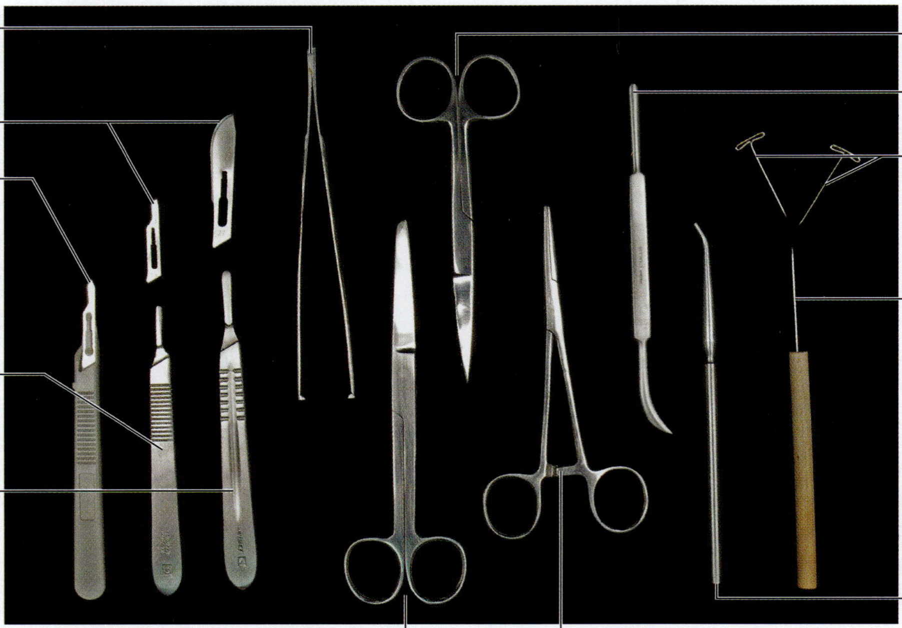

1 ____
2 ____
3 ____
4 ____
5 ____
6 ____
7 ____
8 ____
9 ____
10 ____
11 ____
12 ____

Figure 1.2 Identification of Common Dissection Instruments. Use the terms below to fill in the numbered labels in the figure. Answers may be used more than once.

- ☐ blunt probe
- ☐ dissecting needle
- ☐ dissecting pins
- ☐ forceps
- ☐ hemostat
- ☐ scalpel (disposable)
- ☐ scalpel blade handle (#3)
- ☐ scalpel blade handle (#4)
- ☐ scalpel blades
- ☐ scissors (curved)
- ☐ scissors (pointed)

Hazardous Chemicals

Various chemicals are used in the human anatomy and physiology laboratory. The discussion in this chapter will address chemicals that are used to preserve, or "embalm," animal specimens or human cadavers. Safety precautions for chemicals used in physiology "wet lab" experiments will be covered when these chemicals are encountered within specific laboratory exercises. For embalming chemicals, most are not in their full-strength form in the laboratory. Instead, most tissues and specimens encountered in the laboratory will have been previously injected with solutions containing these chemicals. Thus, safety measures in the laboratory are designed to protect users from the forms of these chemicals that are most likely to be encountered. The most common chemicals used for embalming purposes are formalin, ethanol, phenol, and glycerol.

Table 1.6 summarizes the uses and hazards of these chemicals. The majority of these chemicals are used to fix tissues and prevent the growth of harmful microorganisms, such as bacteria, viruses, and fungi. **Fixation** refers to the ability of the chemical to solidify proteins, thus preventing their breakdown. **Preservatives** both fix tissues and inhibit the growth of harmful microorganisms. Because most preservatives also dehydrate tissues, **humectants** are added to embalming solutions. Humectants, such as glycerol, attract water. When humectants act alongside preservatives, they help keep tissues moist. Other chemicals that may be added to embalming solutions are pigments, which either make the tissues look more natural or mask the odors of the preservative chemicals. **Formalin** and **phenol** are the most toxic and odoriferous preservative chemicals. Luckily, exposure to them in the anatomy laboratory will be very low. Although it may smell as if the concentrations of these chemicals are high, the odor is often misleading because these chemicals can be detected by odor in extremely small quantities. Although the concentrations of formalin and phenol that users are exposed to may be very low, if these chemicals have been used to preserve specimens, then protective clothing is required to prevent the chemicals from contacting the user's skin or eyes. Use gloves whenever handling the specimens, and use protective eyewear whenever there is a risk of chemicals getting into your eyes. If your skin is exposed, rinse it immediately. If your eyes are exposed, use the eyewash station in the laboratory to rinse your eyes thoroughly. If you have experienced contact exposure to these chemicals and your skin or eyes continue to be irritated after rinsing, consult a medical doctor.

Table 1.6 Preservative Chemicals Encountered in the Human Anatomy & Physiology Laboratory

Chemical	Description	Use	Hazard	Preventing Exposure	Disposal
Ethanol	Inhibits growth of bacteria and fungi	Preservative	Flammable, so requires storage in a fire-safe cabinet. Generally safe in small quantities.	Gloves and eye protection. Rinse tissues immediately if exposed, particularly eyes. Seek medical attention if irritation persists.	Small amounts may be flushed down the sink along with plenty of water to dilute the solution.
Formalin	Fixes tissues by causing proteins to cross-link (solidify). Destroys autolytic enzymes, which initiate tissue decomposition. Inhibits growth of bacteria, yeast, and mold.	Preservative	Flammable, so requires storage in a fire-safe cabinet. Toxic at full strength. Penetrates skin. Corrosive. Burns skin. Damages lungs if inhaled. May be carcinogenic.	Gloves and eye protection. Rinse tissues immediately if exposed, particularly eyes. Seek medical attention if irritation persists.	Do not pour into sinks.
Glycerine (glycerol)	Helps control moisture balance in tissues. When used with formalin, it counteracts the dehydrating effects of formalin.	Humectant	Flammable, so requires storage in a fire-safe cabinet. Generally safe. Can pose a slipping hazard if spilled on the floor.	Gloves and eye protection. Rinse tissues immediately if exposed, particularly eyes. Seek medical attention if irritation persists.	Small amounts may be flushed down the sink along with plenty of water to dilute the solution.
Phenol	Assists formalin in fixing tissues through protein solidification. Inhibits growth of bacteria, yeast, and mold.	Preservative	Flammable, so requires storage in a fire-safe cabinet. Extremely toxic at full strength. Rapidly penetrates the skin. Corrosive. Burns skin. Damages lungs if inhaled. NOTE: When used as embalming preservative, concentration (and thus toxicity) is extremely low.	Gloves and eye protection. Rinse tissues immediately if exposed. Use an eye wash station if solution gets in the eyes. Seek medical attention if irritation persists.	Do not pour into sinks.

Proper Disposal of Laboratory Waste

There are several types of waste that must be disposed of in the human anatomy laboratory. Much of this waste is "normal" waste, such as tissues, paper towels, or rubber gloves. Such waste should be disposed of in the regular garbage/waste container found in the classroom. However, any potentially **hazardous waste** must be disposed of in a special container. The general rule for determining if something is potentially hazardous or not is this: If you think someone else may be injured *in any way* from handling this waste, it is hazardous. Follow this rule, and be sure to ask the instructor how to properly dispose of something any time there is a question as to whether it is hazardous or not. It is always better to err on the side of caution.

What is hazardous waste?

1. Any sort of fresh tissue and/or blood
2. Laboratory chemicals
3. Broken glass, scalpel blades, or any other sharp item that may cut an individual who handles the waste

Sharps Containers

Sharps containers (figure 1.3) are plastic containers (often red or orange) that are used to dispose of anything "sharp," such as needles, scalpel blades, broken glass, pins, or anything else that has the potential to cut or puncture a person who handles it. Such items should NEVER go in the garbage, because they may injure anyone who handles the garbage thereafter. When in doubt, put it in the sharps container.

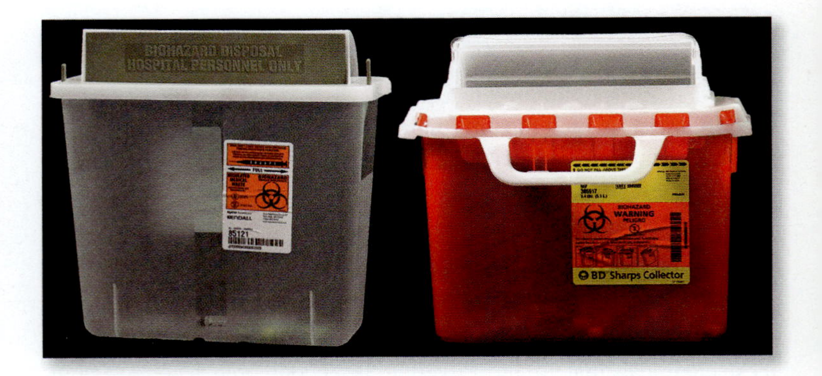

Figure 1.3 Sharps Containers. Samples of two different models of sharps containers. Such containers allow one to place sharp objects into the container, but the objects cannot be removed once placed inside. Note the biohazard warning symbol on the containers.

Biohazard Bags

Special **biohazard bags** may be available in the laboratory. These are used for biological material such as blood or other fresh animal tissue that requires special disposal. When it comes to human blood, an item containing a small amount of blood (such as a Band-aid™) can be disposed of in a normal wastebasket. However, if a towel is soaked with blood, then it must be disposed of in a biohazard bag. A biohazard bag is usually red or clear and has the symbol shown in **figure 1.4** on it. When dealing with tissues that must be disposed of in a biohazard bag, the instructor generally will provide disposal instructions. Again, when in doubt, always ask before disposing of something potentially hazardous. Important note: Human cadaveric tissues do *not* go into biohazard bags. They must be kept with the cadaver. Any piece of human tissue removed from a cadaver must eventually be returned to the cadaver to be cremated with the entire body.

Figure 1.4 Biohazard Waste Symbol.

EXERCISE 1.5

PROPER DISPOSAL OF LABORATORY WASTE

Circle the letter (a, b, or c) of the correct waste receptacle (shown in **figure 1.5**) for each item listed below.

1.	broken scissors	A	B	C
2.	cotton swab	A	B	C
3.	dissecting pins	A	B	C
4.	glass slide	A	B	C
5.	paper towel	A	B	C
6.	rubber glove	A	B	C
7.	scalpel blades	A	B	C
8.	fresh tissue	A	B	C
9.	hypodermic needle	A	B	C

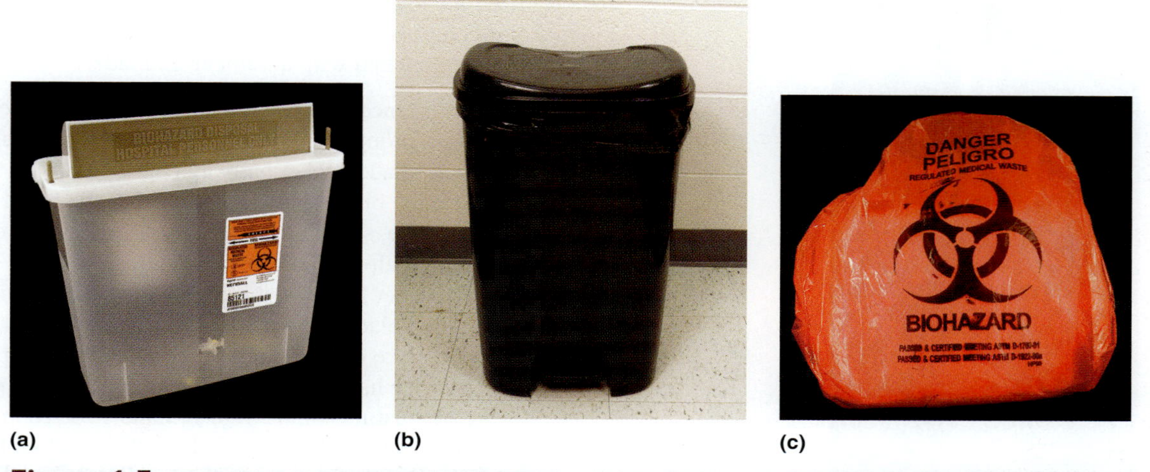

(a) (b) (c)

Figure 1.5 Common Waste Receptacles in the Laboratory. (a) Sharps container. (b) Wastebasket. (c) Hazardous waste bag.

Dissection Techniques

The word *dissect* literally means to cut something up. Most individuals have been led to think that the first thing a surgeon or anatomist does when planning to dissect is to pick up a scalpel and cut. However, skilled dissection does not always involve actually cutting tissues. In fact, the dissector's best friend is a technique called "blunt dissection." Blunt dissection specifically involves separation of tissues without using sharp instruments (hence the term *blunt*). When dissecting tissues, always try using blunt dissection before picking up sharp instruments such as scissors and scalpels. Sharp instruments are very handy—as they are good at cutting things. However, often students will end up cutting many things they do not wish to cut, purely by accident. Thus, being sparse and prudent in the use of sharp tools is one of the most important tips for performing a good dissection.

For this exercise, the demonstration of techniques will be shown using a fresh chicken purchased from a grocery store. However, the instructor may choose another specimen for you to practice on. For now, the goal of the dissection is to separate the skin from the underlying tissues such as bones and muscle (the "meat") of the specimen.

Sharp Dissection Techniques

Sharp dissection techniques are the techniques most familiar to most individuals. These techniques involve the use of sharp instruments such as scissors and scalpels. They are "cutting" techniques. They are advantageous in that tough tissues may be easily separated from each other, or tissue pieces may be removed from a dissection specimen. The danger in using sharp techniques is that novice and experienced dissectors alike will often end up cutting things they do not wish to cut, such as blood vessels and nerves. Thus, sharp dissection techniques should be used with care.

EXERCISE 1.6

PLACING A SCALPEL BLADE ON A SCALPEL BLADE HANDLE

Scalpels come in many forms. Some are of the disposable type, which typically means that the handle and blade come as one unit and the handle is made out of plastic **(figure 1.6).** Often the blades and handles are separate items. Such items allow the blade to be replaced whenever it becomes dull from use. This exercise covers how to properly place a scalpel blade on a scalpel blade handle, and how to properly remove the blade once finished.

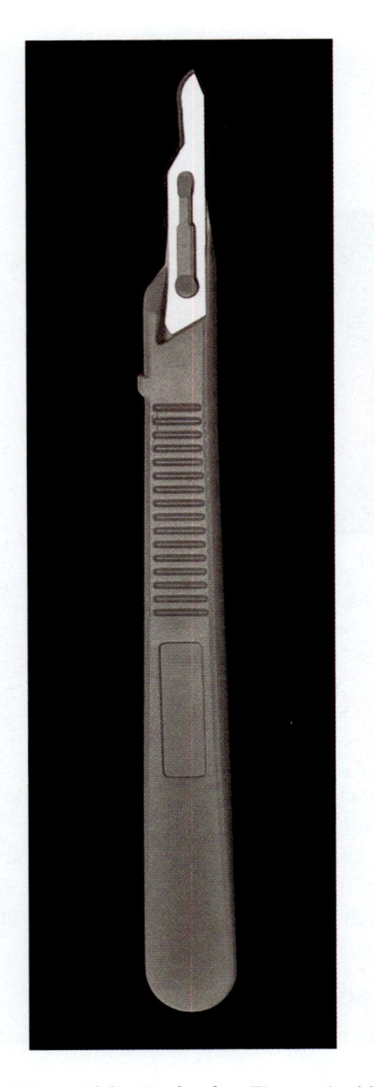

Figure 1.6 Disposable Scalpel. The scalpel blade and scalpel blade handle are both disposable. The entire unit must be disposed of in a sharps container.

1. Obtain a **scalpel blade** and **scalpel blade handle** from the instructor. Scalpel blades and handles come in various sizes, and it is important to match the size of the blade to the size of the blade handle. Observe the scalpel handle and look for a number stamped on it, which will be a 3 or a 4 **(figure 1.7*a*).** Next, observe the blade packet and note the number on it (figure 1.7*b*). A number 3 handle is used to fit number 10, 10A, 11, 12, 12D, and 15 blades. A number 4 handle is used to fit number 18, 20, 21, 22, 23, 24, 24D, and 25 blades. Larger handles and blades are generally used for making bigger, deeper cuts, whereas the smaller handles and blades are generally used for finer dissection. One of the most commonly used combinations in anatomy laboratories is a number 4 handle matched with a number 22 blade.

2. Once the scalpel handle and blade size are properly paired, carefully open the scalpel blade packet halfway **(figure 1.8*a*-1).** Note the bevel on the blade. This bevel matches the bevel on the blade handle, so that there is only one way to properly place the scalpel blade on the handle. The blade handle has a bayonet fitting that is matched to the opening on the scalpel blade (figure 1.8*a*-2), which will lock the blade in place on the handle. The safest way to place the blade on the handle is to first grasp the end of the blade using **hemostats** (figure 1.8*a*-2; table 1.5). Then, while matching the bevel on the blade to the bevel on the handle, slide the blade onto the handle until it clicks, indicating it is locked in place (figure 1.8*a*-3, *a*-4). If it does not go on easily, check to make sure that the blade has not been placed on the handle incorrectly (example: figure 1.8*b*). Now it is ready for use!

3. The safest way to remove a blade from a handle is to use a device that is both a **blade remover** and a sharps container all in one (an example is shown in **figure 1.9**).

4. If a blade remover is not available, remove the blade using hemostats. Obtain a pair of hemostats. Pointing the blade away from you (but not toward someone else), clamp the part of the blade nearest the handle with the hemostats **(figure 1.10-1).** Once you have a firm grip on the blade, slide it over the bayonet on the handle and away from you until the blade comes off the handle (figure 1.10-2). Using the hemostats, transport the blade to a sharps container and dispose of it in the sharps container (figure 1.10-3).

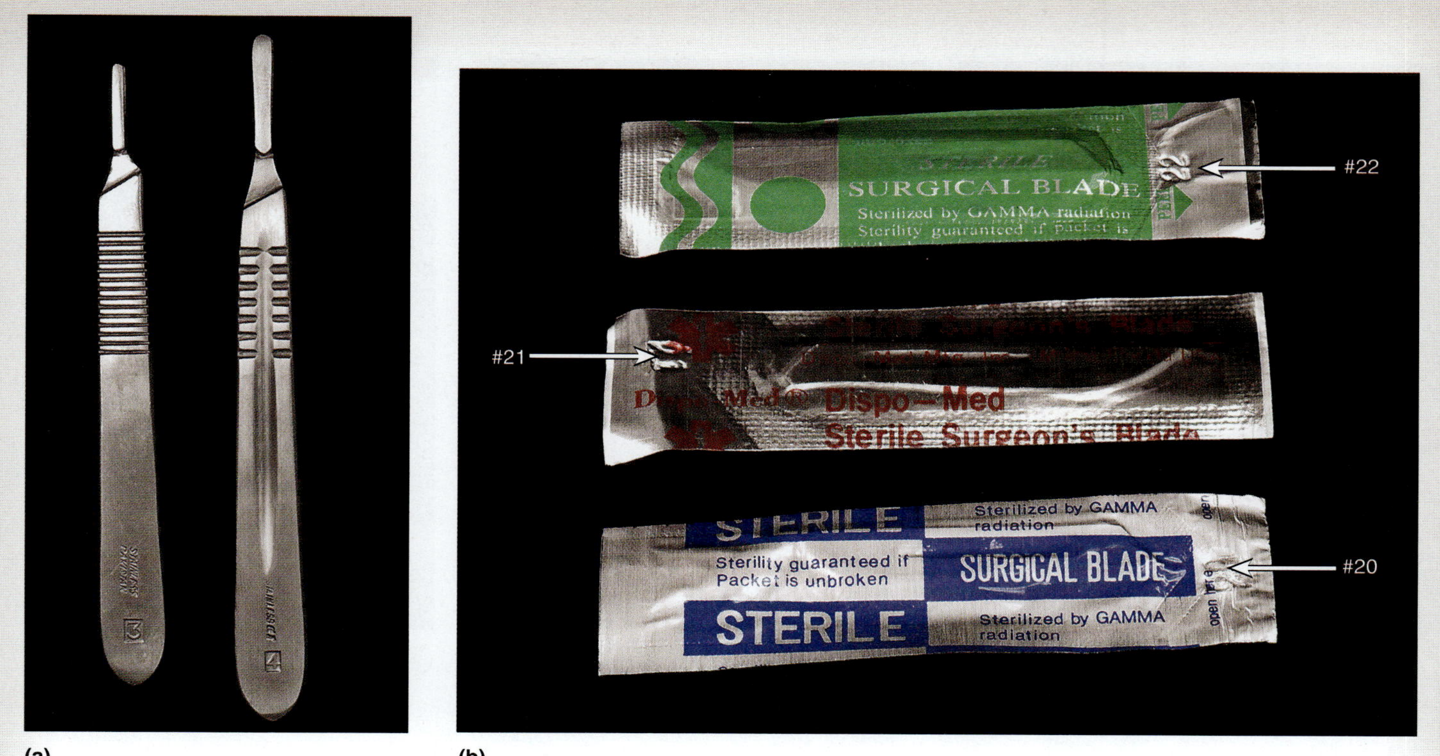

Figure 1.7 **Scalpel Blade Handles and Blades.** (a) The number on the scalpel blade handle indicates what size blades will fit on the handle. (b) The number on the blade wrapper indicates the size of the blade. See text for description of what size blades fit on what size blade handles.

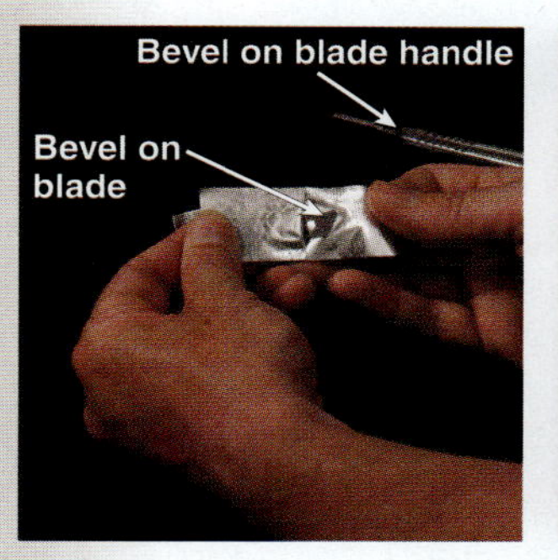

① Open the foil packet and note the bevel on the blade.

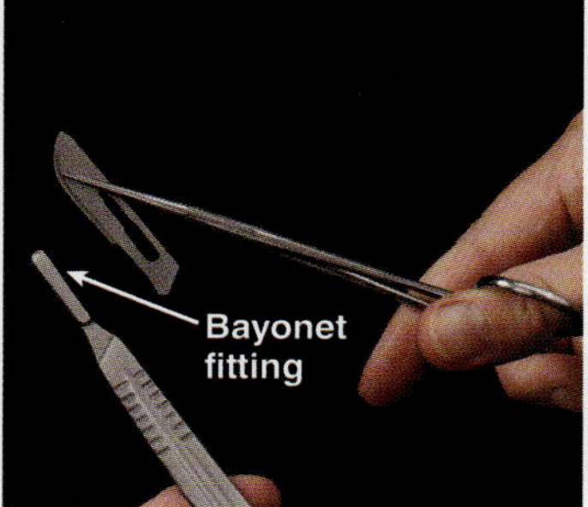

② Grasp the blade firmly using hemostat and line the blade up so that it matches the bevel on the blade handle.

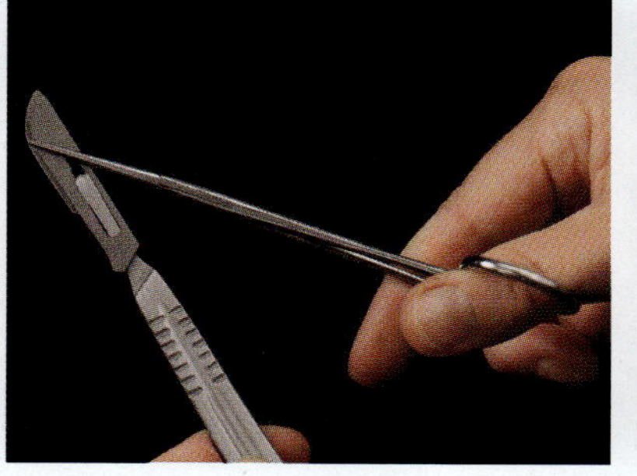

③ Slide the blade onto the bayonet of the blade handle.

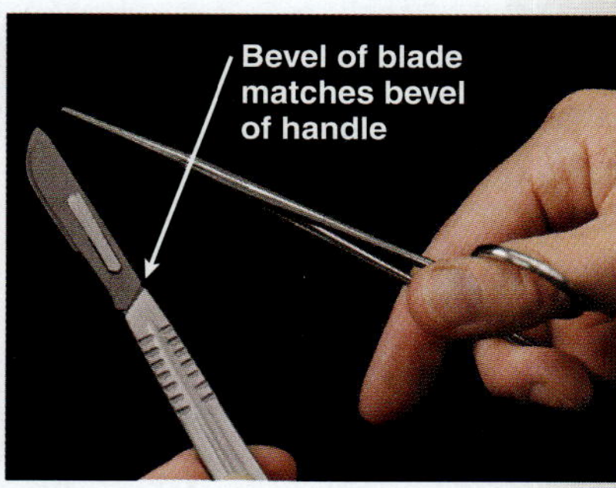

④ The blade should "click" as it locks in place on the blade handle.

(a)

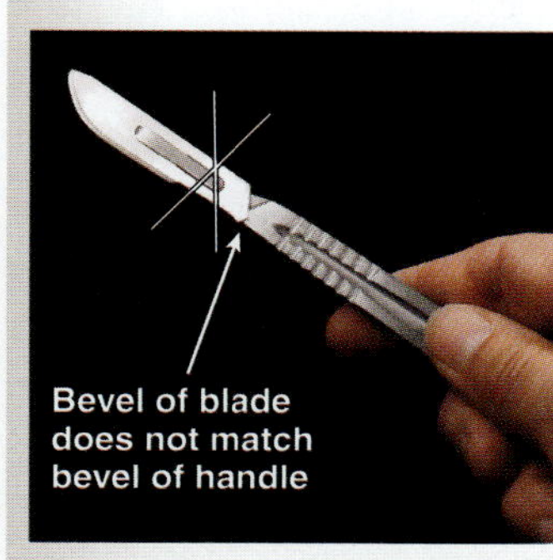

(b)

Figure 1.8 **Scalpel Blade Placement.** (a) Correct procedure. (b) Incorrect placement of a blade on a blade handle. Notice that the bevel on the blade does not match up with the bevel on the blade handle. If placed in this fashion, the blade will not be secure on the handle and may slip off the handle and injure someone.

(continued on next page)

(continued from previous page)

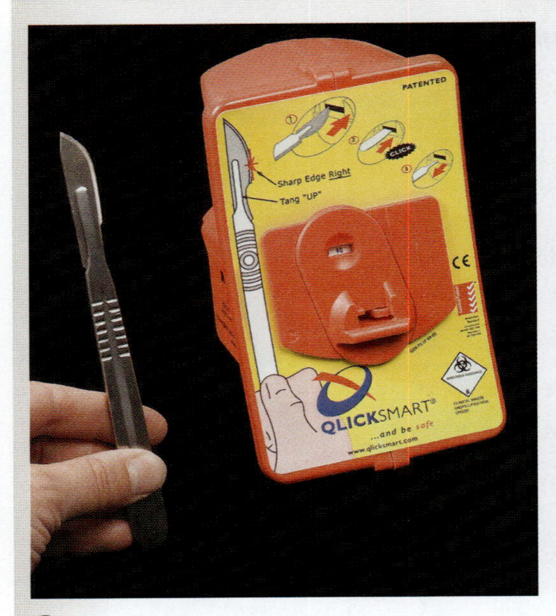

① Orient the blade and blade handle with sharp edge of the blade pointed to the right, as shown on the front of the device.

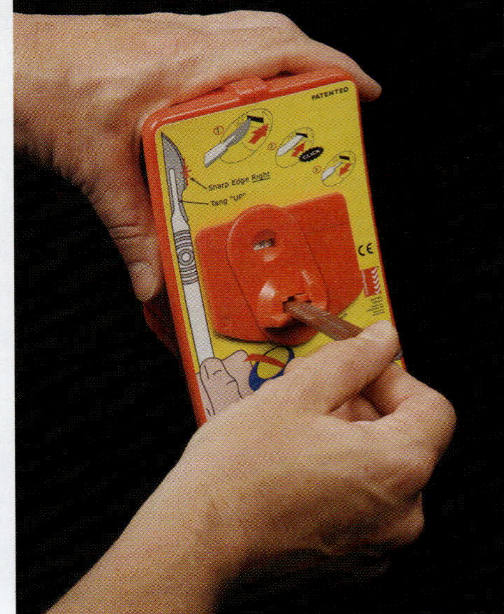

② Push the blade into the slot on the device until a distinct "click" is both heard and felt.

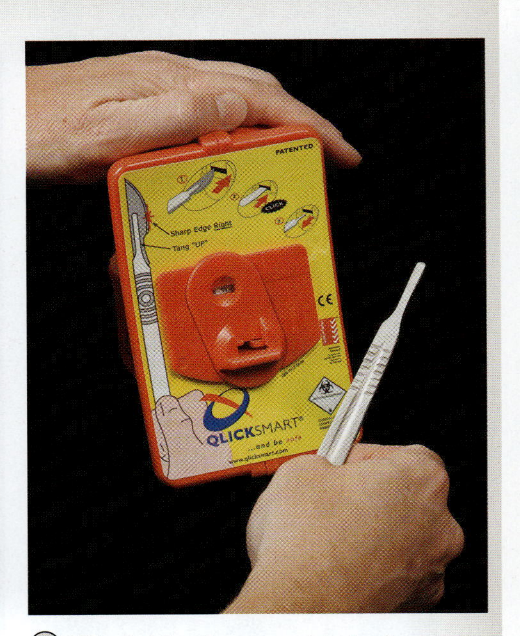

③ While holding the removal device firmly with your free hand, pull the blade handle out of the device.

Figure 1.9 Removal of a Scalpel Blade from Handle Using All-in-One Blade Remover/Sharps Container.

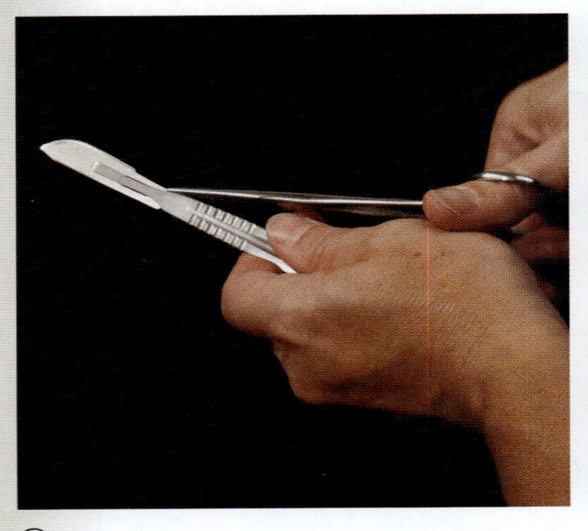

① With the blade pointed away from your body and the bayonet surface of the handle also directed away from your body, grasp the base of the blade with hemostats and lock the hemostats firmly to the blade.

② Slide the blade off of the bayonet on the blade handle. Again, push it away from your body (and away from others in your vicinity as well).

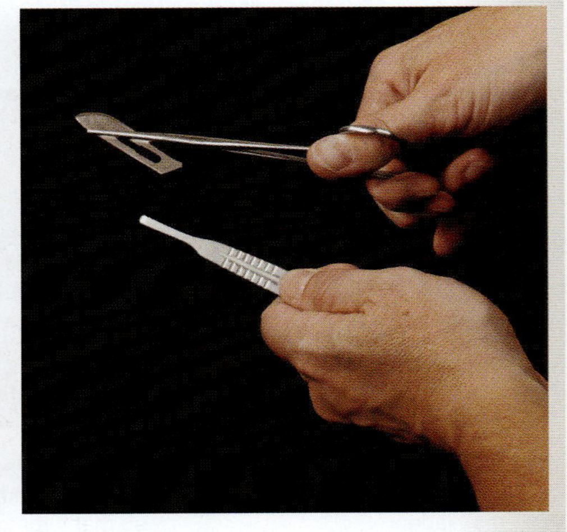

③ Once the blade has been removed from the handle, continue to grasp it firmly with the hemostats. Dispose of the scalpel blade in a sharps container.

Figure 1.10 Removal of a Scalpel Blade from Handle Using Hemostats.

EXERCISE 1.7

DISSECTING WITH A SCALPEL

1. Obtain a dissection specimen and place it on a dissecting tray.
2. Obtain a scalpel with a blade (see exercise 1.4) and some **tissue forceps** (see table 1.5). If the tissue is difficult to grasp use **hemostats** instead of forceps. Hemostats allow you to "lock" on to the tissue so the tissue is not dropped when you release your grip on the handle.
3. Using the forceps or hemostats, pull the skin away from the muscle on the dissection specimen **(figure 1.11-1).** Carefully cut into the skin, using the tip of the scalpel blade (figure 1.11-2). Note how easily a new blade cuts into the tissue. When cutting with a scalpel, take care not to cut too deep, or too aggressively, or underlying tissues will be damaged. Once a small slit has been cut in the skin, observe the stringy tissue that lies between the skin and the muscle. This tissue is a loose connective tissue called fascia (figure 1.11-3), which is discussed further in chapter 5. Because the goal is to separate the skin from the muscle, it is necessary to loosen the "grip" of the fascia that holds the skin and muscle together. One way to do this is to cut into the fascia using the scalpel.
4. Next, *without* holding the skin away from the muscle with forceps, cut into the skin using a considerable amount of pressure. Note how easy it is to cut through the skin directly into the muscle. This is not desirable. To avoid damaging the underlying tissues, push a blunt probe or scalpel handle (*without* blade attached) into the space between the skin and muscle, thus protecting the underlying tissues. Then cut with the scalpel superficial to the probe **(figure 1.12).** This way the probe limits the depth at which the scalpel blade can cut, thus protecting the underlying tissues.
5. Once enough skin has been pulled back for it to be easily grasped with forceps or hemostats, put as much tension on the skin as possible, thus stretching out the fibers in the fascia **(figure 1.13).** Once the fascia is stretched, use the scalpel to cut the fascia and remove the skin from the specimen. When cutting with the scalpel, always point the sharp end of the blade toward the skin, not toward the underlying tissues, so as to protect those underlying tissues.
6. Practice using the forceps, hemostats, blunt probe, and scalpel to remove the skin from part of the specimen. Note areas where this is more difficult than others. When practicing, consider carefully whether the scalpel is the best instrument for the job, or if using it is causing damage to tissues.

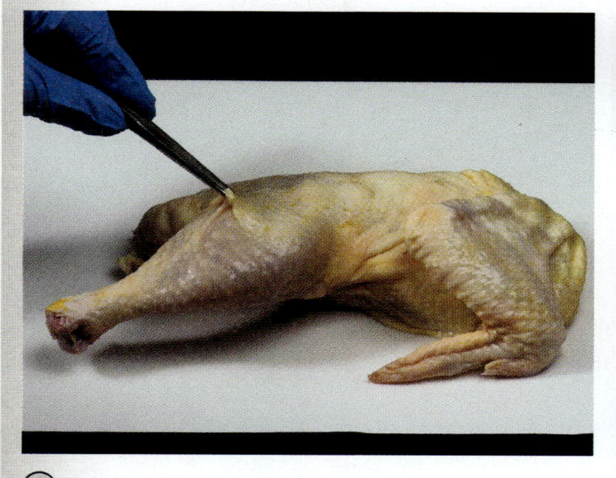

(1) Pull the skin away from the underlying tissues using tissue forceps.

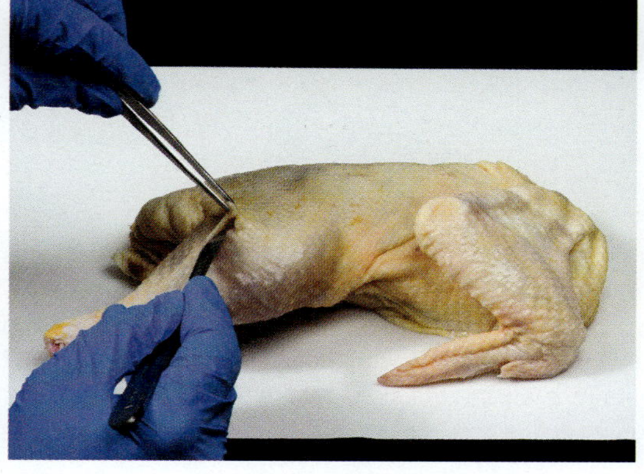

(2) Begin cutting the skin with the scalpel, taking care not to cut delicate tissues deep to the skin.

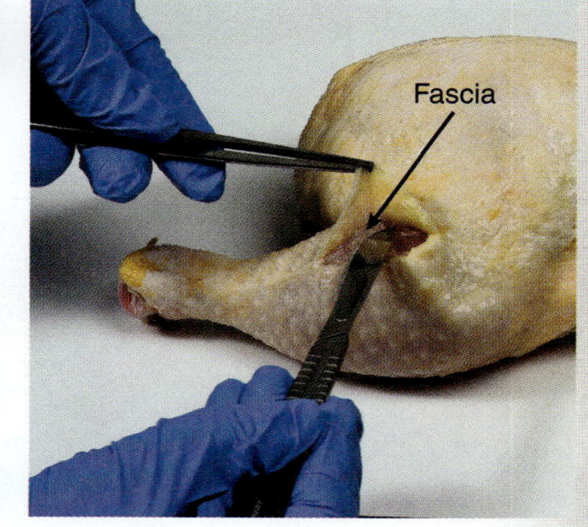

(3) To assist with removal of the skin, use the scalpel to gently cut away the fascia that loosely holds the skin to the muscle. Maintain as much tension on the skin as possible and always keep the sharp end of the blade pointed toward the skin, not the underlying tissues.

Figure 1.11 **Dissecting with a Scalpel.**

(continued on next page)

(continued from previous page)

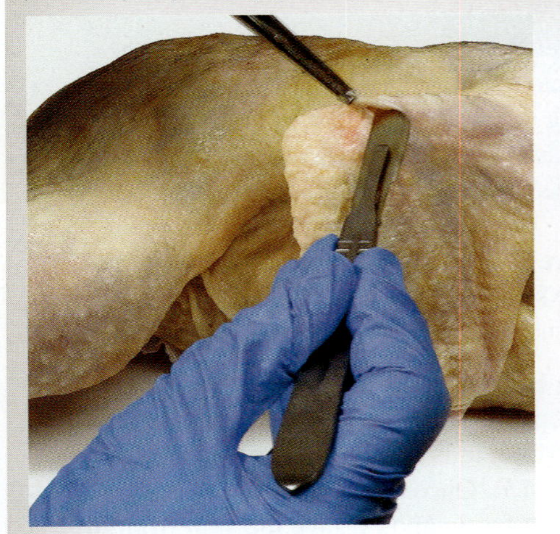

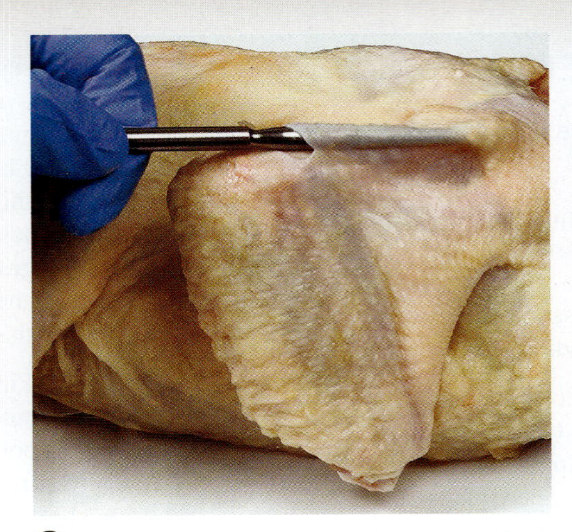

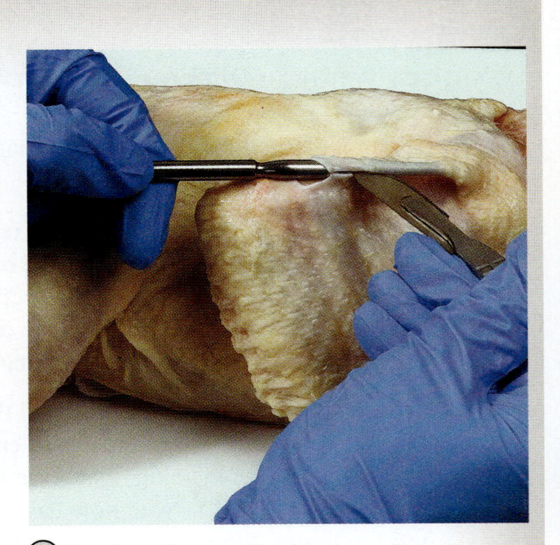

(1) Pull the skin away from the underlying tissues using tissue forceps and cut a small slit in the skin with the scalpel, taking care not to cut delicate tissues deep to the skin.

(2) Push the probe under the skin along the line where the cut will be made.

(3) Cut the skin superficial to the probe with the scalpel. Notice how the blunt probe limits the depth at which the scalpel can cut, thus protecting underlying tissues.

Figure 1.12 **Protecting Underlying Tissues with a Probe When Dissecting with a Scalpel.**

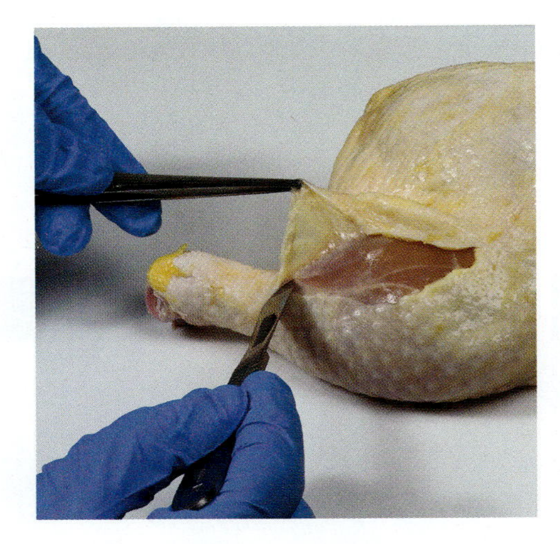

Figure 1.13 **Removing the Rest of the Skin with the Scalpel.** Use the forceps to pull the skin away from the underlying tissues and keep as much tension as possible on the fascia. Cut the fascia with the scalpel, always keeping the sharp part of the scalpel blade pointed toward the skin. This way, if the blade slips accidentally, it will cut the skin, not the underlying tissues.

EXERCISE 1.8

DISSECTING WITH SCISSORS

1. Using the same dissection specimen used in exercise 1.7, practice using scissors to cut tissues.
2. Obtain a pair of **pointed scissors** and **forceps** (table 1.5). Using the forceps, grasp part of the skin covering part of the specimen that has not already been dissected and pull it away from the muscle. Next, cut a small slit into the skin until the fascia beneath it is visible. Continue to lengthen the cut until it is about 2 inches long **(figure 1.14).**
3. **Open scissors technique:** There are a lot of tissues within the fascia that may need to be preserved, such as nerves and blood vessels. When using "sharp" techniques, these structures may accidentally get cut. For this reason, "blunt" dissection technique is preferred to preserve potentially important structures. One blunt dissection technique is called an "open scissors" technique, so named because the dissecting action of the scissors is performed by starting with the scissors closed and then actively opening them. This is the opposite of how most scissors are used.

4. With the scissors closed, push the tip of the scissors into the space between the skin and the muscle so that it pierces the fascia **(figure 1.15-1).** Once the tip of the scissors is within the fascia, open the scissors (figure 1.15-2). Notice how this action causes the fibers within the fascia to separate from each other and loosens the hold between the skin and the fascia. When using the open scissors technique, be sure to keep the scissors open within the specimen. Remove the scissors completely prior to closing the scissors to prevent any unwanted damage to surrounding tissues and structures.

5. Continue to loosen the fascia using the open scissors technique. While doing this, observe small structures such as blood vessels and nerves that may run in the space between the skin and the underlying tissues. Notice how the fibers in the fascia easily separate from each other without damaging the vessels and nerves when using the open scissors technique. At times, the hold of the fascia is too tight, and "open scissors" technique will no longer work effectively. At those times, switch to "normal" scissors technique and simply cut away the tough tissue.

6. Practice using both open and normal scissors techniques to continue to remove skin from underlying tissues.

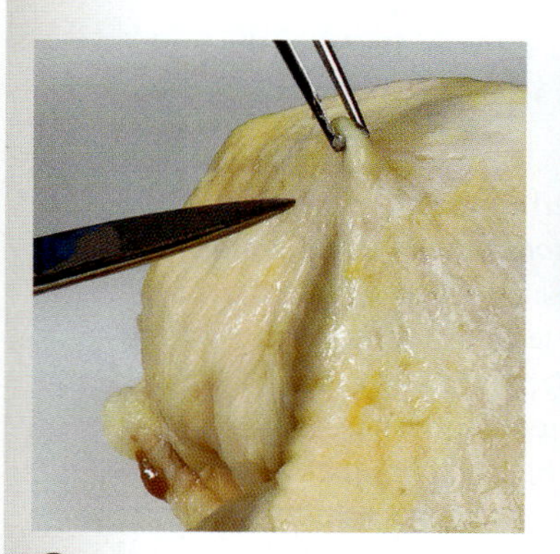

(1) Use tissue forceps to pull the skin away from underlying tissues.

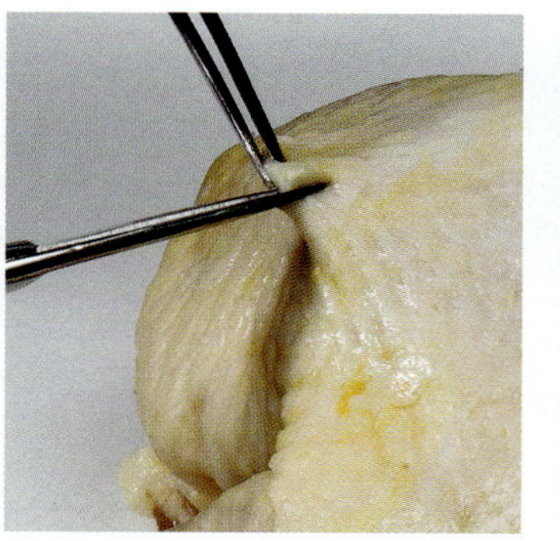

(2) Make a small cut in the skin with the scissors.

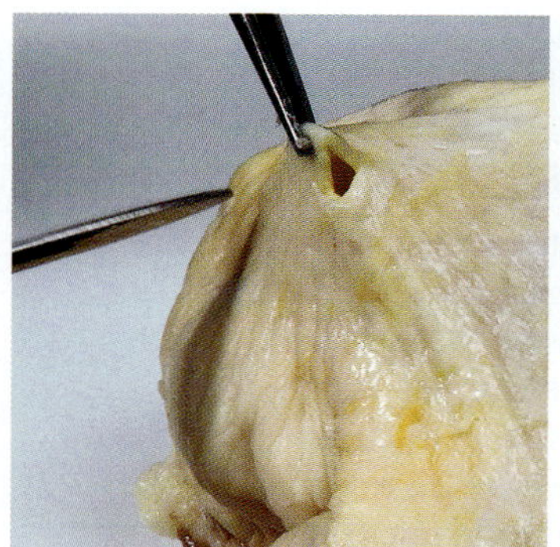

(3) A small hole has now been created in the skin. Insert the tip of the scissors into this hole and begin cutting directionally along the skin.

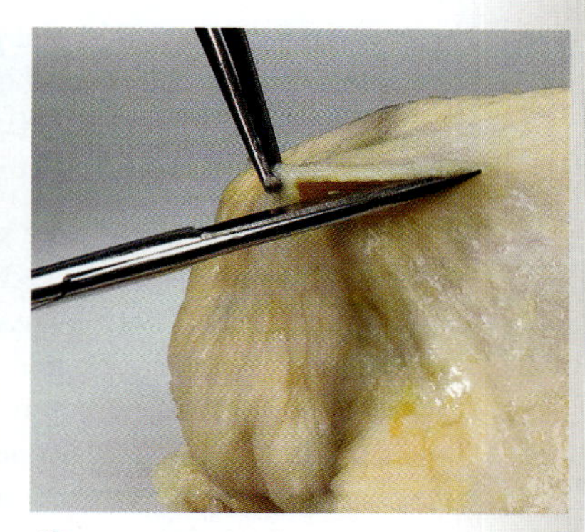

(4) Continue the cut along the skin. Continue to use the tissue forceps to pull the skin away from underlying tissues before making each cut to avoid damaging underlying structures.

Figure 1.14 Dissecting with Scissors.

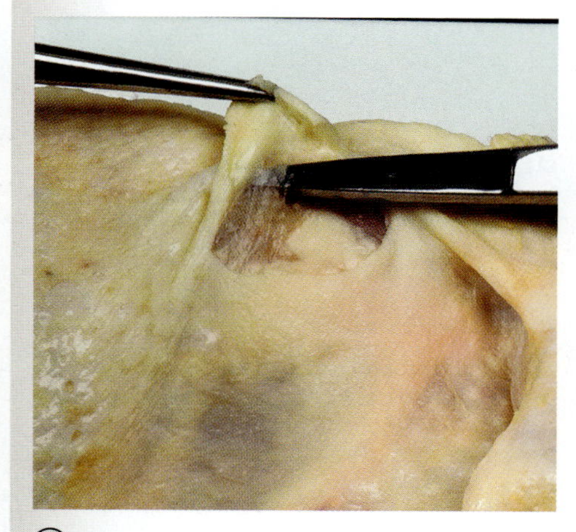

(1) Make a small cut in the skin. Using forceps, pull the skin away from the underlying fascia. Push the tip of the *closed* scissors into the fascia that lies deep to the skin.

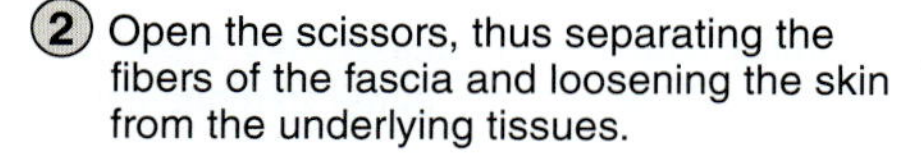

(2) Open the scissors, thus separating the fibers of the fascia and loosening the skin from the underlying tissues.

Figure 1.15 Open Scissors Technique.

Blunt Dissection Techniques

There are times when sharp dissection technique is undesirable, because tissues might be damaged if sharp instruments are used. At these times it is best to switch to blunt dissection techniques. Blunt dissection is designed to separate tissues without damaging delicate structures.

EXERCISE 1.9

BLUNT DISSECTION TECHNIQUES

1. Practice using blunt dissection techniques using the same dissection specimen used in exercises 1.7 and 1.8.
2. Obtain a pair of **pointed scissors** and **forceps** (table 1.5). Using the forceps, grasp part of the skin of the specimen not previously dissected and pull it away from the muscle. Next, make a small cut into the skin until the fascia beneath it is visible **(figure 1.16-1).** This is the "sharp" dissection technique previously described.
3. When attempting to preserve structures such as nerves and blood vessels, "sharp" techniques may damage these delicate structures. For this reason, use "blunt" dissection techniques whenever possible to prevent damage to potentially important structures. "Open scissors," described in step 3 of exercise 1.8, is one blunt dissection technique. Use the open scissors technique to loosen the hold between the skin and the fascia of the specimen.
4. Once the space between the skin and muscle is large enough for a finger to be pushed in, set the scissors down. Proceed to separate the skin from the muscle using only your fingers (figure 1.16-2). Because sharp instruments are not used to perform this part of the dissection, it is referred to as "blunt" dissection.
5. Obtain a blunt probe (table 1.5). A blunt probe can be used in place of fingers to separate structures when fingers are too large (figure 1.16-3). Because the probe does not cut the tissue, this is also a "blunt" dissection technique.
6. Other items that can be used for blunt dissection are scalpel blade handles (*without* the blades on them!), or the rounded ends of forceps. Practice using these tools to separate skin from muscle in different regions of the dissection specimen. The general rule of thumb is to start with sharp dissection techniques to cut slits in the skin, but then transition to blunt techniques whenever possible to prevent accidental damage to underlying structures.

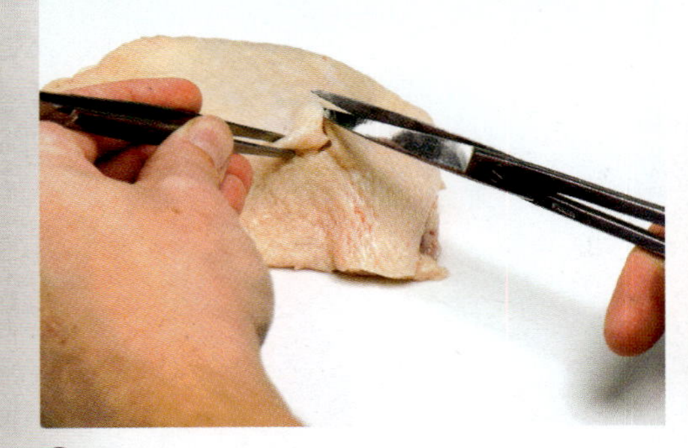

(1) Using tissue forceps and scissors, pull the skin away from the underlying tissues and make a cut in the skin. Use open scissors technique to loosen the fascia and to create a space where a blunt probe or fingers may be pushed in.

(2) Using fingers, pull the skin away from the underlying tissues. When necessary, use a sharp instrument to cut any fascia that is very tough and won't separate using blunt techniques.

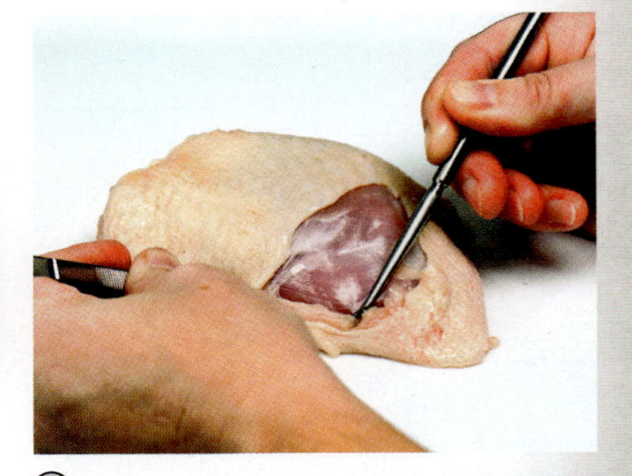

(3) A blunt probe can be moved around under the skin to gently separate the connective tissue without damaging underlying structures.

Figure 1.16 Blunt Dissection. Blunt dissection techniques involve separating tissues with fingers or a blunt instrument such as a probe. When handling fresh tissue such as this chicken thigh, either use gloves or make sure to wash your hands thoroughly when the dissection is complete.

Chapter 1: The Laboratory Environment

Name: ______________________

Date: ____________ Section: ____________

POST-LABORATORY WORKSHEET

The ❶ corresponds to the Learning Objective(s) listed in the chapter opener outline.

Do You Know the Basics?

Exercise 1.1: The Scientific Method

1. List the 6 steps involved in the scientific method, in the correct order. ❶

   a. ______________________ d. ______________________

   b. ______________________ e. ______________________

   c. ______________________ f. ______________________

2. Define the following terms: ❷

   a. Independent variable:

   ____

   b. Dependent variable:

   ____

   c. Confounding variable:

   ____

3. The purpose of an experimental control is to limit as many confounding variables as possible. ____________ (True/False) ❸

4. Given the set of numbers to the left of the box, calculate the mean and the range. Show your work in the box. ❹

25
98
32
46
22
87
34
26
15

Mean: Range:

Exercise 1.2: Presenting Data

5. Match the components of a typical graph listed in column A with what each component represents, listed in column B. ❺

Column A	*Column B*
____ 1. label	a. axis where the independent variable is plotted
____ 2. units	b. axis where the dependent variable is plotted
____ 3. values	c. description (name) of the variable in question
____ 4. X-axis	d. description of what the numbers on the graph indicate (e.g., ml)
____ 5. Y-axis	e. numerical representation of the data points

EXERCISE 2.1

ANATOMIC PLANES AND SECTIONS

EXERCISE 2.1A Human Brain Sections

1. **Figure 2.2** shows photographs of a brain that has been sectioned along different planes. Determine which plane the brain was sectioned along for each photo, then enter the information in the appropriate spaces in figure 2.2. Answer choices are provided in the figure legend, and each answer may be used only once. Refer to the chapter on the brain in the main textbook for assistance in getting oriented.

2. *Optional Activity:* **AP|R 1: Body Orientation**—Review all dissections in this module to become familiar with with general anatomic terminology.

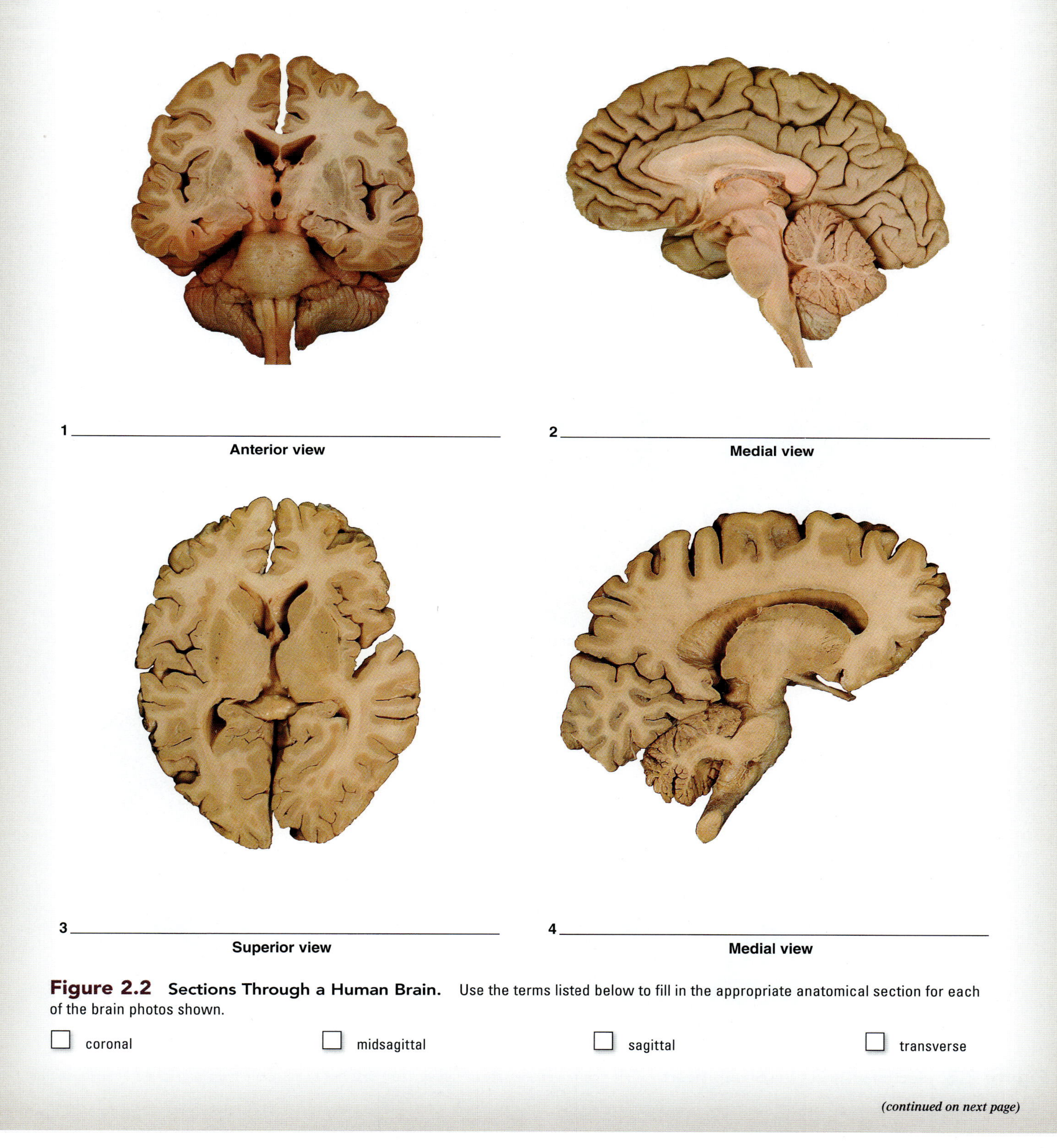

Figure 2.2 Sections Through a Human Brain. Use the terms listed below to fill in the appropriate anatomical section for each of the brain photos shown.

- ☐ coronal
- ☐ midsagittal
- ☐ sagittal
- ☐ transverse

(continued on next page)

(continued from previous page)

EXERCISE 2.1B Sectioning a Specimen

Obtain the Following:

- dissecting tray
- knife and/or scalpel
- specimen (e.g., sheep brain, sheep heart, sheep kidney)

1. **Figure 2.3** demonstrates a sheep heart that has been sectioned along coronal, sagittal, and transverse planes. Using figure 2.3 as a guide, section the specimen along the following planes:

   ☐ coronal ☐ sagittal ☐ transverse

2. Sketch the appearance of each of the resulting portions of the specimen in the spaces provided.

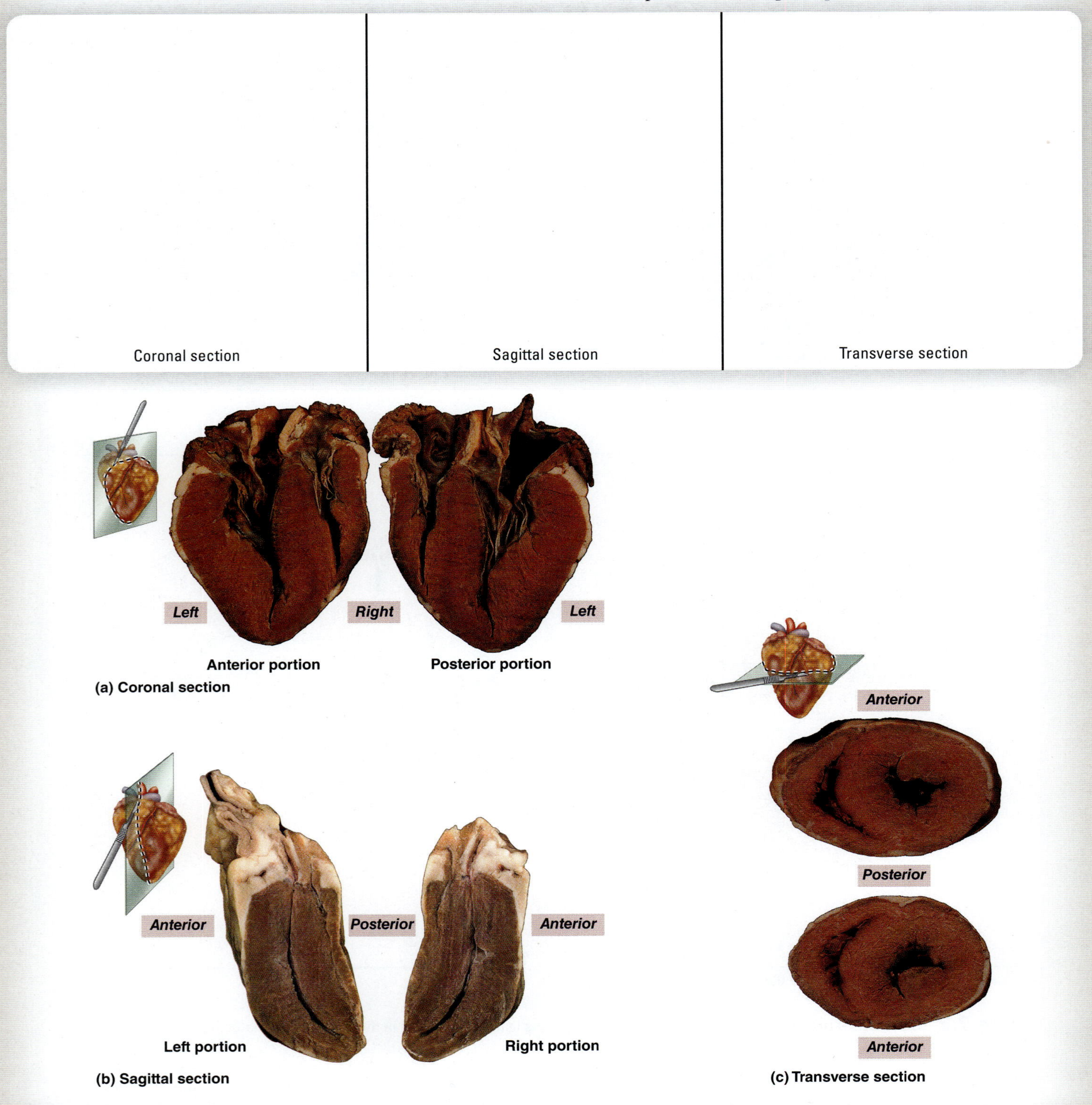

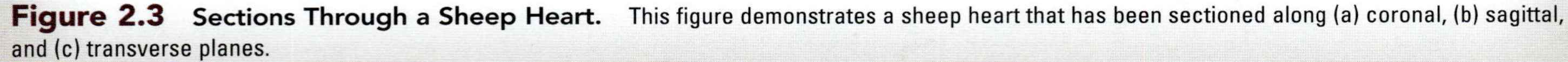

Figure 2.3 Sections Through a Sheep Heart. This figure demonstrates a sheep heart that has been sectioned along (a) coronal, (b) sagittal, and (c) transverse planes.

Directional Terms

Everyday terms like "front" and "back," and "on top of" or "on the bottom of," are often used to give directions. While this is perfectly appropriate in everyday language, it can cause confusion when referring to directions in the human body. For instance, when saying "on top of," we need to know to what part of the body "on top of" refers. Different people might use the term in different ways or in reference to different structures. Furthermore, an individual may be thinking that the direction "on top of" can change, and is relative to different body positions. This becomes problematic in a medical setting because it leads to confusion, which has the potential to create severe consequences for a patient. Thus, an agreed-upon set of directional terms is used in anatomy and medicine to be as specific as possible when describing directions, but also to ensure that everyone is speaking the same language. **Table 2.2** lists these directional terms and gives definitions of each of them. Exercise 2.2 involves practicing the use of these directional terms.

Table 2.2 Directional Terms

Directional Term	Definition
Anterior (ventral)	Toward the front of the body (the belly side)
Posterior (dorsal)	Toward the back of the body (the back side)
Superior	Above; closer to the head
Inferior	Below; closer to the feet
Cranial (cephalic)	At the head end of the body
Caudal	At the tail end of the body
Medial	Toward the midline of the body
Lateral	Away from the midline of the body
Superficial	Toward the surface of the body; on the outside
Deep	Beneath the surface of the body; on the inside
Proximal	Near; closer to the attachment point of a limb to the trunk
Distal	Far; farther from the attachment point of a limb to the trunk

EXERCISE 2.2

DIRECTIONAL TERMS

Figure 2.4 shows a posterior view of a human. Three locations marked on the body are marked with the circled numbers 1–3. Describe the locations of markings 1–3 as specifically as possible using correct anatomic terminology. When finished, compare your answers to those of other students in the class to see how similar the answers are. Note that there is more than one correct answer for each of these.

Location 1

____

____

____

Location 2

____

____

____

Location 3

____

____

____

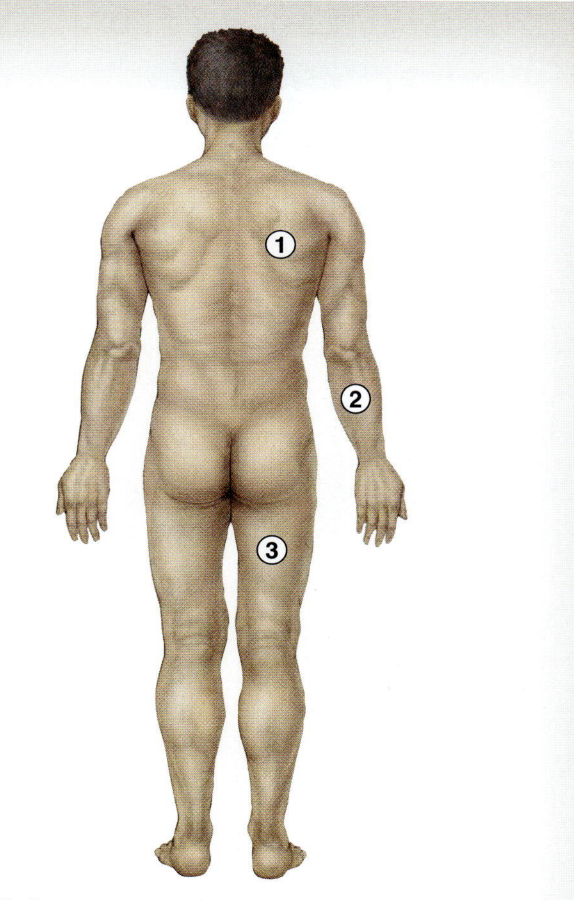

Figure 2.4 Posterior View of an Individual with Three Reference Locations (1–3) Marked. Number the terms below to match the numbered reference locations in the figure.

☐ antebrachial ☐ femoral ☐ thoracic

Regional Terms

Just as with directional terms, there are common, everyday terms that are often used to describe regions of the body such as *arm* or *back*. The correct anatomic terms to describe regions of the body are basically synonyms for these terms, and are closer to the Latin or Greek derivatives of the terms. **Table 2.3** correlates some commonly used regional terms in anatomy with words typically used to describe the same region using everyday language. The main textbook has a much more inclusive table of regional terms, as well as a figure describing these regions. Use the main textbook as a reference when completing exercise 2.3.

INTEGRATE

LEARNING STRATEGY

The directional terms *superior* and *inferior* are used when describing one structure with respect to another structure in the trunk of the body. The directional terms *proximal* and *distal* are used when describing the position of one structure with respect to another structure on the limbs. Thus, it is more appropriate to say the elbow is located *proximal* to the wrist, rather than to say it is superior to the wrist. On the other hand, it is quite appropriate to say the thorax is located *superior* to the abdomen.

Table 2.3 Selected Regional Terms

Regional Term	Description
Axillary	Armpit
Brachial	Arm (between the shoulder and elbow)
Buccal	Cheek
Carpal	Wrist
Cephalic	Head
Cervical	Neck
Digital/phalangeal	Fingers or toes
Femoral	Thigh
Gluteal	Buttock
Inguinal	Groin
Lumbar	Lower back
Mental	Chin
Orbital	Eye
Plantar	Sole of the foot
Umbilical	Navel
Vertebral	Spinal column

EXERCISE 2.3

REGIONAL TERMS

1. Identify the body regions listed in **figure 2.5** on your own body.
2. Label figure 2.5 with the appropriate regional terms, using the main textbook as a guide.

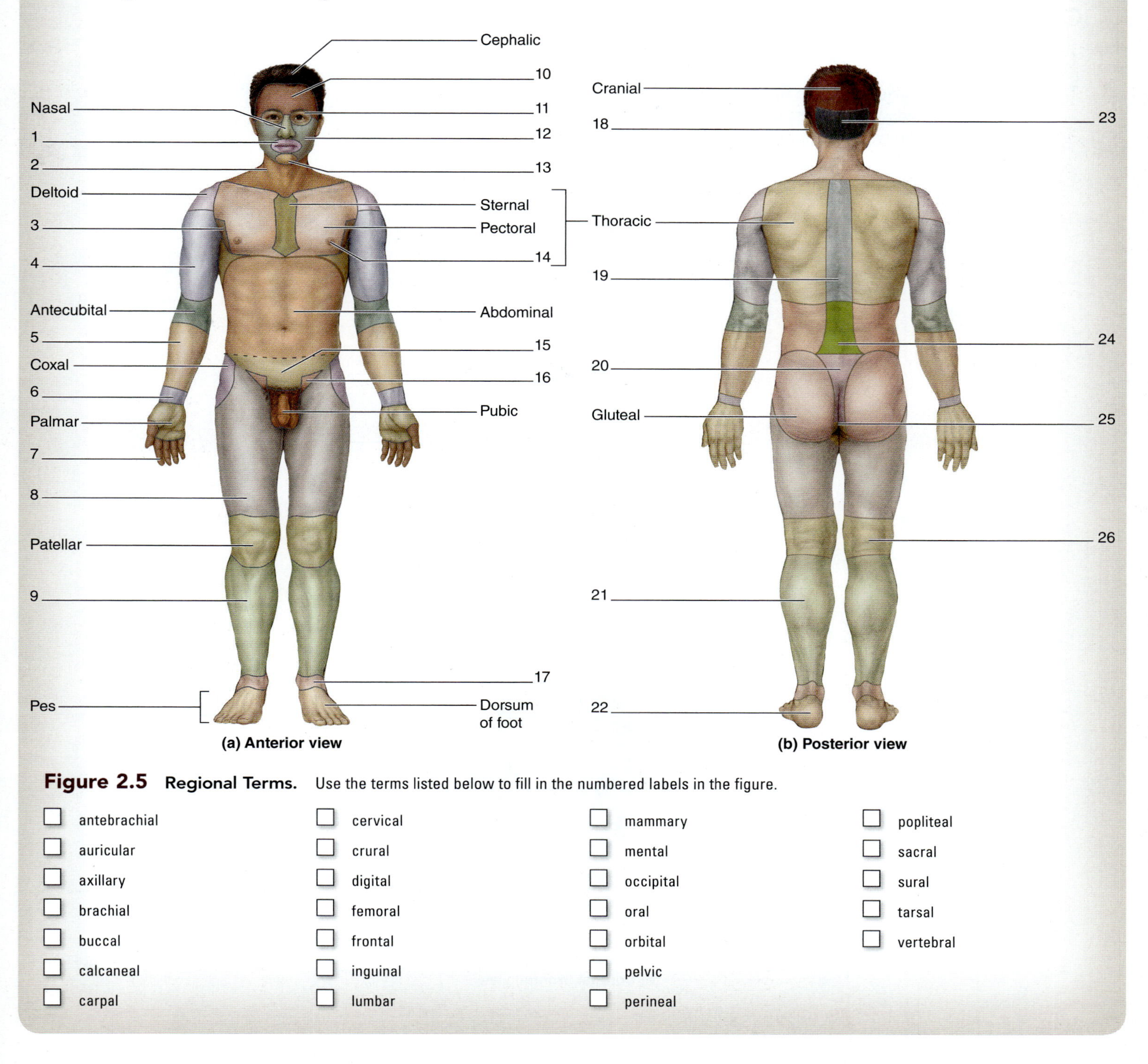

Figure 2.5 **Regional Terms.** Use the terms listed below to fill in the numbered labels in the figure.

- ☐ antebrachial
- ☐ auricular
- ☐ axillary
- ☐ brachial
- ☐ buccal
- ☐ calcaneal
- ☐ carpal
- ☐ cervical
- ☐ crural
- ☐ digital
- ☐ femoral
- ☐ frontal
- ☐ inguinal
- ☐ lumbar
- ☐ mammary
- ☐ mental
- ☐ occipital
- ☐ oral
- ☐ orbital
- ☐ pelvic
- ☐ perineal
- ☐ popliteal
- ☐ sacral
- ☐ sural
- ☐ tarsal
- ☐ vertebral

Body Cavities and Membranes

Many organs within the body are compartmentalized and separated from each other by a *body cavity*. Compartmentalizing the organs this way allows the separate organs to perform their functions without interfering with the functioning of other organs. For example, the pumping action of the heart does not interfere with the expansion and contraction of the lungs because each organ is enclosed in its own cavity. In addition, the encasement of organs within separate cavities helps to prevent the spread of infection from one cavity to another.

EXERCISE 2.4

BODY CAVITIES

1. Observe a human torso model or a human cadaver.
2. Identify the body cavities listed in **figure 2.6** on the torso model or human cadaver, using the textbook as a guide. Then label figure 2.6 with the appropriate terms.

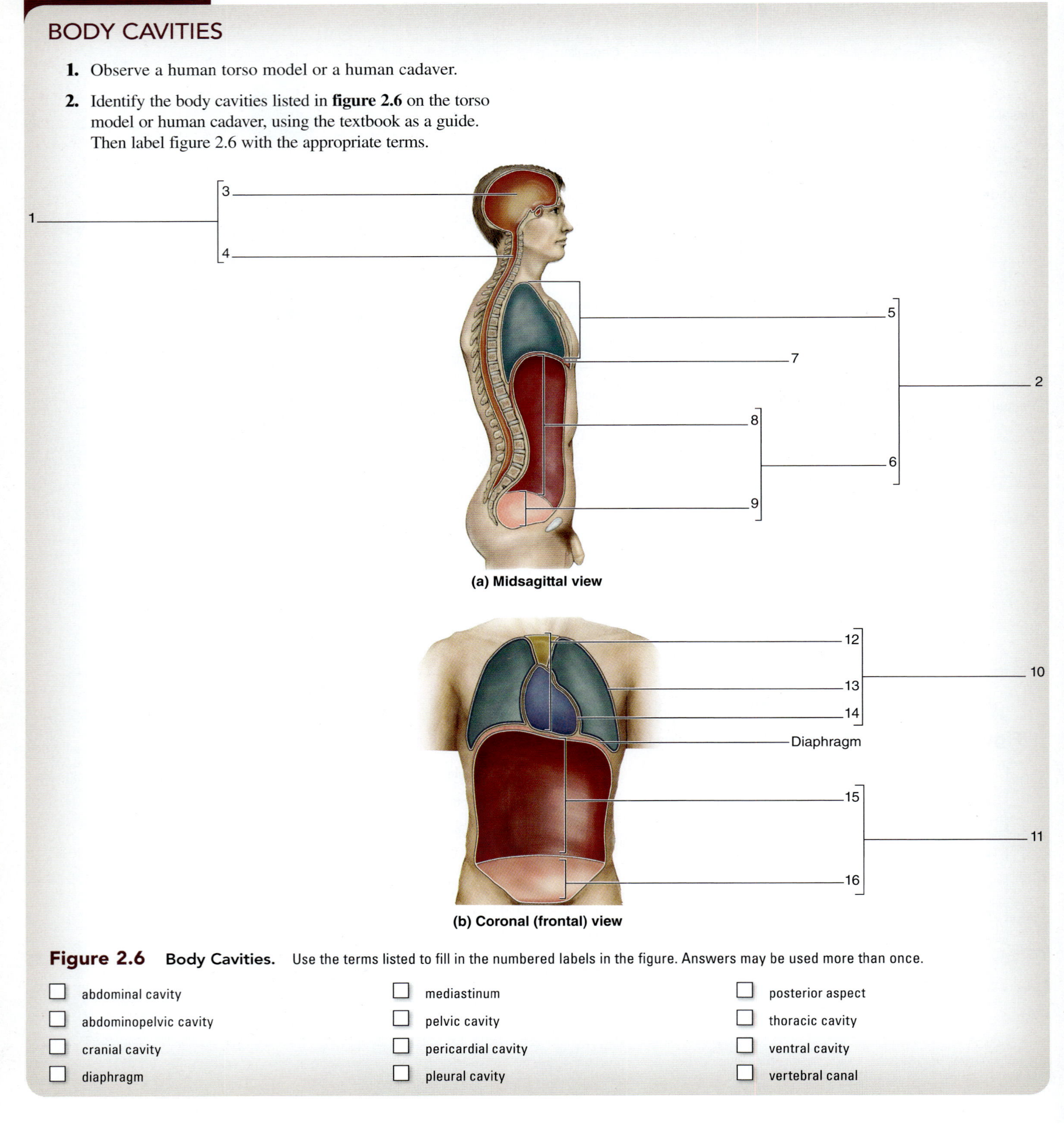

Figure 2.6 Body Cavities. Use the terms listed to fill in the numbered labels in the figure. Answers may be used more than once.

- ☐ abdominal cavity
- ☐ abdominopelvic cavity
- ☐ cranial cavity
- ☐ diaphragm
- ☐ mediastinum
- ☐ pelvic cavity
- ☐ pericardial cavity
- ☐ pleural cavity
- ☐ posterior aspect
- ☐ thoracic cavity
- ☐ ventral cavity
- ☐ vertebral canal

INTEGRATE

CONCEPT CONNECTION

Serous (*serosus*, watery fluid) **membranes** cover organs within the body cavities. These membranes are composed of two layers: visceral and parietal. The visceral (*viscus,* internal organ) layer lies adjacent to the organ, whereas the parietal (*paries*, wall) layer attaches to the wall of the body cavity. The cells that compose the serous membrane secrete a slippery substance called **serous fluid** in the space between the two layers of membrane to act as a lubricant. When organs such as the heart and lungs expand and contract, serous fluid reduces friction between the organs and surrounding structures. The tissues that compose serous membranes will be explored further in chapter 5. The structure of specific organ coverings will be explored in the "gross anatomy" section of subsequent chapters. Note that a specific serous membrane that surrounds an organ generally has a unique name that relates to the organ it encases. For example, the *pericardium* surrounds the heart and *pleurae* surround the lungs.

INTEGRATE

CLINICAL VIEW

Inflammation of Serous Membranes

Serous membranes are membranes that line body cavities, such as the pericardial cavity (which encases the heart). Inflammation of serous membranes can be a serious health risk. Clinical terms related to inflammation of serous membranes reference the specific serous membrane affected. For example, *pericarditis* (*peri-*, around + *cardio*, heart + *itis*, inflammation) is inflammation of the pericardium surrounding the heart; *pleurisy* is inflammation of the pleurae surrounding the lungs; *peritonitis* is inflammation of the peritoneum surrounding the abdominopelvic cavity. One consequence of inflammation of a serous membrane is swelling, which can impede the function of surrounding organs within the body cavity. For example, pericarditis may interfere with the heart's ability to pump blood. Pleurisy may prevent adequate gas exchange in the lungs.

Abdominopelvic Regions and Quadrants

Describing the locations of organs in the abdominopelvic cavity can be complicated because it is such a large cavity. Having a set of terms that divides the larger cavity into smaller sections allows for more specific descriptions of where within the cavity an organ or tissue is located. There are two approaches used to do this. The first approach is to divide the abdominopelvic cavity into **quadrants** (*quad-*, four). This is done by passing one imaginary line vertically through the *umbilicus* (belly button) and another horizontally through the umbilicus. The four resulting quadrants are the right upper, left upper, right lower, and left lower (**figure 2.7*a***). Because this approach is simple, it is the approach used most often in a clinical setting. The second approach is to divide the abdominopelvic cavity into **regions.** This is done by drawing one vertical line to the left of the umbilicus and another to the right of the umbilicus (at the *midclavicular line,* a vertical line that passes through the midpoint of the clavicle), then drawing one horizontal line superior to the umbilicus and another inferior to the umbilicus. The result is a grid similar to a tic-tac-toe layout with nine regions formed (figure 2.7*b*). This approach allows greater specificity in describing the locations of organs and tissues within the abdominopelvic cavity. The resulting regions are the umbilical, epigastric (*epi-*, above, + *gastēr*, belly), hypogastric (*hypo-*, below, + *gastēr*, belly), right and left hypochondriac (*hypo-*, below, + *chondro-*, cartilage, as in the cartilages that attach ribs to sternum), right and left lumbar (*lumbus*, a loin), and right and left iliac (*ilium*, flank). Exercise 2.5 is designed as an introduction to the locations of major organs within the abdominopelvic cavity and to help students become comfortable with the terms used to describe the quadrants and regions.

EXERCISE 2.5

LOCATING MAJOR BODY ORGANS USING ABDOMINOPELVIC REGION AND QUADRANT TERMINOLOGY

1. Identify the following structures on a human torso model or on a human cadaver, using figure 2.7 as a guide.
   - ☐ left kidney
   - ☐ liver
   - ☐ pancreas
   - ☐ small intestine
   - ☐ spleen
   - ☐ stomach
   - ☐ urinary bladder

2. Based on observations of a cadaver or human torso model, complete the chart on p. 39 by indicating both the quadrant(s) and region(s) in which each organ is located.

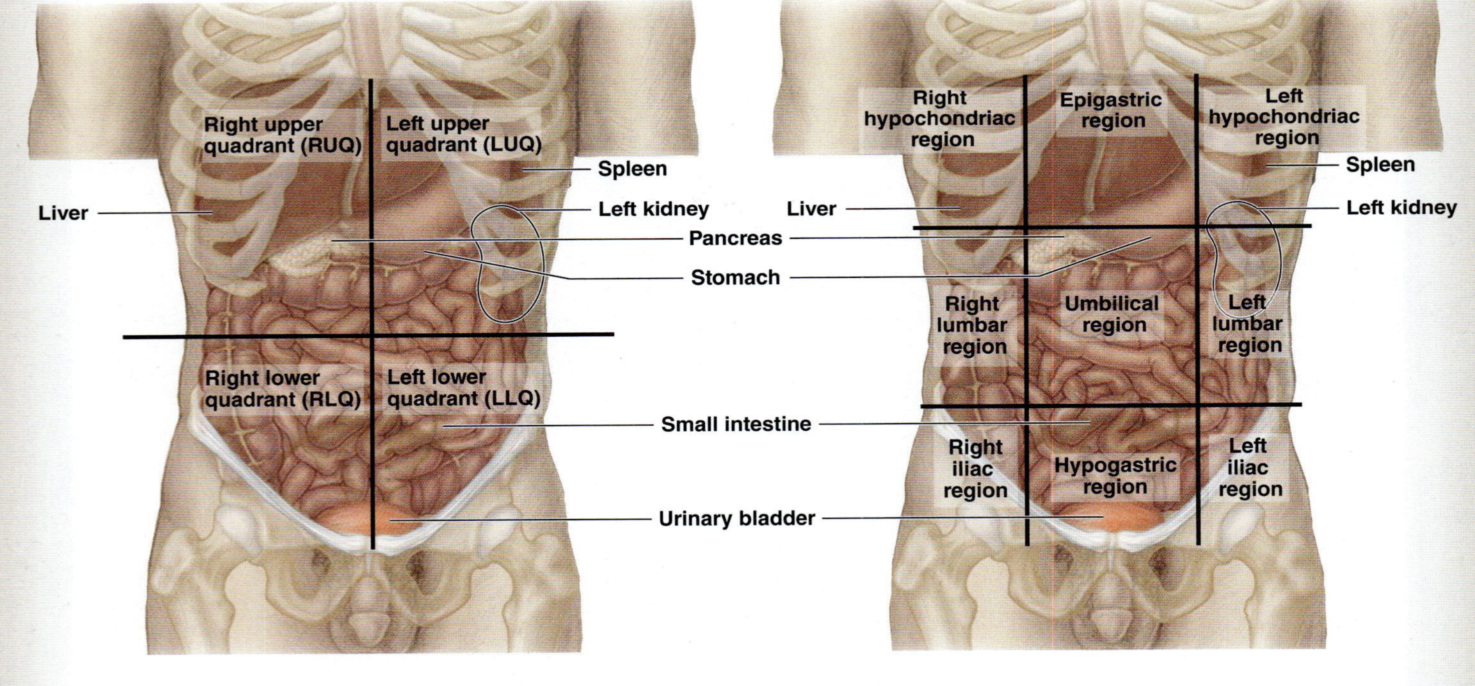

Figure 2.7 The Abdominopelvic Cavity. The abdominopelvic cavity can be subdivided into (a) four abdominopelvic quadrants, or (b) nine abdominopelvic regions. An outline of the left kidney is shown because the actual view of the kidney is obscured by other organs within the abdominopelvic cavity.

Organ	Quadrant(s)	Region(s)
Left kidney		
Liver		
Pancreas		
Small Intestine		
Spleen		
Stomach		
Urinary Bladder		

Chapter 2: Orientation to the Human Body

Name: ______________________

Date: ____________ Section: ____________

POST-LABORATORY WORKSHEET

The ❶ corresponds to the Learning Objective(s) listed in the chapter opener outline.

Do You Know the Basics?

Exercise 2.1: Anatomic Planes and Sections

1. Which of the following correctly describes the anatomic position? (Check all that apply.) ❶
   _______ a. one can be sitting down or standing up
   _______ b. the feet are directed forward (anterior)
   _______ c. no two bones of the body cross each other
   _______ d. the palms of the hands are facing backward (posterior)

2. A ______________ (plane/section) is an imaginary two-dimensional flat surface. A ______________ (plane/section) is a slice made along one of these two-dimensional flat surfaces. ❷

3. To divide this image of a Valentine heart ♡ into equal sections, a ______________ (coronal/midsagittal) section could be used. ❸

4. Match each definition listed in column A with the anatomic plane listed in column B. ❹

Column A	Column B
_______ 1. separates the right and left portions equally	a. coronal (frontal)
_______ 2. separates anterior portions from posterior portions	b. midsagittal (median)
_______ 3. separates superior portions from inferior portions	c. oblique
_______ 4. runs at an angle to any of the three main planes of the body	d. transverse (horizontal)

5. Which of the following shapes would end up with identical portions when sectioned along all of the following planes: midsagittal, transverse, and coronal? (Circle one.) ❹
   a. pyramid
   b. rectangular box
   c. egg
   d. square box

Exercise 2.2: Directional Terms

6. For each of the following, insert the most appropriate directional term. ❺
   a. The elbow is located ______________________ to the wrist.
   b. The mouth is located ______________________ to the ears.
   c. The lungs are located ______________________ to the ribs.
   d. The umbilicus is located ______________________ to the sternum.
   e. The nose is located ______________________ and ______________________ to the ears.
   f. A scratch wound, which does not penetrate the skin, is said to be a ______________________ wound. In contrast, a stab wound, which penetrates the skin, is referred to as a ______________________ wound.

Exercise 2.3: Regional Terms

7. Which of the following statements matches the regional term with the appropriate common term (example: the *tarsal* bones are in the ankle)? (Circle one.) ❻
   a. The *brachial* artery is in the wrist.
   b. The *carpal* bones are in the ankle.
   c. The *cervical* vertebrae are in the lower back.
   d. The *femoral* nerve is in the arm.
   e. The *popliteal* artery is in the knee.

Exercise 2.4: Body Cavities

8. Match the description listed in column A with the appropriate body cavity listed in column B. ❼

Column A		***Column B***
______	1. encases the brain	a. cranial cavity
______	2. encases the heart	b. pelvic cavity
______	3. contains most of the reproductive and urinary systems	c. pericardial cavity
______	4. contains the cardiovascular and respiratory systems	d. pleural cavity
______	5. encases the lungs	e. thoracic cavity

Exercise 2.5: Locating Major Body Organs Using Abdominopelvic Region and Quadrant Terminology

9. Fill in the figure provided with the names of the appropriate abdominopelvic regions. Note that the terms "Right" and "Left" refer to the body, *not* to the right and left side of the page. ❽

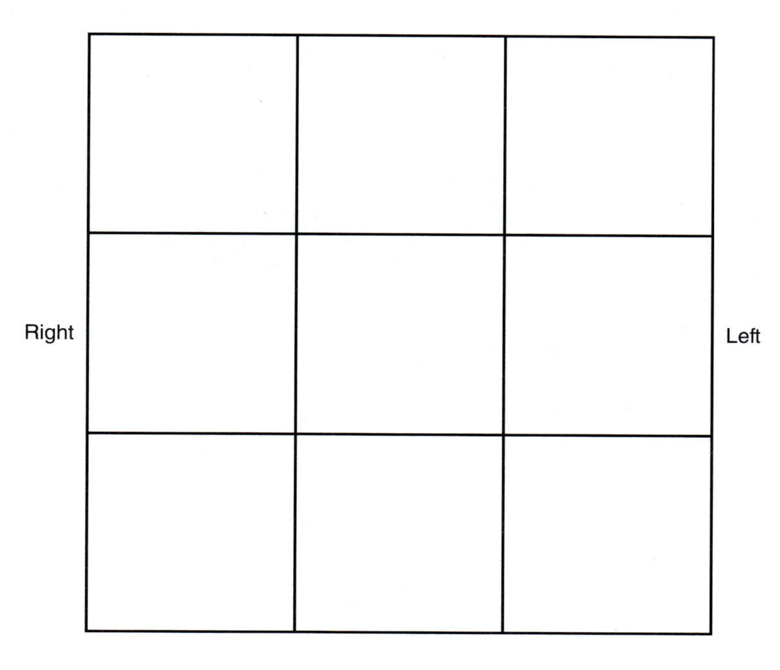

10. The left kidney is located in the ______________________________ abdominopelvic quadrant and the ______________________________ abdominopelvic region. ❾

Can You Apply What You've Learned?

11. The anatomic position requires that the forearm be rotated so the palm of the hand faces anteriorly. Why is the forearm rotated this way in the anatomic position?

____

12. A physician would like to obtain a radiographic view of both the liver and the spleen of her patient, so she orders a CT scan. What plane of section does the physician ask the radiology department to scan so she may view both of these organs in the same image?

____

13. The same physician would like to obtain a radiographic view of both the right lung and the liver of her patient, so she orders a CT scan. What plane of section does the physician ask the radiology department to scan so she may view both of these organs in the same image?

____

14. A teenage boy is diagnosed with appendicitis (inflammation of the appendix) and must undergo an operation. To remove the appendix, the surgeon will operate on the boy's ____ abdominopelvic quadrant.

Can You Synthesize What You've Learned?

15. The abdominopelvic region located directly lateral to the umbilical region is the ______________________ region.

16. A horizontal section through the tarsus would separate the ______________________ from the ______________________.

17. A patient was in a knife fight and suffered a 2-inch stab wound to his back. The knife entered the posterior thorax along the midsagittal (median) plane and entered the posterior aspect (posterior body cavity). What organ in the posterior aspect is likely to have been injured by this wound? ______________________

18. A man who had been in a car accident arrived at the emergency department. He was awake and alert and was able to tell the physician that he was experiencing severe pain near his lower ribs on the left side of his body and in his abdomen. Upon palpation, the patient's abdomen was rigid, indicating the possibility of internal bleeding. The physician suspected that broken ribs may have injured an organ within the abdominal cavity. Which organ would most likely be injured in this case? ______________________ What abdominopelvic quadrant is this organ located in? ______________________ What abdominopelvic region is it located in? ______________________

The Microscope

OUTLINE AND LEARNING OBJECTIVES

INTRODUCTION

The study of anatomy involves observation of both gross (large) and microscopic (small) structures. Viewing the details of body tissues and cells is possible using a microscope. Both light microscopes and electron microscopes allow detailed observation of the ultrastructure of tissues. Importantly, microscopes allow the user to view specific cellular components of tissues that make up the body. Such components include the shapes of the cells, intracellular components such as the nucleus of the cells, and modifications of the plasma membrane (e.g., cilia or microvilli). Observing specific cellular features provides a visual image that assists with the process of integrating structure and function.

To get the most from the experience of observing microscopic structures, it is important to know how to use a microscope properly. Although you may have used a microscope before, pay very careful attention to the instructions in this chapter. Poor technique when using microscopes causes frustration and is an impediment to the learning process. This laboratory exercise is an opportunity to refine and improve upon your technique.

This laboratory exercise covers both how to care for and how to use a compound microscope. After performing these exercises, you will be prepared to make observations of histology (tissues) in later chapters. Also covered in this chapter is the proper use and care of a dissecting microscope (also called a stereomicroscope). A dissecting microscope is used to view larger specimens such as hair and muscle fibers.

List of Reference Tables

Chapter 3: The Microscope

Name: ______________________

Date: __________ Section: __________

These Pre-Laboratory Worksheet questions may be assigned by instructors through their connect course.

PRE-LABORATORY WORKSHEET

1. Which of the following are proper procedures for holding, transporting, and storing a microscope? (Check all that apply.)
   ______ a. Always use both hands when carrying the microscope.
   ______ b. Place one hand on the base of the microscope and the other on the arm of the microscope.
   ______ c. Use care and move deliberately when carrying the microscope.
   ______ d. Slides may be left on the stage when storing a microscope.
   ______ e. Move the stage to its highest position when storing a microscope.
   ______ f. Wrap the power cord around the base and replace the dust cover after using the microscope.

2. Broken slides should be placed in a sharps container. __________ (True/False)

3. When holding a slide, be sure to hold it by the edges so as to not leave smudges and fingerprints on the slide. __________ (True/False)

4. Match the definition listed in column A with the microscope part listed in column B.

Column A	*Column B*
______ 1. everything that is visible when looking through the eyepiece	a. arm
______ 2. the part of the microscope that connects the head to the base	b. coarse adjustment knob
______ 3. the platform that a slide is placed upon	c. field of view
______ 4. lens that is attached to the nosepiece	d. fine adjustment knob
______ 5. a knob that moves the mechanical stage up and down in small increments	e. mechanical stage
______ 6. a knob that moves the mechanical stage up and down in large increments	f. objective lens

5. When viewing structures using the compound microscope, which of the following statements describes an appropriate action to take when light is visible in the field of view, but there is no clear specimen in the field of view? (Circle one.)
   a. Check the power switch and make sure it is turned on.
   b. Adjust the condenser.
   c. Set the microscope to the scanning power objective and lower the stage; adjust the coarse adjustment knob until the specimen comes into view.
   d. Make sure the microscope is plugged into a working power outlet.
   e. Move the eyepieces either closer together or farther apart until a clear image is visible.

6. Which of the following correctly describes how to properly calculate the total magnification of a microscope? (Circle one.)
   a. Add the ocular lens magnification and objective lens magnification.
   b. Subtract the ocular lens magnification from the objective lens magnification.
   c. Multiply the ocular lens magnification by the objective lens magnification.
   d. Divide the objective lens magnification by the ocular lens magnification.

7. As total magnification power increases, depth of field __________ (increases/decreases).

8. As total magnification power increases, diameter of the field of view __________ (increases/decreases).

9. A dissecting microscope is used to view __________ (microscopic/macroscopic) structures.

10. Which of the following statements is an accurate comparison between a dissecting microscope and a compound microscope? (Check all that apply.)
    ______ a. A dissecting microscope can view smaller structures than a compound microscope.
    ______ b. An image viewed with a dissecting microscope is not inverted, as it is with a compound microscope.
    ______ c. Tissues must be prepared prior to viewing with a compound microscope, but no such preparation is needed for a dissecting microscope.
    ______ d. Cells and tissues can be viewed with either a dissecting microscope or a compound microscope.

HISTOLOGY

The Compound Microscope

A compound microscope is used to view structures that are not visible to the naked eye. Most compound microscopes can magnify images anywhere from 40 to 1000 times (40× to 1000×) normal size. The total magnification of a compound microscope is achieved through a combination of objective and ocular lenses (see **table 3.1** for definitions). The term "compound" comes from the fact that total magnification is "compounded" by the use of more than one lens.

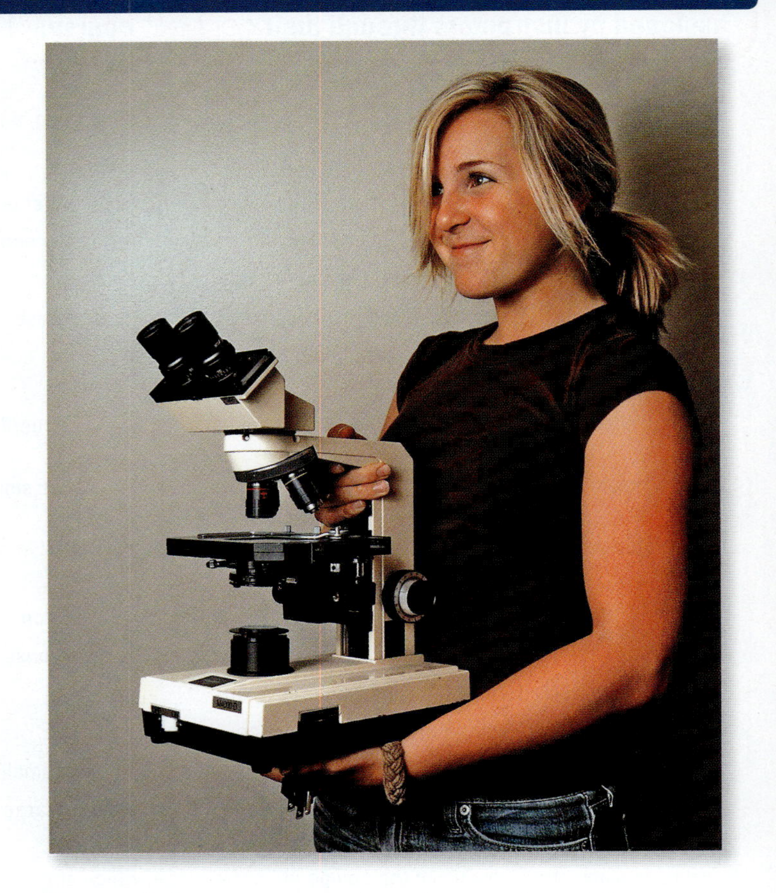

Figure 3.1 Proper Technique for Carrying a Microscope. Note that one hand is supporting the base and another is holding the arm of the microscope.

Caring for the Compound Microscope

Microscopes are very expensive instruments. Use great care when transporting, using, and storing microscopes.

- *Workspace:* Keep the workspace clear of all materials except for the microscope, laboratory manual, lens paper, and microscope slides. Remove or restrict any loose articles of clothing or jewelry. Long necklaces that dangle can damage the microscope. Long hair should be tied back.
- *Transport: Always* use both hands when carrying the microscope. Place one hand on the base of the microscope and the other on the arm. Use care to always move deliberately when carrying the microscope **(figure 3.1).**
- *Lens care:* Lenses are some of the most expensive and delicate parts of the microscope and should be treated with great care. Special lens paper is the only thing that should ever be used to clean lenses. This paper is designed so it will not scratch the lenses. *Never* use facial tissues, paper towels, articles of clothing, or anything else to clean the lenses because they may scratch the lenses. A special cleaning solution can be used with the lens paper, although lens paper generally does a fine job of cleaning when used alone. *Never* use saliva, water, or other fluids to moisten the lens paper.
- *Storage:* When finished using the microscope, move the stage to its lowest position. Make sure that no slides are left on the stage, and then rotate the nosepiece so the lowest-power objective lens is over the stage. Wrap the power cord around the base of the microscope (or around a structure that holds the power cord, if available) and replace the dust cover.
- *Handling slides:* Always hold a slide by the edges to prevent smudges and fingerprints, which will interfere with the ability to view the slide clearly. If the slide is dirty, clean it using lens paper before placing it on the microscope stage. If a slide becomes cracked or broken, notify the instructor so that it can be disposed of properly. Broken slides should be placed in a special broken glass container or sharps container, never in the garbage can.

⚠ When removing the microscope cover, be careful not to pull the ocular lens off with the cover. When using the coarse adjustment knob, be careful not to allow the objective lens to smash into the slide, especially when using the high power objectives.

EXERCISE 3.1

PARTS OF A COMPOUND MICROSCOPE

1. Obtain a compound microscope.
2. Identify all the parts listed in table 3.1 on the compound microscope, using **figures 3.2** and **3.3** and table 3.1 as guides.
3. The ocular lens magnification is engraved on the eyepiece, and the objective lens magnification is etched on the silver tube that holds each objective lens. Record the magnifications of the lenses on the microscope in the spaces below.

   Magnification of ocular lenses: ______

   Magnification of objective lenses:

   Scanning ______ Low ______ High ______
4. To determine the total magnification of an image, multiply the magnification of the ocular lens with that of the objective lens. For example, if the magnification of the ocular lens is 10× and the magnification of the objective lens is 4×, then the total magnification is 40×. Calculate the total magnification for each of these objective lenses—scanning, low-power, and high-power—for the microscope you are using. Write the answer in the spaces provided.

   Total magnification (ocular × objective):

   Scanning ______ Low ______ High ______
5. Use the information from steps 3 and 4 to answer question 4 of the Post-Laboratory Worksheet (p. 55).

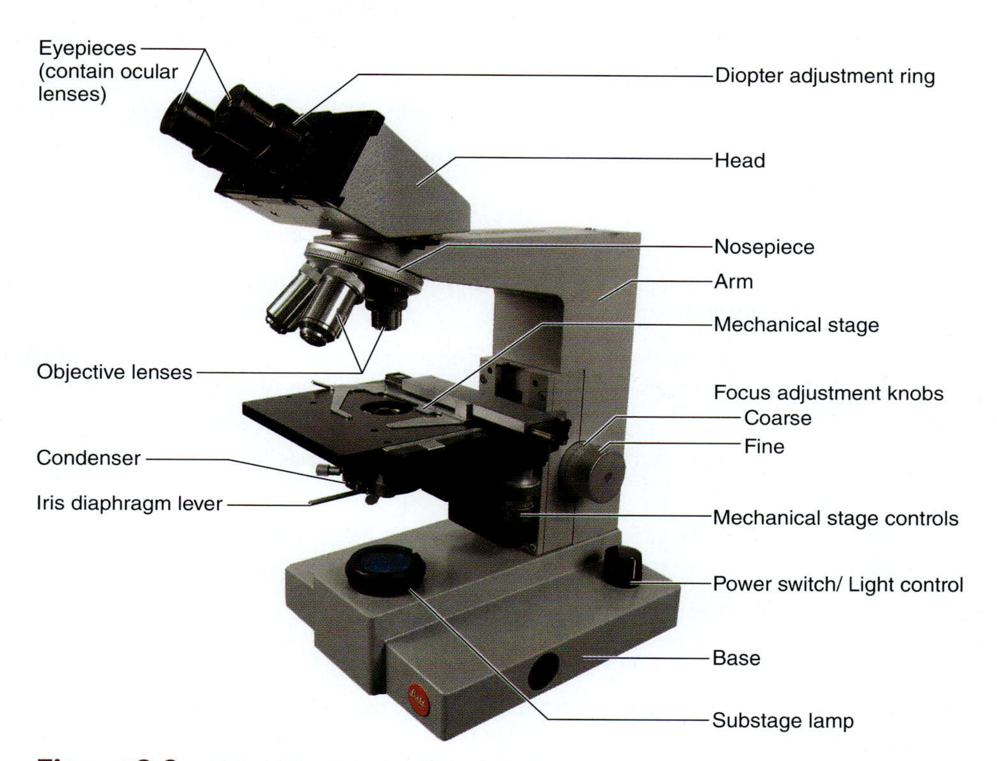

Figure 3.2 Parts of a Compound Microscope.

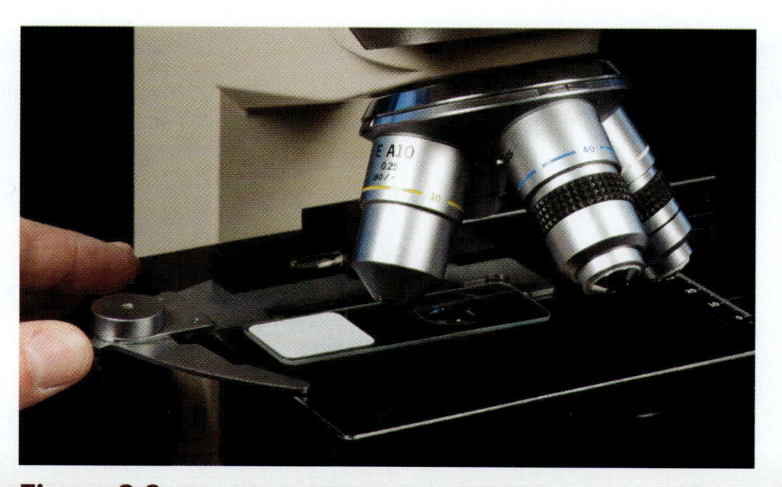

Figure 3.3 Loading a Microscope Slide.

(continued on next page)

(continued from previous page)

Table 3.1 Parts of the Microscope

Term	Description and Usage
Arm	Connects the head to the base. When transporting the microscope, always grasp the arm with one hand.
Base	The bottom part of the microscope. It supports the entire microscope and encases much of the electrical wiring. When transporting the microscope, one hand should always be placed under the base for support.
Condenser	A device used to focus the light coming from the illuminator/substage lamp. It generally works best in the position closest to the mechanical stage.
Coarse Adjustment Knob	Moves the mechanical stage up and down in relatively large increments. It is used to position a specimen into the field of view when scanning a slide at low power. It should never be used with the higher-power objectives.
Depth of Field	How much of the total thickness of the specimen is in focus
Diopter Adjustment Ring	Changes the focus of the eyepiece. Used to make adjustments when the user has better vision in one eye and is not using corrective lenses. To use this device, begin by closing the eye that looks through the ocular lens containing the diopter adjustment ring. Looking through the *other* ocular lens, bring the sample into focus. Then, open the eye that was closed and bring the sample into focus for that eye using the diopter adjustment ring.
Eyepieces	The parts of the microscope that the user looks into. Always use both eyepieces. The eyepieces are movable so they can be adjusted to the distance between each user's eyes. Each eyepiece holds an ocular lens.
Field of View	Everything visible when looking through the eyepieces
Fine Adjustment Knob	Moves the mechanical stage up and down in small increments. Once a specimen is brought into view using the coarse adjustment knob, the fine adjustment knob is used to bring the specimen into focus.
Head	The part of the microscope that provides attachment and support for the objective and ocular lenses. It also serves as the support for the eyepieces.
Iris Diaphragm	A part of the condenser that can be opened or closed to control the amount of light passing through the condenser. More light (iris open) decreases contrast between structures. Less light (iris closed) increases contrast.
Light Control (voltage regulator)	Typically, a rotating or sliding knob that alters the voltage going to the substage lamp to regulate the brightness of the light
Mechanical Stage	Platform that holds the specimen. It contains clips to hold a slide in place and knobs for positioning the slide on the stage.
Mechanical Stage Controls	Knobs used to position the slide on the stage
Nosepiece	Device that connects the objective lenses to the head of the microscope, and rotates to change objective lenses
Objective Lenses	A typical microscope has three objective lenses, though some have four. Each objective lens has a label that tells how much it magnifies the image. For instance, a 4× objective lens magnifies the specimen four times normal size. The lowest-power objective lens is also called the "scanning" objective.
Ocular Lenses	Lenses located within the eyepieces, which typically magnify the specimen ten times normal size
Power Switch	Usually located on the base of the microscope; used to turn the power on or off
Substage Lamp (illuminator)	The light source located in the base of the microscope. When it is turned on, light passes through the specimen on the stage, through the lenses of the microscope, and it ultimately hits the user's eye, allowing the user to see the specimen.
Working Distance	The distance between the mechanical stage (and the slide on it) and the tip of the objective lens

Focus and Working Distance

The purpose of this exercise is to familarize students with the microscope to develop an appreciation for the relationship between what is visible when looking directly at a slide and what is visible when looking at a slide through the microscope lenses. If you experience difficulties and cannot see anything through the microscope or cannot focus the lenses, refer to **table 3.2,** "Troubleshooting the Compound Microscope."

Table 3.2 Troubleshooting the Compound Microscope

Problem	Solution
"No light is coming from my illuminator."	Make sure the microscope is plugged into a working power outlet.
	Check the power switch and make sure it is turned on.
	Check the light control/voltage regulator and make sure it isn't turned all the way down.
	If the first 3 steps don't solve the problem, see the instructor. The bulb may have burned out.
"I can't find anything on my slide."	Go back to scanning power. Lower the stage as far as it will go. Look at the slide on the stage and position it so that the specimen is illuminated and the lower power objective is in the lowest position (position closest to the stage). Look through the eyepiece and, using the coarse adjustment knob, slowly bring the objective up until the specimen comes into view.
"I can't use both eyes to view the slide." or *"I have to close one eye to view the slide."*	Move your head back from the eyepieces slightly. If you are too close, it is more difficult to see a single image.
	Move the eyepieces (closer together or farther apart) until a single image is visible through the microscope.
	If the first 2 steps don't work, get a classmate to help you measure the distance between your pupils using a ruler. Then use the ruler to move the eyepieces apart that same distance.
"I see a dark crescent in the view."	Make sure the objective lens is clicked into place.
	Adjust the condenser.
"I can't get the specimen in clear focus."	Make sure the slide is not upside down on the mechanical stage.

EXERCISE 3.2

VIEWING A SLIDE OF THE LETTER *e*

EXERCISE 3.2A Focusing the Microscope

1. Obtain a compound microscope and a slide of the letter *e*.
2. *Always begin observations with the lowest possible total magnification (scanning objective in place).* Make sure the mechanical stage is in its lowest position, closest to the base of the microscope, then place the slide on the stage (see figure 3.3) and turn on the illuminator. Adjust the light control/voltage regulator so it is somewhere near the middle or low end of its range.
3. After checking to be sure that light is coming out of the illuminator, position the microscope stage so the letter *e* on the slide is over the opening in the stage where light comes through. Do this by looking directly at the microscope stage (not through the eyepiece[s]). This ensures that the specimen will be in the field of view, or at least very close to it, when viewed through the eyepieces and lenses.
4. Check to see if the scanning objective is in place over the stage, and look into the eyepiece(s). Using the coarse adjustment knob, slowly move the stage up until the specimen comes into view. Then use the fine adjustment knob to bring the specimen into focus. Next, adjust the iris diaphragm and light control/voltage regulator to see how each affects both the clarity (the sharpness of the image) and contrast (the ability to distinguish the specimen from the background) of the image.

(continued on next page)

(continued from previous page)

5. Draw what is visible within the field of view in the space provided. Record the total magnification.

Total magnification = ______ ×

6. How does the image seen through the microscope differ from what is seen when looking directly at the slide?

____

____

____

____

____

7. Without changing the position of the slide or the stage, rotate the nosepiece so the low-power objective is in place. Because the slide was in focus with the scanning objective, use only the fine adjustment knob to bring the specimen into focus with the low-power objective. Each time the magnification increases, use only the fine adjustment knob to focus the specimen.

8. Repeat step 7 using the high-power objective. It may be necessary to adjust the light control to increase illumination, because light intensity decreases when changing from a low-power to a high-power objective.

EXERCISE 3.2B Working Distance

The **working distance** is the distance between the mechanical stage (and the slide upon it) and the tip of the objective lens **(figure 3.4).** The shorter the working distance, the more likely it is that the objective lens will touch the slide or the stage. Because of this, only the fine adjustment knob should be used when viewing specimens using high-power objective lenses.

1. Begin with the microscope stage at its lowest position. Place the slide of the letter *e* on the stage and bring it into focus using the scanning objective.
2. Using a millimeter ruler, measure the working distance: the distance between the top of the slide on the stage and the bottom of the objective lens. Record the distance here: ______ mm.
3. Change to the low-power objective lens and repeat the process in step 2. Record the distance here: ______ mm.
4. Change to the high-power objective and repeat the process in step 2. Record the distance here: ______ mm.
5. Use the results from steps 2–4 to answer question 4 of the Post-Laboratory Worksheet (p. 55).

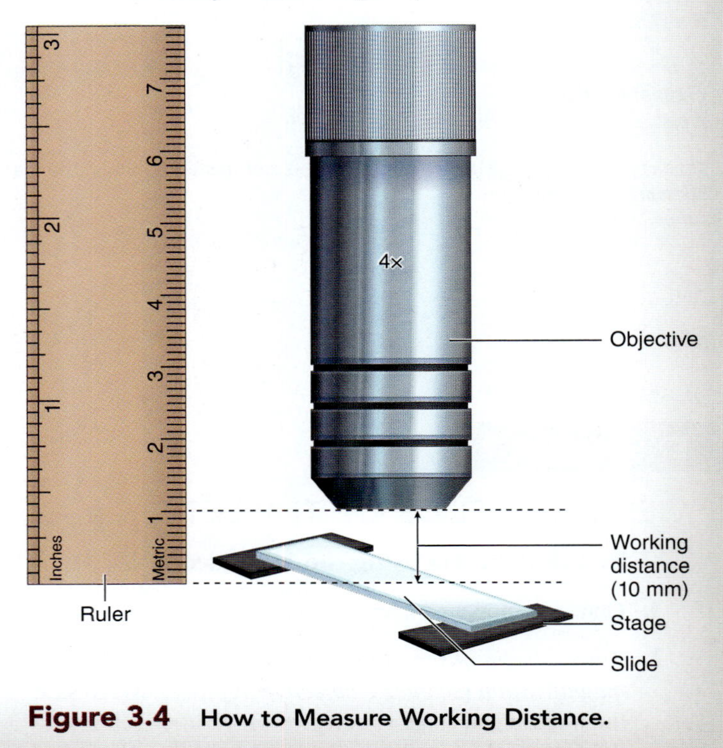

Figure 3.4 How to Measure Working Distance.

Diameter of the Field of View

The **field of view** is everything that is visible when looking through the eyepiece. As magnification increases, the field of view decreases. In this activity, the diameter of the field of view will be determined for the various magnifications of the microscope. The diameter of the field of view is simply the distance (usually given in millimeters [mm], and sometimes in micrometers [μm]) across the widest part of the visible field.

EXERCISE 3.3

MEASURING THE DIAMETER OF THE FIELD OF VIEW

1. Obtain a compound microscope and a stage micrometer or a clear ruler with mm increments.
2. Begin with the scanning objective in place. Place the stage micrometer slide on the stage and position it so that the markings are visible within the field of view when looking through the eyepiece. Then line up the first marking with the left-hand side of the field of view at the widest part. If using a ruler, line up one of the mm lines with the left-hand side of the field of view at the widest part.
3. Count the number of markings visible across the widest part of the field of view and record the number: ______ mm. This is the diameter of the field of view at scanning power.

INTEGRATE

LEARNING STRATEGY

A *millimeter* (mm) is 10^{-3} meters. A *micrometer* (μm) is 10^{-6} meters. At higher magnifications, the size of structures should be recorded using micrometers instead of millimeters. To convert 0.2 mm to micrometers, simply move the decimal point three positions to the right. Thus, a distance of 0.2 mm is a distance of 200 μm.

Practice: If the diameter of the field of view at a magnification of 400× is 0.48 mm, what is the diameter in micrometers? ______ μm

4. Repeat steps 1 and 2 with the low-power objective in place. Record the diameter of the field of view at low power: ______ mm.
5. It is often difficult to use the method from steps 1 and 2 to determine the diameter of the field of view for the high-power objective. However, the diameter can be calculated based on the measurements taken for the scanning or low-power objectives using the following formula (LP = low-power objective, HP = high-power objective):

   diameter at LP × total magnification at LP = diameter at HP × total magnification at HP

Example:

- Diameter of the field of view using low-power objective: 4.8 mm
- Total magnification using low-power objective: 40 ×
- Diameter of field of view using high-power objective: unknown (this is what will be calculated)
- Total magnification using high-power objective: 400 ×
- Calculation: 4.8 mm · 40× = unknown · 400×
  192 = unknown · 400×
  192/400 = 0.48 mm (round off the number if necessary)

The diameter of the field of view at a magnification of 400× is 0.48 mm.

EXERCISE 3.4

ESTIMATING THE SIZE OF A SPECIMEN

Once the diameter of the field of view for each of the objective lenses of the microscope is known, the approximate size of a specimen can be determined by comparing the size of the specimen to the diameter of the field of view. Use calculations shown in exercise 3.3 to determine the size of the same specimen viewed at three different magnifications.

1. **Figure 3.5** shows a specimen (outlined in red) at various magnifications. First, calculate the diameter of each field of view before estimating the length of the object (0.48 is the diameter of the field of view for a total magnification of 40×.).
2. Next, calculate the diameter of the object.

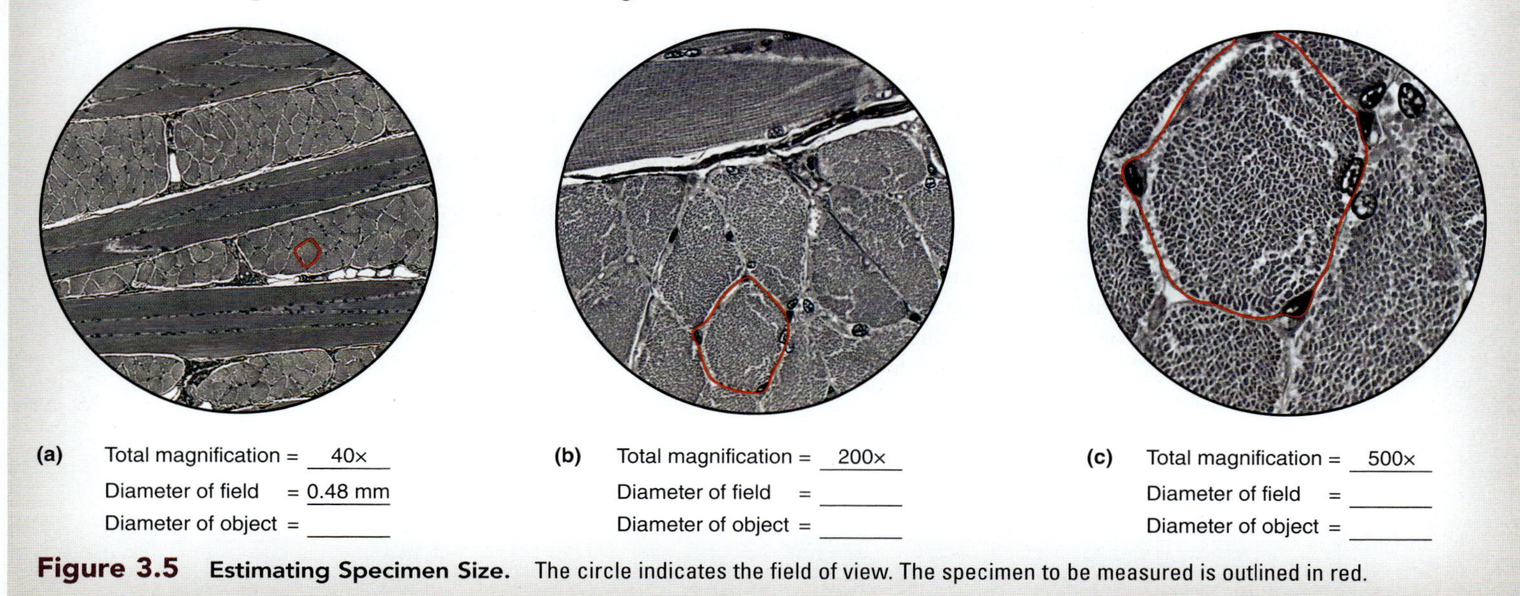

Figure 3.5 Estimating Specimen Size. The circle indicates the field of view. The specimen to be measured is outlined in red.

(continued on next page)

(continued from previous page)

3. Is it possible to measure or estimate the diameter of the object in figure 3.5*c*? ______

Explain your answer. ______________________________ ______________________________ ______________________________ ______________________________

Depth of Field

The **depth of field** is how much of the total thickness of a specimen is in focus at each magnification. As total magnification increases, the depth of field decreases and a smaller portion of the specimen will be in focus. **Figure 3.6** demonstrates what happens to the depth of field as total magnification changes.

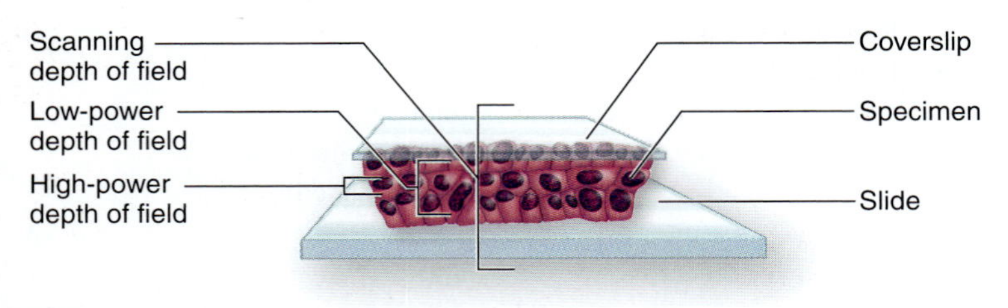

Figure 3.6 Depth of Field. Depth of field narrows as total magnification increases.

EXERCISE 3.5

DETERMINING DEPTH OF FIELD

1. Obtain a compound microscope and a slide of three crossed, colored threads.
2. Place the slide on the microscope stage and observe the slide with the scanning objective in place. Once three crossed threads are visible in the field of view, use the fine adjustment knob to focus up and down and determine which of the threads is on top, which is in the middle, and which is on the bottom. Then record the answers:

   Color of top thread: ______________________________

   Color of middle thread: ______________________________

   Color of bottom thread: ______________________________
3. Now observe the slide with the low-power objective in place. To see all the threads, does the fine adjustment knob need to be moved more or less at this magnification than at the magnification with the scanning objective in place? ____________________
4. Finally, observe the slide with the high-power objective in place. At this magnification, each individual thread will take up almost the entire diameter of the field of view. Is the entire thickness of an individual thread in focus at this magnification? __________. What does this indicate about the depth of field at this magnification?

   ____

   ____

   ____

INTEGRATE

CONCEPT CONNECTION

In each chapter of this laboratory manual, there are photographs of slides demonstrating cells, tissues, and organs. These slides will be used to determine both structure and function of the viewed specimens. When viewing these images, look for consistent structural patterns among tissue types and relate these patterns to the function of the cells, tissues, or organs. For example, there are cilia associated with the epithelium of the respiratory tract. There are also cilia associated with the epithelium that lines the uterine tube (see chapter 27). In both locations, cilia aid in movement of a substance along the epithelium. Cilia are responsible for clearing the airways and in moving a fertilized ovum (egg) toward the uterus for implantation. Similarly, microvilli are present in the epithelial tissue found in part of the kidney (see chapter 25) and small intestine (see chapter 26). In both cases, microvilli increase the overall surface area of the epithelium to facilitate absorption. When viewing histological slides, make every attempt to make connections among similar structures, their associated functions, and the body systems in which the structures are located.

Finishing Up

When you are finished, follow proper cleanup procedures.

1. Remove the slide from the microscope stage and put it back where it belongs.
2. Turn the power switch to the "off" position, unplug the power cord, and wrap the cord neatly around the base of the microscope.
3. Rotate the nosepiece so the scanning power objective clicks into place.
4. Lower the microscope stage to its lowest position.
5. Put the dust cover back on the microscope.
6. Return the microscope to its proper storage location.

GROSS ANATOMY

The Dissecting Microscope

A dissecting microscope (also called a stereomicroscope) is used to view structures that are larger than those viewed with a compound microscope, but that are often too small to see with the naked eye. The term "dissecting microscope" comes from the fact that these microscopes are useful for observing small dissections. The total magnification of a dissecting microscope is lower than that of a compound microscope. Use a dissecting microscope to view something like a human hair close up to see its gross structure in more detail. To see the actual cells and tissues that compose a human hair, use a compound microscope. Recall that a compound microscope is used to view tissues that have been prepared on microscope slides. In contrast, objects viewed with a dissecting microscope need not be prepared in this way. Instead, the objects are simply placed on the stage of the microscope and illuminated with a light from above (and sometimes below). The image is not inverted as it is with a compound microscope. For example, when the letter *e* is placed on a dissecting microscope and viewed, the letter *e* will appear the same as it does with the naked eye (not upside down as it does when viewed with a compound microscope).

Dissecting microscopes and compound microscopes have distinct features. Like a compound microscope, a dissecting microscope has both ocular and objective lenses. However, the objective lens(es) are much lower power, and there are two of them—one for each eye. Thus, each of the eyes looks through the microscope independently of the other, which allows the user to have a "stereo" or three-dimensional view of the specimen. In many cases, the objective lenses are also of varying powers, with typical magnification ranges of 1× to 3×.

While working through this exercise, note that many of the parts of a dissecting microscope have the same names as those of a compound microscope. **Figure 3.7** demonstrates an example of a dissecting microscope. This microscope uses an external illuminator that inserts into the back of the base of the microscope (not visible in the figure). The mirror below the stage then reflects the light through the glass on the stage to illuminate a specimen on the stage. Although it is often useful to illuminate specimens from below, specimens may also be illuminated from above. Many dissecting microscopes have both above- and below-stage illuminators for this very purpose. The microscope shown in figure 3.7 has only a below-stage illuminator.

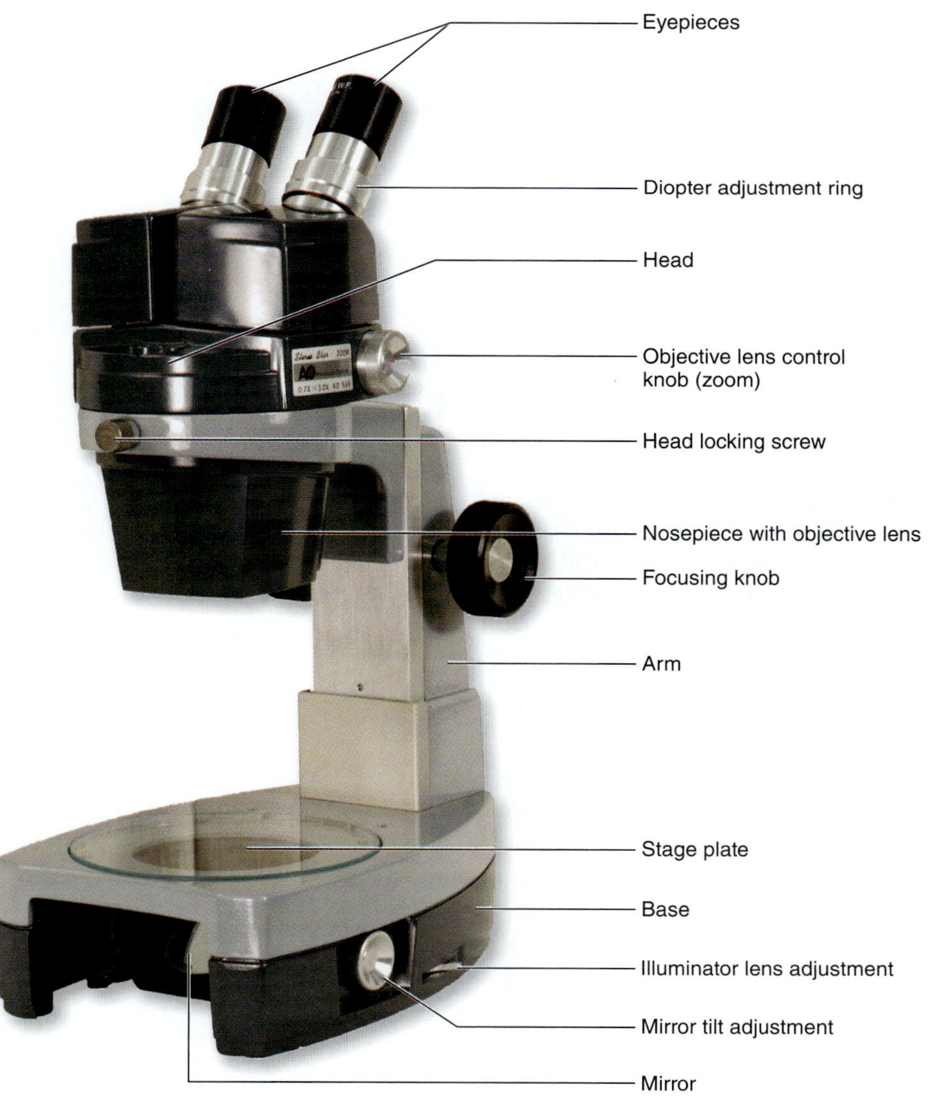

Figure 3.7 Parts of a Dissecting Microscope.

EXERCISE 3.6

PARTS OF A DISSECTING MICROSCOPE

1. Obtain a dissecting microscope.

2. Identify the following structures on the dissecting microscope. Use figure 3.7 and table 3.1 as guides.

- ☐ arm
- ☐ base
- ☐ diopter adjustment ring
- ☐ eyepieces
- ☐ focusing knob
- ☐ head
- ☐ head locking screw
- ☐ illuminator lens adjustment
- ☐ mirror
- ☐ mirror tilt adjustment
- ☐ nosepiece with objective lens
- ☐ objective lens control knob
- ☐ stage plate

3. Record the magnifications of the lenses on the microscope in the spaces below.

Magnification of ocular lenses: ______

Magnification of objective lenses:

Lowest ______ Highest (if applicable) ______

4. What is the total magnification of the microscope? If there is more than one objective lens, list the lowest and highest total magnifications possible.

Total magnification:
Lowest ______ Highest (if applicable) ______

5. Place a pen or pencil on the microscope stage and adjust the controls (focus, zoom, illuminator, etc.) while observing the object through the ocular lenses. Place a piece of paper with text written on it on the microscope stage and view that through the ocular lenses. Is there any change in the orientation of the letters as seen through the microscope as compared to viewing the letters without the microscope? ______________________
If so, describe what differences are observed. __________
____
____
____
____
____
____
____

INTEGRATE

CLINICAL VIEW

Nail Fungus

Knowing how to view structures using a microscope is an essential skill for physicians to possess. Many diseases are accurately diagnosed only after viewing structures using a microscope. For example, a dermatologist may use a microscope to determine if a patient is suffering from nail fungus. The dermatologist first scrapes under the patient's nail, then places the debris on a slide. After adding a drop or two of an isotonic wetting solution, the dermatologist looks for the presence of fungi under the microscope before the patient even leaves the examination room. All types of these microorganisms (i.e., bacteria and fungi) require a microscope for viewing, because none of these microorganisms are visible with the naked eye. Proper and rapid diagnosis of the microorganism causing the infection allows the dermatologist to determine the best course of treatment for the patient. This saves the patient and the physician both expense and time as compared to the alternative of having to send the tissue sample out to an external laboratory for testing.

Chapter 3: The Microscope

Name: ____________________

Date: __________ Section: __________

POST-LABORATORY WORKSHEET

The ❶ corresponds to the Learning Objective(s) listed in the chapter opener outline.

Do You Know the Basics?

Exercise 3.1: Parts of a Compound Microscope

1. Label the parts of the compound microscope. ❶

1 ____
2 ____
3 ____
4 ____
5 ____
6 ____
7 ____
8 ____
9 ____
10 ____
11 ____
12 ____
13 ____
14 ____
15 ____
16 ____

2. Describe how to perform each of the following tasks: ❶
   a. Transport the microscope. ____________________
   b. Position the microscope on a laboratory workstation. ____________________
   c. Clean the microscope lenses. ____________________
   d. Prepare the microscope for storage. ____________________

Exercise 3.2: Viewing a Slide of the Letter *e*

3. What is the total magnification of a microscope set up with an ocular lens magnification of 10× and an objective lens magnification of 43×? ______ ❷

Exercise 3.3: Measuring the Diameter of the Field of View

4. Complete the following table with the numbers observed or calculated in exercises 3.1, 3.2, and 3.3. ❸ ❹ ❺

Power	Ocular Magnification	Objective Magnification	Total Magnification	Diameter of the Field of View	Working Distance
Scanning					
Low					
High					

Exercise 3.4: Estimating the Size of a Specimen

5. Refer back to the calculations on the sample "specimens" in figure 3.5. Enter information from those calculations in the spaces provided. Pay attention to the units specified next to the answers, because they are not all the same. ❻

A.
Total magnification = 40×
Diameter of field = ______ mm
Length of object = ______ mm

B.
Total magnification = 200×
Diameter of field = ______ mm
Length of object = ______ mm

C.
Total magnification = 500×
Diameter of field = ______ µm
Length of object = ______ µm

Exercise 3.5: Determining Depth of Field

6. Record the answers to the questions about colored threads on page 52 in the spaces provided. ❼

   Color of top thread: ____________________ Color of middle thread: ____________________ Color of bottom thread: ____________________

7. Explain why proper microscope technique requires always viewing a slide with the scanning objective first before moving to higher-power objectives. Use the concept of *depth of field* in the explanation. ❼

   ____

   ____

   ____

   ____

   ____

Exercise 3.6: Parts of a Dissecting Microscope

8. Identify two differences between a dissecting microscope and compound microscope. ❽

   a. ______________________________

   b. ______________________________

9. List three types of specimens a dissecting microscope might be used to view: ❾

   a. ______________________________

   b. ______________________________

   c. ______________________________

Can You Apply What You've Learned?

10. What microscope structures are used to control the amount of light illuminating the specimen?

    ____

11. What happened to the light intensity when switching from low to high power?

    ____

12. What adjustment will typically have to be made to the light after changing from the low-power to the high-power objective?

    ____

13. a. How does working distance change as total magnification increases?

    ____

    b. What are the practical consequences of this change in working distance?

    ____

14. If four cells are visible within the field of view at the field's maximum diameter, and the total magnification is 200×, how many cells will be visible at a total magnification of 500×?

    ____

Can You Synthesize What You've Learned?

15. Describe why images viewed with a compound microscope are two-dimensional, whereas images viewed with a dissecting microscope are three-dimensional.

    ____

    ____

16. A patient presented to his physician complaining of an unusual growth on the skin of his upper back. The physician was unable to identify the growth, so she decided to perform a biopsy (take a tissue sample). After obtaining a sample of the unusual growth on the patient's back, the physician sent the sample to the pathology lab. Discuss how the pathologist would make use of a compound microscope to correctly diagnose the identity of the unusual growth.

    ____

    ____

Cell Structure and Membrane Transport

OUTLINE AND LEARNING OBJECTIVES

MODULE 2: CELLS & CHEMISTRY

INTRODUCTION

The cell is the basic unit of life. Organisms can be unicellular or multicellular, but they must be composed of cells to be considered living entities. Human beings are, of course, multicellular organisms. Our bodies are composed of **tissues:** groups of similar cells and associated extracellular materials that function together as a unit. The study of tissues is called **histology** [*histos*, web (tissue), + *logos*, study]. Understanding the study of histology first requires identifying cells and cellular organelles under the microscope. Most cells are easily seen with a light microscope, but most cellular organelles are too small to be seen without the use of a more powerful electron microscope.

Most animal cells are transparent. Because of this, when tissue samples are prepared for use in the anatomy and physiology laboratory, the slides are stained so cellular details will be visible when viewed under a microscope. Different parts of a cell attract biological stains to different degrees, which makes some parts of the cell appear darker in color, and others appear lighter in color, or even transparent. The nucleus of the cell has a high attraction for most biological stains, so it is often the most recognizable part of a cell.

Most of the slides that are viewed in the anatomy and physiology laboratory have been stained with hematoxylin and eosin, and are labeled "H and E." This stain makes the **cytoplasm** of the cell appear pink in color and makes visible the outline of the cell where the **plasma membrane** (the boundary of the cell) is located. The **nucleus** of the cell appears dark purple in color, and the **nucleolus** often appears as a dark spot within the nucleus. Note that these structures have been described using references to colors that result from the use of hematoxylin and eosin stains (i.e., pink and purple). It is important to remember that other types of stains may be used on slides prepared for the laboratory. Use of stains other than H and E will cause the same structures to have different colors. For this reason, do not use color alone as an identifying feature when viewing slides. Instead, learn to recognize cells and cellular organelles based on *shape*.

The exercises in the "histology" and "gross anatomy" sections of this chapter are designed to aid in the identification of cellular organelles that are visible using a light microscope, and with identification of the stages of mitosis in a whitefish embryo. The exercises in the "physiology" section of this chapter are mainly "wet lab" activities that demonstrate the processes of diffusion, osmosis, and filtration.

One of the most important components of a cell is the plasma membrane, which is **selectively permeable.** The selective permeability of the plasma membrane allows the cell to control both its internal and external environment through the physiological processes of membrane transport. Mechanisms of membrane transport include **simple diffusion, facilitated diffusion, osmosis, active transport,** and **vesicular transport.** The exercises in this chapter explore the effect of factors such as temperature on the rate of diffusion and the process of osmosis across an artificial membrane. A simulation activity (Ph.I.L.S.) covers observation of the effect of placing erythrocytes (red blood cells) in solutions that mimic changes in extracullular concentration. Although filtration is not a cellular process, the process of filtration will be observed. An additional activity covers the effect of changing fluid (hydrostatic) pressure on filtration rate.

List of Reference Tables

Chapter 4: Cell Structure and Membrane Transport

Name: ______________________

Date: __________ Section: __________

These Pre-Laboratory Worksheet questions may be assigned by instructors through their connect course.

PRE-LABORATORY WORKSHEET

1. Match the function listed in column A with the cell structure listed in column B.

Column A

_____ 1. provides a selectively permeable barrier between the intracellular and extracellular environment of the cell

_____ 2. contains the cell's genetic material (DNA)

_____ 3. synthesizes new proteins destined for the plasma membrane, for lysosomes, or for secretion from the cell

_____ 4. a stack of flattened membranes that are the site where proteins from the ER are modified, packaged, and sorted for delivery to other organelles or to the plasma membrane of the cell

_____ 5. often referred to as the "powerhouse" of the cell, these organelles are the site of cellular respiration

Column B

a. Golgi apparatus

b. mitochondria

c. nucleus

d. plasma membrane

e. rough endoplasmic reticulum (RER)

2. Number the following stages of mitosis in the correct order.

_____ a. anaphase

_____ b. metaphase

_____ c. prophase

_____ d. telophase

3. The stage of the cell cycle when cells are *not* undergoing mitosis is ______________. During this phase, individual chromosomes ______________ (are/are not) visible within the nucleus of the cell.

4. Which of the following factors increase the rate of diffusion of a substance? (Check all that apply.)

_____ a. decreased temperature

_____ b. decreased viscosity of the solvent

_____ c. increased molecular weight of the substance

_____ d. increased permeability of the membrane

5. When a red blood cell is placed in a hypertonic solution, which of the following may occur? (Check all that apply.)

_____ a. crenation

_____ b. intracellular volume will decrease

_____ c. intracellular volume will increase

_____ d. lysis

6. Match the definition listed in column A with the term listed in column B.

Column A		*Column B*
______	1. the dissolved substance in a solution	a. hypertonic
______	2. a solution into which another substance dissolves	b. hypotonic
______	3. a solution that has a lower osmotic pressure than intracellular fluid	c. isotonic
______	4. a solution that has a higher osmotic pressure than intracellular fluid	d. osmosis
______	5. a solution that has the same osmotic pressure as intracellular fluid	e. solute
______	6. the movement of water across a semipermeable membrane	f. solvent

7. Which of the following properties that influence the rate of diffusion is studied when measuring the diffusion distance of potassium permanganate in both water and agar? (Circle one.)
   a. molecular weight of the substance
   b. permeability of the membrance
   c. temperature of the solvent
   d. viscosity of the solvent

8. Which of the following properties that influence the rate of diffusion is studied when measuring the diffusion distance of both methylene blue and potassium permanganate in agar? (Circle one.)
   a. molecular weight of the substance
   b. permeability of the membrance
   c. temperature of the solvent
   d. viscosity of the solvent

9. A ____________________ (concentration/pressure) gradient drives the movement of solutes in diffusion, whereas a ____________________ (concentration/pressure) gradient drives the movement of fluid in filtration.

HISTOLOGY

Structure and Function of a Generalized Animal Cell

Table 4.1 lists the parts of an animal cell and gives descriptions of the functions and microscopic features of each. The parts of an animal cell that are most readily visible under a light microscope are the nucleus, nucleolus, and plasma membrane (which is the boundary of the cell). While observing animal cells under the light microscope, focus on finding these parts of a typical animal cell. Learning to recognize what parts of an animal cell are typically visible under the light microscope serves as preparation for observing different cell types that will be presented in future laboratory exercises.

Table 4.1 Parts of a Generalized Animal Cell

Organelle / Structure	Function	Microscopic Features	Word Origin	Appearance
Centrioles	Paired organelles that are used to organize the spindle microtubules that attach to chromosomes during mitosis. The area next to the nucleus that contains the centrioles is called the *centrosome*.	Visible only when a cell is actively undergoing nuclear division (mitosis)	*kentron*, center	Centrosome; Centriole
Chromatin	Genetic material within the nucleus; consists of uncoiled chromosomes and associated proteins	Most of the colored material visible in the nucleus (with exception of the nucleolus) consists of chromatin.	*chroma*, color	Nucleus; Nuclear pores; Nuclear envelope; Nucleolus; Chromatin
Cytoplasm	Includes cellular organelles and cytosol. Cytosol contains enzymes that mediate cytosolic reactions, such as glycolysis and fermentation.	Clear and homogeneous in appearance; may contain granular substances such as glycogen in certain cells (e.g., hepatocytes)	*kytos*, a hollow (cell), + *plasma*, something formed	Cytoplasm; Cytosol; Organelles; Inclusions
Cytoskeleton	Provides the main structural support for the cell and is composed of microtubules, intermediate filaments, and microfilaments	Not generally visible under the light microscope	*kytos*, a hollow (cell), + *skeletos*, dried	Cytoskeleton; Intermediate filament; Microfilament; Microtubule
Endoplasmic Reticulum (ER)	Site of lipid synthesis and detoxification of drugs and alcohol (smooth ER). Additionally, rough ER synthesizes proteins destined for the cell membrane, for lysosomes, or for secretion.	Not generally visible under the light microscope. In neurons, the rough ER stains very dark and is called chromatophilic substance (Nissl bodies).	*endon*, within, + *plasma*, something formed, + *rete*, a net	

(continued on next page)

Table 4.1 Parts of a Generalized Animal Cell *(continued)*

Organelle / Structure	Function	Microscopic Features	Word Origin	Appearance
Golgi Apparatus	A stack of flattened membranes that receive proteins from the rough ER and then modify, package, and sort them for delivery to other organelles or to the plasma membrane of the cell	Not generally visible under a light microscope	*Golgi*, Camillo, Italian histologist and Nobel laureate, 1843–1926	
Lysosomes	Membrane-enclosed sacs that contain digestive enzymes; function in the breakdown of intracellular debris	Not generally visible under a light microscope	*lysis*, a loosening, + *soma*, body	
Mitochondria	Often referred to as the "powerhouse" of the cell. These organelles are the site of cellular respiration; the metabolic pathway that utilizes oxygen in the breakdown of food molecules to produce ATP.	Not generally visible under a light microscope	*mitos*, thread, + *chondros*, granule	
Nucleolus	Synthesizes rRNA and assembles ribosomes in the nucleus	Recognized as a small, dark, circular structure within the nucleus	*nucleus*, a little nut	Nucleus; Nuclear pores; Nuclear envelope; Nucleolus; Chromatin
Nucleus	Contains the cell's genetic material (DNA)	The most noticeable feature of a cell; typically stains very dark	*nucleus*, a little nut	
Peroxisomes	Membrane-enclosed sacs that contain catalase and other oxidative enzymes. The enzymes break down lipids and toxic substances by first converting them into hydrogen peroxide and then breaking down the hydrogen peroxide into water and oxygen.	Not generally visible under a light microscope	*peroxi*, relating to hydrogen peroxide, + *soma*, body	
Plasma Membrane	Provides a selectively permeable barrier between the intracellular and extracellular environments of the cell	Visible only using an electron microscope. However, the outer border of the cell, where the cell membrane is located, is often visible under the light microscope.	*plasma*, something formed, + *membrane*, a membrane	Plasma membrane
Ribosomes	Sites of protein synthesis: may be bound to the ER ("fixed") or within the cytoplasm ("free")	Not generally visible under a light microscope	*ribose*, the sugar in RNA, + *soma*, body	Free ribosomes; Fixed ribosomes

EXERCISE 4.1

OBSERVING CELLULAR ANATOMY WITH A COMPOUND MICROSCOPE

Preparing a Wet Mount of Human Cheek Cells

This exercise involves taking a sample of cells from the inside of the cheek and preparing a wet mount. A wet mount is a procedure that involves placing a tissue sample in a wet medium onto a microscope slide. The "wet" medium is typically an isotonic saline solution. Why is it important for the solution to be isotonic? ______________________________

____

The cells on the inside of a human cheek are squamous cells, which are flattened cells. The inside of the cheek is lined with multiple layers of these cells. Therefore, a few cells may be gently scraped off with a toothpick without causing damage to the entire epithelium (lining) of the inside of the mouth. Details regarding the structure of cheek cells are described in the histology section covering epithelial tissues in chapter 5.

After obtaining cheek cells, they will be placed on a microscope slide, a stain (methylene blue) will be applied, and they will be covered with a coverslip. The stain is necessary to visualize the cells because normal cells are nearly transparent. Methylene blue is basophilic (base-loving) and is attracted to eosinophilic (acid-loving) components of the cell. The most eosinophilic part of the cell is the nucleus, which contains the nucleic acids DNA and RNA. Thus, the nucleus of the cell stains more intensely than other parts of the cell.

Obtain the Following:

- **compound microscope**
- **microscope slide and coverslip**
- **toothpick or wood applicator stick**
- **methylene blue solution with eyedropper**
- **fine tissue paper or KimWipes®**

1. Place the microscope slide on a piece of white paper on the lab bench. The piece of white paper will make observations easier.
2. Place a small drop of normal saline on the microscope slide **(figure 4.1*a*).** Very *gently* scrape the toothpick along the inside of your mouth to pick up a few cells. This process should not be painful, and most definitely should not draw blood!
3. Next, place the tip of the toothpick in the drop of saline on the slide (figure 4.1*b*). Roll the toothpick around gently so the cells detach from the toothpick and fall into the drop of saline.

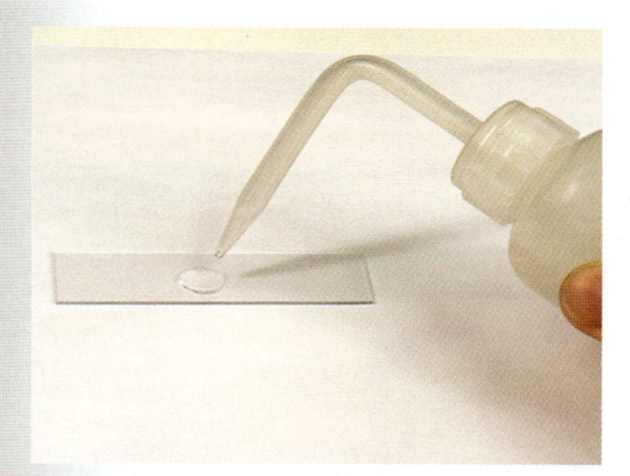

(a) Place a drop of normal saline on the slide.

(b) After collecting cheek cells, gently roll the tip of the toothpick in the drop of saline.

(c) Place a drop of methylene blue on the slide.

(d) Slowly lower a cover slip over the drop of liquid containing cheek cells.

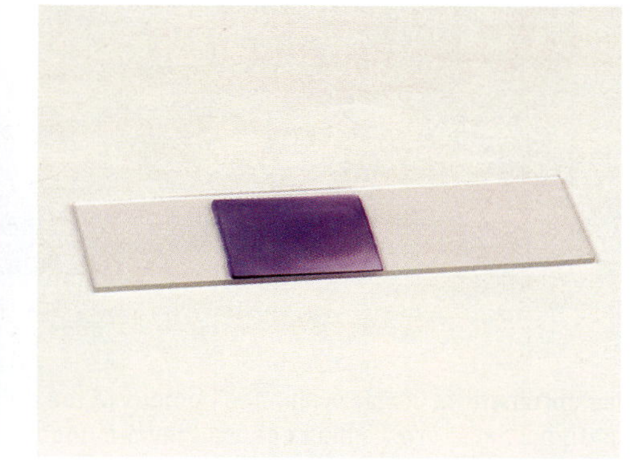

(e) Make sure there are no air bubbles between the cover slip and the slide. The slide is now ready to view with the microscope.

Figure 4.1 Preparing a Wet Mount of Human Cheek Cells.

(continued on next page)

(continued from previous page)

4. Obtain a vial of methylene blue solution. Place a single drop of methylene blue on the drop of saline containing cheek cells (figure 4.1*c*).
5. Obtain a coverslip and place it on the edge of the liquid on the microscope slide as shown in figure 4.1*d*. Carefully and slowly lower the coverslip onto the drop of liquid. The goal is to place the coverslip over the drop of saline without introducing air bubbles. Simply dropping the coverslip on the slide may cause large air bubbles to form between the slide and the coverslip, which will interfere with the ability to see the cells on the slide. To prevent air bubbles from forming, carefully and slowly lower the coverslip, starting on one side and lowering it down at an angle (figure 4.1*d*). Most air bubbles will be pushed out of the way as the coverslip is placed. Obtain a piece of tissue paper or a KimWipe® and use it to dab any excess liquid on the sides of the coverslip (if necessary). A successful wet mount of cheek cells should resemble that shown in figure 4.1*e*.
6. Observe the slide with your naked eye. Can you see anything on the slide? In particular, can you see any cheek cells?
7. Arrange the objective lens on the microscope so it is set to use the scanning objective. Place the slide containing the cheek cells on the microscope stage and bring the tissue sample into focus using the scanning objective.

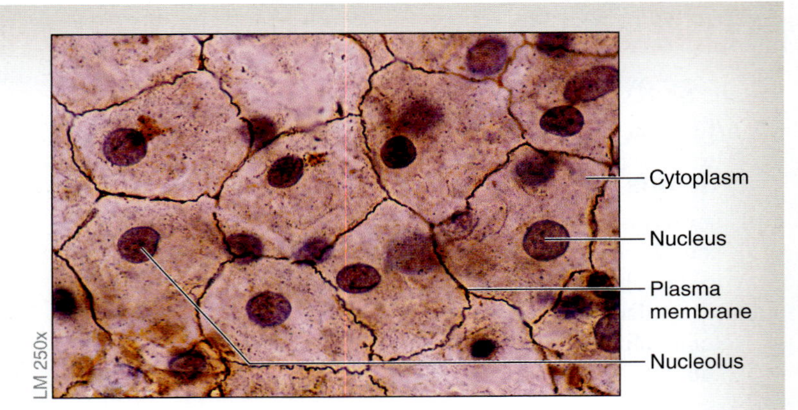

Figure 4.2 **Human Cheek Cells.**

Next, change to a higher power and bring the tissue sample into focus once again. The cells observed should somewhat resemble those in **figure 4.2,** although they will be isolated from one another rather than being in a sheet as in figure 4.2.

8. Scan the slide until cheek cells are visible in the field of view. Sketch the cheek cells as seen through the microscope in the space provided. Label the following on the sketch:

☐ **cytoplasm** ☐ **nucleolus**
☐ **nucleus** ☐ **plasma membrane**

Table 4.2 Appearance of Whitefish Embryo Cells Undergoing Mitosis During Phases of the Cell Cycle

Stage	Interphase	Prophase
Histological View	Nucleus with chromatin LM 450x	Nucleus with dispersed chromosomes LM 450x
Recognizable Features	Loose chromatin is visible within the nucleus of the cell. No chromosomes are visible because they are uncoiled (i.e., chromatin) at this stage.	Chromosomes become visible as the chromatin coils. If the cell is in early prophase, the fibers of the mitotic spindle may be visible adjacent to the nucleus.
Word Origins	*inter*, between, + *phasis*, an appearance	*pro*, before + *phasis*, an appearance

Troubleshooting:

If the slide appears to consist mostly of cellular debris instead of intact cells, it is likely that distilled water was mistakenly used in place of saline when making the slide. Distilled water is a hypotonic solution. When cells are placed in a hypotonic solution, they will lyse, leaving only cellular debris on the slide.

Sometimes the view through the microscope appears to vibrate or shake, which makes it impossible to visualize the cells. If this happens, there is too much fluid between the coverslip and the slide. Use a KimWipe® to draw some of the excess fluid out from under the coverslip. Then, observe the slide again.

9. *Optional Activity:* **APR 2: Cells & Chemistry**—Examine the "Generalized cell" dissection and test yourself on cell structures in the Quiz area.

Mitosis

The cell cycle describes the events that occur during the process of forming a new cell. As a cell passes through the stages of the cell cycle, two identical daughter cells are formed from one original parent cell. The cell cycle is divided into two main phases: interphase and mitotic phase. Interphase is the time in which the genetic material is uncoiled as chromatin. Mitotic phase (*mitos*, thread) includes the processes by which cells reproduce and includes mitosis (nuclear division, which is divided into four stages) and cytokinesis (*cyto-*, cell, + *kinesis*, movement), which is division of the cytoplasm. Note that casual usage of the term "mitosis" usually implies that the cytoplasm and cellular organelles have also divided. **Table 4.2** describes the microscopic appearance of cells in interphase and each of the four stages of mitosis. The four stages are (in order): prophase, metaphase, anaphase, and telophase. (The stages of mitosis can be remembered with the acronym P-MAT.) In this laboratory exercise the goal is to locate cells in interphase and each of the stages of mitosis by observing whitefish embryos (blastulas). Whitefish embryos are very small and are rapidly developing, which makes them ideal specimens for observing cells undergoing mitosis.

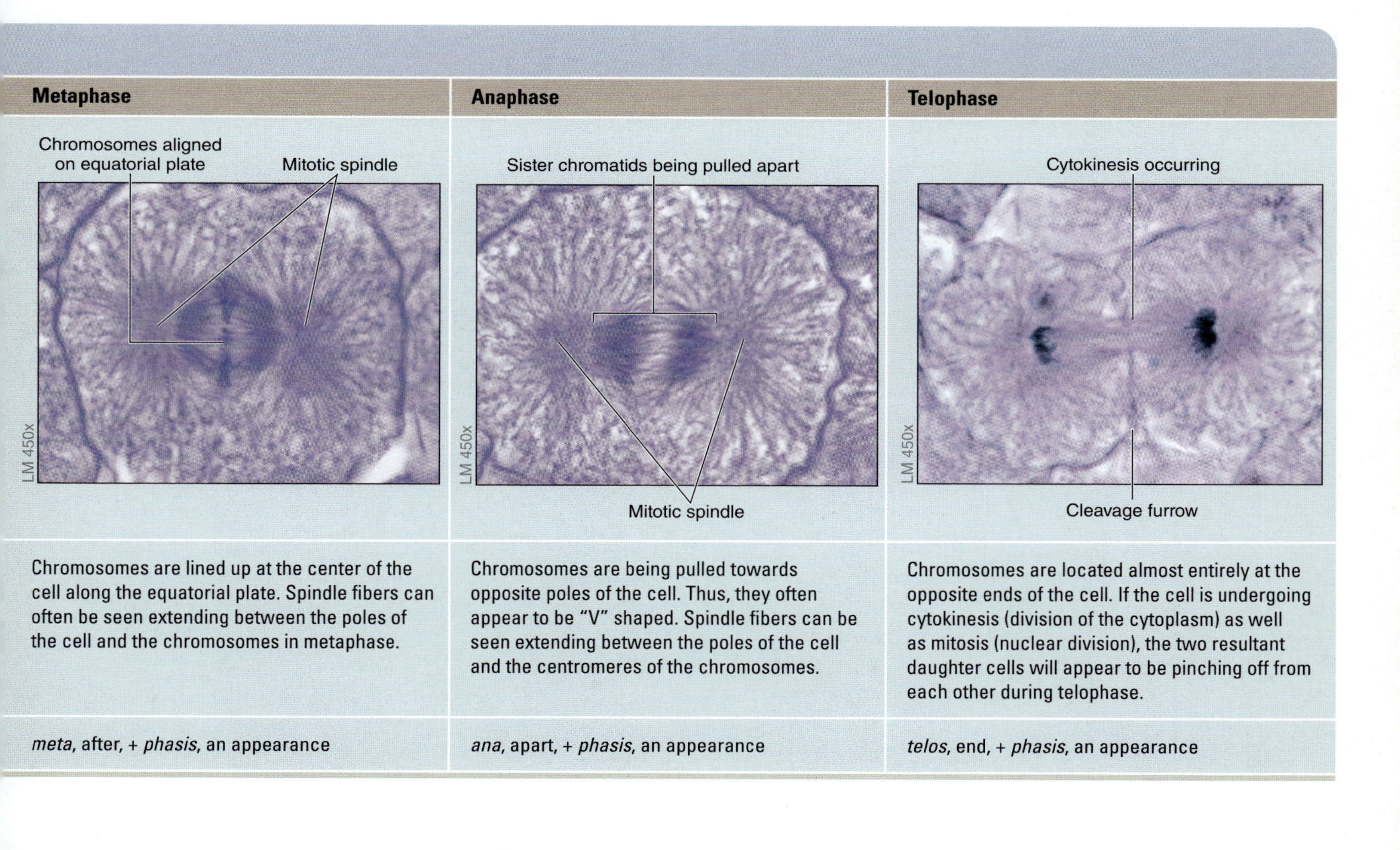

Metaphase	Anaphase	Telophase
Chromosomes are lined up at the center of the cell along the equatorial plate. Spindle fibers can often be seen extending between the poles of the cell and the chromosomes in metaphase.	Chromosomes are being pulled towards opposite poles of the cell. Thus, they often appear to be "V" shaped. Spindle fibers can be seen extending between the poles of the cell and the centromeres of the chromosomes.	Chromosomes are located almost entirely at the opposite ends of the cell. If the cell is undergoing cytokinesis (division of the cytoplasm) as well as mitosis (nuclear division), the two resultant daughter cells will appear to be pinching off from each other during telophase.
meta, after, + *phasis*, an appearance	*ana*, apart, + *phasis*, an appearance	*telos*, end, + *phasis*, an appearance

EXERCISE 4.2

OBSERVING MITOSIS IN A WHITEFISH EMBRYO

1. Obtain a compound microscope and a prepared slide of a whitefish embryo (blastula) or a slide of a different type of cell undergoing mitosis.
2. Scan the slide and locate cells in interphase and the four stages of mitosis that are listed in table 4.2. Once a cell in a particular phase has been located, switch to a higher-power objective to see the cell more clearly. Note that not all cells undergo mitosis simultaneously. Therefore, cells in various stages of mitosis and interphase may all be observed on the same slide.

   ☐ **anaphase** ☐ **prophase**
   ☐ **interphase** ☐ **telophase**
   ☐ **metaphase**

3. Sketch the appearance of cells in each of the phases listed in step 2 in the spaces provided.

Stage: Interphase

Stage: Anaphase

Stage: Prophase

Stage: Telophase

Stage: Metaphase

GROSS ANATOMY

Models of a Generalized Animal Cell

Exercises in this section involve observing classroom models demonstrating a generalized animal cell and observing classroom models demonstrating the stages of mitosis. **Figure 4.3** is a photograph of a classroom model of a generalized animal cell, which has the organelles labeled for reference.

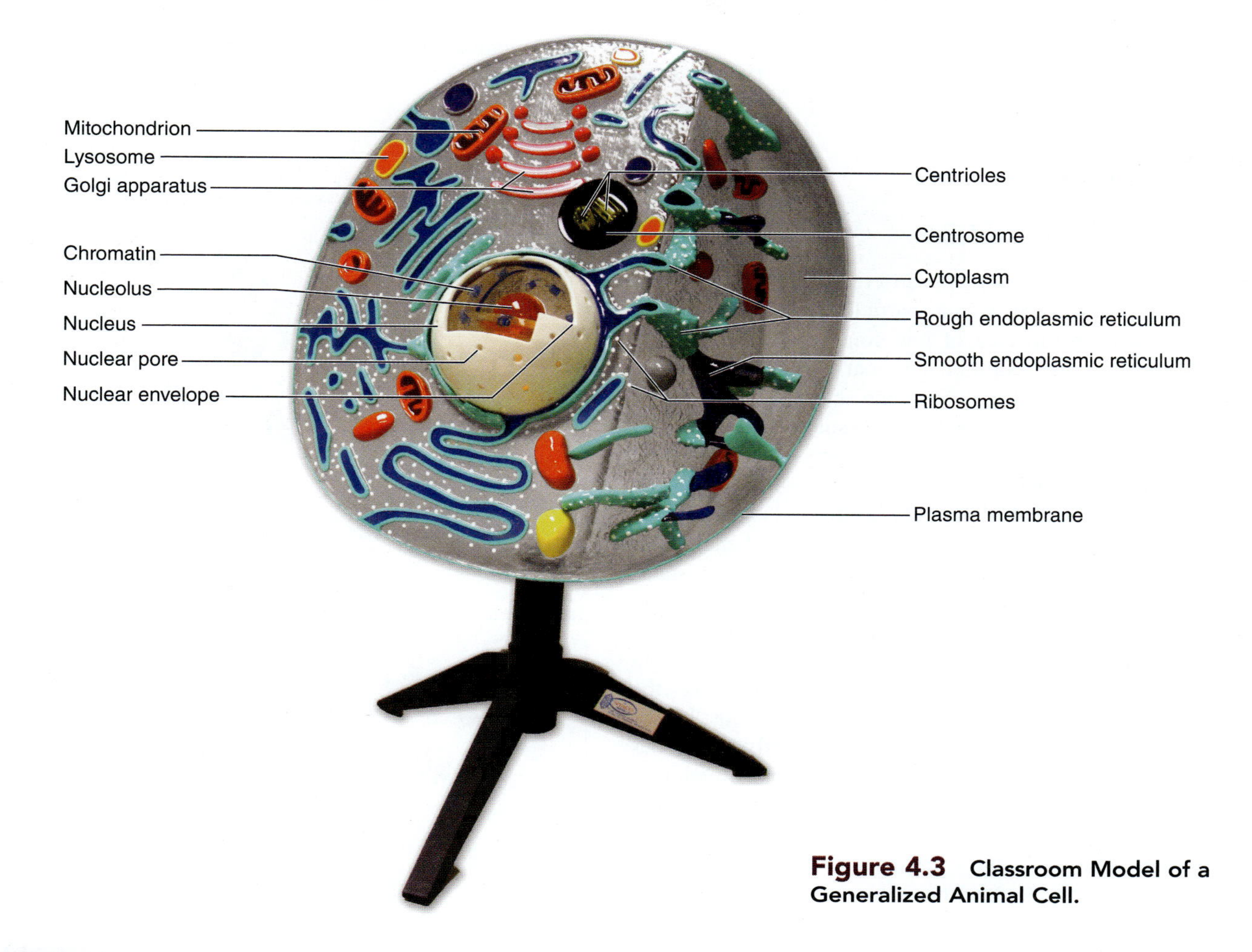

Figure 4.3 **Classroom Model of a Generalized Animal Cell.**

EXERCISE 4.3

OBSERVING CLASSROOM MODELS OF CELLULAR ANATOMY

1. Obtain a classroom model demonstrating a generalized animal cell.
2. Identify the listed structures on the classroom model of a generalized animal cell. Use figure 4.3, table 4.1, and the textbook as guides.

- ☐ **centrioles**
- ☐ **chromatin**
- ☐ **cytoplasm**
- ☐ **Golgi apparatus**
- ☐ **lysosome**
- ☐ **mitochondria**
- ☐ **nuclear envelope**
- ☐ **nucleolus**
- ☐ **nucleus**
- ☐ **peroxisome**
- ☐ **plasma membrane**
- ☐ **nuclear pore**
- ☐ **ribosomes**
- ☐ **rough ER**
- ☐ **smooth ER**

3. Sketch the appearance of a generalized cell, including all visible organelles, in the space provided.

PHYSIOLOGY

Mechanisms of Passive Membrane Transport

Passive transport mechanisms are mechanisms substances use to cross the cell membrane that do not require an input of energy. Because of this, such mechanisms can be studied using artificial membranes and/or simulations of membranes. The following exercises explore several of the factors that affect the rates of passive transport across cell membranes. Before beginning these exercises, read the textbook to become familiar with the following concepts: kinetic energy, passive transport, molecular weight, simple diffusion, osmosis, hypertonic solution, hypotonic solution, isotonic solution, and membrane impermeable solute.

EXERCISE 4.4

DIFFUSION (WET LAB)

Diffusion (*diffundo*, to pour in different directions) refers to the net movement of solute particles from an area of high concentration to an area of low concentration. Diffusion occurs whenever a concentration gradient exists. As concerns a solution, the substance dissolved in the solution is the **solute,** and the substance in which the solute is dissolved is the **solvent.** When working with solutions in the anatomy and physiology laboratory, the solvent will typically be water.

The **second law of thermodynamics** (also called the **law of entropy**) is the law governing the natural behavior of molecules, including molecules in solution. This law states that, over time, all molecules move at random toward a state of increasing disorder. **Entropy** (*entropia*, a turning toward) is a measure of the disorder of a system. Thus, for a system to become more disorderly, no energy input is required (as the saying goes, "entropy happens"). Similarly, energy must be put into a system for the system to become more orderly. How does the law of entropy apply to diffusion? Consider the situation in **figure 4.4.** The purple dots represent solute particles in solution. On the left of figure 4.4*a* the solution is more concentrated, and on the right of figure 4.4*a* the solution is more dilute. Observe each particle in relation to the other solute particles. Notice that where the solution is more concentrated, the particles are packed close together, and thus are more orderly. Where the solution is more dilute, the solute particles are farther away from each other and thus are more disorderly. When following the movement, or **flux** (*fluxus,* to flow), of the solute particles over time, the random movement of solute particles in solution is expected to result in a situation where the particles become as disorderly as possible. This means each particle will be as far away from another particle as is possible. Figure 4.4*b* shows what happens after diffusion has taken place.

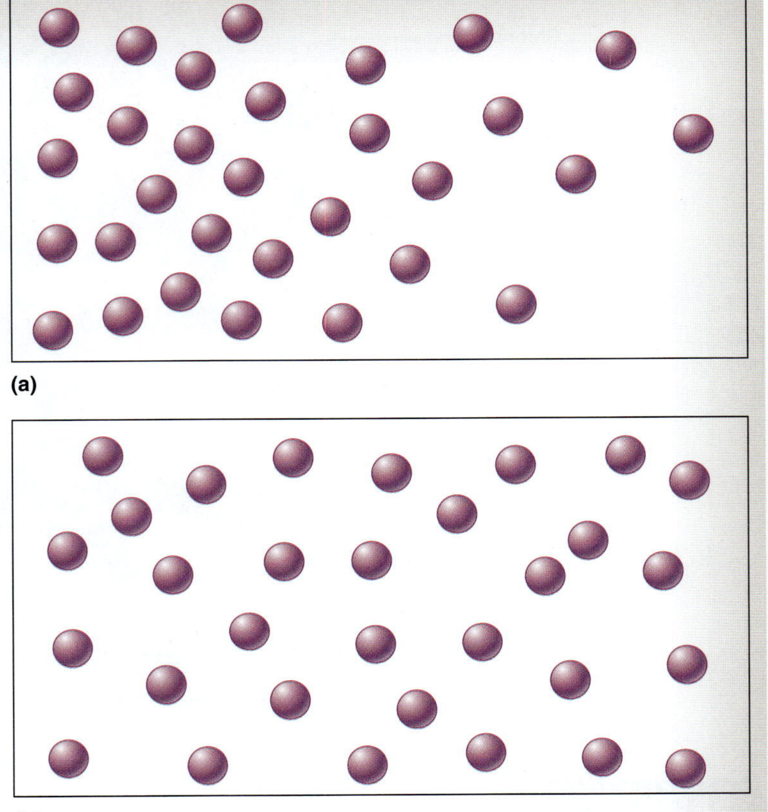

Figure 4.4 Diffusion. (a) Initially concentrated molecules. (b) Molecule distribution after diffusion takes place.

When two solutions of differing concentrations are separated from each other by a membrane that is permeable to the solute in question, solute particles will move, at random, both with and against the concentration gradient because of their inherent kinetic energy (kinetic energy is energy of *motion*). Thus, there will be two one-way fluxes: A → B and B → A **(figure 4.5).** However, the *net* movement of particles (the difference between the two one-way fluxes: A → B and B → A) will be from an area of high concentration to an area of low concentration. Once the concentration of solute is the same on both sides of the membrane, the solution is said to be at *equilibrium.* It is important to note that although equilibrium has been reached at that point, this does not mean that solute particles have stopped moving across the membrane. On the contrary, particles continue to move at random and will still cross the membrane. However, the rate at which particles move from container A to container B will be equal in magnitude but opposite in direction to the rate at which particles move from container B to container A. Thus, *no net movement* of solute will occur. This is how the point of equilibrium is defined.

Some factors that affect the rate at which a substance diffuses include

1. viscosity of the solvent
2. temperature of the solvent
3. molecular weight of the solute
4. permeability of the membrane

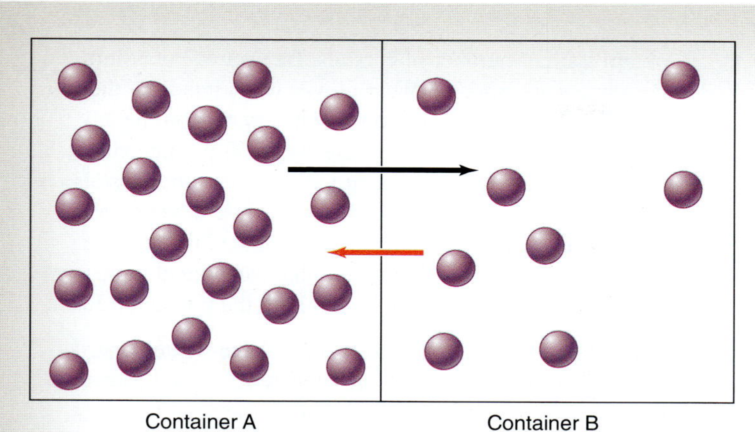

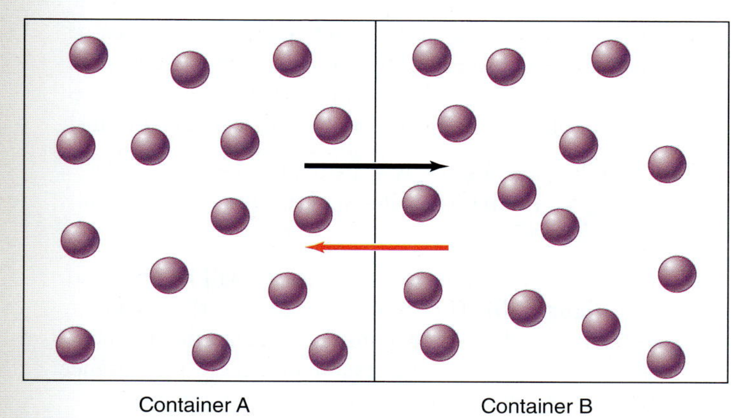

Figure 4.5 Diffusion Through a Semipermeable Membrane. (a) Diffusion of solute from A → B is greater than the diffusion of solute from B → A. Thus, the net movement is from A → B. (b) Once equilibrium has been reached, the movement from A → B is equal in magnitude but opposite in direction to the movement from B → A. Thus, there is no net movement at equilibrium, and the concentration of solute in both containers is the same.

INTEGRATE

CONCEPT CONNECTION

The movement of substances into and out of the cell is an important concept that provides the foundation for many physiological processes. For example, passive and active transport are the basic processes that make our neurons and muscle cells "excitable." These cells, as with all cells, contain a plasma membrane that is selectively permeable, allowing water and small, mainly lipid-soluble substances to pass freely into and out of the cell by passive transport. Protein pumps embedded within the plasma membrane actively pump ions (active transport) such as Na^+ and K^+, maintaining the appropriate concentrations of solutes in the intracellular fluid (ICF) and extracellular fluid (ECF). When protein channels open, these ions can flow along their concentration gradients from an area of high concentration to an area of low concentration (passive transport). In doing so, they also carry positive charges across the membrane. The movement of charges across the membrane is responsible for generating the electrical signals called action potentials. This process will be described in much more detail in chapter 14.

In the following exercises the effect of each of these factors on rates of diffusion will be observed.

EXERCISE 4.4A Effect of Viscosity on Rate of Diffusion

This exercise involves observing the diffusion of a solute (potassium permanganate) in media of differing viscosities (water and agar). Agar is a gel-like substance that is made from algae. Agar is 98% water. Thus, molecules will diffuse through it. However, agar is more viscous (thicker) than water. In this experiment one crystal of potassium permanganate will be placed in a petri dish containing agar. Another crystal of potassium permanganate will be placed in a petri dish containing water. The rate of diffusion will then be observed in each dish. Before beginning the experiment, state a hypothesis regarding the effect of solvent viscosity on the rate of diffusion.

Hypothesis: ____

____

____

____

Obtain the Following:

- **letter-size (8.5" × 11") piece of white paper**
- **petri dish containing distilled water**
- **petri dish containing agar**
- **fine-tissue forceps**
- **potassium permanganate crystals**
- **one large (~30 cm) and 2 small (~15 cm) metric rulers**
- **a stopwatch or other device for recording the time**

1. Obtain a small weigh boat containing potassium permanganate crystals **(figure 4.6).** Only two moderate-sized crystals are needed for this experiment, one for the agar dish and one for the water dish. Thus, only

Pour a small amount of potassium permanganate into a weigh boat or other small container. Obtain a single moderate-sized crystal using tissue forceps.

Figure 4.6 Obtaining Potassium Permanganate Crystals.

(continued on next page)

(continued from previous page)

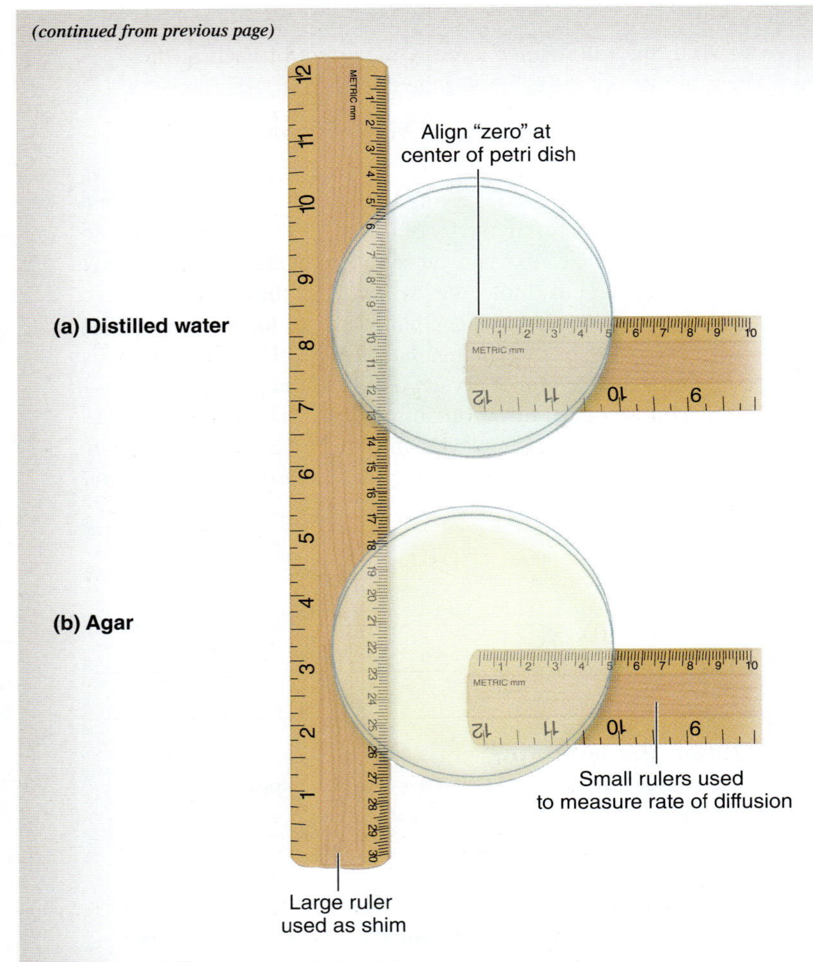

Figure 4.7 Setup of Petri Dishes and Rulers for Exercise 4.4A.

obtain a small number of crystals. Be aware! Potassium permanganate is a dye. Wear gloves when handling the crystals and take care not to spill the crystals on the lab table, books, or clothing because they will permanently dye those items. If a spill occurs, first sweep the crystals up with a dry brush or cloth. Try not to use a wet towel until most of the crystals have been cleaned up because the water will create a liquid dye that can spread more easily (and thus stain more items) than the powdered form.

2. Place the rulers and the petri dishes on the white paper as shown in **figure 4.7.** The large ruler will be used as a shim to keep the petri dishes level, whereas the small rulers will be used to measure rates of diffusion. If using very thin rulers, a shim will not be necessary. Place the large ruler on the left side of the paper and orient it vertically. Place the small rulers at right angles to the large ruler, approximately 8 cm apart from each other, and with the "zero" point approximately at the center of the petri dish.
3. Label the petri dishes "A – distilled water" and "B – agar" on the paper in the space next to each dish (see figure 4.7).
4. The diffusion of dye crystals in **water** will be observed first **(figure 4.8).** This process happens fairly quickly, so be prepared before beginning the experiment. One lab partner will place the crystal and record the distance diffused. Another student will start the stopwatch and record time intervals. To begin: Carefully place a single potassium permanganate crystal in the center of the petri dish containing water (dish A) at the "0" mark on the ruler. The moment the crystal is placed in the water, start the stopwatch and observe diffusion of the dye. Record the radius of the dye spot in 30-second intervals for a total of at least 5 minutes (see **figure 4.9** for a description

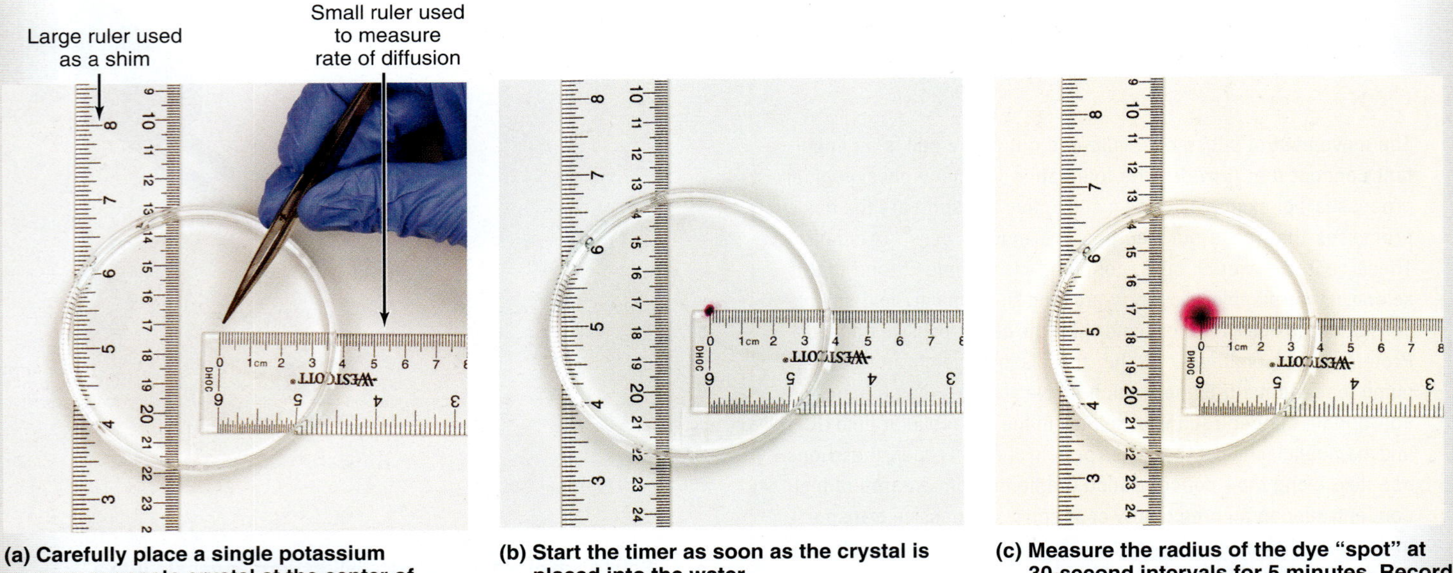

(a) Carefully place a single potassium permanganate crystal at the center of the petri dish at the "0" mark on the ruler.

(b) Start the timer as soon as the crystal is placed into the water.

(c) Measure the radius of the dye "spot" at 30-second intervals for 5 minutes. Record your data in table 4.3.

Figure 4.8 Observing Diffusion of Potassium Permanganate Crystals in Water.

Table 4.3 Effect of Solvent Viscosity on Rate of Diffusion

Water		Agar	
Time	**Distance Diffused (mm)**	**Time**	**Distance Diffused (mm)**
30 sec		*15 min*	
1 min		*30 min*	
1.5 min		*45 min*	
2 min		*1 hour*	
2.5 min		*1 hour 15 min*	
3.0 min		*1 hour 30 min*	
3.5 min		*1 hour 45 min*	
4.0 min		*2 hours*	
4.5 min			
5.0 min			

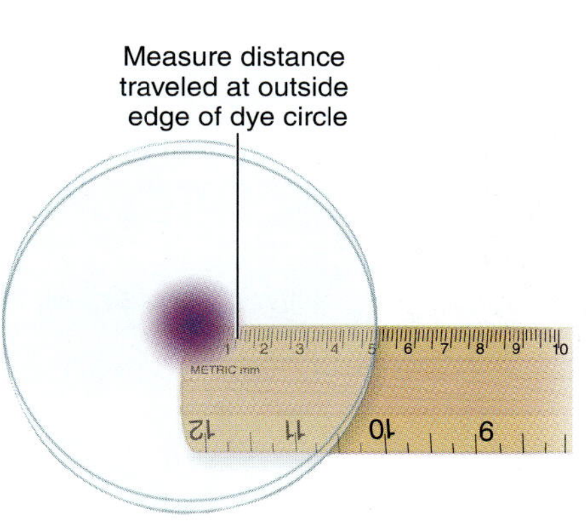

Figure 4.9 Measuring the Diffusion Distance.

of how to measure diffusion distance). Record the data in **table 4.3.** Note any pertinent observations in the space provided:

____

____

____

____

5. Observe diffusion of the dye crystals in **agar.** This process happens fairly slowly, so you need not be as "at the ready" as for the water diffusion exercise. To begin, carefully place a single potassium permanganate crystal in the center of the petri dish containing agar (dish B) at the "0" mark on the ruler. The moment the crystal is placed on the agar, start the stopwatch and observe diffusion of the dye. Record the radius of the dye spot in 15-minute intervals for a total of 2 hours. Record the data in table 4.3.

6. Make a note regarding conclusions made from this experiment (i.e., did it support or refute the hypothesis?) in the space provided:

____

____

____

____

EXERCISE 4.4B Effect of Solvent Temperature on Rate of Diffusion

This exercise repeats the method of observing diffusion of potassium permanganate crystals in water from exercise 4.4A, but this time, the rates of diffusion in water at different temperatures will be compared. Recall that molecules move at random in solution because of their inherent kinetic energy. When molecules are heated, the amount of kinetic energy increases and the molecules move faster. Likewise, when molecules are cooled, their kinetic energy decreases and they move more slowly. In this exercise potassium permanganate crystals will be placed in petri dishes containing water of differing temperatures. The rates of diffusion of each solute over time will be measured. Before beginning the experiment, state a hypothesis regarding the effect that temperature has on the rate of diffusion.

(continued on next page)

(continued from previous page)

Hypothesis: ____

____

____

____

____

Obtain the Following:

- **letter-size (8.5" × 11") piece of white paper**
- **three empty petri dishes**
- **water from one of three flasks: (1) ice water, (2) boiling water, and (3) room-temperature water**
- **fine-tissue forceps**
- **potassium permanganate crystals**
- **one large (~30 cm) and 3 small (~15 cm) metric rulers**
- **a stopwatch or other device for recording the time**

1. Place the rulers and the petri dishes on the white paper as was done in exercise 4.4A, only this time, set up three petri dishes instead of two. Use the large ruler as a shim to keep the petri dishes level, and the small rulers to measure rates of diffusion. Place the large ruler on the left side of the paper and orient it vertically. Place the small rulers at right angles to the large ruler, approximately 8 cm from each other, and with the "zero" point approximately at the center of the petri dish.
2. Label the petri dishes "A – ice water," "B – room-temperature water," and "C – boiling water" on the paper in the space next to each dish.
3. Obtain a small container of potassium permanganate crystals (figure 4.6). Only three moderate-sized crystals are needed for this experiment; one for each dish.
4. Observe diffusion of dye crystals in ice water. Before obtaining the ice water, measure the temperature of the water and record it here: ________________. Pour some of the ice water into petri dish "A." As observed in exercise 4.4A, the diffusion of potassium permanganate happens fairly quickly in water, so be prepared before beginning this exercise. One lab partner will place the crystal and record the distance diffused, while another will start the stopwatch and record time intervals. To begin: Place a single potassium permanganate crystal in the center of the petri dish containing ice water (dish A) at the "0" mark on the ruler. The moment the crystal is dropped in the water, start the stopwatch and observe diffusion of the dye. Record the radius of the dye in 30-second intervals for a total of 5 minutes. Record the data in **table 4.4.**
5. Make a note regarding conclusions made from this experiment (i.e., did it support or refute the hypothesis?) in the space provided:

____

____

____

____

Table 4.4 Effect of Temperature on Rate of Diffusion

	Ice Water Temp: ______	**Room-Temp. Water Temp: ______**	**Hot/boiling Water Temp: ______**
Time	**Distance Diffused (mm)**	**Distance Diffused (mm)**	**Distance Diffused (mm)**
30 sec			
1 min			
1.5 min			
2 min			
2.5 min			
3.0 min			
3.5 min			
4.0 min			
4.5 min			
5.0 min			

6. Next, observe diffusion of dye crystals in room-temperature water. Before obtaining the water, measure the temperature of the water and record it here: ________________. Pour some of the room-temperature water into petri dish "B." Place a single potassium permanganate crystal in the center of the petri dish containing room-temperature water (dish B) at the "0" mark on the ruler. The moment the crystal is dropped in the water, start the stopwatch and observe diffusion of the dye. Record the radius of the dye in 30-second intervals for a total of 5 minutes. Record the data in table 4.4. Note any pertinent observations in the space provided:

   ____
   ____
   ____
   ____

7. Next, observe diffusion of dye crystals in **boiling water.** Before obtaining the boiling water, measure the temperature of the water and record it here: ________________. Pour some of the boiling water into petri dish "C." Place a single potassium permanganate crystal in the center of the petri dish containing boiling water (dish C) at the "0" mark on the ruler. The moment the crystal is dropped in the water, start the stopwatch and observe diffusion of the dye. Record the radius of the dye in 30-second intervals for a total of 5 minutes. Record the data in table 4.4.

8. Make a note regarding conclusions of this experiment (i.e., did it support or refute the hypothesis?) in the space provided.

   ____
   ____
   ____
   ____

INTEGRATE

CONCEPT CONNECTION

Calculating the molecular weight of a substance requires knowledge of the molecular formula for the substance. Once the molecular formula is known, multiply the atomic mass of each element in the formula by the number of atoms of that element. Then add the products (answers from each) together to determine the total. For example, the molecular weight of table salt, with molecular formula NaCl, is calculated as shown:

Atomic mass of Na = 23; Contribution of Na to molecular weight of NaCl = 23 × 1 (only 1 atom of Na in NaCl) = 23

Atomic mass of Cl = 35; Contribution of Cl to molecular weight of NaCl = 35 × 1 (only 1 atom of Cl in NaCl) = 35

Molecular weight of NaCl = 23 + 35 = 58

The molecular weights of the two substances used in the exercises in this chapter (potassium permanganate and methylene blue) have already been provided. Try now to calculate them on your own in the space provided. Show your work so that if you obtain the wrong answer you can ask the instructor for assistance. The molecular formula for potassium permanganate is $KMnO_4$. The molecular formula for methylene blue is $C_{16}H_{18}ClN_3S$. Use the textbook or a periodic table of the elements as a guide for locating the atomic mass of each element.

(continued on next page)

(continued from previous page)

EXERCISE 4.4C Effect of Solute Molecular Weight on Rate of Diffusion

The **molecular weight** of a substance is the mass of a single molecule of the substance. Molecular weight is calculated by taking the sum of the atomic weights of the consituent atoms that make up the molecule. In this exercise, drops of solutions composed of molecules of differing molecular weights will be placed in a petri dish containing agar. The rate of diffusion of each solute over time will then be measured. The two solutes that will be compared are potassium permanganate (molecular weight 158) and methylene blue (molecular weight 319). Before beginning the experiment, state a hypothesis regarding the effect that molecular weight has on the rate of diffusion.

Hypothesis: ____

____

____

____

____

Obtain the Following:

- **petri dish filled with agar**
- **drinking straw or medicine dropper**
- **dropper vials containing 0.1M solutions of potassium permanganate and methylene blue**
- **metric ruler**

1. In this exercise there is no need to place the metric ruler underneath the petri dish because it is difficult to read through the agar. Thus, the additional metric ruler is not needed as a shim. Instead, diffusion distance will be measured by holding the ruler above the diffusion circle (without letting it touch the agar).

2. Obtain a drinking straw or medicine dropper and make a well in the agar dish as shown in **figure 4.10.** If using a drinking straw, press the end of the straw into the agar (figure 4.10*a*) and then lift it out at a slight angle to take agar with the straw as it is removed, leaving a well in the agar (figure 4.10*b*). If using a medicine dropper, compress the bulb before pushing it into the agar, then release the bulb to suction the agar into the dropper, leaving a well in the agar. Form two wells of equal size in the agar, approximately two inches apart.

3. Obtain dropper bottles of 0.1M potassium permanganate and 0.1M methylene blue. Using a medicine dropper, carefully fill the first well with a drop of potassium permanganate as shown in figure 4.10*c*. Take care not to allow the dye to spill over the edges of the well.

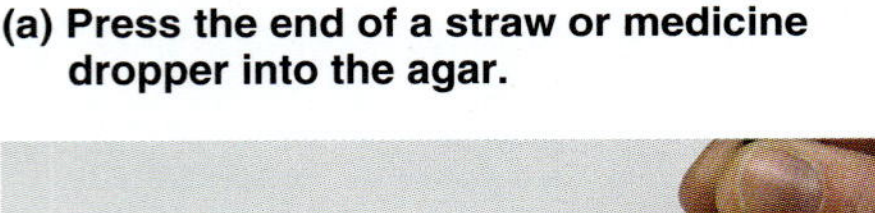

(a) Press the end of a straw or medicine dropper into the agar.

(b) Carefully withdraw the straw so as to take the agar with it to form a well in the agar.

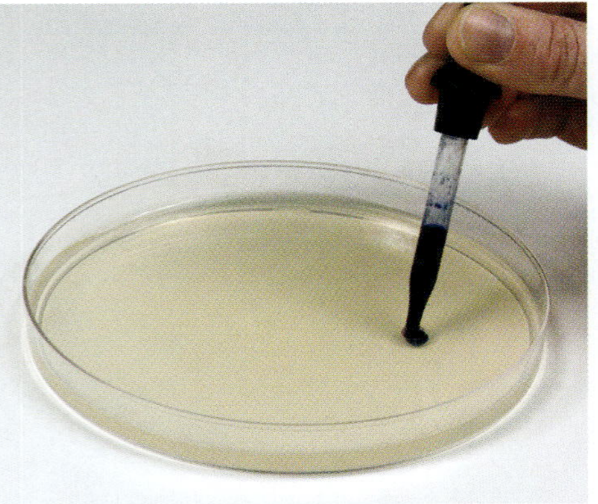

(c) Fill the well with dye, taking care not to let it spill over the edge of the well onto the surface of the agar.

(d) Form a second well at least two inches away from the first well, and fill it with the other dye.

Figure 4.10 **Observing the Effect of Solvent Molecular Weight on Diffusion Rates.**

Table 4.5	Effect of Solute Molecular Weight on Rate of Diffusion	
Solute	**Methylene Blue**	**Potassium Permanganate**
Molecular Weight	**319**	**158**
Time	**Distance Diffused (mm)**	**Distance Diffused (mm)**
15 min		
30 min		
45 min		
1 hour		
1 hour 15 min		
1 hour 30 min		
1 hour 45 min		
2 hours		

Next, using another medicine dropper, carefully fill the second well with a drop of methylene blue (figure 4.10*d*). Record the time: ___________________

4. Measure the diffusion radius of the dye every 15 minutes for 1.5–2.0 hours. Record the data in **table 4.5.**

5. Make a note regarding conclusions of this experiment (i.e., did it support or refute your hypothesis?) in the space provided.

____

____

____

____

EXERCISE 4.4D Effect of Membrane Permeability on Rate of Diffusion

Recall that the plasma membrane of cells is a semipermeable membrane. This means it allows certain molecules to pass through it, while preventing others. The permeability of a cell's plasma membrane is imparted by characteristics of the lipid bilayer and various proteins inserted into the bilayer. In this exercise, a semipermeable membrane is used to determine how size of a substance influences its ability to diffuse across a membrane. Dialysis tubing, which is an artificial membrane, is used to simulate the plasma membrane of a cell.

A starch solution will be placed into the dialysis tubing; the tubing will be sealed, and then placed into a beaker containing water and iodine. Therefore, starch and iodine solutions will be separated by a semipermeable membrane. The goal of this experiment is to determine if the membrane is permeable to starch, iodine, or both.

How is it possible to determine if any of these substances pass through the dialysis tubing? Indicator solutions are used to test for the presence of each of the different substances. The indicator substances that will be used in this exercise are:

- **Iodine,** which is an indicator for starch. Iodine is yellow-brown in color. When Iodine comes in contact with starch it turns purple/black in color.
- **Benedict's solution** is an indicator for glucose. It is light blue in color. When Benedict's solution comes in contact with glucose (and is heated) it turns yellow/orange in color.

Glucose test strips can also be used to test for glucose.

Before beginning the experiment, state a hypothesis regarding diffusion of substances into and out of the dialysis bag.

Hypothesis (What substance(s) will diffuse into/out of the dialysis bag? How will this be determined?): ___________________

____

____

____

____

Obtain the Following:

- **piece of 1" diameter dialysis tubing about 8" long**
- **~150 mL of 10% uncooked starch solution**
- **500 mL glass beaker filled with ~300 mL distilled water**
- **dropper bottle of Lugol's solution (iodine potassium iodide)**
- **funnel**

1. Soak the dialysis tubing in warm water for at least one minute to soften it up.

(continued on next page)

(continued from previous page)

2. Open the dialysis tubing by gently rubbing it between your fingers. Tie off one end of the dialysis tubing or close it with a dialysis tubing clamp **(figure 4.11*a*).**
3. Using a funnel, pour starch solution into the dialysis bag. Leave enough room so the open end can be tied or clamped off after it is filled (figure 4.11*b*).
4. Rinse the dialysis bag with water to ensure no starch remains on the outside of the tubing. Be sure to carefully rinse the tied-off ends of the bag because starch can become trapped there.
5. Place the dialysis bag into a beaker. Add distilled water to the beaker until it just covers the tubing containing the starch solution (figure 4.11*c*).
6. Add 3–4 drops of Lugol's solution (iodine potassium iodide) to the beaker containing the dialysis bag and water (figure 4.11*d*). Note any observations made upon adding the iodine to the water here:

____

____

____

____

7. Allow the beaker to sit while continuing to work on other exercises in this chapter. Observe the setup again after 30 minutes to one hour, and note any observations here:

____

____

____

____

(a) After soaking the dialysis tubing in water to soften it up, tie off one end.

(b) Using a funnel, pour starch solution into the dialysis tubing. Next, tie off or clamp the open end.

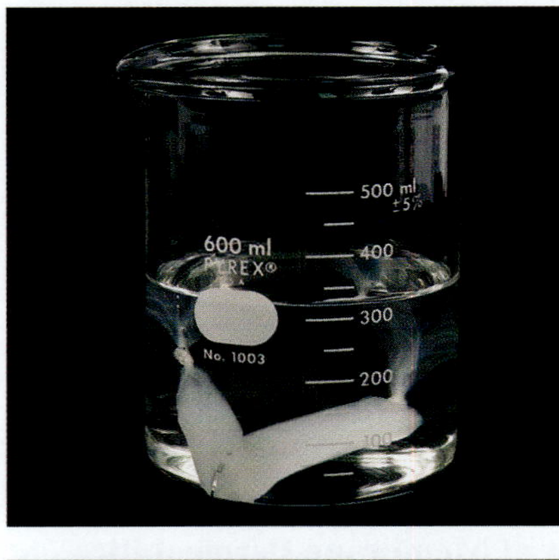

(c) After rinsing the tubing to ensure no starch is on the outside of the tubing, place it in a beaker containing distilled water.

(d) Add 3–4 drops of iodine solution to the beaker. Let it sit for 30 minutes to 1 hour to allow diffusion to take place.

Figure 4.11 Exercise Demonstrating the Effect of Membrane Permeability on Diffusion.

EXERCISE 4.5

OSMOSIS (WET LAB)

Osmosis (*osmos,* a thrusting) is a type of passive transport that involves the net movement of **water** from an area of high concentration of water (low solute concentration) to an area of low concentration of water (high solute concentration). Osmosis occurs when solutions with different concentrations of solutes are separated by a selectively permeable (or semipermeable) membrane through which water can pass, but solutes cannot pass. Such solutes are referred to as "membrane impermeable" or "nonpenetrating" solutes. If the solutes cannot pass, then water will move to obtain equilibrium. In this exercise, either an animal membrane or dialysis tubing will be used because both block the passage of large molecules (e.g., sugar molecules).

Place a hypertonic solution (a sugar solution containing molasses) in a thistle tube (**figure 4.12**). Cover the opening of the thistle tube with dialysis tubing, and place it in a beaker of distilled water (a hypotonic solution). If water moves out of the tube into the beaker, the level of sugar solution in the thistle tube will decrease. If water moves into the tube from the beaker, the level of sugar solution in the thistle tube will increase. This exercise requires two people working together, so partner up with another student in the lab to do the exercise. Before beginning the experiment, state a hypothesis regarding the effect of selective permeability of the membrane on diffusion and osmosis.

Hypothesis: ____

____

____

____

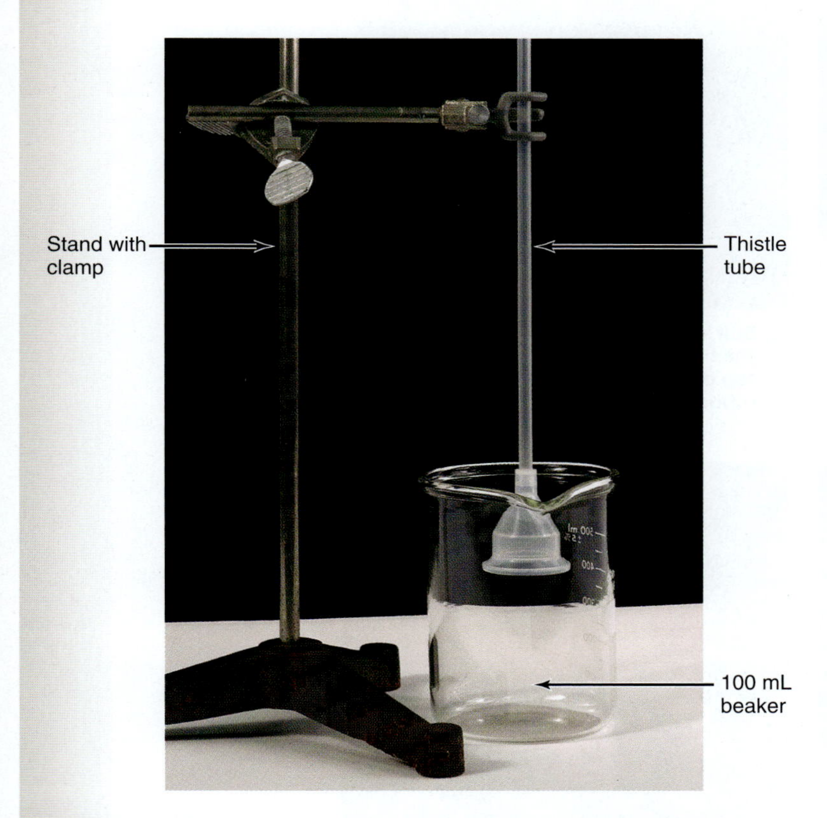

Figure 4.12 Setup for Osmosis Experiment with Thistle Tube. Make sure to have the stand, clamp, and beaker set up so the end of the thistle tube is low enough to be covered by water in the beaker before rotating the thistle tube and proceeding to fill it with molasses.

Obtain the Following:

- **stand and clamp**
- **molasses**
- **3″ diameter disc of animal membrane (alternative: 2″ long piece of 1.5″ wide dialysis tubing)**
- **100 mL glass beaker**
- **distilled water**
- **thistle tube**
- **stopwatch or other timing device**
- **wax pencil**

1. Soak the animal membrane or dialysis tubing in warm water for a few minutes to soften it up. If using dialysis tubing, loosen and open it by rubbing it between your thumb and index finger. Cut along one side of the tubing to open it and create a flat sheet of tubing. Place the tubing back into the water to soak until it is needed.
2. Fill the glass beaker about halfway with distilled water.
3. Set up the thistle tube, beaker, stand, and clamp as shown in figure 4.12. Next, turn the thistle tube so the funnel side is up. Have one lab partner place a finger over the bottom end of the tube to form a seal.
4. With one person holding the thistle tube with his/her finger covering the bottom end of the tube, have the other person pour molasses into the funnel part of the thistle tube until it begins to leave the funnel part and enter the tube **(figure 4.13*a*).** Try to fill the tube ~1/2" because the molasses will run back into the funnel part of the thistle tube when it is turned back over.

(continued on next page)

(continued from previous page)

5. Obtain the animal membrane or piece of dialysis tubing from the beaker of water and place it over the top of the thistle tube. Secure the membrane/tubing with a rubber band (figure 4.13*b*).

6. Carefully turn the tube over so the funnel part is facing the beaker and table. Let the molasses settle into the funnel for a couple of minutes, then mark the starting level with a wax pencil (figure 4.13*c*).

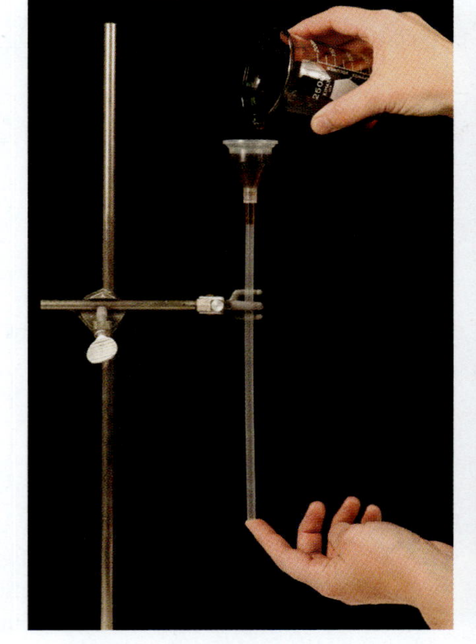

(a) Keeping a finger over the bottom of the tube, pour molasses into the funnel until it fills the bulb completely and runs into the thin part of the thistle tube.

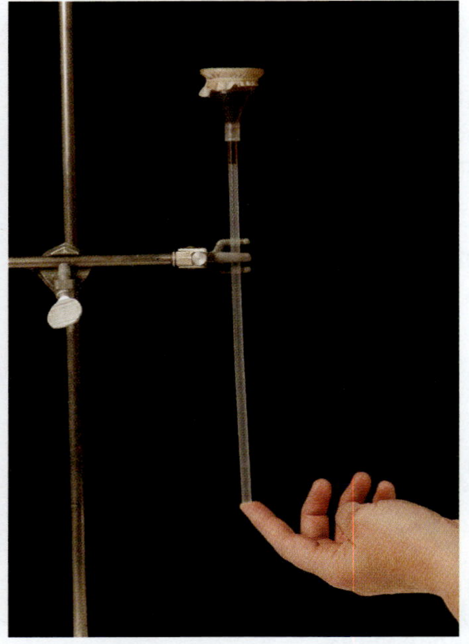

(b) Still keeping a finger on the bottom of the tube, wrap the membrane over the top of the funnel and secure it with a rubber band.

(c) Turn the tube over and affix it to the stand clamp. Mark the starting level of the molasses with a wax pencil.

(d) Lower the thistle tube into the water. Record the level of the molasses every 15 minutes for at least 2 hours.

Figure 4.13 Osmosis Exercise Using an Animal Membrane and Thistle Tube.

7. Lower the thistle tube into the glass beaker containing water so it is suspended in the water as shown in figure 4.13*d*. Tighten the clamp on the tube so it will remain suspended. Record the start time:

____

8. After lowering the thistle tube into the beaker of water, start a timer. Record the change in the level of the meniscus of molasses in **table 4.6** at 15-minute intervals from 30 minutes to several hours.

9. Note any conclusions of this experiment in the space provided (i.e., did it support or refute the hypothesis?).

____

____

____

____

Table 4.6 Osmosis

Time (min)	Height of Molasses Column (mm)
0	0
15	
30	
45	
60	
75	
90	
105	
120	
135	
150	
165	
180	

EXERCISE 4.6 Mc Graw Hill Ph.I.L.S.

Ph.I.L.S. LESSON 1: OSMOSIS AND DIFFUSION: VARYING EXTRACELLULAR CONCENTRATION

The purpose of this laboratory exercise is to demonstrate the principle of osmosis and observe the effect of osmosis on human cells. This will be accomplished by observing the effect of placing an erythrocyte (red blood cell) into solutions of differing concentrations of sodium chloride [NaCl].

Before beginning the experiment, become familiar with the following concepts (use the textbook as a guide):

- Normal intracellular and extracellular concentrations of sodium (Na^+) for a generalized cell
- The mechanisms of transport used to move Na^+ into and out of a generalized cell
- Differences between hypotonic, hypertonic, and isotonic solutions

1. Open Ph.I.L.S. Lesson 1: Osmosis and Diffusion: Varying Extracellular Concentration.
2. Read the objectives, introduction, and wet lab sections for the exercise. The objectives and introduction contain several hyperlinks to videos that may help to explain the experimental process. After reading through the introduction, complete the pre-laboratory quiz.
3. The laboratory exercise will open after clicking on "Submit Quiz" **(figure 4.14).**
4. Click the power switch on the front of the spectrophotometer to turn it on.

(continued on next page)

INTEGRATE

LEARNING STRATEGY

To distinguish active and passive transport mechanisms, think of padding a canoe on a river. In this analogy, water naturally flows in the river from an area of high "concentration" (upstream) to an area of low "concentration" (downstream). Attempting to move the canoe "upstream," or *against* the current, requires an input of energy to paddle the canoe—just as it takes energy in the form of ATP to move a substance across a membrane from an area of low concentration to an area of high concentration. On the other hand, to move the canoe "downstream," or *with* the current, requires no input of energy (the canoe just "goes with the flow . . .")—just as no energy is required to move a substance across a membrane from an area of high concentration to an area of low concentration.

(continued from previous page)

5. Using the arrow controls, set the wavelength to 510 nm (the wavelength at which an intact cell membrane absorbs light).
6. Click and drag the green pipette to the stock solution of blood. Click the "up" arrow to draw 1 mL of stock blood solution into the pipette.
7. Drag the pipette to an empty test tube. Release the blood into the tube by pressing the "down" arrow. If experiencing difficulty moving the pipette, click on the bottom of the pipette to get it to move.
8. Repeat steps 6 and 7 until all of the test tubes are filled with blood.
9. To calibrate the spectrophotometer, adjust the transmittance value to "0" using the arrows above the "zero" buttons.
10. Open the lid of the "holder" of the spectrophotometer (if unsure, click on the word "spectrophotometer holder" in the instructions at the bottom of the screen). Click the "blank" test tube, which is located on the far left of the rack and is labeled "0," and drag it to the spectrophotometer. The blank contains no salt solution and therefore serves as the control. Close the lid and set the transmittance to 100 using the "Calibrate" arrows.
11. Click the "Journal" in the lower right corner to record the transmittance value. Data will then be displayed in a table and in graphic form. After viewing the data, close the journal by clicking on the X in the upper right corner of the graph.
12. Click the lid of the spectrophotometer to open it. Then click on the test tube to remove it. The test tube will pop straight up out of the spectrophotometer. Click on the test tube again to drag it back to the rack.
13. Click on the next test tube and drag it to the spectrophotometer. Close the lid. A transmittance reading will be taken once the lid is closed. Click on the "Journal" to record and view the data. Close the journal.
14. Click on the lid of the spectrophotometer to open it, then click on the test tube to remove it. Drag the test tube back to the rack.
15. Repeat steps 13 and 14 until readings from all test tubes have been taken. Record the data in **table 4.7**.
16. Click on "graph data" to view the results. With the graph still on the screen, transfer the data points from the graph to figure 4.15 (p. 81). If the program is accidentally closed before graphing the data, the data can be regraphed using the data recorded in table 4.7.
17. Open the "post-lab quiz" and the "lab report" by clicking on them. Answer the post-lab quiz questions that appear on the computer screen.
18. Read the conclusion on the computer screen. You may either click "Print Lab" to print a hard copy of the report or click "Save PDF" to save an electronic copy of the report.
19. Make note of pertinent observations in the space provided:

____

____

____

____

____

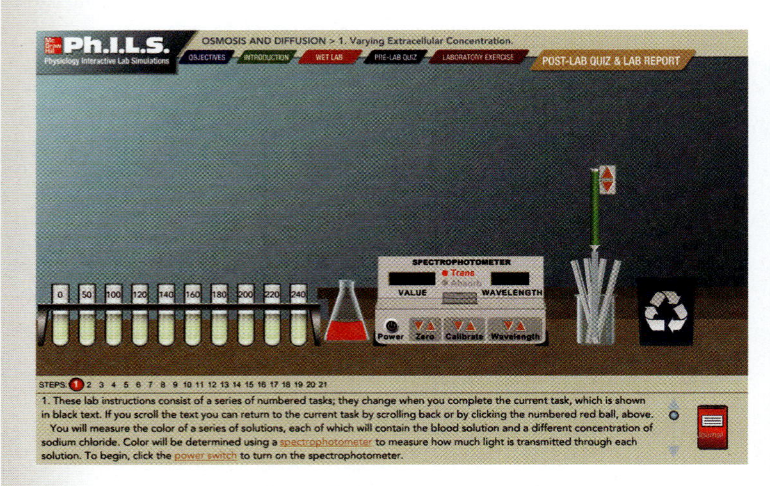

Figure 4.14 Opening Screen for the Ph.I.L.S. Exercise "Osmosis and Diffusion: Varying Extracellular Concentration."

Table 4.7	Ph.I.L.S. Lesson 1: Varying Extracellular Concentration
[NaCl] (mM)	**Transmittance (%)**
0	100
50	
100	
120	
140	
160	
180	
200	
220	
240	

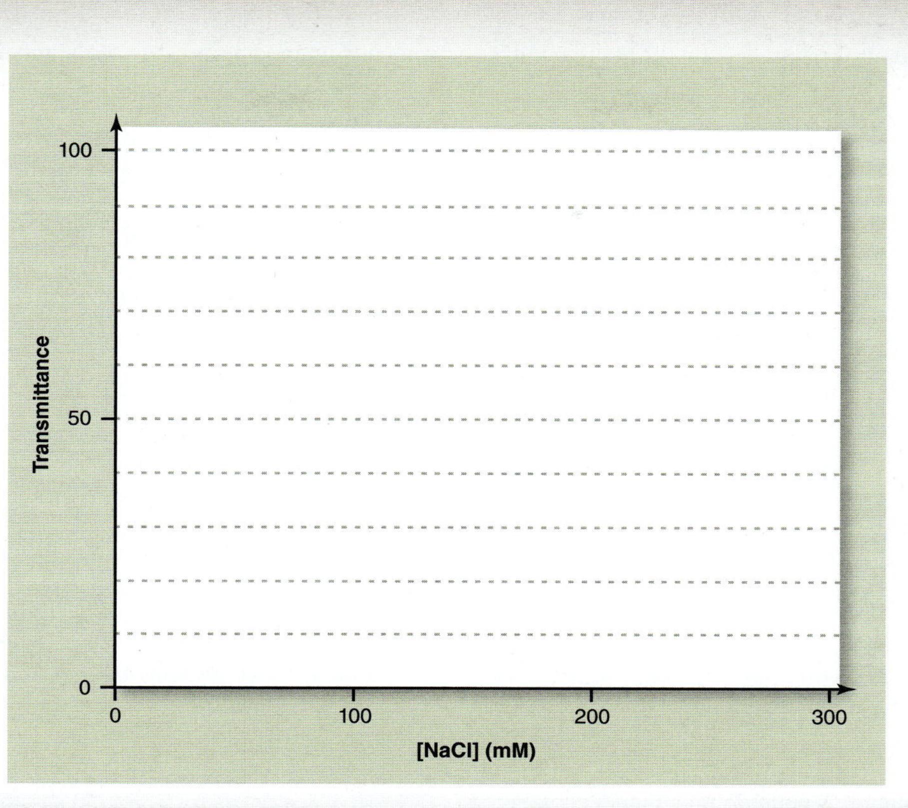

Figure 4.15 Graph of Transmittance vs. Salt Concentration from Ph.I.L.S. Lesson 1: Osmosis and Diffusion: Varying Extracellular Concentration.

INTEGRATE

CLINICAL VIEW
Hemodialysis

Patients with reduced kidney function have a decreased ability to remove wastes and excess fluid from the blood. Consequently, toxic substances can increase in the blood, and the volume of interstitial fluid can also increase, which causes edema, or swelling, of the tissues. In addition, urinary output may also decline. When kidney function becomes critically low, these patients must undergo a procedure called **hemodialysis** (*hemo-*, blood + *dialysis*, a separation). Hemodialysis is a procedure that takes advantage of the same properties of membrane transport discussed in this chapter. In hemodialysis, the patient's arterial blood is passed through a series of tubes that contain semipermeable membranes. The membranes allow substances such as metabolic wastes and electrolytes to pass from the blood into a filtrate that is contained within the tubes, but prevent the passage of larger substances such as plasma proteins. This process mimics the function of the filtration membrane that is normally functional in the kidneys. A dialysis fluid (called dialysate) surrounds the tubes. This fluid matches the composition of the plasma (fluid) portion of blood, but contains no metabolic wastes. As a result, metabolic wastes flow along their individual concentration gradients from the blood to the surrounding dialysate. Equally important, there is no net movement (and hence, loss) of electrolytes or larger proteins from the blood into the dialysate. The filtered blood is returned to the patient's circulatory system through one of the patient's veins. This process must be repeated several times weekly, depending on the extent of kidney damage.

Filtration

Filtration (*filtro,* to strain through) is a process that is likely familiar if a person has ever brewed coffee using a filtration brewing system. Filtration is a process by which solutes are separated from a solvent by being passed through a filter. The size of pores in the filter prevents solute particles that are larger than the size of the pores from passing through, while allowing any solute particles that are smaller than the size of the pores to pass. At times, the charge on a filter can also prevent substances from passing through. For example, if the substance is small and negatively charged and the filter is also negatively charged, even if the substance is small enough to fit through the pores in the filter, it will likely be prevented from passing because of the repellent forces of the negative charges. Notably, if enough fluid pressure, or **hydrostatic pressure** (*hydro,* water + *statikos,* causing to stand), is exerted on the fluid, the pressure can overcome the influence of like charges repelling each other. In the case of making coffee using a filtration system, coffee grounds are placed in a filter and hot water is passed over the grounds. The extracts of the coffee, such as caffeine, readily pass through the filter, but the coffee grounds do not pass because they are too large. The resulting coffee that ends up in the coffee pot is a **filtrate.**

EXERCISE 4.7

FILTRATION (WET LAB)

Although there are multiple factors that can affect filtration, this exercise explores only two of those factors:

1. *Particle size*—particles smaller than the size of the pores in the filter will pass through. Particles larger than the size of the pores in the filter will remain in the filter.
2. *Hydrostatic pressure*—a higher column of fluid exerts more pressure and thus increases filtration rate.

In this exercise a solution containing substances of varying sizes will be made. The solution is poured into a funnel containing a paper filter. The rate at which fluid passes through the filter and fills up a graduated cylinder will be measured. The rate of filtration when the filter is completely full will then be compared to the rate of filtration when the filter is only 50% full. The solution will contain the following:

a. distilled water

b. 10% uncooked starch solution

c. 10% dark corn syrup solution

d. charcoal or ground black pepper

Before beginning the experiment, state a hypothesis regarding the effect of particle size and hydrostatic pressure on filtration rate. Will there be a difference in filtration rate when the filter is 100% vs. 50% full?

Hypothesis: ____

____

____

____

____

Obtain the Following:

- **a ring stand and ring clamp to hold a glass funnel**
- **one piece of filter paper (18.5 cm diameter)**
- **glass funnel (10 cm diameter)**
- **100 mL glass beaker**
- **100 mL graduated cylinder**
- **uncooked dry starch**
- **dark corn syrup**
- **charcoal or black pepper**
- **stopwatch or other timing device**
- **Lugol's iodine (to test for starch)**
- **glucose test strips (to test for glucose)**

1. Set up the ring stand with the funnel and graduated cylinder as shown in **figure 4.16.** To prepare the filter paper, first fold it in half. Then fold it in half once again and open it up into a cone that will fit into the glass funnel. Place the filter in the funnel.
2. Mix the following solution:

   a. 100 mL distilled water

   b. 1 teaspoon charcoal or black pepper

   c. 1 Tablespoon dark corn syrup

   d. 1 Tablespoon dry uncooked corn starch
3. One lab partner will be in charge of timing; the other will be in charge of pouring the solution into the funnel

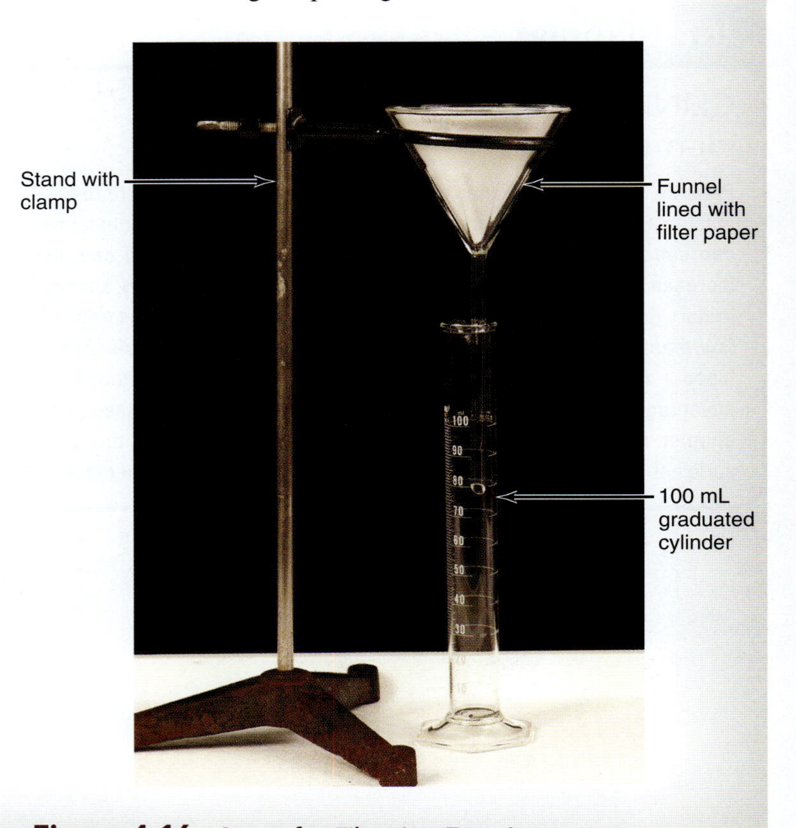

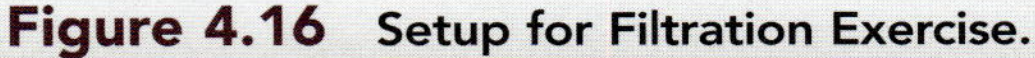

Figure 4.16 Setup for Filtration Exercise.

and recording the times in **table 4.8.** Before pouring the solution into the funnel, give it a quick stir so that solids are not settled at the bottom of the beaker. Then, quickly pour the solution into the filter-lined funnel. The liquid should come up to just below the lip of the filter paper. Start the timer. Note the time it takes to fill the cylinder in increments of 5 mL, and record these times in table 4.8. NOTE THE TIME WHEN THE COLUMN OF FLUID IN THE FUNNEL IS AT 50 mL. When that happens, record the time here: __________________. This will allow comparison of filtration rates.

Table 4.8	Filtration Rate
Volume (mL)	**Time (sec)**
5	
10	
15	
20	
25	
30	
35	
40	
45	
50	
55	
60	
65	
70	
75	
80	
85	
90	

4. Allow filtration to continue until the level of filtrate in the cylinder reaches 80 or 90 mL. When filtration is complete, remove the funnel containing the filter. Note any substances retained by the filter in the space provided:

____

____

____

____

5. Next, perform the following tests on the filtrate in the graduated cylinder:

a. Observe: Are there **charcoal** or black pepper flecks present in the filtrate? __________________

b. Dip a glucose test strip into the filtrate and look for a color change. Is **glucose** present in the filtrate?

____

c. Place a couple of drops of Lugol's iodine into the filtrate and look for a color change. Is there **starch** in the filtrate? __________________

6. Make a note regarding any conclusions of this experiment (i.e., did it support or refute your hypothesis?) in the space provided.

____

____

____

____

Chapter 4: Cell Structure and Membrane Transport

Name: ______________________

Date: __________ Section: __________

POST-LABORATORY WORKSHEET

The (1) corresponds to the Learning Objective(s) listed in the chapter opener outline.

Do You Know the Basics?

Exercise 4.1: Observing Cellular Anatomy with a Compound Microscope

1. When preparing a wet mount of human cheek cells, methylene blue is a ______________ (basophilic/eosinophilic) dye that is most attracted to ______________ (basophilic/eosinophilic) components of the cell. (1)

2. Which of the following components of a human cheek cell will be stained most intensely? (Circle one.) (2)

   a. cytoplasm
   b. nucleus
   c. plasma membrane
   d. rough endoplasmic reticulum

3. Which of the following types of solutions should be used when preparing a wet mount? (Circle one.) (3)

   a. hypertonic
   b. hypotonic
   c. isotonic

4. Match the functions listed in column A with the appropriate cell structures listed in column B. (4)

	Column A	*Column B*
______	1. includes cellular organelles and cytosol; cytosol contains enzymes that mediate many cytosolic reactions such as glycolysis and fermentation	a. centrioles
______	2. composed of both protein and RNA; forms ribosomes	b. chromatin
______	3. genetic material within the nucleus; consists of DNA and associated proteins in an "uncoiled" form	c. cytoplasm
______	4. sites of protein synthesis; may be bound to the ER ("bound") or found within the cytoplasm ("free")	d. cytoskeleton
______	5. membrane–enclosed sacs that contain digestive enzymes and function in the breakdown of intracellular molecules	e. lysosomes
______	6. membrane–enclosed sacs that contain oxidative enzymes (e.g., catalase)	f. nucleolus
______	7. paired organelles composed of microtubules that are used to organize the mitotic spindles that attach to chromosomes during mitosis	g. peroxisomes
______	8. composed of protein filaments called microtubules, intermediate filaments, and microfilaments; provides structural support for the cell	h. ribosomes

Exercise 4.2: Observing Mitosis in a Whitefish Embryo

5. Match the features listed in column A with the appropriate stage of mitosis or interphase listed in column B. (5) (6)

	Column A	*Column B*
______	1. chromosomes are located at opposite ends of the cell; a cleavage furrow may be visible	a. anaphase
______	2. chromosomes are being pulled toward the opposite poles of the cell	b. interphase
______	3. chromation coils into chromosomes, which become visible when viewed with a light microscope	c. metaphase
______	4. chromosomes line up at the center of the cell along the equatorial plate	d. prophase
______	5. uncoiled chromatin exists within the nucleus of the cell; chromosomes are not visible when viewed with a light microscope	e. telophase

Exercise 4.3: Observing Classroom Models of Cellular Anatomy

6. Label the structures in this diagram of a generalized cell: 7

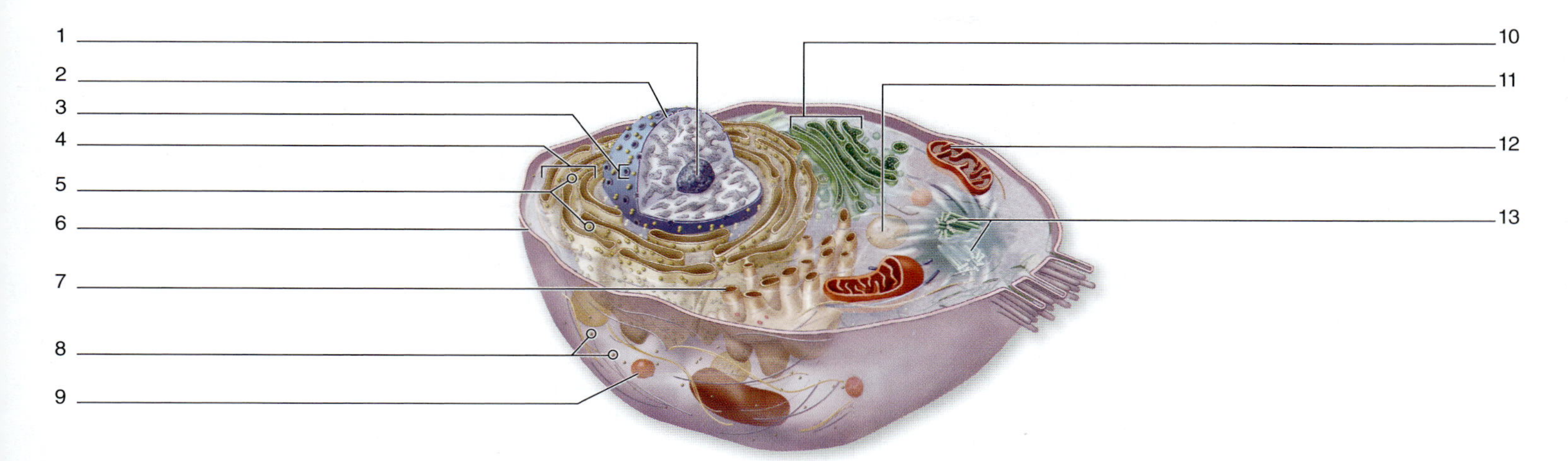

Exercise 4.4: Diffusion (Wet Lab):

7. Graph the results from table 4.3: Effect of **Solvent Viscosity** on Rate of Diffusion in the space provided. There should be two data lines, one for each medium (agar and water). Remember to label the axes with the appropriate units.

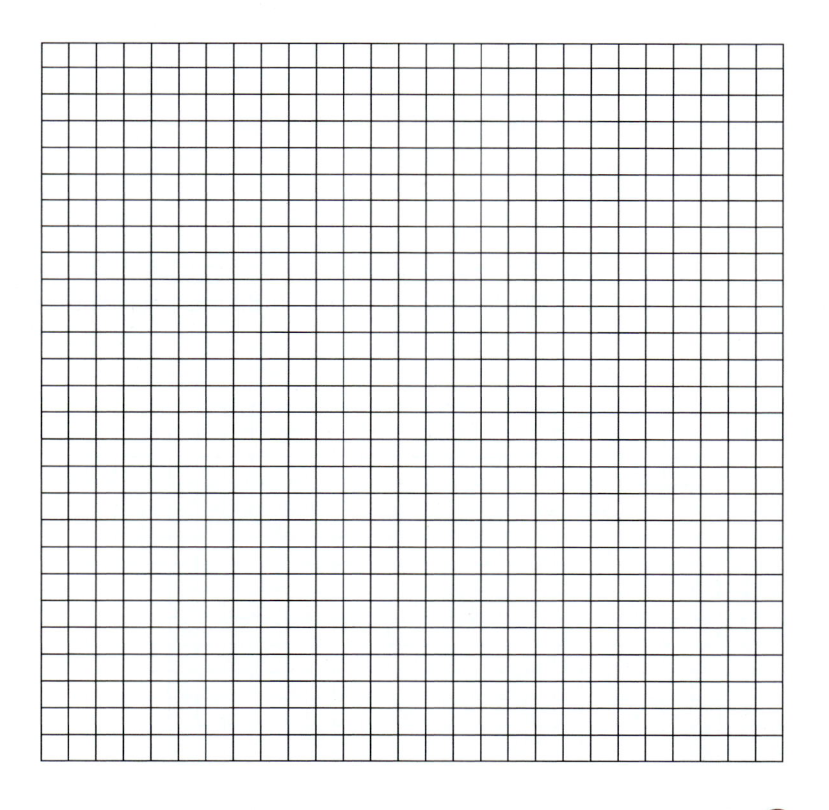

8. Based on the results of this exercise, describe how and why the **solvent viscosity** effects the rate of diffusion. 8

____

____

____

9. Graph the results from table 4.4: Effect of **Temperature** on Rate of Diffusion in the space provided. There should be three data lines, one for each temperature. Remember to label the axes with the appropriate units. 8

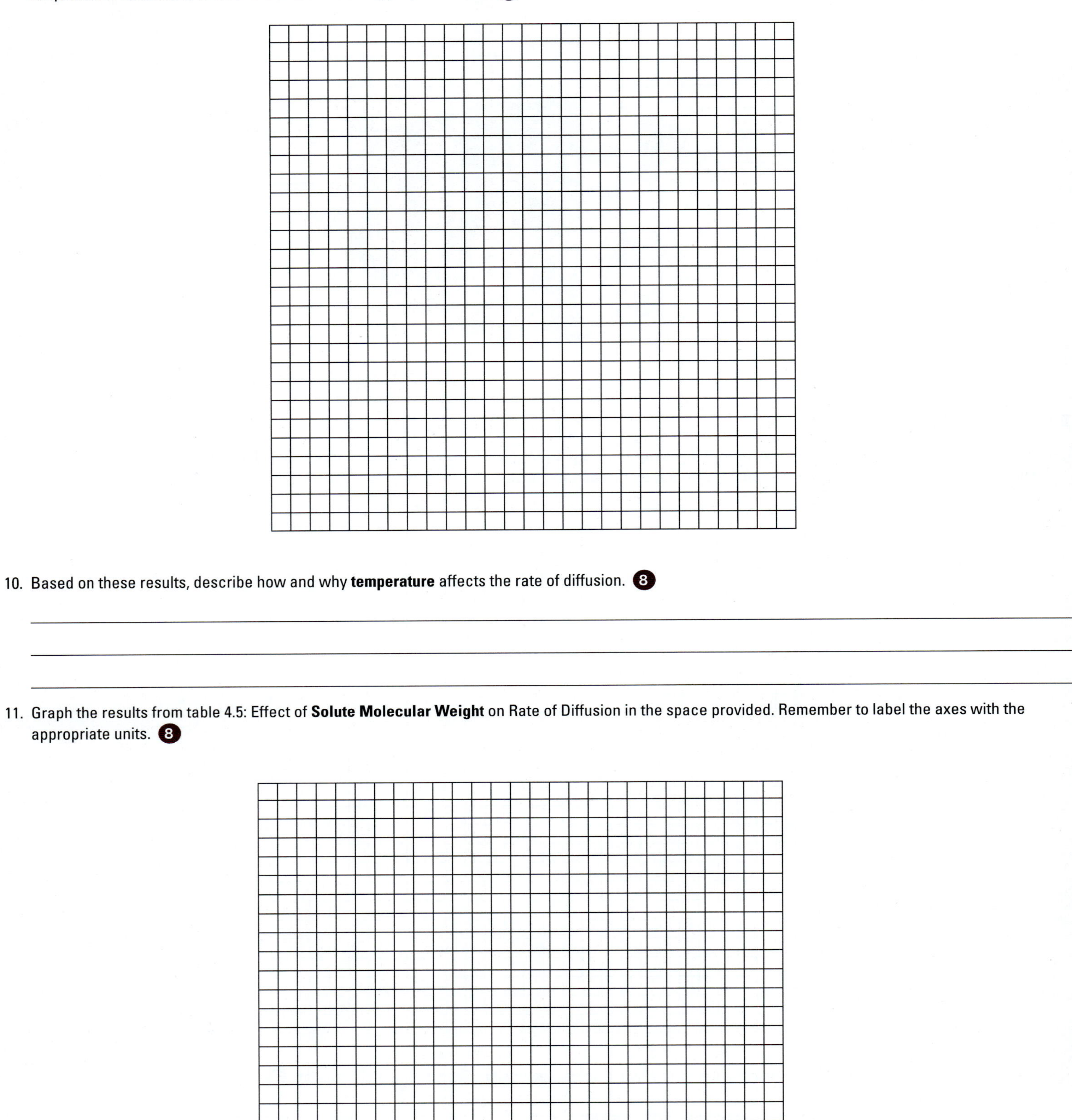

10. Based on these results, describe how and why **temperature** affects the rate of diffusion. 8

____

____

____

11. Graph the results from table 4.5: Effect of **Solute Molecular Weight** on Rate of Diffusion in the space provided. Remember to label the axes with the appropriate units. 8

12. Based on these results, describe how and why **solute molecular weight** effects the rate of diffusion. 8

____

____

____

13. Was there a change in the color of the water in the beaker when iodine was added? ____________ Based on the answer to this question, did starch diffuse from inside the dialysis tubing to the water in the beaker? 8 ____________

14. Was there a change in the color of the starch solution in the dialysis bag after the bag was immersed in the iodine solution for a while? ____________ Based on the answer to this question, did iodine diffuse into the dialysis bag from the water in the beaker? 8 ____________

15. Based on the results of this exercise, identify the substance(s) to which the dialysis tubing was permeable. 8

____

Exercise 4.5: Osmosis (Wet Lab)

16. Which of the following will occur if the concentration of membrane-impermeable solutes in the intracellular fluid of a cell is higher than the concentration in the extracellular fluid? (Check all that apply.) 9 10

____ a. There will be a net movement of water INTO the cell.

____ b. There will be a net movement of water OUT OF the cell.

____ c. There will be a net movement of solutes INTO the cell.

____ d. There will be a net movement of solutes OUT OF the cell.

____ e. There will be a net movement of solutes OUT OF the cell and water INTO the cell.

____ f. There will be a net movement of solutes INTO the cell and water OUT OF the cell.

Exercise 4.6: Ph.I.L.S. Lesson 1: Osmosis and Diffusion: Varying Extracellular Concentration

17. a. Identify the *independent variable* in this virtual experiment: 11 ____________

b. Identify the *dependent variable* in this virtual experiment: 11 ____________

18. a. Describe why the same volume of blood is placed in each vial. 12 ____________

____

____

b. Identify the control in this experiment. ____________

____

19. Using the graph created in figure 4.15, explain the relationship between transmittance and [NaCl]. In your explanation, be sure to note the status of the erythrocytes at the various concentrations of salt solution. 13

____

____

____

____

20. What concentration of sodium chloride solution is isotonic to erythrocytes? ____________ (Fill in the blank.) 14

Exercise 4.7: Filtration (Wet Lab)

21. Graph the results from table 4.8: **Filtration Rate** in the space provided. Remember to label the axes with the appropriate units. After graphing the results, put a star (*) on the graph to indicate the time at which the volume of fluid within the graduated cylinder reached 50 mL. 15

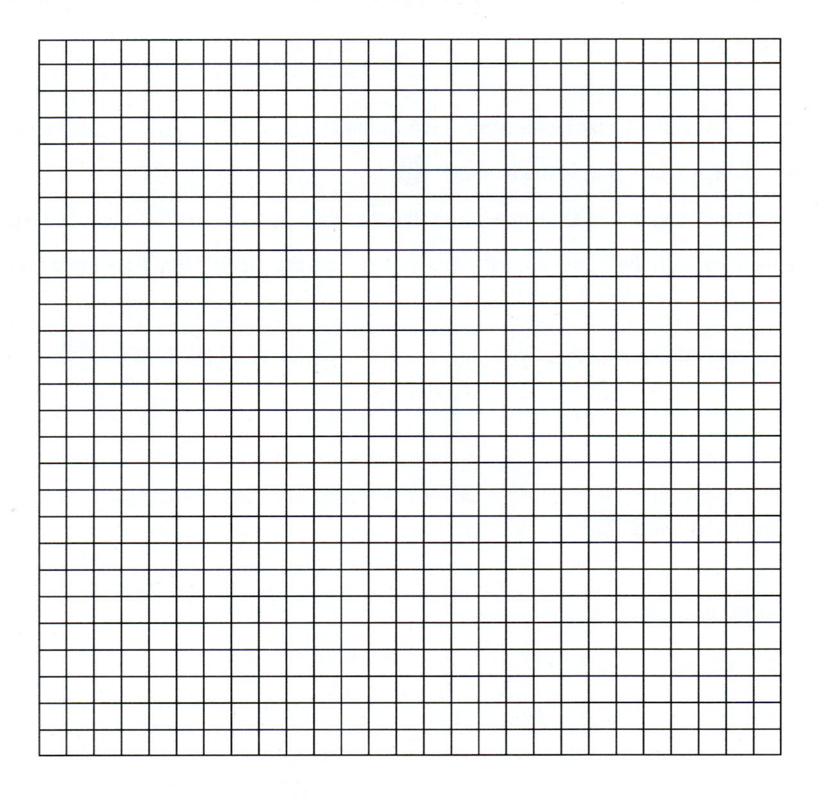

22. The slope (X/Y) of the line on the graph from question 21 is an estimate of the **filtration rate** (mL/sec). Observe the slope of the line before and after the point at which the volume of fluid within the graduated cylinder reached 50 mL. Is there a difference in these two slopes? ____________ (yes/no) If "yes," explain the difference in the two slopes. 15 16

____

23. The results indicate that an increase in hydrostatic pressure (fluid height) will ____________ (increase/decrease) filtration rate. 16

Can You Apply What You've Learned?

24. One major function of liver cells (hepatocytes) is to detoxify alcohol. Based on this function, what organelle(s) do you predict hepatocytes would contain large numbers of? Why?

____

____

____

25. Certain cells within the pancreas function in the synthesis and secretion of the hormone insulin, which is a protein. Based on this function, what organelle(s) do you predict these cells would contain large numbers of? Why?

____

____

____

26. A patient enters the emergency department with considerable blood loss from blunt force trauma. The physician attending to the patient needs to increase the volume of the patient's bodily fluids so the patient's blood pressure and systemic perfusion of tissues can be maintained. The physician quickly administers an intravenous solution of NaCl. To avoid causing more harm to the patient, the tonicity of the intravenous solution should be ____________.

27. While on vacation your brother runs out of contact lens solution. In an attempt to improvise, he uses tap water to hydrate his contact lenses. Immediately upon application of the contact lenses to his eyes, he starts complaining of pain in his eye. The pain is bad enough that he removes the contact lenses and is afraid he will not be able to wear them for the rest of the vacation. Explain why tap water caused such significant discomfort to your brother's eyes.

28. You are in the laboratory observing a wet mount of fresh erythrocytes under the microscope. In an attempt to make the erythrocytes easier to see, you decide to dilute the sample. Following dilution you make another wet mount, but you are unable to find any cells when you observe the slide under the microscope. What happened to the erythrocytes? (Hint: You may have grabbed the wrong solution to dilute the sample.)

Can You Synthesize What You've Learned?

29. Chemotherapy treatments are given to cancer patients in an attempt to halt or slow the growth of a tumor, which is composed of rapidly dividing cells. Certain chemotherapy drugs exert their actions by interfering with mitosis. For example, some drugs act to prevent microtubules from lengthening or shortening. Microtubules are protein filaments that attach to chromosomes and centrioles, forming the mitotic spindle, which moves the chromosomes during mitosis. Based on this role of microtubules during mitosis, with which stage(s) of mitosis would these drugs most likely interfere?

30. What is the function of rough endoplasmic reticulum? Why do you think neurons might contain large amounts of rough endoplasmic reticulum? (Hint: Neurons are cells that need to be able to transport numerous ions into and out of the cell. They accomplish this task by synthesizing many ion channels to insert in the plasma membrane.)

31. Exercise 4.4D (Effect of Membrane Permeability on Rate of Diffusion) may have made you curious to know if dialysis tubing is permeable to glucose. In the space provided, describe an experiment that could determine if the dialysis tubing is permeable to glucose.

CHAPTER 5

Histology

OUTLINE AND LEARNING OBJECTIVES

Anatomy & Physiology REVEALED® aprevealed.com **MODULE 3: TISSUES**

INTRODUCTION

This chapter of the laboratory manual introduces the practical study of **histology,** tissue biology (*histo*, tissue, + *logos*, the study of). **Tissues** consist of multiple cells that function together as a unit. An understanding of histology is important in the health sciences because many times the first manifestation of disease is seen at the tissue level of organization. For example, when a patient presents to his or her doctor with a tumor (*tumere*, to swell), one of the primary methods for determining the type of tumor and for determining whether the tumor is cancerous is to do a biopsy (take a tissue sample) and look at the tissue through a microscope. Thus, understanding normal histology is the first step in understanding histopathology (*histo*, tissue + *pathos*, disease).

This chapter introduces the key features of the four basic tissue types—epithelial, connective, muscle, and nervous. The text and figures in this chapter provide descriptions of each tissue type and examples of where each type of tissue is located in the body.

The exercises in this chapter involve looking at tissues through the microscope. At first it may seem as if every slide is simply a slide with a lot of "pink and purple stuff" on it. It may be difficult to identify particular tissue types and structures, particularly at first. However, with practice this will get easier. The exercises in this chapter address characteristics and classifications of each of the four basic tissue types. The exercises need not be covered in any particular order. However, be sure to complete all the tasks for one particular tissue type (e.g., epithelial tissues) before moving on to another tissue type.

List of Reference Tables

Chapter 5: Histology

Name: ______________________

Date: __________ Section: __________

These Pre-Laboratory Worksheet questions may be assigned by instructors through their connect course.

PRE-LABORATORY WORKSHEET

1. Which of the four basic tissue types is/are excitable? (Check all that apply.)
   ____ a. connective
   ____ b. epithelial
   ____ c. muscle
   ____ d. nervous
2. Epithelial tissue is classified by the number of layers of cells and the shape of the cells on the apical surface. ______________ (True/False)
3. Bone is classified as which of the following tissue types? (Circle one.)
   a. connective
   b. epithelial
   c. muscle
   d. nervous
4. Which of the four tissue types contains an extensive extracellular matrix (ECM)? (Circle one.)
   a. connective
   b. epithelial
   c. muscle
   d. nervous
5. Which of the three types of muscle tissue is/are striated? (Check all that apply.)
   ____ a. cardiac
   ____ b. skeletal
   ____ c. smooth
6. Identify the basic tissue type that exhibits polarity (has both apical and basal surfaces). (Circle one.)
   a. connective
   b. epithelial
   c. muscle
   d. nervous
7. Bone is a unique connective tissue that is avascular (lacking blood vessels). ______________ (True/False)
8. Match the description of connective tissue listed in column A with the tissue type listed in column B.

Column A	*Column B*
____ 1. dense connective tissue with fibers oriented in many directions	a. adipose
____ 2. loose connective tissue characterized by long, thin, dark-staining fibers	b. areolar
____ 3. dense connective tissue characterized by thick fibers oriented in one direction	c. dense irregular
____ 4. loose connective tissue characterized by a network of short, dark-staining fibers	d. dense regular
____ 5. loose connective tissue characterized by large cells that appear "empty"	e. reticular

9. Which of the following are classified as supportive connective tissues? (Check all that apply.)
   ____ a. adipose
   ____ b. bone
   ____ c. cartilage
   ____ d. muscle
   ____ e. reticular
10. Fluid connective tissue includes both blood and lymph. ______________ (True/False)

HISTOLOGY

Epithelial Tissue

Epithelial tissues are tissues that cover body surfaces, line body cavities, and form the majority of glands. As such, they will have a free surface (see Learning Strategy on p. 96). Epithelial tissues are characteristically highly **cellular** (mostly composed of cells, with little extracellular material) and **avascular** (no blood vessels). Epithelial cells exhibit **polarity;** they have a distinct *basal* (bottom) and *apical* (top) surface. On their basal surface, they have a specialized extracellular structure called a **basement membrane,** which anchors the epithelium to the underlying tissues. The characteristics used to classify epithelial tissue include (a) the number of layers of cells (simple, stratified, or pseudostratified) **(table 5.1)**, (b) the shape of the cells on the *apical* surface of the epithelium **(table 5.2)**, and (c) presence of any surface modifications **(table 5.3)**.

Figure 5.2 is a flowchart for classification of epithelial tissues that can be used as a tool when attempting to identify an unknown slide containing epithelial tissue. Following the flowchart will assist in the process of deciding how to classify an epithelial tissue.

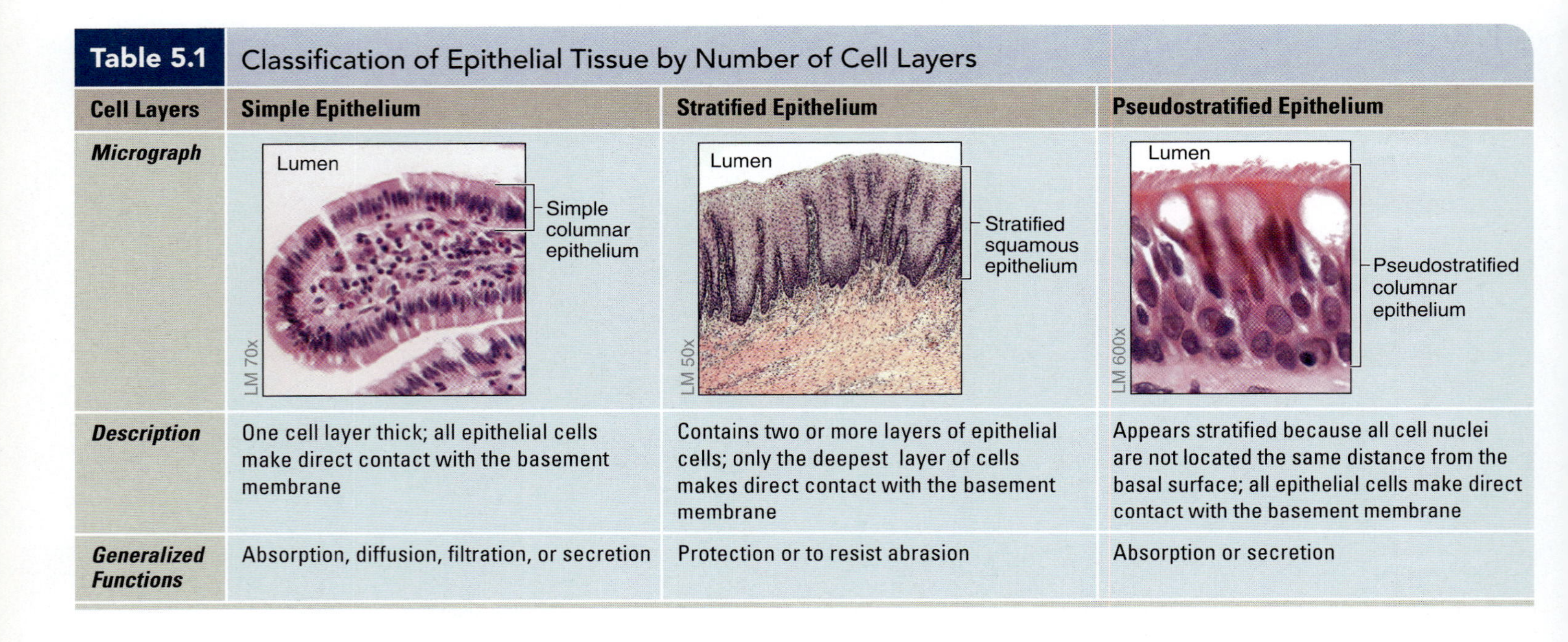

Table 5.1 Classification of Epithelial Tissue by Number of Cell Layers

Cell Layers	Simple Epithelium	Stratified Epithelium	Pseudostratified Epithelium
Micrograph	(micrograph)	(micrograph)	(micrograph)
Description	One cell layer thick; all epithelial cells make direct contact with the basement membrane	Contains two or more layers of epithelial cells; only the deepest layer of cells makes direct contact with the basement membrane	Appears stratified because all cell nuclei are not located the same distance from the basal surface; all epithelial cells make direct contact with the basement membrane
Generalized Functions	Absorption, diffusion, filtration, or secretion	Protection or to resist abrasion	Absorption or secretion

INTEGRATE

CLINICAL VIEW
Histopathology

Knowledge of the microscopic structure of tissues is critical for health professionals so they may be able to communicate with other medical professionals about tissue-level structures. Although most health professionals will rarely view slides of tissues in practice, nearly all health professionals will need to be able to interpret histopathology reports that are pertinent to their patients' diagnoses. A *histopathologist* is a physician and/or scientist who analyzes tissue samples that have been taken from a patient via biopsy (*bi*, two + *-opsy*, inspection). Once the histopathologist receives the tissue sample in the laboratory, he or she goes through the following process to create a microscopic slide containing a slice of the tissue. This process is very similar to the process used to create the slides that are viewed in the anatomy & physiology laboratory.

Consider briefly how the slide shown in **figure 5.1** may have been prepared. The process of making a histology slide involves five general steps:

1. Obtain a tissue sample.
2. Prepare the tissue sample for slicing.
3. Cut thin slices of the tissue using a special knife called a microtome.
4. Transfer the tissue slices to a microscope slide.
5. Stain the slide.

After a tissue sample has been obtained, it must be made rigid so that it will be easy to slice. This is done either by freezing the tissue or by embedding

Table 5.2 Classification of Epithelial Tissue by Cell Shapes

Cell Shape	Squamous	Cuboidal	Columnar	Transitional
Micrograph	Squamous cells; Lumen; LM 205x	Cuboidal cells; Lumen; LM 165x	Columnar cells; Lumen; LM 500x	Transitional cell; Lumen; LM 180x
Description	Cells are flattened and have irregular borders	Cells are as tall as they are wide	Cells are taller than they are wide	Cells change shape depending on the stress on the epithelial tissue. The cells change between a cuboidal shape and a more flattened, squamous shape.
Generalized Functions	If the epithelium is only one cell layer thick, it provides a very thin barrier for *diffusion.* If the epithelium is several layers thick, the cells specialize in *protection* (as in epidermal cells of the skin).	The shape of the cell allows more room for cellular organelles (e.g., mitochondria, endoplasmic reticulum). Cuboidal cells generally function in *secretion* and/or *absorption.*	The large size of the cell allows even more room for cellular organelles (e.g., mitochondria, endoplasmic reticulum). Columnar cells generally function in *secretion* and/or *absorption.*	The fact that these cells change shape means that they are good at *resisting stretch* without being torn apart from each other. Transitional cells are only found lining structures of the urinary tract (e.g., ureters, urinary bladder).
Identifying Characteristics	In cross section, the nucleus is the most visible structure. The nucleus will be very flattened. In a surface view of the epithelium, the cell borders will be irregular in shape.	Generally, cuboidal cells are identified by their very round, plump nucleus, and by equal amounts of cytoplasm in the spaces between the nucleus and the plasma membrane on all sides.	The nuclei of columnar cells can be either oval or round in shape, and they generally line up in a row. If the nuclei are round, more cytoplasm will be visible between the nucleus and the plasma membrane on the apical side of the nucleus than on the other three sides.	Transitional cells are located on the apical surface of the epithelium. However, the transitional cells appear much more rounded or dome-shaped than typical cuboidal cells, and they are sometimes binucleate.

the tissue in a block of paraffin wax. The next step in preparing the slide involves slicing the frozen sample (or wax block) into very thin slices (on the order of micrometers—1 μm is 10^{-6} meters) using a special knife called a microtome (*micro,* small, + *tome* or *temmein,* to cut). The slices are cut so thin that often only a single layer of cells is contained in the slice. Once the slices are made, they are transferred to a microscope slide, which is then covered with a cover slip. Finally, the samples are stained to make intracellular and extracellular structures visible. The most common method of staining, hematoxylin and eosin (H and E), makes most structures appear pink or purple and makes the nucleus of the cell, in particular, easily visible.

Once a slide of a patient's tissue has been made, a histopathologist analyzes the sample to determine if the tissue appears as expected. Any variations from expected structure are then characterized and described. These observations are analyzed in conjunction with lab tests to aid in making a clinical diagnosis of the patient's condition.

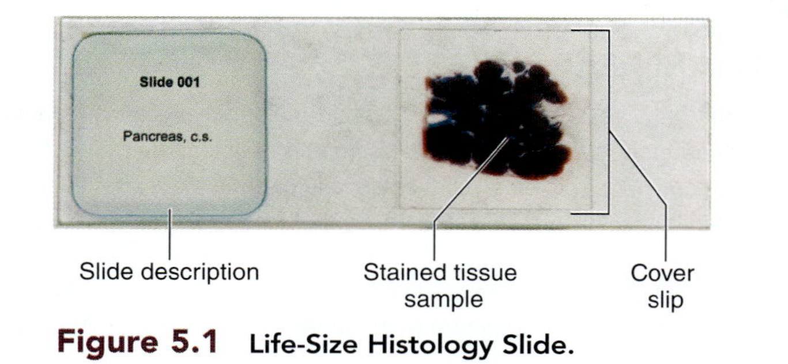

Figure 5.1 Life-Size Histology Slide.

Table 5.3 Cell Surface Modifications and Specialized Cells of Epithelial Tissues

Surface Modification	Cilia	Goblet Cells	Keratinization	Microvilli
Micrograph	Cilia; Lumen; LM 1000x	Goblet cell; Columnar epithelial cell; Mucin within goblet cell; Lumen; Goblet cell nucleus; Location of basement membrane; LM 100x	Keratinization; Lumen; LM 25x	Brush border of microvilli; Lumen; LM 130x
Description and Function	Cilia are small, hairlike structures that extend from the apical surface of epithelial cells. Cilia actively move to *propel substances along the apical surface of an epithelial sheet.* Cilia move substances in only one direction.	Goblet cells are named for their shape. They are rounded near the apical surface and they narrow toward their basal surface. Goblet cells contain many small mucin granules and function in the *production of mucus.* The mucus is used to *assist in transport* of substances along an epithelial sheet, to *provide a protective barrier* along the apical surface of the epithelium, or to provide *lubrication.*	Stratified squamous epithelial cells of the skin contain keratin (an intermediate filament). Bundles of keratin fill up entire cells and bind to desmosomes, which firmly anchor the dead squamous epithelial cells together. The layers of cells appear to be a single homogeneous unit. Keratin imparts *strength* and *protection* to dead skin epithelial cells.	Microvilli are extremely small extensions of the plasma membrane of the apical surface of cells. Microvilli *increase the surface area* of the cell to enhance the process of *absorption.*
Identifying Characteristics	When cilia are present, and the slide is viewed at sufficient magnification, what appear to be individual "hairs" are visible on the apical surface of the epithelial cells.	Goblet cells are named for their shape. They are rounded near the apical surface and they narrow toward their basal surface. The shape is similar to the shape of a wine glass. The mucin inside the cells does not typically take up biological stains, so the cells often appear white or "empty." If the slide is stained specifically for mucin, then the goblet cells will appear dark.	Keratinization is recognized as a homogeneous, acellular-looking portion of a stratified squamous epithelium.	Individual microvilli can be seen only when the specimen is viewed with an electron microscope. Thus, individual microvilli will *not* be visible with a light microscope. Instead, the apical surface of the epithelium will appear to be "fuzzy." For this reason, epithelia containing microvilli are often said to have a "brush border."

INTEGRATE

LEARNING STRATEGY

When viewing a slide for the purpose of identifying an epithelial tissue, remember that epithelial tissues form linings and coverings of organs. Most histology slides typically contain more tissues than just epithelial tissues. To locate the epithelial tissue, first look for any white space, or "empty" space, on the slide. This space typically will be the outside of an organ or the lumen (inside) of the organ. The tissue that lies directly adjacent to the empty space will usually be an epithelial tissue.

INTEGRATE

LEARNING STRATEGY

A **simple** epithelium is only one cell layer thick. A **stratified** epithelium is two or more cell layers thick. Be aware that cell shape can vary in stratified epithelium. To avoid confusion, always identify the shapes of cells on the *apical* surface when classifying stratified epithelium.

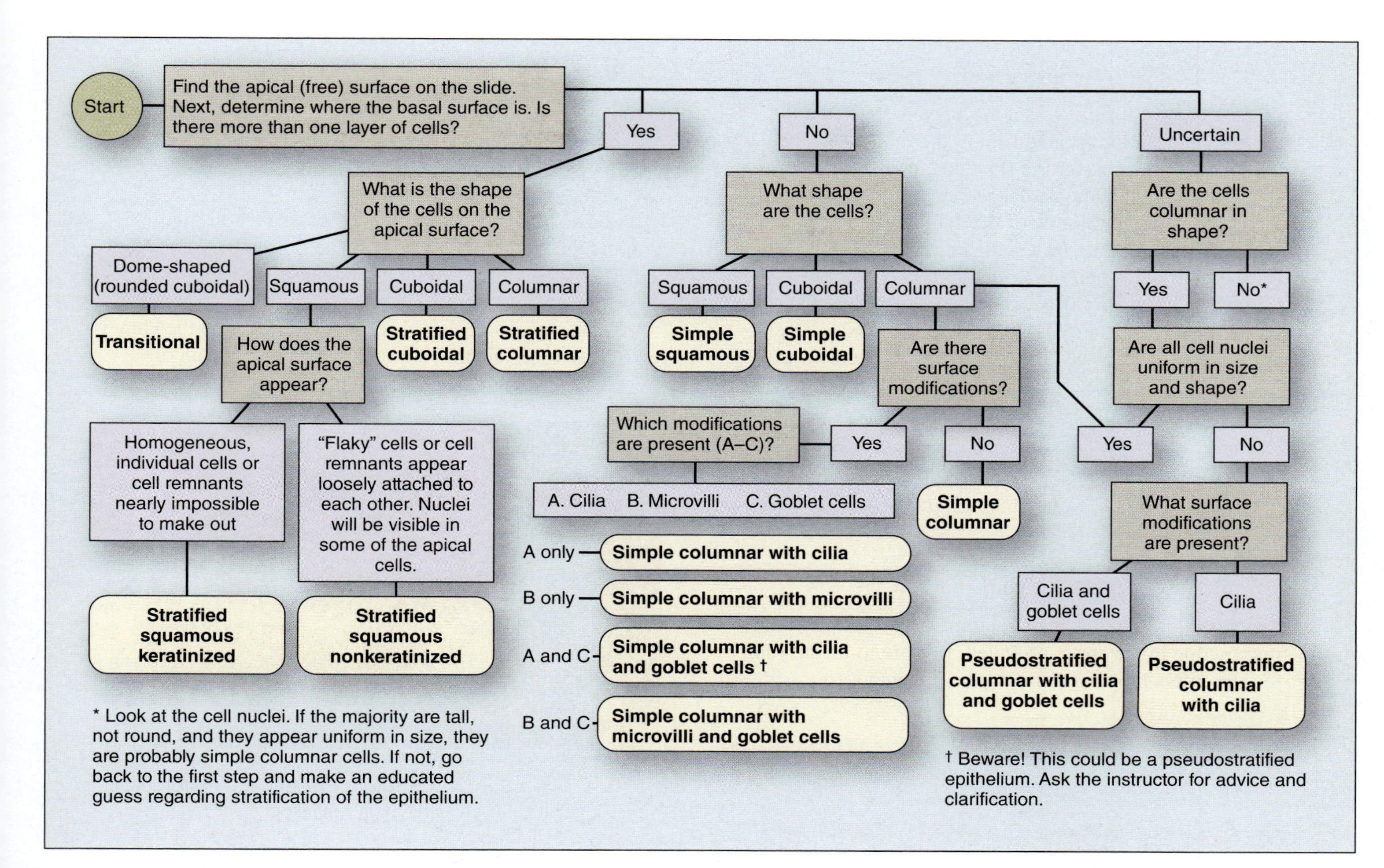

Figure 5.2 Flowchart for Classifying Epithelial Tissues.

EXERCISE 5.1

IDENTIFICATION AND CLASSIFICATION OF EPITHELIAL TISSUE

EXERCISE 5.1A Simple Squamous Epithelium

1. Obtain a slide of a small vein in cross section **(figure 5.3).**
2. Place the slide on the microscope stage and bring the tissue sample into focus on low power.
3. Look for any "empty" space on the slide. The empty space on the slide will be either the inside of the vessel (the **lumen** of the vessel) or the outside edge of the tissue sample.
4. Locate the lumen of the vessel and move the microscope stage so the lumen is at the center of the field of view. The lumen of all blood vessels, lymphatic vessels, and the heart, is lined with a simple squamous epithelium called *endothelium.* This type of epithelium is always somewhat difficult to see because it is extremely thin.

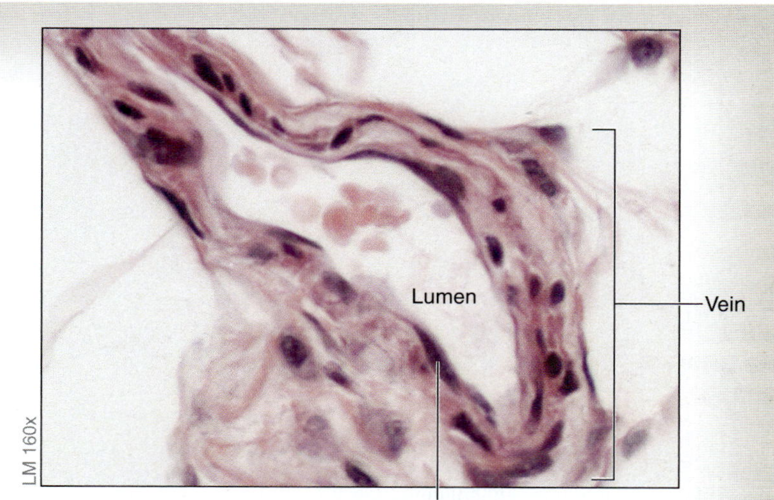

Figure 5.3 **Simple Squamous Epithelium.** Cross section through a small vein lined with epithelial cells, which are simple squamous epithelial cells.

(continued on next page)

(continued from previous page)

5. Once the lumen of the vessel has been identified and the epithelial tissue located (lining the lumen), change to high power. Look for the flattened nuclei of the squamous epithelial cells that line the lumen. It is unlikely that much, if any, of the cellular cytoplasm will be visible because the cells are extremely thin. Because simple squamous epithelium is extremely thin, it functions in diffusion—a process that occurs over very short distances (distances of approximately 10 μm). Diffusion (see chapter 4) is the movement of particles from an area of high concentration to an area of low concentration, and is one mechanism by which substances are transported in the body.

6. Identify the following structures on the slide, using figure 5.3 and tables 5.2 and 5.3 as guides:

   ☐ **lumen of vein**

   ☐ **nucleus of squamous epithelial cell**

7. Sketch simple squamous epithelium as seen through the microscope in the space provided. Be sure to identify all the structures listed in step 6 in the drawing.

__________ ×

INTEGRATE

CONCEPT CONNECTION

Simple squamous epithelium lining the cardiovascular system is called **endothelium.** Simple squamous epithelium lining body cavities is called **mesothelium.** Thus, these terms (endothelium and mesothelium) indicate not only the *type* of epithelium (simple squamous), but also the *location* of the epithelium.

EXERCISE 5.1B Simple Cuboidal Epithelium

1. Obtain a slide of the kidney **(figure 5.4).**

2. Place the slide on the microscope stage and bring the tissue sample into focus on low power.

3. Locate the lumen of a tubule in cross section (figure 5.4), and then identify the cells that lie next to the lumen. These cells should have plump, round nuclei and approximately equal amounts of cytoplasm surrounding each nucleus. These are cuboidal epithelial cells, which line the kidney tubules and function in secretion and absorption of substances across the epithelium (table 5.2).

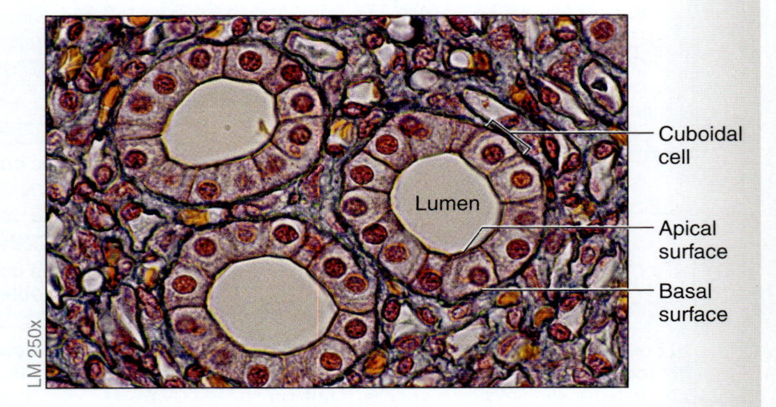

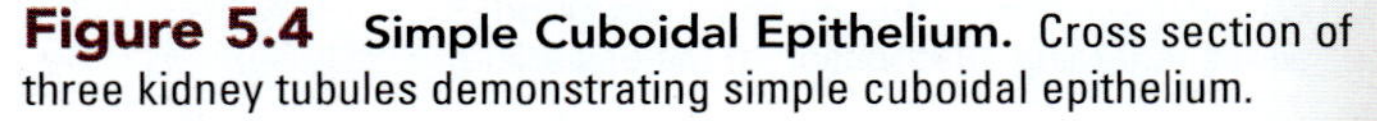

Figure 5.4 Simple Cuboidal Epithelium. Cross section of three kidney tubules demonstrating simple cuboidal epithelium.

4. Identify the following structures on the slide, using figure 5.4 and table 5.2 as guides:

   ☐ **apical surface** ☐ **cuboidal cell**

   ☐ **basal surface** ☐ **lumen of tubule**

5. Sketch simple cuboidal epithelium as seen through the microscope in the space provided. Be sure to identify all the structures listed in step 4 in the drawing.

__________ ×

EXERCISE 5.1C Simple Columnar Epithelium (nonciliated)

1. Obtain a slide of the small intestine **(figure 5.5).**
2. Place the slide on the microscope stage and bring the tissue sample into focus on low power.
3. This slide will show only a part of the wall of the intestine, so find some empty space on the slide first and then look for epithelium next to that empty space. Once the epithelium is in the center of the field of view, switch to high power. Look for epithelial cells with oval, elongated nuclei that have most of their cytoplasm on the apical side of the nucleus. Columnar cells are taller than they are wide, and their nuclei generally appear to be lined up in a row. The nuclei can be either elongated or round in shape.

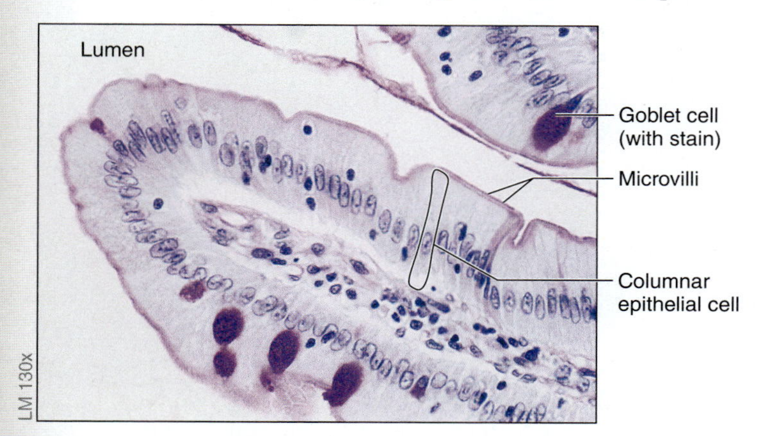

Figure 5.5 Simple Columnar Epithelium with Microvilli. Simple columnar epithelium with microvilli lining the lumen of the small intestine.

4. This epithelium also demonstrates **goblet cells,** which secrete mucin, and **microvilli,** which increase the surface area of the epithelial cells for absorption (table 5.3).
5. Identify the following structures on the slide, using figure 5.5 and tables 5.2 and 5.3 as guides:

- ☐ **columnar epithelial cell**
- ☐ **goblet cell**
- ☐ **lumen of intestine**
- ☐ **microvilli**

6. Sketch simple columnar epithelium as seen through the microscope in the space provided. Be sure to identify all the structures listed in step 5 in the drawing.

_________ ×

EXERCISE 5.1D Simple Columnar Epithelium (ciliated)

1. Obtain a slide of a uterine tube **(figure 5.6).**
2. Place the slide on the microscope stage and bring the tissue sample into focus on low power.
3. Look for a tubular structure cut in cross section (figure 5.6), and identify its lumen. Look for columnar epithelial cells next to the lumen. Once the epithelium is in the field of view, switch to high power. When cilia are present, the individual "hair"-like cilia should be visible on the apical surface of the epithelial cells.

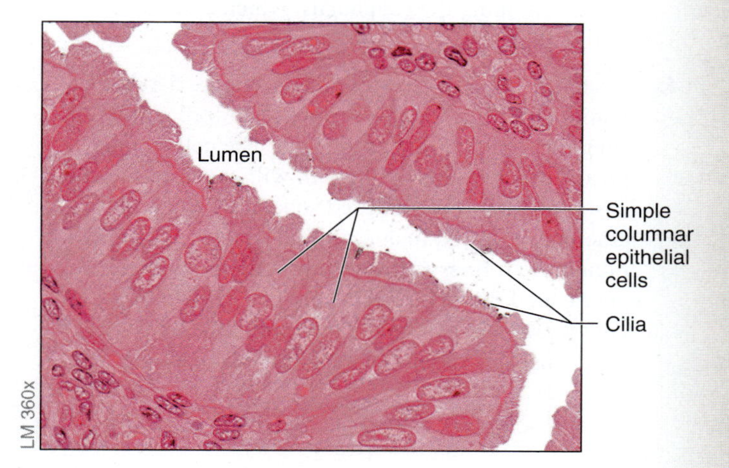

Figure 5.6 Simple Columnar Epithelium with Cilia. Simple columnar ciliated epithelium lining the lumen of the uterine tube.

4. Identify the following structures on the slide, using figure 5.6 and tables 5.2 and 5.3 as guides:

- ☐ **cilia**
- ☐ **lumen of uterine tube**
- ☐ **simple columnar epithelial cell**

5. Sketch simple columnar epithelium with cilia as seen through the microscope in the space provided. Be sure to identify all the structures listed in step 4 in the drawing.

_________ ×

(continued on next page)

(continued from previous page)

EXERCISE 5.1E Stratified Squamous Epithelium (nonkeratinized)

1. Obtain a slide of the trachea and esophagus **(figure 5.7*a*,*b*).**
2. Place the slide on the microscope stage and bring the tissue sample into focus on low power.
3. The slide shown contains two different organs in cross section, the trachea and the esophagus. Look for the epithelium lining the esophagus, which consists of multiple layers of cells.
4. Locate the lumen of the esophagus (figure 5.7*a*), then identify stratified squamous nonkeratinized epithelium lining the lumen of the esophagus. The cells on the apical surface of this epithelium will appear flattened. This nonkeratinized stratified squamous epithelium is sometimes referred to as stratified squamous "moist" epithelium. It lines surfaces within the body that experience friction and abrasion, but where water loss is not a problem (for example, lining the oral cavity, esophagus, and vagina).
5. Identify the following structures on the slide, using figure 5.7*b* and tables 5.2 and 5.3 as guides:
   - ☐ **lumen of esophagus**
   - ☐ **squamous epithelial cells**
   - ☐ **stratified squamous nonkeratinized epithelium**
6. Sketch stratified squamous nonkeratinized epithelium as seen through the microscope in the space provided. Be sure to identify the structures listed in step 5 in the drawing.

_______________ ×

7. Keep the slide on the microscope stage and proceed to exercise 5.1F to focus on the epithelium lining the trachea.

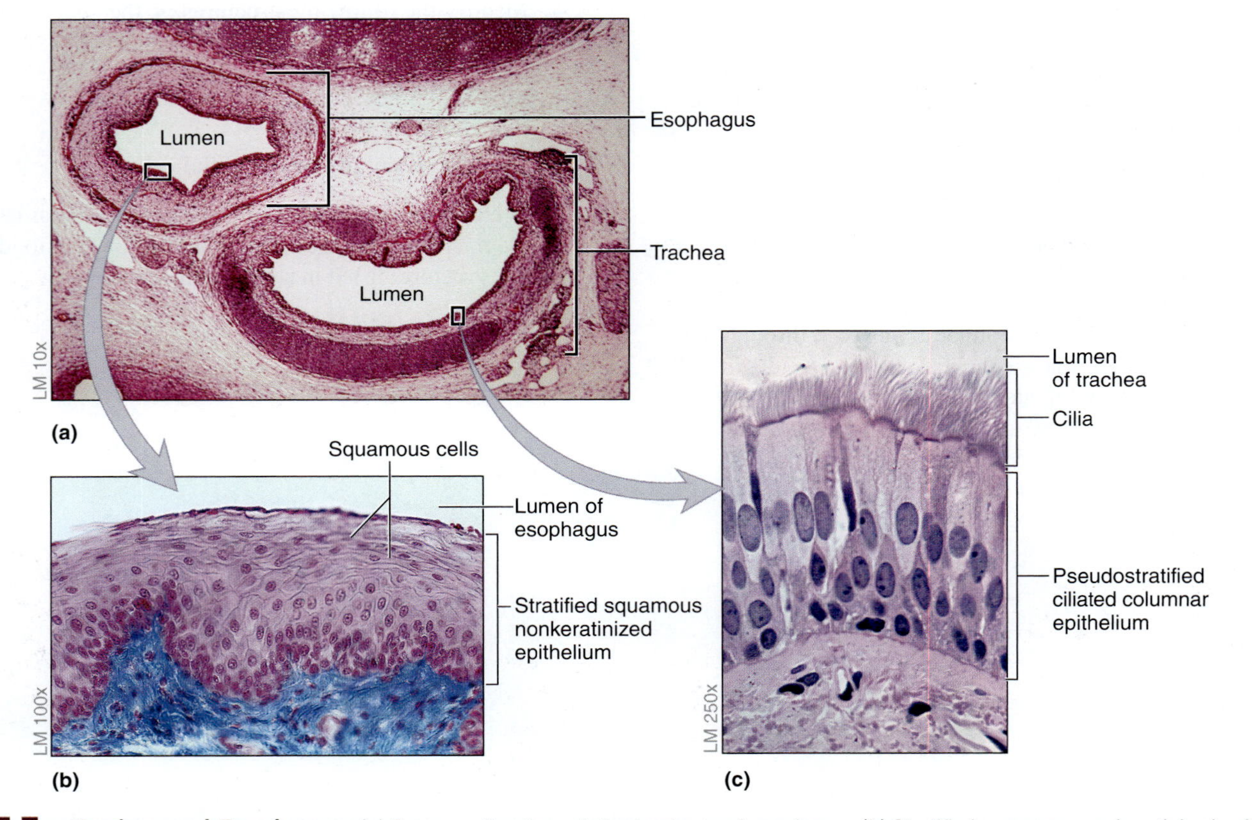

Figure 5.7 Trachea and Esophagus. (a) Cross section through the trachea and esophagus. (b) Stratified squamous nonkeratinized epithelium lining the esophagus. (c) Pseudostratified ciliated columnar epithelium lining the trachea.

EXERCISE 5.1F Pseudostratified Columnar Epithelium

1. Obtain a slide of the trachea and esophagus (figure 5.7*a*).
2. Place the slide on the microscope stage and bring the tissue sample into focus on low power.
3. This slide will contain two organs in cross section, the trachea and the esophagus. Look for the epithelium lining the trachea, which contains ciliated cells.
4. Locate the lumen of the trachea (figure 5.7*a*), then identify pseudostratified columnar epithelium (figure 5.7*c*) lining the trachea. Once the epithelium is in the field of view, switch to high power. This epithelium is characterized by the presence of columnar cells that are not all the same height. All cells contact the basement membrane, but not all reach the apical surface. Because not all cells reach the apical surface of the epithelium, the nuclei will *not* be nicely lined up in rows as with simple columnar epithelium. Instead, the nuclei will appear to be layered. Hence, the name of this epithelium: **pseudostratified** (*pseudo-*, false). Most, but not all, pseudostratified epithelia also contain cilia and goblet cells.
5. Identify the following structures on the slide, using figure 5.7*c* and tables 5.2 and 5.3 as guides:
   - ☐ **cilia**
   - ☐ **columnar epithelial cell**
   - ☐ **lumen of trachea**
6. Sketch pseudostratified columnar epithelium as seen through the microscope in the space provided. Be sure to label the structures listed in step 5 in the drawing.

_______ ×

EXERCISE 5.1G Stratified Cuboidal or Stratified Columnar Epithelium

1. Obtain a slide of a merocrine sweat gland **(figure 5.8).**
2. Place the slide on the microscope stage and bring the tissue sample into focus on low power.
3. Locate the lumen of a duct of the sweat gland (figure 5.8). Identify stratified cuboidal epithelium next to the lumen. Once the epithelium is in the field of view, switch to high power. Stratified cuboidal epithelium is found lining the ducts of merocrine sweat glands, which are located in the dermis of the skin. Stratified cuboidal epithelium is generally only two cell layers thick. How many layers are visible on the slide? The laboratory may have other slides available that demonstrate stratified *columnar* epithelium lining the ducts of other exocrine glands. As with stratified cuboidal epithelium, stratified columnar epithelium is also rarely more than two cell layers thick.
4. Identify the following structures on the slide, using figure 5.8 and tables 5.2 and 5.3 as guides:
   - ☐ **basement membrane**
   - ☐ **cuboidal epithelial cell**
   - ☐ **lumen of the duct**

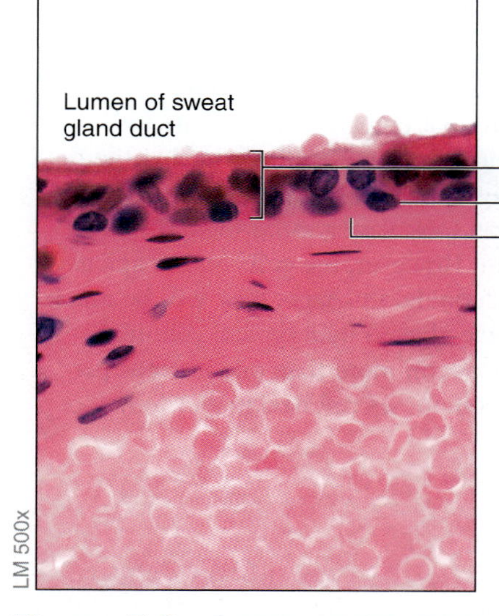

Figure 5.8 Stratified Cuboidal Epithelium. Epithelium lining the duct of a merocrine sweat gland.

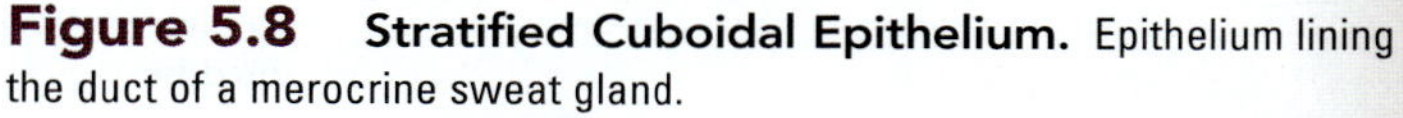

INTEGRATE

LEARNING STRATEGY

If the slide being viewed appears to be stratified columnar/cuboidal, but also has cilia, it *cannot* be stratified columnar/cuboidal. Only *pseudostratified* columnar epithelium can have cilia. That said, an epithelia *can* be pseudostratified and *not* have cilia, so the distinction only goes one way.

(continued on next page)

(continued from previous page)

5. Sketch stratified cuboidal (or stratified columnar) epithelium as seen through the microscope in the space provided. Be sure to label the structures listed in step 4 on the drawing.

_____________ ×

EXERCISE 5.1H Transitional Epithelium

1. Obtain a slide of the urinary bladder **(figure 5.9).**
2. Place the slide on the microscope stage and bring the tissue sample into focus on low power.
3. This slide will show only a part of the wall of the bladder, so first locate the empty space and then look for epithelium next to that empty space. Locate the transitional epithelium. Once the epithelium is in the field of view, switch to high power. Transitional epithelium is stratified, and it can look very similar to stratified squamous epithelium. However, the cells on the apical surface will appear cuboidal in shape, they will be much more rounded or dome-shaped than typical cuboidal cells, and they may contain more than one nucleus per cell (these cells are sometimes referred to as "dome" or "umbrella" cells).

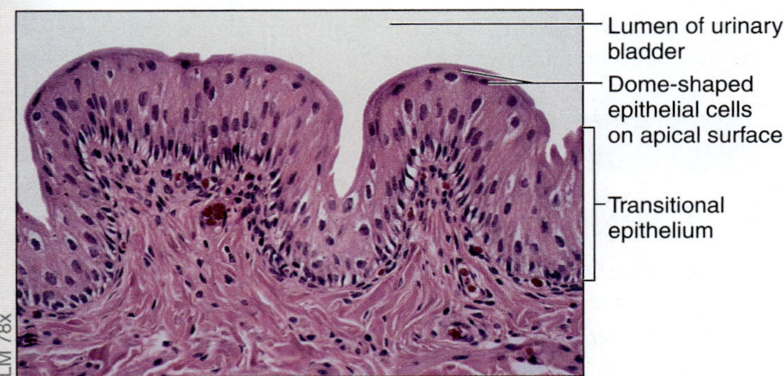

Figure 5.9 Transitional Epithelium. Epithelium lining the urinary bladder.

4. Observe the cells on the basal surface of the transitional epithelium. These cells are usually columnar in shape, in contrast to the cells on the basal surface of a stratified squamous epithelium, which tend to be more cuboidal in shape. Transitional epithelium is found lining the urinary bladder and other urine-draining structures. Its structure allows the epithelium to stretch easily to accommodate the passage or storage of urine without causing the epithelial cells to tear apart.
5. Identify the following structures on the slide, using figure 5.9 and tables 5.2 and 5.3 as guides:
   - ☐ **dome-shaped epithelial cells**
   - ☐ **lumen of urinary bladder**
   - ☐ **transitional epithelium**
6. Sketch transitional epithelium as seen through the microscope in the space provided. Be sure to label the structures listed in step 5 in the drawing.

_____________ ×

7. *Optional Activity:* **AP|R 3: Tissues**—Watch the "Epithelial Tissue Overview" animation.

INTEGRATE

LEARNING STRATEGY

Students often confuse the terms *basement membrane* and *basal surface*. To clarify, the **basement membrane** is a connective tissue structure that lies at the **basal surface** of the epithelium. The two terms are not synonymous. That is, the basement membrane is a *structure;* the basal surface is a *location.* Also note that the basement membrane is an extracellular structure on which the epithelium rests. The function of the basement membrane is to anchor the epithelium to the underlying connective tissue, provide physical support for the epithelium, and act as a barrier to regulate the passage of large molecules between the epithelium and the underlying tissues.

INTEGRATE

CONCEPT CONNECTION

The presence of epithelial modifications such as cilia, microvilli, and goblet cells in epithelial tissues reveals the overall function of the epithelium in question. In the exercises in this chapter, representative examples of each type of epithelial tissue are being observed. These examples will be revisited in the histology sections of subsequent chapters, where they will be covered in more detail. Rather than memorizing the locations of these tissues at this time, focus on identifying how the modification relates to tissue function. Cilia aid in *movement/transport,* microvilli aid in *absorption,* and goblet cells aid in *lubrication.* For example, consider the epithelial tissue lining the upper respiratory structures: pseudostratified ciliated columnar epithelium. This epithelium contains goblet cells, which produce mucus, and cilia, which trap and transport mucus and debris along the epithelial surface. Together, these modifications form a "mucus escalator" that prevents foreign objects and pathogens from entering lower respiratory tract structures. Damage to this epithelium, specifically the cilia, is detrimental, because it renders the lower respiratory tract more prone to infection.

Connective Tissue

All connective tissues share three basic components: cells, protein fibers, and ground substance. The different types of protein fibers include collagen, elastic, and reticular (see **table 5.4**). Connective tissues are derived from an embryonic tissue called **mesenchyme (figure 5.10).** There are three broad categories of mature connective tissues: connective tissue proper, supporting connective tissues, and fluid connective tissues. **Connective tissue proper** is the "glue" that holds things together (such as tendons and ligaments) and the "stuffing" that fills in spaces (such as the fat that fills in spaces between muscles). Subcategories of connective tissue proper include both loose connective tissue (areolar, adipose, and reticular) and dense connective tissue (dense regular, dense irregular, and elastic) as described in **table 5.5. Supporting connective tissues** (cartilage and bone) are specialized connective tissues that provide support and protection for the body. **Fluid connective tissues** (blood and lymph) are specialized connective tissues that function to transport substances throughout the body. The details of bone and blood are covered in chapters 7 and 20 of this laboratory manual.

Connective Tissue Proper

Connective tissue proper is a kind of "grab bag" category that contains all of the unspecialized connective tissues (that is, any connective tissue other than supporting or fluid connective tissues). These tissues are used either to hold things together (as with tendons and ligaments) or to fill up space (as with adipose tissue). Loose connective tissues have a loose association of cells and fibers, whereas dense connective tissues have cells and fibers that are densely packed together, which makes them much tougher than loose connective tissues. Table 5.5 summarizes the characteristics of the different types of connective tissue proper. **Figure 5.11** is a flowchart for classification of connective tissue proper that can be used when attempting to identify an unknown slide containing connective tissue.

Table 5.4 Connective Tissue Fibers

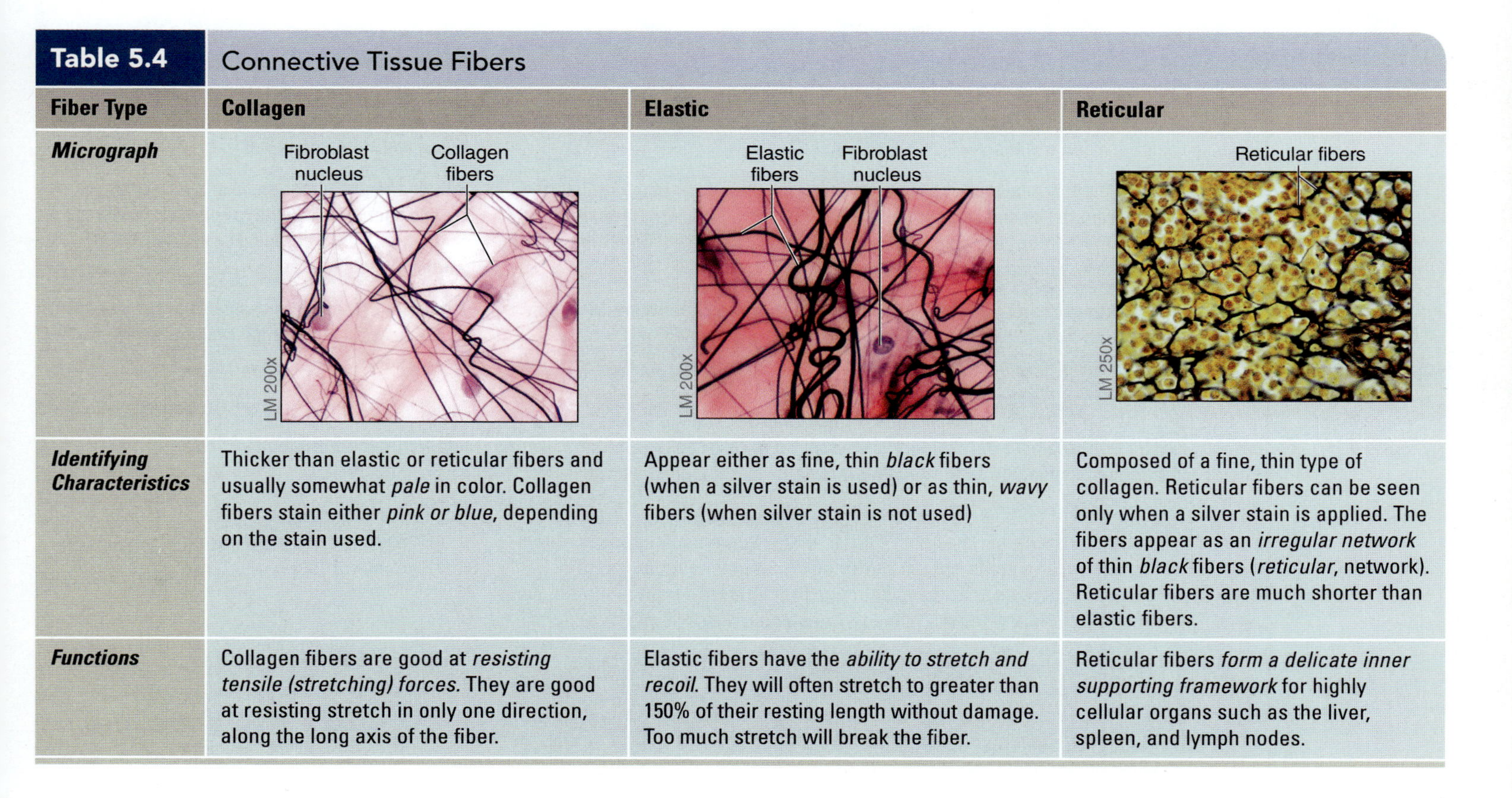

Fiber Type	**Collagen**	**Elastic**	**Reticular**
Micrograph	Fibroblast nucleus; Collagen fibers; LM 200x	Elastic fibers; Fibroblast nucleus; LM 200x	Reticular fibers; LM 250x
Identifying Characteristics	Thicker than elastic or reticular fibers and usually somewhat *pale* in color. Collagen fibers stain either *pink or blue,* depending on the stain used.	Appear either as fine, thin *black* fibers (when a silver stain is used) or as thin, *wavy* fibers (when silver stain is not used)	Composed of a fine, thin type of collagen. Reticular fibers can be seen only when a silver stain is applied. The fibers appear as an *irregular network* of thin *black* fibers (*reticular,* network). Reticular fibers are much shorter than elastic fibers.
Functions	Collagen fibers are good at *resisting tensile (stretching) forces.* They are good at resisting stretch in only one direction, along the long axis of the fiber.	Elastic fibers have the *ability to stretch and recoil.* They will often stretch to greater than 150% of their resting length without damage. Too much stretch will break the fiber.	Reticular fibers *form a delicate inner supporting framework* for highly cellular organs such as the liver, spleen, and lymph nodes.

EXERCISE 5.2

IDENTIFICATION OF EMBRYONIC CONNECTIVE TISSUE

1. Obtain a slide of mesenchyme (figure 5.10).
2. Place the slide on the microscope stage and bring the tissue sample into focus on low power, then switch to high power.
3. Notice that there are no visible fibers within the extracellular matrix (mature fibers do not yet exist within mesenchyme). The **mesenchymal cells** are recognized by their large oval nuclei. Mesenchyme has the ability to differentiate into any of the mature connective tissue cell types. That is, mesenchymal cells can differentiate into any of the adult connective tissue cell types (e.g., fibroblasts, chondroblasts, osteoblasts).
4. Identify the following structures on the slide, using figure 5.10 as a guide:

☐ **ground substance**
☐ **mesenchymal cell**

5. Sketch mesenchyme as seen through the microscope in the space provided. Be sure to label the structures listed in step 4 in the drawing.

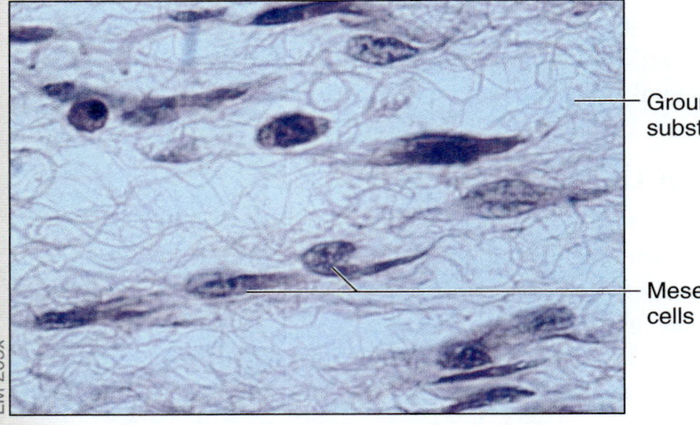

Figure 5.10 **Mesenchyme.** Mesenchyme is embryonic connective tissue.

_______ ×

Table 5.5 Connective Tissue Proper

Classification	Description and Function	Location(s)
Loose Connective Tissue		
Areolar (Figure 5.12)	Consists of a loose arrangement of collagen and elastic fibers with numerous fibroblasts. Areolar connective tissue loosely anchors structures to each other or fills in spaces between organs.	Located in the superficial fascia below the skin, which anchors skin to underlying muscle. It is also found surrounding many organs.
Adipose (Figure 5.13)	Characterized by adipocytes, which appear to be large, "empty" cells because the process of preparing the tissue removes all lipid within the cells. Collagen fibers located between the adipocytes hold the tissue together. Adipose tissue functions in insulation, protection, and energy storage.	Subcutaneous (under the skin). Surrounds organs such as the kidneys, where it provides protection. Fills in potential spaces such as within the popliteal fossa.
Reticular (Figure 5.14)	Composed of reticular fibers, which form an inner supporting framework for highly cellular organs such as the liver	The inner stroma (*stroma*, bed) of organs such as the spleen, liver, and lymph nodes
Dense Connective Tissue		
Dense Regular (Figure 5.15)	Composed of regular bands of collagen fibers all oriented in the same direction. The flattened nuclei of fibroblasts can be seen between bundles of collagen fibers. Dense regular connective tissue is good at resisting tensile forces in one direction only.	Tendons (connect muscle to bone) and ligaments (connect bone to bone)
Dense Irregular (Figure 5.16)	Composed of bundles of collagen fibers arranged in many directions. Fibroblast nuclei can be seen between bundles of collagen fibers. Some nuclei appear round (if cut in cross section) and some appear flattened (if cut in longitudinal section). Many of the collagen fibers appear wavy because they are not all cut along the same plane. This tissue is tough and resists tensile forces applied in multiple directions.	Organ capsules, dermis of the skin, periosteum (outer covering of bone), perichondrium (outer covering of cartilage)
Elastic (Figure 5.17)	Consists of both collagen and elastic fibers all oriented in the same direction. Collagen fibers are thick and typically stain light pink or purple. Elastic fibers appear thin and black (if stained) or wavy. Fibroblast nuclei can be seen between the densely packed fibers. Elastic connective tissue is extensible and allows structures to stretch and recoil back to their original shape.	Walls of large arteries such as the aorta and some ligaments

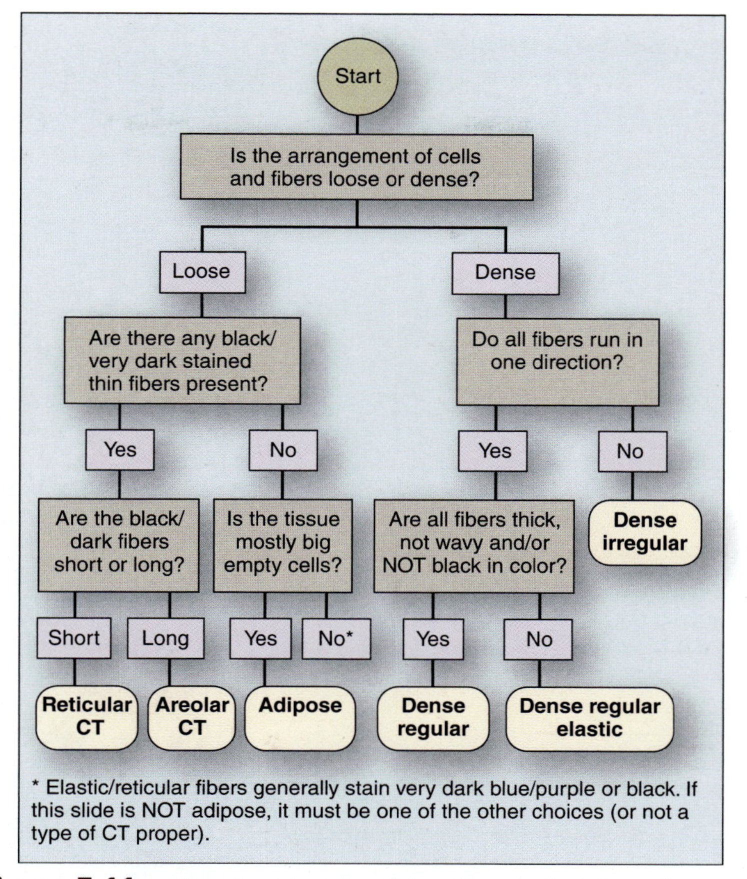

Figure 5.11 Flowchart for Classifying Connective Tissue Proper.

EXERCISE 5.3

IDENTIFICATION AND CLASSIFICATION OF CONNECTIVE TISSUE PROPER

EXERCISE 5.3A Areolar Connective Tissue

1. Obtain a slide of areolar connective tissue **(figure 5.12).**
2. Place the slide on the microscope stage and bring the tissue sample into focus on low power. Then change to high power.

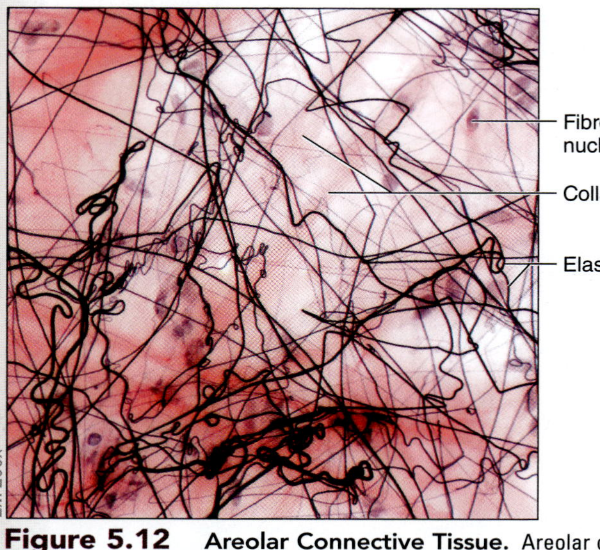

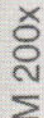

Figure 5.12 **Areolar Connective Tissue.** Areolar connective tissue contains predominantly fibroblasts, collagen fibers, and elastic fibers.

3. Several small cells are scattered throughout the slide. Most of these cells are **fibroblasts,** which secrete the elastic and collagen fibers (table 5.5). Collagen fibers will generally be light pink in color and somewhat thick. Elastic fibers will generally be long and thin, and will stain very dark or black.
4. Identify the following structures on the slide, using figure 5.12 and tables 5.4 and 5.5 as guides:

☐ **collagen fibers** ☐ **fibroblasts**
☐ **elastic fibers**

5. Sketch areolar connective tissue as seen through the microscope in the space provided. Be sure to label the structures listed in step 4 on the drawing.

_____________ ×

(continued on next page)

(continued from previous page)

EXERCISE 5.3B Adipose Connective Tissue

1. Obtain a slide of adipose connective tissue **(figure 5.13).**
2. Place the slide on the microscope stage and bring the tissue sample into focus on low power. Then change to high power.

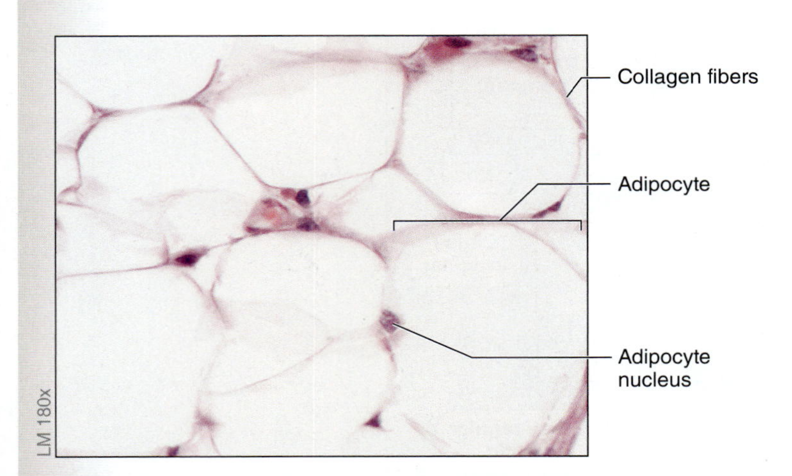

Figure 5.13 Adipose Connective Tissue. Adipose connective tissue is characterized by large adipocytes held together with a loose arrangement of collagen fibers.

3. Identify the following structures on the slide, using figure 5.13 and tables 5.4 and 5.5 as guides:

- ☐ **adipocyte**
- ☐ **collagen fibers**
- ☐ **adipocyte nucleus**

4. Sketch adipose connective tissue as seen through the microscope in the space provided. Be sure to label all the structures listed in step 3 in the drawing.

_______ ×

EXERCISE 5.3C Reticular Connective Tissue

1. Obtain a slide of reticular connective tissue **(figure 5.14).**
2. Place the slide on the microscope stage and bring the tissue sample into focus on low power. Then change to high power.

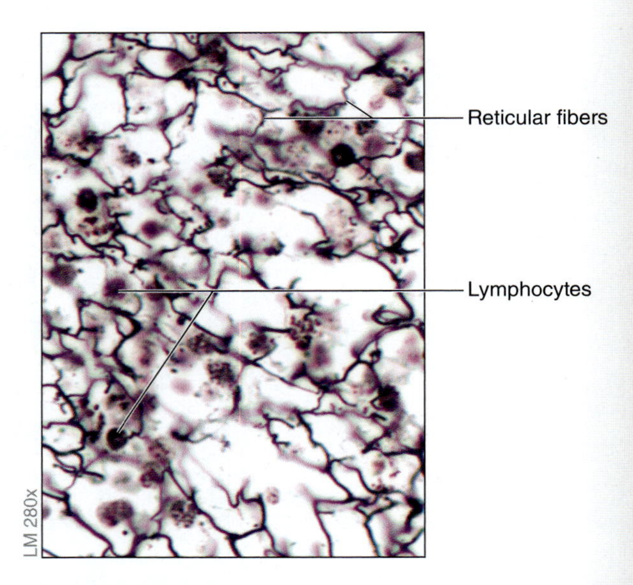

Figure 5.14 Reticular Connective Tissue. Reticular connective tissue consists of reticular fibers (a fine, thin form of collagen) that are visible only when a special stain is used, which makes them appear black.

3. Identify the following on the slide, using figure 5.14 and tables 5.4 and 5.5 as guides:

- ☐ **lymphocytes**
- ☐ **reticular fibers**

4. Sketch reticular connective tissue as seen through the microscope in the space provided. Be sure to label all the structures listed in step 3 in the drawing.

_______ ×

INTEGRATE

LEARNING STRATEGY

Though a variety of stains are used for visualization of cells and tissues, certain tissues are visible only when special stains are used. In connective tissues, elastic and reticular fibers can be seen only when a special stain is used that stains the fibers black. Thus, if black fibers are visible, these are either elastic or reticular fibers. If the black fibers are very long and thin or wavy, they are elastic fibers. If they are short and form a network (*rete*, network), they are reticular fibers.

EXERCISE 5.3D Dense Regular Connective Tissue

1. Obtain a slide of a tendon or ligament **(figure 5.15).**
2. Place the slide on the microscope stage and bring the tissue sample into focus on low power. Then change to high power.
3. Identify the following structures on the slide, using figure 5.15 and tables 5.4 and 5.5 as guides:

☐ **collagen fibers** ☐ **fibroblast nuclei**

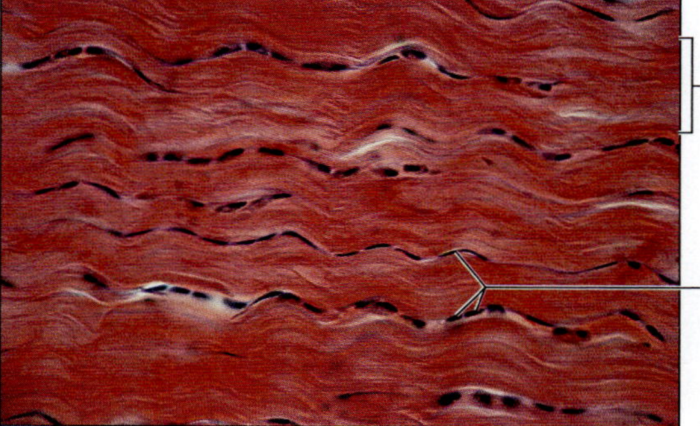

Figure 5.15 **Dense Regular Connective Tissue.** Dense regular connective tissue consists of bundles of collagen fibers all oriented in the same direction, with fibroblasts located between the bundles of collagen fibers.

INTEGRATE

LEARNING STRATEGY

Sometimes it helps to relate what is seen through the microscope to something that is already familiar. For example, the regular arrangement of collagen fibers in dense regular connective tissue often resembles uncooked lasagna noodles stacked upon each other, whereas the irregular arrangement of collagen and elastic fibers in areolar connective tissue often resembles a piece of abstract art.

4. Sketch dense regular connective tissue as seen through the microscope in the space provided. Be sure to label the structures listed in step 3 in the drawing.

_______ ×

EXERCISE 5.3E Dense Irregular Connective Tissue

1. Obtain a slide of skin **(figure 5.16).** Skin consists of two major layers, an outer epidermis, composed of stratified squamous epithelial tissue, and an inner dermis, composed of connective tissue (areolar and dense irregular).

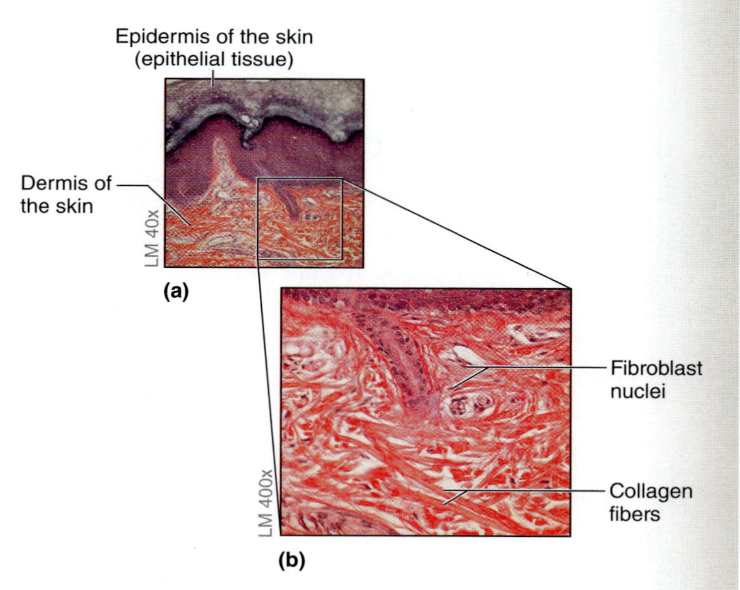

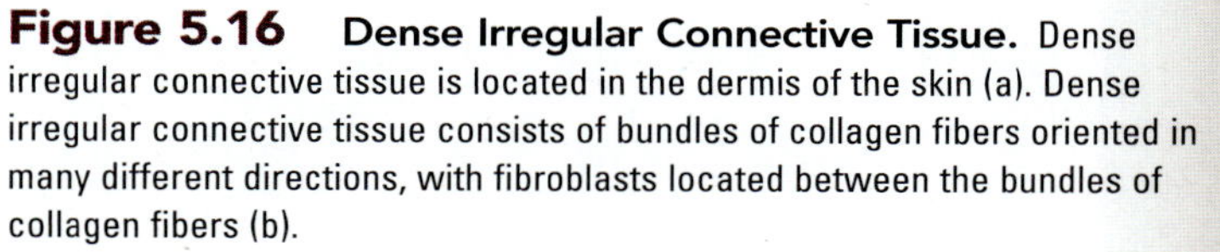

Figure 5.16 **Dense Irregular Connective Tissue.** Dense irregular connective tissue is located in the dermis of the skin (a). Dense irregular connective tissue consists of bundles of collagen fibers oriented in many different directions, with fibroblasts located between the bundles of collagen fibers (b).

(continued on next page)

(continued from previous page)

2. Place the slide on the microscope stage and bring the tissue sample into focus using the scanning objective. Identify the epithelial tissue in the epidermis (figure 5.16*a*), then move the stage so the lens focuses on the underlying connective tissue in the dermis (figure 5.16*b*). Switch to a higher-power objective.
3. Identify the following structures on the slide, using figure 5.16 and tables 5.4 and 5.5 as guides:

☐ **collagen fibers** ☐ **fibroblast nucleus**

4. Sketch dense irregular connective tissue as seen through the microscope in the space provided. Be sure to label the structures listed in step 3 in the drawing.

_________ ×

EXERCISE 5.3F Elastic Connective Tissue

1. Obtain a slide of the aorta or an elastic artery **(figure 5.17).**
2. Place the slide on the microscope stage and bring the tissue sample into focus on low power. Then switch to high power.
3. Identify the following structures on the slide, using figure 5.17 and tables 5.4 and 5.5 as guides:

☐ **collagen fibers** ☐ **fibroblasts**
☐ **elastic fibers**

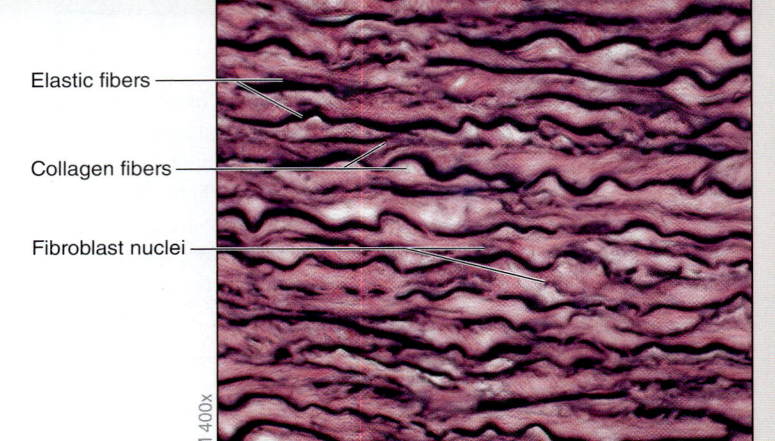

Figure 5.17 Elastic Connective Tissue. The wall of the aorta consists of dense regular elastic connective tissue, which contains both collagen and elastic fibers (which stain black). All the fibers are oriented in the same direction. In this slide, the collagen fibers are dark pink and the elastic fibers are black and wavy. Some fibroblast nuclei are visible between the bundles of fibers.

4. Sketch elastic connective tissue as seen through the microscope in the space provided. Be sure to label the structures listed in step 3 in the drawing.

_________ ×

Supporting Connective Tissue

Cartilage

Cartilage is a specialized connective tissue whose function is to provide strong, yet flexible, support. Cartilage is unique as a connective tissue in that it is avascular (*a-*, without + *vasa*, vessel). A dense irregular connective tissue covering, called the perichondrium, surrounds all types of cartilage except fibrocartilage. The innermost part of the perichondrium contains immature cartilage cells called chondroblasts. The function of the chondroblasts is to secrete the fibers and ground substance that compose the extracellular matrix of cartilage. As a chondroblast secretes extracellular matrix, it eventually becomes completely surrounded by the matrix; at this point it is considered a mature cell, a chondrocyte. The space in the matrix where a chondrocyte sits is a lacuna (*lacus*, a lake). The function of a chondrocyte is to maintain the matrix that has already been formed. All types of cartilage contain chondrocytes in lacunae and have a ground substance that consists largely of glycosaminoglycans (GAGs). The three types of cartilage differ mainly in the type and arrangement of fibers within the matrix. **Table 5.6** summarizes the characteristics of the different types of cartilage.

Bone

Bone (table 5.6) is a specialized connective tissue whose function is to provide strong support. It protects vital organs and provides strong attachment points for skeletal muscles. Similar to cartilage, bone is surrounded by a dense irregular connective tissue covering, called the periosteum. The innermost part of the periosteum contains precursor bone cells called osteoprogenitor cells, which develop into immature bone cells called osteoblasts. Osteoblasts secrete the extracellular matrix (fibers and ground substance) of bone. When osteoblasts become completely enveloped by the bony matrix, they become mature osteocytes.

There are two types of bone tissue: compact bone (dense) and spongy bone (cancellous). The structural and functional unit of compact bone is an osteon, which consists of concentric layers of bony matrix (lamellae). Along the lamellae are lacunae that contain osteocytes. The details of the two types of bone tissue will be covered in chapter 7.

Figure 5.18 is a flowchart for classification of supporting connective tissues that can be used when attempting to identify an unknown slide of connective tissue.

Table 5.6 Supporting Connective Tissue: Cartilage and Bone

Tissue Type	Description	Functions and Locations	Identifying Characteristics
Hyaline Cartilage	Chondrocytes are located within lacunae. Contains a perichondrium of dense irregular connective tissue. Extracellular matrix consists of diffuse collagen fibers spread throughout a semirigid ground substance, which is composed mainly of glycoproteins and water.	Hyaline cartilage provides *strong, semiflexible support* for structures such as the nasal septum, costal cartilages, articular cartilages, larynx, and tracheal "C" rings.	Hyaline cartilage is recognized by the chondrocytes in lacunae and the fact that no fibers are visible in the extracellular matrix.
Fibrocartilage	Very similar to hyaline cartilage except there is no perichondrium and the collagen fibers form thick visible bundles	The organization and density of the collagen fibers make fibrocartilage particularly effective at *resisting compressive forces.* Located in areas where compressive forces are high, such as intervertebral discs, the pubic symphysis, and the menisci of the knee joint.	Bands of fibers are easily visible. In addition, chondrocytes will appear to be lined up in rows because the thick bands of collagen fibers force them into this configuration.
Elastic Cartilage	Very similar to hyaline cartilage in all aspects. However, the addition of elastic fibers to the matrix makes elastic cartilage much more flexible than hyaline cartilage.	The addition of elastic fibers within the cartilage provides *flexible support.* Locations include the epiglottis, the lining of the auditory tube, and the external ear.	Elastic fibers are stained black in most preparations. The elastic fibers are generally higher in concentration near the lacunae.
Compact Bone	Composed of osteons (Haversian systems), which are concentric layers of bony matrix (lamellae) surrounding a central canal. Lamellae have lacunae located along them, with osteocytes inside the lacunae. Canaliculi connect adjacent lacunae, and perforating canals run perpendicular to the central canals.	Provides strong, rigid support. Compact bone is thickest in the diaphysis of long bones, but is also found as a thin layer forming the peripheral component of all bone.	Multiple osteons packed tightly together. No marrow spaces.

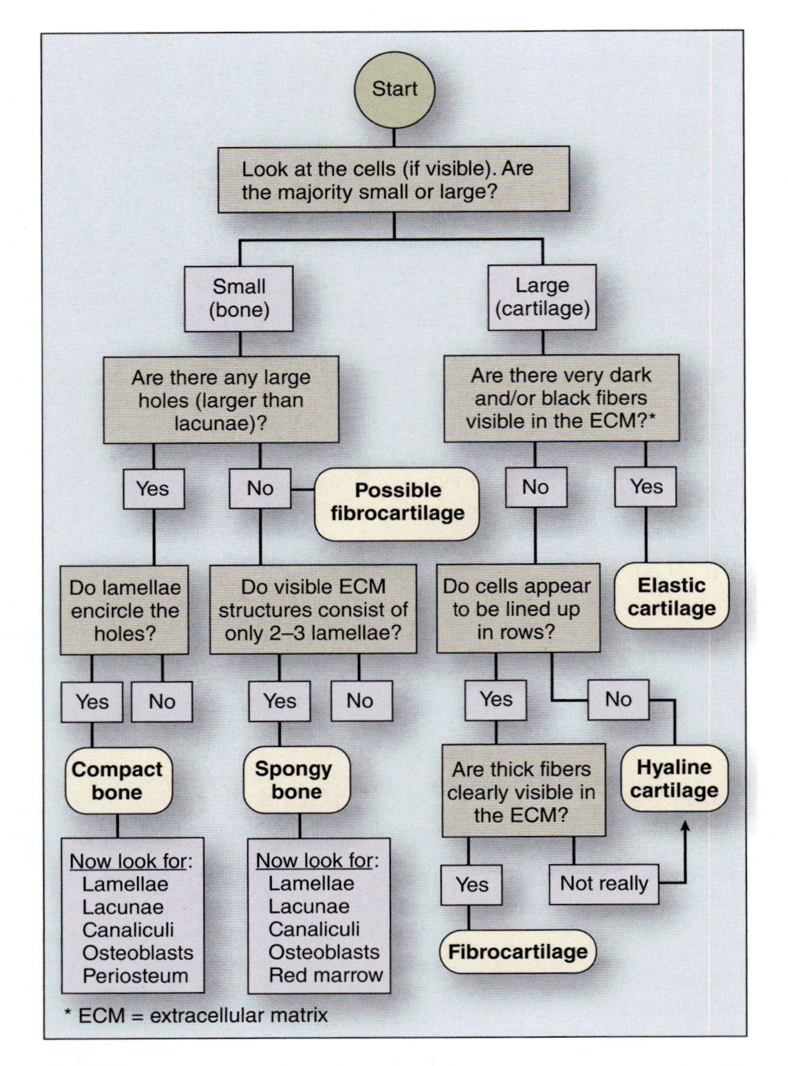

Figure 5.18 Flowchart for Classifying Supporting Connective Tissues.

EXERCISE 5.4

IDENTIFICATION AND CLASSIFICATION OF SUPPORTING CONNECTIVE TISSUE

EXERCISE 5.4A Hyaline Cartilage

1. Obtain a slide of the trachea and esophagus (see exercise 5.1F, p. 101).
2. Place the slide on the microscope stage and bring the tissue sample into focus on low power.
3. Find the lumen of the trachea and identify the epithelial tissue that lines the trachea (see exercise 5.1F, p. 101).
4. Move the microscope stage so that the tissue deep to the epithelium of the trachea is in the center of the field of view. A plate of hyaline cartilage will be visible here **(figure 5.19).**
5. Observe the slide on the lowest magnification possible. Identify the perichondrium, which surrounds the cartilage will be visible here. plate.
6. Look for small nuclei on the inner surface of the perichondrium. These are the nuclei of chondroblasts.

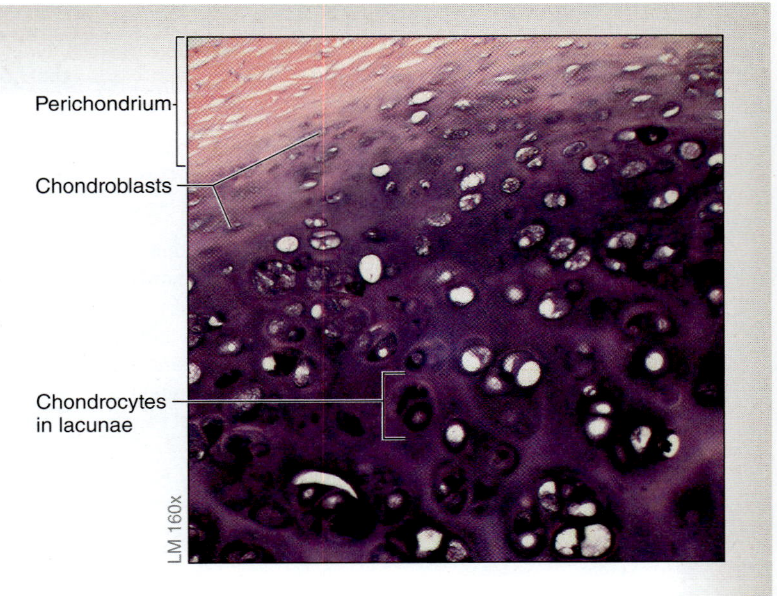

Figure 5.19 **Hyaline Cartilage.** Hyaline cartilage contains prominent lacunae with chondrocytes. No fibers are visible in the extracellular matrix. At the top of this photograph the perichondrium (light pink) with small, flattened chondroblasts on its inner surface is visible.

7. Next, identify the chondrocytes located within lacunae. In hyaline cartilage there are no visible fibers in the matrix because the fibers are spread very diffusely throughout the matrix. Instead, the matrix will appear uniform and smooth.
8. Identify the following structures on the slide, using figure 5.19 and table 5.6 as guides:
   - ☐ **chondroblasts**
   - ☐ **lacunae**
   - ☐ **chondrocytes**
   - ☐ **perichondrium**
9. Sketch hyaline cartilage as seen through the microscope in the space provided. Be sure to label all the structures listed in step 8 in the drawing.

______ ×

EXERCISE 5.4B Fibrocartilage

1. Obtain a slide of an intervertebral disc **(figure 5.20)**.
2. Place the slide on the microscope stage and bring the tissue sample into focus on low power. Then change to high power.

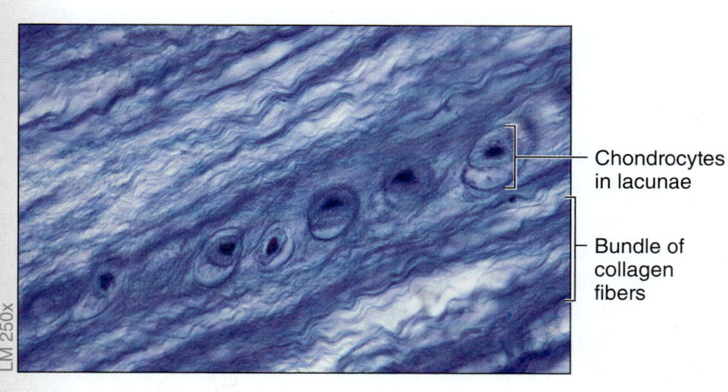

Figure 5.20 Fibrocartilage. Fibrocartilage contains visible bundles of collagen fibers within its matrix. The chondrocytes (in their lacunae) often appear to line up in rows.

3. Identify the following structures on the slide, using figure 5.20 and table 5.6 as guides:
   - ☐ **chondrocytes**
   - ☐ **bundle of collagen fibers**
4. Sketch fibrocartilage as seen through the microscope in the space provided. Be sure to label all the structures listed in step 3 in the drawing.

______ ×

EXERCISE 5.4C Elastic Cartilage

1. Obtain a slide of elastic cartilage **(figure 5.21)**.
2. Place the slide on the microscope stage and bring the tissue sample into focus on low power. Then change to high power.

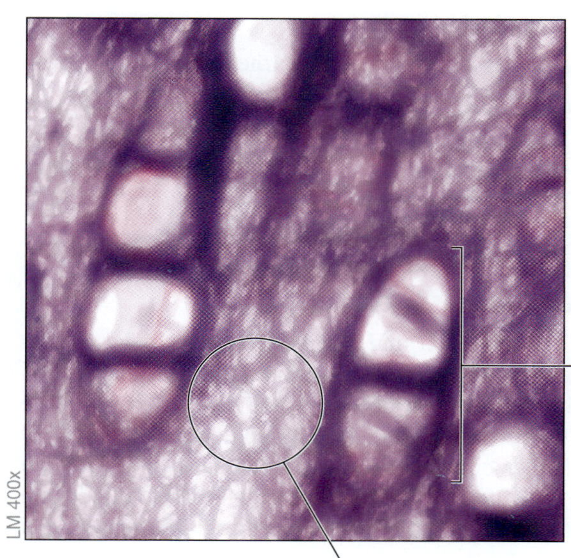

Figure 5.21 Elastic Cartilage. Elastic cartilage contains chondrocytes in lacunae and visible elastic fibers (which stain dark blue/purple or black) in its extracellular matrix.

(continued on next page)

(continued from previous page)

3. Identify the following structures on the slide, using figure 5.21 and table 5.6 as guides:

- ☐ **chondroblasts**
- ☐ **chondrocytes**
- ☐ **elastic fibers**
- ☐ **lacunae**
- ☐ **perichondrium**

4. Sketch elastic cartilage as seen through the microscope in the space provided. Be sure to label all the structures listed in step 3 in the drawing.

_____________ ×

EXERCISE 5.4D Bone

1. Obtain a slide of compact bone **(figure 5.22).**
2. Place the slide on the microscope stage and bring the tissue sample into focus on low power. Then change to high power.
3. Identify the following structures on the slide, using figure 5.22 and table 5.6 as guides:

- ☐ **canaliculus**
- ☐ **central canal**
- ☐ **lacuna**
- ☐ **lamella**
- ☐ **osteocyte**
- ☐ **osteon**

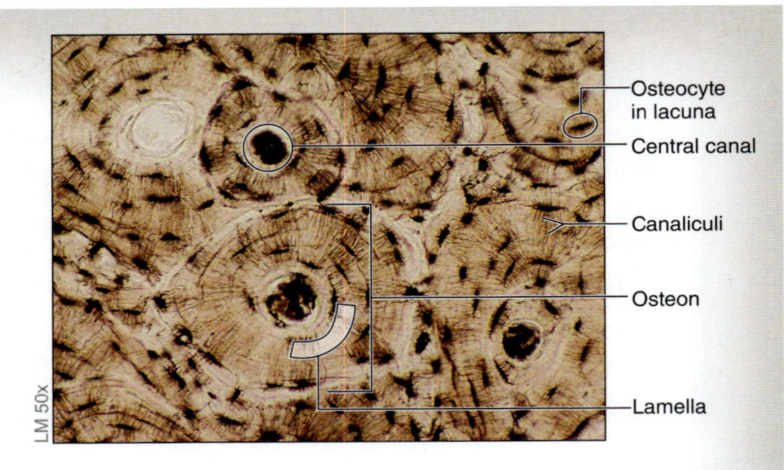

Figure 5.22 Compact Bone. Compact bone is composed of multiple osteons. Each osteon is characterized by a central canal surrounded by concentric layers of bony matrix called lamellae.

4. Sketch compact bone as seen through the microscope in the space provided. Be sure to label all the structures listed in step 3 in the drawing.

_____________ ×

Fluid Connective Tissue

Fluid connective tissue (blood and lymph) are specialized in that the extracellular matrix of these tissues consists of a liquid ground substance and soluble fibers that become insoluble only in response to tissue injury. These tissues will be covered in detail in chapters 20 and 23. In this chapter only the basic characteristics of blood as a connective tissue will be considered. The cell types in blood are **erythrocytes** (red blood cells), **leukocytes** (white blood cells), and **platelets** (thrombocytes). Platelets are not actually cells. Instead they are cytoplasmic fragments of cells called **megakaryocytes,** which are found in the bone marrow. The extracellular matrix of blood consists of blood **plasma.** The ground substance is liquid, composed mainly of water and a number of dissolved substances. In addition, blood contains soluble proteins, some of which are called clotting proteins, such as **fibrinogen.** Fibrinogen becomes insoluble **fibrin** and forms fibers in response to tissue injury.

EXERCISE 5.5

IDENTIFICATION AND CLASSIFICATION OF FLUID CONNECTIVE TISSUE

1. Obtain a slide of a blood smear **(figure 5.23)**.
2. Place the slide on the microscope stage and bring the tissue sample into focus on low power. Change to medium power, and then bring the sample into focus using the oil immersion lens. Consult the instructor if assistance is needed using the oil immersion lens.

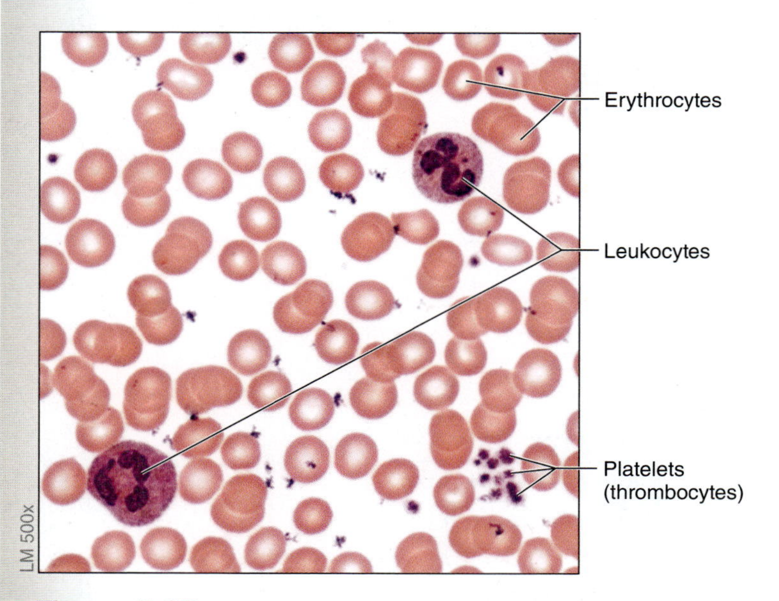

Figure 5.23 Blood. Blood is a fluid connective tissue containing erythrocytes (red blood cells), leukocytes (white blood cells), platelets (thrombocytes), and an extracellular matrix called plasma.

3. Identify the following structures on the slide, using figure 5.23 as a guide:
   - ☐ **erythrocytes**
   - ☐ **leukocytes**
   - ☐ **platelets (thrombocytes)**
4. Sketch blood as seen through the microscope in the space provided. Be sure to label all the structures listed in step 3 in the drawing.

__________ ×

5. *Optional Activity:* **AP|R 3: Tissues**—Watch the "Connective Tissue Overview" animation.

INTEGRATE

CLINICAL VIEW

Carcinomas and Sarcomas

Clinically, a tumor derived from epithelial tissues is called a *carcinoma* (*karkinos*, crab + *-oma*, tumor), and a tumor derived from connective tissues is called a *sarcoma* (*sarc-*, flesh + *-oma*, tumor). Carcinomas are considered noninvasive when they do not penetrate the basement membrane that lies between the epithelial and connective tissue layers. Once rapidly dividing cells penetrate the basement membrane, the cancer is considered invasive. While epithelial tissue is avascular, the underlying connective tissue is not. Thus, invading cancer cells can easily metastasize (*meta-*, change + *-ize*, an action) to other locations of the body through blood and lymphatic vessels. Sarcomas, which arise from connective tissues, pose a similar risk. However, they tend to grow and metastasize much more readily than carcinomas because of the highly vascular nature of connective tissues.

Muscle Tissue

Muscle tissue is both excitable and contractile. Excitable tissues are able to generate and propagate electrical signals called action potentials. As a contractile tissue, muscle has the ability to actively shorten and produce force. There are three types of muscle tissue: skeletal muscle, cardiac muscle, and smooth (visceral) muscle. The three types of muscle tissue are distinguished by the presence or absence of visible striations, the shape of the cells, and the number and location of nuclei. Skeletal muscle is found in the voluntary muscles that move the skeleton and the facial skin. Cardiac muscle is found in the heart, and smooth muscle is found in the walls of soft viscera, such as the blood vessels, stomach, urinary bladder, intestines, and uterus. **Table 5.7** compares the three types of muscle tissue, and the flowchart in **figure 5.24** explains steps that can be used to identify muscle tissues.

Table 5.7 Muscle Tissue

Type of Muscle	Description	Generalized Functions	Identifying Characteristics
Skeletal	Elongate, cylindrical cells with multiple nuclei. Nuclei are peripherally located. Tissue appears striated (light and dark bands along the length of the cell).	Provides voluntary movement	Length of cells (extremely long), striations, multiple peripheral nuclei
Cardiac	Short, branched cells with 1 to 2 nuclei. Nuclei are centrally located. Dark bands (intercalated discs) are seen where two cells join. Tissue appears striated (light and dark bands along the length of each cell).	Responsible for pumping action of the heart	Branched, 1–2 nuclei/cell, striations, intercalated discs
Smooth	Elongate, spindle-shaped cells (fatter in the center, narrowing at the ends) with single, "cigar-shaped" or "spiral" nuclei. Nuclei are centrally located. No striations are apparent.	Creates movement within viscera such as intestines, bladder, uterus, and stomach. Moves blood through blood vessels.	Spindle-shaped cells, no striations, cigar-shaped nuclei that are centrally located

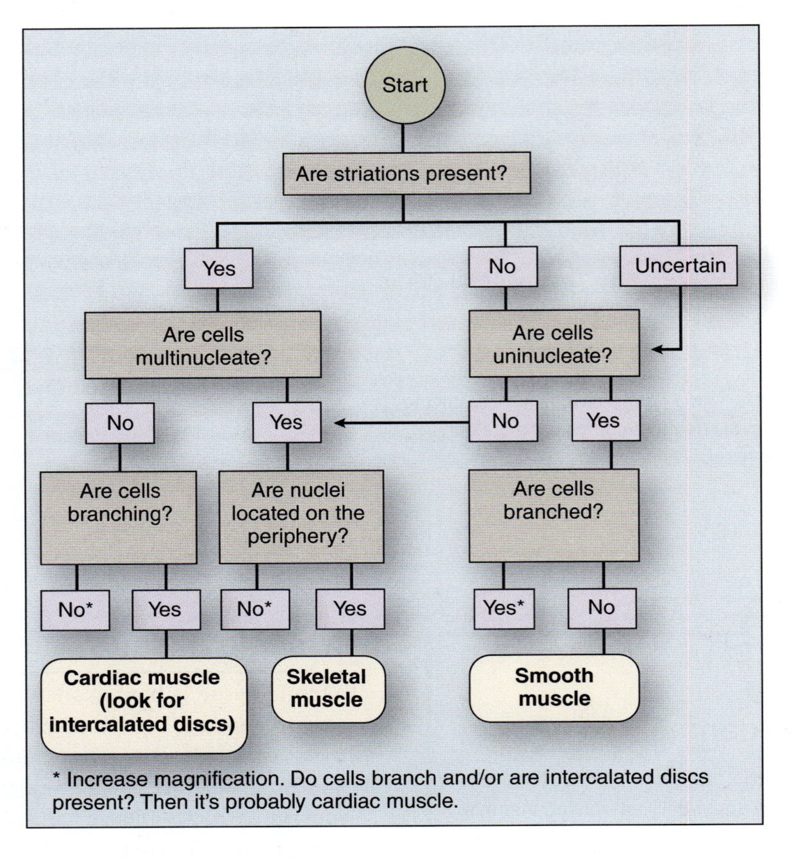

Figure 5.24 Flowchart for Classifying Muscle Tissues.

EXERCISE 5.6

IDENTIFICATION AND CLASSIFICATION OF MUSCLE TISSUE

EXERCISE 5.6A Skeletal Muscle Tissue

1. Obtain a slide of skeletal muscle **(figure 5.25).**
2. Place the slide on the microscope stage and bring the tissue sample into focus on low power. Then change to high power.

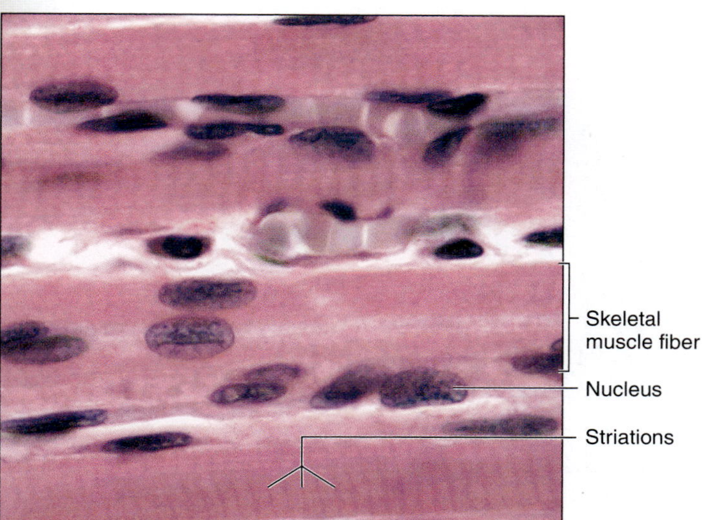

Figure 5.25 Skeletal Muscle. Skeletal muscle fibers are elongated and striated, and contain multiple peripherally located nuclei.

3. Identify the following structures on the slide, using figure 5.25 and table 5.7 as guides:

   - ☐ **nucleus**
   - ☐ **striations**
   - ☐ **skeletal muscle fiber**

4. Sketch skeletal muscle as seen through the microscope in the space provided. Be sure to label all the structures listed in step 3 in the drawing.

_______ ×

EXERCISE 5.6B Cardiac Muscle Tissue

1. Obtain a slide of cardiac muscle **(figure 5.26).**
2. Place the slide on the microscope stage and bring the tissue sample into focus on low power. Then change to high power.

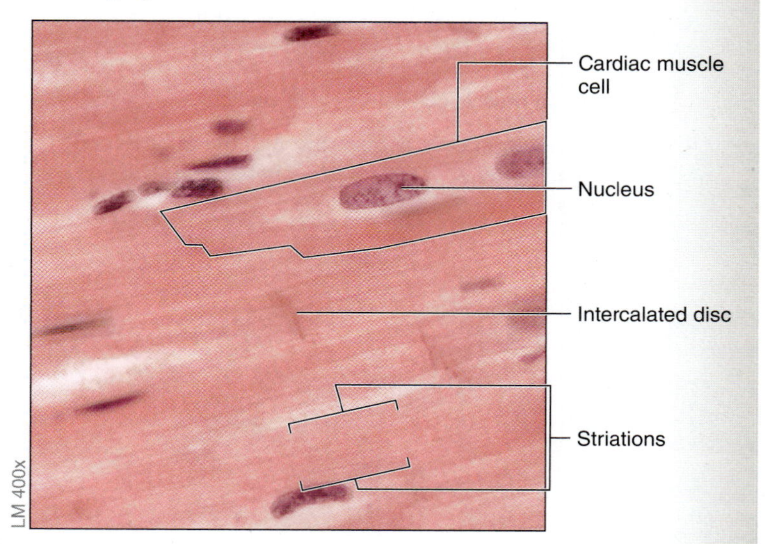

Figure 5.26 Cardiac Muscle. Cardiac muscle cells are short and branched, striated, and generally contain only one centrally located nucleus.

3. Identify the following structures on the slide, using figure 5.26 and table 5.7 as guides:

   - ☐ **cardiac muscle cell**
   - ☐ **nucleus**
   - ☐ **intercalated disc**
   - ☐ **striations**

4. Sketch cardiac muscle as seen through the microscope in the space provided. Be sure to label all the structures listed in step 3 in the drawing.

_______ ×

(continued on next page)

(continued from previous page)

EXERCISE 5.6C Smooth Muscle Tissue

1. Obtain a slide of smooth muscle **(figure 5.27)**.
2. Place the slide on the microscope stage and bring the tissue sample into focus on low power. Then change to high power.
3. Identify the following structures on the slide, using figure 5.27 and table 5.7 as guides:

   ☐ smooth muscle cell ☐ nucleus

4. Sketch smooth muscle as seen through the microscope in the space provided. Be sure to label all the structures listed in step 3 in the drawing.

_____________ ×

5. *Optional Activity:* **AP|R 3: Tissues**—Watch the "Muscle Tissue Overview" animation.

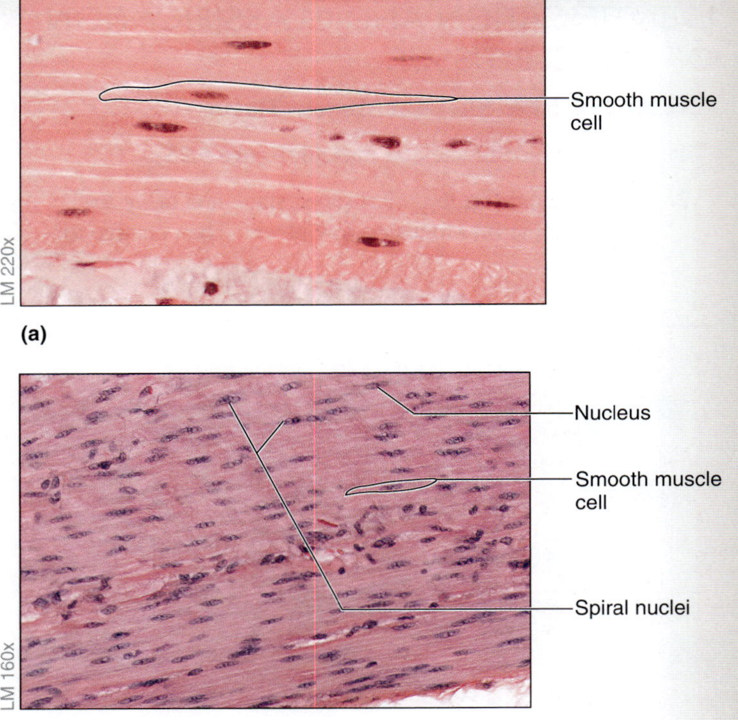

Figure 5.27 Smooth Muscle. Smooth muscle fibers are short and spindle-shaped, not striated, and contain only one centrally located nucleus (a). In figure 5.27*b* some nuclei that have taken on a "spiral" shape are visible. This happens when the muscle fibers contract. As the fibers shorten, the nuclei start to coil up, or "spiral."

Nervous Tissue

Nervous tissue is characterized by its *excitability*: the ability to generate and propagate electrical signals called action potentials. Nervous tissue is composed of two basic cell types **(table 5.8)**. **Neurons** are excitable cells that send and receive electrical signals. They have a limited ability to divide and multiply in the adult brain. **Glial cells** are supporting cells that support and protect neurons. Glial cells maintain the ability to divide and multiply in the adult brain. Glial cells constitute over 60% of the cells found in neural tissue.

Table 5.8 Nervous Tissue

Cell Type	Description	Generalized Functions
Neurons	Though varied in shape, most neurons appear to have numerous branches coming off the cell body (soma). Neurons contain large amounts of rough endoplasmic reticulum (ER). The ribosomes and rough ER stain very dark and are collectively called chromatophilic substance.	These cells are responsible for *generating and transmitting information via electrical impulses* within the nervous system. Thus, they are "excitable" cells.
Glial Cells	Even more varied in shape than neurons, glial cells are generally much smaller than neurons with fewer (if any) branching processes.	These cells are the *general supporting cells* of the nervous system. They protect, nourish, and support the excitable cells, the neurons.

EXERCISE 5.7

IDENTIFICATION AND CLASSIFICATION OF NERVOUS TISSUE

1. Obtain a slide of nervous tissue **(figure 5.28).**

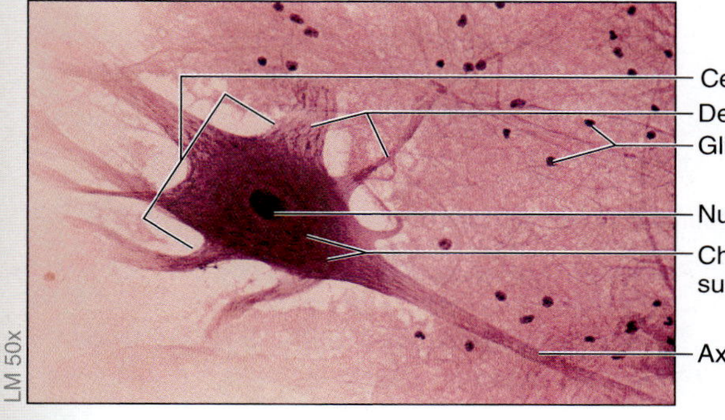

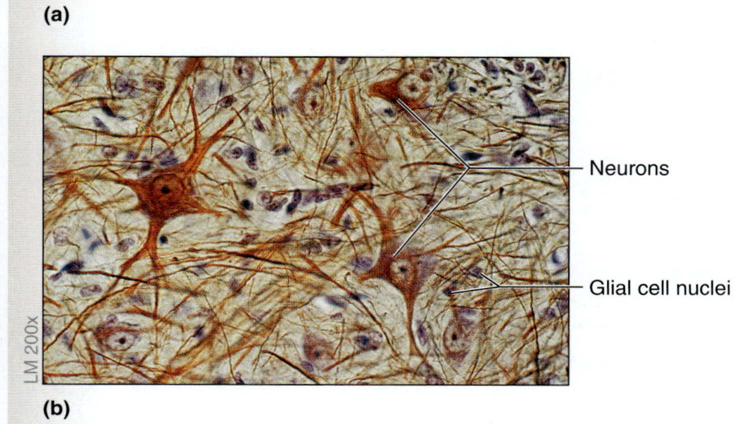

Figure 5.28 Nervous Tissue. (a) Neurons are very large cells with a prominent nucleus and nucleolus, and dark-staining endoplasmic reticulum (chromatophilic substance). (b) Glial cells are much smaller than neurons, and are also more abundant.

2. Place the slide on the microscope stage and bring the tissue sample into focus on low power. Then change to high power.

3. Identify the following structures on the slide, using figure 5.28 and table 5.8 as guides:

☐ **axon of neuron**	☐ **dendrite of neuron**
☐ **cell body of neuron**	☐ **nucleus of glial cell**
☐ **chromatophilic substance**	☐ **nucleus of neuron**

4. Sketch nervous tissue as seen through the microscope in the space provided. Be sure to label all the structures listed in step 3 in the drawing.

_______ ×

5. *Optional Activity:* **AP|R 3: Tissues**—Watch the "Nervous Tissue Overview" animation.

Chapter 5: Histology

Name: ____________________

Date: __________ Section: __________

POST-LABORATORY WORKSHEET

The ❶ corresponds to the Learning Objective(s) listed in the chapter opener outline.

Do You Know the Basics?

Exercise 5.1: Identification and Classification of Epithelial Tissue

1. The epithelial type that protects against abrasion is ____________________. ❶
2. Endothelium is a ____________________ epithelium that lines the walls of blood vessels and the heart. ❷
3. Match the function listed in column A with the surface modifications and specialized cells found in epithelial tissues listed in column B. ❸ ❺

Column A	*Column B*
____ 1. provide lubrication	a. cilia
____ 2. increase surface area for enhanced absorption	b. goblet cells
____ 3. aid in movement of substances in one direction	c. keratinization
____ 4. imparts strength and protection	d. microvilli

4. When observing a cross section of a tube that is lined with epithelial tissue (e.g., the gut tube), the epithelial surface that faces the lumen of the tube is the ____________________ (apical/basal) surface. ❹

Exercise 5.2: Identification of Embryonic Connective Tissue

5. Embryonic connective tissue is called ____________________. ❻

Exercise 5.3: Identification and Classification of Connective Tissue Proper

6. The extracellular matrix of connective tissue is composed of cells and fibers. ____________________ (True/False) ❼
7. For each category of connective tissue listed in the following table, write in the major cell types, the fiber types, and the characteristics of the ground substance of the tissue. Refer to the textbook for assistance. ❼–⓭

	Connective Tissue Proper	Cartilage	Bone	Fluid Connective Tissue
Cells				
Fibers				
Ground Substance				

Exercise 5.4: Identification and Classification of Supporting Connective Tissue

8. The three types of cartilage are differentiated from one another by the type and arrangement of ____________________. ❿
9. Fibrocartilage can be found in intervertebral discs, whereas elastic cartilage can be found in the ear. ____________________ (True/False) ⓫
10. Which of the following is/are a type of bone tissue? (Check all that apply.) ⓬
    ____ a. areolar
    ____ b. compact
    ____ c. elastic
    ____ d. reticular
    ____ e. spongy

Exercise 5.5: Identification and Classification of Fluid Connective Tissue

11. Which of the following is a type of fluid connective tissue? (Check all that apply.) 13
    ____ a. blood
    ____ b. bone
    ____ c. cartilage
    ____ d. lymph
    ____ e. mesenchyme

12. Which of the following are possible cells present in fluid connective tissue? (Check all that apply.) 14
    ____ a. chrondrocytes
    ____ b. erythrocytes
    ____ c. leukocytes
    ____ d. megakaryocytes
    ____ e. osteocytes

Exercise 5.6: Identification and Classification of Muscle Tissue

13. In the following table, compare and contrast the characteristics of the three types of muscle tissue. 15

Characteristic	Skeletal Muscle	Cardiac Muscle	Smooth Muscle
Location and Number of Nuclei			
Cell Shape			
Presence or Absence of Striations			

14. List a location in the body where smooth muscle tissue might be found. ____________________ 16

Exercise 5.7: Identification and Classification of Nervous Tissue

15. The two main cell types found in nervous tissue are ____________________ and ____________________. 17

Can You Apply What You've Learned?

16. In the following table, compare and contrast epithelial and connective tissues with respect to the following:

Characteristic	Epithelial Tissues	Connective Tissues
Cell Number and Arrangement		
Polarity		
Extracellular Matrix		
Vascularity		

17. Match the description listed in column A with the specific type of connective tissue listed in column B.

Column A

____ 1. collagen fibers are not visible; ground substance appears clear and glassy
____ 2. collagen fibers arranged in concentric rings; hard ground substance; osteocytes in lacunae
____ 3. fluid ground substance; many biconcave disc-shaped pink cells and larger, dark-staining cells
____ 4. chondrocytes lined up in rows in between bundles of collagen fibers
____ 5. densely packed elastic fibers; chondrocytes in lacunae

Column B

a. blood
b. bone
c. elastic cartilage
d. fibrocartilage
e. hyaline cartilage

18. What does the presence of cilia indicate about the function of an epithelial tissue?

____

____

19. Identify which statements about connective tissue are accurate. (Check all that apply.)

____ a. conducts neural impulses
____ b. contains bundles of nerve fibers and muscle fibers
____ c. forms bones, cartilage, ligaments, and tendons
____ d. forms deep layers of skin (dermis)
____ e. forms glands (e.g., salivary, pancreas)
____ f. forms a structural framework for organs
____ g. highly vascular (with some exceptions)
____ h. includes blood and components of blood vessel walls

20. Match the description listed in column A with the name of the connective tissue listed in column B.

Column A

____ 1. makes up the cartilage discs between the vertebrae and between the pubic bones
____ 2. transports materials within blood vessels
____ 3. inner supporting framework of spleen, liver, and lymph nodes
____ 4. tough tissue that connects bone to bone and muscle to bone
____ 5. cartilage that retains its original shape after being deformed, such as in the cartilage of the ear
____ 6. allows the aorta (a large blood vessel) to stretch with each pulse of blood pumped into it from the heart and then return to its original size
____ 7. helps keep the body warm and stores excess fuel (energy)
____ 8. forms the ends of long bones, the larynx, the costal cartilages, and the embryonic skeleton

Column B

a. adipose connective tissue
b. blood
c. dense regular connective tissue
d. elastic cartilage
e. elastic connective tissue
f. fibrocartilage
g. hyaline cartilage
h. reticular connective tissue

21. Explain the characteristics that make blood a connective tissue. (Hints: From what embryonic connective tissue is blood derived? What is the composition of the extracellular matrix? What is the cellular component? What are the fibers?)

____

____

____

____

____

____

22. Write the name of the specific tissue type below each photo.

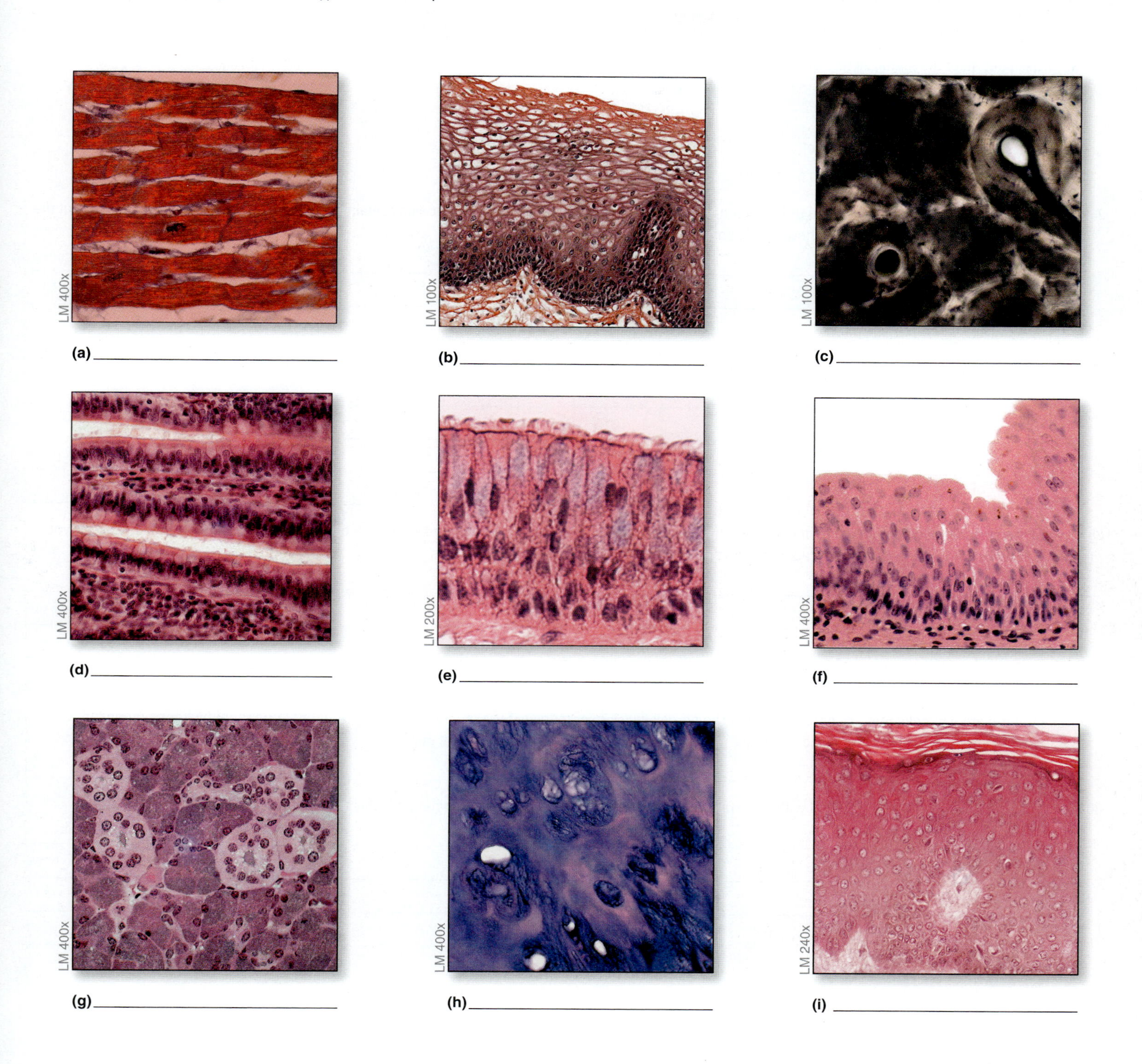

23. A friend was recently diagnosed with a tumor called a glioma. What basic tissue type is the tumor derived from? ____________________

24. The biceps brachii, deltoid, and pectoralis muscles are composed of what type of muscle tissue? ____________________

Can You Synthesize What You've Learned?

25. For each of the examples given below, is the structure likely to fit in its entirety on a microscope slide? (Circle yes or no for each.)

a. cross section of the humerus (the arm bone)	Yes	No
b. cross section of the large intestine	Yes	No
c. cross section of a pinky finger	Yes	No
d. cross section of a small blood vessel	Yes	No
e. cross section of a ureter (~0.5 cm in diameter)	Yes	No

26. A scientist is interested in developing a tissue that can resist stretch in multiple directions, yet also remains very flexible. Given what is known about the properties of human tissues, describe how the scientist might design such a tissue.

27. Cartilage is unique as a connective tissue in that it is avascular. What limitations does avascularity impose on cartilage as a tissue?

28. Two of the surface modifications of epithelial cells, microvilli and cilia, are commonly confused with each other. In the space below, describe how you would teach a fellow student to distinguish between these two surface modifications.

29. Most tumors arise in cells that are constantly undergoing cell division, or mitosis (for example, skin cells). A patient has recently been diagnosed with a brain tumor. What type of cell did the tumor most likely arise from—a neuron or a glial cell?

30. A patient has recently been diagnosed with an osteosarcoma. In the space below, explain what cells this tumor most likely arose from and why.

Integument

OUTLINE AND LEARNING OBJECTIVES

MODULE 4: INTEGUMENTARY SYSTEM

INTRODUCTION

The exercises in this chapter explore the body's largest organ system, the **integument** (*integumentum*, a covering). The integument is often mistakenly referred to as the "skin." However, the two terms do not refer to the same thing. The **skin** is the outer covering of the body and is composed of an **epidermis** (*epi*, upon, + *derma*, skin) and **dermis** (*derma*, skin). The integument is an organ system that includes the skin plus all of the accessory structures of skin, such as sensory organs, glands, hair, and nails. The integument is a truly remarkable system, and its functions may be underestimated. For example, damage to large areas of the integument (as might happen to a burn victim) easily can be life-threatening.

The materials used in the exercises in this chapter include histological slides of the skin and its accessory structures, and models of integument. While it is important to identify the structures of the integument both as viewed through the microscope and as present on classroom models, do not forget to use your own body as an example while working through this chapter of the laboratory manual. For example, while viewing slides of thin skin and thick skin, be sure to also find locations on your own body that contain thin skin and thick skin. See and feel the differences between the two, and try to correlate what is seen and felt on your own body with what is seen through the microscope. Finally, be sure to compare the appearance and texture of your own skin and hair with that of another student in the lab who has skin or hair of a different color or texture from your own.

List of Reference Tables

Chapter 6: Integument

Name: ____________________

Date: __________ Section: __________

These Pre-Laboratory Worksheet questions may be assigned by instructors through their connect course.

PRE-LABORATORY WORKSHEET

1. Which of the following are functions of the integument? (Check all that apply.)
   ____ a. mineral storage
   ____ b. movement
   ____ c. protection
   ____ d. secretion
   ____ e. structural support

2. Identify the type of tissue that composes the epidermis of the skin. (Circle one.)
   a. adipose connective tissue
   b. areolar connective tissue
   c. dense irregular connective tissue
   d. simple squamous epithelial tissue
   e. stratified squamous epithelial tissue

3. Dense irregular connective tissue is found in the *dermis* of the skin. ____________ (True/False)

4. Number the following five layers of the epidermis of thick skin, starting with 1 for the layer closest to the basal surface.
   ____ a. stratum basale
   ____ b. stratum corneum
   ____ c. stratum granulosum
   ____ d. stratum lucidum
   ____ e. stratum spinosum

5. Indicate which of the following are among the layers of the dermis. (Check all that apply.)
   ____ a. papillary layer
   ____ b. reticular layer
   ____ c. subcutaneous layer (hypodermis)

6. Identify the main protein that composes a hair. (Circle one.)
   a. collagen
   b. elastin
   c. fibrin
   d. keratin

7. The subcutaneous layer (hypodermis) of the skin is mainly composed of *adipose* connective tissue. ____________ (True/False)

8. Thick skin is located *only* on the palms of the hands and tops of the feet. ____________ (True/False)

9. Sweat glands are composed of which type of tissue? (Circle one.)
   a. connective tissue
   b. epithelial tissue
   c. muscle tissue
   d. nervous tissue

10. Which of the following is an accessory structure of the skin located within the reticular layer of the dermis? (Check all that apply.)
    ____ a. dermal papilla
    ____ b. hair follicle
    ____ c. sebaceous gland
    ____ d. sweat gland

HISTOLOGY

The Epidermis

The **skin** is composed of an **epidermis** (*epi,* upon, + *derma,* skin) and **dermis** (*derma,* skin). The epidermis is the outermost layer of the skin and consists of keratinized stratified squamous epithelial tissue. Thus there are no blood vessels in the epidermal layers. The main cell type composing the epithelial tissue in the skin is a **keratinocyte,** so named because of its role in synthesizing the protein **keratin.** Keratin is an insoluble protein that imparts strength to the skin and makes it almost completely waterproof. Other cell types in the epidermis include melanocytes that produce the pigment melanin, and tactile (Merkel) cells that release chemicals to stimulate nerve endings when compressed. The epidermis of *thin skin* (for example, see figure 6.3*b*), which is located on most of the body's surface, has four distinct layers, whereas the epidermis of *thick skin* (see figure 6.1), which is located only on the palms of the hands and soles of the feet, has five distinct layers. **Table 6.1** summarizes the layers of the epidermis and the characteristics of each.

INTEGRATE

LEARNING STRATEGY

The terms *thick skin* and *thin skin* refer only to the thickness of the epidermal layer. Technically, the thickest skin on the body is located on the back. However, that thickness takes into account not only the thickness of the epidermis, but that of the dermis as well. The thick skin on the palms of the hands and the soles of the feet is actually rather thin when considering the total thickness of the epidermis and the dermis combined. To understand this a little better, compare the total thickness of the skin on your back (by feel) to the total thickness of the skin on the middle arch of the sole of your foot or the palm of your hand.

Table 6.1 Layers of the Epidermis

Epidermal Layer	Description	Word Origin
Stratum Basale	The deepest layer; composed of a single layer of cells. At higher magnification, the cells appear large, with round nuclei. Cells undergoing cell division (*mitosis*) may be visible. In addition, melanin granules are most concentrated in this layer of the skin. Melanin granules may only be visible in a sample of pigmented skin.	*stratum,* layer, + *basalis,* situated near the base
Stratum Spinosum	Named for the spiny appearance of the keratinocytes when viewed at high magnification. Consists of several layers of keratinocytes that appear "prickly" or "spiny." The "spines" are artifacts that form during tissue preparation. During this process the cytoplasm of the cells shrinks, while the cytoskeletal elements and desmosomes remain intact, keeping neighboring cells attached.	*stratum,* layer, + *spina,* a thorn
Stratum Granulosum	Consists of three to five layers of cells that appear granular and darker in color than those of the underlying stratum spinosum. The graininess of the cells in this layer makes it distinct. Keratinocytes begin to die within the stratum granulosum as organelles and nuclei degenerate and the cells accumulate bundles of keratin.	*stratum,* layer, + *granulum,* a small grain
Stratum Lucidum	Present only in thick (glabrous) skin, which is located on the palm of the hand and sole of the foot. Consists of 2 to 3 layers of dead cells containing the translucent protein eleidin, which is an intermediate product in keratin formation. This clear layer sometimes stains darker than the remainder of the stratum corneum, depending on the stain used.	*stratum,* layer, + *lucidus,* clear
Stratum Corneum	Consists of multiple layers (20 to 30) of dead, scaly, interlocking keratinocytes. A few layers of dead cells that are in the process of **desquamation** (*squamosus,* scaly, fr. *squama,* scale) may be visible on the outer (apical) surface of the stratum corneum.	*stratum,* layer, + *cornu,* horn, hoof

INTEGRATE

LEARNING STRATEGY

Follow these steps to identify the epidermal layers of the skin under the microscope:

1. **Determine if the layer is more superficial or deep.** Remember the stratum corneum is the most superficial layer (next to the free surface), whereas the stratum basale is the deepest epidermal layer.
2. **Examine the shape of the cells.** The stratum corneum and stratum lucidum both contain squamous cells, the stratum spinosum contains polygonal cells, and the stratum basale contains cuboidal to low columnar cells.
3. **Determine if the keratinocytes have a nucleus or not.** Living keratinocytes (as found in the stratum granulosum, stratum spinosum, and stratum basale) have prominent nuclei. Dead keratinocytes (as found in the stratum lucidum and stratum corneum) no longer have a nucleus.
4. **Count the number of cells in the layer.** The stratum corneum contains 20 to 30 layers of cells. The stratum lucidum, stratum granulosum, and stratum spinosum contain two to five layers of cells. The stratum basale contains only one layer of cells.
5. **Determine if the cytoplasm of the cells contains visible granules.** The only layer with visible (and generally dark) cytoplasmic granules is the stratum granulosum.

EXERCISE 6.1

LAYERS OF THE EPIDERMIS

1. Obtain a histology slide of **thick skin** (from the palm of the hand or sole of the foot) **(figure 6.1).** Thick skin is also called *glabrous* skin, meaning it lacks hair (*glaber*, smooth). It does, however, contain all five epidermal layers, so it is an ideal slide for demonstrating characteristics of the epidermis.
2. Place the slide on the microscope stage and scan the slide at low power. Look for any empty or clear space present in the slide, and try to find the apical surface of the skin epithelium lying next to the clear space.
3. Once the skin epithelium is visible in the field of view, switch to a higher power. Scan the slide until the basal surface of the epithelium is in the center of the field of view. Look for the junction between the epidermis and the dermis. Notice that the basal layer of the epidermis is thrown into folds that dip into the dermis. These folds of the epidermis are called **epidermal ridges.** Epidermal ridges are thickened layers of the epidermis that are located between upward extensions of the papillary layer of the dermis, which are called **dermal papillae** (*papilla*, nipple). The increased area of adhesion between the epidermis and the dermis in the areas of epidermal ridges and dermal papillae helps prevent the epidermis and dermis from coming apart in areas where the skin experiences large frictional forces. Epidermal ridges are deeper in the tips of the fingers and are the structures that form fingerprints.
4. Identify the following structures on the slide of thick skin, using figure 6.1 and table 6.1 as guides:

   ☐ **stratum basale** ☐ **stratum spinosum**
   ☐ **stratum granulosum** ☐ **stratum lucidum**
   ☐ **stratum corneum** ☐ **epidermal ridges**

5. Sketch the layers of the epidermis of thick skin as seen through the microscope in the space provided. Be sure to label all of the structures listed in step 4 in the drawing.

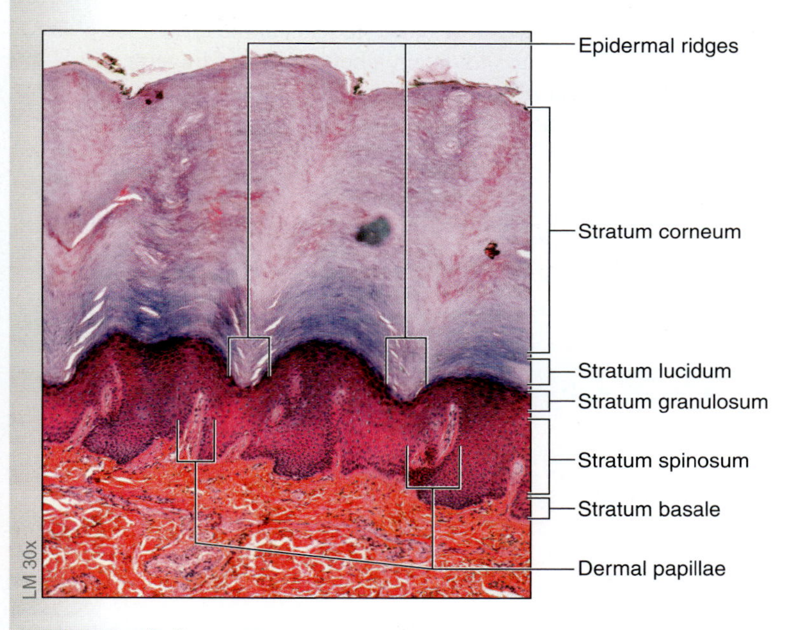

Figure 6.1 **Thick Skin.** Epidermal layers in thick skin (found on the palms of the hands and soles of the feet).

EXERCISE 6.2

FINGERPRINTING

Obtain the Following:

- **ink pad**
- **towelettes for removing ink**
- **magnifying glass**

1. Prior to beginning the laboratory exercise, wash your hands with soap and warm water. Dry your hands completely.
2. Choose a lab partner to be the subject. Open the ink pad and place the ink pad and laboratory notebook (open to this page) on the laboratory bench within easy reach.
3. Creating an accurate fingerprint requires making a print of the entire area of the finger bulb, from the crease of the distal interphalangeal joint to the tip of the finger, and extending from the right to the left edges of the nail bed. For best results, have the subject relax, look away, and not provide any assistance during the fingerprinting process.
4. Stand to the *left* of the subject and grasp the subject's *right* hand with both of your hands. Place the *left* edge of the subject's index finger (digit II) on the ink pad and roll the finger away from the subject's midline to the opposite edge of the finger. Carefully apply steady, even pressure while rolling the subject's finger on the ink pad. Lift the finger away from the ink pad. Repeat the rolling motion to create the print in the space. When you have finished making the print, lift the finger carefully to avoid smudging the newly inked print.
5. Remove the ink from the subject's finger with a towelette.
6. If desired, repeat the fingerprinting process using the subject's right thumb (pollex). To make a thumb print, place the *right* edge of the subject's thumb on the ink pad and roll the subject's thumb *toward* the subject's midline. Lift the thumb away from the ink pad, then repeat the rolling motion to create the print in the space provided. When you have finished making the print, lift the thumb carefully to avoid smudging the newly inked print.

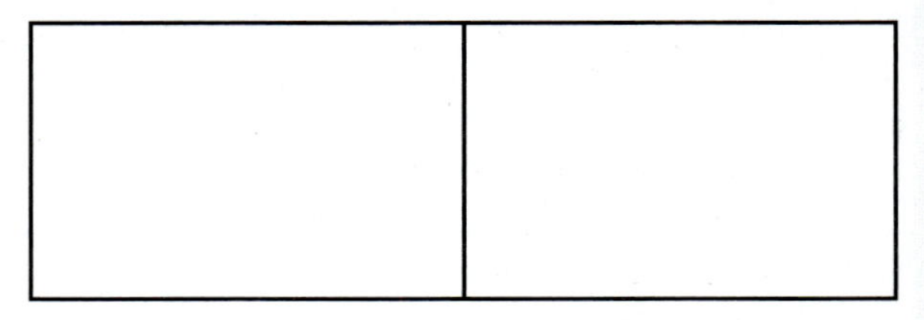

7. Inspect the finger and thumb prints to ensure that clear patterns are visible. If the prints are unclear (e.g., too pale, too dark, or blurry), repeat steps 1 through 6.
8. Compare your subject's fingerprint patterns with the fingerprint patterns of other students in the class. Use a magnifying glass to obtain a better view of fingerprint patterns and ridges. Briefly note observations in the space provided.

____

____

____

____

INTEGRATE

CLINICAL VIEW

Fingerprinting

Epidermal ridges serve not only to increase the adhesion of the dermal and epidermal layers, but also as personal identification markers by forming fingerprints. Fingerprints may be recorded using the methods described in this chapter, whereby the bulb of the finger is rolled onto an inking device and a printed impression is obtained. Experts may also record digital fingerprints using similar techniques. To make digital fingerprints, the impression is made on a glass plate or clear surface and stored digitally.

The presence of dirt, oil, or sweat on the surface of the skin may cause a person to leave a *latent fingerprint* when he or she touches a surface with his or her bare hands. Flexibility of the skin and the amount of pressure applied to the surface may lead to variations in the quality of latent prints. Latent fingerprints degrade with time, and children's fingerprints degrade faster than those of adults because children lack the thick, oily sweat that is produced by sweat glands after puberty. Fewer oils on the surface of the skin cause more rapid degradation of latent prints.

Latent fingerprints are often not visible to the naked eye. Thus, forensic experts must often use chemical reagents or powders to make such prints visible. This "dusting for prints" is particulary important in forensic science, and may aid in determining the events of a crime. Because fingerprints are unique to an individual, latent prints may provide very accurate information leading to the identification of a suspect or suspects present at a crime scene.

EXERCISE 6.3

PIGMENTED SKIN

The color of a person's skin is partially due to the pigment **melanin** (*melas*, black). Melanin is produced by cells called **melanocytes,** which are found along the base of the stratum basale of the epidermis. Melanocytes have long dendritic (*dendrites*, relating to a tree) processes that extend between keratinocytes. Melanocytes transfer the melanin granules they produce to adjacent keratinocytes. The melanin then accumulates on the apical side of the nuclei of the keratinocytes, which protects keratinocytes from harmful ultraviolet light that can damage their DNA. A single melanocyte produces melanin for several adjacent keratinocytes (the melanocyte plus the keratinocytes served by it are referred to as an *epidermal-melanin unit*). Some areas of the body have a higher density of melanocytes than others (e.g., the skin of the areola of the breast has more melanocytes than the skin on the rest of the breast). However, the *total number of melanocytes in light- and dark-skinned individuals is the same.* Differences in skin color have to do with the *amount of melanin* produced by melanocytes. The melanocytes of dark-skinned individuals simply produce greater amounts of melanin than those of light-skinned individuals.

1. Obtain a slide of pigmented skin and place it on the microscope stage. Observe at low power to locate the epidermis, and then change to a higher-power lens.
2. Focus on the stratum spinosum of the pigmented skin. Look for distinct black/brown melanin granules within the keratinocytes **(figure 6.2).** Melanocytes themselves are located in the stratum basale and are characterized by small, round nuclei and pale-staining cytoplasm. Try to locate a melanocyte on the slide, keeping in mind that melanocytes are not easily identified on the standard slides found in the laboratory.

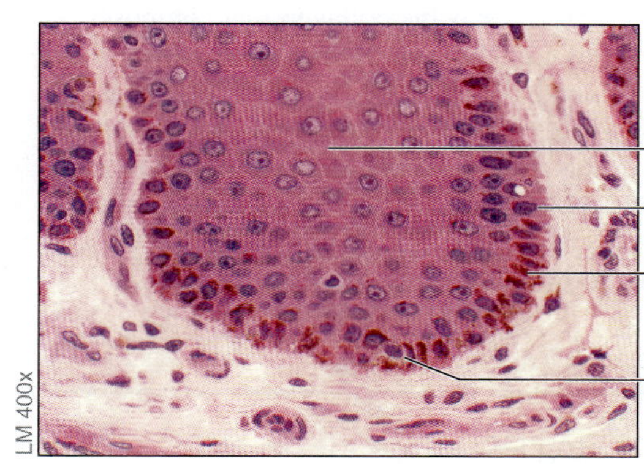

Figure 6.2 Pigmented Skin. Pigmented skin containing melanin granules (brown) within keratinocytes, particularly in the stratum basale of the epidermis. Melanocytes themselves are distinguished by a halo of clear cytoplasm that surrounds the nucleus.

INTEGRATE

CLINICAL VIEW
Melanoma

Melanoma (*melano-*, black + *-oma*, tumor) is an aggressive and sometimes deadly skin cancer that involves uncontrolled proliferation of melanocytes in the stratum basale layer of the epidermis. Melanoma usually presents as a mole on the surface of the skin that exhibits the following "ABCD" characteristics: **asymmetry** (A), **irregular borders** (B), **varied colors** (C), and a **diameter of greater than 6 mm** (D) **(figure 6.3*a*).** Doctors test the suspicious lesion by taking a biopsy of the affected tissue to confirm the diagnosis (figure 6.3*b*). Once the lesion has been identified as melanoma, doctors classify the severity of the case by identifying the "stage" of cancer. Tumors that remain in the outer portions of the epidermis are classified as stage I. These are the simplest types of melanomas to treat. Stage II occurs when the tumor invades the underlying dermis, but there are no signs of spread to local lymph nodes. Because the dermis contains both lymph vessels and blood vessels, there is the risk of metastasis, or spreading of the cancer cells from one part of the body to another. When the cancer spreads to nearby lymph nodes or tissues, it is classified as stage III. Once there is evidence of metastasis to distant organs, it is classified as stage IV.

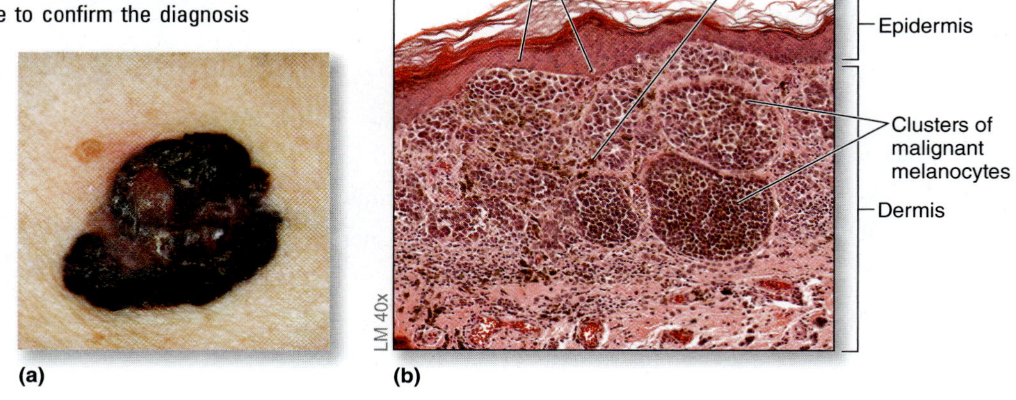

Figure 6.3 Malignant Melanoma. (a) This mole demonstrates the A, B, C, Ds of melanoma: A = asymmetry, B = irregular borders, C = color, D = diameter >6 mm. (b) Histopathology of a malignant melanoma. Note the large clusters of melanocytes in the dermis, and the brownish melanin pigment granules scattered throughout.

The Dermis

The **dermis** of the skin is more complex than the epidermis because it contains numerous *skin appendages* such as hair follicles, glands, blood vessels, and nerves. The dermis consists of two layers: an outer **papillary layer,** and an inner **reticular layer (figure 6.4).** The papillary layer is the part of the dermis that contains "nipplelike" extensions that project upward into the epidermis (the dermal papillae). The papillary layer is generally quite thin and composed of areolar connective tissue. The reticular layer is named for its "networked" appearance (*rete*, a net), not because it contains reticular fibers. In fact, the major fiber type found in this layer is the collagen that composes the dense irregular connective tissue. These collagen fibers are interwoven into a meshwork that surrounds structures of the dermis such as hair follicles, sweat glands, sebaceous glands, blood vessels, and nerves. The slide descriptions in this section of the laboratory manual serve as a guide for both general observation of the structure and function of the dermis, and for identification of the various skin appendages found within the dermis.

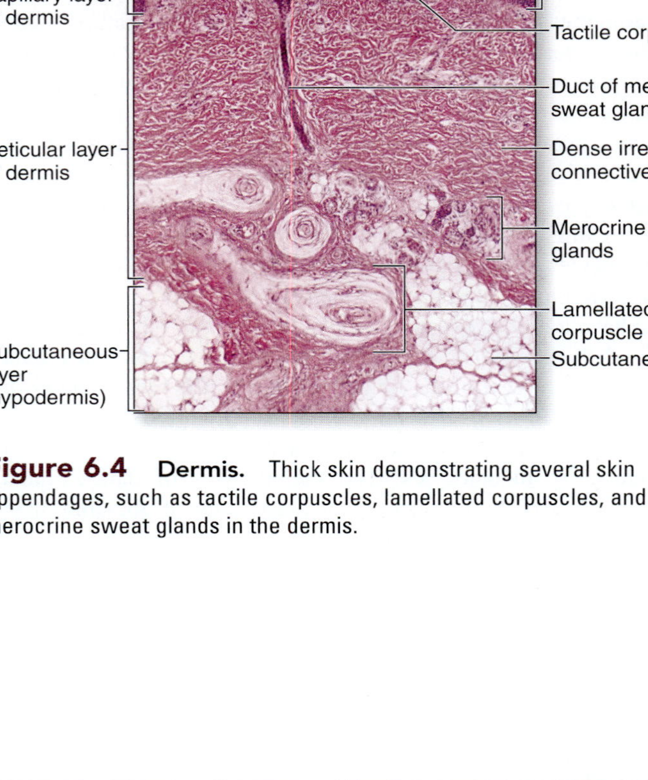

Figure 6.4 Dermis. Thick skin demonstrating several skin appendages, such as tactile corpuscles, lamellated corpuscles, and merocrine sweat glands in the dermis.

INTEGRATE

LEARNING STRATEGY

The next time you see a piece of leather, look at it carefully and think about your own skin. The very tough tissue that composes most of the leather is dense irregular connective tissue (collagen fibers) from the dermis of the animal's skin.

EXERCISE 6.4

LAYERS OF THE DERMIS

1. Obtain a slide of thick skin or thin skin and place it on the microscope stage.
2. Scan the slide at low power to identify the skin tissue, and then bring the dermal layer of the skin into the center of the field of view.
3. Switch to high power and distinguish the papillary dermis from the reticular dermis **(figure 6.5)**. Locate a **tactile (Meissner) corpuscle** within a dermal papilla. Tactile corpuscles are sensory receptors for fine touch (see table 6.3 for a description).
4. Identify the following structures of the dermis on the slide, using figures 6.4 and 6.5 as guides:
   - ☐ **dense irregular connective tissue**
   - ☐ **dermal papillae**
   - ☐ **papillary dermis**
   - ☐ **reticular dermis**
   - ☐ **tactile (Meissner) corpuscle**

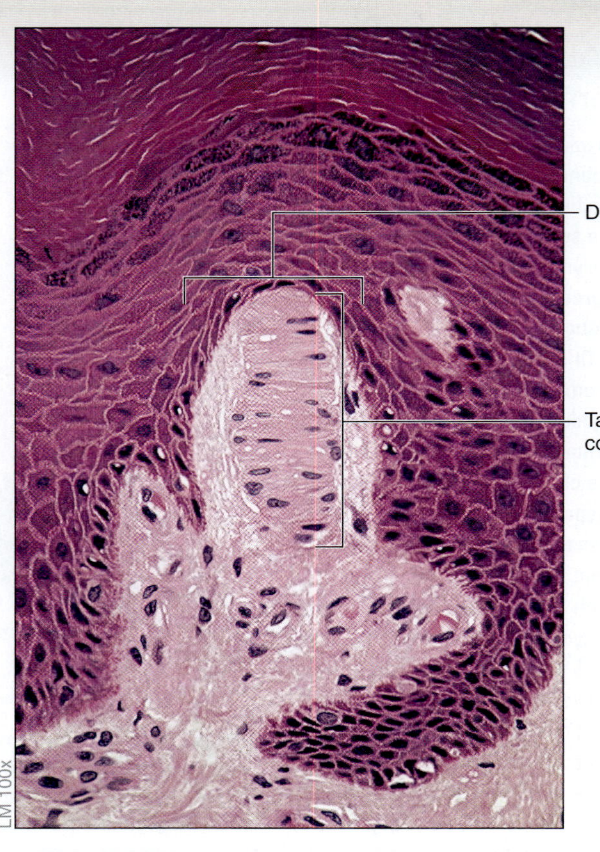

Figure 6.5 Tactile Corpuscle. High-magnification view of a dermal papilla containing a tactile corpuscle.

5. Sketch the dermis as seen through the microscope in the space provided. Be sure to label all the structures listed in step 4 in the drawing.

_______ ×

EXERCISE 6.5

MEROCRINE (ECCRINE) SWEAT GLANDS AND SENSORY RECEPTORS

1. Obtain a slide of thick skin or thin skin and place it on the microscope stage. Begin by observing the epidermal and dermal layers of the skin. When viewing the current slide, focus on the numerous skin appendages found within the dermis.

2. Locate the coiled, tubular glands located deep within the reticular dermis (these glands open to the skin surface, and a duct traveling to the surface of the skin may be visible). These are **merocrine (eccrine) sweat glands** (*meros*, to share, + *krino*, to separate) (figure 6.4). Merocrine sweat glands are located in skin covering nearly the entire body (as opposed to apocrine sweat glands, which are located only in the axillary region, pubic region, and areola). **Table 6.2** lists the types of glands located in the dermis and describes their locations and functions.

3. Focus on sensory receptors found in the dermis. **Table 6.3** summarizes the characteristics of sensory

Table 6.2 Glands in the Dermis

Gland	Location	Description	Mode of Secretion	Function	Word Origin
Apocrine Sweat Glands	Axilla, areola of the breast, pubic, and anal regions	Coiled tubular glands located next to hair follicles. Ducts open into the hair follicle.	*Merocrine*—exocytosis of vesicles containing product into the duct of the gland	Produce a thick, slightly oily sweat that may have pheromone-like properties	*apo-*, away from, + *krino*, to separate or secrete
Merocrine (eccrine) Sweat Glands	Most of the surface of the body	Coiled tubular glands whose main secretory portions are found deep within the reticular layer of the dermis. Ducts open to the surface of the skin.	*Merocrine*—exocytosis of vesicles containing product into the duct of the gland	Produce the thin, watery sweat that cools the body	*meros*, share, + *krino*, to separate or secrete
Sebaceous Glands	Wherever hair follicles are found. Particularly abundant on the scalp.	Glands located next to hair follicles. Ducts commonly open into the hair follicle.	*Holocrine*—disintegrated whole cells filled with product are discharged into the duct of the gland	Produce sebum, an oily substance that lubricates the skin surface, keeps it from drying out, and inhibits the growth of bacteria	*sebaceous*, relating to sebum; oily; *holos*, whole, + *krino*, to separate or secrete

(continued on next page)

(continued from previous page)

Table 6.3 Sensory Receptors in the Dermis

Sensory Receptor	Location	Structure/Appearance	Function
Free Nerve Ending	At epidermal/dermal junction	Dendritic endings of sensory neurons. There is no special structure at the end of the nerve (hence "free" nerve ending).	Light touch, temperature, pain, and pressure
Lamellated (Pacinian) Corpuscle	Deep in the dermis and hypodermis	Dendritic endings of sensory neurons are ensheathed with an inner core of neurolemmocytes and outer concentric layers of connective tissue. Lamellated corpuscles resemble an onion in cross-section.	Sensation of deep pressure and high-frequency vibration
Tactile (Merkel) Cell	At epidermal/dermal junction	Tactile cells are round cells located in the stratum basale of the epidermis. A tactile cell associates with a sensory nerve ending in the dermis (a tactile disc).	Sensation of fine touch, textures, and shapes
Tactile (Meissner) Corpuscle	In dermal papillae	Highly intertwined dendritic endings enclosed by modified neurolemmocytes and connective tissue. Oval-shaped structure with cells that appear almost layered on top of each other.	Sensation of fine, light touch and texture

receptors in the dermis. Many of the receptors are difficult to identify histologically, so refer to the gross anatomy section of this manual to identify them on classroom models of integument. Observe the many dermal papillae found in the slide. Center one or more papillae in the field of view and increase the magnification.

4. Using figure 6.5 as a guide, once again try to locate the tiny sensory receptors called **tactile (Meissner) corpuscles** (*tactus*, to touch, + *corpus*, body) within the papillae. These sensory receptors, appropriately located near the surface of the skin, are responsible for sensing fine touch.
5. After identifying the tactile corpuscles, change back to a lower magnification and scan the lower reticular dermis and subcutaneous regions of the skin. Look for a large onion-shaped organ. This is a sensory receptor called a **lamellated (Pacinian) corpuscle** (*lamina,* plate, + *corpus,* body) (figure 6.4). Such sensory organs, located deep within the dermis, are responsible for sensing deep pressure applied to the skin. The other main sensory receptors found within the dermis, the **tactile (Merkel) cells** and **free nerve endings,** are not easily identifiable through the light microscope, so classroom models may be necessary for their identification.
6. Finally, identify cross sections through the numerous small blood vessels located in the dermis.
7. Sketch the skin appendages observed through the microscope in the space provided.

_______ ×

INTEGRATE

CONCEPT CONNECTION

The skin is an organ that is composed of both epithelial and connective tissues. Recall that epithelial tissues cover surfaces and are avascular, whereas connective tissues provide support and vary in degree of vascularity. The epidermis of the skin, the outermost layer, is composed of keratinized stratified squamous epithelial tissue. The epidermis aids in temperature regulation, helps prevent excessive water loss, and protects underlying tissues from abrasion and invasion from foreign pathogens. The dermis of the skin is composed of two types of connective tissue. The papillary layer contains areolar connective tissue, whereas the reticular layer contains dense irregular connective tissue. It is critical to relate structure to function. That is, the epidermis contains *stratified* epithelium; therefore, it provides *protection* for underlying structures. The dermis contains largely dense irregular connective tissue, which is composed of large quantities of collagen. Therefore, the dermis provides support and resistance to stress in multiple directions.

EXERCISE 6.6

THE SCALP—HAIR FOLLICLES AND SEBACEOUS GLANDS

1. Obtain a slide of the **scalp** and place it on the microscope stage. Begin by observing the epidermal and dermal layers of the scalp. The scalp epithelium is thin, but the dermis is thick and contains numerous **hair follicles** (**figure 6.6** and **table 6.4**).

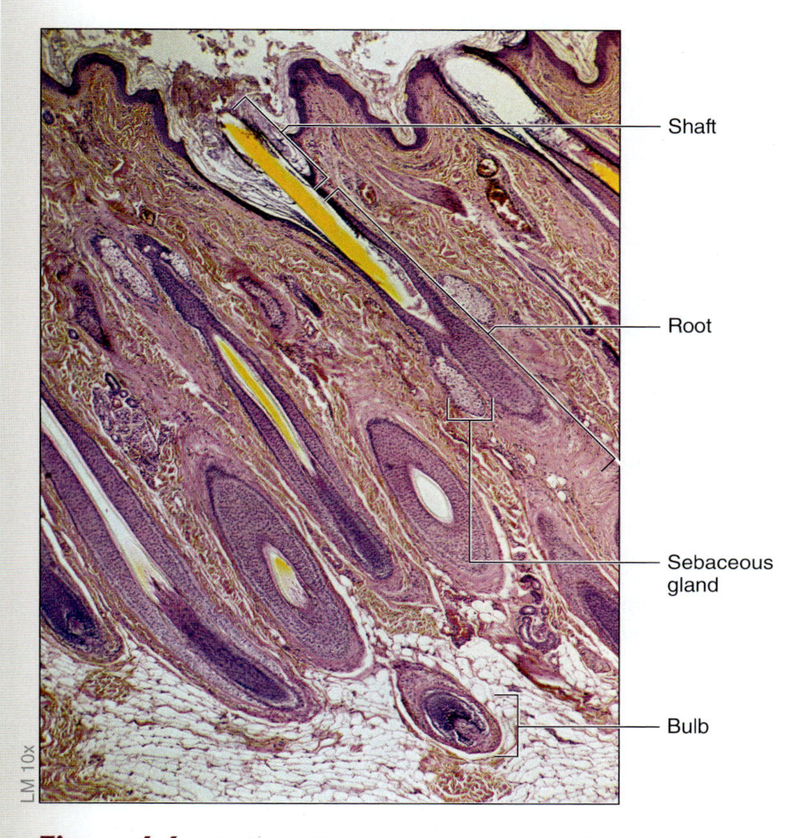

Figure 6.6 Scalp. Skin of the scalp, demonstrating several hair follicles. Arrector pili muscles are not visible in this photomicrograph.

2. Scan the slide until a hair follicle that is sliced longitudinally is visible in the field of view. The entire hair, from the base of the hair follicle to where it exits the skin, should be visible. Notice that the color of the cells lining the hair follicle is similar to the color of the cells within the epidermis. This is because hair follicles are derivatives of the epidermis and they develop as downgrowths of the stratum basale. There are three distinct regions to a hair: (1) the **shaft,** which is the portion of the hair that exits the skin surface; (2) the **root,** which is the portion of the hair within the skin itself; and (3) the **bulb,** which is the swelled base of the hair.
3. Observe the bulb of the hair at higher magnification **(figure 6.7).** Notice the **papilla,** a cone-shaped structure in the middle of the base of the follicle. The papilla is part of the dermis, and is separated from the hair follicle by the basement membrane of the hair follicle epithelium (this basement membrane continues external to the hair follicle as the "glassy membrane"). The papilla contains sensory nerve endings and numerous blood vessels, which are important in supplying nutrients to the developing hair.
4. Return to a lower magnification and look for the oil-secreting **sebaceous glands** (figure 6.6) that connect to the hair follicles in the region of the hair roots. Sebaceous glands secrete an oily substance, **sebum,** into the hair follicle.
5. Carefully observe several hair follicles to locate the small **arrector pili** muscles (*arrector*, that which raises, + *pili*, hair) that attach to the base of the hair follicle (not visible in figure 6.6; see figure 6.11). When these smooth muscles contract, they pull at the base of the hair follicle. This causes the hair to stand up straight, rather than lie flat against the surface of the skin. Muscle contraction pulls down on the epidermis of the skin, while the area where the hair shaft exits the epidermis remains elevated. Thus, in humans it gives the appearance of "goose bumps" or "goose pimples."

Table 6.4 Parts of a Hair Follicle

Structure	Description and Function
Connective Tissue (Dermal) Root Sheath	The connective tissue of the dermis (mainly dense collagen fibers) that surrounds the entire hair follicle
Cortex	Constitutes the bulk of the hair; composed predominantly of keratin
Cuticle Layer	The outer portion of the hair itself; composed of several layers of hard plates of keratin that surround the cortex of the hair
External Root Sheath	The outer layers of the hair follicle, which are continuous with the stratum basale and stratum spinosum of the epidermis
Glassy Membrane	A specialized basement membrane located external to the external root sheath and internal to the connective tissue that surrounds the hair follicle (the connective tissue root sheath)
Internal Root Sheath	A sheath derived from epithelial tissue that lies between the external root sheath and the hair itself

(continued on next page)

(continued from previous page)

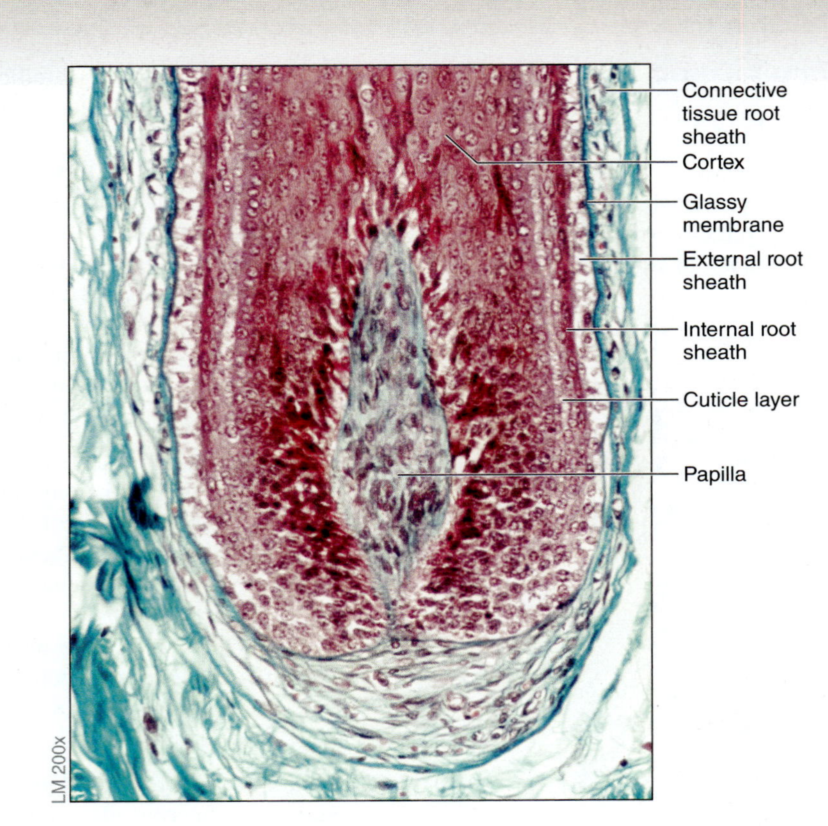

Figure 6.7 Close-up View of a Hair Bulb.

6. Identify the following structures related to the hair follicle, using figures 6.6 and 6.7 and tables 6.2 and 6.3 as guides:
   - ☐ **arrector pili muscle**
   - ☐ **bulb of hair follicle**
   - ☐ **hair follicle**
   - ☐ **papilla of hair follicle**
   - ☐ **root of hair follicle**
   - ☐ **sebaceous gland**
   - ☐ **shaft of hair follicle**

7. Sketch a hair follicle and associated skin appendages as viewed through the microscope in the space provided. Be sure to label all the structures listed in step 6 in the drawing.

_______________ ×

8. Next, scan the slide until a hair follicle in cross section is visible within the field of view **(figure 6.8).**

9. Identify the following structures, using figure 6.8 and table 6.4 as guides:

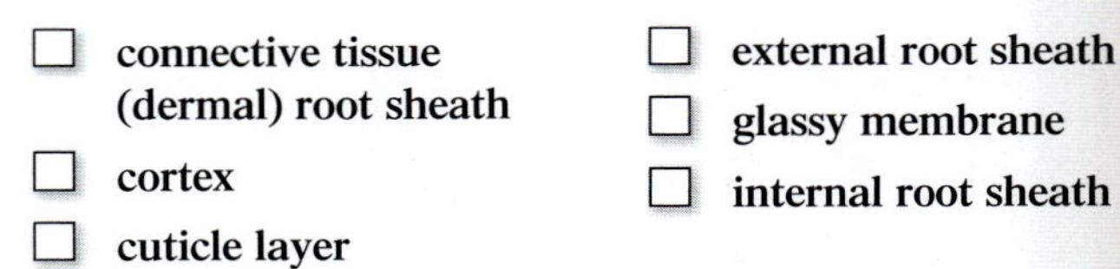

   - ☐ **connective tissue (dermal) root sheath**
   - ☐ **cortex**
   - ☐ **cuticle layer**
   - ☐ **external root sheath**
   - ☐ **glassy membrane**
   - ☐ **internal root sheath**

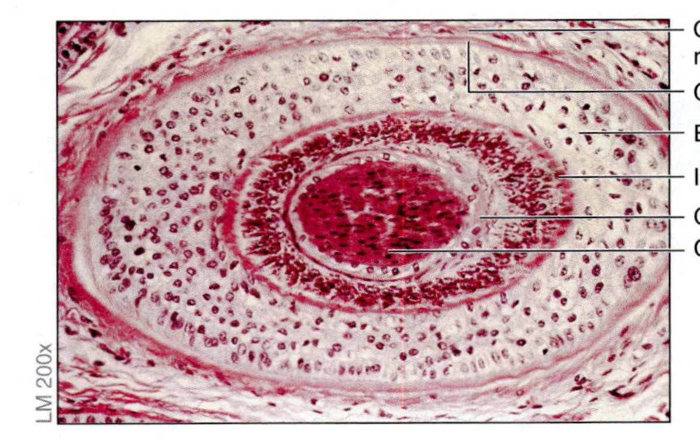

Figure 6.8 Cross Section of a Hair Follicle.

EXERCISE 6.7

AXILLARY SKIN—APOCRINE SWEAT GLANDS

1. Obtain a slide of **axillary skin** and place it on the microscope stage. Locate a hair follicle on the slide. Notice that the hair follicles in this skin are oriented at a fairly steep angle with respect to the apical surface of the epithelium. This, in part, is what makes the hairs in the axillary region curly.

2. Look for glands that open into the hair follicle. These are **apocrine sweat glands** (*apo*-, away from, + *krino*, to separate or secrete) **(figure 6.9).** Apocrine sweat glands are predominantly located in the axillary, pubic, and anal regions of the body, though they are also located in the areola of the nipple and in men's facial hair. These glands produce their secretions via exocytosis, the same mechanism used by merocrine glands. However, apocrine glands produce a secretion that is thicker and oilier than that of the merocrine glands. Although the term *apocrine* historically referred to a different mode of secretion as compared to merocrine glands, these glands are still referred to as apocrine glands.*

3. Sketch a hair follicle from axillary skin and its associated apocrine sweat gland in the space provided.

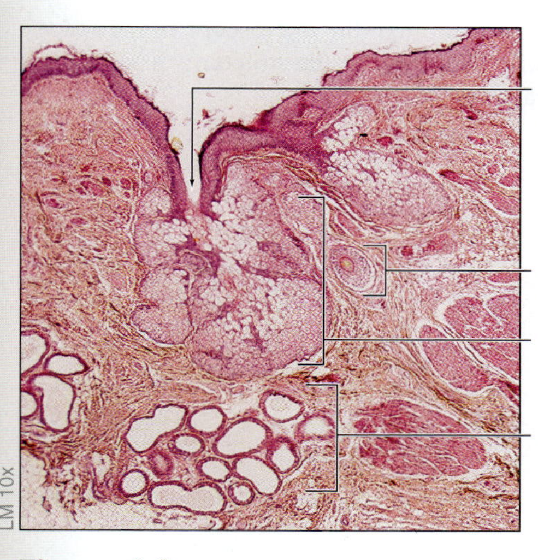

Figure 6.9 Apocrine Sweat Glands. Axillary skin demonstrating apocrine sweat glands.

*Traditionally, *apocrine* referred to a process by which the apical part of the cell discharged into the duct of the gland along with the secretion. However, we now know that this is not how apocrine glands produce their secretions.

INTEGRATE

CONCEPT CONNECTION

The sweat produced by apocrine sweat glands is unique. Similar to merocrine sweat glands, apocrine sweat glands produce sweat that contains water, salt, and nitrogenous wastes. In contrast to merocrine sweat glands, the products of apocrine sweat glands also include lipids and pheromones. A **pheromone** is a type of chemical messenger similar to a hormone. However, whereas hormones act as chemical messengers *within* an individual's own body, pheromones act as chemical messengers *between* two individuals of the same species.

Interestingly, researchers have found that the pheromones produced by apocrine sweat glands are responsible for phenomena such as the coordination of menstrual cycles that occur when two or more women live in the same space for a period of time (such as in college dorm rooms). Pheromones produced by apocrine sweat glands have also been found to influence mating behavior. This discovery came from a study in which researchers had several male subjects sleep in the same t-shirt two nights in a row. Researchers then had female subjects smell the t-shirts and rate them using an "attraction" scale. Finally, the researchers analyzed the DNA that codes for part of the individual's immune system for both the male and female subjects. The results showed that females consistently rated t-shirts worn by males whose DNA was the most *different* from their own as most attractive. What does this mean? In essence, the better a potential mate smells, the higher the likelihood that mating will produce offspring with a more diverse, hence robust, immune system. Thus, if a potential date smells a little "off," it just might mean that he or she isn't a good genetic match.

EXERCISE 6.8

STRUCTURE OF A NAIL

1. Obtain a slide of a **nail.** A nail is a very specialized skin appendage that arises from epithelial tissue. **Table 6.5** lists the parts of a nail, and **figure 6.10** shows a longitudinal section of a nail.
2. Identify the following structures on the slide, using figure 6.10 and table 6.5 as guides:
   - ☐ **body**
   - ☐ **hidden border**
   - ☐ **eponychium**
   - ☐ **hyponychium**

Table 6.5 Parts of a Nail

Structure	Description	Word Origins
Body	The main portion of the nail	
Eponychium	The fold of skin at the root of the nail that folds over the body of the nail	*epi*, upon, + *onyx*, nail
Hidden Border	The portion of the nail that lies beneath the eponychium	
Hyponychium	The skin underneath the free border of the nail	*hypo*, under, + *onyx*, nail
Lunula	A white, curved area at the base of the nail	*luna*, moon

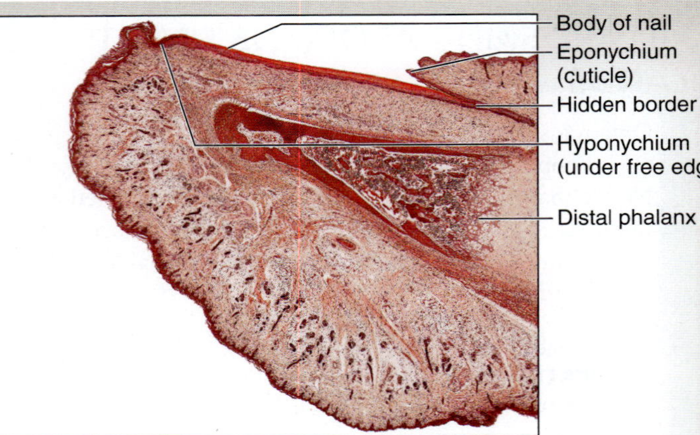

Figure 6.10 Longitudinal Section of a Nail.

3. Sketch a nail as seen through the microscope in the space provided. Be sure to label all of the structures listed in step 2 in the drawing.

_______ ×

GROSS ANATOMY

Integument Model

Viewing laboratory models or charts of integument allows for the identification of structures that are not easily visible through the light microscope. **Figure 6.11** demonstrates structures that are typically visible in classroom models of the integument.

EXERCISE 6.9

OBSERVING CLASSROOM MODELS OF INTEGUMENT

1. Locate models or posters of the integument in the laboratory.
2. Identify the following structures on a model of the integument, using figure 6.11 as a guide. Pay particular attention to tactile corpuscles, lamellated corpuscles, and free nerve endings, because these structures are not easily seen through the microscope.

Epidermis

- ☐ **epidermal ridges**
- ☐ **stratum basale**
- ☐ **stratum corneum**
- ☐ **stratum granulosum**
- ☐ **stratum lucidum**
- ☐ **stratum spinosum**

3. *Optional Activity:* AP|R **4: Integumentary System—** Use the Quiz feature to review integumentary system structures.

Dermis

- ☐ **blood vessels**
- ☐ **dense irregular connective tissue**
- ☐ **free nerve ending**
- ☐ **hair follicle**
- ☐ **lamellated (Pacinian) corpuscle**
- ☐ **merocrine sweat gland**
- ☐ **papillary layer**
- ☐ **reticular layer**
- ☐ **sebaceous gland**
- ☐ **tactile (Meissner) corpuscle**

Subcutaneous layer (hypodermis)

- ☐ **adipose connective tissue**
- ☐ **lamellated (Pacinian) corpuscle**

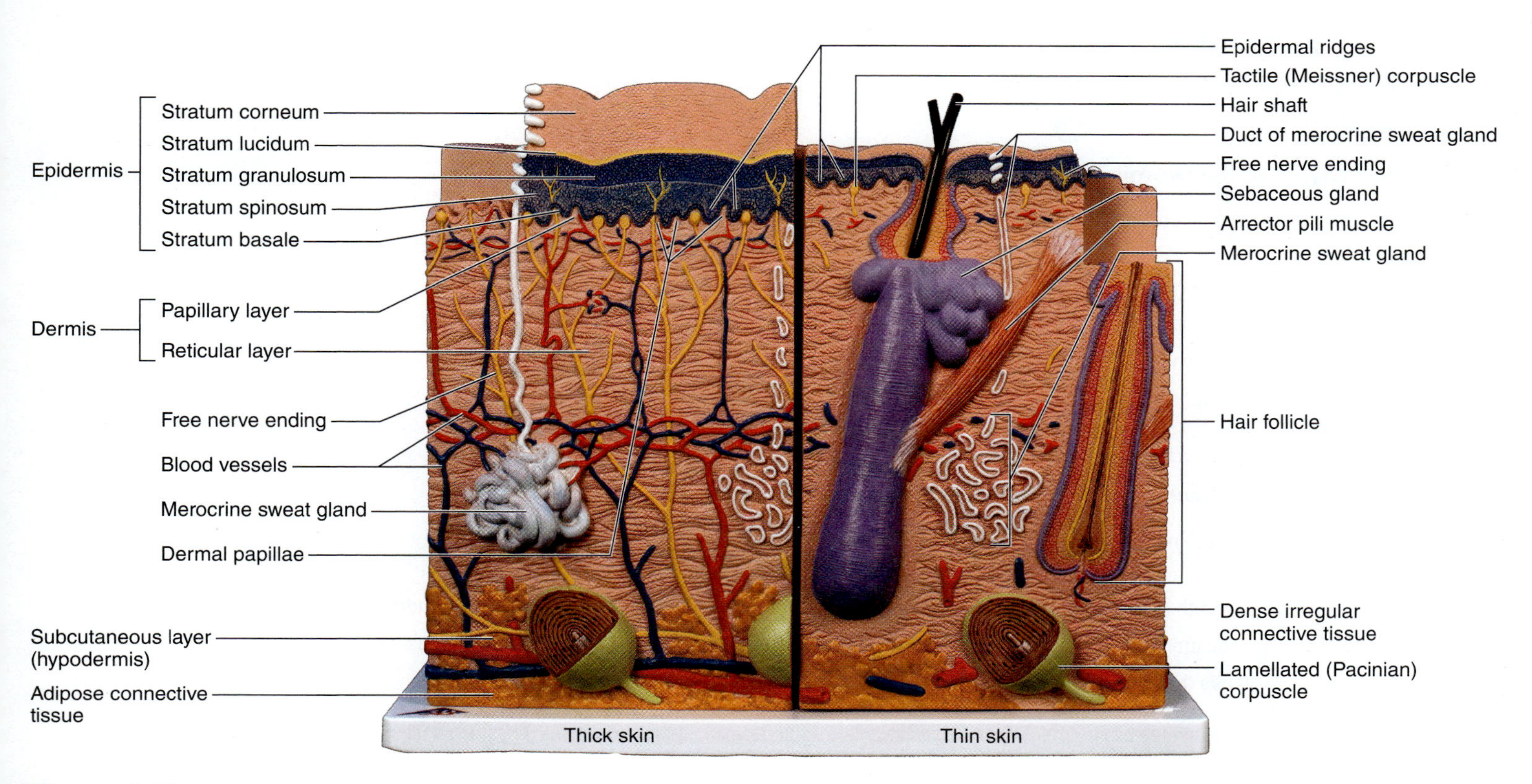

Figure 6.11 **Classroom Model of the Integument.** The left side of this model demonstrates thick (glabrous) skin, which has five epidermal layers. The right side of this model demonstrates thin skin, which has only four epidermal layers and contains hair follicles.

Chapter 6: Integument

Name: ____________________

Date: __________ Section: __________

POST-LABORATORY WORKSHEET

The (1) corresponds to the Learning Objective(s) listed in the chapter opener outline.

Do You Know the Basics?

Exercise 6.1: Layers of the Epidermis

1. Which of the following is the type of tissue found in the most superficial layer of thick skin? (Circle one.) (1)
   a. adipose connective tissue
   b. areolar connective tissue
   c. dense irregular connective tissue
   d. simple squamous epithelial tissue
   e. stratified squamous epithelial tissue

2. Identify the location(s) on the body where skin is glabrous and contains five layers in the epidermis. (Check all that apply.) (2)
   ____ a. back
   ____ b. face
   ____ c. palms of the hands
   ____ d. scalp
   ____ e. soles of the feet

Exercise 6.2: Fingerprinting

3. Identify structures of the epidermis that are responsible for the formation of fingerprints. (Check all that apply.) (3)
   ____ a. dermal papillae
   ____ b. epidermal ridges
   ____ c. keratinocytes
   ____ d. melanocytes
   ____ e. tactile corpuscles

Exercise 6.3: Pigmented Skin

4. Although melanocytes are the cells that produce melanin, the cells that actually *concentrate* the melanin in the apical region of the cell to protect the cell nucleus are ____________________. (4)

Exercise 6.4: Layers of the Dermis

5. Match the definition listed in column A with the corresponding term listed in column B. (5)

Column A	Column B
____ 1. a cell located in the epidermis that produces melanin	a. dermal papilla
____ 2. an epithelial cell composing the majority of the epidermis	b. epidermal ridge
____ 3. a fold of the epidermis that interdigitates with dermal papillae	c. keratinocyte
____ 4. a "nipplelike" extension of the dermis into the epidermis	d. melanocyte

6. The ____________________ (papillary/reticular) layer of the dermis is the thickest layer of the dermis. (5)

Exercise 6.5: Merocrine (Eccrine) Sweat Glands and Sensory Receptors

7. Match the description listed in column A with the corresponding gland listed in column B.

Column A	Column B
____ 1. *merocrine* glands that empty into hair follicles	a. apocrine sweat glands
____ 2. *holocrine* glands that empty into hair follicles	b. eccrine sweat glands
____ 3. *merocrine* glands that open to the surface of the skin	c. sebaceous glands

8. Lamellated (Pacinian) corpuscles are located in the *papillary* layer of the dermis, whereas tactile (Meissner) corpuscles are located in the *reticular* layer of the dermis. ____________________ (True/False) (6)

9. Match the function listed in column A with the corresponding sensory receptor listed in column B. 7

Column A	Column B
____ 1. sensation of fine touch, textures, and shapes	a. free nerve ending
____ 2. sensation of light touch, temperature, pain, and pressure	b. lamellated (Pacinian) corpuscle
____ 3. sensation of deep pressure and high-frequency vibration	c. tactile (Meissner) corpuscle
____ 4. sensation of fine, light touch and texture	d. tactile (Merkel) cell

Exercise 6.6: The Scalp—Hair Follicles and Sebaceous Glands

10. Match the description and function listed in column A with the corresponding structure listed in column B. 8

Column A	Column B
____ 1. the swelled base of the hair	a. arrector pilli
____ 2. the portion of the hair that exits the skin surface	b. bulb
____ 3. the portion of the hair contained within the skin	c. root
____ 4. small muscles attached to the base of the hair follicle	d. sebaceous gland
____ 5. oil-secreting glands that connect to the hair follicle	e. shaft

11. Which of the following correctly identifies the product of sebaceous glands and the corresponding function of this product? (Circle one.) 9
    a. sebum; lubrication
    b. sebum; temperature regulation
    c. thick, oily sweat; pheromone-like properties
    d. thin, watery sweat; temperature regulation
    e. thin, watery sweat; lubrication

Exercise 6.7: Axillary Skin—Apocrine Sweat Glands

12. Identify the regions of the body where apocrine glands are located. (Check all that apply.) 10
    ____ a. anal
    ____ b. axillary
    ____ c. femoral
    ____ d. frontal
    ____ e. pubic

13. Apocrine sweat glands open to hair follicles, whereas merocrine (eccrine) sweat glands open to the surface of the skin. ____________________ (True/False) 11

Exercise 6.8: Structure of a Nail

14. Match the description listed in column A with the structure listed in column B. 12

Column A	Column B
____ 1. skin underneath the free border of the nail	a. body
____ 2. main portion of the nail	b. eponychium
____ 3. portion of the nail that lies beneath the eponychium	c. hidden border
____ 4. white curved area at the base of the nail	d. hyponychium
____ 5. fold of skin at the root of the nail	e. lunula

Exercise 6.9: Observing Classroom Models of Integument

15. Label the following diagram of the integument. 13

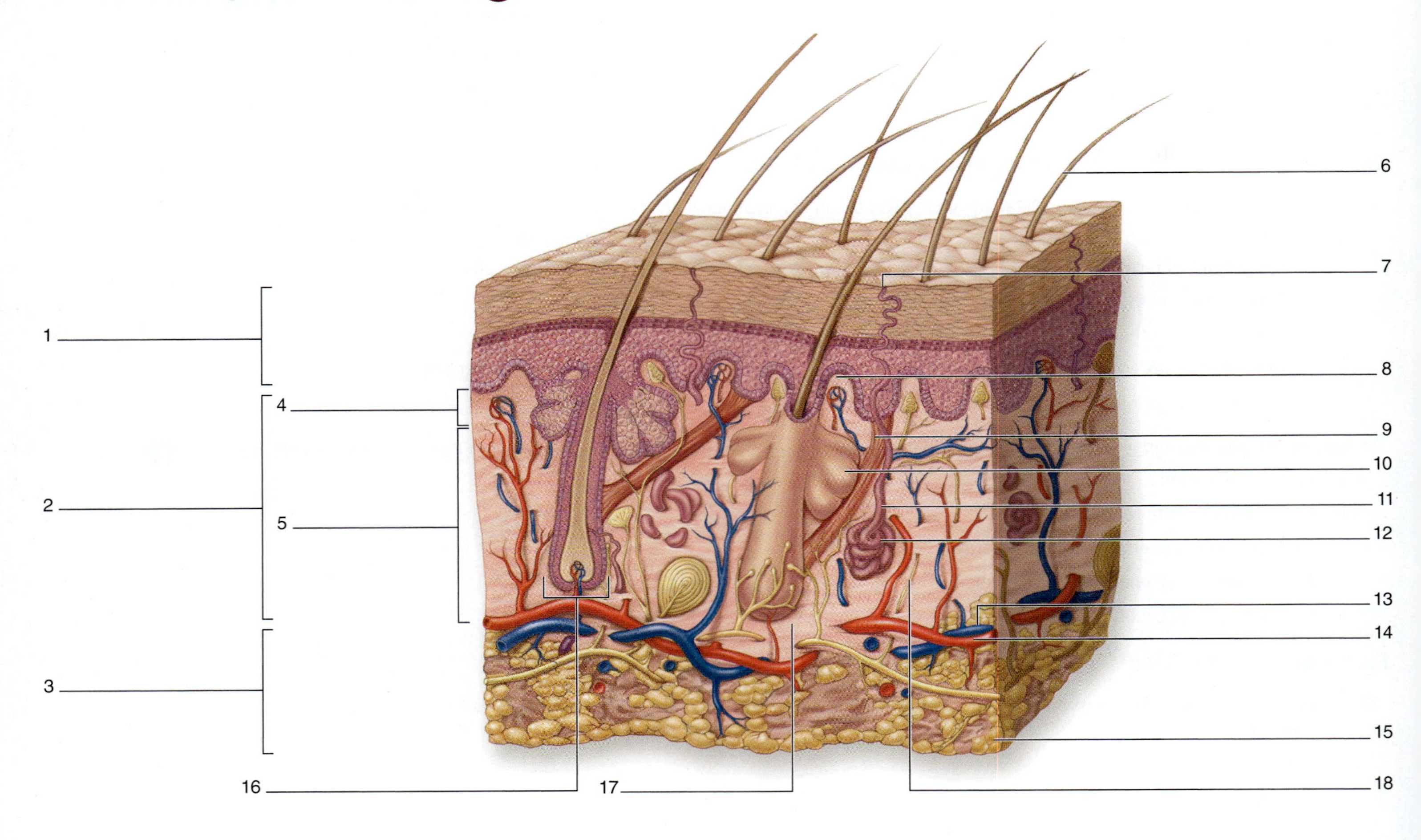

Can You Apply What You've Learned?

16. While sitting on the couch reading your anatomy and physiology book, you realize that you left a window open because you feel a light breeze flowing across your arm. What type of sensory receptor in the skin is responsible for detecting this sensation? ____________________

17. When you sit down on a bench to wait for a bus to arrive, you notice that there is a rock on the bench that you didn't see before you sat down on it. What type of sensory receptor in the skin is responsible for detecting this sensation? ____________________

18. Susan accidentally tripped on the sidewalk and scraped her knee. The scrape was superficial and did not bleed. What is the deepest layer of her skin that could have been damaged without causing her to bleed? ____________________

19. Label the following photomicrograph of the skin in part (a), then answer parts (b) and (c) of this question (below the diagram).

a.

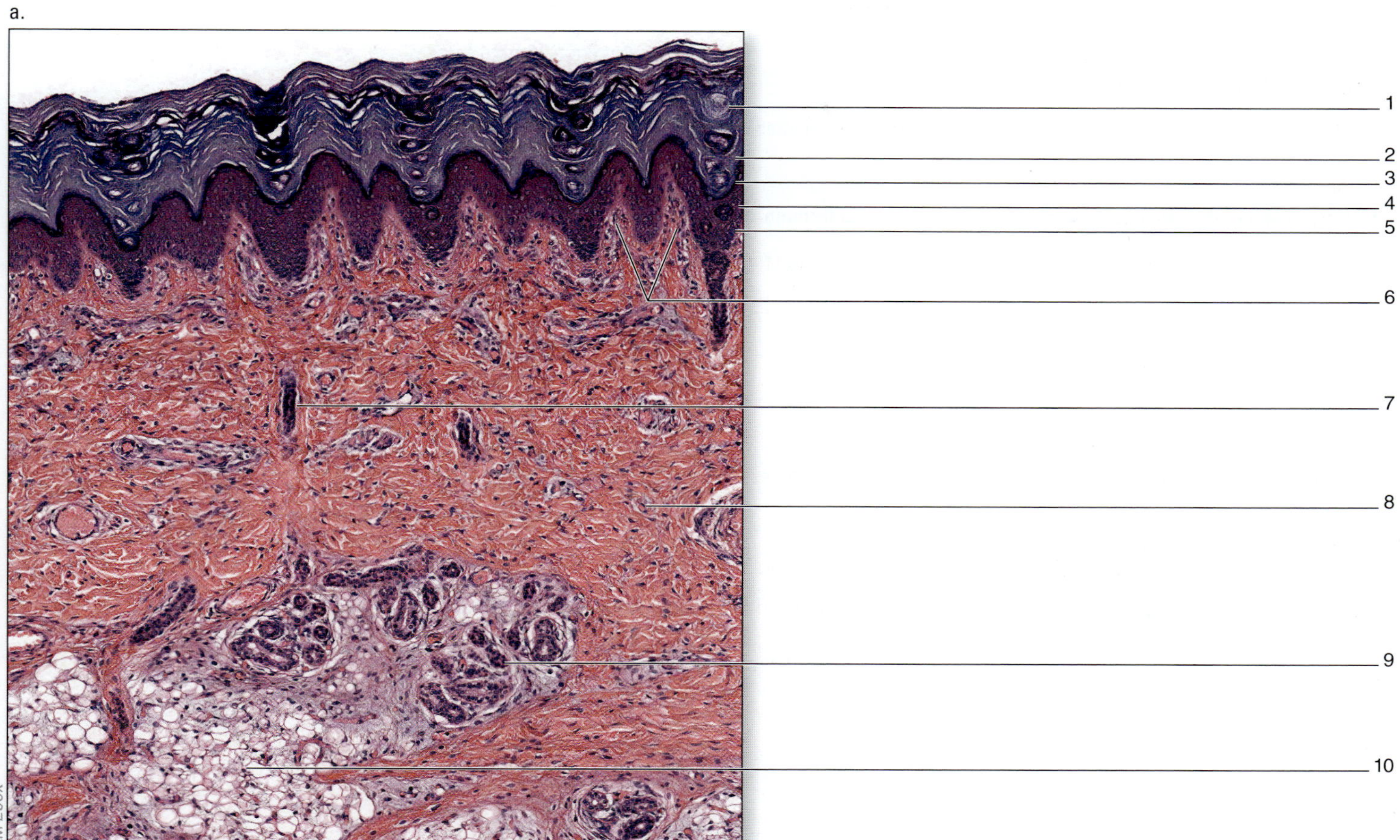

b. This section of skin likely came from what area of the body? ______________________________

____

c. Explain how you came up with the answer to part (b) of this question. That is, the presence of what specific structures in this tissue led to your answer?

____

____

____

____

Can You Synthesize What You've Learned?

20. What areas of the body lack sebaceous glands?

____

____

21. When part of the skin is exposed to a great deal of friction, such as when part of the foot rubs against a shoe, a **blister** often forms.

a. A blister is a collection of fluid that accumulates between two layers of the skin as they separate from each other. Which two layers of the skin are these? ______________________________ and ______________________________

b. When the outer layers of skin are pulled off a blister, exposure of the underlying tissue causes a great deal of pain. What type of sensory receptor is responsible for sensation of this pain?

____

c. Why is the pain worse after removal of the outer layer of the blister?

22. A hypodermic needle is used to give certain types of injections.

    a. Based on the name of the needle, what space is the tip of the needle usually directed into?

    b. What layers of the skin must the hypodermic needle pass through, in order to get to this space? (In your answer, include all sublayers of the dermis or epidermis that may apply. Assume the needle is passing through thin skin.)

    c. What structures do you think are the likely targets of these needles?

23. a. What layer of the epidermis represents the transition from living to dead epithelial cells?

    b. Why do keratinocytes begin to die within this layer?

24. Contrast the epidermal/dermal junction in thick skin with that of thin skin. Specifically note the structure of the epidermal/dermal junction in thick vs. thin skin. Is there a difference in the number of dermal papillae? What function do you think such differences serve?

25. Why is it important that melanin is present in its highest concentration in the keratinocytes at or near the basal layer of cells? (Hint: Were cells that were undergoing cell division observed in this layer?)

26. a. What is the advantage of having tactile (Meissner) corpuscles located near the surface of the skin?

    b. Do you think there would be more of these sensory receptors per unit area in the skin on the palm of the hand or in the skin on the back (or would they be the same)?

The Skeletal System: Bone Structure and Function

OUTLINE AND LEARNING OBJECTIVES

Anatomy & Physiology REVEALED apreveaed.com **MODULE 5: SKELETAL SYSTEM**

INTRODUCTION

The human skeletal system is composed of bones, as well as cartilage, ligaments, and other connective tissues that stabilize or connect the bones. Bones are organs that are composed primarily of bone tissue, which serves as the rigid framework to support the body. Bone is a highly specialized connective tissue that is both rigid and flexible. A common misconception about bones is that they are static, nonliving tissues. This misconception most likely occurs because the type of bones commonly observed in the anatomy laboratory are the preserved skeletons. In reality, bone is a dynamic, versatile tissue that is in a constant state of turnover and is one of the most metabolically active kinds of tissue within the body. Bone has a rich blood supply and an amazing ability to alter its structure in response to the changing stresses placed upon it. Indeed, prolonged absence of stress will cause a loss of bone density and strength, whereas increased stress will cause an increase in bone density and strength.

The focus of the exercises in this chapter is to explore the general micro- and macrostructure of bone, and to provide a brief overview of the structures and functions of the human skeleton. The detailed structures and functions of the bones making up the human skeleton are covered in chapters 8 and 9.

List of Reference Tables

Chapter 7: The Skeletal System: Bone Structure and Function

Name: ______________________

Date: ____________ Section: ____________

These Pre-Laboratory Worksheet questions may be assigned by instructors through their connect course.

PRE-LABORATORY WORKSHEET

1. Which of the following is a function of the skeletal system? (Check all that apply.)
   ____ a. hemopoiesis
   ____ b. movement
   ____ c. storage of minerals
   ____ d. structural support

2. The type of bone that contains osteons, characterized by lamellae surrounding central canals, is called ____________________ (compact/spongy) bone.

3. Match the description and function listed in column A with the type of bone cell listed in column B.

Column A	*Column B*
____ 1. immature cell that lays new bone	a. osteoblast
____ 2. mature bone cell that maintains the matrix	b. osteoclast
____ 3. phagocytic cell that breaks down bone	c. osteocyte
____ 4. stem cell that differentiates into an osteoblast	d. osteoprogenitor cell

4. Number the five functional layers of the epiphyseal plate in the correct order, from diaphyseal side to epiphyseal side.
   ____ a. calcified cartilage
   ____ b. hypertrophic cartilage
   ____ c. ossification
   ____ d. proliferating cartilage
   ____ e. resting cartiliage

5. Match the descriptions of the components of a long bone listed in column A with the appropriate term listed in column B.

Column A	*Column B*
____ 1. the area where the epiphysis of a developing long bone meets the shaft	a. diaphysis
____ 2. the end of a long bone	b. epiphysis
____ 3. the shaft of a lone bone	c. metaphysis

6. In *mature* skeleton, ____________________ (red/yellow) bone marrow fills the medullary cavity of a long bone, and ____________________ (red/yellow) bone marrow is located within the proximal epiphyses of long bones such as the humerus and femur.

7. The outer covering of bone, the periosteum, is composed of which type of connective tissue? (Circle one.)
   a. cartilage
   b. dense irregular connective tissue
   c. elastic connective tissue
   d. reticular connective tissue

8. The ribs are classified as bones of the axial skeleton. ____________________ (True/False)

9. The bones composing the appendicular skeleton include the upper limb bones, lower limb bones, and vertebrae.____________________ (True/False)

10. Match the bone listed in column A with its appropriate classification listed in column B.

Column A	*Column B*
____ 1. carpal bone	a. flat
____ 2. ethmoid bone	b. irregular
____ 3. femur	c. long
____ 4. frontal bone	d. short

HISTOLOGY

The exercises in this chapter explore the histology of compact bone tissue, spongy bone tissue, and endochondral bone development. **Table 7.1** summarizes the characteristics of the types of bone cells that will be observed during histologic studies of bone tissue.

Bone Tissue

There are two types of bone histology preparations generally observed in the laboratory. In the first type of preparation, **ground bone** (see figure 7.1), a solid bone sample, is ground down into a section that is thin enough to be viewed under a microscope. This type of preparation destroys living cells. Thus, no osteocytes, osteoblasts, or other cells are visible. In the second type of preparation, **decalcified bone** (see figure 7.2), the rigid, mineralized (calcified) matrix of the bone is dissolved away so the tissue is soft enough to be sectioned in the traditional manner. This type of preparation preserves living cells such as osteocytes and osteoblasts, but the fine structure of the bony matrix cannot be visualized because of the removal of the rigid portion of the matrix. However, details of the bony matrix—such as central (Haversian) canals, perforating (Volkmann) canals, and canaliculi—are preserved.

Table 7.1 Types of Bone Cells

Cell Name	Description and Function	Drawing	Word Origins
Osteoprogenitor Cell	A stem cell derived from mesenchyme that differentiates into an osteoblast. These cells are located in both the endosteum and the inner layer of the periosteum.	Nuclei	(*osteon*, bone + *pro*, before + *genesis*, origin)
Osteoblast	A small, immature bone cell derived from an osteoprogenitor cell that functions to lay down new bone for bone growth, remodeling, and repair. These cells often appear lined up in rows next to a trabecula of spongy bone.	Nucleus	(*osteon*, bone + *blastos*, germ)
Osteocyte	A mature bone cell derived from an osteoblast that functions to maintain the matrix surrounding it. These cells are located within lacunae.	Nucleus	(*osteon*, bone + *kytos*, a hollow; a cell)
Osteoclast	A very large, multinucleate, phagocytic cell derived from bone marrow cells that also produce monocytes. Osteoclasts break down bone and are often found on the side of a trabecula of spongy bone that is opposite to the layer of osteoblasts.	Nuclei	(*osteon*, bone + *klastos*, broken)

EXERCISE 7.1

COMPACT BONE

1. Obtain a slide of **ground compact bone (figure 7.1)** and place it on the microscope stage. Bring the tissue sample into focus on low power and locate an **osteon.** Remember, living cells are not visible in this tissue sample.

2. With the osteon at the center of the field of view, increase the magnification. The **central (Haversian) canals** will be the largest holes visible in the sample. (The blood vessels and nerves that are housed within these canals can be viewed in the slide of decalcified compact bone.) In contrast, **lacunae** will appear to be much smaller holes because they contain only a single cell in living bone. (Osteocytes present in lacunae can be viewed in the slide of decalcified compact bone.) **Concentric lamellae** extend outward from the central canals in concentric rings. **Lacunae** that house osteocytes are found at the border between two adjacent lamellae. **Interstitial lamellae** can be seen between osteons. **Circumferential lamellae** can be viewed by scanning the outer border of the bone sample, because circumferential lamellae surround the entire diaphysis of the bone. Lacunae appear very dark in color, and at very high magnification they may appear to be empty because of the lack of living osteocytes in the tissue.

   Observe this slide at the highest magnification possible (without using the oil immersion lens). Note what appear to be tiny little "cracks" or fractures that run perpendicular to the central canals. These are the tiny **canaliculi,** which contain cytoplasmic extensions of osteocytes in living bone tissue. Within the canaliculi osteocytes connect to each other via gap junctions, which allow them to exchange nutrients and other substances with each other.

3. Scan the slide to look for large canals that run perpendicular to the central canals. These canals are **perforating (Volkmann) canals,** which convey blood vessels from the outer periosteum into the central canals.

4. Obtain a slide of **decalcified compact bone (figure 7.2)** and place it on the microscope stage. Bring the tissue sample into focus on low power and locate an osteon. This slide will demonstrate the remnants of living structures (cells, blood vessels, and nerves) within the bone.

5. With an osteon at the center of the field of view, increase the magnification. The **central (Haversian) canals** will be the largest holes visible in the sample. The central canals will likely appear to have "junk" inside their lumens. This "junk" consists of blood vessels and nerves. In contrast, **lacunae** will appear to be much smaller holes and there will only be one cell (an osteocyte) inside each lacuna. **Lamellae** are sometimes difficult to make out. First locate the lacunae, and then search for the lamellae. Lacunae are located at the border between two adjacent lamellae. Scan toward the outer border of the bone sample to identify **circumferential lamellae,** which surround the entire diaphysis of the bone.

6. Scan the outer edge of the bone tissue on the slide and identify the **periosteum** of the bone. The periosteum contains an outer layer of dense irregular connective tissue and an inner layer of **osteoprogenitor cells,** which may or may not be visible.

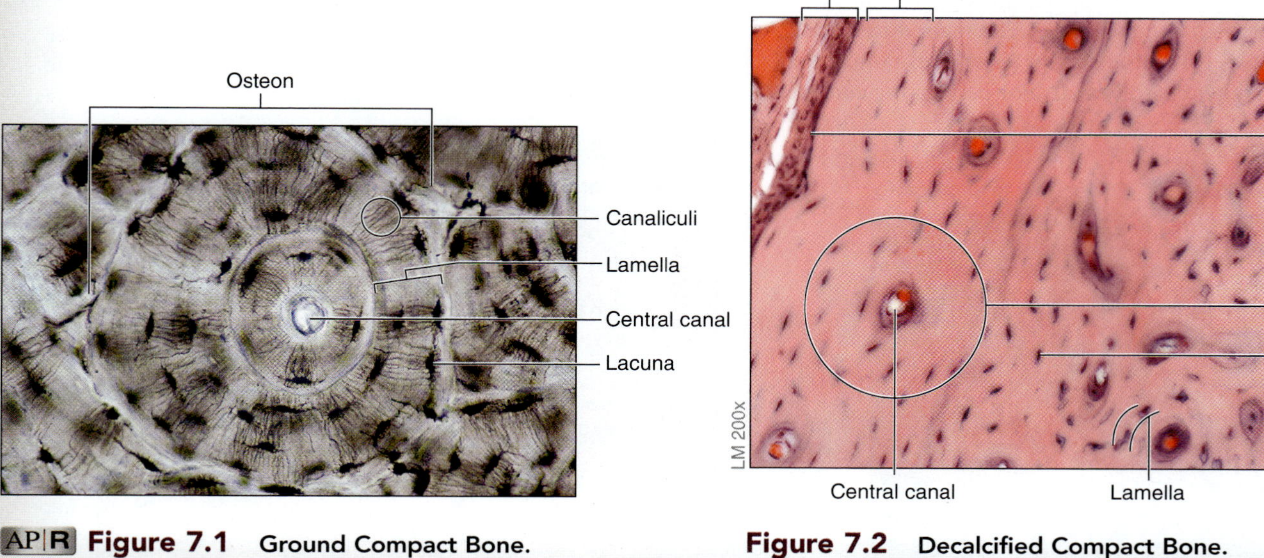

APR **Figure 7.1** Ground Compact Bone. No perforating canals are visible in this section.

Figure 7.2 Decalcified Compact Bone.

(continued on next page)

(continued from previous page)

7. Make sketches of ground compact bone and decalcified compact bone as viewed through the microscope in the spaces provided. Label the following on the sketches, using table 7.1 and figures 7.1 and 7.2 as guides:

- ☐ **central canal**
- ☐ **circumferential lamellae**
- ☐ **concentric lamellae**
- ☐ **interstitial lamellae**
- ☐ **lacuna**
- ☐ **osteoblast**
- ☐ **osteocyte**
- ☐ **osteon**
- ☐ **osteoprogenitor cell**
- ☐ **perforating canal**
- ☐ **periosteum**

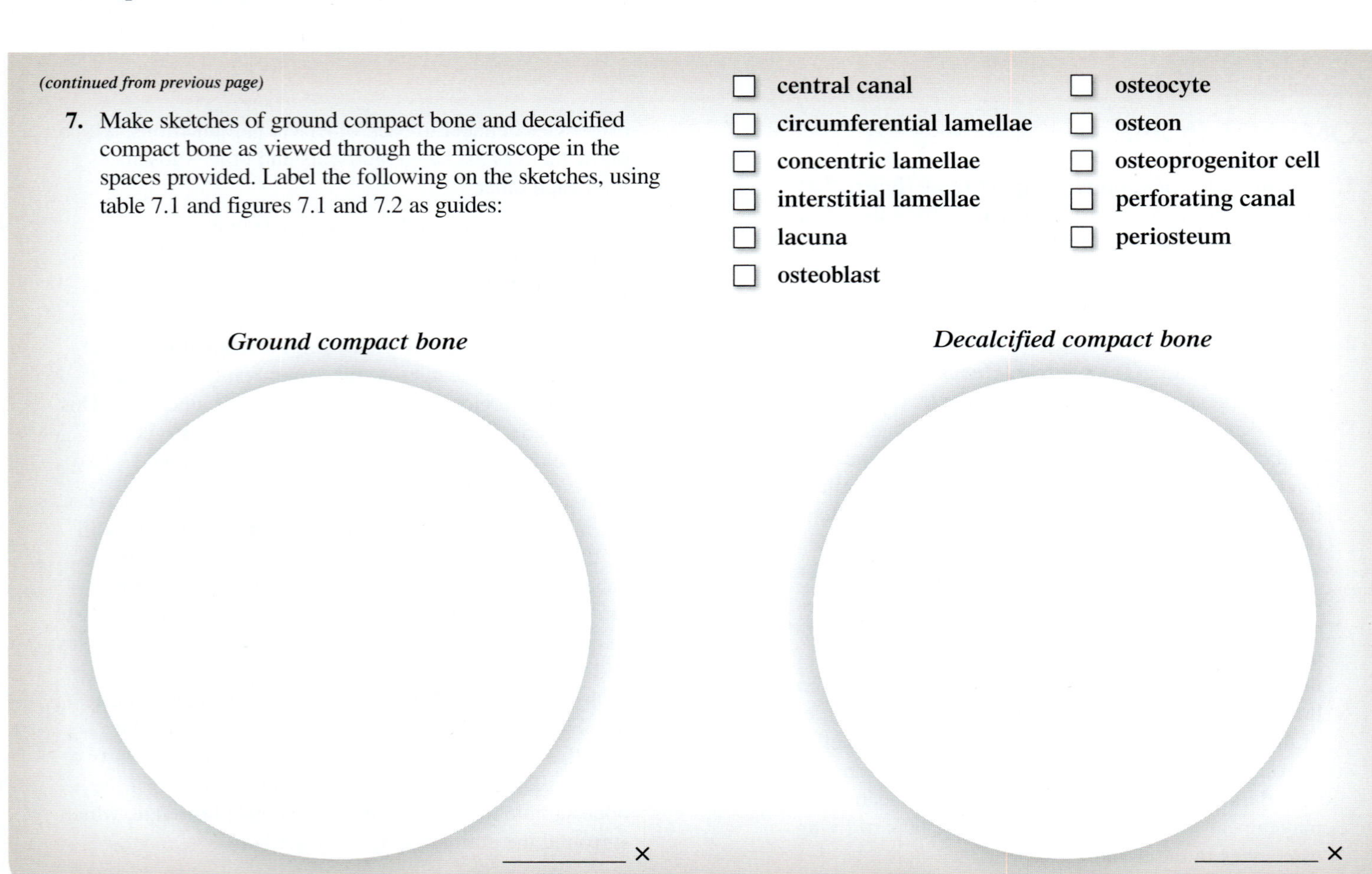

EXERCISE 7.2

SPONGY BONE

1. Obtain a slide of decalcified **spongy (cancellous) bone (figure 7.3)** and place it on the microscope stage. The sample of spongy bone will appear most similar to the slide of decalcified compact bone. The main difference between the two is that the spongy bone does *not* contain osteons, so neither central canals nor perforating canals are visible.

2. Observe the slide at high magnification and look for areas where several small cells are lined up next to each other on the edge of a trabecula. These are **osteoblasts** (figure 7.3*a*) that actively secrete new bony matrix. Are any very large, multinucleate cells visible on the slide? If so, these are bone-resorbing **osteoclasts** (figure 7.3*b*). Both osteoclasts and osteoblasts are actively involved in the process of bone remodeling.

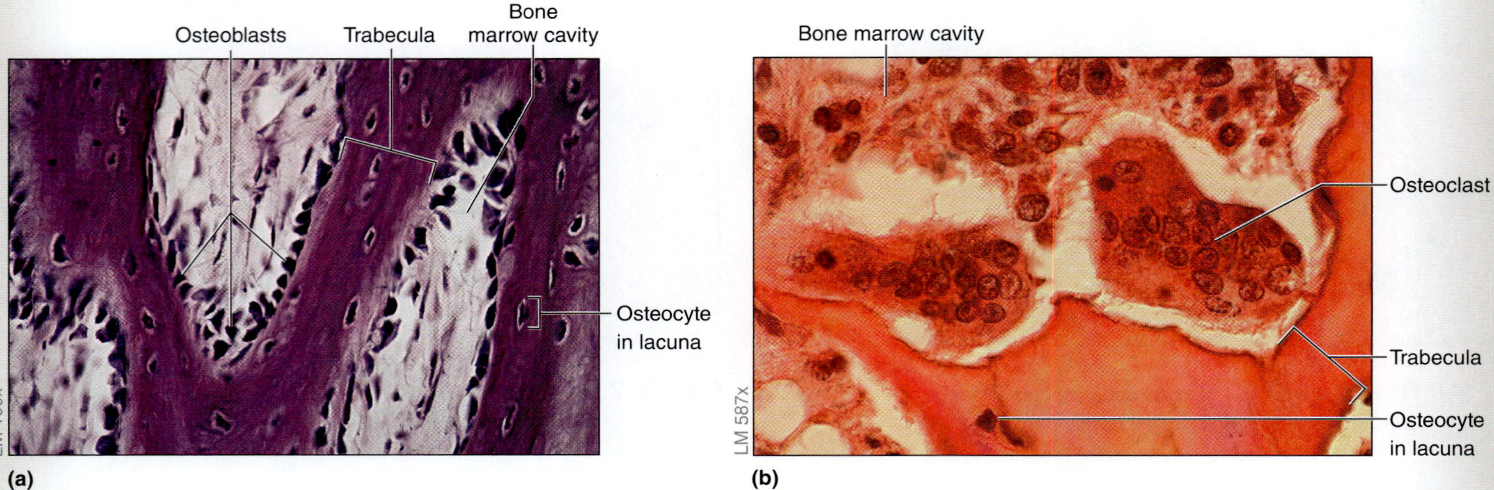

APR Figure 7.3 Spongy Bone. Decalcified sections of spongy bone showing (a) osteoblasts and (b) osteoclasts. Note how the osteoblasts appear to be lined up in rows along the lacuna in (a). The osteoclast in (b) is very large compared to adjacent cells and is multinucleate.

3. Sketch spongy bone as viewed through the microscope in the space provided. Label the following on the sketch, using table 7.1 and figure 7.3 as guides:

- ☐ bone marrow cavity
- ☐ lacuna
- ☐ osteoblast
- ☐ osteoclast
- ☐ osteocyte
- ☐ trabecula

________ ×

INTEGRATE

CONCEPT CONNECTION

Recall that bone is a supporting connective tissue that protects organs and allows body movement by providing attachment sites for muscles. Bone matrix is composed of both organic and inorganic components. *Osteoid,* the organic portion of bone, contains a dense collection of collagen fibers and proteoglycans that provide strength and flexibility. This is the "soft" portion of bone that allows for the penetration of blood vessels during development. The "hard," inorganic portion of bone contains crystals of the mineral compound hydroxyapatite. Hydroxyapatite is formed through a reaction between calcium phosphate and calcium hydroxide. The presence of hydroxyapatite in the mineral matrix of bone allows the bone to store calcium and phosphate, which can then be released into the blood as necessary. This mineralized matrix also provides rigidity and incompressiblility to bone. As bone tissue is loaded by mechanical stresses, osteoblasts are stimulated to secrete osteoid, which leads to calcification of the bone matrix. When mechanical stresses are reduced (for example, when a limb is immobilized in a cast for some time), osteoclasts are stimulated to secrete substances to break down the osteoid and hydroxyapatite. The result of the breakdown of the bony matrix is the release of calcium and phosphorus into the blood. Maintenance of blood calcium levels is so important that regulatory mechanisms prioritize breaking down bone—even to the point of sacrificing the structural integrity of the bone—just to be able to maintain blood calcium levels. For example, a female may experience decreased blood calcium levels during pregnancy due, in part, to increased demand from the developing fetus. Therefore, women are at risk for bone density loss due to blood calcium depletion during pregnancy.

EXERCISE 7.3

ENDOCHONDRAL BONE DEVELOPMENT

1. Obtain a slide of **developing long bone (figure 7.4)** and place it on the microscope stage. This slide will typically contain the developing femur of a young mammal. The femur develops using **endochondral ossification,** a process by which a hyaline cartilage model of the bone is gradually replaced by bone tissue.

2. Scan the slide at the lowest magnification and identify the parts of the bone—specifically, the epiphysis, diaphysis, and metaphysis. In the metaphysis of a developing long bone is an **epiphyseal (growth) plate**. Name the type of tissue found at this location: ____

(continued on next page)

(continued from previous page)

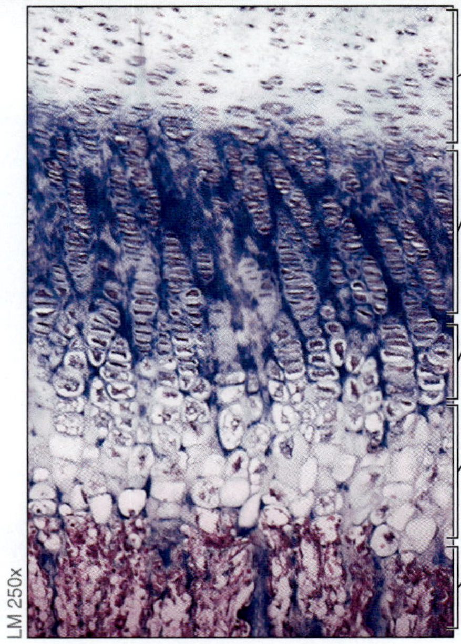

Zone 1: Resting cartilage	Consists of typical hyaline cartilage. Chondrocytes within lacunae are small and the matrix is very light in color.
Zone 2: Proliferating cartilage	Chondrocytes are lined up in rows. They are actively undergoing cell division (mitosis), and the matrix is very dark in color.
Zone 3: Hypertrophic cartilage	Chondrocytes hypertrophy (increase in size). Cell division ceases and the chondrocytes mature.
Zone 4: Calcified cartilage	Chondrocytes begin to calcify the matrix, and then die.
Zone 5: Ossification	Osteoprogenitor cells and blood vessels enter the spaces left behind by the degenerated cartilage and new bone is deposited.

LM 250x

Figure 7.4 Epiphyseal Plate. The five functional layers of the epiphyseal plate. In this photomicrograph, the epiphyseal side of the plate is located at the top of the photo, whereas the diaphyseal side of the plate is located at the bottom of the photo.

3. Position the metaphysis at the center of the field of view and increase the magnification. Identify the five functional layers within the epiphyseal plate, using figure 7.4 as a guide.

4. Sketch the epiphyseal growth plate as viewed through the microscope in the space provided. Label the following on the sketch, using figure 7.4 and the textbook as guides:
   - ☐ **zone of calcified cartilage**
   - ☐ **zone of hypertrophic cartilage**
   - ☐ **zone of ossification**
   - ☐ **zone of proliferating cartilage**
   - ☐ **zone of resting cartilage**

_____________ ×

INTEGRATE

CONCEPT CONNECTION

Many hormones (endocrine system, chapter 19) influence normal growth of bone tissue. Growth hormone, thyroid hormone, calcitonin, and sex steroid hormones (estrogen and testosterone) all help promote bone growth, whereas glucocorticoids (e.g., cortisol) may either inhibit bone growth or stimulate an increase in bone resorption by osteoclasts. Thus, disorders of the endocrine system may often manifest themselves as disorders of the skeletal system.

INTEGRATE

LEARNING STRATEGY

A developing long bone has five functional layers in the epiphyseal plate. As the bone develops, cartilage is gradually replaced by bone tissue. While the overall length of the bone increases, the width of the epiphyseal plate remains unchanged. Once the bone has grown to its maximum length, the growth of new cartilage ceases, allowing the epiphyseal plate to fully ossify, which leaves the epiphysis completely fused with the diaphysis of the bone.

GROSS ANATOMY

Classification of Bones

Bones of the human skeleton appear in various shapes and sizes, depending on their function(s). The four categories of bone classified by shape are flat, irregular, long, and short. The characteristics of this classification scheme are summarized in **table 7.2** and an example of each is shown in **figure 7.5.**

Table 7.2 Classification of Bones Based on Shape

Class of Bone	Description	Examples
Flat	Have thin, flat surfaces; may be slightly curved	Many skull bones (e.g., frontal, parietal)
Irregular	Complex shape that does not fit into other classifications	Vertebrae, some skull bones (e.g., ethmoid, sphenoid)
Long	Greater in length than width	Most limb bones (e.g., humerus, femur, metacarpals)
Short	Nearly equal in length and width	Wrist and ankle bones (e.g., carpals, tarsals)

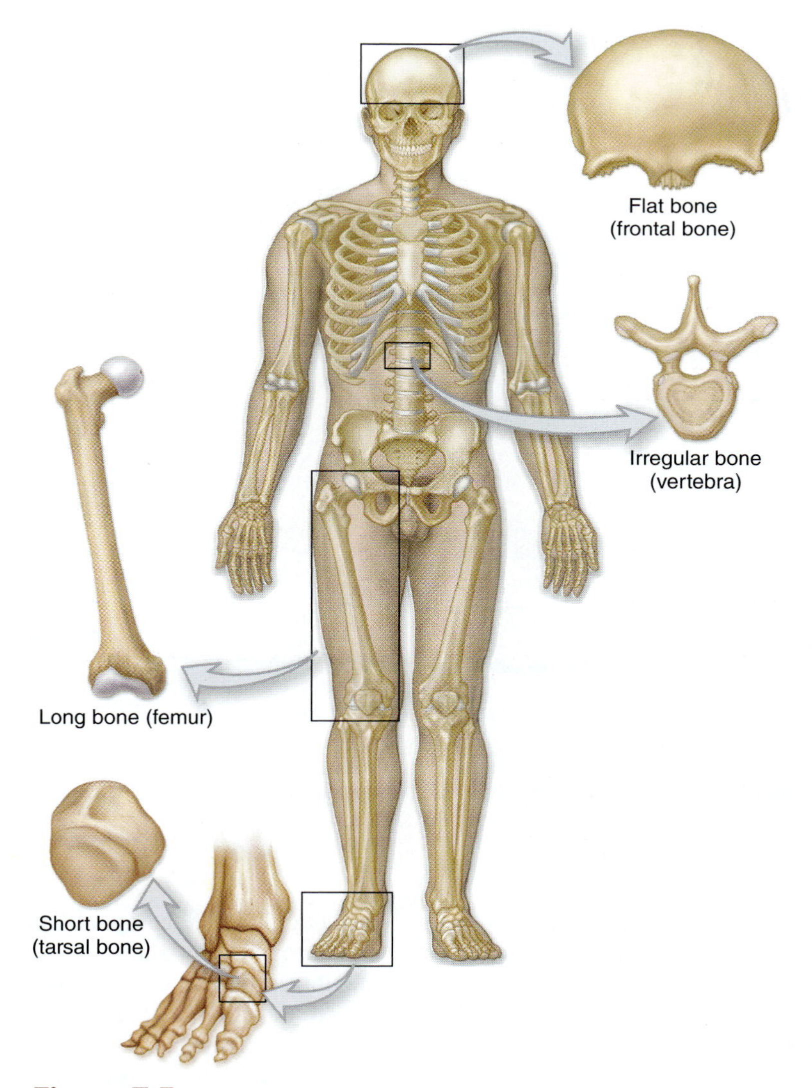

Figure 7.5 Structural Classifications of Bones.

EXERCISE 7.4

IDENTIFYING CLASSES OF BONES BASED ON SHAPE

1. Obtain a box containing disarticulated human bones. Pull each bone out of the box and classify the bone according to its shape. Note that there is a fair amount of variability in the shape of bones within each classification.

2. **Figure 7.6** contains photographs of several disarticulated human bones. In the space next to each bone, name the category to which each bone belongs: flat, irregular, long, and short.

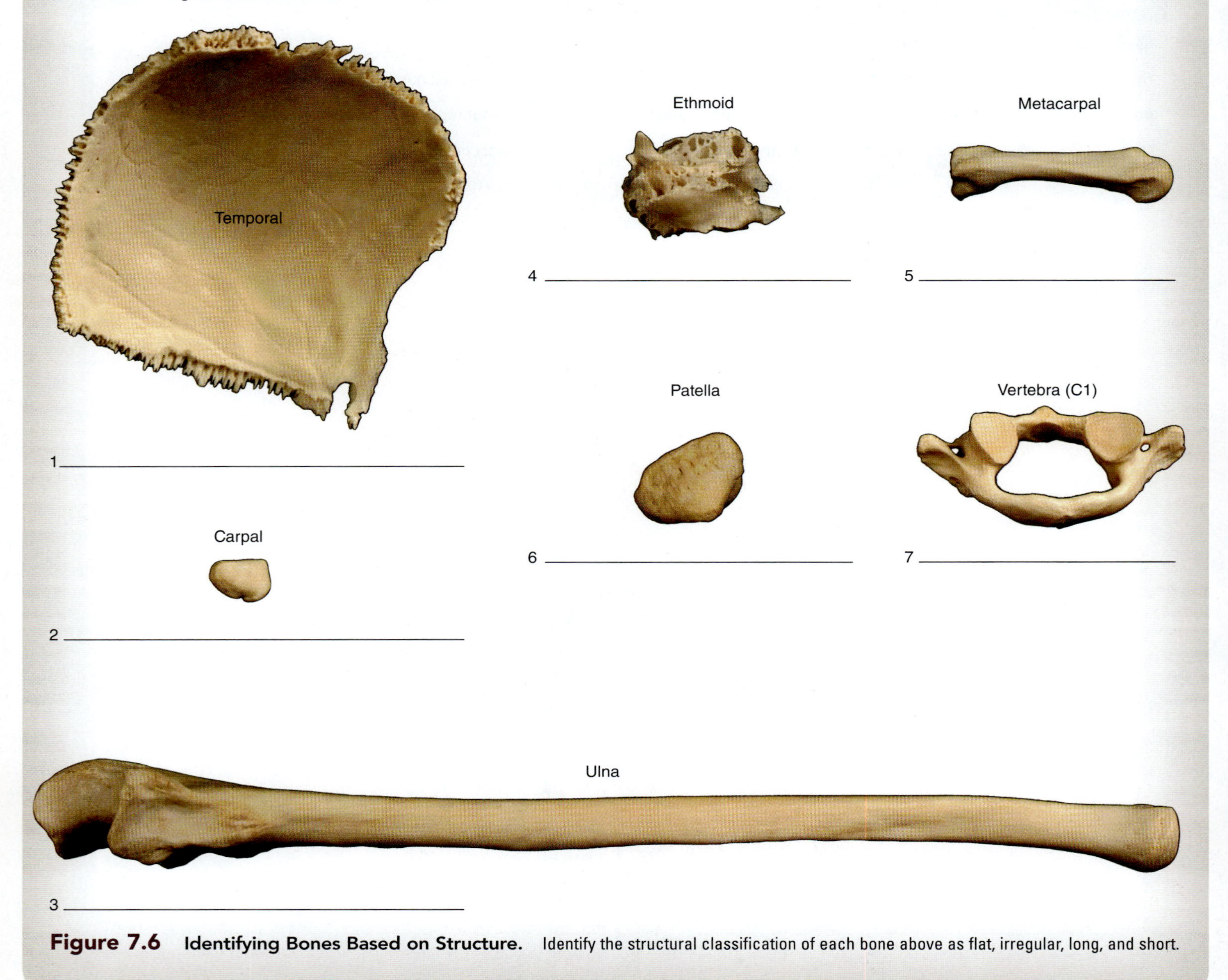

Figure 7.6 Identifying Bones Based on Structure. Identify the structural classification of each bone above as flat, irregular, long, and short.

Structure of a Typical Long Bone

A typical long bone such as the femur **(figure 7.7)** is composed of a long shaft, called the **diaphysis** (*dia,* through, + *physis,* growth); rounded ends, called **epiphyses** (*epi,* upon, + *physis,* growth); and articulation points between the two, called **metaphyses** (*meta,* between, + *physis,* growth). Within the shaft is a large cavity called the **medullary cavity,** which is filled with **yellow bone marrow** (adipose tissue) in the adult. The walls of the diaphysis are composed of a thick layer of **compact bone** tissue. The epiphyses of the bone are surrounded by a thin layer of compact bone and have **articular cartilages** on the ends. **Spongy bone** tissue is found within the epiphyses of the bone. In the fetus, the marrow spaces between the trabeculae of spongy bone are composed of red bone marrow. However, in the adult they are composed mainly of yellow bone marrow because of the conversion of red marrow to yellow marrow that occurs as the skeleton matures. In the adult, red bone marrow is primarily limited to the proximal epiphyses of the humerus and femur, the sternum, and the iliac crest. Observing a fresh bone specimen allows observation of many of the tissues that normally associate with bones, such as periosteum, articular cartilages, muscles, tendons, ligaments, marrow, and blood vessels. The following exercises involve observing a fresh specimen of a long bone from a cow, and an articulated human skeleton. The goal of these exercises is to familiarize students with the major bones of the human body.

EXERCISE 7.5

COMPONENTS OF A LONG BONE

1. Obtain both an intact femur (not cut) and a femur that has been cut along its longitudinal axis.

2. Identify the following components of a long bone, using figure 7.7 as a guide:

- ☐ **compact bone**
- ☐ **diaphysis**
- ☐ **epiphyseal line**
- ☐ **epiphysis**
- ☐ **medullary cavity**
- ☐ **spongy bone**

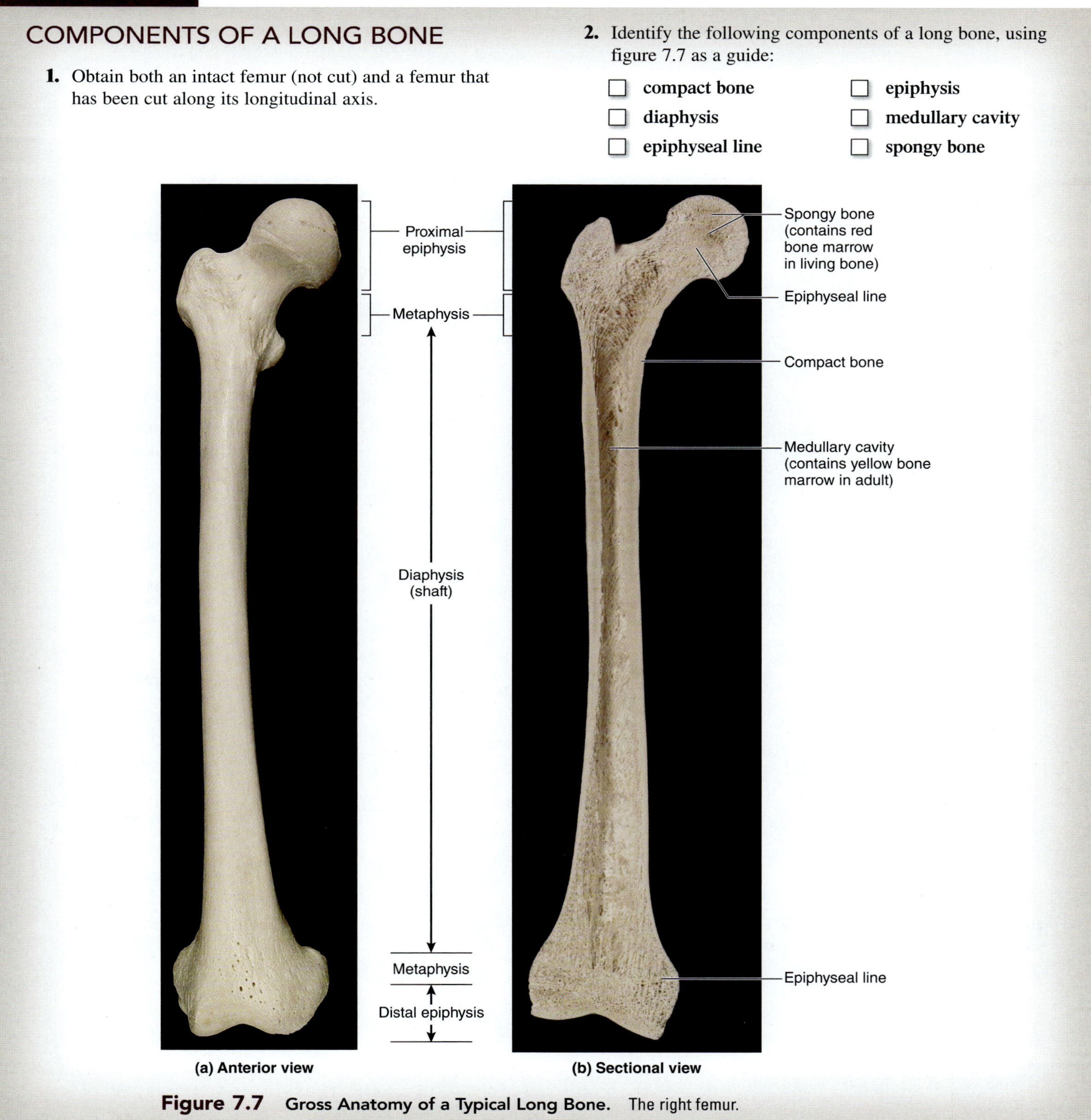

Figure 7.7 **Gross Anatomy of a Typical Long Bone.** The right femur.

EXERCISE 7.6

COW BONE DISSECTION

1. Obtain a dissecting pan, a blunt probe, forceps, and a fresh cow bone cut in a longitudinal section. Place the bone in the dissecting pan and begin your observations. Find the large medullary cavity that is filled with yellow bone marrow **(figure 7.8).** Notice the thick layer of compact bone that surrounds the medullary cavity.

2. If using a dissecting microscope or magnifying glass, focus in on the compact bone tissue. Notice how "bloody" it appears. Are there any tiny dots of blood within the compact bone? These result from the rupture of tiny blood vessels in the tissue when the bone was sectioned. This observation should reaffirm that living bone is a highly vascular, metabolically active tissue—quite different from the appearance of preserved bones.

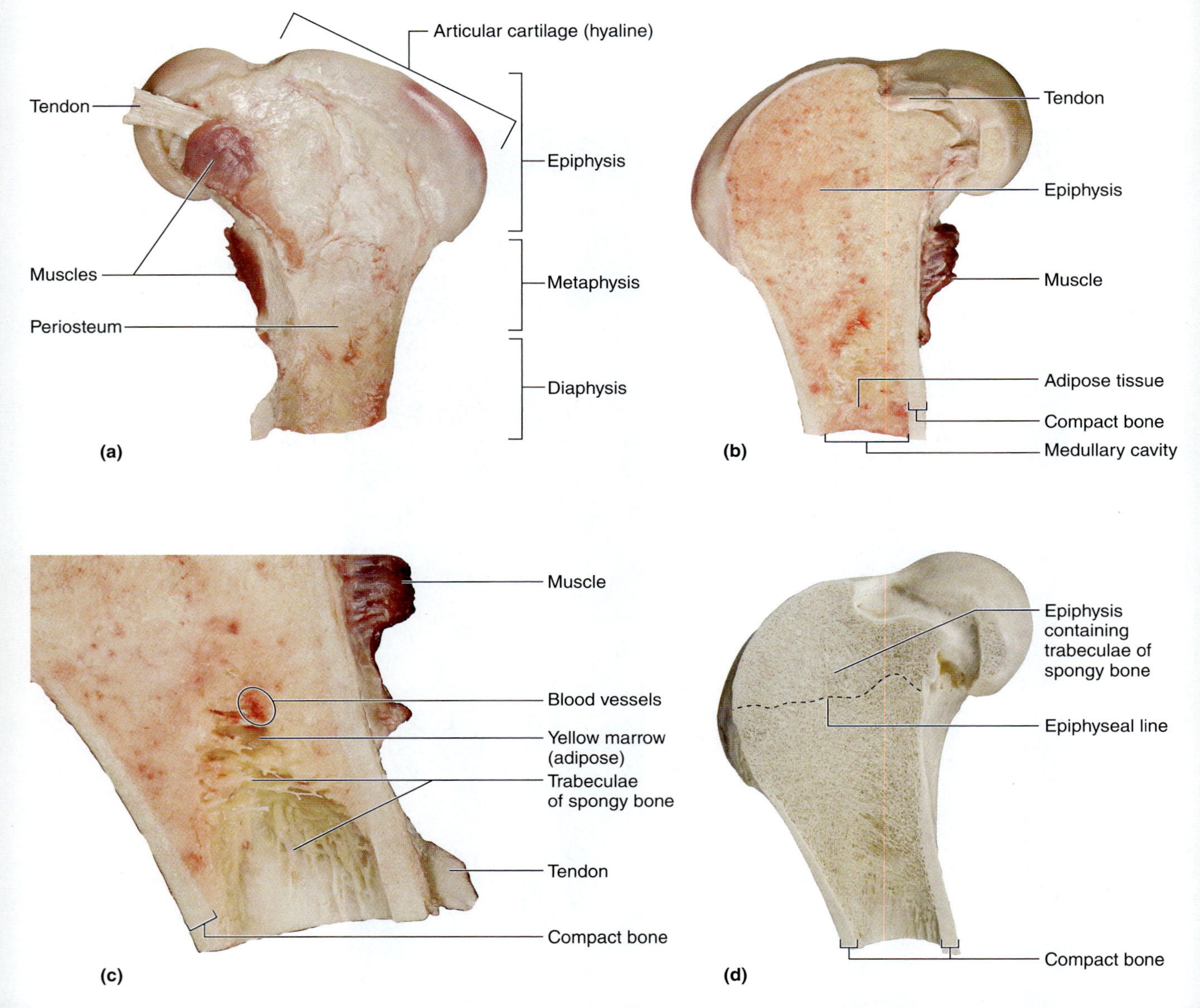

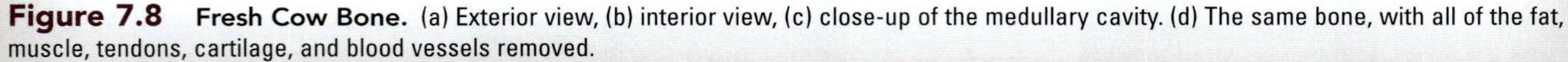

Figure 7.8 Fresh Cow Bone. (a) Exterior view, (b) interior view, (c) close-up of the medullary cavity. (d) The same bone, with all of the fat, muscle, tendons, cartilage, and blood vessels removed.

3. Using the blunt probe, carefully begin to clean out the adipose tissue from the medullary cavity (figure 7.8*b*). Gently probe for medium- to large-size blood vessels embedded within the adipose tissue. If there is a blood vessel that travels from the outer surface of the diaphysis into the medullary cavity, this is most likely the **nutrient artery,** an artery that grows into the diaphysis of the bone during the initial stages of ossification of the bone.
4. Continue to clean out the inside of the diaphysis of the bone and progress toward the inside of the epiphysis. Locate the small, hard "strings" of bony tissue. These are the **trabeculae** (*trabs,* beam) of spongy bone, which are located within the epiphysis and lining the inside of the diaphysis. These trabeculae will make it difficult to clean out the adipose tissue in the epiphysis, but spend some time poking around to get a better understanding of the arrangement of the trabeculae within. Some parts of the epiphysis may appear much more "bloody" than the rest of the inside of the bone. These areas consist of **red bone marrow**—a **hemopoietic** (*hemo-*, blood, + *poiesis,* a making) tissue that produces blood cells. Because these are adult cow bones, there should be very little red marrow; most of the spaces will be filled with yellow marrow.
5. Observe the structures on the outside of the bone. Using forceps, pick away at the dense connective tissue on the outside of the diaphysis (figure 7.8*a*). This is the **periosteum,** or outer covering of the bone. The periosteum acts as an attachment point for tendons and ligaments and as an anchoring point for blood vessels and nerves that enter the bone.
6. Observe the outside of the epiphysis and look for the cut portion of a tendon or ligament where it attaches to the periosteum (figure 7.8*a,c*). Tendons and ligaments consist of a regular arrangement of shiny, white fibers (i.e., dense regular connective tissue). These are the tough collagen fibers that give the tendon or ligament its great tensile strength.
7. Observe the shiny, white cartilage on the ends of the bone. This is the **articular cartilage,** which is composed of hyaline cartilage. Notice that the periosteum ends where the articular cartilage begins.
8. If observing the tibia of a cow, try to identify the C-shaped pads of fibrocartilage located on top of the articular cartilages. These are the **menisci** (s. *meniscus*) of the knee joint **(figure 7.9).** If no meniscus is present, find another group of students in the laboratory whose cow bone has a meniscus, to observe its gross structure. Within the meniscus there are more of the shiny, white collagen fibers observed in the tendons and ligaments. Recall from chapter 5 that fibrocartilage contains thick bundles of collagen fibers in its extracellular matrix. Note that the meniscus is somewhat "tied in to" the connective tissues surrounding the joint.

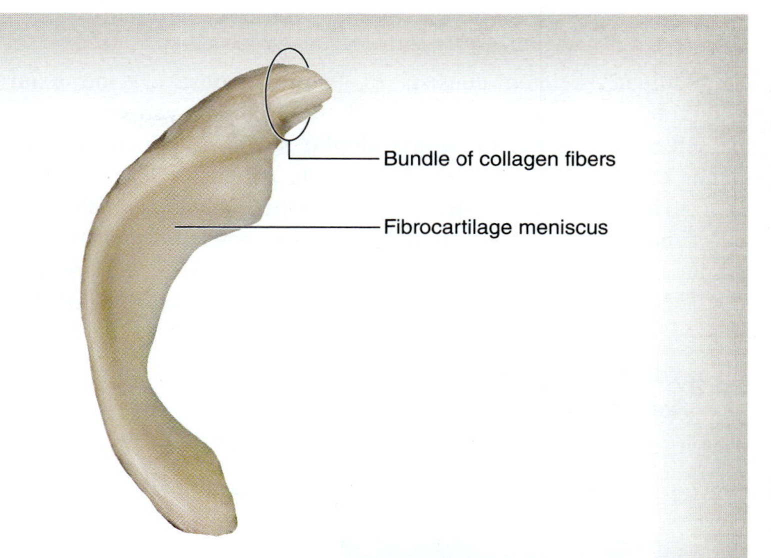

Figure 7.9 **Meniscus.** One of the C-shaped meniscal cartilage pads removed from the end of the tibia of a fresh cow bone.

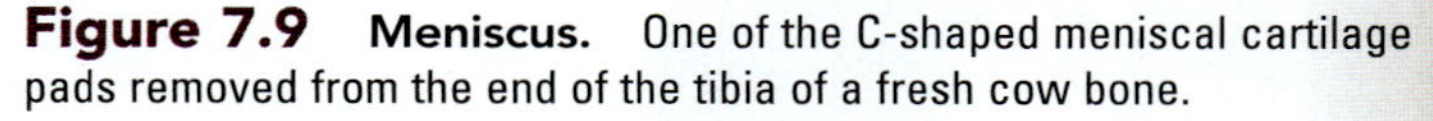

9. Sketch the dissected fresh cow bone in the space provided. Label the following on the sketch, using figure 7.8 as a guide:

- ☐ **articular cartilage**
- ☐ **compact bone**
- ☐ **medullary cavity**
- ☐ **periosteum**
- ☐ **spongy bone**
- ☐ **tendon/ligament**

10. When the dissection is complete, dispose of the cow bones according to the laboratory instructor's directions, and clean all dissection instruments and the workspace.

Survey of the Human Skeleton

The human skeleton consists of two divisions: (1) the **axial skeleton,** which includes bones of the cranium, vertebrae, ribs, and sternum, and (2) the **appendicular skeleton,** which includes bones of the pectoral girdle, upper limb, pelvic girdle, and lower limb. The **pectoral girdle** consists of the scapula and clavicle. These bones provide support and attachment points for muscles that connect the limbs to the axial skeleton. The **pelvic girdle** consists of the bones composing the **os coxae** (*os,* bone, + *coxa,* hip): the ilium, ischium, and pubis. These bones protect contents of the pelvic cavity and provide support and attachment points for muscles that connect the lower limb to the axial skeleton. When learning the names of all the bones in the human body and their features, it is best to break the larger task into many small tasks that can be tackled individually in short segments of time. An excellent "first task" is to learn the names of the bones that compose the skeleton. The exercises in chapters 8 and 9 will move beyond this and require learning the names of specific features unique to each bone of the skeleton.

EXERCISE 7.7

THE HUMAN SKELETON

1. Observe an articulated human skeleton.
2. Identify the bones listed in **figure 7.10** on the articulated skeleton, using the textbook as a guide. Then label them in figure 7.10.

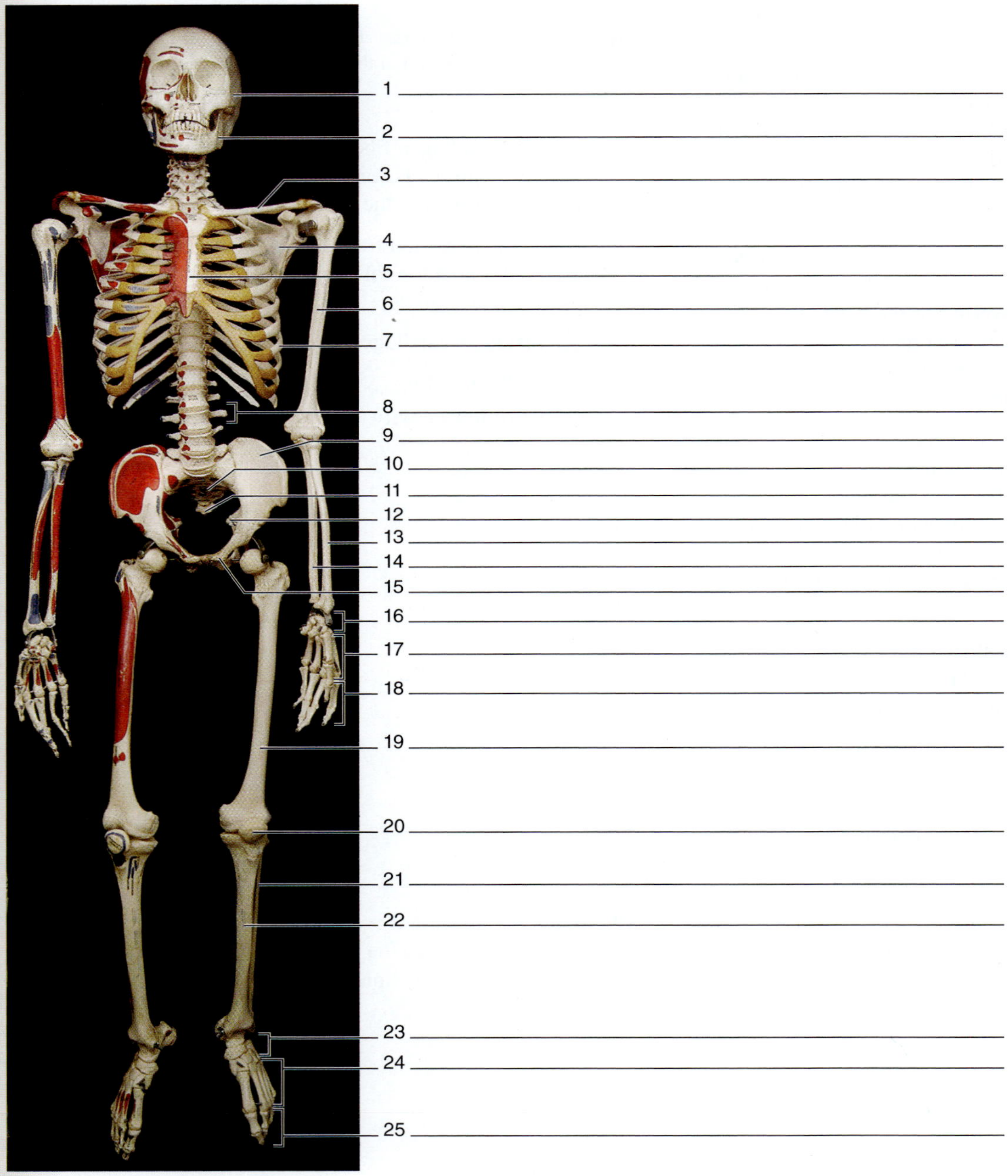

1 ____
2 ____
3 ____
4 ____
5 ____
6 ____
7 ____
8 ____
9 ____
10 ____
11 ____
12 ____
13 ____
14 ____
15 ____
16 ____
17 ____
18 ____
19 ____
20 ____
21 ____
22 ____
23 ____
24 ____
25 ____

- ☐ carpals
- ☐ clavicle
- ☐ coccyx
- ☐ femur
- ☐ fibula
- ☐ humerus
- ☐ ilium
- ☐ ischium
- ☐ mandible
- ☐ metacarpals
- ☐ metatarsals
- ☐ patella
- ☐ phalanges (of the foot)
- ☐ phalanges (of the hand)
- ☐ pubis
- ☐ radius
- ☐ rib
- ☐ sacrum
- ☐ scapula
- ☐ skull
- ☐ sternum
- ☐ tarsals
- ☐ tibia
- ☐ ulna
- ☐ vertebra

(a) Anterior view

Figure 7.10 **The Human Skeleton.** Use the terms listed to fill in the numbered labels in the figure.

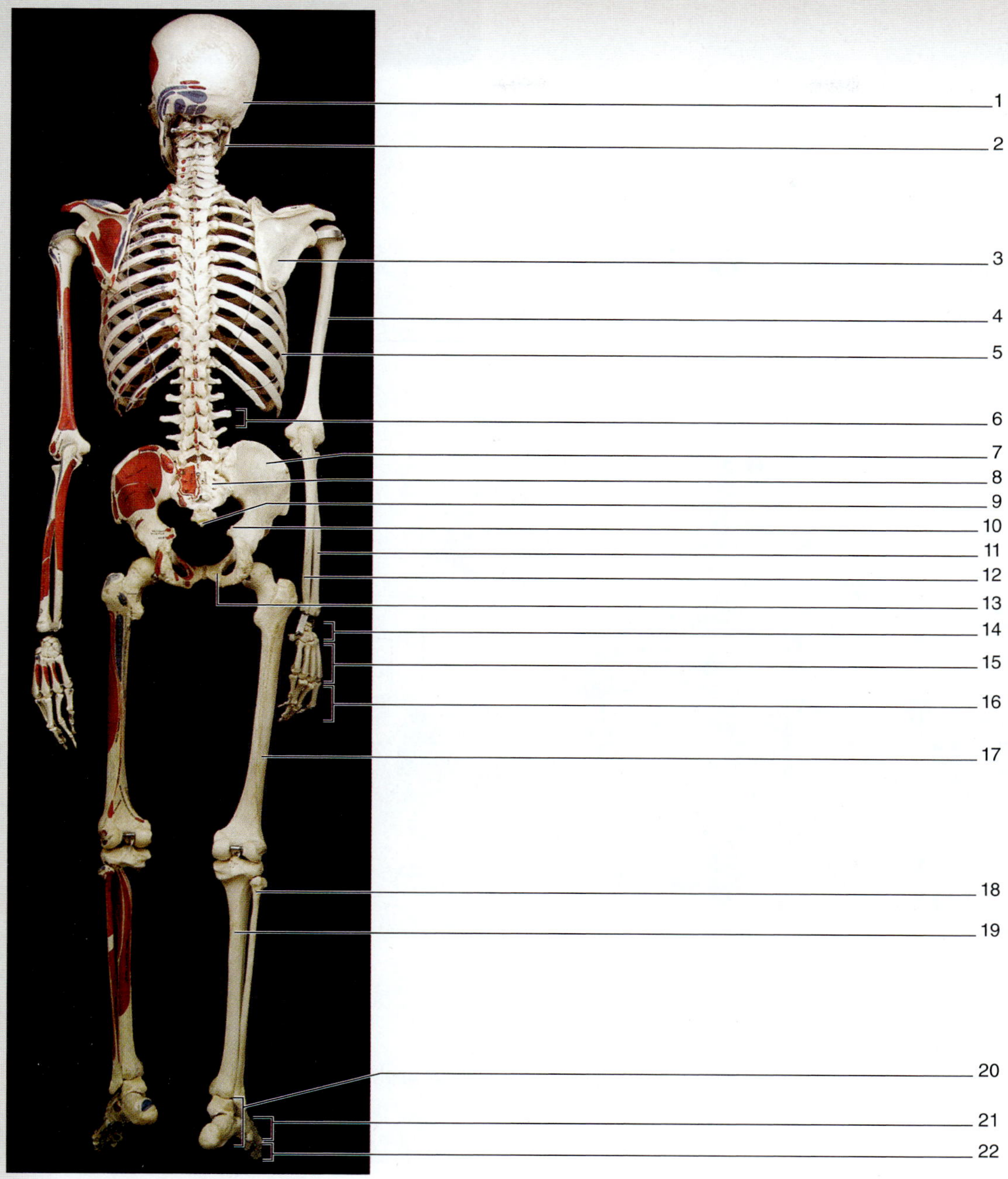

1
2
3
4
5
6
7
8
9
10
11
12
13
14
15
16
17
18
19
20
21
22

- ☐ carpals
- ☐ coccyx
- ☐ femur
- ☐ fibula
- ☐ humerus
- ☐ ilium
- ☐ ischium
- ☐ mandible
- ☐ metacarpals
- ☐ metatarsals
- ☐ phalanges (of the hand)
- ☐ phalanges (of the foot)
- ☐ pubis
- ☐ radius
- ☐ rib
- ☐ sacrum
- ☐ scapula
- ☐ skull
- ☐ tarsals
- ☐ tibia
- ☐ ulna
- ☐ vertebra

(b) Posterior view

Figure 7.10 The Human Skeleton *(continued).* Use the terms listed to fill in the numbered labels in the figure.

INTEGRATE

CLINICAL VIEW

Bones and Mechanical Stress

Bone remodels itself in response to applied mechanical stresses that are placed upon it. In this instance, "stress" refers to force per unit area of tissue. For example, weight-bearing exercises "load" bone, thereby increasing stress on the tissue. The increased force directly impacts osteoblast activity. That is, there is an increase in the deposition of bone matrix and calcification with increasing force placed on the bone. Bones are loaded on a daily basis as an individual maintains an upright posture because of the constant downward pull of gravity. However, the bone remodeling process is disrupted when gravity is greatly reduced. Consider, for example, quadriplegics, who are wheelchair-bound. These patients experience decreased mechanical loading on their bones due to decreased mobility and decreased muscle mass. Research suggests that the reduction in mechanical loading inhibits osteoblast activity and stimulates osteoclast activity in these patients' bones. The net result is an increase in bone resorption without bone matrix deposition. To counteract these changes, patients with quadriplegia undergo gait training, sometimes with the assistance of direct muscle stimulation, to simulate the mechanical loading they normally would experience by standing upright and walking. Unfortunately, such exercise has not been able to completely reverse the reduction in new bone formation that these patients experience.

Chapter 7: The Skeletal System: Bone Structure and Function

Name: ______________________

Date: ____________ **Section:** ____________

POST-LABORATORY WORKSHEET

The ❶ corresponds to the Learning Objective(s) listed in the chapter opener outline.

Do You Know the Basics?

Exercise 7.1: Compact Bone

1. Label the components of compact bone on the following diagram. ❶ ❷

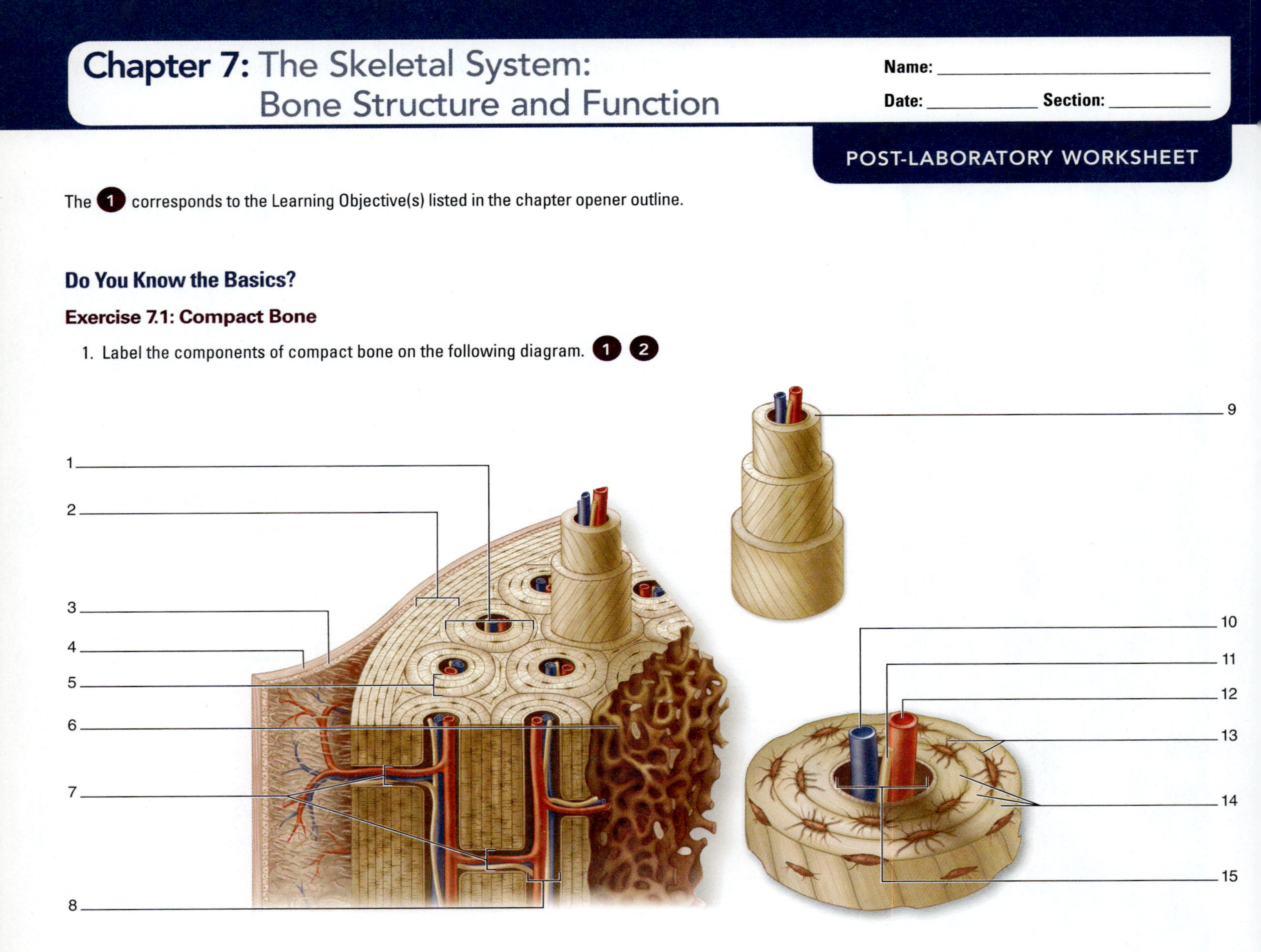

Exercise 7.2: Spongy Bone

2. Which of the following is/are visible on a slide of decalcified spongy (cancellous) bone? (Check all that apply.) ❸

____ a. bone marrow cavity

____ b. central canal

____ c. osteocyte

____ d. osteon

____ e. perforating canal

____ f. trabecula

Exercise 7.3: Endochondral Bone Development

3. Which of the following bones undergo endochondral bone development (ossification)? (Check all that apply.) ❹

____ a. femur

____ b. frontal bone

____ c. radius

____ d. temporal bone

____ e. tibia

4. The figure below is a light micrograph of developing endochondral bone. The dotted line divides the epiphyseal plate into two major regions, a and b. One region is **new bone tissue** and the other is **hyaline cartilage.** Label these two components of the developing bone on the following diagram. ❺

   a. ______________________

   b. ______________________

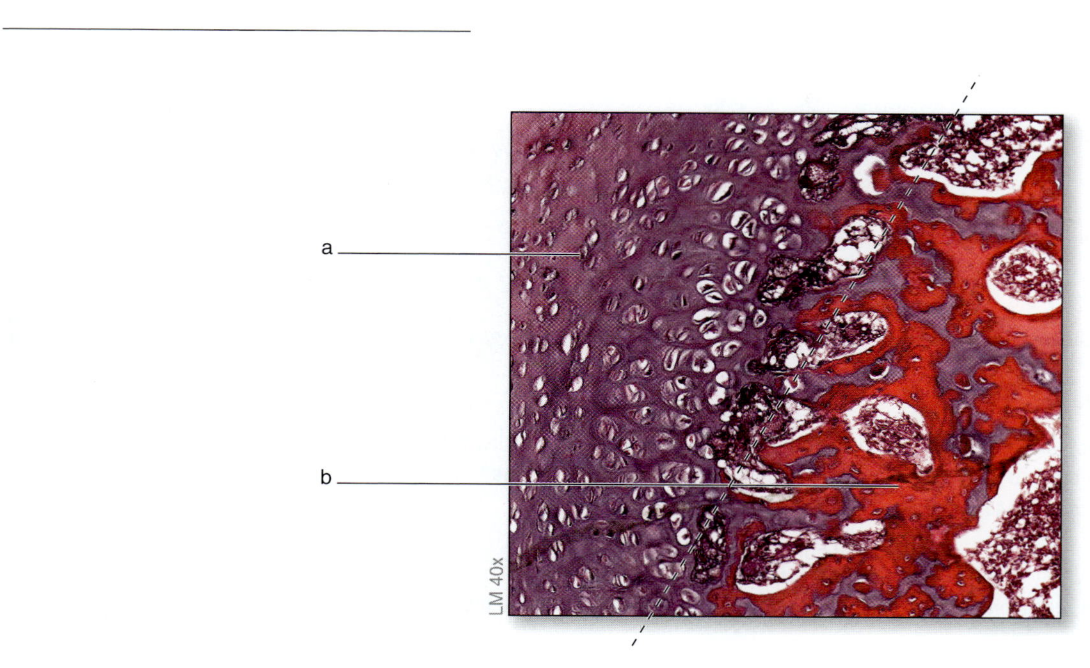

Exercise 7.4: Identifying Classes of Bones Based on Shape

5. Which of the following is the correct classification for a vertebra based on shape? (Circle one.) ❻

   a. flat b. irregular c. long d. short

Exercise 7.5: Components of a Long Bone

6. Label the parts of a typical long bone on the following diagram. ❼

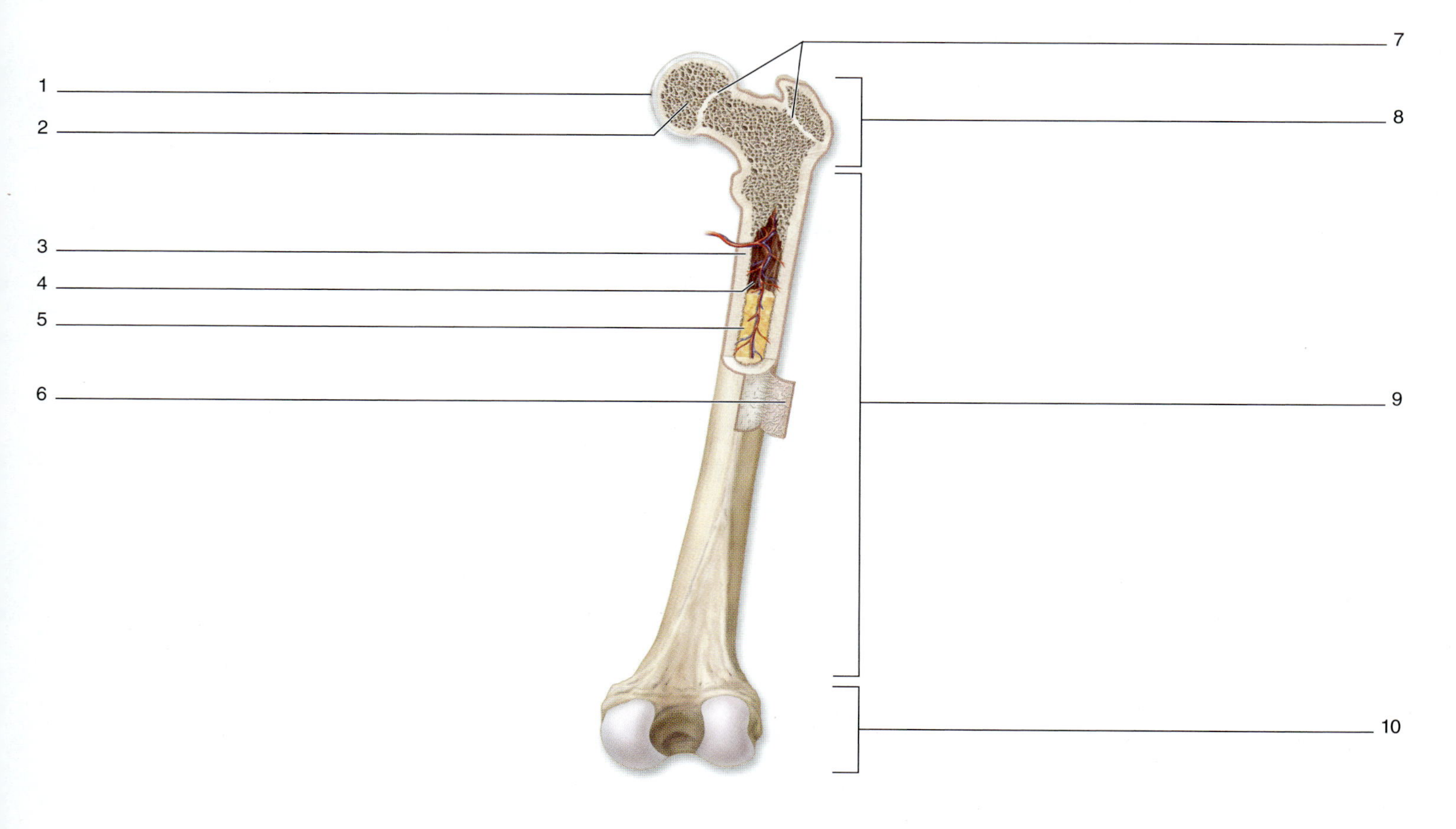

Exercise 7.6: Cow Bone Dissection

7. Which of the following is the outer covering of bone that serves as an attachment point for tendons and ligaments? (Circle one.) 8

   a. articular cartilage   b. medullary cavity   c. periosteum   d. trabeculae

8. The thick layer of bone surrounding the medullary cavity is composed of ____________________ (compact/spongy) bone tissue. 8

Exercise 7.6: The Human Skeleton

9. Which of the following bone(s) is/are part of the axial skeleton? (Check all that apply.) 9

   ______ a. pubis   ______ b. rib   ______ c. scapula   ______ d. skull   ______ e. sternum

10. Which of the following bones compose the os coxae? (Check all that apply.) 9

    ______ a. coccyx   ______ b. ilium   ______ c. ischium   ______ d. pubis   ______ e. sacrum

11. The pectoral girdle consists of the scapula and clavicle, and the pelvic girdle consists of the os coxae. ____________________ (True/False) 9

Can You Apply What You've Learned?

12. You are a biomedical researcher who is interested in designing a drug to increase bone density. You have decided you can take one of two approaches: (1) design a drug that stimulates bone to be built faster, or (2) design a drug that prevents bone from breaking down. To begin this process, you need to study the cells that would likely respond to either drug 1 or drug 2. What cells are these?

    a. ________________________________ (build bone)   b. ________________________________ (break down bone)

13. What process must stop in order for the epiphyseal plate to close? (That is, which layer of the plate must stop its development first?)

    ____

    ____

14. A forensic anthropologist identifies a bone as a left femur with epiphyses that are detached from the diaphysis. Does this bone belong to a juvenile or an adult? Justify your answer.

    ____

    ____

Can You Synthesize What You've Learned?

15. One effect of growth hormone (GH) is the stimulation of cartilage growth. Discuss how decreased GH production during development might impact a person's height.

    ____

    ____

    ____

16. Why might astronauts, who spend only a few days in outer space under microgravity (reduced gravity), experience a significant loss in bone density?

    ____

    ____

    ____

17. Vitamin D promotes increased dietary calcium and phosphorus absorption in the small intestine. Describe why doctors may recommend vitamin D supplements to patients diagnosed with osteoporosis (weak/brittle bones).

    ____

    ____

    ____

The Skeletal System: Axial Skeleton

OUTLINE AND LEARNING OBJECTIVES

Anatomy & Physiology REVEALED® aprevealed.com **MODULE 5: SKELETAL SYSTEM**

INTRODUCTION

The skeletal system, which is typically composed of 206 bones in an adult, is organized into two divisions: the axial skeleton and the appendicular skeleton. The term **axial** is a derivative of the term *axis*. The axial skeleton consists of bones that form the main axis of the body—the skull, vertebrae, ribs, and sternum, which collectively form the body's core structural foundation. These bones are also critical for protection of the body's most vital organs, such as the brain, heart, and lungs.

While completing the exercises in this chapter, be sure to view both individual disarticulated bones (e.g., the humerus) and the same bones on an articulated skeleton. This will increase understanding of how individual bones fit in with the rest of the axial skeleton (e.g., how the humerus articulates with the scapula, radius, and ulna).

The exercises in this chapter involve observing the bones that compose the axial skeleton on articulated human skeletons, on disarticulated bones of the human skeleton, or on bone models. The goal is to identify the major landmarks and identifying features of each bone, and associate the observed structures with associated functions.

List of Reference Tables

Chapter 8: The Skeletal System: Axial Skeleton

Name: ____________________

Date: __________ **Section:** __________

These Pre-Laboratory Worksheet questions may be assigned by instructors through their connect course.

PRE-LABORATORY WORKSHEET

1. Match the description of the bone feature listed in column A with the appropriate name listed in column B.

Column A	*Column B*
____ 1. smooth, grooved, pulley-like process	a. condyle
____ 2. small, flat, shallow, articulating surface	b. facet
____ 3. large, smooth, and round projection	c. head
____ 4. prominent, rounded epiphysis of a bone	d. trochlea

2. Which of the following is/are bony features that function as attachment points for tendons and ligaments? (Check all that apply.)
   ____ a. fossa
   ____ b. process
   ____ c. sinus
   ____ d. tuberosity

3. A foramen is a round hole that passes through a bone, whereas a fissure is a narrow, slit-like opening that passes through a bone. ____________________ (True/False)

4. In a typical human, how many vertebrae (individual or fused) does each section of the vertebral column contain? Write the number in the space provided.
   ____ a. cervical
   ____ b. thoracic
   ____ c. lumbar
   ____ d. sacral
   ____ e. coccygeal

5. The sella turcica is a feature of which bone? (Circle one.)
   a. ethmoid bone
   b. frontal bone
   c. occipital bone
   d. sphenoid bone

6. All ribs articulate with which of the following vertebrae? (Circle one.)
   a. cervical
   b. coccygeal
   c. lumbar
   d. sacral
   e. thoracic

7. Arrange the parts of the sternum from superior (1) to the inferior (3) by placing the numbers 1–3 next to the parts.
   ____ a. body
   ____ b. manubrium
   ____ c. xiphoid

8. Which is the only skull bone that is mobile (movable)? ____________________

9. Which of the following bone(s) contain(s) a paranasal sinus? (Check all that apply.)
   ____ a. ethmoid bone
   ____ b. frontal bone
   ____ c. mandible
   ____ d. maxilla
   ____ e. temporal bone

10. The suture that forms between the frontal and parietal bones is called the coronal suture. ____________________ (True/False)

GROSS ANATOMY

Bone Markings

Prior to learning the specific bones of the axial and appendicular skeleton, it is important to become familiar with general terminology that describes the features of each of the bones. Recall from chapter 7 that all bones undergo a process of ossification, whereby osteoblasts lay down new bony matrix (osteoid) during bone development. At the same time, muscles, tendons, and ligaments attach to these bones as they begin to develop, and blood vessels and nerves pass through or between bones as the vessels grow into their target organs. Continual movement during development, particularly when bones are soft and pliable, leads to the formation of distinguishing features on each bone. Once a bone completely ossifies, the markings become solid, recognizable features of the bone. In general, smooth surfaces are found on articulating surfaces (surfaces that form joints), projections represent points of muscle, ligament, and tendon attachment, and foramina are passageways for blood vessels and nerves. **Table 8.1** lists each of the major bone markings and provides a description and general definition for the marking. This information will assist in understanding what kind of structure to look for when identifying individual bones and their respective features in subsequent exercises.

INTEGRATE

LEARNING STRATEGY

When learning the processes, projections, foramina (holes), and other markings of the bones, view each structure, study it closely, and contemplate its function. The process may be an attachment point for a ligament, a tendon, or a muscle. The opening or hole may serve as a passageway for a nerve, artery, or vein. The smooth surface may be where the bone articulates with another bone. For each structure, relate form to function.

Table 8.1 Bone Markings

General Structure	Anatomic Term	Description	Word Origin
Articulating Surfaces	Condyle	Large, smooth, round articulating structure	*kondylos*, a knuckle
	Facet	Small, flat, shallow articulating surface	*facet,* a small face
	Head	Prominent, rounded epiphysis	NA
	Trochlea	Smooth, grooved, pulley-like articular process	*trochlea,* a pulley block
Depressions	Alveolus (pl., *alveoli*)	Deep pit or socket in the maxillae or mandible	*alveus,* a hollow cavity
	Fossa (pl., *fossae*)	Flattened or shallow depression	*fossa,* a pit/cavity
	Sulcus	Narrow groove	*sulcus,* a furrow/groove
Projections for Tendon and Ligament Attachment	Crest	Narrow, prominent, ridge-like projection	*crista,* a crest
	Epicondyle	Projection adjacent to a condyle	*epi,* above + *kondylos,* a knuckle
	Line	Low ridge	*linea,* line
	Process	Any marked bony prominence	*processus,* a projection
	Ramus (pl., *rami*)	Angular extension of a bone relative to the rest of the structure	*ramus,* a branch
	Spine	Pointed, slender process	*spinosus,* a spine
	Trochanter	Massive, rough projection found only on the femur	*trokhanter,* to run
	Tubercle	Small, round projection	*tuberculum,* a small bump
	Tuberosity	Large, rough projection	*tuberosus,* knobby
Openings and Spaces	Canal	Passageway through a bone	*canalis,* channel
	Fissure	Narrow, slit-like opening through a bone	*fissura,* cleft
	Foramen (pl., *foramina*)	Rounded passageway through a bone	*foramen,* a hole
	Meatus	Passageway through a bone	*meatus,* a channel
	Sinus	Cavity or hollow space in a bone	*sinus,* a curve or bay

INTEGRATE

LEARNING STRATEGY

Learning the names of bones and the bone markings is like learning a new language. In fact, most of the names are derived from Latin or Greek words. It is helpful to learn the meanings of each word as a way to remember each bone. For example, *femur* is the Latin word for thigh and mandible comes from the Latin word *mandere,* which means to chew. The names of the bones, then, often indicate location and/or function. The same is true for bone markings. *Foramen* comes from the Latin word *forare,* which means to bore. It should be no surprise, then, to discover that a foramen is a hole. For each bone and bone marking, the word origins listed in the tables in this chapter serve as a guide to learning their names.

The Skull

The bones that make up the skull are separated into two functional categories: **cranial bones** (frontal, parietal, temporal, occipital, sphenoid, and ethmoid) and **facial bones** (maxilla, mandible, zygomatic, nasal, lacrimal, palatine, inferior nasal conchae, and vomer). The roof of the cranium—the **calvaria,** or skullcap—is the dome-shaped part of the skull that protects the brain. In an adult, all of the skull bones, with the exception of the mandible, are fused to each other via synarthrotic (immovable) joints called **sutures.**

Table 8.2 describes each individual bone of the skull and lists the best view(s) for observing its features. An organized approach to learning the bones of the skull begins with viewing the skull from six points of reference: anterior view, lateral view, posterior view, superior view, inferior view, and superior view of the cranial floor. For each view of the skull, first identify the individual bones that are visible. Second, identify all processes, foramina, and major features (often formed from multiple bones) that are visible in each view of the skull. Always relate the bony processes, fossae, and foramina to the individual bone(s) from which they are formed.

Table 8.2 The Axial Skeleton: Skull Bones and Important Bony Landmarks

Major Bone	Bone Features	Description and Related Structures of Importance	Best View	Word Origins
Ethmoid	Cribriform plate	Forms roof of nasal cavity and part of cranial floor	Cranial floor	*cribrum, sieve* + *forma,* form + *platus,* flat
	Cribriform foramina	The olfactory nerve (CN I) passes through to the brain	Cranial floor	*cribrum,* sieve + *forma,* form + *foramen,* a hole
	Crista galli	Projection that serves as attachment point for falx cerebri	Cranial floor	*crista galli,* cockscomb
	Superior nasal concha	Forms superior lateral wall of nasal cavity; causes turbulent air flow	Anterior	*superus,* upper + *nasus,* nose + *concha,* shell
	Middle nasal concha	Forms middle lateral wall of nasal cavity; causes turbulent air flow	Anterior	*middle,* middle + *nasus,* nose + *concha,* shell
	Perpendicular plate	Forms superior part of nasal septum	Anterior	*perpendiculum,* plumb line + *platus,* flat
Frontal	Frontal sinus	A cavity within frontal bone	Cranial floor	*frontellum,* forehead + *sinus,* a hollow
	Supraorbital foramen (notch)	A hole or notch on the superior ridge of orbit	Anterior	*supra-,* above + *orbit,* eye socket + *foramen,* a hole
	Superciliary arch	Process that forms brow ridges; more pronounced in males than in females	Anterior	*superus,* upper + *cilium,* the eyelid + *arcus,* a bow
Inferior Nasal Conchae	NA	Forms inferior part of lateral wall of nasal cavity; causes turbulent air flow	Anterior	*inferus,* lower + *nasus,* nose + *concha,* shell
Lacrimal	Lacrimal groove	Forms medial, inferior aspect of orbit of eye. Groove connects orbital and nasal cavities.	Lateral	*lacrima,* tear + *groove,* a pit

(continued on next page)

Table 8.2 The Axial Skeleton: Skull Bones and Important Bony Landmarks *(continued)*

Major Bone	Bone Features	Description and Related Structures of Importance	Best View	Word Origins
Mandible	Alveolar processes	Cavities that form tooth "sockets"	Lateral	*alveolus,* a trough + *processus,* a projection
	Angle	Portion of mandible connecting the body to the ramus, forming a right angle	Lateral	*angulus,* a corner
	Body	Anterolateral portion of mandible	Lateral	NA
	Coronoid process	Insertion point for temporalis muscle	Lateral	*corona,* crown + *eidos,* resembling + *processus,* a projection
	Head	Forms a joint with the mandibular fossa of temporal bone (temporomandibular joint)	Lateral	NA
	Mandibular foramen	Passageway for mandibular branch of the trigeminal nerve (CN V_3)	Lateral	*mandere,* to chew + *bula,* a means + *foramen,* a hole
	Mental foramen	Passageway for mental artery and nerve (CN V_3)	Anterior	*mental,* chin + *foramen,* a hole
	Mental protuberance	Anterior projection of mandible that forms the chin	Lateral	*mental,* chin + *protuberare,* to swell
	Ramus	Part of bone that forms an angle with body of mandible	Lateral	*ramus,* branch
Maxilla	Infraorbital foramen	Passageway for infraorbital artery and nerve	Lateral	*infra-,* below *orbit,* eye socket + *foramen,* a hole
	Incisive foramen (fossa)	Contains arteries and nerves passing from nasal cavity into oral cavity	Inferior	*incidere,* to cut into + *foramen,* a hole
	Palatine process	Forms anterior floor and part of lateral wall of nasal cavity	Inferior	*palatin,* the palate + *processus,* a projection
Nasal	NA	Forms most of the bridge of nose	Frontal	
Occipital	External occipital protuberance	Large projection palpated on the posterior aspect of the head; muscle attachment point	Posterior	*externus,* outside + *occipital,* occipital bone + *protuberare,* to swell
	Foramen magnum	Large hole for passage of spinal cord	Cranial floor	*foramen,* a hole + *magnus, great*
	Hypoglossal canal	Passageway for hypoglossal nerve (CN XII)	Cranial floor	*hypo,* under + *glossus,* tongue + *canalis,* channel
	Jugular foramen	Passageway for internal jugular vein and several nerves (CN IX, X, and XI)	Cranial floor	*jugal,* throat + *foramen,* a hole
	Occipital condyle	Smooth surface for articulation with atlas (first cervical vertebra)	Inferior	*occipital,* occipital bone + *kondylos,* knuckle
Palatine	NA	Forms posterior floor of nasal cavity; part of orbit and hard palate	Inferior	*palatin,* the palate
Parietal	NA	L-shaped bone that forms the lateral, superior wall of cranial cavity	Lateral	*paries,* wall
Sphenoid	Foramen ovale	Passageway for mandibular branch of the trigeminal nerve (CN V_3)	Cranial floor	*foramen,* a hole + *ovalis,* oval
	Foramen rotundum	Passageway for maxillary branch of the trigeminal nerve (CN V_2)	Cranial floor	*foramen,* a hole + *rotundum,* round

Table 8.2 The Axial Skeleton: Skull Bones and Important Bony Landmarks *(continued)*

Major Bone	Bone Features	Description and Related Structures of Importance	Best View	Word Origins
Sphenoid (continued)	Foramen spinosum	Passageway for middle meningeal artery and vein and a branch of the trigeminal nerve (CN V)	Cranial floor	*foramen,* a hole + *spinosus,* spine-like
	Greater wing	Forms parts of posterior orbit and middle cranial fossa	Cranial floor	NA
	Inferior orbital fissure	Passageway for maxillary branch of the trigeminal nerve (CN V_2) and infraorbital artery and vein	Cranial floor	*inferus,* lower + *orbit,* eye socket + *fissura,* cleft
	Lesser wing	Forms part of anterior cranial fossa	Cranial floor	NA
	Optic foramen	Passageway for optic nerve (CN II)	Cranial floor	*optikos,* eye + *foramen,* a hole
	Sella turcica	"Turkish saddle"-shaped depression housing the pituitary gland	Cranial floor	*sella,* saddle + *turcica,* Turkish
	Superior orbital fissure	Passageway for oculomotor (CN III), trochlear (CN IV), trigeminal (CN V_1), and abducens (CN VI) nerves	Cranial floor	*superus,* upper + *orbit,* eye socket + *fissura,* cleft
Temporal	Carotid canal	Passageway for internal carotid artery and associated nerves	Inferior	*karotides,* arteries of the neck + *canalis,* channel
	External acoustic (auditory) meatus	Opening into external auditory canal	Lateral	*externus,* outside + *auditorius,* related to hearing + *meatus,* channel
	Foramen lacerum	Largely covered by cartilage in living human; no structures pass entirely through it	Cranial floor	*foramen,* a hole + *lacer,* mangled
	Internal acoustic (auditory) meatus	Passageway for facial (CN VII) and vestibulocochlear nerves (CN VIII)	Cranial floor	*internus,* inside + *auditorius,* related to hearing + *meatus,* channel
	Mandibular fossa	Point of articulation with head of mandible, forming the temporomandibular joint	Lateral	*mandere,* to chew + *bula,* a means + *fossa,* a pit/cavity
	Mastoid process	Attachment point for muscles of neck	Lateral	*mastos,* breast + *oideos,* resembling + *processus,* a projection
	Petrous part	Houses structures for hearing and equilibrium; separates middle and posterior cranial cavities	Superior	*petrosus,* like a rock
	Squamous part	Forms inferior, posterior part of temporal fossa	Lateral	*squamosus,* scale-like
	Styloid process	Serves as attachment point for muscles controlling tongue	Lateral	*stylus,* stylus + *oideos,* resembling + *processus,* a projection
	Zygomatic process	Projection that articulates with temporal process of the zygomatic bone	Lateral	*zygoma,* a yoke + *processus,* a projection
Vomer	NA	Forms inferior and posterior part of nasal septum	Inferior	
Zygomatic	Frontal process	Articulates with frontal bone	Lateral	*frontellum,* forehead + *processus,* a projection
	Maxillary process	Articulates with zygomatic process of maxillary bone	Lateral	*maxilla,* jawbone + *processus,* a projection
	Temporal process	Articulates with zygomatic process of temporal bone	Lateral	*temporalis,* temple/time + *processus,* a projection

EXERCISE 8.1

ANTERIOR VIEW OF THE SKULL

EXERCISE 8.1A Anterior View of the Skull

1. Obtain a skull and observe it from an anterior view **(figure 8.1).** An anterior view of the skull reveals much of the detail of the facial bones. Facial bones play a role in mastication (chewing) and in the protection and support of special sensory organs such as the eye. **Table 8.3** describes bony structures of the face and lists the bones that compose each structure.
2. Identify the structures that are listed in figure 8.1 on a skull or model, using table 8.3 and the textbook as guides. Then label figure 8.1.
3. *Optional Activity:* AP|R **5: Skeletal System—**Watch the "Skull" animation, which demonstrates how the bones of the skull fit together. This animation also facilitates understanding of difficult concepts such as the location of the sphenoid and ethmoid bones with respect to the rest of the skull.

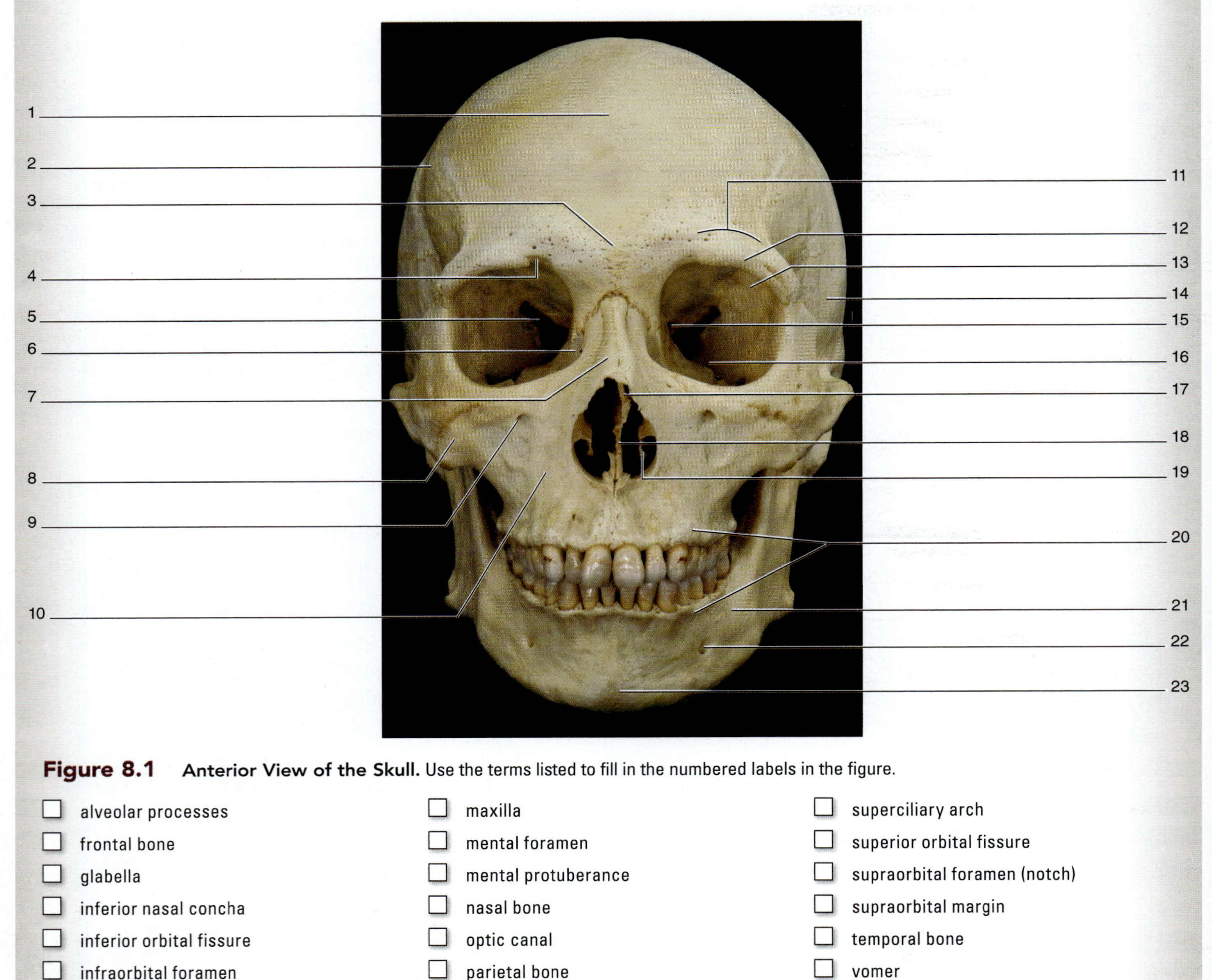

Figure 8.1 **Anterior View of the Skull.** Use the terms listed to fill in the numbered labels in the figure.

- ☐ alveolar processes
- ☐ frontal bone
- ☐ glabella
- ☐ inferior nasal concha
- ☐ inferior orbital fissure
- ☐ infraorbital foramen
- ☐ lacrimal bone
- ☐ mandible
- ☐ maxilla
- ☐ mental foramen
- ☐ mental protuberance
- ☐ nasal bone
- ☐ optic canal
- ☐ parietal bone
- ☐ perpendicular plate of ethmoid
- ☐ sphenoid bone
- ☐ superciliary arch
- ☐ superior orbital fissure
- ☐ supraorbital foramen (notch)
- ☐ supraorbital margin
- ☐ temporal bone
- ☐ vomer
- ☐ zygomatic bone

Table 8.3 The Axial Skeleton: Anterior View of the Skull

Facial Structure	Major Bone	Bone Feature	Description and Related Structures of Importance	Word Origins
Forehead	Frontal	Coronal suture	Suture between frontal and parietal bones	*corona,* crown + *sutura,* a sewing together
		Metopic suture	Suture between two parts of frontal bone; named in adult if suture persists	*metopa,* the space between two hollows + *sutura,* a sewing together
		Squamous suture	Suture between frontal and temporal bones	*squama, scale-like* + *sutura,* a sewing together
		Glabella	Prominent bony ridge located immediately superior to nasal bone	*glabellus,* smooth
		Superciliary arch	"Brow" ridges, located superior to supraorbital margin	*superus,* upper + *cilium,* an eyelid + *arcus,* a bow
Orbit	Frontal	Supraorbital margin	Bony support and protection of superior border of orbit	*supra-,* above + *orbit,* eye socket + *margo,* a border
		Supraorbital foramen	Passageway for supraorbital artery and nerve; sometimes a notch	*supra-,* above + *orbit,* eye socket + *foramen,* a hole
	Sphenoid	Optic foramen	Passageway for optic nerve (CN II)	*optikos,* eye + *foramen,* a hole
		Superior orbital fissure	Passageway for oculomotor (CN III), trochlear (CN IV), trigeminal (CN V_1), and abducens (CN VI) nerves	*superus,* upper + *orbit,* eye socket + *fissura,* cleft
		Inferior orbital fissure	Passageway for maxillary branch of trigeminal nerve (CN V_2) and infraorbital artery and vein	*inferus,* lower + *orbit,* eye socket + *fissura,* cleft
	Ethmoid		Forms medial wall and part of posterior wall of orbit	*ethmos,* sieve + *oideos,* resembling
	Lacrimal	Lacrimal fossa	Drains tears from surface of the eye into nasal cavity	*lacrima,* tear + *fodere,* to dig
	Maxilla		Forms medial and inferior walls of orbit	*maxilla,* lower jaw
		Infraorbital foramen	Passageway for infraorbital nerve (a branch of CN V) and artery	*infra-,* below + *orbit,* eye socket + *foramen,* a hole
	Zygomatic		Forms lateral border and wall of orbit	*zygos,* yolk
Nose	Nasal		Forms most of bridge and anterior portion of bony skeleton of nose	*nasalis,* the nose
	Maxilla	Frontal processes	Forms lateral aspect of bony skeleton of nose	*frontellum,* the forehead + *processus,* a projection
	Frontal	Nasal spine	Forms superior aspect of bony skeleton of nose	*nasalis,* the nose + *spinosus,* spine-like
Nasal Septum	Ethmoid	Perpendicular plate	Forms superior portion of nasal septum	*perpendiculum,* plumb line + *platus,* flat
	Vomer		Forms posterior-inferior portion of nasal septum	*vomer,* plowshare
Nasal Cavity	Ethmoid	Cribriform foramina	Passageway for olfactory nerves (CN I) to travel through to get to the CNS	*cribrum,* sieve + *forma,* form + *foramen,* a hole
		Cribriform plate	Forms roof of nasal cavity	*cribrum,* sieve + *platus,* flat
		Superior and middle conchae	Curved bony structures that form superior part of lateral wall; cause turbulent air flow	*superus,* upper + *concha,* shell; *middle,* middle + *concha,* shell
	Inferior nasal concha		Curved bone that forms inferior part of lateral wall; causes turbulent air flow	*infero,* lower + *concha,* shell
	Maxilla	Palatine process	Forms anterior floor and part of lateral wall of nasal cavity	*palatin,* the palate + *processus,* a projection
	Palatine		Forms posterior floor of nasal cavity	*palatin,* the palate
Oral Cavity (Buccal)	Palatine		Forms posterior roof of oral cavity	*palatin,* the palate
	Maxilla	Palatine process	Forms anterior roof of oral cavity	*palatin,* the palate
		Incisive foramen	Contains arteries and nerves passing from nasal cavity into oral cavity	*incidere,* to cut into + *foramen,* a hole
		Alveolar processes	Form joints with teeth	*alveolus,* a hollow
	Mandible	Alveolar processes	Form joints with teeth	*alveolus,* a hollow
		Body	Forms anterior portion of inferior border of oral cavity	NA
		Ramus	Forms lateral portion of inferior border of oral cavity	*ramus,* a branch

(continued on next page)

(continued from previous page)

Table 8.3 The Axial Skeleton: Anterior View of the Skull *(continued)*

Facial Structure	Major Bone	Bone Feature	Description and Related Structures of Importance	Word Origins
Chin	Mandible	Body	Anterolateral portion of mandible	NA
		Angle	Portion of mandible connecting the body to the ramus, forming a right angle	*angulus,* a corner
		Mental foramen	Passageway for mental artery and nerve (CN V_3)	*mental,* chin + *foramen,* a bore/hole
		Ramus	Part of bone that forms an angle with body of mandible	*ramus,* a branch
		Alveolar processes	Cavities that form tooth "sockets"	*alveolus,* a concave vessel, + *processus,* an advance
		Mental protuberance	Anterior projection of mandible, forming anterior projection of chin	*mental,* chin + *protubero,* to swell or bulge out

EXERCISE 8.1B The Orbit

1. Observe the **orbit (figure 8.2)** on a skull or model.
2. The orbit is the bony casing that supports and protects the eyeball. Parts of the frontal, zygomatic, maxillary, ethmoid, and lacrimal bones form the anterior border of the orbit. The ethmoid and lacrimal bones form most of the medial wall of the orbit, and the sphenoid bone forms most of the posterior wall of the orbit. Identify the walls and borders of the orbit on a skull.
3. Identify the features of the orbit that are listed in figure 8.2 on a skull, using table 8.3 and the textbook as guides. Then label figure 8.2.

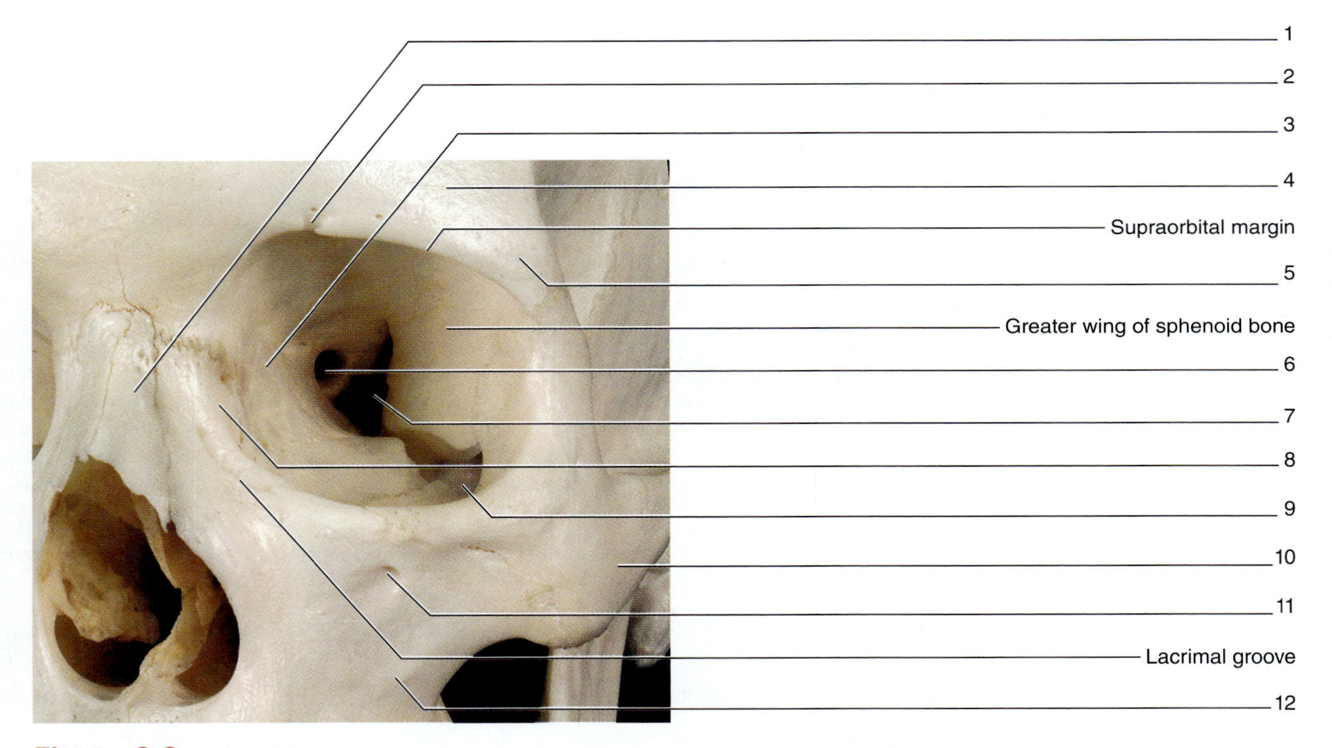

Figure 8.2 The Orbit. Anterior view. Use the terms listed to fill in the numbered labels in the figure.

- ☐ ethmoid bone
- ☐ frontal bone
- ☐ inferior orbital fissure
- ☐ infraorbital foramen
- ☐ lacrimal bone
- ☐ maxilla
- ☐ nasal bone
- ☐ optic foramen
- ☐ superior orbital fissure
- ☐ supraorbital foramen (notch)
- ☐ zygomatic bone
- ☐ zygomatic process of frontal bone

EXERCISE 8.1C The Nasal Cavity

1. Observe the **nasal cavity (figure 8.3)** on a skull or model.
2. The nasal cavity is a large, complex cavity that is separated into two halves by a **nasal septum** (*saeptum*, a partition). The ethmoid bone forms parts of the roof, septum, and lateral walls. The **cribriform** (*cribrum*, a sieve, + *forma*, a form) **plate** of the ethmoid bone forms most of the roof. The **palatine processes of the maxillary bones** and the **palatine bones** form the floor of the nasal cavity (and the roof of the oral cavity). The **nasal bones** form most of the bridge of the nose. Finally, the bony portion of the nasal septum is formed from the **perpendicular plate of the ethmoid bone** and the **vomer.**
3. Identify the features of the nasal cavity listed in figure 8.3 on a skull, using table 8.3 and the textbook as guides. Then label figure 8.3.

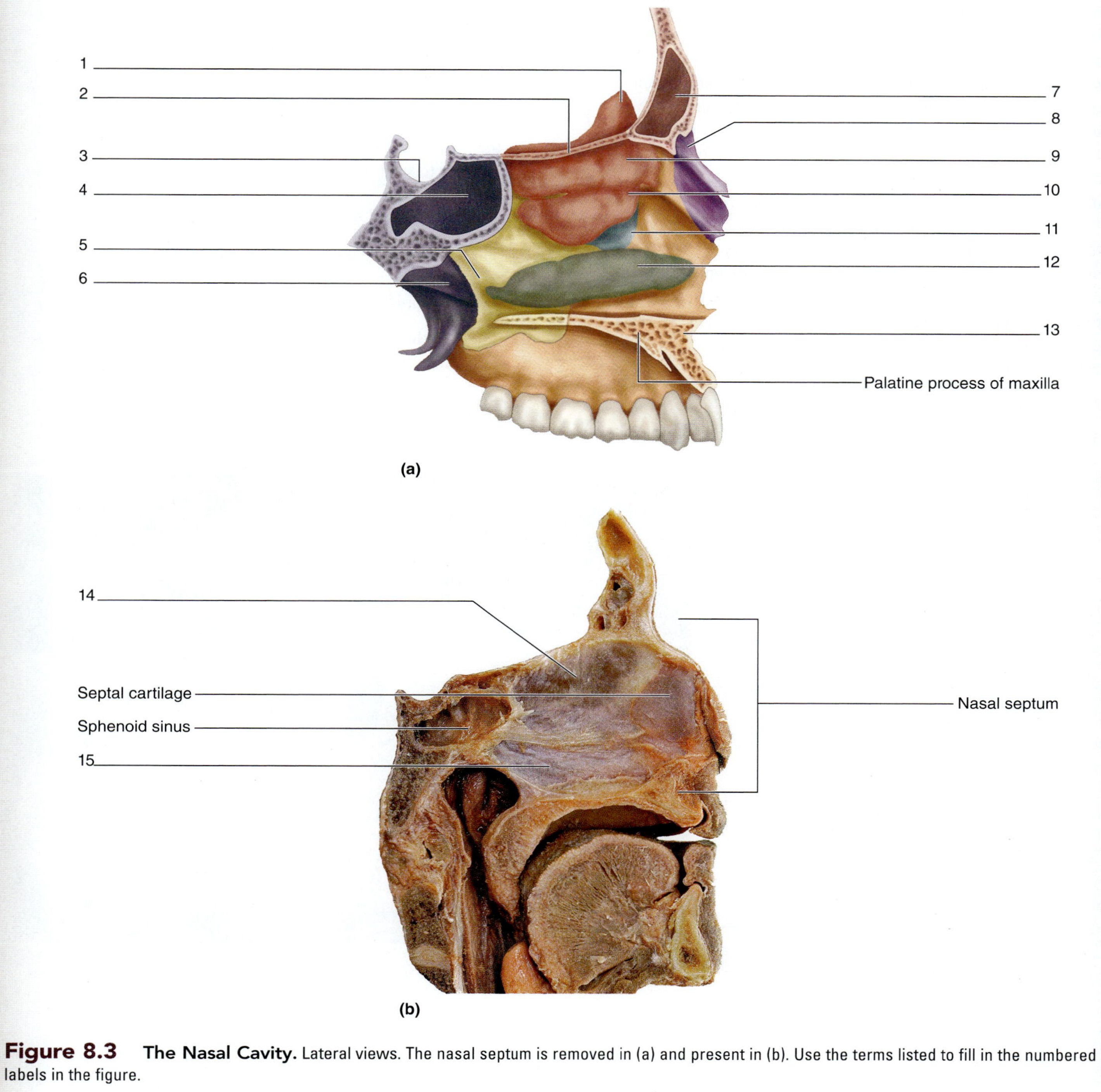

Figure 8.3 **The Nasal Cavity.** Lateral views. The nasal septum is removed in (a) and present in (b). Use the terms listed to fill in the numbered labels in the figure.

- ☐ cribriform plate of ethmoid bone
- ☐ crista galli (of ethmoid bone)
- ☐ frontal sinus
- ☐ inferior nasal concha
- ☐ lacrimal bone
- ☐ maxilla
- ☐ middle nasal concha
- ☐ nasal bone
- ☐ palatine bone
- ☐ perpendicular plate of ethmoid bone
- ☐ sella turcica
- ☐ sphenoid bone
- ☐ sphenoid sinus
- ☐ superior nasal concha
- ☐ vomer

(continued on next page)

(continued from previous page)

INTEGRATE

CONCEPT CONNECTION

The nasal cavity plays an important role in respiration and olfaction (the sense of smell). The role of the nasal cavity in respiration is to moisten, filter, and warm the air before it enters lower respiratory structures such as the lungs. The role of the nasal cavity in olfaction is to house the olfactory epithelium, which is responsible for detecting odorants in the air. Olfactory receptor cells are embedded within this epithelium, and their nerve processes project through the cribriform foramina of the ethmoid bone to reach the olfactory nerves. This pathway relays sensory input regarding the sense of smell to the brain. As odorants enter the nasal cavity, they dissolve in mucus in the nasal cavity and then bind to receptors on olfactory "hairs," thereby exciting these sensory neurons. Nerve signals, or action potentials, are then transmitted along the olfactory nerves to the olfactory bulb. From there, the signals are transmitted to the brain for the conscious perception of smell. This topic will be revisited in chapter 18 in the discussion of special senses.

EXERCISE 8.1D The Mandible

1. Obtain an isolated mandible or observe the mandible on an articulated skeleton or complete skull.
2. The **mandible (figure 8.4)** is unique among skull bones because it is the only bone that is movable. It shares an articulation with the temporal bone, forming the **temporomandibular joint.** Here, the **head of the mandible** articulates with the mandibular fossa of the temporal bone. The **mandibular condyle** is the rounded projection on the head of the mandible that actually forms the joint with the temporal bone. Place your fingers just anterior to your ears and then open and close your mouth to feel the movement of the joint formed between the mandible and the temporal bone (the temporomandibular joint).
3. The mandible also has a prominent **coronoid process** (*corona*, crown, + *eidos*, resemblance) that serves as the insertion point for the temporalis muscle.
4. The mandible contains two prominent paired foramina. The first, found on the inner (medial) surface of the ramus of the mandible, is the **mandibular foramen.** A continuation of the mandibular branch of the

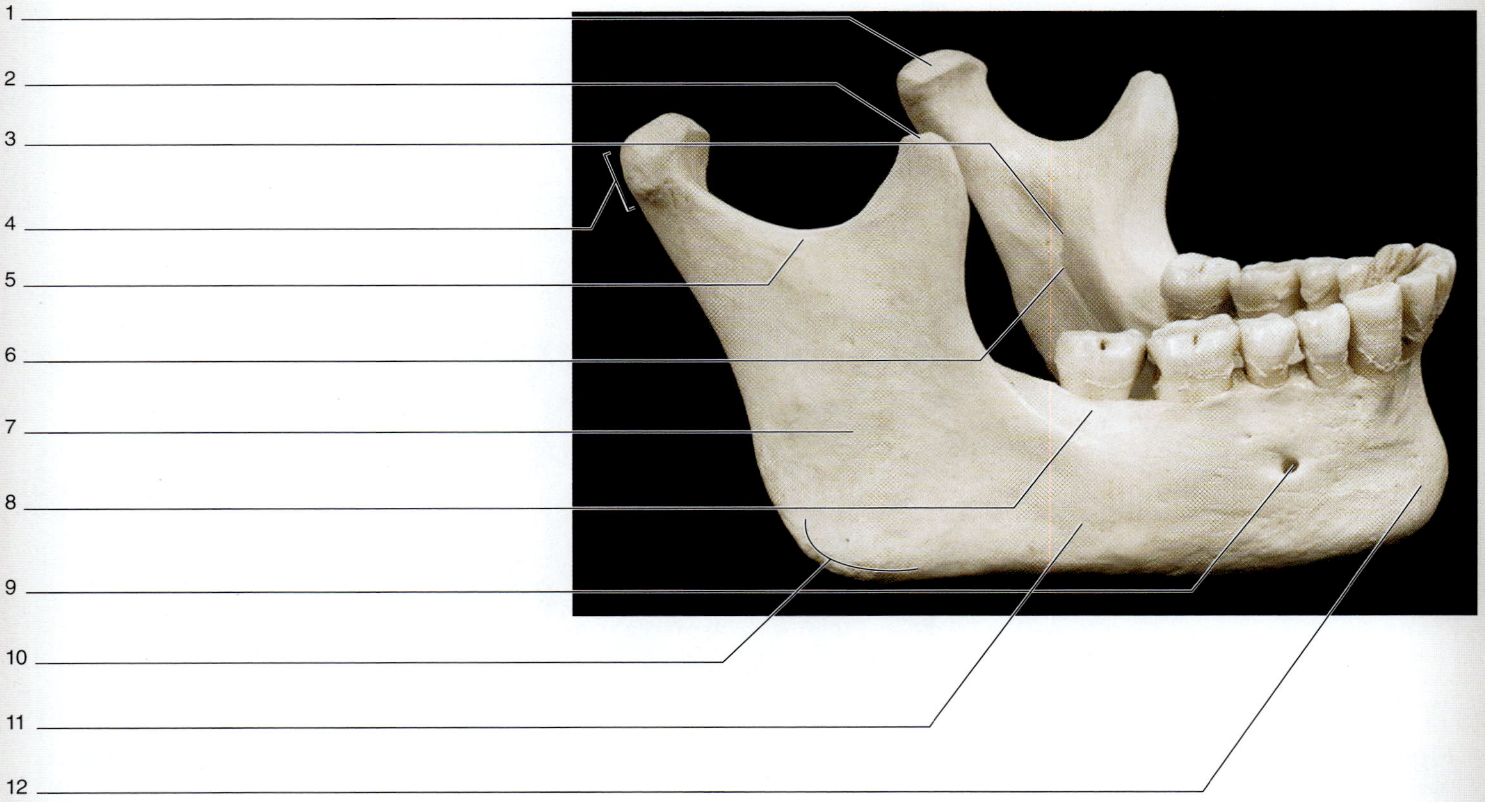

1 ______
2 ______
3 ______
4 ______
5 ______
6 ______
7 ______
8 ______
9 ______
10 ______
11 ______
12 ______

Figure 8.4 The Mandible. Lateral view. Use the terms listed to fill in the numbered labels in the figure.

☐ alveolar process	☐ coronoid process	☐ mental foramen
☐ angle	☐ head of mandible	☐ mental protuberance
☐ body	☐ mandibular foramen	☐ mylohyoid line
☐ condylar process	☐ mandibular notch	☐ ramus

trigeminal nerve (CN V_3) passes through this foramen into the interior of the bone. The nerve then sends branches to the **alveolar processes** of the mandible to innervate the roots of the teeth in a living individual. Dental work done on the teeth of the lower jaw involves directing a needle containing anaesthetic at the soft tissues surrounding this foramen to bathe the nerve branches that travel into the mandible. Finally, the **mental foramen** (*mental*, chin) is located just superior to the inferior border of the mandible at about the midpoint of the body of the mandible. The mental artery and nerve travel through the mental foramen in a living individual.

5. Identify the structures listed in figure 8.4 on a mandible, using table 8.3 and the textbook as guides. Then label figure 8.4.

EXERCISE 8.2

ADDITIONAL VIEWS OF THE SKULL

EXERCISE 8.2A Lateral View of the Skull

1. Obtain a skull and observe it from a lateral view **(figure 8.5).**
2. The most notable feature in a lateral view of the skull is the **zygomatic arch,** which is the bony structure that forms the superior part of a person's cheek. It is formed from the zygomatic process of the temporal bone and the temporal process of the zygomatic bone.
3. Identify the structures listed in figure 8.5 on a skull, using table 8.2 and the textbook as guides. Then label figure 8.5.

Figure 8.5 **Lateral View of the Skull.** Use the terms listed to fill in the numbered labels in the figure.

- ☐ body of mandible
- ☐ coronal suture
- ☐ ethmoid bone
- ☐ external acoustic meatus
- ☐ frontal bone
- ☐ greater wing of sphenoid bone
- ☐ head of mandible
- ☐ inferior temporal line
- ☐ lacrimal bone
- ☐ lacrimal groove
- ☐ lambdoid suture
- ☐ mastoid process
- ☐ maxilla
- ☐ mental foramen
- ☐ mental protuberance
- ☐ nasal bone
- ☐ occipital bone
- ☐ parietal bone
- ☐ parietal eminence
- ☐ pterion
- ☐ squamous part of temporal bone
- ☐ squamous suture
- ☐ styloid process
- ☐ superior temporal line
- ☐ temporal process of zygomatic bone
- ☐ zygomatic bone
- ☐ zygomatic process of temporal bone

(continued on next page)

(continued from previous page)

EXERCISE 8.2B **Posterior View of the Skull**

1. Obtain a skull and observe it from a posterior view **(figure 8.6).**
2. The most notable feature in a posterior view of the skull is the **lambdoid suture** (*lambda*, the Greek letter λ, + *eidos*, resemblance), which is named for its resemblance to the Greek letter lambda.
3. Identify the structures listed in figure 8.6 on a skull, using table 8.2 and the textbook as guides. Then label figure 8.6.

EXERCISE 8.2C **Superior View of the Skull**

1. Obtain a skull (with the skullcap intact) and observe it from a superior view **(figure 8.7).**
2. The most notable feature in a superior view of the skull is the **sagittal suture.**
3. Identify the structures listed in figure 8.7 on a skull, using table 8.2 and the textbook as guides. Then label figure 8.7.

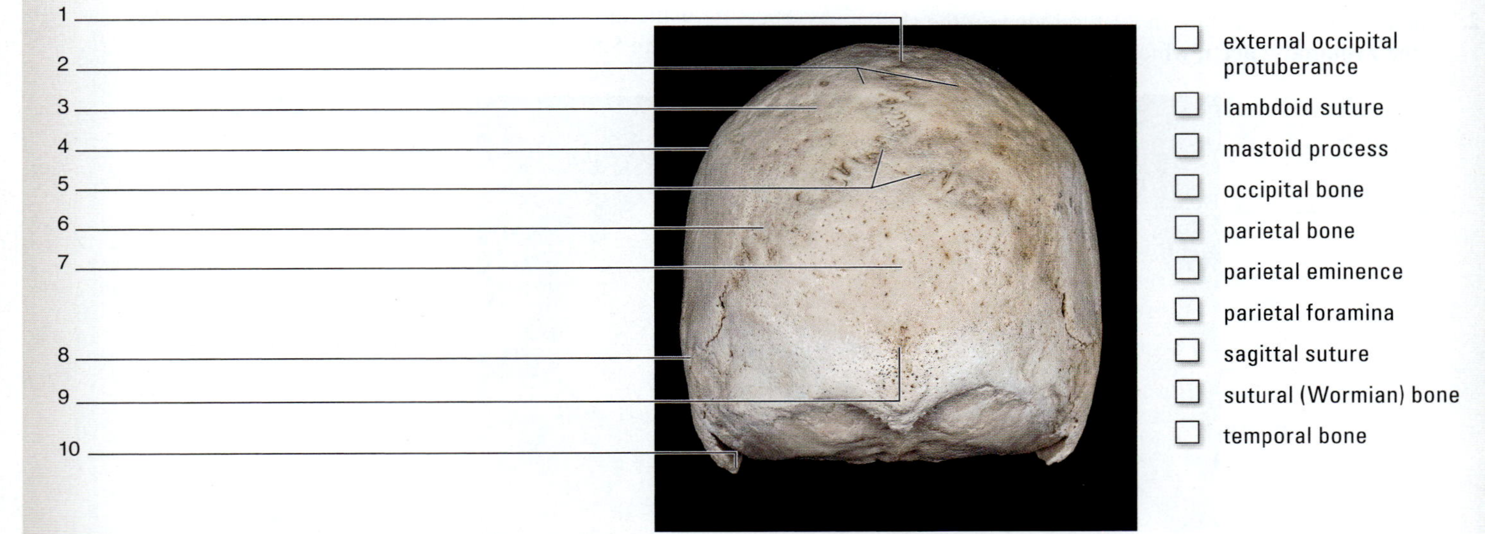

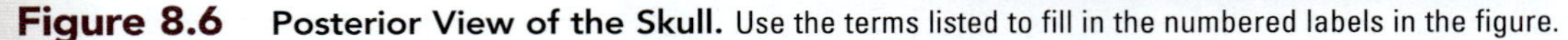

Figure 8.6 **Posterior View of the Skull.** Use the terms listed to fill in the numbered labels in the figure.

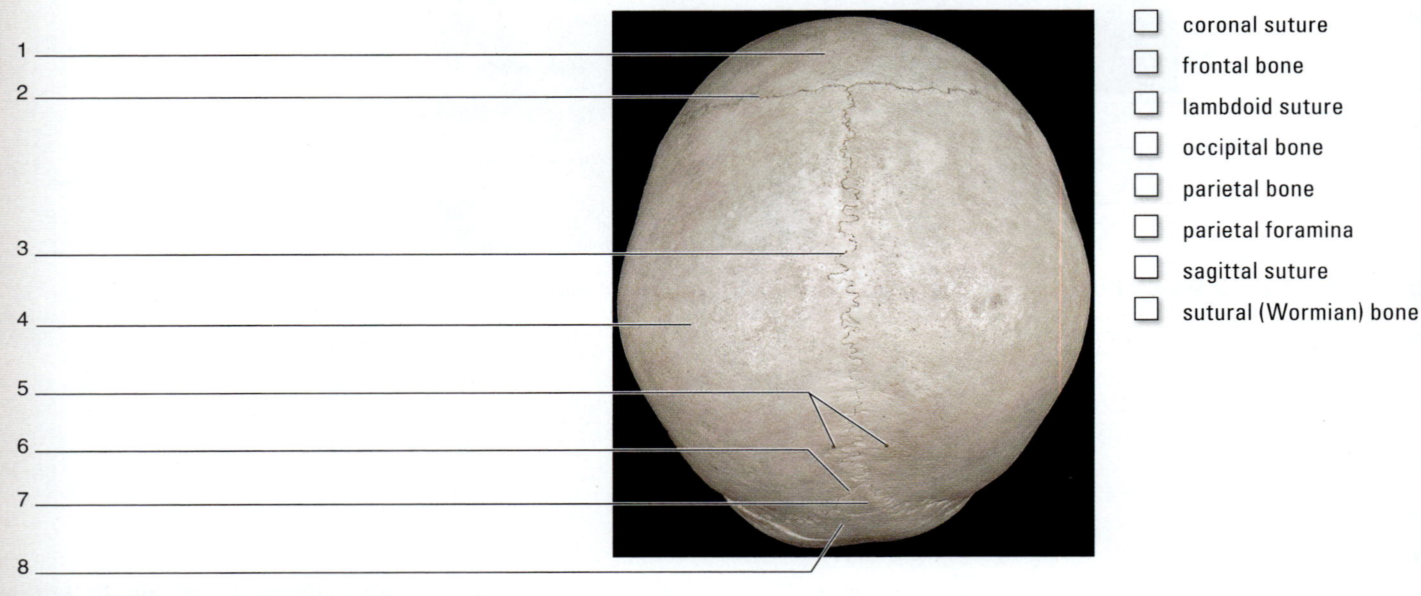

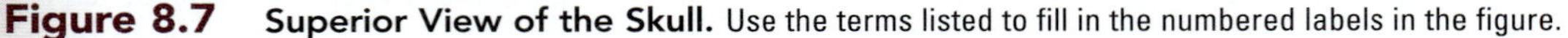

Figure 8.7 **Superior View of the Skull.** Use the terms listed to fill in the numbered labels in the figure.

EXERCISE 8.2D Inferior View of the Skull

1. Turn the skull over and observe it from an inferior view **(figure 8.8).** Many of the structures visible in this view are foramina that nerves and blood vessels pass through to get into and out of the cranial cavity.

2. The **foramen lacerum** (*lacero*, to tear to pieces) is unique among cranial foramina. It is one of the longest canals in the skull (about a centimeter in length). However, no single structure passes completely through it from one opening to the other. Instead, several structures pass through small portions of the canal. Such structures include a number of nerves as well as the internal carotid artery. As the internal carotid artery travels superiorly from the thorax into the cranial cavity, it first passes through the **carotid canal** and then enters the superior portion of the foramen lacerum as it proceeds toward the brain.

3. Identify the structures listed in figure 8.8 on a skull, using table 8.2 and the textbook as guides. Then label figure 8.8.

INTEGRATE

LEARNING STRATEGY

Obtain a broom straw, coffee stir stick (unused), or other nonmarking pointing device (NO pens or pencils). When identifying each foramen in the inferior view of the skull, pass the broom straw through the foramen and see where it comes out in the cranial floor. Visualize the pathways that structures take when traveling through the foramina to get into or out of the cranial cavity.

1 ______
2 ______
3 ______
4 ______
5 ______
6 ______
7 ______
8 ______
Stylomastoid foramen
9 ______
10 ______
11 ______
12 ______
13 ______
14 ______
Medial pterygoid plate
15 ______
16 ______
17 ______
18 ______
19 ______
20 ______
21 ______
22 ______
23 ______
24 ______
25 ______

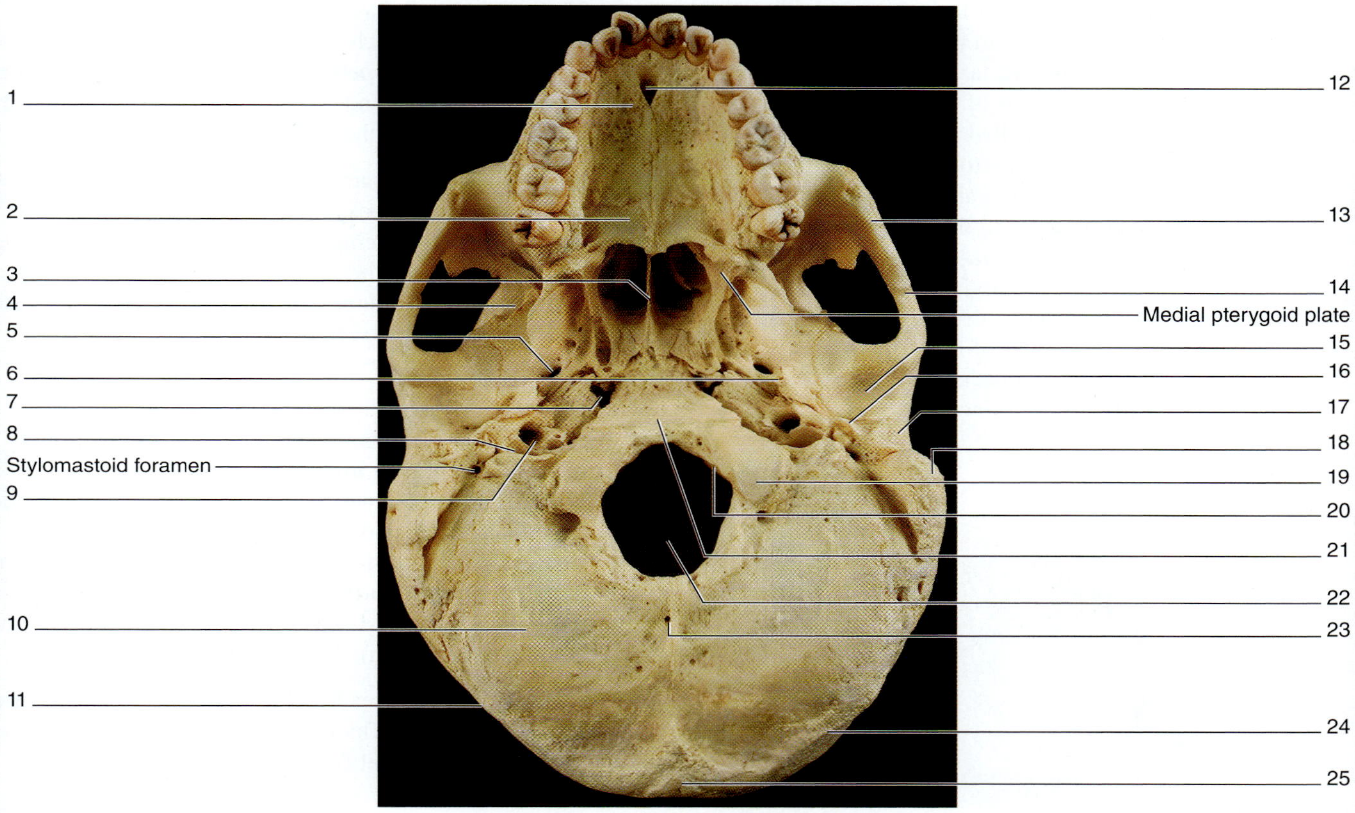

Figure 8.8 **Inferior View of the Skull.** Use the terms listed to fill in the numbered labels in the figure.

- ☐ basilar region of occipital bone
- ☐ carotid canal
- ☐ external occipital crest
- ☐ external occipital protuberance
- ☐ foramen lacerum
- ☐ foramen magnum
- ☐ foramen ovale
- ☐ foramen spinosum
- ☐ hypoglossal canal
- ☐ incisive foramen
- ☐ inferior nuchal line
- ☐ jugular foramen
- ☐ mandibular fossa
- ☐ mastoid process
- ☐ maxilla
- ☐ occipital bone
- ☐ occipital condyle
- ☐ palatine bone
- ☐ sphenoid bone
- ☐ styloid process
- ☐ superior nuchal line
- ☐ temporal bone
- ☐ temporal process of zygomatic bone
- ☐ vomer
- ☐ zygomatic process of temporal bone

EXERCISE 8.3

SUPERIOR VIEW OF THE CRANIAL FLOOR

1. Obtain a skull and remove the superior part of the cranium so the cranial floor is visible **(figure 8.9).**
2. Notice how the floor of the cranium is separated into three fossae: anterior, middle, and posterior. The **lesser wing of the sphenoid bone** forms the border between anterior and middle cranial fossae. The **petrous part of the temporal bone** (*petrosus*, a rock) forms the "rocky" border between the middle and posterior cranial fossae.
3. Using colored pencils, color and label the anterior, middle, and posterior cranial fossae in figure 8.9. Take note of the structures that form the natural divisions between these fossae.
4. Observe the **sella turcica** (*sella*, saddle, + *turcica*, Turkish) in the central portion of the sphenoid bone. This structure gets its name from its resemblance to a Turkish saddle, which contains large, prominent horns **(figure 8.10).** The anterior part of the sella turcica contains a slight projection called the **tuberculum sellae** (*tuber*, a knob). Posterior to that is the **hypophyseal fossa** (*hypophysis*, an undergrowth), which houses the pituitary gland in a living human. The pituitary gland is a small pea-shaped endocrine gland that connects to the brain via a small stalk called the infundibulum (*infundibula*, a funnel). The larger projection in the posterior part of the sella turcica is the **dorsum sellae,** which connects laterally to the two **posterior clinoid processes** (*klino*, to slope).

Figure 8.10 **Photograph of a Turkish Saddle.**

5. The temporal bone has two major portions: a lateral **squamous part,** which forms part of the lateral wall of the cranium, and a thick **petrous part,** which forms the border between the middle and posterior cranial fossae (although it is considered part of the middle cranial fossa).
6. Locate the petrous part of the temporal bone. Notice how it forms a sort of "rocky" ridge within the cranial floor. This portion of the temporal bone is very large and bulky because it contains the structures for hearing (the cochlea) and equilibrium/balance (the semicircular canals). These structures cannot be seen from the surface of the bone; instead, the structures are contained within the bone.
7. Find the internal and external acoustic (auditory) meatuses. The **internal acoustic** (auditory) **meatus** is the opening into the **internal auditory canal,** which is a passageway for the nerves that carry sensory information from the cochlea and semicircular canals to the brain. The **external acoustic** (auditory) **meatus** is the opening into the **external auditory canal,** which is a passageway through which sound waves travel to reach the tympanic membrane (eardrum).
8. Identify the structures listed in **figure 8.11** on a superior view of the cranial floor, using table 8.2 and the textbook as guides. Then label figure 8.11.

INTEGRATE

LEARNING STRATEGY

It is useful to observe one fossa at a time when identifying bony structures within the cranial floor. This is a natural way to divide the features of the cranial floor into manageable pieces of material. For example, when asked to identify one of the many foramina in the cranial floor, first narrow down the choices by identifying the cranial fossa where the foramen is located.

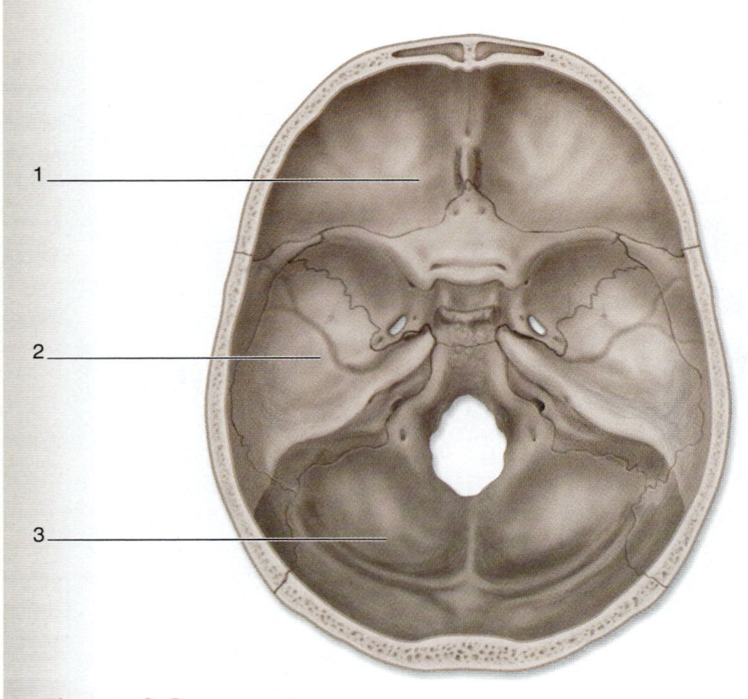

Figure 8.9 **Cranial Fossae.** Use the terms listed to fill in the numbered labels in the figure. Use colored pencils to color in each cranial fossa to further differentiate them from each other.

- ☐ anterior cranial fossa
- ☐ middle cranial fossa
- ☐ posterior cranial fossa

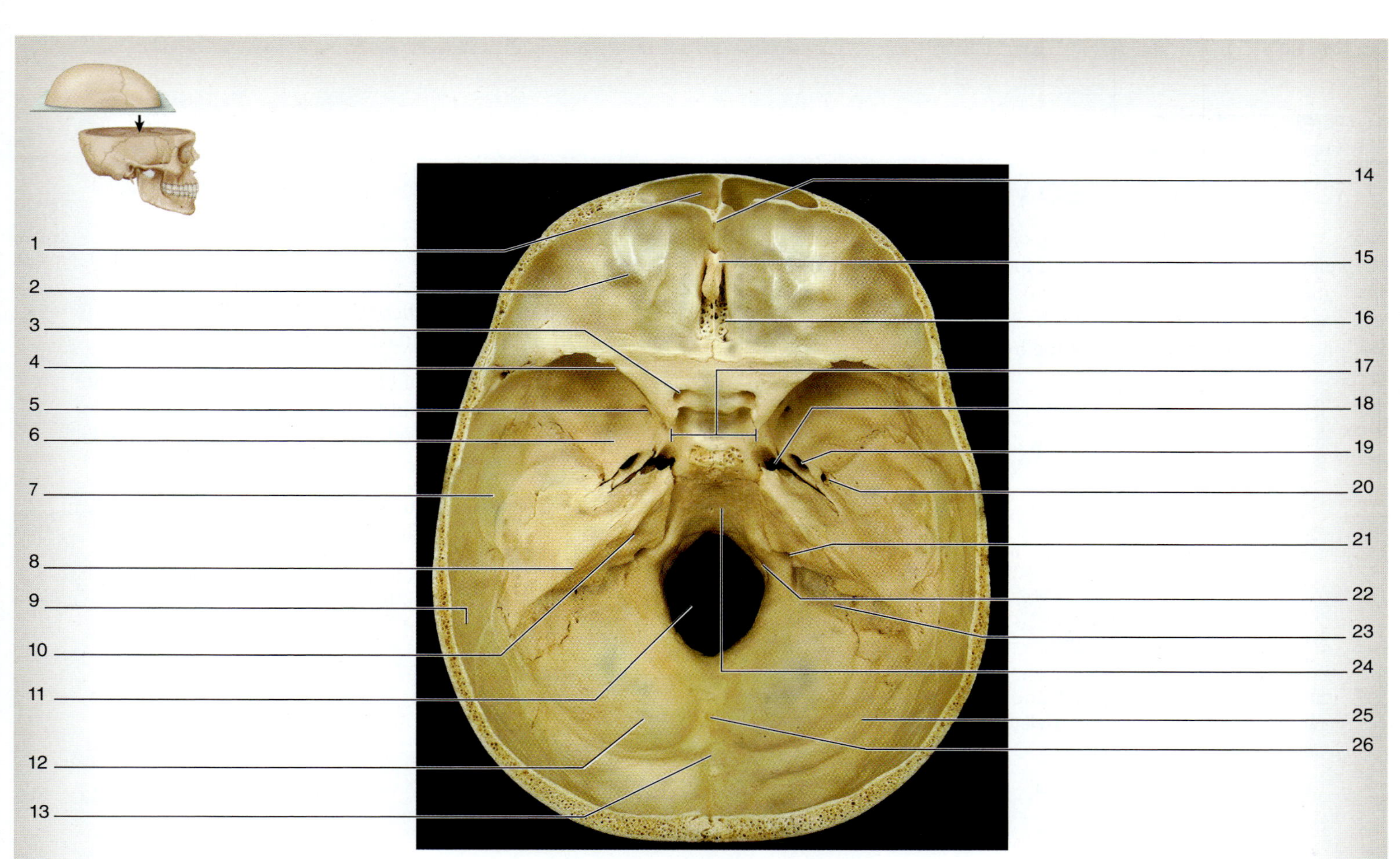

(a) Superior view

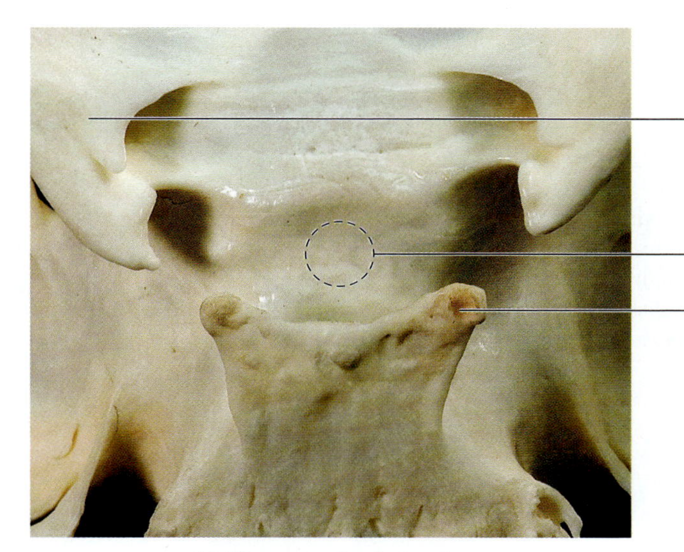

(b) Close-up of sella turcica

Figure 8.11 **Superior View of the Cranial Floor.** Use the terms listed to fill in the numbered labels in the figure.

Anterior Cranial Fossa

- ☐ cribriform plate of ethmoid bone
- ☐ crista galli
- ☐ frontal bone
- ☐ frontal crest
- ☐ frontal sinus
- ☐ lesser wing of sphenoid bone

Middle Cranial Fossa

- ☐ anterior clinoid process
- ☐ foramen lacerum
- ☐ foramen ovale
- ☐ foramen rotundum
- ☐ foramen spinosum
- ☐ greater wing of sphenoid bone
- ☐ hypophyseal fossa
- ☐ optic canal
- ☐ petrous part of temporal bone
- ☐ posterior clinoid process
- ☐ sella turcica
- ☐ temporal bone

Posterior Cranial Fossa

- ☐ basilar part of occipital bone
- ☐ foramen magnum
- ☐ groove for sigmoid sinus
- ☐ groove for transverse sinus
- ☐ hypoglossal canal
- ☐ internal acoustic meatus
- ☐ internal occipital crest
- ☐ internal occipital protuberance
- ☐ jugular foramen
- ☐ occipital bone
- ☐ parietal bone

EXERCISE 8.4

BONES ASSOCIATED WITH THE SKULL

The hyoid bone and the auditory ossicles are bones of the axial skeleton that are associated with the skull, but they are not part of the skull proper. The auditory ossicles are part of the hearing apparatus and will be covered in chapter 18. This exercise will concentrate on the features of the hyoid bone.

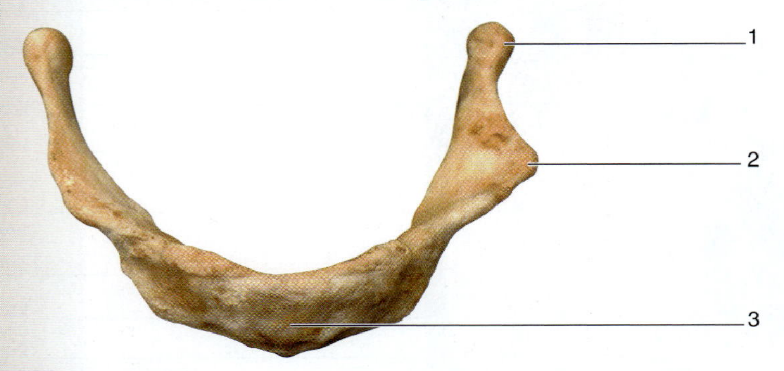

Figure 8.12 **The Hyoid Bone.** The hyoid bone is not in direct contact with any other bone of the skeleton. Use the terms listed to fill in the numbered labels in the figure.

☐ body ☐ greater cornu ☐ lesser cornu

1. Observe the hyoid bone **(figure 8.12)** on an articulated skeleton.
2. The **hyoid bone** (*hyoeidēs*, shaped like the Greek letter upsilon, υ) is the only bone in the body with no direct articulation with another bone. Muscles that move the tongue and pharynx attach to the hyoid.
3. Palpate your own hyoid bone by placing your thumb and index finger just medial to the angle of the mandible on either side. Then move them from side to side. Can you feel the rigid structure that is moving?
4. Observe an articulated skeleton and look at the placement of the hyoid bone with respect to the mandible.
5. Identify the structures listed in figure 8.12 on a hyoid bone, using the textbook as a guide. Then label figure 8.12.

The Fetal Skull

Like all bones in the body, the skull bones of the fetus are still developing. Recall that the skull bones develop via intramembranous ossification, which involves replacing a connective tissue membrane with bone tissue. Thus, when a fetus is born, the sutures between skull bones have not yet formed. This allows the head to distort and "squish" as it moves through the birth canal. The spaces between the plates of bone in the developing skull still consist of connective tissue membranes, which are largest in places where more than two bones come together. These membranes are **fontanels** (*fontaine*, a small fountain). The fontanels can be felt as "soft spots" on a baby's head. The fontanels will close and form sutures between the ages of 2 and 3.

EXERCISE 8.5

THE FETAL SKULL

1. Observe a fetal skull or a model of a fetal skull **(figure 8.13).**
2. Based on observations of the adult skull, locate the major cranial bones on the fetal skull (e.g., frontal, parietal, occipital).
3. Identify the structures listed in figure 8.13 on a fetal skull or model of a fetal skull, using the textbook as a guide. Then label them in figure 8.13.

1 ______
2 ______
3 ______
4 ______
5 ______
6 ______
7 ______
8 ______

(a) Lateral view

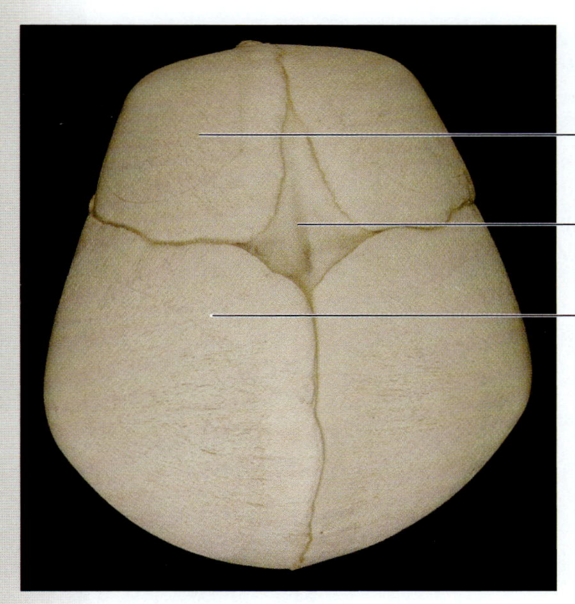

9 ______
10 ______
11 ______

(b) Superior view

Figure 8.13 The Fetal Skull. Use the terms listed to fill in the numbered labels in the figure. Some answers may be used more than once.

- ☐ anterior fontanel
- ☐ frontal bone
- ☐ mandible
- ☐ mastoid fontanel
- ☐ occipital bone
- ☐ parietal bone
- ☐ sphenoid bone
- ☐ sphenoidal fontanel
- ☐ temporal bone

The Vertebral Column

The vertebral column lies at the core of the human skeleton. It quite literally is the "backbone" that anchors nearly every major component of the skeletal support system. The vertebral column is divided into five major regions: **cervical, thoracic, lumbar, sacral,** and **coccygeal.** The vertebrae themselves change size and shape rather drastically from the cervical region to the coccygeal region. These changes reflect the different weight-bearing demands placed on the vertebrae in each region. **Cervical vertebrae** are small and light because they are not supporting a lot of weight (relatively speaking), and they are specialized to allow a lot of movement of the neck, particularly rotation. **Thoracic vertebrae** are specialized to provide articulation points for the ribs. **Lumbar vertebrae** are very large, bulky vertebrae that are specialized for supporting the weight of the entire vertebral column and body structures above them. They do not allow much movement, but instead are designed to keep the vertebral column stable. The **sacrum,** which consists of fused vertebrae, is specialized to provide a stable anchoring point for the bones of the pelvic girdle. Finally, the **coccyx** consists of 3–5 small vertebrae, which have fused together during development. It serves as an attachment point for several ligaments and for muscles of the pelvic floor. **Table 8.4** summarizes the characteristics of each type of vertebra.

Table 8.4 The Axial Skeleton: Vertebral Column

Vertebrae	Number of Vertebrae	Bone Features	Description and Related Structures of Importance	Word Origins
Typical Vertebra	32	Lamina	Connects transverse process to spinous process on either side of each vertebra	*lamina,* a saw (or the flap of the ear)
		Pedicle	Connects body to transverse process	*ped,* foot
		Transverse processes	Processes directed laterally (one on each side)	*trans,* across + *versus,* a line + *processus,* a projection
		Spinous process	Process directed posteriorly	*spina,* spine + *processus,* a projection
		Inferior articular process	Contains a facet that forms a joint with the superior articular process of inferior vertebra	*inferus,* lower + articulo, a joint + processus, a projection
		Superior articular process	Contains a facet that forms a joint with the inferior articular process of superior vertebra	*superus,* above + articulo, a joint + *processus,* a projection
		Vertebral foramen	Large within each vertebra; spinal cord extends through "stacked" vertebral foramina	*vertebra,* vertebra of the spine + *foramen,* a hole
		Body	Largest part of vertebra. Intervertebral discs are found between bodies of adjacent vertebrae.	NA
		Intervertebral foramina	Formed when two vertebrae come together; passageway for spinal nerves	*inter,* between + *vertebra,* of the spine + *foramen,* a hole
Cervical (C)	7	Body	Small body, oval/kidney bean shape	NA
		Spinous process	Horizontal, bifid (forked) spine of cervical vertebrae 3–6	*spina,* spine + *processus,* a projection
		Vertebral foramen	Large (especially with respect to size of the body); slight oval shape	*vertebra,* vertebra of the spine + *foramen,* a hole
		Transverse processes	Each contains transverse foramina	*trans,* across + *versus,* a line + *processus,* a projection
		Transverse foramen	Passageway for vertebral artery	*trans,* across + *versus,* a line + *foramen,* a hole
Atlas (C_1)		Body	Has no body; body has become dens (odontoid process) of axis	NA
		Arch	Contains articular surface for dens of axis and posterior tubercle (no spinous process)	arus, *a bow*

Table 8.4	The Axial Skeleton: Vertebral Column *(continued)*			
Vertebrae	**Number of Vertebrae**	**Bone Features**	**Description and Related Structures of Importance**	**Word Origins**
Axis (C_2)		Body	Has odontoid process (dens), which is the fused body of C_1	NA
Vertebra Prominens (C_7)		Spinous process	Very large and blunt, not bifid, not covered by ligamentum nuchae; first spinous process easily felt under skin	*spina*, spine + *processus*, a projection
Thoracic (T)	12	Body	Heart-shaped, contains demifacets for articulation of head of rib	NA
		Spinous process	Points inferiorly	*spina*, spine + *processus*, a projection
		Vertebral foramen	Relatively small; circular in shape; houses spinal cord	*vertebra*, vertebra of the spine + *foramen*, a hole
		Transverse processes	Contain facets for articulation with tubercle of rib	*trans*, across + *versus*, a line + *processus*, a projection
		Costal facets	Located on lateral surface of body and transverse processes; form joints with ribs	*costa*, a rib + *facet*, a small face
Lumbar (L)	5	Body	Very large, heavy	NA
		Spinous process	Short and blunt, square shaped, points horizontal	*spina*, spine + *processus*, a projection
		Vertebral foramen	Small (especially with respect to size of body), round; houses spinal cord	*vertebra*, vertebra of the spine + *foramen*, a hole
		Transverse processes	Short and tapered at the ends	*trans*, across + *versus*, a line + *processus*, a projection
Sacrum (S)	5 (fused)	Anterior sacral foramina	Passageway for exit of anterior (ventral) rami of sacral spinal nerves	*anterior*, foremost + *sacro*, to hold sacred + *foramen*, a hole
		Posterior sacral foramina	Passageway for exit of posterior (dorsal) rami of sacral spinal nerves	*posterus*, that which comes later + *sacro*, to hold sacred + *foramen*, a hole
		Median sacral crest	Represents fused spinous processes of sacral vertebrae (S_1–S_4)	*medianus*, in the middle + *sacro*, to hold sacred + *cristatus*, tufted
		Auricular processes	Ear-like (*auris*, ear) processes that articulate with iliac bones	*auris, ear*, + *processus*, an advance
		Superior articular processes	Contain facets to form joints with inferior articular processes of L_5	*superus, above*, + *auris, ear*, + *processus*, an advance
		Sacral hiatus	Opening at inferior end of sacral canal; formed by unfused laminae of S_5	*sacro*, to hold sacred + *hiatus*, an opening
		Sacral promontory	Anterosuperior border of the body of S_1	*sacro*, to hold sacred + *promunturium*, a mountain ridge
Coccyx (Co)	3 to 5 (fused)	Cornu (horns)	Small projections that point superiorly (part of Co_1)	*cornu*, horn

EXERCISE 8.6

VERTEBRAL COLUMN REGIONS AND CURVATURES

1. Observe the vertebral column of an articulated skeleton **(figure 8.14).**
2. Using colored pencils, color and label the regions of the vertebral column in figure 8.14. Use a different color for each region. Count the number of vertebrae that make up the region and write that number in the appropriate space in figure 8.14.
3. As the vertebral column develops, it forms several curvatures because of the stresses placed on it. The first curvatures to develop during the fetal period are **primary curvatures.** These form in the thoracic and sacral regions due to growth of the viscera. The second curvatures, which develop after birth, are **secondary curvatures.** These form in the cervical and lumbar regions. The cervical curvature forms when an infant begins to lift its head and the lumbar curvature forms when an infant begins to stand on its feet.
4. Locate all of the curvatures of the vertebral column on an articulated skeleton, and label the curvatures on figure 8.14.

- ☐ cervical curvature
- ☐ cervical vertebrae
- ☐ coccygeal vertebrae
- ☐ lumbar curvature
- ☐ lumbar vertebrae
- ☐ sacral curvature
- ☐ sacrum
- ☐ thoracic curvature
- ☐ thoracic vertebrae

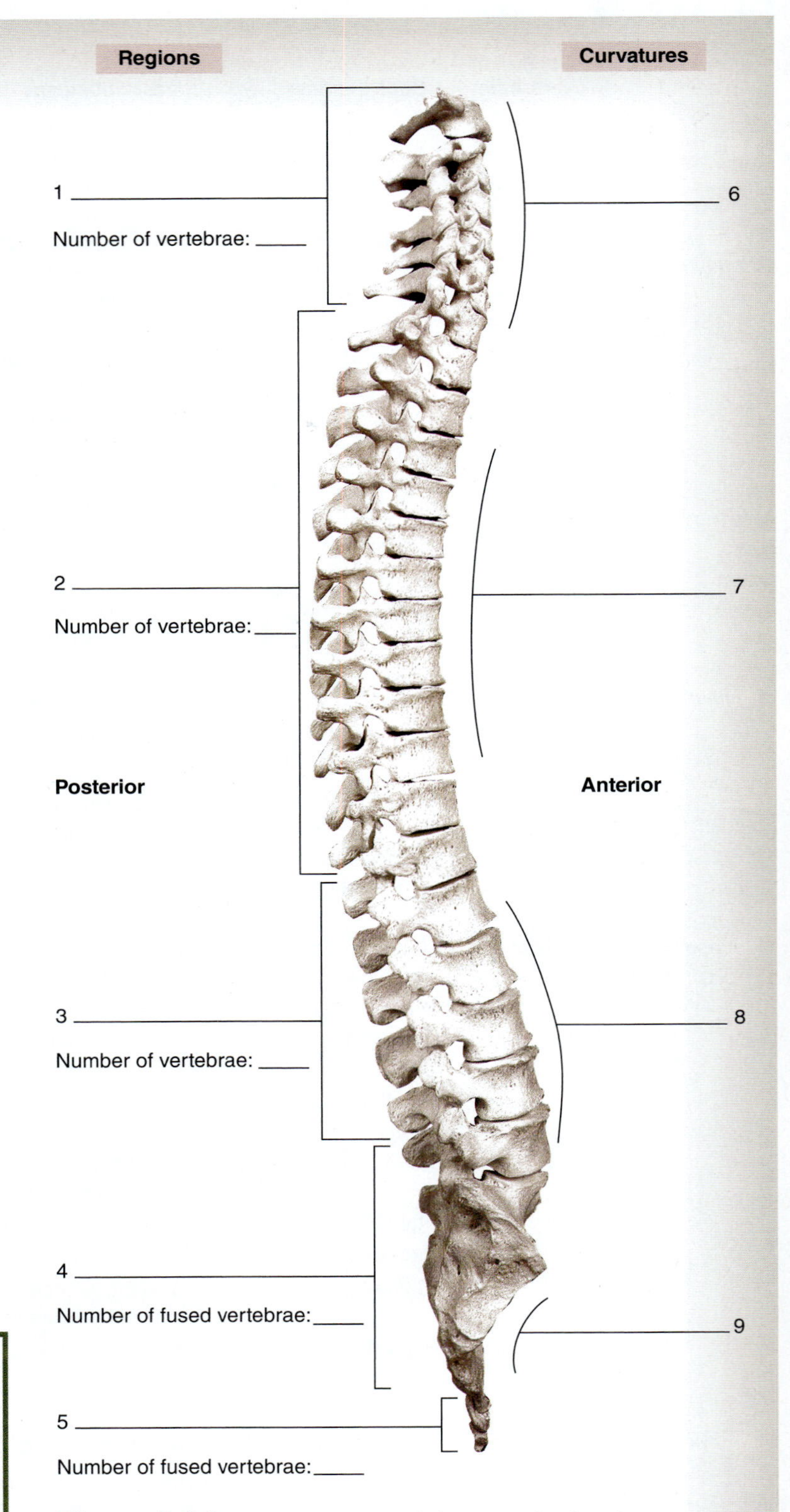

Figure 8.14 Lateral View of the Vertebral Column. Use the terms listed to fill in the numbered labels in the figure. Write the number of vertebrae for each region in the space provided in the figure.

INTEGRATE

LEARNING STRATEGY

Remember the number of vertebrae in each region using mealtimes as an aid. Breakfast is at *seven,* lunch is at *twelve,* dinner is at *five,* and a bedtime snack is at *nine.* That is, there are seven cervical vertebrae (C_1–C_7), twelve thoracic vertebrae (T_1–T_{12}), five lumbar vertebrae (L_1–L_5), five fused primitive vertebrae in the sacrum, and three to five small bones in the coccyx, with an average of four (5 + 4 = 9).

EXERCISE 8.7

STRUCTURE OF A TYPICAL VERTEBRA

1. Obtain a **thoracic vertebra (figure 8.15)**—this will be an example of a "typical" vertebra. It is helpful to begin studying the vertebral column by first taking a "typical" vertebra and identifying its component parts. This will make the study of specific features of vertebrae in the different regions of the vertebral column easier.
2. Looking at the vertebra from a superior view, notice the large **body.** The body is generally the largest part of a vertebra, and connections between adjacent vertebral bodies (with intervertebral discs in between) provide the main support of the vertebral column. Just posterior to the body is the large foramen called the **vertebral (spinal) foramen.** The spinal cord extends through the "stacked" vertebral foramina, which collectively form the vertebral canal. Each vertebral foramen is formed by the **vertebral arch.** The vertebral arch is composed of two sets of processes and the structures that connect them.
3. Observe the vertebral process that projects posteriorly and the processes that project laterally from the vertebral arch. The largest vertebral process is the **spinous process,** which is directed posteriorly, and the **transverse processes,** which are directed laterally.
4. Observe the vertebral arch. Notice the bony connections between the vertebral body and the transverse processes. These structures are called **pedicles** (L. *pediculus*, dim. of *pes*, foot). The word *pedicle* comes from a word meaning "foot." Imagine how the vertebral arch stands upon the body on its "feet." Now notice the bony connections between the transverse processes and the spinous process. These structures are called **laminae** (*lamina*, layer).
5. Next, turn the vertebra to observe it from an anterior view. Notice that there are two prominent structures that project superiorly from the vertebral arch and two that project inferiorly. The projections are respectively called the **superior articular processes** and **inferior articular processes.** Note that on each process there is a smooth, flat surface. These surfaces are called **facets** (*facette*, face). The term *facet* literally means "a little face." (This is the same term used to describe the surfaces on a diamond.) Each vertebra contains a pair of **superior articular facets** and **inferior articular facets** on its superior and inferior processes. These facets are the surfaces that form the joints between vertebrae, as described in the next step (6).
6. Pick up a second vertebra that articulates (forms a joint) with the first vertebra. Put the two together to observe how the superior facets and inferior facets articulate with each other to form a joint. These joints are much more mobile than the intervertebral joints (the joints between the vertebral bodies). They are also the sites where most of the movement is allowed in the vertebral column.
7. Once two vertebrae are articulated with each other, look at them from a lateral view. Notice the foramen that forms between the pedicles of adjacent vertebrae. This is the **intervertebral foramen.** This foramen is the location where spinal nerves (nerves that come off of the spinal cord) exit the vertebral canal to travel to their destinations throughout the body.
8. Identify the structures listed in figure 8.15 on a typical vertebra, using table 8.4 and the textbook as guides. Then label them in figure 8.15.

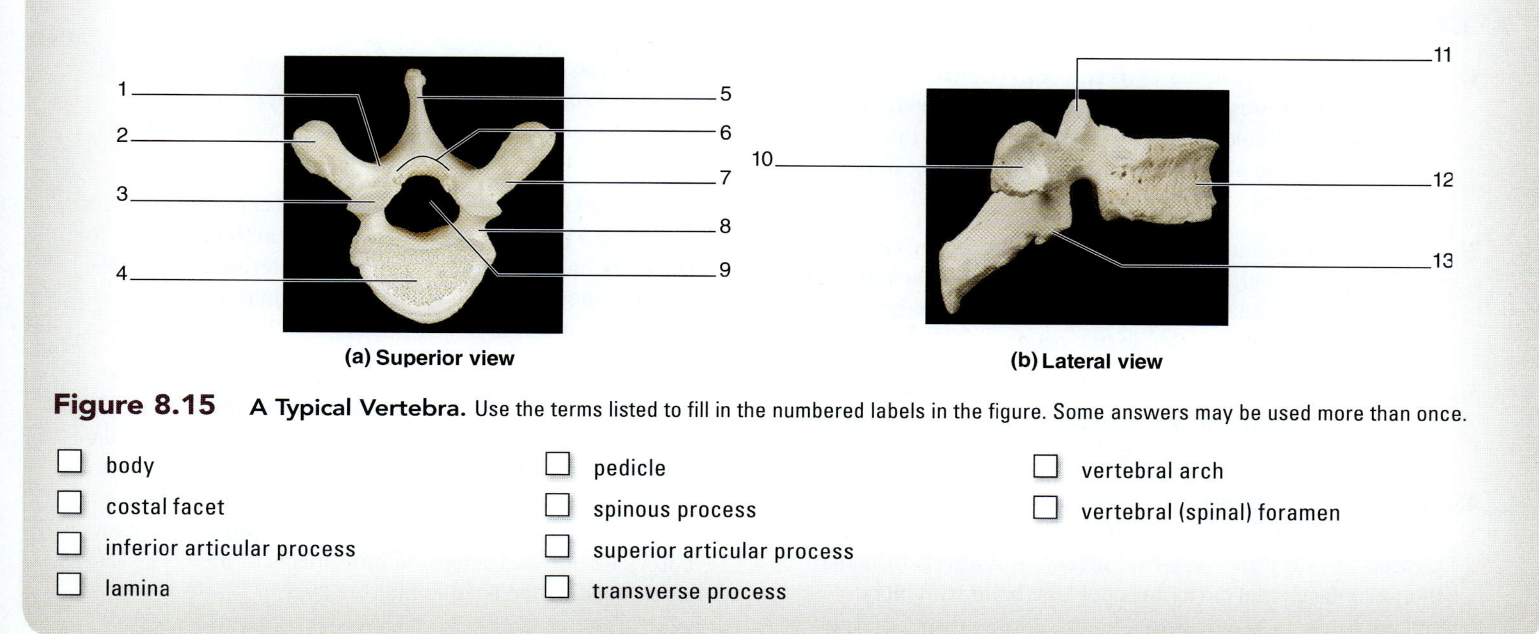

Figure 8.15 **A Typical Vertebra.** Use the terms listed to fill in the numbered labels in the figure. Some answers may be used more than once.

- ☐ body
- ☐ costal facet
- ☐ inferior articular process
- ☐ lamina
- ☐ pedicle
- ☐ spinous process
- ☐ superior articular process
- ☐ transverse process
- ☐ vertebral arch
- ☐ vertebral (spinal) foramen

INTEGRATE

CLINICAL VIEW
Spondylolisthesis

Spondylolisthesis (*spondylos,* vertebra + *olisthesis,* a slipping, falling) is a condition that involves the anterior or posterior displacement of a vertebra, typically in the lumbar region of the vertebral column. Displacement may occur during development, or may be the result of degeneration or fracture of the pars interarticularis (i.e., pars defect), which is the structure located between the superior and inferior articular processes of the vertebrae. The displaced vertebra may compress nearby spinal nerves, resulting in low back pain or sciatica, a condition that can cause pain to radiate down the posterior aspect of the lower limb. A grading system rates the severity of the displacement, with grade 1 being the least severe, and grade 4 being the most severe. In the most severe cases, the body of the superior vertebra may come to rest completely on the body of the inferior vertebra or the sacrum. The incidence of spondylolisthesis is particularly high in young athletes, particularly gymnasts, whose vertebrae experience trauma from repetitive lower back hyperextension and frequent, sudden impacts. Many confuse spondylolisthesis with a herniated disc, another common source of lower back pain. However, a herniated disc involves the bulging of an intervertebral disc rather than displacement of the vertebra itself.

EXERCISE 8.8

CHARACTERISTICS OF INDIVIDUAL VERTEBRAE

INTEGRATE

LEARNING STRATEGY

When observing individual vertebrae, always keep function in mind. Ask questions such as these: Why does this vertebra have such a large/small body? Why does this vertebra have such a large/small vertebral canal? How does this vertebra "fit" with other aspects of the skeletal system? Asking these questions will facilitate identification of each type of vertebra. Finally, when observing an isolated vertebra, always be sure to identify the same vertebra on an articulated skeleton to develop an appreciation for how the complete vertebral column is assembled.

EXERCISE 8.8A Typical Cervical Vertebrae

1. Obtain a cervical vertebra (*cervix*, neck) **(figure 8.16).** Identify the features of each of the individual **cervical vertebrae** and think about how modifications of the cervical vertebrae allow a great deal of movement in the neck region of the vertebral column.
2. Observe the vertebra from a superior view. A typical cervical vertebra has a small, oval body and a large triangular vertebral foramen. The spinous process of some cervical vertebrae is forked, or **bifid** (*bifidus*, cleft in two parts).
3. Observe cervical vertebrae on an articulated skeleton. Notice how the fork on one vertebra fits over the top of the spinous process of the vertebra below it.
4. Observe the transverse processes on a cervical vertebra. Notice that it has a hole, or **transverse foramen,** in it. This foramen protects an artery, the **vertebral artery,** as it travels from the thorax to the cranial cavity to supply the brain with blood.
5. Identify the structures listed in figure 8.16 on a cervical vertebra, using table 8.3 and the textbook as guides. Then label them in figure 8.16.

INTEGRATE

CONCEPT CONNECTION

The Greek Titan, Atlas, held up the heavens on his shoulders. The first cervical vertebra is named the **atlas** because it holds up the head in much the same way. The second cervical vertebra is called the **axis** because it forms an axis of rotation for the first cervical vertebra to rotate about. Both of these vertebrae are specialized to allow for extensive flexion, extension, and rotational movements of the neck. When observing the unique modifications of the atlas and axis, try to visualize how these modifications allow extensive movement of the head and neck.

EXERCISE 8.8B The Atlas (C_1)

1. Obtain an **atlas** (C_1) and an axis (C_2).
2. Notice that the atlas is missing a body **(figure 8.17).** During development, the tissue that would normally become the body of the atlas fuses with the body of the axis (and separates from the atlas), forming the *dens* (odontoid process) of the axis. This modification allows the atlas to rotate around the axis.
3. Instead of laminae and pedicles, the atlas has an **anterior arch** and a **posterior arch.** Notice the **articular facet for the dens** on the inner surface of the anterior arch. Also note that instead of a spinous process there is a smaller **posterior tubercle** on the posterior arch.
4. Observe the **superior articular facets** of the atlas. These facets are oriented horizontally in the atlas, rather than vertically as with the other vertebrae. These facets articulate with the **occipital condyles.** Observe the occipital bone and atlas on an articulated skeleton to see how these structures fit together to form the **atlanto-occipital joint.** This joint allows flexion and extension movements of the neck—as when nodding the head to indicate "yes."
5. Identify the structures listed in figure 8.17 on an atlas (C_1 vertebra), using table 8.4 and the textbook as guides. Then label them in figure 8.17.

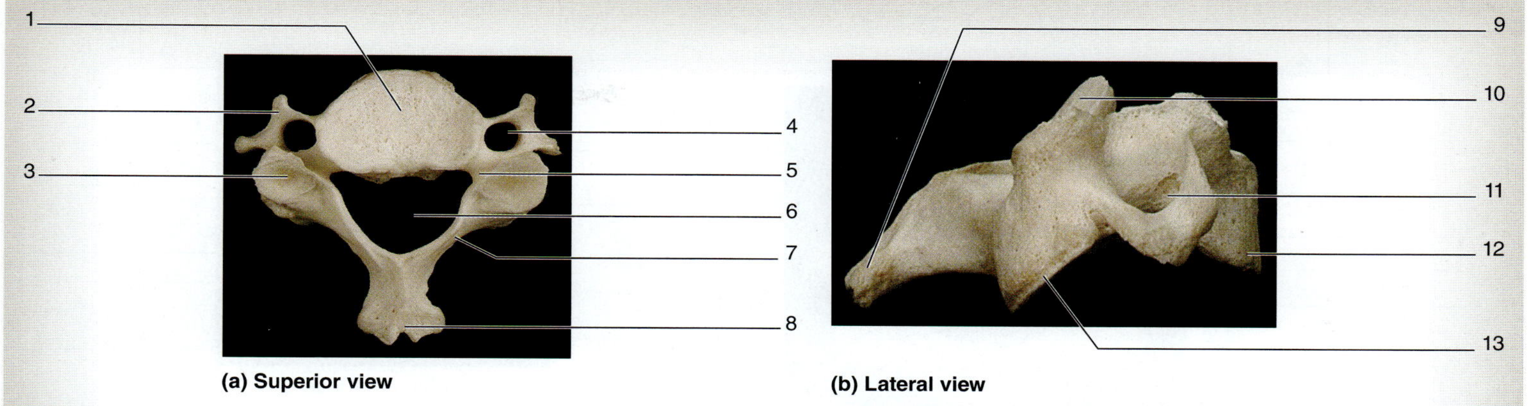

1
2
3
4
5
6
7
8

(a) Superior view

9
10
11
12
13

(b) Lateral view

Figure 8.16 **Cervical Vertebra.** Use the terms listed to fill in the numbered labels in the figure. Some answers may be used more than once.

- ☐ body
- ☐ inferior articular process (and facet)
- ☐ lamina
- ☐ pedicle
- ☐ spinous process
- ☐ superior articular process (and facet)
- ☐ transverse foramen
- ☐ transverse process
- ☐ vertebral (spinal) foramen

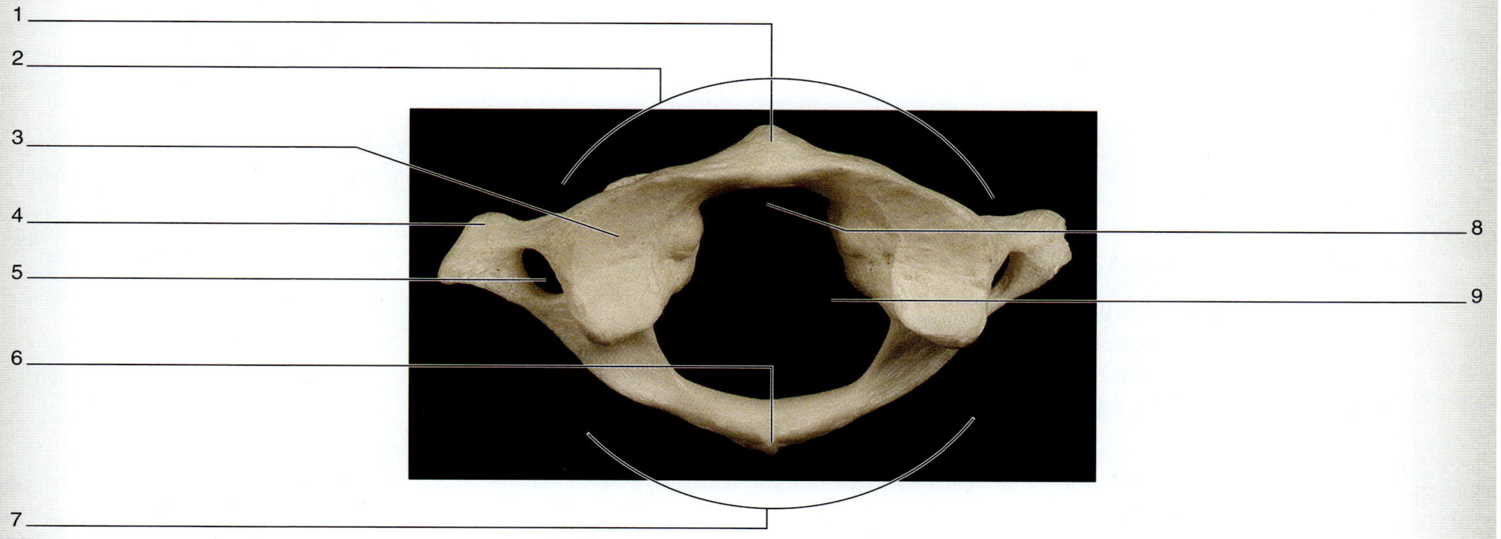

1
2
3
4
5
6
7
8
9

Figure 8.17 **The Atlas (C_1).** Superior view. Use the terms listed to fill in the numbered labels in the figure.

- ☐ anterior arch
- ☐ anterior tubercle
- ☐ articular facet for dens
- ☐ posterior arch
- ☐ posterior tubercle
- ☐ superior articular facet
- ☐ transverse foramen
- ☐ transverse process
- ☐ vertebral foramen

EXERCISE 8.8C The Axis (C_2)

1. Obtain an atlas (C_1) and an **axis** (C_2).
2. The axis **(figure 8.18)** is more similar than the atlas to a typical cervical vertebra. However, it has an extra process that no other vertebra has. This process is the **dens,** or **odontoid process** (*odont-*, tooth). Where did this process come from (developmentally)?

   ____

   ____

3. Place the atlas upon the axis and observe their articulation with each other to form the **atlantoaxial joint.** This joint allows lateral rotation of the neck—as when turning the head from side to side to indicate "no." Holding the atlas (C_2) in place, rotate the axis (C_1) around the dens of the atlas to observe this movement.
4. Similar to the atlas (C_1), the axis (C_2) has superior and inferior articular processes that lie in a horizontal plane. In addition, the axis (C_2) has a large bony surface where the laminae and pedicles come together called the **lateral mass.** The transverse processes connect to the lateral mass.
5. Identify the structures listed in figure 8.18 on an axis (C_2 vertebra), using table 8.4 and the textbook as guides. Then label them in figure 8.18.

(continued on next page)

(continued from previous page)

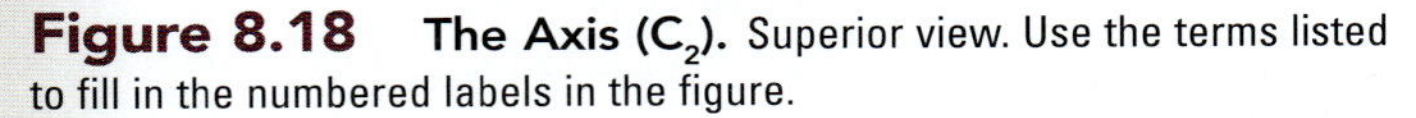

Figure 8.18 **The Axis (C_2).** Superior view. Use the terms listed to fill in the numbered labels in the figure.

- ☐ dens (odontoid process)
- ☐ lamina
- ☐ pedicle
- ☐ spinous process
- ☐ superior articular process
- ☐ transverse foramen
- ☐ transverse process
- ☐ vertebral (spinal) foramen

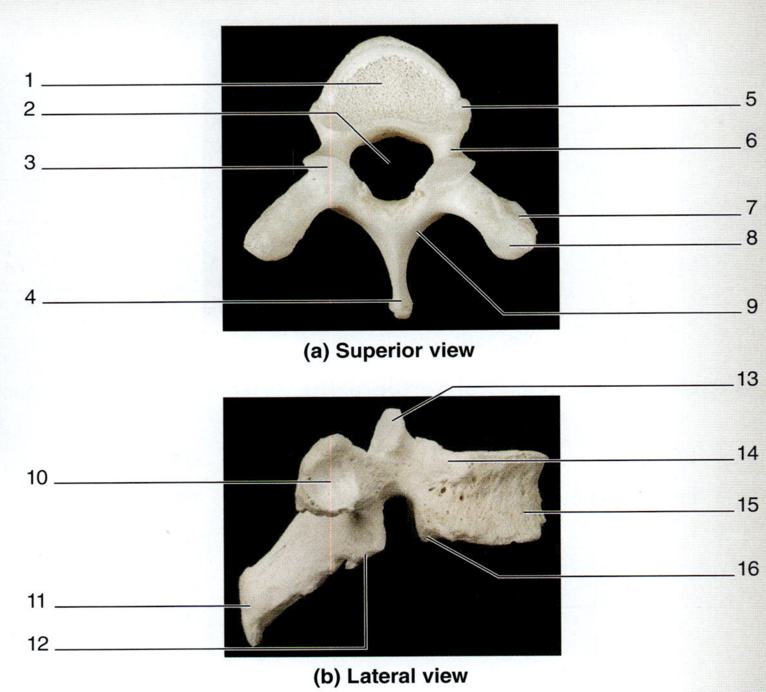

Figure 8.19 **Thoracic Vertebra.** Use the terms listed to fill in the numbered labels in the figure. Some answers may be used more than once.

- ☐ body
- ☐ costal demifacet
- ☐ costal facet
- ☐ inferior articular process
- ☐ lamina
- ☐ pedicle
- ☐ spinous process
- ☐ superior articular process
- ☐ transverse process
- ☐ vertebral (spinal) foramen

EXERCISE 8.8D Thoracic Vertebrae

1. Obtain a **thoracic vertebra (figure 8.19).** Thoracic vertebrae are the only vertebrae that articulate with the ribs. Thus, these vertebrae have special articular surfaces (facets) in locations where the ribs and vertebrae meet and form joints.
2. Observe the thoracic vertebra from a superior view. Thoracic vertebrae typically have a heart-shaped body (medium in size), a round vertebral foramen, a spinous process that projects inferiorly, and superior and inferior articular processes with surfaces that lie in the frontal plane.
3. Look at the relationship between the ribs and vertebrae on an articulated skeleton. Notice that the **tubercle** of a rib articulates with the transverse process of a thoracic vertebra. Notice also that the head of a rib articulates at the junction between two vertebral bodies. Thus, it articulates with the **costal demi facet** of the vertebra superior to it and the **costal facet** of the vertebra inferior to it.
4. Identify the structures listed in figure 8.19 on a thoracic vertebra, using table 8.4 and the textbook as guides. Then label them in figure 8.19.

INTEGRATE

LEARNING STRATEGY

It can be helpful to develop a "gestalt" (broad, generalized) view of vertebrae from each region of the vertebral column to assist with rapid identification. When viewed from an anterior view, thoracic vertebrae often resemble the head of a giraffe. The spinous process is the nose, the superior articular processes are the horns, and the transverse processes are the ears. When viewed from a lateral view, lumbar vertebrae often resemble a moose. The spinous process is the nose, the superior articular processes are the horns, the transverse processes are the ears, and the inferior articular processes are the dewlap—the little flap of tissue that hangs down from the chin of a bull (male) moose.

EXERCISE 8.8E Lumbar Vertebrae

1. Obtain a **lumbar vertebra (figure 8.20).** Lumbar vertebrae have very large, round or oval bodies, small vertebral foramina, and a short and blunt spinous process that projects posteriorly. The superior and inferior articular processes have facets that face medial and lateral, respectively.
2. Identify the structures listed in figure 8.20 on a lumbar vertebra, using table 8.4 and the textbook as guides. Then label them in figure 8.20.

(a) Superior view

(b) Lateral view

Figure 8.20 **Lumbar Vertebra.** Use the terms listed to fill in the numbered labels in the figure. Some answers may be used more than once.

- ☐ body
- ☐ inferior articular process (and facet)
- ☐ lamina
- ☐ pedicle
- ☐ spinous process
- ☐ superior articular process (and facet)
- ☐ transverse process
- ☐ vertebral (spinal) foramen

EXERCISE 8.8F The Sacrum and Coccyx

1. Obtain a sacrum and coccyx **(figure 8.21),** or observe them on an articulated skeleton.
2. The **sacrum** (*sacr-*, sacred) forms by the fusion of five primitive vertebrae that subsequently form a single bony structure. When identifying the features of the sacrum, try to recognize the parts of a typical vertebra within the sacrum.
3. The coccyx is usually composed of three to five small bones. The coccygeal vertebrae have only two prominent structures, the *cornu* (*cornu*, horn) and the *transverse processes*.
4. Identify the structures listed in figure 8.21 on a sacrum and coccyx, using table 8.4 and the textbook as guides. Then label them in figure 8.21.

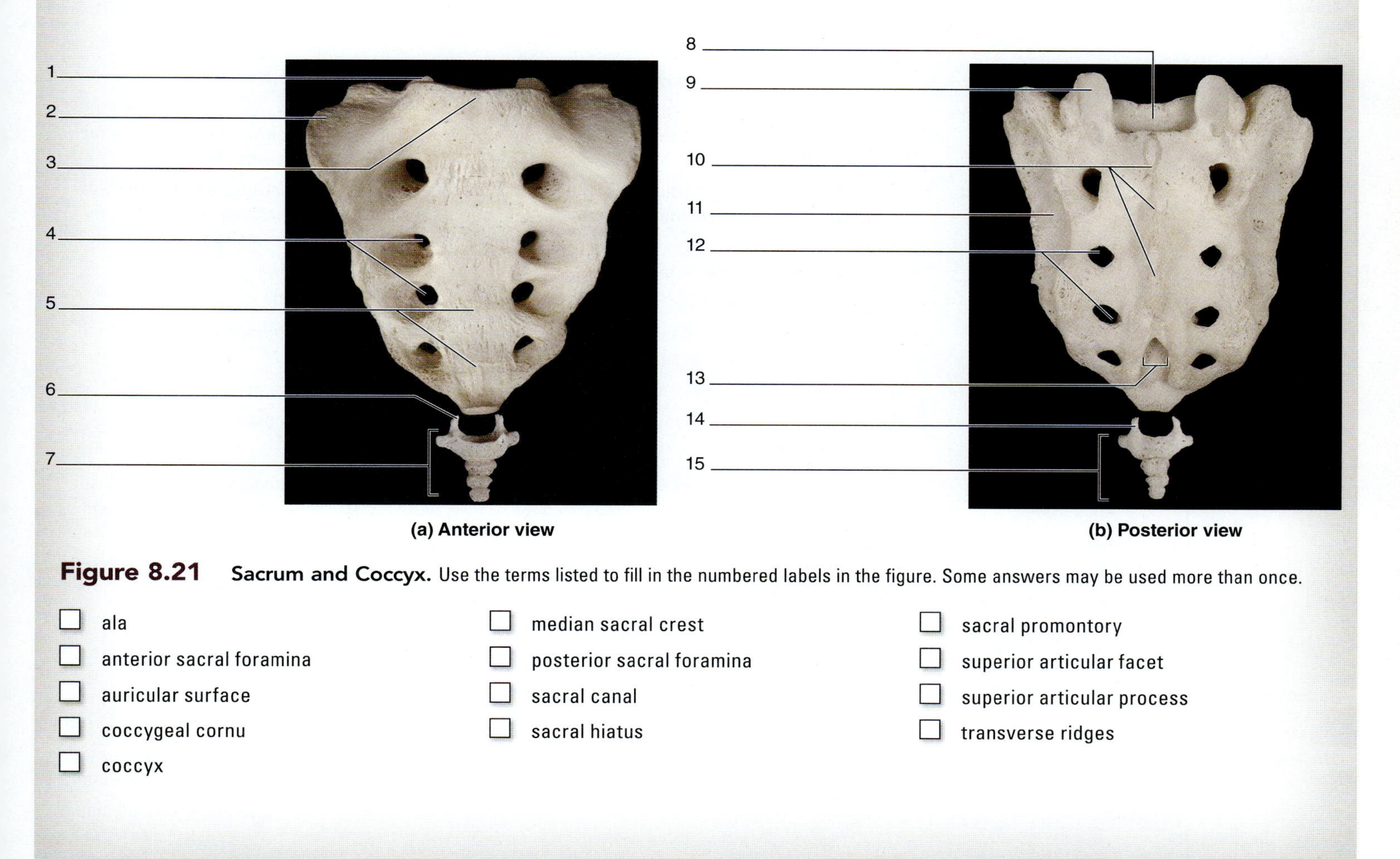

(a) Anterior view

(b) Posterior view

Figure 8.21 **Sacrum and Coccyx.** Use the terms listed to fill in the numbered labels in the figure. Some answers may be used more than once.

- ☐ ala
- ☐ anterior sacral foramina
- ☐ auricular surface
- ☐ coccygeal cornu
- ☐ coccyx
- ☐ median sacral crest
- ☐ posterior sacral foramina
- ☐ sacral canal
- ☐ sacral hiatus
- ☐ sacral promontory
- ☐ superior articular facet
- ☐ superior articular process
- ☐ transverse ridges

The Thoracic Cage

The **thoracic cage** consists of the **sternum, ribs,** and **thoracic vertebrae.** Its main function is to protect vital organs such as the heart and lungs. However, the bones of the thoracic cage also serve as important attachment sites for muscles involved with respiratory movements and muscles involved with movements of the back, chest, and neck. **Table 8.5** summarizes the key features of the sternum and ribs. Refer to table 8.4 for descriptions of the features of the thoracic vertebrae, which articulate with the ribs.

Table 8.5 The Axial Skeleton: Sternum and Ribs

Bone	Bone Features	Description and Related Structures of Importance	Word Origins
Sternum			
Manubrium	Clavicular notch	Point of articulation with clavicle	*manus,* hand *clavicula,* a small key
	Suprasternal notch	Depression at superior border	*supra,* above + *sternuo,* to sneeze (refers to the sternum)
	Notch for rib 1	Location of articulation with costal cartilage of rib 1	NA
	Sternal angle	Joint between manubrium and body; point of articulation with costal cartilage of rib 2	*sternuo,* to sneeze (refers to the sternum) + *angulus,* a corner
Body	Notches for ribs 2–7	Point of articulation for costal cartilages of ribs 2–7; partial notch for rib 2	NA
	Xiphisternal joint	Joint between body and xiphoid; point of articulation with superior part of costal cartilage of rib 7	*xiphion,* sword-shaped + *sternuo,* to sneeze (refers to the sternum)
Xiphoid	Partial notch for rib 7	Point of articulation for inferior part of the costal cartilage of rib 7	*xiphion, sword-shaped*
Ribs			
Typical Rib	Head	Part of rib that articulates with bodies of thoracic vertebrae	NA
	Superior articular facet	Facet on head of rib that articulates with inferior costal facet on body of vertebra that lies one level above it (i.e., superior articular facet of rib 6 with T_5)	*superus,* above + *articularis,* pertaining to joints + *facet,* a small face
	Inferior articular facet	Facet on head of rib that articulates with superior costal facet on body of numerically equivalent thoracic vertebra (i.e., inferior articular facet of rib 6 to T_6)	*inferus,* below + *articularis,* pertaining to joints + *facet,* a small face
	Shaft	Main part (body) of rib; begins at angle of rib; projects anteriorly	NA
	Neck	Narrow region where head meets tubercle of rib	NA
	Tubercle	Projection at the junction between shaft and neck; contains facet for articulation with transverse process of thoracic vertebra	*tuberculum,* a small bump
	Angle	Location where rib curves anteriorly	*angulus,* a corner
	Costal groove	Groove on inferior, deep border of shaft; contains intercostal artery, vein, and nerve	*costa,* a rib + *sulcus,* groove
First Rib	Scalene tubercle	Attachment for anterior scalene muscle	*scalenus,* of unequal sides + *tuberculum,* a small bump
	Groove for subclavian artery	Depression where subclavian artery passes out of thoracic cavity	NA
	Groove for subclavian vein	Depression where subclavian vein passes into thoracic cavity	NA
	Articular facet	Singular facet on head of rib (typical rib has two facets)	*articularis,* pertaining to joints + *facet,* a small face
Second Rib	All markings of a typical rib	Unique features of rib 2 are a rough tuberosity and a shallow costal groove	NA
11th and 12th Ribs	Articular facet	Singular facet on the head of the rib (a typical rib has two facets)	*articularis,* pertaining to joints + *facet,* a small face
	Tubercle	Absent	*tuberculum,* a small bump
	Neck	Absent	NA

EXERCISE 8.9

THE STERNUM

1. Observe the thoracic cage on an articulated skeleton and locate the sternum **(figure 8.22).** The sternum has three portions: the **manubrium,** the **body,** and the **xiphoid process** (*xiphos*, sword). The depression on the superior part of the manubrium is the **suprasternal notch.**
2. Palpate the suprasternal notch on yourself. Keeping your fingers on the manubrium, move your fingers inferiorly until you feel a rough ridge. This is the **sternal angle.** The sternal angle is located where the manubrium meets the body of the sternum. It is an important clinical landmark, because this is where the second rib articulates with the sternum.
3. Identify the structures listed in figure 8.22 on a sternum, using table 8.5 and the textbook as guides. Then label them in figure 8.22.

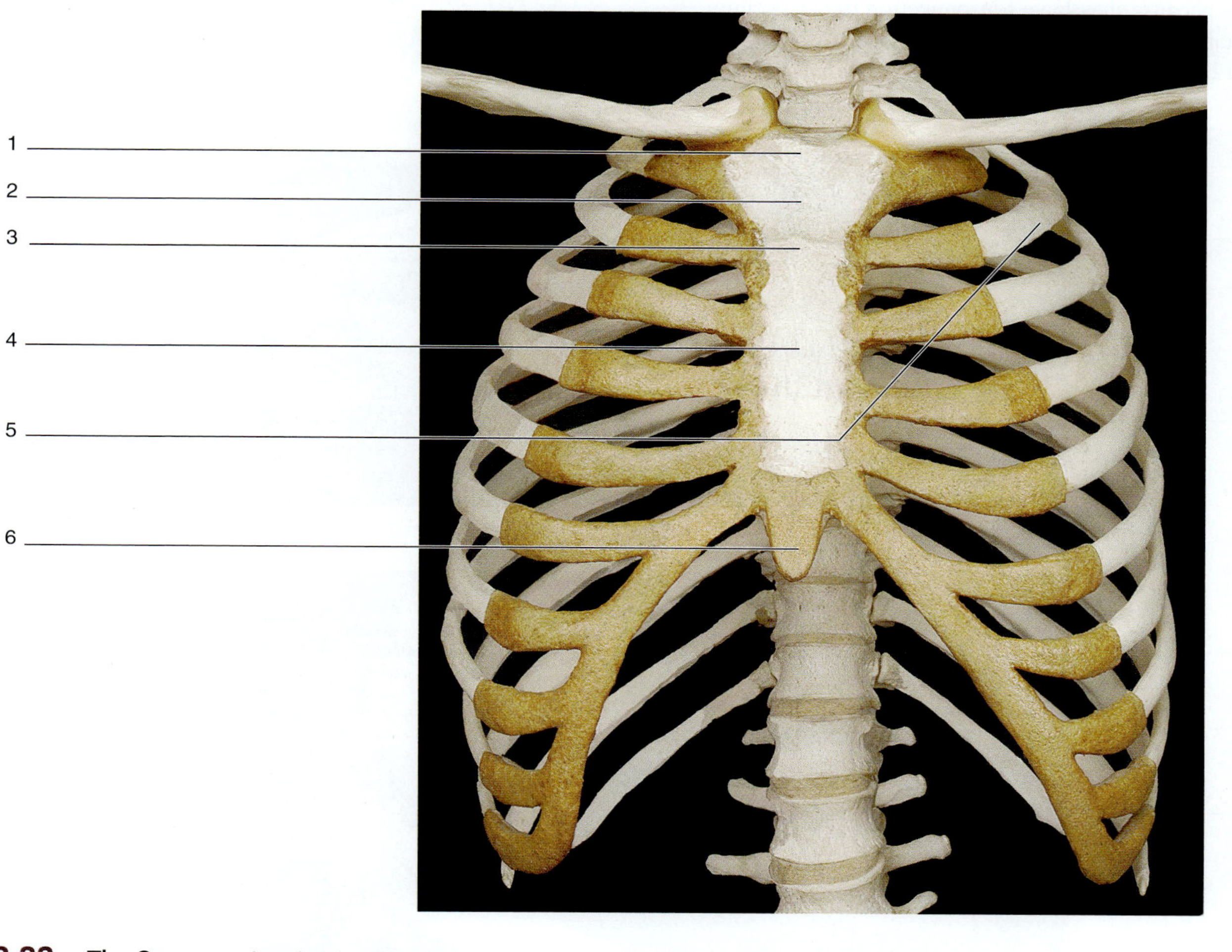

Figure 8.22 **The Sternum.** Anterior view. Use the terms listed to fill in the numbered labels in the figure.

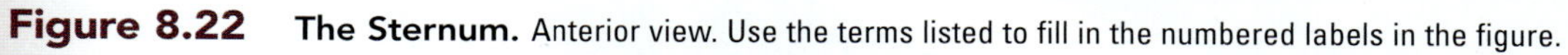

☐ body ☐ second rib ☐ suprasternal notch
☐ manubrium ☐ sternal angle ☐ xiphoid process

EXERCISE 8.10

THE RIBS

EXERCISE 8.10 The Ribs

1. There are twelve pairs of **ribs**, one pair for each thoracic vertebra. Ribs 1–7 are called **true ribs** because each true rib connects individually to the sternum by separate cartilaginous extensions called costal cartilages. Ribs 8–12 are called **false ribs** because their costal cartilages do not attach directly to the sternum. The two pairs of false ribs (ribs 11 and 12) are called **floating ribs** because they have no connection to the sternum. Obtain a typical rib (any rib other than ribs 1, 2, 11, or 12) **(figure 8.23).**
2. When observing the features of a typical rib, pay particular attention to the surfaces of the rib that articulate with the thoracic vertebrae.
3. Observe the articulations between the ribs and the thoracic vertebrae on an articulated skeleton, and review the unique features of thoracic vertebrae that allow them to form articulations with the ribs.
4. Using table 8.5 and the textbook as guides, identify the structures listed in figure 8.23 on a typical rib. Then label them in figure 8.23.

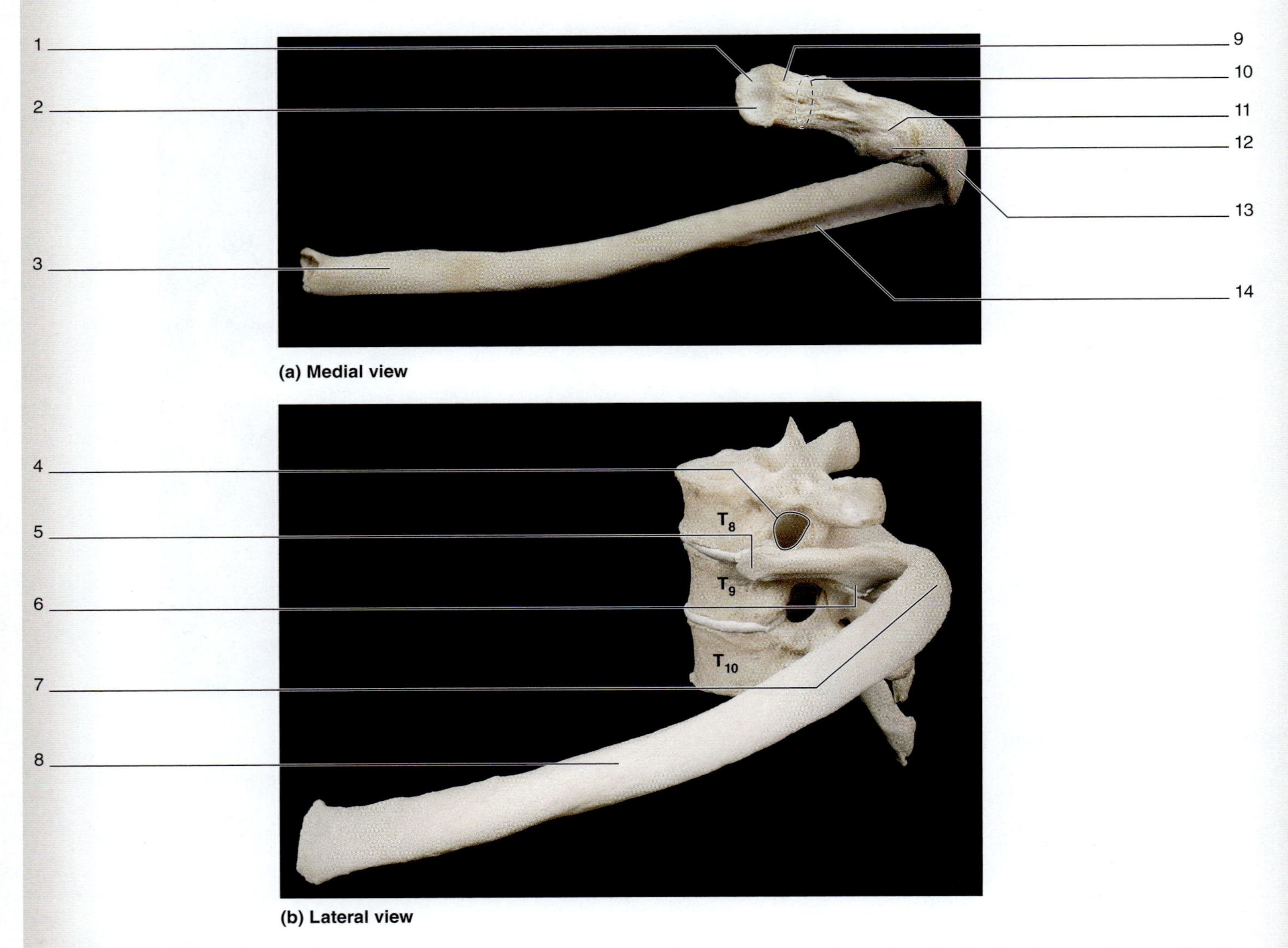

Figure 8.23 **A Typical Rib.** (a) Medial view; (b) lateral view of the 9th rib articulated with vertebrae $T_8 - T_{10}$. Use the terms listed to fill in the numbered labels in the figure. Some answers may be used more than once.

- ☐ angle
- ☐ articular facet for transverse process
- ☐ body
- ☐ costal groove
- ☐ head
- ☐ inferior articular facet
- ☐ intervertebral foramen
- ☐ neck
- ☐ shaft
- ☐ superior articular facet
- ☐ tubercle

Chapter 8: The Skeletal System: Axial Skeleton

Name: ______________________

Date: __________ Section: __________

POST-LABORATORY WORKSHEET

The (1) corresponds to the Learning Objective(s) listed in the chapter opener outline.

Do You Know the Basics?

Exercise 8.1: Anterior View of the Skull

1. Which of the following bone(s) is/are visible from the anterior view of the skull? (Check all that apply.) (1)

   ______ a. ethmoid bone ______ b. frontal bone ______ c. occipital bone ______ d. parietal bone ______ e. sphenoid bone

2. Which of the following skull bones form the orbit of the eye? (Check all that apply.) (2)

   ______ a. ethmoid bone ______ b. frontal bone ______ c. lacrimal bone ______ d. maxilla ______ e. nasal bone ______ f. zygomatic bone

3. The perpendicular plate of the ethmoid bone and the vomer form the nasal septum. ________________ (True/False) (3)

4. The projection of the mandible that forms the anterior part of the chin is known as the ________________. (Circle one.) (4)

   a. alveolar process b. angle c. body d. mental foramen e. mental protuberance

Exercise 8.2: Additional Views of the Skull

5. Match the bone marking listed in column A with the bone it is associated with, listed in column B. Some answers in column B may be used more than once. (5)

Column A	*Column B*
____ 1. foramen ovale	a. mandible
____ 2. mastoid process	b. occipital bone
____ 3. jugular foramen	c. sphenoid bone
____ 4. styloid process	d. temporal bone
____ 5. mental foramen	
____ 6. foramen spinosum	

6. The sagittal suture joins the two parietal bones. ________________ (True/False) (5)

7. The zygomatic arch is composed of which of the following? (Check all that apply.) (6)

   ____ a. frontal process of the zygomatic bone

   ____ b. mental foramen of the mandible

   ____ c. palatine process of the maxilla

   ____ d. temporal process of the zygomatic bone

   ____ e. zygomatic process of the temporal bone

Exercise 8.3: Superior View of the Cranial Floor

8. a. What does the term *petrous* mean? (7) ________________________________

   b. Why is this term used to describe the petrous part of the temporal bone?

   ____

   ____

9. Match the bony landmark listed in column A with the cranial fossa in which the landmark can be found, listed in column B. Some answers in column B may be used more than once. (8)

Column A	*Column B*
____ 1. internal acoustic meatus	a. anterior cranial fossa
____ 2. sella turcica	b. middle cranial fossa
____ 3. lesser wing of the sphenoid bone	c. posterior cranial fossa
____ 4. foramen ovale	
____ 5. crista galli	

10. a. What does the term *lacero* mean? (8) ______

    b. Why is the term *lacero* used to describe the foramen lacerum? ______

    ______

    ______

    c. What is unique about the foramen lacerum (as compared to other cranial foramina)? ______

    ______

    ______

Exercise 8.4: Bones Associated with the Skull

11. The only bone of the axial skeleton that does not articulate with any other bone is the (9) ______.

Exercise 8.5: The Fetal Skull

12. a. What is a fontanel? (10) ______

    b. By what age do most of the fontanels completely close? ______

Exercise 8.6: Vertebral Column Regions and Curvatures

13. The primary curvatures of the vertebral column are the *cervical* and *lumbar* curvatures, whereas the secondary curvatures are the *thoracic* and *sacral* curvatures. ______ (True/False). (11)

Exercise 8.7: Structure of a Typical Vertebra

14. Which of the following compose the vertebral arch? (Check all that apply.) (12)

    ______ a. inferior articular process ______ b. lamina ______ c. pedicle ______ d. spinous process ______ e. transverse process

Exercise 8.8: Characteristics of Individual Vertebrae

15. a. What foramen is present in cervical vertebrae, but is not present in other vertebrae? (13) ______

    b. What structure runs through this foramen in a living human? ______

16. The superior and inferior articular processes of the atlas are oriented along a vertical plane, whereas the inferior articular processes of "typical" vertebrae are oriented along a horizontal plane. ______ (True/False). (14)

17. The dens (odontoid process) is a projection of the ______ (atlas/axis). The dens arose from the body of the ______ (atlas/axis) (14)

18. Identify the locations on a thoracic vertebra where the ribs articulate. (Check all that apply.) (15)

    ______ a. inferior costal facets ______ b. pedicle ______ c. spinous process ______ d. superior costal facets ______ e. transverse process

19. The anterior and posterior sacral foramina are the equivalent of the intervertebral foramina in other regions of the vertebral column. ______ (True/False) (16)

Exercise 8.9: The Sternum

20. Identify the bones that compose the thoracic cage. (Check all that apply.) (17)

    ______ a. cervical vertebrae ______ b. clavicle ______ c. ribs ______ d. sternum ______ e. thoracic vertebrae

21. The sternal angle is the point of articulation between which two structures? (Circle one.) (17)

    a. body and manubrium of the sternum
    b. body and xiphoid of the sternum
    c. body of the sternum and second rib
    d. manubrium of the sternum and clavicle
    e. manubrium and xiphoid of the sternum

22. The clinical significance of the sternal angle is that it serves as the point of articulation of which rib? (Circle one.) (18)

    a. first b. second c. third d. fourth e. fifth

Exercise 8.10: The Ribs

23. Match the description listed in column A with the features of the typical rib, listed in column B. 19

Column A	*Column B*
____ 1. a projection between the shaft and neck	a. costal groove
____ 2. the main part (body) of a rib	b. neck
____ 3. a depression that contains the intercostal artery, vein, and nerve	c. shaft
____ 4. a narrow region where the head meets the tubercle	d. tubercle

24. The ______________________________ of a rib articulates with the bodies of the thoracic vertebrae. 20

Can You Apply What You've Learned?

25. What is a functional consequence of the shape (and arrangement) of the superior and inferior articular processes of the lumbar vertebrae? (Hint: Put two of them together and see what movement is, or is not, allowed.) ______________________________

____

26. The optic nerve extends from the eye toward the brain by traveling through the optic foramen. The optic foramen is a hole in what bone? ______________

____

27. Explain the difference between the vertebral foramen and the intervertebral foramen. ______________________________

____

28. Examine the skeleton and the different types of vertebrae to answer these questions.

    a. In the photos below, circle the structures that differentiate cervical, thoracic, and lumbar vertebrae on the photo of each vertebra. Then describe the feature(s) in the numbered space(s) below each figure.

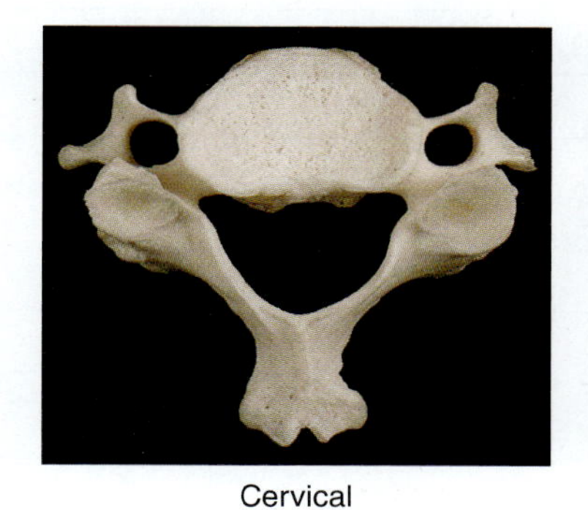

Cervical

1. ____________________
2. ____________________
3. ____________________

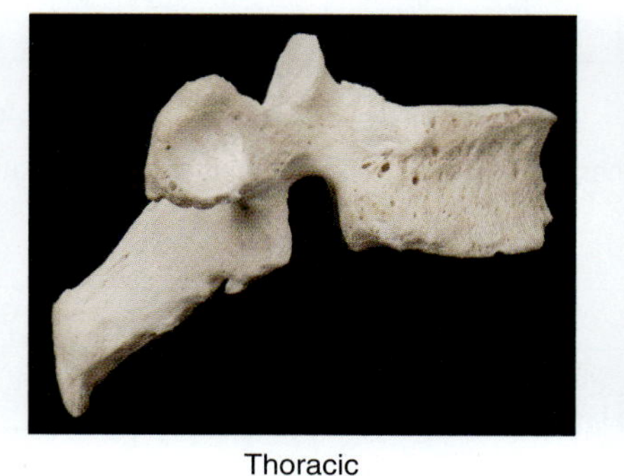

Thoracic

1. ____________________
2. ____________________
3. ____________________

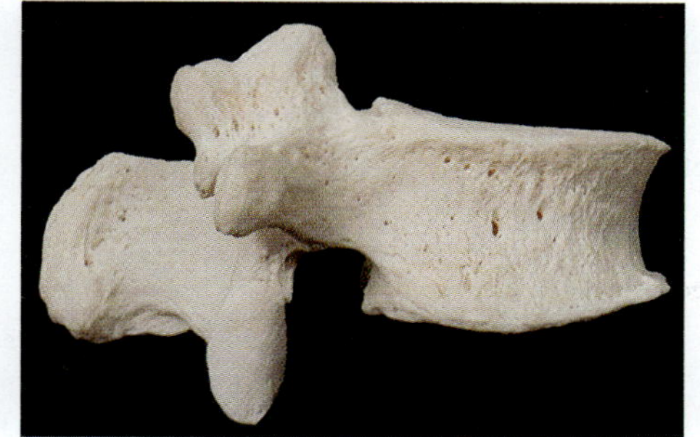

Lumbar

1. ____________________
2. ____________________
3. ____________________

29. Identify the region of the vertebral column where the intervertebral foramina (or their equivalent) project anteriorly and posteriorly to allow for the exit of spinal nerves. (Circle one.)

    a. cervical
    b. coccygeal
    c. lumbar
    d. thoracic
    e. sacral

30. a. The cervical vertebra that does not have a body or a spinous process is the ______________.

    b. The cervical vertebra that does not have a bifid spinous process is ______________.

    c. The dens (odontoid process) is a component of this cervical vertebra: ______________.

31. a. Does the atlantooccipital joint allow the head movement that indicates "yes" or "no"? ______________

    b. Does the atlantoaxial joint allow the head movement that indicates "yes" or "no"? ______________

32. What bone(s) does the sacrum articulate with superiorly? ______________ inferiorly? ______________ laterally? ______________

33. Describe the two points of articulation between a rib and the thoracic vertebrae.

____

____

34. a. The manubrium of the sternum articulates with rib ____________ and the body of the sternum articulates with ribs ____________ to ____________. Does the xiphoid process articulate with the ribs? ____________

    b. True ribs articulate with the sternum by individual costal cartilages. The true ribs are ____________ to ____________. The rest (ribs ____________ to ____________) are false ribs. The two false ribs that are also floating ribs are ribs ____________ and ____________.

Can You Synthesize What You've Learned?

35. Why do the fontanels persist until well after the birth of an infant?

____

____

____

36. How might the bifid spinous processes of cervical vertebrae affect anterior-posterior movement in the cervical region of the vertebral column?

____

____

37. How do the superior and inferior articular processes of the atlas differ from the same processes on a "typical" vertebra? How does this difference contribute to the special movement allowed at the atlanto-occipital and atlantoaxial joints?

____

____

____

38. Notice that when two lumbar vertebrae are put together, little to no lateral rotation is allowed because of the shape of the articulating bones. Why do you think they interact in this manner?

____

____

39. What part of the vertebral column is removed when a laminectomy is performed? ____________

40. Needles inserted into the thoracic cavity must always be placed along the superior border of a rib so as not to injure important structures. What important structures could be damaged by insertion of a needle too close to the inferior border of a rib? (Hint: Refer to table 8.5.)

____

____

The Skeletal System: Appendicular Skeleton

OUTLINE AND LEARNING OBJECTIVES

Anatomy & Physiology REVEALED® aprevealed.com **MODULE 5: SKELETAL SYSTEM**

INTRODUCTION

The appendicular skeleton is composed of the bones that attach to the axial skeleton. The term **appendicular** comes from the word *appendage.* A dictionary definition of an appendage is *something that is added or attached to an item that is larger or more important.* While the appendages of the human body (the upper and lower limbs) *could* be described as simply "added" structures, most would consider the upper and lower limbs to be essential. For, without the limbs, most of the movement essential to the movement of the human body would be absent. The exercises in chapter 8 covered the bones that compose the axial skeleton, which serves as the structural support of the body. The exercises in this chapter cover the appendicular skeleton, which includes the bones of the pectoral girdle, the upper limbs, the pelvic girdle, and the lower limbs. Each "girdle" serves to attach either upper or lower limbs to the axial skeleton. The bones of the **pectoral girdle** (the clavicles and scapulae) attach the **upper limbs,** whereas the bones of the **pelvic girdle** (ossa coxae or hip bones) attach the **lower limbs.** The exercises in this chapter cover identification of the bones of the appendicular skeleton, and the features of each bone.

Bones and their bony features are summarized in separate tables, one table for each major area of the appendicular skeleton. These tables include the name of the bone, the bony feature, a description, and the word origin. Word origins often provide information that make it easier to remember the structures. For example, the *conoid tubercle* of the clavicle gets its name from its conical shape (*konoeides,* cone-shaped). The *coracoid* process of the scapula is named for its resemblance to a crow's beak (*karakodes,* like a crow's beak). The two names look very similar and can be easily confused, so pay close attention to their meanings.

The exercises in this chapter guide the user in studying the bones that compose the appendicular skeleton on an articulated human skeleton, on disarticulated bones of the human skeleton, and on bone models. The goal of these exercises is to identify the major bones, identify features of each bone, and associate the observed structures with their functions. Use the textbook as a reference while completing the labeling exercises.

List of Reference Tables

INTEGRATE

CONCEPT CONNECTION

Learning the bony features of the appendicular skeleton better prepares one to learn attachment points and functions of the appendicular muscles, which are covered in chapter 12. Many bony landmarks (e.g., tuberosities, trochanters, tubercles) exist because of the pulling action of muscles that attach to them and stress them as they develop. For example, without a sternocleidomastoid muscle pulling on the mastoid process of the temporal bone, a mastoid process would not exist. While observing the features of the bones, remember that bony features tell a story about the development of the musculoskeletal system.

Chapter 9: The Skeletal System: Appendicular Skeleton

Name: ______________________________

Date: ____________ Section: ____________

These Pre-Laboratory Worksheet questions may be assigned by instructors through their connect course.

PRE-LABORATORY WORKSHEET

1. The appendicular skeleton is composed of which of the following? (Check all that apply.)
   ____ a. lower limb bones
   ____ b. pectoral girdle
   ____ c. pelvic girdle
   ____ d. thoracic cage
   ____ e. upper limb bones

2. Which of the following bones compose the pectoral girdle? (Check all that apply.)
   ____ a. clavicle
   ____ b. humerus
   ____ c. ribs
   ____ d. scapula
   ____ e. sternum

3. Which of the following bones composes part of the os coxa of the pelvic girdle? (Check all that apply.)
   ____ a. femur
   ____ b. ilium
   ____ c. ischium
   ____ d. pubis
   ____ e. sacrum

4. In the anatomic position, the radius lies ____________________ (medial/lateral) to the ulna.

5. In the anatomic position, the tibia lies ____________________ (medial/lateral) to the fibula.

6. Carpal bones are located in the wrist, whereas the tarsal bones are located in the ankle. ____________________ (True/False)

7. Match the description listed in column A with the appropriate bone listed in column B.

Column A	*Column B*
____ 1. a bone that has two large tubercles on its proximal end	a. carpals
____ 2. a bone that has two large trochanters on its proximal end	b. femur
____ 3. a bone that contains the olecranon process	c. humerus
____ 4. bones that form the wrist	d. patella
____ 5. a sesamoid bone found in the knee	e. tibia
____ 6. the largest bone in the leg	f. ulna

8. The lateral malleolus is a feature of which bone? (Circle one.)
   a. calcaneus
   b. femur
   c. fibula
   d. talus
   e. tibia

9. Identify the bone that has both an acromial and a coracoid process. (Circle one.)
   a. clavicle
   b. humerus
   c. radius
   d. scapula
   e. ulna

10. The calcaneus is located in the wrist, whereas the pisiform bone is located in the heel. ____________________ (True/False)

GROSS ANATOMY

The Pectoral Girdle

The **pectoral girdle** (*girdle*, a belt) is an incomplete bony ring formed by the paired clavicles and scapulae. The function of these bones is to act as the bony support to attach the upper limb to the axial skeleton. Each upper limb is attached to the axial skeleton by one bony joint (the sternoclavicular joint) and numerous ligaments and muscles. **Table 9.1** lists the bones of the pectoral girdle and describes the key features of each bone.

Table 9.1 The Appendicular Skeleton: Pectoral Girdle

Bone	Bony Landmark	Description	Word Origin
Clavicle clavicula, a small key	Acromial end	The flattened, lateral end of the clavicle; articulates with scapula superior to the shoulder joint	*akron*, tip, + -omos, shoulder
	Conoid tubercle	A small "cone-shaped" projection on the lateral, inferior end of the clavicle; attachment for conoid ligament	*konoeides*, cone-shaped
	Costal tuberosity	A rough impression on the inferior surface of the sternal end; serves as the attachment point for the costoclavicular ligament	*costa*, rib
	Sternal end	The triangular medial end of the clavicle; articulates with sternum	*sternon*, chest
Scapula scapula, the shoulder blade	Acromion	The large process at the lateral tip of the scapular spine, which projects laterally and slightly anteriorly; articulates with acromial end of clavicle	*akron*, tip, + *-omos*, shoulder
	Coracoid process	The smaller approximately C-shaped process that projects anteriorly; attachment site for several ligaments and muscles	*korakodes*, like a crow's beak
	Glenoid fossa	A shallow depression on the superior, lateral border; articulates with humerus	*glenoeides*, resembling a socket
	Inferior angle	The angle between the medial and lateral borders	*inferior*, lower
	Infraglenoid tubercle	A rough projection at the inferior border of the glenoid fossa; attachment point for the long head of the triceps brachii muscle	*infra*, below, + *glenoeides*, resembling a socket
	Infraspinous fossa	A large depression inferior to the scapular spine; attachment for the infraspinatus and teres minor muscles	*infra*, below, + *spina*, spine
	Lateral (axillary) border	The border of the scapula that has the glenoid fossa on its superior part	*axilla*, armpit
	Medial (vertebral) border	The longest border of the scapula; contains very few notable features	*medialis*, middle
	Spine	A long "spiny" process on the posterior surface; attachment point for trapezius and deltoid muscles	*spina*, spine
	Subscapular fossa	A large depression on the anterior surface of the bone; origin of the subscapularis muscle	*sub*, under, + *spina*, spine
	Superior angle	The angle between the superior and medial borders	*superus*, above
	Superior border	The border from which the acromial and coracoid processes project	*superus*, above
	Supraglenoid tubercle	A rough projection at the superior border of the glenoid fossa; attachment point for the long head of the biceps brachii muscle	*supra*, on the upper side, + *glenoeides*, resembling a socket
	Suprascapular notch	A small, deep notch just medial to the coracoid process; the suprascapular nerve, artery, and vein pass through the notch	*supra*, on the upper side, + *scapula*, shoulder blade
	Supraspinous fossa	A large depression superior to the scapular spine; attachment of the supraspinatus muscle	*supra*, on the upper side, + *spina*, spine

EXERCISE 9.1

BONES OF THE PECTORAL GIRDLE

EXERCISE 9.1A The Clavice

1. Observe the clavicle **(figure 9.1)** on an articulated skeleton (or see figure 7.10, which shows a full skeleton). The **clavicle** is an S-shaped bone that extends between the sternum and the scapula and has very few muscular attachments compared to other bones. Rather than acting as a rigid attachment point for muscles, the clavicle functions more like a strut that pushes the shoulders laterally and keeps them from collapsing anteriorly toward the sternum. Notice the location where the clavicle joins with the sternum, which is called the **sternoclavicular joint**. This is the only point of attachment between the pectoral girdle (and upper limb) and the axial skeleton.
2. Palpate your own clavicle. The superficial location of the clavicle allows this to be done relatively easily. Locate the two ends of the clavicle and name the bones that articulate with each end of the clavicle.
3. Obtain a disarticulated clavicle. Identify the bony landmarks of a clavicle that are listed in table 9.1.
4. Obtain both a right and a left disarticulated clavicle. What distinguishes a right clavicle from a left clavicle?

   ____

   ____

5. Label the structures on the clavicle shown in figure 9.1, using table 9.1 and the textbook as guides.

EXERCISE 9.1B The Scapula

1. Observe the scapula **(figure 9.2)** on an articulated skeleton (or see figure 7.10, which shows a full skeleton). The **scapula** (*scapula*, shoulder blade) is a large, irregular bone that is not directly attached to the axial skeleton. Notice that the scapula articulates with the sternum only at the acromion. (Recall that the clavicle articulates with the axial skeleton only at the sternum.) Having only one point of articulation between the bones of the pectoral girdle and the axial skeleton makes the shoulder somewhat unstable, but at the same time it allows for a great deal of flexibility in movements of the upper limb.
2. Palpate your own scapula. Locate the spine, medial border, and inferior angle. Palpate the joint between the acromion and the scapula. To do this, begin by palpating the clavicle. Then "walk" your fingers laterally until you reach the tip of the shoulder, where you will feel the acromion of the scapula. If you raise and lower your upper limb while keeping your fingers on the bones, you will be able to feel the joint between the clavicle and the scapula, the **acromioclavicular joint**.
3. Obtain a disarticulated scapula. Identify the bony landmarks of the scapula listed in table 9.1.
4. Obtain both a right and a left disarticulated scapula. What distinguishes a right scapula from a left scapula?

   ____

   ____

   ____

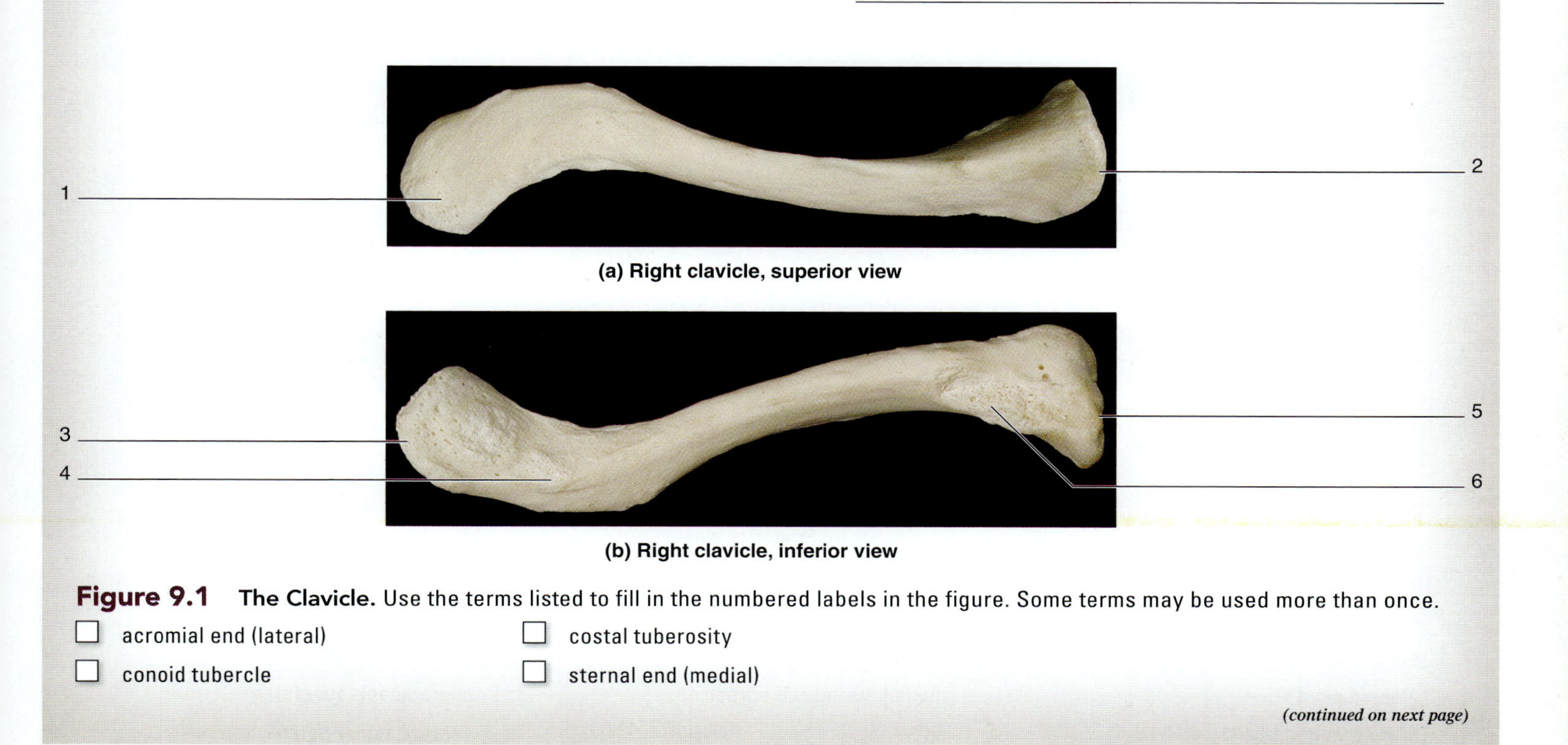

Figure 9.1 The Clavicle. Use the terms listed to fill in the numbered labels in the figure. Some terms may be used more than once.

- ☐ acromial end (lateral)
- ☐ conoid tubercle
- ☐ costal tuberosity
- ☐ sternal end (medial)

(continued on next page)

(continued from previous page)

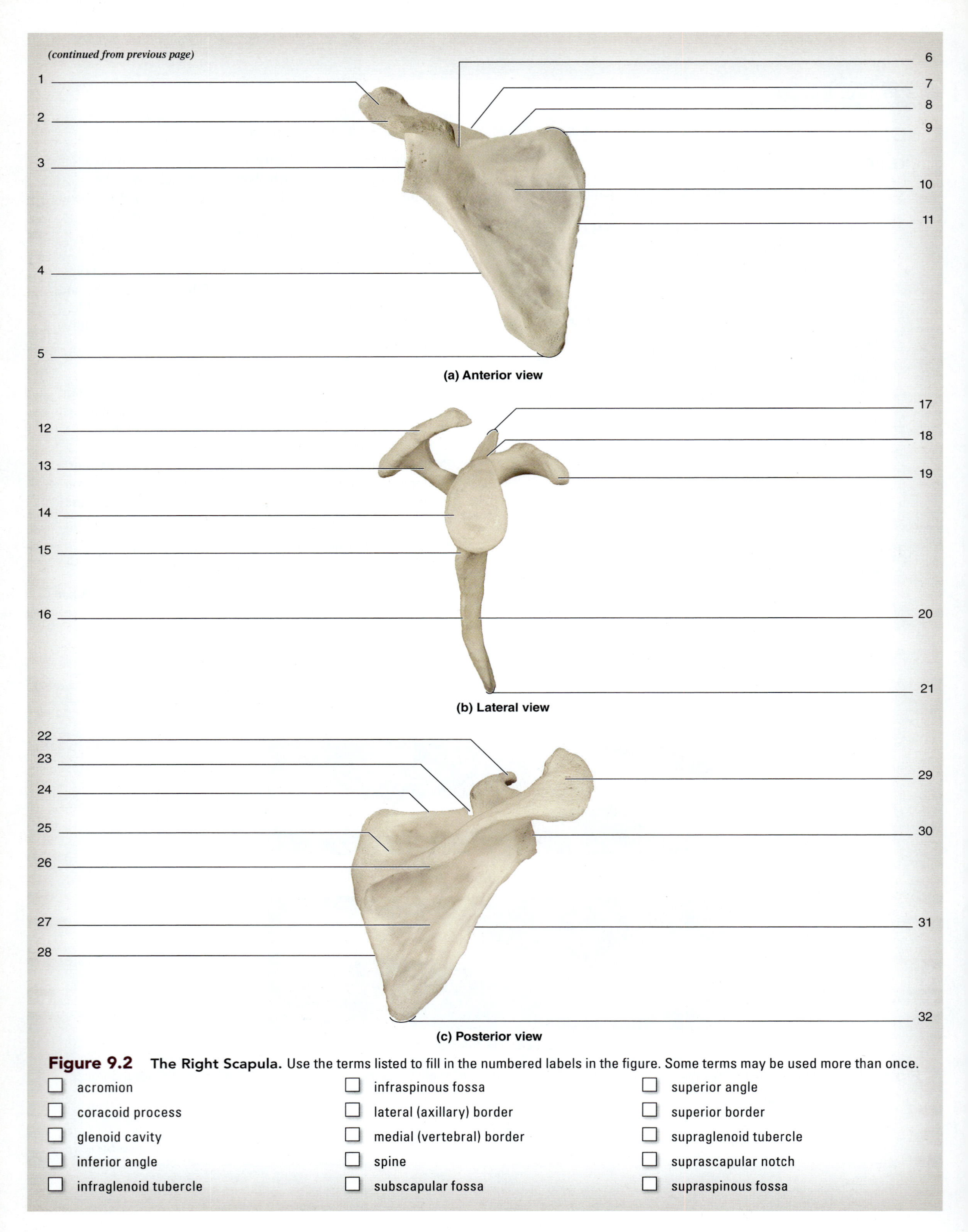

(a) Anterior view

(b) Lateral view

(c) Posterior view

Figure 9.2 **The Right Scapula.** Use the terms listed to fill in the numbered labels in the figure. Some terms may be used more than once.

- ☐ acromion
- ☐ coracoid process
- ☐ glenoid cavity
- ☐ inferior angle
- ☐ infraglenoid tubercle
- ☐ infraspinous fossa
- ☐ lateral (axillary) border
- ☐ medial (vertebral) border
- ☐ spine
- ☐ subscapular fossa
- ☐ superior angle
- ☐ superior border
- ☐ supraglenoid tubercle
- ☐ suprascapular notch
- ☐ supraspinous fossa

5. Identify the structures listed in figure 9.2 on the scapula, using table 9.1 and the textbook as guides. Then label them in figure 9.2.

6. Obtain a right clavicle and a right scapula and articulate them with each other. Sketch the relationship between the clavicle and the scapula in the space provided.

7. What is the name of the joint that forms between the clavicle and scapula?

____

INTEGRATE

CLINICAL VIEW
Clavicular Fracture

A fracture of the clavicle (or broken collarbone) is a common injury that occurs either from a direct blow to the clavicle (e.g., while playing contact sports) or from force applied to an outstretched arm (e.g., when reaching out to brace oneself in a fall). Babies are also susceptible to the injury during childbirth if the shoulders are overly compressed during passage through the birth canal. A clavicular fracture may occur at any location along the clavicle. The most common site of fracture is the middle of the bone. The clavicle may also break at or near its attachment point with the scapula (acromioclavicular joint), or the sternum (sternoclavicular joint).

Signs and symptoms of a clavicular fracture include a "sagging" shoulder, limited range of motion, pain, swelling, and sometimes bruising. Nonsurgical interventions involve placing the arm in a sling to limit movement of the shoulder joint. Healing time can be lengthy due to the risk of reinjury. Severe fractures require surgical intervention, which involves the placement of plates and screws in the clavicle to realign the bone and allow it to heal. In either case, physical therapy is recommended to strengthen the muscles acting at the shoulder and restore range of motion.

The Upper Limb

The bones of the upper limb consist of the humerus, radius, ulna, carpals, metacarpals, and phalanges. Nearly all of the projections on these bones serve as attachment points for the muscles that move the upper limb. **Table 9.2** lists the bones of the upper limb and describes their key features.

Table 9.2 The Appendicular Skeleton: Upper Limb

Bone	Bony Landmark	Description	Word Origin
Humerus **humerus,** ***shoulder***	Anatomical neck	The narrow part between the head and the tubercles; location of the former epiphyseal (growth) plate	NA
	Capitulum	The rounded surface (condyle) on the distal end of the humerus; articulates with the radius	*caput*, head
	Coronoid fossa	A depression for the coronoid process of the ulna	*corona*, crown, + *eidos*, resemblance, + *fossa*, trench
	Deltoid tuberosity	A rough projection on the proximal diaphysis; attachment for deltoid	*delta*, triangle, + *eidos*, resemblance
	Greater tubercle	A large lateral projection on the proximal epiphysis; attachment for rotator cuff muscles	*tuber*, a knob
	Head	A rounded projection on the medial side of the proximal epiphysis; articulates with the glenoid cavity of the scapula	*head*, the rounded extremity of a bone
	Intertubercular sulcus	A groove between the greater and lesser tubercles; also called bicipital groove	*inter-*, between, + *tuber*, knob
	Lateral epicondyle	A rough ridge proximal to the capitulum; attachment for radial collateral ligament	*epi-*, above, + *kondylos*, knuckle
	Lesser tubercle	A small, rough projection on the anterior side of the proximal epiphysis; attachment for subscapularis	*tuber*, a knob
	Medial epicondyle	A rough ridge proximal to the trochlea; attachment for ulnar collateral ligament	*epi-*, above, + *kondylos*, knuckle
	Olecranon fossa	A depression on the posterior surface at the distal end of humerus; articulates with olecranon process of the ulna	*olecranon*, the head of the elbow, + *fossa*, trench

(continued on next page)

Table 9.2 The Appendicular Skeleton: Upper Limb *(continued)*

Bone	Bony Landmark	Description	Word Origin
***Humerus* (continued)**	Radial fossa	A depression on the anterolateral surface of the distal end of the humerus; articulates with the head of the radius	*radius*, spoke of a wheel, + *fossa*, trench
	Supracondylar ridges	Sloped ridges (medial and lateral) proximal to the epicondyles; attachment points for several muscles	*supra-*, on the upper side, + *kondylos*, knuckle
	Surgical neck	The narrowing of the shaft immediately distal to the tubercles; the most common site of fracture of the humerus	NA
	Trochlea	The rounded surface (condyle) that articulates with the ulna	*trochileia*, a pulley
Ulna ulna, elbow	Coronoid process	A projection on the proximal, anterior surface of the ulna; articulates with the humerus	*corona*, crown, + *eidos*, resemblance
	Olecranon	A large projection on the proximal, posterior surface of the ulna; forms the point of the elbow and is an attachment for the triceps brachii muscle	*olecranon*, the head of the elbow
	Radial notch	A depression on the lateral, proximal surface of the ulna; articulates with the head of the radius	*radius*, spoke of a wheel
	Styloid process	A pointed process on the distal ulna; forms the medial aspect of the wrist	*stylos*, pillar, + *eidos*, resemblance
	Trochlear notch	A C-shaped groove on the proximal end of the ulna; articulates with the trochlea of the humerus	*trochileia*, a pulley
	Tuberosity of ulna	A process on the anterior surface of the proximal end of the ulna; serves as an attachment point for the brachialis muscle	*ulna*, elbow
Radius radius, spoke of a wheel	Head	The disc-shaped proximal end of the radius; articulates with capitulum of the humerus	NA
	Neck	Narrow region where the head of the radius meets the shaft of the bone	NA
	Radial tuberosity	A large projection on the medial surface distal to the proximal epiphysis of the radius; serves as an attachment point for the biceps brachii muscle	*radio-*, ray
	Styloid process of the radius	A small, pointed projection on the distal radius; forms the lateral aspect of the wrist	*stylos*, pillar, + *eidos*, resemblance
Carpals carpus, wrist — ***Proximal row***	Scaphoid	A large, "boat-shaped" bone; articulates with the radius	*skaphe*, boat, + *eidos*, resemblance
	Lunate	A "moon-shaped" bone; articulates with the radius	*luna*, moon
	Triquetrum	A pyramid-shaped bone located between the pisiform, lunate, and hamate bones	*triquetrus*, three-cornered
	Pisiform	A "pea-shaped" bone; positioned on the medial, palmar surface of the wrist	*pisum*, pea, + *forma*, appearance
Distal row	Trapezium	A "table-shaped" bone located at the base of the first metacarpal (base of the thumb)	*trapezion*, a table
	Trapezoid	A "table-shaped" bone located at the base of the second metacarpal	*trapezion*, a table, + *eidos*, resemblance
	Capitate	A "head-shaped" bone located in the center of the wrist, at the base of the third metacarpal	*caput*, head
	Hamate	A "hook-shaped" bone located at the base of the fifth metacarpal	*hamus*, a hook
Metacarpals		Each bone consists of a head, body, and base	meta-, *after*, + carpus, *wrist*
Phalanges	NA	Bones of the fingers and thumb	*phalanx*, line of soldiers
II through V	Proximal	The phalanx closest to the palm of the hand	*proximus*, nearest
	Middle	The middle phalanx	NA
	Distal	The small, cone-shaped phalanx forming the tips of the fingers	*distalis*, away
Pollex (I) pollex, thumb	Proximal	The phalanx closest to the palm of the hand	*proximus*, nearest
	Distal	The small, cone-shaped phalanx forming the tip of the thumb	*distalis*, away

EXERCISE 9.2

BONES OF THE UPPER LIMB

EXERCISE 9.2A The Humerus

1. Observe the humerus **(figure 9.3)** on an articulated skeleton (or see figure 7.10, which shows a full skeleton). Locate the point of articulation with the scapula to form the glenohumeral joint. Locate the points of articulation with the ulna and radius to form the elbow joint.
2. Palpate your own humerus both at the shoulder joint and at the medial epicondyle, just proximal from the elbow joint.
3. Obtain a disarticulated humerus. Identify the bony landmarks of the humerus that are listed in table 9.2.
   a. What distinguishes the proximal end of the humerus from the distal end?

   ____

   ____

   ____

   b. What distinguishes the anterior surface of the humerus from the posterior surface?

   ____

   ____

   ____

4. Obtain both a right and a left disarticulated humerus. What distinguishes a right humerus from a left humerus?

   ____

   ____

   ____

5. Label the features of the humerus shown in figure 9.3, using table 9.2 and the textbook as guides.
6. Obtain a scapula and a humerus and articulate them with each other. Sketch the relationship between the scapula and the humerus in the space provided.

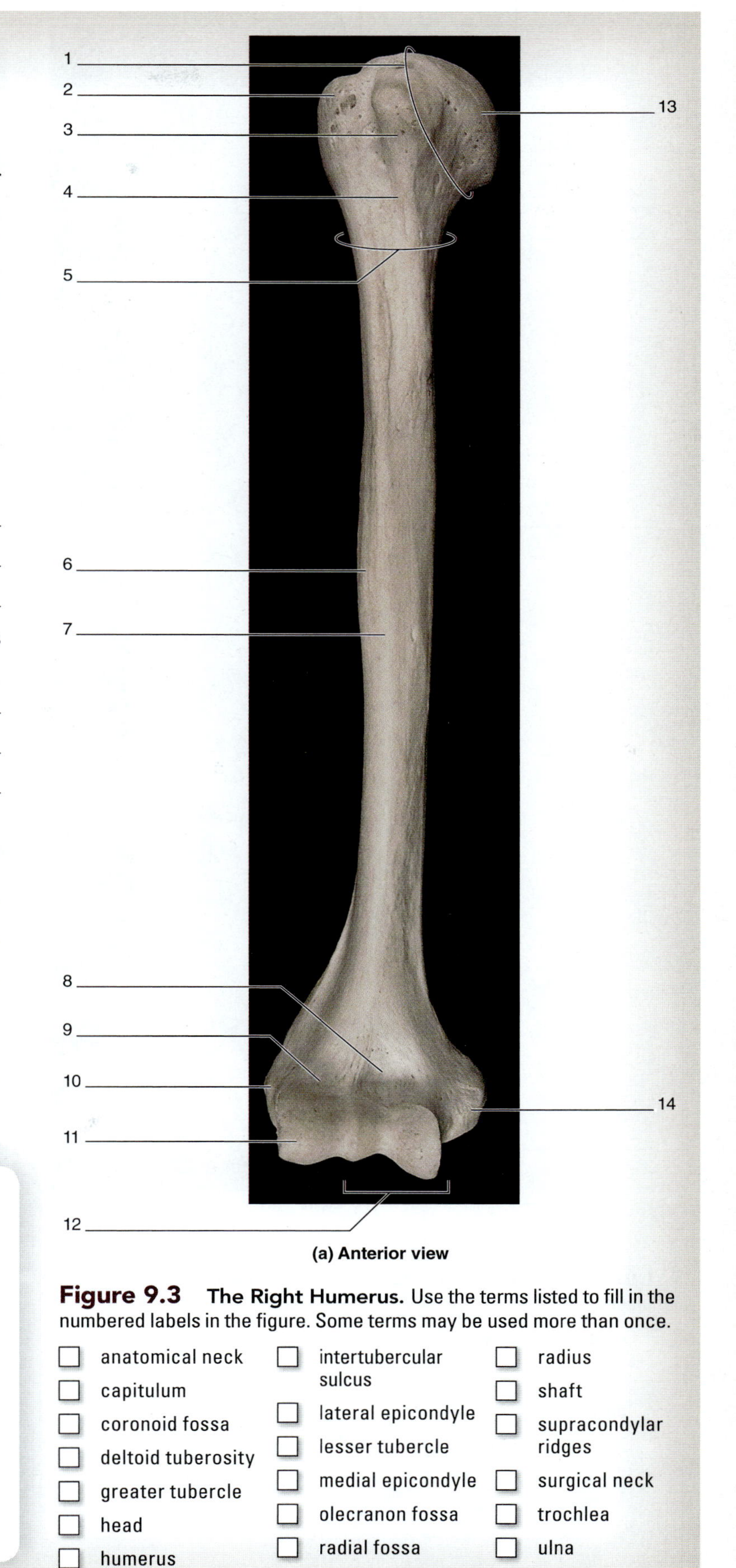

Figure 9.3 **The Right Humerus.** Use the terms listed to fill in the numbered labels in the figure. Some terms may be used more than once.

- ☐ anatomical neck
- ☐ capitulum
- ☐ coronoid fossa
- ☐ deltoid tuberosity
- ☐ greater tubercle
- ☐ head
- ☐ humerus
- ☐ intertubercular sulcus
- ☐ lateral epicondyle
- ☐ lesser tubercle
- ☐ medial epicondyle
- ☐ olecranon fossa
- ☐ radial fossa
- ☐ radius
- ☐ shaft
- ☐ supracondylar ridges
- ☐ surgical neck
- ☐ trochlea
- ☐ ulna

(continued on next page)

(continued from previous page)

15 ____
16 ____
17 ____
18 ____
19 ____
20 ____
21 ____
22 ____
23 ____

(b) Posterior view

24 ____
25 ____
26 ____
27 ____
28 ____
29 ____

(c) Bones of the elbow, anterior view

30 ____
31 ____
32 ____
33 ____
34 ____

(d) Bones of the elbow, posterior view

Figure 9.3 **The Right Humerus *(continued).*** Use the terms listed to fill in the numbered labels in the figure. Some terms may be used more than once.

- ☐ anatomical neck
- ☐ capitulum
- ☐ coronoid fossa
- ☐ deltoid tuberosity
- ☐ greater tubercle
- ☐ head
- ☐ humerus
- ☐ intertubercular sulcus
- ☐ lateral epicondyle
- ☐ lesser tubercle
- ☐ medial epicondyle
- ☐ olecranon fossa
- ☐ radial fossa
- ☐ radius
- ☐ shaft
- ☐ surgical neck
- ☐ trochlea
- ☐ ulna

EXERCISE 9.2B The Radius

1. Observe the radius **(figure 9.4)** on an articulated skeleton (or see figure 7.10, which shows a full skeleton). Locate its articulation with the humerus at the elbow. Also locate its articulation with the carpals at the wrist.
2. Palpate your own radius at the styloid process, which is on the lateral side of the wrist and is aligned with the thumb.
3. Obtain a disarticulated radius. Identify the bony landmarks of the radius that are listed in table 9.2.
   a. What distinguishes the proximal end of the radius from the distal end?

   ____

   ____

   ____

   b. What distinguishes the anterior surface of the radius from the posterior surface?

   ____

   ____

   ____

4. Obtain both a right and a left disarticulated radius. What distinguishes a right radius from a left radius?

   ____

   ____

   ____

   ____

5. Label the structures on the radius shown in figure 9.4, using table 9.2 and the textbook as guides.

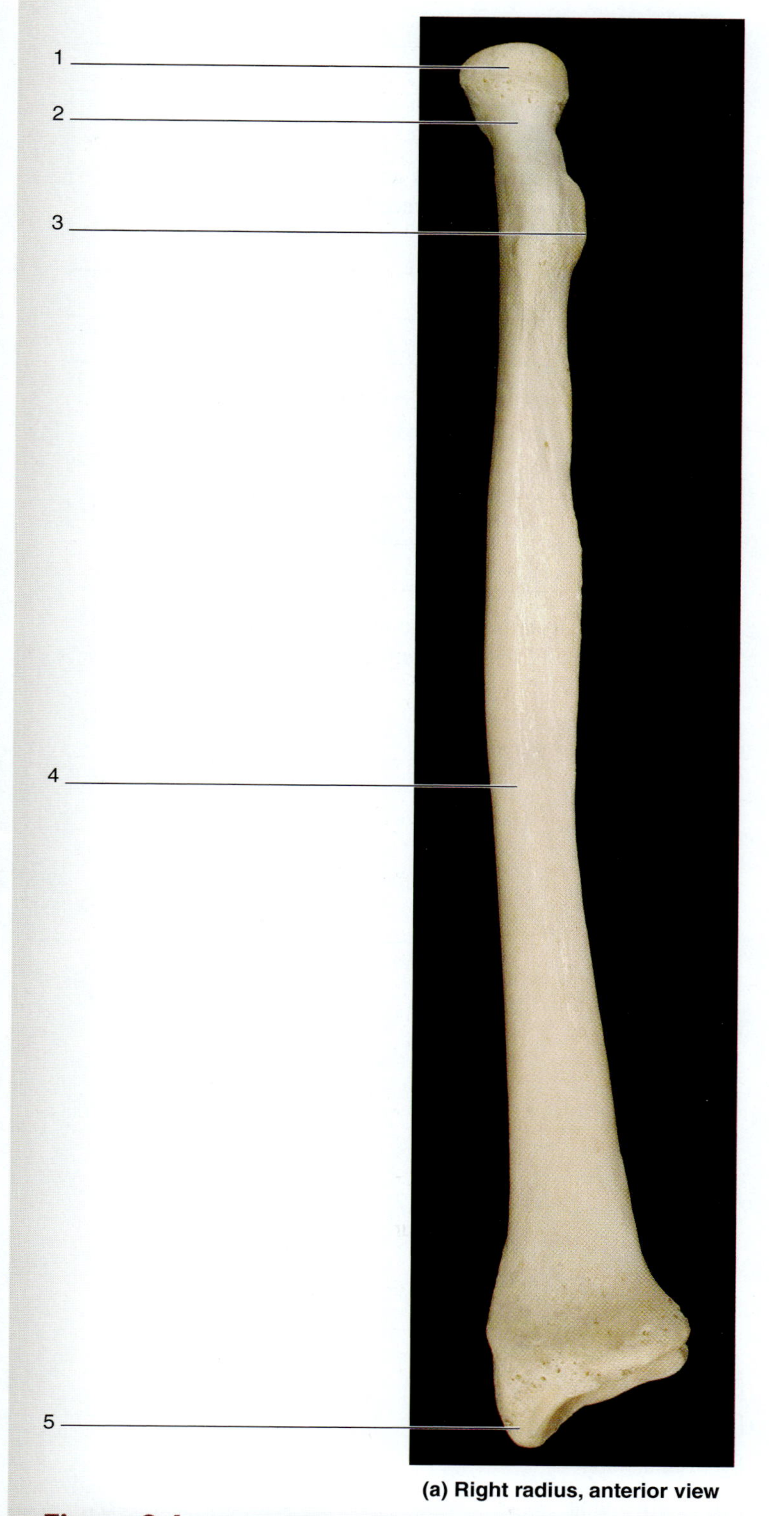

(a) Right radius, anterior view

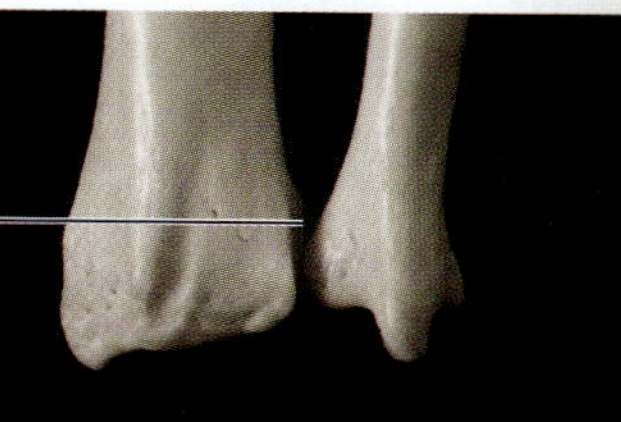

(b) Distal radius

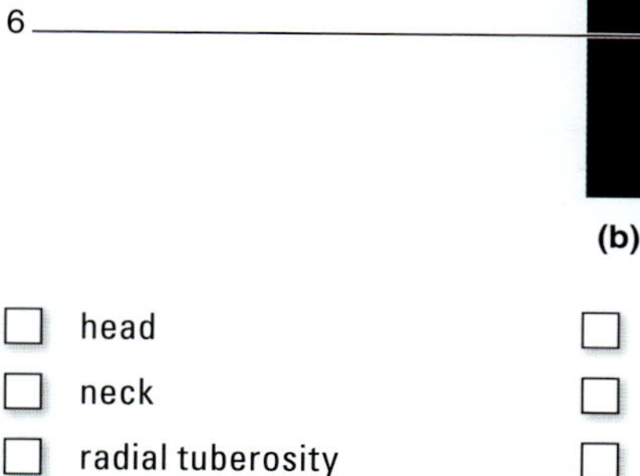

- ☐ head
- ☐ neck
- ☐ radial tuberosity
- ☐ shaft
- ☐ styloid process of radius
- ☐ ulnar notch

Figure 9.4 The Radius. Use the terms listed to fill in the numbered labels in the figure.

(continued on next page)

(continued from previous page)

EXERCISE 9.2C The Ulna

1. Observe the ulna **(figure 9.5)** on an articulated skeleton (or see figure 7.10, which shows a full skeleton). Locate its articulation with the humerus at the elbow and its articulation with the carpals at the wrist.

2. Palpate your own ulna both at the olecranon at the elbow and at the styloid process on the medial side of the wrist, which is aligned with the "little finger."

3. Obtain a disarticulated ulna. Identify the bony landmarks of the ulna that are listed in table 9.2.

   a. What distinguishes the proximal end of the ulna from the distal end?

   ____

   ____

   ____

   b. What distinguishes the anterior surface of the ulna from the posterior surface?

   ____

   ____

   ____

4. Obtain both a right and a left disarticulated ulna. What distinguishes a right ulna from a left ulna?

   ____

   ____

   ____

5. Label the structures on the ulna shown in figure 9.5, using table 9.2 and the textbook as guides.

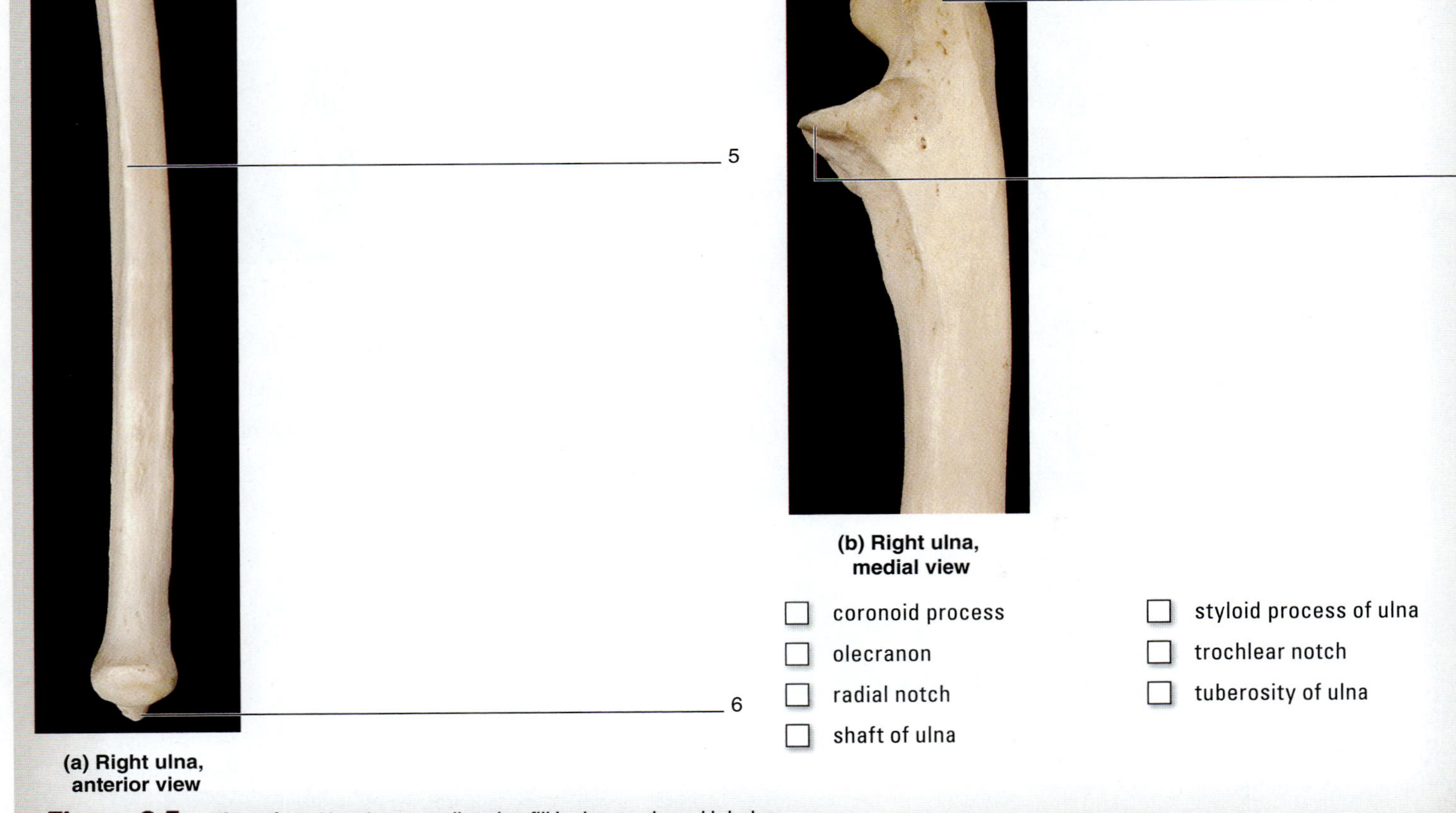

Figure 9.5 **The Ulna.** Use the terms listed to fill in the numbered labels in the figure. Some terms may be used more than once.

6. Obtain a right humerus, right radius, and right ulna and articulate them with each other. Sketch the relationship between the right humerus, radius, and ulna from an anterior view in the space provided.

EXERCISE 9.2D The Carpals

1. Obtain articulated bones of the wrist (carpals), or observe the **carpal** bones on an articulated skeleton **(figure 9.6).** The term *carpus* means "wrist." Thus, the carpal bones are the bones of the wrist.
2. The eight carpal bones are arranged in two rows of four bones each. The proximal row, which is adjacent to the radius and ulna, consists of the scaphoid, lunate, triquetrum, and pisiform. The distal row, which is adjacent to the metacarpals, consists of the trapezium, trapezoid, capitate, and hamate. Draw and label the carpal bones as they would be positioned in an anterior view of the wrist in the spaces provided.

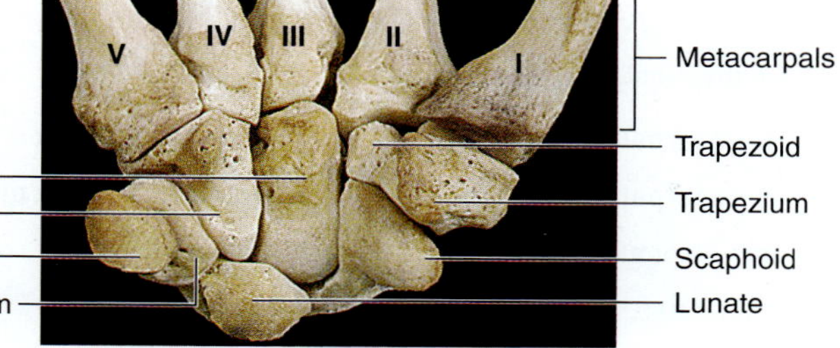

(a) Right wrist and hand, anterior view

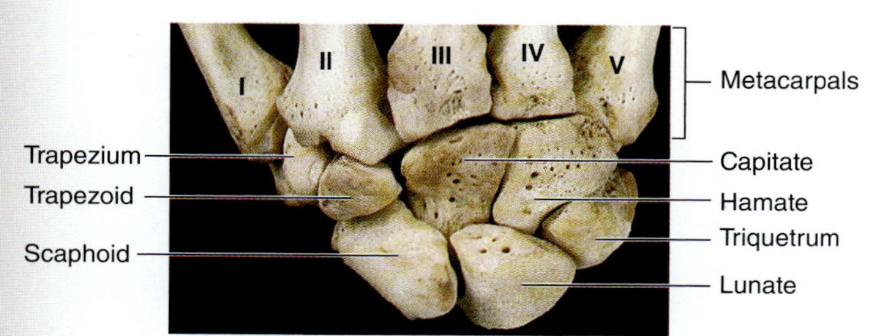

(b) Right wrist and hand, posterior view

Figure 9.6 The Carpals.

Proximal row

Distal row

3. Complete the following chart for the carpal bones, using table 9.2 as a reference.

Carpal Bone	Bone Shape/Appearance	Word Origin
Scaphoid		
Lunate		
Triquetrum		
Pisiform		
Trapezium		
Trapezoid		
Capitate		
Hamate		

4. *Optional Activity:* **AP|R 5: Skeletal System**—Visit the Quiz area for focused drill and practice on the bones of the appendicular skeleton.

(continued on next page)

(continued from previous page)

EXERCISE 9.2E The Metacarpals and Phalanges

1. Obtain articulated bones of the hand (metacarpals and phalanges), or observe bones of the hand on an articulated skeleton **(figure 9.7).**
2. **Metacarpals:** The prefix *meta-* means "after." Thus, the metacarpal bones are the bones that come after the carpals, and they are located in the palm of the hand. There are five metacarpal bones, numbered I through V. The first metacarpal is on the lateral surface of the hand and forms the base of the thumb, or **pollex**. Palpate the palm of your hand to feel the metacarpal bones.
3. **Phalanges:** The term *phalanx* means "a line of soldiers." Palpate the three phalanges that compose each of the four fingers. While typing on a keyboard writing a term paper or some other assignment, think about how these little "soldiers" march along the keyboard.
4. How do the phalanges (and metacarpal I) of the pollex differ from the phalanges of digits II–V?

____

____

____

____

5. Identify the metacarpals and phalanges on an articulated skeleton. Then label them in figure 9.7, using the textbook as a guide.

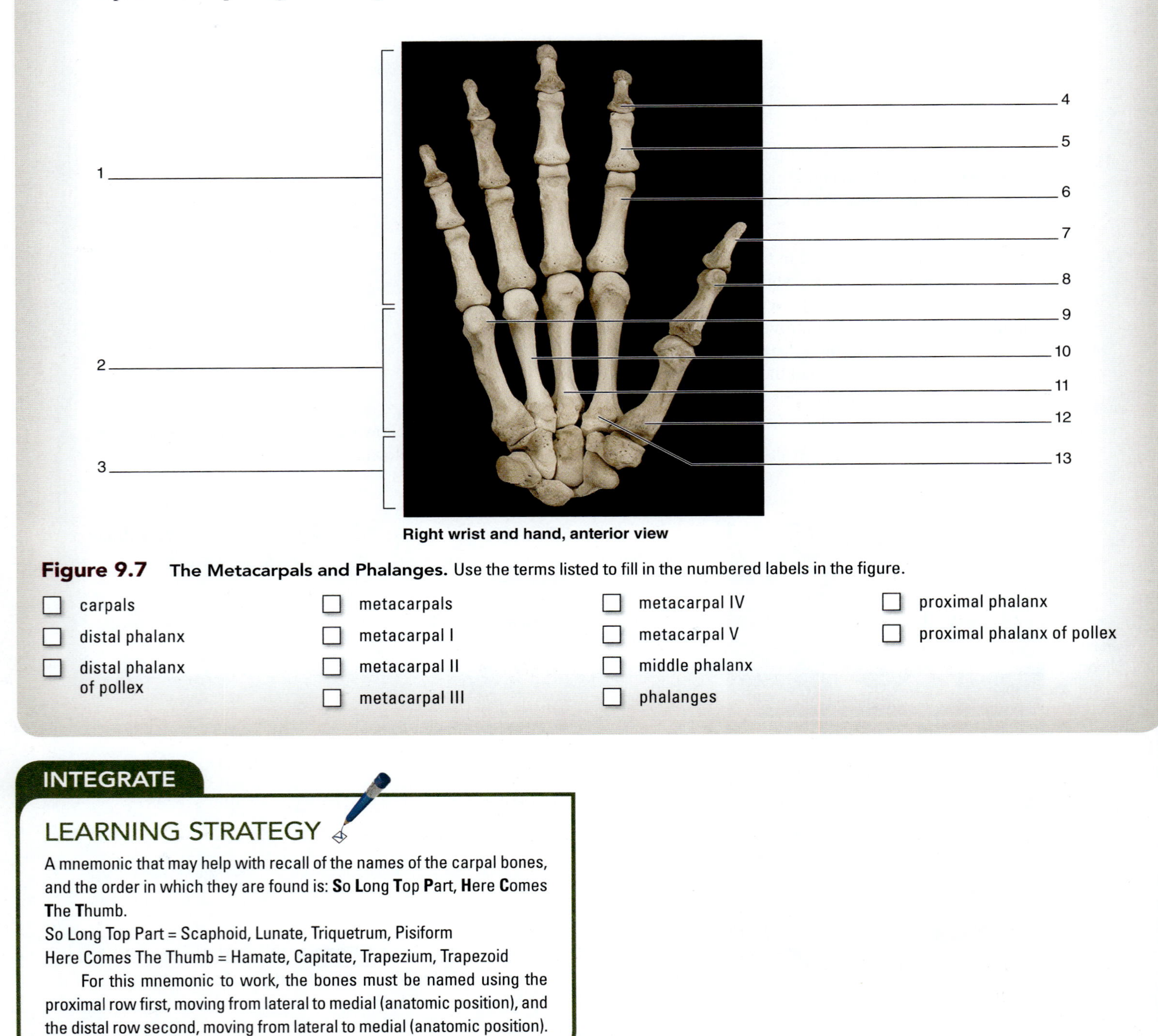

Figure 9.7 The Metacarpals and Phalanges. Use the terms listed to fill in the numbered labels in the figure.

- ☐ carpals
- ☐ distal phalanx
- ☐ distal phalanx of pollex
- ☐ metacarpals
- ☐ metacarpal I
- ☐ metacarpal II
- ☐ metacarpal III
- ☐ metacarpal IV
- ☐ metacarpal V
- ☐ middle phalanx
- ☐ phalanges
- ☐ proximal phalanx
- ☐ proximal phalanx of pollex

INTEGRATE

LEARNING STRATEGY

A mnemonic that may help with recall of the names of the carpal bones, and the order in which they are found is: **So Long Top Part, Here Comes The Thumb**.

So Long Top Part = Scaphoid, Lunate, Triquetrum, Pisiform

Here Comes The Thumb = Hamate, Capitate, Trapezium, Trapezoid

For this mnemonic to work, the bones must be named using the proximal row first, moving from lateral to medial (anatomic position), and the distal row second, moving from lateral to medial (anatomic position).

EXERCISE 9.3

SURFACE ANATOMY REVIEW—PECTORAL GIRDLE AND UPPER LIMB

In previous exercises, certain landmarks on the bones of the pectoral gridle and upper limbs were palpated. This exercise serves as an opportunity to review these surface landmarks.

1. Palpate the manubrium of the sternum and suprasternal (jugular) notch on yourself. Move your fingers just lateral from the sternal notch to palpate the joint between the manubrium and the proximal end of the clavicle: the **sternoclavicular joint.** Recall that the sternoclavicular joint is the only bony attachment between the pectoral girdle and the axial skeleton.
2. Palpate along the **clavicle** and make note of the curvatures of the clavicle while moving your fingers from medial to lateral. At the tip of the shoulder, palpate the joint between the lateral aspect of the clavicle and the **acromion** of the scapula: the **acromioclavicular joint.**
3. Continue to palpate along the acromion as it curves posteriorly and becomes the **spine of the scapula.**
4. Palpate the inferior, lateral border of the deltoid muscle, which covers much of the shoulder. You should be able to feel part of the diaphysis of the humerus where the deltoid attaches to the humerus at the **deltoid tuberosity** because there is very little muscle between the bone and the skin at that point.
5. Moving your fingers distally to the elbow, palpate the large **olecranon** of the ulna. This is the bony process that rests on a table when leaning on the elbows.
6. Just proximal from the olecranon on the medial aspect of the elbow, palpate the **medial epicondyle of the humerus.** Place your thumb in the hollow on the posterior part of the elbow between the olecranon of the ulna and the medial epicondyle of the humerus to feel the cable-like **ulnar nerve.** This nerve is what causes the pain or tingly sensations that are felt when hitting the "funny bone."
7. Palpate the olecranon once again. Continue to palpate along the ulna distally until reaching the wrist joint. The bump on the medial aspect of the wrist is the **styloid process of the ulna.** Palpate the corresponding location on the lateral aspect of the wrist to feel the **styloid process of the radius.**
8. Finally, palpate the small metacarpal and phalangeal bones of the hand (see figure 9.7) while reviewing the names of the bones.
9. Label the surface anatomy structures in **figure 9.8,** using the textbook as a guide.

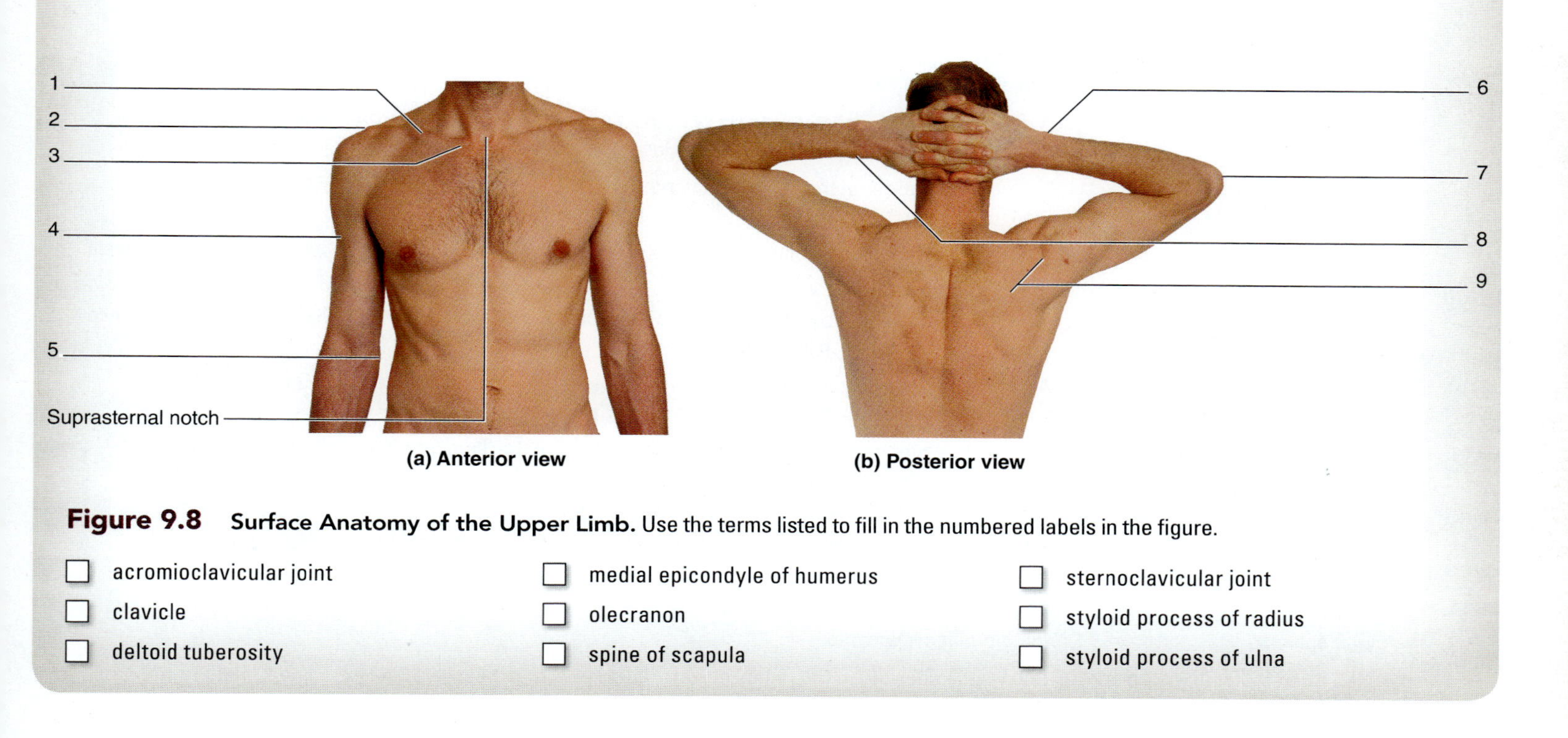

Figure 9.8 Surface Anatomy of the Upper Limb. Use the terms listed to fill in the numbered labels in the figure.

- ☐ acromioclavicular joint
- ☐ clavicle
- ☐ deltoid tuberosity
- ☐ medial epicondyle of humerus
- ☐ olecranon
- ☐ spine of scapula
- ☐ sternoclavicular joint
- ☐ styloid process of radius
- ☐ styloid process of ulna

The Pelvic Girdle

The **pelvic girdle** consists of the paired **os coxae** (os, bone, + *coxa*, hip) bones. Each os coxa (singular) is composed of three bones: the **ilium,** the **ischium,** and the **pubis** bones. Together, the three bones compose the *os coxae* (*os*, bone, + *coxa*, hip), a structure that forms between the ages of 13 and 15 years. Unlike the bones of the pectoral girdle, the bones of the pelvic girdle fuse together during development to become one solid structure that is joined anteriorly at the pubic symphysis. **Table 9.3** lists the bones composing the os coxae and describes their key features. A complete pelvis is formed from four bones: the right os coxa, the left os coxa, the sacrum, and the coccyx.

In the following exercises, the features of the os coxa as a whole are described first. The features of the individual bones that compose the os coxa are described next. Finally, the structural differences between the male pelvis and the female pelvis are described.

Table 9.3 The Appendicular Skeleton: Pelvic Girdle

Bone	Bony Landmark	Description	Word Origin
Os Coxae ***os*, bone, + *coxa*, hip**	Acetabulum	A deep bony socket; articulates with the head of the femur	*acetabula*, a shallow cup
	Linea terminalis (pelvic brim)	An oblique ridge on the inner surface of the ilium and pubic bones that consists of the pubic crest, pectineal line, and arcuate line; separates the true pelvis (below) from the false pelvis (above)	*linea*, line, + *terminalis*, ending
	Lunate surface	The half-moon-shaped (curved) smooth surface on the superior border of the acetabulum; articulates with the head of the femur	*luna*, moon
	Obturator foramen	A large, oval hole in the inferior part of the os coxae; covered by a membrane in a living individual	*obturo*, to occlude
Ilium **ilium, *flank***	Ala	Concave, curved anterior region of the ilium	*ala*, wing
	Anterior gluteal line	A rough line running obliquely on the lateral surface of the ilium from the iliac crest to the greater sciatic notch; attachment site for gluteal muscles	*gloutos*, buttock
	Anterior inferior iliac spine	A process inferior to the anterior superior iliac spine; attachment site for rectus femoris muscle	*spina*, a spine
	Anterior superior iliac spine	A projection at the anteriormost part of the iliac crest; attachment site for the inguinal ligament and sartorius muscle	*spina*, a spine
	Arcuate line	An oblique line between the ilium and ischium that composes the iliac part of the linea terminalis (pelvic brim) of the bony pelvis	*arcuatus*, bowed
	Auricular surface	An "ear-like" rough surface on the medial aspect of the ilium; articulates with the sacrum	*auris*, ear
	Greater sciatic notch	A deep notch on the posterior surface of the ilium inferior to the posterior iliac spine; the sciatic nerve is located adjacent to this notch	*sciaticus*, the hip joint
	Iliac crest	The superior border of the ilium, beginning at the sacrum and ending on the lateral aspect of the hip	*crista*, a ridge
	Iliac fossa	A large fossa on the anteromedial (internal) surface of the ilium inferior to the iliac crest	*ilium*, flank, + *fossa*, a trench
	Iliac tuberosity	A large projection on the posterior, superior aspect of the ilium; attachment for sacroiliac ligaments	*ilium*, flank, + *tuber*, a knob
	Inferior gluteal line	A rough line running transversely on the lateral surface of the ilium just superior to the acetabulum; attachment site for gluteal muscles	*gloutos*, buttock
	Posterior gluteal line	A rough line running vertically on the lateral surface of the ilium from the iliac crest to the posterior rim of the greater sciatic notch; attachment site for gluteal muscles	*gloutos*, buttock
	Posterior inferior iliac spine	A small projection on the posterior inferior point of the ilium; attachment site for the rectus femoris	*spina*, a spine
	Posterior superior iliac spine	A projection on the posterior superior point of the ilium; attachment site for sacroiliac ligaments	*spina*, a spine
Ischium **ischion, *hip***	Body of ischium	Bulky bone superior to the ramus of ischium	*ischion*, hip, + *spina*, a spine
	Ischial spine	A small, sharp spine on the posterior aspect of the ischium; attachment site for sacrospinous ligaments	*ischion*, hip
	Lesser sciatic notch	A notch located immediately inferior to the ischial spine on the posterior surface of the ilium	*sciaticus*, the hip joint
	Ramus of ischium	The inferior part of the ischium that connects to the pubis anteriorly; forms the inferior part of the obturator foramen	*ramus*, branch
	Ischial tuberosity	A large, rough projection on the posterior, inferior surface of the ischium; attachment site for hamstring muscles	*ischion*, hip, + *tuber*, a knob

Table 9.3 The Appendicular Skeleton: Pelvic Girdle *(continued)*

Bone	Bony Landmark	Description	Word Origin
Pubis **pubis, *pubic bone***	Inferior pubic ramus	The inferior part of the pubis that joins with the ischium	*ramus*, branch
	Pectineal line	Rough ridge on the medial surface of the superior ramus of the pubis; attachment site of pectineus muscle	*pectineal*, relating to the pubis
	Pubic crest	A ridge on the lateral part of the superior ramus of the pubis	*crista*, crest
	Pubic tubercle	A projection composing the anteriormost point of the bone; attachment site for the inguinal ligament	*tuber*, a knob
	Superior pubic ramus	The superior part of the pubis that joins with the ilium	*ramus*, branch
	Symphysial surface	Site of articulation for pubic bones at the pubic symphysis	*symphysis*, a growing together

EXERCISE 9.4

BONES OF THE PELVIC GIRDLE

EXERCISE 9.4A The Os Coxae

1. Observe the os coxae **(figure 9.9)** on an articulated skeleton (or see figure 7.10, which shows a full skeleton). Locate the point of articulation of the os coxae with the axial skeleton. With what major bone does the os coxa articulate?

   ____

2. Palpate your own os coxae at the iliac crest by placing your hands on your hips.
3. Obtain both a disarticulated pelvis and a disarticulated os coxa. Identify the bony landmarks that are listed in table 9.3.
   - **a.** What distinguishes the superior portion of the os coxa from the inferior portion?

     ____

     ____

     ____

   - **b.** What distinguishes the anterior surface of the os coxa from the posterior surface?

     ____

     ____

     ____

4. Obtain both right and left disarticulated os coxae. What distinguishes the right os coxa from the left os coxa?

   ____

   ____

   ____

5. Label the structures on the os coxae shown in figure 9.9, using table 9.3 and the textbook as guides.

EXERCISE 9.4B Male and Female Pelves

1. Obtain a male pelvis and a female pelvis and lay them next to each other on the workspace with the anterior surfaces facing toward you. **Figure 9.10** demonstrates the features of male and female pelves.
2. There are numerous features that help distinguish a **male pelvis** from a **female pelvis.** For example, a female pelvis is generally wider and more flared; it has a broader subpubic angle; a smaller, triangular obturator foramen; a wide, shallow greater sciatic notch; and ischial spines that rarely project into the pelvic outlet. All of these are adaptations that facilitate childbirth. However, no method of sexing a pelvis is completely foolproof. At the very least, use the guidelines to determine if a pelvis is more male-like than female-like, or vice versa. Observe both male and female pelves to develop a method of differentiating the male pelvis from the female pelvis.

INTEGRATE

LEARNING STRATEGY

There are several methods for determining if a pelvis belonged to a male or a female. One involves estimating the subpubic angles by comparing the angles on the pelves with the angles formed between the digits on the hand with the fingers spread apart (abducted). For example, the angle between the thumb and index finger when spread apart approximates the wide subpubic angle of a female pelvis, whereas the angle formed between the index and middle fingers when spread apart approximates the narrower subpubic angle of a male pelvis.

The width of the greater sciatic notch may also provide clues. Females typically have a larger greater sciatic notch. Three fingers can be placed in the greater sciatic notch of a female pelvis, whereas only one finger may fit in the greater sciatic notch of a male.

(continued on next page)

(continued from previous page)

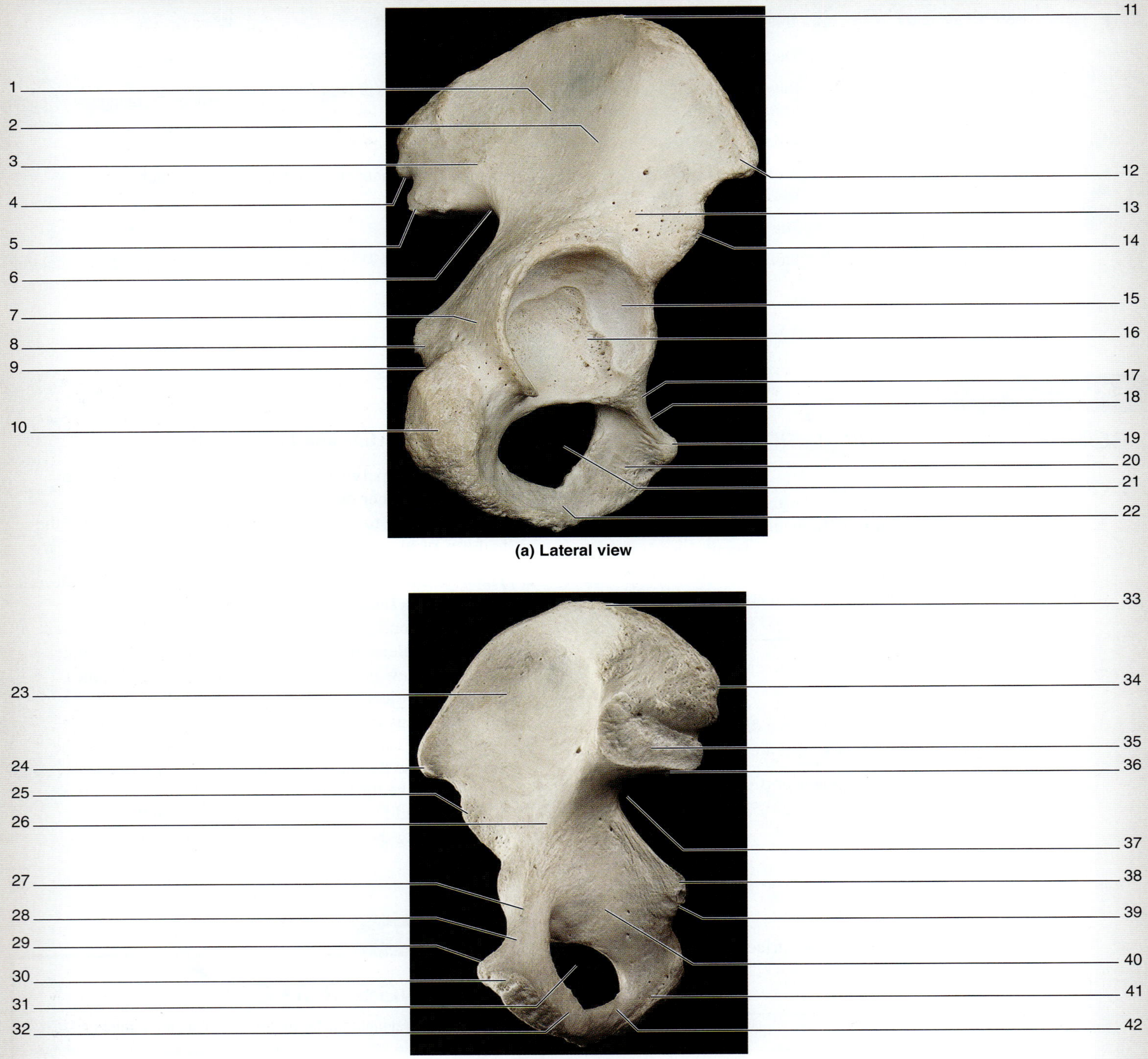

1 ____
2 ____
3 ____
4 ____
5 ____
6 ____
7 ____
8 ____
9 ____
10 ____
11 ____
12 ____
13 ____
14 ____
15 ____
16 ____
17 ____
18 ____
19 ____
20 ____
21 ____
22 ____

(a) Lateral view

23 ____
24 ____
25 ____
26 ____
27 ____
28 ____
29 ____
30 ____
31 ____
32 ____
33 ____
34 ____
35 ____
36 ____
37 ____
38 ____
39 ____
40 ____
41 ____
42 ____

(b) Medial view

Figure 9.9 The Right Os Coxa. Use the terms listed to fill in the numbered labels in the figure. Some terms may be used more than once.

- ☐ acetabulum
- ☐ ala
- ☐ anterior gluteal line
- ☐ anterior inferior iliac spine
- ☐ anterior superior iliac spine
- ☐ arcuate line
- ☐ auricular surface
- ☐ body of ischium
- ☐ greater sciatic notch
- ☐ iliac crest
- ☐ iliac fossa
- ☐ inferior gluteal line
- ☐ inferior pubic ramus
- ☐ ischial spine
- ☐ ischial tuberosity
- ☐ lesser sciatic notch
- ☐ lunate surface
- ☐ obturator foramen
- ☐ pectineal line
- ☐ posterior gluteal line
- ☐ posterior inferior iliac spine
- ☐ posterior superior iliac spine
- ☐ pubic crest
- ☐ pubic tubercle
- ☐ ramus of ischium
- ☐ superior pubic ramus
- ☐ symphysial surface of pubic bone

3. Sketch male and female pelves in the spaces provided. Then list the features that distinguish the male pelvis from the female pelvis. Use the photos in figure 9.10 as guides if there are not samples of both a male and a female pelvis in the laboratory.

Male Pelvis

Female Pelvis

Female

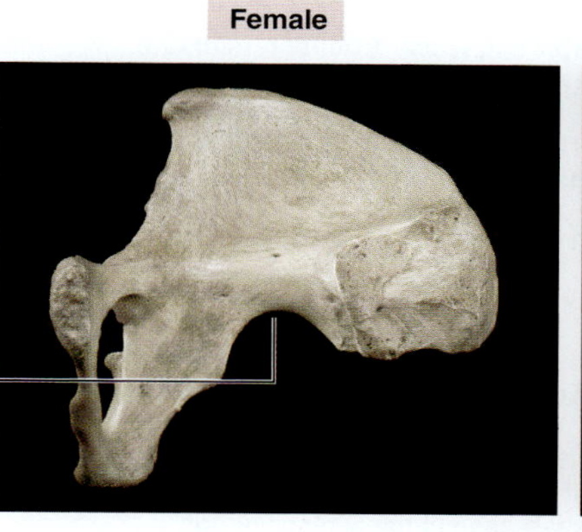

Male

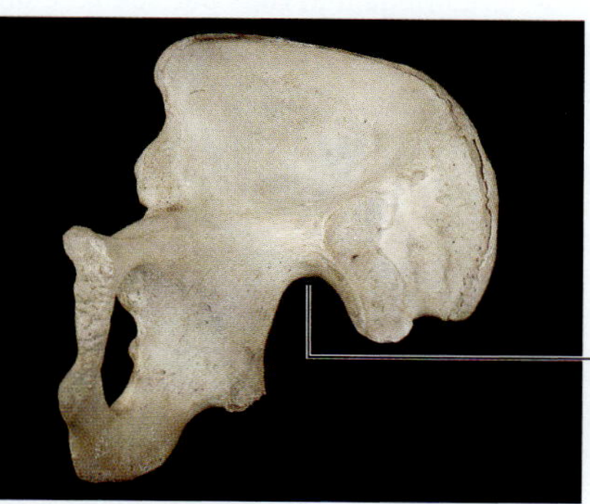

Medial views

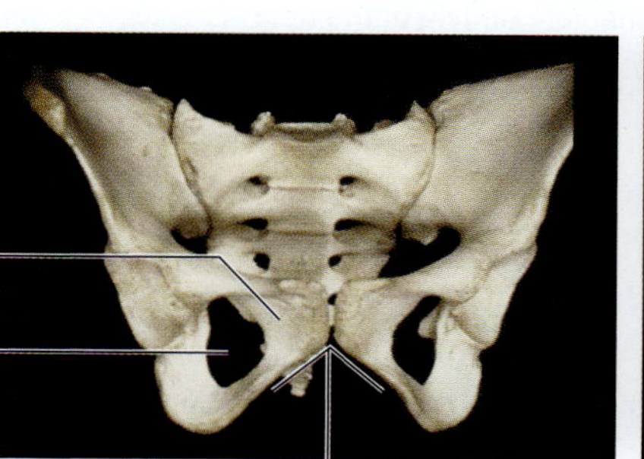

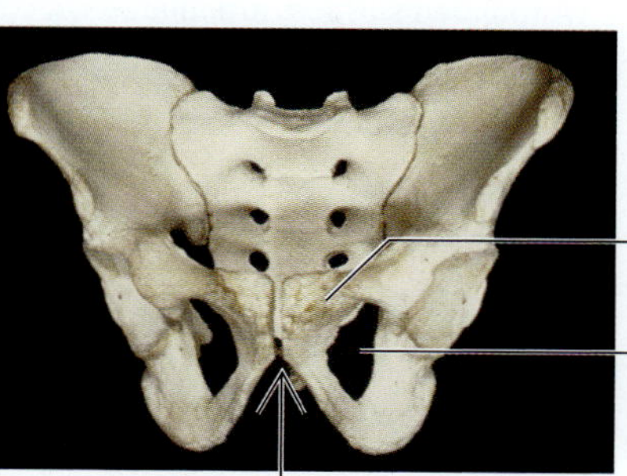

Anterior views

Figure 9.10 **Male and Female Pelves.**

INTEGRATE

CLINICAL VIEW

Pregnancy and Childbirth

Figure 9.10 demonstrates several of the features that distinguish the female pelvis from the male pelvis, and that facilitate childbirth. Specifically, females generally have a wider, flared pelvis and greater subpubic angle. These features allow a baby's head to pass through the pelvic outlet during childbirth. Females also usually have ischial spines that remain clear of the pelvic outlet, so as not to impede the passage of the baby through the pelvic outlet. In addition to these anatomic differences between female and male pelves there are hormones that play an important role in creating a "roomy" birth canal. Specifically, the hormone *relaxin*, which is primarily secreted by the ovaries during pregnancy, promotes remodeling of connective tissue. Relaxin increases the flexibility of the pubic symphysis, the slightly movable fibrocartilaginous joint that connects the right and left pubic bones. Overall, these factors increase separation of the pubic bones in the latest stages of pregnancy. In rare cases, the pubic bones separate more than the physiologic limit (10 mm), a condition called *diastasis symphysis pubis*. Factors that predispose a woman to this condition are the baby's position during birth, a rapid delivery, or excessive relaxation of the pubic symphysis joint. Symptoms of diastasis symphysis pubis include pain and decreased hip mobility. Treatment involves bracing the pelvic region.

The Lower Limb

The bones of the lower limb are the femur, patella, tibia, fibula, tarsals, metatarsals, and phalanges. Nearly all of the projections on these bones are attachment points for the muscles that move the limb. **Table 9.4** lists the bones of the lower limb and describes their key features.

Table 9.4 The Appendicular Skeleton: Lower Limb

Bone	Bony Landmark	Description	Word Origin
Femur **femur, *thigh***	Adductor tubercle	A small projection proximal or superior to a medial condyle; attachment point for adductor magnus	*tuber*, a knob
	Fovea	A circular depression within the head of the femur; attachment site for the ligament of the head of the femur	*fovea*, a dimple
	Gluteal tuberosity	A projection on the proximal aspect of the linea aspera; attachment site for the gluteus maximus	*gloutos*, buttock, + *tuber, a knob*
	Greater trochanter	A very large projection on the lateral surface of the proximal epiphysis; attachment site for both gluteal and thigh muscles	*trochanter*, a runner
	Head	Large, spherical structure on the proximal end of the femur; articulates with the acetabulum of os coxa	NA
	Intercondylar fossa	A depression on the distal end of the femur between the two condyles; attachment site for the cruciate ligaments of the knee	*inter-*, between, + *kondylos*, knuckle
	Intertrochanteric crest	A large ridge that runs between the greater and lesser trochanters on the posterior surface; landmark separating shaft and neck of femur	*inter-*, between, + *trochanter*, a runner
	Intertrochanteric line	A shallow ridge that runs between the greater and lesser trochanters on the anterior surface; landmark separating shaft and neck of femur	*inter-*, between, + *trochanter*, a runner
	Lateral condyle	A large, rounded surface; articulates with the lateral condyle of the tibia	*kondylos*, knuckle
	Lateral epicondyle	A rough surface superior to the lateral condyle; attachment site for fibular collateral ligament	*epi*, above, + *kondylos*, knuckle
	Lesser trochanter	A large projection on the medial surface of the proximal epiphysis; attachment site of iliopsoas muscle	*trochanter*, a runner
	Linea aspera	A "rough line" that runs along the posterior surface of the diaphysis; attachment site for both the adductor longus and the short head of the biceps femoris	*linea*, line, + *aspera*, rough
	Medial condyle	A rounded surface; articulates with the medial condyle of the tibia	*kondylos*, knuckle
	Medial epicondyle	A rough surface superior to the medial condyle; attachment site for tibial collateral ligament	*epi*, above, + *kondylos*, knuckle
	Neck	The narrow portion where the head meets the shaft of the femur; this is the part of the femur that is fractured in a "broken hip"	NA
	Patellar surface	A smooth depression on the anterior surface of the distal epiphysis; articulates with the patella	(patella) *patina*, a shallow disk
	Pectineal line	A line on the posterior, superior aspect of the femur; attachment site for the pectineus muscle	*pectineal*, ridged or comb-like
	Popliteal surface	A triangular region on the posterior aspect of the distal femur	*popliteal*, the back of the knee
	Shaft	The diaphysis of the bone	NA

Table 9.4 The Appendicular Skeleton: Lower Limb *(continued)*

Bone	Bony Landmark	Description	Word Origin
Tibia **tibia, *the large shin bone***	Anterior border	A ridge on the anterior surface extending distally from the tibial tuberosity; commonly referred to as the "shin"	NA
	Fibular articular surface	Smooth region on the proximal, posterolateral surface of the femur; articulates with the head of the fibula	NA
	Intercondylar eminence	A prominent projection between the two condyles on the proximal epiphysis	*eminentia*, a raised area on a bone
	Lateral condyle	A large, flat surface on the lateral aspect of the proximal epiphysis; articulates with the lateral condyle of the femur	*kondylos*, knuckle
	Medial condyle	A large, flat surface on the medial aspect of the proximal epiphysis; articulates with the medial condyle of the femur	*kondylos*, knuckle
	Medial malleolus	A projection on the medial surface of the distal epiphysis	*malleus*, hammer
	Shaft	The diaphysis of the bone	NA
	Tibial tuberosity	A projection on the anterior surface of the proximal epiphysis; attachment site for the patellar ligament (quadriceps femoris muscle attachment)	*tibia*, shin bone
Fibula **fibula, *a clasp or buckle***	Head	The rounded proximal end of the bone; articulates with lateral tibial condyle	NA
	Lateral malleolus	A projection on the lateral surface of the distal epiphysis; attachment site for ligaments	*malleus*, hammer
	Neck	The narrow portion where the head meets the diaphysis	NA
	Shaft	The diaphysis of the bone	NA
Tarsals **tarsus, *a flat surface***	Calcaneus	Bone that forms the heel of the foot; attachment site for calcaneal tendon	*calcaneus*, the heel
	Cuboid	A cube-shaped bone; located at the base of the fourth metatarsal	*kybos*, cube, + *eidos*, resemblance
	Intermediate cuneiform	A wedge-shaped bone; located at the base of the second metatarsal	*cuneus*, wedge, + *forma*, shape
	Lateral cuneiform	A wedge-shaped bone; located at the base of the third metatarsal	*cuneus*, wedge, + *forma*, shape
	Medial cuneiform	A wedge-shaped bone; located at the base of the first metatarsal	*cuneus*, wedge, + *forma*, shape
	Navicular	A bone shaped like a boat ("ship"); located just anterior to the talus	*navis*, ship
	Talus	The major weight-bearing bone of the ankle; articulates with the tibia and fibula	*talus*, ankle
Metatarsals		Each bone consists of a head, body, and base	meta-, *after*, + tarsus, a *flat surface*
Phalanges **phalanx, *line of soldiers***			
II through V	Proximal	The phalanx closest to the metatarsal bones	*proximus*, nearest
	Middle	The middle phalanx	NA
	Distal	The small, cone-shaped phalanx forming the ends of the toes	*distalis*, away
Hallux **hallux, *the big toe***	Proximal	The phalanx closest to the sole of the foot	*proximus*, nearest
	Distal	The phalanx forming the end of the big toe	*distalis*, away

INTEGRATE

LEARNING STRATEGY

When attempting to identify individual bones, determine the different aspects of the bone (e.g., anterior vs. posterior), and determine if a bone is from the right side or left side of the body. Think about what features *you* find most distinctive for each bone. For example, you may look for a large head to distinguish the femur from other bones. You may look for the linea aspera to distinguish the posterior surface from the anterior surface (the linea aspera is on the posterior surface). Both of these features may be considered together to determine if the femur is a right or a left femur. Follow this procedure for each bone of the appendicular skeleton with an understanding that the features that help *you* identify a bone may not be the same features somone else uses to identify that same bone.

EXERCISE 9.5

BONES OF THE LOWER LIMB

EXERCISE 9.5A The Femur

1. Observe the femur (**figure 9.11**) on an articulated skeleton (or see figure 7.10, which shows a full skeleton). Locate the points of articulation with the os coxa, tibia, and patella.
2. Palpate the femur at the hip by placing your hand about 4 inches below the iliac crest. The proximal end of the femur can be felt when bending laterally at the hip.
3. Obtain a disarticulated femur. Identify the bony landmarks that are listed in table 9.4.
   a. What distinguishes the proximal end of the femur from the distal end?

      ____

      ____

      ____

   b. What distinguishes the anterior surface of the femur from the posterior surface?

      ____

      ____

      ____

4. Obtain both a right and a left disarticulated femur. What distinguishes a right femur from a left femur?

   ____

   ____

   ____

5. Label the structures on the femur shown in figure 9.11, using table 9.4 and the textbook as guides.
6. Obtain a right os coxa and a right femur and articulate them with each other. Sketch the relationship between the bones of the os coxa and the femur in the space provided.
7. What is the name of the joint formed between the bones of the os coxa and the femur?

   ____

1 ____
2 ____
3 ____
4 ____
5 ____
6 ____
7 ____

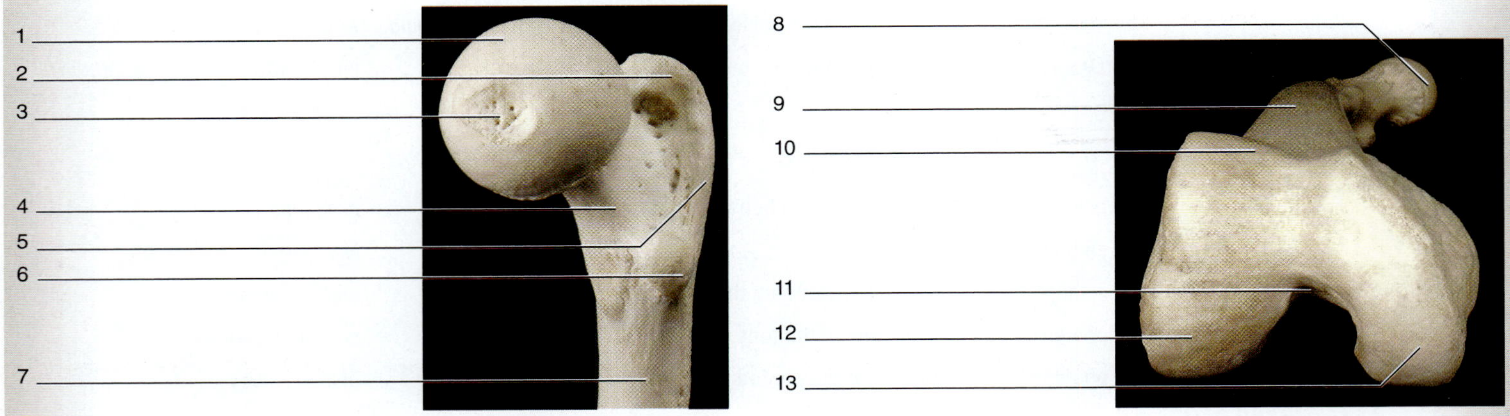

(a) Right femur, medial view

8 ____
9 ____
10 ____
11 ____
12 ____
13 ____

(b) Right femur, inferior view

Figure 9.11 **The Right Femur.** Use the terms listed to fill in the numbered labels in the figure. Some terms may be used more than once.

- ☐ adductor tubercle
- ☐ fovea
- ☐ gluteal tuberosity
- ☐ greater trochanter
- ☐ head
- ☐ intercondylar fossa
- ☐ intertrochanteric crest
- ☐ intertrochanteric line
- ☐ lateral condyle
- ☐ lateral epicondyle
- ☐ lateral supracondylar line
- ☐ lesser trochanter

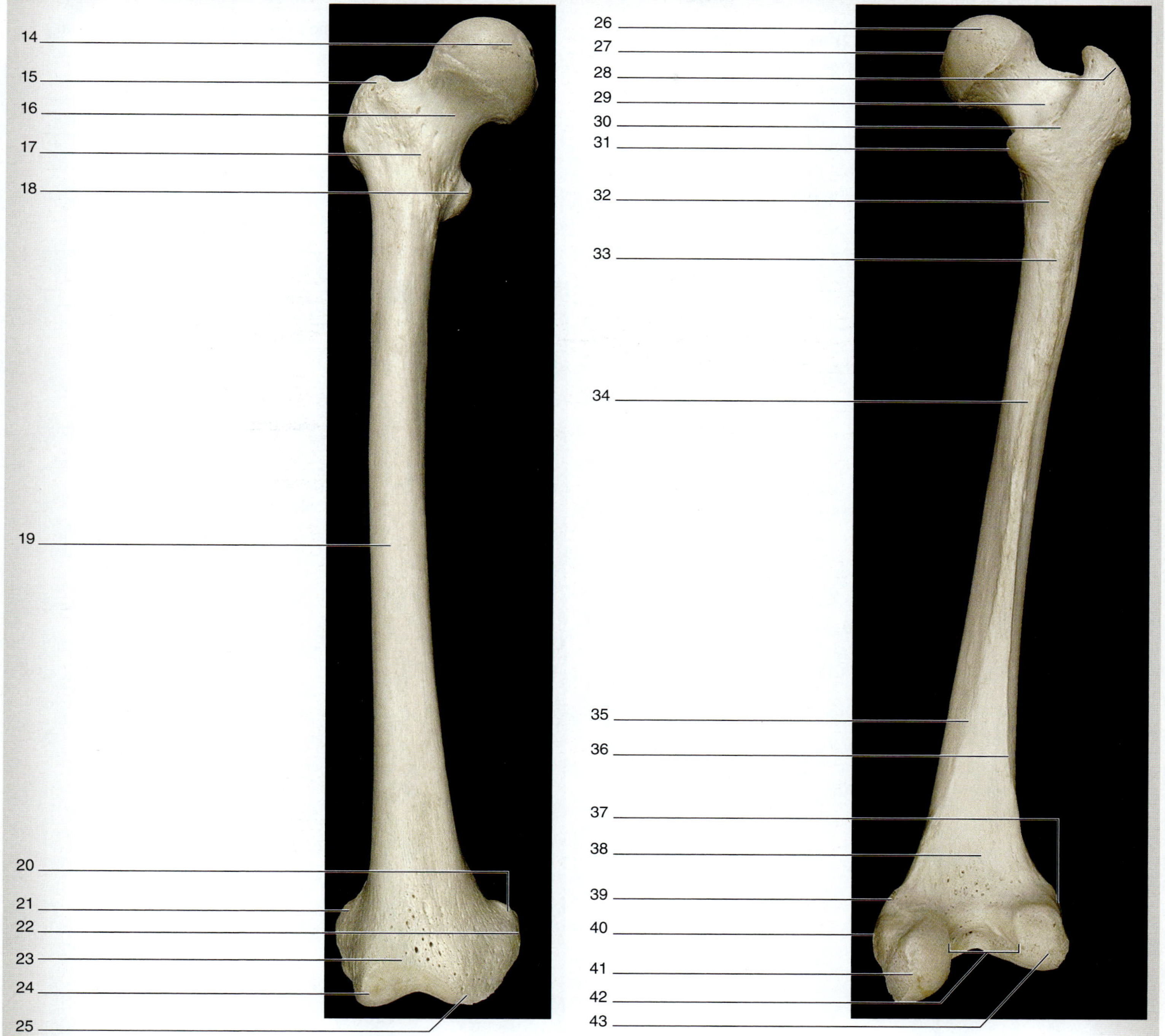

14
15
16
17
18
19
20
21
22
23
24
25

(c) Right femur, anterior view

26
27
28
29
30
31
32
33
34
35
36
37
38
39
40
41
42
43

(d) Right femur, posterior view

Figure 9.11 **The Right Femur *(continued)*.** Use the terms listed to fill in the numbered labels in the figure. Some terms may be used more than once.

- ☐ linea aspera
- ☐ medial condyle
- ☐ medial epicondyle
- ☐ medial supracondylar line
- ☐ neck
- ☐ patellar surface
- ☐ pectineal line
- ☐ popliteal surface
- ☐ shaft

(continued on next page)

(continued from previous page)

EXERCISE 9.5B The Tibia

1. Observe the tibia **(figure 9.12)** on an articulated skeleton (or see figure 7.10, which shows a full skeleton). Locate the point of articulation between the femur and the talus.
2. Palpate your own tibia on the anterior border (or shin). Walk your fingers down the anterior border of the tibia to the medial malleolus. Note that the tibia forms the medial aspect of the ankle, whereas the fibula forms the lateral aspect of the ankle.
3. Obtain a disarticulated tibia. Identify the bony landmarks that are listed in table 9.4.

   a. What distinguishes the proximal end of the tibia from the distal end?

   ____

   ____

   ____

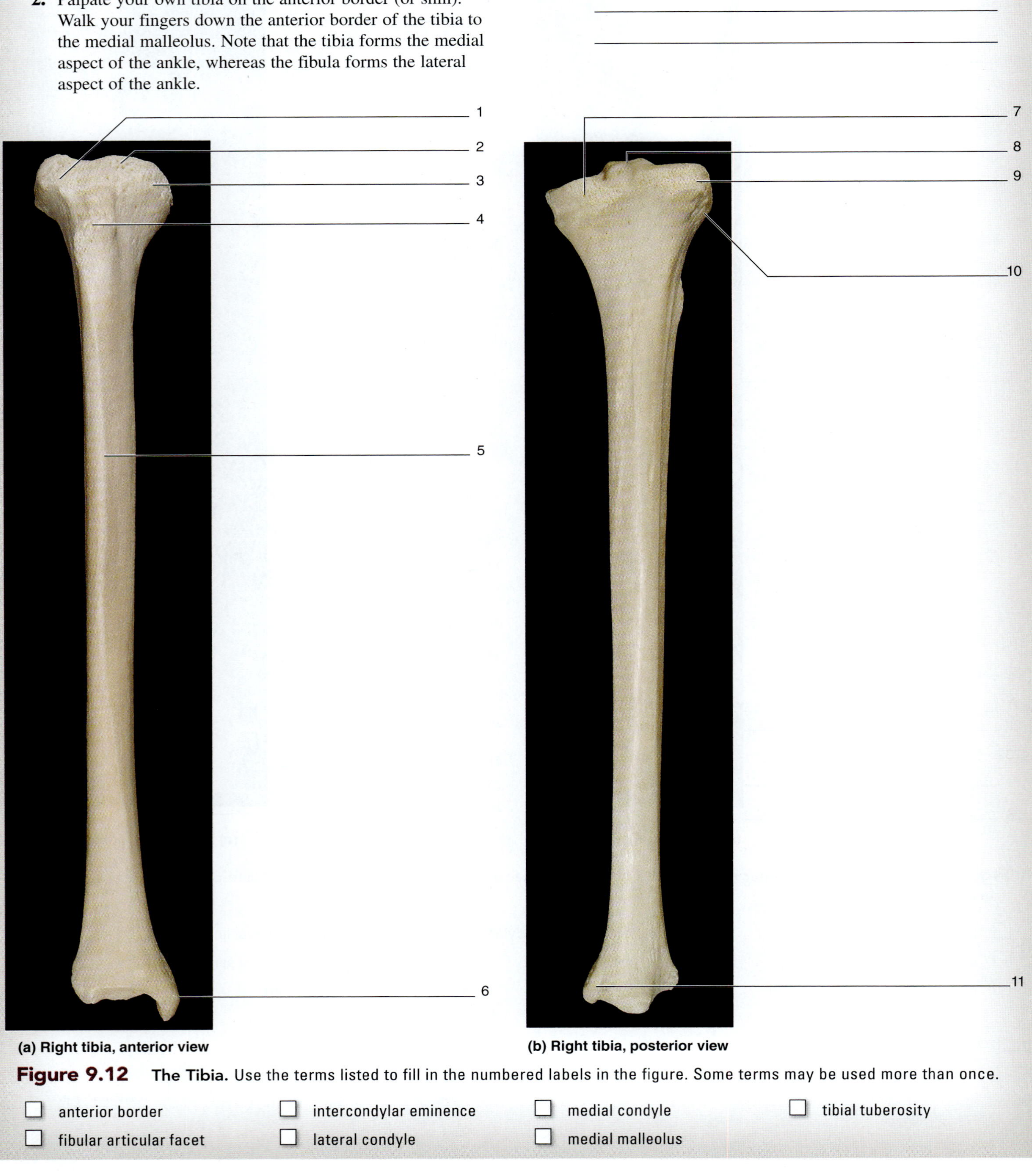

(a) Right tibia, anterior view

(b) Right tibia, posterior view

Figure 9.12 **The Tibia.** Use the terms listed to fill in the numbered labels in the figure. Some terms may be used more than once.

- ☐ anterior border
- ☐ fibular articular facet
- ☐ intercondylar eminence
- ☐ lateral condyle
- ☐ medial condyle
- ☐ medial malleolus
- ☐ tibial tuberosity

b. What distinguishes the anterior surface of the tibia from the posterior surface?

____

____

____

4. Obtain both a right and a left disarticulated tibia. What distinguishes a right tibia from a left tibia?

____

____

____

5. Label the structures on the tibia shown in figure 9.12, using table 9.4 and the textbook as guides.

EXERCISE 9.5C The Fibula

1. Observe the fibula **(figure 9.13)** on an articulated skeleton. (or see figure 7.10, which shows a full skeleton). Locate the points of articulation between the fibula and the tibia at both proximal and distal ends.

    a. Does the fibula form part of the knee joint?

    ____

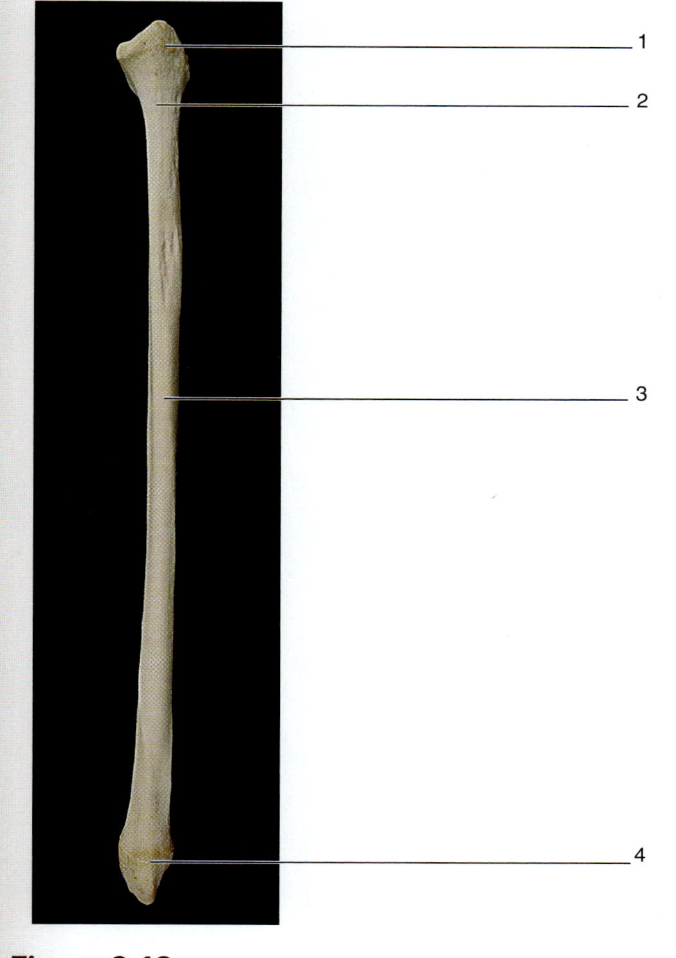

Figure 9.13 **The Right Fibula.** Lateral view. Use the terms listed to fill in the numbered labels in the figure.

☐ head ☐ neck
☐ lateral malleolus ☐ shaft

b. Does the fibula form part of the ankle joint?

____

2. Palpate the head of the fibula on the lateral aspect of the knee. Next, palpate the fibula at the lateral malleolus.

3. Obtain a disarticulated fibula. Identify the bony landmarks that are listed in table 9.4.

a. What distinguishes the proximal end of the fibula from the distal end?

____

____

____

b. What distinguishes the anterior surface of the fibula from the posterior surface?

____

____

____

4. Obtain both a right and a left disarticulated fibula. What distinguishes a right fibula from a left fibula?

____

____

____

5. Label the structures on the fibula shown in figure 9.13, using table 9.4 and the textbook as guides.

6. Obtain a right femur, right tibia, and right fibula and articulate them with each other. Sketch the relationship between the right femur, tibia, and fibula from an anterior view in the space provided. Include the location of the patella.

(continued on next page)

(continued from previous page)

EXERCISE 9.5D The Tarsals

1. Obtain bones of the ankle (tarsals) or observe the **tarsal** bones on an articulated skeleton (**figures 9.14** and **9.15**).
2. The term *tarsal* means "flat surface," as in the sole of the foot. The tarsal bones are much larger than their counterparts in the wrist (the carpal bones). The largest differences in structure are seen in the talus and calcaneus, two bones that form a major part of the weight-bearing ankle joint.
3. Identify the tarsal bones on an articulated skeleton, using table 9.4 and the textbook as guides. Then label them in figure 9.14.
4. The seven tarsal bones are arranged in two rows. The proximal row consists of the talus, calcaneus, and navicular bone. The distal row consists of the medial cuneiform, intermediate cuneiform, lateral cuneiform, and cuboid. The bones of the distal row articulate with the metatarsal bones. Sketch the tarsal bones of the proximal and distal rows as they would be positioned in a superior view in the spaces provided. Then label the bones in each row.

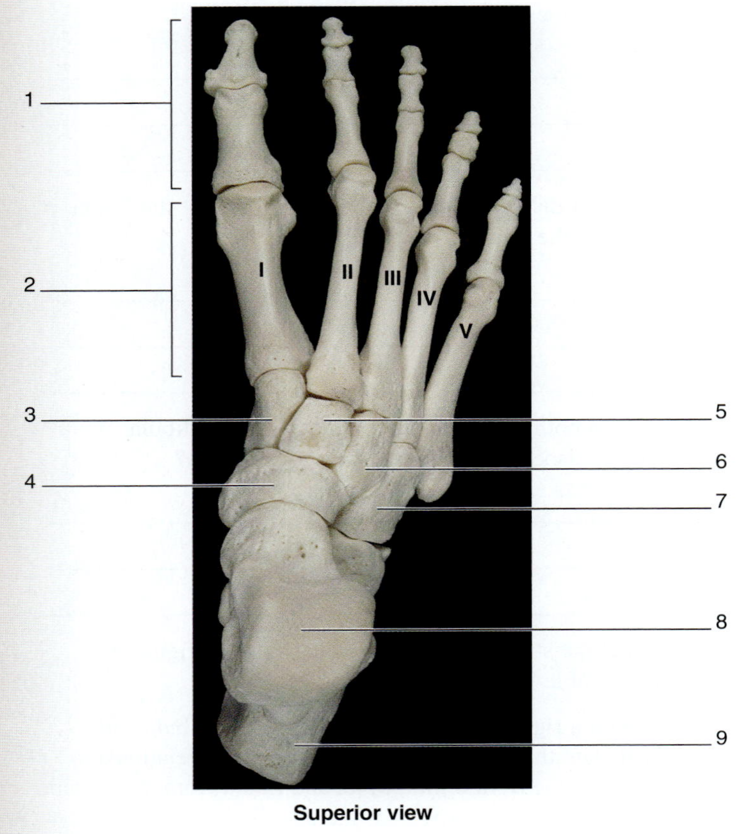

Figure 9.14 **The Tarsals.** Use the terms listed below to fill in the numbered labels in the figure.

- ☐ calcaneus
- ☐ cuboid
- ☐ intermediate cuneiform
- ☐ lateral cuneiform
- ☐ medial cuneiform
- ☐ metatarsals
- ☐ navicular
- ☐ phalanges
- ☐ talus

Proximal Row

INTEGRATE

LEARNING STRATEGY

The following mnemonic may help facilitate recall of the names of the tarsal bones and the order in which they are found:

Taking **C**lass **N**otes **M**akes **I**ntelligent **L**iterate **C**hampions.

Taking Class Notes = **T**alus, **C**alcaneus, **N**avicular
Makes Intelligent Literate Champions = **M**edial cuneiform, **I**ntermediate cuneiform, **L**ateral cuneiform, **C**uboid

The mnemonic only works when starting with the proximal row, moving from lateral to medial (anatomic position), followed by the distal row, moving from lateral to medial (anatomic position).

Distal Row

5. Complete the following chart for the tarsal bones, using table 9.4 as a reference.

Tarsal Bone	Bone Shape/Appearance	Word Origin
Talus		
Calcaneus		
Navicular		
Medial Cuneiform		
Intermediate Cuneiform		
Lateral Cuneiform		
Cuboid		

6. Obtain a right tibia, right fibula, right talus, and right calcaneus (or an articulated foot with tarsals) and articulate them with each other. Sketch the relationship between the tibia, fibula, calcaneus, and talus in the space provided.

7. Which bones or parts of bones compose the ankle joint?

____

____

____

(continued on next page)

(continued from previous page)

EXERCISE 9.5E The Metatarsals and Phalanges

1. Obtain bones of the foot (metatarsals and phalanges) or observe the bones of the foot on an articulated skeleton (figures 9.14 and 9.15).
2. **Metatarsals:** The term *meta-* means "after." Thus, the metatarsal bones are the bones that come after the tarsus. These bones are found in the sole of the foot. Palpate the sole of your own foot and feel for the metatarsal bones. There are five metatarsal bones, numbered I through V. The first metatarsal is on the medial surface of the foot and articulates distally with the base of the big toe, or **hallux**.
3. How do the phalangeal bones of the hallux differ from the phalanges of digits II–V?

   ____

   ____

   ____
4. Identify the metatarsal bones and phalanges on an articulated skeleton, using table 9.4 and the textbook as guides. Then label them in figure 9.15.

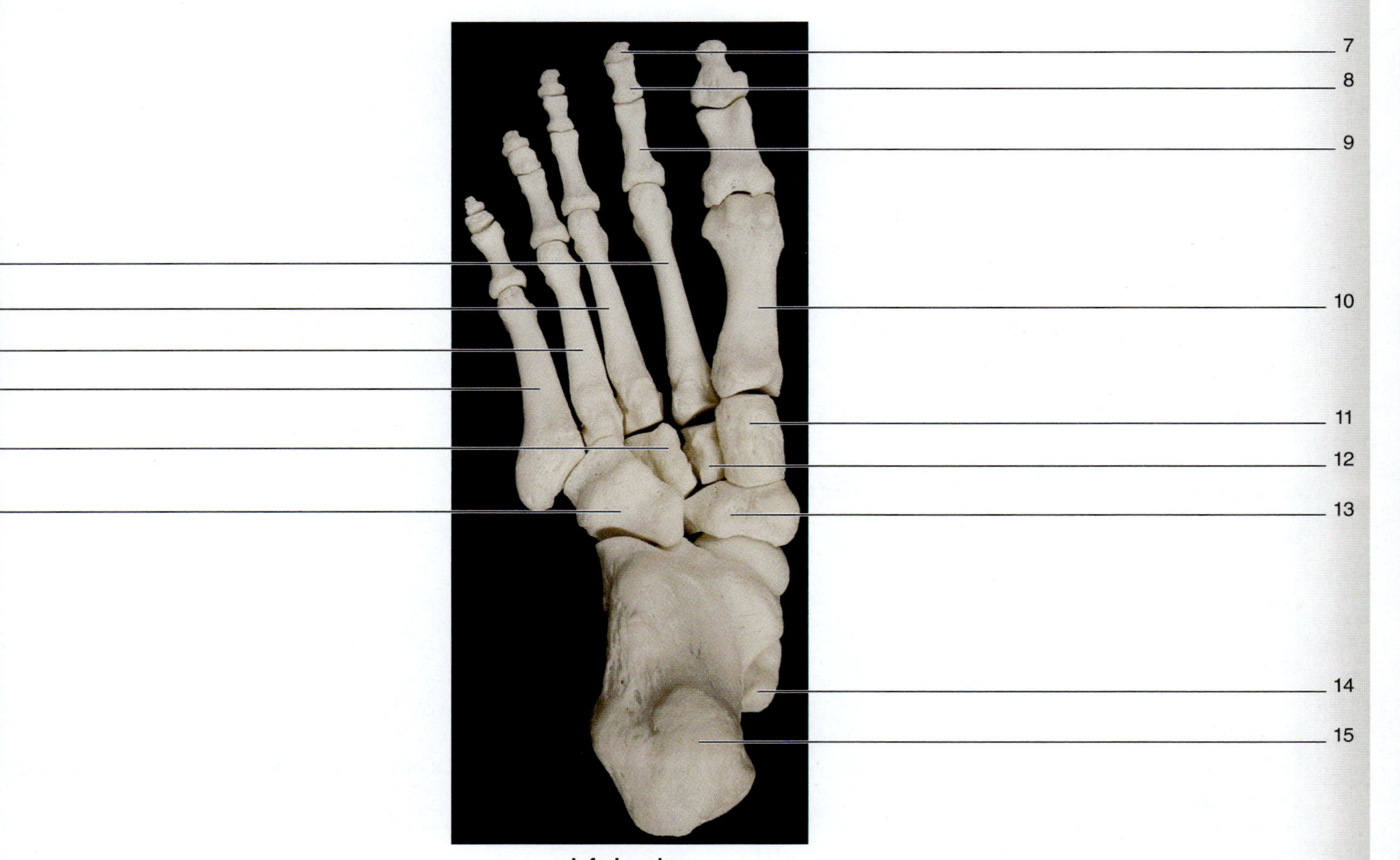

Figure 9.15 **The Metatarsals and Phalanges (inferior view).** Use the terms listed to fill in the numbered labels in the figure.

Metatarsals
- ☐ metatarsal I
- ☐ metatarsal II
- ☐ metatarsal III
- ☐ metatarsal IV
- ☐ metatarsal V

Phalanges
- ☐ proximal phalanx
- ☐ middle phalanx
- ☐ distal phalanx

Tarsals
- ☐ calcaneus
- ☐ cuboid
- ☐ intermediate cuneiform
- ☐ lateral cuneiform
- ☐ medial cuneiform
- ☐ navicular
- ☐ talus

EXERCISE 9.6

SURFACE ANATOMY REVIEW—PELVIC GIRDLE AND LOWER LIMB

In previous exercises, certain landmarks were palpated on the bones of the pelvic girdle and lower limbs. This exercise provides an opportunity to review these surface landmarks.

1. Place your hands on your hips. The bony ridge you feel under your skin is the **iliac crest.** The iliac crest is located at vertebral level L_4.
2. Palpate anteriorly along the iliac crest until you come to the large bony projection on the anterior surface, the **anterior superior iliac spine.**
3. Palpate laterally along the iliac crest until your fingers are on the lateral aspect of the hip. Palpate the soft tissue of the gluteus medius muscle just inferior to the lateral portion of the iliac crest.
4. Continue to palpate inferiorly to feel another bony projection. This is the **greater trochanter of the femur.** You can feel the greater trochanter more easily by bending laterally at the hip. Depending upon the amount of fat and muscle in your gluteal region, you may also be able to palpate the **ischial tuberosities** (or "sit bones") deep within the posterior region of the buttocks.
5. Palpate the **patella** on the anterior aspect of the knee.
6. Place your thumb and pinky finger (fifth phalanx) on the medial and lateral aspects of the knee. Here you will feel the large **medial** and **lateral epicondyles of the femur.**
7. Palpate the lateral epicondyle of the femur. Now move your fingers distally until you feel the knob-like **head of the fibula.** As you continue to palpate distally, you will be unable to feel the shaft of the fibula because of the fibularis muscles that overlie it.
8. As you palpate distally along the fibula toward the ankle joint, you will eventually feel the **lateral malleolus** of the fibula in the ankle joint.
9. Palpate the patella once again. Distal to the patella on the anterior surface of the leg, you will feel the **tibial tuberosity.** Continue to palpate distally along the tibia. Notice that you can feel the entire shaft of the bone because it is covered only with skin and a little bit of fat. The subcutaneous part of the tibia here is commonly referred to as the shin.
10. Palpate the most distal end of the tibia to feel the **medial malleolus** of the tibia in the ankle joint.
11. Palpate the large **calcaneus** in your heel. The talus and other tarsal bones are more difficult to palpate. However, if you wiggle your toes, you can see and palpate the **metatarsal bones** and the **phalanges.** As you palpate these bones, review their names. Recall that the first metatarsal is the metatarsal that lines up with the base of the big toe, or hallux, and the numbering continues, II through V, as you move toward the lateral aspect of the foot.
12. Label the surface anatomy structures in **figure 9.16**, using the textbook as a guide.

(continued on next page)

INTEGRATE

CONCEPT CONNECTION

Red bone marrow is the site of hemopoiesis, or blood cell formation. In young children, red bone marrow is found within most bones. However, by adulthood most red bone marrow is replaced by yellow bone marrow, which is mainly adipose tissue. In an adult, red bone marrow is limited to the flat bones of the skull, the vertebrae, the ribs, the sternum, the os coxa, and proximal epiphyses of the femur and humerus. Red bone marrow contains stem cells known as hemocytoblasts, which differentiate into *erythrocytes* (red blood cells), *leukocytes* (white blood cells), and *megakaryocytes*, which produce platelets. Chemical "factors," including hormones released in the blood, determine the ultimate fate of these stem cells. Bone marrow donation involves harvesting red bone marrow from either the iliac crest or the sternum (locations where red bone marrow is abundant and relatively easy to access). As with blood transfusions, donor and recipient bone marrow must "match" to prevent rejection. Properties of erythrocytes and leukocytes will be explored in chapters 20 (blood) and 23 (lymphatic and immune systems), respectively.

(continued from previous page)

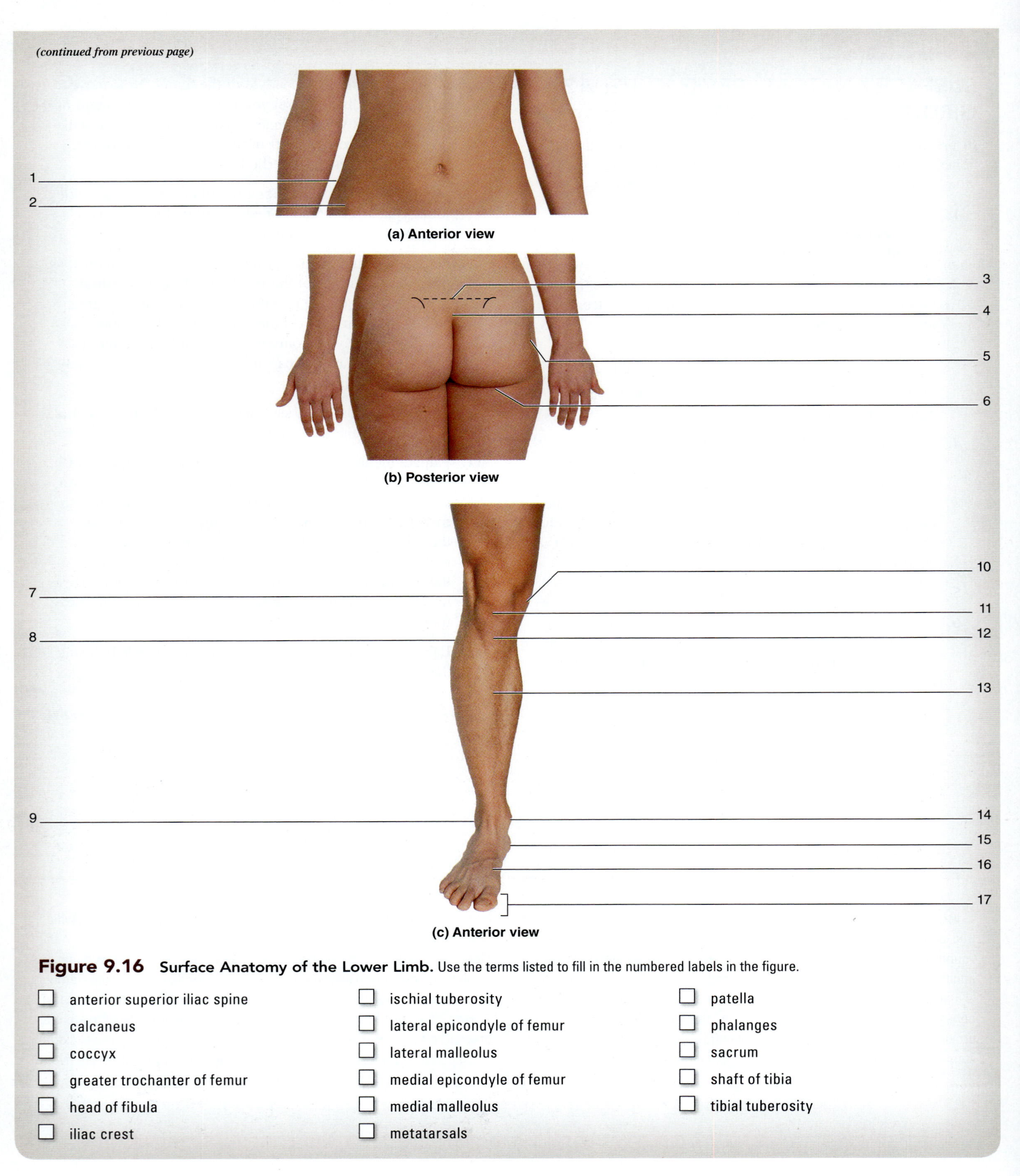

Figure 9.16 Surface Anatomy of the Lower Limb. Use the terms listed to fill in the numbered labels in the figure.

- ☐ anterior superior iliac spine
- ☐ calcaneus
- ☐ coccyx
- ☐ greater trochanter of femur
- ☐ head of fibula
- ☐ iliac crest
- ☐ ischial tuberosity
- ☐ lateral epicondyle of femur
- ☐ lateral malleolus
- ☐ medial epicondyle of femur
- ☐ medial malleolus
- ☐ metatarsals
- ☐ patella
- ☐ phalanges
- ☐ sacrum
- ☐ shaft of tibia
- ☐ tibial tuberosity

Chapter 9: The Skeletal System: Appendicular Skeleton

Name: ______________________________

Date: ____________ **Section:** ____________

POST-LABORATORY WORKSHEET

The ❶ corresponds to the Learning Objective(s) listed in the chapter opener outline.

Do You Know the Basics?

Exercise 9.1: Bones of the Pectoral Girdle

1. The head of the humerus articulates with which bony landmark of the scapula? (Circle one.) ❶
   a. acromion
   b. coracoid process
   c. glenoid fossa
   d. spine
   e. suprascapular notch

2. The only point of articulation between the pectoral girdle and the axial skeleton is the sternoclavicular joint. ____________________ (True/False) ❷

3. Match the bones listed in column A with the corresponding joint listed in column B. ❶ ❷ ❸

Column A	*Column B*
____ 1. clavicle and scapula	a. acromioclavicular joint
____ 2. humerus and ulna	b. elbow joint
____ 3. scapula and humerus	c. glenohumeral joint
____ 4. ulna, radius, and carpal bones	d. wrist joint

4. Label the following diagram of an articulated shoulder girdle using the terms listed: ❶ ❷

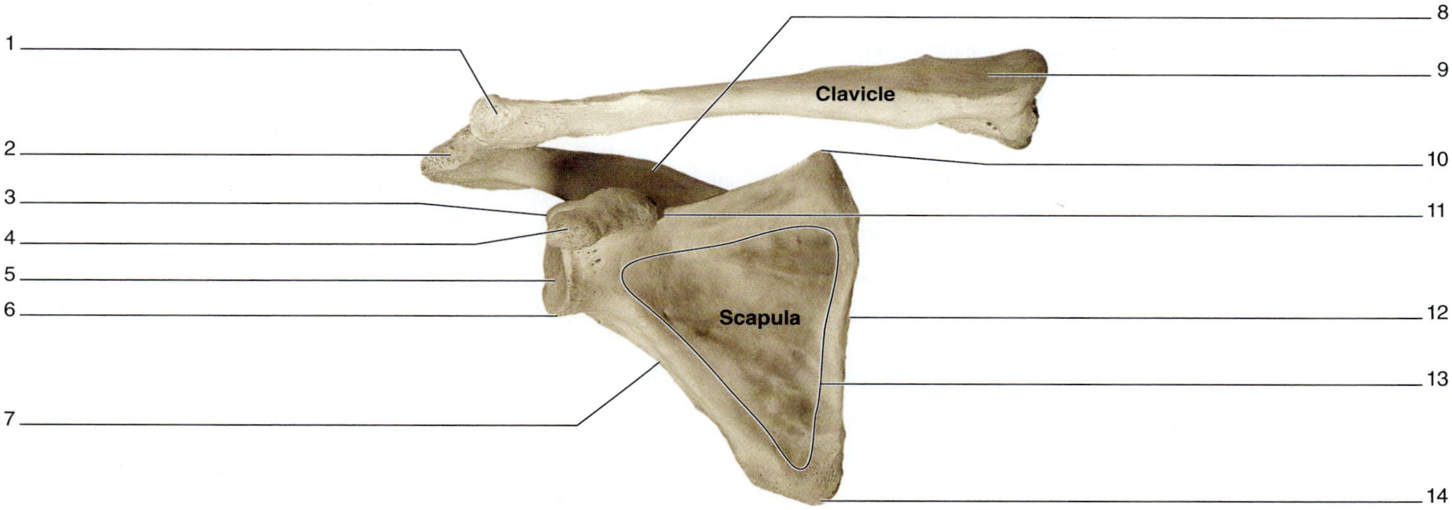

Right scapula and clavicle articulation, anterior view

Scapula	*Clavicle*
acromion	acromial end
coracoid process	sternal end
glenoid cavity	
inferior angle	
infraglenoid tubercle	
lateral border	
medial border	
spine	
subscapular fossa	
superior angle	
supraglenoid tubercle	
suprascapular notch	

Exercise 9.2: Bones of the Upper Limb

5. Match the description listed in column A with the bone or marking listed in column B. ❹

Column A	Column B
____ 1. bone of the arm	a. acromion
____ 2. bone of the forearm that aligns with the thumb	b. carpals
____ 3. bone of the forearm that aligns with digit V (pinky finger)	c. clavicle
____ 4. middle bone of the index finger	d. coracoid process
____ 5. part of the scapula that articulates with the humerus	e. glenoid cavity
____ 6. part of the scapula that articulates with the clavicle	f. humerus
____ 7. anterior projection of the scapula that resembles a crow's beak	g. metacarpal I
____ 8. bone that articulates with both the sternum and scapula	h. middle phalanx II
____ 9. first bone composing the palm of the hand	i. middle phalanx IV
____ 10. long bony ridge on the posterior aspect of the scapula	j. radius
____ 11. bones of the wrist	k. spine of scapula
____ 12. middle bone of the ring finger	l. ulna

6. The anatomic name for the thumb is the pollex. ____________________ (True/False) ❹

7. Match the description listed in column A with the carpal bone listed in column B. ❺

Column A	Column B
____ 1. shaped like a half moon	a. hamate
____ 2. shaped like a "table" and located at the base of the *first* metacarpal (thumb)	b. lunate
____ 3. shaped like a "table" and located at the base of the *second* metacarpal (index finger)	c. pisiform
____ 4. smallest carpal bone; shaped like a "pea"	d. trapezium
____ 5. shaped like a hook and located at the base of the *fifth* metacarpal (pinky)	e. trapezoid

Exercise 9.3: Surface Anatomy Review—Pectoral Girdle and Upper Limb

8. Which process of the scapula can be palpated at the tip of the shoulder? (Circle one.) ❻

a. acromion   b. coracoid   c. infraglenoid tubercle   d. spine   e. supraglenoid tubercle

9. Match the description listed in column A with the bony landmark listed in column B. ❻

Column A	Column B
____ 1. forms the tip of the elbow	a. medial epicondyle
____ 2. process of the radius palpated at the wrist	b. olecranon
____ 3. process of the ulna palpated on the medial aspect of the elbow	c. styloid process

Exercise 9.4: Bones of the Pelvic Girdle

10. Match the description listed in column A with the bone or marking listed in column B. ❼

Column A	Column B
____ 1. three bones fuse to form this bone of the pelvis	a. acetabulum
____ 2. bone that the pelvis rests on when sitting	b. ilium
____ 3. most anterior bone of the os coxa	c. ischium
____ 4. "hip" bone	d. obturator foramen
____ 5. large hole in the os coxa	e. os coxa
____ 6. structure that articulates with the femur at the hip	f. pubic bone

11. Identify the bones that form the os coxae. (Check all that apply). 8

____ a. coccyx

____ b. ilium

____ c. ischium

____ d. pubis

____ e. sacrum

Exercise 9.5: Bones of the Lower Limb

12. Match the bones listed in column A with the corresponding joint listed in column B. 9

Column A	*Column B*
____ 1. os coxa and femur	a. ankle
____ 2. femur and tibia	b. hip
____ 3. tibia, fibula, and talus	c. knee

13. The anatomic name for the big toe is talus. ____________________ (True/False) 9

14. Match the description listed in column A with the tarsal bone listed in column B. 10

Column A	*Column B*
____ 1. shaped like a "ship" and located anterior to the talus	a. calcaneus
____ 2. major weight-bearing bone that articulates with the tibia and fibula to form the ankle joint	b. cuneiforms
____ 3. shaped like a "wedge" and located at the base of the first, second, and third metatarsals	c. navicular
____ 4. forms the heel	d. talus

15. There are ____________________ (seven/eight) carpal bones and ____________________ (seven/eight) tarsal bones. 11

Exercise 9.6: Surface Anatomy Review—Pelvic Girdle and Lower Limb

16. Match the description listed in column A with the bone or bony landmark listed in column B. 12

Column A	*Column B*
____ 1. ridge of bone palpated when placing the hands on the hips	a. iliac crest
____ 2. palpated on the anterior part of the knee joint	b. lateral epicondyle of the femur
____ 3. palpated on the lateral aspect of the knee joint	c. lateral malleolus of the fibula
____ 4. palpated on the anterior leg, distal to the knee joint	d. medial epicondyle of the femur
____ 5. palpated on the medial aspect of the ankle joint	e. medial malleolus of the tibia
____ 6. palpated on the lateral aspect of the ankle joint	f. patella
____ 7. palpated on the medial aspect of the knee joint	g. tibial tuberosity

Can You Apply What You've Learned?

17. What are the structural and functional differences between the anatomical and surgical necks of the humerus?

____

____

____

18. Construct a unique mnemonic to assist with recall of the carpal bones. Write the mnemonic here:

____

____

____

19. Compare and contrast both the appearance and the location of the carpal and tarsal bones.

____

____

____

20. Construct a unique mnemonic device to assist with recall of the tarsal bones. Write the mnemonic here:

____

____

____

21. List three of the features that help differentiate a male pelvis from a female pelvis.

Male Pelvis	***Female Pelvis***
a. ____________	a. ____________
b. ____________	b. ____________
c. ____________	c. ____________

22. What bone or bony process of the upper limb serves the same function as the patella in the lower limb? (Hint: The patella's function is to act as a lever. It forces the tendons of the muscles on the anterior surface of the thigh farther away from the center of rotation of the joint in order to give the muscles a greater mechanical advantage—greater leverage.) ____________

23. Explain why one end of the clavicle is called the sternal end, and the other is called the acromial end.

____

____

____

Can You Synthesize What You've Learned?

24. What bony landmark of the upper limb is used as a reference point for locating the ulnar nerve at the elbow?

____

25. Which of the two necks of the humerus do you think is more likely to fracture in an accident? Why?

____

____

26. Observe the relationship between the carpal bones and the distal portion of the radius and ulna on an articulated skeleton. Which of the carpal bones do you think is most likely to fracture when someone falls on an outstretched hand?

____

27. Why do you think it is functionally important that the bones of the os coxae fuse together rather than remain independent bones?

____

____

28. A fracture to which of the leg bones (tibia or fibula) would result in the greatest loss of function of the lower limb? Why?

____

____

CHAPTER 10

Articulations

OUTLINE AND LEARNING OBJECTIVES

MODULE 5: SKELETAL SYSTEM

INTRODUCTION

When asked to name joints of the body, most people think only of the ankle, knee, hip, wrist, elbow, and shoulder joints. However, the sutures between the skull bones are also joints, as are the connections between the ribs and sternum. A joint, or **articulation** (*articulatio*, a forming of vines), is formed wherever one bone comes together with another bone. The study of joints is called **arthrology** (*arthron*, a joint, + *logos*, the study of). Joints are classified both by anatomic structure and by function. Joint classification based on structure reflects (1) whether a space called a joint cavity exists between the articulating bones, and (2) the type of connective tissue that holds the articulating surfaces of the bones together. The three *structural* categories of joints include fibrous joints, cartilaginous joints, and synovial joints.

Joint classification based upon function reflects the extent of movement permitted at the joint. The three *functional* categories of joints include synarthroses, which are immovable joints; amphiarthroses, which are slightly movable joints; and diarthroses, which are freely movable joints. **Table 10.1** summarizes the functional (movement) classification of joints.

The early exercises in this chapter provide an overview of fibrous and cartilaginous joints, and later exercises focus on the details of synovial joints. The final laboratory exercises take an in-depth look at the structure and function of a representative synovial joint: the knee joint. All of the exercises in this chapter can be performed either on cadaver specimens or on joint models. The text is written as a guide to finding structures associated with each of the joints using resources that are available in the anatomy and physiology laboratory. Use the tables and figures provided in this manual, which contain detailed information about each type of joint, to guide your observations.

List of Reference Tables

Table 10.1 Functional (Movement) Classification of Joints

Type of Joint	Description	Examples	Word Origin
Synarthrosis	An immobile joint	Skull suture and tooth gomphosis	*syn,* together, + *arthron-*, a joint
Amphiarthrosis	A slightly mobile joint	Pubic symphysis	*amphi,* on both sides, + *arthron,* a joint
Diarthrosis	A freely mobile joint	All synovial joints	*di-*, two, + *arthron,* a joint

Chapter 10: Articulations

Name: ______________________

Date: ____________ Section: ____________

These Pre-Laboratory Worksheet questions may be assigned by instructors through their connect course.

PRE-LABORATORY WORKSHEET

1. Which of the following are categories describing *structural* classifications of joints? (Check all that apply.)
   ____ a. cartilaginous
   ____ b. diarthrosis
   ____ c. fibrous
   ____ d. synarthrosis
   ____ e. synovial

2. Which of the following terms is used in the *functional* classification of an immobile joint? (Circle one.)
   a. cartilaginous
   b. diarthrosis
   c. fibrous
   d. synarthrosis
   e. synovial

3. Match the description of the fibrous joint listed in column A with the fibrous joint classification listed in column B.

Column A	***Column B***
____ 1. between skull bones	a. gomphosis
____ 2. distal radioulnar or tibiofibular joints	b. suture
____ 3. between alveolar processes of maxilla or mandible and the root of a tooth	c. syndesmosis

4. Which of the following is classified as a synchondrosis joint? (Check all that apply.)
   ____ a. epiphyseal plate
   ____ b. intervertebral disc
   ____ c. public symphysis
   ____ d. sternocostal joint

5. Fibrous and cartilaginous joints may be classified based on movement as synarthrotic joints, amphiarthrotic joints, or diarthrotic joints.

   ____________________ (True/False)

6. Based on movement allowed, all synovial joints are classified as ____________________. (Circle one.)
   a. amphiarthroses
   b. diarthroses
   c. synarthoses

7. There is a(n) ____________________ (direct/indirect) relationship between mobility and stability of a joint. Therefore, an increase in mobility leads to a(n) ____________________ (increase/decrease) in stability of the joint.

8. The knee joint is an example of a ____________________ joint. (Circle one.)
   a. ball-and-socket
   b. condylar
   c. hinge
   d. pivot
   e. plane

9. Circumduction is a movement allowed at the hip joint. ____________________ (True/False)

10. Synovial fluid ____________________ (increases/decreases) friction in a joint.

GROSS ANATOMY

Fibrous Joints

Fibrous joints are characterized by having no joint cavity. Rather, fibrous connective tissue binds the neighboring bones. Functionally, fibrous joints are classified into two types that are immovable joints **(gomphoses** and **sutures)** and one type that is slightly movable **(syndesmoses; table 10.2).** When working through the structural classifications of joints in these exercises, practice using the functional terms from table 10.1 to describe the movement allowed at each joint.

Table 10.2 Classification of Fibrous Joints

Fibrous Joints	Structure and Description	Examples	Word Origin
Gomphosis	Consists of a cone-shaped peg fitting into a socket and anchored by the periodontal membrane; immovable joint	Teeth articulating with alveolar processes of mandible or maxilla	*gomphos*, nail, + *-osis*, condition
Suture	Found exclusively between skull bones; consists of a small amount of connective tissue (the sutural ligament) holding the bone surfaces together; immovable joint	Lambdoid suture, sagittal suture	*sutura*, a seam
Syndesmosis	Consists of large surfaces of bones that are anchored together by a connective tissue membrane called an interosseous membrane; slightly movable joint	Distal radioulnar joint; distal tibiofibular joint	*syn-*, together, + *desmos*, a band

EXERCISE 10.1

FIBROUS JOINTS

1. ***Gomphosis***—(a) Observe the **gomphosis** (*gomphos*, nail, + *-osis*, condition) between a tooth and the mandible in **figure 10.1.** Although a tooth is not a bone, the connection of a tooth into the mandible or maxilla is considered a joint. Each tooth normally fits very tightly into a socket in a living individual because of both the deep socket and the fibrous tissue (called a **periodontal membrane**) that binds the tooth. (b) Obtain a skull and observe a gomphosis. The teeth might be very loose (or absent) in the skull because the periodontal membranes were destroyed when the skull was prepared. If the skull has some empty tooth sockets (alveolar processes), observe the shape of the inside of the socket that the cone-shaped root of a tooth fits into. How is a gomphosis classified based on the amount of normal *movement* allowed?

   ____

2. ***Suture***—(a) Observe the **suture** (*sutura*, a seam) between skull bones in figure 10.1. The skull bones fit together tightly in a living individual. This is due to both the interlocking shapes of the articulating skull bones and the membranous connective tissue that binds them together. Recall that the flat bones of the skull form by **intramembranous ossification.** The fibrous connective tissue within the sutures is a remnant of the original membrane that served as the structural framework for the developing bones. (b) Obtain a skull and observe the numerous sutures between the cranial bones. The sutural joints may be somewhat loose in the preserved skull because the membranous connective tissue that normally holds the bones together was destroyed when the skull was prepared. How is a sutural joint classified based on the amount of *movement* allowed?

   ____

3. ***Syndesmosis***—(a) Observe the **syndesmosis** (*syn-*, together, + *desmos*, a band) joint between the radius and ulna in figure 10.1. Notice the fibrous connective tissue between these two bones, which is called an **interosseus membrane**. This fibrous connective tissue composes the syndesmosis between these two bones. (b) Observe the radius and ulna on an articulated skeleton. The articulated skeleton no longer contains the fibrous tissue that forms the syndesmosis joint because it was destroyed during preparation of the skeleton. The movement at the distal radioulnar joint can still be observed by moving the radius and ulna to mimic the action of turning a doorknob. How is a syndesmosis classified based on the amount of *movement* allowed?

   ____

4. Label the types of fibrous joints in figure 10.1, using the textbook as a guide.

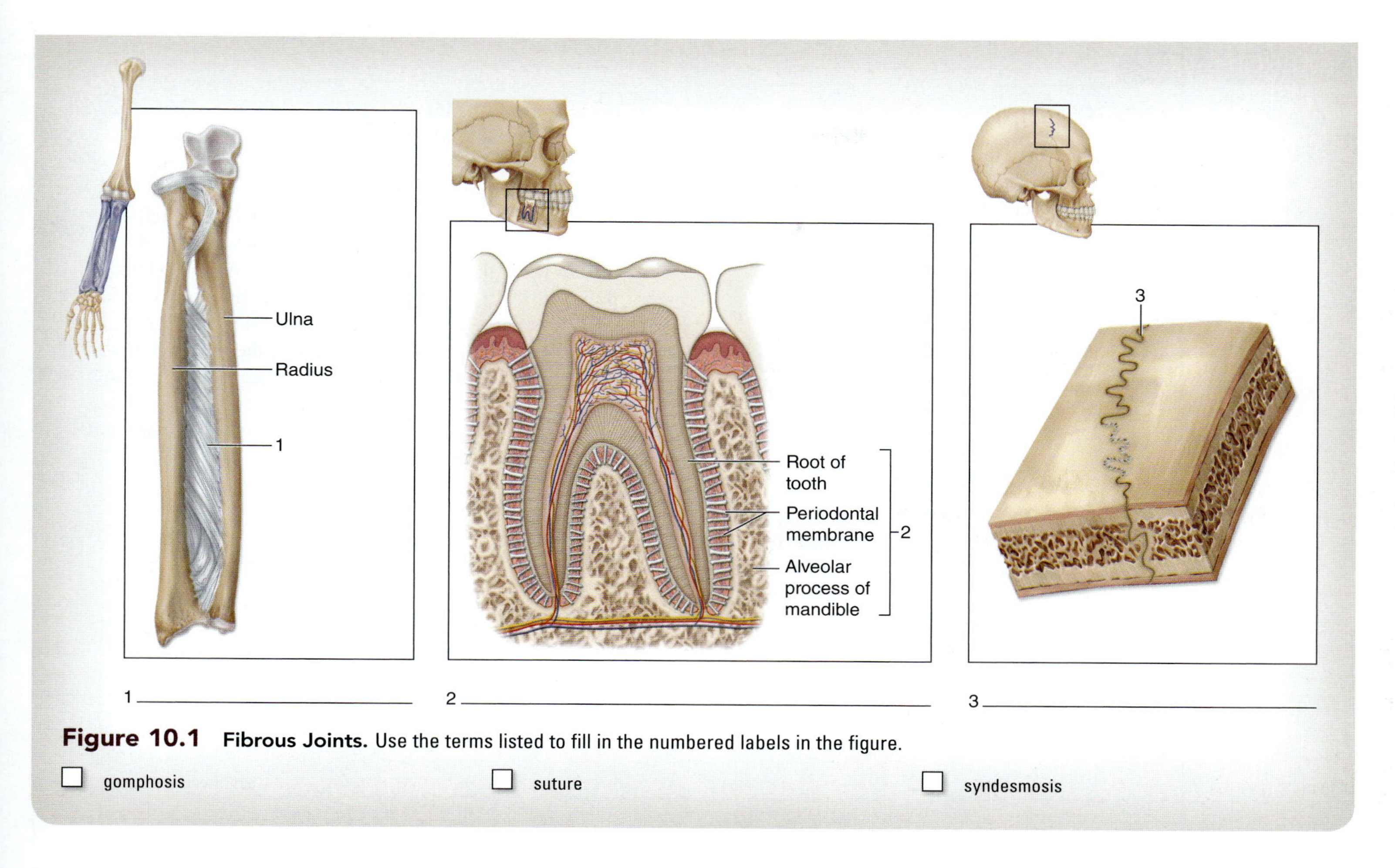

Figure 10.1 **Fibrous Joints.** Use the terms listed to fill in the numbered labels in the figure.

☐ gomphosis ☐ suture ☐ syndesmosis

Cartilaginous Joints

Cartilaginous joints are characterized by having no joint cavity. Rather, cartilage connects the neighboring bones. Cartilaginous joints are classified based on the type of cartilage found in the joint. **Synchondroses** are composed of hyaline cartilage, and **symyphyses** are composed of fibrocartilage **(table 10.3).**

Table 10.3 Classification of Cartilaginous Joints

Cartilaginous Joints	Structure and Description	Examples	Word Origin
Synchondrosis	Bone connected to bone by hyaline cartilage; immovable joint	Sternocostal joints; epiphyseal plates	*syn-*, together, + *chondrion*, cartilage
Symphysis	Bone connected to bone by fibrocartilage; slightly movable joint	Pubic symphysis; intervertebral discs	*symphysis*, a growing together

EXERCISE 10.2

CARTILAGINOUS JOINTS

1. ***Synchondrosis***—(a) Observe the **synchondrosis** (*syn-*, together, + *chondrion*, cartilage) between a rib and costal cartilage and at an epiphyseal plate in **figure 10.2.** The joint is composed of **hyaline cartilage** in a living individual.
(b) Observe the articulations between the ribs and sternum on an articulated skeleton. The "cartilage" between the ribs and sternum is some sort of replacement material such as plastic or rubber on the articulated skeleton. Attempt to move the bones at one of the synchondroses. How are the sternocostal joints classified based on the amount of *movement* allowed?

____

2. ***Symphysis***—(a) Observe the **symphysis** (*symphysis*, a growing together) joints both at the **pubic symphysis** (between the pubic bones) and at an **intervertebral disc** (between two vertebrae) in figure 10.2. The joint is composed of **fibrocartilage** in a living individual. The intervertebral discs have a more complex structure than the pubic symphysis. They consist of an outer band of fibrocartilage, called the **anulus fibrosus,** which surrounds a gel-like interior, called the **nucleus pulposus.** A "ruptured," "herniated," or "slipped" disc occurs when the anulus fibrosus tears and the nucleus pulposus leaks out. The leaked nucleus pulposus can compress nerve fibers, causing pain and numbness.
(b) Observe the pubic symphysis and the intervertebral discs on an articulated skeleton. The "cartilage" is some sort of replacement material on the articulated skeleton. How is a symphysis joint classified based on the amount of *movement* allowed?

____

3. Label the types of cartilaginous joints in figure 10.2, using the textbook as a guide.

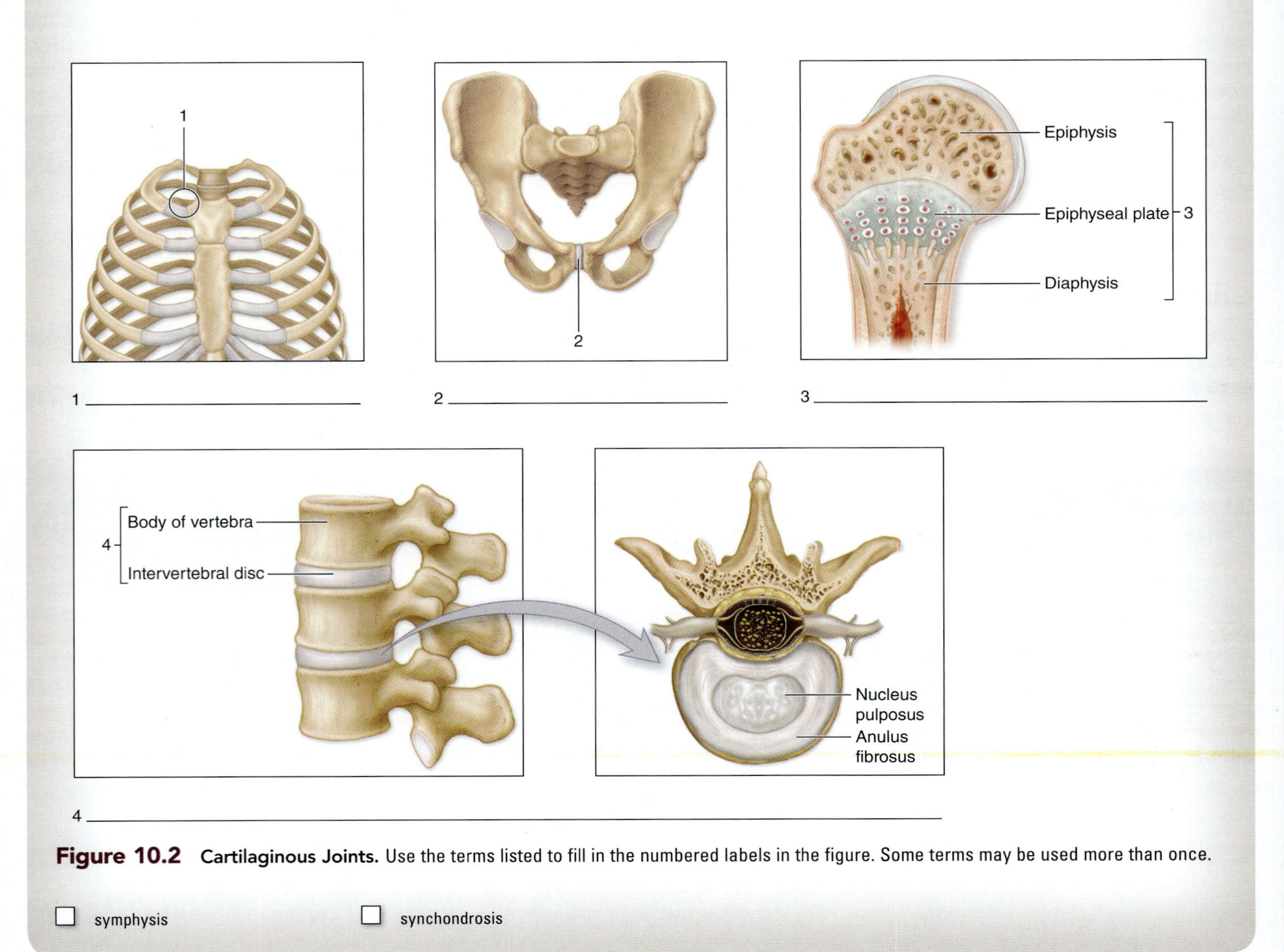

Figure 10.2 **Cartilaginous Joints.** Use the terms listed to fill in the numbered labels in the figure. Some terms may be used more than once.

☐ symphysis ☐ synchondrosis

Synovial Joints

Synovial joints have a complex structure that includes a joint cavity filled with fluid. The term *synovial* literally means "together with egg" (*syn*, together, + *ovum*, egg). This term refers to the fluid inside the joint (the synovial fluid), which has the consistency and appearance of egg white. The joint cavity filled with synovial fluid allows the articulating bones to move easily past one another with very little friction between the bones. The features of synovial joints are covered in detail in the next several exercises. These include exercises covering the general structure of a synovial joint, categories of synovial joints, and the types of movements allowed by synovial joints.

EXERCISE 10.3

GENERAL STRUCTURE OF A SYNOVIAL JOINT

1. Observe a model of a synovial joint, preferably a model of the knee joint. Identify the features of a typical synovial joint listed in **table 10.4.**

2. Label the components of a synovial joint in figure 10.3, using table 10.4 and the textbook as guides.

3. *Optional Activity:* AP|R **5: Skeletal System**—Watch the "Synovial Joint" animation for a summary of synovial joint structure and types.

Table 10.4 Components of Synovial Joints

Structure	Description	Word Origin
Articular Capsule	Consists of two layers: an outer fibrous capsule and an inner synovial membrane	*arthron*, a joint, + *capsa*, a box
Articular Cartilage	Hyaline cartilage found on the epiphyses of the articulating bones	*arthron*, a joint
Fibrous Layer of Articular Capsule	Dense irregular connective tissue that anchors the periosteum of the two articulating bones to each other; thickenings of the fibrous capsule form several joint ligaments	*fibra*, fiber
Synovial Cavity	A cavity within the joint that is lined by a synovial membrane and filled with synovial fluid	*syn*, together, + *ovum*, egg, + *cavus*, hollow
Synovial Fluid	A viscous, oily fluid located within the synovial joint; functions as a lubricant, to nourish the articular cartilage, and as a shock absorber	*syn*, together, + *ovum*, egg, + *fluidus*, to flow
Synovial Membrane	Composed primarily of areolar connective tissue that forms the inner lining of the articular capsule and covers internal joint surfaces not covered by cartilage; responsible for the formation of synovial fluid	*syn*, together, + *ovum*, egg, + *membrana*, a skin

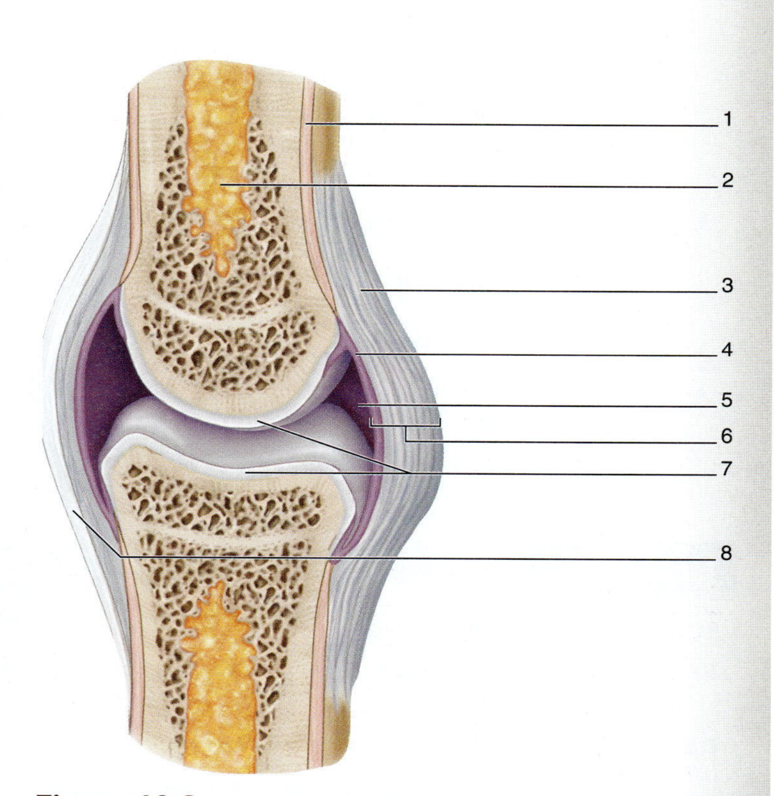

Figure 10.3 **Diagram of a Representative Synovial Joint.** Use the terms listed to fill in the numbered labels in the figure.

- ☐ articular capsule
- ☐ articular cartilage
- ☐ fibrous layer of articular capsule
- ☐ ligament
- ☐ periosteum
- ☐ synovial (joint) cavity
- ☐ synovial membrane
- ☐ yellow bone marrow

EXERCISE 10.4

CLASSIFICATION OF SYNOVIAL JOINTS

Synovial joints are classified using several different methods, most of which involve a description of the shape of the bones and the type of movement allowed at the joint **(table 10.5).** For example, in a ball-and-socket joint, the "ball" (rounded) part of one bone fits into the "socket" (concave) part of another bone.

1. **Ball-and-Socket Joint**—Observe the ball-and-socket joint between the femur and the os coxa in **figure 10.4.** Now observe how the head of the femur articulates with the os coxa on an articulated skeleton. Notice that the head of the femur ("the ball") fits into the acetabulum ("the socket"). What other joint in the body is also a ball-and-socket joint? ________________.
2. **Condylar (ellipsoid) Joint**—Observe the condylar joint between a metacarpal and proximal phalanx in figure 10.4. Now observe this joint on an articulated skeleton. Notice the shape of the two articulating bones. The bone with the convex oval surface is the ________________ and the bone with the elliptical concavity is the ________________.
3. **Hinge Joint**—Observe the hinge joint between the humerus and ulna in figure 10.4. Now observe this joint on an articulated skeleton. The bone with the convex oval surface is the ________________ and the bone with the concave surface is the ________________.
4. **Pivot Joint**—Observe the pivot joint between the dens of the axis and the atlas in figure 10.4. Now observe this joint on an articulated skeleton. The bone with the round surface is the ________________ and the bone with the "ring" is the ________________.
5. **Plane Joint**—Observe the plane joint between two carpal bones in figure 10.4. Now observe this joint on an articulated skeleton. What other joints in the foot are plane joints? ________________
6. **Saddle Joint**—Observe the saddle joint at the base of the thumb in figure 10.4. Now observe this joint on an articulated skeleton. The two bones that have both a concave and a convex surface are the ________________ and ________________.
7. Identify the synovial joints in figure 10.4, using table 10.5 and the textbook as guides.

INTEGRATE

LEARNING STRATEGY

Both synchondroses and synovial joints contain cartilage within the joint. However, remember that synchondroses do not have a joint cavity and the bones are tightly joined together (making them either immovable or slightly movable). In comparison, synovial joints contain cartilage on the ends of the bones (the articular cartilage) but the bones are separated by a joint cavity, which allows them to be freely movable.

Table 10.5 Classification of Synovial Joints

Surface Shape	Structure and Description	Examples	Word Origin
Ball-and-Socket	A spherical head fits into a concave socket; multiaxial	Hip joint; shoulder joint	*ball*, a round mass, + *soccus*, a shoe or sock
Condylar (ellipsoid)	A convex oval surface fits into an elliptical concavity; biaxial	Radiocarpal joints; metacarpophalangeal joints	*kondylos*, knuckle, or *ellips*, oval, + *eidos*, form
Hinge (gynglymoid)	A convex surface fits into a concave surface; uniaxial	Knee joint; elbow joint	*gynglymos*, a hinge joint
Pivot	A round surface fits into a ring formed by a ligament and a depression in another bone; uniaxial	Atlantoaxial joint	*pivot*, a post upon which something turns
Plane (gliding)	Two flat surfaces come together; uniaxial	Intertarsal joints; some intercarpal joints	*planus*, flat
Saddle (sellar)	Both bones have a concave and convex surface with the bones positioned at right angles to each other; biaxial	Carpometacarpal joint of the thumb; ankle joint; calcaneocuboid joint	*sella*, saddle

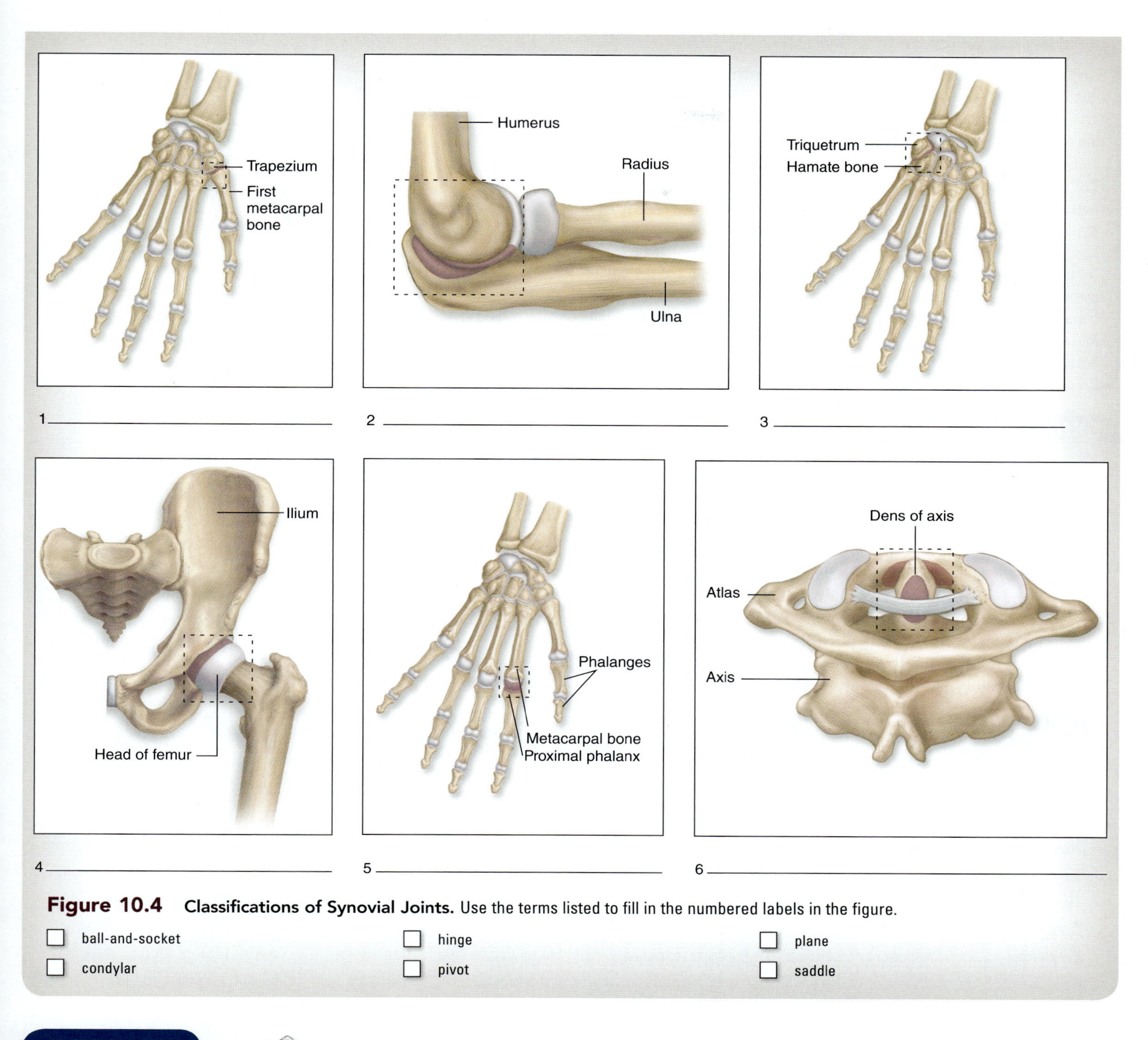

Figure 10.4 **Classifications of Synovial Joints.** Use the terms listed to fill in the numbered labels in the figure.

- ☐ ball-and-socket
- ☐ condylar
- ☐ hinge
- ☐ pivot
- ☐ plane
- ☐ saddle

INTEGRATE

CLINICAL VIEW

Hip Replacement Surgery

What comes to mind when thinking of someone's "hip"? Many people think this is the part of the body that lies superficial to the iliac crest. In fact, the term *ilium* literally means "hip." When you "put your hands on your hips," you are putting them on your iliac crests. However, when someone falls and suffers a fractured hip, it has little to do with the iliac bones at all. Instead, it concerns a fracture of another bone that composes part of the hip joint: the femur. Specifically, a fractured hip refers to a fracture of the neck of the femur.

The hip joint is the joint formed between the acetabulum of the os coxae and the head of the femur. This joint suffers wear and tear with age, and is a common site of osteoarthritis. In the elderly, particularly elderly women, the bones that compose the hip joint may become less dense over time. These individuals are at increased risk of developing a femoral or "hip" fracture if they fall. Unfortunately, hip fractures do not heal particularly well, because the fracture often disrupts the blood supply to the head of the femur. Therefore, instead of repairing the fractured bones, a surgeon might opt to completely replace the hip joint.

Hip replacement surgery involves cleaning out the acetabulum. Previous techniques involved reconstructing the acetabulum with bone grafts. More modern approaches include bone augmentation with platelet-rich plasma (PRP), polymethyl methacrylate (PMMA, bone cement), or Cortoss, a composite that consists of resins and reinforcing glass particles. The purpose of reconstruction or augmentation is to create a more efficient and complete socket that will be a better fit for the two parts of the prosthetic hip.

(continued on next page)

(continued from previous page)

Augmentation techniques allow a "forced fit," which does away with the need for fixation screws. Within the acetabulum, the surgeon places an artificial cup to compose the "socket" of this "ball-and-socket" joint. Next, the fractured head and neck of the femur are removed, and a prosthetic is placed on the proximal end of the femur, thus composing the new "ball" of the "ball-and-socket" joint. **Figure 10.5** shows a few different types of prosthetics. Figure 10.5*a* is an example of a very old hip prosthetic. Notice how large the ball is. Figure 10.5*b* is an example of a prosthetic with a much smaller head, and figure 10.5*c* shows a similar prosthetic that is still within the bone. Notice in figure 10.5*c* that the greater and lesser trochanters of the femur are still intact. This is necessary to maintain the connections between the bone and the gluteal muscles and other muscles that act about the hip. Such muscles are separated during the surgery so the surgeon can access the hip joint. However, they must be reconnected to the bone after the surgery.

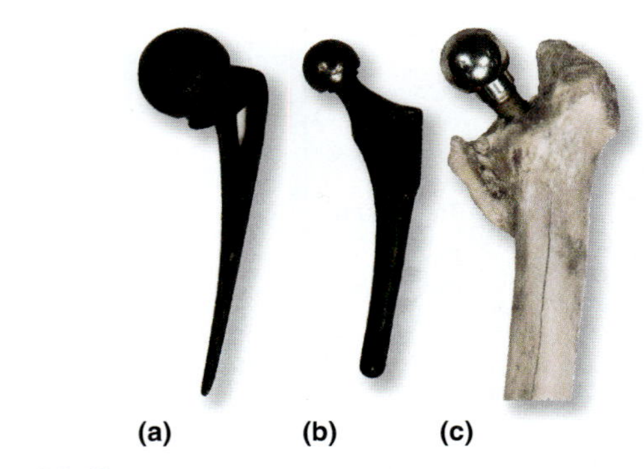

Figure 10.5 Three Different Examples of Prostheses Used for Hip Replacements.

EXERCISE 10.5

PRACTICING SYNOVIAL JOINT MOVEMENTS

1. All synovial joints are classified as **diarthrotic** (*di-*, two, + *arthron*, a joint) because they are freely movable. There are different types of movement possible with the various types of synovial joints. **Table 10.6** summarizes the types of movements possible at synovial joints. Practice all of the movements listed in table 10.6 with a laboratory partner to ensure that you can demonstrate or describe them all. It may be helpful to view figures of these movements in the textbook. This knowledge is critically important for success in subsequent chapters, which cover muscle actions. Muscle actions are described in terms of joint movements (e.g., flexor digitorum = muscle that flexes the digits).
2. *Optional Activity:* APIR **5: Skeletal System**—Review the series of joint movement animations to see examples of multiple movements at each joint.

Table 10.6 Movements of Synovial Joints

Movement	Description	Opposite Movement	Word Origin
Abduction	Movement of a body part away from the midline	Adduction	*ab*, from, + *duco*, to lead
Adduction	Movement of a body part toward the midline	Abduction	*ad-*, toward, + *duco*, to lead
Circumduction	Movement of the distal part of an extremity in a circle	NA	*circum-*, around, + *duco*, to lead
Depression	Movement of a body part inferiorly (especially mandible and scapula)	Elevation	*depressio*, to press down
Dorsiflexion	Movement of the ankle such that the foot moves toward the dorsum (back)	Plantar flexion	*dorsum*, the back, + *flexus*, to bend
Elevation	Movement of a body part superiorly (especially mandible and scapula)	Depression	*e-levo-atus*, to lift up
Eversion	Movement of the ankle such that the plantar surface of the foot faces laterally	Inversion	*e-*, out, + *versus*, to turn
Extension	An increase in joint angle	Flexion	*extensio*, a stretching out
Flexion	A decrease in joint angle	Extension	*flexus*, to bend
Inversion	Movement of the ankle such that the plantar surface of the foot faces medially	Eversion	*in-*, inside, + *versus*, to turn
Opposition	Placement of the thumb (pollex) such that it crosses the palm of the hand and can touch all of the remaining digits	Reposition	*op -*, against, + *positio*, a placing
Plantar Flexion	Movement of the ankle such that the foot moves toward the plantar surface	Dorsiflexion	*plantaris*, the sole of the foot, + *flexus*, to bend
Pronation	Movement of the palm from anterior to posterior	Supination	*pronatus*, to bend forward
Protraction	Movement of a body part anteriorly (especially mandible and scapula)	Retraction	*pro-*, before + *tractio*, to draw
Reposition	Movement of the thumb to anatomic position after opposition	Opposition	*re-*, backward, + *positio*, a placing
Retraction	Movement of a body part posteriorly (especially mandible and scapula)	Protraction	*re-*, backward, + *tractio*, to draw
Rotation	Movement of a body part around its axis	NA	*rotatio*, to rotate
Supination	Movement of the palm from posterior to anterior	Pronation	*supinatus*, to bend backward

INTEGRATE

LEARNING STRATEGY

When studying the structural classification of synovial joints, take a close look at how each bone fits with surrounding bones to allow movement. For example, a ball-and-socket joint involves the *ball* of one bone, (e.g., the head of the humerus or femur) fitting into the *socket* of another bone (e.g., the glenoid fossa of the scapula or the acetabulum of the os coxae). Once you have an idea of how two bones fit together, try to predict the movement allowed at each joint by forming each joint using a disarticulated skeleton. Then, predict what movements are allowed at the joint. For example, the ball-and-socket joint between the head of the humerus and the glenoid fossa allows flexion, extension, rotation, and circumduction. This is quite different from a hinge joint, such as the elbow joint, which only allows flexion and extension movements. Perform this exercise with each synovial joint that you study to further understand both structural classifications of joints and synovial joint movements.

INTEGRATE

CLINICAL VIEW
Bursitis

Bursae are small round sacs lined with synovial membrane and filled with synovial fluid. They are found both within synovial joints and in other areas of the body that experience a great deal of friction. Although they are structures designed to reduce friction, they themselves may become inflamed when frictional forces are severe and long-lasting. *Bursitis* refers to inflammation of a bursa, and can be a common source of joint pain. Repetitive movement of the joint and subsequent inflammation of the synovial membrane within the bursa causes bursitis. The inflammation causes the production of excessive synovial fluid. The swelling of the bursa can put pressure on surrounding structures and induce pain. Bursitis can also alter the mobility of the joint, significantly reducing the range of motion. Individuals suffering from bursitis may reduce their joint movements to avoid this pain. The muscles surrounding the joint may also stiffen as a result of bursitis. Common sites of bursitis are the shoulder, elbow, knee, and ankle joints. Treatment typically involves rest, ice, and elevation along with a course of anti-inflammatory drugs to reduce localized swelling and pain. In severe cases, inflamed bursae may be surgically removed.

EXERCISE 10.6

THE KNEE JOINT

This exercise involves observation of the structure of the knee joint, a representative synovial joint. The knee joint is complex, as are most synovial joints. It contains several modifications, such as bursae, tendon sheaths, and menisci. The knee joint also contains several strong ligaments, which help to stabilize the joint. **Table 10.7** summarizes the structures that compose the knee joint.

Most individuals have some peripheral knowledge of the knee joint, having known individuals who have suffered a ruptured ACL, torn meniscus, or other knee injury, even without knowledge of the anatomic structure of the knee. The ACL (anterior cruciate ligament) is one of two **cruciate ligaments** (*cruciatus*, resembling a cross) found in the knee joint, and a ruptured ACL is a knee injury that is common in football players, downhill skiers, and others involved in contact sports.

1. Observe a model of the knee joint or the knee joint of a cadaver. Notice that the knee joint is actually two joints: the **tibiofemoral joint,** which is classified as a synovial hinge joint that acts in flexion and extension (though it allows for some rotational movement as well), and the **patellofemoral joint,** which is classified as a planar (gliding) joint. The bony structure of the tibiofemoral joint includes the medial and lateral femoral condyles, which sit on top of the medial and lateral tibial condyles. The bony structure of the patellofemoral joint includes the patellar surface of the femur, which consists of the smooth anterior surface between the femoral condyles, and the medial and lateral facets of the patella.
2. Identify the structures listed in **figure 10.6** on a model of the knee joint or on a cadaver, using table 10.4 and the textbook as guides. Then, label them in figure 10.6.

(continued on next page)

(continued from previous page)

Table 10.7 Structures of the Knee Joint

	Description	Word Origin
Ligaments		
Anterior Cruciate Ligament	Connects the anterior intercondylar eminence of the tibia to the medial surface of the lateral condyle of the femur	*ante-*, in front of, + *cruciatus*, shaped like a cross
Fibular (lateral) Collateral Ligament	Connects the lateral epicondyle of the femur to the head of the fibula	*co*, together, + *latus*, side
Patellar Ligament	Connects the patella to the tibial tuberosity	*patina*, a shallow disk
Posterior Cruciate Ligament	Connects the posterior intercondylar eminence of the tibia to the anterior part of the lateral surface of the medial condyle of the femur	*post-*, behind, + *cruciatus*, shaped like a cross
Tibial (medial) Collateral Ligament	Connects the medial epicondyle of the femur to the medial surface of the tibia; its deep surface is anchored to the medial meniscus	*co*, together, + *latus*, side
Menisci		
Lateral Meniscus	A crescent-shaped pad of fibrocartilage located between the lateral condyle of the femur and the lateral condyle of the tibia	*meniskos*, crescent
Medial Meniscus	A crescent-shaped pad of fibrocartilage located between the medial condyle of the femur and the medial condyle of the tibia	*meniskos*, crescent
Bursae		
Infrapatellar Bursa	Located between the proximal tibia and the patellar ligament	*infra*, below, + *patina*, a shallow disc, + *bursa*, a purse
Prepatellar Bursa	Located between the patella and the overlying skin	*pre-*, before, + *patina*, a shallow disc, + *bursa*, a purse
Suprapatellar Bursa	Located between the distal femur and the quadriceps femoris tendon; communicates with the synovial cavity of the knee joint	*supra*, on the upper side, + *patina*, a shallow disc, + *bursa*, a purse
Tendon		
Quadriceps Femoris Tendon	Connects the quadriceps femoris muscle to the patella	*quad,* four + *femoris,* femur

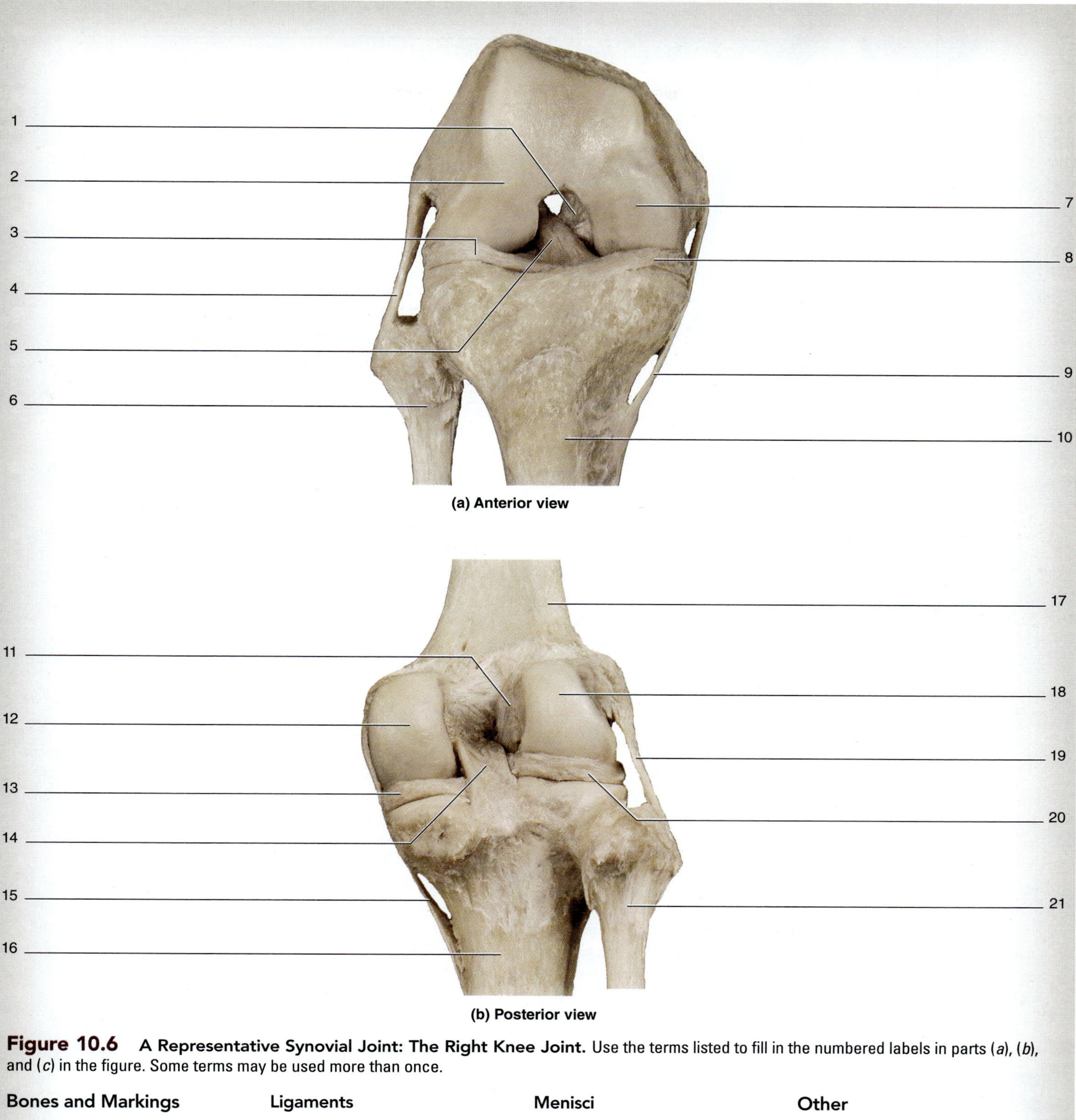

Figure 10.6 A Representative Synovial Joint: The Right Knee Joint. Use the terms listed to fill in the numbered labels in parts (*a*), (*b*), and (*c*) in the figure. Some terms may be used more than once.

Bones and Markings

- ☐ femur
- ☐ fibula
- ☐ lateral condyle of femur
- ☐ medial condyle of femur
- ☐ patella
- ☐ tibia

Ligaments

- ☐ anterior cruciate ligament
- ☐ fibular collateral ligament
- ☐ patellar ligament
- ☐ posterior cruciate ligament
- ☐ tibial collateral ligament

Menisci

- ☐ lateral meniscus
- ☐ medial meniscus

Bursae and tendon sheaths

- ☐ infrapatellar bursa
- ☐ prepatellar bursa
- ☐ suprapatellar bursa

Other

- ☐ articular cartilage
- ☐ quadriceps femoris tendon

(continued on next page)

(continued from previous page)

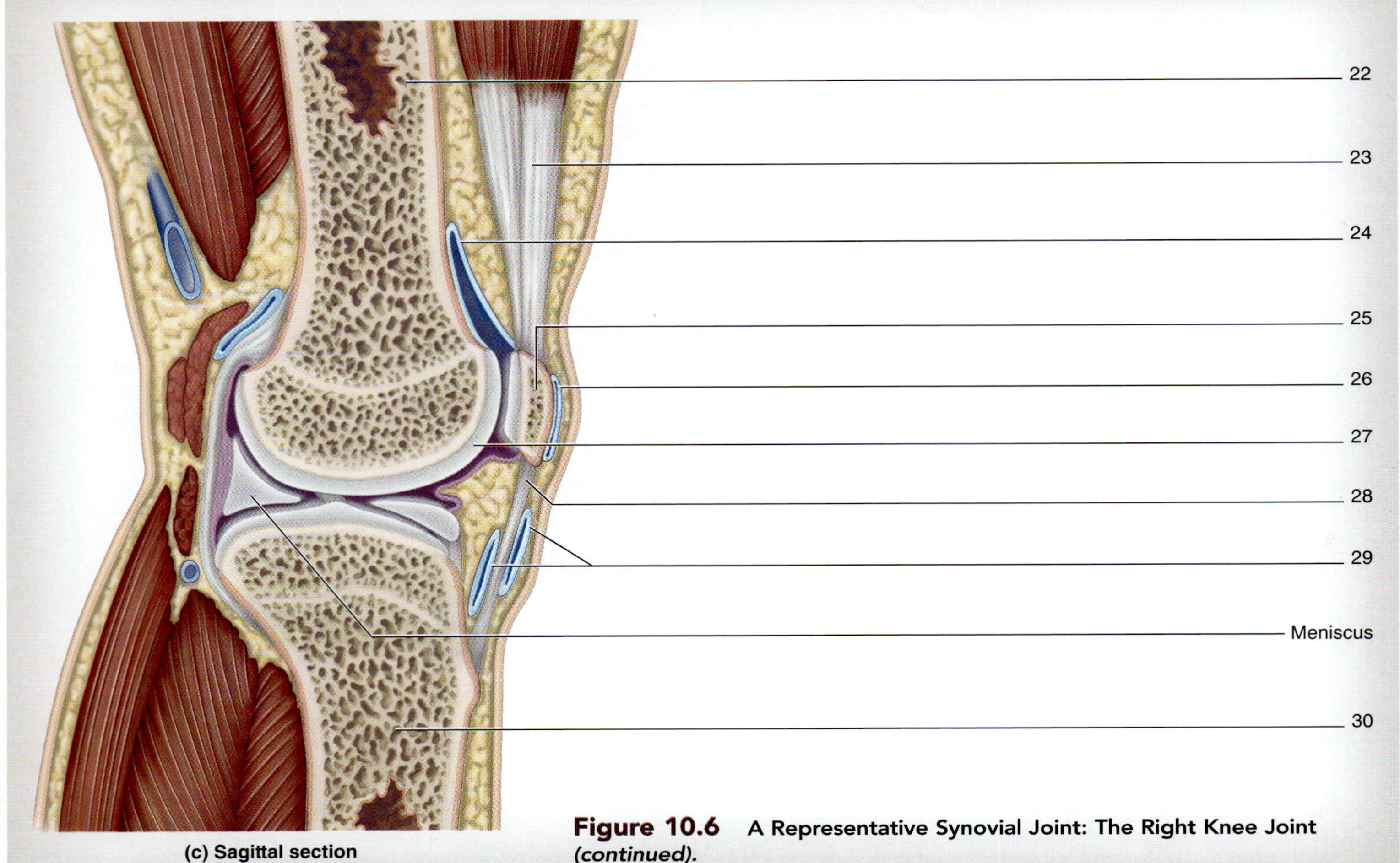

Figure 10.6 A Representative Synovial Joint: The Right Knee Joint *(continued).*

INTEGRATE

CONCEPT CONNECTION

Movement of synovial joints requires the contraction of muscles that surround the joint. Recall that the knee joint is a hinge joint, allowing only flexion and extension movements. Extending the knee joint (increasing the joint angle) requires contraction of the quadriceps muscles (anterior compartment of thigh). Flexing the knee joint (decreasing the joint angle) requires contraction of the hamstring muscles (posterior compartment of thigh). A muscle that causes the given action about a joint is called an **agonist** (*agon*, a contest), or **prime mover**. A muscle whose action opposes the action of the agonist is an **antagonist** (*anti*, against, + *agon*, a contest). In the example of movement about the knee joint, the quadriceps muscle group acts as the agonist, and the hamstring muscles act as antagonists. The actions of these two muscle groups oppose each other and cause the "hinge" movements at the joint. The action of both agonist and antagonist muscles is also important for stabilization of synovial joints.

More complex joint movements may also be possible using additional muscles surrounding a joint. Recall that the glenohumeral joint (shoulder joint) is an example of a ball-and-socket joint, allowing flexion, extension, abduction, adduction, rotation, and circumduction. The shoulder joint is more mobile than a hinge joint, but is subsequently more unstable. In fact, the shoulder joint is the most mobile joint in the body, and therefore susceptible to injury due to its corresponding instability. The "rotator cuff" muscles stabilize the glenohumeral joint and create movements of the joint such as adduction, medial rotation, and extension of the arm.

The individual muscles that compose the quadriceps, hamstrings, and rotator cuff muscle groups will be discussed in more detail in chapter 13.

INTEGRATE

CLINICAL VIEW
Low Back Pain

The sacroiliac (SI) joint has recently been identified as a possible common source of low back pain. Each SI joint (right and left) is a condylar joint, with the oval surface of the sacrum fitting into the concavity of the ilium. The sacral surface is lined with hyaline cartilage, whereas the ilial surface is lined with fibrocartilage. Because there are two SI joints, they are collectively considered bicondylar joints, because the movement of one joint affects the movement of the other. The SI joints are classified as synovial joints; however, mobility is typically limited to small **excursions** of the surrounding ligaments. Changes in mobility of these joints can result in sacroiliac joint dysfunction (SIJD), or even sacroiliitis, inflammation of the SI joint. The result is unilateral or bilateral low back pain that worsens with activities such as prolonged sitting or exercise. Increased lumbar lordosis and the effects of hormones such as relaxin, both of which are conditions that occur during pregnancy, stretch the ligaments, thereby inducing hypermobility of the SI joints. In contrast, arthritis can lead to hypomobility (reduced motion) of the SI joints. Both hypermobility and hypomobility can induce low back pain and be a source of SIJD.

Chapter 10: Articulations

Name: ______________________________

Date: ____________ Section: ____________

POST-LABORATORY WORKSHEET

The ❶ corresponds to the Learning Objective(s) listed in the chapter opener outline.

Do You Know the Basics?

Exercise 10.1: Fibrous Joints

1. Which of the following joints contains an interosseous membrane? (Circle one.) ❶
   a. gomphosis
   b. suture
   c. syndesmosis

2. Which of the following is a structural classification for fibrous joints? (Circle one.) ❷
   a. suture
   b. symphysis
   c. synchrondosis
   d. synovial

3. Which of the following correctly pairs the classification of a fibrous joint with an example of its proper location in the body? (Circle one.) ❸
   a. gomphosis; between skull bones
   b. gomphosis; distal tibiofibular joint
   c. suture; distal radioulnar joint
   d. syndesmosis; between skull bones
   e. syndesmosis; distal radioulnar joint

Exercise 10.2: Cartilaginous Joints

4. Symphysis joints consist of bones connected to bones by ________________________ (fibrocartilage/hyaline cartilage), whereas synchondrosis joints consist of bones connected to bones by ________________________ (fibrocartilage/hyaline cartilage). ❹

5. Which of the following terms are used to classify a synchrondrosis joint based on the extent of movement? (Check all that apply.) ❺

   ______ a. amphiarthrosis ______ b. diarthrosis ______ c. synarthrosis

6. Which of the following is an example of a cartilaginous joint? (Circle all that apply.) ❻
   a. epiphyseal plates
   b. intervertebral discs
   c. pubic symphysis
   d. sternocostal joints

Exercise 10.3: General Structure of a Synovial Joint

7. Of the four main tissue types, which tissue is the main component of the synovial membrane? (Circle one.) ❼
   a. connective tissue
   b. epithelial tissue
   c. muscle tissue
   d. nervous tissue

8. Match the descriptions of synovial joint structures listed in column A with the corresponding structure listed in column B. ❽

Column A	*Column B*
______ 1. crescent-shaped pad of fibrocartilage found within the joint	a. articular cartilage
______ 2. slippery fluid containing hyaluronic acid and glycoproteins	b. bursa
______ 3. small, round sac filled with fluid	c. meniscus
______ 4. lining within the joint	d. synovial fluid
______ 5. hyaline cartilage found on the ends of long bones	e. synovial membrane

Exercise 10.4: Classification of Synovial Joints

9. Match the description of the synovial joint listed in column A with the corresponding classification listed in column B. **9**

Column A	*Column B*
_______ 1. biaxial; oval, convex and concave	a. ball-and-socket
_______ 2. multiaxial; head fits into a socket	b. condylar
_______ 3. multiaxial; resembles the shape of a saddle	c. hinge
_______ 4. uniaxial; convex and concave surfaces	d. pivot
_______ 5. uniaxial; round surface in ring	e. plane
_______ 6. uniaxial; two flat surfaces	f. saddle

10. Which of the following correctly pairs the classification of a synovial joint with an example of its location? (Circle one.) **10**
    a. ball-and-socket; elbow joint
    b. hinge; shoulder joint
    c. pivot; atlantoaxial joint
    d. plane; radiocarpal joint
    e. saddle; knee joint

Exercise 10.5: Practicing Synovial Joint Movements

11. For the following synovial joints, list the type(s) of movement allowed at the joint (e.g., flexion, extension, abduction, adduction, rotation). **11**

Synovial Joint	Type of Movement
Shoulder	
Knee	
Carpals	
Wrist	
Ankle	
Atlas/axis	
Interphalangeal	
Metacarpal/phalangeal	
Hip	
Thumb	

12. Match the description of the joint movement listed in column A with the appropriate definition listed in column B. **11**

Column A	*Column B*
_______ 1. angle of the joint decreases	a. abduction
_______ 2. movement of limb away from the midline	b. adduction
_______ 3. angle of the joint increases	c. circumduction
_______ 4. movement of limb toward the midline	d. extension
_______ 5. limb pivots around its long axis	e. flexion
_______ 6. distal part of a limb moves in a circle	f. rotation

13. Match the description of the joint movement listed in column A with the appropriate term listed in column B. **11**

Column A	*Column B*
______ 1. palm turns from posterior to anterior	a. depression
______ 2. palm turns from anterior to posterior	b. dorsiflexion
______ 3. inferior movement of the mandible or scapula	c. elevation
______ 4. body part is moved anterior in the horizontal plane	d. eversion
______ 5. sole of foot faces medial (inward)	e. inversion
______ 6. dorsum of foot moves superiorly	f. plantar flexion
______ 7. superior movement of the mandible or scapula	g. pronation
______ 8. body part is moved posterior in horizontal plane	h. protraction
______ 9. sole of foot faces lateral (outward)	i. retraction
______ 10. dorsum of foot moves inferiorly	j. supination

Exercise 10.6: The Knee Joint

14. Match the description listed in column A with its corresponding structure listed in column B. **12**

Column A	*Column B*
______ 1. ligament connecting femur to tibia	a. ACL and PCL
______ 2. ligaments that cross in the middle of the knee joint	b. bursae
______ 3. sesamoid bone located in the knee joint	c. fibular (lateral) collateral ligament
______ 4. ligament connecting femur to fibula	d. meniscus
______ 5. structure connecting patella to tibial tuberosity	e. patella
______ 6. structure connecting quadriceps muscles to patella	f. patellar ligament
______ 7. fibrocartilaginous structure within knee joint	g. patellar tendon
______ 8. synovial sacs located above, below, and anterior to the patella	h. tibial (medial) collateral ligament

Can You Apply What You've Learned?

15. Where do the articular cartilages arise developmentally?

____

____

____

16. Define *bursitis*.

____

____

____

17. How does a symphysis joint differ from a synchondrosis in terms of both structure and extent of movement?

____

____

____

18. Explain why the knee joint is *not* classified as a synchondrosis, even though it has hyaline cartilage.

19. Fully classify the glenohumeral joint both structurally and functionally (degree of movement).

Can You Synthesize What You've Learned?

20. Why do you think the anterior cruciate ligament is so often injured during contact sports?

21. Injuries to the tibial collateral ligament are often accompanied by a torn medial meniscus. After observing the fibrous tissues that form the joint capsule of the knee, why do you think this pattern of injury is so common?

CHAPTER 11

The Muscular System: Muscle Structure and Function

OUTLINE AND LEARNING OBJECTIVES

Anatomy & Physiology REVEALED® aprevealed.com **MODULE 6: MUSCULAR SYSTEM**

INTRODUCTION

While reading the text on this page, skeletal muscles connected to the eyeballs (the *extrinsic* eye muscles) are contracting to produce very fine movements necessary for the eyes to track the words on the paper. At the same time, smooth muscles within the ciliary bodies of the eyes are contracting to alter the shape of the lens so the image on the page is focused clearly upon the retina. In addition, contraction of cardiac muscle in the heart is generating the force necessary to propel blood through the arteries to deliver oxygen and glucose to the working tissues within the eyes, the brain, and the rest of the body. Indeed, proper function of all three types of muscle tissue is essential for survival.

There are three types of muscle tissue: skeletal muscle, smooth muscle, and cardiac muscle. **Skeletal muscle** comprises the voluntary muscles that move the skin of the face and the skeleton, **smooth muscle** is found mainly in the walls of the viscera (such as the blood vessels, stomach, urinary bladder, intestines, and uterus), and **cardiac muscle** is found in the heart. These three types of muscle tissue are distinguished from each other based on location, neural control (voluntary vs. involuntary), the presence or absence of visible striations (striped appearance), the shape of the cells, and the number of nuclei per cell.

The exercises in this chapter guide the user in studying muscle from several different perspectives, including (1) histology of the three types of muscle, (2) gross anatomy of skeletal muscle, with a focus on naming of skeletal muscles, and (3) physiology of skeletal muscle.

List of Reference Tables

Chapter 11: The Muscular System: Muscle Structure and Function

Name: ______________________

Date: __________ Section: __________

These Pre-Laboratory Worksheet questions may be assigned by instructors through their connect course.

PRE-LABORATORY WORKSHEET

1. Match the location in the body listed in column A with the corresponding muscle tissue type listed in column B.

Column A	*Column B*
____ 1. face and skeleton	a. cardiac muscle
____ 2. heart	b. skeletal muscle
____ 3. visceral organs (e.g., bladder, intestine, stomach, uterus)	c. smooth muscle

2. Excitable tissues are able to propagate action potentials. __________________ (True/False)

3. Which of the following are characteristics of muscle tissue? (Check all that apply.)
   ____ 1. contractile
   ____ 2. distensible
   ____ 3. elastic
   ____ 4. excitable
   ____ 5. extensible

4. Identify the structural and functional unit of skeletal muscle. (Circle one.)
   a. actin
   b. myofibril
   c. myofilament
   d. myosin
   e. sarcomere

5. Which type of tissue composes a tendon, the structure that attaches a muscle to a bone? (Circle one.)
   a. areolar connective tissue
   b. dense irregular connective tissue
   c. dense regular connective tissue
   d. elastic connective tissue
   e. reticular connective tissue

6. Intercalated discs are found in which type of muscle tissue? (Circle one.)
   a. cardiac
   b. skeletal
   c. smooth

7. The cylindrical bundles of contractile proteins located inside skeletal muscle fibers are called __________________. (Circle one.)
   a. myofibers
   b. myofibrils
   c. myofilaments
   d. myosin
   e. sarcomeres

8. Skeletal muscles are given names that reflect location, shape, attachments, or other features related to the muscles. These names are based on Latin and Greek word roots. Match the meaning listed in column A with the word root listed in column B.

Column A	*Column B*
____ 1. around	a. *caput-*
____ 2. belly	b. *endo-*
____ 3. between	c. *epi-*
____ 4. head	d. *gastro-*
____ 5. upon	e. *inter-*
____ 6. within	f. *peri-*

HISTOLOGY

Muscle tissue is one of the most metabolically active tissues in the body and is the only tissue capable of creating movement of the body or body organs. All types of muscle tissue are characterized by the following properties: *excitability, contractility, elasticity*, and *extensibility*. **Excitable** tissues are able to respond to a stimulus, and generate an electrical signal called an action potential. **Contractile** tissues actively shorten and produce force. **Elastic** tissues return to their original shape following either contraction or stretching. **Extensible** tissues are able to be lengthened by the pull of an external force (such as an external weight or the action of an opposing muscle). **Conductive** tissues allow electrical signals such as action potentials to travel along the plasma membrane of the cell.

The exercises in this section guide the user through observations of the histology of skeletal, cardiac, and smooth muscle tissues. **Table 11.1** lists the characteristics of the three types of muscle tissue and serves as a reference for these exercises.

Skeletal Muscle Tissue

Skeletal muscle cells are some of the largest cells in the body. Their enormous size results from the fusion of hundreds of *myoblasts* (embryonic muscle cells) during development into a single muscle cell, or fiber. A mature muscle fiber is multinucleated, containing 200 to 300 nuclei per millimeter of fiber length. The nuclei of skeletal muscle fibers are located directly beneath the *sarcolemma* (plasma membrane of the muscle fiber).

A muscle, such as the biceps brachii muscle of the arm, consists of several bundles, or **fascicles** (*fascis*, bundle) **(figure 11.1*a*).** Each fascicle consists of hundreds of long, cylindrical **muscle fibers** (figure 11.1*b*). Each muscle fiber contains within it a number of cylindrical bundles of contractile proteins called **myofibrils** (figure 11.1*c*). The myofibrils contain the **myofilaments** actin and myosin (the main contractile proteins of muscle), which are arranged into **sarcomeres** (the structural and functional unit of skeletal muscle, figure 11.1*d*). The regular arrangement of actin and myosin into sarcomeres gives each myofibril a striated or banded appearance, with visible **A bands** (dark bands), **I bands** (light bands), and **Z discs** (a dark line in the middle of an I band) (figure 11.1*d,e*). An adult skeletal muscle fiber typically contains about 2000 myofibrils per cell. Because intermediate filaments within the muscle fiber anchor and align the Z discs of adjacent myofibrils, the entire muscle fiber takes on the same regular striated appearance of A bands and I bands found in the myofibrils when viewed through the light microscope. Both thick filaments, composed of myosin, and Z discs appear dark, whereas thin filaments, composed primarily of actin, appear light when skeletal muscle tissue is viewed through a light microscope.

Figure 11.1 demonstrates the relationships between the gross structure of a skeletal muscle and the microstructure of a skeletal muscle fiber. Though myofibrils and myofilaments are not visible when viewing muscle tissue through the microscope, A bands, I bands, and Z discs are usually visible. While viewing the banding pattern of skeletal muscle cells, try to relate the bands to the corresponding arrangement of myofilaments into sarcomeres as shown in the drawing of a sarcomere in figure 11.1.

Table 11.1 Muscle Tissues

Type of Muscle	Description	Functions	Identifying Characteristics
Skeletal	Elongated, cylindrical cells with multiple nuclei that are peripherally located. Tissue appears striated (alternating light and dark bands along the length of the cell).	Produces voluntary movement of the skin and the skeleton	Length of cells (extremely long); striations; multiple, peripheral nuclei
Cardiac	Short, branched cells with a single nucleus that is centrally located. Dark lines (intercalated discs) are seen where two cells come together. Tissue appears striated (light and dark bands along the length of each cell).	Performs the contractile work of the heart. Responsible for the pumping action of the heart.	Branched cells connected by intercalated discs; striations; one or two central nuclei
Smooth	Elongated, spindle-shaped cells (fatter in the center, narrowing at the tips) with single, "cigar-shaped" nuclei that are centrally located. No striations.	Produces movement within the walls of visceral organs such as intestines, bladder, uterus, and stomach; produces other specialized involuntary muscle movement (e.g., focusing the lens in the eye)	Spindle shape of the cells; lack of striations; single, central, cigar-shaped or spiral nuclei

INTEGRATE

CONCEPT CONNECTION

The study of muscle tissue invites a comparison between the three types of muscle tissue: skeletal, cardiac, and smooth. The mode of activation for these muscle types differs. Skeletal muscle is voluntary, whereas cardiac and smooth muscle are involuntary. Involuntary contraction of cardiac and smooth muscle is advantageous, as it is inconvenient to "think" about contracting the heart every time blood circulation is required, or to "think" about propelling food through the gastrointestinal tract when digestion is required. Instead, cardiac muscle contracts ("beats") *intrinsically* (on its own) and smooth muscle in the gut contracts when food is present. The rate of contraction of cardiac and smooth muscle is *extrinsically* regulated by the autonomic nervous system. The electrical conduction system of the heart is explored further in chapter 21.

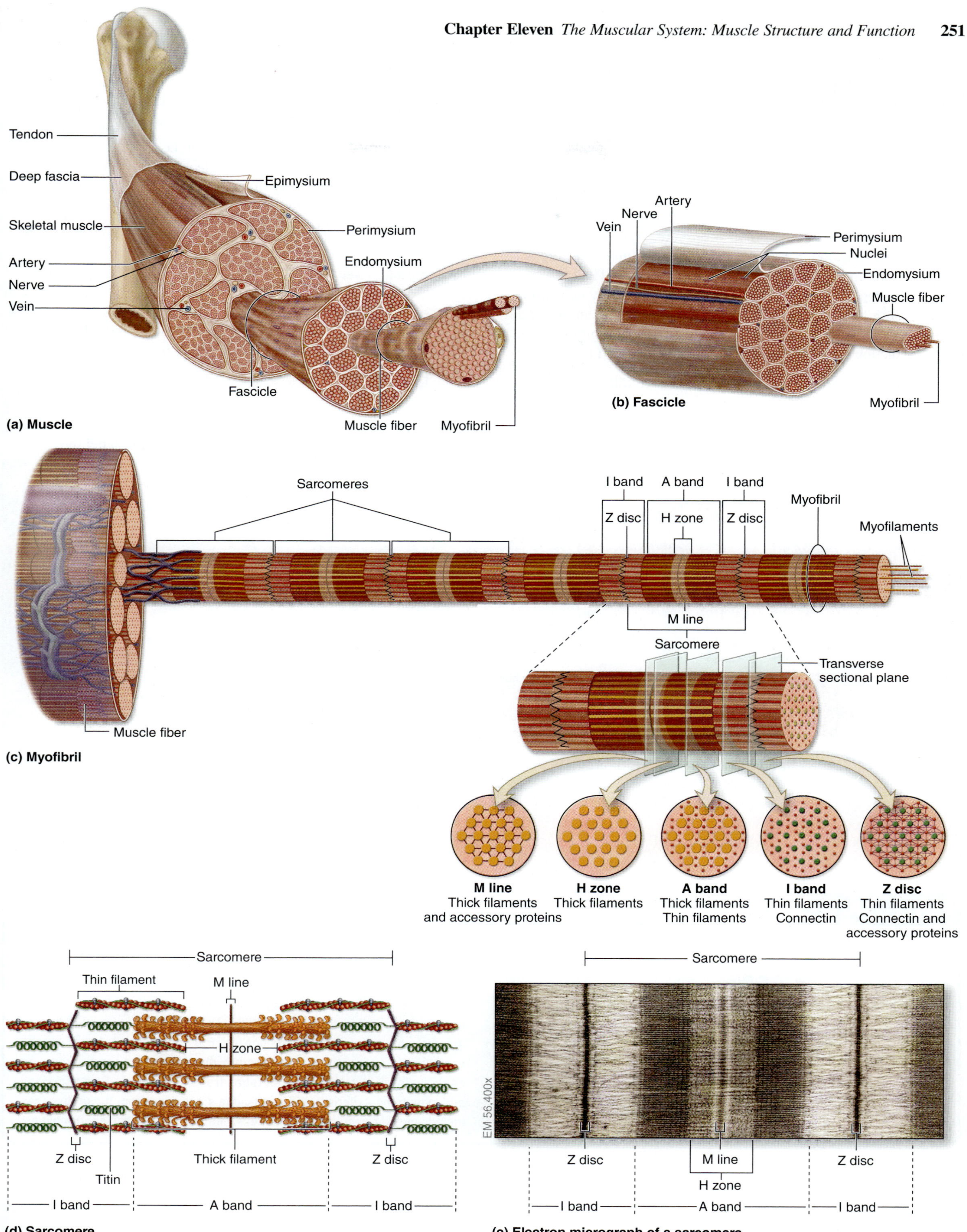

Figure 11.1 **Levels of Structural Organization of Skeletal Muscle.** (a) A whole muscle, (b) a muscle fascicle, (c) a myofibril, (d) a sarcomere, (e) an electron micrograph of a sarcomere. AP|R

EXERCISE 11.1

HISTOLOGY OF SKELETAL MUSCLE FIBERS

1. Obtain a slide of skeletal muscle tissue and place it on the microscope stage.
2. Bring the tissue into focus using the scanning objective. Switch to low power, bring the tissue sample into focus once again, and then switch to high power. The slide of skeletal muscle contains muscle fibers usually shown in both longitudinal section and cross section. Scan the slide and identify muscle fibers shown in both longitudinal section and cross section **(figures 11.2 and 11.3).**
3. Focus on muscle fibers shown in longitudinal section (figure 11.2) and observe them at high power. Identify an individual muscle fiber. Notice the numerous peripherally located nuclei. Most of the nuclei that are visible on the slide belong to the muscle fibers. However, about 5% to 15% of the visible nuclei are those of **satellite cells,** myoblast-like cells located between the muscle fibers. Satellite cells give skeletal muscle a limited ability to repair itself after injury. It is not possible to tell which nuclei belong to satellite cells when viewing the tissue through the light microscope.
4. Identify the following structures on the slide of skeletal muscle tissue, using table 11.1 and figure 11.2 as guides:

- ☐ **A band**
- ☐ **I band**
- ☐ **nucleus**
- ☐ **sarcomere (if visible)**
- ☐ **Z disc (if visible)**

5. Sketch skeletal muscle fibers as viewed through the microscope in the space provided. Be sure to label the structures listed in step 4 in the drawing.

_____________ ×

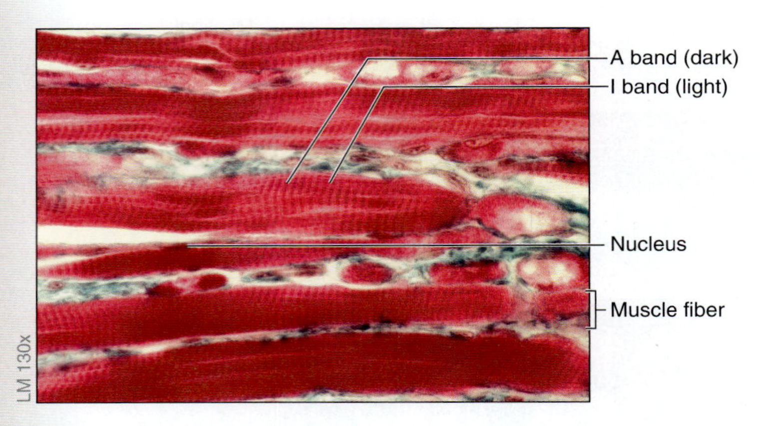

Figure 11.2 Skeletal Muscle in Longitudinal View.
Z discs and sarcomeres are not visible in this micrograph.

INTEGRATE

CLINICAL VIEW

Muscular Dystrophies

Muscular dystrophies are genetic diseases that involve abnormalities in the structural proteins and enzymes of skeletal muscle cells. Several forms of muscular dystrophy exist. Most involve disruption of a protein called **dystrophin** (*dys*, bad + *trophe*, nourishment). **Duchenne muscular dystrophy (DMD)** is the most common and severe form. A milder and less common form is **Becker muscular dystrophy (BMD)**. Both forms involve a mutation in dystrophin, a protein that links cytoskeletal elements of the skeletal muscle fiber (or cell) to the extracellular matrix (ECM). More specifically, dystrophin connects actin to the sarcolemma and to the surrounding endomysium. Disruption in this protein results in an inability to transmit force from the sarcomeres to the surrounding ECM, thereby preventing coordinated contraction of the skeletal muscle fiber. This leads to an overall reduction in force output of muscle. In addition, contraction of muscle fibers causes extensive damage to the fibers. Structural disruptions also result in reduced capacity for skeletal muscle cell regeneration and repair. Common signs and symptoms experienced by individuals suffering from these diseases include muscle atrophy and weakness, clumsiness, and gait abnormalities.

EXERCISE 11.2

CONNECTIVE TISSUE COVERINGS OF SKELETAL MUSCLE

After observing the structure of individual muscle fibers, it is time to investigate the relationship between individual muscle fibers and whole muscles. A whole skeletal muscle, such as the biceps brachii muscle, consists of many individual muscle fibers bundled together with connective tissue (figure 11.3). Each individual muscle fiber is covered by a thin layer of connective tissue called the **endomysium** (*endo,* within, + *mys,* muscle). Several muscle fibers are bundled together into **fascicles** (*fascis,* bundle) by a surrounding layer of connective tissue called the **perimysium** (*peri,* around, + *mys,* muscle). Finally, the entire skeletal muscle is surrounded by a layer of connective tissue called the **epimysium** *(epi,* upon, + *mys,* muscle). The epimysium is an extension of the **deep fascia,** which will be discussed shortly.

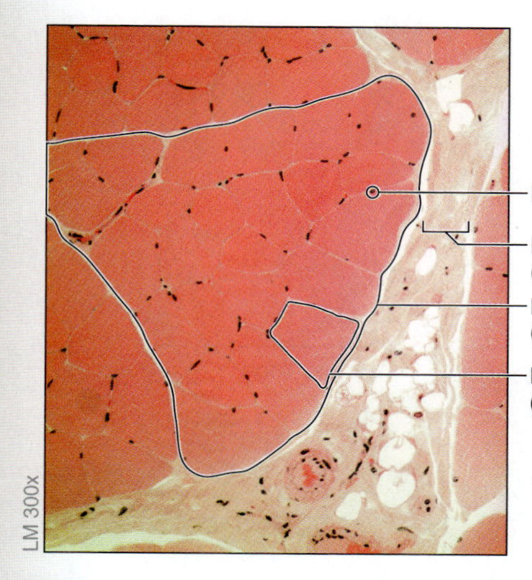

Figure 11.3 **Connective Tissue Coverings of Skeletal Muscle.** Muscle fibers are seen in a cross-sectional view.

1. View the slide of skeletal muscle tissue at low magnification. Move the stage until a cross section of skeletal muscle is visible. Try to identify the connective tissue that surrounds (a) a muscle fiber, (b) a fascicle, and (c) the whole skeletal muscle.
2. Identify the following structures, using figure 11.3 as a guide:*

   ☐ **endomysium** ☐ **fascicle** ☐ **nucleus**
   ☐ **epimysium** ☐ **muscle fiber** ☐ **perimysium**

3. Sketch the connective tissue coverings of skeletal muscle as viewed through the microscope in the space provided. Be sure to label the structures listed in step 2 in the drawing.

_______ ×

*All three layers of connective tissue should be visible if the tissue on the slide is skeletal muscle from a small mammal such as a mouse. However, only endomysium and perimysium may be visible if the tissue on the slide is from a larger mammal.

The Neuromuscular Junction

Each skeletal muscle fiber is innervated by a somatic motor neuron. Each motor neuron branches to innervate several muscle fibers, which are dispersed throughout the muscle. A single motor neuron and all muscle fibers innervated by that motor neuron constitute a **motor unit.** When a nerve signal (an action potential) travels down the motor neuron to stimulate the muscle, all fibers within that motor unit contract. The point of interaction between an axonal branch and an individual skeletal muscle fiber is called a **neuromuscular junction (figure 11.4).** Each neuromuscular junction is composed of an oval-shaped **synaptic knob** of a motor neuron, a **motor end plate** of a skeletal muscle fiber, and a **synaptic cleft** (the space between the two). This exercise involves observing a neuromuscular junction in a histology slide that has been stained to show axonal branches and synaptic knobs of a motor neuron, and skeletal muscle fibers.

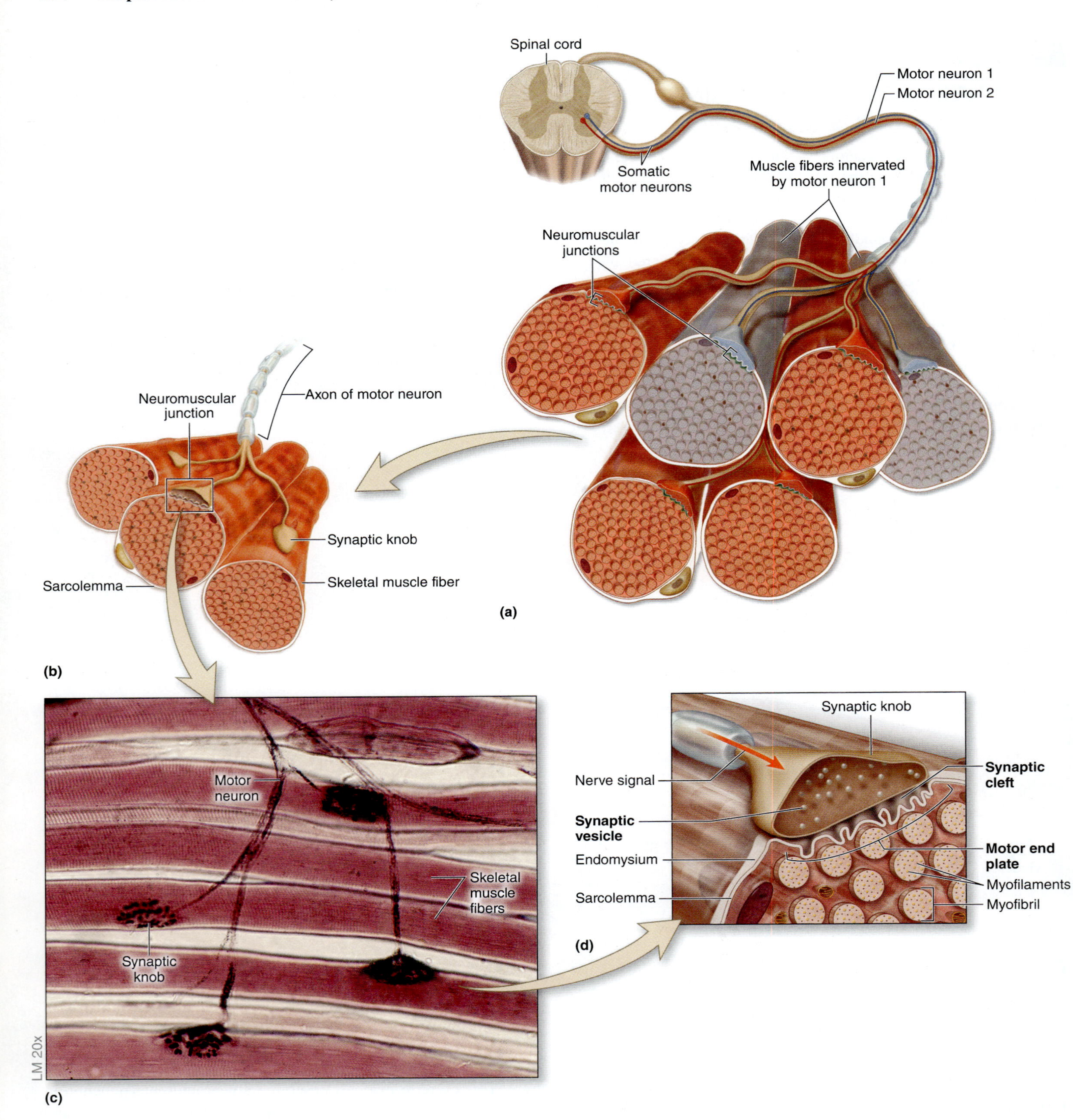

Figure 11.4 **The Neuromuscular Junction.** (a) Diagram of somatic motor neurons that extend from the spinal cord to several skeletal muscle fibers. (b) Each of the fibers within the motor unit has an area where the synaptic knob of the motor neuron contacts the muscle fiber; the neuromuscular junction. (c) Light micrograph of a motor neuron and skeletal muscle fibers in a motor unit. Synaptic knobs are the darkly stained structures associated with skeletal muscle fibers. (d) At each synaptic knob, there are synaptic vesicles that contain the neurotransmitter acetylcholine (ACh).

EXERCISE 11.3

THE NEUROMUSCULAR JUNCTION

1. Obtain a slide of a neuromuscular junction and place it on the microscope stage.
2. Bring the tissue sample into focus using the scanning objective. Switch to low power and bring the tissue sample into focus once again. Then switch to high power. Scan the slide until skeletal muscle fibers are visible. Next, look for darkly staining thin fibers with "knobby" ends. The "fibers" are the **axonal branches** of a somatic motor neuron and the oval-shaped "knobs" are the neuromuscular junctions that contain the synaptic knobs of the somatic motor neuron (figure 11.4).
3. Identify the following structures on the slide of the neuromuscular junction, using figure 11.4 as a guide.

☐ **axonal branches of the somatic motor neuron**
☐ **skeletal muscle fibers**
☐ **synaptic knobs**

4. Sketch the neuromuscular junction as viewed through the microscope in the space provided. Be sure to label the structures listed in step 3.

_______ ×

Cardiac Muscle Tissue

Branching cells with centrally located nuclei are characteristic of **cardiac muscle tissue (figure 11.5).** Like skeletal muscle fibers, cardiac muscle fibers are striated. The striations appear because cardiac muscle fibers also contain sarcomeres within myofibrils. Unlike skeletal muscle fibers, the myofibrils in cardiac muscle fibers are not anchored to each other with intermediate filaments at the Z discs. The resulting slight offset of myofibrils within the muscle fiber means that the banding pattern isn't always as clear in cardiac muscle as it is in skeletal muscle.

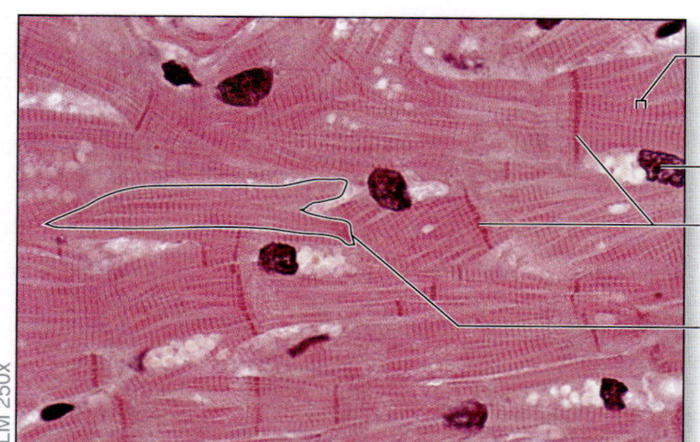

Figure 11.5 Cardiac Muscle Tissue.

EXERCISE 11.4

CARDIAC MUSCLE TISSUE

1. Obtain a slide of cardiac muscle tissue and place it on the microscope stage.
2. Bring the tissue sample into focus using the scanning objective. Switch to low power and bring the tissue sample into focus once again. Then switch to high power. Notice that the cells contain only one or two nuclei, and that the cells are short and branched. Where two cells come together, there should be a darkly stained line. This is an **intercalated disc.** Intercalated discs contain numerous **desmosomes,** which function to hold the fibers together, and **gap junctions,** which allow electrical signals to be transmitted very rapidly from one fiber to the next.
3. Identify the following structures on the slide of cardiac muscle tissue, using table 11.1 and figure 11.5 as guides.

☐ **A band**
☐ **branching fibers**
☐ **I band**
☐ **intercalated disc**
☐ **nucleus**

4. Sketch cardiac muscle fibers as viewed through the microscope in the space provided. Be sure to label the structures listed in step 3.

_______ ×

Smooth Muscle Tissue

Spindle-shaped cells with cigar-shaped or spiral nuclei and an absence of striations are characteristic of smooth muscle tissue. Individual cells have tapered ends, and there is only one centrally located nucleus per cell. Most of the nuclei will appear to be somewhat cigar shaped. However, in cells that have contracted, the nuclei take on a corkscrew or spiral appearance **(figure 11.6*a*)**, which can be a key identifying feature.

Smooth muscle is generally found in two layers around tubular organs such as the small intestine. The most common arrangement is an *inner circular layer* and an *outer longitudinal layer* of smooth muscle cells (figure 11.6*b*).

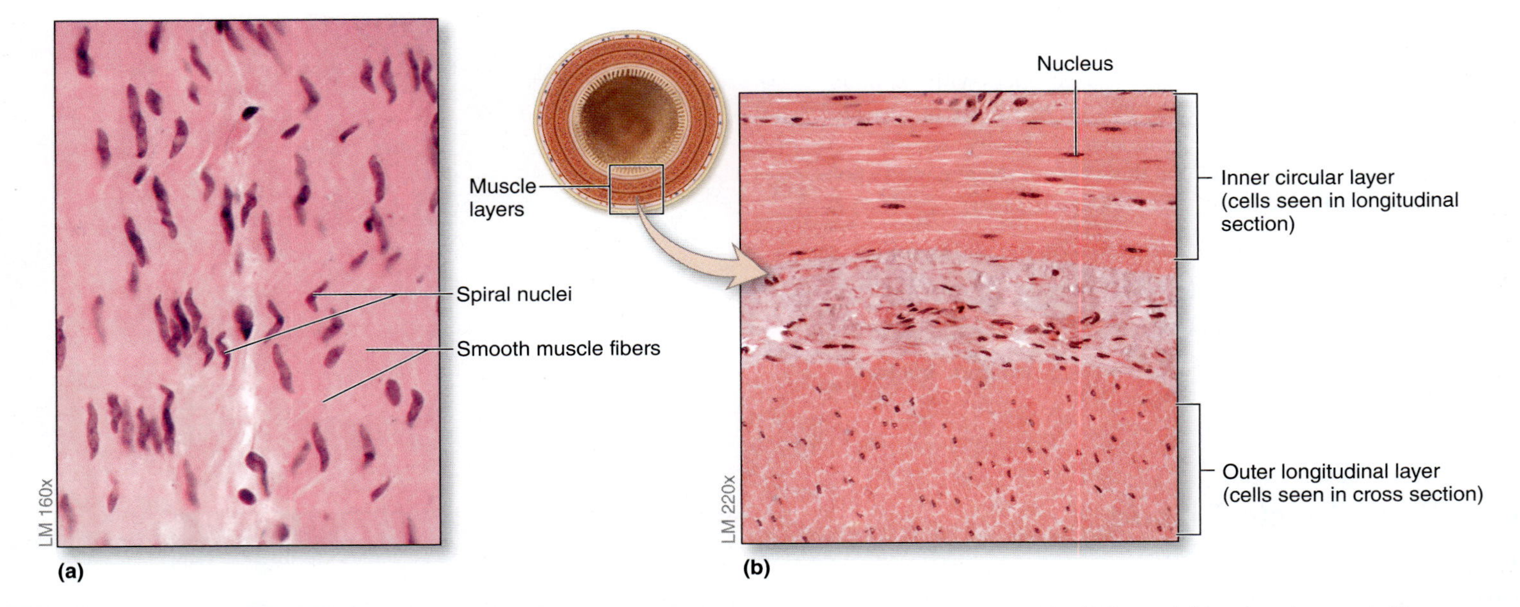

Figure 11.6 Smooth Muscle Tissue. (a) Close-up view of smooth muscle demonstrating spiral nuclei, which are visible when the muscle fibers are contracted; (b) circular and longitudinal layers of smooth muscle tissue.

EXERCISE 11.5

SMOOTH MUSCLE TISSUE

1. Obtain a slide containing smooth muscle tissue (figure 11.6) and place it on the microscope stage.
2. Bring the tissue into focus using the scanning objective. Switch to low power and bring the tissue sample into focus once again. Then switch to high power. If viewing a slide of the small intestine, identify smooth muscle cells that have been cut in both longitudinal section and cross section. Note that the cells do not appear to be of uniform diameter when viewed in cross section. This is because some cells are sectioned through the tapered ends, while others are sectioned through the thickest part of the fiber, which contains the nucleus.
3. Identify the following structures on the slide of smooth muscle tissue, using table 11.1 and figure 11.6 as guides.

- ☐ inner circular layer
- ☐ muscle cell in cross section
- ☐ muscle cell in longitudinal section
- ☐ nucleus
- ☐ outer longitudinal layer

4. Sketch smooth muscle cells as viewed through the microscope in the space provided. Be sure to label the structures listed in step 3 in the drawing.

INTEGRATE

LEARNING STRATEGY

A common error is to confuse smooth muscle tissue and dense regular connective tissue. A few key features help distinguish the two tissues from each other. First, smooth muscle cell nuclei are plumper and can be seen *within* the cells, whereas fibroblast nuclei (found in dense regular connective tissue) appear flattened and are located *in between* fibers. In addition, there are typically more nuclei per unit area in smooth muscle tissue than in dense regular connective tissue. Finally, the appearance of spiral or corkscrew nuclei is a good clue that the tissue is smooth muscle, not dense regular connective tissue.

GROSS ANATOMY

Gross Anatomy of Skeletal Muscles

The names of skeletal muscles often seem overly complex. However, understanding the basis of their names makes the task of identifying skeletal muscles of the body seem far less daunting. The name of a muscle often gives a clue to its location, size, shape, action, or attachment points. The exercises in this section include an introduction to the logic behind the naming of skeletal muscles (Exercise 11.6) and an overview of the typical architectures found in skeletal muscles (Exercise 11.7).

EXERCISE 11.6

NAMING SKELETAL MUSCLES

Table 11.2 summarizes some of the common ways skeletal muscles are named and gives word origins for the muscle names. Plan to spend some time mastering these word origins to be better prepared to handle the material to come in chapters 12 and 13. The efforts to learn the Latin and Greek word roots of anatomical terms will become even more valuable when working through the next three chapters.

Table 11.2 Common Methods for Naming Skeletal Muscles

Name	Meaning	Word Origin	Example
Naming Skeletal Muscles Based on Shape			
Deltoid	Triangular	*delta*, the Greek letter delta (a triangle), + *eidos*, resemblance	Deltoid
Gracilis	Slender	*gracilis*, slender	Gracilis
Lumbrical	Wormlike	*lumbricus*, earthworm	Lumbricals
Rectus	Straight	*rectus*, straight	Rectus abdominis
Rhomboid	Diamond-shaped	*rhombo-*, an oblique parallelogram with unequal sides, + *eidos*, resemblance	Rhomboid major
Teres	Round	*teres*, round	Teres major
Trapezius	A four-sided geometrical figure having no two sides parallel	*trapezion*, a table	Trapezius

(continued on next page)

(continued from previous page)

Table 11.2 Common Methods for Naming Skeletal Muscles *(continued)*

Name	Meaning	Word Origin	Example
Naming Skeletal Muscles Based on Size			
Brevis	Short	*brevis*, short	Adductor brevis
Latissimus	Broadest	*latissimus*, widest	Latissimus dorsi
Longissimus	Longest	*longissimus*, longest	Longissimus capitis
Longus	Long	*longus*, long	Adductor longus
Major	Bigger	*magnus*, great	Teres major
Minor	Smaller	*minor*, smaller	Teres minor
Naming Skeletal Muscles Based on the Number of Heads and/or Bellies			
Biceps	2 heads	*bi*, two, + *caput*, head	Biceps brachii
Digastric	2 bellies	*bi*, two, + *gastro*, belly	Digastric
Quadriceps	4 heads	*quad*, four, + *caput*, head	Quadriceps femoris
Triceps	3 heads	*tri*, three, + *caput*, head	Triceps brachii
Naming Skeletal Muscles Based on Position			
Abdominis	Abdomen	*abdomen*, the greater part of the abdominal cavity	Rectus abdominis
Anterior	On the front surface of the body	*ante-*, before, in front of	Serratus anterior
Brachii	Arm	*brachium*, arm	Biceps brachii
Dorsi	Back	*dorsum*, back	Latissimus dorsi
Femoris	Thigh	*femur*, thigh	Rectus femoris
Infraspinatus	Below the scapular spine	*infra-*, below, + *spina*, spine	Infraspinatus
Interosseous	In between bones	*inter*, between + *osseus*, bone	Interossei
Oris	Mouth	*oris*, mouth	Orbicularis oris
Pectoralis	Chest	*pectus*, chest	Pectoralis major
Posterior	On the back surface of the body	*posterus*, following	Serratus posterior
Supraspinatus	Above the scapular spine	*supra-*, on the upper side, + *spina*, spine	Supraspinatus
Naming Skeletal Muscles Based on Depth			
Externus	External	*external*, on the outside	Obturator externus
Internus	Internal	*internal*, away from the surface	Obturator internus
Profundus	Deep	*pro*, before, + *fundus*, bottom	Flexor digitorum profundus
Superficialis	Superficial	*super*, above, + *facies*, face	Flexor digitorum superficialis

Table 11.2 Common Methods for Naming Skeletal Muscles *(continued)*

Name	Meaning	Word Origin	Example
Naming Skeletal Muscles Based on Action			
Abductor	Moves a body part away from the midline	*ab*, from, + *ductus*, to bring toward	Abductor pollicis brevis
Adductor	Moves a body part toward the midline	*ad*, toward, + *ductus*, to bring toward	Adductor pollicis
Constrictor	Acts as a sphincter and closes an orifice	*cum*, together, + *stringo*, to draw tight	Superior pharyngeal constrictor
Depressor	Flattens or lowers a body part	*de-*, away, + *pressus*, to press	Depressor anguli oris
Dilator	Causes an orifice to open, or dilate	*dilato*, to spread out	Dilator pupillae
Extensor	Causes an increase in joint angle	*ex-*, out of, + *-tensus*, to stretch	Extensor carpi ulnaris
Flexor	Causes a decrease in joint angle	*flectus*, to bend	Flexor carpi ulnaris
Levator	Raises a body part superiorly	*levo* + *atus*, a lifter	Levator scapulae
Pronator	Turns the palm of the hand from anterior to posterior	*pronatus*, to bend forward	Pronator teres
Supinator	Turns the palm of the hand from posterior to anterior	*supino* + *atus*, to bend backward	Supinator

INTEGRATE

LEARNING STRATEGY

Recall from chapter 10 that synovial joints are highly movable joints. Activation of the muscles surrounding a joint creates movement of the joint. For example, flexion of the forearm (or elbow joint) requires contraction of the brachialis muscle and concurrent relaxation of the triceps brachii muscle. When learning the naming conventions for muscles, it is often helpful to observe the location of the muscle, including the attachment points (i.e., origin and insertion), and use those to predict the motion that is allowed at a particular joint. For example, when the flexor digitorum longus shortens, the digits "flex." It is no surprise, then, that "flex" appears in the name of this muscle based on its action. Be sure to review the movements of synovial joints (table 10.6) in preparation for learning names of axial and appendicular muscles in chapters 12 and 13.

EXERCISE 11.7

ARCHITECTURE OF SKELETAL MUSCLES

The overall architecture of a skeletal muscle affects how the muscle functions. When a whole muscle is observed, the individual fibers and fascicles are visible, making it relatively easy to see how the fascicles are arranged within the muscle. Recall that when skeletal muscle contracts, it generally gets shorter and brings its attachment points closer to each other. Thus, the orientation of the muscle fascicles compared to the attachment points of the muscle will directly affect the force produced by the muscle and the complexity of the muscle's actions. For example, **pennate** architecture (*penna,* feather) allows a muscle to produce greater force per distance shortened than **parallel** architecture. In addition, muscles with more than two attachments (for example, biceps and triceps) produce more complex movements than muscles with only two attachments (one proximal attachment and one distal attachment). **Table 11.3** summarizes the common patterns of fascicle arrangement that contribute to skeletal muscle architecture.

1. Using classroom models of skeletal muscles or a prosected human cadaver, observe the arrangement of fascicles in several different skeletal muscles of the body, using table 11.3 as a guide.
2. Locate the muscles that are listed in **figure 11.7** on classroom models or on a human cadaver, using the textbook as a guide.
3. Complete the table provided by listing the architecture of each of the listed muscles.

Table 11.3 Common Architectures of Skeletal Muscles

Diagram						
Name	**Unipennate**	**Bipennate**	**Multipennate**	**Circular**	**Convergent**	**Parallel**
Word Origin	*uni*, one, + *penna*, feather	*bi*, two, + *penna*, feather	*multi*, many, + *penna*, feather	*circum*, around	*cum*, together, + *vergo*, to incline	*para*, alongside

INTEGRATE

LEARNING STRATEGY

It is helpful to observe the orientation of muscle fibers to predict the action of a muscle. For example, circular muscle fibers, when shortened, will decrease the diameter of any opening they surround (e.g., the orbicularis oculi surrounds the eye and contraction of this muscle closes the eye). Other muscle fibers may be oriented in multiple directions (i.e., multipennate), such as those of the deltoid muscle. The variation in fiber orientation allows the muscle to pull in multiple directions. Contraction of the deltoid muscle creates flexion, abduction, and extension of the arm. The multipennate architecture of the deltoid allows it to create this variety of movements. When discerning the action of each muscle, observe the orientation of the muscle fibers on anatomically correct models, and use this information to predict the corresponding muscle action.

Muscle Number	Muscle Name	Architecture
1	Deltoid	
2	Extensor digitorum	
3	Gastrocnemius	
4	Orbicularis oculi	
5	Pectoralis major	
6	Rectus femoris	
7	Sartorius	
8	Trapezius	
9	Triceps brachii	

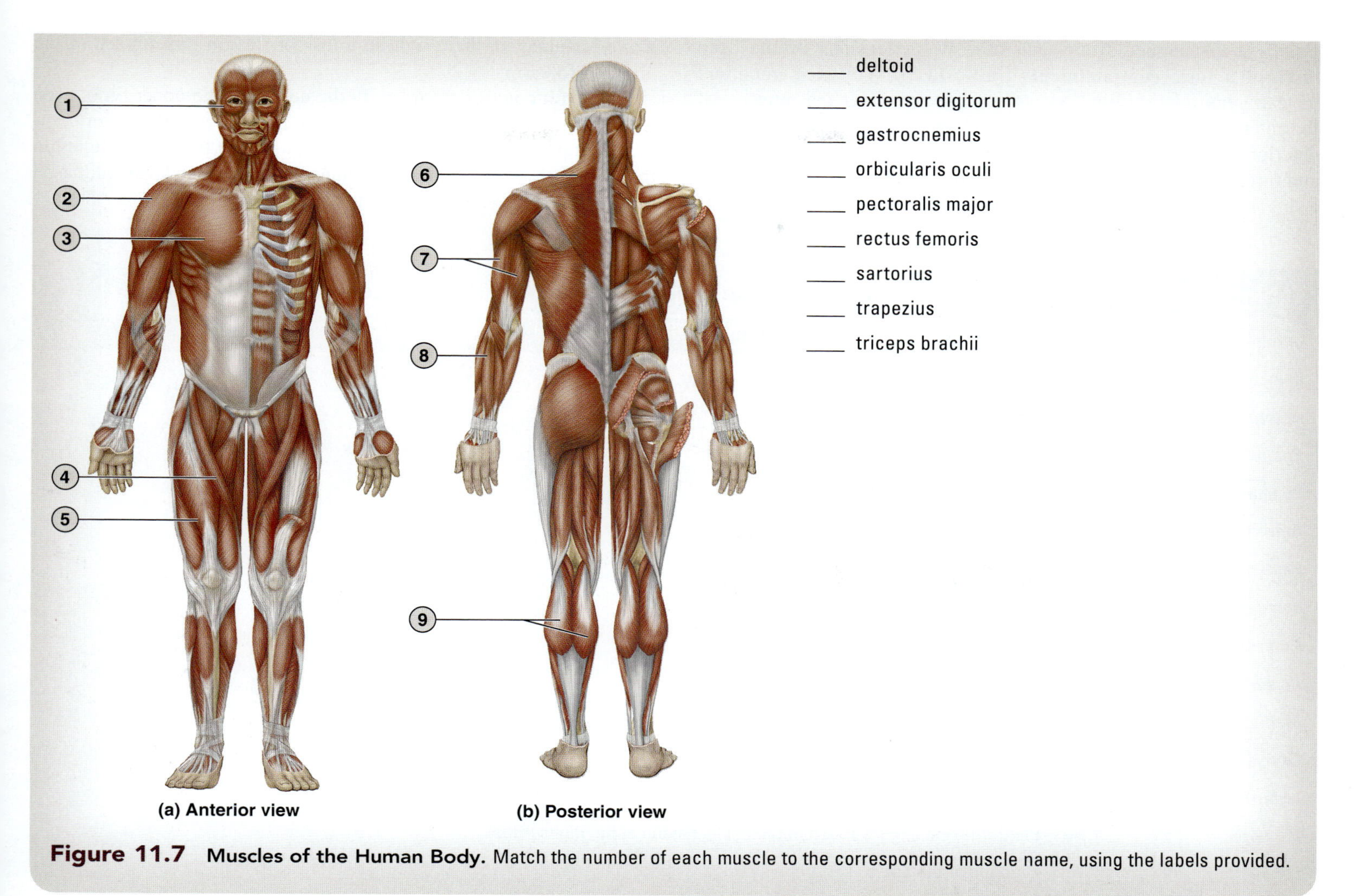

Figure 11.7 **Muscles of the Human Body.** Match the number of each muscle to the corresponding muscle name, using the labels provided.

Organization of the Human Musculoskeletal System

Table 11.4 lists the major muscle groups of the body and the common actions of each group of muscles. Each individual muscle within a group of muscles has its own specific actions. However, skeletal muscles do not act alone, so it makes sense to learn the muscles as groups and to learn the major actions of each group. Muscles contained within a group commonly act as synergists (muscles that facilitate the prime mover) and often have a common nerve and blood supply. Thus, damage to the nerve and blood supply often affects a whole group of muscles, not just an individual muscle.

In the limbs, muscles are not only *functionally* organized into groups; they are also *anatomically* organized into groups. Muscle

Table 11.4 Major Muscle Groups and Their Actions

Major Muscle Group	General Description of Muscle Actions
Muscles of facial expression	Produce facial expressions
Muscles of mastication	Used for chewing
Suprahyoid and infrahyoid muscles	Move the tongue and pharynx
Muscles of the neck	Rotate, flex, and extend the head
Muscles of respiration	Involved in breathing movements
Muscles of the abdomen	Flex, bend, and rotate the spine
Muscles of the back and spine	Extend, bend, and rotate the spine
Muscles of the pelvic floor	Support pelvic contents and form sphincters around structures such as the urethra and the anus
Muscles that act about the pectoral girdle	Stabilize the pectoral girdle and anchor the upper limb to the pectoral girdle
Muscles that act about the pelvic girdle	Stabilize the pelvic girdle and anchor the lower limb to the pelvic girdle

groups are separated from each other by extensions of the deep fascia that form **compartments (table 11.5).** Why is it important to learn this relationship? As an example, consider the muscles of the anterior compartment of the arm. All of the muscles located in the anterior compartment of the arm act primarily in flexion of the arm and forearm. All of the muscles in the compartment are innervated by a single nerve, the musculocutaneous nerve. If the musculocutaneous nerve is damaged, there will be a loss of flexion movements of the arm and forearm.

When it comes time to study the nerves of the upper limb, even if you forget the names of the individual muscles in the anterior compartment of the arm, you will be able to predict the loss of function that would result if the musculocutaneous nerve was damaged, based on your knowledge of compartments and muscle groups.

Table 11.5 Fascial Compartments of the Limbs and Their General Muscle Actions

Compartment	General Description of Muscle Actions
Compartments of the Arm	
Anterior	Flexion of shoulder and elbow
Posterior	Extension of shoulder and elbow
Compartments of the Forearm	
Anterior	Flexion of the wrist and digits (fingers)
Posterior	Extension of the wrist and digits (fingers)
Compartments of the Thigh	
Anterior	Extension of the knee, flexion of the hip
Posterior	Extension of the hip, flexion of the knee
Medial	Adduction of the thigh
Compartments of the Leg	
Anterior	Dorsiflexion and inversion of the ankle, extension of the digits (toes)
Posterior	Plantarflexion and inversion of the ankle, flexion of the digits (toes)
Lateral	Eversion of the ankle

EXERCISE 11.8

MAJOR MUSCLE GROUPS AND FASCIAL COMPARTMENTS OF THE LIMBS

1. Identify each of the major muscle groups and compartments of the limbs, using tables 11.4 and 11.5 as guides.
2. When identifying each muscle group or compartment, practice performing the actions listed in the tables. Try to feel the contraction of the muscles by palpating the skin overlying the muscle groups. If necessary, refer back to chapter 10 for descriptions of joint actions.
3. **Figures 11.8** and **11.9** represent cross sections of the arm (brachium) and thigh, respectively. The details in cross-sectional views such as these may be difficult to negotiate the first time through. Thus, for now just focus on the organization of the muscles into compartments. Visualize how the cross-sectional diagram relates to your own arm or thigh. When performing the actions listed for each compartment, correlate the muscles used to perform each action with a specific compartment of the arm or thigh.

INTEGRATE

LEARNING STRATEGY

Terminology related to the action of muscle fibers can sometimes be confusing. Recall from chapter 10 that the term *flexion* refers to any joint movement in which the angle between two bones is decreased. However, people often use the term *flex* (as in "flex your bicep") when they are asking you to *contract* a muscle. The term *contract* is the term scientists prefer to use to describe the action of a muscle fiber producing force. Even so, the term *contract* can also be misleading at times. Literally, to contract means to shorten. All muscle fibers have the ability to shorten when they produce force. However, skeletal muscle fibers can also produce force while remaining at a fixed length or while lengthening.

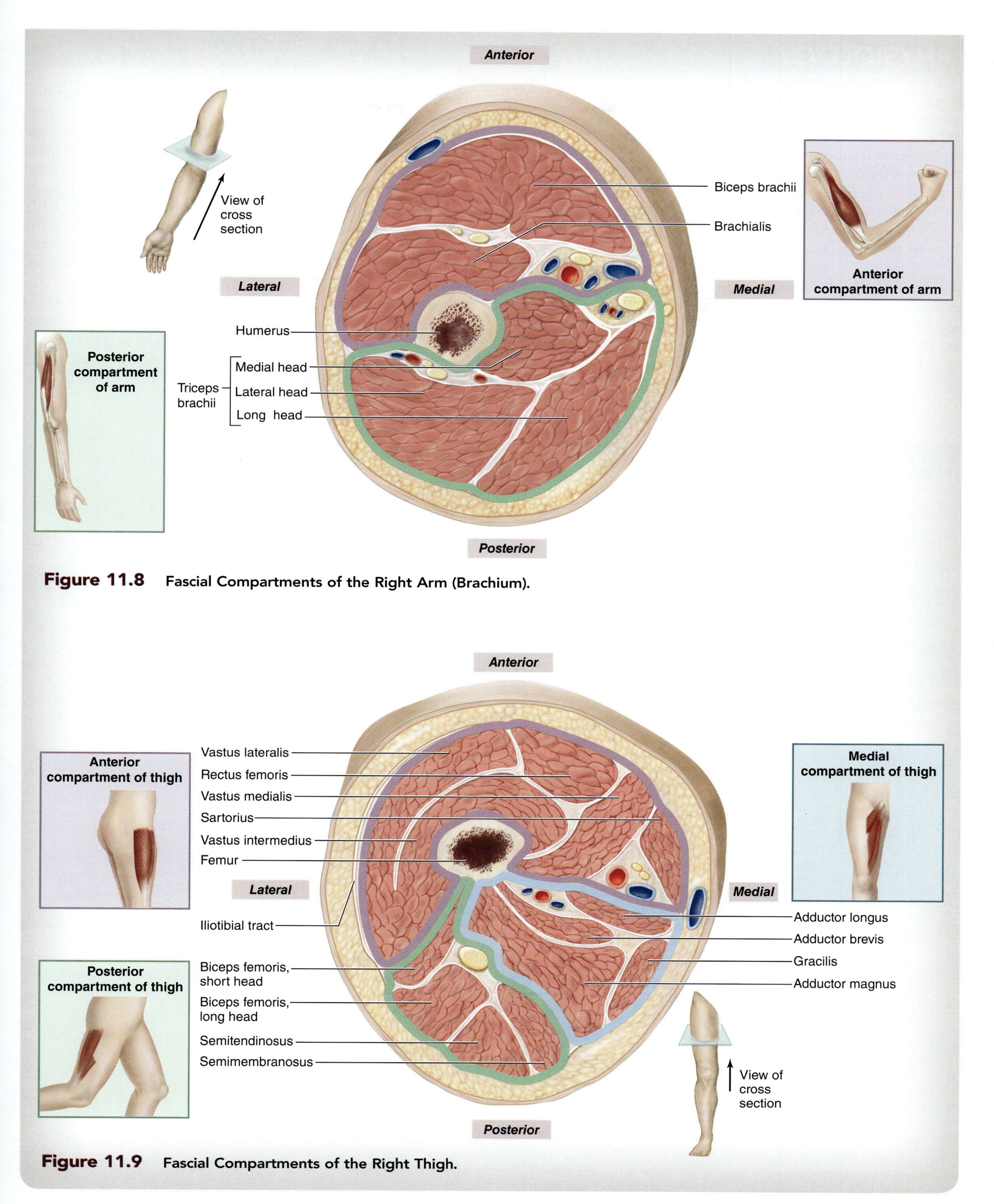

Figure 11.8 Fascial Compartments of the Right Arm (Brachium).

Figure 11.9 Fascial Compartments of the Right Thigh.

PHYSIOLOGY

Force Generation of Skeletal Muscle

Skeletal muscles are composed of three types of skeletal muscle fibers, whose features are summarized in **table 11.6**. Two features are significant for this discussion. (1) Skeletal muscle fiber types vary in size. Type I fibers are small, Type IIa fibers are medium, and Type IIb are large. (2) Skeletal muscle fiber types vary in force of contraction. Type I fibers exert a small force, Type IIa fibers a medium force, and Type IIb fibers a large force.

Typically, the skeletal muscle fibers within a motor unit are of the same type. There is also a relationship between the size of a motor unit and the type of skeletal muscle fibers that compose the motor unit. The smallest motor units consist of Type I fibers, medium-sized motor units consist of Type IIa fibers, and the largest motor units consist of Type IIb fibers. In addition, for most activities, the pattern of motor unit recruitment occurs such that the smallest motor units are recruited first. Successive increases in muscle force are then achieved by recruitment of increasingly larger motor units. Because each motor unit consists of muscle fibers that are all of the same type, this also means that recruitment follows the pattern of Type I fibers first, then Type IIa, and finally Type IIb. This order of recruitment (I → IIa → IIb) ensures that successive addition of larger and larger motor units will cause incremental increases in total force produced by the muscle (**figure 11.10**).

Observe the properties of Type IIb fibers in table 11.6. These fibers compose the largest motor units, which produce the greatest force. At the same time, these fibers are also the least fatigue-resistant, which means they fatigue easily. This is why it is impossible to sustain a maximal contraction for an extended period of time (such as when lifting a very heavy weight). In contrast, Type I fibers compose the smallest motor units and the fibers are the most fatigue-resistant. Thus, activities that require only tonic contraction of small motor units (e.g., standing) can be maintained for very long periods of time.

Table 11.6 Properties of Skeletal Muscle Fiber Types

Fiber Type	Type I (SO; Slow Oxidative)	Type IIa (FOG; Fast Oxidative-glycolic)	Type IIb (FG; Fast-glycolic)
Fiber cross-sectional area	Small	Medium	Large
Order of recruitment	First	Second	Third
Resistance to fatigue	Large	Medium	Small
Size of motor units	Small	Medium	Large
Speed of contraction	Slow	Fast	Fast
Twitch force	Small	Medium	Large
Twitch speed	Slow	Fast	Fast

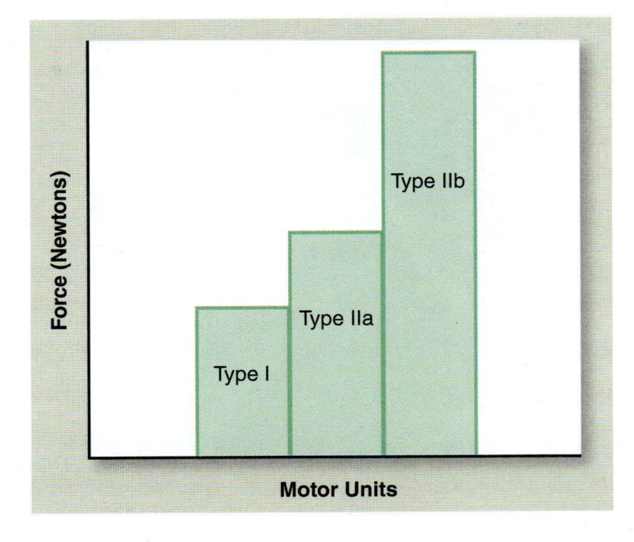

Figure 11.10 **Recruitment of Motor Units.** The first motor units recruited consist of Type I fibers, which produce very little force. Further recruitment brings in Type IIa fibers, which produce more force. Type IIb fibers are recruited last. Maximal force generation by a muscle requires recruitment of all fibers in the muscle.

EXERCISE 11.9

MOTOR UNITS AND MUSCLE FATIGUE (HUMAN SUBJECT)

This exercise involves observing the relationship between motor unit recruitment and fatigue by successively working motor units to the point of fatigue. The exercise begins by having a study subject perform a maximal contraction holding a heavy weight until the point of fatigue and measuring the time it takes to reach fatigue. Next, the subject will hold successively lighter weights until the point of fatigue. The time to fatigue will be measured for each of these successive weights as well.

In this experiment, what is the independent variable?

____

What is the dependent variable? ______________________

Before beginning, state a hypothesis regarding the relationship between force generation and muscle fatigue.

____

____

____

____

Obtain the Following:

- **hand barbells of the following weight increments: 1 lb, 2 lb, 5 lb, 10 lb, 25 lb**
- **stopwatch (or other mechanism for keeping track of time)**

1. Choose a volunteer to be the subject. The subject will need to initially guess the maximum weight that he or she will be able to hold in an outstretched arm for a period of time. Begin with the heaviest weight the subject can comfortably hold (even if only for a few seconds). To start the first trial, have the subject pick up the weight with one hand, extend his or her arm laterally so that his or her arm is positioned parallel to the floor, and then hold the weight in that position for as long as possible.
2. As soon as the subject gets the upper limb into position with the weight, start a stopwatch.
3. Observe the subject's upper limb for signs of fatigue (shaking, trembling, and an inability to hold the weight steady). As soon as the subject can no longer hold the weight in position (i.e., the upper limb falls below 90 degrees), stop the timer and have the subject lay the weight down. Record both the weight used and the time to fatigue in **table 11.7.**
4. Trial #2: Have the subject repeat steps 1–3 using a lighter weight. For example, if a 25-lb weight was used previously, now use a 10-lb weight for trial #2. Again, record the weight and time to fatigue in table 11.7.
5. Repeat the process with successive trials until all available weights have been used. At that time, do one last trial having the subject hold his or her arm out without holding any weight at all. For each trial, record the weight used and the time to fatigue. The weight for the last trial should be zero.
6. Note any observations and/or initial thoughts about the outcome of this experiment in the space provided.

____

____

____

____

____

Table 11.7 Results for Exercise 11.9 (Motor Units and Muscle Fatigue)

Trial #	Weight (lb)	Time to Fatigue (min)
1		
2		
3		
4		
5		

Note: It is possible to have fewer than five trials.

INTEGRATE

CLINICAL VIEW
Amyotrophic Lateral Sclerosis (ALS)

Amyotrophic lateral sclerosis (ALS), or Lou Gehrig's Disease, is named for the famous baseball player, Lou Gehrig, who was diagnosed with this fatal disease at the age of 36 after suffering a dramatic decline in the quality of his game. ALS impacts somatic motor neurons, whose cell bodies are located in the posterior (ventral) horn of the spinal cord (see chapter 16). These neurons innervate skeletal muscle fibers. While the etiology of ALS is still unclear, it is known that ALS leads to the destruction of these motor neurons. As motor neurons die and retract from their respective muscle fibers, there is a loss in function in the nerve–muscle connection, or neuromuscular junction. With this loss comes an inability to excite the skeletal muscle fibers. Without excitation, there is no contraction, and if you don't use it, you lose it! That is, muscle fibers will begin to atrophy (*a-*, without + *-trophe*, nourishment) due to disuse. This also impacts motor unit recruitment, because a loss of motor neurons results in atrophy of all of the fibers that the motor neurons innervate. As a compensatory mechanism, healthy motor neurons begin sprouting collateral processes to make connections with the fibers that are no longer receiving electrical signals from the central nervous system. Often, muscle fiber types will begin to "group" together, such that Type I fibers may be concentrated in one particular compartment of a muscle. This is in contrast to a healthy muscle, where muscle fiber types are often distributed throughout a compartment. As the disease progresses, patients may begin to experience slurred speech, muscle fatigue, and clumsiness. Unfortunately, ALS is an aggressive disease that attacks all somatic motor neurons in the body, including those responsible for innervating skeletal muscles that aid in respiration, such as the diaphragm. Over time, often in as little as 3 to 5 years, there is so much denervation and muscle atrophy that patients are no longer able to effectively contract the diaphragm, and they die from respiratory failure.

EXERCISE 11.10

CONTRACTION OF SKELETAL MUSCLE (WET LAB)

This exercise involves observing muscle contraction using a preparation of glycerinated muscle fibers. The process of preparing these fibers (glycerination) removes ions (e.g., Na^+, K^+, Cl^-, Ca^{2+}) and ATP from the tissue. It also disrupts the troponin/tropomyosin complex in such a way that tropomyosin no longer covers the myosin binding sites on actin **(figure 11.11).** Thus, if the fibers are given the appropriate materials (particularly ATP), actin and myosin will be able to interact, crossbridge cycling will take place, and the fibers will contract. The salt solutions added contain two specific salts: KCl (potassium chloride) and $MgCl_2$ (magnesium chloride). Myosin has a high affinity for these salts. Ironically, magnesium interferes with contraction of muscle *in situ* (within the body), but has been observed to *increase* the strength of contraction in glycerinated muscle. This exercise tests this directly.

Before beginning, become familiar with the following concepts:

- Structure of actin and myosin
- Role of calcium in inducing muscle contraction
- Events of the crossbridge cycle

Figure 11.12 demonstrates the events of a single crossbridge cycle. Using the textbook as a guide, fill in the descriptions of the events that happen at each stage of the crossbridge cycle.

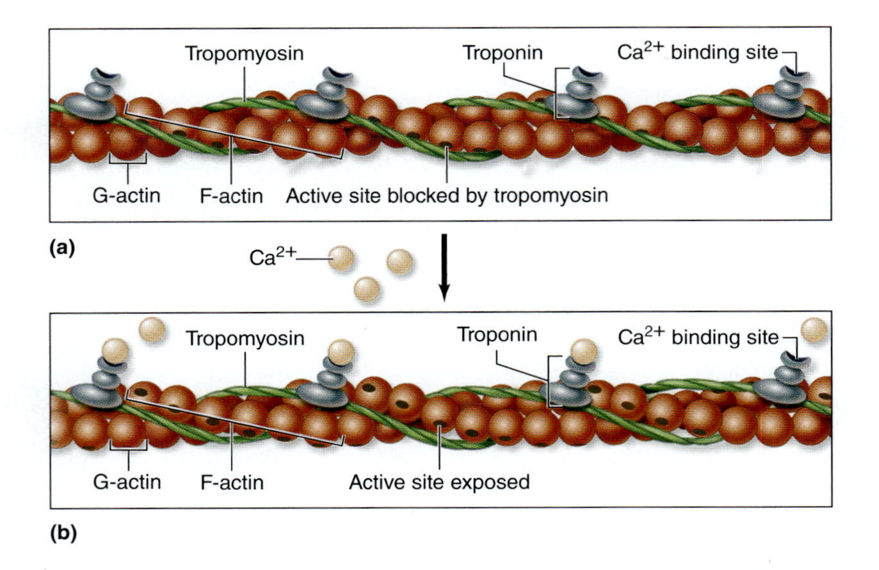

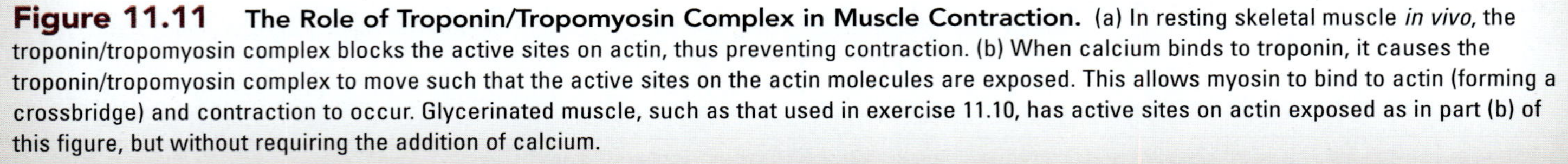
Figure 11.11 The Role of Troponin/Tropomyosin Complex in Muscle Contraction. (a) In resting skeletal muscle *in vivo*, the troponin/tropomyosin complex blocks the active sites on actin, thus preventing contraction. (b) When calcium binds to troponin, it causes the troponin/tropomyosin complex to move such that the active sites on the actin molecules are exposed. This allows myosin to bind to actin (forming a crossbridge) and contraction to occur. Glycerinated muscle, such as that used in exercise 11.10, has active sites on actin exposed as in part (b) of this figure, but without requiring the addition of calcium.

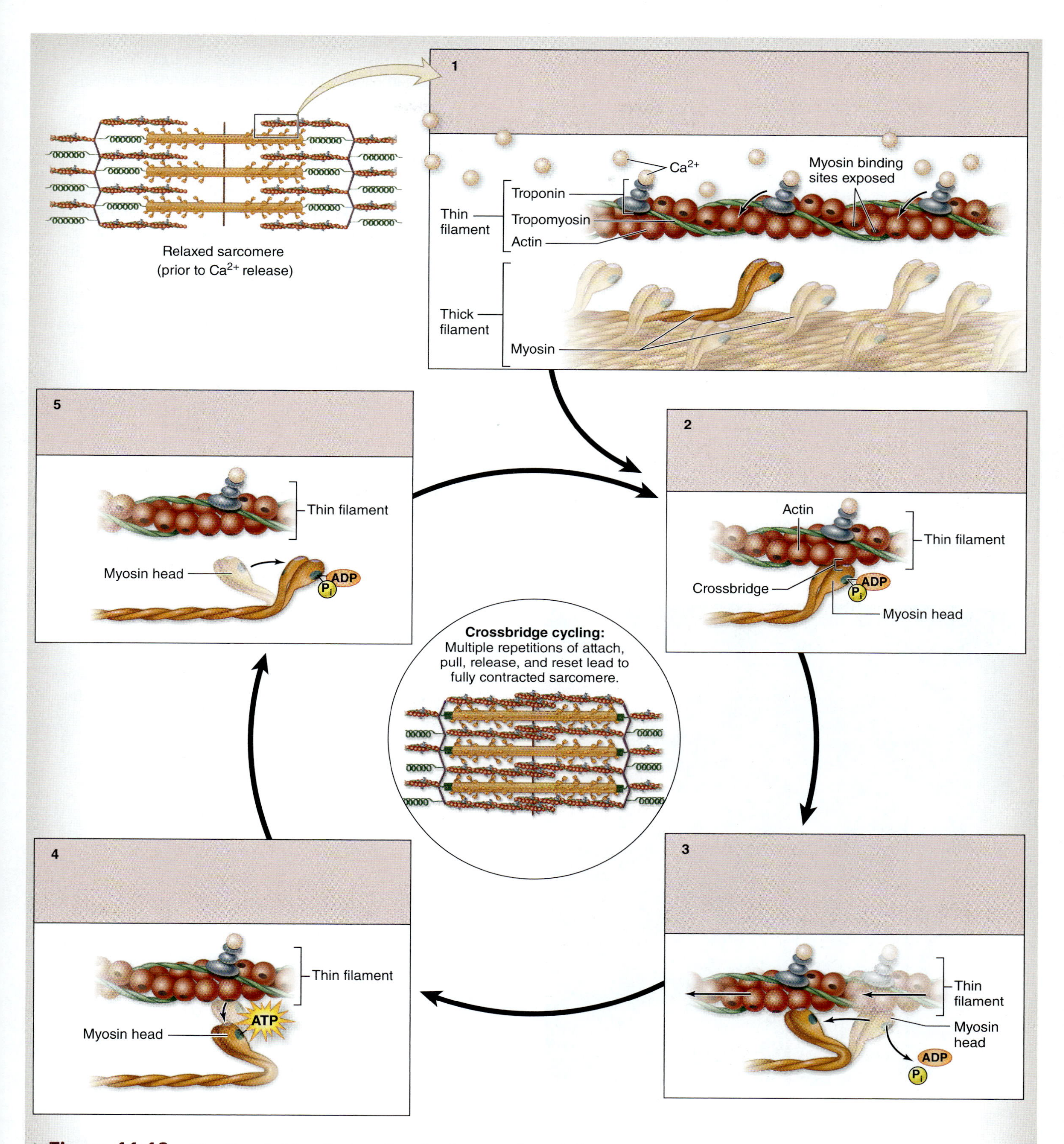

Figure 11.12 **The Crossbridge Cycle.** Describe the steps of crossbridge cycling, using the textbook as a guide.

(continued on next page)

(continued from previous page)

EXERCISE 11.10A Observation of Muscle Fibers

This exercise involves observing the effect of adding ATP alone, ATP plus salts (KCl and $MgCl_2$), and salts alone (KCl and $MgCl_2$) on contraction of preserved (glycerinated) skeletal muscle. Contraction of the muscle fibers will be observed using a microscope. The contraction can also be observed grossly without the use of a microscope; however, the bands will not be visible.

Obtain the Following:

- **dissecting needles (2)**
- **fine-tissue forceps ("watchmaker" forceps)**
- **~ 2-cm-long section of muscle tissue in petri dish covered in glycerol**
- **solutions: (1) ATP only, (2) KCl/$MgCl_2$ only, (3) ATP + KCl/$MgCl_2$**
- **medicine dropper or pipette**
- **5 glass microscope slides**
- **5 glass microscope slide cover slips**
- **millimeter ruler**
- **compound microscope**
- **dissecting microscope**
- **wax pencil**

1. Obtain the petri dish with the muscle tissue. Using fine forceps and the dissecting needles, tease the fibers apart longitudinally until strands that are as small (thin) as possible are visible, as shown in **figure 11.13.** Tease the sample apart enough to get at least 5 separate bundles of fibers because each of the bundles will be used to perform a separate exercise in this series. Keep the petri dish covered, and in a cool spot away from any heat source, while performing these exercises.

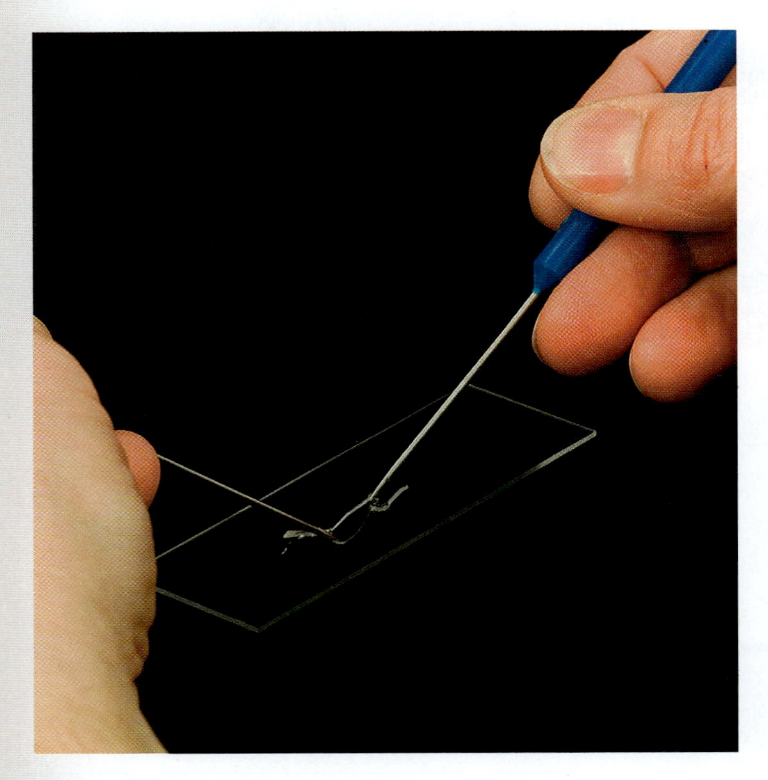

Figure 11.13 Glycerinated Muscle Fibers. Using forceps and a dissecting needle, or two dissecting needles (shown here), gently tease apart the muscle fiber bundle in an attempt to obtain the thinnest possible strands for exercise 11.10.

2. Place the compound microscope on the laboratory bench and obtain the glass slides and cover slips. Transfer one small bundle of muscle fibers from the petri dish to a glass slide. Cover the slide with a cover slip, and place it on the microscope stage.
3. Bring the tissue sample into focus using the scanning objective. Because the tissue has not been stained, it may be somewhat difficult to see. Closing the iris diaphragm on the microscope may allow for better tissue contrast (see chapter 3, The Microscope). Next, change to a higher-power objective and bring the tissue sample into focus once again. Look for the following on the slide:

   ☐ **nuclei** ☐ **striations**
   ☐ **skeletal muscle fibers**

4. Observe the striations carefully and compare them to the drawing in figure 11.1. Which of the following are visible in the sample (check any and all that are visible)?

   ☐ **A band** ☐ **I band**
   ☐ **H zone** ☐ **Z disc**

5. Sketch the muscle fibers as viewed through the microscope in the space provided. This sketch will serve as an aid for subsequent laboratory exercises.

_______ ×

EXERCISE 11.10B Observing the Effect of ATP and Salts on Contraction of Glycerinated Muscle Fibers

This exercise involves observing the effect of adding solutions containing ATP, salts ($MgCl_2$ and KCl) only, and ATP plus salts on contraction of glycerinated muscle fibers. Refer back to the exercise introduction for assistance in formulating hypotheses.

State a hypothesis regarding how each of the following influence muscle contraction:

a. Addition of ATP: ______________________________

____

b. Addition of salts (KCl/$MgCl_2$) only: ______________________________

____

c. Addition of ATP plus salts (MgCl): ______________________________

____

1. Obtain three microscope slides and cover slips. Label with a wax pencil three of the slides "A," "B," and "C" to correspond to the substances to be added as listed in (a), (b), and (c) above. Then obtain the solutions and three separate eyedroppers (one for each solution so as to not cross-contaminate the solutions).

Observe the effect of adding ATP

2. Obtain the slide labeled "A," a millimeter ruler, and the solution labeled "ATP Alone."
3. Obtain a bundle of fibers from the petri dish and place it on the microscope slide labeled "A."
4. Using the ruler, measure the starting length of the muscle fibers and record it in **table 11.8.**
5. Using a clean eyedropper, flood the muscle fibers with the ATP-only solution. Record your observations:

   ____

   ____

6. After approximately 1 minute, measure the ending length of the muscle fibers and record it in table 11.8.
7. Obtain another clean glass slide. Tease a few fibers from the bundle of fibers treated with ATP and place them on the clean slide. Place a cover slip over the fibers and then observe them with the compound microscope. Record your observations: ______________________________

   ____

Observe the effect of adding salts only (KCl/$MgCl_2$)

8. Obtain the slide labeled "B," a millimeter ruler, and the solution labeled "Salt Solution."
9. Obtain a bundle of fibers from the petri dish and place it on the microscope slide labeled "B."
10. Measure the starting length of the muscle fibers and record it in table 11.8.
11. Using a clean eyedropper, flood the muscle fibers with the salt solution. Record your observations: ______________

    ____

12. After approximately 1 minute, measure the ending length of the muscle fibers and record it in table 11.8.

Observe the effect of adding ATP + salts (KCl/MgCl)

13. Obtain the slide labeled "C," a millimeter ruler, and the solution labeled "ATP Plus Salt Solution."
14. Obtain a bundle of fibers from the petri dish and place it on the microscope slide labeled "C."
15. Measure the starting length of the muscle fibers and record it in table 11.8.
16. Using a clean eyedropper, flood the muscle fiber with the ATP plus salt solution. Record your observations:

    ____

    ____

17. After approximately 1 minute, measure the ending length of the muscle fibers and record in table 11.8.

Table 11.8 Results for Exercise 11.10 (Contraction of Skeletal Muscle)

Slide	Solution	Starting Length (mm)	Ending Length (mm)
A	ATP only		
B	Salts only (KCl/$MgCl_2$)		
C	ATP + Salts (MgCl)		

EXERCISE 11.11 Ph.I.L.S.

Ph.I.L.S. LESSON 5: STIMULUS-DEPENDENT FORCE GENERATION

The purpose of this exercise is to demonstrate how stimulus intensity relates to maximum isometric twitch force generated in a skeletal muscle. Electrical stimuli will be applied over a range of voltages to a frog gastrocnemius muscle. Each stimulus mimics an action potential delivered to a muscle by a motor neuron. In this Ph.I.L.S. exercise, increases in voltage will serve as the increased stimulus. As the voltage is increased from 0 volts up to 1.6 volts, observe how an increase in stimulus intensity relates to increases in maximal isometric twitch force produced by the muscle. A **muscle twitch** is the response of a muscle to a single action potential. In this simulated Ph.I.L.S. experiment, the muscle is being held at a fixed length; hence, the contraction is an **isometric** contraction. In reality, muscles *in vivo* are stimulated by a series of action potentials delivered by the motor neuron (hence, they do not "twitch"). However, studying the twitch response by muscles *in vitro* or through a simulated *in vitro* experiment furthers the understanding of the basic physiological properties of skeletal muscle.

The lowest stimulus that generates a muscle twitch is known as the *threshold stimulus.* As stimulus intensity increases above the threshold stimulus, the maximum isometric twitch force generated by the skeletal muscle also increases. This occurs as more motor units are recruited. Subsequent recruitment of motor units will continue to increase the total muscle force until the muscle reaches a maximal contraction (i.e., all motor units have been recruited). The smallest stimulus that elicits a maximal twitch contraction in the muscle is called the *maximal stimulus.* Stimulating the muscle with voltages greater than the maximal stimulus does not increase the overall force generation. These stimuli are termed *supramaximal stimuli.*

Before beginning, become familiar with the following concepts, using the textbook as a reference:

- Phases of an isometric muscle twitch (latent period, contraction period, relaxation period)
- Innervation of skeletal muscle fibers by motor neurons
- Motor units
- The differences between threshold, submaximal, maximal, and supramaximal stimuli

State a hypothesis regarding the influence of voltage on maximal isometric twitch force.

____

____

____

____

1. Open Lesson 5: Stimulus-Dependent Force Generation.
2. Read the objectives, introduction, and wet lab. Complete the pre-lab quiz.
3. The lab exercise will open once the pre-lab quiz has been completed **(figure 11.14).**
4. Turn on the power of the virtual computer screen and then the power of the data acquisition unit (DAQ).
5. Connect the force transducer to the DAQ by dragging the blue plug to recording input 1.
6. Connect the stimulating electrodes to the stimulator outputs by dragging the black plug to the negative output and the red plug to the positive output.
7. Set the stimulus voltage to 1.0 volts. This voltage will give a baseline reading.
8. To stimulate the muscle to contract, click the "start" button in the control panel (a graph will appear with a blue line indicating the force generated by the muscle and a red line indicating the stimulus applied).
9. Measure the muscle force (magnitude of contraction) by positioning the crosshairs (using the mouse) at the top of the wave. Click the "journal" panel (red rectangle at the bottom of screen) to enter the value into the journal. A graph with applied voltage (V) and muscle tension (g) will appear. Note the range in values for volts from 0 to 1.6 volts. Close the journal window by clicking the "X" in the upper right corner.
10. Adjust the voltage by one-tenth of a volt and repeat the procedure in steps 7–10 to record the data. Continue to repeat this exercise, until the muscle tension that corresponds to applied voltages of 0 to 1.6 volts has been recorded. Record the data from the Ph.I.L.S. "journal" in **table 11.9**.

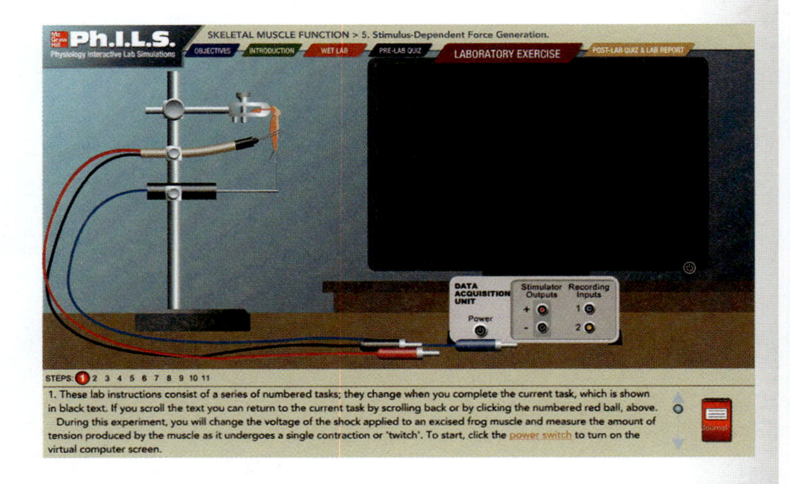

Figure 11.14 Opening Screen for Ph.I.L.S. Lesson 5: Stimulus-Dependent Force Generation.

Table 11.9 Results of Ph.I.L.S. Lesson 5: Stimulus-Dependent Force Generation

Volts	Amplitude/Force
0.0	
0.1	
0.2	
0.3	
0.4	
0.5	
0.6	
0.7	
0.8	
0.9	
1.0	
1.1	
1.2	
1.3	
1.4	
1.5	
1.6	

11. Graph data recorded in table 11.9 in the space provided.

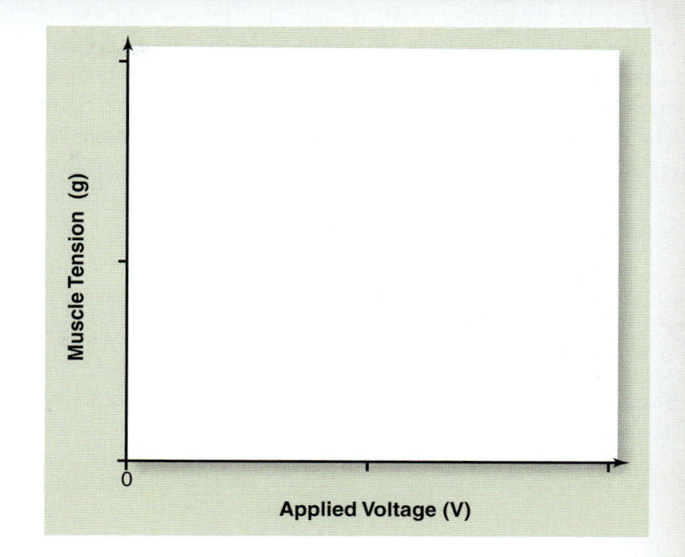

12. Click the "Post-lab Quiz & Lab Report" to complete the post-lab quiz.
13. Make note of any pertinent observations here:

____

____

____

____

____

EXERCISE 11.12 Ph.I.L.S.

Ph.I.L.S. LESSON 7: THE LENGTH–TENSION RELATIONSHIP

The sliding filament theory of muscle contraction states that muscle contraction occurs as thick and thin filaments slide past each other. This process occurs as myosin heads interact with binding sites on actin. The degree of overlap between thick and thin filaments determines how many crossbridges are formed and consequently how much force the muscle will be able to produce during an isometric contraction. The "optimal length" of a muscle is the length at which the maximum number of myosin crossbridges are allowed to interact with the binding sites on actin without interference. It just so happens that a skeletal muscle at its normal *in vivo* resting length is also at its "optimal length," and thus is able to produce the maximal amount of tension when stimulated to contract isometrically. As the length of the muscle (and hence, of the sarcomeres within) gets longer or shorter than the optimal/resting length, force decreases because fewer crossbridges are formed between myosin and actin.

This experiment will use a virtual frog gastrocnemius muscle. The muscle will be held at a fixed length and then stimulated to induce a maximum contraction (hence, the

(continued on next page)

contraction will be isometric, producing maximum isometric force). The experiment will be repeated by varying the length of the muscle (making it longer and/or shorter than the starting length), stimulating it to contract maximally once again, measuring the force generated, and recording it in the "journal." The data will be used to construct a graph that demonstrates the effect of muscle length on maximum isometric force production.

Before beginning, become familiar with the following concepts. Use the textbook as a reference:

- Sliding filament theory of muscle contraction
- Events of the crossbridge cycle
- Optimal length of a skeletal muscle

State a hypothesis regarding the influence of muscle length on maximum isometric force.

____

____

____

____

____

____

____

1. Open Ph.I.L.S. Lesson 7, The Length–Tension Relationship **(figure 11.15).**
2. Read the objectives, introduction, and wet lab. Take the pre-lab quiz. The lab exercise will open when the pre-lab quiz has been completed.
3. Click the "power" switch on the virtual computer screen and data acquisition unit (DAQ) to begin the experiment.

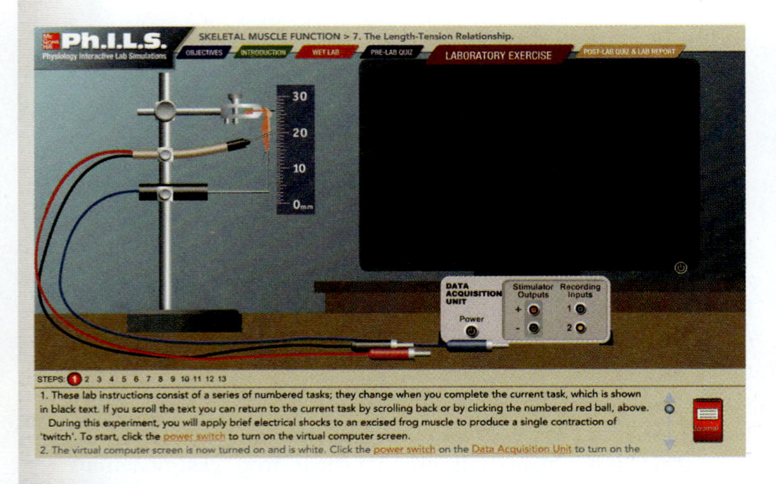

Figure 11.15 Opening Screen for Ph.I.L.S. Lesson 7: The Length–Tension Relationship.

4. Connect the force transducer to the DAQ by dragging the blue plug to recording input 1.
5. Connect the stimulating electrodes to the stimulator outputs by dragging the black plug to the negative output and the red plug to the positive output.
6. Click the "zoom" button in the middle of the computer screen to get an enlarged view of the muscle preparation. The muscle is already stretched to 26 mm at the beginning of the experiment (notice the measurement on the blue ruler next to the muscle).
7. Set the voltage on the control panel by clicking the up or down arrows on the control panel (right side of the screen). Set the voltage at 1.6 volts.
8. Click the "start" button on the control panel. This will elicit a contraction (the blue line on the screen is the muscle tension and the red line is the stimulus).
9. Position the crosshair at the tallest portion of the tension peak. Click the "journal" panel (red rectangle at bottom right of screen) to enter the data into the on-screen table. Then, enter the data in **table 11.10**. Close the journal by clicking the "X" in the upper right-hand corner of the journal window.

Table 11.10 Results from Ph.I.L.S. Lesson 7: The Length–Tension Relationship

Length (mm)	Tension (g)
26.0	
26.5	
27.0	
27.5	
28.0	
28.5	
29.0	
29.5	
30.0	

10. Click the up arrow on the muscle clamp to increase the length of the skeletal muscle by 0.5 mm.
11. Repeat steps 8–9 until the muscle has been fully stretched.

12. Graph data recorded in table 11.10 in the space provided.

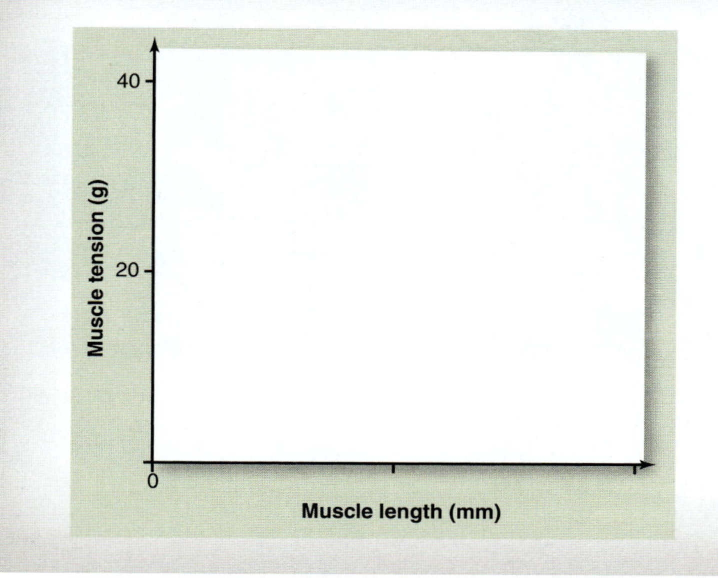

13. Click the "Post-lab Quiz & Lab Report" to complete the post-lab quiz.

14. Make note of any pertinent observations here:

____

____

____

____

____

INTEGRATE

CONCEPT CONNECTION

While the mode of activation for skeletal and cardiac muscle may differ, there are some structural similarities between these two muscle types. Both skeletal and cardiac muscle tissues are striated. Therefore, both contain sarcomeres that are responsible for force generation. Recall that sarcomere length determines force output. This is the foundation for the length–tension curve of muscle. There is an optimal sarcomere length for maximal force generation. Typically, the sarcomeres within skeletal muscles operate close to their optimal length when they are at rest within the body. However, the sarcomeres in cardiac muscle "rest" at a shorter-than-optimal length for maximal force production. The sarcomeres must be lengthened in order to generate maximal force. In the heart, blood enters the heart chambers and provides the force necessary to stretch the cardiac muscle fibers. When more blood enters the heart (as when activity increases), the blood stretches sarcomeres to their optimal length, which results in an increase in force production by cardiac muscle fibers. Therefore, as increased demands are placed on cardiac muscle, cardiac muscle responds by contracting more forcefully.

EXERCISE 11.13 Ph.I.L.S.

Ph.I.L.S. LESSON 8: PRINCIPLES OF SUMMATION AND TETANUS

Action potential propagation along the sarcolemma of skeletal muscle fibers (cells) results in the release of calcium from the sarcoplasmic reticulum into the surrounding sarcoplasm (cytoplasm within the muscle cell). The duration of the electrical event is rapid (1–2 ms) when compared to the duration of the mechanical contraction (100–200 ms), principally because of the time that it takes to actively pump the calcium back into the sarcoplasmic reticulum. The presence of calcium in the sarcoplasm prolongs the period of crossbridge formation. As action potentials increase in frequency, more calcium is available in the sarcoplasm; therefore, muscle tension increases.

The purpose of this laboratory exercise is to observe the change in muscle tension as the frequency of action potentials increases. As the frequency increases, muscle twitches occur more rapidly, ultimately resulting in summation of muscle tension.

Before beginning, become familiar with the following concepts. Use the textbook as a reference:

- Voltage-gated calcium channels in skeletal muscle
- Crossbridge cycling
- A single muscle twitch
- Summation of muscle twitches
- Incomplete and complete tetanus

State a hypothesis regarding the effect that action potential frequency has on calcium concentration in the sarcoplasm and resulting muscle tension.

____

____

____

____

____

____

____

1. To begin the experiment, open Ph.I.L.S. Lesson 8: Principles of Summation and Tetanus (see **figure 11.16**).
2. Read the objectives, introduction, and wet lab. Take the pre-lab quiz. The lab exercise will open when the pre-lab quiz has been completed.

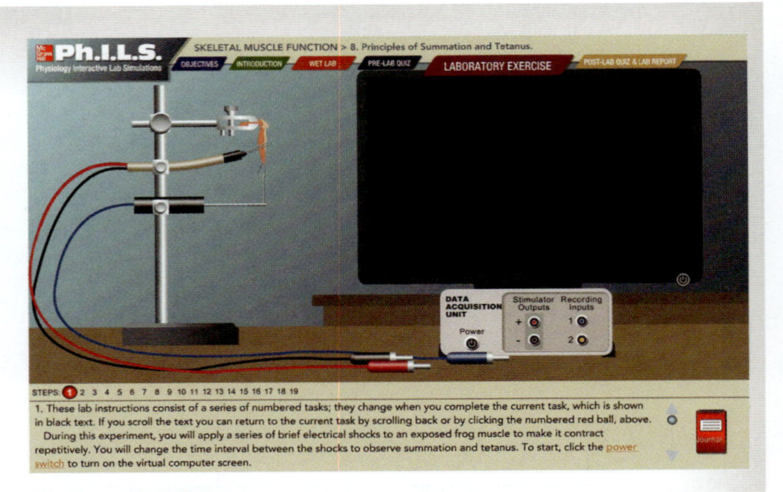

Figure 11.16 Opening Screen for Ph.I.L.S. Lesson 8: Principles of Summation and Tetanus.

3. Click the "power" button on the virtual computer screen and the data acquisition unit (DAQ).
4. Connect the force transducer to the data acquisition unit by clicking and dragging the blue plug to recording input 1.
5. Connect the stimulating electrodes to the stimulator outputs on the DAQ by clicking and dragging the black plug to the negative terminal and clicking and dragging the red plug to the positive terminal.
6. On the control panel, set the voltage to 1.6 volts by clicking the up arrow.
7. Click the "start" button to begin delivering stimuli to the muscle. The default time interval is 500.
8. At a time interval of 500, there are single muscle twitches. To determine the interval that will elicit summation, decrease the interval between stimuli by clicking the down arrow on the control panel. Observe the changes seen in the twitches after decreasing the time interval.
9. Continue the above procedure until summation is achieved, where the blue trace does not always return to the red baseline. Click the stop button once summation is achieved.
10. Click the "journal" in the lower right corner of the screen. Record the time interval where summation is first observed in **table 11.11.** Click the "X" in the upper right corner of the journal window to continue the experiment.
11. Click the "start" button to begin stimulating the muscle again.

Table 11.11 Results from Ph.I.L.S. Lesson 8: Principles of Summation and Tetanus

Type of Contraction	Time Interval of Stimulation
Summation	
Incomplete Tetanus	
Complete Tetanus	

12. Continue to decrease the time interval by clicking the down arrow until incomplete tetanus is observed, where the blue muscle tension trace begins to extend beyond the gray line. Click the "journal" to record the time interval where the muscle reached incomplete tetanus. Click the "X" to close the journal window. Record this time interval in table 11.11.
13. Click the "start" button to begin stimulating the muscle again.
14. Continue to decrease the time interval by clicking the "down" arrow until complete tetanus is observed, where individual twitches are no longer visible. Click the "journal" to record the time interval where the muscle reaches incomplete tetanus. Click the "X" to close the journal window. Record this time interval in table 11.11.
15. Click the "Post-lab Quiz & Lab Report" to complete the post-lab quiz.
16. Make note of any pertinent observations here:

____

____

____

EXERCISE 11.14 McGraw Hill **Ph.I.L.S.**

Ph.I.L.S. LESSON 9: EMG AND TWITCH AMPLITUDE

The purpose of this laboratory exercise is to record the electrical activity from skeletal muscles using virtual surface electrodes, to generate an electromyogram (EMG). Pressure generated from a contracting muscle and EMG readings will be obtained in a virtual subject, and this data will be used to estimate the number of muscle fibers recruited to generate muscle tension.

Before beginning, become familiar with the following concepts. Use the textbook as a reference:

- Physiological principles of the neuromuscular junction
- Anatomical and physiological properties of a motor neuron
- Depolarization and repolarization of a muscle cell
- Contractile properties of skeletal muscle
- The "all-or-none" principle of skeletal muscle contraction

State a hypothesis regarding the relationship between muscle force and EMG.

____

____

____

____

____

____

____

1. Open Ph.I.L.S Lesson 9: EMG and Twitch Amplitude (see **figure 11.17**).
2. Read the objectives, introduction, and wet lab. Take the pre-lab quiz. The lab exercise will open when the pre-lab quiz has been completed.
3. Click the power switch on the virtual computer screen and data acquisition (DAQ) unit to begin the lab.

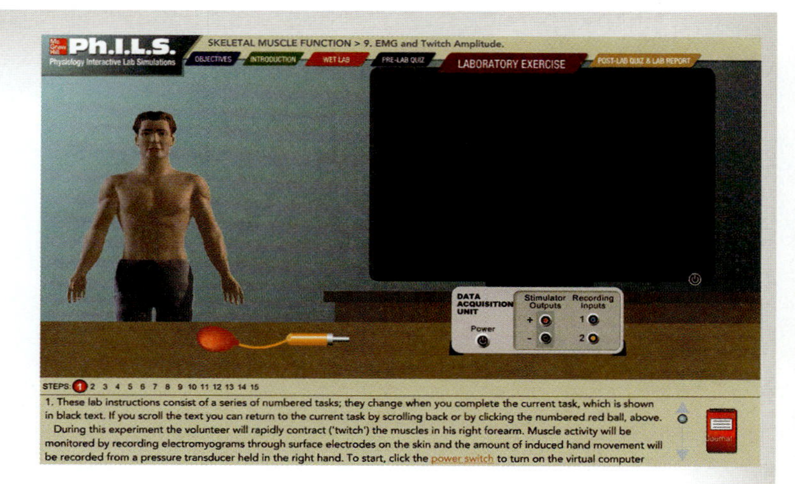

Figure 11.17 Opening Screen for Ph.I.L.S. Lesson 9: EMG and Twitch Amplitude.

4. Connect the pressure transducer by clicking and dragging the orange plug to the recording input number 2.
5. Click and drag the hand dynamometer to the subject's right hand.
6. To set up the EMG, connect the electrodes to the DAQ by clicking and dragging the blue plug to recording input number 1.
7. Click and drag the green electrode to a position just above the subject's right wrist. Click and drag the red electrode to the upper right forearm of the subject (just inside the elbow joint). Click and drag the black electrode to a position between the red and green electrodes on the forearm.
8. Just below the subject's right hand, a slider bar with arrows will appear. This is used to control the amount of forearm contraction. To increase the contraction strength, click the up arrow. To decrease the contraction strength, click the down arrow.
9. To begin the experiment, adjust the grip strength by clicking the down arrow until the setting on the slider bar is set at 1.

(continued on next page)

(continued from previous page)

10. Click "start" on the control panel. The EMG is the upper blue tracing and is recorded in millivolts (mV). The pressure is the lower orange tracing. Position the crosshairs on the highest point on the orange tracing. Click on "journal" in the lower right-hand corner of the screen. Enter the pressure data in **table 11.12.** Click the "X" in the upper right corner of the journal window to continue the experiment.

11. To record the EMG, move the mouse pointer to the highest peak in the EMG recording (blue tracing) and click the "journal." Enter this EMG data in table 11.12.

12. Repeat steps 10–11 by increasing the forearm contraction from 2 to 6. Record data for all points in table 11.12.

Table 11.12 Results from Ph.I.L.S. Lesson 9: EMG and Twitch Amplitude

Recording (Value)	Pressure Amplitude	EMG Amplitude
1		
2		
3		
4		
5		
6		

13. Graph the data generated in the space provided.

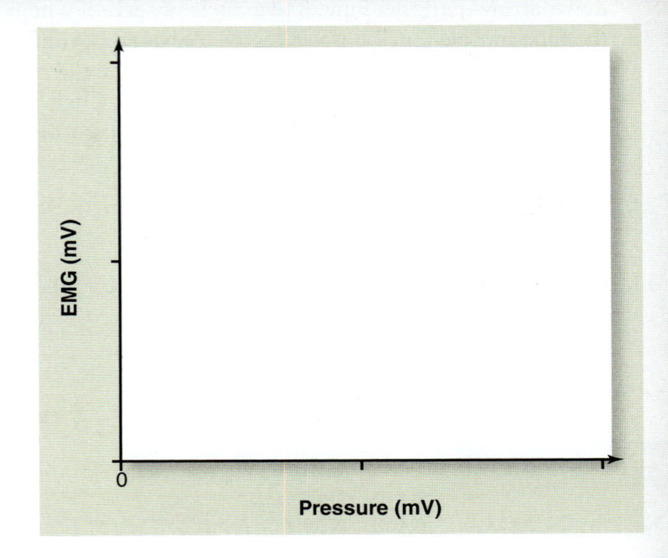

14. Click the "Post-lab Quiz & Lab Report" to complete the post-lab quiz.

15. Make note of any pertinent observations here:

____

____

____

____

EXERCISE 11.15

BIOPAC ELECTROMYOGRAPHY (EMG)

This exercise involves observing the electrical activity of muscle at rest and at maximum exertion. A subject will perform the activity (clenching a fist) and observe the corresponding electromyography (EMG) tracing to determine the relationship between motor unit recruitment and whole muscle force production in the subject's dominant versus nondominant forearm.

Obtain the Following:

- **BIOPAC electrode lead set (SS2L)**
- **BIOPAC general purpose electrodes (EL503)**
- **BIOPAC headphones (OUT1)**
- **electrode gel**
- **BIOPAC MP36/35 recording unit**
- **Laptop Computer with BSL 4 software installed**

1. Prepare the BIOPAC equipment by turning the computer ON and the MP36/35 unit OFF. Plug in the following: electrode lead set (SS2L) (CH 1), and headphones (OUT 1; back of unit). Turn the MP36/35 unit ON.

2. Prepare a subject for placement of electrodes by rubbing the skin vigorously with an alcohol swab at each of the designated locations on the subject's *dominant* forearm and allowing the skin to dry completely **(figure 11.18).** Apply a small quantity of electrode gel to the center of the electrodes. Next, place the electrode firmly on the skin of the subject's *dominant* forearm using the color code described in figure 11.18.

3. Have the subject sit in a comfortable position facing the computer monitor. Start the BIOPAC Student Lab Program. Select "L01-Electromyography (EMG) I" from the drop-down menu, enter the filename, and click "OK."

4. Click "calibrate" in the upper left corner. Wait two seconds; then have the subject clench his/her fist as hard as he/she can, and then relax. Once the calibration has stopped, check the graph to ensure that an EMG was recorded. If so, continue to the next step, "Data Recording." If no recording was made, repeat the calibration process by clicking on, "Redo Calibration." See **figure 11.19** for sample calibration data.

5. OBTAIN DATA: During the recording session, the subject will repeat the clench and release task four times. Each time the subject clenches and releases his/her fist,

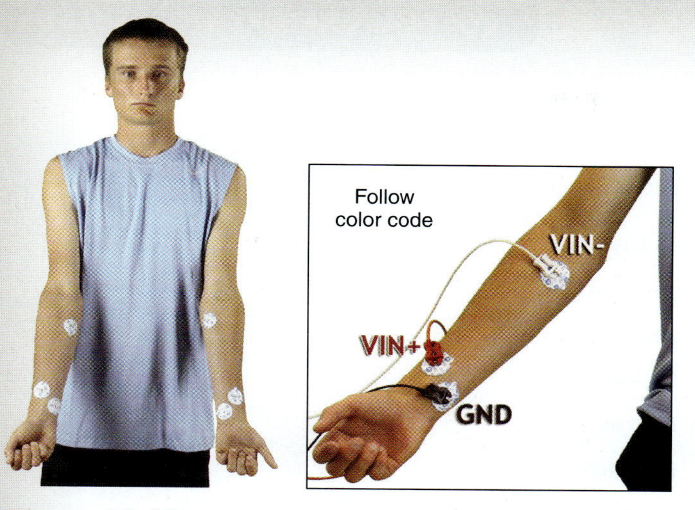

Figure 11.18 BIOPAC EMG Electrode Placement and Lead Attachment.
Courtesy of and © BIOPAC Systems, Inc.

he/she should clench it harder than the previous time. Maximum intensity should be obtained on the fourth repetition. To begin recording EMG data for each trial, click "Record." After the subject has completed a trial, click "Suspend." Review the data recording after each trial to ensure that EMG activity increased in intensity with each trial. If not, click "Redo."

6. Repeat the experiment using the subject's *nondominant* forearm by following the same procedure described in steps 2–5. Place electrodes on the subject's opposite, or *nondominant* forearm (see step 2 and figure 11.18). Click "Resume." Repeat step 5, having the subject complete the task with the opposite fist. Review the data to ensure the EMG increased in intensity with each trial. If so, click "Stop." If not, click "Redo."

7. To listen to the EMG, place the headphones on and click "Listen." Repeat the clench and release task. Adjust the sound as needed to hear the EMG signal through the headphones. Note that data recorded while listening through the headphones is not saved. Click "Stop," and "Redo" to listen again. Click "Done" to conclude the experiment.

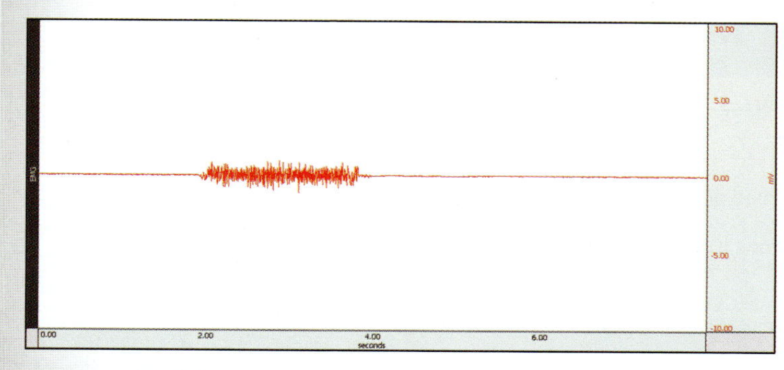

Figure 11.19 Sample Calibration Data for BIOPAC EMG Exercise.
Courtesy of and © BIOPAC Systems, Inc.

Table 11.13	EMG Measurements	
Between Clenches #	**Dominant Arm**	**Nondominant Arm**
	40 Mean	**40 Mean**
1–2		
2–3		
3–4		

8. If analyzing data that was just collected, click on "Analyze Current Data File." If opening data that was collected previously, click "Review Saved Data." Note that CH 1 displays "EMG," and CH40 displays "Integrated EMG." View the first recorded segment.

9. Measurement boxes appear above the marker region in the data window. Note that CH 40 displays "Mean" (average value).

10. Locate the "I-beam" cursor. Select the point on the recorded segment that corresponds to the plateau of the first EMG data cluster. Record the values in **table 11.13.**

11. Repeat step 10 for each EMG data cluster for both the subject's dominant forearm and the subject's nondominant forearm.

12. The resting state between the EMG data clusters represents "Tonus." Use the "I-beam"cursor to highlight each tonus for both the dominant and the nondominant forearms. Record the values in **table 11.14.**

13. Make note of any pertinent observations here:

____

____

____

14. When data analysis is complete, click "Save" or "Print"; then click "Quit" to close the program.

Table 11.14	Tonus Measurements	
Clench #	**Dominant Arm**	**Nondominant Arm**
	40 Mean	**40 Mean**
1		
2		
3		
4		

Chapter 11: The Muscular System: Muscle Structure and Function

Name: ____________________

Date: __________ Section: __________

POST-LABORATORY WORKSHEET

The ❶ corresponds to the Learning Objective(s) listed in the chapter opener outline.

Exercise 11.1: Histology of Skeletal Muscle Fibers

1. The nervous control of skeletal muscle is ________________ (voluntary/involuntary). ❶
2. Match the appropriate description listed in column A with the zone or band listed in column B. ❷

Column A	*Column B*
____ 1. a dark line in the middle of the A band	a. A band
____ 2. a light region in the middle of the A band	b. H zone
____ 3. contains only thin filaments (actin)	c. I band
____ 4. the dark band in skeletal muscle	d. M line
____ 5. edges of a sarcomere are determined by this	e. Z disc

Exercise 11.2: Connective Tissue Coverings of Skeletal Muscle

3. Match the structure listed in column A with the connective tissue covering listed in column B. ❸

Column A	*Column B*
____ 1. covers an entire skeletal muscle	a. endomysium
____ 2. covers a fascicle	b. epimysium
____ 3. covers a skeletal muscle fiber	c. perimysium

Exercise 11.3: The Neuromuscular Junction

4. A somatic motor neuron and all the muscle fibers innervated by that motor neuron comprises a motor unit. ________________ (True/False) ❹

Exercise 11.4: Cardiac Muscle Tissue

5. Which of the following are properties of cardiac muscle tissue? (Check all that apply.) ❺

____ 1. involuntary

____ 2. spindle-shaped cells

____ 3. striated

____ 4. uninucleate

____ 5. voluntary

6. For the following, fill in the blank with the name of the muscle tissue that is described (skeletal, cardiac, or smooth). ❻

a. Muscle tissue of the heart ________________

b. A component of the iris that changes the size of the pupil in the eye ________________

c. Muscle tissue that attaches to the skeleton ________________

d. Changes the size of the opening (lumen) of an air passageway (bronchiole) ________________

e. Responsible for forcing a baby from the uterus during childbirth ________________

f. Changes the diameter of the lumen of blood vessels and helps regulate blood pressure ________________

g. Helps (through involuntary control) to expel urine from the bladder and feces from the digestive tract ________________

Exercise 11.5: Smooth Muscle Tissue

7. Which of the following are properties of smooth muscle tissue? (Check all that apply.) 7

____ 1. involuntary

____ 2. spindle-shaped cells

____ 3. striated

____ 4. uninucleate

____ 5. voluntary

8. In smooth muscle tissue within the wall of a tubular organ (e.g., small intestine), when the inner circular layer of fibers contracts, the diameter of the tube ____________________ (increases/decreases) and the length of the tube ____________________ (increases/decreases). When the outer longitudinal layer of muscle contracts, the diameter of the tube ____________________ (increases/decreases) and the length of the tube ____________________ (increases/decreases). 8

Exercise 11.6: Naming Skeletal Muscles

9. Match the appropriate meaning listed in column A with the word origin listed in column B. 9

Column A	*Column B*
____ 1. smaller	a. abductor
____ 2. round	b. biceps
____ 3. a lifter	c. gracilis
____ 4. around	d. internus
____ 5. to move away	e. latissimus
____ 6. widest	f. levator
____ 7. straight	g. minor
____ 8. two-headed	h. peri-
____ 9. slender	i. rectus
____ 10. internal	j. teres

Exercise 11.7: Architecture of Skeletal Muscles

10. The orbicularis oris muscle has circular architecture. ____________________ (True/False) 10

Exercise 11.8: Major Muscle Groups and Fascial Compartments of the Limbs

11. Match the muscle actions listed in column A with the appropriate compartment and/or major muscle group listed in column B. 11

Column A	*Column B*
____ 1. adduct the thigh	a. anterior compartment of the arm
____ 2. allow one to chew	b. anterior compartment of the leg
____ 3. create facial expressions	c. anterior compartment of the thigh
____ 4. dorsiflex and invert ankle	d. lateral compartment of the leg
____ 5. evert the ankle	e. medial compartment of the thigh
____ 6. extend the knee; flex the hip	f. muscles of the abdomen
____ 7. extend the wrist and digits	g. muscles of facial expression
____ 8. flex, bend, and rotate the spine	h. muscles of mastication
____ 9. flex shoulder and elbow	i. muscles of the neck
____ 10. rotate, flex, and extend the head	j. posterior compartment of the forearm

12. Flexor muscles are generally found in the ____________________ (anterior/posterior) compartment of the arm. 12

Exercise 11.9: Motor Units and Muscle Fatigue (Human Subject)

13. Match the characteristic listed in column A with the muscle fiber type listed in column B. Answers may be used more than once. 13

Column A	*Column B*
____ 1. fast-twitch with high fatigue resistance	a. Type I
____ 2. fast-twitch with moderate fatigue resistance	b. Type IIa
____ 3. largest fiber cross-sectional area	c. Type IIb
____ 4. slow-twitch with very low fatigue resistance	
____ 5. smallest fiber cross-sectional area	

14. Rank the motor units in the order of recruitment as increased force is demanded of skeletal muscle. 14

____ a. Type I

____ b. Type IIa

____ c. Type IIb

15. If a muscular contraction requires the recruitment of Type IIb muscle fibers, the force produced will be ____________________ (high/low), the speed of contraction will be ____________________ (fast/slow), and the contraction ____________________ (will/will not) be able to be sustained for a significant length of time. 15

Exercise 11.10: Contraction of Skeletal Muscle (Wet Lab)

16. Adding which of the following solutions to the glycerinated muscle preparation resulted in contraction? (Check all that apply.) 16

____ a. ATP only

____ b. salts only

____ c. ATP + salts

17. When the supply of ATP in a muscle runs out, myosin heads remain detached from actin. ____________________ (True/False) 17

Exercise 11.11: Ph.I.L.S. Lesson 5: Stimulus-Dependent Force Generation

18. Put the following phases of a muscle twitch in the correct order. 18

____ a. period of contraction

____ b. latent period

____ c. period of relaxation

19. Match the description of the stimulus applied to a muscle fiber listed in column A with the corresponding definition listed in column B. 19 20

Column A	*Column B*
____ 1. not strong enough to produce a muscle twitch	a. subthreshold
____ 2. strong enough to produce a muscle twitch but no summation	b. suprathreshold
____ 3. stronger than necessary to cause summation	c. threshold

Exercise 11.12: Ph.I.L.S. Lesson 7: The Length–Tension Relationship

20. Match the statements with the corresponding numbered region of the length tension curve shown in the figure below. 21 22

____ a. The length of the sarcomere is too great, such that fewer crossbridges may form because the overlap of thin and thick filaments has decreased.

____ b. The sarcomere is at its normal resting length prior to stimulation; therefore, maximal force generation is achieved.

____ c. The sarcomere is shortened prior to stimulation; therefore, the force generated in the muscle is less than its maximum.

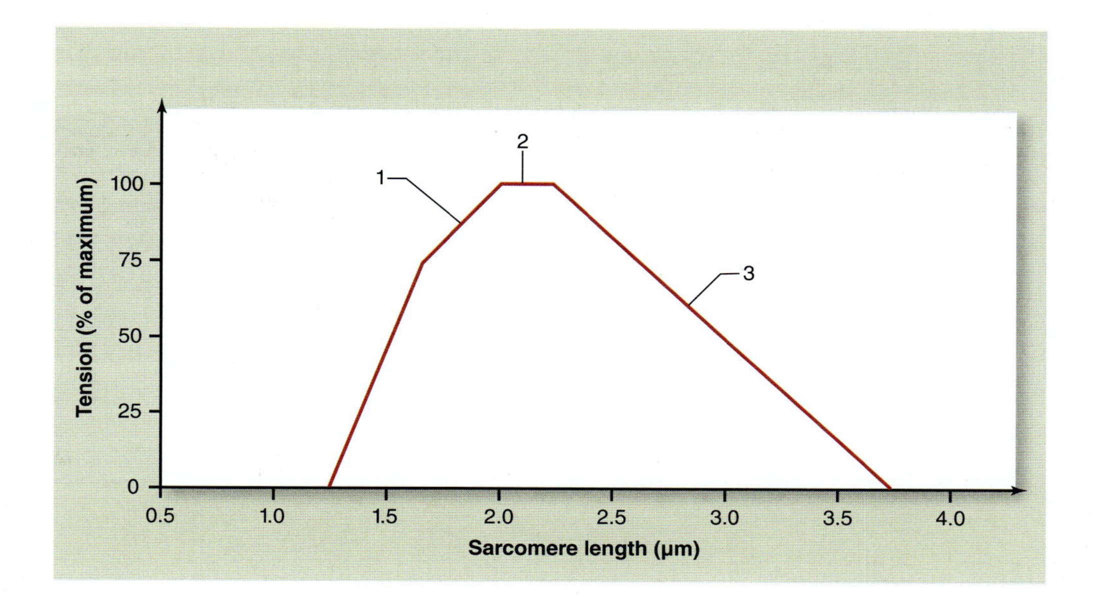

Exercise 11.13: Ph.I.L.S. Lesson 8: Principles of Summation and Tetanus

21. An increase in the frequency of action potentials leads to a(n) ____________________(decrease/increase) in muscle force. 23

22. When the overall amount of force in a muscle increases but muscle twitches are still visible in the force trace, this is known as ____________________ (complete/incomplete) tetanus. 24

Exercise 11.14: Ph.I.L.S. Lesson 9: EMG and Twitch Amplitude

23. An increase in overall EMG amplitude typically corresponds to an increase in muscle tension. ____________________ (True/False) 25

Exercise 11.15: BIOPAC Electromyography (EMG)

24. An increase in motor unit recruitment typically leads to a(n) ____________________ (increase/decrease) in EMG amplitude. 26

Can You Apply What You've Learned?

25. Place the three types of muscle tissues (cardiac, skeletal, smooth) in order of size (smallest to largest). ____________________

____

____

26. The proximal joint of the front limb of a dog contains two compartments of muscles, an anterior and a posterior compartment. If the muscles in the anterior compartment are all flexors, what is the common action of all the muscles in the posterior compartment? ____________________

____

27. A new muscle has been discovered in the body that extends along the tibia. The muscle is a long muscle that contains four heads. Based on knowledge of word origins, suggest a logical name for this muscle. ____________________

28. Describe the events that must take place for a weight lifter to generate enough tension to lift a 100-lb load.

____

____

____

Can You Synthesize What You've Learned?

29. Tetrodotoxin, also known as "zombie powder," is synthesized by the puffer fish as a defense mechanism. Tetrodotoxin's mechanism of action is to block voltage-gated sodium (Na+) channels in motor neurons, thus preventing them from transmitting action potentials. What effect will tetrodotoxin have on the generation of force by a muscle?

____

____

____

30. Using knowledge of the crossbridge cycle, explain why muscles remain in a contracted state (rigor) after death, even though there is no longer a supply of ATP.

____

____

____

CHAPTER 12

The Muscular System: Axial Muscles

OUTLINE AND LEARNING OBJECTIVES

MODULE 6: MUSCULAR SYSTEM

INTRODUCTION

Imagine sleeping on a very uncomfortable mattress and waking up the next morning with a stiff back and neck. For the next few days, every action seems to cause pain. The classic comment on such an experience is, "I discovered muscles I never knew I had!" Indeed, it often seems that muscles of the back and neck only become apparent to us when the muscles are sore or injured, as opposed to when they are being actively used. At such a time it is impossible *not* to be consciously aware of these muscles—they constantly hurt. This example is a reminder of the importance of the muscles of the back, neck, and thorax for creating the most basic motions of the human body.

The exercises in this chapter involve identifying, naming, and exploring the structure and function of muscles of the head, neck, and vertebral column. Muscles responsible for such vital bodily functions as breathing and coughing are also explored. Before beginning the exercises in this chapter, find out which muscles you are required to learn. Then, highlight or star those muscles on the tables throughout this chapter. In addition, before studying the muscles in this chapter, make sure to complete the Gross Anatomy section in chapter 11. Those exercises introduce major muscle groups and the common actions of those muscle groups (see table 11.4). Refer to images in the textbook while working through the exercises in this chapter.

List of Reference Tables

Chapter 12: The Muscular System: Axial Muscles

Name: ______________________

Date: ____________ Section: ____________

These Pre-Laboratory Worksheet questions may be assigned by instructors through their connect course.

PRE-LABORATORY WORKSHEET

1. Which of the following muscles are classified as muscles of facial expression? (Check all that apply.)
   ____ a. buccinator
   ____ b. masseter
   ____ c. orbicularis oris
   ____ d. temporalis
   ____ e. zygomaticus major

2. Which of the following muscles are classified as muscles of mastication? (Check all that apply.)
   ____ a. epicranius
   ____ b. masseter
   ____ c. platysma
   ____ d. risorius
   ____ e. temporalis

3. Match the definition listed in column A with the corresponding term listed in column B.

Column A	*Column B*
____ 1. a broad, flat tendon	a. aponeurosis
____ 2. the "groin" region	b. extension
____ 3. an increase in joint angle	c. extrinsic
____ 4. outside of	d. inguinal
____ 5. pertaining to the cheek	e. buccal

4. Identify the largest and most important respiratory muscle that separates the thoracic cavity from the abdominal cavity. (Circle one.)
   a. anterior scalene
   b. diaphragm
   c. middle scalene
   d. posterior scalene
   e. transversus thoracis

5. Which of the following muscles compose the anterior abdomial wall? (Check all that apply.)
   ____ a. external oblique
   ____ b. internal oblique
   ____ c. quadratus lumborum
   ____ d. rectus abdominis
   ____ e. transversus abdominis

6. Which of the following is *not* one of the three attachment points for the sternocleidomastoid muscle? (Circle one.)
   a. mastoid process of the temporal bone
   b. medial third of the clavicle
   c. spine of the scapula
   d. sternum

7. Match the muscle listed in column A with the corresponding muscle group to which it belongs, listed in column B.

Column A	*Column B*
____ 1. capitis	a. erector spinae
____ 2. digastric	b. laryngeal elevators
____ 3. levator veli palatini	c. semispinalis muscles
____ 4. longissimus	d. suprahyoid muscles

8. The diaphragm changes from dome-shaped to flat when it contracts. ____________________ (True/False)

9. The actions of abdominal wall muscles include extension and rotation of the trunk. ____________________ (True/False)

GROSS ANATOMY

Muscles of the Head and Neck

Muscles of the head and neck include the muscles of the face (muscles of facial expression and muscles of mastication), extrinsic eye muscles, muscles that move the tongue, and muscles that move the neck. (Note that extrinsic eye muscles are covered in chapter 18 with the exercises on the eye.)

INTEGRATE

LEARNING STRATEGY

Use the following procedure when studying each major muscle group to help build knowledge and understanding in logical steps. Once the information from step 1 is mastered, move on to subsequent steps. Each successive step requires a more detailed level of understanding. Following these steps will prepare you to deal with the increased level of detail required at each knowledge level.

Stepwise Approach to Learning Muscles of the Body

1. Describe the general location of the muscle group.
2. Describe the general actions that all muscles of the group have in common.
3. List the names of all muscles belonging to that group (or just the ones required by the laboratory instructor).
4. Identify the muscles on a model or a cadaver (this can be done concurrently with step 3).
5. Learn the outliers—the muscles that DO NOT share the common actions of the group.
6. Learn specific points of attachment and actions of the muscles, as required by the laboratory instructor.

Follow these suggestions: Study one muscle group at a time, take a lot of breaks, and study using frequent, short time intervals. The results may be pleasantly surprising.

EXERCISE 12.1

MUSCLES OF FACIAL EXPRESSION

Muscles of the face are separated into two groups based on function and innervation: muscles of facial expression and muscles of mastication (chewing). **Muscles of facial expression (table 12.1)** allow us to express emotions such as fright, delight, confusion, and surprise. These muscles are unique in that they have distal attachments on skin instead of bone. Thus, when the muscles contract, they pull on the skin. This movement is easily seen on the surface of the face as a facial expression. Over time, the pulling of these muscles on the face causes a characteristic wrinkling of the skin.

1. Observe muscles of the face on a human cadaver or a classroom model.
2. Identify the muscles of facial expression listed in **figure 12.1** on the cadaver or model of the face, using table 12.1 and the textbook as guides.
3. Label the muscles of facial expression in figure 12.1.
4. Each of the seemingly endless variety of facial expressions is performed by contracting specific muscles of the face. Determine the muscles involved in the facial expressions listed by observing the facial expression either on yourself (in a mirror) or by observing the expressions on the face of a classmate (use table 12.1 as a guide).

   ☐ **anger or doubt**
   ☐ **happiness (smiling or laughter)**
   ☐ **kissing (close mouth, purse cheeks, close eyes)**
   ☐ **sadness (frowning)**
   ☐ **surprise or delight**

5. *Optional Activity:* APIR **6: Muscular System**—Watch the muscle action animations to review the actions of many of the muscles mentioned in chapters 12 and 13. Also try the action, origin, and insertion questions found in the quiz area for challenging drill and practice.

Table 12.1 Muscles of Facial Expression*

Muscle	Origin	Insertion	Action	Innervation	Word Origin
Buccinator	Mandible, molar region of mandible and maxilla	Orbicularis oris (corners of the lips)	Presses cheek against molar teeth, as in chewing, whistling, or playing a wind instrument	Facial (CN VII)	*bucca*, cheek
Corrugator Supercilii	Superciliary arch	Skin of eyebrow	Creates vertical wrinkles in medial forehead, as in frowning	Facial (CN VII)	*corrugo*, to wrinkle, + *superus*, above, + *cilium*, eyelid

Table 12.1	Muscles of Facial Expression* *(continued)*				
Muscle	**Origin**	**Insertion**	**Action**	**Innervation**	**Word Origin**
Depressor Anguli Oris	Mandible (antero-lateral surface of the body)	Muscles and skin in the lower lip near the angle of the mouth	Pulls corners of the mouth inferior, as in frowning	Facial (CN VII)	*depressus*, to press down, + *angulus*, angle, + *oris*, mouth
Depressor Labii Inferioris	Mandible (between the midline and the mental foramen)	Oribicularis oris and skin of the lower lip	Depresses the lower lip, as in expressions of doubt and sadness	Facial (CN VII)	*depressus*, to press down, + *labia*, lip, + *inferior*, lower
Epicranius (Occipitofrontalis)*	Epicranial aponeurosis	Skin of the forehead (frontalis); superior nuchal line (occipitalis)	Elevates the eyebrows and causes horizontal wrinkles in the forehead, as in expressions of surprise or delight	Facial (CN VII)	*occiput*, the back of the head, + *frontalis*, in front
Levator Anguli Oris	Maxilla (lateral portion)	Skin at the superior corner of the mouth	Elevates the corners of the mouth and pulls them laterally, as in smiling	Facial (CN VII)	*levatus*, to lift, + *labia*, lip, + *superus*, above
Levator Labii Superioris	Maxilla (inferior to infraorbital foramen)	Orbicularis oris and skin of the upper lip	Elevates the upper lip, as in expressions of sadness or seriousness	Facial (CN VII)	*levatus*, to lift, + *anguli*, angle, + *oris*, mouth
Mentalis	Mandible (incisive fossa)	Skin of the chin	Wrinkles the skin of the chin and elevates and protrudes the lower lip, as in expressions of doubt	Facial (CN VII)	*mentum*, the chin
Nasalis	Maxilla	Alar cartilages of the nose	Flares the nostrils, widens the anterior nasal aperture	Facial (CN VII)	*nasus*, nose
Orbicularis Oculi	Skin around the margin of the orbit of the eye	Skin surrounding the eyelids	Closes the eyelids as in blinking	Facial (CN VII)	*orbiculus*, a small disk, + *oculus*, eye
Orbicularis Oris	Deep surface of skin of maxilla and mandible	Mucous membrane of the lips	Purses and protrudes the lips, closes the mouth	Facial (CN VII)	*orbiculus*, a small disk, + *oris*, mouth
Platysma	Fascia superficial to the deltoid and pectoralis major muscles at ribs 1 and 2	Mandible (lower border) and skin of the cheek	Stretches the skin of the anterior neck, depresses the lower lip, as in expressions of fright	Facial (CN VII)	*platys*, flat, broad
Procerus	Nasal bones and nasal cartilages	Aponeurosis at the bridge of the nose and the skin of the forehead	Depresses the eyebrows and elevates the nose producing wrinkles in the skin of the nose, as in frowning and squinting the eyes	Facial (CN VII)	*procerus*, long or stretched out
Risorius	Fascia overlying the masseter muscles	Orbicularis oris and skin of the corner of the mouth	Pulls the corners of the mouth laterally, as in expressions of laughter or smiling	Facial (CN VII)	*risus*, to laugh
Zygomaticus (Major and Minor)	Zygomatic bone	Skin and muscle at the corner of the mouth	Pulls the corners of the mouth posteriorly and superiorly, as in smiling	Facial (CN VII)	*zygon*, yoke

*All muscles in this table are innervated by the facial nerve (CN VII).

**The epicranius consists of the epicranial aponeurosis and the occipitofrontalis muscle, which has two bellies: frontal belly of occipitofrontalis and occipital belly of occipitofrontalis.

(continued on next page)

(continued from previous page)

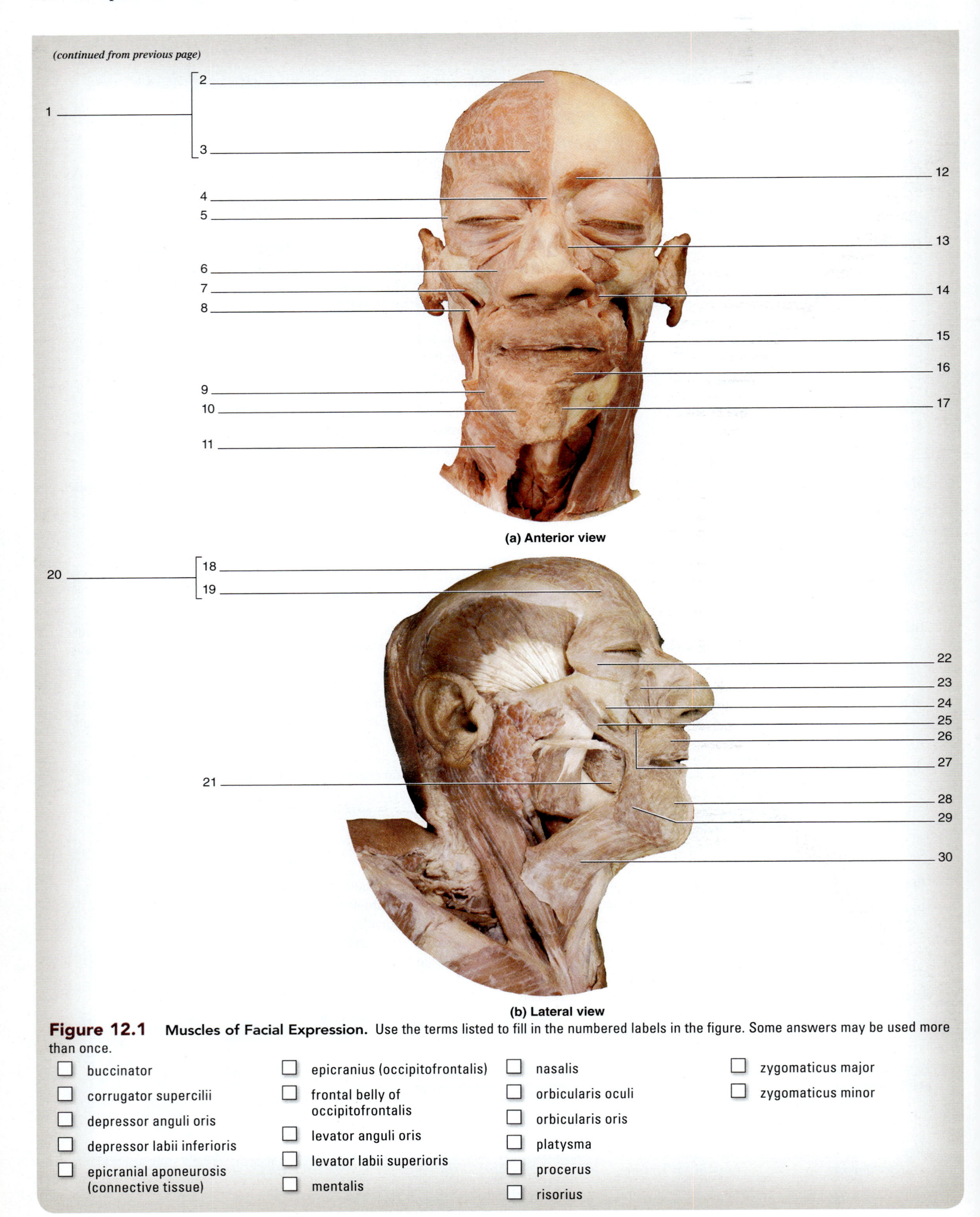

Figure 12.1 Muscles of Facial Expression. Use the terms listed to fill in the numbered labels in the figure. Some answers may be used more than once.

- ☐ buccinator
- ☐ corrugator supercilii
- ☐ depressor anguli oris
- ☐ depressor labii inferioris
- ☐ epicranial aponeurosis (connective tissue)
- ☐ epicranius (occipitofrontalis)
- ☐ frontal belly of occipitofrontalis
- ☐ levator anguli oris
- ☐ levator labii superioris
- ☐ mentalis
- ☐ nasalis
- ☐ orbicularis oculi
- ☐ orbicularis oris
- ☐ platysma
- ☐ procerus
- ☐ risorius
- ☐ zygomaticus major
- ☐ zygomaticus minor

EXERCISE 12.2

MUSCLES OF MASTICATION

Muscles of mastication (*masticate*, to chew) are used in chewing **(table 12.2).** These muscles attach to the only mobile bone of the skull, the mandible.

1. Observe muscles of the face on a human cadaver or a classroom model demonstrating facial muscles.
2. The two most powerful muscles of mastication are the **masseter** and **temporalis** muscles. To palpate these on yourself, place your fingers over the angle and ramus of the mandible (just below the cheek) and close your jaw forcefully (elevate the mandible) to feel contraction of the masseter. Repeat the process, only this time, place your fingers over your temples to feel the contraction of the temporalis muscle. Muscles that depress the mandible (open the mouth) are called infrahyoid muscles. The infrahyoid muscles are covered in exercise 12.5.
3. Identify the muscles of mastication listed in **figure 12.2** on a cadaver or the model demonstrating facial musculature, using table 12.2 and the textbook as guides.
4. Label the muscles of mastication in figure 12.2.

Table 12.2 Muscles of Mastication*

Muscle	Origin	Insertion	Action	Innervation	Word Origin
Lateral Pterygoid*	Sphenoid (greater wing and lateral pterygoid plate)	Mandible (neck)	Protracts the mandible and depresses the chin; produces a side-to-side motion of the mandible	Trigeminal (CN V)	*pteryx*, wing, + *eidos*, resemblance
Masseter	Zygomatic arch	Mandible (lateral surface of the ramus)	Elevates and protracts the mandible	Trigeminal (CN V)	*masetér*, chewer
Medial Pterygoid*	Sphenoid (lateral pterygoid plate) and maxilla	Mandible (medial surface of the ramus and neck)	Elevates and protracts the mandible; produces a side-to-side motion of the mandible	Trigeminal (CN V)	*pteryx*, wing, + *eidos*, resemblance
Temporalis	Temporal line	Mandible (coronoid process)	Elevates and retracts the mandible	Trigeminal (CN V)	*tempus*, temple

*All muscles in this table are innervated by the trigeminal nerve (CN V).

**Lateral and medial pterygoids, when acting alone (alternating one side at a time), produce a side-to-side grinding motion.

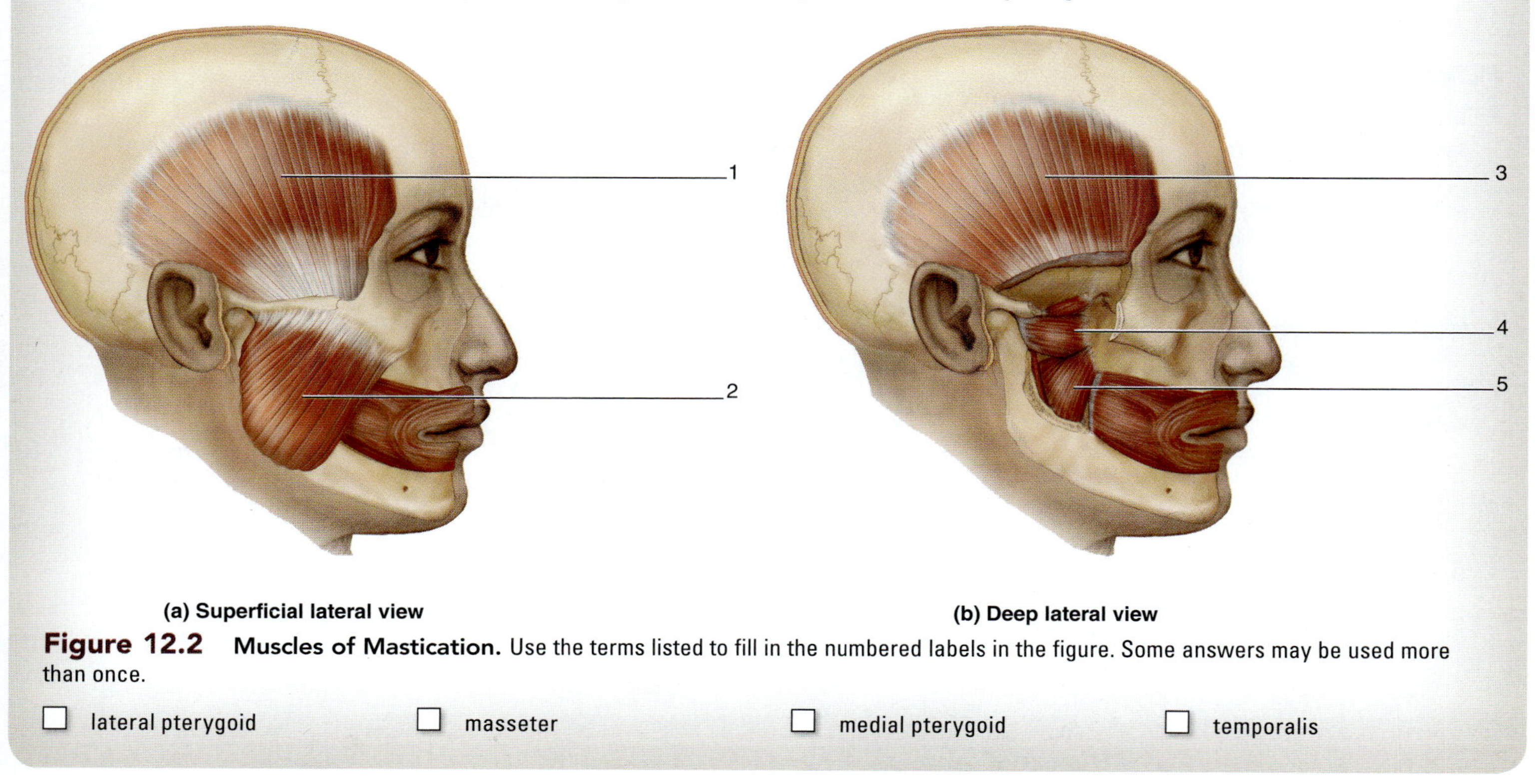

Figure 12.2 Muscles of Mastication. Use the terms listed to fill in the numbered labels in the figure. Some answers may be used more than once.

☐ lateral pterygoid ☐ masseter ☐ medial pterygoid ☐ temporalis

INTEGRATE

CONCEPT CONNECTION

The muscles of facial expression are innervated by the **facial nerves,** which are the paired seventh cranial nerves (CN VII). A person with damage to a facial nerve will have impaired ability to demostrate facial expression on the affected side. The muscles of mastication are innervated by branches of the **trigeminal nerves,** which are the paired fifth cranial nerves (CN V). A person with damages to the trigeminal nerve will have impaired ability to chew on the affected side.

EXERCISE 12.3

MUSCLES THAT MOVE THE TONGUE

The tongue has both intrinsic and extrinsic muscles **(table 12.3).** The *intrinsic* muscles, within the tongue, change the shape of the tongue. The *extrinsic* muscles, which connect the tongue to bony structures of the head and neck, cause fine movements of the tongue (e.g., to form speech and manipulate food).

1. Observe a model of a head that has been cut in a midsagittal plane so the muscles of the tongue are visible.
2. Identify the **muscles that move the tongue** listed in **figure 12.3** on a cadaver or on a model of the head and neck, using table 12.3 and the textbook as guides. (Note: These are all extrinsic muscles of the tongue.)
3. Label the muscles of the tongue in figure 12.3.
4. Stick your tongue out, and manipulate it in various ways (e.g., curl it, flip it). When performing these actions, decide if the muscles used are intrinsic or extrinsic muscles of the tongue.

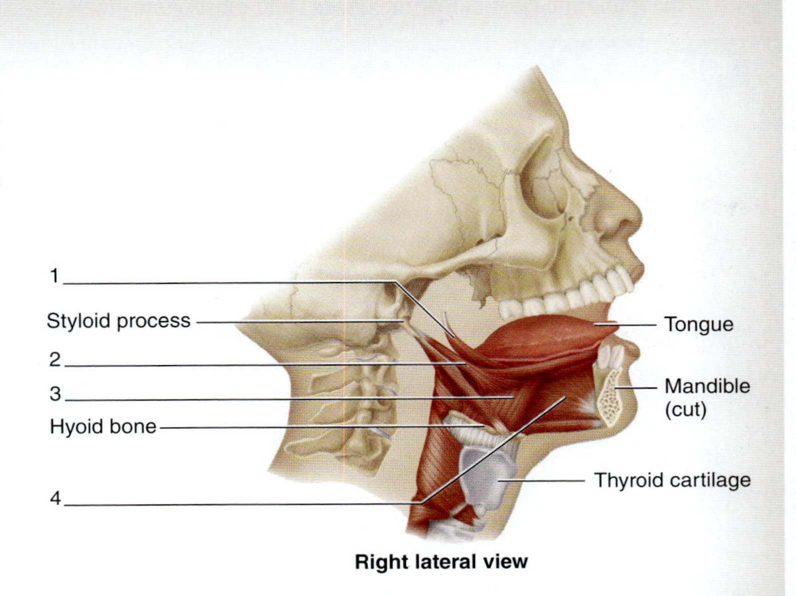

Figure 12.3 **Muscles That Move the Tongue.** Use the terms listed to fill in the numbered labels in the figure.

- ☐ genioglossus
- ☐ hyoglossus
- ☐ palatoglossus
- ☐ styloglossus

Table 12.3 Muscles That Move the Tongue

Muscle	Origin	Insertion	Action	Innervation	Word Origin
Genioglossus	Mandible (mental spine)	Hyoid bone (body) and inferior portion of the tongue	Depresses and protrudes the tongue	Hypoglossal (CN XII)	*geneion*, chin, + *glossa*, tongue
Hyoglossus	Hyoid bone (body and greater horn)	Inferior and lateral aspects of the tongue	Depresses and retracts the tongue	Hypoglossal (CN XII)	*Hyo-*, hyoid bone, + *glossa*, tongue
Palatoglossus	Soft palate (palatine aponeurosis)	Inferiolateral aspect of the tongue	Depresses the soft palate and elevates the posterior aspect of the tongue	Vagus (CN X)	*palatum*, palate, + *glossa*, tongue
Styloglossus	Styloid process of the temporal bone	Inferior and lateral aspects of the tongue	Retracts and elevates the tongue during swallowing	Hypoglossal (CN XII)	*stylo*, styloid process, + *glossa*, tongue

EXERCISE 12.4

MUSCLES OF THE PHARYNX

The pharyngeal muscles **(table 12.4)** are used during the swallowing process.

1. Observe a prosected cadaver, a model of the larynx, or a model of the head and neck. (The models should also show the muscles of the pharynx.)
2. Identify the **muscles of the pharynx** listed in **figure 12.4** on a cadaver, on a model of the head and neck, or on a model of the larynx, using table 12.4 as a guide.
3. Label the muscles of the pharynx in figure 12.4.

Table 12.4 Muscles of the Pharynx

Muscle	Origin	Insertion	Action	Innervation	Word Origin
Laryngeal Elevators					
Levator Veli Palatini	Temporal bone (petrous portion) and cartilage of the auditory tube	Soft palate (palatine aponeurosis)	Elevates the soft palate, as in swallowing and yawning	Vagus (CN X)	*levatus*, to lift, + *velum*, veil, + *palatum*, palate
Tensori Veli Palatini	Sphenoid bone, region around auditory tube	Soft palate	Tenses soft palate and opens auditory tube when swallowing or yawning	Trigeminal (CN V)	*tensus,* to stretch, + *velum,* veil, + *palatum,* palate
Palate Muscles					
Palatopharyngeus	Soft palate (palatine aponeurosis)	Lateral wall of the pharynx	Elevates the larynx and pharynx	Vagus (CN X)	*palatum*, palate, + *pharyngo-*, pharynx
Stylopharyngeus	Styloid process of temporal bone	Larynx (thyroid cartilage)	Elevates the larynx and pharynx	Glossopharyngeal (CN IX)	*stylo-*, styloid process, + *pharyngo-*, pharynx
Tensor Veli Palatini	Sphenoid bone (pterygoid process)	Soft palate (palatine aponeurosis)	Elevates the soft palate	Trigeminal (CN V)	*tensus*, to stretch, + *velum*, veil, + *palatum*, palate
Pharyngeal Constrictors					
Inferior Constrictor	Larynx (thyroid and cricoid cartilages)	Posterior median raphe	Constricts the pharynx	Vagus (CN X)	*inferior*, lower, + *constringo*, to draw together
Middle Constrictor	Hyoid bone	Posterior median raphe	Constricts the pharynx	Vagus (CN X)	*middle*, middle, + *constringo*, to draw together
Superior Constrictor	Sphenoid bone (pterygoid process)	Posterior median raphe	Constricts the pharynx	Vagus (CN X)	*superus*, above, + *constringo*, to draw together

INTEGRATE

CLINICAL VIEW
Dysphagia

Swallowing is a complex process that involves the coordination of many muscles associated with the mouth, tongue, pharynx, larynx, and esophagus. The muscles of mastication are used for chewing to create a bolus of food. The intrinsic muscles of the tongue and the mylohyoid muscle, which elevates the floor of the mouth and the tongue, are used to push the bolus of food into the pharynx. Laryngeal elevators prevent the bolus of food from entering the larynx. Palatine muscles elevate the pharynx in preparation for receiving the bolus. Finally, pharyngeal constrictors sequentially contract to narrow the opening of the pharynx, thereby pushing the bolus into the **esophagus** (*oisein-*, to carry + *phagos*, to eat). Once the bolus enters the esophagus, peristaltic contraction of skeletal muscle in the wall of upper portions of the esophagus and smooth muscle in the walls of the lower portions of the esophagus propel the bolus into the stomach.

Disruptions in the swallowing process can lead to **dysphagia,** or difficulty swallowing. Dsyphagia can be classfied as or *opharyngeal,* which is characterized by difficulty moving food through the oral cavity and pharynx, or *esophageal,* which is characterized by difficulty moving food through the esophagus. Some cases of dysphagia are neurogenic (*neuro*, nervous + *genesis*, birth), meaning they arise secondary to conditions such as Amyotrophic Lateral Sclerosis (ALS), stroke, or traumatic brain injury. Other causes of dysphagia include obstruction as may occur with an esophageal tumor. Patients suffering from dysphagia are at a greater risk for aspirating substances into the airways, and malnutrition from an inability to ingest food.

(continued on next page)

(continued from previous page)

4. To get a feel for how these muscles function, swallow some saliva or water and pay attention to the role these muscles play during swallowing. Notice that during swallowing, the larynx is moved superiorly by the action of suprahyoid muscles (exercise 12.5). (The larynx is best felt by palpating the thyroid cartilage, or "Adam's apple," which is located in the upper portion the of anterior neck.) The muscles of focus in this exercise are the ones felt at the back of the throat at the very end of the swallowing process. The pharyngeal constrictors move swallowed materials (e.g., food, saliva) into the esophagus.

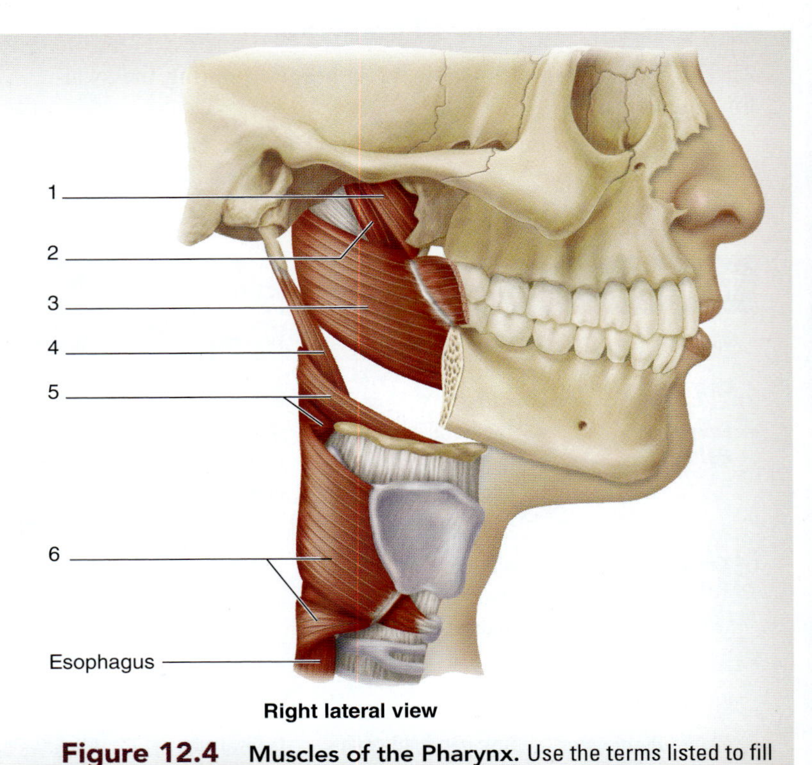

- ☐ inferior constrictor
- ☐ levator veli palatini
- ☐ middle constrictor
- ☐ stylopharyngeus
- ☐ superior constrictor
- ☐ tensor veli palatini

Figure 12.4 **Muscles of the Pharynx.** Use the terms listed to fill in the numbered labels in the figure.

EXERCISE 12.5

MUSCLES OF THE NECK

This exercise explores both the muscles of the anterior neck and the muscles that move the head and neck. The muscles of the anterior neck are organized into the suprahyoid muscles, which are associated with the floor of the mouth, and the infrahyoid muscles, which alter the position of the hyoid bone and larynx. The muscles of the head and neck originate on the vertebral column, the thoracic cage, and the pectoral girdle, and insert on bones of the cranium.

1. Observe a prosected cadaver or a classroom model demonstrating muscles of the neck.
2. Identify the muscles of the neck listed in **figure 12.5** on a cadaver or a model of the head and neck, using **table 12.5** and the textbook as guides.
3. Label the muscles of the neck in figure 12.5.
4. The most prominent muscle in the anterior neck is the **sternocleidomastoid** muscle. The attachment of

INTEGRATE

CONCEPT CONNECTION

The sternocleidomastoid and splenius capitis are muscles that provide a good example of how muscles can act as either **synergists** or **antagonists** (see chapter 10), depending upon the movement required. These muscles can act as synergists for the action of lateral rotation of the neck, whereas they act as antagonists (see chapter 10) depending upon the movement required. These muscles can act as synergists for the action of lateral rotation of the neck, whereas they act as antagonists for the actions of flexion and extension of the neck. Palpate the sternal head of the sternocleidomastoid on your right side, and then laterally rotate your head to the right. Do you feel tension in the muscle? ________________ While palpating the muscle, laterally rotate your head to the left. Do you feel tension in the muscle? ________________ Based on your observations, in which direction does the right sternocleidomastoid rotate the neck? ________________

The most superficial muscle on the posterior side of the neck is the trapezius, a muscle that can move the neck, but is more of a prime mover of the scapula (see table 13.1). If you palpate the posterior side of your neck as you rotate your neck, you can feel the trapezius muscle and a smaller muscle located deep to the trapezius, the splenius capitis, which is a prime mover of the neck (see table 12.5 for reference to its actions). Place your right hand over the posterior aspect of the right side of your neck and rotate your head to the right. Keeping your hand in the same location, rotate your head to the left. In which direction of rotation do you feel more tension in the muscles on the back of the neck? ________________ Based on your observations, in which direction does the right splenius capitis muscle rotate the neck? ________________

In summary: To rotate your neck to the right, you use the sternal head of the sternocleidomastoid on the ________________ side of the neck, and the splenius capitis muscle on the ________________ side of the neck.

this muscle can be palpated easily through the skin. To do this, place your fingers just behind your ears to palpate the **mastoid process,** the insertion of the sternocleidomastoid. Next, palpate your sternum, just lateral to the suprasternal notch as you rotate your neck from right to left, and feel for the tendons of the **sternal head** of the sternocleidomastoid. Finally, palpate just superior to the medial third of the clavicle while laterally flexing your neck to see if you can feel contraction of the **clavicular head** of the sternocleidomastoid. Understanding the locations of the attachments, bellies, and borders of the two heads of this muscle is important because the sternocleidomastoid is a major clinical landmark in the neck.

Table 12.5 Muscles of the Anterior Neck and Muscles That Move the Head and Neck

Muscle	Origin	Insertion	Action	Innervation	Word Origin
Muscles of the Anterior Neck: Suprahyoid Muscles					
Digastric	Mastoid process (digastric groove on medial aspect)	Mandible (lower border near the midline)	When the mandible is fixed, it elevates the hyoid (posterior belly). When the hyoid is fixed, it depresses the mandible (anterior belly).	Posterior belly: facial (CN VII) Anterior belly: mandibular branch of the trigeminal (CN V)	*di-*, two, + *gastro-*, belly
Geniohyoid	Mandible (mental spine)	Hyoid bone	Elevates the hyoid	Hypoglossal (CN XII)	*geneion*, chin, + *hyoides*, shaped like the letter U
Mylohyoid	Mandible (mylohyoid line)	Hyoid bone	Elevates the floor of the mouth and tongue. When the hyoid is fixed it depresses the mandible.	Mandibular branch of the trigeminal (CN V)	*myle*, a mill, + *hyoides*, shaped like the letter U
Stylohyoid	Styloid process of temporal bone	Hyoid bone	Elevates the hyoid	Facial (CN VII)	*stylos*, pillar, + *hyoides*, shaped like the letter U
Muscles of the Anterior Neck: Infrahyoid Muscles					
Omohyoid	Scapula (between the superior angle and the scapular notch)	Hyoid bone	Depresses the hyoid	Cervical spinal nerves C1–C3 through ansa cervicalis	*omos*, shoulder, + *hyoides*, shaped like the letter U
Sternohyoid	Manubrium of sternum (posterior surface) and first costal cartilage	Hyoid bone	Depresses the hyoid	Cervical spinal nerves C1–C3 through ansa cervicalis	*sternon*, chest, + *hyoides*, shaped like the letter U
Sternothyroid	Manubrium of sternum (posterior surface) and first costal cartilage	Thyroid cartilage	Depresses the larynx	Cervical spinal nerves C1–C3 through ansa cervicalis	*sternon*, chest, + *thyroid*, shaped like a shield
Thyrohyoid	Thyroid cartilage	Hyoid bone	Moves hyoid toward the larynx	First cervical spinal nerve C1 via hypoglossal (CN XII)	*thyro-*, thyroid, + *hyoides*, shaped like the letter U
Muscles That Move the Head and Neck					
Longissimus Capitis	T_1–T_4 (transverse processes) and C_4–C_7 (articular processes)	Mastoid process of temporal bone	Bilateral: extends the neck Unilateral: laterally rotates the neck to the same side	Cervical and thoracic spinal nerves	*longissimus*, longest, + *caput*, head
Semispinalis Capitis	T_1–T_5 (spinous processes), C_4–C_7 (articular processes)	Occipital bone (between superior and inferior nuchal lines)	Bilateral: extends the neck Unilateral: laterally flexes the neck to the same side	Dorsal rami of cervical spinal nerves	*semi*, half, + *spina*, spine, + *caput*, head

(continued on next page)

(continued from previous page)

Table 12.5 Muscles of the Anterior Neck and Muscles That Move the Head and Neck *(continued)*

Muscle	Origin	Insertion	Action	Innervation	Word Origin
Splenius Capitis	Ligamentum nuchae and T_1–T_6 (spinous processes)	Superior nuchal line of occipital bone (lateral aspect) and mastoid process of temporal bone	Bilateral: extends the neck Unilateral: laterally rotates and laterally flexes the neck to the same side	Dorsal rami of spinal nerves	*splenion*, a bandage, + *caput*, head
Sterno-cleidomastoid	Sternal head: manubrium of the sternum Clavicular head: clavicle (medial third)	Mastoid process of temporal bone	Bilateral: flexes the neck Unilateral: laterally rotates the neck to the opposite side	Accessory (CN XI)	*sterno-*, sternum, + *cleido-*, clavicle, + *mastoid*, resembling a breast

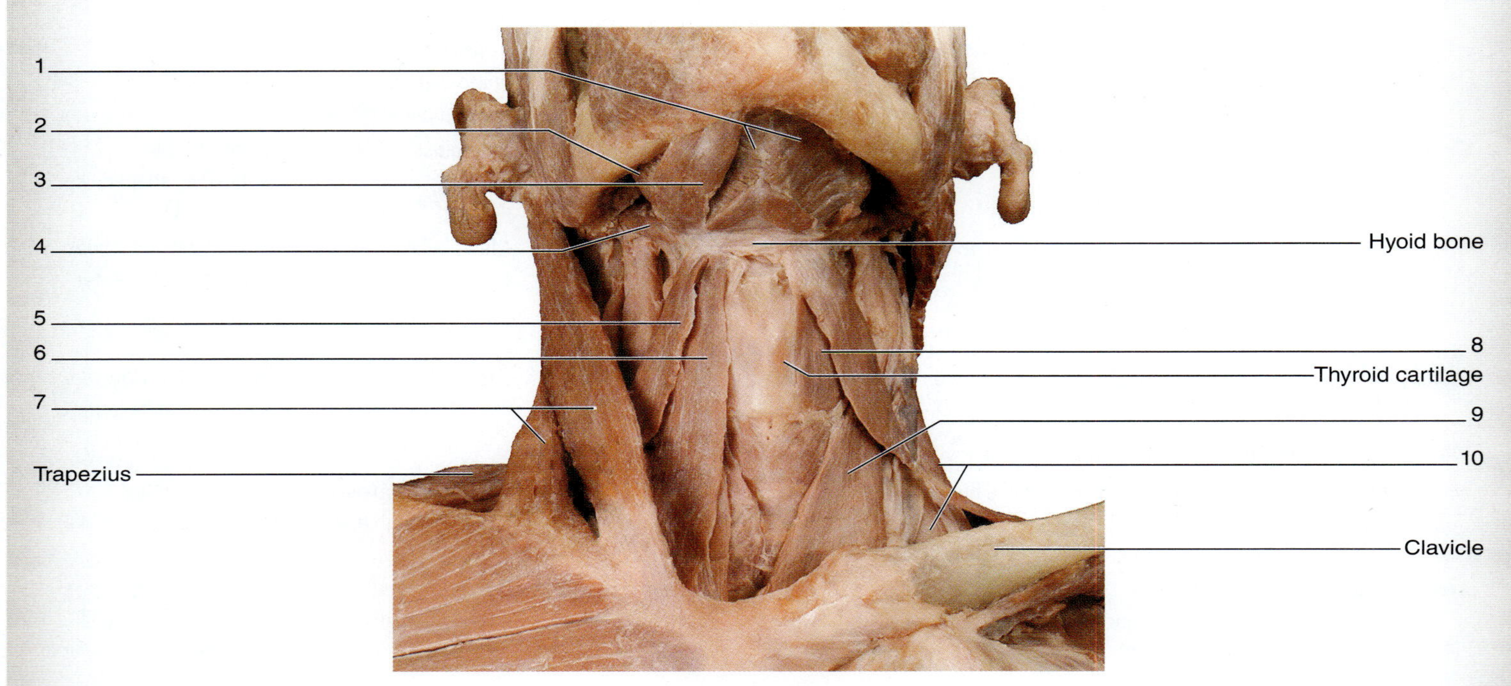

(a) Anterior view

Figure 12.5 Muscles of the Anterior Neck and Muscles That Move the Head and Neck. Use the terms listed to fill in the numbered labels in the figure. Some answers may be used more than once.

- ☐ digastric (anterior belly)
- ☐ digastric (posterior belly)
- ☐ mylohyoid
- ☐ omohyoid (superior belly)
- ☐ scalenes
- ☐ sternocleidomastoid
- ☐ sternohyoid
- ☐ sternothyroid
- ☐ stylohyoid
- ☐ thyrohyoid

INTEGRATE

LEARNING STRATEGY

When studying muscles, or any other anatomical structure for that matter, remember that directional terms like "anterior/posterior" or "medial/lateral" are used in the naming of muscles only when necessary. That is, they are generally used only when there are two or more similar muscles with the same name. For example: If there is a "major" (e.g., zygomaticus major), then there must also be a "minor" (e.g., zyogmaticus minor); if there is a "superior" (e.g., superior oblique), there must also be an "inferior" (e.g., inferior oblique). This may seem simple, but it is not always obvious to the beginning student of anatomy.

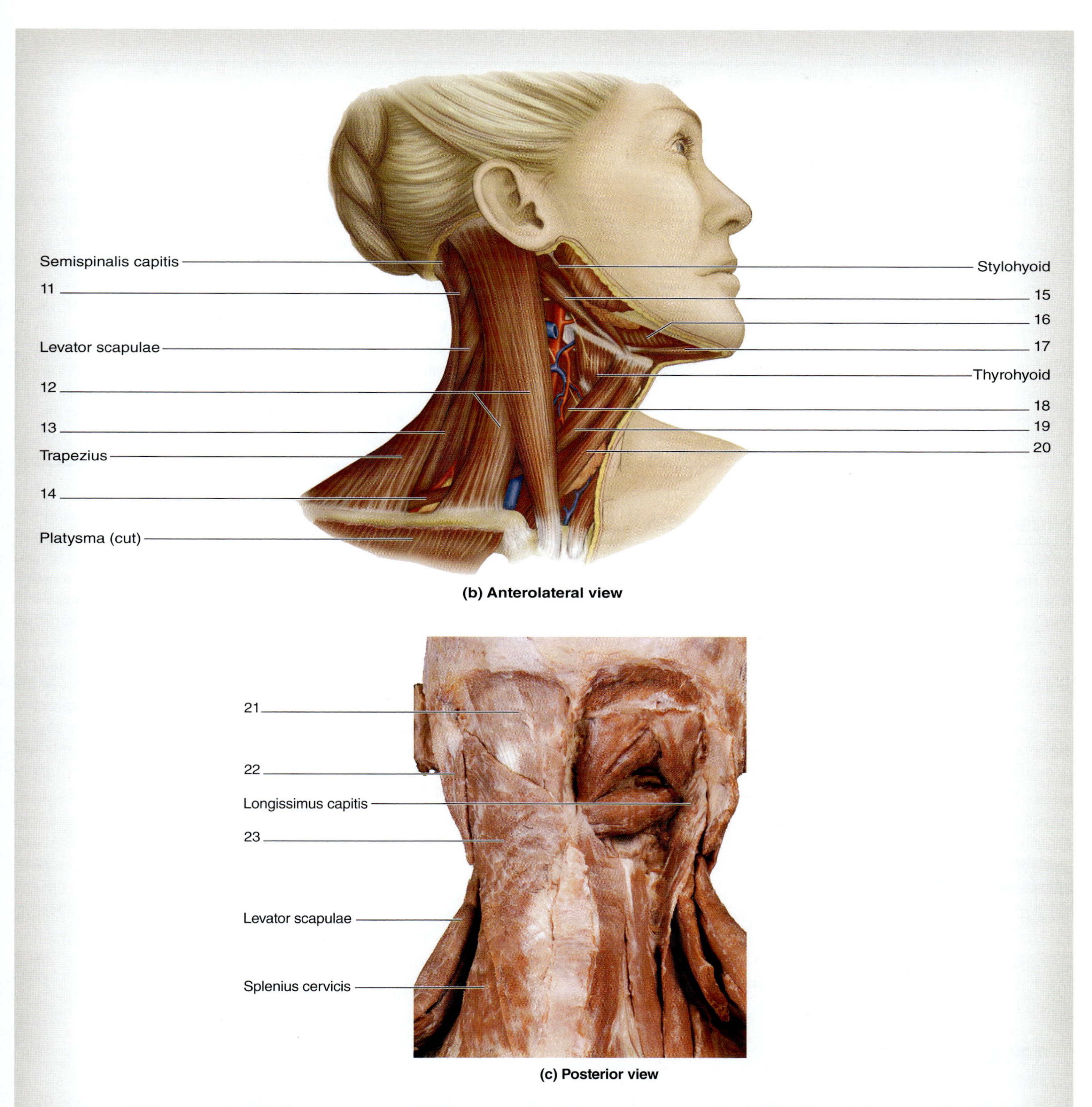

Figure 12.5 Muscles of the Anterior Neck and Muscles That Move the Head and Neck *(continued).* Use the terms listed to fill in the numbered labels in the figure. Some answers may be used more than once.

- ☐ digastric (anterior belly)
- ☐ digastric (posterior belly)
- ☐ mylohyoid
- ☐ omohyoid (inferior belly)
- ☐ omohyoid (superior belly)
- ☐ scalenes
- ☐ semispinalis capitis
- ☐ splenius capitis
- ☐ sternocleidomastoid
- ☐ sternohyoid

INTEGRATE

CLINICAL VIEW
Stroke

A **stroke** occurs when one or more parts of the brain suffer(s) a lack of oxygen either from a reduction in blood flow—as can happen with a blood clot or narrowed artery (ischemic stroke)—or from rupture of a blood vessel in the brain (hemorrhagic stroke). The consequences of a stroke can be great, depending on the portion of the brain affected. Chapter 15 covers topics related to stroke such as the parts of the brain that function in planning movements, perceiving sensations, and maintaining homeostasis. Loss of oxygen or rupture of a blood vessel to the part of the brain responsible for regulating bodily functions such as heart rate can have very serious implications. Patients who suffer from a stroke often experience symptoms such as gross motor impairments, slurred speech, an inability to swallow, and a "droopy" face. These symptoms often indicate damage to the specific areas of the brain responsible for each modality. For example, the command for voluntary movement of skeletal muscles arises from the frontal lobe. Therefore, damage to the frontal lobe may result in an inability to stimulate skeletal muscles in various regions of the body. The nuclei for the cranial nerves innervate skeletal muscles of the head and neck. For example, paired facial nerves (CN VII) innervate the muscles of the face responsible for creating facial expressions. The cell bodies of the facial nerves are located within the brainstem. If the supply of oxygen is reduced in this critical part of the brain, nerve signals will not be sent from the cranial nerve nuclei out to the skeletal muscles. This results in muscle paralysis so the patient is unable to demonstrate facial expressions on the affected side. The patient may also experience an inability to elevate the right upper lip. Likewise, damage to the nuclei for the paired trigeminal (CN V) nerves may cause a patient to experience difficulty with mastication (chewing). Damage to the hypoglossal (CN XII) nerve may make swallowing or speech difficult. Such symptoms give clinicians some clues regarding what structures or areas of the brain have been damaged by a stroke. Typically, a magnetic resonance imaging (MRI) scan is performed on the brain to confirm the exact location of damage.

Muscles of the Vertebral Column

Muscles that move the vertebral column are complex in both location and function. The ability to identify all of the muscles listed here will depend on the degree to which the cadaver in the laboratory is dissected or the type(s) of models available in the laboratory. To simplify the process of learning these muscles, initially focus on learning how the groups of muscles are arranged from superficial to deep, and from medial to lateral. Then focus on specifically identifying the individual muscles belonging to each group.

The largest muscles of the back that move the vertebral column are collectively referred to as the **erector spinae** (*erector,* to make erect or straight, + *spina,* spine). The erector spinae consist of three muscle groups that form long columns along both sides of the vertebral column. The muscle groups, from medial to lateral, are the spinalis, longissimus, and iliocostalis. Deep to the erector spinae is the **transversospinal** group of muscles, so named because they attach to transverse and spinous processes of adjacent vertebrae, and several smaller muscles that create fine movements of the vertebral column: interspinales and intertransversarii. The deeper muscles can be difficult to identify on cadavers or models. In general, the **semispinalis** muscles are the most superficial of the transversospinal muscles, span 5–6 vertebrae, and are most highly developed in the cervical and upper thoracic regions of the vertebral column. The **multifidus** lie deep to the semispinalis muscles, span 3–4 vertebrae, and are most highly developed in the lumbar region of the vertebral column. Finally, the **rotatores** are the deepest muscles of the transversospinal group, span 1–2 vertebrae, and are most highly developed in the lower thoracic region of the vertebral column.

EXERCISE 12.6

MUSCLES OF THE VERTEBRAL COLUMN

1. Observe a prosected cadaver or a classroom model of the thorax/abdomen that demonstrates muscles of the vertebral column **(table 12.6).**
2. Identify the **muscles of the vertebral column** listed in **figure 12.6** on a cadaver or on models of the thorax and abdomen, using table 12.6 and the textbook as guides.
3. Label the muscles of the vertebral column in figure 12.6.

Table 12.6 Muscles of the Vertebral Column

Muscle Group	Individual Muscles	Origin	Insertion	Action	Innervation	Word Origin
Superficial Layer — Splenius Muscles						
Splenius Muscles	The splenius muscles are thick, flat muscles on the lateral and posterior aspect of the neck.	Midline	Cervical vertebrae and skull	Hold the deep neck muscles in position; extend, laterally flex, and laterally rotate the neck	Dorsal rami of spinal nerves	*splenius,* bandage

Table 12.6 Muscles of the Vertebral Column *(continued)*

Muscle Group	Individual Muscles	Origin	Insertion	Action	Innervation	Word Origin
REGIONS	Capitis (splenius capitis)	Ligamentum nuchae and T_1–T_6 vertebrae (spinous processes)	Superior nuchal line (lateral aspect)	Bilateral: extends the neck Unilateral: laterally rotates and laterally flexes the neck to the same side Bilateral: extends the neck	Dorsal rami of spinal nerves	*splenion*, a bandage, + *caput*, head
	Cervicis (splenius cervicis)	Nuchal ligament and C_7–T_4 (spinous processes)	C_1–C_3 vertebrae (posterior tubercles)	Unilateral: laterally rotates and laterally flexes the neck to the same side Bilateral: extends the neck	Dorsal rami of spinal nerves	*cervix*, neck
Intermediate Layer — Erector Spinae (Sacrospinalis) Muscles						
Erector Spinae	The erector spinae muscles compose the intermediate layer of back muscles. They are arranged into groups.	Broad tendon covering the posterior iliac crest, the lumbar vertebrae, and the sacrum	Vertebrae and ribs	Bilateral: extends the vertebral column and the head/neck Unilateral: laterally bends the vertebral column	Dorsal rami of spinal nerves	*erector*, to make erect, + *spina*, spine
GROUPS	Iliocostalis (lateral group)	Broad tendon covering the posterior iliac crest, the lumbar vertebrae, and the sacrum	Ribs (angles of lower ribs) and cervical vertebrae (transverse processes)	Extends the vertebral column	Dorsal rami of spinal nerves	*ilium*, groin, + *costal*, rib
	Longissimus (intermediate group)	Broad tendon covering the posterior iliac crest, the lumbar vertebrae, and the sacrum	Ribs (between tubercles and angles), cervical and thoracic vertebrae (transverse processes), and mastoid process of temporal bone	Extends the neck and vertebral column and laterally rotates the head	Dorsal rami of spinal nerves	*longissimus*, longest
	Spinalis (medial group)	Broad tendon covering the posterior iliac crest, the lumbar vertebrae, and the sacrum	Vertebrae (spinous processes of upper thoracic), and skull	Extends the neck and vertebral column and laterally rotates the head	Dorsal rami of spinal nerves	*spina*, spine
Spinal Flexors — Quadratus Lumborum						
	Quadratus lumborum	Iliac crest and transverse processes of lower lumbar vertebrae	Rib 12, transverse processes of upper lumbar vertebrae	Abducts the trunk	Ventral rami of lumbar spinal nerves	*quadratus*, square, + *lumbus*, loin

(continued on next page)

(continued from previous page)

Table 12.6 Muscles of the Vertebral Column (continued)

Muscle Group	Individual Muscles	Origin	Insertion	Action	Innervation	Word Origin
Deep Layer—Transversospinalis Muscles						
Transversospinal Group	The transversospinal muscles are the deepest muscles of the back and lie between transverse and spinous processes of adjacent vertebrae.	Transverse processes of inferior vertebrae	Spinous process of cervical and thoracic vertebrae 1–3 levels above the vertebra of origin and/or the posterior aspect of the occipital bone	Extends and rotates the vertebral column; stabilizes the vertebrae during local movements of the vertebral column	Dorsal rami of spinal nerves	*transverse*, across, + *spina*, spine
Multifidus	NA	T_1–T_3 vertebrae (transverse processes), C_4–C_7 (articular processes), ilium and sacrum	Spinous process of vertebra located 2–4 segments superior to vertebra of origin	Assists with local extension and rotation of the vertebral column	Dorsal rami of spinal nerves	*multus*, much, + *findo*, to cleave
Rotatores	NA	Transverse processes of all vertebrae (most developed in the thoracic region)	Vertebral arch (between lamina and transverse process) of vertebra superior to the vertebra of origin	Assists with local extension and rotation of the vertebral column	Dorsal rami of spinal nerves	*rotatus*, to rotate
Semispinalis Group	The semispinalis muscles are the deepest muscles of the back and lie between transverse and spinous processes of adjacent vertebrae.	Transverse processes of inferior vertebrae	Spinous process of vertebra above and/or the posterior aspect of the occipital bone	Extends and rotates the vertebral column; stabilizes the vertebrae during local movements of the vertebral column	Dorsal rami of spinal nerves	*semis*, half, + *spina*, spine
REGIONS	Capitis (semispinalis capitis)	Inferior cervical and superior thoracic vertebrae (spinous and transverse processes)	Occipital bone (between superior and inferior nuchal lines)	Extends and rotates the vertebral column; stabilizes the vertebrae during local movements of the vertebral column	Dorsal rami of spinal nerves	*caput*, head
	Cervicis (semispinalis cervicis)	T_1–T_6 vertebrae (transverse processes)	C_2–C_3 vertebrae (spinous processes)	Extends and rotates the vertebral column; stabilizes the vertebrae during local movements of the vertebral column	Dorsal rami of spinal nerves	*cervix*, neck
	Thoracis (semispinalis thoracis)	T_6–T_{10} vertebrae (transverse processes)	C_5–T_4 vertebrae (spinous processes)	Extends and rotates the vertebral column; stabilizes the vertebrae during local movements of the vertebral column	Dorsal rami of spinal nerves	*thoracis*, thorax

Table 12.6 Muscles of the Vertebral Column *(continued)*

Muscle Group	Individual Muscles	Origin	Insertion	Action	Innervation	Word Origin
Minor Deep Back Muscles						
	Interspinales	Cervical and lumbar vertebrae (superior surfaces of spinous processes)	Spinous process of the vertebra superior to the vertebra of origin (inferior surface)	Extends and rotates the vertebral column	Dorsal rami of spinal nerves	*inter*, between, + *spina*, spine
	Intertransversarii	Cervical and lumbar vertebrae (transverse processes)	Transverse processes of vertebra above or below vertebra of origin	Laterally flexes the vertebral column	Dorsal rami of spinal nerves	*inter*, between, + *transversarii*, relating to the transverse process

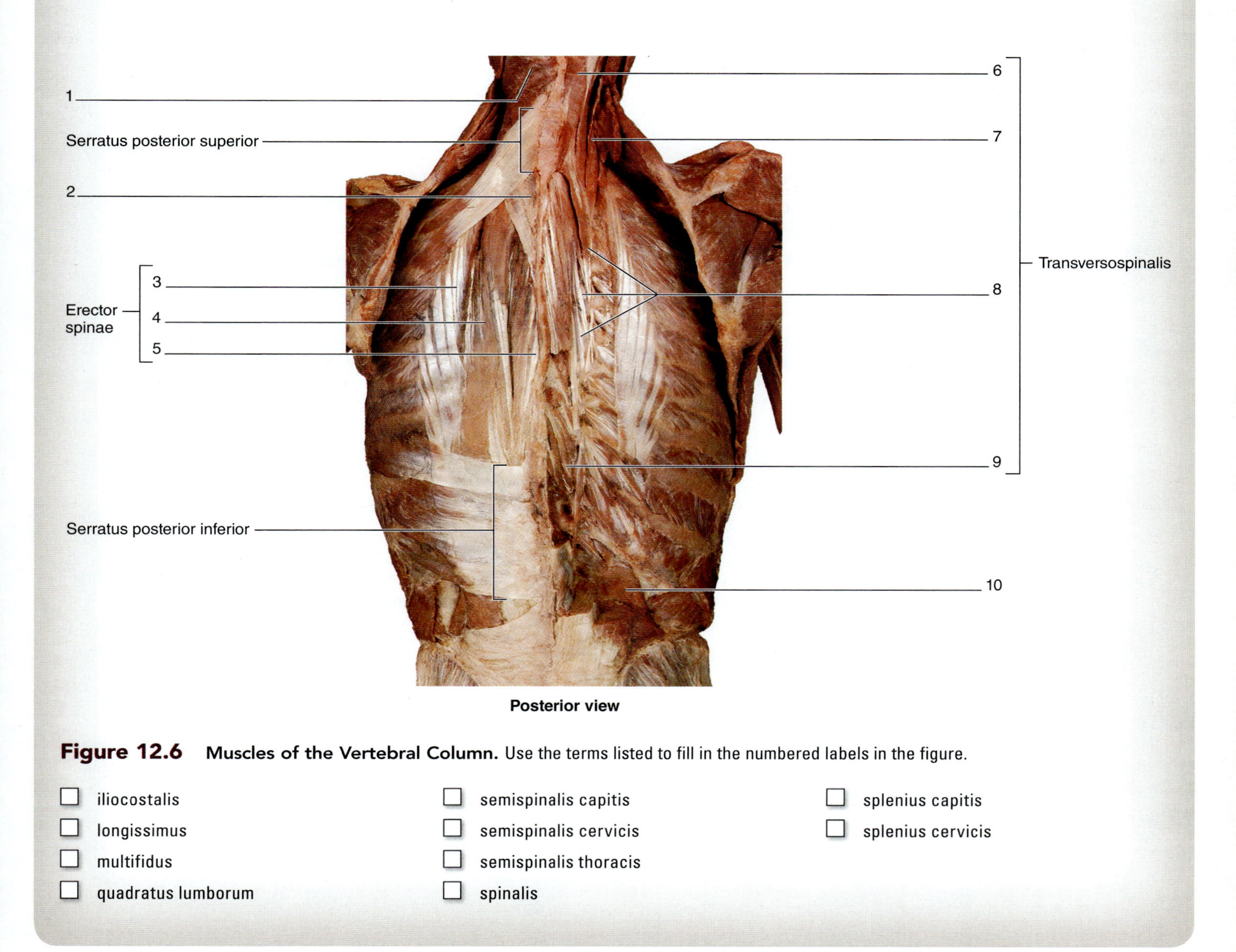

Figure 12.6 **Muscles of the Vertebral Column.** Use the terms listed to fill in the numbered labels in the figure.

- ☐ iliocostalis
- ☐ longissimus
- ☐ multifidus
- ☐ quadratus lumborum
- ☐ semispinalis capitis
- ☐ semispinalis cervicis
- ☐ semispinalis thoracis
- ☐ spinalis
- ☐ splenius capitis
- ☐ splenius cervicis

Muscles of Respiration

The diaphragm and muscles of the thoracic cage, which include the external intercostals, internal intercostals, the transversus thoracis, and the scalenes, are the primary **muscles of respiration (table 12.7).**

Table 12.7 Muscles of Respiration

Muscle	Origin	Insertion	Action	Innervation	Word Origin
Diaphragm	Inferior borders of rib 12, sternum, and the xiphoid process; costal cartilages of ribs 6–12, and lumbar vertebrae	Central tendon	Prime mover for inspiration; flattens when contracted, and increases intra-abdominal pressure and the size of the thoracic cavity	Phrenic nerves	*diaphragma*, a partition wall
External Intercostals	Inferior border of superior rib	Superior border of the rib below	Elevates the ribs	Intercostal nerves (continuations of thoracic spinal nerves)	*externus*, on the outside, + *inter*, between, + *costal*, rib
Internal Intercostals	Superior border of inferior rib	Inferior border of the rib above	Depresses the ribs	Intercostal nerves (continuations of thoracic spinal nerves)	*internus*, away from the surface, + *inter*, between, + *costal*, rib
Transversus Thoracis	Posterior surface of the lower half of the body of the sternum	Costal cartilages of ribs 2–6 (posterior surface)	Depresses the ribs	Intercostal nerves (continuations of thoracic spinal nerves)	*transversus*, crosswise, + *thoracis*, thorax
Anterior Scalene	C_3–C_6 (transverse processes)	First rib (scalene tubercle)	Elevates the first rib	Cervical plexus	*skalenos*, uneven
Middle Scalene	C_2–C_6 (transverse processes)	First rib (posterior to groove for subclavian artery)	Elevates the first rib	Cervical plexus	*skalenos*, uneven
Posterior Scalene	C_4–C_6 (transverse processes)	Second rib (lateral surface)	Elevates the second rib	Cervical and brachial plexuses	*skalenos, uneven*

INTEGRATE

CONCEPT CONNECTION

Pulmonary ventilation (*ventulus,* a breeze) is the process of moving air between the atomsphere and the lungs. The movement of air occurs because of pressure gradients between the lungs and the atomsphere. Pressures within the lungs change as the volume of the lungs changes (see Boyle's Law, chapter 24). There is an inverse relationship between pressure and volume. As the volume of the lungs increases, the pressure within the lungs decreases. When pressure within the lungs drops below atmospheric pressure, air flows into the lungs (inspiration). As the volume of the lungs decreases, the pressure within the lungs increases. When pressure within the lungs exceeds atomspheric pressure, air flows out of the lungs (expiration).

The changes in lung volume that occur during pulmonary ventilation are created by contractions of skeletal muscles. The **diaphragm** is the primary muscle of respiration because it causes the largest changes in volume of the thoracic cavity. Contraction of the diaphragm, external intercostal, and scalene muscles causes the volume of the thoracic cavity to increase, thereby causing inspiration. The natural recoil of the lungs and subsequent relaxation of these inspiratory muscles causes expiration. The internal intercostals and transversus thoracis muscles are used for forced expiration (e.g., sneezing or coughing). These muscles depress the ribs, which decreases the volume of the thoracic cavity.

EXERCISE 12.7

MUSCLES OF RESPIRATION

1. Observe a prosected cadaver or a classroom model of the thorax/abdomen that demonstrates muscles of the thoracic cage.
2. Identify the muscles of respiration listed in **figure 12.7,** using table 12.7 and the textbook as guides.
   a. ***Diaphragm***—The primary muscle involved in breathing is the diaphragm. This dome-shaped muscle forms a partition between the thoracic and abdominal cavities. The diaphragm contracts and moves inferiorly during inspiration. The diaphragm relaxes and moves superiorly during expiration. Remove the breastplate from the cadaver or model and observe the diaphragm. The structures that pass through the diaphragm are best seen from an inferior view. If possible, try to observe the diaphragm from both superior and inferior points of view. Observe the broad origin along the bones that constitutes the lower border of the thoracic cage. The muscle has a unique central attachment, the central tendon of the diaphragm.
   b. ***External Intercostals***—The majority of the muscle mass of the **external intercostals** is located on the posterior and lateral thorax, extending from the vertebral column to the **midclavicular line** (a vertical line that passes through the middle of the clavicle (see **figure 12.8*a***). The **external intercostal membrane** lies in place of the external intercostal muscles in the space between the midclavicular line and the sternum. Notice that the muscle fibers of the external intercostals are arranged obliquely, pointing in an inferomedial direction. The external intercostals elevate the ribs during inspiration.

      The external intercostal membrane can be identified on a cadaver because the connective tissue fibers parallel the direction of the muscle fibers of the external intercostals. If observing classroom models, this membrane is not identifiable.
   c. ***Internal Intercostals***—In contrast to the external intercostals, the majority of the muscle mass of the **internal intercostals** is located on the anterior surface of the thorax. These muscles extend from the sternum to the **scapular line** (a vertical line that passes through the inferior angle of the scapula; **figure 12.8*b***) on the posterior thorax. The **internal intercostal membrane** lies in place of the internal intercostal muscles on the posterior thorax between the scapular line and the vertebral column. The muscle fibers of the internal intercostals are arranged obliquely, pointing in an inferolateral direction, at right angles to the fibers of the external intercostals The internal intercostal muscles depress the ribs during forced expiration.
   d. ***Transversus thoracis***—If possible, remove the breastplate on the cadaver (or on a model of the thorax) and observe its interior surface. Lying adjacent to the inferior part of the sternum is the **transversus thoracis** muscle, which consists of several muscle bellies running obliquely. The transversus thoracis assists in depression of the ribs during forced expiration.
3. Label the muscles of respiration in figure 12.7.

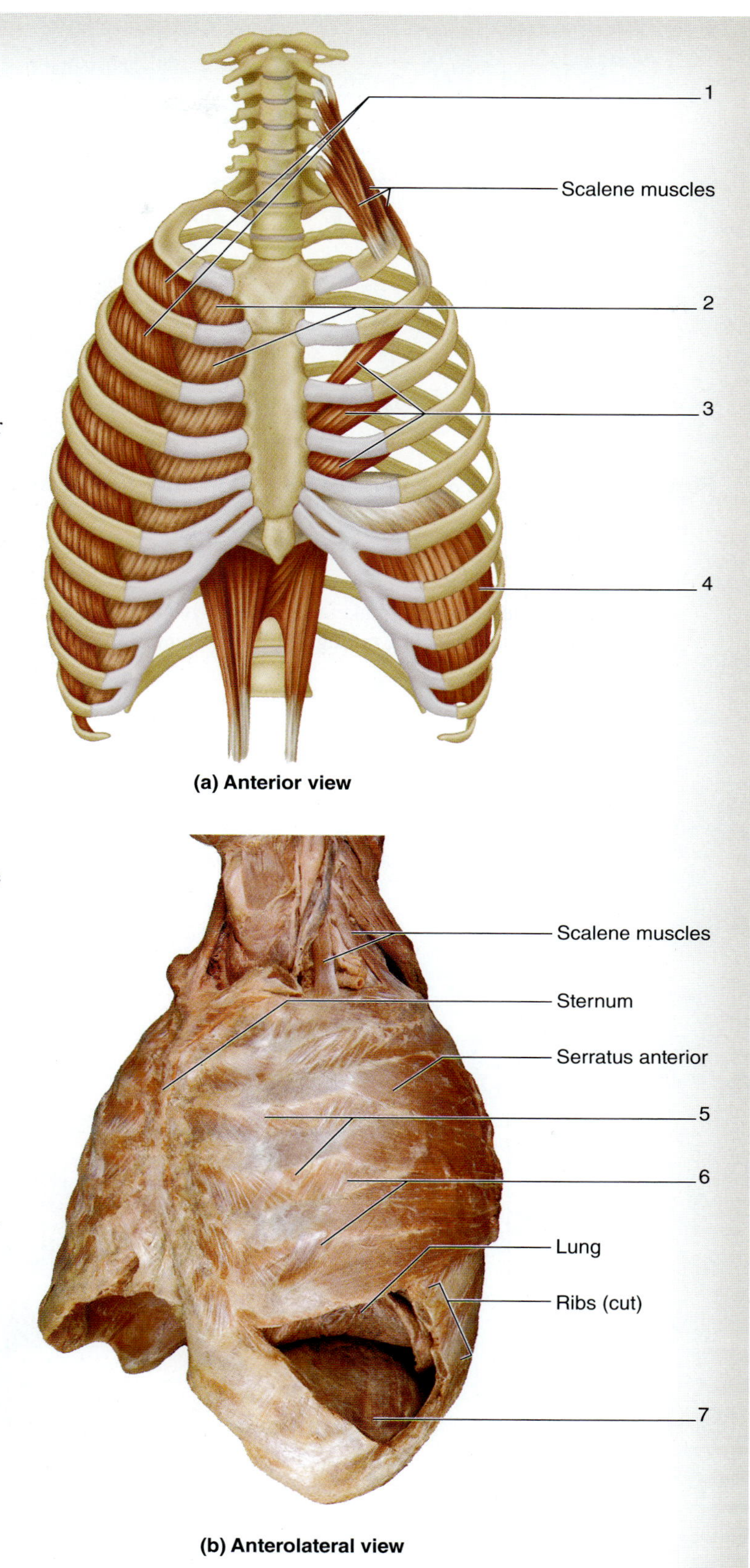

Figure 12.7 Muscles of Respiration. Use the terms listed to fill in the numbered labels in the figure. Some answers may be used more than once.

- ☐ diaphragm
- ☐ external intercostals
- ☐ internal intercostals
- ☐ transversus thoracis

(continued on next page)

(continued from previous page)

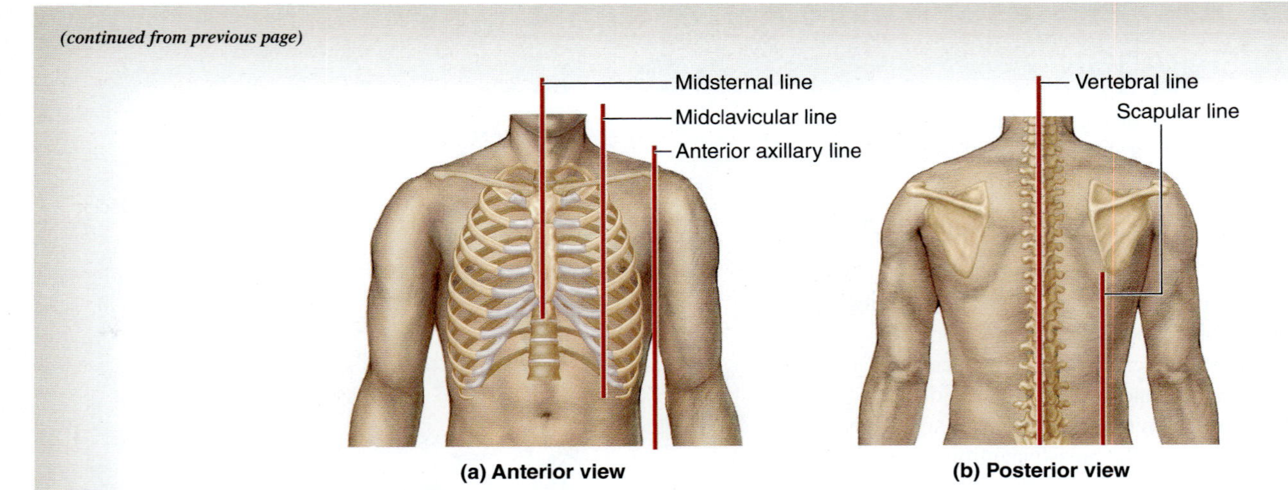

Figure 12.8 Points of Reference on the Anterior and Posterior Thorax.

4. In the space provided, make a representative drawing of the intercostal muscles, noting the fiber orientation for each of them.

5. ***Optional Activity***—Several important structures pass through the diaphragm, including the aorta, inferior vena cava, and esophagus. The aorta and inferior vena cava have passages through the central tendon, whereas the passage for the esophagus is surrounded by the muscle fibers of the diaphragm. Very practical consequences result from this arrangement. As the diaphragm contracts during inspiration, it changes from its dome-shape to a more flattened position. The muscle fibers of the diaphragm squeeze the esophagus and act as a sphincter to prevent stomach contents from being pushed back into the esophagus. In contrast, because the aorta and inferior vena cava pass through the central tendon, they are not constricted during inspiration. Label **figure 12.9,** using the textbook as a guide.

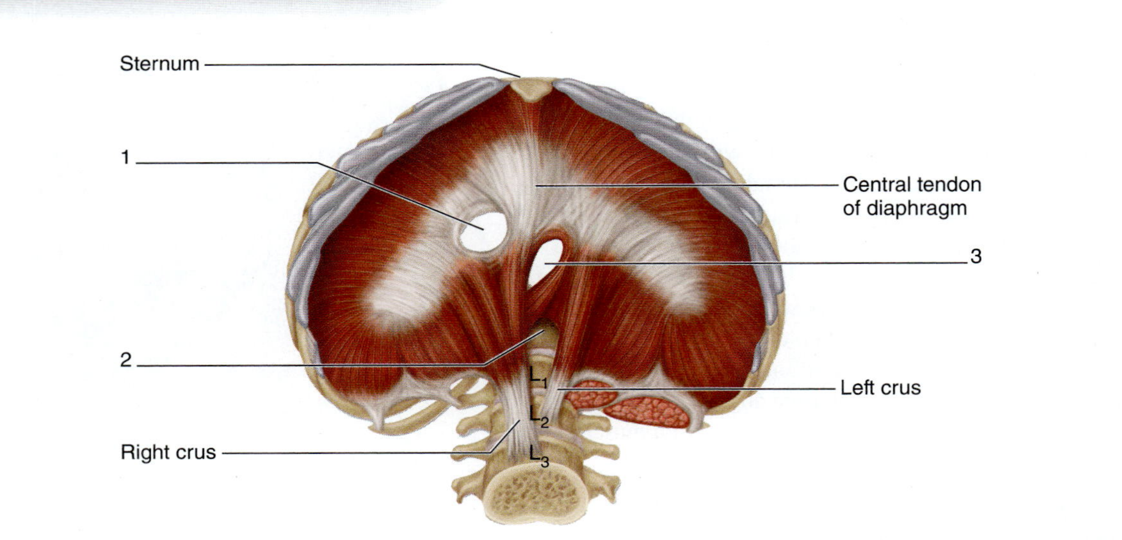

Figure 12.9 **Diaphragm.** Inferior View. Use the terms listed to fill in the numbered labels in the figure. Some answers may be used more than once.

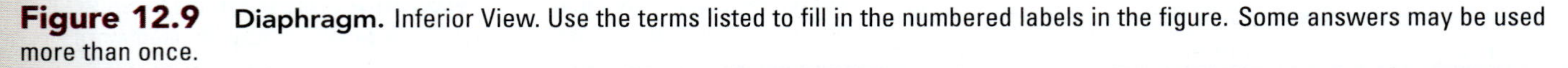

- ☐ aortic opening (hiatus)
- ☐ caval opening (for inferior vena cava)
- ☐ esophageal opening (hiatus)

Muscles of the Abdominal Wall

Muscles of the abdominal wall include four pairs of muscles that collectively compress and hold the abdominal organs in place. The abdominal muscles include the external obliques, internal obliques, transversus abdominis, and rectus abdominis. The muscle fiber orientation of the abdominal muscles parallels the muscle fiber orientation of the intercostal muscles. Like the external intercostals, the **external obliques** have muscle fibers that point in an inferomedial direction. Like the internal intercostals, the **internal obliques** have muscle fibers that point in an inferolateral direction. Deep to the internal obliques, the **transversus abdominis** has muscle fibers that run in a transverse, or horizontal, direction. All three of these muscles have broad, flat tendons called **aponeuroses** (*apo-*, from, + *neuron*, sinew). The aponeuroses of these muscles begin at the midclavicular line and extend to a central, tendon-like structure called the **linea alba**. In addition, the aponeuroses of these muscles form the rectus sheath, which surrounds the fourth abdominal muscle, the **rectus abdominis**.

EXERCISE 12.8

MUSCLES OF THE ABDOMINAL WALL

1. Observe a prosected cadaver or a model of the thorax/abdomen that demonstrates muscles of the abdominal wall (**table 12.8** and figures 12.10–12.12).
2. Identify the **abdominal muscles** listed in **figure 12.10** on a cadaver or on a classroom model of the abdomen, using table 12.8 and the textbook as guides.
3. Label the muscles of the abdominal wall in figure 12.10.

Table 12.8 Muscles of the Abdominal Wall

Muscle	Origin	Insertion	Action	Innervation	Word Origin
External Oblique	Anterior surface of inferior 8 ribs	Linea alba and the anterior iliac crest	Flexes and rotates the trunk; compresses the abdominal viscera	Spinal nerves T8–T12, L1	*externus*, on the outside, + *obliquus*, slanting
Internal Oblique	Thoracolumbar fascia, lateral half of inguinal ligament, and iliac crest	Linea alba, iliac crest, pubic tubercle, and the inferior border of last 4 ribs; costal cartilages of ribs 8–10	Flexes and rotates the trunk; compresses the abdominal viscera	Spinal nerves T8–T12, L1	*internus*, away from the surface, + *obliquus*, slanting
Rectus Abdominis	Pubic symphysis and crest	Xiphoid process and the costal cartilages of ribs 5–7	Flexes and rotates the trunk; compresses the abdominal viscera	Spinal nerves T7–T12	*rectus*, straight, + *abdominis*, the abdomen
Transversus Abdominis	Lateral third of inguinal ligament, iliac crest, costal cartilages of inferior 6 ribs	Linea alba and pubic crest	Compresses the abdominal viscera	Spinal nerves T8–T12, L1	*transversus*, crosswise, + *obliquus*, slanting
Structures Related to Abdominal Musculature					
Structure	**Attachment 1**	**Attachment 2**	**Description**	**Innervation**	**Word Origin**
Aponeurosis	NA	NA	A broad, flat tendon, such as those connecting abdominal muscles to the linea alba	NA	*apo-*, from, + *neuron*, sinew
Inguinal Canal	NA	NA	An oblique passage in the anterior abdominal wall located superior to the inguinal ligament; it is formed from the aponeuroses of the external and internal oblique muscles	NA	*inguen*, groin

(continued on next page)

(continued from previous page)

Table 12.8 Muscles of the Abdominal Wall *(continued)*

Structure	Attachment 1	Attachment 2	Description	Innervation	Word Origin
Inguinal Ligament	Anterior superior iliac spine	Pubic tubercle	A structure formed from the aponeurosis of the external oblique muscle; an important anatomical landmark in the inguinal region	NA	*inguen*, groin
Linea Alba	Xiphoid process of the sternum	Pubic symphysis	Literally, the "white line"; a tendinous structure that acts as the insertion point for the oblique and transversus abdominis muscles	NA	*linea*, line, + *alba*, white
Rectus Sheath	NA	NA	A connective tissue sheath that surrounds the rectus abdominis muscle and is formed from the aponeuroses of the external oblique, internal oblique, and transversus abdominis muscles	NA	*rectus*, referring to the rectus abdominis muscle
Tendinous Intersections	NA	NA	Tendinous bands that run across a muscle. In this case, the tendinous intersections are the structures that separate the parts of the rectus abdominis muscle.	NA	*tendo-*, to stretch out, + *inscriptio*, to write on

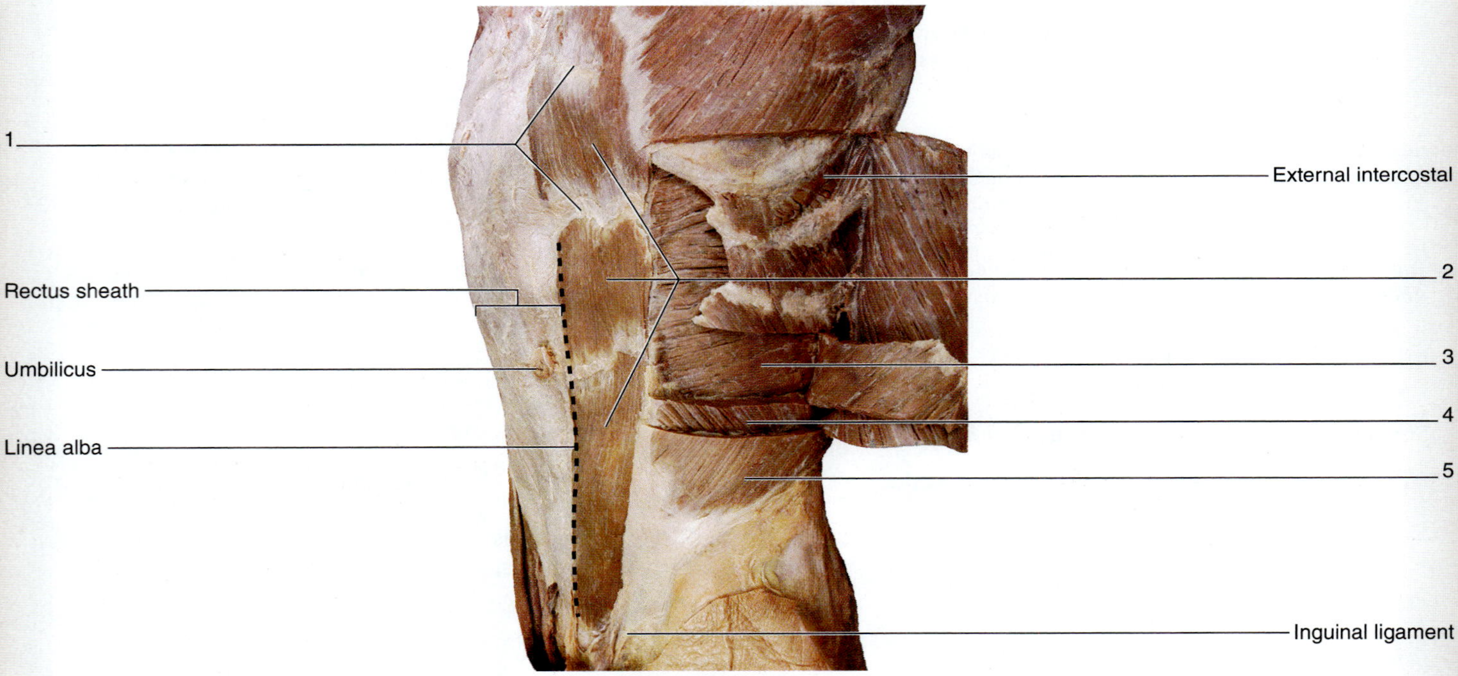

Figure 12.10 **Muscles of the Abdominal Wall.** Use the terms listed to fill in the numbered labels in the figure.

- ☐ external oblique
- ☐ internal oblique
- ☐ rectus abdominis
- ☐ tendinous intersections
- ☐ transversus abdominis

The Rectus Sheath, Inguinal Ligament, and Inguinal Canal

The rectus sheath, inguinal ligament, and inguinal canal are important structures associated with the abdominal musculature.

EXERCISE 12.9

THE RECTUS SHEATH, INGUINAL LIGAMENT, AND INGUINAL CANAL

1. Observe a prosected cadaver or a classroom model of the thorax/abdomen that demonstrates muscles of the abdominal wall. Identify the structures listed in **figure 12.11** on a cadaver or on a classroom model, using table 12.8 and your textbook as guides.
2. Identify these structures associated with abdominal muscles shown in figures 12.11 and 12.12 on a cadaver or on a classroom model of the abdomen.
   - **a.** ***Rectus Sheath***—The **rectus sheath** (figure 12.11; see table 12.8) is a structure formed from the aponeuroses of the external obliques, the internal obliques, and the transversus abdominis muscles (this relationship is accurate only for the sheath superior to the umbilicus, so observations should be made in that location). The rectus abdominis muscle is located within the rectus sheath, and is divided into four sections by connective tissue partitions called **tendinous intersections.** The tendinous intersections effectively divide one very long muscle into four smaller muscles, arranged in series. This allows the muscle as a whole to make greater overall change in length, in addition to increasing its force of contraction.
   - **b.** ***Inguinal Ligament***—The **inguinal ligament** (**figure 12.12;** table 12.8) is formed by the *aponeurosis of the external oblique* muscle. This ligament extends from the anterior superior iliac spine (ASIS) laterally, to the pubic tubercle medially. Instead of being a straight ligament, the ligament folds back upon itself, forming a trough. The inguinal ligament is an important landmark of the abdomen and thigh. In addition, the trough formed by the aponeurosis of the external oblique forms part of the *inguinal canal.*
   - **c.** ***Inguinal Canal***— **The inguinal canal** (figure 12.12; table 12.8) is a tube-like passageway in the inferior abdominal wall. Its floor is formed by the trough of the **aponeurosis of the external oblique.** Its roof is formed by fibers of the **aponeurosis of the internal oblique.** Within this canal, structures pass from the abdomen into the subcutaneous tissues of the **perineum** (*perneon*, the area between the thighs below the pelvic diaphragm). In males, structures that pass through compose the **spermatic cord,** whereas in females, the **round ligament of the uterus** passes through. The **superficial inguinal ring** is located lateral to the pubic symphysis. On a male cadaver or on a classroom model of the abdomen, identify the spermatic cord as it passes through the superficial inguinal ring. Attempt to identify the round ligament of the uterus on a female cadaver, keeping in mind that identification of the round ligament of the uterus is often quite difficult.

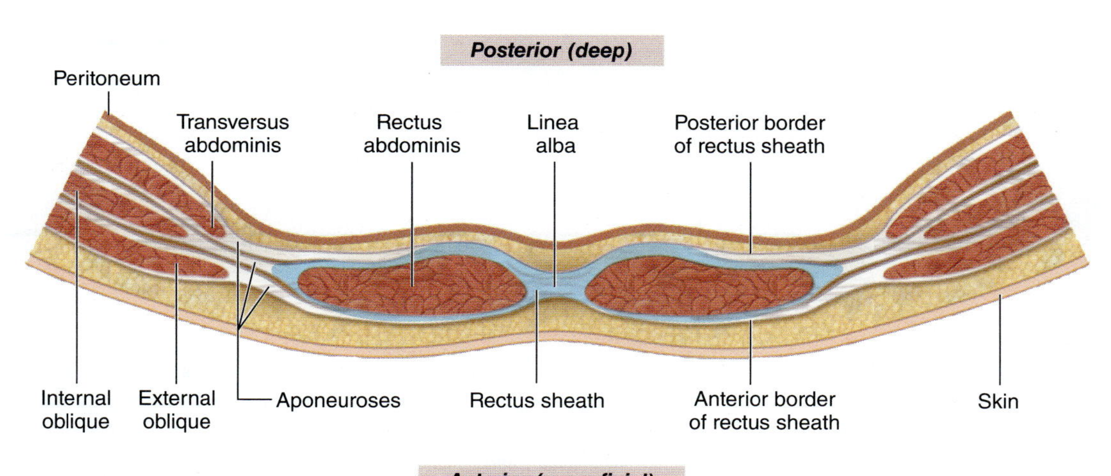

Figure 12.11 **The Rectus Sheath.** A transverse section through the abdominal wall demonstrates components of the rectus sheath.

(continued on next page)

(continued from previous page)

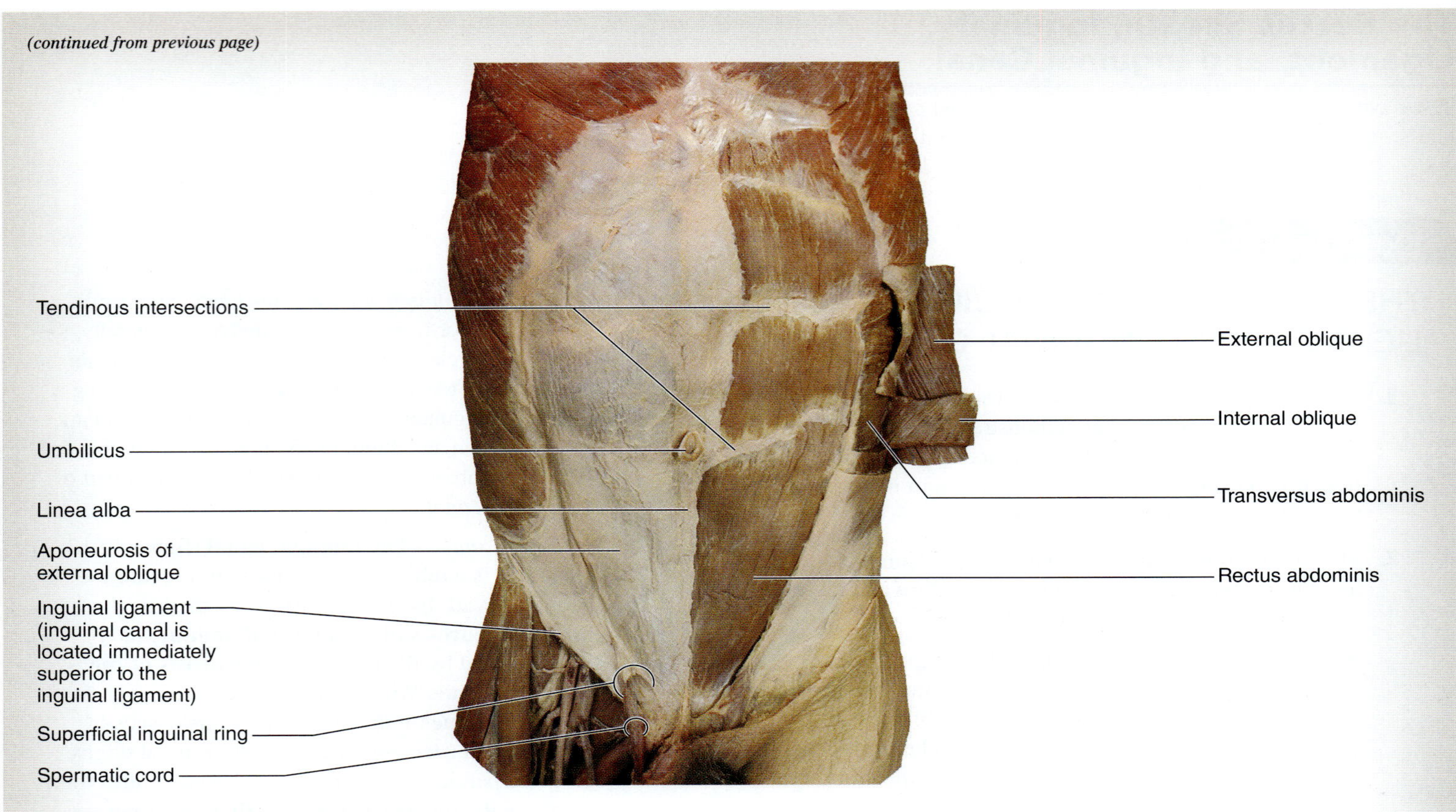

Figure 12.12 The Anterior Abdominal Wall.

INTEGRATE

LEARNING STRATEGY

When learning the external and internal oblique muscles, it is helpful to note the orientation of the fibers. A learning aid to help with this involves placing your thumbs inside the belt loops on the front of a pair of jeans while resting your hands on your hips. The position of the thumbs (inside the belt loops) represents the orientation of the muscle fibers of the *internal* obliques. That is, muscle fibers within the *internal* oblique muscles are directed toward the midline and point superiorly. The position of the fingers (outside the belt loops) represents the orientation of the muscle fibers of the *external* obliques. That is, muscle fibers within the external obliques are directed toward the midline and point inferiorly. This same convention can be used for external and internal intercostal because the direction of muscle fibers of the internal intercostals is the same as that of the internal obliques and vice versa.

INTEGRATE

CLINICAL VIEW
Athletic Pubalgia

Athletic pubalgia (*pubis,* groin + *algia,* pain) (AP), previously referred to as a "sports hernia," is a syndrome affecting many athletes, and is particularly common among hockey players due either to overuse (i.e., repetitive turning and twisting) or direct injury to the groin region. Symptoms include pain in the groin, medial thigh, lower abdomen, or pubic region. Causes of AP are varied, and include tendon or fascial damage in the specified regions, or nerve entrapment of the ilioinguinal or genitofemoral nerves, which traverse the inguinal canal as they travel to the genitals and medial thigh. Pain may be reproduced in these patients by having them contract muscles composing the abdominal wall (e.g., have the patient perform abdominal crunches). While AP is difficult to diagnose, scientists are coming up with promising new clinical diagnostic tools, such as magnetic resonance imaging (MRI), to both improve diagnosis and monitor treatment of AP.

Chapter 12: The Muscular System: Axial Muscles

Name: ____________________

Date: __________ Section: __________

POST-LABORATORY WORKSHEET

The ❶ corresponds to the Learning Objective(s) listed in the chapter opener outline.

Do You Know the Basics?

Exercise 12.1: Muscles of Facial Expression

1. Match the movements listed in column A with the corresponding muscle that causes the movement, listed in column B. ❶

Column A	*Column B*
____ 1. blink or close eyes	a. buccinator
____ 2. close mouth	b. epicranius
____ 3. compress cheeks	c. orbicularis oculi
____ 4. tense skin of neck	d. orbicularis oris
____ 5. wrinkle forehead, raise eyebrows	e. platysma

2. Muscles of facial expression are innervated by which of the following cranial nerves? (Circle one.) ❶
   a. abducens (CN VI)
   b. facial (CN VII)
   c. hypoglossal (CN XII)
   d. trigeminal (CN V)
   e. trochlear (CN IV)

Exercise 12.2: Muscles of Mastication

3. Match the movements listed in column A with the corresponding muscle of mastication listed in column B. ❷

Column A	*Column B*
____ 1. depresses the chin	a. lateral pterygoid
____ 2. elevate mandible (3 answers)	b. masseter
____ 3. grind food side to side (2 answers)	c. medial pterygoid
____ 4. protract mandible, depress chin (3 answers)	d. temporalis

4. Which of the following is classified as a muscle of mastication? (Circle one.) ❷
   a. hypoglossus
   b. masseter
   c. platysma
   d. risorius
   e. zygomaticus major

Exercise 12.3: Muscles That Move the Tongue

5. Extrinsic muscles of the tongue connect the tongue to bony structures of the head and neck and cause fine movements of the tongue. ____________________ (True/False). ❸

6. The genioglossus muscle is a(n) ____________________ (intrinsic/extrinsic) muscle of the tongue. ❹

Exercise 12.4: Muscles of the Pharynx

7. Which of the following muscles acts to narrow the diameter of the pharynx during swallowing? (Circle one) ❺
   a. laryngeal elevators
   b. palate muscles
   c. pharyngeal constrictors
   d. pharyngeal sphincters

Exercise 12.5: Muscles of the Neck

8. Match the muscle group listed in column A with the appropriate muscles listed in column B. Answers may be used more than once. 6

Column A	*Column B*
____ 1. infrahyoid muscle	a. longissimus capitis
____ 2. muscle that moves the neck	b. mylohyoid
____ 3. suprahyoid muscle (2 answers)	c. omohyoid
	d. stylohyoid

9. The sternocleidomastoid and splenius muscles act as ____________________ (antagonists/synergists) for lateral rotation of the neck. 7

10. The sternocleidomastoid and splenius muscles act as ____________________ (antagonists/synergists) for neck extension and flexion. 7

Exercise 12.6: Muscles of the Vertebral Column

11. The erector spinae is a collective term for ____________________ (posterior/anterior) groups of muscles that function to straighten the spine. 8

12. Identify the three muscle groups that compose the erector spinae. (Check all that apply.) 8

____ a. iliocostalis

____ b. longissimus

____ c. semispinalis

____ d. spinalis

____ e. transversospinal

Exercise 12.7: Muscles of Respiration

13 There are two sets of intercostal muscles (muscles between the ribs). The external intercostals are ____________________ (deep/superficial) to the internal intercostals. 9

14 Match the description listed in column A with the appropriate muscle listed in column B. 9

Column A	*Column B*
____ 1. compresses the abdomen and assists with forced expiration	a. diaphragm
____ 2. depresses the ribs and assists with forced expiration	b. external intercostals
____ 3. elevates the ribs and assists with inspiration	c. internal intercostals
____ 4. primary muscle used for breathing	d. transversus thoracis

15. Identify the major structures that pass through the diaphragm. (Check all that apply.) 10

____ a. aorta

____ b. esophagus

____ c. hepatic vein

____ d. inferior vena cava

____ e. superior vena cava

Exercise 12.8: Muscles of the Abdominal Wall

16. Internal oblique abdominal muscles are ____________________ (deep/superficial) to the external oblique abdominal muscles. 11

17. Match the direction of the muscle fibers listed in column A with the corresponding muscle of the abdominal wall listed in column B. 11

Column A	*Column B*
_____ 1. horizontal	a. external obliques
_____ 2. inferolateral (down and away from middle)	b. internal obliques
_____ 3. inferomedial (down and toward middle)	c. rectus abdominis
_____ 4. vertical	d. transversus abdominis

18. The inguinal ligament is formed from the aponeurosis of which abdominal muscle(s)? (Circle one.) 12
    a. external obliques
    b. internal obliques
    c. rectus abdominis
    d. transversus abdominis

Exercise 12.9: The Rectus Sheath, Inguinal Ligament, and Inguinal Canal

19. The rectus sheath comes together at the midline to form a vertical fibrous strip called the linea alba, which sometimes ____________________ (darkens/lightens) during pregnancy. 13

20. What structures run through the inguinal canal? 14

    Males: ____________________

    Females: ____________________

Can You Apply What You've Learned?

21. If the facial nerve is severed, which of the following muscles will *not* be paralyzed? (Check all that apply.)

    ____ a. buccinator

    ____ b. epicranius

    ____ c. frontalis

    ____ d. masseter

    ____ e. zygomaticus

22. Explain the major difference in muscular movement produced by the masseter and the temporalis during mastication (chewing). ____________________

23. Explain how the suprahyoid and infrahyoid muscles are named. ____________________

24. a. Describe the action of the sternocleidomastoid if both heads are contracted simultaneously. ____________________

    b. Describe the action of the sternocleidomastoid if only the right clavicular head contracts. ____________________

25. Explain how the external obliques, internal obliques, and transversus abdominis muscles relate to the rectus sheath.

Can You Synthesize What You've Learned?

26. In an **indirect inguinal hernia**, abdominal contents, such as a loop of small intestine, pass into the inguinal canal. Why are such hernias more common in males than in females?

27. In people with gastroesophageal reflux disease (GERD), acidic stomach contents enter into the lower part of the esophagus, thus damaging the lining of the esophagus. GERD is often caused by weakness in part of the diaphragm. Using knowledge of the diaphragm and the locations where important structures pass through, hypothesize a mechanism by which a weak diaphragm could contribute to the development of GERD.

CHAPTER 13

The Muscular System: Appendicular Muscles

OUTLINE AND LEARNING OBJECTIVES

MODULE 6: MUSCULAR SYSTEM

INTRODUCTION

The **appendicular muscles** are familiar muscles to most people. Anyone who has gone to a gym and lifted weights, or who has admired the muscular body of a basketball player or gymnast, has at least some familiarity with these muscles. They are the muscles used to perform active tasks such as typing, walking, running, and lifting. They are the bulging muscles seen in the arms and legs of elite athletes. When lifting a heavy weight, these muscles are visible in our upper limbs, so it is fairly easy to determine which muscles are used to perform the particular action. If this is not immediately clear from the fatigue or pain experienced immediately in the working muscle, it is definitely clear in the next two days as delayed-onset muscle soreness (DOMS) sets in.

All of these activities give us a fascination and curiosity about the muscles responsible for causing movement. When working through the task of learning names, attachments, and actions for the appendicular muscles, try to identify the muscles on your own body. Practice using them. An excellent way to do this is to go to a gym that has weight machines. Observe the illustrations on the weight machines that demonstrate what muscle or muscles the exercise is meant to work on. Then practice the movement *without using any weights*. When performing the stated action of the muscle(s), feel the muscle(s) produce tension under your skin. Once the connection is made between the muscles in your own body and the muscles seen on the cadaver, on models, or in photographs, you will begin to truly appreciate their functions.

The exercises in this chapter involve identifying, naming, and exploring the structure and function of muscles that move the appendicular skeleton on a model, photo, or cadaver. Prior to beginning, find out from the laboratory instructor which muscles will be assessed on practical exams. Then, place check marks in the summary tables of this chapter so as to focus on those muscles that are required for the course. In addition, prior to beginning the more detailed study of the appendicular musculature in this chapter, be sure to complete the exercises related to appendicular musculature in chapter 11. Those exercises introduced the major muscle groups and the common actions of those groups. Refer to tables 11.4 and 11.5 to review the muscle compartments of the upper and lower limbs, and the major actions of the muscles in each compartment.

List of Reference Tables

Chapter 13: The Muscular System: Appendicular Muscles

Name: ______________________

Date: __________ Section: __________

These Pre-Laboratory Worksheet questions may be assigned by instructors through their connect course.

PRE-LABORATORY WORKSHEET

1. Match the action(s) described in column A with the compartment listed in column B. Actions are those performed by the majority of muscles within the listed compartment.

Column A	***Column B***
____ 1. extension of arm and forearm	a. anterior compartment of the arm
____ 2. extension of the wrist and digits	b. anterior compartment of the forearm
____ 3. flexion of arm and forearm	c. posterior compartment of the arm
____ 4. flexion of the wrist and digits	d. posterior compartment of the forearm

2. Match the action(s) described in column A with the compartment listed in column B. Actions are those performed by the majority of muscles within the listed compartment.

Column A	***Column B***
____ 1. eversion of the ankle	a. anterior compartment of the leg
____ 2. extension of the leg, flexion of the thigh	b anterior compartment of the thigh
____ 3. extension of the thigh, flexion of the leg	c. lateral compartment of the leg
____ 4. dorsiflexion and inversion of the ankle, extension of the toes	d. medial compartment of the thigh
____ 5. adduction of the thigh	e. posterior compartment of the leg
____ 6. plantar flexion and inversion of the ankle, flexion of the toes	f. posterior compartment of the thigh

3. Which of the following terms describes a muscle that is primarily responsible for causing a given action about a joint? (Circle one.)

   a. agonist b. antagonist c. protagonist d. synergist

4. The upper limb includes the arm, forearm, and wrist. ____________________ (True/False)

5. The portion of the lower limb from the hip to the knee is the ____________________ (leg/thigh).

6. The portion of the lower limb from the knee to the ankle is the ____________________ (leg/thigh).

7. Which of the following muscles compose the rotator cuff? (Check all that apply.)

   ____ a. infraspinatus
   ____ b. subscapularis
   ____ c. supraspinatus
   ____ d. teres major
   ____ e. teres minor

8. Which of the following muscles compose the hamstring muscles? (Check all that apply.)

   ____ a. biceps femoris
   ____ b. gracilis
   ____ c. rectus femoris
   ____ d. semimembranosus
   ____ e. semitendinosus

9. Identify the most superficial muscle of the chest. (Circle one.)

   a. deltoid b. pectoralis major c. pectoralis minor d. rhomboid major e. trapezius

10. Which of the following muscles moves both the pectoral girdle and the glenohumeral joint? (Circle one.)

   a. coracobrachialis b. latissimus dorsi c. levator scapulae d. pectoralis minor

GROSS ANATOMY

Muscles That Act About the Pectoral Girdle/Glenohumeral Joint

Recall from chapter 9 that the pectoral girdle consists of the clavicle and scapula, and that there is only one bony attachment between the clavicle and the sternum (sternoclavicular joint). This anatomic arrangement requires very strong muscular attachments between the pectoral girdle and the axial skeleton. The muscles that act about the pectoral girdle both stabilize the scapula and move the pectoral girdle, and are classified as either anterior or posterior thoracic muscles **(table 13.1).**

The glenohumeral joint (or shoulder joint) consists of the head of the humerus articulating with the relatively shallow glenoid cavity of the scapula. This anatomic arrangement allows for a great deal of flexibility, but requires numerous muscular attachments. The muscles acting about the glenohumeral joint both stabilize and create movement of this joint and are classified as muscles originating on the axial skeleton and muscles originating on the scapula **(table 13.2).**

Table 13.1 Muscles That Act About the Pectoral Girdle

Muscle	Origin	Insertion	Action*	Innervation	Word Origin
Anterior Muscles					
Pectoralis Minor	Ribs 3–5 (anterior surface)	Coracoid process of scapula	Protracts and depresses the scapula; elevates the ribcage	Medial pectoral nerve	*pectus*, chest, + *minor*, smaller
Serratus Anterior	Ribs 1–8 (outer surface)	Scapula (medial border)	Protracts the scapula and rotates it superiorly; most important for holding the scapula flat against the rib cage; damage causes a "winging" of the scapula	Long thoracic nerve	*serratus*, a saw, + *anterior*, the front surface
Posterior Muscles					
Trapezius	Occipital bone, ligamentum nuchae, spinous processes of C_7–T_{12}	Clavicle (lateral third) and scapula (acromial process and spine)	Superior fibers: Elevate and superiorly rotate scapula Middle fibers: Retract scapula Inferior fibers: Depress scapula	Accessory nerve (CN XI)	*trapeza*, a table
Levator Scapulae	Transverse processes of C_1–C_4	Scapula (superior vertebral border)	Elevates the scapula and tilts the glenoid inferiorly	Dorsal scapular nerve	*levatus*, to lift, + *scapula*, the shoulder blade
Rhomboid Minor	Spines of C_7–T_1	Scapula (medial border)	Retracts, adducts, and stabilizes the scapula; tilts the glenoid inferiorly	Dorsal scapular nerve	*rhomboid*, resembling an oblique parallelogram, + *minor*, smaller
Rhomboid Major	Spines of T_2–T_5	Scapula (medial border)	Retracts, adducts, and stabilizes the scapula; tilts the glenoid inferiorly	Dorsal scapular nerve	*rhomboid*, resembling an oblique parallelogram, + *major*, larger

*Only actions that apply to movement of the pectoral girdle are listed.

EXERCISE 13.1

MUSCLES THAT ACT ABOUT THE PECTORAL GIRDLE/GLENOHUMERAL JOINT

EXERCISE 13.1A Muscles That Act About the Pectoral Girdle

1. Observe a prosected human cadaver or a classroom model demonstrating muscles of the thorax and upper limb.
2. Identify the **muscles that act about the pectoral girdle** listed in **figure 13.1** on the cadaver or on a classroom model, using table 13.1 and the textbook as guides. Then label them in figure 13.1.
   a. *Anterior Muscles*—The **pectoralis minor** is a small muscle that can have multiple actions, depending upon what other muscles of the thorax and shoulder are contracting at the same time. If the scapula is fixed, the pectoralis minor assists in elevation of the rib cage (as in inspiration). If the scapula is not fixed, the pectoralis minor acts to pull the scapula anteriorly. The **serratus anterior** is a large, flat, fan-shaped muscle positioned between the ribs and the scapula; it is named based on the saw-toothed (serrated) appearance of its origin on the ribs. This muscle helps to both stabilize and move the scapula.
   b. *Posterior Muscles*—The largest posterior muscle that moves the pectoral girdle is the superficial **trapezius** muscle, which acts to anchor the scapula to the entire

superior two-thirds of the vertebral column and to the back of the head. As noted in table 13.1, this muscle performs multiple actions on the scapula, depending upon which part of the muscle contracts at a given time. Deep to the trapezius are smaller, more numerous muscles that act as **synergists** of the trapezius muscle. These include the **rhomboids** (major and minor) and the **levator scapulae.** When identifying these muscles on the cadaver or models, think about the actions they share with the trapezius. Practice performing the actions listed in table 13.1 to correlate the actions with the location and fiber orientation of the muscles.

EXERCISE 13.1B Muscles That Act About the Glenohumeral Joint

1. Observe a prosected human cadaver or a classroom model demonstrating muscles of the thorax and upper limb.
2. Identify muscles that act about the glenohumeral joint listed in figure 13.1 on the cadaver or on a classroom model, using table 13.2 and the textbook as guides. Then label them in figure 13.2.
   a. *Muscles Originating on Axial Skeleton*—The posteriorly located **latissimus dorsi** is a broad, triangular muscle located on the inferior part of the back originating on the axial skeleton and inserting on the humerus. It is often called the "swimmer's muscle" because its actions are required for many swimming strokes. Contraction of this muscle extends, adducts, and medially rotates the arm. The anteriorly located **pectoralis major** is the large, thick, fan-shaped muscle that covers the superior part of the chest, and acts to flex and adduct the arm.
   b. *Muscles Originating on the Scapula*—Numerous muscles originate on the scapula and insert on the humerus (table 13.2). One important group of muscles is the rotator cuff muscles. The rotator cuff is a musculotendinous cuff formed by four muscles that impart strength and stability to the glenohumeral joint. These muscles include the **subscapularis** (located in the subscapular fossa) and the **supraspinatus, infraspinatus,** and **teres minor**. More than any other factor, the strength of these muscles determines how stable the joint is. Weaknesses in these muscles contribute to many musculoskeletal problems in the glenohumeral joint. Rotator cuff injuries are common in baseball players and in older adults who fall on an outstretched arm. The term *rotator cuff* comes from the fact that these muscles act to medially or laterally rotate the humerus, in addition to stabilizing the glenohumeral joint.
3. Make a representative drawing of the three rotator cuff muscles (located on the external surface of the scapula) that are seen in a posterior view in the space provided.

INTEGRATE

LEARNING STRATEGY

The following mnemonic is helpful for remembering the names of the rotator cuff muscles. "A baseball pitcher who tears his rotator cuff **SITS** out the season." (**SITS: S** = supraspinatus, **I** = infraspinatus, **T** = teres minor, **S** = subscapularis.) To remember it is the teres **minor** (not the teres major) that forms part of the rotator cuff, know that this injured pitcher will then be relegated to the **minor** leagues.

INTEGRATE

CLINICAL VIEW

Winged Scapula

One of the most important muscles for stabilization of the scapula is the serratus anterior. This muscle's main action is to prevent the scapula from pulling away from the rib cage ("winging"). To test the action of the serratus anterior muscle, stand and face a wall. With your arms held straight out horizontally in front of you, place your palms flat against the wall in front of you and then push against the wall. If the serratus anterior is working properly, the scapula will remain flat against your rib cage. If not, the scapula will "wing" and its medial border will be pushed away from your rib cage.

Damage to the serratus anterior or the long thoracic nerve may lead to a winged scapula. Due to the action of the muscle, the person suffering from a winged scapula may also be unable to lift his/her arm above his/her head. Physical therapy is prescribed for those with some recoverable muscle function. Surgical interventions may be required for those with more serious injuries.

(continued on next page)

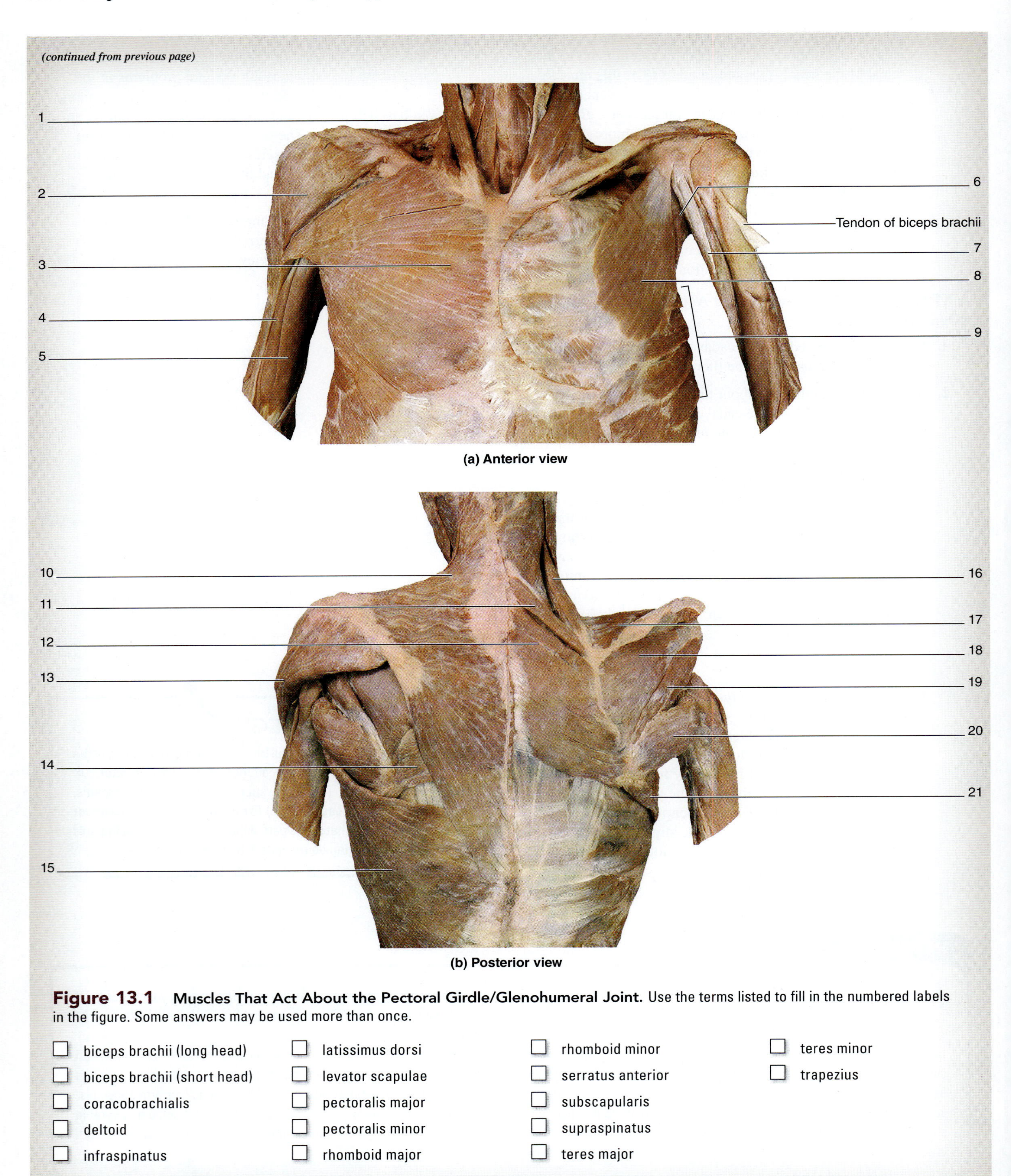

Figure 13.1 Muscles That Act About the Pectoral Girdle/Glenohumeral Joint. Use the terms listed to fill in the numbered labels in the figure. Some answers may be used more than once.

- ☐ biceps brachii (long head)
- ☐ biceps brachii (short head)
- ☐ coracobrachialis
- ☐ deltoid
- ☐ infraspinatus
- ☐ latissimus dorsi
- ☐ levator scapulae
- ☐ pectoralis major
- ☐ pectoralis minor
- ☐ rhomboid major
- ☐ rhomboid minor
- ☐ serratus anterior
- ☐ subscapularis
- ☐ supraspinatus
- ☐ teres major
- ☐ teres minor
- ☐ trapezius

Table 13.2 Muscles That Act About the Glenohumeral Joint

Muscle	Origin	Insertion	Action*	Innervation	Word Origin
Muscles Originating on Axial Skeleton					
Latissimus Dorsi	Spinous process of T_6–L_5, iliac crest, and ribs 10–12	Humerus (intertubercular groove)	Extends, adducts, medially rotates arm	Thoracodorsal nerve	*latissimus*, widest, + *dorsum*, the back
Pectoralis Major	Sternum, medial clavicle, and costal cartilages 2–6	Lateral part of intertubercular groove of humerus	Flexes and adducts arm	Medial pectoral nerve	*pectus*, chest, + *major*, larger
Muscles Originating on Scapula					
Pectoralis Minor	Anterior surfaces of ribs 3–5	Coracoid process of scapula	Flexes, adducts, medially rotates arm	Lateral pectoral nerve	*pectus*, chest, + *minor*, smaller
Deltoid	Clavicle (lateral third), acromion and spine of scapula	Humerus (deltoid tuberosity)	Anterior fibers: Flex and medially rotate arm Middle fibers: Prime movers of arm abduction Posterior fibers: Extend and laterally rotate arm	Axillary nerve	*deltoid*, resembling the Greek letter delta (a triangle)
Coracobrachialis	Coracoid process of scapula	Midhumerus (medial surface)	Flexes and adducts arm	Musculocutaneous nerve	*coraco*, referring to the coracoid process, + *brachium*, arm
Teres Major	Posterior surface of scapula at inferior angle	Crest of lesser tubercle on anterior humerus	Extends, adducts, medially rotates arm	Lower subscapular nerve	*teres*, round, + *major*, larger
Triceps Brachii (Long Head)	Infraglenoid tubercle of scapula	Olecranon process of ulna	Extends, adducts arm	Radial nerve	*tri*, three, + *caput*, head, + *brachium*, arm
Biceps Brachii (Long Head)	Supraglenoid tubercle	Radial tuberosity	Flexes arm (weak)	Musculocutaneous nerve	*bi*, two, + *caput*, head, + *brachium*, arm
Biceps Brachii (Short Head)	Coracoid process of scapula	Radial tuberosity	Flexes arm (weak)	Musculocutaneous nerve	*bi*, two, + *caput*, head, + *brachium*, arm
Rotator Cuff Muscles					
Subscapularis	Subscapular fossa	Lesser tubercle of humerus	Medially rotates arm	Upper and lower scapular nerves	*sub*, under, + *scapula*, the shoulder blade
Supraspinatus	Supraspinous fossa	Humerus (greater tubercle)	Stabilizes arm, assists in abduction	Suprascapular nerve	*supra*, above, + *spina*, referring to the spine of the scapula
Infraspinatus	Infraspinous fossa of scapula	Greater tubercle of humerus	Adducts and laterally rotates arm	Suprascapular nerve	*infra*, below, + *spina*, referring to the spine of the scapula
Teres Minor	Inferior angle of scapula	Greater tubercle of humerus	Adducts and laterally rotates arm	Axillary nerve	*teres*, round, + *minor*, smaller

*Only actions that apply to movement of the glenohumeral joint are listed.

Upper Limb Musculature

The upper limb consists of the **arm** and the **forearm.** Each is divided into anterior and posterior compartments. The **anterior compartment** is collectively referred to as a **flexor compartment** because its muscles act to flex the arm, forearm, or wrist and fingers. The **posterior compartment** is collectively referred to as an **extensor compartment** because its muscles act to extend the arm, forearm, or wrist and fingers. Understanding the compartmental nature of the muscles not only is important for simplifying the task of learning the muscles within each compartment—it also helps when learning the distribution of peripheral nerves. In most cases, one nerve innervates one compartment. Understanding the common function of the muscles in an entire compartment helps to more easily determine the deficits an individual will suffer when a specific nerve is damaged.

EXERCISE 13.2

COMPARTMENTS OF THE ARM

EXERCISE 13.2A Anterior Compartment of the Arm

1. Observe a prosected human cadaver or a classroom model demonstrating muscles of the upper limb.
2. Identify the **muscles of the anterior compartment of the arm** listed in **figure 13.2** on the cadaver or on a classroom model, using table 13.3 and the textbook as guides. Then label them in figure 13.2. When identifying these muscles, make note of the following:

(continued on next page)

(continued from previous page)

Table 13.3	Anterior (Flexor) Compartment of the Arm				
Muscle	**Origin**	**Insertion**	**Action**	**Innervation**	**Word Origin**
Biceps Brachii (Long Head)	Supraglenoid tubercle of scapula	Radial tuberosity	Supinates forearm and flexes forearm (weak)	Musculocutaneous nerve	*bi*, two, + *caput*, head, + *brachium*, arm
Biceps Brachii (Short Head)	Coracoid process of scapula	Radial tuberosity	Supinates forearm and flexes forearm (weak)	Musculocutaneous nerve	*bi*, two, + *caput*, head, + *brachium*, arm
Brachialis	Humerus (lower half of anterior surface)	Ulna (coronoid process and ulnar tuberosity)	Flexes forearm	Musculocutaneous nerve	*brachium*, arm
Coracobrachialis	Coracoid process of scapula	Midhumerus (medial surface)	Flexes and adducts arm (weak)	Musculocutaneous nerve	*coraco*, referring to the coracoid process, + *brachium*, arm

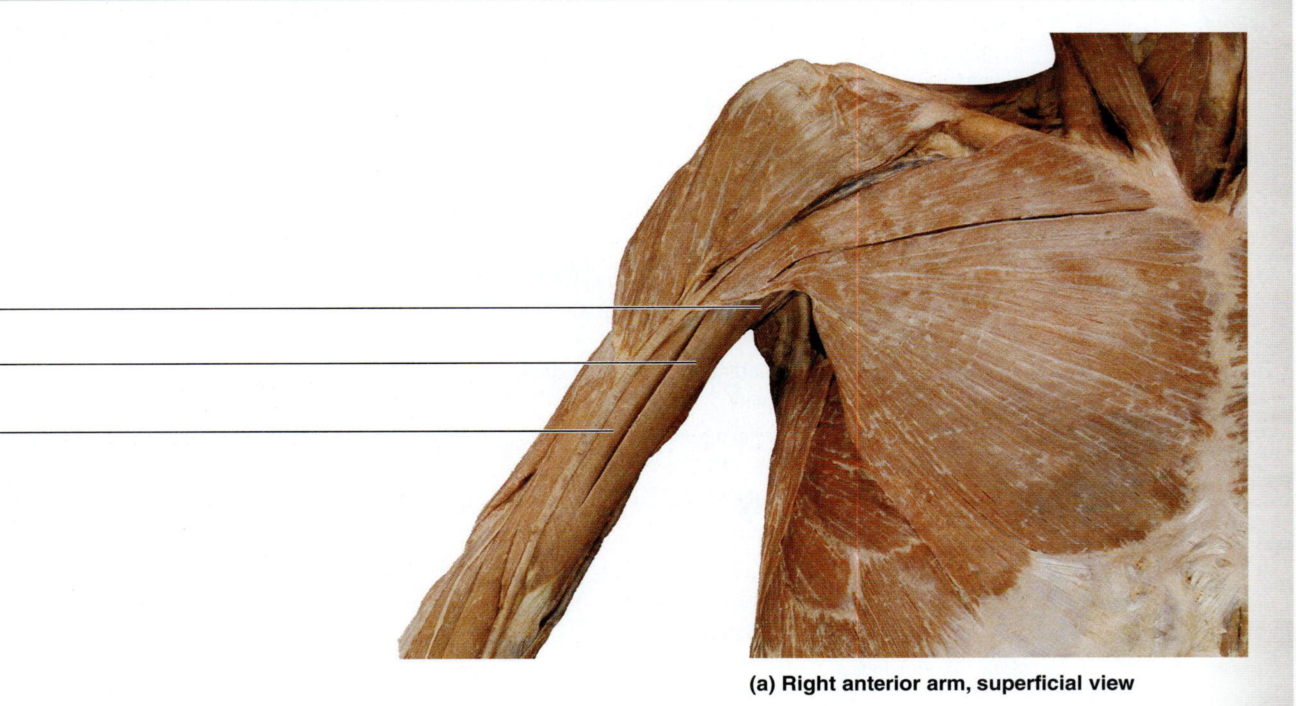

(a) Right anterior arm, superficial view

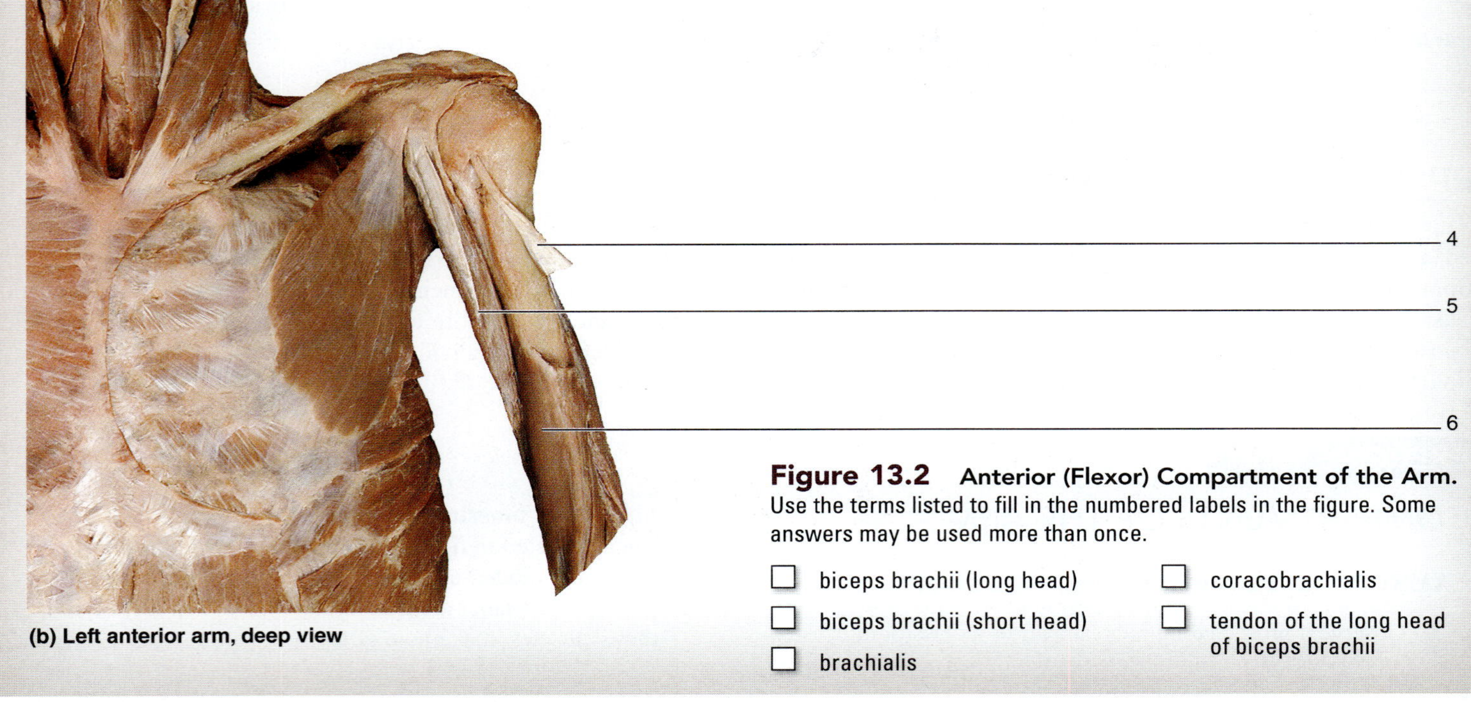

(b) Left anterior arm, deep view

Figure 13.2 Anterior (Flexor) Compartment of the Arm. Use the terms listed to fill in the numbered labels in the figure. Some answers may be used more than once.

- ☐ biceps brachii (long head)
- ☐ biceps brachii (short head)
- ☐ brachialis
- ☐ coracobrachialis
- ☐ tendon of the long head of biceps brachii

Common Actions: Muscles in this compartment flex the arm or forearm.

Exceptions: The coracobrachialis muscle acts only about the glenohumeral joint.

The brachialis muscle acts only about the elbow joint.

a. Note the following when identifying the biceps brachii muscle on models or cadavers. The **long head of the biceps brachii** disappears into the intertubercular groove of the humerus on its way to its attachment on the supraglenoid tubercle; only a small portion of its tendon is visible. On the other hand, the entire tendon of the **short head of the biceps brachii**, which attaches to the coracoid process of the scapula, is visible. Thus, on models and cadavers, the short head of the biceps brachii will generally appear to be longer.

b. *Brachialis*—A common misconception is that the biceps brachii muscle is the most powerful flexor of the forearm (elbow). It is not. Test this by performing the following activity. With your right wrist pronated, place the palm of your left hand over the belly of the biceps brachii on your right arm. Flex your elbow and note the degree to which the biceps brachii muscle produces tension. Next, supinate your wrist and perform the flexion action again. Did the biceps brachii produce more tension when the wrist was pronated or supinated? __________ You should have noticed a large difference in the amount of tension produced under the two conditions. In fact, the **prime mover,** or **agonist,** for the action of forearm flexion is the **brachialis** muscle, not the biceps brachii muscle. The biceps brachii is a **synergist** for this action. The biceps brachii is a very powerful **supinator** of the wrist and forearm. Thus, actions that require both flexion of the elbow and supination of the wrist and forearm use the biceps brachii to its full capacity. Think about this the next time you are trying to twist off a stuck bottle lid.

EXERCISE 13.2B Posterior Compartment of the Arm

The posterior compartment of the arm consists of one muscle, the triceps brachii, with three heads (long, lateral, and medial).

1. Observe a prosected human cadaver or a classroom model demonstrating muscles of the upper limb.
2. Identify the **muscles of the posterior compartment of the arm** listed in **figure 13.3** on the cadaver or on a classroom model, using table 13.4 and the textbook as guides. Then label figure 13.3. While identifying these muscles, make note of the following:

Common Actions: Extend the forearm (elbow).

Exception: The long head also extends the arm.

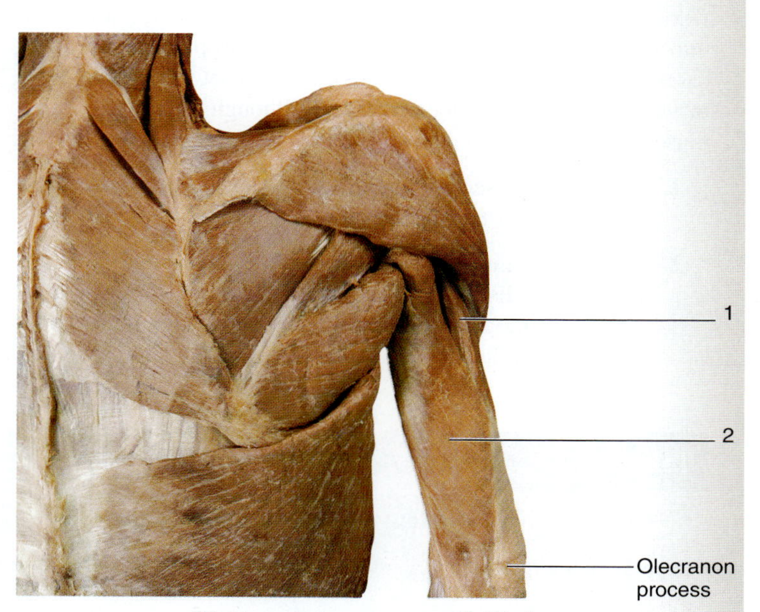

Figure 13.3 **Posterior (Extensor) Compartment of the Arm.** Use the terms listed to fill in the numbered labels in the figure.

- ☐ lateral head of triceps brachii
- ☐ long head of triceps brachii

INTEGRATE

LEARNING STRATEGY

The *long* head of the *triceps brachii* attaches to the *infraglenoid tubercle* (of the scapula), while the *long* head of the *biceps brachii* attaches to the *supraglenoid tubercle* (of the scapula). Also, the medial head of the triceps brachii can be visualized only if the lateral head is bisected. It is a deep muscle, and it attaches directly to the humerus.

Table 13.4 Posterior (Extensor) Compartment of the Arm

Muscle	Origin	Insertion	Action	Innervation	Word Origin
Triceps brachii					*tri,* three, + *caput,* head, + *brachium,* arm
Lateral Head	Proximal half of posterior humerus	Olecranon of ulna	Extends forearm	Radial nerve	
Long Head	Infraglenoid tubercle of scapula	Olecranon of ulna	Extends forearm, extends arm	Radial nerve	
Medial Head	Distal half of posterior humerus	Olecranon of ulna	Extends forearm	Radial nerve	

EXERCISE 13.3

COMPARTMENTS OF THE FOREARM

EXERCISE 13.3A Anterior Compartment of the Forearm

The muscles of the **anterior compartment of the forearm** cause flexion of the wrist and the digits (fingers). They are arranged into two layers—a superficial group and a deep group. In most cases, the name of the muscle tells exactly what the muscle does or where the muscle is located. Thus, although the names of the muscles seem long, the meanings of the names are very useful.

1. Observe a prosected human cadaver or a classroom model demonstrating muscles of the upper limb.
2. Identify the muscles of the anterior compartment of the forearm listed in **figure 13.4** on the cadaver or on a classroom model, using **table 13.5** and the textbook as guides. Then label them in figure 13.4.

   a. *Palmaris Longus*—Approximately 15% to 20% of humans do not have a palmaris longus muscle. The palmaris longus passes *superficial* to the **flexor retinaculum** (a band of connective tissue that wraps around the wrist) and inserts onto the **palmar aponeurosis.** Perform the following exercise to determine whether you have a palmaris longus muscle. With your forearm supinated, touch the tips of your first and fifth digits (thumb and pinky finger) together and flex the wrist *slightly*. If the palmaris longus is present, you will see its tendon passing longitudinally in the middle of your wrist. If no tendon is visible, the muscle is probably absent.

   b. *Flexor Digitorum*—There are two muscles that flex the digits, the **flexor digitorum superficialis** (superficial) and the **flexor digitorum profundus** (deep) **(figure 13.5).** These muscles, as their names

Table 13.5 Anterior (Flexor) Compartment of the Forearm

Muscle	Origin	Insertion	Action	Innervation	Word Origin
Superficial Group					
Flexor Carpi Radialis	Humerus (medial epicondyle)	Metacarpals II and III (base)	Flexes and abducts wrist	Median nerve	*flex*, to bend, + *carpus*, wrist, + *radialis*, radius
Flexor Carpi Ulnaris	Humerus (medial epicondyle)	Pisiform, hamate, and metacarpal V (base)	Flexes and adducts wrist	Ulnar nerve	*flex*, to bend, + *carpus*, wrist, + *ulnaris*, ulna
Flexor Digitorum Superficialis	Humerus (medial epicondyle)	Middle phalanx of digits 2–5	Flexes the metacarpophalangeal and proximal interphalangeal joints of digits 2–5	Median nerve	*flex*, to bend, + *digit*, finger, + *superficialis*, surface
Palmaris Longus	Humerus (medial epicondyle)	Palmar aponeurosis	Flexes wrist	Median nerve	*palmaris*, the palm, + *longus*, long
Pronator Teres	Humerus (medial epicondyle)	Radius (lateral shaft)	Pronates forearm and flexes elbow (weak)	Median nerve	*pronatus*, to bend forward, + *teres*, round
Deep Group					
Flexor Digitorum Profundus	Ulna (anteromedial surface) and interosseous membrane	Distal phalanx of digits 2–5	Flexes the distal phalanx of digits 2–5	Ulnar and median nerves	*flexus*, to bend, + *digitorum*, finger, + *profundus*, deep
Flexor Pollicis Longus	Radius (anterior surface) and interosseous membrane	Distal phalanx of the thumb	Flexes the distal phalanx of the thumb	Median nerve	*flexus*, to bend, + *pollex*, the thumb, + *longus*, long
Pronator Quadratus	Ulna (distal anterior shaft)	Radius (distal, anterior shaft)	Pronates the forearm	Median nerve	*pronatus*, to bend forward, + *quadratus*, square

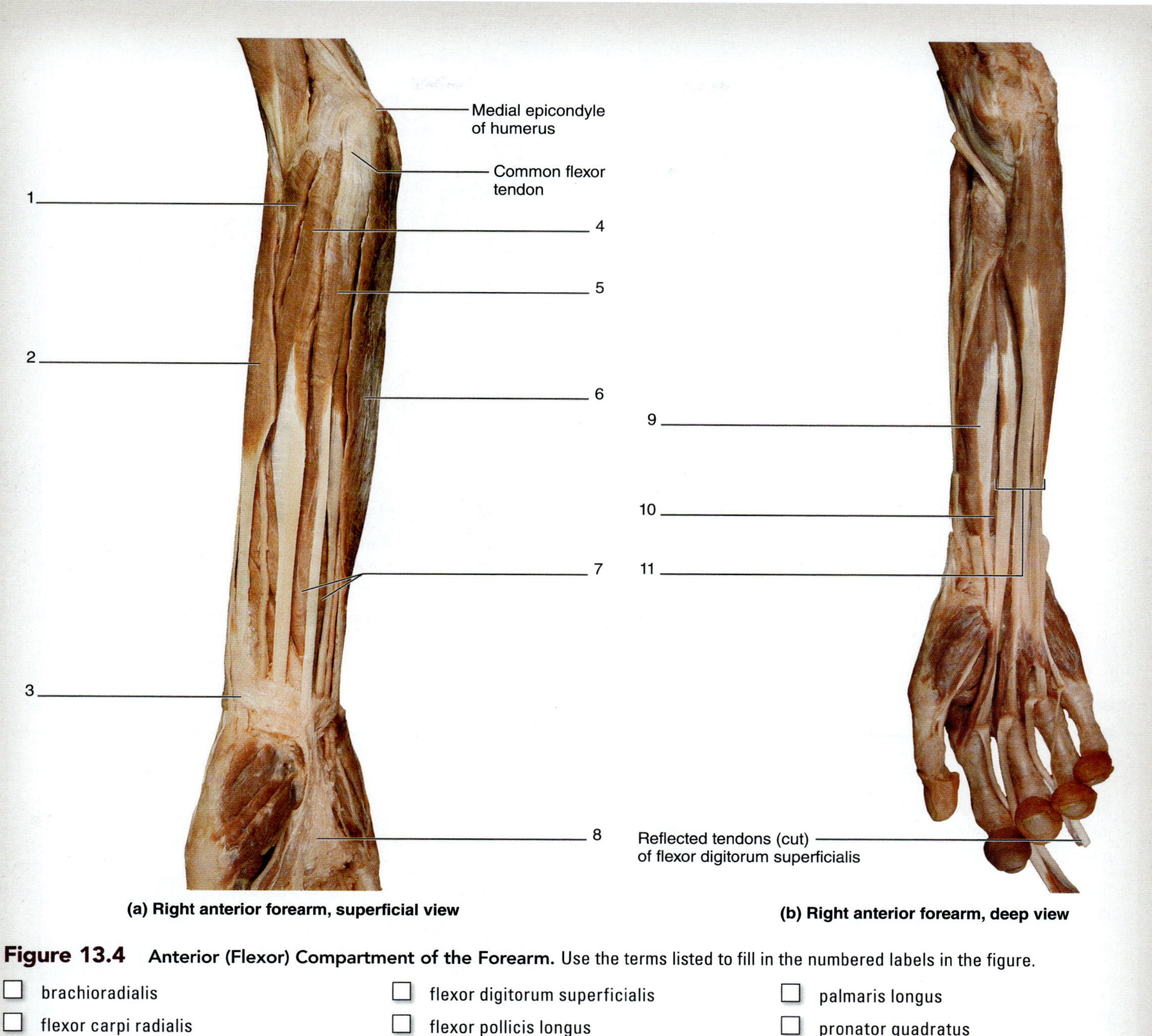

(a) Right anterior forearm, superficial view

(b) Right anterior forearm, deep view

Figure 13.4 Anterior (Flexor) Compartment of the Forearm. Use the terms listed to fill in the numbered labels in the figure.

- ☐ brachioradialis
- ☐ flexor carpi radialis
- ☐ flexor carpi ulnaris
- ☐ flexor digitorum profundus
- ☐ flexor digitorum superficialis
- ☐ flexor pollicis longus
- ☐ flexor retinaculum
- ☐ palmar aponeurosis
- ☐ palmaris longus
- ☐ pronator quadratus
- ☐ pronator teres

suggest, flex the fingers. They do so by attaching to the phalanges. These muscles take somewhat unique routes to get to their respective attachments on the phalanges. The **flexor digitorum profundus** tendon attaches to the **distal phalanx.** Thus, it must pass through the tendon of the more superficial muscle (the tendon of the flexor digitorum superficialis) as it travels to the distal phalanx. Observe a cadaver or a classroom model of the hand, and locate the middle and distal phalanges of one digit. The tendon of **flexor digitorum superficialis** inserts onto the **middle phalanx** of the digit. Before it reaches its insertion point, it has a slit through which the tendon of flexor digitorum profundus passes. Thus, though both muscles flex the digits, they do so at different joints. Perform an experiment to test the function of these two muscles. With your right hand, pull the tip of the third digit (middle finger) of your left hand posteriorly. This stretches the flexor digitorum profundus muscle beyond the point at which it can contract. Now try to flex the other digits of your hand. Notice that you can flex the joint between your proximal and middle phalanges, but you cannot flex the joint between your middle and distal phalanges. Once you release your

(continued on next page)

(continued from previous page)

hold on the third digit, you will be able to flex all the interphalangeal joints together.

3. *Optional Activity:* **AP|R 6: Muscular System**—Visit the quiz area for drill and practice on muscles of the upper limb.

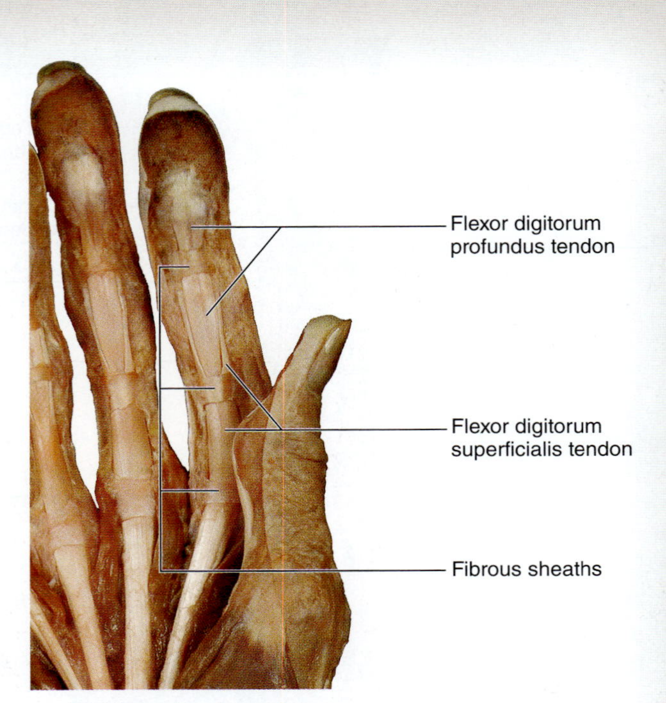

Figure 13.5 Flexor Tendons Within the Hand. The tendon of the flexor digitorum profundus pierces the tendon of the flexor digitorum superficialis as it travels to its attachment point on the distal phalanx.

INTEGRATE

LEARNING STRATEGY

To identify superficial muscles and tendons, place your left palm on the medial epicondyle of your right humerus. In this position, the order of the muscles on your right forearm, from lateral to medial, is:

Index finger—pronator teres (PT)

Middle finger—flexor carpi radialis (FCR)

Ring finger—palmaris longus (PL)

Pinky finger—flexor carpi ulnaris (FCU)

While performing this exercise, flex your wrist and digits to identify the tendons, from lateral to medial, of the flexor carpi radialis, palmaris longus, and flexor carpi ulnaris.

Pronator teres
Flexor carpi radialis
Palmaris longus
Flexor carpi ulnaris

(Left hand covers medial epicondyle)

EXERCISE 13.3B Posterior Compartment of the Forearm

1. Observe a prosected human cadaver or a classroom model demonstrating muscles of the upper limb.
2. Identify the **muscles of the posterior compartment of the forearm** listed in **figure 13.6** on the cadaver or on a classroom model, using table 13.6 and the textbook as guides. Then label them in figure 13.6.

Table 13.6 Posterior (Extensor) Compartment of the Forearm

Muscle	Origin	Insertion	Action	Innervation	Word Origin
Abductor Pollicis Longus	Ulna and radius (posterior surface) and interosseous membrane	Metacarpal I (base)	Abducts the thumb	Radial nerve	*abduct*, to move away from the midline, + *pollex*, thumb, + *longus*, long
Brachioradialis	Humerus (lateral supracondylar ridge)	Radius (styloid process)	Flexes the forearm	Radial nerve	*brachium*, arm, + *radialis*, radius
Extensor Carpi Radialis Brevis	Humerus (lateral epicondyle)	Metacarpal III (base)	Extends and abducts the wrist	Radial nerve	*extendo*, to stretch out, + *carpus*, wrist, + *radialis*, radius, + *brevis*, short
Extensor Carpi Radialis Longus	Humerus (lateral supracondylar ridge)	Metacarpal II (base)	Extends and abducts the wrist	Radial nerve	*extendo*, to stretch out, + *carpus*, wrist, + *radialis*, radius, + *longus*, long
Extensor Carpi Ulnaris	Humerus (lateral epicondyle) and proximal ulna	Metacarpal V (base)	Extends and adducts the wrist	Radial nerve	*extendo*, to stretch out, + *carpus*, wrist, + *ulnaris*, ulna
Extensor Digiti Minimi	Humerus (lateral epicondyle)	Proximal phalanx of digit 5 (little finger)	Extends the little finger at the metacarpophalangeal and interphalangeal joints	Radial nerve	*extendo*, to stretch out, + *digit*, finger, + *minimi*, smallest
Extensor Digitorum	Humerus (lateral epicondyle)	Distal and middle phalanges of digits 2–5	Extends digits 2–5 at the metacarpophalangeal joint, extends the wrist	Radial nerve	*extendo*, to stretch out, + *digit*, finger
Extensor Indicis	Ulna (posterior surface) and interosseous membrane	Tendon of extensor digitorum of digit 2 (index finger)	Extends the index finger	Radial nerve	*extendo*, to stretch out, + *indicis*, the forefinger
Extensor Pollicis Brevis	Radius (posterior surface) and interosseous membrane	Proximal phalanx of the thumb	Extends the proximal phalanx of the thumb at the metacarpophalangeal joint	Radial nerve	*extendo*, to stretch out, + *pollex*, thumb, + *brevis*, short
Extensor Pollicis Longus	Ulna (middle third, posterior surface) and interosseous membrane	Distal phalanx of the thumb	Extends the distal phalanx of the thumb at the metacarpophalangeal and interphalangeal joints	Radial nerve	*extendo*, to stretch out, + *pollex*, thumb, + *longus*, long
Supinator	Humerus (lateral epicondyle) and proximal ulna	Radius (proximal third)	Supinates the forearm	Radial nerve	*supinatus*, to move backward

(continued on next page)

(continued from previous page)

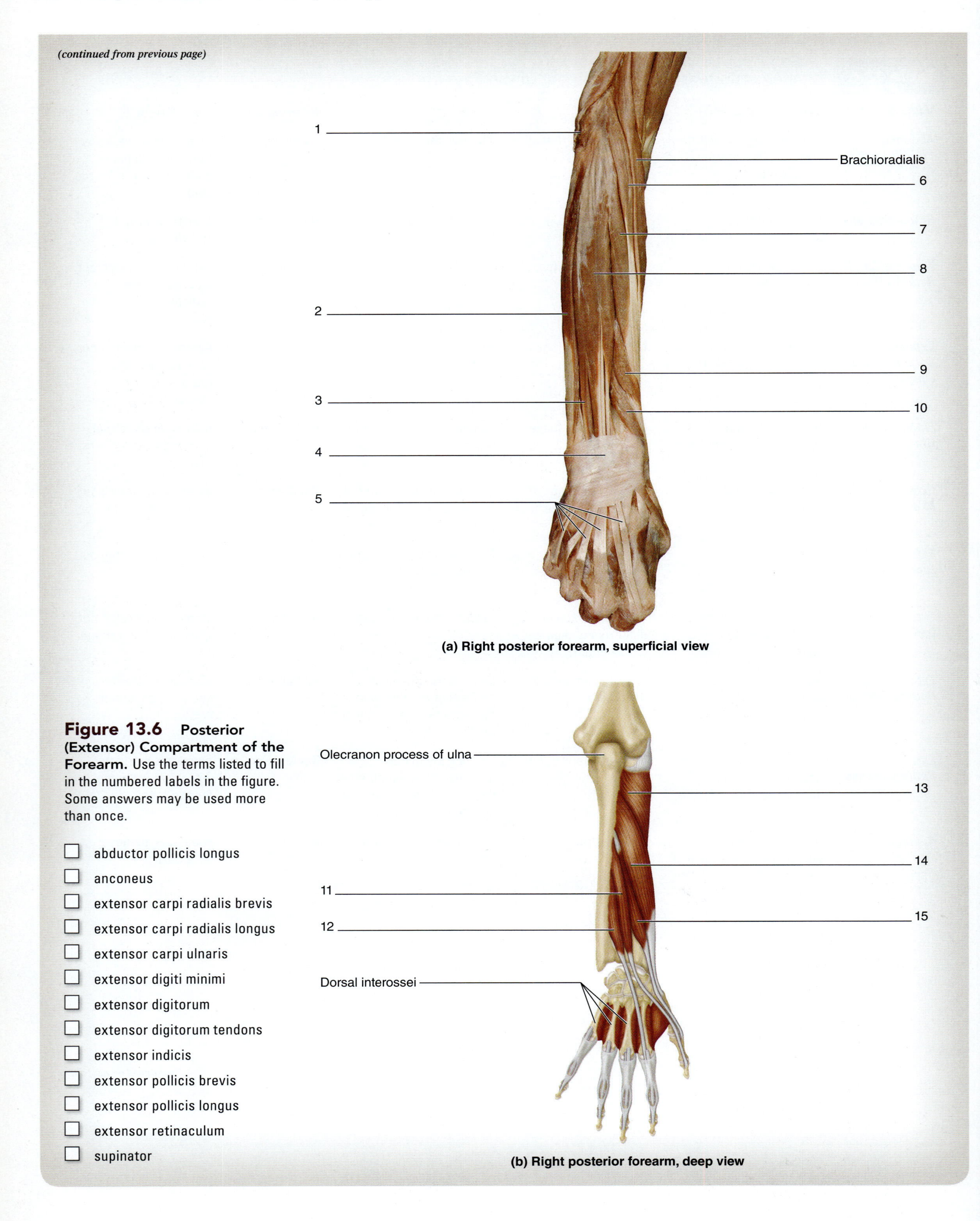

(a) Right posterior forearm, superficial view

(b) Right posterior forearm, deep view

Figure 13.6 Posterior (Extensor) Compartment of the Forearm. Use the terms listed to fill in the numbered labels in the figure. Some answers may be used more than once.

- ☐ abductor pollicis longus
- ☐ anconeus
- ☐ extensor carpi radialis brevis
- ☐ extensor carpi radialis longus
- ☐ extensor carpi ulnaris
- ☐ extensor digiti minimi
- ☐ extensor digitorum
- ☐ extensor digitorum tendons
- ☐ extensor indicis
- ☐ extensor pollicis brevis
- ☐ extensor pollicis longus
- ☐ extensor retinaculum
- ☐ supinator

EXERCISE 13.4

INTRINSIC MUSCLES OF THE HAND

The **intrinsic muscles of the hand (table 13.7)** consist of the thenar and hypothenar groups of muscles and the midpalmar group. The **thenar** group of muscles forms the large pad at the base of the thumb, the **thenar eminence.** Muscles in this group create special movements of the pollex (thumb), the most important of which is **opposition.** The **hypothenar** group of muscles forms a pad at the base of the fifth digit, the **hypothenar eminence.** Muscles in this group create special movements of the fifth digit (little finger). Observe both the thenar and hypothenar eminences on one of your hands. The **midpalmar** group is a group of muscles that lies between the thenar and hypothenar muscle groups. The midpalmar group consists of the interossei, lumbricals, and adductor pollicis. The **lumbricals** and **interossei** muscles are small muscles, located very deep in the hand, that create small, intricate movements of the fingers.

Table 13.7 Intrinsic Muscles of the Hand

Muscle	Origin	Insertion	Action	Innervation	Word Origin
Thenar Group					*thenar*, the palm of the hand
Abductor Pollicis Brevis	Flexor retinaculum, scaphoid and trapezium (tubercles)	Base of the proximal phalange of the thumb (lateral side)	Abducts the thumb	Recurrent branch of the median nerve	*abduct*, to move away from the midline, + *pollex*, thumb, + *brevis*, short
Flexor Pollicis Brevis	Flexor retinaculum and trapezium (tubercles)	Base of the proximal phalange of the thumb (lateral side)	Flexes the thumb	Recurrent branch of the median nerve	*flexus*, to bend, + *pollex*, thumb, + *brevis*, short
Opponens Pollicis	Flexor retinaculum and trapezium (tubercles)	Metacarpal I (lateral side)	Assists in opposition and medial rotation of the thumb	Recurrent branch of the median nerve	*oppono*, to place against, + *pollex*, thumb
Hypothenar Group					*hypo-*, under, + *thenar*, the palm of the hand
Abductor Digiti Minimi	Pisiform bone and tendon of flexor carpi ulnaris	Base of the proximal phalanx of digit 5 (medial side)	Abducts digit 5	Ulnar nerve	*abduct*, to move away from the median plane, + *digitus*, a finger, + *minimi*, smallest
Flexor Digiti Minimi Brevis	Hamate (hook) and flexor retinaculum	Base of the proximal phalanx of digit 5 (medial side)	Flexes the proximal phalanx of digit 5	Ulnar nerve	*flex*, to bend, + *digitus*, a finger, + *minimi*, smallest
Opponens Digiti Minimi	Hamate (hook) and flexor retinaculum	Metacarpal V (medial border)	Medially rotates and opposes digit 5 toward the thumb	Ulnar nerve	*oppono*, to place against + *digitus*, a finger, + *minimi*, smallest
Midpalmar Group					
Adductor Pollicis	Capitate bone, metacarpals II–III	Medial side of proximal phalanx of thumb	Adducts thumb	Ulnar nerve	*adduct*, to move toward the midline, + *pollex*, thumb
Dorsal Interossei	Metacarpals II–IV (medial and lateral surface)	Tubercle of the proximal phalanx and dorsal aponeurosis	Abducts digits 2–4	Ulnar nerve	*dorsal*, the back, + *inter-*, between, + *os*, bone
Lumbricals	Flexor digitorum profundus tendon	Lateral side of the dorsal expansion of digits 2–5	Flexes metacarpal-phalangeal joint	Ulnar and median nerves	*lumbricus*, earthworm
Palmar Interossei	Metacarpals (medial and lateral surface)	Tubercle of the proximal phalanx and the dorsal aponeurosis	Adducts digits 2–5	Ulnar nerve	*palmar*, the palm of the hand, + *inter-*, between, + *os*, bone

(continued on next page)

(continued from previous page)

INTEGRATE

LEARNING STRATEGY

The **D**orsal interossei **AB**duct the digits (mnemonic is **DAB**), whereas the **P**almar interossei **AD**duct the digits (mnemonic is **PAD**).

1. Observe a prosected human cadaver or a model demonstrating muscles of the hand.
2. Identify the intrinsic muscles of the hand listed in **figure 13.7** on a cadaver or on a classroom model, using table 13.7 and the textbook as guides. Then label them in figure 13.7.
   - **a.** *Interossei*—The two groups of interossei muscles act to abduct and adduct the digits. There are two groups of interossei: dorsal and palmar. The *dorsal interossei* abduct the digits, and the *palmar interossei* adduct the digits. To feel the belly of the first dorsal interosseous muscle, forcibly press (adduct) your thumb and index finger together. Then palpate between metacarpals I and II using your other hand to feel the first dorsal interosseous muscle.
   - **b.** *Lumbricals*—The lumbricals are relatively easy to identify because they appear slim and "worm-like," as their name implies. There is a lumbrical muscle located between every metacarpal bone. The lumbricals flex the metacarpophalangeal joints and extend the interphalangeal joints.
3. Make a representative drawing of the intrinsic muscles of the hand in the space provided.

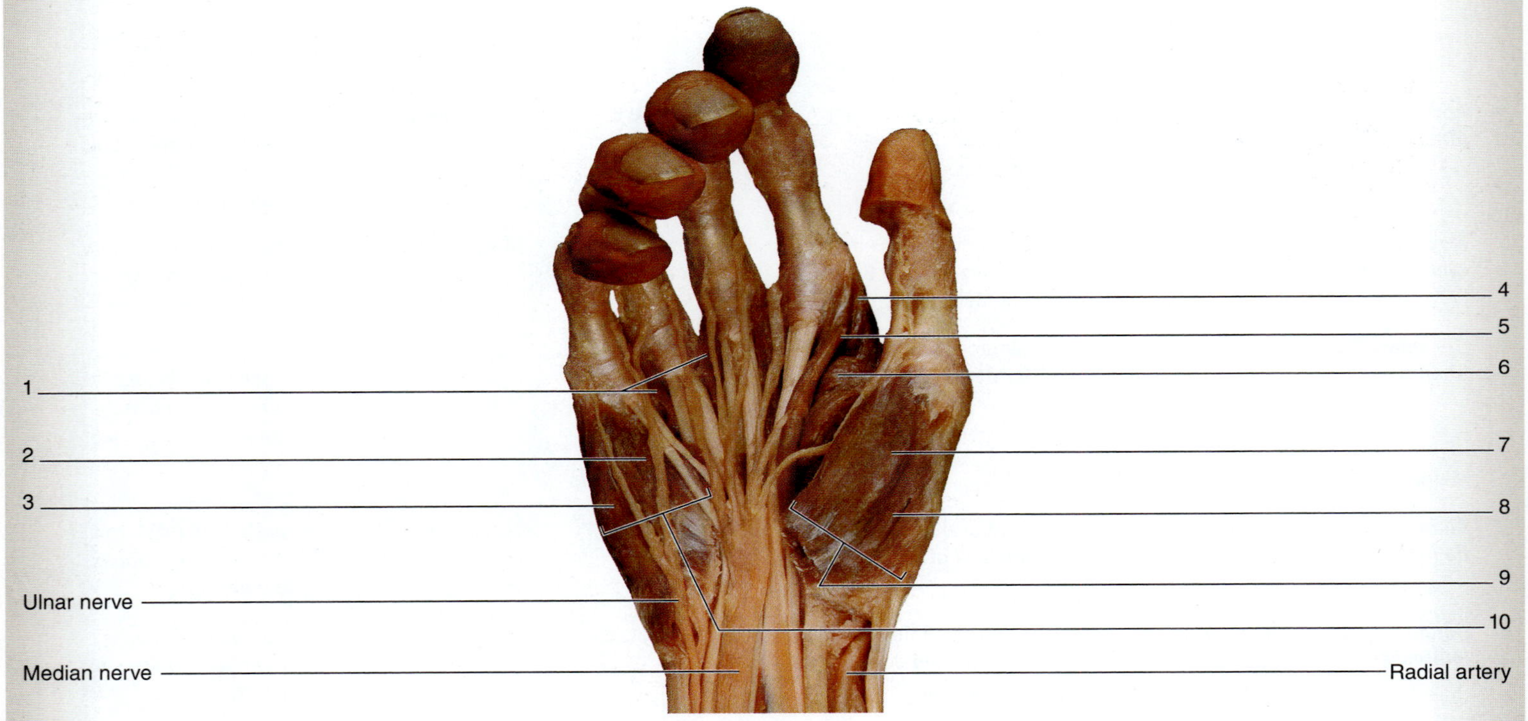

Figure 13.7 **Intrinsic Muscles of the Hand.** Use the terms listed to fill in the numbered labels in the figure.

- ☐ abductor digiti minimi
- ☐ abductor pollicis brevis
- ☐ adductor pollicis
- ☐ first dorsal interosseous
- ☐ flexor digiti minimi brevis
- ☐ flexor pollicis brevis
- ☐ hypothenar group
- ☐ lateral lumbricals
- ☐ medial lumbrical
- ☐ thenar group

Muscles That Act About the Hip Joint/Thigh

The pelvic girdle is composed of the paired os coxae, which articulate with the axial skeleton at the sacrum. Each lower limb is attached to an os coxa at the hip joint, which consists of the head of the femur articulating with the relatively deep acetabulum of the os coxa. Numerous muscles cross this joint to both stabilize and allow for movement—including standing, walking, running, and cycling—at this joint.

Deep fascia, called the fascia lata, both encircles the muscles of the thigh like a support stocking, and extends inward between the muscles to divide the thigh muscles into compartments—each with its own blood and nerve supply. The compartments include the anterior, medial, and posterior thigh compartments **(table 13.8).** Observe in table 13.8 how muscles within a compartment generally overlap in the types of movement that they produce (e.g., lateral rotation of muscles in the medial thigh compartment). Thickening of the deep fascia of the thigh also forms the iliotibial tract (or IT band), which extends from the anterior iliac crest to the lateral tibia.

INTEGRATE

LEARNING STRATEGY

- All the muscles in the medial compartment of the thigh attach to the pubic bone and, except for the gracilis, to the linea aspera of the femur.
- Think of these muscles as "the short one" (adductor brevis), "the long one" (adductor longus), "the really big one" (adductor magnus), and "the graceful one" (gracilis) to make them easier to identify relative to each other.
- As you view the anterior thigh, you will see these muscles medial to the sartorius. From superior to inferior, the order of muscles is pectineus, adductor longus, and adductor magnus. The adductor brevis is located deep to the pectineus and adductor longus muscles, so you will have to move them aside to see it clearly.

Table 13.8 Muscles That Act About the Hip Joint/Thigh

Muscle	Origin	Insertion	Action	Innervation	Word Origin
Anterior Thigh Compartment (Thigh Flexors)					
Psoas Major	T_{12}–L_5 (bodies and transverse processes)	Femur (lesser trochanter)	Flexes the hip joint	Lumbar plexus (L2–L3)	*psoa,* the muscles of the loins
Iliacus	Iliac bone (iliac fossa)	Femur (lesser trochanter)	Flexes the hip joint	Femoral nerve (L2–L3)	*ilium,* groin
Sartorius	ASIS (anterior superior iliac spine)	Medial tibia	Crosses legs, flexes the hip and knee, abducts and laterally rotates the thigh	Femoral nerve (L2–L4)	*sartor,* a tailor
Rectus Femoris	Anterior inferior iliac spine	Tibial tuberosity	Extends the knee	Femoral nerve (L2–L4)	*rectus,* straight, + *femur,* thigh
Medial Thigh Compartment (Thigh Adductors)					
Adductor Longus	Pubis	Femur (linea aspera)	Adducts and laterally rotates the thigh	Obturator nerve (L2–L4)	*adduct,* to bring toward the median plane, + *longus,* long
Adductor Brevis	Pubis	Femur (linea aspera)	Adducts and laterally rotates the thigh	Obturator nerve (L2–L 3)	*adduct,* to bring toward the median plane, + *brevis,* short
Gracilis	Pubis	Tibia (medial condyle)	Adducts the thigh	Obturator nerve (L2–L4)	*gracilis,* slender
Pectineus	Pubis	Femur (pectineal condyle)	Adducts and laterally rotates the thigh	Obturator or femoral nerve (L2–L4)	*pectineal,* a ridged or comb-like structure
Adductor Magnus	Pubis	Femur (linea aspera)	Adducts and laterally rotates the thigh	Obturator nerve (L2–L4)	*adduct,* to bring toward the median plane, + *magnus,* large
Obturator Externus	Margin of obturator foramen and obturator membrane	Femur (trochanteric fossa)	Laterally rotates thigh	Obturator nerve (L3–L4)	*obturator,* structure that occludes opening, + *externus,* outside

(continued on next page)

Table 13.8 Muscles that Act About the Hip Joint/Thigh (*continued*)

Muscle	Origin	Insertion	Action	Innervation	Word Origin
Posterior Compartment of the Thigh (Hamstring Group)					
Biceps Femoris					*bi*, two, + *caput*, head, + *femur*, thigh
Long Head	Ischial tuberosity	Head of fibula	Extends the hip	Tibial nerve (L4–S1)	
Short Head	Linea aspera of femur	Head of fibula	Extends the hip	Common fibular nerve (L5 and S1)	
Semimembranosus	Ischial tuberosity	Medial tibia	Extends the hip	Tibial nerve (L4–S1)	*semi*, half, + *membrana*, a membrane
Semitendinosus	Ischial tuberosity	Medial tibia	Extends the hip	Tibial nerve (L4–S1)	*semi*, half, + *tendinosus*, tendon
Lateral Thigh (Thigh Abductor)					
Tensor Fasciae Latae	Iliac crest (anterior aspect), anterior superior iliac spine (ASIS)	Iliotibial tract	Abducts the thigh	Superior gluteal nerve (L4 –S1)	*tensus*, to stretch, + *fascia*, a band, + *latus*, side
Gluteal Group					
Gluteus Maximus	Dorsal ilium, sacrum, and coccyx	Gluteal tuberosity of the femur, iliotibial tract	Extends the thigh, assists in lateral rotation	Inferior gluteal nerve (L5–S2)	*gloutos*, buttock, + *maximus*, greatest
Gluteus Medius	Ilium (between the anterior and posterior gluteal lines)	Greater trochanter of femur	Abducts and medially rotates the thigh	Superior gluteal nerve (L4–S1)	*gloutos*, buttock, + *medius*, middle
Gluteus Minimus	Ilium (between the anterior and inferior gluteal lines)	Greater trochanter of femur	Abducts and medially rotates the thigh	Superior gluteal nerve (L4–S1)	*gloutos*, buttock, + *minimus*, smallest
Deep Muscles of the Gluteal Region (Lateral Thigh Rotators)					
Piriformis	Anterior sacrum, sacro-tuberous ligament	Greater trochanter of femur	Laterally rotates the thigh	Nerve to piriformis (S1–S2)	*pirum*, pear, + *forma*, form
Superior Gemellus	Ischial spine	Greater trochanter of the femur (medial surface)	Laterally rotates the thigh	Nerve to obturator internus (L5 and S1)	*geminus*, twin, + *superus*, above
Obturator Internus	Obturator membrane (posterior surface)	Margins of obturator foramen	Laterally rotates the thigh	Nerve to obturator internus (L5 and S1)	*obturo*, to occlude, + *internus*, away from the surface
Inferior Gemellus	Ischial tuberosity	Greater trochanter of femur (medial surface)	Laterally rotates the thigh	Nerve to quadratus femoris (L5 and S1)	*geminus*, twin, + *inferior*, lower
Quadratus Femoris	Ischial tuberosity (lateral border)	Intertrochanteric crest of femur	Laterally rotates the thigh	Nerve to quadratus femoris (L5 and S1)	*quad*, four (sided), + *femur*, thigh

EXERCISE 13.5

MUSCLES THAT ACT ABOUT THE HIP JOINT/THIGH

1. Observe a prosected human cadaver or a model demonstrating muscles of the hip.
2. Identify the muscles that move the hip listed in **figures 13.8** and **13.9** on the cadaver or on a classroom model, using table 13.8 and the textbook as guides. Then label them in figures 13.8 and 13.9.

Anterior Muscles

a. *Anterior Compartment of the Thigh*—Two important hip flexors are the iliacus and psoas major. They have separate origins, but have a common insertion on the lesser trochanter of the femur, so together are called the iliopsoas. These muscles are difficult to identify on the cadaver and models if you look at the anterior thigh to find them, because only a small portion of the insertion is visible there. The bellies of these muscles are located deep within the pelvis on the posterior abdominopelvic wall. Thus, when identifying them on the cadaver or on models, be sure to view the inside of the abdominopelvic cavity where the muscles originate. In the anterior thigh, the iliopsoas muscle forms the floor of the femoral triangle. You must look deep to the structures in the femoral triangle (the femoral nerve, artery, and vein) to see the iliopsoas muscle here. Both the sartorius and the rectus femoris,which also move the leg, are included in exercise 13.6.

b. *Medial Compartment of the Thigh*—All six muscles of the medial thigh compartment adduct the thigh, and some perform additional functions (see figure 13.10).

c. *Lateral Thigh*—The tensor fascia latae muscle inserts on the iliotibial tract. This muscle abducts the thigh.

Posterior Muscles

d. *Gluteal Group*—The gluteus maximus (along with the tensor fasciae latae muscle) inserts on the iliotibilal tract. When these two muscles become stiff from overuse, they put excessive tension on the iliotibial tract. Most commonly this causes pain in the knee joint because the tight iliotibial tract compresses structures in the lateral knee and the underlying vastus lateralis muscle. The gluteus maximus extends the hip and *laterally* rotates the thigh. The gluteus medius and minimus muscles abduct the hip and *medially* rotate the thigh. Consider the action of **hip abduction.** It may not seem that we perform this action often, except perhaps in a kickboxing class. In fact, the gluteus medius and minimus muscles are rarely used for extreme abduction of the hip, as in kicking the lower limb out to the side. Instead, when attempting to abduct the hip against resistance, as when standing on one leg, these muscles prevent the hip from *adducting,* which would cause the hip on the side of the stance (standing) leg to protrude laterally (figure 13.8). To test the function of the gluteus medius and minimus muscles, first take a standing position. Next, place the palm of your right hand flat against your right lateral hip (superficial to the location of the gluteus medius and minimus muscles). Now flex the knee on your left lower limb, bringing your left foot off the ground so you are balancing on your right limb only. Do you feel tension in the muscles deep to your palm? If you are holding your hip vertical, you will. If you then relax the gluteus medius and minimus muscles, you will find that your hip protrudes laterally and the body tilts to the unsupported side. The gluteus medius and minimus hold the hip in the vertical position every time the limb is in its stance position. By holding the hip in such a position, they allow the foot on the swing limb to move forward without scraping against the ground. It may seem like a minor action, but it can be quite problematic for an individual whose muscles are not functioning properly due to nerve injury or other disease.

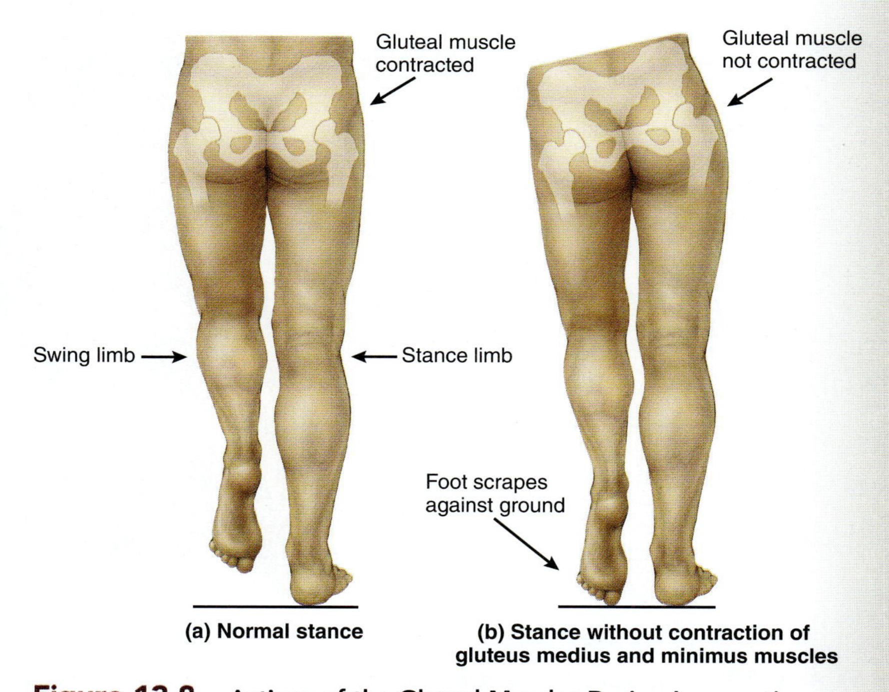

Figure 13.8 **Actions of the Gluteal Muscles During Locomotion.** The gluteus medius and minimus support the hip on the right side of the body when the right leg is the stance (supporting) leg. This allows the foot on the swing leg to clear the ground.

(continued on next page)

(continued from previous page)

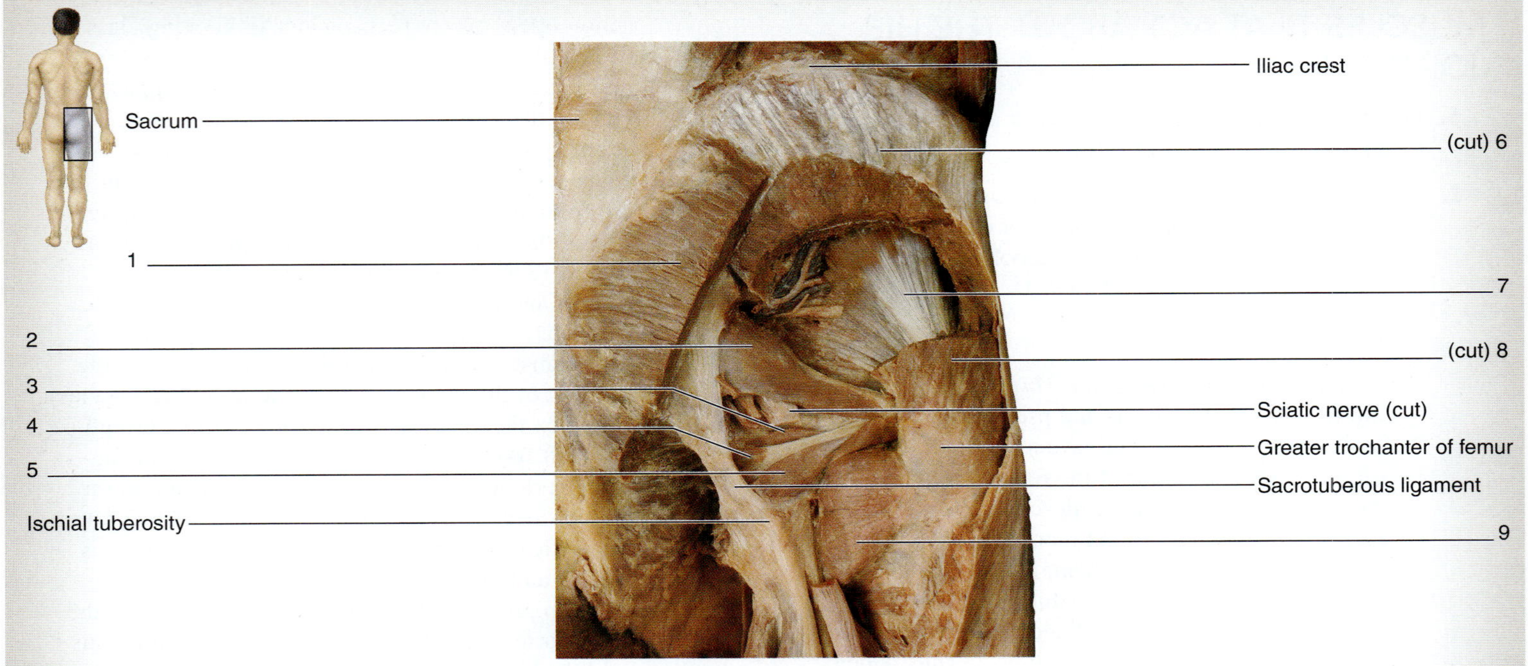

Iliac crest
Sacrum
(cut) 6
1
7
2
(cut) 8
3
Sciatic nerve (cut)
4
Greater trochanter of femur
5
Sacrotuberous ligament
Ischial tuberosity
9

(a) Right thigh and hip, deep posterior view

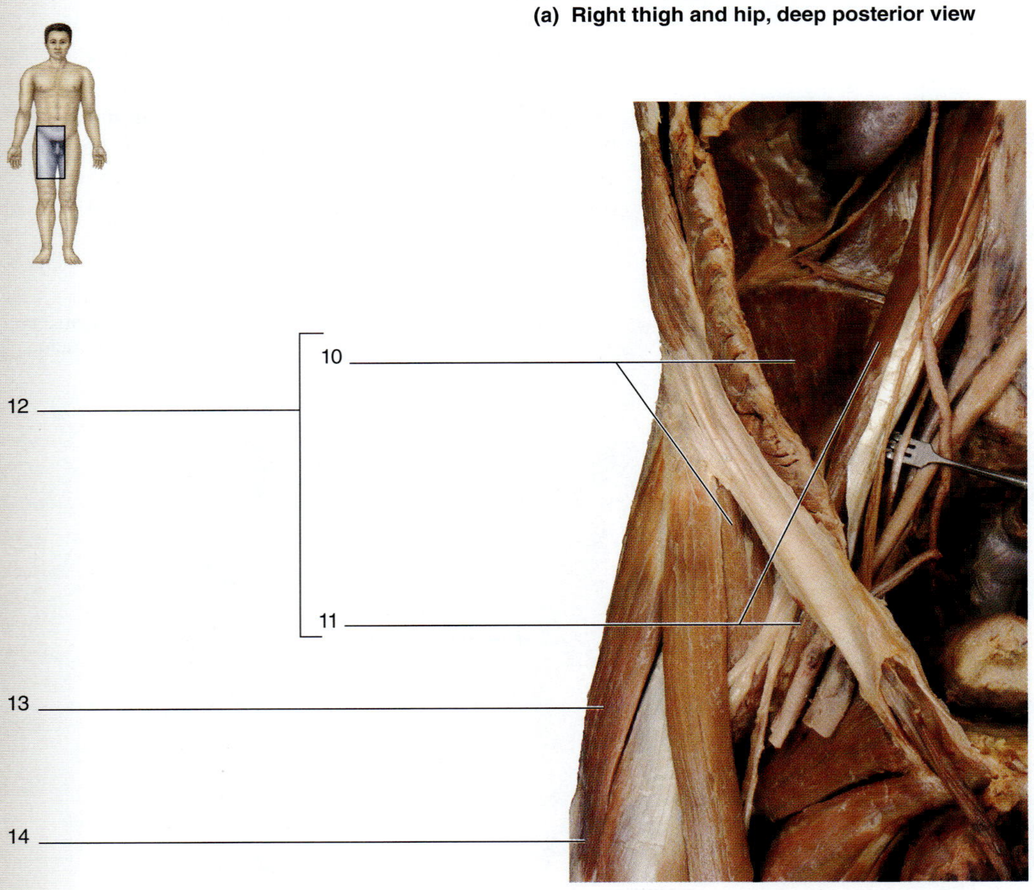

10
12
11
13
14

(b) Right thigh and hip, deep anterior view

Figure 13.9 Muscles That Act About the Hip Joint/Thigh. Use the terms listed to fill in the numbered labels in the figure. Some answers may be used more than once.

- ☐ gluteus maximus
- ☐ gluteus medius
- ☐ gluteus minimus
- ☐ iliacus
- ☐ iliopsoas
- ☐ iliotibial tract
- ☐ inferior gemellus
- ☐ obturator internus
- ☐ piriformis
- ☐ psoas
- ☐ quadratus femoris
- ☐ superior gemellus
- ☐ tensor fasciae latae

e. *Deep Muscles of the Gluteal Region*—The muscles in this group collectively laterally rotate the thigh as their primary action. However, they are also very important in stabilizing the hip joint, in much the same way that the rotator cuff muscles stabilize the shoulder joint. The **piriformis** muscle is a major landmark in the gluteal region. For example, the sciatic nerve exits inferior to this muscle, and the superior and inferior gluteal arteries and nerves are named for their passages above and below the piriformis, respectively.

f. *Posterior Compartment of the Thigh (Hamstring Group)*—The hamstring group, which is composed of the biceps femoris, semimembranosus, and semiteninosus, have a common origin on the ischial tuberosity of the os coxa, and insert on the bones of the leg. These muscles are collectively referred to as the hamstring muscles because of their association with the analogous muscles in pigs. The meat we call ham comes from these posterior thigh muscles of a pig. In a slaughterhouse, pig thighs are hung by the long tendons of these muscles—hence the term hamstrung. In fact, when we say that we feel "hamstrung," we say that we feel (figuratively) incapacitated. If our hamstring muscles were severed, we literally would be incapacitated. Their primary function at the hip is extension. These muscles are described in more detail in exercise 13.6.

3. Observe a prosected cadaver or a model in which the gluteus maximus muscle has been reflected or cut away. First identify the gluteus medius and minimus muscles. Inferior to the gluteus medius is the pear-shaped piriformis muscle. Note the tough **sacrotuberous ligament** (figure 13.9*a*) that overlies the medial attachment of the piriformis, and the large **sciatic nerve,** which exits inferior to the piriformis. Finally, locate the lateral rotators, from superior to inferior:

- ☐ **superior gemellus**
- ☐ **obturator internus** (in this location, only its tendon is visible)
- ☐ **inferior gemellus**
- ☐ **quadratus femoris**

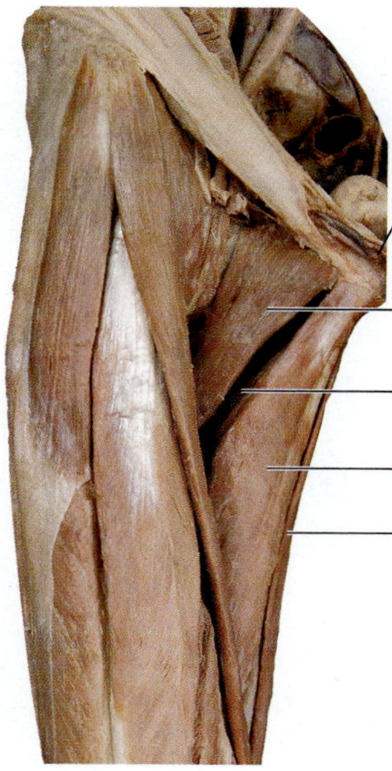

Pubic symphysis

1

2

3

4

Right thigh, anterior view

Figure 13.10 **Medial Compartment of the Thigh.** Use the terms listed to fill in the numbered labels in the figure. The adductor magnus is not visible in this photo.

- ☐ adductor brevis
- ☐ gracilis
- ☐ adductor longus
- ☐ pectineus

INTEGRATE

CLINICAL VIEW
Piriformis Syndrome

The piriformis muscle is a "pear-shaped" muscle that lies in close proximity to important structures within the gluteal region, such as the sciatic nerve, and the gluteal arteries and nerves. Piriformis syndrome is a painful condition that results from inflammation or overuse of the piriformis muscle. The incidence of piriformis syndrome is relatively common in athletes such as runners and cyclists, who may develop an imbalance in the strength of the piriformis muscle as compared to the gluteal muscles. Specifically, the syndrome occurs when the piriformis muscle (which laterally rotates the thigh) is stronger than the gluteus medius and gluteus minimus muscles (which are responsible for medial rotation of the thigh). As the piriformis muscle becomes inflamed or experiences spasms, it may also compress the underlying sciatic nerve, resulting in sciatica. **Sciatica** is a tingling, painful, or even numbing sensation that travels down the path of the sciatic nerve. Patients complain of shooting pain that runs from the gluteal region down the lateral aspect of the thigh, and toward the leg. Often the pain may be exacerbated when the body is held in certain positions, such as prolonged sitting or standing. The symptoms of piriformis syndrome can be reduced with the administration of anti-inflammatory drugs and through stretching exercises.

Lower Limb Musculature

The lower limb is composed of the **thigh** (from hip to knee), the **leg** (from knee to ankle), and the **foot.** The thigh and the leg each consist of three compartments. As with the upper limb, each compartment consists of muscles that perform similar actions, and each compartment is served by one peripheral nerve branch.

EXERCISE 13.6

COMPARTMENTS OF THE THIGH

EXERCISE 13.6A Anterior Compartment of the Thigh

1. Observe a prosected human cadaver or a classroom model demonstrating muscles of the thigh.
2. Identify the **leg extensor muscles** listed in **figure 13.11** on the cadaver or on a classroom model, using **table 13.9** and the textbook as guides. Then label them in figure 13.11. When identifying these muscles, make note of the following:

   Common Actions: Extend the knee.

   Exceptions: The rectus femoris and sartorius muscles flex the hip joint. The sartorius also flexes the knee joint.

   a. *Quadriceps femoris*—This large group of muscles includes the rectus femoris, vastus lateralis, vastus medialis, and vastus intermedius. All four muscles converge on a single patellar tendon, which extends to the patella and then continues as the patellar ligament to insert on the tibial tuberosity.

   b. *Sartorius*—The word *sartorius* literally means "tailor." Remember this and think about the typical cross-legged position tailors take to mend something by hand. It should

Table 13.9 Anterior Compartment of the Thigh

Muscle	Origin	Insertion	Action	Innervation	Word Origin
Sartorius	ASIS	Medial tibia	Flexes the leg	Femoral nerve (L2-L4)	*sartor,* a tailor
Quadriceps Femoris Group					
Rectus Femoris	Anterior inferior iliac spine	Tibial tuberosity	Extends the knee	Femoral nerve (L2–L4)	*rectus*, straight, + *femur*, thigh
Vastus Intermedius	Femur (medial and lateral)	Tibial tuberosity	Extends the knee	Femoral nerve (L2–L4)	*vastus*, great, + *intermedius*, in between
Vastus Lateralis	Femur (medial and posterior)	Tibial tuberosity	Extends the knee	Femoral nerve (L2–L4)	*vastus*, great, + *lateralis*, to the side
Vastus Medialis	Femur (inferior and posterior)	Tibial tuberosity	Extends the knee	Femoral nerve (L2–L4)	*vastus*, great, + *medialis*, medial

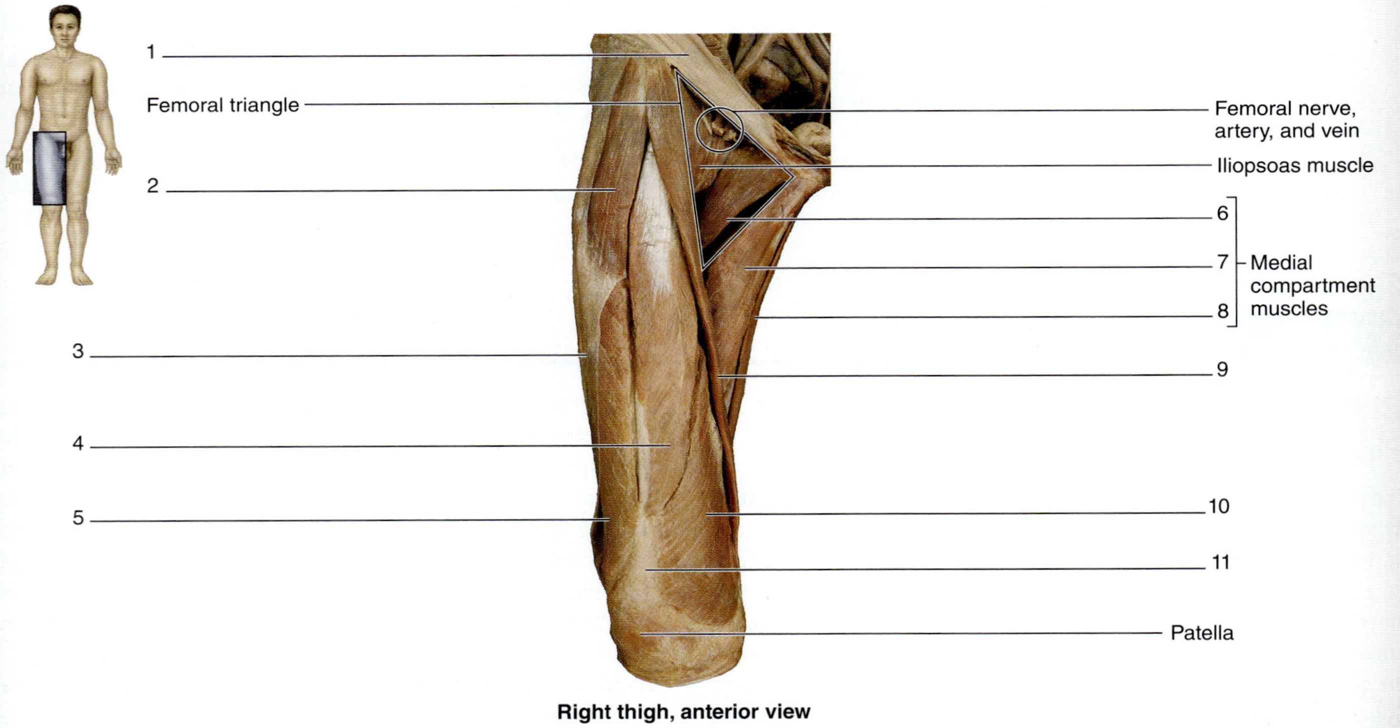

1 ____
Femoral triangle
2 ____
3 ____
4 ____
5 ____
Femoral nerve, artery, and vein
Iliopsoas muscle
6 ____
7 ____ Medial compartment muscles
8 ____
9 ____
10 ____
11 ____
Patella

Right thigh, anterior view

Figure 13.11 **Anterior Compartment of the Thigh.** Use the terms listed to fill in the numbered labels in the figure.

- ☐ adductor longus
- ☐ gracilis
- ☐ iliotibial tract
- ☐ inguinal ligament
- ☐ pectineus
- ☐ patellar tendon
- ☐ rectus femoris
- ☐ sartorius
- ☐ tensor fasciae latae
- ☐ vastus lateralis
- ☐ vastus medialis

be easy to remember the actions of the sartorius muscle. Try this out. Sit in a chair and cross your right leg over your left so that your right ankle is resting on your left knee. Next, note the positions of your right hip and knee joints. The hip is flexed and laterally rotated, and the knee is also flexed. These are the actions of the sartorius muscle.

3. Locate the femoral triangle. The **femoral triangle** is a triangular space in the upper, anterior thigh. Its borders are the **sartorius,** the **adductor longus,** and the **inguinal ligament.** It is clinically relevant because of the structures located within the space, which include the femoral nerve, artery, and vein. The only structures superficial to these are fat and skin; thus, it is a location where the vascular system can be accessed with relative ease.
4. Make a representative drawing of the femoral triangle in the space provided. Be sure to draw and label the structures that form its boundaries, and draw and label the contents found within the triangle.

INTEGRATE

LEARNING STRATEGY

The femoral triangle contains some very important structures. The mnemonic for remembering the contents of the femoral triangle is **NAVEL,** which lists the contents of the femoral triangle from lateral to medial:

N—femoral **N**erve

A—femoral **A**rtery

V—femoral **V**ein

E—**E**mpty space

L—**L**ymphatic vessel

EXERCISE 13.6B Medial Compartment of the Thigh

The muscles of the medial compartment were first discussed in exercise 13.5 because collectively they adduct the hip. The only muscle of this group that acts at the knee is the gracilis. This is because instead of inserting on the femur (like the other muscles of this group), the gracilis inserts on the tibia.

1. Observe a prosected human cadaver or a classroom model demonstrating muscles of the thigh.
2. Review the muscles of the medial compartment of the thigh, which are listed in figure 13.10.
3. Locate both the origin and the insertion of the gracilis.
4. *Optional Activity:* AP|R **6: Muscular System**—Visit the quiz area for drill and practice on muscles of the lower limb.

INTEGRATE

CLINICAL VIEW

Graciloplasty

In addition to being the outlier in terms of its attachments, the gracilis muscle is a weak adductor of the thigh—so weak, in fact, that it can be removed without a patient's experiencing great loss in function. When a patient needs a muscle graft, the gracilis muscle is often used for this purpose. The superficial location and limited function of the gracilis make it a good candidate for grafting procedures. Grafting may involve removing a portion of the muscle, or removing the muscle in its entirety. A flap of the gracilis muscle may be used in breast reconstruction following mastectomy. **Graciloplasty** is a procedure that involves transplanting the entire muscle to create an external anal sphincter. The procedure is used to treat cases of fecal incontinence.

(continued on next page)

(continued from previous page)

EXERCISE 13.6C Posterior Compartment of the Thigh

The muscles of the posterior compartment were first discussed in exercise 13.5 because they cross the hip joint and function in extension of the hip. In addition, all three of these muscle cross the knee joint and function in flexion of the knee.

1. Observe a prosected human cadaver or a classroom model demonstrating muscles of the thigh. Identify the muscles listed in **figure 13.12,** using **table 13.10** and the textbook as guides. Then label them in figure 13.12.
2. Locate both the origin and the insertion of the three hamstring muscles. Do both heads of the biceps femoris cross the hip? ________________ Which head is responsible for hip extension? ________________

Table 13.10 Posterior Compartment of the Thigh

Muscle	Origin	Insertion	Action	Innervation	Word Origin
Gracilis	Pubis	Tibia (medial condyle)	Flexes the leg	Obturator nerve (L2-L4)	*gracilis,* slender
Biceps Femoris	Ischial tuberosity	Head of fibula	Flexes the leg	Tibial nerve (long head) Common fibular nerve (short head)	*bi,* two, + *caput,* head, + *femur,* thigh
Semimembranosus	Ischial tuberosity	Medial tibia	Flexes the leg	Tibial nerve	*semi,* half, + *membrana*
Semitendinosus	Ischial tuberosity	Medial tibia	Flexes the leg	Tibial nerve	*semi,* half, + tendon

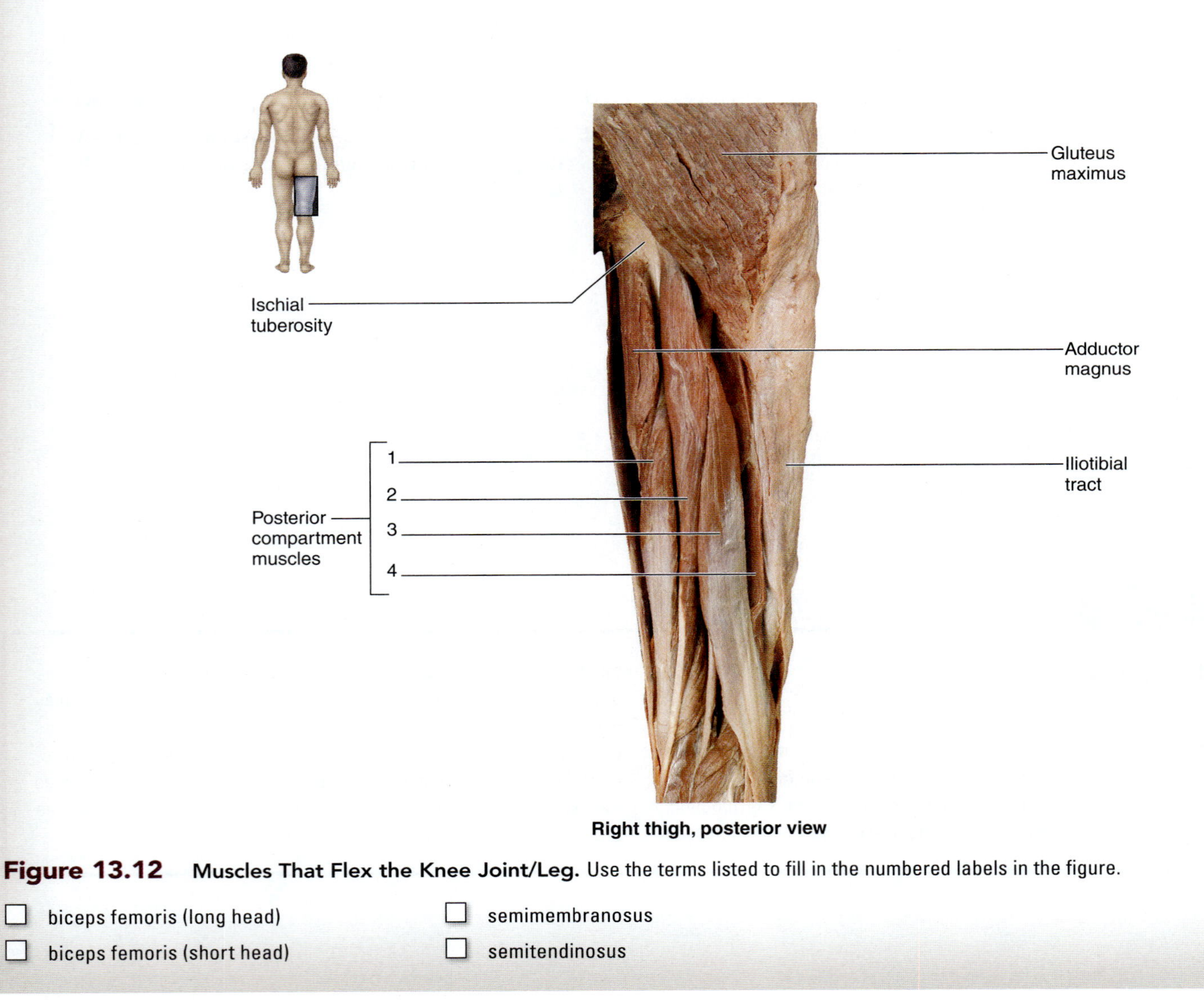

Figure 13.12 **Muscles That Flex the Knee Joint/Leg.** Use the terms listed to fill in the numbered labels in the figure.

- ☐ biceps femoris (long head)
- ☐ biceps femoris (short head)
- ☐ semimembranosus
- ☐ semitendinosus

EXERCISE 13.7

COMPARTMENTS OF THE LEG

EXERCISE 13.7A Anterior Compartment of the Leg

1. Observe a prosected human cadaver or a classroom model demonstrating muscles of the leg.
2. Identify the **muscles of the anterior compartment of the leg** listed in **figure 13.13** on the cadaver or on a classroom model, using **table 13.11** and the textbook as guides. Then label the muscles in figure 13.13. When identifying these muscles, make note of the following:

   Common Actions: Dorsiflex the ankle and extend the digits.

   Exception: The tibialis anterior does not act on the digits. In addition, it inverts the ankle as it dorsiflexes.

EXERCISE 13.7B Lateral Compartment of the Leg

1. Observe a prosected human cadaver or a classroom model demonstrating muscles of the leg.
2. Identify the **muscles of the lateral compartment of the leg** on the cadaver or on classroom models, using **table 13.12** and the textbook as guides.

   Fibularis Longus and Brevis—The fibularis muscles were formerly referred to as peroneus muscles. Thus, be aware that the terms relate to the same muscles. The **fibularis (peroneus) longus** has a very long tendon that lies superficial to the belly of the **fibularis (peroneus) brevis** muscle. The tendons of both muscles pass immediately posterior to the lateral malleolus en route to their attachments on the plantar surface of the foot. Their major action is to evert the ankle.

INTEGRATE

LEARNING STRATEGY

p**E**roneus (fibularis) muscles always **E**vert, whereas t**I**bialis muscles always **I**nvert.

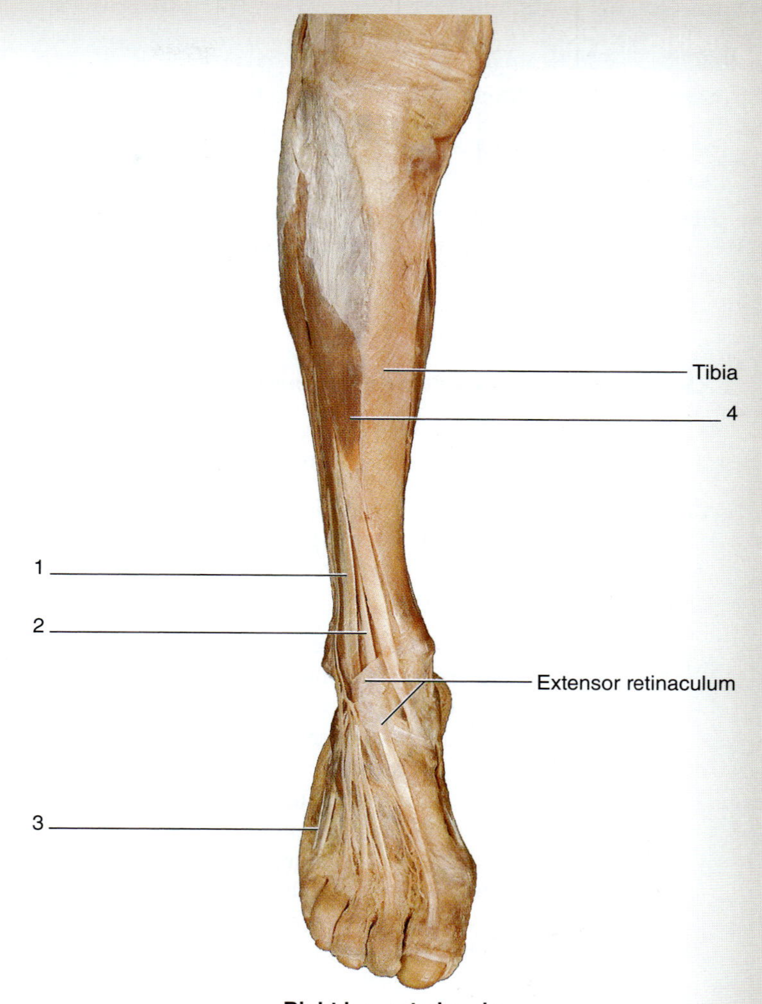

Figure 13.13 **Anterior Compartment of the Leg.** Use the terms listed to fill in the numbered labels in the figure.

- ☐ extensor digitorum longus
- ☐ extensor hallucis longus
- ☐ fibularis tertius tendon
- ☐ tibialis anterior

Table 13.11 Anterior Compartment of the Leg

Muscle	Origin	Insertion	Action	Innervation	Word Origin
Extensor Digitorum Longus	Tibia (lateral condyle and proximal three-fourths)	2nd and 3rd phalanges of digits 2–5	Extends digits 2–5 and dorsiflexes the ankle	Deep fibular nerve	*extendo*, to stretch out, + *digit*, a toe, + *longus*, long
Extensor Hallucis Longus	Fibula (anteromedial surface) and interosseous membrane	Distal phalanx of big toe	Extends the great toe and dorsiflexes the ankle	Deep fibular nerve	*extendo*, to stretch out, + *hallux*, the great toe, + *longus*, long
Fibularis Tertius	Fibula (anterior, distal surface) and interosseous membrane	Base of metatarsal V	Dorsiflexes and weakly everts foot	Deep fibular nerve	*fibularis*, fibula, + *tertius*, third
Tibialis Anterior	Tibia (lateral condyle and upper two-thirds)	Middle cuneiform (inferior surface) and metatarsal I	Dorsiflexes and inverts the foot	Deep fibular nerve	*tibia*, the shinbone, + *anterior*, the front surface

(continued on next page)

(continued from previous page)

Table 13.12	Lateral Compartment of the Leg				
Muscle	**Origin**	**Insertion**	**Action**	**Innervation**	**Word Origin**
Fibularis Brevis	Distal fibula	Metatarsal V	Everts the foot and plantar flexes the ankle	Superficial fibular nerve	*fibularis*, relating to the fibula, + *brevis*, short
Fibularis Longus	Head of fibula	Metatarsal I and middle cuneiform	Everts the foot and plantar flexes the ankle	Superficial fibular nerve	*fibularis*, relating to the fibula, + *longus*, long

EXERCISE 13.7C Posterior Compartment of the Leg

1. Observe a prosected human cadaver or a classroom model demonstrating muscles of the leg.
2. Identify the **muscles of the posterior compartment of the leg** listed in **figure 13.14** on the cadaver or on a classroom model, using **table 13.13** and the textbook as guides. Then label them in figure 13.14.

   Triceps Surae—The **triceps surae** includes the **soleus** and the two-headed **gastrocnemius** muscle. The tendons of these muscles come together to insert onto the calcaneus via the calcaneal (Achilles) tendon. The most superficial of the muscles is the gastrocnemius, though the lateral edges of the soleus can be seen as they poke out beneath the inferior portion of the gastrocnemius. The **plantaris** is a small muscle that can be seen only by reflecting or removing the gastrocnemius and soleus muscles. The plantaris has a very small belly, located just below the popliteal fossa in the back of the knee. It has a very long, flat tendon that often resembles ribbon used to decorate presents. The muscle itself is a very weak flexor of the leg. If you are viewing a cadaver and cannot find the plantaris, this may be because it isn't there. The plantaris is absent in approximately 6% to 8% of humans, and it is more commonly absent in the left leg than in the right leg. Interestingly, because the plantaris' actions are minimal and its tendon is long, the tendon is often removed and used for tendon grafts.
3. Label **figure 13.15** to review muscles from all three compartments of the leg.

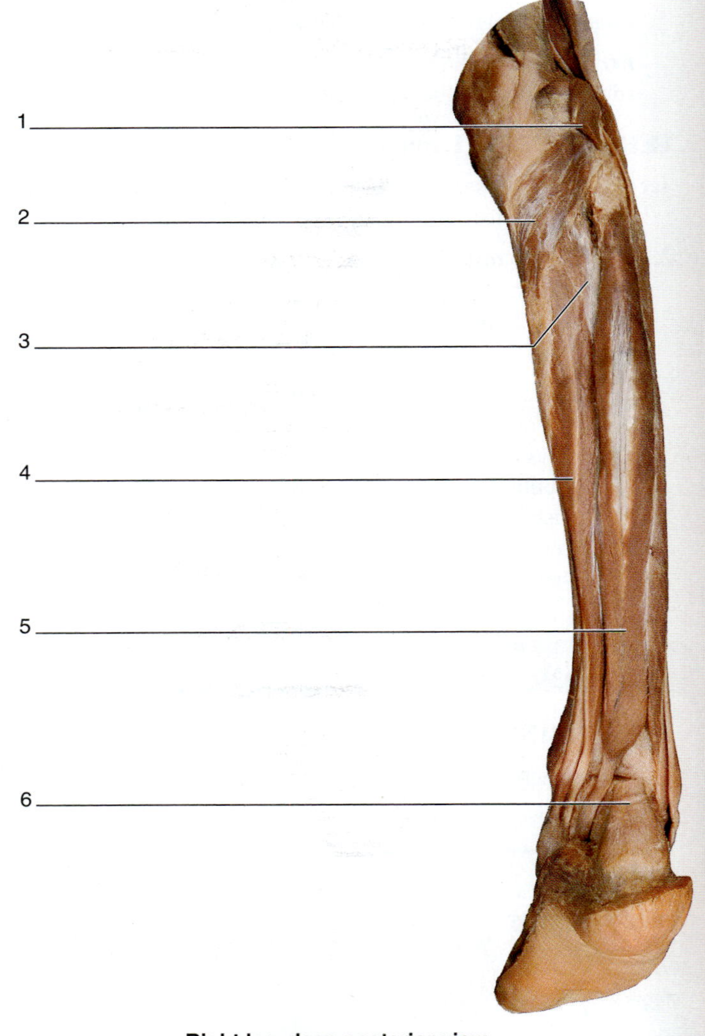

Right leg, deep posterior view

Figure 13.14 Posterior Compartment of the Leg. Use the terms listed to fill in the numbered labels in the figure.

- ☐ calcaneal tendon
- ☐ flexor digitorum longus
- ☐ flexor hallucis longus
- ☐ plantaris
- ☐ popliteus
- ☐ tibialis posterior

Table 13.13 Posterior Compartment of the Leg

Muscle	Origin	Insertion	Action	Innervation	Word Origin
Flexor Digitorum Longus	Posterior tibia	Distal phalanx of digits 2–5	Flexes digits 2–5	Tibial nerve (L5–S1)	*flexus*, to bend, + *digit*, a toe, + *longus*, long
Flexor Hallucis Longus	Inferior two-thirds of fibula	Distal phalanx of the great toe	Flexes the great toe	Tibial nerve (L5–S1)	*flexus*, to bend, + *hallux*, the great toe, + *longus*, long
Gastrocnemius	Femoral condyles	Calcaneus	Plantar flexes the ankle and flexes the leg (weak)	Tibial nerve (L4–S1)	*gaster*, belly, + *kneme*, leg
Plantaris	Femur (supracondylar ridge)	Calcaneus	Plantar flexes the ankle and flexes the leg (weak)	Tibial nerve (L4–S1)	*plantar*, relating to the sole of the foot
Popliteus	Lateral condyle of femur	Posterior, proximal surface of tibia	Flexes leg, unlocks the knee joint	Tibial nerve	*poplit*, back of the knee
Soleus	Proximal tibia, fibula	Calcaneus	Plantar flexes the ankle	Tibial nerve (L4–S1)	*solea*, a sandal, sole of the foot
Tibialis Posterior	Proximal tibia and fibula	Medial cuneiform and navicular	Inverts the foot and plantar flexes the ankle	Tibial nerve (L5–S1)	*tibia*, the shinbone, + *posterior*, the back surface

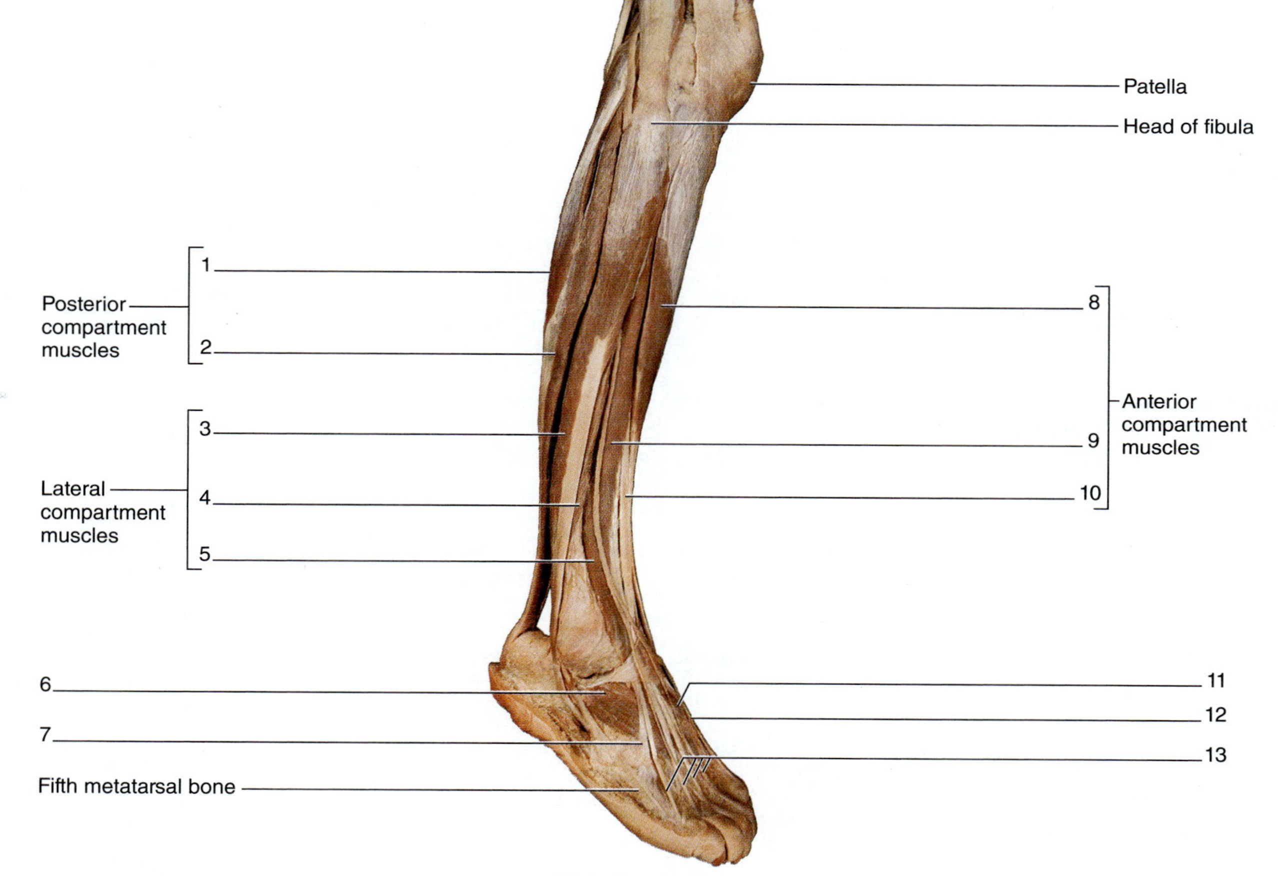

Figure 13.15 Lateral View of the Leg. In a lateral view of the leg, muscles from all three leg compartments can be seen. Use the terms listed to fill in the numbered labels in the figure.

- ☐ extensor digitorum brevis
- ☐ extensor digitorum longus
- ☐ extensor digitorum longus tendons
- ☐ extensor hallucis brevis
- ☐ extensor hallucis longus
- ☐ extensor hallucis longus tendon
- ☐ fibularis brevis
- ☐ fibularis longus
- ☐ fibularis tertius
- ☐ fibularis tertius tendon
- ☐ gastrocnemius
- ☐ soleus
- ☐ tibialis anterior

EXERCISE 13.8

INTRINSIC MUSCLES OF THE FOOT

The intrinsic muscles of the plantar surface of the foot are arranged in four layers, named layer 1 to layer 4 from superficial to deep. The foot also contains two neurovascular planes (a neurovascular plane is a region where the nerves and blood vessels that serve the region are located). The first is located between muscle layers 1 and 2, and the second is located between layers 3 and 4.

1. Observe a prosected human cadaver or a classroom model demonstrating muscles of the foot. Then label them in figure 13.16.
2. Identify the **intrinsic muscles of the foot** listed in **figure 13.16** on the cadaver or on a classroom model, using **table 13.14** and the textbook as guides. Then label them in figure 13.16.
   a. *Adductor Hallucis*—The **adductor hallucis** muscle is located in the third plantar muscle layer. It is a good landmark for the identification of other muscles in the foot. In addition, its two heads (oblique and transverse) appear like the number 7 when viewed together. To identify this muscle and others in the deeper layers of the foot, the tough **plantar aponeurosis** must first be cut and reflected.
   b. *Interossei*—As in the hand, there are two groups of interosseous muscles in the foot, though they are called dorsal and *plantar*, instead of dorsal and *palmar*. The functions of the interossei in the foot are the same as in the hand: The dorsal interossei abduct the digits, whereas the plantar interossei adduct the digits.

INTEGRATE

LEARNING STRATEGY

Conveniently, the mnemonic for remembering the functions of the interossei in the foot is the same as that for the hand: **DAB** and **PAD.** The **D**orsal interossei **AB**duct the digits **(DAB),** whereas the **P**lantar interossei **AD**duct the digits **(PAD).**

Table 13.14 Intrinsic Muscles of the Foot

Muscle	Origin	Insertion	Action	Innervation	Word Origin
Dorsal Musculature					
Extensor Digitorum Brevis	Calcaneus and extensor retinaculum (inferior surface)	Digits 2–4 (dorsal surfaces)	Extends the proximal phalanx of digits 2–4	Deep fibular nerve	*dorsal,* the back, + *inter-,* between, + *os,* bone
Extensor Hallucis Brevis	Calcaneus and extensor retinaculum (inferior surface)	Dorsal surface of the great toe	Extends the proximal phalanx of the great toe	Deep fibular nerve	*palmar,* the palm of the hand (foot), + *inter-,* between, + *os,* bone
Plantar Musculature—Layer 1					
Abductor Digiti Minimi	Calcaneus and plantar aponeurosis	Base of the proximal phalanx of digit 5 (lateral surface)	Abducts and flexes digit 5 (the little toe)	Lateral plantar nerve	*abduct,* to move away from the median plane, + *digitus,* toe, + *minimi,* smallest
Abductor Hallucis	Calcaneus and plantar aponeurosis	Base of the proximal phalanx of digit 1 (medial side)	Abducts the great toe	Medial plantar nerve	*abduct,* to move away from the median plane, + *hallux,* the great toe
Flexor Digitorum Brevis	Calcaneus and plantar aponeurosis	Middle phalanx of digits 2–5	Flexes digits 2–5	Medial plantar nerve	*flex,* to bend, + *digitus,* toe, + *brevis,* short
Plantar Musculature—Layer 2					
Lumbricals	Tendons of the flexor digitorum longus	Tendons of the extensor digitorum longus	Flex the proximal phalanges and extend the distal phalanges of digits 2–5	Medial and lateral plantar nerve	*lumbricus,* earthworm
Quadratus Plantae	Calcaneus (medial surface and lateral margin)	Tendon of the flexor digitorum longus (lateral side)	Flexes digits 2–5	Lateral plantar nerve	*quadratus,* having four sides, + *plantae,* sole of the foot

Table 13.14 Intrinsic Muscles of the Foot *(continued)*

Muscle	Origin	Insertion	Action	Innervation	Word Origin
Plantar Musculature—Layer 3					
Adductor Hallucis (Oblique Head)	Base of metatarsals II–V	Base of the proximal phalanx of the great toe	Adducts the great toe and helps maintain the transverse arch of the foot	Lateral plantar nerve	*adduct*, to move toward the median plane, + *hallux*, the great toe
Adductor Hallucis (Transverse Head)	Capsules of the metatarsophalangeal joints III–V	Base of the proximal phalanx of the great toe	Adducts the great toe and helps maintain the transverse arch of the foot	Lateral plantar nerve	*adduct*, to move toward the median plane, + *hallux*, the great toe
Flexor Digiti Minimi Brevis	Base of metatarsal V	Proximal phalanx of digit 5	Flexes the proximal phalanx of digit 5 (the little toe)	Lateral plantar nerve	*flex*, to bend, + *digitus*, a toe, + *minimi*, smallest, + *brevis*, short
Flexor Hallucis Brevis	Cuboid and lateral cuneiform (plantar surfaces)	Proximal phalanx of the great toe	Flexes the proximal phalanx of the great toe	Medial plantar nerve	*flex*, to bend, + *hallux*, the great toe, + *brevis*, short
Plantar Musculature—Layer 4					
Dorsal Interossei	Metatarsals I–V (adjacent surfaces)	Proximal phalanges of digits 2–4 (sides)	Abduct digits 2–4 and flex the metatarsophalangeal joints	Lateral plantar nerve	*dorsal*, the back, + *inter-*, between, + *os*, bone
Plantar Interossei	Bases of metatarsals III–V	Bases of proximal phalanges of digits 3–5 (medial side)	Adduct digits 2–4 and flex the metatarsophalangeal joints	Lateral plantar nerve	*plantae*, the sole of the foot, + *inter-*, between, + *os*, bone

(continued on next page)

INTEGRATE

CONCEPT CONNECTION

Sensory receptors embedded in muscles and tendons sense precise information regarding muscle length and tension. These specialized receptors, called *proprioceptors* (*proprio-*, one's own), relay this information to the central nervous system, enabling perception of body position. For example, when closing your eyes, you generally still have some awareness of the position of your limbs. Likewise, when walking into a dark room, you are still able to navigate through the space, in part due to proprioception. Two important types of proprioceptors are muscle spindles and Golgi tendon organs, which detect muscle length and tension, respectively. Muscle spindles lie parallel to muscle fibers so that they experience the same changes in length as the muscle fibers. Golgi tendon organs are embedded within the tendons of muscles. As muscle force builds during a muscle contraction, the sensory receptor becomes compressed. Both muscle spindles and Golgi tendon organs also play a role in reflexes, or preprogrammed responses to stimuli. For example, the stretch reflex involves the muscle spindle, and it is often tested clinically with a hammer tap to the patellar ligament (see chapter 16). The tap causes a rapid stretch of the quadriceps muscle group, which also stretches the muscle spindles. The result is a contraction of the entire quadriceps group, thereby causing knee extension, which counteracts the stretch that was applied. It is a protective mechanism to prevent too much force on a muscle from causing injury. Proprioceptors are distributed in all skeletal muscles. However, the number of muscle spindles and Golgi tendon organs varies from one muscle to another.

(continued from previous page)

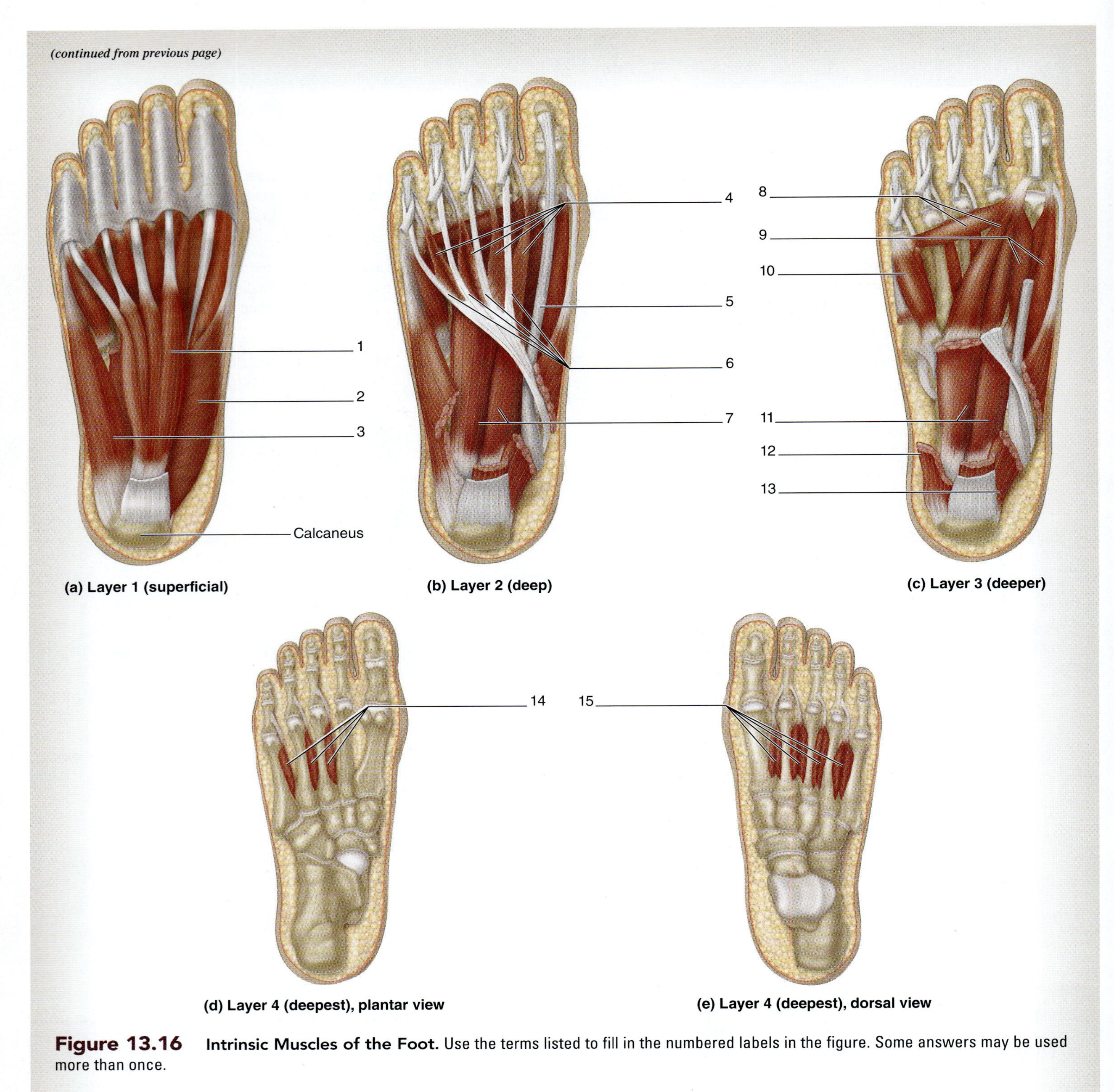

Figure 13.16 **Intrinsic Muscles of the Foot.** Use the terms listed to fill in the numbered labels in the figure. Some answers may be used more than once.

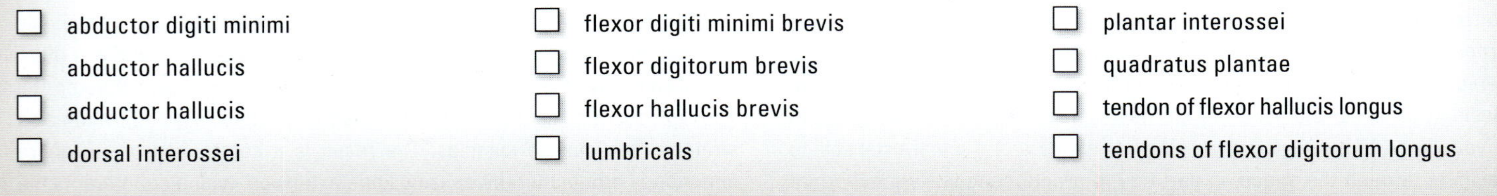

- ☐ abductor digiti minimi
- ☐ abductor hallucis
- ☐ adductor hallucis
- ☐ dorsal interossei
- ☐ flexor digiti minimi brevis
- ☐ flexor digitorum brevis
- ☐ flexor hallucis brevis
- ☐ lumbricals
- ☐ plantar interossei
- ☐ quadratus plantae
- ☐ tendon of flexor hallucis longus
- ☐ tendons of flexor digitorum longus

Chapter 13: The Muscular System: Appendicular Muscles

Name: ____________________

Date: __________ **Section:** __________

POST-LABORATORY WORKSHEET

The (1) corresponds to the Learning Objective(s) listed in the chapter opener outline.

Do You Know the Basics?

Exercise 13.1: Muscles That Act About the Pectoral Girdle/Glenohumeral Joint

1. Match the movements listed in column A with the corresponding muscle that creates the movement listed in column B. (1)

Column A	*Column B*
____ 1. extends, adducts, and medially rotates arm	a. deltoid
____ 2. flexes and medially rotates arm; prime mover of arm abduction; extends and laterally rotates arm	b. latissimus dorsi
____ 3. prime mover of arm extension; adducts and medially rotates arm (big muscle)	c. pectoralis major
____ 4. prime mover of arm flexion; adducts and medially rotates arm (big muscle)	d. teres major

2. A synergist of the trapezius for the action of *adduction* of the scapula is the rhomboid major, whereas a synergist of the trapezius for the action of *elevation* of the scapula is the levator scapulae. ____________________ (True/False). (2)

3. Damage to the serratus anterior muscle is tested by having a patient push against a wall with an outstretched upper limb. If there is weakness or paralysis of the muscle, the scapula will do which of the following? (Circle one.) (3)
   a. adduct excessively
   b. abduct excessively
   c. "wing" (medial border moves posterior)
   d. depress excessively

4. One or more of the rotator cuff muscles cause which of the following movements of the arm? (Check all that apply.) (4)
   ____ a. abduction
   ____ b. adduction
   ____ c. extension
   ____ d. flexion
   ____ e. lateral rotation
   ____ f. medial rotation

Exercise 13.2: Compartments of the Arm

5. The ____________________ (anterior/posterior) compartment of the arm consists largely of forearm flexors, whereas the ____________________ (anterior/posterior) compartment consists largely of forearm extensors. (5)

6. The biceps brachii is a prime mover for the action of forearm____________________ (flexion/supination). The biceps brachii is a synergist with the brachialis muscle for the action of forearm____________________ (flexion/supination). (6)

Exercise 13.3: Compartments of the Forearm

7. Match the muscles listed in column A with the appropriate compartment of the forearm listed in column B. Each compartment may be used more than once. (7)

Column A	*Column B*
____ 1. extensor indicis	a. anterior compartment
____ 2. flexor carpi ulnaris	b. posterior compartment
____ 3. palmaris longus	
____ 4. pronator quadratus	
____ 5. supinator	

8. The palmaris longus is ____________________(deep/superficial) to the flexor retinaculum and inserts onto the palmar aponeurosis. 8

9. The tendons of the flexor digitorum superficialis muscle attach to the (proximal/middle/distal) phalanges, whereas the tendons of the flexor digitorum profundus attach to the (proximal/middle/distal) phalanges. 9

Exercise 13.4: Intrinsic Muscles of the Hand

10. Which of the following is an intrinsic muscle of the hand? (Check all that apply.) 10

____ a. abductor digiti minimi

____ b. abductor pollicis brevis

____ c. dorsal interossei

____ d. flexor digitorum brevis

____ e. lumbricals

11. The large pad at the base of the thumb is called the ____________________ (hypothenar/thenar) eminence and it contains muscles collectively referred to as ____________________ (hypothenar/thenar) muscles. The large pad at the base of the pinky finger (digit V) is called the ____________________ (hypothenar/thenar) eminence, and it contains muscles collectively referred to as ____________________ (hypothenar/thenar) muscles. 11

12. The dorsal interossei ____________________ (abduct/adduct) the digits, whereas the palmar interossei ____________________ (abduct/adduct) the digits. 12

Exercise 13.5: Muscles That Act About the Hip Joint/Thigh

13. Which of the following muscles is a major hip extensor whose antagonist is the psoas major muscle (a major flexor of the hip)? (Circle one.) 13

a. gluteus maximus b. gluteus medius c. iliacus d. quadratus femoris e. tensor fascia latae

14. During normal locomotion, which of the following muscles on the stance limb must function correctly for the foot of the swing limb to clear the ground? (Check all that apply.) 14

____ a. gluteus maximus

____ b. gluteus medius

____ c. piriformis

____ d. psoas major

____ e. quadratus femoris

15. A deep muscle of the posterior hip that is an important clinical landmark due to its association with the gluteal arteries and nerves, and the sciatic nerve, is the piriformis muscle ____________________ (True/False). 15

16. The iliopsoas muscle is composed of two muscles with a common insertion. The two muscles that compose the iliopsoas are the iliacus and psoas minor muscles. ____________________ (True/False). 16

Exercise 13.6: Compartments of the Thigh

17. Match the muscle listed in column A with the appropriate compartment of the thigh in column B. Each compartment may be used more than once. 17

Column A	*Column B*
____ 1. adductor magnus	a. anterior compartment
____ 2. biceps femoris	b. medial compartment
____ 3. gracilis	c. posterior compartment
____ 4. rectus femoris	
____ 5. sartorius	
____ 6. semitendinosus	
____ 7. vastus lateralis	

18. The only two muscles of the anterior compartment of the thigh that cross the hip joint are the rectus femoris and the sartorius muscles. The action of these two muscles about the hip is to ________________ (flex/extend) the hip. 18

19. Identify the structures that form the borders of the femoral triangle. (Check all that apply.) 19

____ a. adductor longus muscle

____ b. inguinal ligament

____ c. sartorius muscle

____ d. sacrotuberous ligament

Exercise 13.7: Compartments of the Leg

20. Match the muscle listed in column A with the appropriate compartment of the leg listed in column B. Each compartment may be used more than once. 20

Column A	*Column B*
____ 1. extensor digitorum longus	a. anterior compartment
____ 2. fibularis (peroneus) brevis	b. lateral compartment
____ 3. fibularis (peroneus) longus	c. posterior compartment
____ 4. flexor hallucis longus	
____ 5. gastrocnemius	
____ 6. soleus	
____ 7. tibialis anterior	

21. The fibularis longus and brevis evert the foot at the ankle joint. Both the tibialis anterior and the tibialis posterior oppose this action by inverting the foot at the ankle joint. ________________ (True/False). 21

22. The triceps surae is composed of which of the following? (Check all that apply.) 22

____ a. flexor hallucis longus

____ b. gastrocnemius

____ c. plantaris

____ d. soleus

____ e. tibialis anterior

Exercise 13.8: Intrinsic Muscles of the Foot

23. Which of the following is an intrinsic muscle of the foot? (Check all that apply.) 23

____ a. abductor hallucis

____ b. dorsal interossei

____ c. flexor digitorum brevis

____ d. soleus

Can You Apply What You've Learned?

24. Name three muscles that attach to the coracoid process of the scapula.

a. ____

b. ____

c. ____

25. One action of the gluteal muscles (gluteus maximus, medius, and minimus) is to rotate the hip.

a. Which gluteal muscle(s) laterally rotate the hip? ________________________________

b. Which gluteal muscle(s) medially rotate the hip? ________________________________

26. Which muscle of the medial compartment of the thigh does *not* insert onto the linea aspera of the femur? ____________________

27. Which muscle of the quadriceps muscle group is the only muscle to cross the hip joint? ____________________ What is the *origin* (proximal attachment) of this muscle? ____________________. What is the *insertion* (distal attachment)? ____________________

28. Describe the difference between the patellar *tendon* and the patellar *ligament*, and describe the location of each. (Hint: Think about the specific definitions of *tendon* and *ligament*.)

29. Describe how to distinguish between the semitendinosus muscle and the semimembranosus muscle.

30. The tendons of which three of the rotator cuff muscles hold the humerus into the glenoid cavity on the posterior surface? ____________________ The tendon of the rotator cuff muscle that holds the humerus into the glenoid cavity on the anterior surface is the tendon of the ____________________ muscle.

31. What two muscles insert onto the iliotibial tract (IT band)?

    a. ____________________ b. ____________________

32. The gluteus maximus muscle is a major extensor of the hip. What muscle acts as the major antagonist to the gluteus maximus muscle? (Hint: The muscle you name would be a prime mover for hip *flexion*.) ____________________

33. Describe the involvement of the biceps in the following movements:

    a. Flexion of the forearm:

    b. Supination of the forearm/wrist:

Can You Synthesize What You've Learned?

34. A patient is told he has a "torn hamstring." List the possible muscles that could have been torn in this case. Then, describe the actions that would demonstrate weakness in the muscle as a result of the tear.

35. Compartment syndrome is a disorder in which inflammation within the compartment of a limb causes pressure such that blood flow is reduced, nerves become damaged, and the muscles start to become nonfunctional (and eventually necrotic, if the situation is not treated). A patient is experiencing compartment syndrome in the anterior compartment of the leg, which has weakened the muscles in this compartment. As a result, what actions would this patient have difficulty performing?

Nervous Tissues

14

OUTLINE AND LEARNING OBJECTIVES

MODULE 7: NERVOUS SYSTEM

INTRODUCTION

Many of the structures that make up the nervous system are not particularly impressive in the absence of any knowledge of what they do, particularly when compared with very dynamic-looking organs such as the heart. Imagine how unimpressed early anatomists must have been with the brain. All they saw was a very dense, oatmeal-colored mass of tissue with no known function. Their observations of brain tissue could hardly have been as fascinating or romantic as their observations of the heart. At first, the most inspiring names they could find for brain tissues were terms like "gray matter" and "white matter." Today we know that the nervous system is much more fascinating than we could ever have imagined, and that scientists are barely beginning to grasp its complexity. Most information about the complexity of the nervous system has been gained through dynamic studies of the brain using methods such as magnetic resonance imaging (MRI), as opposed to static observations of gross structures.

A student of anatomy and physiology beginning to learn about the gross structures of the nervous system must trust what others state about the *functions* of those structures because these functions cannot be observed at the gross level. However, observing nervous tissue through the microscope makes the study of nervous tissue much more interesting and helps one begin to understand some of the relationships between structure and function. When viewing various types of neurons, try to associate the parts of the neuron (e.g., soma, axon, dendrite) with things learned in the lecture about the functions of the nerve cell membrane in those areas and its ability to generate graded or action potentials.

Nervous tissue is composed of two basic cell types: neurons and neuroglia. Neurons are the excitable cells. They have a limited ability to divide and multiply in the adult. Neuroglia (glial cells) are supporting cells whose functions are to support and protect neurons. Glial cells retain the ability to divide and multiply, and they constitute more than half of the matter found in nervous tissue.

The exercises in this chapter involve making observations of the histological structure of both neurons and glial cells.

When working on exercises in subsequent chapters that cover the gross structures of the nervous system, try to relate the microscopic observations with gross observations of nervous system anatomy. Additional exercises within this chapter cover simulated experiments on nervous tissues using the Ph.I.L.S. program that are designed to guide investigation of the functional properties of the neuron plasma membrane (neurolemma).

List of Reference Tables

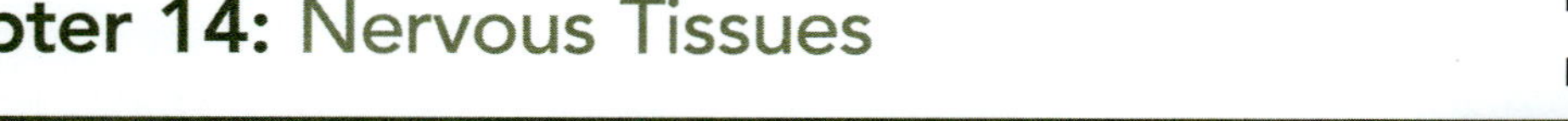

Chapter 14: Nervous Tissues

Name: ______________________

Date: ____________ **Section:** ____________

PRE-LABORATORY WORKSHEET

These Pre-Laboratory Worksheet questions may be assigned by instructors through their connect course.

1. For each structure listed, write the corresponding letter as labeled in the diagram.

____ axon
____ axon hillock
____ cell body (soma)
____ chromatophilic substance
____ dendrite
____ myelin sheath/neurolemmocyte (Schwann cell)
____ neurofibril node (node of Ranvier)
____ nucleolus
____ nucleus
____ synaptic knobs

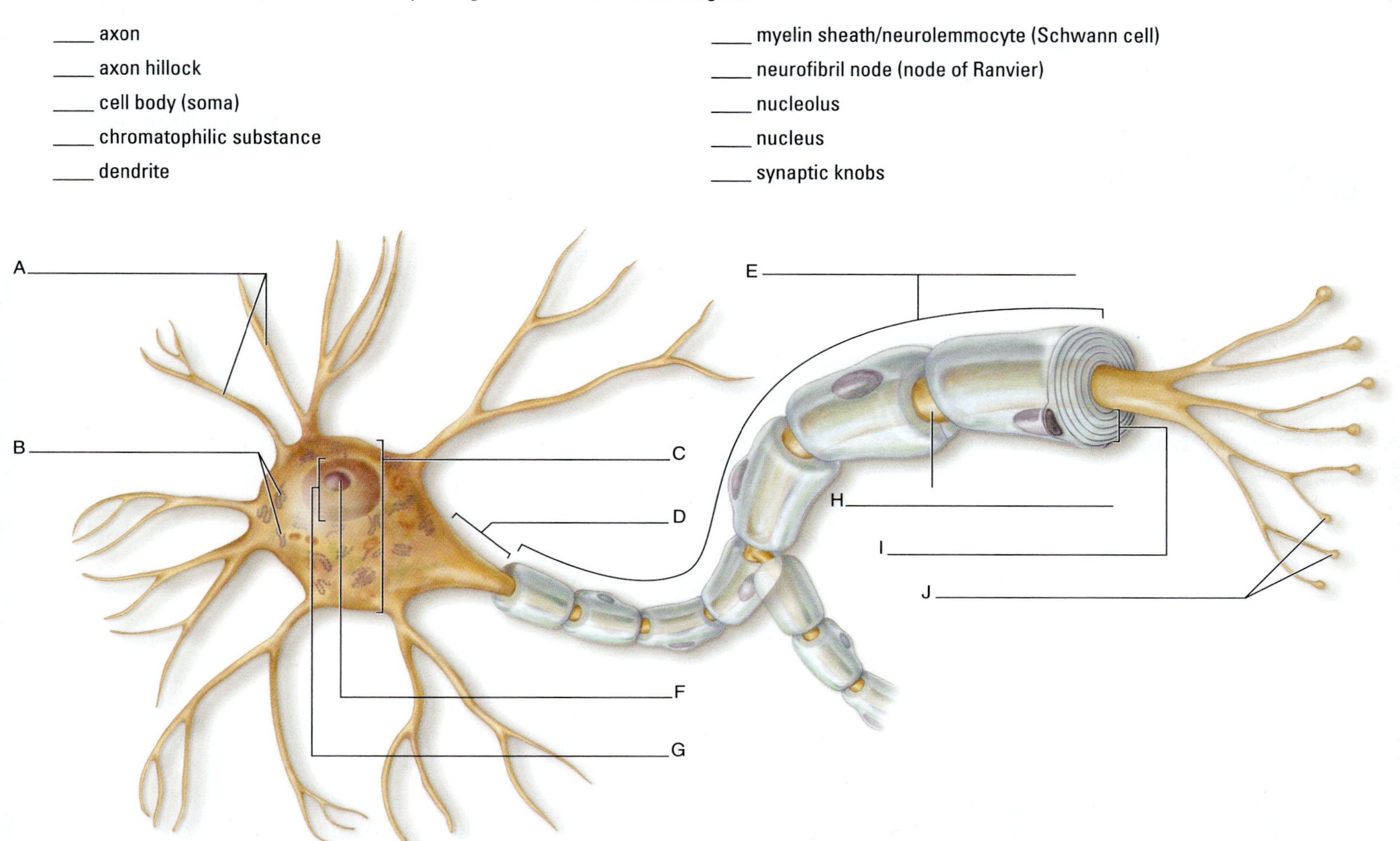

2. Match the description listed in column A with the corresponding cell type listed in column B.

Column A

____ 1. cells that myelinate axons in the central nervous system
____ 2. cells that help reinforce the blood-brain barrier
____ 3. glial cells found within peripheral nerve ganglia
____ 4. cells that myelinate axons in the peripheral nervous system
____ 5. cells that engage in phagocytosis in response to tissue injury
____ 6. cells that line the ventricles of the brain

Column B

a. astrocytes
b. ependymal cells
c. microglia
d. neurolemmocytes (Schwann cells)
e. oligodendrocytes
f. satellite cells

3. Match the description listed in column A with the corresponding connective tissue structure listed in column B.

Column A

____ 1. surrounds a fascicle of axons
____ 2. surrounds an individual axon
____ 3. surrounds the entire nerve

Column B

a. endoneurium
b. epineurium
c. perineurium

HISTOLOGY

At the gross level, nervous tissue is classified as either white matter or gray matter. **White matter** consists mainly of myelinated axons. Bundles of myelinated axons in the central nervous system are called **tracts,** and bundles of axons (either mylelinated or unmyelinated) in the peripheral nervous system are called **nerves.** White matter is white because of the presence of myelin, a fatty substance produced by glial cells that is used to insulate axons to increase the speed of propagating action potentials along the plasma membrane of an axon. **Gray matter** consists mainly of neuron cell bodies, dendrites, and unmyelinated axons. Collections of gray matter within the central nervous system are called **nuclei,** and collections of gray matter within the peripheral nervous system are called **ganglia.** Gray matter is gray due to an absence of myelin and the presence of chromatophilic substance (aggregates of ribosomes within neurons). The exercises in this chapter begin with observations of the histological appearance of white matter and gray matter and then proceed to observations of specific cell types found within nervous tissues.

EXERCISE 14.1

GRAY AND WHITE MATTER

1. Obtain a slide of a cross section of the spinal cord.
2. Observe the slide with the naked eye before placing it on the microscope stage. Notice that the spinal cord consists of an inner, butterfly-shaped core of gray matter surrounded by an outer region of white matter **(figure 14.1).** What neuronal structures are found within the gray matter?

   ____

   What neuronal structures are found within the white matter? ______________________________________

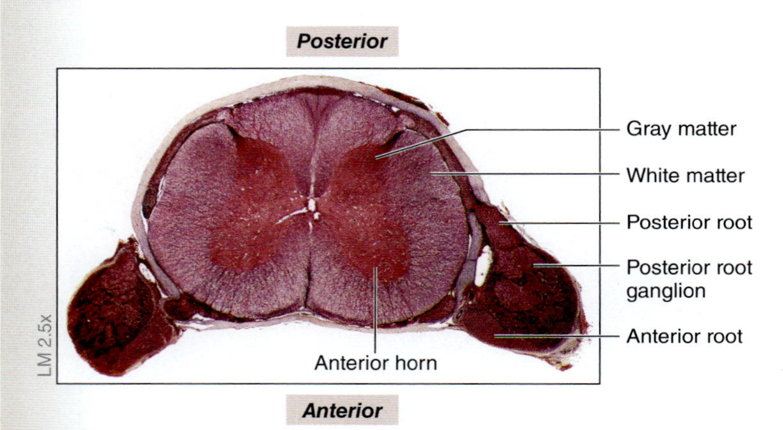

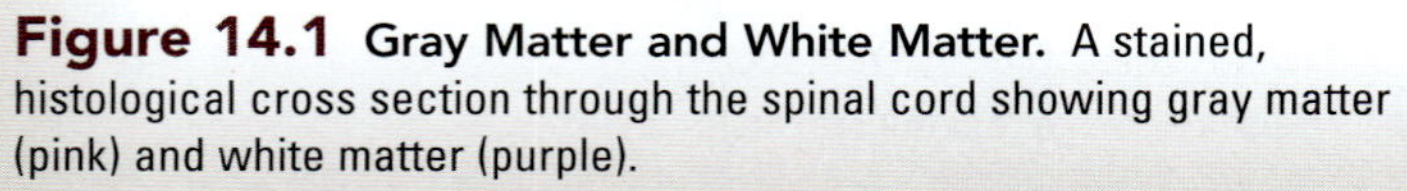

Figure 14.1 Gray Matter and White Matter. A stained, histological cross section through the spinal cord showing gray matter (pink) and white matter (purple).

3. Sketch a cross-section of the spinal cord as seen through the microscope in the space provided. Label the following in the drawing:

- ☐ **gray matter**
- ☐ **white matter**

____________ ×

INTEGRATE

CONCEPT CONNECTION

Recall that the gray matter of the spinal cord contains predominately neuron cell bodies, dendrites, and unmyelinated axons. In addition, gray matter also contains various types of neurons (e.g., motor neurons, somatic sensory neurons, etc.), which are typically contained within a specified area of the gray matter. For example, the posterior horn of the spinal cord contains predominately unmyelinated axons of somatic sensory neurons. The cell bodies for those sensory neurous are located in the posterior root ganglion, and their axons enter the spinal cord via the posterior root. The anterior horn of the spinal cord contains cell bodies for somatic motor neturons. Somatic motor neurons are the neurons that stimulate skeletal muscles to contract. Recall from chapter 11 that a motor unit is a motor neuron and all of the fibers that it innervates. That *somatic* motor neuron becomes "excited" within the anterior horn of gray matter within the spinal cord, and its axon exits the central nervous system through the anterior root. It then joins a spinal nerve and ultimately travels to individual muscle fibers within a skeletal muscle. An action potential traveling along a somatic motor neuron will excite the muscle fibers, causing an increase in force production by the muscle. The lateral horns of gray matter in the thoracic and lumber regions of the spinal cord contain cell bodies of autonomic motor neurons. Autonomic motor neurons innervate cardiac muscle, smooth muscle, and glands (see chapter 17). As with somatic motor neurons, the axons of autonomic motor neurons exit the central nervous system by traveling through the anterior root to a spinal nerve.

Neurons

Neurons are typically the largest cells observed in a slide of nervous tissue. Neurons have very large, light-colored nuclei with prominent nucleoli. The **cell body** (soma) of the neuron contains a large amount of rough endoplasmic reticulum that stains darkly and is called **chromatophilic substance** (or Nissl bodies). Generally **dendrites** are not easily seen, even at high magnification. On the other hand, a single large **axon** can often be identified leaving the cell body. There will be an absence of chromatophilic substance at the location where the axon leaves the cell body, an area of the neuron called the **axon hillock. Table 14.1** summarizes the parts of a typical neuron and their functions.

Neurons are structurally classified based upon the number of cellular processes that extend from the cell body. **Multipolar** neurons have many dendrites and a single axon. **Bipolar** neurons have one dendrite and one axon. Finally, **unipolar** neurons have a single process (*uni*, one). Multipolar neurons are the types of neurons within the anterior horn of the spinal cord, the cerebral cortex, and the cerebellum of the brain **(table 14.2).**

Table 14.1 Parts of a Neuron

Part of the Neuron	Description and Function	Word Origin
Axon	A single large process of a neuron that propagates action potentials away from the cell body of a neuron	*axon*, axis
Axon Hillock	A light-staining region where the axon leaves the cell body of a neuron; devoid of chromatophilic substance; the location where action potentials are initiated by a neuron	*hillock*, a small elevation
Cell Body (Soma)	The part of the neuron that contains the nucleus and cellular organelles	*soma*, body
Chromatophilic (Nissl) Substance	Dark-staining material found within the soma of a neuron; composed of both clusters of ribosomes associated with rough endoplasmic reticulum and free ribosomes that function in the production of proteins	*chroma*, color, + *phileo*, to love
Dendrite	A branching process that extends from the soma of a neuron that transmits graded potentials to the cell body	*dendron*, tree
Neurofibril Node (Node of Ranvier)	A bare region on a myelinated axon where there is an absence of myelin and where action potentials are propagated	*neuron*, nerve, + *fibrilla*, fiber; *Louis Ranvier*, French pathologist
Synaptic Knobs	Swellings on the ends of an axon that form synapses with either another neuron or an effector organ	*syn-*, together, + *hapto*, to clasp, + *knob*, a protuberance
Telodendria (Axon Terminals)	Branches at the end of an axon, with each process containing a synaptic knob at its end	*telos*, end, + *dendron*, tree

Table 14.2 Multipolar Neurons Within the Spinal Cord, Cerebrum, and Cerebellum

Neuron Type	Location	Features
Anterior Horn Cell	Anterior horn of the spinal cord	Very large neuron with prominent chromatophilic substance; cell body is irregularly shaped with multiple dendrites extending from the soma
Purkinje Cell	Cerebellum	Cell bodies appear "basket-like" with a rounded area facing the granular layer of the cerebellum, and a tuft of dendrites extending into the molecular layer of the cerebellum
Pyramidal Cell	Cerebral cortex	Cell body has a triangular shape and multiple dendrites extend from the cell body

INTEGRATE

CONCEPT CONNECTION

Nervous tissue is characterized by **excitability** and **conductivity.** Recall that excitability is the ability of a cell to respond to a stimulus (e.g., chemical, stretch, pressure change). The stimulus causes a local change in the resting membrane potential of the excitable cell. Excitability is due to the opening of either chemically-gated or modality-gated channels that result in graded potentials. Conductivity involves an electrical change that is quickly propagated along the plasma membrane as voltage-gated channels open sequentially during an action potential. Not all portions of the neurolemma (plasma membrane) of a neuron are excitable. The neuron can be divided into segments based on structural features and function. For example, the *receptive segment* contains the dendrites and the soma of a neuron. The neurolemma of these structures contain a large number of ligand-gated ion channels, which are opened or closed by the binding of neurotransmitters ("ligands") released by the axon terminals of neighboring neurons. Therefore, the structures that make up the receptive segment *receive* the signal. Because these structures do not contain a large number of voltage-gated sodium channels, the membrane can only respond by generating graded potentials (called "post-synaptic potentials"). The *intial segment,* or axon hillock, of the neuron contains a high density of voltage-gated sodium ion channels, making the neurolemma very excitable. Thus, action potentials are *initiated* here. The *conductive segment,* or axon, also contains a large number of voltage-gated sodium ion channels along its length. Thus, the axon *conducts* the action potential to the axon terminals.

EXERCISE 14.2

GENERAL MULTIPOLAR NEURONS—ANTERIOR HORN CELLS

1. Obtain a slide of the *spinal cord in cross section.*
2. Place the slide on the microscope stage and bring the tissue sample into focus on low power **(figure 14.2*a*).**

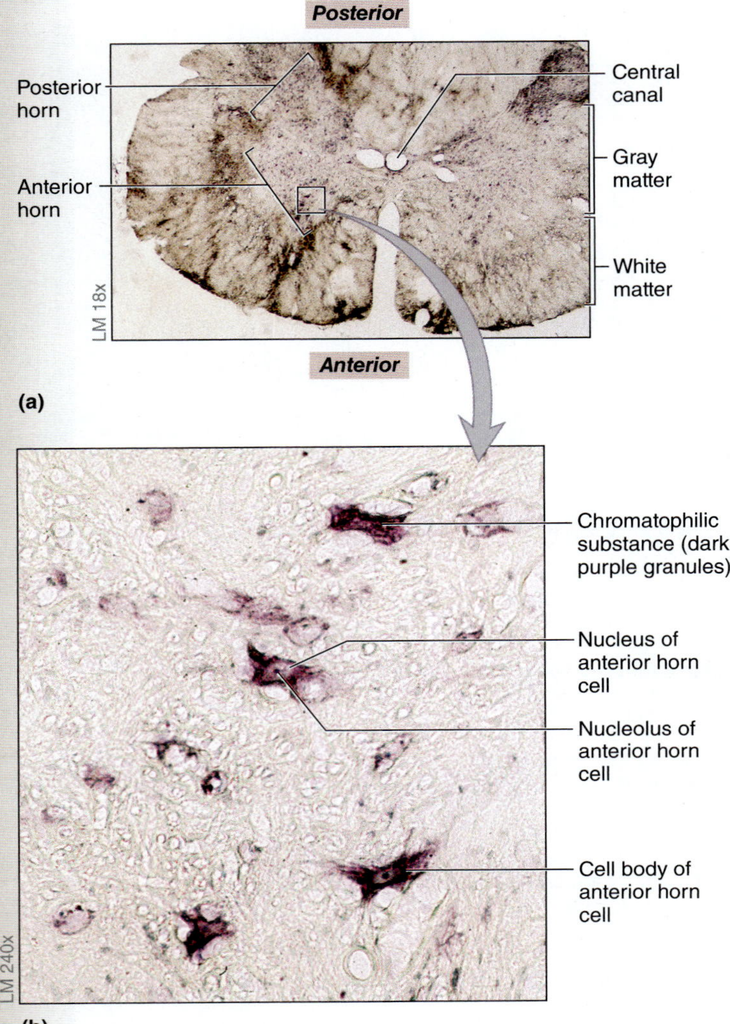

Figure 14.2 Gray Matter of the Spinal Cord. (a) Cross section of spinal cord. (b) Close up showing large motor neurons (anterior horn cells).

Again, note the inner core of gray matter surrounded by an outer region of white matter. Move the microscope stage so the inner gray matter is in the center of the field of view and then switch to high power (figure 14.2*b*). Look for very large, multipolar neurons found in the anterior (ventral) horn of the gray matter. These cells are called **anterior horn cells,** based on their location. They are large somatic motor neurons whose axons exit the spinal cord and travel through peripheral nerves to skeletal muscles in the body.

3. Identify the following structures on the spinal cord slide, using tables 14.1 and 14.2 and figure 14.2 as guides:

- ☐ **cell body of anterior horn cell**
- ☐ **chromatophilic substance**
- ☐ **gray matter**
- ☐ **nucleolus of anterior horn cell**
- ☐ **nucleus of anterior horn cell**
- ☐ **white matter**

4. Sketch anterior horn cells as seen through the microscope in the space provided.

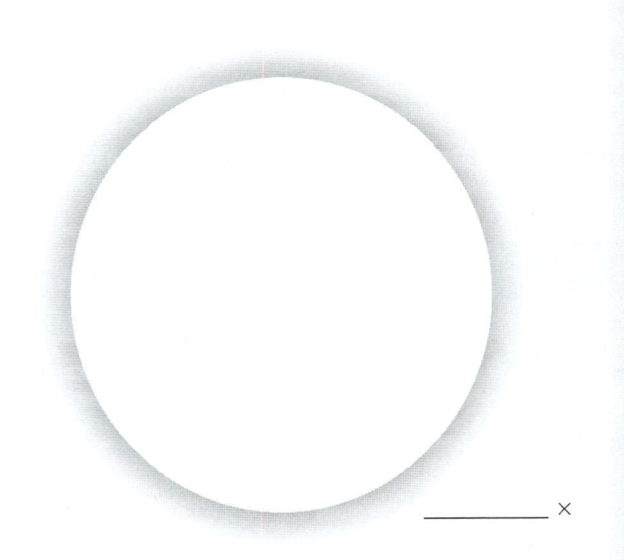

EXERCISE 14.3

CEREBRUM—PYRAMIDAL CELLS

1. Obtain a slide of the cerebrum that has been stained with Nissl stain. Nissl stain colors the **rough endoplasmic reticulum** and free ribosomes (the chromatophilic substance) of the neurons an intense blue color.
2. Place the slide on the microscope stage and bring the tissue sample into focus on low power. Identify areas of gray matter and white matter on the slide **(figure 14.3).**
3. Bring an area of gray matter to the center of the field of view and switch to high power. Note the very large cells located within the gray matter. These cells are **neurons.** Note the very large nuclei of the *neurons*. Neurons are surrounded by much smaller cells called *glial cells*. Locate large neurons whose cell bodies appear triangular in shape. These neurons of the cerebrum are called **pyramidal cells** because the cell body's three-dimensional shape is pyramid-shaped. (figure 14.3, table 14.2).

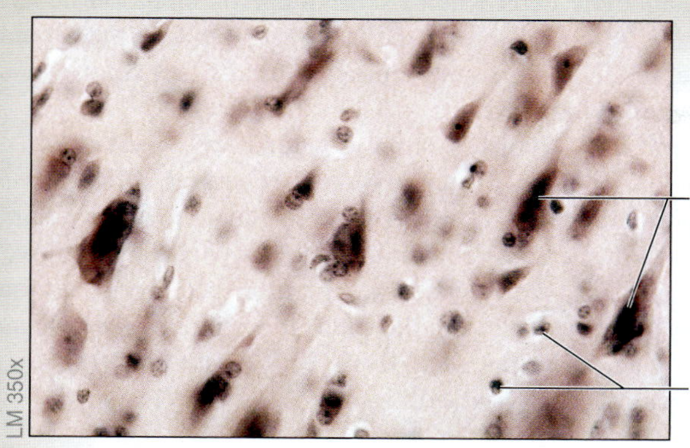

Figure 14.3 Pyramidal Cells in the Cerebral Cortex. The pyramidal cells (neurons) are the large triangular cells. The smaller nuclei visible in the slide are mainly those of glial cells.

4. Sketch pyramidal cells as seen through the microscope in the space provided. Label the following in the drawing:

☐ glial cells ☐ pyramidal cells

_________ ×

EXERCISE 14.4

CEREBELLUM—PURKINJE CELLS

1. Obtain a slide of the **cerebellum.** Place it on the microscope stage and bring the tissue sample into focus on low power **(figure 14.4*a*).**
2. Note the folds of tissue within the outer region, the *molecular layer*, and the inner region, the *granular layer* of the cerebellum (figure 14.4*a*). Deep to the folds is the white matter of the cerebellum. Move the microscope stage to bring the junction between the molecular and granular layers to the center of the field of view. Then switch to high power (figure 14.4*b*). Identify the very large cells located at this juncture. These cells are **Purkinje cells** (table 14.2), also known as basket cells, which are large multipolar neurons of the cerebellum (the layer where they are located is the *Purkinje cell layer*).
3. Sketch Purkinje cells as seen through the microscope at high power in the space provided. Label the following in the drawing:

☐ granular layer ☐ Purkinje cells
☐ molecular layer ☐ white matter

_________ ×

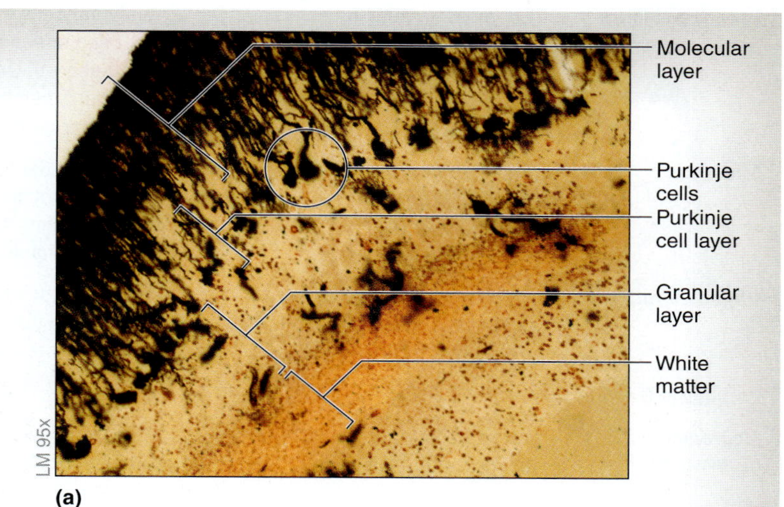

(a)

Dendrites
Cell body of Purkinje cell
LM 300x

(b)

Figure 14.4 Cerebellum—Purkinje Cells. The cerebellum, demonstrating Purkinje cells between the granular and molecular layers. (a) Low power. (b) High power.

Glial Cells

Glial cells, or **neuroglia** (literally, "nerve glue"), are much more abundant than neurons in the nervous system, and they retain the capacity for cell division. Thus, most brain tumors arise from glial cells and are referred to as *gliomas*. The two most abundant glial cells in the central nervous system (CNS) are **astrocytes** and **oligodendrocytes.** Because of their large size, astrocytes and oligodendrocytes are collectively referred to as *macroglia*. The central nervous system also contains resident macrophages called **microglia.** As their name implies, microglia are very tiny cells. Their name is a bit of a misnomer, however, because these cells become very large, phagocytic cells when tissue injury or infection occurs. Microglial cells are not visible in the slides available in the laboratory. Special epithelial cells found lining the fluid-filled ventricles of the brain are **ependymal cells.**

The peripheral nervous system contains only one type of glial cell, but it is named differently—and functions differently—depending upon its location. These glial cells have the same embryonic origin, but not the same function, which is why they are given different names. Thus, we typically refer to the peripheral nervous system (PNS) as having two glial cell types: *neurolemmocytes (Schwann cells)* and *satellite cells*. **Table 14.3** summarizes the characteristics of each type of glial cell.

Table 14.3 Glial Cells

Cell Name	Description	Function(s)	Word Origin
Central Nervous System			
Astrocytes	Star-shaped cells that are very abundant within the central nervous system	General supporting cells in the CNS. Transfer nutrients to neurons from the blood. Reinforce the blood-brain barrier. Maintain the extracellular environment around neurons.	*astron*, star, + *kytos*, cell
Ependymal Cells	Cuboidal to columnar-shaped cells with microvilli and cilia on their apical surfaces	Ependymal cells are epithelial cells that line the ventricles (fluid-filled spaces) of the brain and the central canal of the spinal cord. Unlike other epithelia, they DO NOT have a basement membrane. They play a role in the production and circulation of CSF.	*ependyma*, an upper garment
Microglia	Small cells with oval nuclei and multiple branching processes	Microglia are derived from blood monocytes, and they are the resident macrophages in the CNS. They are normally very small (hence the name), but transform into very large, phagocytic cells (macrophages) when tissues are injured or infected.	*mikros*, small, + *glia*, glue
Oligodendrocytes	Cells with several long processes that wrap around axons in the CNS	Myelinate axons in the central nervous system. Each oligodendrocyte can myelinate multiple axons. Myelination allows for faster nerve signal propagation.	*oligos*, few, + *dendro-*, like a tree, + *kytos*, cell
Peripheral Nervous System			
Neurolemmocytes (Schwann Cells)	Large cells that wrap around axons in the PNS	Myelinate axons in the peripheral nervous system. Each neurolemmocyte can myelinate only part of one axon; typically takes many neurolemmocytes to myelinate an entire axon. Myelination allows for faster nerve signal propagation.	*neuron*, nerve, + *lemma*, husk, + *kytos*, cell; *Theodor Schwann*, German histologist and physiologist
Satellite Cells	Small glial cells that surround the cell bodies within a ganglion (e.g., posterior root ganglion)	Satellite cells surround the cell bodies of somatic sensory neurons, hence the appearance of "satellites" around those neurons. They provide general support for the neurons and are analogous in function to astrocytes in the CNS.	*satelles*, attendant

Glial Cells of the Central Nervous System

EXERCISE 14.5

ASTROCYTES

1. Obtain a slide of the cerebrum or cerebellum that has been stained with silver stain. Silver stain makes the general supporting cells of the central nervous system, the **astrocytes,** stain very dark so they are visible through the microscope **(figure 14.5).**
2. Place the slide on the microscope stage and bring the tissue sample into focus on low power. Then switch to high power and bring the tissue sample into focus once again.
3. The two most prominent cell types visible are the large neurons (e.g., pyramidal cells in the cerebrum or Purkinje cells in the cerebellum) and the smaller astrocytes. Astrocytes, as their name implies, are shaped like stars (*astron-*, star). They have multiple long cellular processes that wrap themselves around neurons and around blood vessels in the central nervous system. These processes allow astrocytes to perform one of their main functions, which is to transport nutrients from the blood to the neurons.

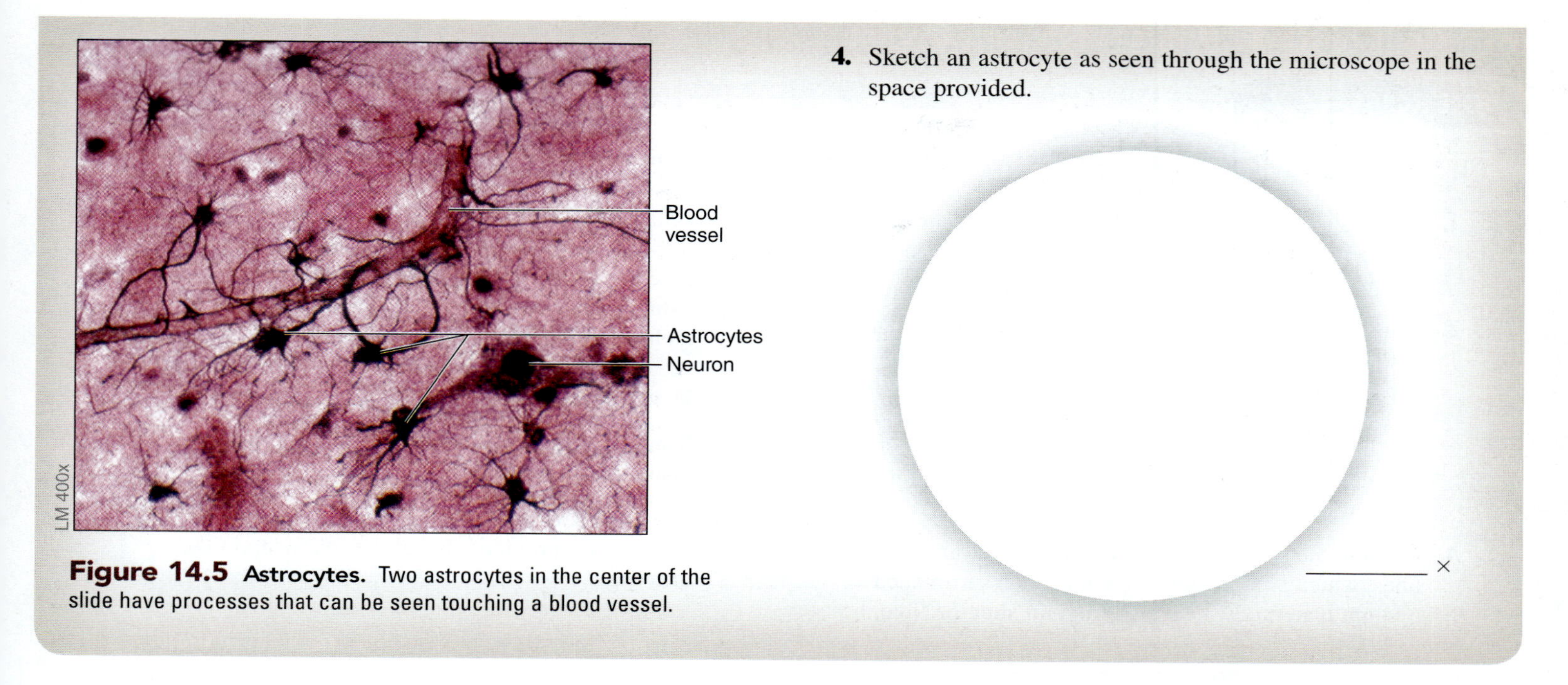

Figure 14.5 Astrocytes. Two astrocytes in the center of the slide have processes that can be seen touching a blood vessel.

4. Sketch an astrocyte as seen through the microscope in the space provided.

______ ×

EXERCISE 14.6

EPENDYMAL CELLS

1. Obtain a slide of the brain and place it on the microscope stage. Bring the tissue sample into focus on low power. Then switch to high power and bring the tissue sample into focus once again. Look for the "empty" spaces on the slide, because those spaces will likely be the ventricular spaces that are lined with ependymal cells.
2. **Ependymal cells (figure 14.6)** are cuboidal to columnar-shaped epithelial cells that line the ventricles of the brain and the central canal of the spinal cord. They also form the outer layer of the choroid plexus. They have both *cilia* and *microvilli* on their apical surfaces, but they are unique as epithelia in that they do not have a basement membrane.
3. Locate ependymal cells on the slide, using figure 14.6 as a guide.
4. Sketch ependymal cells as seen through the microscope in the space provided. Label the following in the drawing:

☐ **ependymal cells** ☐ **ventricular space**

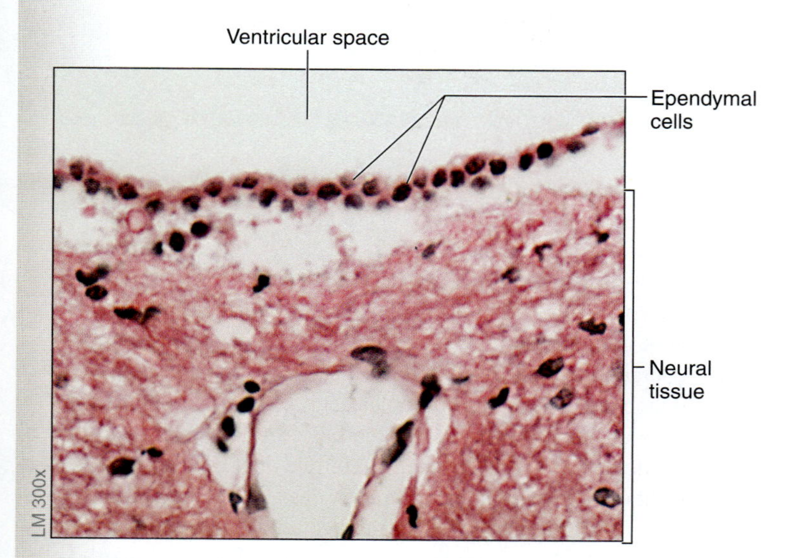

Figure 14.6 Ependymal Cells. Cuboidal ependymal cells lining a ventricular surface of the brain.

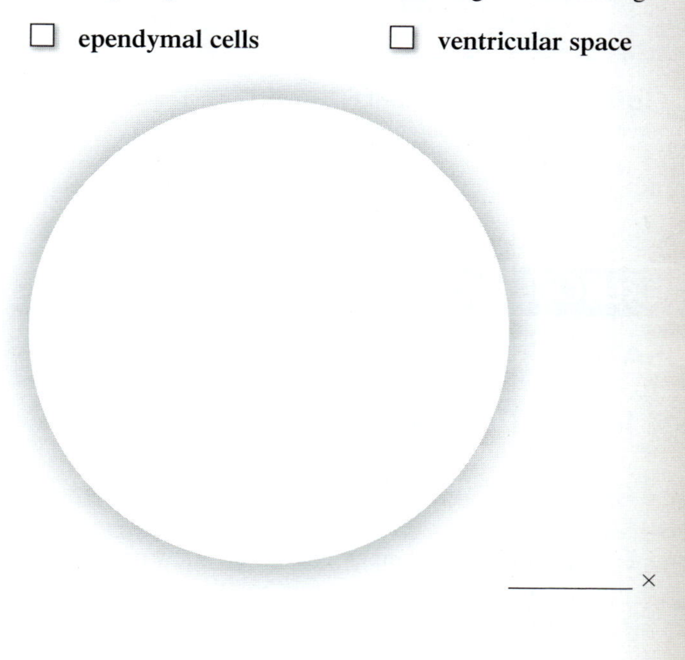

______ ×

Glial Cells of the Peripheral Nervous System

EXERCISE 14.7

NEUROLEMMOCYTES (SCHWANN CELLS)

1. Obtain a slide of a longitudinal section of a myelinated peripheral nerve and place it on the microscope stage **(figure 14.7).**
2. Bring the tissue sample into focus on low power and note the wavy appearance of the axons. The axons appear wavy because the nerve is not stretched. When a nerve is stretched, the axons also stretch out. The fact that peripheral nerves have elasticity is important because it allows the nerves to stretch during movement without being damaged.
3. Observe the slide on high power. Note the very dark lines, which are the **axons**. Each axon is surrounded by light material, which is the **myelin sheath.** Note the elongated nuclei seen throughout the slide. These nuclei belong to **neurolemmocytes (Schwann cells),** the glial cells responsible for myelinating axons in the peripheral nervous system. Each individual neurolemmocyte myelinates only a portion of a single axon, and a single axon typically has hundreds of neurolemmocytes along its length. The bare areas of axon found between myelin sheaths are called **neurofibril nodes,** or **nodes of Ranvier.**
4. Scan the slide to see if one or more neurofibril nodes are visible.
5. Sketch a peripheral nerve as seen through the microscope in the space provided. Label the following in the drawing:

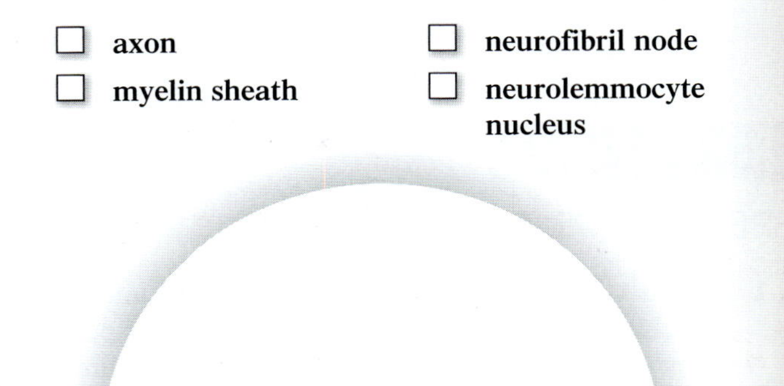

______ ×

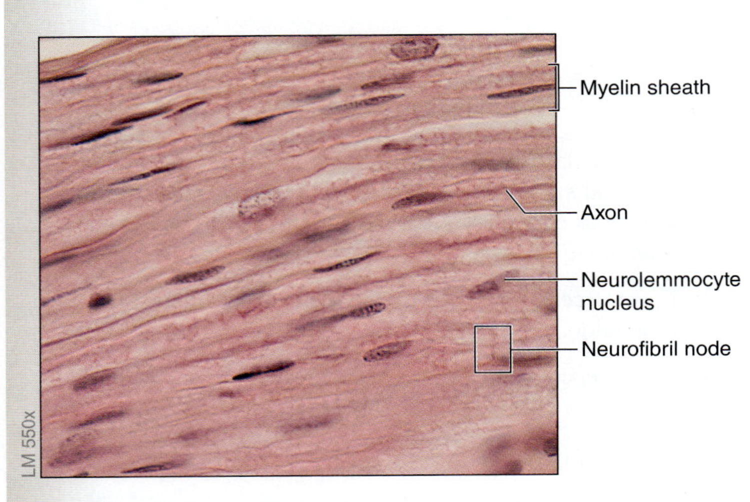

Figure 14.7 Neurofibril Nodes. Longitudinal section of a nerve demonstrating neurofibril nodes.

EXERCISE 14.8

SATELLITE CELLS

1. Obtain a slide of a spinal ganglion (the slide may be labeled dorsal root ganglion, posterior root ganglion, or peripheral nerve ganglion) and place it on the microscope stage.
2. Bring the tissue sample into focus on low power and locate both the nerve *root* and the *ganglion* **(figure 14.8*a*).**
3. Recall that the peripheral nervous system is mainly composed of nerves, which are bundles of myelinated axons. Occasionally collections of neuron cell bodies are found along these nerves. A collection of neuron cell bodies in the peripheral nervous system is called a **ganglion** (*ganglion*, swelling). The name comes from the fact that the area where the neuron cell bodies aggregate appears as a "swelling" on the cord-like nerve. The posterior roots of the spinal cord contain swellings called **posterior (dorsal) root ganglia** (peripheral nerve ganglia). These structures contain the neuron cell bodies of somatic sensory neurons, which are classified as unipolar neurons. Unipolar neurons have only a single process extending from the cell body of the neuron. The absence of multiple dendrites coming right off the cell bodies of the neurons allows the glial cells (which are also found in the ganglion) to lie directly adjacent to the cell bodies of the neurons. They look like small "satellites" surrounding the neuron cell body, and are called **satellite cells.** Satellite cells are general supporting cells for neurons within the posterior root ganglia.

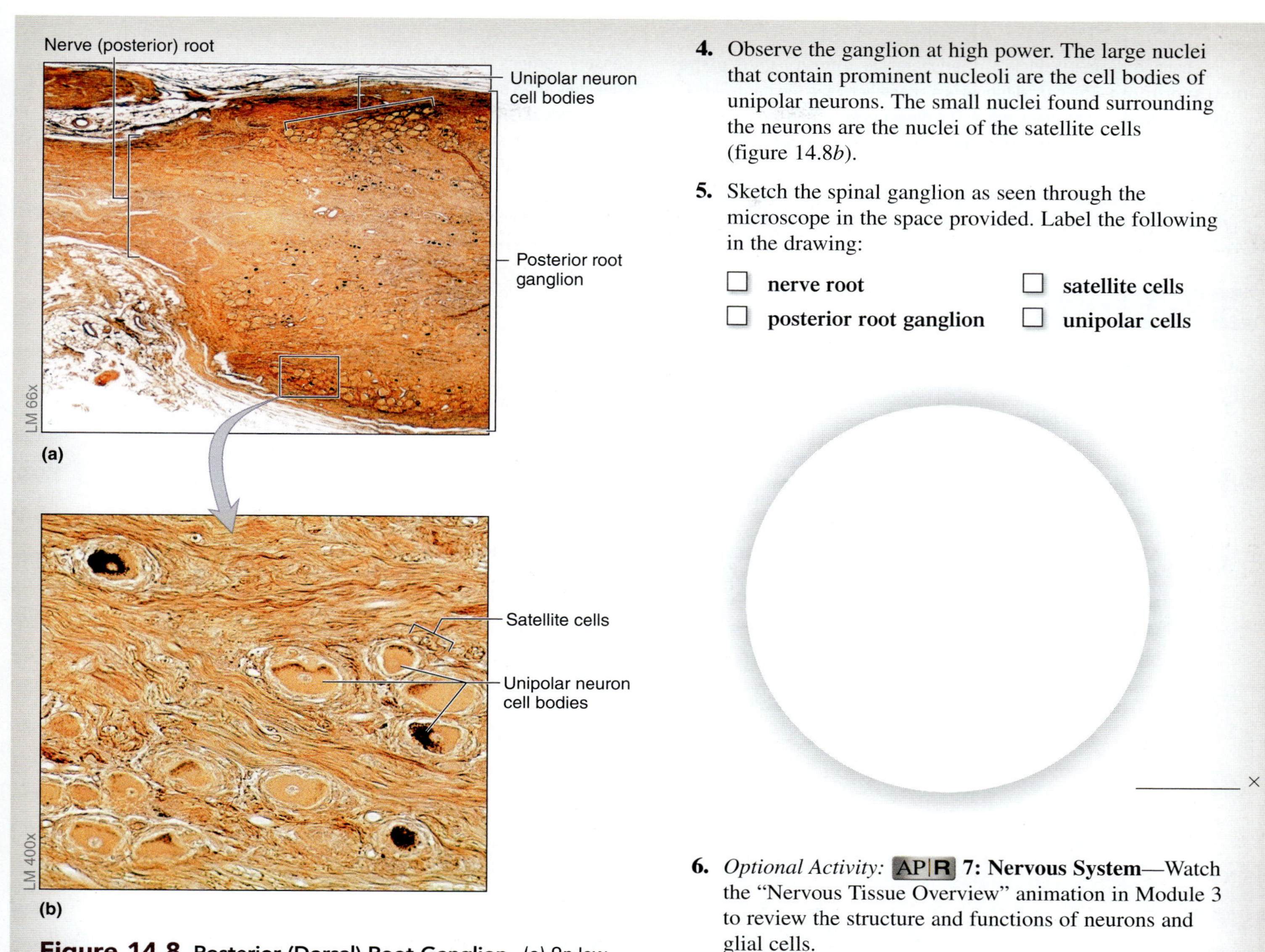

Figure 14.8 Posterior (Dorsal) Root Ganglion. (a) On low power the nerve root and the ganglia are visible. (b) On high power the unipolar neurons surrounded by satellite cells are visible.

4. Observe the ganglion at high power. The large nuclei that contain prominent nucleoli are the cell bodies of unipolar neurons. The small nuclei found surrounding the neurons are the nuclei of the satellite cells (figure 14.8*b*).

5. Sketch the spinal ganglion as seen through the microscope in the space provided. Label the following in the drawing:

   - ☐ **nerve root**
   - ☐ **posterior root ganglion**
   - ☐ **satellite cells**
   - ☐ **unipolar cells**

   _______ ×

6. *Optional Activity:* **AP|R 7: Nervous System**—Watch the "Nervous Tissue Overview" animation in Module 3 to review the structure and functions of neurons and glial cells.

Peripheral Nerves

Both cranial nerves (nerves extending from the brain) and spinal nerves (nerves extending from the spinal cord) are collections of myelinated axons that are bundled together with connective tissue. The bundling pattern is exactly like the bundling pattern seen with skeletal muscle, and the prefixes used for the names of the connective tissue coverings are also the same. An entire peripheral nerve is bundled by connective tissue called **epineurium.** Within an entire nerve, the axons are grouped into bundles called **fascicles.** Each fascicle is surrounded by a connective tissue covering called the **perineurium.** Within each fascicle are individual axons. Each individual axon is surrounded by a connective tissue covering called the **endoneurium.** Between the fascicles are blood vessels that supply nutrients to the nerve.

INTEGRATE

CLINICAL VIEW
Peripheral Nerve Injury

Injury to peripheral nerves can be very disabling. There are varying degrees of severity of peripheral nerve injuries. In the least severe cases, the nerve itself or nearby blood vessels may be compressed (i.e., "crush" injury), which leads to decreased excitability of axons within the nerves. The damage to the nerves is typically temporary, and is relieved when compressive forces are removed. Slightly more severe cases occur when axons within the nerve are severed, but the integrity of surrounding connective tissues and glial cells are maintained. When axons are severed, the propagation of action potentials along the nerve temporarily ceases, causing the patient to experience loss of motor function and sensation along the nerve. However, as long as the surrounding glial cells and connective tissue are intact, nerve regeneration can occur in the weeks and months following injury. The most severe cases of peripheral nerve injury to the neurons and glial cells occurs when connective tissues and glial cells that surround the nerve are also damaged. In such cases the loss of motor function and sensation may be permanent.

(continued on next page)

(continued from previous page)

One common clinical example of a peripheral nerve crush injury is **carpal tunnel syndrome.** The carpal tunnel is a space between the carpal bones and the flexor retinaculum in the wrist. The median nerve, and tendons of the flexor muscles of the forearm, travel through the carpal tunnel. The median nerve is a mixed nerve, carrying both motor and sensory information to the palm of the hand and muscles in the region of the thumb. In minor injuries, a patient may report a tingling sensation, which occurs because of damage to sensory portions of the nerve. In more severe cases, motor output may be impacted, and a patient may experience partial paralysis of muscles that control thumb movements. Typical treatments for carpal tunnel syndrome invlove splinting the wrist (to prevent further injury to the nerve due to movement of the wrist) or local injections of steroids (to decrease inflammation of the nerve). The most severe cases require "carpal tunnel release" surgery, in which the flexor retinaculum is cut or lengthened to relieve compression of the nerve.

EXERCISE 14.9

COVERINGS OF A PERIPHERAL NERVE

1. Obtain a slide of a cross section of a peripheral nerve and place it on the microscope stage.
2. Bring the tissue sample into focus on low power. Identify epineurium, fascicles, and perineurium, using **figure 14.9** as a guide.
3. Observe the slide at high power so the individual *axons* are visible in cross section. Each axon is surrounded by a myelin sheath, which is then surrounded by a connective tissue covering of endoneurium.
4. Sketch a cross section of a peripheral nerve as seen through the microscope in the space provided. Label the following in the drawing:

- ☐ **axon**
- ☐ **blood vessel**
- ☐ **endoneurium**
- ☐ **epineurium**
- ☐ **fascicle**
- ☐ **myelin sheath**
- ☐ **perineurium**

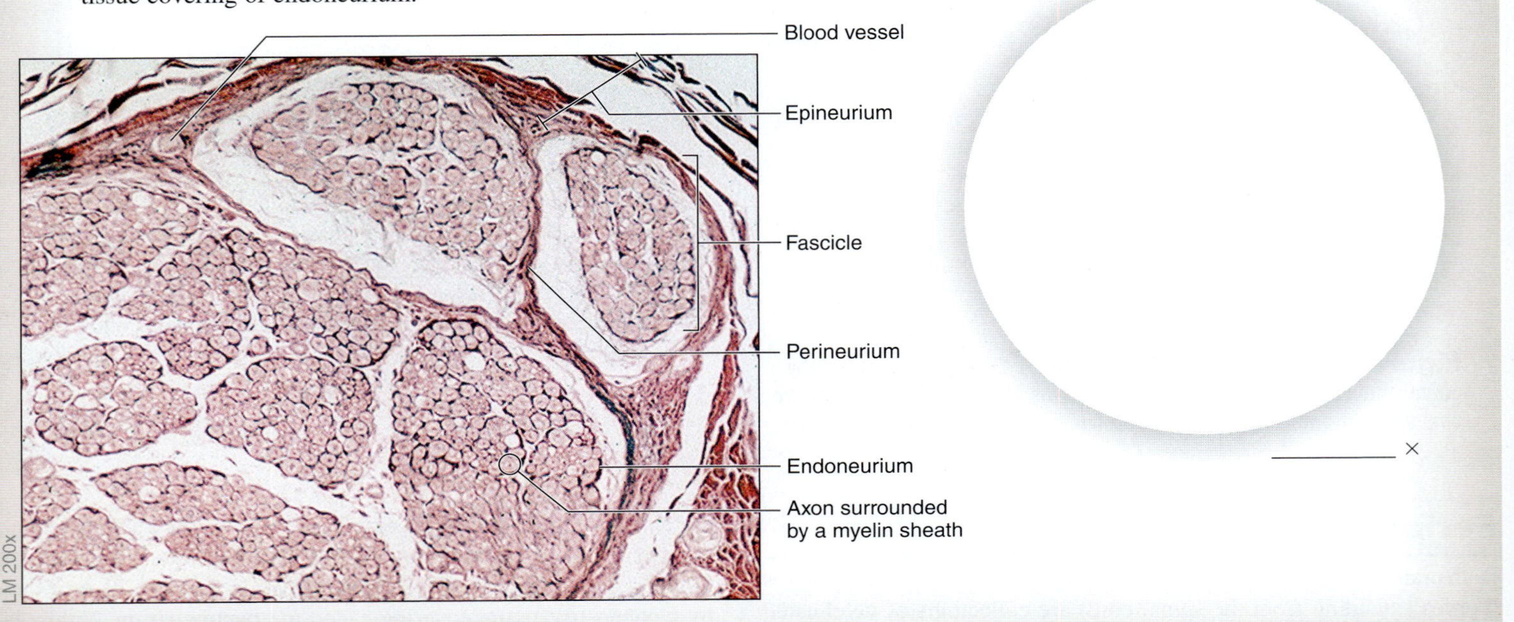

Figure 14.9 Cross Section Through a Peripheral Nerve.

PHYSIOLOGY

Resting Membrane Potential

The plasma membrane of an excitable cell is composed of the same biomolecular structures discussed in chapter 4. That is, the selectively permeable membrane is a phospholipid bilayer, oriented such that the hydrophilic (*hydro-*, water + *-philos*, fond) heads are pointed toward the intracellular and extracellular fluid and the hydrophobic (*hydro-*, water + *-phobos*, fear) tails are pointed toward one another within the interior of the membrane. Proteins, including transport pumps, receptors, and ion channels, provide mechanisms to move ions into and out of the cell.

At rest, the selectively permeable membrane allows for the separation of positive and negative charges. The **resting membrane potential** is negative, meaning that the inside of the membrane is relatively negative compared to the outside of the membrane. That is, there are more negatively charged ions and molecules inside of the membrane than outside of the membrane. Establishing and maintaining the resting membrane potential is dependent on K^+ leak channels, Na^+ leak channels, and Na^+/K^+ pumps. The Na^+/K^+ pump moves sodium out of the cell and potassium into the cell in a ratio of 3 Na^+ out of the cell for every 2 K^+ into the cell, thereby maintaining concentration gradients for both Na^+ and K^+. At rest, there are leak channels for both Na^+ and K^+, although there are many more leak channels for K^+ than for Na^+. Therefore, the cell is more permeable to K^+ than to Na^+ at rest. There is also an electrical gradient across the membrane that is largely due to an excess of negatively charged phosphate ions and proteins in the intracellular fluid, which attract the K^+ ions back into the cell. Eventually, there is no net movement of K^+, as the electrical and chemical gradients are equal in magnitude and opposite in direction.

The typical resting membrane potential for neurons is –70 millivolts (mV). However, rather than simply memorizing a value of the resting membrane potential, know that this number may fluctuate between –40 mV and –90 mV as the membrane's permeability to particular ions changes. Most significantly, the changes in the concentration of extracellular Na^+ or K^+ have the ability to alter the resting membrane potential. The membrane potential deviates from resting when gated channels specific for ions (e.g., Na^+, K^+) are opened. The following laboratory exercises investigate the changes in resting membrane potential that result from changing the concentration of these ions in the extracellular fluid.

INTEGRATE

LEARNING STRATEGY

In discussions of energy (see chapter 4,) we discussed two main forms of energy: **potential energy,** which is stored energy (e.g., water held behind a hydroelectric dam) and **kinetic energy,** which is energy of movement (e.g., water released from a hydroelectric dam). As more potential energy is stored, more energy is available for conversion to kinetic energy, which can then be used to do work. In the example of a hydroelectric dam, when water is released from the dam it moves turbines within the dam. The moving turbines create electrical energy, which can then be used to provide power to homes.

In an excitable cell such as a neuron, the separation of positive and negative charges by the selectively permeable plasma membrane represents a form of potential energy similar to holding water behind a dam. The amount of energy that is *stored* across the membrane is the **membrane *potential*.** The unit of measure for electrical potential energy is a **volt.** Thus, we speak of a resting membrane potential, or **membrane voltage.** The membrane voltage is very small, on the order of millivolts.

EXERCISE 14.10 Ph.I.L.S.

Ph.I.L.S. LESSON 10: RESTING POTENTIAL AND EXTERNAL [K^+]

This exercise involves observing the changes in resting membrane potential of crayfish muscle fibers when they are placed in solutions of increasing concentrations of KCl. This simulates the effect of altering the [K^+] (potassium ion concentration) in normal extracellular fluid. This exercise—along with exercise 14.11 (Resting Potential and External [Na^+])—emphasizes the contribution of K^+ and Na^+ to development of the resting membrane potential (E_m). This exercise explores why it is so important physiologically for the body to maintain concentration gradients of K^+ and Na^+ between intracellular fluid (ICF) and extracellular fluid (ECF) within narrow limits. Large variations in the concentrations of either of these two ions can have drastic consequences for the functioning of excitable cells.

While the study cells examined in this simulation are crayfish muscle fibers, the concepts apply to any excitable cells, including neurons. In fact, early studies that were performed to understand the function of excitable cells were done on squid axons because they are so large that microelectrodes could be placed in them in much the same way that the simulated microelectrodes are placed in crayfish muscle in this exercise.

Before beginning the exercise, become familiar with the following concepts (use the textbook as a reference):

- Function of the Na^+/K^+ ATPase pump
- Normal concentrations of Na^+ and K^+ in the ICF and ECF
- Resting membrane potential
- Permeability of the cell membrane to Na^+ and K^+ ions at rest
- Relative contributions of Na^+ and K^+ to development of the resting membrane potential
- Depolarization, hyperpolarization, and repolarization

Before beginning the experiment, state a hypothesis regarding the influence of extracellular [K^+] on resting membrane potential.

____

____

____

____

1. Open Ph.I.L.S. Lesson 10: Resting Potential and External [K^+] **(figure 14.10).**
2. Read the objectives, introduction, and wet lab exercises. Then, take the pre-lab quiz. The lab exercise will open when the pre-lab quiz has been completed.
3. To start the experiment, click the "power" switch on the virtual computer screen. Then, click the "power" switch on the data acquisition unit and the electrometer.

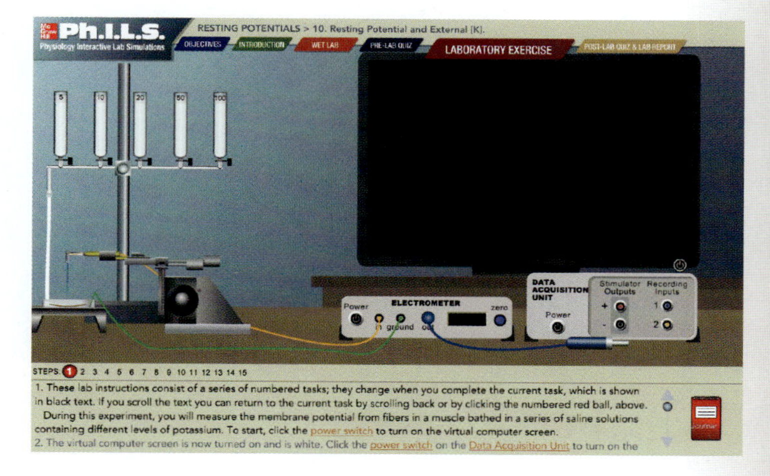

Figure 14.10 Setup of the Wet Lab for Ph.I.L.S. Exercise 10: Resting Potential and External [K^+]. An isolated muscle fiber has been placed in the apparatus and is hooked up to both stimulating and recording electrodes. Test tubes above the setup are arranged to allow for bathing the muscle in solutions containing varying concentrations of K^+.

(continued on next page)

(continued from previous page)

4. Click and drag the "blue plug" to input 1 on the data acquisition unit to connect the electrometer.
5. To record the output of the electrometer, click the "start" button in the upper right-hand corner of the virtual computer screen. A blue line appears on the virtual monitor. This blue line will represent the voltage measured from the crayfish preparation.
6. Zero the electrometer by clicking the "zero" button near the read-out on the electrometer.
7. The microelectrode is a thin blue line above the crayfish preparation. To insert the electrode into the sample, click the "down control" on the micromanipulator (large black wheel on the bottom left of the screen). Watch the blue line as the electrode is inserted into the crayfish muscle fiber. When a change in the voltage is registered on the electrometer, release the down control. It may be required to move the microelectrode laterally for correct positioning. If so, click the "up control" and then the "lateral controls" to correctly position the microelectrode.
8. The electrode has penetrated the cell membrane and is now measuring the membrane potential. Record that value here: E_m (normal saline) = ____________________ Then click the "journal" to enter the data into the Ph.I.L.S. journal. The control part of the experiment is now complete. Subsequent recordings will be experimental recordings where the [K^+] varies in the ECF.
9. To begin the experiment, raise the micromanipulator so the electrode is no longer in the muscle fiber by clicking the "up control" arrow.
10. Perfuse the sample with 5 mEq/L K^+ solution by clicking the "tap" below the appropriately labeled test tube.
11. Reposition the microelectrode by clicking the "lateral controls" to the left, right, anterior, or posterior. Find an intact portion of membrane to make a new recording.
12. Click the "zero" on the electrometer.
13. Insert the microelectrode into the crayfish muscle preparation by clicking the "down control" on the micromanipulator. Remember to watch for the blue line to move, indicating the microelectrode is recording a change in voltage across the membrane. When the blue line changes, release the down control and record the membrane potential in the Ph.I.L.S. journal, and in **table 14.4.**
14. Repeat step 13 until there are five measurements from different portions of the membrane in the 5 mEq/L K^+ solution. Record the data in table 14.4.
15. Repeat steps 7 through 14 with the 10, 20, 50, and 100 mEq/L K^+ solutions.
16. After all of the measurements are complete, press the "CALC" button in the journal to perform the logarithmic calculations for the data that are needed to graph. Enter the calculated values in **table 14.5,** and then plot the data on the graph provided.
17. Open the post-lab quiz and lab report by clicking on it. Answer the post-lab questions that appear on the computer screen. Click "Print Lab" to print a hard copy of the report or click "Save PDF" to save an electronic copy of the report.
18. Make note of any pertinent observations here.

____

____

____

Table 14.4 Raw Data for Ph.I.L.S. Lesson 10: Resting Potential and External [K^+]

Tube	5 mEq/L K^+ E_m	10 mEq/L K^+ E_m	20 mEq/L K^+ E_m	50 mEq/L K^+ E_m	100 mEq/L K^+ E_m
1					
2					
3					
4					
5					

Table 14.5 Log Data for Ph.I.L.S. Lesson 10: Resting Potential and External [K^+]

[K^+]	Av. E_m	#
5		
10		
20		
50		
100		

E (mV)

0

0

Log [K^+] (mEq/L)

EXERCISE 14.11 Ph.I.L.S.

Ph.I.L.S. LESSON 11: RESTING POTENTIAL AND EXTERNAL [Na^+]

This exercise is a logical extension of exercise 14.10, in which there were changes in resting membrane potential of crayfish muscle fibers when they were placed in solutions of increasing concentrations of KCl. This simulated the effect of altering the [K^+] in normal extracellular fluid. This exercise involves simulating the effect of altering the [Na^+] (sodium ion concentration) in normal extracellular fluid by placing the crayfish muscle fibers into solutions of increasing concentrations of NaCl. Because of the enormous contribution of K^+ and Na^+ in forming the resting membrane potential (RMP) of a cell, this exercise, as well as exercise 14.10, are important first steps to take to understand how changes in the concentrations of these ions in the ECF alter membrane potential. This will help in understanding why it is so important physiologically for the body to maintain concentrations of K^+ and Na^+ within such narrow limits in intracellular fluid (ICF) and extracellular fluid (ECF). Subsequent exercises (exercises 14.12 to 14.14) will allow for the investigation of the role these two ions play in causing *changes* in membrane potential: the mechanism neurons and other excitable cells use to create signals such as graded potentials and action potentials. As with exercise 14.10, the study tissue in this simulation consists of crayfish muscle fibers, which are excitable cells.

Before beginning the exercise, become familiar with the following concepts (use the textbook as a reference):

- Function of the Na^+/K^+ ATPase pump
- Normal concentrations of Na^+ and K^+ in the ICF and ECF
- Resting membrane potential
- Permeability of the cell membrane to Na^+ and K^+ ions at rest
- Relative contributions of Na^+ and K^+ to development of the resting membrane potential
- Depolarization, hyperpolarization, and repolarization

Before beginning the experiment, state a hypothesis regarding the effect of extracellular [Na^+] on resting membrane potential.

____

____

____

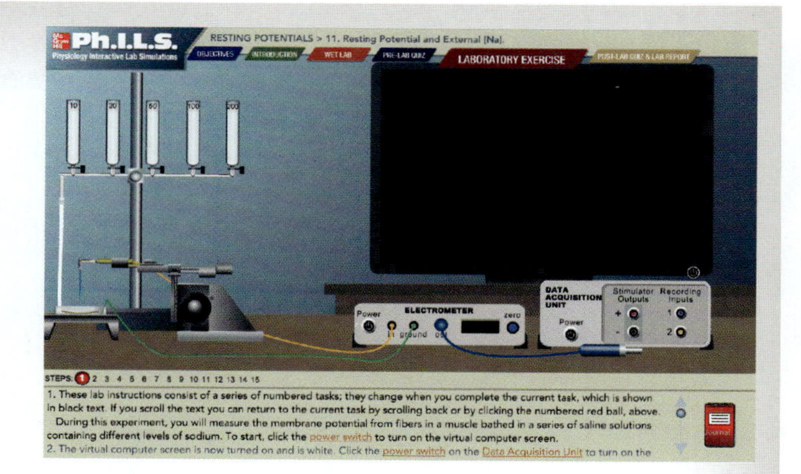

Figure 14.11 **Setup of the Web Lab for Ph.I.L.S. Exercise 11: Resting Potential and External [Na^+].** An isolated muscle fiber has been placed in the apparatus and is hooked up to both stimulating and recording electrodes. Test tubes above the setup are arranged to allow for bathing the muscle in solutions containing varying concentrations of Na^+.

1. Open Ph.I.L.S. Lesson 11: Resting Potential and External [Na^+] **(figure 14.11).**
2. Read the objectives, introduction, and wet lab sections. Then take the pre-lab quiz. The lab exercise will open when the pre-lab quiz has been completed.
3. To start the experiment, click the "power" switch on the virtual computer screen. Then, click the "power" switch on the data acquisition unit and the electrometer.
4. Click and drag the "blue plug" to recording input 1 to connect the electrometer.
5. To record the output of the electrometer, click the "start" button in the upper right-hand corner of the virtual computer screen. A blue line appears on the virtual monitor. This blue line will represent the voltage measured from the crayfish preparation.
6. Zero the electrometer by clicking the "zero" button near the read-out on the electrometer.
7. The microelectrode is a thin blue line above the crayfish preparation. To insert the electrode into the sample, click the "down control" on the micromanipulator (large black wheel on the bottom left of the screen). Watch the blue line as the electrode is inserted into the crayfish muscle fiber. When a change in the voltage is registered on the electrometer, release the down control. It may be required to move the microelectrode laterally for correct positioning. If so, click the "up control" and then the "lateral controls" to correctly position the microelectrode.
8. The electrode has penetrated the plasma membrane and is now measuring the membrane potential. Record that value here: E_m (normal saline) = ______________ Then click the "journal" to enter the data into the Ph.I.L.S. journal. The control part of the experiment is now complete. Subsequent recordings will be experimental recordings when varying the [Na^+] in the ECF.
9. To begin the experiment, raise the micromanipulator so the electrode is no longer in the muscle fiber by clicking the "up control" arrow.
10. Perfuse the sample with 200 mM NaCl solution by clicking the "tap" below the appropriately labeled test tube.

(continued on next page)

(continued from previous page)

Table 14.6 Raw Data for Ph.I.L.S. Lesson 11: Resting Potential and External [Na^+]

Tube	200 mM NaCl E_m	100 mM NaCl E_m	50 mM NaCl E_m	20 mM NaCl E_m	10 mM NaCl E_m
1					
2					
3					
4					
5					

11. Reposition the microelectrode by clicking the "lateral controls" to the left, right, anterior, or posterior. Find an intact portion of membrane to make a new recording.
12. Click the "zero" on the electrometer.
13. Insert the microelectrode into the crayfish muscle preparation by clicking the "down control" on the micromanipulator. Remember to watch for the blue line to move, indicating the microelectrode is recording a change in voltage across the membrane. When the blue line changes, release the down control and record the membrane potential in the Ph.I.L.S. journal, and in **table 14.6.**
14. Repeat step 13 until there are five measurements from different portions of the membrane in the 200 mM NaCl solution.
15. Repeat steps 7 through 14 with the 100, 50, 20 and 10 mM NaCl solutions.
16. After completing all of the measurements, press the "CALC" button in the journal to perform the logarithmic calculations for the data that are needed to graph. Enter the calculated values in table 14.6, and then plot the data on the graph provided.
17. Open the post-lab quiz and lab report by clicking on it. Answer the post-lab questions that appear on the computer screen. Click "Print Lab" to print a hard copy of the report or click "Save PDF" to save an electronic copy of the report.

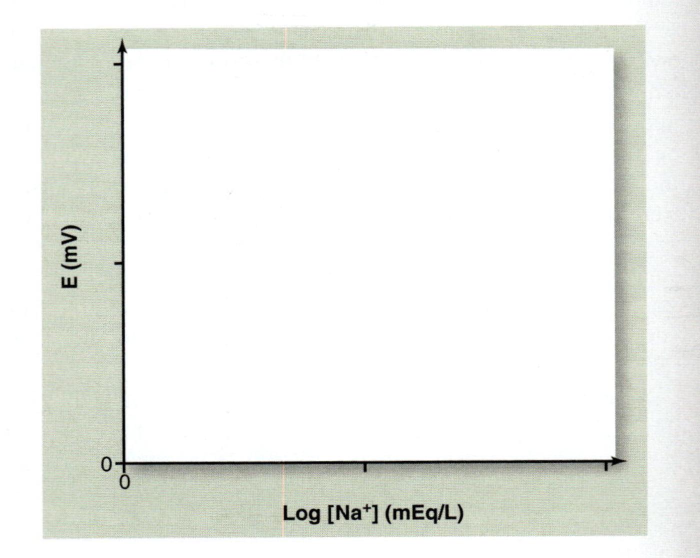

18. Make note of any pertinent observations here.

____

____

____

Action Potential Propagation

Although the plasma membranes of all nerve and muscle cells have a negative resting membrane potential (RMP), not all parts of the membrane have the ability to form the same types of electrical signals. The two main types of electric signals that these cells create are graded potentials and action potentials. **Graded potentials** are small, transient changes in membrane potential that become weaker with time and distance along the plasma membrane. **Action potentials** are very large, specific changes in membrane potential that maintain their intensity with time and distance along the plasma membrane. Instead, action potentials are propagated along the plasma membrane until they reach the end of the membrane or a part of the membrane that is not excitable. Recall from the Concept Connection on p. 349 that the receptive segment of a typical neuron contains ligand-gated channels. This segment allows for the generation of graded potentials. However, the initial and conductive segments contain voltage-gated channels and allow for action potential generation and propagation. Any cell that has a membrane capable of responding to stimuli to generate graded potentials is called an **excitable cell.** Cells that propagate an action potential are referred to as conductive. The two types of cells in the body that are both excitable and conductive are nerve and muscle cells.

Depolarization corresponds to any event that causes the inside of an excitable cell to become more positive. Depolarization first occurs with graded potentials. For example, some ligand-gated channels allow for positive ions, such as Na^+, to flow into the cell. When graded potentials are summed and depolarization of an excitable cell reaches a critical threshold voltage (which is usually around –55 mV for a neuron), the result is the initiation of an **action potential** on the membrane. The action potential involves a specific, sequential opening and closing of fast voltage-gated sodium channels and then slow voltage-gated potassium channels. The voltage-gated sodium channels quickly open in response to membrane depolarization and then rapidly close. The initial membrane depolarization that occurs with an action potential activates voltage-gated potassium channels in addition to the sodium channels. These are slow ion channels that do not completely open until after the peak of the action potential has been reached. When opened, these channels allow potassium ions to diffuse out of the cell, causing repolarization of the membrane. This potassium efflux causes the membrane potential to become relatively more negative than –70 mV, or **hyperpolarize** before returning to resting membrane potential once again **(figure 14.12).** Note that the terms *depolarization* and *repolarization* are labeled on figure 14.12. While these terms are general, meaning that the inside of the cell becomes more positive (depolarization) or less positive (repolarization), when referencing an action potential, these terms also refer to specific phases. That is, the depolarization phase of the action potential is the period where voltage-gated sodium channels open to allow the cell to become more depolarized.

An excitable cell has some limit to its response to external stimuli, in that there are times during the action potential when the membrane is insensitive to another stimulus. For example, during the depolarization phase, rapid closure of the voltage-gated sodium channels causes channel inactivation, thereby preventing the depolarization of the membrane.

In this state, the inactivated sodium channels cannot be opened (activated) again until they reset to the resting state. This resetting process takes time. As a consequence, after an action potential has been generated at the membrane, there is a period of time when the membrane cannot produce another action potential, no matter how great the stimulus that is applied. This period is called the **absolute refractory period.** However, if another stimulus is applied to the membrane during the hyperpolarization period, the membrane potential is essentially starting at a potential that is more negative than normal (e.g., starting at –80 mV instead of the usual –70 mV). Thus, the stimulus must be much greater than normal for the membrane potential to reach threshold. Because an action potential can be generated, but it requires a larger stimulus to do so, this time period is called the **relative refractory period.**

In many nerve cells, action potentials must travel long distances along the axons to reach the target organs. For example, motor neurons that send signals to the muscles in the big toe have cell bodies located in the lumbar region of the spinal cord. The axons of those motor neurons start in the spinal cord and travel along the lower limb nerves for several meters before reaching ("innervating") the muscles in the foot. For action potential propagation to be effective over long distances, **myelination** of the axon by glial cells is required. Myelin forms an insulating sheath around the axon, which allows the action potential to travel much faster than it would along a "bare" axon. It does so by preventing leakage of ions. The speed at which an action potential is propagated along an axon is the **conduction velocity.** When comparing "bare" to myelinated axons, the conduction velocity achieved by the myelinated axon can be up to 200 times faster than that of the "bare" axon—a huge difference!

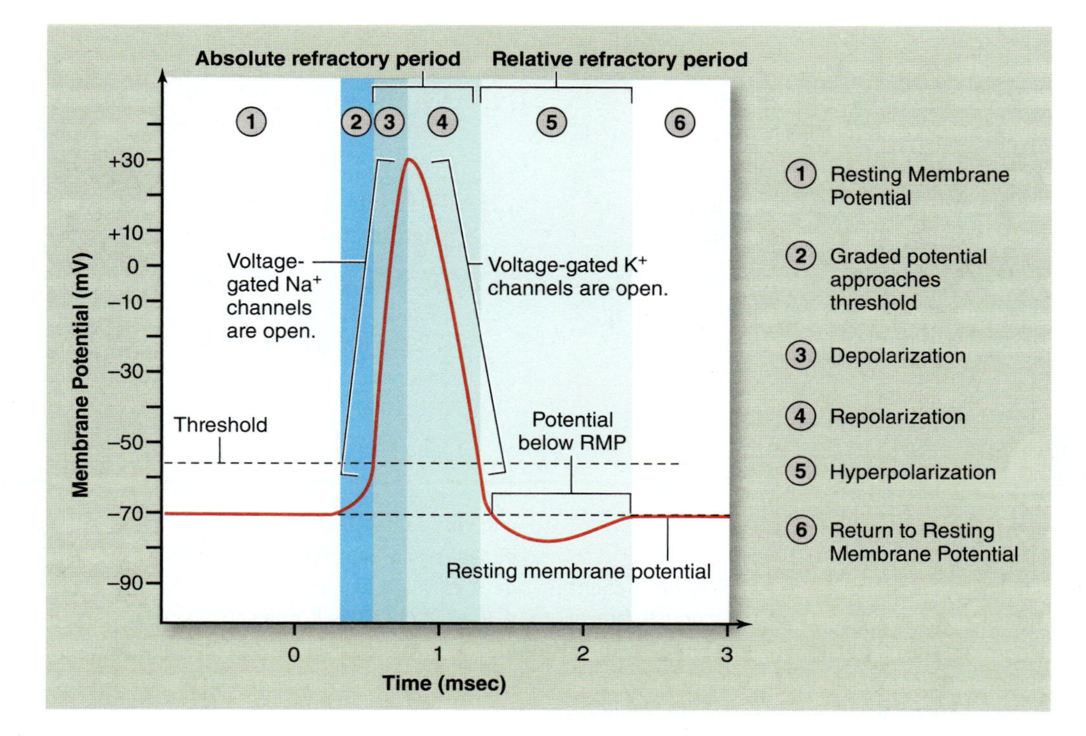

Figure 14.12 Absolute and Relative Refractory Periods and the Ionic Events Associated with Each.

EXERCISE 14.12 Ph.I.L.S.

Ph.I.L.S. LESSON 12:

THE COMPOUND ACTION POTENTIAL

The structure of a nerve is such that there are multiple axons contained within a bundle of connective tissue. An external stimulus (voltage) that allows the membrane of an axon to reach threshold will initiate an action potential. Increasing the stimulus intensity will recruit more neurons, thereby producing compound action potentials. Compound action potentials can be recorded at the postsynaptic membrane. The purpose of this exercise is to vary the intensity of external stimuli on an excised frog sciatic nerve to demonstrate both threshold and recruitment.

Before beginning the exercise become familiar with the following concepts (use the main textbook as a reference):

- The role of sodium (Na^+) and potassium (K^+) in generating and maintaining membrane potential
- How the concept of threshold applies to the generation of an action potential
- The functionality of voltage-gated channels
- The definition of depolarization and hyperpolarization in terms of membrane potential

Before beginning the experiment, state a hypothesis regarding the effect of stimulus intensity on threshold and recruitment.

____

____

____

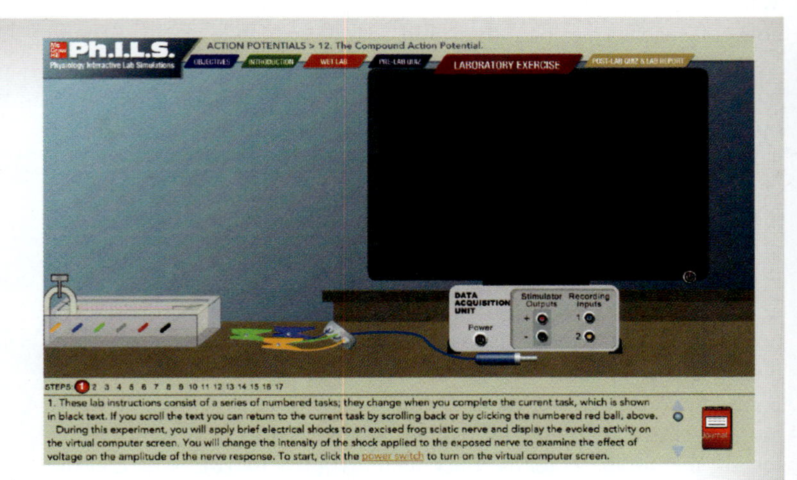

Figure 14.13 Setup of the Wet Lab for Ph.I.L.S. Lesson 12: The Compound Action Potential. An isolated nerve has been placed in the apparatus and is hooked up to both stimulating and recording electrodes.

1. Open Ph.I.L.S. Lesson 12: Action Potentials: The Compound Action Potential **(figure 14.13).**
2. Read the objective, introduction, and wet lab sections. Then, take the pre-lab quiz. The laboratory exercise will open when the pre-lab quiz has been completed.
3. To start the experiment, click the "power switch" on the virtual computer screen. Then, click the "power switch" to turn on the data acquisition unit.
4. Drag the "blue plug" to recording input 1 on the data acquisition unit. Click and drag the "recording cables" to the matching colored post on the nerve chamber.
5. Drag the "red plug" to the positive stimulator output on the data acquisition unit, and drag the "red recording cable" to the matching post on the nerve chamber.
6. Drag the "black plug" to the negative stimulator output on the data acquisition unit, and drag the "black recording cable" to the matching post on the nerve chamber.
7. Click and hold the "tap" until all saline is drained from the chamber.
8. The default setting for the voltage is 1.0 volt. Click the "start" button in the upper right-hand corner of the virtual computer screen to apply a stimulus to the nerve.
9. Measure the voltage of the compound action potential by positioning one of the two cursors (vertical black lines) at the top of the wave on the graph. Now position the second cursor at the bottom of the deflection (0 volts).
10. Click the "journal" panel to enter the value into the journal. A table with volts and amplitude (amp) and a graph will appear. Note the range in values for volts from 0 to 1.6 volts.
11. Close the journal window by clicking on the "X" in the right-hand corner.
12. To complete the table and produce the graph, repeat steps 10 to 12 for the entire range of voltages (0–1.6 mV). Record the data in **table 14.7.**

INTEGRATE

LEARNING STRATEGY

Membrane potential, or voltage, is not all that is required for "action" in an excitable cell. Consider a 9-volt battery, which stores 9 volts of electrical potential energy. To make that energy useful, the battery must be hooked up to wires, which will allow the electrical charges to flow toward each other (e.g., positive moves toward negative and vice versa). Placing the battery in a device allows for such connections and completes the circuit, thereby providing power to the device. This is akin to releasing water from a dam to get energy from it. In an excitable cell, opening ion channels allows stored charges to move (e.g., Na^+ flowing into the cell or K^+ flowing out of the cell). This is a form of *electrical kinetic energy,* and is called a **current** (*currere,* running).

Table 14.7 Membrane Potentials with Applied External Stimuli

Volts	Amplitude
0.0	
0.1	
0.2	
0.3	
0.4	
0.5	
0.6	
0.7	
0.8	
0.9	
1.0	
1.1	
1.2	
1.3	
1.4	
1.5	
1.6	

13. After finishing the laboratory exercise, print the line tracing by clicking on the "P" in the upper left-hand corner of the virtual computer screen.

14. Construct a graph of a compound action potential in the space provided. Be sure to indicate where voltage-gated sodium channels and voltage-gated potassium channels are opened and closed.

15. Open the post-lab quiz & lab report by clicking on it. Answer the post-lab questions that appear on the computer screen. Click "Print Lab" to print a hard copy of the report or click "Save PDF" to save an electronic copy of the report.

16. Make note of any pertinent observations here.

___

___

___

Nerve Response (mV)

0

0 5

Applied Voltage (V)

INTEGRATE

CONCEPT CONNECTION

The membrane potential of an excitable cell is regulated by active and passive **membrane transport** mechanisms. Active pumps move Na^+ out of the cell and K^+ into the cell. The movements of Na^+ and K^+ in this case are caused by active transport because the pump is moving the ions *against* their respective concentration gradients. This leads to a separation of charge across the plasma membrane due to the membrane's selective permeability. The movement of Na^+ or K^+ ions into or out of the cell is driven by passive transport mechanisms. These ions pass freely through their respective leak channels along their individual electrochemical gradients (e.g., K^+ moves out of the cell, from an area of high concentration to an area of low concentration). In addition to leak channels, ions may pass through either chemically-gated or voltage-gated channels. Opening many gates simultaneously creates a large permeability change. Again, passive transport mechanisms allow for these ions to move along their respective electrochemical gradients through these channels. Rather than memorizing the movement of ions into or out of the plasma membrane, simply refer to the governing principles of membrane transport for guidance.

EXERCISE 14.13 Ph.I.L.S.

Ph.I.L.S. LESSON 13: CONDUCTION VELOCITY AND TEMPERATURE

This experiment involves measuring the conduction velocity of action potentials as they propagate along myelinated axons in the frog sciatic nerve. The nerve will be excised from the frog hind limb and placed in a bath containing an isotonic (Ringer's) solution. Electrical shocks will be applied to one end of the excised nerve, which will generate action potentials in the nerve. The currents associated with the action potentials that are generated will be recorded at the other end of the nerve. Subsequently, the stimulating and recording electrodes will be moved closer to each other, another electrical shock will be applied, and the currents will be measured once again. The conduction velocity of the nerve will then be calculated by comparing the stimulus-response times for the two lengths of nerve. This technique will then be used to study the effect of decreasing temperature on the conduction velocity of action potentials within the nerve.

Before beginning the exercise, become familiar with the following concepts (use the textbook as a reference):

- Graded potentials
- Action potentials and excitable cells
- The effect of temperature on membrane transport
- The definition of depolarization and hyperpolarization in terms of membrane potential

Before beginning the experiment, state a hypothesis regarding the effect of temperature on action potential propagation.

1. Open Ph.I.L.S. Lesson 13: Conduction Velocity and Temperature **(figure 14.14).**
2. Read the objectives, introduction, and wet lab sections. Then, take the pre-lab quiz. The laboratory exercise will open once the pre-lab quiz has been completed.
3. Note that the sciatic nerve has already been removed from the frog and placed in the nerve chamber. To begin, click "power" on the temperature probe. Then, click the "power" on the virtual computer screen and data acquisition unit.
4. Drag the "blue plug" to recording input 1 on the data acquisition unit. Click and drag the "recording cables" to the matching colored posts on the nerve chamber.
5. Drag the "red plug" to the positive stimulator output on the data acquisition unit, and drag the "red recording cable" to the matching post on the nerve chamber.
6. Drag the "black plug" to the negative stimulator output on the data acquisition unit, and drag the "black recording cable" to the matching post on the nerve chamber.
7. Before the stimulus can be applied, the saline must be drained from the nerve chamber to prevent the current from passing through the saline instead of the neuron.

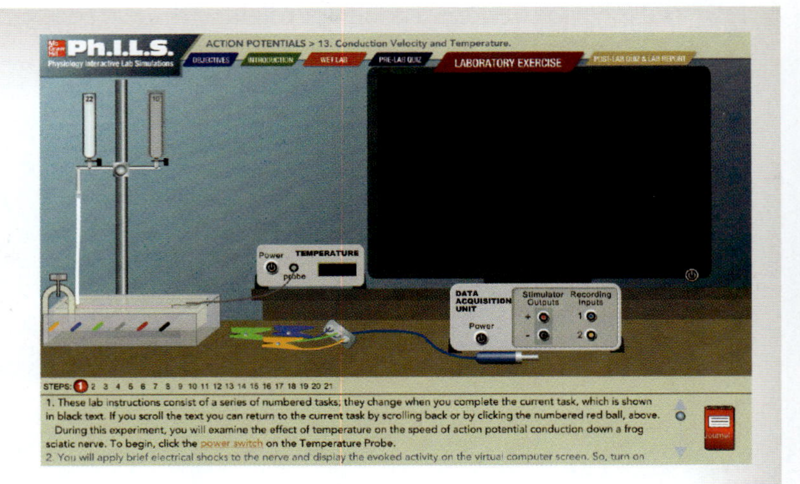

Figure 14.14 Setup of the Wet Lab for Ph.I.L.S. Lesson 13: Conduction Velocity and Temperature. An isolated nerve has been placed in the apparatus and is hooked up to both stimulating and recording electrodes.

To drain the saline from the nerve chamber, click the "tap" and hold it until all the saline has been removed.

8. The initial shock or stimulus voltage is set at the default of 1.0 volt. Click the "start" button in the upper right-hand corner of the virtual computer screen to apply a single stimulus to the nerve. The response from the nerve is displayed on the control panel screen (the blue line trace). This response is a compound action potential (CAP).
9. Next, change the distance the stimulus will travel. Disconnect the red and black clips by clicking on each of them. Then, reconnect the red clip to the gray post and connect the black clip to the red post. This moves the stimulating and recording cables 10 mm closer to each other.
10. Click the "start" button to produce a CAP (at 1.0 volt) to the nerve.
11. There will now be two responses recorded on the display screen. The one on the right is the original CAP, and the one on the left is the CAP that was recorded after moving the stimulating and recording cables closer (by 10 mm). Now measure the difference in time between the two trials by positioning one of the two cursors (vertical black lines) on the highest point on the CAP on the left. Then, position the second cursor on the highest point on the CAP on the right. Press the "journal" icon to have the program calculate the conduction velocity (conduction velocity = distance/time). Record the data in **table 14.8.**
12. Close the journal window by clicking on the "X" in the right-hand corner.
13. Moisten the nerve with 22 °C Ringer's solution by clicking and holding the tap. Then repeat steps 7 through 11.
14. Moisten the nerve with 10 °C Ringer's solution by clicking and holding the tap. Then repeat steps 7 through 11.

15. After finishing the laboratory exercise, print the line tracing by clicking on the "P" in the upper left-hand corner of the virtual computer screen.
16. Open the post-lab quiz & lab report by clicking on it. Answer the post-lab questions that appear on the computer screen. Click "Print Lab" to print a hard copy of the report or click "Save PDF" to save an electronic copy of the report.
17. Make note of any pertinent observations here.

____

____

____

Table 14.8 Raw Data for Ph.I.L.S. Lesson 13: Conduction Velocity and Temperature

Temperature	Distance (mm)	Time (ms)	Conduction Velocity (Distance/Time)
22 °C			
22 °C			
10 °C			
10 °C			

EXERCISE 14.14 Ph.I.L.S.

Ph.I.L.S. LESSON 14: REFRACTORY PERIODS

This experiment involves observing the response of myelinated axons to successive stimuli in a frog sciatic nerve. The nerve will be excised from the frog hind limb and placed in a bath containing an isotonic (Ringer's) solution. Electrical shocks will be applied to one part of the excised nerve, which will generate an action potential in the nerve. Subsequently, a series of two stimuli will be applied to the same region of the neuron, and the amplitude of the second action potential recorded. Look for changes in the peak amplitude of the second action potential to observe both absolute and relative refractory periods in the nerve.

Before beginning the experiment, become familiar with the following concepts (use the textbook as a reference):

- The difference between graded potentials and action potentials
- The meaning of a threshold potential as it applies to excitable cells
- The function of voltage-gated Na^+ and K^+ channels in the generation of an action potential
- Absolute refractory period of a nerve
- Relative refractory period of a nerve

Before beginning the experiment, state a hypothesis regarding the influence of refractory period on action potential propagation:

____

____

____

1. Open Ph.I.L.S. Lesson 14: Refractory Periods **(figure 14.15).**
2. Read the objectives, introduction, and wet-lab sections. Then, take the pre-lab quiz. The laboratory exercise will open once the pre-lab quiz has been completed.
3. When observing the opening screen (similar to that in figure 14.14) note that the sciatic nerve has already been removed from the frog and placed in the nerve chamber. To begin, click "power" on the virtual computer screen. Then, click the "power" on the data acquisition unit.
4. Drag the "blue plug" to recording input 1 on the data acquisition unit. Click and drag the "recording cables" to the matching colored posts on the nerve chamber.

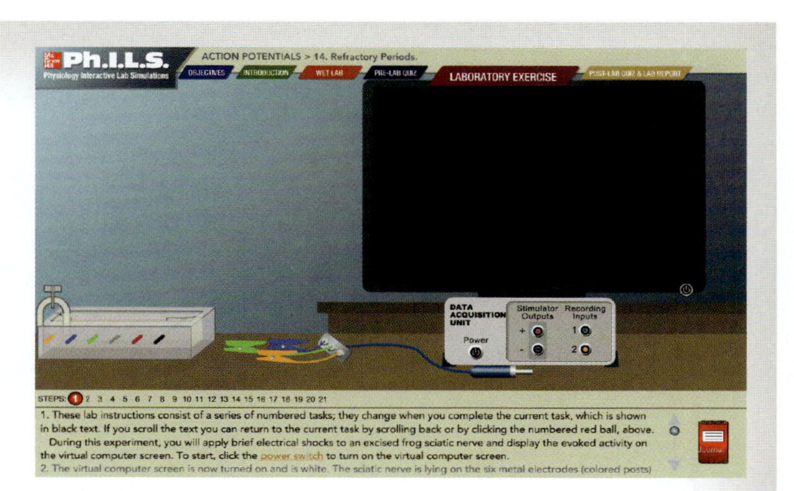

Figure 14.15 Setup of the Wet Lab for Ph.I.L.S. Lesson 14: Refractory Periods. An isolated nerve has been placed in the apparatus and is hooked up to both stimulating and recording electrodes.

5. Drag the "red plug" to the positive stimulator output on the data acquisition unit, and drag the "red recording cable" to the matching post on the nerve chamber.
6. Drag the "black plug" to the negative stimulator output on the data acquisition unit, and drag the "black recording cable" to the matching post on the nerve chamber.
7. Before the stimulus can be applied, the saline must be drained from the nerve chamber to prevent the current from passing through the saline instead of the neurons. To drain the saline from the nerve chamber, click the "tap" and hold it until all the saline has been removed.
8. The voltage is set at the default of 0.7 V. Click the "start" button on the control panel in the upper right-hand corner of the virtual computer screen to see the nerve response. The recording will include a graded potential (larger peak) and a stimulus artifact (smaller peak).
9. To determine the minimal stimulus required to elicit a compound action potential (CAP), increase the voltage by clicking the "up arrow" on the control panel until the voltage reaches 0.8 V. Click the "start" button and observe the change in the amplitude of the recording.

(continued on next page)

(continued from previous page)

10. Repeat step 9 until finding the *minimum* stimulus required to elicit a maximal peak height of the CAP. Record the amplitude of this peak on the line at the top of **table 14.9.**

11. To determine the refractory period of the nerve, deliver two stimuli (shocks) at different time intervals. To begin, increase the number of shocks by clicking the "up arrow." Then, set the time interval to 1 millisecond by clicking the "down arrow."

12. Click "start" to deliver the two stimuli with a 1 millisecond interval between.

13. Two CAPs will be produced. Ensure that one of the cursors remains on the flat part of the tracing. Drag the second cursor to the tallest part of the second CAP. Click the "journal" button in the lower right-hand corner to enter the data in the journal. Then enter the data into table 14.9.

14. Click the "up arrow" in the time interval box to set the interval between stimuli to 1.5 milliseconds. Click "start" and repeat the procedure in step 13 to measure and record the difference in peak amplitudes.

15. Repeat step 14 until the maximum time interval of 9 milliseconds has been reached. Graph the data from table 14.9 on the figure provided.

16. After finishing the laboratory exercise, print the line tracing by clicking on the "P" in the upper left-hand corner of the virtual computer screen.

17. Open the post-lab quiz & lab report by clicking on it. Answer the post-lab questions that appear on the computer screen. Click "Print Lab" to print a hard copy of the report or click "Save PDF" to save an electronic copy of the report.

18. Make note of any pertinent observations here.

____

____

____

Table 14.9 Raw Data for Ph.I.L.S. Lesson 14: Refractory Periods

Maximal Peak Amplitude: ____________

Interval	Amplitude
1.0	
1.5	
2.0	
2.5	
3.0	
3.5	
4.0	
4.5	
5.0	
5.5	
6.0	
6.5	
7.0	
7.5	
8.0	
8.5	
9.0	

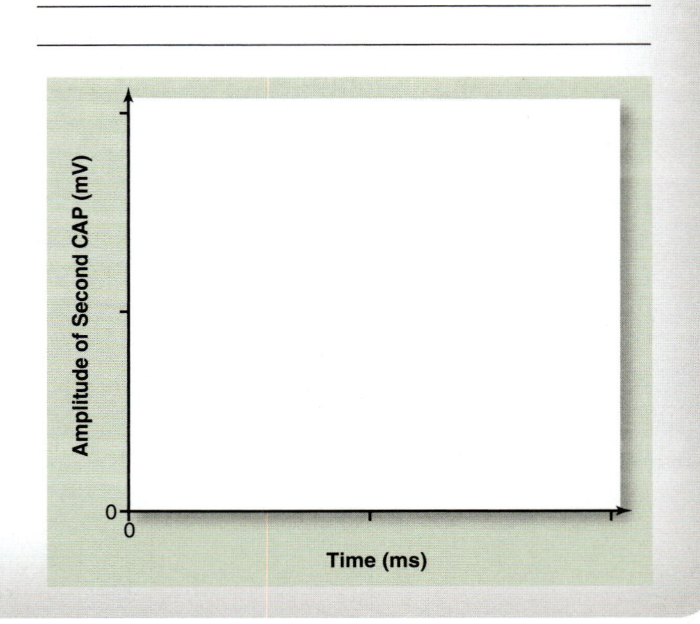

INTEGRATE

CLINICAL VIEW
Potassium Channel Blockers

Demyelinating diseases, such as multiple sclerosis, result in the loss of myelin that surrounds axons. Myelin insulates axons, thereby preventing the leakage of ions. Losing myelin leads to slower nerve conduction velocities, as more positive charge (K^+) can "leak" out of the cell along the length of the axon. *Dalfampridine*, a drug that blocks voltage-gated potassium channels, is now FDA approved for patients suffering from multiple sclerosis (MS). The drug has been shown to increase nerve conduction velocities due to its mechanism of prolonging the action potential. That is, blocking voltage-gated potassium channels blocks the rapid repolarization phase of the action potential, thereby prolonging its duration. This blockage keeps the cell depolarized for longer, thereby increasing the conduction velocity. MS patients report an improvement in sensory and motor function while taking dalfampridine. Incidentally, potassium channel blockers are also used to treat arrhythmias in the heart. In this application, the drug targets the rapid repolarization phase of cardiac muscle cells rather than neurons. However, the result is the same: The excitable cell remains depolarized for longer. In cardiac muscle cells, prolonging the repolarization phase of the action potential allows for a slower heart rate, therefore allowing the heart more time to fill with blood. Regulation of heart rate and cardiac output will be explored further in chapter 21.

Chapter 14: Nervous Tissues

Name: ______________________

Date: __________ Section: __________

POST-LABORATORY WORKSHEET

The ❶ corresponds to the Learning Objective(s) listed in the chapter opener outline.

Do You Know the Basics?

Exercise 14.1: Gray and White Matter

1. Nervous tissue that contains predominately neuron cell bodies, dendrites, and unmyelinated axons is known as ______________ (gray/white) matter, whereas nervous tissue that consists mainly of myelinated axons is known as ______________ (gray/white) matter. ❶
2. Which of the following segments of a neuron contain chemically-gated ion channels, and are therefore considered excitable? (Check all that apply.) ❷

    ____ a. conductive segment

    ____ b. initial segment

    ____ c. receptive segment

    ____ d. transmissive segment

3. Match the description listed in column A with the corresponding term listed in column B. ❷

Column A	*Column B*
____ 1. branches at the end of an axon	a. axon hillock
____ 2. branching processes that transmit graded potentials to the cell body	b. axon terminals
____ 3. location where action potentials are generated by a neuron	c. chromatophilic (Nissl) substance
____ 4. rough endoplasmic reticulum within the soma of a neuron	d. dendrites
____ 5. swellings at the end of an axon	e. synaptic knobs

Exercise 14.2: General Multipolar Neurons—Anterior Horn Cells

4. Large, multipolar neurons found in the anterior horn of the spinal cord are known as anterior horn cells. These cells are also which of the following? (Circle one.) ❸
    a. autonomic motor neurons
    b. interneurons
    c. somatic motor neurons
    d. somatic sensory neurons
    e. visceral sensory neurons

Exercise 14.3: Cerebrum—Pyramidal Cells

5. Which of the following are large, triangularly-shaped neurons found in the cerebral cortex? (Circle one.) ❹
    a. astrocytes
    b. ependymal cells
    c. neurolemmocytes
    d. Purkinje cells
    e. pyramidal cells

Exercise 14.4: Cerebellum—Purkinje Cells

6. Rank the following layers of the cerebellum, from outermost to innermost. ❺

    ____ a. granular layer

    ____ b. molecular layer

    ____ c. Purkinje cell layer

    ____ d. white matter

Exercise 14.5: Astrocytes

7. Which of the following is a function of astrocytes? (Check all that apply.) 6

_____ a. produce and circulate CSF

_____ b. maintain extracellular environment around neurons

_____ c. reinforce the blood-brain barrier

_____ d. transfer nutrients to neurons from the blood

_____ e. transform into phagocytic cells following tissue injury

Exercise 14.6: Ependymal Cells

8. Ependymal cells are unique as epithelia in that they do not have a basement membrane. __________________ (True/False). 7

Exercise 14.7: Neurolemmocytes (Schwann Cells)

9. Neurolemmocytes (Schwann cells) are located within the __________________ (CNS/PNS), whereas oligodendrocytes are located within the __________ (CNS/PNS). 8

10. Each neurolemmocyte can myelinate an entire axon; therefore, an axon has only one neurolemmocyte along its length. __________________ (True/False) 8

Exercise 14.8: Satellite Cells

11. Satellite cells are glial cells that surround the cell bodies of __________________ (motor/sensory) neurons. 9

Exercise 14.9: Coverings of a Peripheral Nerve

12. Rank the following connective tissue coverings of a peripheral nerve, from superficial to deep. 10

____ a. endoneurium

____ b. epineurium

____ c. perineurium

Exercise 14.10: Ph.I.L.S. Lesson 10: Resting Potential and External [K^+]

13. In a typical excitable cell, the concentration of K+ in the ECF is __________________ (greater/less) than the concentration of K+ in the ICF. 11

Exercise 14.11: Ph.I.L.S. Lesson 11: Resting Potential and External [Na^+]

14. In a typical excitable cell, the concentration of Na^+ in the ECF is __________________ (greater/less) than the concentration of Na^+ in the ICF. 12

Exercise 14.12: Ph.I.L.S. Lesson 12: The Compound Action Potential

15. Place the characteristics of a compound action potential in the correct order. 13

____ a. depolarization

____ b. hyperpolarization

____ c. repolarization

16. The type of signal that has the ability to travel a great distance without losing its strength is a(n) __________________ (action/graded) potential. 13

17. As stimulation to a nerve increases, __________________ (fewer/more) neurons are recruited. 14

Exercise 14.13: Ph.I.L.S. Lesson 13: Conduction Velocity and Temperature

18. Raising the temperature from 22 °C to 37 °C should __________________ (increase/decrease) conduction velocity. 15

Exercise 14.14: Ph.I.L.S. Lesson 14: Refractory Periods

19. The __________________ (absolute/relative) refractory period is associated with voltage-gated potassium channels. During this refractory period, another action potential __________________ (can/cannot) be generated. 16

20. The __________________ (absolute/relative) refractory period is associated with voltage-gated sodium channels. During this refractory period, another action potential __________________ (can/cannot) be generated. 16

Can You Apply What You've Learned?

21. Describe the structure, function, and specific location of satellite cells.

22. Describe the structure and function of microglia.

23. Give the location where each of the neuron types listed is found.
    a. anterior horn cell __________
    b. Purkinje cell __________
    c. pyramidal cell __________

24. Compare and contrast the structure, location, and function of neurolemmocytes and oligodendrocytes.

25. Is it more common for brain tumors to arise from neurons or from glial cells? __________

26. Using knowledge of electrochemical gradients, describe the mechanism that explains why the resting membrane potential changes as the concentration of potassium in the ECF increases.

27. Describe whether sodium or potassium has a greater effect on resting membrane potential.

28. Describe how conduction velocity is impacted when temperature decreases. In your explanation, refer to the effect of temperature on membrane transport mechanisms.

29. Explain the roles of Na^+ and K^+ in action potential generation.

Can You Synthesize What You've Learned?

30. A patient with an *astrocytoma* (*-oma,* tumor) has a tumor that is derived from ________________. This tumor would be located in what part of the nervous system? ________________

31. Multiple sclerosis (MS) is a disease in which the immune system attacks myelin sheaths. Often a patient suffering from MS will experience a bout of illness, which involves muscle weakness and sensory alterations, followed by a period of recovery, in which normal function partially or completely returns. Using knowledge of glial cells, explain the biological basis for the partial or complete recovery of nerve function in an MS patient.

32. Which type of neural tissue is more involved with the integration of information (rather than the transmission or conduction of information)—white matter or gray matter? Explain.

33. When a peripheral nerve is compressed, blood vessels that run between fascicles of axons are also compressed. What might be a consequence of this?

34. In the United States, some states have legalized capital punishment (the death penalty). If this is done via a "lethal injection," the substance that is injected is a very high dose of potassium chloride (KCl), which will stop the heart from beating. This is due to the effect of the change in extracellular $[K^+]$ on the resting membrane potential of the cardiac muscle cells, which are excitable cells. Explain how a high dose of KCl will stop the beating of the heart.

35. Clinically, many medications are administered intravenously. These medications are often dissolved in a solution of NaCl. Explain why medications can safely be delivered in a NaCl solution, but would be deadly if they were administered in a KCl solution. ________________

36. Novocaine is used by dentists as an anaesthetic so that a tooth can be extracted or a cavity filled. Novocaine works to block action potentials along sensory neurons typically responsible for relaying information for perception of pain. Propose a possible mechanism for this drug. ________________

37. An ice fisherman has fallen through the ice and is under water for 15 minutes. He is rescued and is slowly warmed back up to normal body temperature. Upon recovery, he is found to have little to no brain damage. Explain how the cold water may have helped protect his brain function.

38. If a muscle fiber has a very long refractory period, what consequence would this have on the ability of the muscle to achieve a tetanic contraction?

The Brain and Cranial Nerves

15

OUTLINE AND LEARNING OBJECTIVES

MODULE 7: NERVOUS SYSTEM

INTRODUCTION

Recall a time when solving word problems in math class was difficult. At the time someone may have said, "Come on, use your gray matter!"—meaning "Use your brain! Think!" Having completed the exercises in chapter 14, it is now clear that the brain consists not only of gray matter (dendrites and cell bodies of neurons), but also of white matter (fiber tracts). The activities in this chapter showcase specific parts of the brain that are composed of gray and white matter.

To help make sense of the structures observed in this laboratory session, the tables in this chapter list the names of the structures along with brief summaries of associated functions and information about the derivation of the names of the structures. Keep in mind that brain structures were named well before their functions were known. Most of the names do not relate to function; instead, they derive from the general appearance of the structure. For instance, "mammillary bodies" have little or nothing to do with the female reproductive system, but to early anatomists they appeared to be "little breasts" on the inferior surface of the brain. The area known as the thalamus apparently looked like a "bed" or "bedroom." In reality, the thalamus has very little, if anything, to do with the bedroom or sleep.

When learning the structures of the brain, try to imagine how and why each structure was given its name. This will make it easier to recall the names of structures. In addition, pay attention to the function(s) of each structure identified. Knowledge of function is even more important than knowledge of structure, because when an area of the brain is damaged, there will be deficiencies directly related to the functions of that area of the brain. The same holds true for the cranial nerves, which transmit information between the brain and structures of the head, neck, and thoracic and abdominal viscera.

This laboratory session involves first examining the protective coverings of the brain—the meninges—then observing the ventricular system of the human brain and relating the development of ventricular structures to the development of the brain. These exercises will be followed with the study of preserved human brains or models of the human brain, and dissection of a representative mammalian brain: the sheep brain.

What follows is an investigation of the *cranial nerves*, which are nerves that arise from the brain (as contrasted with *spinal nerves*, which arise from the spinal cord). The cranial nerve exercises in this laboratory session are organized so that if completed in the proper sequence, knowledge of the cranial nerves will become increasingly detailed. The first exercises establish a base knowledge about the locations and names of the nerves. Subsequent exercises build on that base knowledge, exploring the detailed functions, locations, and common disorders of the cranial nerves.

List of Reference Tables

Chapter 15: The Brain and Cranial Nerves

Name: ______________________

Date: __________ **Section:** __________

These Pre-Laboratory Worksheet questions may be assigned by instructors through their connect course.

PRE-LABORATORY WORKSHEET

1. A fold of brain tissue is known as a ______________ (gyrus/sulcus), whereas a groove between folds of brain tissue is known as a ______________ (gyrus/sulcus).

2. Number the three meninges that cover the brain and spinal cord in order, from superficial (1) to deep (3).

   _____ a. arachnoid mater _____ b. dura mater _____ c. pia mater

3. Match the ventricular structure listed in column A with the corresponding associated adult brain structure listed in column B.

Column A	***Column B***
_____ 1. cerebral aquaduct	a. cerebral hemispheres
_____ 2. fourth ventricle	b. diencephalon
_____ 3. lateral ventricle	c. midbrain
_____ 4. third ventricle	d. pons

4. Dural venous sinuses carry only venous blood ____________________ (True/False)

5. Match the description listed in column A with the corresponding brain structure listed in column B.

Column A	***Column B***
_____ 1. deep, horizontal groove between the frontal, parietal, and temporal lobes	a. central sulcus
_____ 2. fold of brain tissue that serves as the primary somatic motor area of the brain	b. lateral sulcus
_____ 3. fold of brain tissue that serves as the primary somatic sensory area of the brain	c. postcentral gyrus
_____ 4. separates the cerebral hemispheres from the cerebellar hemispheres	d. precentral gyrus
_____ 5. separates the frontal lobe from the parietal lobe of the brain	e. transverse fissure

6. Number the three parts of the brainstem in order, from superior (1) to inferior (3).

   _____ a. medulla oblongata _____ b. midbrain _____ c. pons

7. Match the function listed in column A with the corresponding lobe of the brain listed in column B.

Column A	***Column B***
_____ 1. controls conscious movement of skeletal muscle	a. frontal lobe
_____ 2. primary auditory and auditory association area	b. occipital lobe
_____ 3. primary visual area	c. parietal lobe
_____ 4. receives sensory input from the skin and proprioceptors	d. temporal lobe

8. A cranial nerve is a ______________ (central/peripheral) nervous system structure that originates from the brain.

9. Which of the following structures are innervated by visceral motor neurons? (Check all that apply.)

   _____ a. cardiac muscle _____ b. glands _____ c. skeletal muscle _____ d. smooth muscle

10. Which of the following cranial nerve(s) is/are involved with voluntary eye movement? (Check all that apply.)

    _____ a. abducens (CN VI) _____ b. oculomotor (CN III) _____ c. optic (CN II) _____ d. trochlear (CN IV)

GROSS ANATOMY

The Meninges

The **meninges** are connective tissue coverings of the brain and spinal cord and consist of the dura mater, arachnoid mater, and pia mater. These coverings perform many functions and their structure varies slightly depending on whether they are covering the brain or the spinal cord. Exercise 15.1 explores meningeal structures as they apply to the brain, and exercises in chapter 16 consider meningeal structures as they apply to the spinal cord.

EXERCISE 15.1

CRANIAL MENINGES

1. Obtain a model of the dura mater of the brain or a cadaveric specimen of the head with intact dural structures.
2. Identify meningeal structures listed in **figure 15.1** on a model or cadaveric specimen, using **table 15.1** and the textbook as guides.
   - **a.** Observe the dura mater (figure 15.1). The **dura mater** (*dura*, hard, + *mater*, mother) is composed of two layers: the outer **periosteal layer** (which is simply the periosteum that lines the internal portion of the cranial bones), and the inner **meningeal layer.** The meningeal layer folds inward to form cranial dural septae and dural venous sinuses (figure 15.1*b*). **Cranial dural septae** are partitions that stabilize and support the brain. **Dural venous sinuses** are modified veins that transport venous blood from the brain to the internal jugular vein. The largest dural venous sinus is the **superior sagittal sinus.** Eventually all sinuses drain into the **sigmoid** (*sigmoid*, S-shaped) **sinus,** which drains into the **internal jugular vein.** Table 15.1 includes a list of dural venous sinuses and describes their general locations.
   - **b.** If a cadaveric specimen is available, locate the superior sagittal sinus and note the small granular structures located in and around the sinus. These structures are called **arachnoid villi (granulations)** and serve as sites for absorption of cerebrospinal fluid from the subarachnoid space into the dural venous sinuses.
   - **c.** Obtain a human brain with the arachnoid mater and pia mater intact. If one is not available, this activity will be completed when performing the sheep brain dissection. On the human brain, note the thin, transparent covering that lies over the surface of the brain and does not follow the sulci (shallow grooves). This is the **arachnoid mater** (*arachnoid*, shaped like a spiderweb). Deep to the arachnoid mater is the finest, thinnest meningeal layer, the **pia mater** (*pia*, soft or tender). The pia mater is in direct contact with the neural tissue of the brain and follows all of the contours (gyri and sulci) on its surface. The space between the arachnoid mater and the pia mater is the **subarachnoid space.** What fluid is normally found in the subarachnoid space? ______________________

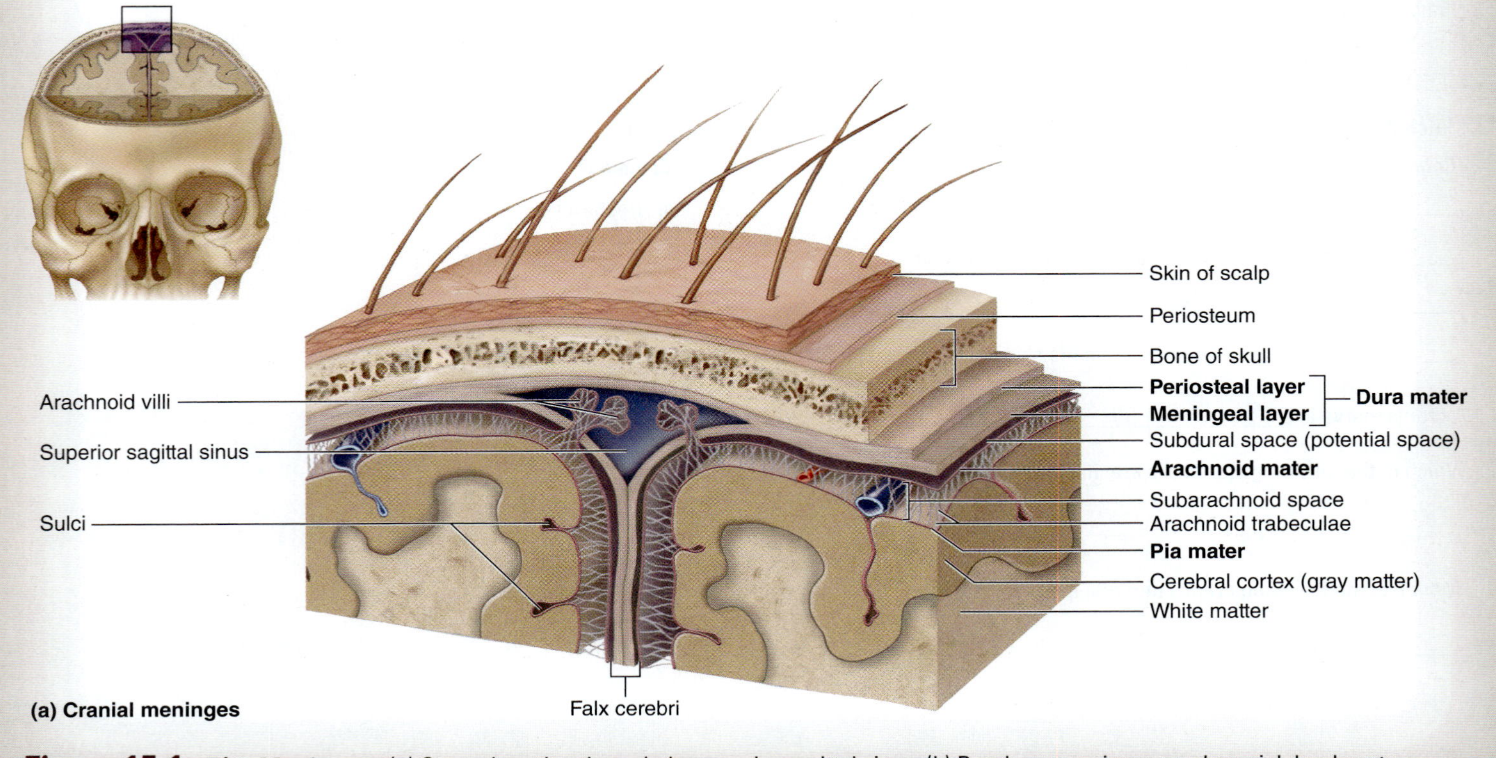

Figure 15.1 The Meninges. (a) Coronal section through the superior sagittal sinus. (b) Dural venous sinuses and cranial dural septa.

Table 15.1 Meninges, Dural Septa, and Dural Venous Sinuses

Structure	Description and Function	Word Origin
Meningeal Layer		
Dura Mater	Very tough, durable membrane composed of dense irregular connective tissue that protects CNS structures within the cranial cavity and vertebral canal; composed of two layers	*durus*, hard, + *mater*, mother
Periosteal Layer	Outer layer of dura mater that composes the inner periosteum of the cranial bones and anchors the dura mater tightly to the cranial bones (not present within the vertebral canal). In most places within the cranial cavity it is anchored to, or continuous with, the meningeal layer.	*periosteal*, relating to the periosteum
Meningeal Layer	Inner layer of dura mater that forms cranial dural septa and dural venous sinuses within the cranial cavity	*meninx*, membrane
Arachnoid Mater	Thin, loose connective tissue membrane that lies adjacent to the dura mater; contains numerous web-like extensions called arachnoid trabeculae, which are composed of collagen and elastic fibers that anchor it to the pia mater; forms a space (the subarachnoid space) for cerebrospinal fluid (CSF) to circulate around the brain and spinal cord	*arachno*, spider cobweb, + *eidos*, resemblance, + *mater*, mother
Pia Mater	A thin, highly vascular, areolar connective tissue membrane in direct contact with the brain and spinal cord. It follows all the surface contours (gyri and sulci) of the brain and is generally inseparable from brain tissue.	*pia*, soft or tender, + *mater*, mother
Cranial Dural Septa	Folds of the meningeal layer that extend inward to form partitions within the cranial cavity that stabilize and support the brain	*durus*, hard, + *septa*, fold
Diaphragma Sellae	Forms a "roof" over the sella turcica of the sphenoid bone and contains a hole for the passage of the infundibulum	*diaphragm*, diaphragm, + *sella*, saddle

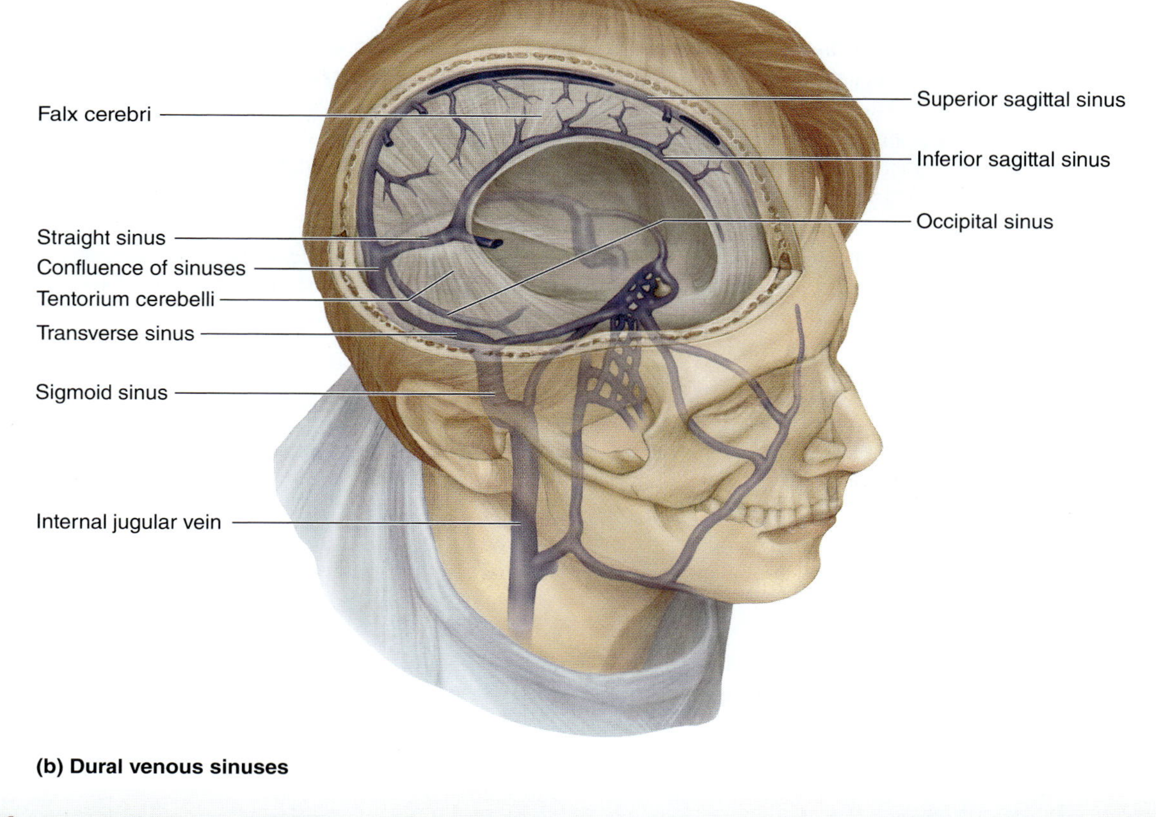

(b) Dural venous sinuses

Figure 15.1 The Meninges *(continued).*

(continued on next page)

(continued from previous page)

Table 15.1 Meninges, Dural Septa, and Dural Venous Sinuses *(continued)*

Structure	Description and Function	Word Origin
Falx Cerebelli	Located between the two cerebellar hemispheres	*falx*, sickle, + *cerebelli*, relating to the cerebellum
Falx Cerebri	Located between the two cerebral hemispheres, within the longitudinal fissure of the brain; anchored anteriorly to the crista galli of the ethmoid bone	*falx*, sickle, + *cerebri*, relating to the cerebrum
Tentorium Cerebelli	Drapes across the cerebellar hemispheres horizontally within the transverse fissure of the brain between the cerebellum and the cerebrum	*tentorium*, a tent, + *cerebelli*, relating to the cerebellum
Dural Venous Sinuses	Spaces formed between meningeal and periosteal layers of the dura mater within the cranial cavity; site of CSF reabsorption from the subarachnoid space and venous blood transport from the brain to the internal jugular vein	*durus*, hard, + *vena*, a blood vessel, + *sinus*, a channel
Inferior Sagittal Sinus	Located within the inferior part of the falx cerebri; drains into the straight sinus	*inferior*, below, + *sagitta*, an arrow (relating to the sagittal plane), + *sinus*, a channel
Occipital Sinus	Smallest sinus within the cranial cavity; located in the margin of the tentorium cerebelli; drains into the confluence of sinuses	*sinus,* a channel + *occipital,* inner surface of occipital bone
Sigmoid Sinuses	Located in the posterior cranial fossa just posterior to the petrous part of the temporal bone and extending into the jugular foramen. Transports venous blood from the transverse sinuses to the internal jugular vein.	*sinus*, a channel, + *sigmoid*, shaped like an S
Confluence of Sinuses	Located within the posterior cranial cavity deep to the external occipital protruberance. Transmits venous blood from the superior sagittal sinus and the straight sinus to the transverse sinuses.	*confluens,* to flow together, + *sinus,* a channel
Straight Sinus	Located at the junction between the falx cerebri, falx cerebelli, and tentorium cerebelli. Transmits venous blood from the inferior sagittal sinus to the confluence of sinuses.	*sinus,* a channel, + straight
Superior Sagittal Sinus	Largest sinus within the cranial cavity; located within the superior portion of the falx cerebri. Transports venous blood from the subarachnoid space and the brain (respectively) to the confluence of sinuses.	*superior*, above, + *sagitta*, an arrow (relating to the sagittal plane), + *sinus*, a channel
Transverse Sinuses	Located posterior to the tentorium cerebelli. Runs from the confluence of sinuses, along the posterior aspect of the occipital bone, to the posterior cranial fossa just posterior to the petrous part of the temporal bone. Transports venous blood from the confluence of sinuses to the sigmoid sinuses.	*transversus*, across, + *sinus*, a channel
Meningeal Spaces		
Epidural Space	Space between the dura mater and the walls of the vertebral canal (there is no epidural space within the cranial cavity)	*epi*, above, + *dura*, relating to the dura mater
Subarachnoid Space	Space between the arachnoid mater and the pia mater where CSF flows as it circulates around the brain and spinal cord. Blood vessels are located within this space.	*sub*, under, + *arachnoid*, relating to the arachnoid mater
Subdural Space	Potential space between the dura mater and the arachnoid mater. In a healthy individual, this space does not exist, but traumatic injury may cause bleeding into the subdural space (subdural hematoma).	*sub*, under, + *dural*, relating to the dura mater

3. Label the meningeal structures in **figure 15.2.**

4. *Optional Activity:* **AP|R 7: Nervous System—** Watch the "Meninges" and "Dural Sinus Blood Flow" animations to reinforce your understanding of these structures and their relationships.

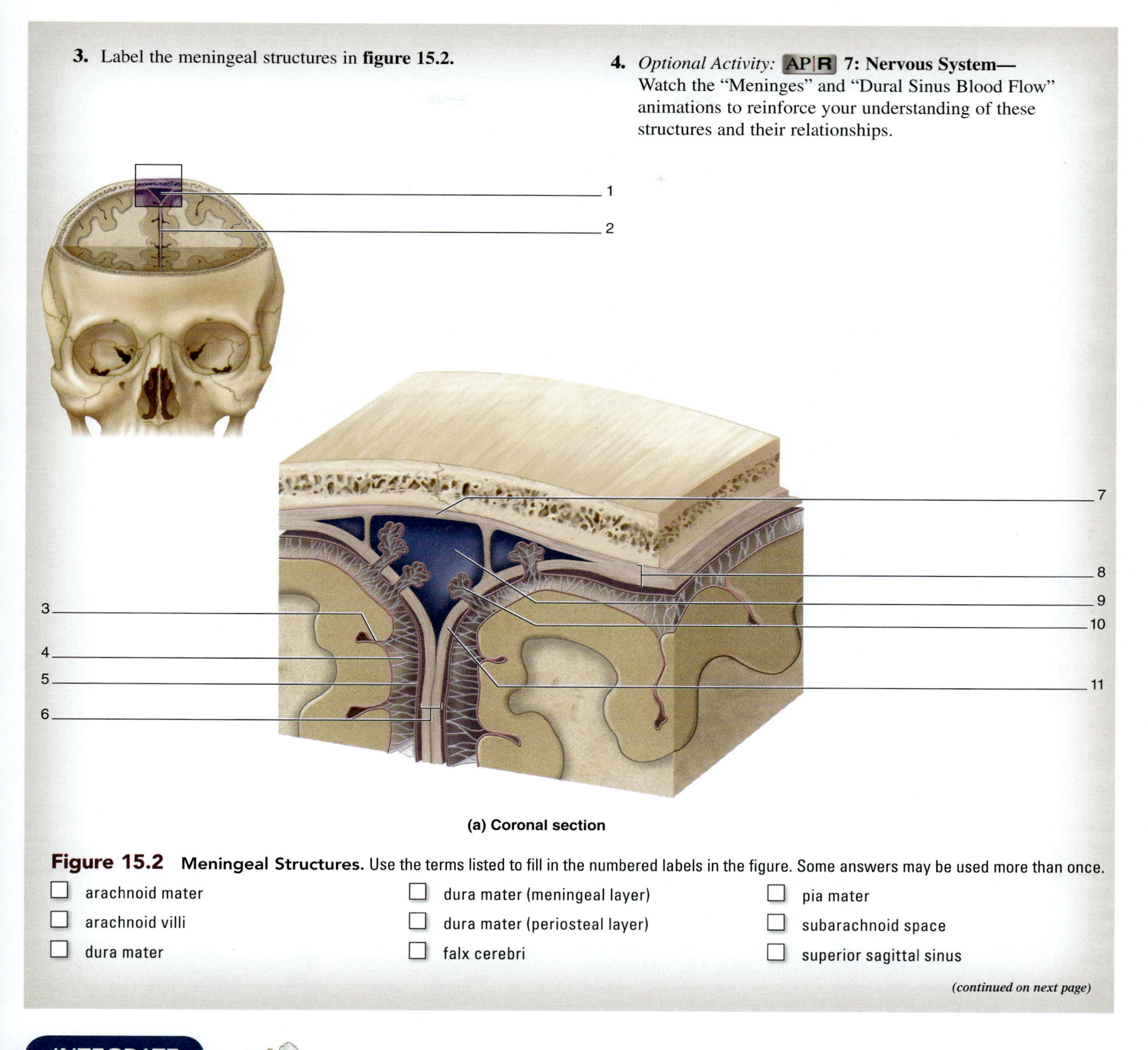

(a) Coronal section

Figure 15.2 Meningeal Structures. Use the terms listed to fill in the numbered labels in the figure. Some answers may be used more than once.

- ☐ arachnoid mater
- ☐ arachnoid villi
- ☐ dura mater
- ☐ dura mater (meningeal layer)
- ☐ dura mater (periosteal layer)
- ☐ falx cerebri
- ☐ pia mater
- ☐ subarachnoid space
- ☐ superior sagittal sinus

(continued on next page)

INTEGRATE

CLINICAL VIEW
Meningiomas

Meningiomas are brain tumors that originate in the meninges of the central nervous system. These slow-growing tumors develop from cells within the arachnoid villi and project into the venous sinuses within the brain. Incidence is higher in women than in men, although the reason is unknown. Most commonly, meningiomas develop either superior to the frontal and parietal lobes near the falx cerebri (parasagittal meningioma) or between the cerebral hemispheres (falcine meningiomas). Some may grow directly under the skull (convexity meningiomas), and some develop within the ventricles (ventricular meningiomas). Other tumors are considered skull base meningiomas due to their location. Subtypes of skull base meningiomas include those located near the sphenoid bone (sphenoid wing meningioma), near the posterior fossa (petrous meningioma), or near the ethmoid bone (paranasal/olfactory meningioma). Meningiomas are dome-shaped, encapsulated tumors that typically attach to the dura mater. In some cases, the tumor may project into and infiltrate surrounding bone tissue. Many meningiomas are asymptomatic and benign. As a tumor grows in size, it may cause symptoms such as headaches, seizures, muscle weakness, sensory and visual disturbances, and increased intracranial pressure. Surgery is required to remove tumors that become symptomatic or malignant (this means that the tumor has spread to other tissues). Some meningiomas are associated with genetic disorders, some are linked to radiation exposure, and some have no known cause at all. Because many meningiomas are asymptomatic, they may not be discovered unless an autopsy is performed.

(continued from previous page)

Cranium
12
13
14
15
16
17
18
19

(b) Midsagittal section

Figure 15.2 **Meningeal Structures *(continued).*** Use the terms listed to fill in the numbered labels in the figure.

- ☐ confluence of sinuses
- ☐ diaphragma sellae
- ☐ falx cerebelli
- ☐ falx cerebri
- ☐ inferior sagittal sinus
- ☐ straight sinus
- ☐ superior sagittal sinus
- ☐ tentorium cerebelli

Ventricles of the Brain

The **ventricles** are fluid-filled spaces within the central nervous system that are complex in shape. Exercise 15.2 involves viewing a *cast* of the ventricles, which is produced by filling the ventricular spaces with plastic, allowing the plastic to harden, and then removing the brain tissue so only the cast is left. A cast allows one to visualize the three-dimensional structure of the ventricles without the brain literally "getting in the way." If casts of the brain ventricles are not available, this exercise can be performed using **table 15.2** and figures in the textbook.

The central nervous system initially develops as a neural tube. As it grows, the neural tube begins to change size and shape. The cephalic end develops into the brain, while the rest develops into the spinal cord. Both the brain and the spinal cord contain fluid-filled spaces inside. The pattern of growth of the neural tissue surrounding the neural tube changes the size and shape of the fluid-filled spaces within. Thus, because the spinal cord remains mostly a tubular structure as it grows, the fluid-filled space inside, the **central canal,** remains tubular. On the other hand, because the cephalic (brain) end of the neural tube undergoes extensive folding as it grows, the fluid-filled spaces within develop into irregular shapes. These shapes tell a story about how the parts of the brain developed.

The cephalic end of the neural tube first develops into three **primary vesicles** (prosencephalon, mesencephalon, and rhombencephalon) and then into five **secondary vesicles** (telencephalon, diencephalon, mesencephalon, metencephalon, and myelencephalon). **Figure 15.3** lists the secondary vesicles of the brain and the parts of the ventricular system that develop from each of them. For instance, the telencephalon undergoes extensive growth as it develops into the cerebral hemispheres. This growth results in the horseshoe-shaped structure of the **lateral ventricles** in the adult brain. On the other hand, the mesencephalon does not undergo such extensive growth as it develops into the midbrain. Hence the fluid-filled space inside, the **cerebral aqueduct,** remains tubular in shape within the adult brain.

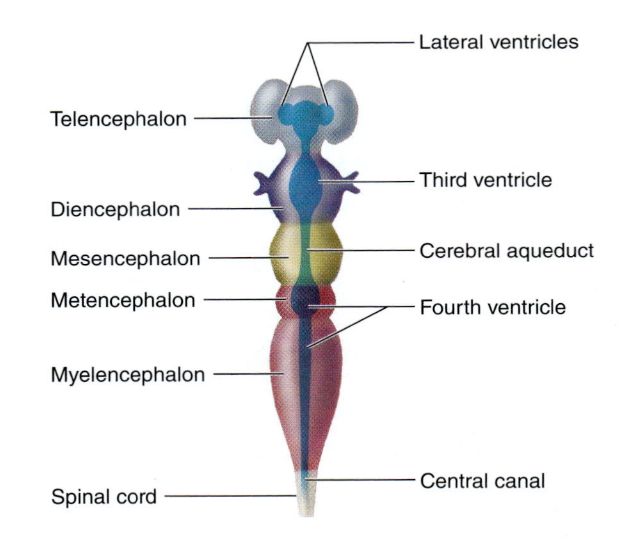

Figure 15.3 **Secondary Brain Vesicles and Associated Ventricular Structures of the Brain.**

EXERCISE 15.2

BRAIN VENTRICLES

1. Obtain a cast of **the ventricles of the brain (figure 15.4).**
2. Identify the ventricles of the brain listed in figure 15.4 on the cast of the ventricles, using table 15.2 and the textbook as guides. Then label them in figure 15.4. When identifying each of the ventricles, relate each ventricle to the **secondary brain vesicle** from which it developed (table 15.2 and figure 15.3).

INTEGRATE

LEARNING STRATEGY

When observing whole brains or brain models in the laboratory, always begin by locating the ventricular spaces and associating each ventricular space with a secondary brain vesicle (table 15.2). Next, identify the adult brain structures that surround the ventricular spaces. Finally, correlate each adult brain structure to the secondary brain vesicle from which it formed. For example, the lateral ventricles (ventricular space), which are part of the telencephalon (secondary brain vesicle), are surrounded by the cerebral hemispheres (adult brain structures). Thus, the cerebral hemispheres (brain structure) are derived from the telencephalon (secondary brain vesicle).

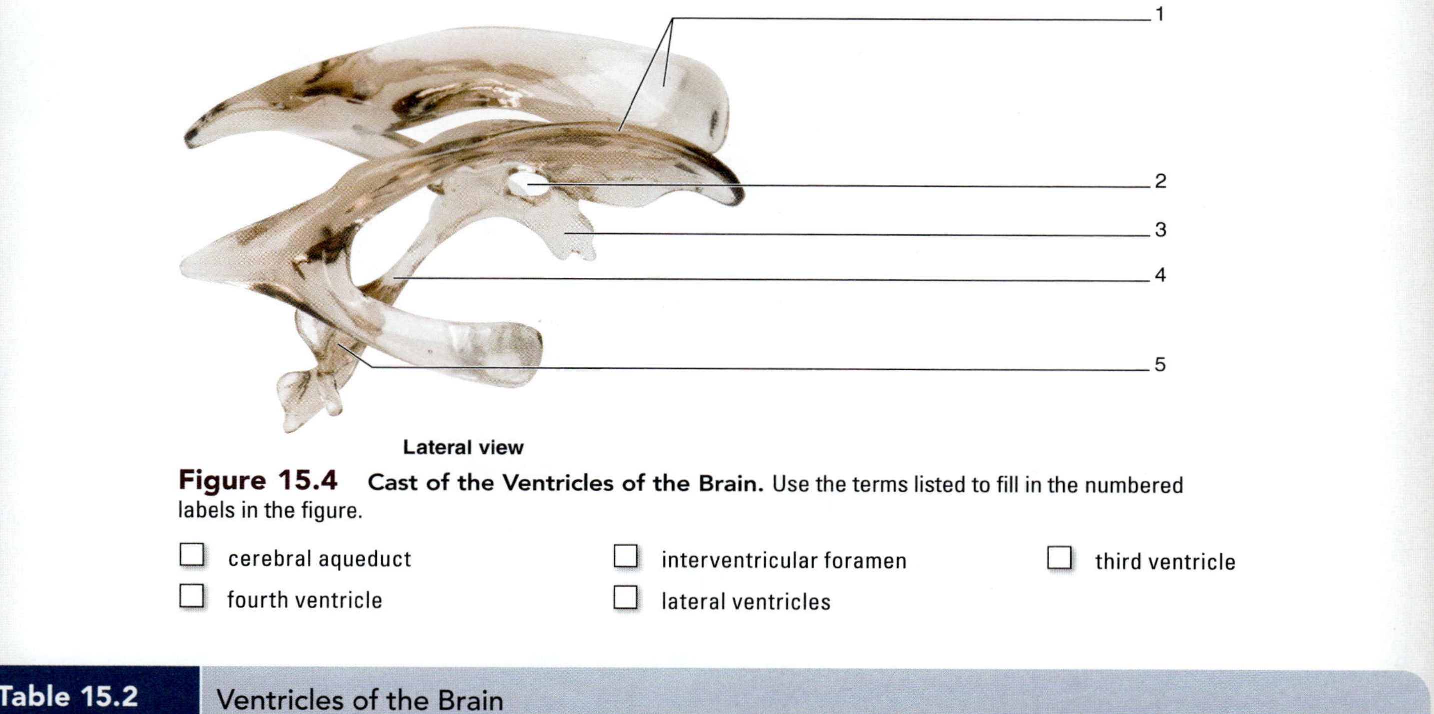

Figure 15.4 Cast of the Ventricles of the Brain. Use the terms listed to fill in the numbered labels in the figure.

- ☐ cerebral aqueduct
- ☐ fourth ventricle
- ☐ interventricular foramen
- ☐ lateral ventricles
- ☐ third ventricle

Table 15.2 Ventricles of the Brain

Structure	Location and Description	Secondary Brain Vesicle Derivative	Word Origin
Lateral Ventricles	Horseshoe-shaped ventricles within the cerebral hemispheres containing anterior and posterior horns whose shape follows the developmental shape of the cerebral hemispheres	Telencephalon	*latus*, to the side, + *ventriculus*, belly
Third Ventricle	Narrow, quadrilateral-shaped ventricle located in the midsagittal plane inferior to the corpus callosum and medial to the thalamic nuclei; surrounded by structures of the diencephalon	Diencephalon	*ventriculus*, belly
Cerebral Aqueduct	Narrow channel that lies in the midbrain between the cerebral peduncles and the tectal plate (corpora quadrigemina)	Mesencephalon	*aquaeductus*, a canal
Fourth Ventricle	Diamond-shaped ventricle located anterior to the cerebellum and posterior to the pons	Metencephalon (superior part) Myelencephalon (inferior part)	*ventriculus*, belly

EXERCISE 15.3

CIRCULATION OF CEREBROSPINAL FLUID (CSF)

Cerebrospinal fluid (CSF) is a clear, colorless fluid produced from blood plasma, which is filtered by the choroid plexus within each brain ventricle. Approximately 500 mL of CSF is produced each day. Following its production within the lateral ventricles, CSF moves through the following structures: the interventricular foramen, the third ventricle, the cerebral aqueduct, and the fourth ventricle. CSF exits the fourth ventricle through the median and lateral apertures to enter the subarachnoid space, which surrounds both the brain and the spinal cord. CSF is returned

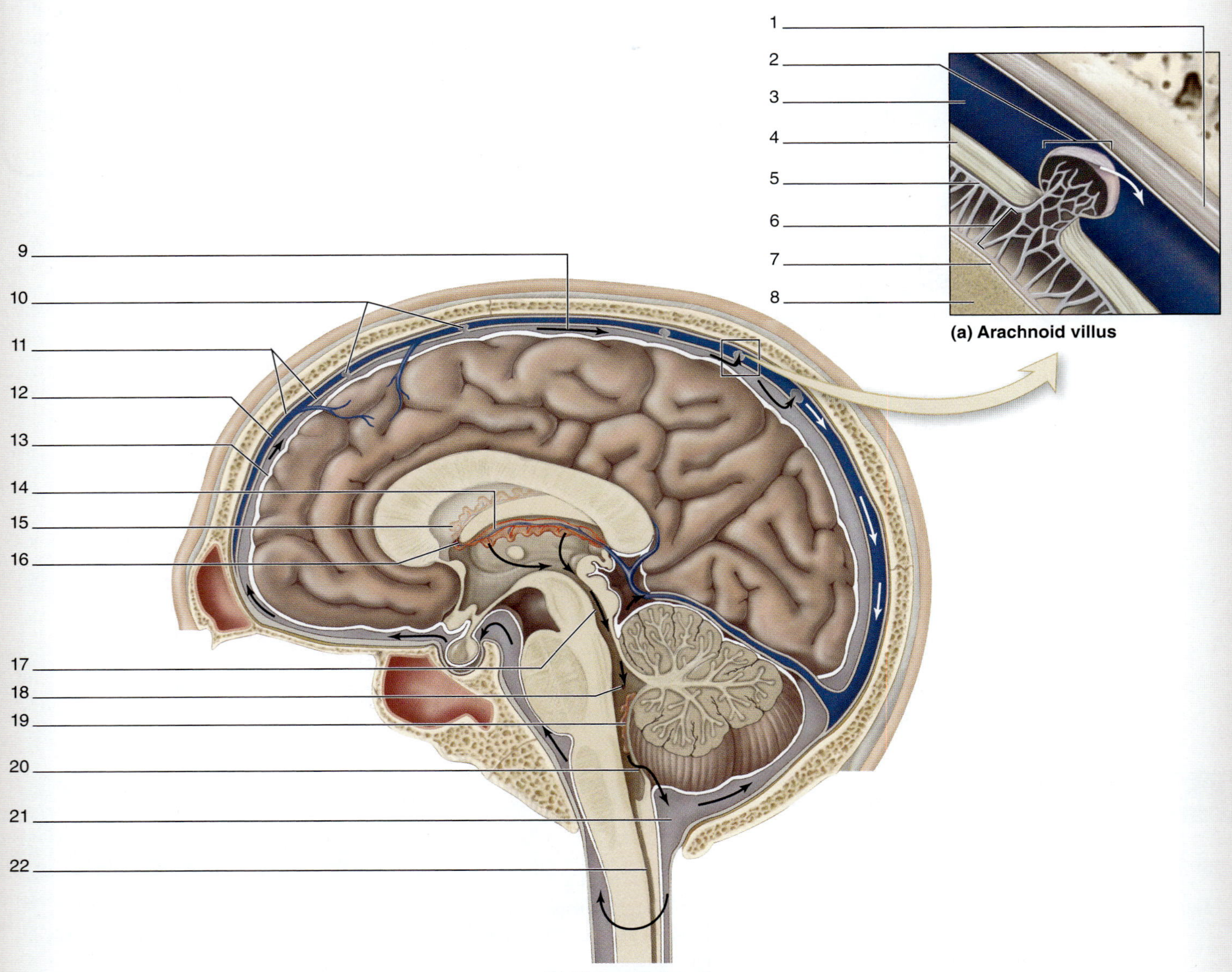

(b) Midsagittal section

Figure 15.5 Cerebrospinal Fluid (CSF) Production and Circulation. Note that arrows show the directional flow of CSF. Use the terms listed to fill in the numbered labels in the figure. Some terms may be used more than once.

- ☐ arachnoid mater
- ☐ arachnoid villi
- ☐ arachnoid villus
- ☐ central canal
- ☐ cerebral aqueduct
- ☐ cerebral cortex
- ☐ choroid plexus of fourth ventricle
- ☐ choroid plexus of lateral ventricle
- ☐ choroid plexus of third ventricle
- ☐ CSF flow
- ☐ dura mater
- ☐ interventricular foramen
- ☐ lateral aperture
- ☐ median aperture
- ☐ meningeal dura
- ☐ periosteal dura
- ☐ pia mater
- ☐ subarachnoid space
- ☐ superior sagittal sinus

to the blood through arachnoid villi within the dural venous sinuses. CSF provides buoyancy to the brain, protecting both the brain and the spinal cord from sudden movements. CSF also provides environmental stability for cells in both the brain and the spinal cord by maintaining extracellular fluid composition within narrow limits.

1. Using table 15.2, figure 15.4, and the textbook as guides, label the structures involved in CSF production and circulation in **figure 15.5.**
2. Trace the flow of CSF from the choroid plexus within the ventricles to the arachnoid villi. Use the terms listed in column A to complete the pathway for CSF circulation in column B, in the spaces provided.

Column A	*Column B*
☐ **cerebral aqueduct**	**1.** lateral ventricles
☐ **fourth ventricle**	**2.** ________________
☐ **interventricular foramen**	**3.** ________________
☐ **median and lateral apertures**	**4.** ________________
☐ **subarachnoid space**	**5.** ________________
☐ **third ventricle**	**6.** ________________
	7. ________________
	8. arachnoid villi

3. *Optional Activity:* **AP|R 7: Nervous System**—View the "Brain Ventricles" and "CSF Flow" animations to see how cerebrospinal fluid flows through the brain ventricles.

The Human Brain

The next series of exercises involves identifying structures that are visible in four views of the human brain: superior, lateral, inferior, and midsagittal. **Table 15.3** lists the main brain structures, the views in which each is visible, and a description of its function.

When identifying parts of the brain, follow these steps to make things easier:

- Name the structures in a logical order, such as the order in which the structures appear from anterior to posterior. Learning the structures in an orderly fashion will allow for improved information recall later on.
- Do not think of brain regions as isolated structures. The structures are easier to identify within the context of their surroundings. Think about this the next time you are traveling to your anatomy & physiology class—would you know how to get to the classroom if all of the buildings on campus (with the exception of the one to which you are traveling) were suddenly moved around?
- Always associate a function with each structure identified. Use table 15.3 as a reference.

INTEGRATE

LEARNING STRATEGY

Because structure/function relationships are not easily visualized when it comes to the brain, consider making flashcards using table 15.3 as a reference. List the brain structure on one side of the card and its function on the other side. Once the names and locations of the listed structures have been mastered, quiz others on the structure and function of each item.

Table 15.3 Brain Structures Visible in Superficial Views of Whole or Sagittally Sectioned Brains

Brain Structure	Description	Function(s)	Word Origin	Views Where Visible
Cerebrum				
Frontal Lobe	Lies deep to the frontal bone	Controls conscious movement of skeletal muscle; contains Broca's area, which controls motor speech. Controls conjugate eye movement (the ability to move the eyes together). Higher-level functions include judgment and foresight (the ability to think before acting).	*frontal*, in the front, + *lobos*, lobe	Superior, lateral, inferior, and midsagittal
Precentral Gyrus	Fold of brain tissue located immediately anterior to the central sulcus	Primary somatic motor area of the brain. Neurons from this gyrus are somatic motor neurons that initiate motor signals to control voluntary muscle activity.	*pre*, before, + *central*, relating to the central sulcus, + *gyros*, circle	Superior, lateral, and midsagittal
Occipital Lobe	Lies deep to the occipital bone	Primary visual area (the first area of the cerebral cortex where visual information synapses, after the thalamus). Visual association area (the ability to interpret visual information).	*occiput*, the back of the head, + *lobos*, lobe	Superior, lateral, and midsagittal
Parietal Lobe	Lies deep to the parietal bone	Receives sensory input from the skin and proprioceptors. Higher-level functions include logical reasoning (math, problem solving).	*parietal*, a wall, + *lobos*, lobe	Superior, lateral, and midsagittal
Postcentral Gyrus	Fold of brain tissue located immediately posterior to the central sulcus	Primary somatic sensory area of the brain. Sensory information that comes in from the body travels to this area of the cerebral cortex.	*post*, after, + *central*, relating to the central sulcus, + *gyros*, circle	Superior, lateral, and midsagittal

(continued on next page)

Table 15.3 Brain Structures Visible in Superficial Views of Whole or Sagitally Sectioned Brains *(continued)*

Brain Structure	Description	Function(s)	Word Origin	Views Where Visible
Cerebrum *(continued)*				
Septum Pellucidum	A thin membrane located between the corpus callosum (above) and fornix (below)	Contains neurons and glial cells, and forms a thin connection between the corpus callosum above and the fornix below; also forms a thin wall between the two anterior horns of the lateral ventricles	*saeptum*, a partition, + *pellucidus*, allowing the passage of light	Midsagittal
Temporal Lobe	Lies deep to the temporal bone	Primary auditory and auditory association area of the brain; conscious perception of smell	*tempus*, time	Lateral
Diencephalon				
Epithalamus	A small projection extending posteriorly from the superior portion of the third ventricle	Contains the pineal body (pineal gland) along with other structures	*epi*, above, + *thalamos*, a bed or bedroom	Midsagittal
Pineal Body (Gland)	Small gland found within the epithalamus. It is not possible to establish the difference between the epithalamus and the pineal gland on gross observation alone.	Secretes the hormone melatonin from its precursor molecule, serotonin, in response to *decreased* light levels. Melatonin has an effect on circadian rhythms. May also play a role in establishing the onset of puberty.	*pineal*, shaped like a pine cone	Midsagittal
Hypothalamus	Located deep to the walls of the inferior part of the third ventricle	Regulates body temperature, metabolism (hunger/thirst), sleep, sex, and emotional control (limbic system functions). It is also a "master" endocrine gland, controlling hormone secretion from the pituitary gland.	*thalamos*, a bed or bedroom	Midsagittal
Mammillary Bodies	Two small bump-like ("breast-shaped") structures of the hypothalamus located immediately posterior to the infundibulum	Involved in short-term memory processing; part of the limbic system (the emotional brain). Also involved with suckling and chewing reflexes.	*mammillary*, shaped like a breast	Inferior and midsagittal
Thalamus	Paired nuclei located deep to the lateral walls of the third ventricle. A pin pierced through the lateral wall of the third ventricle adjacent to the intermediate mass will pass into the thalamic nuclei.	Primary relay center for all sensory information coming into the brain (except olfaction)	*thalamos*, a bed or bedroom	Midsagittal
Intermediate Mass (Interthalamic Adhesion)	A fiber tract that crosses the third ventricle. The cut end of this structure is visible in a midsagittal section of the brain in the middle of the third ventricle.	A fiber tract that connects the two thalamic nuclei to each other. It is absent in about 20% of human brains.	*intermediate*, in the middle, + *mass*, a mass	Midsagittal
Brainstem				
Medulla Oblongata	The most inferior aspect of the brainstem, forming the transition zone between the brain and spinal cord	Contains the centers for regulation of respiration and cardiac function, and contains nuclei of the *reticular activating system*, which is a group of nuclei that are important in regulating wakefulness and selective attention	*medius*, middle, + *oblongus*, rather long	Lateral, inferior, and midsagittal
Midbrain (Mesencephalon)	Found superior to the pons	Contains external structures, inferior and superior colliculi, and is responsible for auditory and visual processing. Internal structures include substantia nigra (movement and emotional responses) and tegmentum (integrate motor information from cerebrum and cerebellum).	*mesos*, middle + *enkephalos*, brain	Lateral, inferior, and midsagittal
Tectal Plate (Corpora Quadrigemina)	Consists of four twin bodies, the superior and inferior colliculi	Control center for visual and auditory reflexes	*corpus*, body, + *quad*, four, + *geminus*, twin	Midsagittal
Inferior Colliculus	A pair of oval projections that make up the inferior part of the tectal plate (corpora quadrigemina)	Controls *auditory reflexes*, such as the sudden turning of the head toward the source of a very loud sound	*inferior*, lower, + *colliculus*, a mound or hill	Midsagittal
Superior Colliculus	A pair of rounded projections that make up superior part of the corpora quadrigemina (tectal plate)	Controls visual reflexes, such as the sudden turning of the head toward the source of a flashing light	*superus*, above, + *colliculus*, a mound or hill	Midsagittal
Pons	Appears as a large mass just superior to the medulla oblongata	A "bridge" of nerve tracts that connect the cerebral hemispheres to the cerebellar hemispheres. Contains centers for control of respiration.	*pons*, bridge	Lateral, inferior, and midsagittal
Cerebellum				
Cerebellum	The second largest part of the brain	Regulation of muscle tone (a low-level muscle contraction), coordination of motor activity, and maintenance of balance and equilibrium	*cerebellum*, little brain	Lateral, inferior, and midsagittal

Table 15.3 Brain Structures Visible in Superficial Views of Whole or Sagitally Sectioned Brains *(continued)*

Brain Structure	Description	Function(s)	Word Origin	Views Where Visible
Limbic System				
Cingulate Gyrus	A gyrus located just superior to the corpus callosum	This area of the brain is not well understood. It is predominantly motor and may play a role in the limbic system (such as controlling motor functions with a strong emotional component).	*cingo*, to surround, + *gyros*, circle	Midsagittal
Fornix	An arching fiber tract located inferior to the septum pellucidum	Connects limbic system structures to each other	*fornix*, arch	Midsagittal
Olfactory Bulbs	Swellings connected to the anterior end of the olfactory tracts that lie on the inferior surface of the frontal lobes of the brain lateral to the longitudinal fissure	Location where cranial nerve I (CN I), the olfactory nerves, first synapse after passing through the cribriform plate of the ethmoid bone	*olfactus*, to smell, + *bulbus*, a globular structure	Inferior
Olfactory Tracts	Nerve fibers that extend from the olfactory bulbs posteriorly to the junction where the frontal lobes meet the optic chiasm	Carry the axons of neurons from the olfactory bulbs toward structures in other areas of the brain involved with olfaction	*olfactus*, to smell, + *tractus*, a drawing out	Inferior
Fissures/Sulci				
Central Sulcus	A deep groove that extends along the coronal plane	Separates the frontal lobe from the parietal lobe	*central*, in the center, + *sulcus*, a furrow	Superior, lateral, and midsagittal
Lateral Sulcus	A horizontal groove between the frontal/parietal lobes and the temporal lobe	Separates the frontal and parietal lobes from the temporal lobe	*latus*, the side, + *sulcus*, a furrow	Lateral
Longitudinal Fissure	A deep fissure between the two cerebral hemispheres	Separates the two cerebral hemispheres; the falx cerebri occupies this fissure in a living human	*longus*, long, + *fissure*, a deep furrow	Superior and inferior
Parieto-occipital Sulcus	Small groove that runs along the coronal plane	Separates the parietal lobe from the occipital lobe	*parieto-occipital*, between the parietal and occipital lobes, + *sulcus*, a furrow	Superior, lateral, and midsagittal
Transverse Fissure	A deep fissure between the cerebrum and the cerebellum	Separates the cerebral hemispheres from the cerebellar hemispheres. The tentorium cerebelli lies in this fissure.	*transversus*, across, + *fissure*, a deep furrow	Lateral
Fibers/Tracts				
Cerebral Peduncles	Tracts located between the midbrain and the pons	The fibers connect the forebrain (cerebral hemispheres and diencephalon) to the hindbrain (medulla oblongata, pons, and cerebellum)	*cerebrum*, brain, + *pedunculus*, a little foot	Inferior, midsagittal
Corpus Callosum	A fiber tract located superior to the lateral ventricles	Contains axons that connect the two cerebral hemispheres	*corpus*, body, + *callosus*, thick-skinned	Midsagittal
Infundibulum	A funnel-shaped inferior extension of the brain located immedately posterior to the optic chiasm	Consists of tracts that connect the hypothalamus to the posterior pituitary (pars nervosa)	*infundibulum*, a funnel	Inferior and midsagittal
Optic Chiasm	The X-shaped structure formed where the two optic nerves join, with most fibers crossing to the opposite side; located just anterior to the infundibulum	Location where fibers from both optic nerves cross over and travel in the optic tract on the opposite side. Not all fibers from the optic nerves cross over.	*optikos*, relating to the eye or vision, + *chiasma*, two crossing lines	Inferior and midsagittal
Optic Nerves	Anterior to the optic chiasm	Sensory neurons carrying visual information from the retina to the optic chiasm, where most fibers cross to the opposite side	*optikos*, relating to the eye or vision, + *nevus*, a white, cord-like structure	Inferior
Optic Tracts	Posterior to the optic chiasm	Sensory neurons carrying visual information from the optic chiasm to the lateral geniculate nucleus of the thalamus	*optikos*, relating to the eye or vision, + *tractus*, a drawing out	Inferior

INTEGRATE

CONCEPT CONNECTION

The hypothalamus is the homeostatic center of the brain, as it regulates such fundamental processes in the body as metabolism (thirst/hunger), body temperature, sleep, and sex. The hypothalamus serves as the master control center for the autonomic nervous system and the endocrine system. That is, there are connections between the hypothalamus and nuclei in the brainstem that regulate heart rate, respiration rate, and digestion. Although the hypothalamus is composed of nervous tissue, it is also considered an endocrine organ, as it plays a critical role in the release of hormones. It has connections with the pituitary gland, which lies inferior to the hypothalamus. The pituitary gland is composed of two lobes, called the posterior pituitary gland and the anterior pituitary gland. The hypothalamus contains neurosecretory cells, which are neurons that secrete hormones. A hormone is a chemical messenger that travels to a distant target via the bloodstream. The axons of the neurosecretory cells located in the paraventricular or supraoptic nuclei extend inferiorly through the infundibulum to the posterior pituitary. Upon stimulation, these neurosecretory cells release a hormone by exocytosis from their axon terminals in the posterior pituitary gland. This occurs much like acetylcholine being released from somatic motor neurons at a neuromuscular junction. Two hormones are released from the posterior pituitary gland: antidiuretic hormone (ADH) and oxytocin.

Neurosecretory cells arising from other nuclei within the hypothalamus release hormones into capillaries in the median eminence. These hormones enter a portal vein and are transported to the anterior pituitary gland. There, the hypothalamic hormones stimulate cells in the anterior pituitary to release their hormones. As was the case with the posterior pituitary gland, these hormones then enter the systemic circulation and are carried to distant targets. The specific hormones released and the cells responsible for hormone production will be covered in detail in chapter 19.

The hypothalamus also plays a role in regulating emotional behavior due to its involvement with the limbic system. The hypothalamus is located at the center of this system, connected to surrounding limbic system structures, such as the hippocampus and amygdala, via the fornix. Therefore, the hypothalamus influences aggression, fear, and pleasure. There are nuclei within the hypothalamus that regulate body temperature (preoptic area), circadian rhythms (suprachiasmatic nucleus), and hunger and satiety (ventromedial nucleus). The central location of the hypothalamus and its proximity to surrounding brain structures will be covered later in this chapter.

EXERCISE 15.4

SUPERIOR VIEW OF THE HUMAN BRAIN

1. Obtain a human brain or models of a human brain and observe the superior surface.
   - Numerous grooves are associated with the brain surface. A large groove is a fissure, whereas a small groove is a sulcus. The most prominent feature in this view is the **longitudinal fissure,** which is a deep groove that separates the two cerebral hemispheres from each other. Observe the many sulci on the brain surface. One major sulcus to identify is the **central sulcus.** Identification of the central sulcus is difficult, though not impossible, on a real human brain. The following are two features to look for:
   - The **precentral gyrus** and the **postcentral gyrus** (two raised areas approximately in the middle of the brain's superior surface) should become continuous with each other on the lateral aspect of the central sulcus just above the lateral sulcus (a groove that separates the temporal lobe from the frontal and parietal lobes). This means the central sulcus will not enter the lateral sulcus.
   - The central sulcus will dip down into the longitudinal fissure.
   - Three of the five lobes of the cerebrum are visible on the superior surface view.
2. Identify the structures listed in **figure 15.6** on the superior view of the brain, using table 15.3 and the textbook as guides. Then label them in figure 15.6.

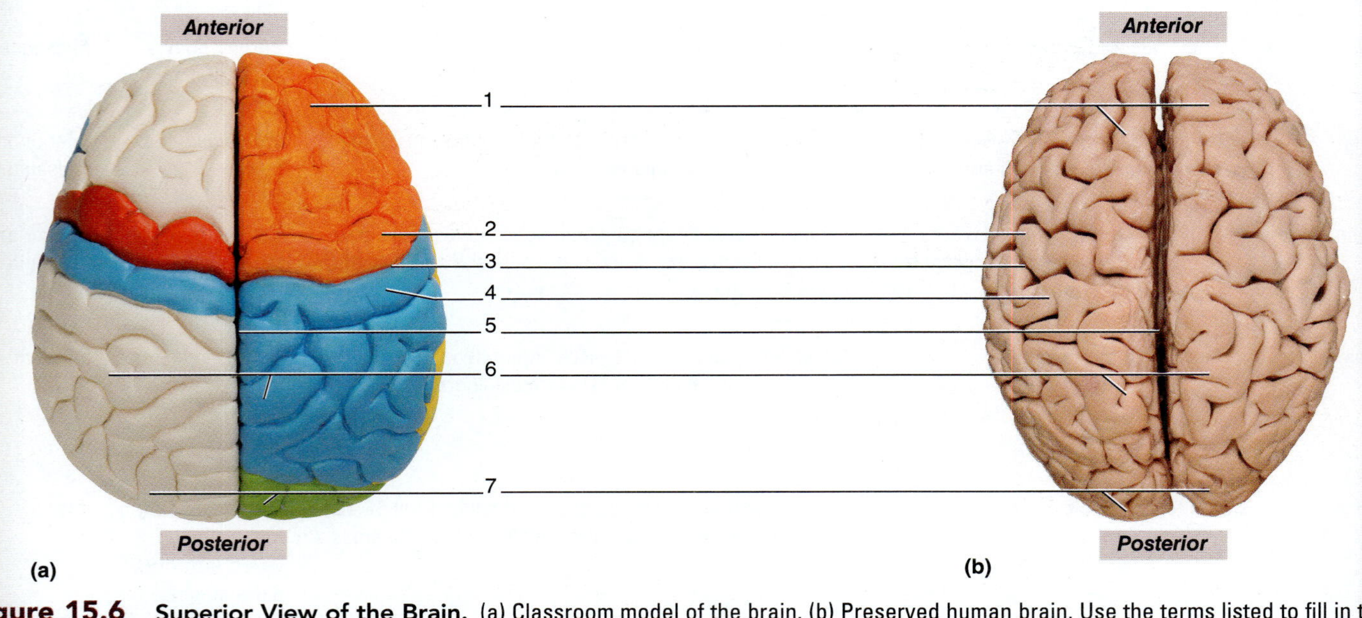

Figure 15.6 Superior View of the Brain. (a) Classroom model of the brain. (b) Preserved human brain. Use the terms listed to fill in the numbered labels in the figure.

- ☐ central sulcus
- ☐ frontal lobe
- ☐ longitudinal fissure
- ☐ occipital lobe
- ☐ parietal lobe
- ☐ postcentral gyrus
- ☐ precentral gyrus

EXERCISE 15.5

LATERAL VIEW OF THE HUMAN BRAIN

1. Obtain a human brain or models of a human brain and observe the lateral surface.
   - As with the superior view, the **central sulcus** should be visible by identifying the location where the pre- and postcentral gyri become continuous with each other just above the **lateral sulcus**.
2. Identify the structures listed in **figure 15.7** on the lateral view of the brain, using table 15.3 and the textbook as guides. Then label them in figure 15.7.
3. *Optional Activity:* **AP|R 7: Nervous System**—Watch the "Divisions of Brain" animation for an overview of the regions of the brain and their general functions.

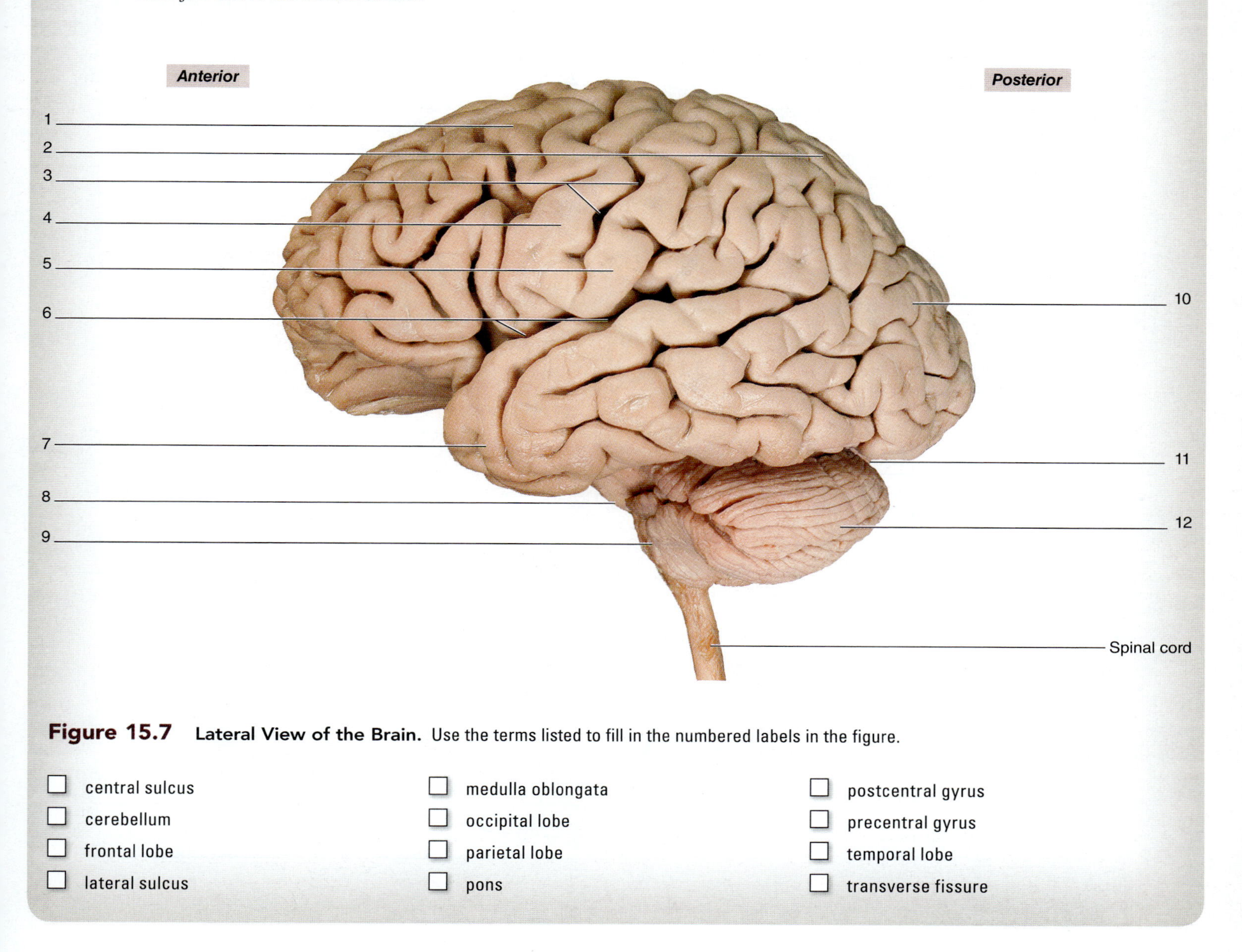

Figure 15.7 Lateral View of the Brain. Use the terms listed to fill in the numbered labels in the figure.

- ☐ central sulcus
- ☐ cerebellum
- ☐ frontal lobe
- ☐ lateral sulcus
- ☐ medulla oblongata
- ☐ occipital lobe
- ☐ parietal lobe
- ☐ pons
- ☐ postcentral gyrus
- ☐ precentral gyrus
- ☐ temporal lobe
- ☐ transverse fissure

EXERCISE 15.6

INFERIOR VIEW OF THE HUMAN BRAIN

1. Obtain a human brain or models of a human brain and observe the inferior surface. This view is considerably more complicated than the superior or lateral views because of the cranial nerves that arise from the brain. Cranial nerve identification is covered later in this chapter.
   - Both the brainstem and the cerebellum are visible from this view.
   - Prominent features associated with the cerebrum are the **optic chiasm** and the **optic tracts**, which extend from it into the brain.
   - The **mammillary bodies** are two small projections posterior to the optic chiasm.
   - One of the more problematic structures to identify in this view on a real brain is the infundibulum. When a brain is removed from the cranium, the pituitary gland almost always gets removed from the brain. The only structure left connected to the brain is the stalk of tissue that connects the pituitary gland to the hypothalamus, which is the **infundibulum,** or **pituitary stalk.** Observation of a model of the brain shows the pituitary gland to be intact. The infundibulum can be identified as a small strand of tissue that is located directly posterior to the optic chiasm and directly anterior to the mammillary bodies.
2. Identify the structures listed in **figure 15.8** on the inferior view of the brain, using table 15.3 and the textbook as guides. Then label them in figure 15.8.

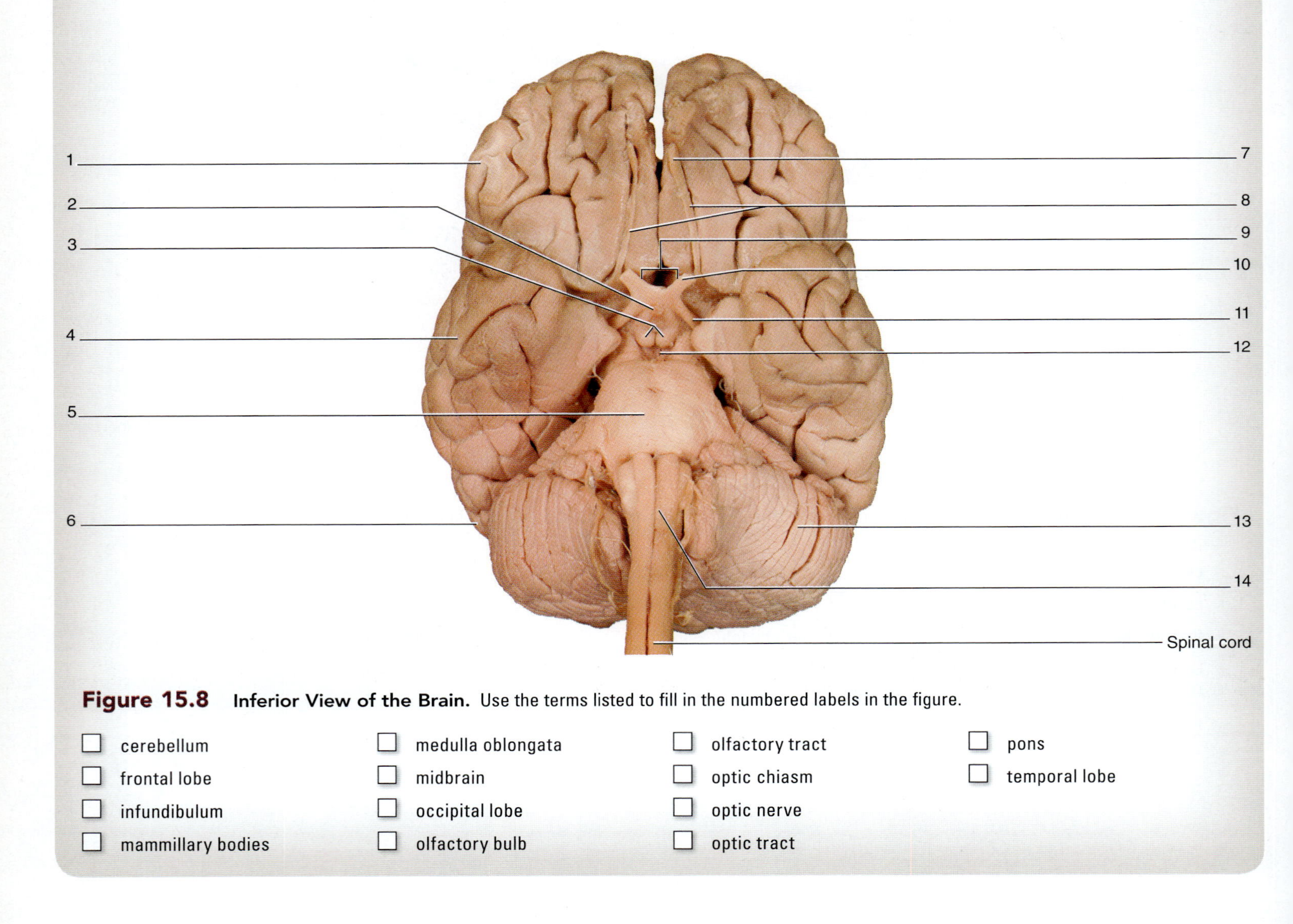

Figure 15.8 Inferior View of the Brain. Use the terms listed to fill in the numbered labels in the figure.

- ☐ cerebellum
- ☐ frontal lobe
- ☐ infundibulum
- ☐ mammillary bodies
- ☐ medulla oblongata
- ☐ midbrain
- ☐ occipital lobe
- ☐ olfactory bulb
- ☐ olfactory tract
- ☐ optic chiasm
- ☐ optic nerve
- ☐ optic tract
- ☐ pons
- ☐ temporal lobe

EXERCISE 15.7

MIDSAGITTAL VIEW OF THE HUMAN BRAIN

1. Obtain a human brain or brain model that has been sectioned along the midsagittal plane and observe its medial surface (**figure 15.9).**

   - In the very center of view, notice the **third ventricle.** The third ventricle is the central depressed area that appears to have a cut nerve in the center. The "cut nerve" isn't really a nerve, but it is similar. It is a fiber tract called the **interthalamic adhesion** (or **intermediate mass**), which connects the two thalamic nuclei to each other. Use the interthalamic adhesion, the thalamus, and the third ventricle as reference points for identification of other structures in this view.
   - Many of the structures that are located around the third ventricle of the brain belong to a system called the **limbic system** (*limbus*, border), so named because structures of the limbic system are located at the border of the third ventricle and the brainstem. The limbic system is referred to as the "emotional brain" because its structures play a role in our emotions.

2. Identify the structures listed in figure 15.9 on the midsagittal view of the brain, using table 15.3 and the textbook as guides. Then label them in figure 15.9.

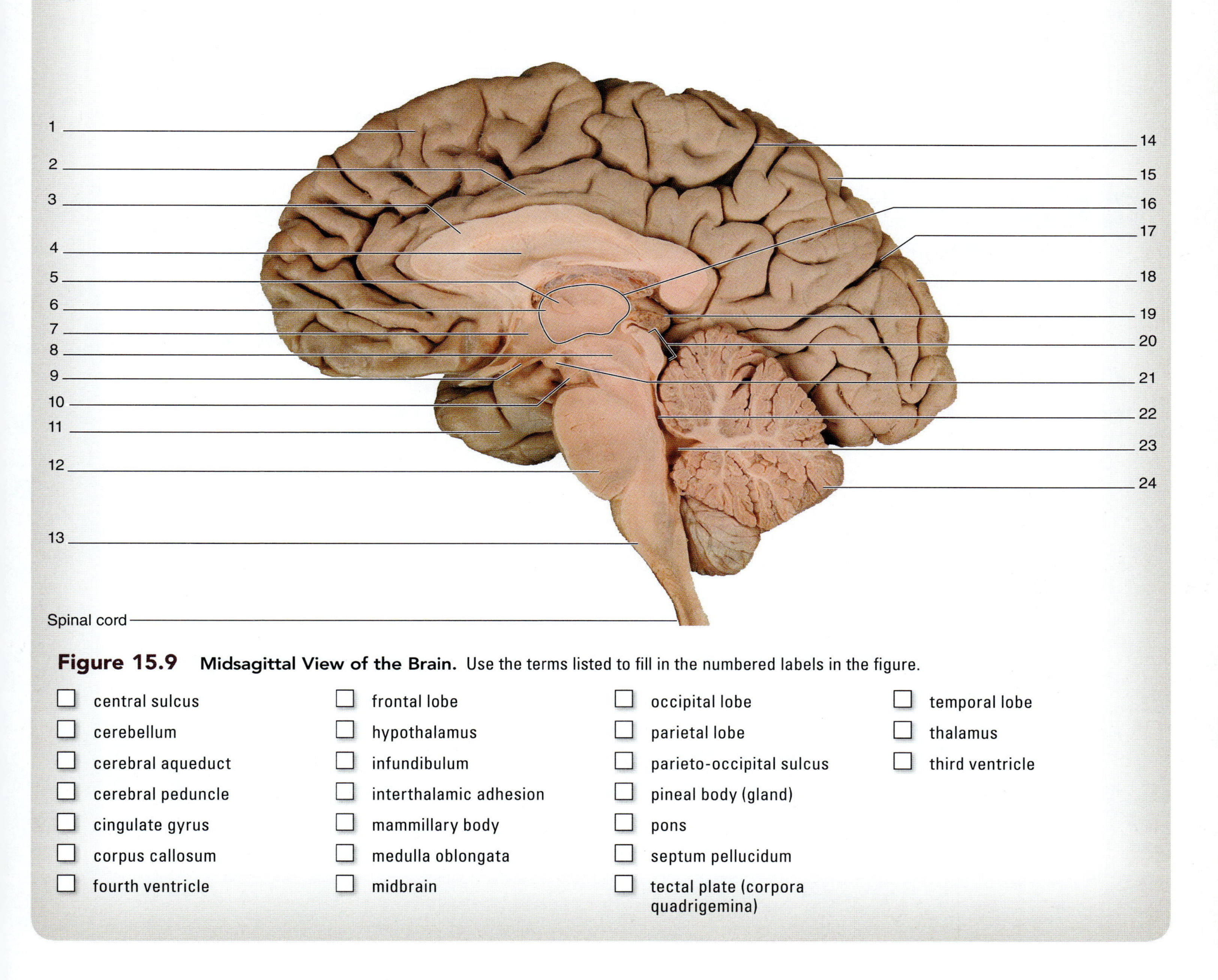

Figure 15.9 Midsagittal View of the Brain. Use the terms listed to fill in the numbered labels in the figure.

- ☐ central sulcus
- ☐ cerebellum
- ☐ cerebral aqueduct
- ☐ cerebral peduncle
- ☐ cingulate gyrus
- ☐ corpus callosum
- ☐ fourth ventricle
- ☐ frontal lobe
- ☐ hypothalamus
- ☐ infundibulum
- ☐ interthalamic adhesion
- ☐ mammillary body
- ☐ medulla oblongata
- ☐ midbrain
- ☐ occipital lobe
- ☐ parietal lobe
- ☐ parieto-occipital sulcus
- ☐ pineal body (gland)
- ☐ pons
- ☐ septum pellucidum
- ☐ tectal plate (corpora quadrigemina)
- ☐ temporal lobe
- ☐ thalamus
- ☐ third ventricle

Cranial Nerves

An inferior view of the brain allows for visualization of the cranial nerves at the location where they arise from the brain. Cranial nerves are numbered, starting from the anterior (rostral) part of the brain and moving posterior (caudal), using Roman numerals I through XII. **Figure 15.10** shows the inferior surface of the brain and the cranial nerves. The olfactory bulbs (where the olfactory nerves (CN I) synapse) and the optic nerves (CN II) are very large, easily identifiable structures located on the inferior surface of the frontal lobes of the cerebrum. The remainder of the cranial nerves (CN III through CN XII) are generally smaller and are located closer together in the region of the midbrain, pons, and medulla that form the brainstem, thus making their identification a little more challenging. Exercise 15.7 will involve identifying the cranial nerves on a brain or on a model of the brain.

EXERCISE 15.8

IDENTIFICATION OF CRANIAL NERVES ON A BRAIN OR BRAINSTEM MODEL

1. Obtain a human brain or a model of a human brain or brainstem.
2. Turn the brain over and observe its inferior surface (figure 15.10). Note all of the small nerves exiting the brain from various locations. These are the cranial nerves **(table 15.4).**
3. Identify the twelve cranial nerves on the inferior surface of the brain, using table 15.4 and figure 15.10 as guides.
4. Complete the chart for cranial nerves III through XII by listing the specific nerves that extend from each of the three areas of the brainstem. Use figure 15.10 as a guide. Keep in mind that knowledge of the general area where a specific nerve emerges can be helpful in identifying the nerves, even if by process of elimination.

Table 15.4 Names of Cranial Nerves

Nerve Number	Name	Foramina of Exit	Word Origin
I	Olfactory	Olfactory foramina in the cribiform plate of the ethmoid bone	*olfacio*, to smell
II	Optic	Optic canal	*optikos*, relating to the eye or vision
III	Oculomotor	Superior orbital fissure	*oculo-*, the eye, + *motorius*, moving
IV	Trochlear	Superior orbital fissure	*trochileia*, a pulley
V	Trigeminal	Superior orbital fissure (V_1—ophthalmic) Foramen rotundum (V_2—maxillary) Foramen ovale (V_3—mandibular)	*tri-*, three, + *geminus*, twins
VI	Abducens	Superior orbital fissure	*abductio-*, to move away from the median plane
VII	Facial	Internal acoustic meatus (exits via the stylomastoid foramen)	*facialis*, relating to the face
VIII	Vestibulocochlear	Internal acoustic meatus	*vestibulum*, entrance, + *cochlea*, snail shell
IX	Glossopharyngeal	Jugular foramen	*glossus*, tongue, + *pharyngeus*, pharynx
X	Vagus	Jugular foramen	*vagus*, wanderer
XI	Accessory	Jugular foramen (accessory division) Foramen magnum (spinal division)	*spina*, spine, + *accessory*, an extra structure
XII	Hypoglossal	Hypoglossal canal	*hypo*, beneath, + *glossus*, tongue

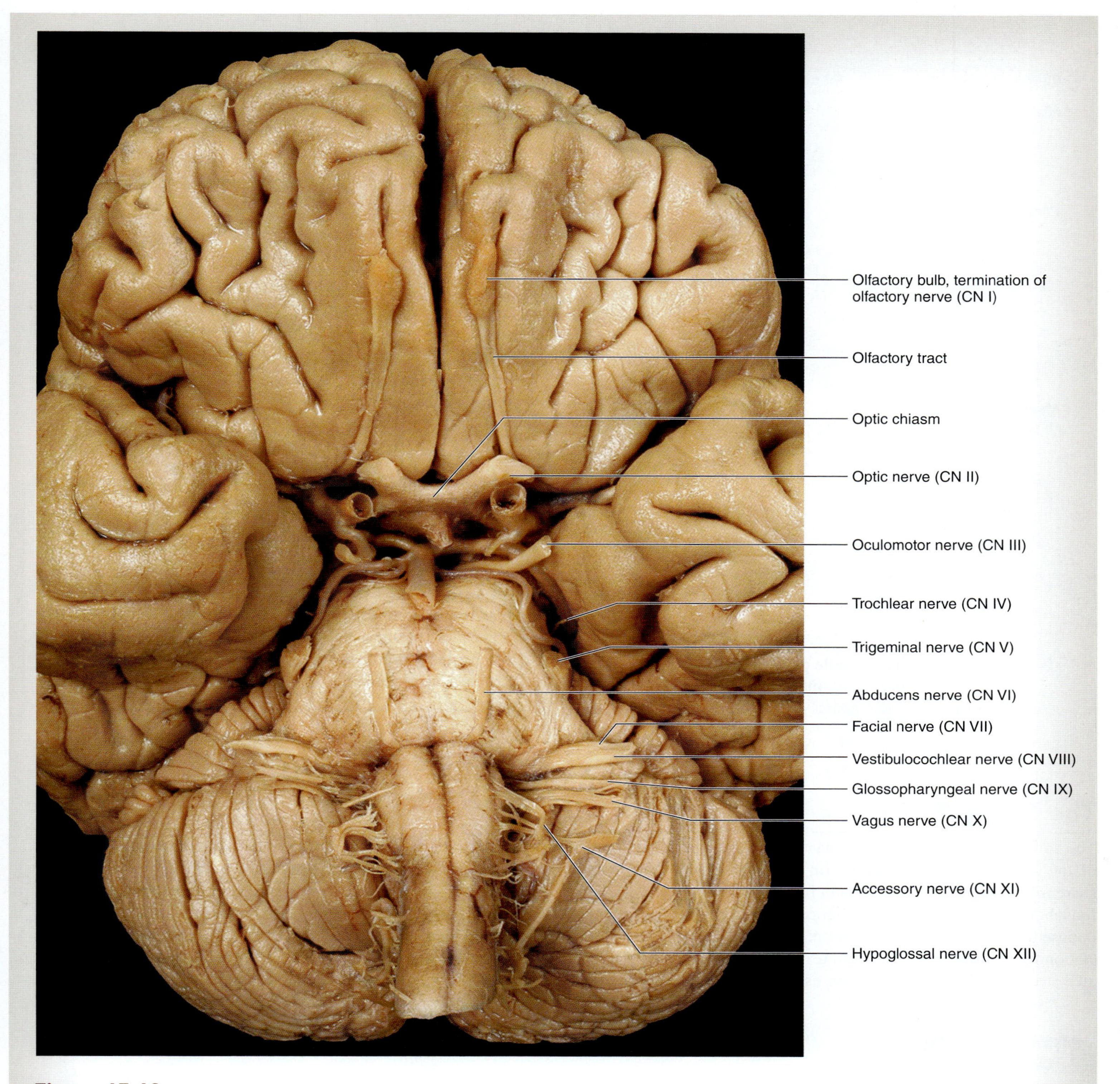

Figure 15.10 **Cranial Nerves on the Inferior Surface of the Brain.**

(continued on next page)

(continued from previous page)

Point of Exit	Cranial Nerve
Midbrain	
Pons	
Medulla Oblongata	

5. **Mnemonic devices,** simple phrases or plays on words that aid in memory recall, make remembering long strings of anatomical names easier. One popular technique is to associate a word with the first letter of the corresponding word you want to remember. A mnemonic that will help you recall the names of the cranial nerves in numerical order is

   Oh **O**nce **O**ne **T**akes **T**he **A**natomy **F**inal, **V**ery **G**ood **V**acations **A**re **H**eavenly.

   Mnemonics work best when they mean something to you. Develop your own mnemonic for remembering the cranial nerves. ______________________________

   ____

   ____

   ____

6. *Optional Activity:* **AP|R 7: Nervous System**—Visit the quiz area for cranial nerve location, composition, and related foramina.

INTEGRATE

LEARNING STRATEGY

When observing a real brain instead of a model, the trochlear nerve (CN IV) may not be visible. It is very small, and often detaches from the brain when the brain is removed from the skull. If this is the case, make sure to identify the trochlear nerve on a model of the human brain. This nerve is unique because it is the only cranial nerve that originates from the posterior part of the brainstem, rather than the anterior or medial part.

The Sheep Brain

The following exercises involve identifying on a sheep brain many of the same structures identified on a human brain. Sheep brains share many similarities with human brains, and are readily available for laboratory studies. The experience of dissecting a real brain (as opposed to observing a plastic model of a brain) will give an appreciation for the true appearance and texture of the brain and its associated structures. Note that the brain tissue itself is relatively delicate. Even so, the tissue is much more solid than it would be if observing a fresh brain because it has been fixed with chemicals. Living brain tissues are *extremely* delicate, and have the consistency of firm gelatin. These dissection exercises involve identifying both brain structures *and* cranial nerves on the sheep brain. Observation of the structure and function of cranial nerves and identification of them on a human brain comes later in this chapter.

EXERCISE 15.9

SHEEP BRAIN DISSECTION

EXERCISE 15.9A Dura Mater

1. Obtain a sheep brain, dissecting tray, and dissecting tools (forceps, scissors, a scalpel, and gloves). Place the sheep brain in the dissecting tray and take turns with a laboratory partner(s) observing its gross structure. This exercise requires a sheep brain with dura mater intact. If the dura mater is missing, proceed to the section "Inferior View of the Sheep Brain" (p. 392). Otherwise begin here.

2. Using hands or blunt forceps, feel the toughness of the dura mater. Notice how it surrounds the entire brain. What type of tissue is the dura mater composed of?

   ____

3. Observe the dura mater on the superior surface of the brain **(figure 15.11*b*),** and locate the following structures:

   ☐ **confluence of sinuses** ☐ **transverse sinuses**
   ☐ **superior sagittal sinus**

4. Rotate the brain so it is resting on its superior surface and the inferior surface is visible (figure 15.11*a*). Notice the relatively large **pituitary gland** projecting inferior to the dura mater. The goal in this part of the dissection will be to cut away the dura mater without disconnecting the pituitary from the rest of the brain. Notice the capillary tufts found just posterior and lateral to the pituitary gland. Just lateral to these capillaries on both sides are the large **trigeminal nerves** (CN V). Using a blunt probe, feel the dura mater surrounding the base of the pituitary gland. This dural membrane is the **diaphragma sellae,** which lies between the pituitary and the rest of the brain (except where the pituitary stalk exits the sella turcica of the sphenoid bone).
5. To free the connections between the dura and the rest of the brain without breaking off the pituitary gland, first cut around the trigeminal nerves and capillary tufts. **Figure 15.12*a*** shows where to make the initial incision. Cut *around* (lateral to) the optic chiasm, diaphragma sellae, pituitary gland, and trigeminal nerves to make a complete circle, which will free the dura mater from its attachments.
6. Next, make an anterior cut in the dura mater along the midsagittal plane between the olfactory bulbs and olfactory tracts (figure 15.12*b*). Once the dura mater has been freed from its connections to the diaphragma sellae, gently pull the dura away from the brain. Pull in a posterior, superior direction so the falx cerebri and tentorium cerebelli slip out of their respective fissures without damaging the delicate brain tissues. When pulling the dura mater away from the brain, gently tease away any remaining connections. Figure 15.12*b* shows what the dura mater should look like after it has been cut away and removed from the brain.
7. Once the dura mater is completely freed from the brain, observe it closely and compare the dural septa and sinuses in the sheep brain to those identified in the human brain. What dural septa is missing in a sheep brain that is present in a human brain?

____

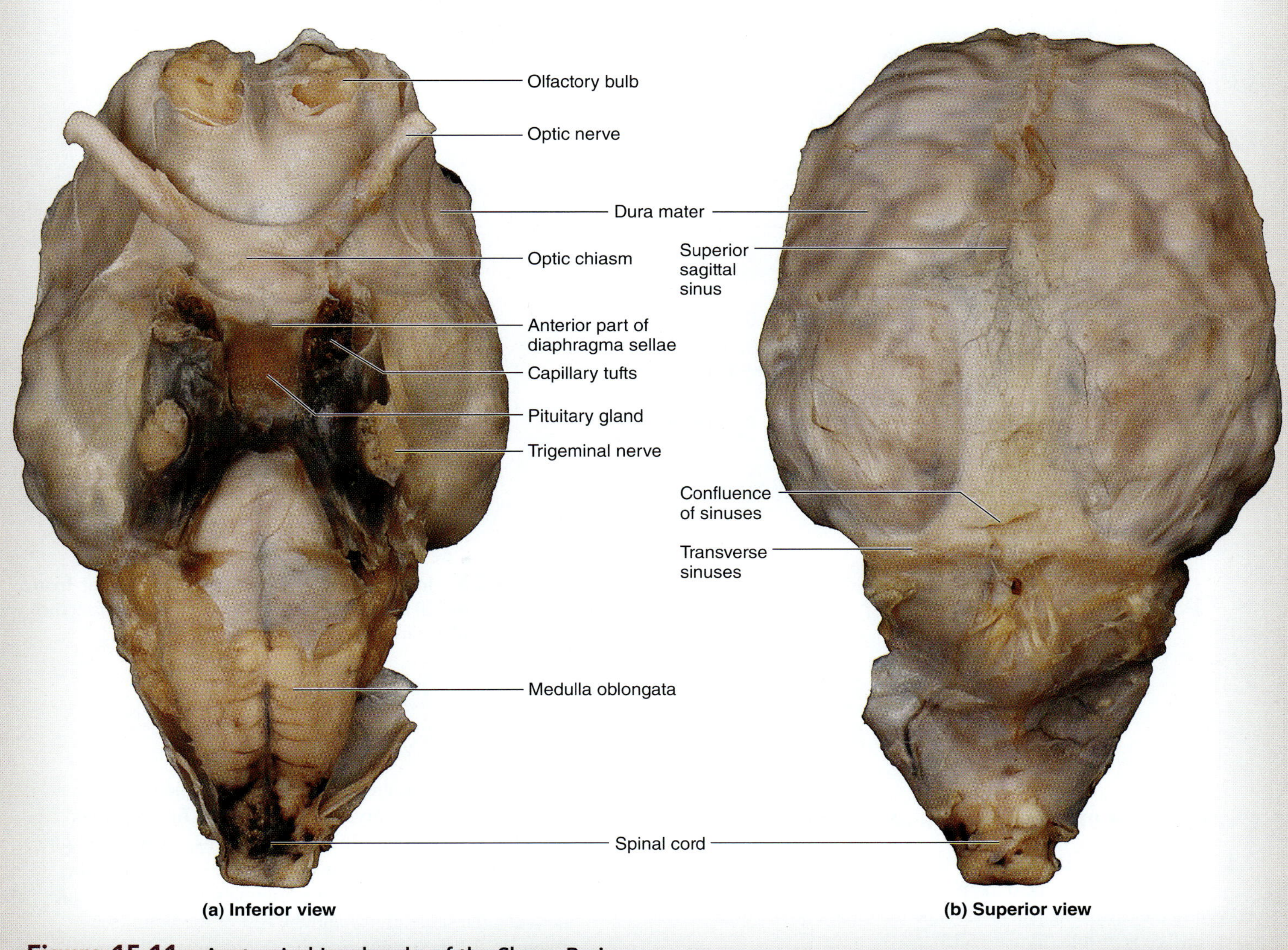

Figure 15.11 **Anatomical Landmarks of the Sheep Brain.**

(continued on next page)

(continued from previous page)

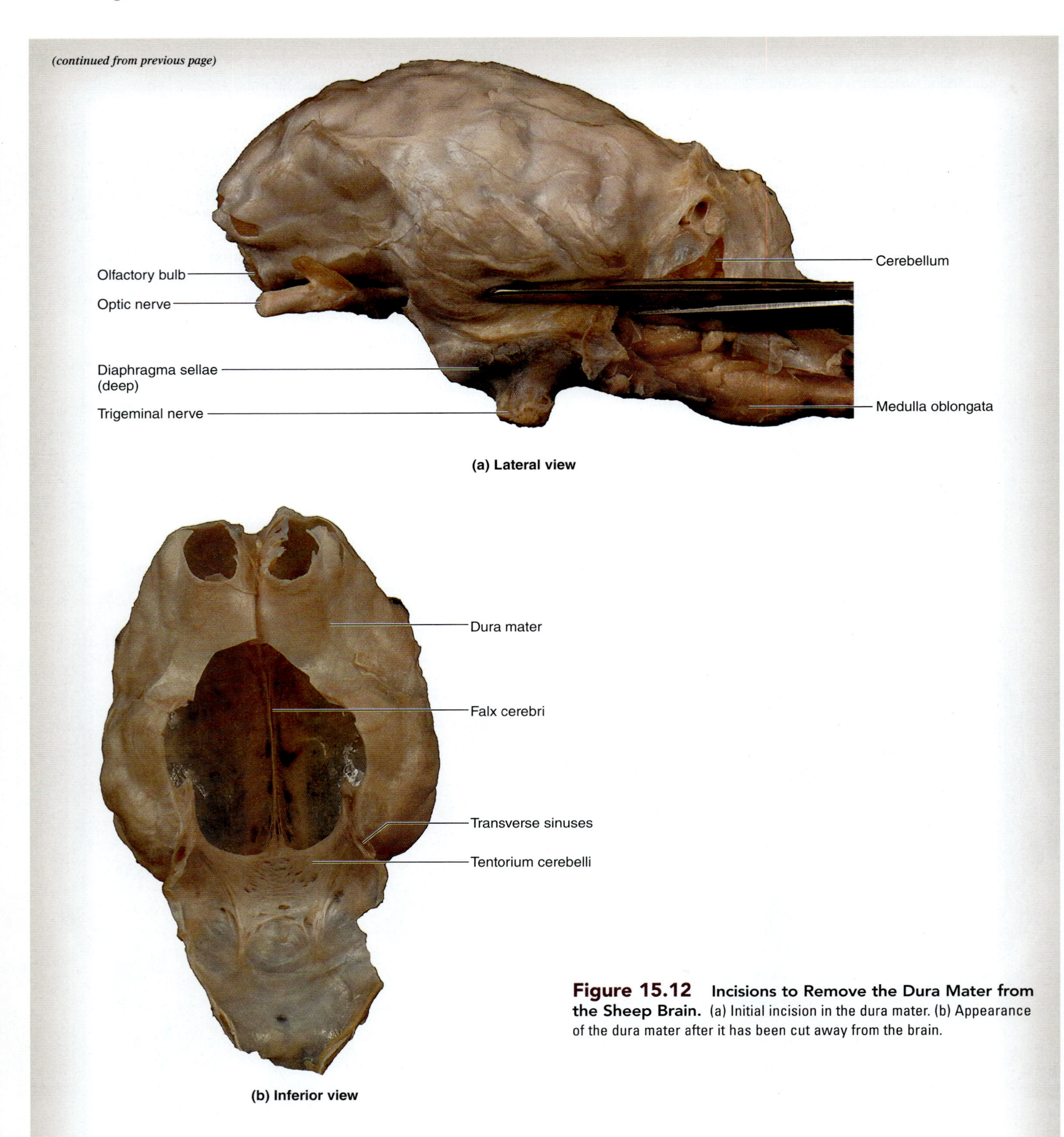

Figure 15.12 Incisions to Remove the Dura Mater from the Sheep Brain. (a) Initial incision in the dura mater. (b) Appearance of the dura mater after it has been cut away from the brain.

EXERCISE 15.9B Inferior View of the Sheep Brain

1. Obtain a sheep brain without the dura mater intact, or use the brain from which the dura mater has been removed. Place it in the dissecting pan on its superior surface so the inferior surface is visible **(figure 15.13).**
2. When sheep brains are collected by a commercial vendor for use in the laboratory, the dura mater is first separated from the cranial bones. In such specimens, most of the dura mater has been dissected away from the cranium and the only part remaining is the diaphragma sellae, a membrane between the sella turcica and the rest of the brain (see figure 15.12*a*). Surrounding the diaphragma sellae are

some capillary tufts and large cranial nerves, the trigeminal nerves (CN V).

3. Identify the following on the sheep brain, using figure 15.13 as a guide:
   - ☐ **capillary tufts**
   - ☐ **diaphragma sellae**
   - ☐ **olfactory bulb**
   - ☐ **olfactory tract**
   - ☐ **optic chiasm**
   - ☐ **pituitary gland**
   - ☐ **trigeminal nerves (CN V)**
4. Next, dissect the diaphragma sellae and the capillary tufts away from the pituitary gland without damaging the cranial nerves, without detaching the pituitary from the infundibulum, and without detaching the trigeminal nerves from the brain. Dissect carefully, because it is very easy to accidentally detach these structures from the brain if there is too much tension on the diaphragma sellae while attempting its removal.
5. Gently lift the dura mater posterior to the pituitary gland to see the small nerves that enter the dura mater on its deep surface **(figure 15.14).**
6. Using scissors or a scalpel, detach the nerves where they enter the dura mater. Cut the nerves where they attach to the dura (not where they attach to the brain!) and then cut the dura and bony material away, removing as much of it as possible while keeping the pituitary intact. Be careful, because the connection between the pituitary and the rest of the brain is delicate.

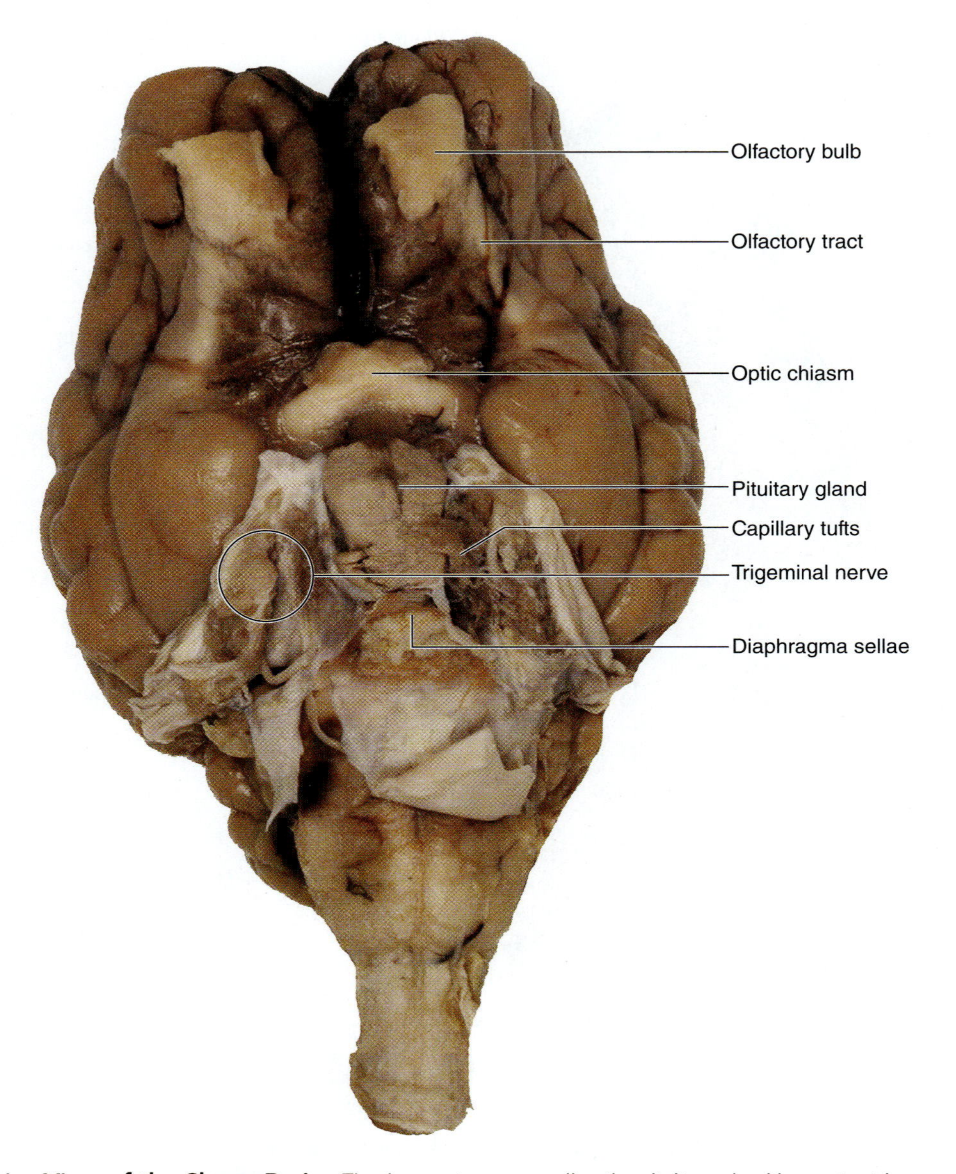

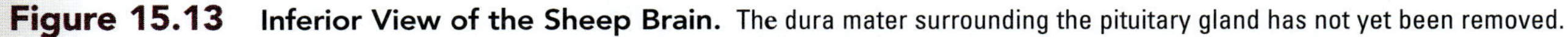

Figure 15.13 Inferior View of the Sheep Brain. The dura mater surrounding the pituitary gland has not yet been removed.

(continued on next page)

(continued from previous page)

7. Identify the following structures in the inferior view of the sheep brain, using **figure 15.15*a*** as a guide.
   - ☐ cerebellum
   - ☐ cerebral peduncle
   - ☐ frontal lobe
   - ☐ longitudinal fissure
   - ☐ medulla oblongata
   - ☐ olfactory bulb
   - ☐ olfactory tract
   - ☐ optic chiasm
   - ☐ optic nerve (CN II)
   - ☐ pituitary gland
   - ☐ pons
   - ☐ spinal cord
   - ☐ temporal lobe
   - ☐ transverse fissure

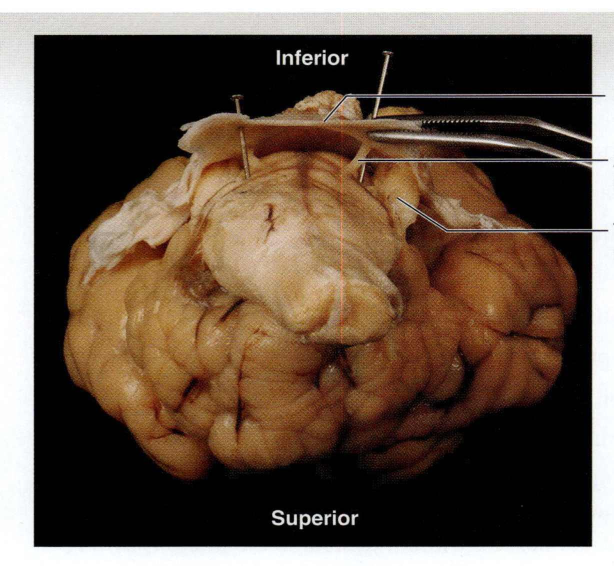

Figure 15.14 **Cranial Nerves Entering the Dura Mater.** The abducens and trigeminal nerves can be seen exiting the inferior surface of the sheep brain and piercing the dura mater.

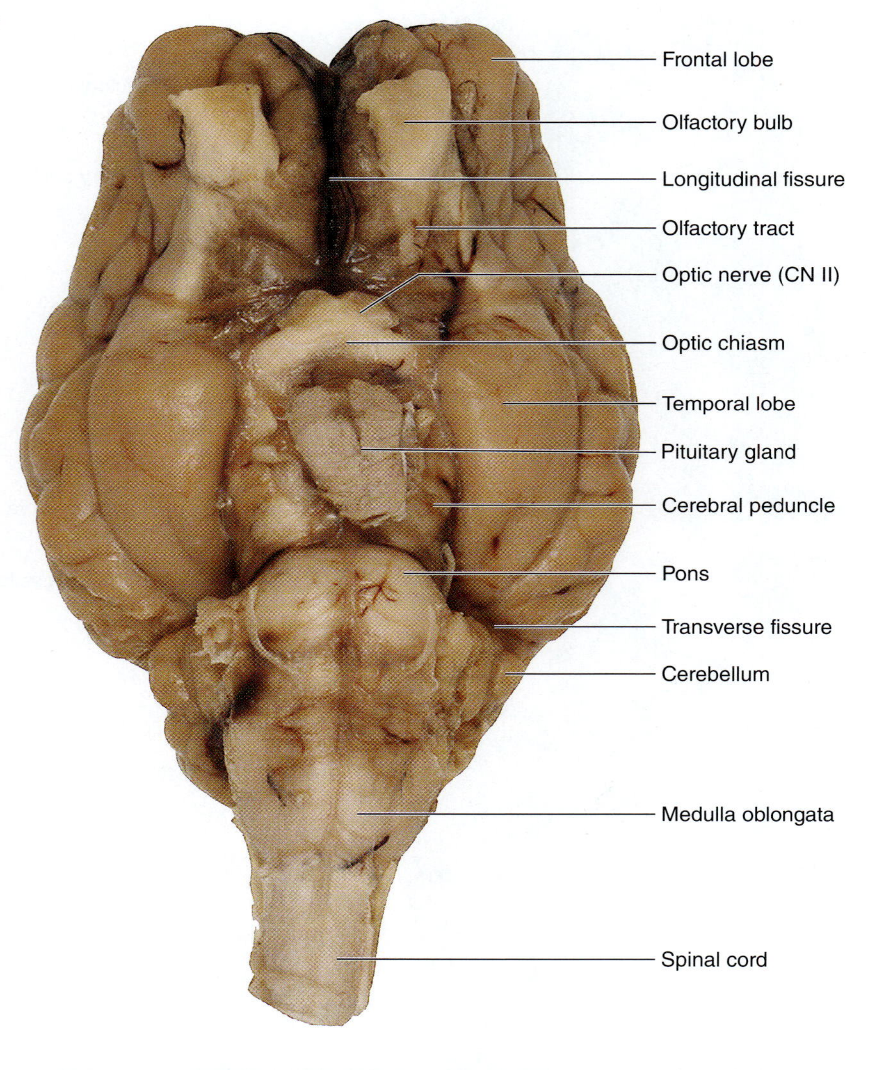

(a) Inferior view, pituitary gland intact; dura mater removed

Figure 15.15 **Inferior Views of the Sheep Brain.**

8. Gently lift the pituitary to observe the **mammillary body** (figure 15.15*b*). Note that the sheep brain has only a single mammillary body, whereas the human brain has two.

9. Finally, attempt to identify the cranial nerves listed in figure 15.15*b*. Identification of cranial nerves IX (glossopharyngeal nerve) through XII (hypoglossal nerve) may not be possible because these nerves are very small and may have been damaged or torn off the brain as the dura mater was removed from the brain, or as the brain was removed from the cranium.

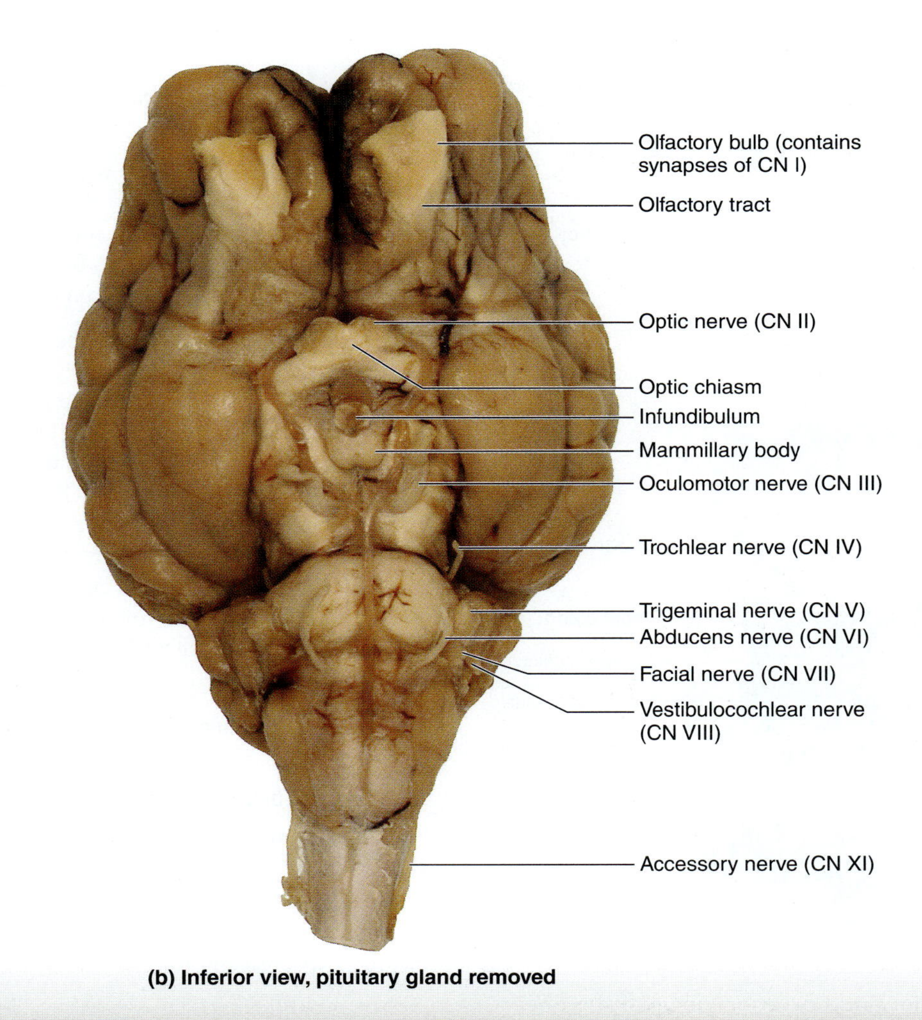

(b) Inferior view, pituitary gland removed

Figure 15.15 **Inferior Views of the Sheep Brain *(continued)*.**

(continued on next page)

(continued from previous page)

EXERCISE 15.9C Superior View of the Sheep Brain

1. Place the brain in the dissecting tray with the inferior side facing down **(figure 15.16*a*).** Note the thin, transparent **arachnoid mater** that covers the entire surface of the brain without dipping into the **sulci** (grooves) between the **gyri** (folds) of the brain. Note the numerous **blood vessels** that lie between the arachnoid mater and the **pia mater.** The space occupied by the blood vessels is also a space where cerebrospinal fluid flows in the living animal. What is the name of this space?

   ____

2. Identify the following structures in the superior view of the sheep brain, using figure 15.16*a* as a guide.

   - ☐ arachnoid mater
   - ☐ blood vessels
   - ☐ cerebellum
   - ☐ cerebrum
   - ☐ gyrus
   - ☐ longitudinal fissure
   - ☐ spinal cord
   - ☐ sulcus
   - ☐ transverse fissure

3. Pick up the brain and gently pull the cerebellum away from the cerebrum so the transverse fissure is visible. Identify the following structures, using figure 15.16*b* as a guide:

   - ☐ cerebellum
   - ☐ cerebrum
   - ☐ inferior colliculus
   - ☐ pineal gland
   - ☐ superior colliculus

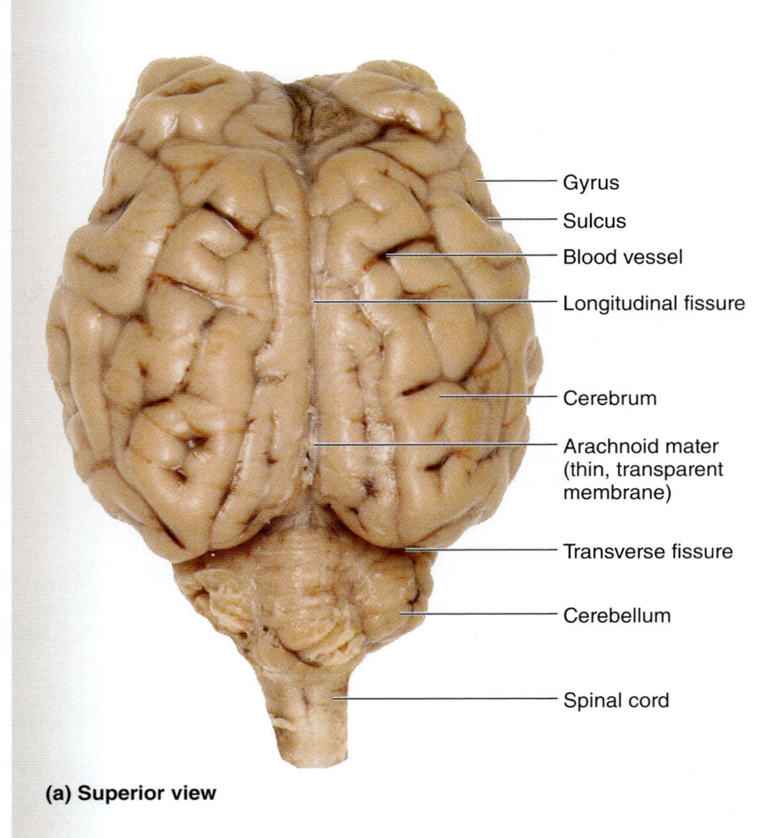

(a) Superior view

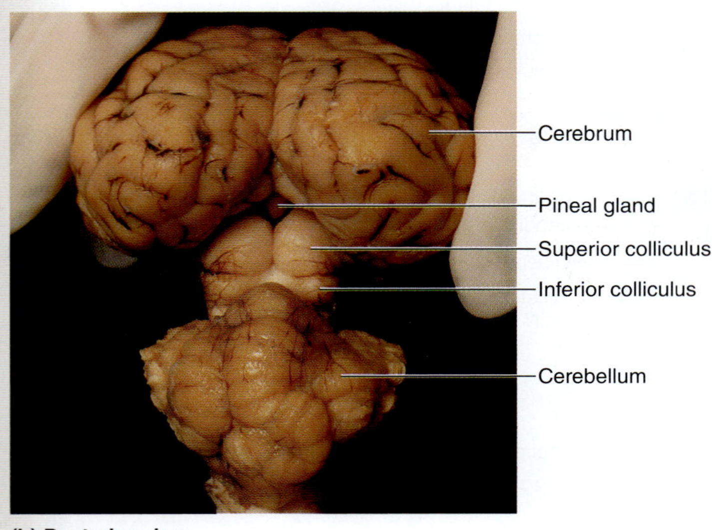

(b) Posterior view

Figure 15.16 Superior and Posterior Views of the Sheep Brain. (a) Superior view of the sheep brain. (b) Posterior view; cerebral hemispheres are pulled away from the cerebellum to reveal deeper structures.

EXERCISE 15.9D Midsagittal and Coronal Sections of the Sheep Brain

1. Some lab members may perform a midsagittal section of the sheep brain; others will perform a coronal section. Ask the instructor which section to make before initiating a cut. Be sure to observe a brain that has been cut along a midsagittal plane and a brain cut along a coronal plane.

2. *Midsagittal Section:* Place the sheep brain in a dissecting tray with its superior surface facing up. Using a scalpel, cut the brain in half along the midsagittal plane. Start a cut on the anterior end of the brain by placing the scalpel blade within the longitudinal fissure. What is the first structure the scalpel blade will cut through?

   ____

3. Once the brain has been cut in half, observe its medial surface. Identify the following structures on the sheep brain, using **figure 15.17** as a guide.

   - ☐ central canal (of spinal cord)
   - ☐ cerebellum
   - ☐ cerebral aqueduct
   - ☐ cerebral peduncle
   - ☐ cerebrum
   - ☐ corpus callosum
   - ☐ fornix
   - ☐ fourth ventricle
   - ☐ mammillary body
   - ☐ medulla oblongata
   - ☐ optic chiasm
   - ☐ pineal gland
   - ☐ pituitary gland
   - ☐ pons
   - ☐ spinal cord
   - ☐ superior colliculus
   - ☐ thalamus

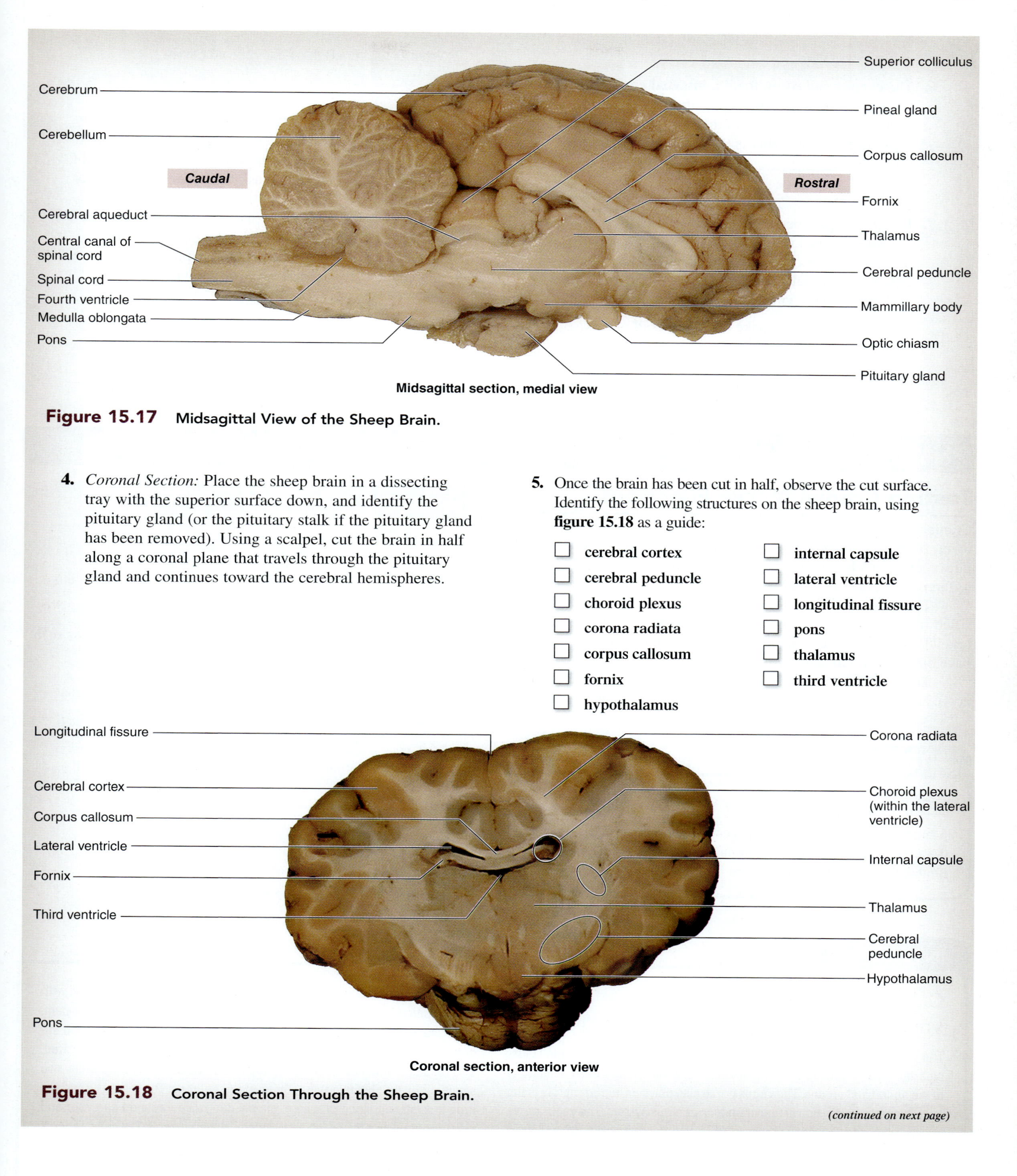

Figure 15.17 Midsagittal View of the Sheep Brain.

4. *Coronal Section:* Place the sheep brain in a dissecting tray with the superior surface down, and identify the pituitary gland (or the pituitary stalk if the pituitary gland has been removed). Using a scalpel, cut the brain in half along a coronal plane that travels through the pituitary gland and continues toward the cerebral hemispheres.

5. Once the brain has been cut in half, observe the cut surface. Identify the following structures on the sheep brain, using **figure 15.18** as a guide:

- ☐ cerebral cortex
- ☐ cerebral peduncle
- ☐ choroid plexus
- ☐ corona radiata
- ☐ corpus callosum
- ☐ fornix
- ☐ hypothalamus
- ☐ internal capsule
- ☐ lateral ventricle
- ☐ longitudinal fissure
- ☐ pons
- ☐ thalamus
- ☐ third ventricle

Figure 15.18 Coronal Section Through the Sheep Brain.

(continued on next page)

(continued from previous page)

6. The **corpus callosum, fornix, internal capsule, corona radiata,** and **cerebral peduncles** are all fiber tracts. Recall from chapter 14 that a tract is a bundle of myelinated axons. What is the function of these fiber tracts? ______________________________

____

7. Using forceps, open the lateral ventricles a bit to find the **choroid plexus** in the wall of the ventricle. Upon gross observation, the choroid plexus kind of looks like "junk" inside the ventricle, but it is really a tuft of capillaries covered by ependymal cells.

   What two structures make up the choroid plexus?

   ____

   ____

   What is the function of the choroid plexus?

   ____

   ____

8. When the dissection has been completed, collect all the organic material from the dissecting pan and dispose of it in the proper containers (ask the laboratory instructor what these are). Dispose of the scalpel blades in the sharps container, and throw used paper towels and gloves into the garbage. Clean all dissecting tools and the dissecting pan and return them to the proper storage area, and disinfect the laboratory workstation.

PHYSIOLOGY

Testing Cranial Nerve Functions

Cranial nerves carry different types of information from the brain to or from the target organ(s) for each nerve. **Sensory** (afferent) information travels from the target organ to the brain, whereas **motor** (efferent) information travels from the brain to the target organ.

The types of information carried by cranial nerves are classified in **table 15.5**. Note that sensory may be somatic, sensory, or special. Motor may be somatic or visceral. Some cranial nerves also have both motor and sensory function.

Table 15.6 lists the type of information carried by each cranial nerve and lists the specific functions of each cranial nerve.

Table 15.5	Modalities Within Cranial Nerves
Somatic Motor	Motor output to skeletal muscle
Somatic Sensory	Sensation from cutaneous touch, pain, temperature, position, and pressure receptors
Special Sensory	Somatic sensation from special sensory organs such as the eye and ear
Visceral Motor	Motor output to cardiac muscle, smooth muscle, and glands
Visceral Sensory	Sensation from sensory receptors within visceral organs (e.g., stretch receptors in bladder wall)

INTEGRATE

LEARNING STRATEGY

A mnemonic device to remember if a cranial nerve is sensory (S), motor (M), or both (B) is this mnemonic device: **S**ome **S**ay **M**arry **M**oney **B**ut **M**y **B**rother **S**ays **B**ig **B**rains **M**atter **M**ore.

INTEGRATE

CLINICAL VIEW

Vasovagal Syncope

The vagus nerve (CN X) innervates not only laryngeal and pharyngeal muscles, but also smooth muscle and glands of the heart, lungs, and the majority of the abdominal organs. The vagus nerve also plays an important role in the parasympathetic division of the autonomic nervous system. The vagus nerve lowers heart rate and blood pressure while also stimulating digestive organs. It carries both sensory and motor information, and it has been implicated in a common form of syncope (or fainting) known as **vasovagal syncope.** Patients that suffer from vasovagal syncope typically experience recurrent episodes triggered by a repeatable stimulus. Common stimuli may be associated with body position (i.e., sitting or standing), stress, dehydration, hunger, or pain. Regardless of the stimulus, sensory information is carried to neurons in the brainstem, where motor and sensory nuclei for the cranial nerve are located. This results in stimulation of motor output along the vagus nerve while also simultaneously inhibiting sympathetic motor neurons. The result is a sudden drop in systemic blood pressure, mainly due to decreased heart rate and cardiac output and vasodilation of systemic blood vessels. Typically, patients are instructed to avoid stimuli that may trigger an event. In severe cases, drugs may be prescribed to either stimulate the central nervous system or increase systemic blood pressure.

Table 15.6 Detailed Functions and Divisions of the Cranial Nerves

Nerve Number	Name	Type of Nerve	Functions
I	Olfactory	Special sensory	Sensory nerve of smell
II	Optic	Special sensory	Sensory nerve of vision
III	Oculomotor	Somatic motor	Motor to levator palpebrae superioris (raises the upper eyelid) and all extrinsic eye muscles except for the superior oblique and the lateral rectus
		Visceral motor (parasympathetic)	Motor to the ciliary body and iris (pupillary sphincter muscles) of the eye
IV	Trochlear	Somatic motor	Motor to superior oblique extrinsic muscle of the eye
V	Trigeminal	Mixed	Sensation from the cornea, scalp, forehead, face, and teeth; motor to the muscles of mastication (chewing)
V_1	Ophthalmic branch	Somatic sensory	Sensory from the cornea, scalp, and forehead
V_2	Maxillary branch	Somatic sensory	Sensory from the face, cheeks, and maxillary (upper) teeth
V_3	Mandibular branch	Somatic motor	Motor to the muscles of mastication
		Somatic sensory	Sensory from the chin, mandibular (lower) teeth, and tongue
VI	Abducens	Somatic motor	Motor to the lateral rectus extrinsic muscle of the eye
VII	Facial	Mixed	Sensory for taste from the anterior two-thirds of the tongue; motor to facial muscles, lacrimal glands, and salivary gland
		Somatic motor	Motor to the muscles of facial expression
		Special sensory	Sensory for taste from the anterior two-thirds of the tongue
		Visceral motor (parasympathetic)	Motor to the lacrimal glands, and the submandibular and sublingual salivary glands
VIII	Vestibulocochlear	Special sensory	Sensory nerve of hearing and balance
IX	Glossopharyngeal	Mixed	Sensory for taste from the posterior one-third of the tongue, sensation from the ear and pharynx, motor to the stylopharyngeus muscle and the parotid salivary gland
		Somatic motor	Motor to the stylopharyngeus muscle
		Somatic sensory	Cutaneous sensation from the external ear; general sensation from posterior one-third of the tongue
		Special sensory	Sensory for taste from the posterior one-third of the tongue
		Visceral motor (parasympathetic)	Motor to the parotid salivary gland
		Visceral sensory	Sensory from the carotid sinus and carotid body, mastoid air cells, pharynx, middle ear, and tympanic cavity
X	Vagus	Mixed	Motor to the pharynx and thoracic and abdominal viscera, sensation from the pharynx and ear, and sense of taste from the pharynx and epiglottis (portion of larynx)
		Somatic motor	Motor to muscles of the pharynx, larynx, and palate (except stylopharyngeus and tensor veli palatini)
		Somatic sensory	Sensory from the external acoustic meatus, tympanic membrane, dura mater of the posterior cranial fossa, and auricle of the ear
		Special sensory	Sensory fibers for taste from the palate and epiglottis
		Visceral motor (parasympathetic)	Motor to glands of the pharynx and larynx, and smooth muscle of the heart, lungs, and abdominal viscera
		Visceral sensory	Sensory from the pharynx, larynx, bronchi, aorta, and abdominal viscera
XI	Accessory	Somatic motor	Motor to the trapezius and sternocleidomastoid muscles
XII	Hypoglossal	Somatic motor	Motor to the intrinsic and extrinsic muscles of the tongue

EXERCISE 15.10

TESTING SPECIFIC FUNCTIONS OF THE CRANIAL NERVES

Neurological tests similar to some of those described in this exercise are performed by physicians when testing for damage to one or more cranial nerves. These tests can indicate if a cranial nerve is damaged. However, they are not infallible. For instance, an inability to hear could indicate damage to the vestibulocochlear nerve. However, the damage could also reside in the auditory cortex of the brain. Each of the tests described here was chosen because it is both easy and quick to perform. **Table 15.7** lists the common disorders of the cranial nerves along with potential signs and symptoms, and causes of the disorders.

Table 15.7 Common Disorders of the Cranial Nerves

Nerve Number	Name	Signs and Symptoms of Damage	Potential Cause of the Disorder
I	Olfactory	Inability to smell (anosmia)	A fracture of the cribriform plate of the ethmoid can damage the olfactory nerves
II	Optic	Blindness on the affected side (hemianopia)	Intracranial tumor or stroke that damages the nerve or tract prevents visual information from reaching the brain
III	Oculomotor	Pupil dilation (mydriasis). Eye deviates down and out (strabismus) from muscle paralysis resulting in double vision (diplopia). Eyelid droops (ptosis).	Increased intracranial pressure is a common cause of compression of the nerve. The parasympathetic fibers that innervate the pupillary sphincter muscle are located on the surface of the nerve, so pupil dilation is often the first sign of nerve damage or increased intracranial pressure. Nerve damage results in paralysis of all extraocular muscles except the superior oblique and lateral rectus muscles. Paralysis of the levator palpebrae superioris muscle causes ptosis.
IV	Trochlear	Difficulty turning the eye inferior and lateral, which leads to double vision (diplopia)	Nerve damage results in paralysis of the superior oblique muscle
V	Trigeminal	Trigeminal neuralgia (tic douloureux), a sudden, intense pain along the course of one of the divisions of the nerve	Pressure on the nerve from the artery that courses alongside it stimulates sensory fibers within the nerve. Pain is often triggered by touching structures inside the mouth.
VI	Abducens	Eye deviates medially (adducts), causing double vision (diplopia)	Any disorder that increases intracranial pressure (for example, a stroke) can cause this nerve to be crushed against the clivus (sloped portion) of the sphenoid bone
VII	Facial	Bell palsy—paralysis of the muscles of facial expression on the side of the face with the affected nerve. Loss of taste sensation on the anterior two-thirds of the tongue (ageusia). Decreased salivation (hypoptyalism).	A viral infection that causes inflammation of the facial nerve is the most likely source. This problem often resolves itself within a couple of months.
VIII	Vestibulocochlear	Loss of balance and equilibrium, nausea, vomiting, and dizziness or inability to hear (anacusis)	Acoustic neuroma—a tumor originating in neurolemmocytes within the internal acoustic meatus—causes compression of the nerve
IX	Glossopharyngeal	Difficulty swallowing (dysphagia). Loss of taste sensation on the posterior one-third of the tongue (ageusia). Decreased salivation (hypoptyalism).	Nerve damage interrupts the sensory component of the swallowing reflex
X	Vagus	Difficulty swallowing (dysphagia) or hoarseness (dysphonia)	Nerve damage interrupts the motor component of the swallowing reflex. Hoarseness results from paralysis of the muscles of the larynx.
XI	Accessory	Difficulty elevating the scapula or rotating the head	Nerve damage results in paralysis of the sternocleidomastoid and/or trapezius muscles
XII	Hypoglossal	When sticking out the tongue, it moves in the direction of the damaged nerve	Compression of nerve from increased intracranial pressure

EXERCISE 15.10A Olfactory (CN I)

The olfactory nerves (**figure 15.19**) are unique as cranial nerves in that there are more than two of them (the rest of the cranial nerves are paired—a right and left for each), and they are constantly being replaced. The nerves lie within the nasal epithelium, and their axons project through the **olfactory foramina** within the cribriform plate of the ethmoid bone (table 15.4). The olfactory neurons then synapse with neurons within the **olfactory bulbs** and the signals are sent to the brain via the **olfactory tracts.**

1. Obtain vials of peppermint, lemon, and vanilla oils.
2. While the subject's eyes are closed, pass an open vial of peppermint oil just under a laboratory partner's nose. Was the subject able to, identify the smell?
3. Repeat this process with the vials of lemon and vanilla oils. Allow some time between applications of the different oils. Damage to the olfactory nerves results in an inability to identify odors. Excessive smoking or inflammation of the nasal mucosa as a result of a viral infection can inhibit the sense of smell, and the sense of smell also declines with age.

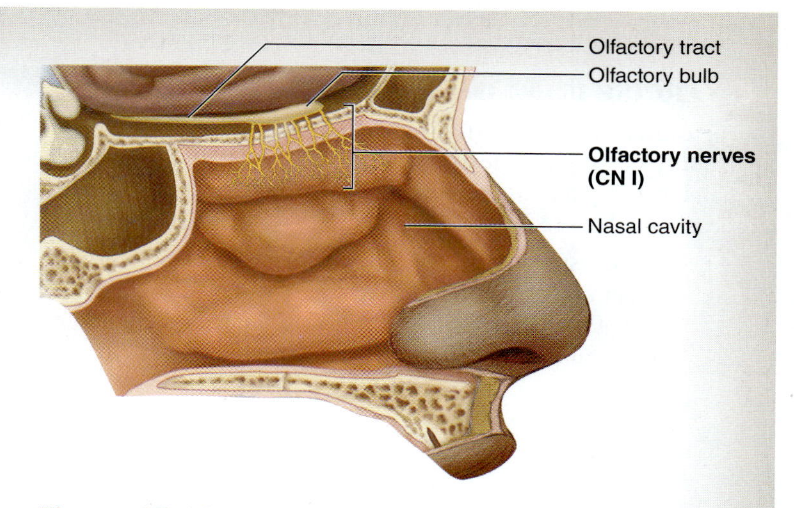

Figure 15.19 Location of the Olfactory Nerves (CN I).

EXERCISE 15.10B Optic (CN II)

Each of the optic nerves (**figure 15.20**) begins as axons of ganglion cells located within the **retina** of the eye. Different parts of the retina correspond to different portions of the visual field. The axons of these nerves exit the eye at the optic disc, or "blind spot," of the eye, at which point they become the **optic nerve,** which then travels posteriorly toward the diencephalon. Anterior to the infundibulum, most of the fibers cross over at the prominent **optic chiasm** (*chiasma*, a crossing of two lines) to the other side of the brain. The fibers then continue to travel posteriorly to reach the visual cortex in the occipital lobe of the brain. Figure 15.20 demonstrates the pattern of flow of visual information from the retina to the brain. When damage to the retina or optic nerve is suspected, visual field tests are performed to discover the location of the damage. These tests of visual function are beyond the scope of this course and will not be performed in this laboratory session.

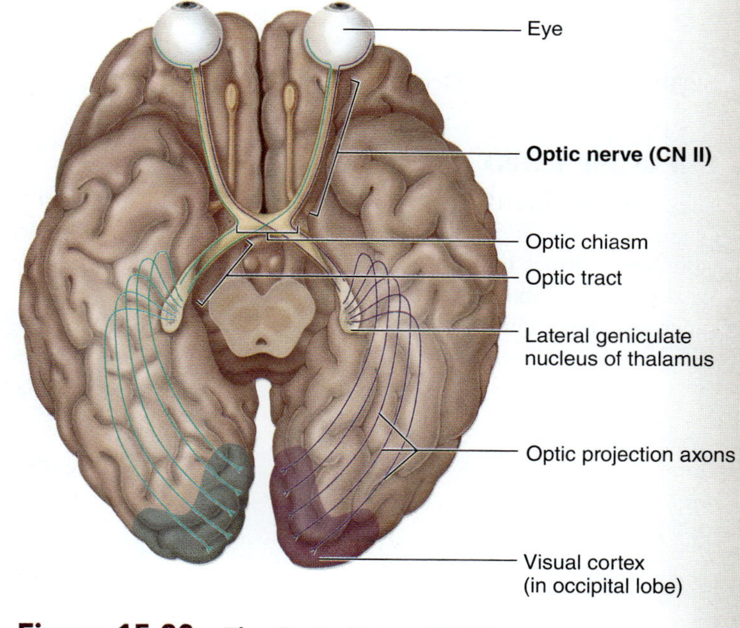

Figure 15.20 The Optic Nerve (CN II).

(continued on next page)

(continued from previous page)

EXERCISE 15.10C Oculomotor (CN III)

The oculomotor nerves **(figure 15.21)** send motor fibers to the majority of the extrinsic eye muscles as well as to the muscles that control pupil diameter and ciliary muscles involved in focusing (tables 15.6 and 15.7).

1. Obtain a small flashlight. Look into one of a laboratory partner's eyes and observe the size of the pupil. While looking into the subject's eye, gently shine the light into the eye (if it is a bright light, just bring it near the eye so that more light enters the eye—the goal here is not to blind the individual with the light).

   Was there a change in pupil diameter? ________________

   If so, what happened? ______________________________

2. Repeat the above activity, but this time observe the pupil of the other eye.

   Was there a change in pupil diameter? ________________

   If so, what happened? ______________________________

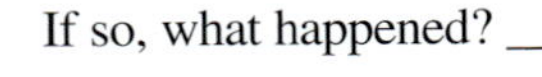

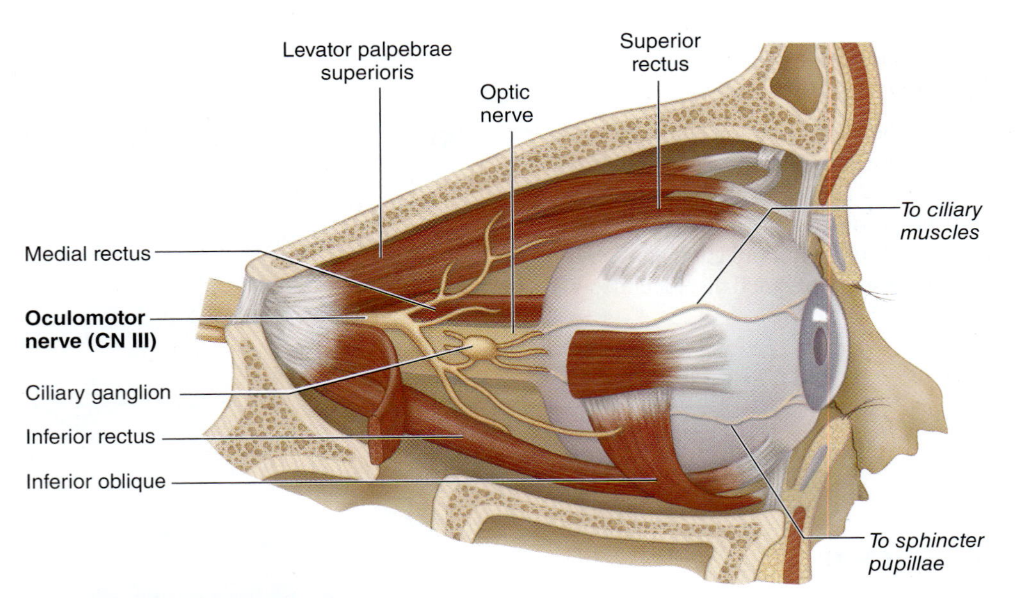

Figure 15.21 The Oculomotor Nerve (CN III).

EXERCISE 15.10D Trochlear (CN IV)

The trochlear nerve **(figure 15.22)** controls only one extraocular eye muscle—the superior oblique (tables 15.6 and 15.7). Ask a laboratory partner to look down and out (inferior and lateral). Weakness or an inability to perform this action indicates a weak superior oblique muscle or damage to the trochlear nerve.

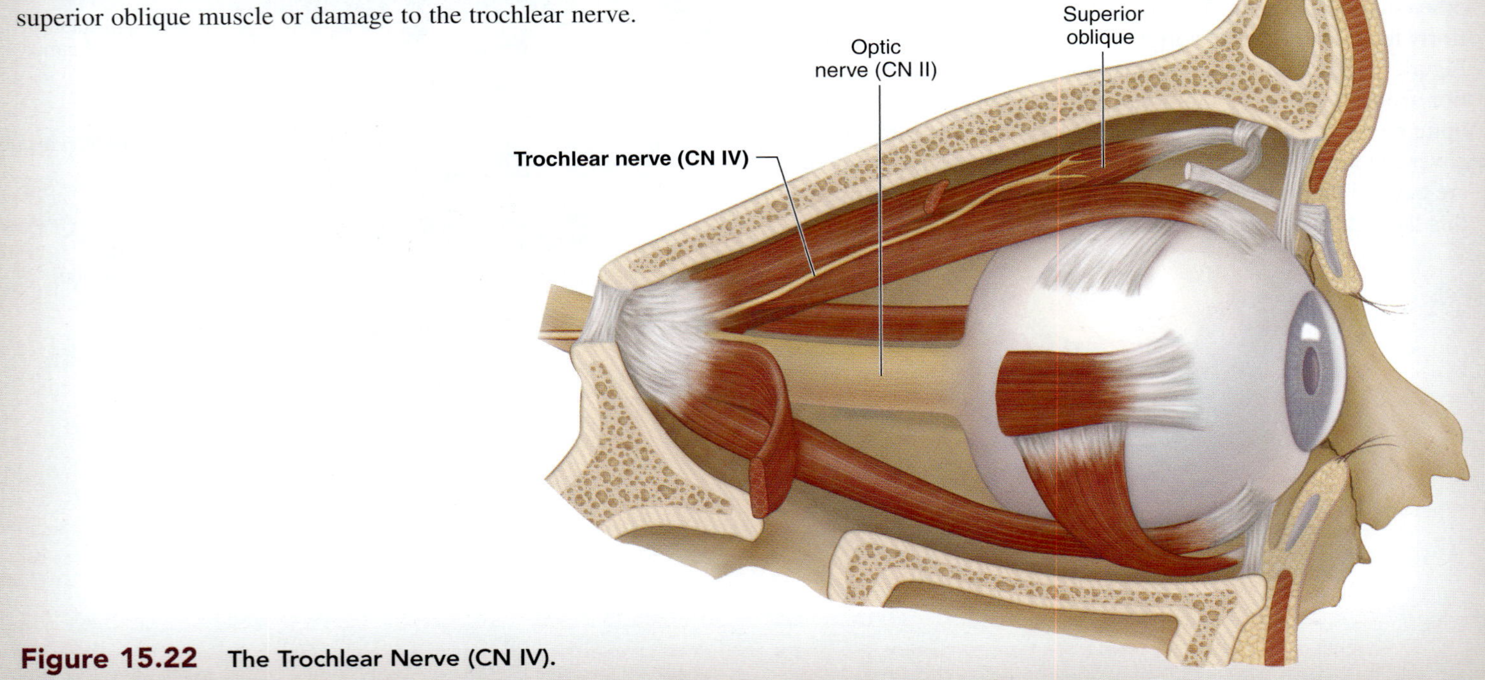

Figure 15.22 The Trochlear Nerve (CN IV).

EXERCISE 15.10E Trigeminal (CN V)

The trigeminal nerve **(figure 15.23)** is the largest and most complex of the cranial nerves. It has three branches: the ophthalmic, maxillary, and mandibular branches, which are named V_1, V_2, and V_3, respectively. It is the predominant nerve carrying sensory information from the face, but it also carries motor output to the muscles of mastication.

1. Obtain a feather and a cotton ball.
2. With the subject's eyes closed, proceed to gently touch the subject's face with the feather in the sensory distribution areas of the trigeminal nerve shown in figure 15.23*b*. An inability to feel this sensation in one or more locations indicates damage to a branch of the trigeminal nerve.
3. While the subject's eyes are open, have the individual look up and away. *Very gently and lightly* touch a few strands of the fibers from the cotton ball to the subject's cornea. This is a test for the *corneal reflex*, whose sensory component is carried by a branch of the trigeminal nerve. Touching the cornea with the cotton should cause the subject to blink. An absent corneal reflex indicates damage to the ophthalmic branch (V_1) of the trigeminal nerve (contact lens wearers may also have a diminished or absent corneal reflex).

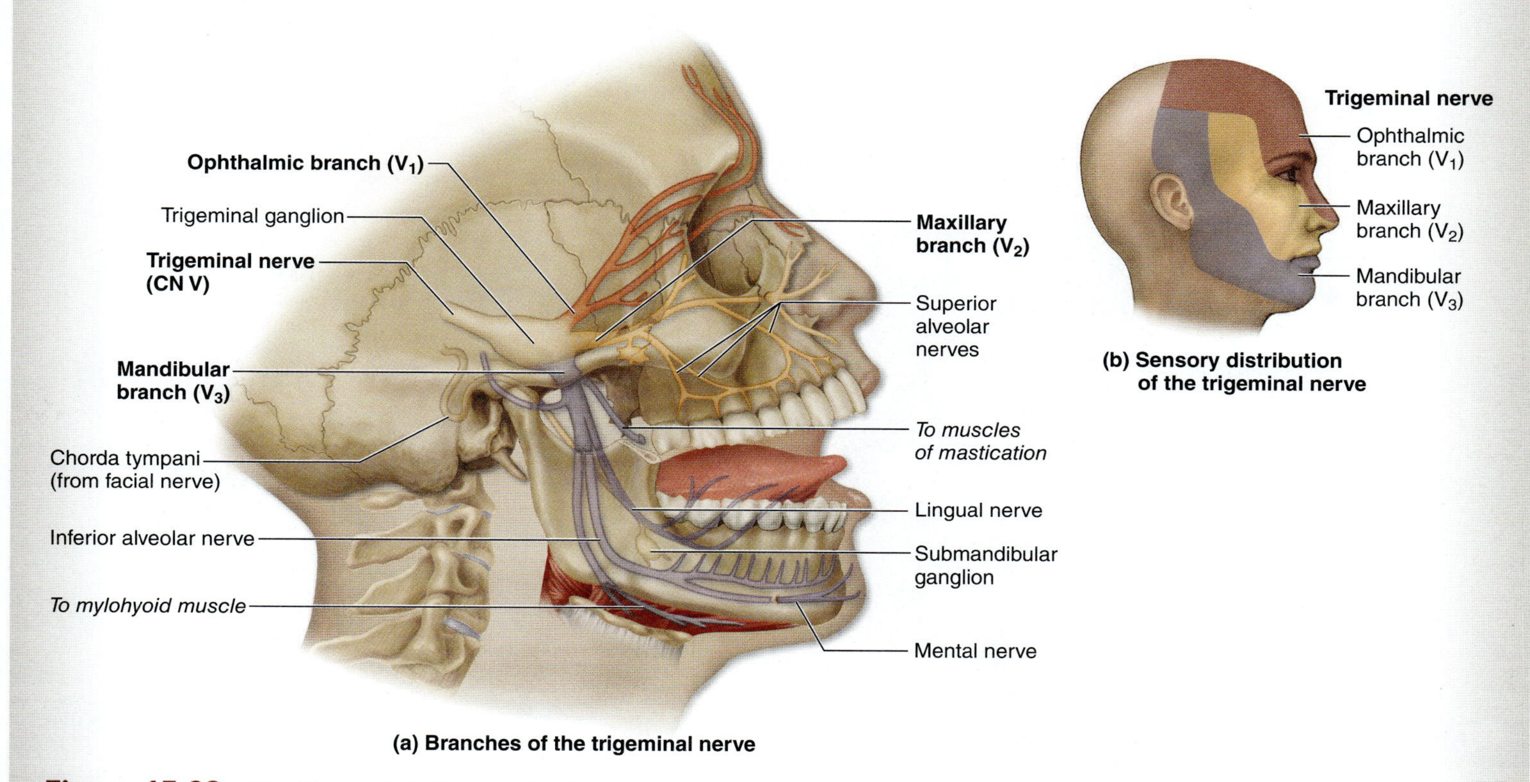

Figure 15.23 The Trigeminal Nerve (CN V).

EXERCISE 15.10F Abducens (CN VI)

The abducens nerve **(figure 15.24)** controls only one extrinsic eye muscle—the lateral rectus (tables 15.6 and 15.7). Ask a laboratory partner to look laterally to the right. Weakness or an inability to do so indicates a weak lateral rectus muscle in the right eye or damage to the right abducens nerve.

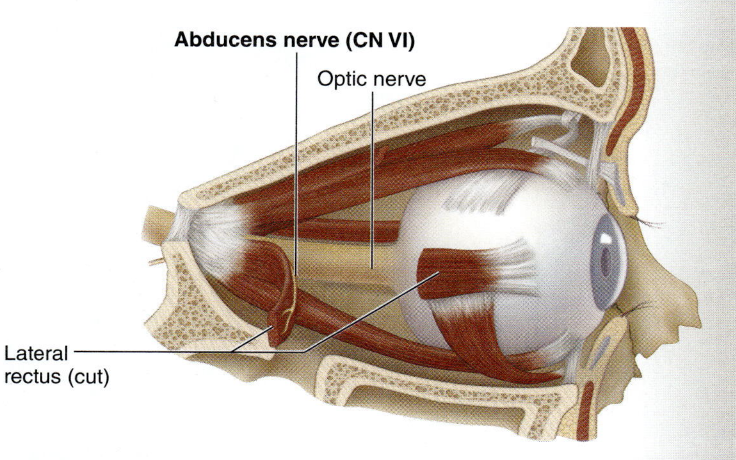

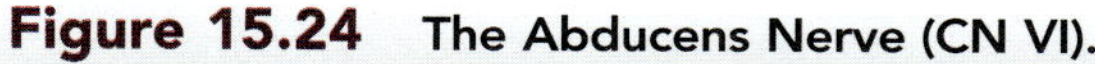
Figure 15.24 The Abducens Nerve (CN VI).

(continued on next page)

(continued from previous page)

EXERCISE 15.10G Facial (CN VII)

The facial nerve **(figure 15.25)** has several functions. Two major functions are to carry motor output to the muscles of facial expression and to carry sensory information to the brain from taste buds on the anterior two-thirds of the tongue **(figure 15.26)**.

1. Obtain vials of salt and sugar, and a cup of drinking water.
2. While a subject has eyes closed, place a few grains of salt on the protruded tongue. Was the individual able to positively identify the taste as salty?
3. Have the subject take a drink of water to refresh the taste buds before performing the next test. Repeat step 2, but this time place a few grains of sugar on the tip of the tongue. Was the individual able to positively identify the taste as sweet? An inability to identify the salty or sweet tastes may indicate damage to the facial nerve.
4. Ask the subject to demonstrate facial expressions, such as surprise, happiness, sadness, and confusion. An inability to express these emotions facially may indicate paralysis of the muscles of facial expression, a common consequence of damage to the facial nerve.

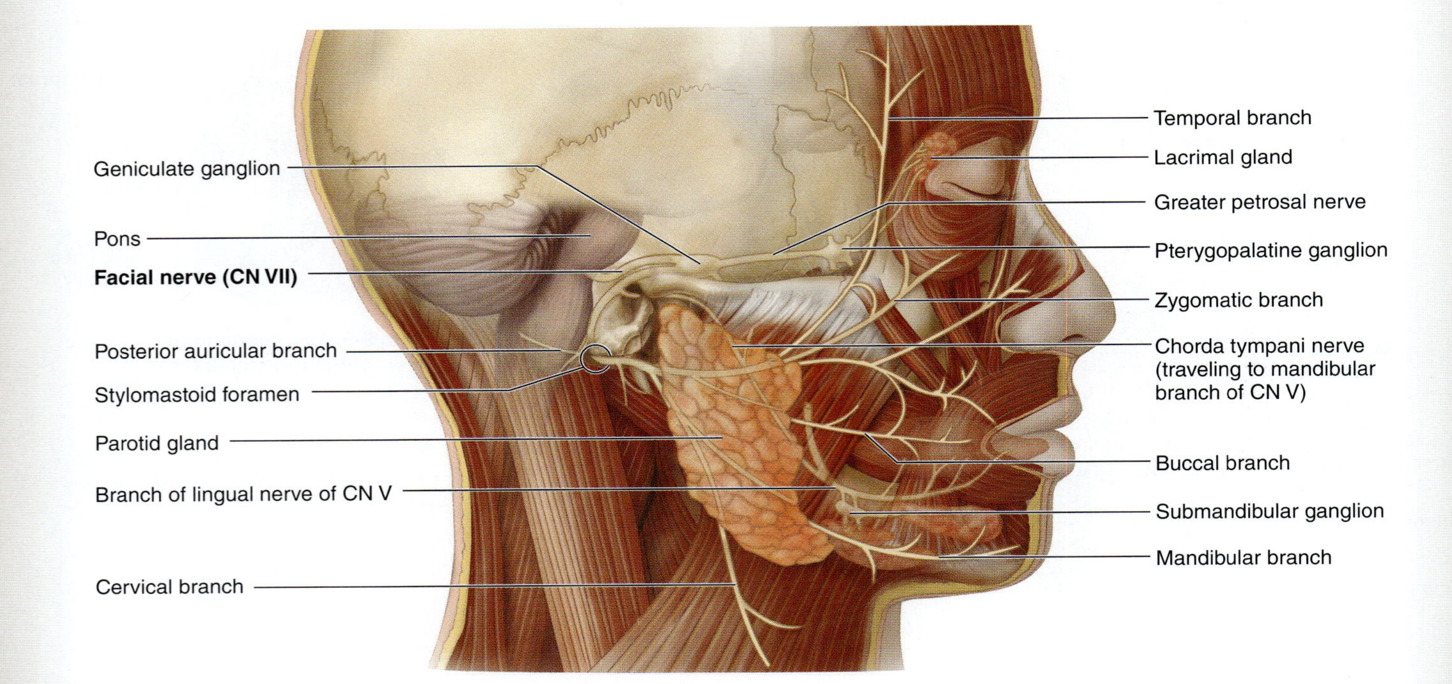

Figure 15.25 The Facial Nerve (CN VII).

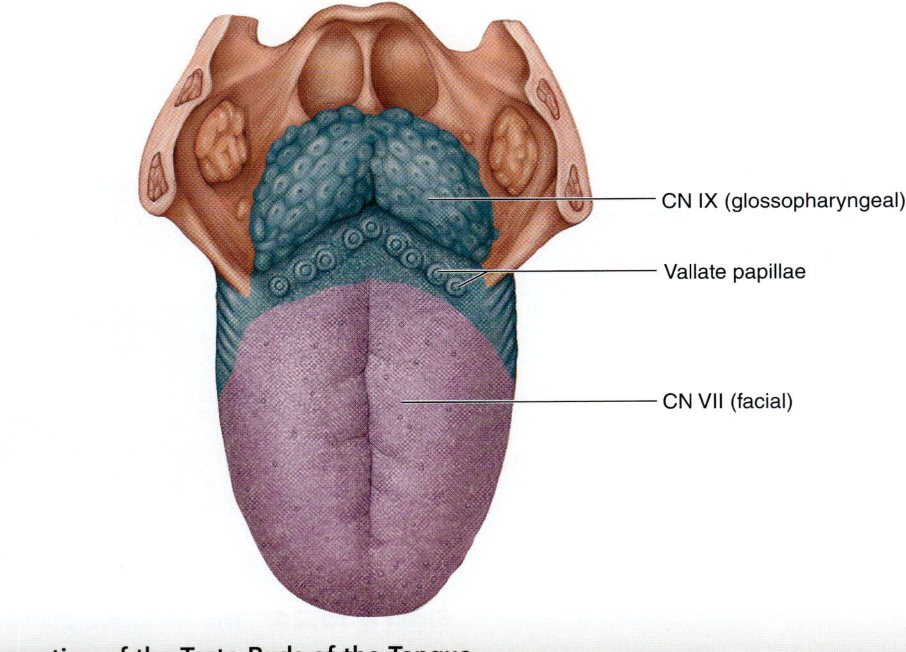

Figure 15.26 Innervation of the Taste Buds of the Tongue.

EXERCISE 15.10H Vestibulocochlear (CN VIII)

The vestibulocochlear nerve **(figure 15.27)** is two nerves: the **vestibular nerve,** which transmits nerve signals from the vestibule and semicircular canals to the brain regarding balance and equilibrium and the **cochlear nerve,** which transmits nerve signals from the cochlea to the brain regarding sound. Both nerves enter the petrous part of the temporal bone through the **internal auditory canal.** Figure 15.27 demonstrates the special sensory structures that are innervated by the vestibulocochlear nerve.

1. Obtain a tuning fork.
2. With the subject's eyes closed, gently strike the "fork" end of the tuning fork on the table, and then hold it near the individual's ear. Is the subject able to detect the sound? An inability to hear the vibrations caused by the tuning fork can indicate damage to the vestibulocochlear nerve.

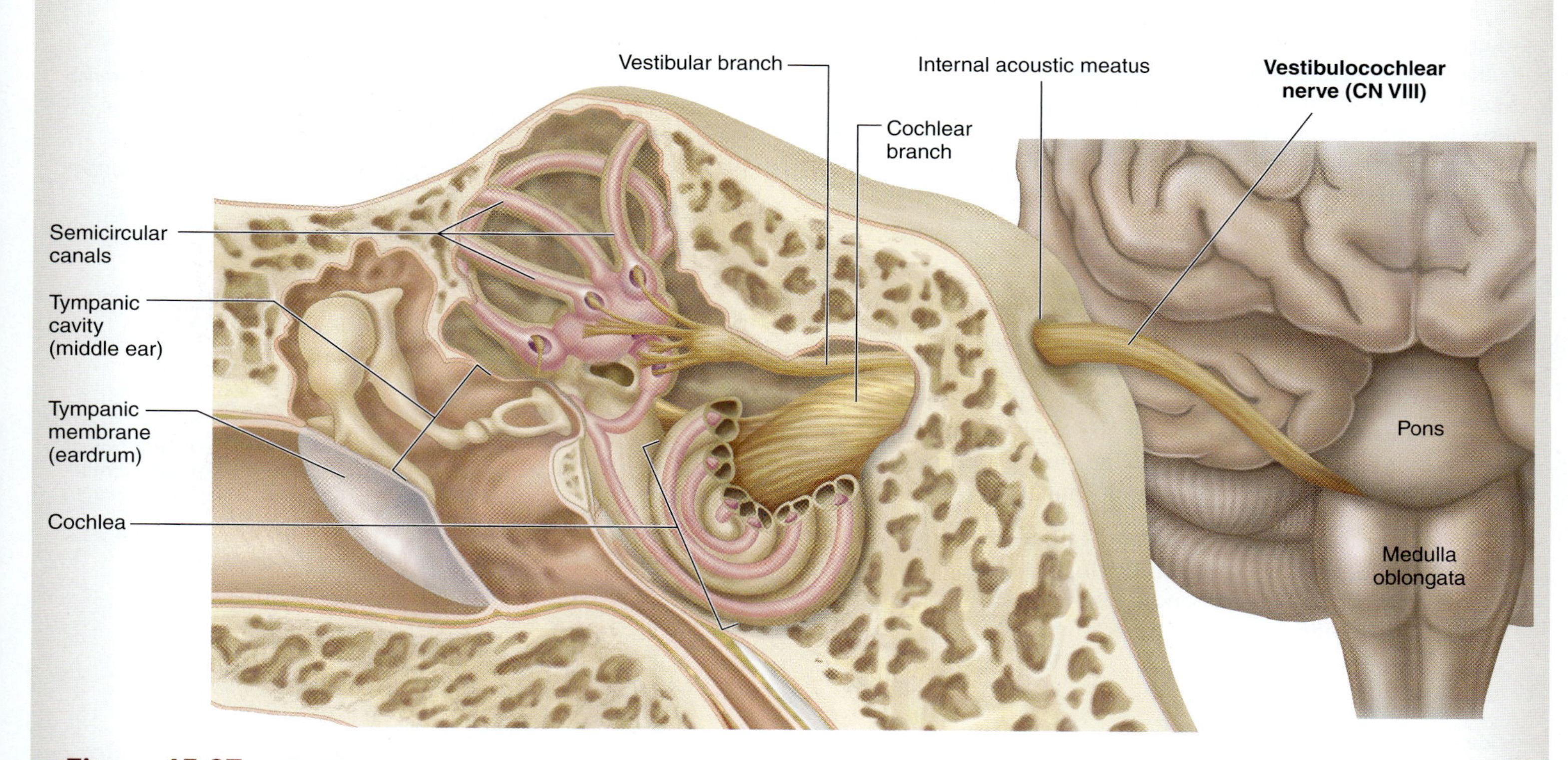

Figure 15.27 The Vestibulocochlear Nerve (CN VIII).

INTEGRATE

CLINICAL VIEW
Cranial Nerve Assessment

Assessing cranial nerve function is standard for any comprehensive neurological examination. Each cranial nerve carries either sensory, motor, or both sensory and motor information, and relays nerve signals regarding the special senses such as equilibrium, olfaction, and vision. Cranial nerves also carry somatic motor information to skeletal muscles in the head and neck regions. Visual acuity and pupillary eye reflex tests assess optic nerve (CN II) function; eye movement and tracking assess oculomotor nerve (CN III), trochlear nerve (CN IV), and abducens nerve (CN VI) function; and facial expressions assess facial nerve (CN VII) function. Touching the chin, cheeks, and forehead with cotton swabs, sharp objects, and cold objects assesses trigeminal nerve (CN V) function. Hearing and vestibular tests assess vestibulocochlear nerve (CN VIII) function, and taste tests assess glossopharyngeal nerve (CN IX) and facial nerve CN VII) function. The gag reflex assesses vagus nerve (CN X) function, tongue movement assesses cranial accessory nerve (CN XI) function, and head and shoulder movement assesses cranial hypoglossal nerve (CN XII) function. Clinicians observe any deficits, whether slight, severe, unilateral, or bilateral. This assessment in combination with imaging techniques, such as magnetic resonance imaging (MRI) and computed tomography (CT), may reveal whether deficits are due to lesions or damage to central or peripheral nervous system structures.

(continued on next page)

(continued from previous page)

EXERCISE 15.10I Glossopharyngeal (CN IX) and Vagus (CN X)

The glossopharyngeal and vagus nerves (**figures 15.28** and **15.29,** tables 15.6 and 15.7) both carry sensory and motor information to and from the soft palate, pharynx, and larynx as part of the coughing and gagging reflexes. In addition, the glossopharyngeal nerve carries sensory information from the taste buds on the posterior one-third of the tongue (see figure 15.26). The vagus nerve is unique as a cranial nerve because it innervates many structures within the thoracic and abdominal cavities. It is the predominant pathway for parasympathetic information to travel from the brain to the visceral organs of the body. Though tests for the functioning of the glossopharyngeal and vagus nerves are not easy to perform, try to observe at least some of the functions of these nerves with this exercise.

1. Have a subject, with the mouth open, say "Ah" while you observe the soft palate and uvula.
2. Unilateral drooping of the soft palate or deviation of the uvula to one side may indicate damage to either the glossopharyngeal or the vagus nerve. The glossopharyngeal nerve carries sensory information from the pharynx to the brain, while the vagus nerve carries motor information back out to the muscles that raise the palate and that are used in swallowing.
3. Another test for the functioning of these two nerves is to test for the gag reflex. This will not be attempted in the lab because inexperienced testing of this reflex could cause choking or, in some cases, vomiting.

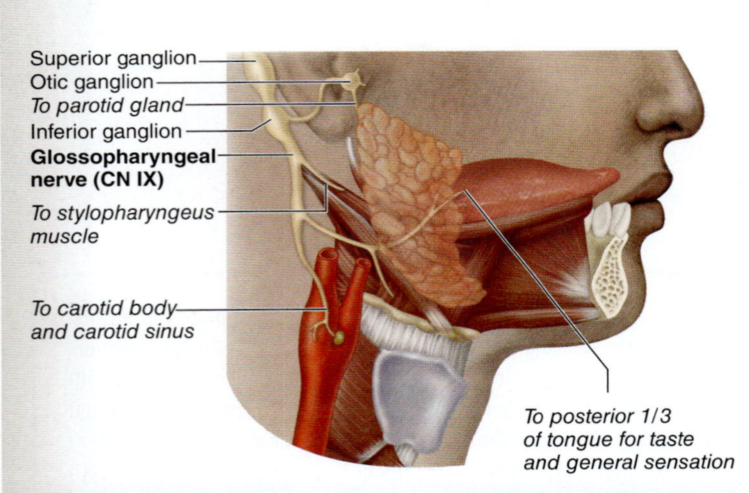

Figure 15.28 The Glossopharyngeal Nerve (CN IX).

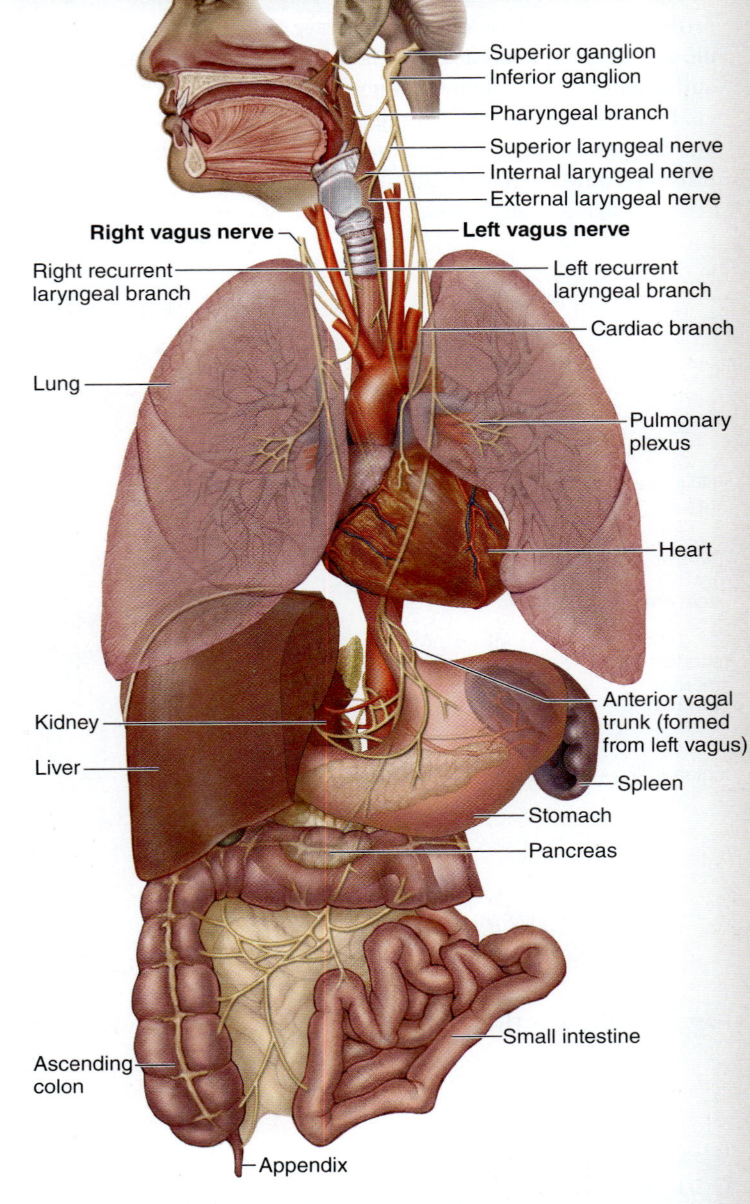

Figure 15.29 The Vagus Nerve (CN X).

EXERCISE 15.10J Accessory (CN XI)

The accessory nerve **(figure 15.30)** carries motor information to the trapezius and sternocleidomastoid muscles (tables 15.5 and 15.6).

1. Have a subject elevate the scapula ("shrug" the shoulders) to test the function of the trapezius muscle.
2. Next, have the subject rotate the head first to the right and then to the left to test the function of the sternocleidomastoid muscle. Damage to the accessory nerve would cause both of these actions to be weak or impossible due to paralysis of the muscles.

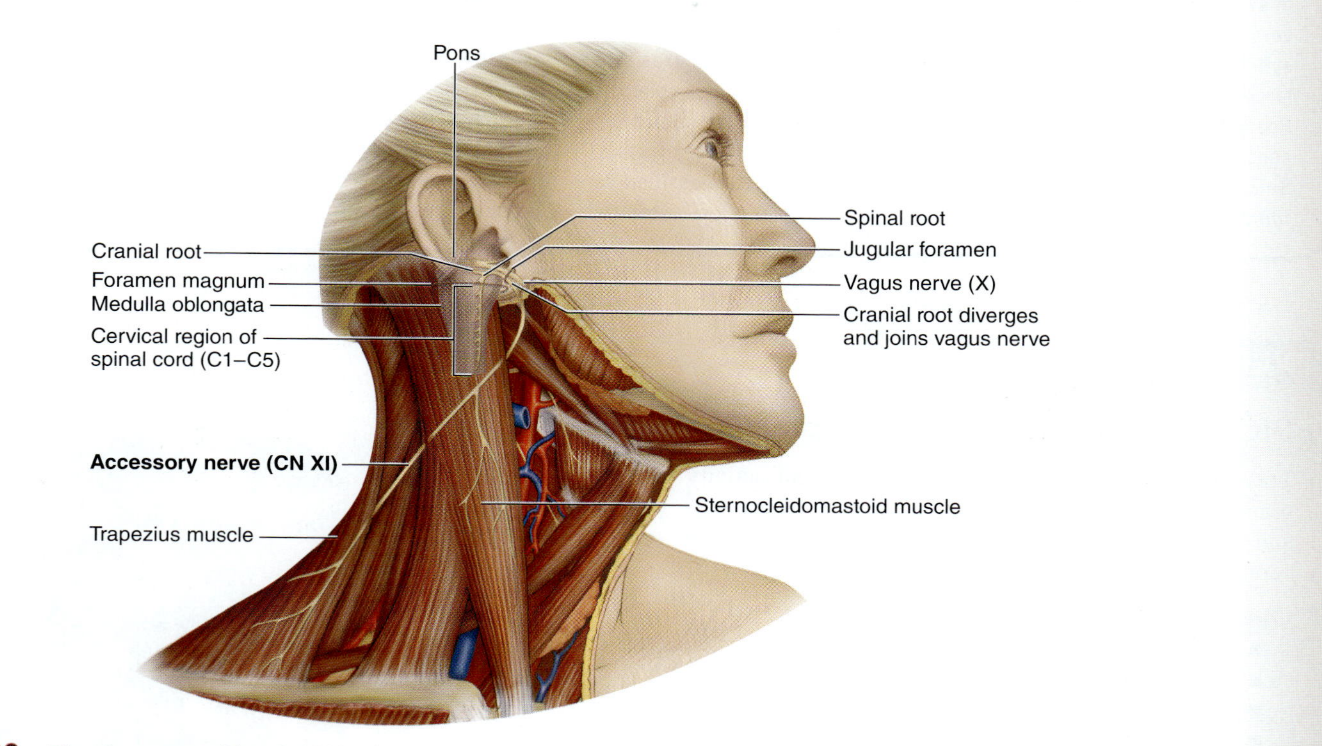

Figure 15.30 The Accessory Nerve (CN XI).

EXERCISE 15.10K Hypoglossal (CN XII)

The hypoglossal nerve **(figure 15.31)** innervates intrinsic and extrinsic muscles of the tongue (tables 15.6 and 15.7). Have the subject stick out the tongue. The tongue should protrude straight out. If the hypoglossal nerve is damaged, the tongue will deviate to the side of the damaged nerve.

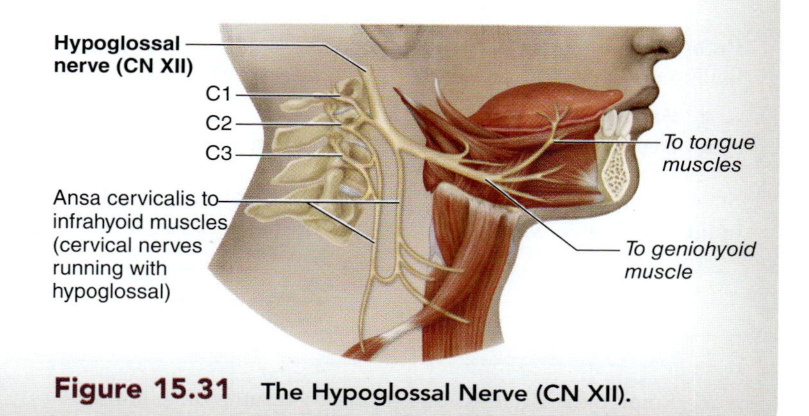

Figure 15.31 The Hypoglossal Nerve (CN XII).

Testing Brain Function

Previous exercises in this chapter involve identifying brain structures on models and/or cadavers, and relating structure to function. An **electroencephalogram (EEG)** is used to record electrical activity of the brain using surface electrodes placed on the skull. Clinically, an EEG is used to assess brain function. The technique is similar to recording electrical activity of skeletal muscles with an electromyogram (EMG), or recording electrical activity of the heart with an electrocardiogram (ECG). Changes in the electrical activity of the brain observed on an EEG may reveal abnormalities in the frontal lobe or cerebellum of a patient who is experiencing motor dysfunction (e.g., gait disturbances). Similarly an EEG may reveal abnormalities in the parietal lobe of a patient experiencing sensory dysfunction (e.g., slow reactions to painful stimuli).

INTEGRATE

CLINICAL VIEW

Epilepsy

Epilepsy is a neurological condition characterized by excessive patterns of synchronous electrical activity of the brain. The resulting **seizure** may vary in severity depending upon the strength, duration, and location of synchronous electrical activity. Seizures may impact a local area or a large portion of the brain. An EEG is used clinically to diagnose epilepsy. Possible symptoms that occur during a seizure include confusion, sensitivity to light and sounds, slurred speech, and a glazed stare. Seizures that impact large portions of the brain may be classified as *tonic-clonic* (formerly *grand mal*) seizures. In a tonic-clonic seizure, abnormal electrical activity in the brain causes a patient's muscles to contract, leading to whole-body stiffening (tonic phase) and loss of consciousness. The muscles then begin to rapidly relax and contract, causing whole-body convulsions (clonic phase). Once patients regain consciousness, they often have no recollection of the event. Treatment depends upon the severity of the seizures. Drugs that influence the excitability of neurons may be prescribed. In patients where the electrical activity is localized to one portion of the brain and there is no response to drug therapy, doctors may implant stimulators that can apply a "resetting" current to nerves or brain tissue. This is similar to a pacemaker that resets the electrical activity in the heart. In the most severe cases, doctors may surgically remove affected brain tissue.

EXERCISE 15.11

BIOPAC Electroencephalography (EEG)

This exercise involves observing the electrical activity of the brain with a subject's eyes open and closed. Alpha, beta, delta, and theta waves will be compared in the subject's electroencephalogram (EEG) for two states.

Obtain the Following:

- **BIOPAC electrode lead set (SS2L)**
- **cot and pillow**
- **eletrode gel**
- **surface eletrodes**
- **swim cap or wrap**

Before beginning the experiment, become familiar with the following concepts (use the textbook as a reference):

- **Brain structure and function**
- **Depolarization and repolarization of an excitable cell**

State a hypothesis regarding how the electrical activity of the brain will change with a subject's eyes open and closed.

____

____

____

____

1. Prepare the BIOPAC equipment by turning the computer ON and the MP3X Data Acquisition Unit (DAQ) OFF. Plug in the following: electrode lead set (SS2L) (CH 1). Turn the MP3X DAQ ON.
2. Allow the subject to assume a relaxed position. It is recommended that the subject lie supine on a cot with the subject's head resting comfortably on a pillow (or rolled-up jacket or sweater) with eyes closed.
3. Apply a small quantity of electrode gel to the center of the electrodes, and place the electrode firmly on the subject's scalp in the designated locations using the color code, as illustrated in **figure 15.32.** Be sure that hair has been kept away from the electrodes as much as possible. A swim cap or wrap may be placed around the subject's head to hold the electrodes firmly in place.

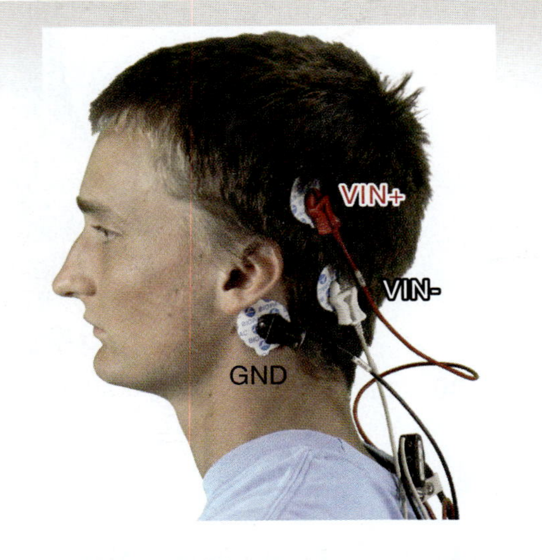

Figure 15.32 BIOPAC Equipment and Electrode Placement for EEG. Place electrodes and leads on the subject's scalp in the designated locations using the color code illustrated. Courtesy of and © BIOPAC Systems.

4. Start the BIOPAC Student Lab Program. Select "L03-EEG-1" from the drop-down menu, enter the file name, and click "OK".
5. Click "calibrate" in the upper left corner of the Setup window. Ensure proper placement of the subject's electrodes and click "OK". Once the calibration has stopped after 8 seconds of recording, check to ensure that the baseline remains unchanged at approximately 0 μV. If spikes appear in the trace, "Redo Calibration."
6. During the 30-second recording session, the subjecct will remain in a relaxed position. The subject will begain with eyes closed (0–10 seconds), then eyes open (11–20 seconds), and then eyes closed (21–30 seconds). During the period when the eyes are open, the subject should try not to blink. One lab member should serve as the **director,** informing the subject when to open and close the eyes, and one lab member will serve as the **recorder,** marking the events on the EEG trace.

7. The subject will close his/her eyes and remain in a relaxed position. To begin recording EEG data, click "Record." After 10 seconds, the director will inform the subject to open his/her eyes. The recorder will press "F9" to add an event marker to the data trace. After 10 seconds with the eyes open, the director will inform the subject to close his/her eyes once again. The recorder will press "F9" to add another event marker to the data trace. After the subject has completed all three conditions, click "Stop." Review the data to ensure that the EEG resembles **figure 15.33.** If not, click "Redo."

8. To observe the component of the EEG trace that corresponds to the various frequency components, click on each of the frequency buttons in the following order: "alpha," "beta," "delta," and "theta." The amplitude should decrease during the "eyes open" condition (see **figure 15.34**). If not, be sure that the electrodes make good contact with the scalp and click "Redo." If the data resembles that in figure 15.34, click "Done" to conclude the experiment. Remove the subject's electrodes, and clean the skin using soap and water.

9. To analyze saved data, click "Review Saved Data." Note that CH 1 displays EEG and CH 2–5 display alpha, beta, delta, and theta waves, respectively. Measurement boxes appear above the marker region in the data window. To set up, indicate the channel number and measurement type for each of four boxes: CH 2–5, "stddev."

10. Locate the "I-beam" tool to the left of the magnifying glass. Select the portion of the EEG that corresponds to the first "eyes closed" condition, from time 0 to the first event marker. Repeat this for the "eyes open" condition (time 11–20 seconds) and the second "eyes closed" condition (time 21–30 seconds). Record the standard deviation values for each condition in **table 15.8.**

11. Measurement boxes appear above the marker region in the data window. Set up the boxes by indicating the channel number and measurement type as follows: CH 2, "Freq", CH 3–5, "none."

12. Obtain values for each of the frequencies by zooming in on 3–4 seconds of data and selecting the first "eyes closed" data for each of the frequency waves. Use the "I-beam" tool to highlight one cycle in the alpha wave, beta wave, delta wave, and theta wave. Record the frequency values in **table 15.9.** Save or print the data and exit the program.

13. Make note of any pertinent observation here: ________
____
____
____

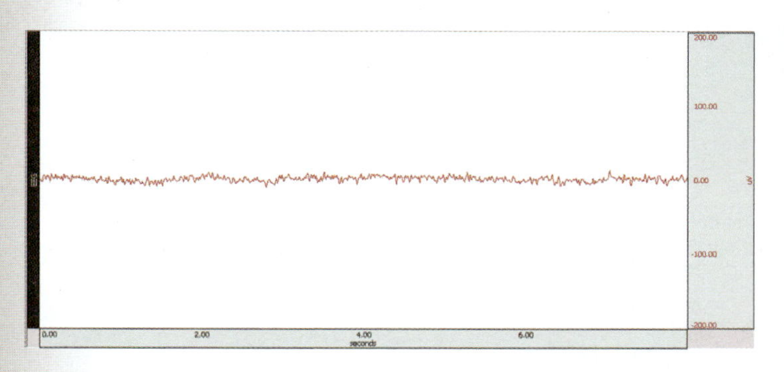

Figure 15.33 EEG Sample Data. Courtesy of and © BIOPAC Systems.

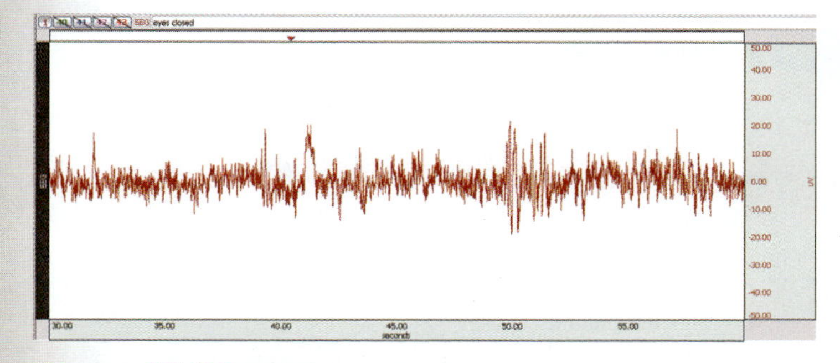

Figure 15.34 EEG Sample Data. Alpha, beta, delta, theta. Courtesy of and © BIOPAC Systems.

Table 15.8 Standard Deviations (μV) in Each Condition

Rhythm	Channel	Eyes Closed (0–10 seconds)	Eyes Open (11–20 seconds)	Eyes Closed (21–30 seconds)
Alpha	CH 2			
Beta	CH 3			
Delta	CH 4			
Theta	CH 5			

Source: Reprinted by permission from BIOPAC Student Lab Manual Revision 4, © BIOPAC Systems, Inc., Goleta, CA.

Table 15.9 Frequency (Hz)

Rhythm	Channel	Cycle 1	Cycle 2	Cycle 3	Mean
Alpha	CH 2				
Beta	CH 3				
Delta	CH 4				
Theta	CH 5				

Source: Reprinted by permission from BIOPAC Student Lab Manual Revision 4, © BIOPAC Systems, Inc., Goleta, CA.

Chapter 15: The Brain and Cranial Nerves

Name: ____________________

Date: __________ **Section:** __________

POST-LABORATORY WORKSHEET

The ❶ corresponds to the Learning Objective(s) listed in the chapter opener outline.

Do You Know the Basics?

Exercise 15.1: Cranial Meninges

1. Match the description listed in column A with the corresponding cranial dural septum listed in column B. ❶

Column A	*Column B*
______ 1. diaphragma sellae	a. drapes across the cerebellar hemispheres horizontally within the transverse fissure
______ 2. falx cerebelli	b. located between the two cerebellar hemispheres
______ 3. falx cerebri	c. located between the two cerebral hemispheres
______ 4. tentorium cerebelli	d. superior to the sella turcica of the sphenoid bone

2. The dura mater is composed of which type of connective tissue? (Circle one.) ❶

   a. areolar
   b. dense irregular
   c. dense regular
   d. reticular

3. Place the following structures through which CSF and venous blood flow in the correct order as they drain from the subarachnoid space into the internal jugular vein. ❷

   ______ a. confluence of sinuses
   ______ b. sigmoid sinuses
   ______ c. superior sagittal sinus
   ______ d. transverse sinuses

Exercise 15.2: Brain Ventricles

4. Match the description listed in column A with the appropriate structure listed in column B. ❸

Column A	*Column B*
______ 1. capillaries that aid in production of CSF	a. cerebral aqueduct
______ 2. a channel located between the third and fourth ventricles	b. choroid plexus
______ 3. a hole between the lateral ventricle and the third ventricle	c. fourth ventricle
______ 4. a membrane that separates the two lateral ventricles	d. interventricular foramen
______ 5. the two most superior ventricles	e. lateral ventricle
______ 6. the ventricle located between the brainstem and cerebellum	f. septum pellucidum
______ 7. the ventricle surrounded by the diencephalon	g. third ventricle

5. Match the ventricular space listed in column A with the secondary brain vesicle from which it developed listed in column B. ❹

Column A	*Column B*
______ 1. cerebral aqueduct	a. diencephalon
______ 2. fourth ventricle	b. mesencephalon
______ 3. lateral ventricles	c. myelencephalon
______ 4. third ventricle	d. telencephalon

Exercise 15.3: Circulation of Cerebrospinal Fluid (CSF)

6. Place the following structures through which CSF flows in the correct order as CSF flows through the ventricular system of the brain. (5)

_____ a. cerebral aquaduct

_____ b. fourth ventricle

_____ c. interventricular foramen

_____ d. lateral ventricles

_____ e. third ventricle

Exercise 15.4: Superior View of the Human Brain

7. Match the description listed in column A with the appropriate structure listed in column B. (6)

Column A	Column B
_____ 1. a fiber tract that connects the right and left cerebral hemispheres	a. central sulcus
_____ 2. a groove between the frontal lobe and the remaining lobes	b. corpus callosum
_____ 3. a groove between the right and left cerebral hemispheres	c. frontal lobe
_____ 4. a groove between the temporal and parietal lobes	d. gyri
_____ 5. a groove that separates the cerebrum from the cerebellum	e. lateral sulcus
_____ 6. the most anterior lobe of the brain	f. longitudinal fissure
_____ 7. the most lateral lobe(s) of the brain	g. parietal lobe
_____ 8. the most superior lobe(s) of the brain	h. sulci
_____ 9. ridges of brain tissue	i. temporal lobe
_____ 10. shallow grooves on the surface of the brain	j. transverse fissure

Exercise 15.5: Lateral View of the Human Brain

8. Which of the following are major lobes visible on the surface of the cerebrum? (Check all that apply.) (7)

_____ a. frontal _____ b. insula _____ c. occipital _____ d. parietal _____ e. temporal

Exercise 15.6: Inferior View of the Human Brain

9. Which of the following brain structures are visible on the inferior surface of the brain? (Check all that apply.) (8)

_____ a. brainstem _____ b. cerebellum _____ c. frontal lobe _____ d. parietal lobe _____ e. temporal lobe

Exercise 15.7: Midsagittal View of the Human Brain

10. Match the description listed in column A with the corresponding structure listed in column B. (9)

Column A	Column B
_____ 1. controls hormone secretion from the pituitary gland	a. corpus callosum
_____ 2. fiber tract that connects the left cerebral hemisphere to the right cerebral hemisphere	b. hypothalamus
_____ 3. involved in suckling reflex and chewing	c. mammillary bodies
_____ 4. primary relay center for sensory information coming into the brain	d. pineal body (gland)
_____ 5. secretes the hormone melatonin from its precursor molecule, serotonin	e. thalamus

Exercise 15.8: Identification of Cranial Nerves on a Brain or Brainstem Model

11. Complete the table by listing the cranial nerves that extend from the three regions of the brainstem. **10**

Region of Brainstem	Cranial Nerves Extending from This Region of Brainstem
Medulla Oblongata	
Pons	
Midbrain	

12. Write the names of the numbered structures shown on the illustration of the inferior brain in the spaces provided. **11**

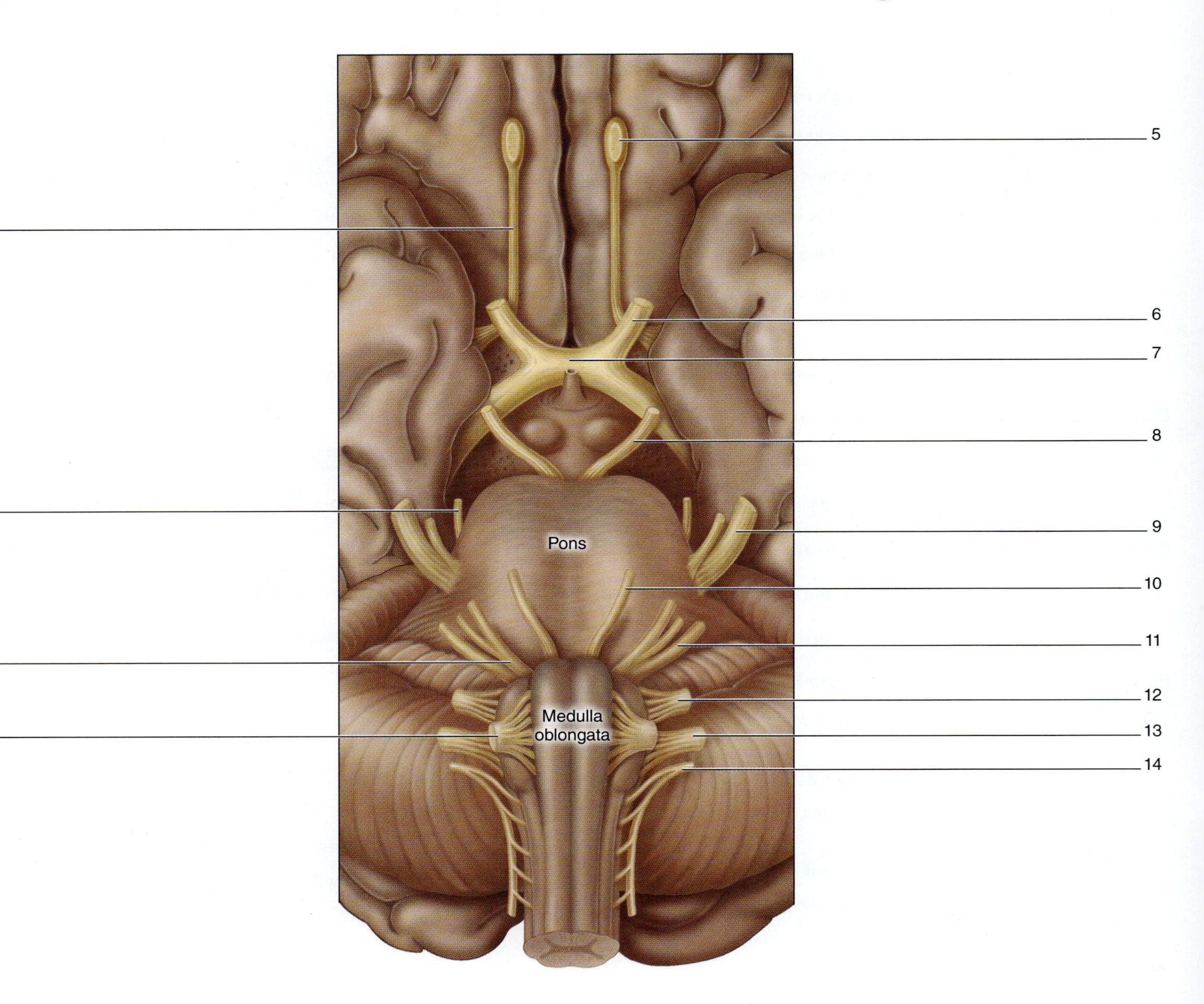

13. Label the following figure of a midsagittal section of a sheep brain. 12

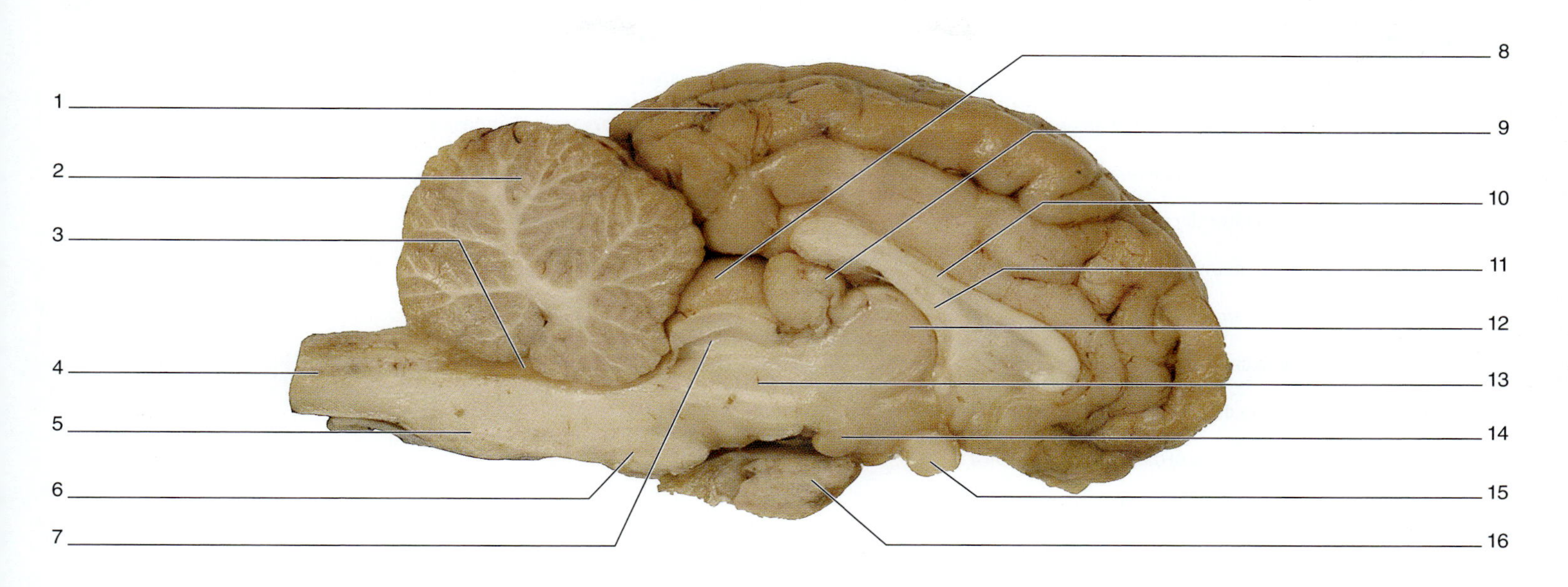

Exercise 15.9: Sheep Brain Dissection

14. Label the following figure of the inferior view of a sheep brain. 12

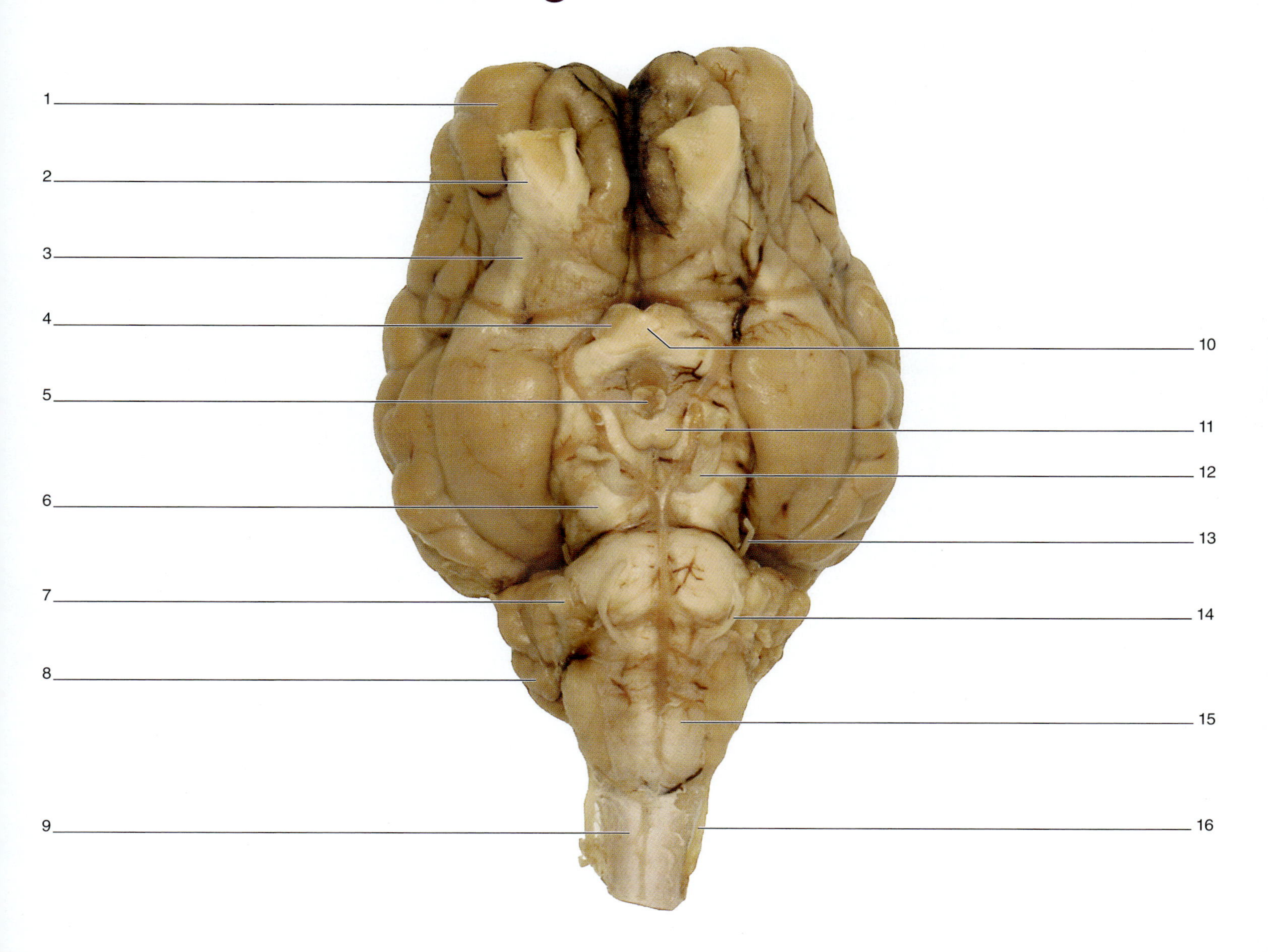

Exercise 15.10: Testing Specific Functions of the Cranial Nerves

15. Match the disorder listed in column A with the cranial nerve associated with that disorder listed in column B. Some answers may be used more than once. 13

Column A		*Column B*
______	1. blindness	a. abducens
______	2. corneal reflex is absent (2 answers)	b. accessory (spinal accessory)
______	3. difficulty turning the eye inferior and lateral	c. facial
______	4. inability to laterally rotate the eye	d. glossopharyngeal
______	5. inability to maintain balance and equilibrium	e. hypoglossal
______	6. inability to pucker the lips	f. oculomotor
______	7. inability to smell	g. olfactory
______	8. inability to taste bitter (sensed by posterior taste buds of the tongue)	h. optic
______	9. pupillary reflexes are absent (2 answers)	i. trigeminal
______	10. soft palate droops on one side (2 answers)	j. trochlear
______	11. tongue deviates to one side when it is stuck out of the mouth	k. vagus
______	12. weakness in elevation of the scapula	l. vestibulocochlear

16. Three of the twelve cranial nerves carry somatic motor fibers to extrinsic muscles of the eye. Complete the table by listing the nerves that carry somatic motor fibers to the extrinsic eye muscles, and then name the muscle(s) innervated by each nerve. 13

Cranial Nerve	Muscles Innervated by the Nerve

17. Several cranial nerves innervate structures of the tongue. Complete the table by listing the nerves that innervate structures of the tongue, and then list the functions of each nerve. 13

Cranial Nerve	Tongue Structures Innervated by the Nerve

18. Four of the twelve cranial nerves carry parasympathetic motor output. Complete the table by listing the nerves that carry parasympathetic motor output, and then give the parasympathetic function(s) of each nerve. 13

Cranial Nerve	Parasympathetic Motor Function(s) of the Nerve

Exercise 15.11: BIOPAC Encephalography (EEG)

19. Which of the following frequency waves are associated with an awake, alert state? (Circle one.) 14

    a. alpha

    b. beta

    c. delta

    d. theta

Can You Apply What You've Learned?

20. Match the following brain structures with the main region of the brain they are associated with by placing checks in the appropriate columns.

Area of the Brain	Brainstem	Cerebellum	Cerebrum	Diencephalon
Arbor Vitae				
Cerebellar Cortex				
Cerebral Cortex				
Corpus Callosum				
Fourth Ventricle				
Hypothalamus				
Intermediate Mass				
Lateral Ventricles				
Medulla Oblongata				
Midbrain				
Pineal Gland				
Pons				
Tectal Plate				
Thalamus				
Third Ventricle				
Vermis				

21. Describe the structural relationship between the optic nerves, optic chiasm, optic tract, hypothalamus, infundibulum, pituitary gland, and mammillary bodies.

____

____

____

____

22. You should have noticed that the tectal plate (corpora quadrigemina) is much larger in the sheep brain, relative to total brain size, than in the human brain. Using this information, answer the following questions:

    a. What is the function of the superior colliculus?

    ____

    ____

    b. What is the function of the inferior colliculus?

    ____

    ____

    c. What does the difference in size between the superior and inferior colliculi tell you about the influence this region of the brain has on the overall functioning of a sheep versus a human? (That is, compare how much influence this area of the brain has on control over body functions.)

    ____

    ____

    ____

23. Compare and contrast structures of the human brain and those of the sheep brain. Complete the table with information about the relative size of the structure compared to the size of the entire brain. Then, based on function, explain why the structure might be more important for survival of the human or the sheep.

Brain Structure	Human Brain	Sheep Brain	
Frontal Lobe			
Inferior Colliculi			
Mammillary Bodies			
Medulla Oblongata			
Olfactory Bulbs			
Pineal Body (Gland)			
Superior Colliculi			

24. What would be the effect of severing the corpus callosum?

Can You Synthesize What You've Learned?

25. If the passage of fluid is blocked at the confluence of sinuses, into which sinuses will fluid back up?

26. An **acoustic neuroma** is a tumor that arises from neurolemmocytes (Schwann cells) surrounding the vestibular portion of the vestibulocochlear nerve (CN VIII). The tumor is benign, but generally grows within the confined space of the petrous part of the temporal bone, thus compressing the nerve and creating problems with balance and hearing loss. What nerve other than CN VIII would you expect to be affected by this tumor (due to its close proximity)?

27. When a light is shined into a patient's right eye, an examiner expects to see a change in pupil diameter in both eyes. The response, called the *consensual light reflex,* is used to test the function of two cranial nerves. The reflex involves one cranial nerve sending the afferent (sensory) signal toward the brain, and another cranial nerve sending the efferent (motor) signal out to the pupil.
    a. Which cranial nerve carries the afferent (sensory) signal to the brain?
    b. Which cranial nerve carries the efferent (motor) signal from the brain?

28. An 8-year-old male suffers from epilepsy. EEG results reveal epileptiform spikes recorded from the left temporal lobe. Doctors performed a left temporal lobe resection, and symptoms resolved. Discuss the consequences of a left temporal lobe resection, and state how these consequences may differ for an adult.

The Spinal Cord, Spinal Nerves, and Reflexes

16

OUTLINE AND LEARNING OBJECTIVES

MODULE 7: NERVOUS SYSTEM

INTRODUCTION

When learning that someone has fractured a vertebra in an accident, people probably think the worst-case scenario: "They are going to be paralyzed!" Though paralysis is common when the spinal cord is severed, the amount of paralysis and subsequent loss of function is highly dependent upon what part of the spinal cord is injured.

The spinal cord transmits nerve signals between the body and the brain. An understanding of where nerve signals enter and exit the spinal cord is important for understanding the degree of paralysis that might result from trauma. For instance, the most common vertebral fractures occur in the lower lumbar region (L_3–L_5). A fracture here cannot sever the spinal cord and rarely results in paralysis. The exercises in this chapter cover concepts that will explain why this is so. In contrast, a fracture high in the vertebral column, such as between the atlas (C_1) and axis (C_2), is commonly fatal. This is not because of paralysis or loss of sensation from the limbs, but because the nerve that controls the diaphragm (the phrenic nerve) can no longer stimulate the muscle to contract, and breathing ceases.

Injury to a spinal nerve or the peripheral nerves that branch from them are most often the result of peripheral nerve compression or irritation. Some familiar examples are *carpal tunnel syndrome,* which causes pain, weakness, or numbness in the hand, and *sciatica,* a condition in which pain or numbness occurs in the lower back and may radiate to the buttock, posterior thigh, leg, and foot. Understanding these clinical conditions requires an appreciation for the organization of the spinal cord and spinal nerves.

The exercises in this chapter explore the structure and function of the spinal cord and spinal nerves, with a focus on major nerves of the body and the structures they serve.

It may be necessary to look at more than one model to be able to see all of the anatomic structures of the spinal cord and spinal nerves. For instance, seeing all the parts of the brachial plexus may require both observing a model of the head, neck, and thorax and then viewing a model of the upper limb. As far as the spinal cord is concerned, understanding the cross-sectional anatomy of the spinal cord is critical for understanding the links between the central and peripheral nervous systems.

Following a detailed study of the spinal cord and spinal nerves, further exercises explore somatic reflexes. Reflexes are preprogrammed responses that the nervous system uses to detect and respond to sensory input. A classic example is the reflex that causes a person to immediately pull his or her hand away from a hot stove. The pain is only noticeable after pulling the hand away. That is because the movement of the hand away from the stimulus is part of a reflex. In this example, the flow of information is only through the spinal cord, making it very fast.

The exercises in this chapter involve not only learning more about the anatomy of these reflexes, but also performing physiological tests to elicit these reflexes in test subjects. Such tests are important in a clinical situation because they allow a clinician to test the region of the nervous system involved in a reflex for signs of dysfunction. When performing the tests, be sure to relate the components of each reflex to anatomical structures covered in the spinal cord histology and gross anatomy sections of this chapter.

List of Reference Tables

Chapter 16: The Spinal Cord, Spinal Nerves, and Reflexes

Name: ______________________

Date: ____________ Section: ____________

These Pre-Laboratory Worksheet questions may be assigned by instructors through their connect course.

PRE-LABORATORY WORKSHEET

1. Which of the following meningeal layers are found in both the brain and the spinal cord? (Check all that apply.)

   _____ a. arachnoid mater _____ b. dura mater _____ c. pia mater

2. Cerebrospinal fluid bathes the spinal cord by circulating within which of the following? (Check all that apply.)

   _____ a. central canal

   _____ b. epidural space

   _____ c. subarachnoid space

   _____ d. subdural space

3. Myelinated axons are found in the ________________ (gray/white) matter of the spinal cord.

4. Both cranial and spinal nerves are part of the peripheral nervous system. ________________ (True/False)

5. Which of the following sections of the spinal cord does not form a nerve plexus? (Circle one.)

   a. cervical b. lumbar c. sacral d. thoracic

6. Match the common action listed in column A with the corresponding compartment listed in column B.

Column A	*Column B*
_____ 1. adduction of the thigh	a. anterior compartment of leg
_____ 2. dorsiflexion and inversion of the ankle	b. anterior compartment of thigh
_____ 3. eversion of the ankle	c. lateral compartment of leg
_____ 4. extension of the hip, flexion of the knee	d. medial compartment of thigh
_____ 5. extension of the knee, flexion of the hip	e. posterior compartment of leg
_____ 6. plantarflexion and inversion of the ankle	f. posterior compartment of thigh

7. Which of the following plexuses is composed of rami, trunks, divisions, and cords? (Circle one.)

   a. brachial b. cervical c. lumbar d. sacral

8. Match the representative nerve listed in column A with the corresponding plexus listed in column B.

Column A	*Column B*
_____ 1. femoral nerve	a. brachial plexus
_____ 2. median nerve	b. cervical plexus
_____ 3. phrenic nerve	c. lumbar plexus
_____ 4. tibial nerve	d sacral plexus

9. Somatic motor neurons are found in the ________________ (anterior/posterior) horn of the spinal cord.

10. Match the description listed in column A with the corresponding component of a reflex arc listed in column B.

Column A	*Column B*
_____ 1. carries out the response	a. effector
_____ 2. detects the stimulus	b. interneuron
_____ 3. neuron that sends impulses away from CNS	c. motor neuron
_____ 4. neuron that sends impulses to the CNS	d. receptor
_____ 5. neuron within the CNS	e. sensory neuron

HISTOLOGY

Spinal Cord Organization

The spinal cord is organized with an outer cortex of white matter surrounding an inner core of gray matter. The gray matter of the spinal cord is organized into horns, so named because of their appearance. Sensory neurons extend from a receptor to the posterior horn. Posterior horns contain the dendrites and cell bodies of interneurons (association neurons). **Anterior horns** contain dendrites and cell bodies of somatic motor neurons, whose axons exit the spinal cord through the anterior roots and extend to skeletal muscle. **Lateral horns,** located only in the thoracic region and first two lumbar segments of the spinal cord, contain the dendrites and cell bodies of visceral motor neurons, with axons that exit the spinal cord through the anterior roots and extend to visceral effectors (smooth muscle, cardiac muscle, and glands). The **gray commissure** is a horizontal bar of gray matter that surrounds the narrow central canal.

The white matter on each side of the spinal cord is partitioned into three **funiculi** (*funis,* cord), which are identified based on their anatomic position: posterior funiculus, lateral funiculus, and anterior funiculus. Each funiculus is composed of myelinated axons that transmit nerve signals between the brain and the body. The anterior funiculi are interconnected by the **white commissure.**

The spinal cord varies in diameter in different regions. It is the largest in the cervical region (called the **cervical enlargement**) and lumbar region (called the **lumbar enlargement**). These enlargements exist because of the increased number of neuron cell bodies in these regions that make the additional connections to the upper and lower limbs, respectively. Make note of similarities and differences between the organization of gray matter and white matter in the spinal cord as compared to the brain.

EXERCISE 16.1

HISTOLOGICAL CROSS SECTIONS OF THE SPINAL CORD

1. Obtain slides containing cross sections of the spinal cord.
2. First observe each slide with the naked eye and attempt to identify the different parts of the spinal cord (cervical, thoracic, lumbar, and sacral) based on cross-sectional area and relative amounts of gray versus white matter.
3. Place the slide on the microscope stage and bring the tissue sample into focus on low power.
4. Identify the following structures on the histology slide of the spinal cord, using **figure 16.1, table 16.1,** and **table 16.2** as guides.

- ☐ **anterior funiculus**
- ☐ **anterior horn**
- ☐ **anterior median fissure**
- ☐ **anterior root**
- ☐ **central canal**
- ☐ **gray commissure**
- ☐ **gray matter**
- ☐ **lateral funiculus**
- ☐ **lateral horn**
- ☐ **nerve roots of cauda equina**
- ☐ **posterior funiculus**
- ☐ **posterior horn**
- ☐ **posterior median sulcus**
- ☐ **posterior root**
- ☐ **posterior root ganglion**
- ☐ **white commissure**
- ☐ **white matter**

INTEGRATE

LEARNING STRATEGY

One way to correctly distinguish anterior from posterior horns is to remember that *posterior* horns reach all the way to the *back.* That is, in the posterior horns the gray matter extends all the way to the edge of the spinal cord, whereas in the anterior horns the gray matter does not extend to the edge of the spinal cord.

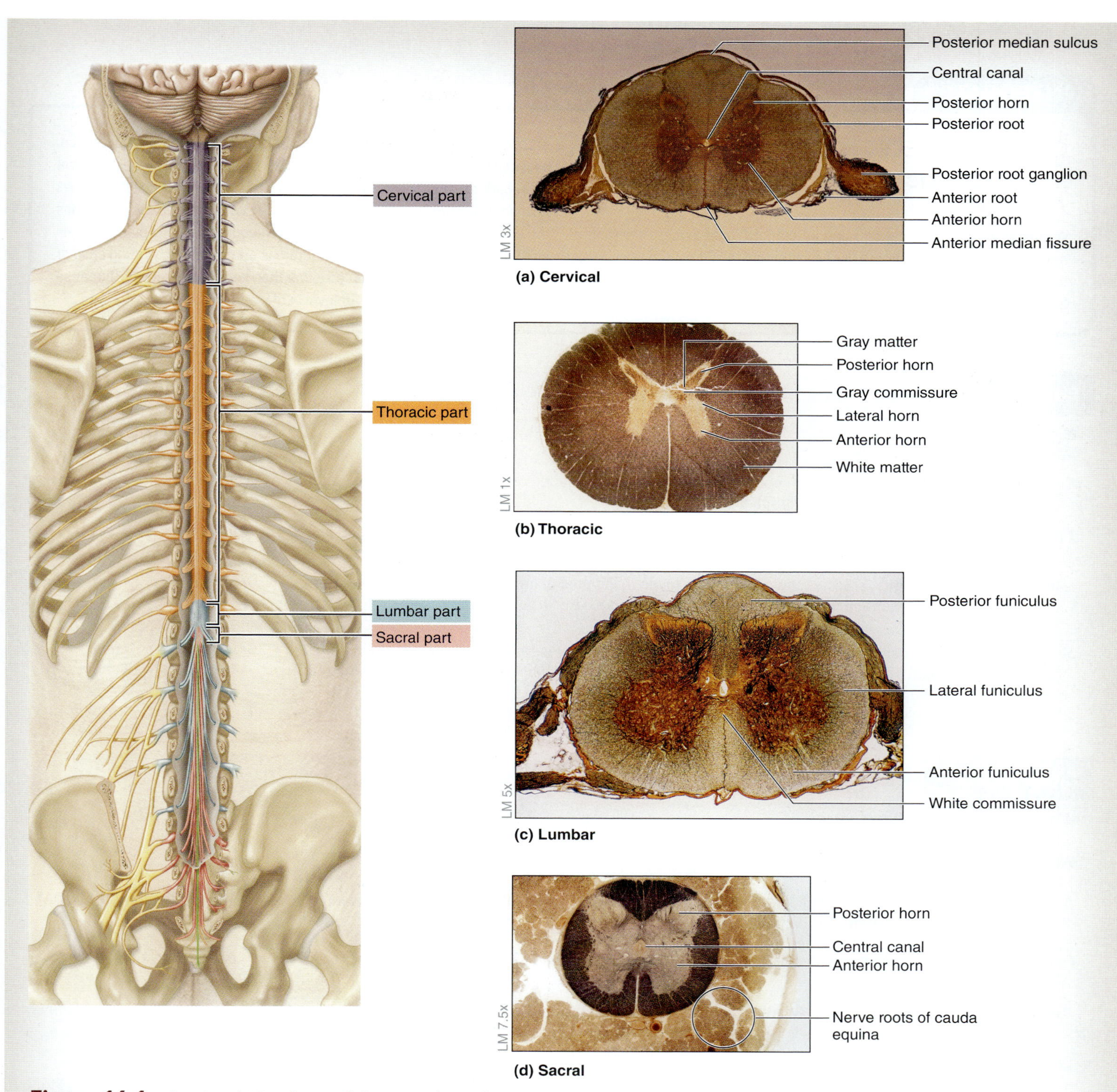

Figure 16.1 Regional Histology of the Spinal Cord in Cross Section.

(continued on next page)

(continued from previous page)

INTEGRATE

CLINICAL VIEW
Lumbar Puncture versus Epidural Anesthesia

A **lumbar puncture** (spinal tap) and an epidural both involve inserting a needle between two vertebrae (L_3–L_4) in the region of the cauda equina of the spinal cord (below where the spinal cord typically ends in an adult at L_1 vertebra). In a lumbar puncture, the needle pierces the dura mater and arachnoid mater within the vertebral canal to enter the subarachnoid space, allowing the withdrawal of cerebrospinal fluid (CSF). In contrast, an **epidural** (a procedure to administer an anesthetic to a woman during childbirth) does not penetrate any meningeal layers within the vertebral canal. Medications are injected into the epidural space, which contains mostly blood vessels and fat. Of note, when a lumbar puncture is performed on a young child or infant, the needle must be placed lower than vertebral level L_3–L_4 because the spinal cord ends at a lower level in these individuals.

Table 16.1 Regional Characteristics of the Spinal Cord

Region of the Spinal Cord	Relative Size	Predominant Tissue	Word Origin
Cervical	Relatively large cross-sectional area due to an abundance of nerves entering and exiting from the cervical and brachial plexuses	White matter	*cervix*, neck
Thoracic	Relatively small cross-sectional area. A distinguishing feature is the presence of lateral horns, which contain the cell bodies of sympathetic motor neurons.	White matter	*thorax*, chest
Lumbar	The largest cross-sectional area, due to an abundance of nerves entering and exiting from the lumbar and sacral plexuses. The anterior horns are large.	Gray matter	*lumbus*, loin
Sacral	The smallest cross-sectional area. Cross sections of numerous nerve roots surrounding the spinal cord are usually visible if the section goes through both the spinal cord and the roots of the cauda equina, which run adjacent to the spinal cord.	Gray matter	*sacer*, sacred

Table 16.2 Histology of the Spinal Cord in Cross Section

Structure	Description	Word Origin
White Matter	Outer portion of the spinal cord consisting of bundles of myelinated axons organized into tracts	literally, a substance that appears white in color
Anterior Funiculus	Tracts of white matter that occupy the anterior region of the spinal cord between the anterior gray horn and the anterior median fissure	*anterior*, the front, + *funis*, cord
Lateral Funiculus	Tracts of white matter that occupy each lateral side of the spinal cord	*latus*, to the side, + *funis*, cord
Posterior Funiculus	Tracts of white matter that occupy the posterior region of the spinal cord between the posterior gray horn and the posterior median sulcus	*posterior*, the back, + *funis*, cord
White Commissure	White matter that interconnects the anterior funiculi	*commissura*, a seam
Gray Matter	Butterfly-shaped inner region of the spinal cord. It contains cell bodies of motor neurons or interneurons, depending upon the location.	literally, a substance that appears gray in color
Anterior Horns	Gray matter on the anterolateral part of the spinal cord that primarily houses the cell bodies of somatic motor neurons (neurons that extend to skeletal muscle)	*anterior*, the front, + *horn*, resembling a horn in shape
Lateral Horns	Small lateral extensions of gray matter found in the T_1–L_2 region of the spinal cord, and containing cell bodies of visceral motor neurons (neurons that extend to smooth muscle, cardiac muscle, and glands)	*latus*, to the side, + *horn*, resembling a horn in shape

Table 16.2 Histology of the Spinal Cord in Cross Section *(continued)*

Structure	Description	Word Origin
Gray Matter		
Posterior Horns	Gray matter on the posterolateral part of the spinal cord that primarily houses the axons of sensory neurons and the cell bodies of interneurons (neurons contained completely within the central nervous system)	*posterior*, the back, + *horn*, resembling a horn in shape
Gray Commissure	A horizontal bar of gray matter that surrounds the central canal	*commissura*, a seam
Other		
Anterior Median Fissure	Deep groove on the anterior surface of the spinal cord	*anterior*, the front, + *median*, the middle, + *fissure*, a deep furrow
Central Canal	Small channel within the gray commissure of the spinal cord that is continuous with the ventricles of the brain; appears as a hole in a microscopic slide of the spinal cord	*central*, in the center, + *canalis*, a duct or channel
Posterior Median Sulcus	Shallow groove located on the posterior surface of the spinal cord	*posterior*, the back, + *median*, the middle, + *sulcus*, a furrow

GROSS ANATOMY

The Spinal Cord

The spinal cord, which is part of the central nervous system, is surrounded by the same three meninges that surround the brain: the dura mater, arachnoid mater, and pia mater (figures 16.2 and 16.3). The dura mater surrounding the spinal cord consists of only a single tissue layer, whereas the dura mater surrounding the brain consists of two layers (periosteal layer and meningeal layer). In addition, there are extensions of pia mater for two structures that help stabilize the spinal cord within the vertebral canal: the denticulate ligaments and the filum terminale. **Denticulate ligaments** are extensions of the pia mater that anchor the spinal cord to the arachnoid mater and dura mater at regular intervals all along the length of the spinal cord. The filum terminale is an extension of the pia mater at the caudal end of the spinal cord that anchors it inferiorly to the coccyx. **Table 16.3** lists some of the key features of structures associated with the spinal cord identified upon gross observation, and describes their functions.

Table 16.3 Gross Anatomy of the Spinal Cord

Structure	Description	Word Origin
Anterior Root*	A nerve root extending from the ventrolateral surface of the spinal cord that contains axons of somatic motor neurons and visceral motor neurons	*anterior*, the front, + *root*, the beginning part
Cauda Equina	A collection of anterior and posterior roots that extend inferiorly from the lumbar and sacral parts of the spinal cord and lie within the vertebral canal. It is named for the fact that the bundle of nerve roots resembles a horse's tail.	*cauda*, tail, + *equinus*, horse
Conus Medullaris	The cone-shaped distal tip of the spinal cord	*konos*, cone, + *medius*, middle
Denticulate Ligaments	Extensions of the pia mater located between anterior and posterior roots. These "ligaments" anchor the spinal cord to the arachnoid and dura mater at intervals between the locations where anterior and posterior roots pierce the dura mater.	*denticulus*, a small tooth, + *ligamentum*, a bandage
Filum Terminale	An extension of pia mater beyond the distal end of the spinal cord. It begins at the conus medullaris and attaches to the distal end of the dura mater at the coccyx.	*filum*, thread, + *terminatio*, ending
Posterior Root*	A nerve root extending from the posterolateral surface of the spinal cord that contains axons of sensory neurons	*posterior*, the back, + *root*, the beginning part
Posterior Root Ganglion*	A swelling on the posterior root that contains cell bodies of sensory neurons and satellite cells (glial cells)	*posterior*, the back, + *root*, the beginning part, + *ganglion*, a swelling
Rootlets (Radicular Fila)	Small branches of nerve fibers coming off the spinal cord that come together to form the anterior and posterior roots	*radicula*, a spinal nerve root, + *filum*, thread
Spinal Nerves	Nerves that form when anterior and posterior roots converge. A spinal nerve exits the vertebral canal through an intervertebral foramina, which forms between adjacent vertebrae.	*spinal*, relating to the spinal cord, + *nevus*, a white, cord-like structure

*The anterior root is also known as the ventral root. The posterior root is also known as the dorsal root.

EXERCISE 16.2

GROSS ANATOMY OF THE SPINAL CORD

1. Observe a whole spinal cord from a cadaver (meninges intact) or a model of the spinal cord.
2. Identify the structures listed in **figure 16.2** on a human spinal cord or a model of the spinal cord, using table 16.3 and the textbook as guides. Then label them in figure 16.2.

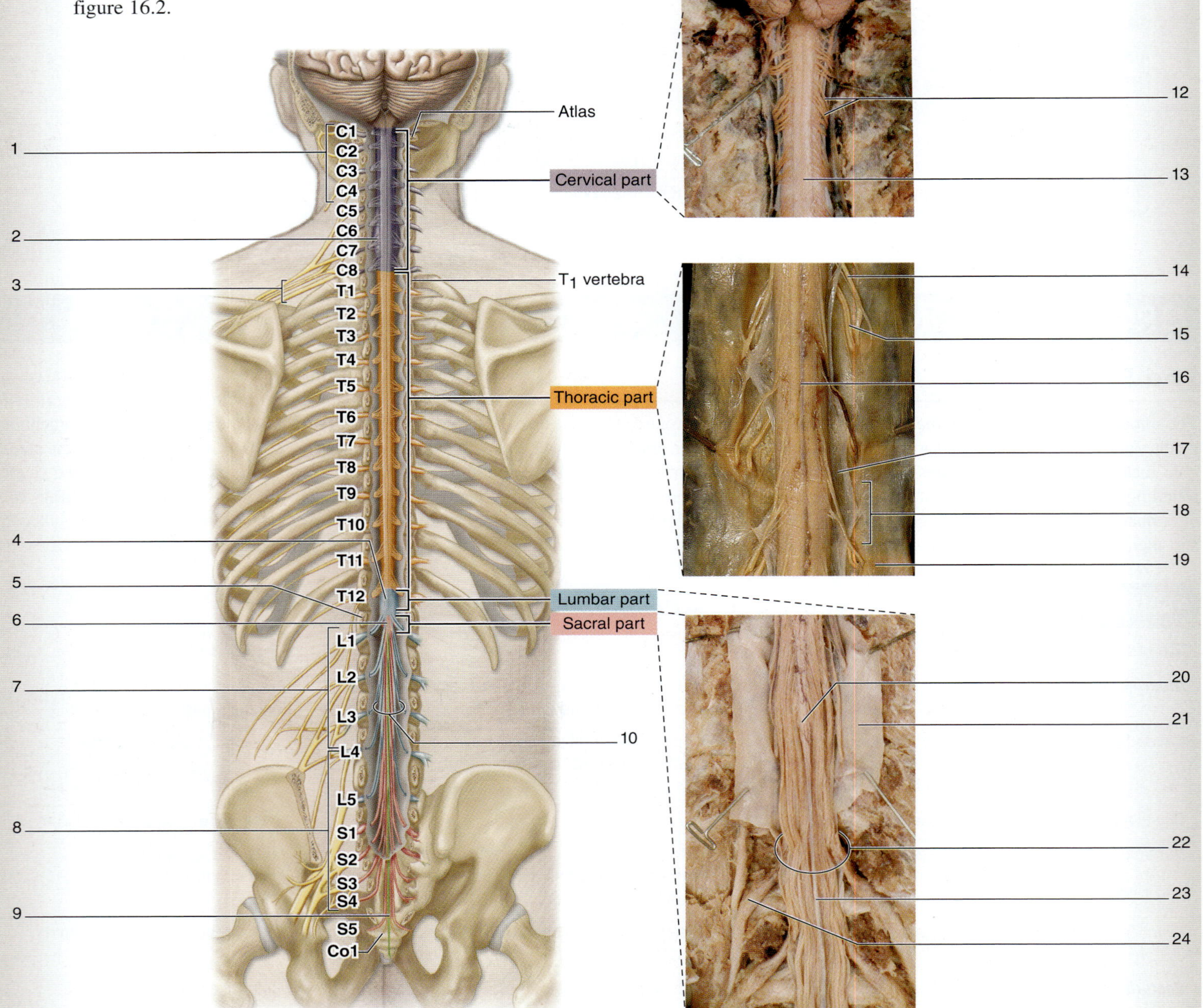

Figure 16.2 Regional Gross Anatomy of the Spinal Cord. Use the terms listed to fill in the numbered labels in the figure. Some answers may be used more than once.

- ☐ anterior roots
- ☐ brachial plexus
- ☐ brain (cerebellum)
- ☐ cauda
- ☐ cervical enlargement
- ☐ cervical plexus
- ☐ conus medullaris
- ☐ denticulate ligament
- ☐ dura mater
- ☐ filum terminal
- ☐ L_1 vertebra
- ☐ lumbar plexus
- ☐ lumbosacral enlargement
- ☐ pia mater
- ☐ posterior median sulcus
- ☐ posterior root ganglion
- ☐ posterior rootlets
- ☐ posterior roots
- ☐ sacral plexus

4. Observe a model of a cross section of the spinal cord.
5. Identify the structures listed in **figure 16.3** on the spinal cord model, using tables 16.2 and 16.3 and the textbook as guides. Then label them in figure 16.3.
6. Describe key features on the anterior and posterior surfaces of the spinal cord (as seen in cross section) that distinguish the anterior surface from the posterior surface.

____

____

____

____

7. *Optional Activity:* APIR **7: Nervous System**—Watch the "Typical Spinal Nerve" animation to help visualize how spinal nerves relate to the spinal cord.

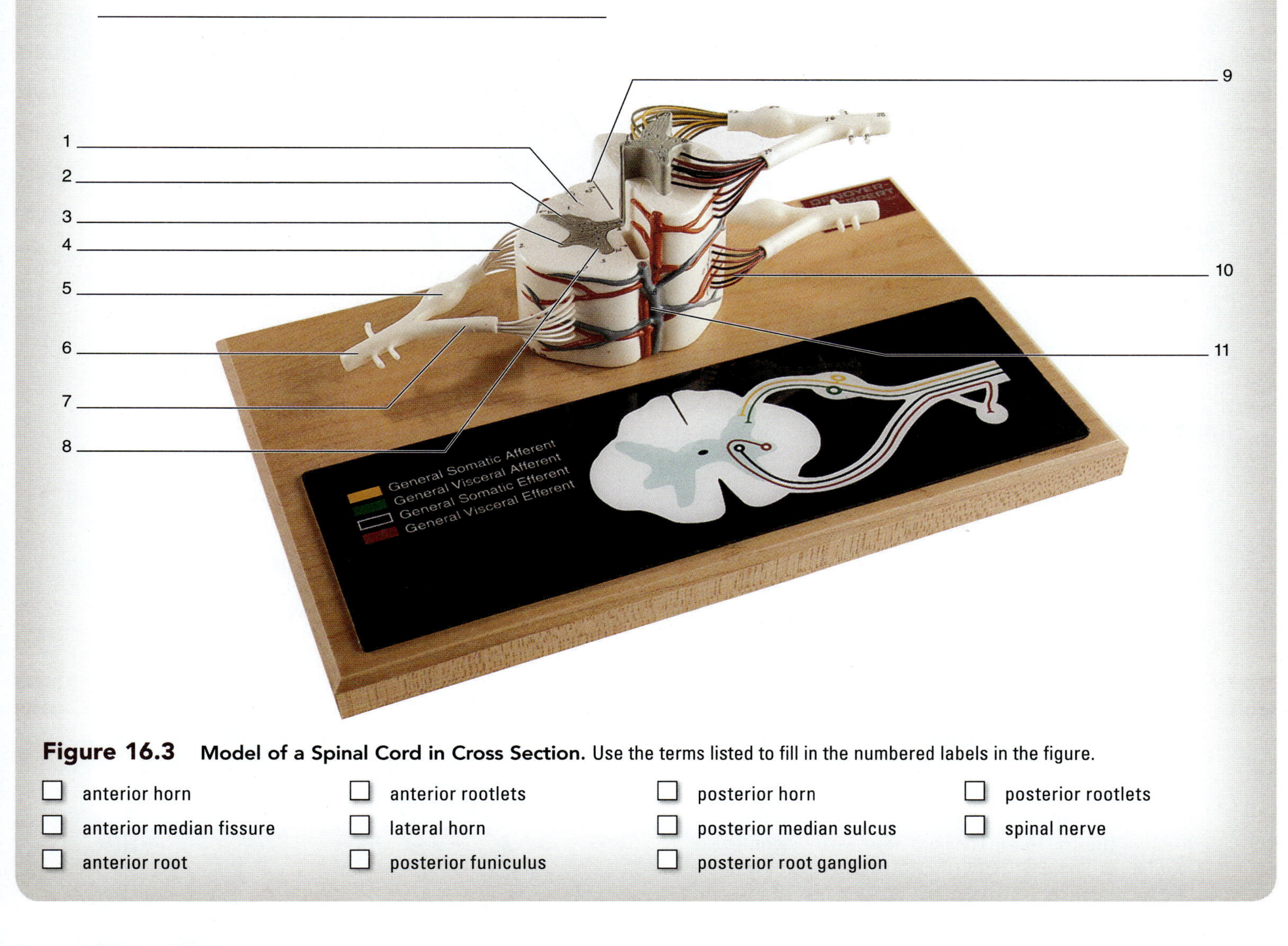

Figure 16.3 **Model of a Spinal Cord in Cross Section.** Use the terms listed to fill in the numbered labels in the figure.

- ☐ anterior horn
- ☐ anterior median fissure
- ☐ anterior root
- ☐ anterior rootlets
- ☐ lateral horn
- ☐ posterior funiculus
- ☐ posterior horn
- ☐ posterior median sulcus
- ☐ posterior root ganglion
- ☐ posterior rootlets
- ☐ spinal nerve

Peripheral Nerves

The peripheral nervous system consists of cranial nerves and spinal nerves. The structure and function of the cranial nerves was covered in chapter 15. This laboratory exercise investigates the structure and function of the **spinal nerves.** After spinal nerves exit the vertebral canal through intervertebral foramina, they immediately branch into posterior and anterior **rami.** The **posterior rami** innervate both skin and muscles of the back that move the vertebral column (this excludes muscles located on the back that move the pectoral girdle and upper limb). In comparison, the structures innervated by the anterior rami depend upon the location at which these rami extend from the vertebral column. The **anterior rami** become **intercostal nerves** that supply skin, bone, and muscle of the thoracic cage in the thoracic region of the vertebral column. The anterior rami form complex networks called **plexuses** (*plexus*, a braid), which then branch to form peripheral nerves that innervate skin, muscle, and bones of the limbs in the cervical, lumbar, and sacral regions of the vertebral column. There are four major nerve plexuses: cervical, brachial, lumbar, and sacral. This laboratory exercise involves investigating each of these plexuses and exploring the major peripheral nerves that arise from each plexus.

EXERCISE 16.3

THE CERVICAL PLEXUS

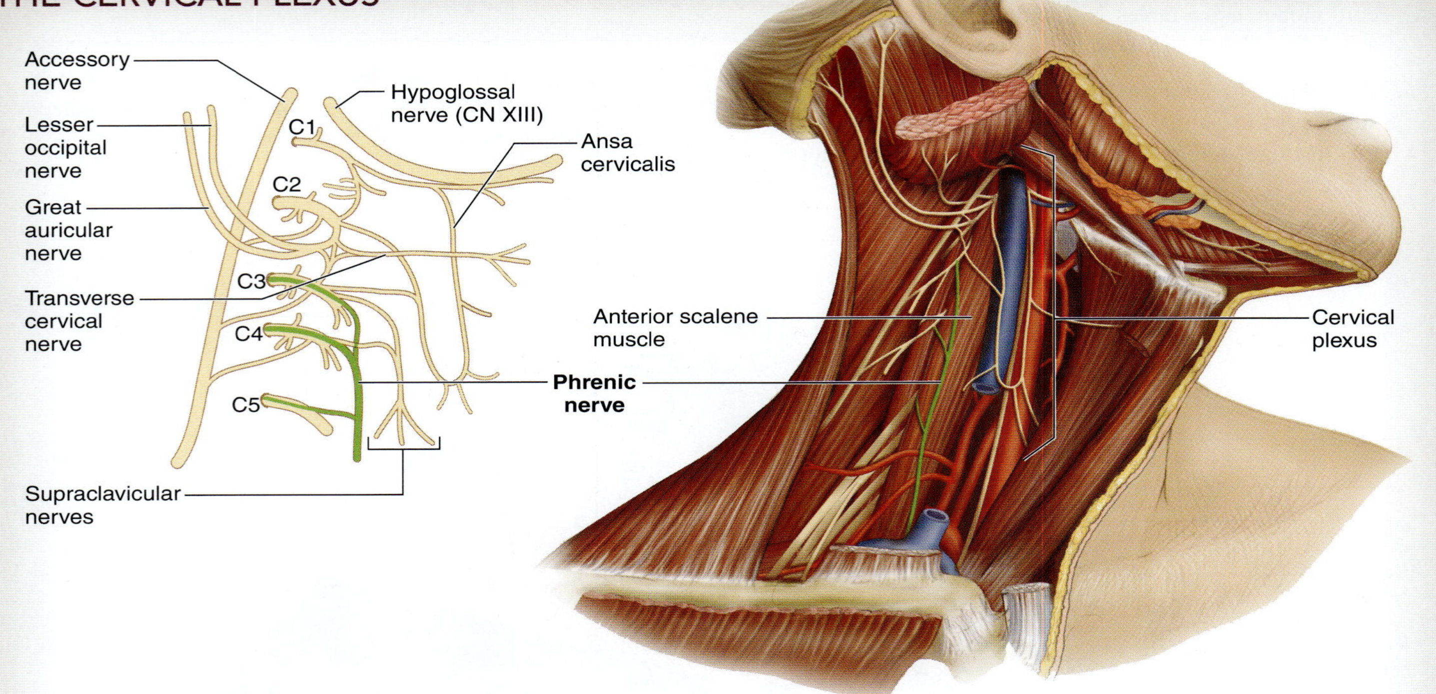

Figure 16.4 The Cervical Plexus and Phrenic Nerve in the Posterior Triangle of the Neck.

1. Observe a prosected human cadaver or a model of the head, neck, and thorax demonstrating nerves of the cervical plexus.

   The cervical plexus arises from the anterior rami of spinal nerves C1–C4. Most of the nerves extending from the cervical plexus transmit nerve signals from skin of the neck and portions of the head and shoulders, and motor information to anterior neck muscles. Perhaps the single most important nerve branching from the cervical plexus is the **phrenic nerve,** which innervates the diaphragm. The phrenic nerves can be seen on the inferior part of the anterior scalene muscle within the neck region, but they are difficult to identify in this location **(figure 16.4).** They are most easily identified within the thoracic cavity **(figure 16.5).** Here they travel within the mediastinum, between the pleural and pericardial cavities. If the heart or lungs have been removed from the cadaver, the phrenic nerves will be easy to identify (if they are still intact). If a model of the thorax is used, look at the location where the pleural and pericardial cavities meet to locate the phrenic nerves.

2. Identify the structures listed in figure 16.5 on a human cadaver or a human torso model, using the textbook as a guide. Then label them in figure 16.5.

INTEGRATE

LEARNING STRATEGY

The phrenic nerve arises from the anterior rami of spinal nerves C3–C5, with the majority of fibers coming from C4. A handy way to remember this is "C3, C4, and C5 keep the body alive." If the phrenic nerves are unable to stimulate the diaphragm, breathing will cease and death will follow.

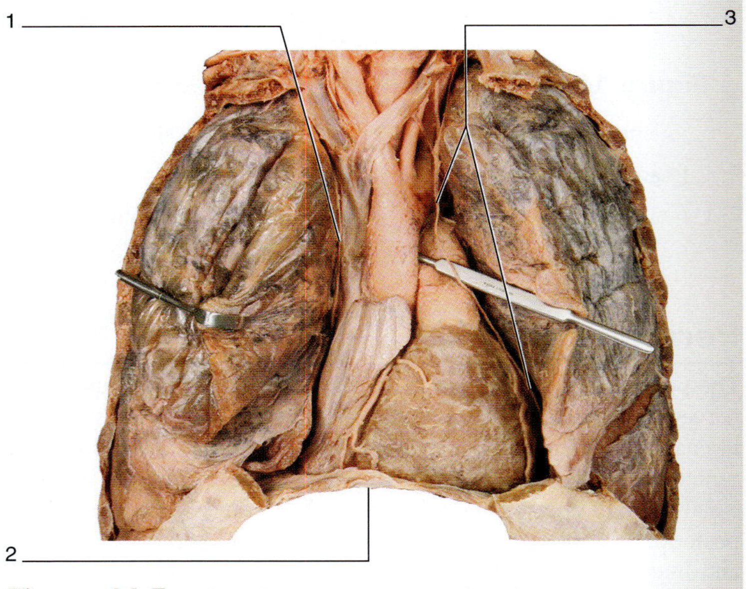

Figure 16.5 The Phrenic Nerves in the Thoracic Cavity. Use the terms listed to fill in the numbered labels in the figure.

- ☐ diaphragm
- ☐ right phrenic nerve
- ☐ left phrenic nerve

EXERCISE 16.4

THE BRACHIAL PLEXUS

1. Observe a prosected human cadaver or a model of the axilla and upper limb demonstrating nerves of the brachial plexus. The organization of the brachial plexus is summarized in **table 16.4.**

   The **brachial plexus** arises from the anterior rami of spinal nerves C5–T1. The overall organization of the brachial plexus follows an organized pattern. The branching pattern of the brachial plexus will be explored in detail in this laboratory exercise. However, note that the branching pattern of the brachial plexus is unique to the brachial plexus. Other plexuses (cervical, lumbar, and sacral) have their own branching patterns, which will not be explored in detail in this laboratory exercise.

   The focus of this exercise is on identification of the **trunks, cords,** and **branches** of the brachial plexus. There are three **trunks,** which pass between the anterior and middle scalene muscles of the neck. The **superior trunk** forms from the anterior rami of C5 and C6, the **middle trunk** is a continuation of the anterior ramus of C7, and the **inferior trunk** forms from the anterior rami of C8 and T1. Each trunk divides into two **divisions,** anterior and posterior. All posterior divisions come together to form the **posterior cord,** and the anterior divisions come together to form either the **medial cord** or the **lateral cord.** The cords are named for their location relative to the axillary artery.

2. First locate the medial and lateral cords around the axillary artery and then spread apart the terminal branches that arise from the cords. Notice that the connections between the cords and the branches appear to form a letter 'M' **(figure 16.6).**

   - *The Posterior Cord:* There are only two major nerves **(table 16.5)** that arise from the posterior cord: the **axillary nerve,** which remains in the axillary region to innervate the deltoid and teres minor muscles; and the **radial nerve,** a large nerve that continues to travel posteriorly along the arm and forearm and innervate skin and muscle along the way (thus, a nerve from the *posterior* cord innervates all structures in the *posterior* compartments of both the arm and the forearm, including the triceps brachii).
   - *The Medial and Lateral Cords:* The medial and lateral cords form three main terminal nerves: the musculocutaneous, median, and ulnar nerves. The **musculocutaneous nerve** forms from the lateral cord and innervates muscles of the anterior compartment of the arm (biceps brachii, coracobrachialis, and brachialis). It is most easily identified where it pierces through the coracobrachialis muscle. The **ulnar nerve** forms from the medial cord and innervates muscles

INTEGRATE

LEARNING STRATEGY

The segments of the brachial plexus, from proximal to distal, are **R**ami, **T**runks, **D**ivisions, **C**ords, and **B**ranches. A mnemonic for remembering this is, "**R**eally **T**ired, **D**rink **C**offee—**B**lack."

Table 16.4 Organization of the Brachial Plexus

Structure	Description	Number	Names
Rami	These are the anterior rami of cervical spinal nerves. The rami combine to form trunks as they pass between the anterior and medial scalene muscles of the neck.	5	C5, C6, C7, C8, T1
Trunks	Located between the anterior and middle scalene muscles of the neck	3	Superior, middle, inferior
Divisions	Each trunk divides into an anterior and a posterior division. The posterior divisions of each trunk come together to form the posterior cord. The anterior divisions of each trunk come together to form the medial and lateral cords.	6	Anterior and posterior division (for each trunk)
Cords	Located in the axilla and named for their location relative to the axillary artery	3	Medial, lateral, and posterior
Branches (Terminal Nerves)	Each terminal nerve innervates a compartment of the arm or forearm, with the exception of the median and ulnar nerves, which both innervate the anterior compartment of the forearm	5	Axillary, radial, musculocutaneous, median, ulnar

(continued on next page)

(continued from previous page)

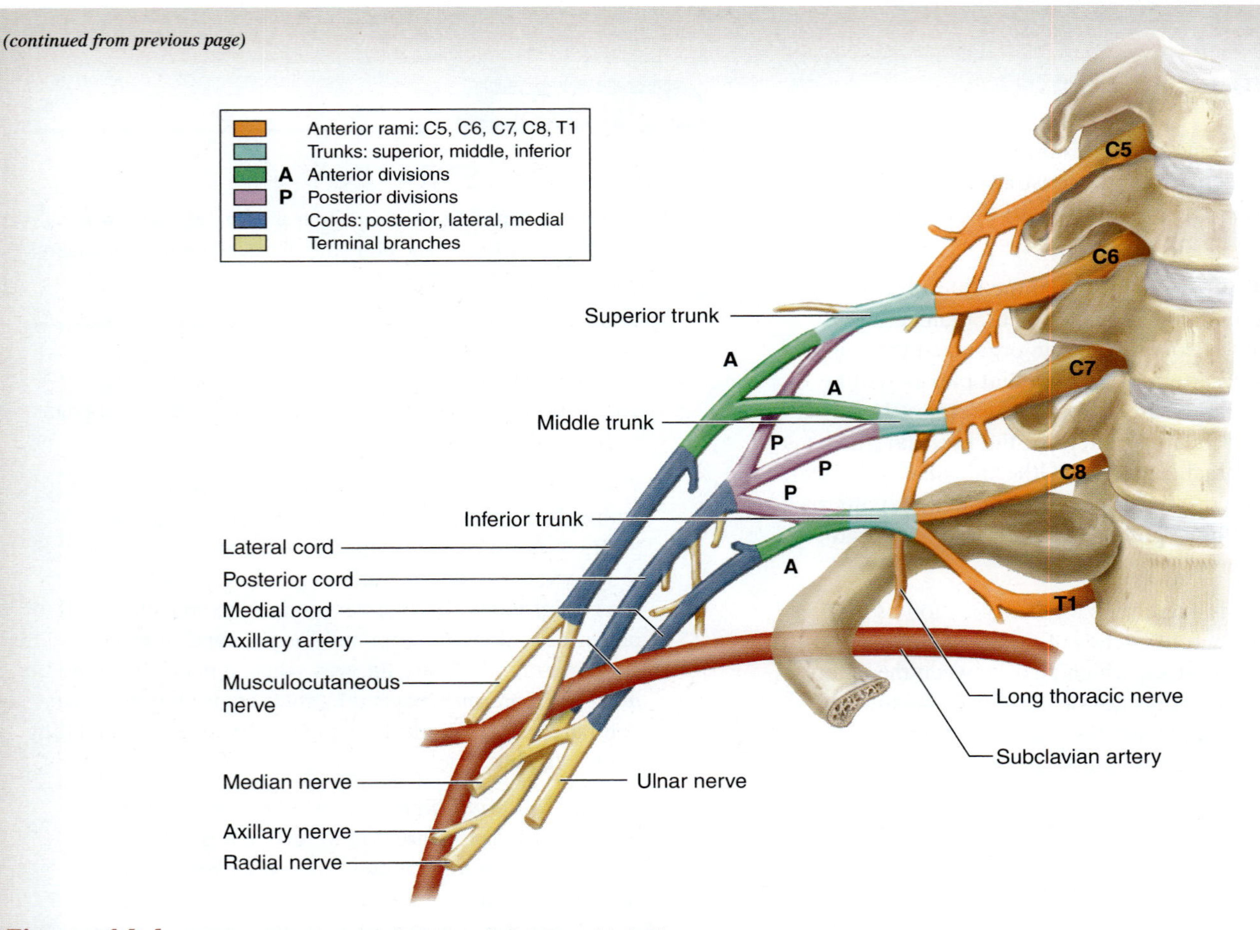

Figure 16.6 Organizational Scheme of the Brachial Plexus.

on the ulnar surface of the forearm along with most intrinsic muscles of the hand. The ulnar nerve is most easily identified where it passes superficially behind the medial epicondyle of the humerus. In this location it is vulnerable to injury; this region of the elbow supplied by the ulnar nerve is commonly referred to as the "funny bone." The **median nerve** forms from branches of both medial and lateral cords. The median nerve is most easily identified in the distal, anterior compartment of the wrist, where it passes through the **carpal tunnel** into the hand. If observing the brachial plexus on a cadaver, try to pass a blunt probe through the carpal tunnel alongside the median nerve. The median nerve is easily irritated by repetitive motions of the wrist, because its passage through the carpal tunnel is very narrow. As the flexor muscles of the forearm contract, their tendons can rub against the median nerve, causing inflammation that results in **carpal tunnel syndrome.**

3. Identify the structures of the brachial plexus listed in **figure 16.7** on a human cadaver or on models of the upper limb, using tables 16.4 and 16.5 and the textbook as guides. Then label them in figure 16.7.

INTEGRATE

CLINICAL VIEW

Additional Nerves of the Brachial Plexus

The **long thoracic** nerve arises from the anterior rami of C5–C7 and innervates the **serratus anterior muscle.** Unlike most nerves, which lie deep to the muscles they innervate, the long thoracic nerve lies superficial to the muscle, which makes it particularly susceptible to injury. Injury can result in paralysis of the serratus anterior muscle. Review the discussion of the serratus anterior muscle in chapter 13, which describes how to diagnose a paralyzed or nonfunctional serratus anterior muscle.

Table 16.5 Major Nerves of the Brachial Plexus

Nerve	Description	Motor Innervation	Sensory Innervation	Word Origin
Posterior Cord				
Axillary (C5–C6)	Arises from the posterior cord and runs around the surgical neck of the humerus en route to the shoulder.	Deltoid and teres minor	Skin on the dorsolateral aspect of the shoulder	*axilla*, the armpit
Radial (C5–T1)	Arises from the posterior cord and extends deep within the posterior compartment of the arm adjacent to the humerus	Posterior (extensor) compartments of the arm and forearm	Skin overlying the posterior compartments of the arm and forearm and the dorsum of the hand (except the tips of the fingers and the entire fifth digit)	*radialis*, relating to the radius
Medial and Lateral Cords				
Musculocutaneous (C5–C7)	Arises from both the medial and lateral cords of the brachial plexus. It pierces through the coracobrachialis muscle and then extends between the brachialis and biceps brachii muscles.	Anterior compartment of the arm; biceps brachii, brachialis, and coracobrachialis	Skin overlying the lateral surface of the forearm	*muscus*, a mouse (muscle), + *cutaneous*, relating to the skin
Median (C5–T1)	Arises from both the medial and lateral cords. Extends along the anteriomedial region of the arm, passes deep to the bicipital aponeurosis in the cubital fossa, then enters the forearm. After passing through the forearm, it courses deep to the flexor retinaculum through the carpal tunnel and into the hand.	Anterior compartment of the forearm except for the flexor carpi ulnaris and ulnar half of the flexor digitorum profundus	Skin of the lateral three and one-half digits on the palmar surface of the hand (and the distal aspect of the same digits on the dorsal surface), skin of the medial palm proximal to its innervation of the digit	*median*, in the middle
Ulnar (C8–T1)	Arises from the medial cord, and continues along the medial aspect of the arm. It travels superficially immediately posterior to the medial epicondyle of the humerus.	Anterior forearm muscles on the medial side including the flexor carpi ulnaris and ulnar half of flexor digitorum profundus, as well as most intrinsic hand muscles	Skin of the medial one and one-half digits on the palmar surface of the hand, skin of the lateral surface of the hand on both palmar and dorsal surfaces	*ulnar*, relating to the ulna
Long Thoracic (C5–C7)	Arises from the roots of C5, C6, and C7 and continues superficially to the serratus anterior muscle	Serratus anterior	None	*longus*, long, + *thoracis*, relating to the thorax

(continued on next page)

(continued from previous page)

5. Make a simple line drawing of the brachial plexus in the space provided. Then label the structures using the terms listed in figure 16.7.

INTEGRATE

LEARNING STRATEGY

The *cord* of the brachial plexus is *medial*, as in located medial to the axillary artery; the *nerve* is *median*, as in located in the middle (of the branches that form from medial and lateral cords).

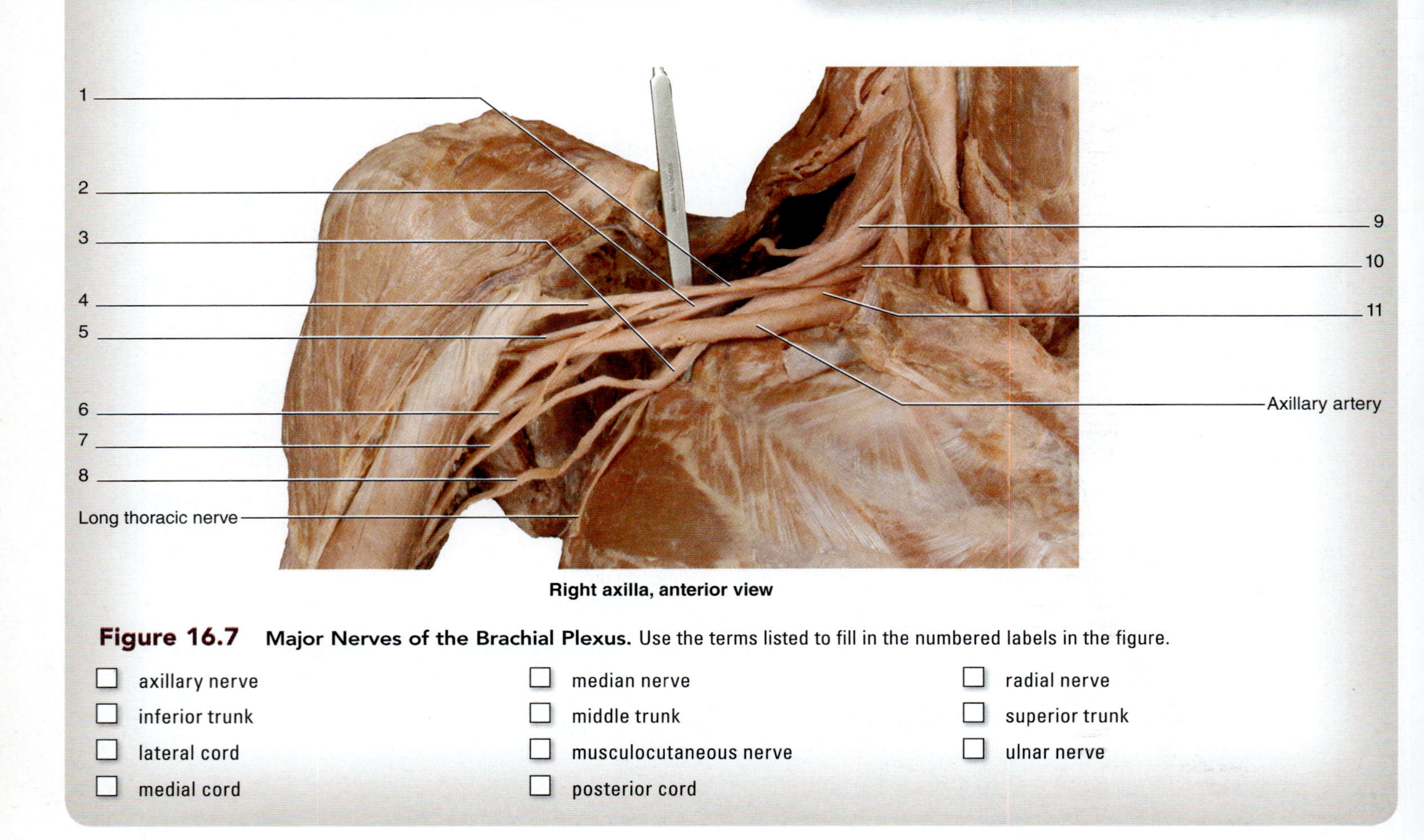

Right axilla, anterior view

Figure 16.7 Major Nerves of the Brachial Plexus. Use the terms listed to fill in the numbered labels in the figure.

- ☐ axillary nerve
- ☐ inferior trunk
- ☐ lateral cord
- ☐ medial cord
- ☐ median nerve
- ☐ middle trunk
- ☐ musculocutaneous nerve
- ☐ posterior cord
- ☐ radial nerve
- ☐ superior trunk
- ☐ ulnar nerve

EXERCISE 16.5

THE LUMBAR AND SACRAL PLEXUSES

1. Observe a prosected human cadaver or a model of the abdomen and lower limb demonstrating nerves of the lumbar plexus. The purpose of this exercise is to identify the major nerves that arise from the lumbar plexus, which forms from the anterior rami of spinal nerves L1–L4 **(table 16.6),** and the major nerves that arise from the sacral plexus, which forms from the anterior rami of spinal nerves L4–S4) **(table 16.7).**

2. *Nerves of the Lower Limb Arising from the Lumbar Plexus:* It will be easiest to learn the nerves of the lower limb when relating each nerve to a single compartment. Recall the limb compartments from chapter 13; there are three compartments of the thigh (anterior, posterior, and medial) and three compartments of the leg (anterior, posterior, and lateral). Each compartment, for the most part, receives innervation from a single nerve from either the lumbar or the sacral plexus. First associate one nerve with one compartment. Then, learn the "exceptions" to

Table 16.6 Major Nerves of the Lumbar Plexus

Nerve	Description	Motor Innervation	Sensory Innervation	Word Origin
Femoral (L2–L4)	Runs along the lateral border of the psoas major muscle, travels under the inguinal ligament into the femoral triangle to innervate the anterior compartment of the thigh	Anterior compartment of the thigh	Skin of the anterior thigh and leg	*femoral*, relating to the femur or thigh
Obturator (L2–L4)	Runs along the medial border of the psoas major muscle, travels through the obturator foramen into the medial compartment of the thigh	Medial compartment of the thigh	Skin on the medial surface of the thigh	*obturatus*, to occlude or stop up

Table 16.7 Major Nerves of the Sacral Plexus

Nerve	Description	Motor Innervation	Sensory Innervation	Word Origin
Common Fibular (L4–S2)	Begins at the bifurcation of the sciatic nerve proximal to the popliteal fossa, passes superficially near the head of the fibula, wraps around the neck of the fibula, and then divides into superfcial and deep branches	Biceps femoris muscle, short head	Skin on the proximal posterolateral surface of the leg	*fibular*, relating to the fibula
Deep Fibular (L4–S1)	Arises from the common fibular nerve at the neck of the fibula, passes through the extensor digitorum longus muscle into the anterior compartment of the leg	Anterior compartment of the leg, dorsal musculature of the foot	Skin on the first interdigital cleft of the foot	*deep*, situated at a deeper level than a corresponding structure, + *fibular*, relating to the fibula
Inferior Gluteal (L5–S2)	Exits the pelvis through the greater sciatic foramen inferior to the piriformis	Gluteus maximus	NA	*inferior*, lower, + *gloutos*, the buttock
Posterior Femoral Cutaneous (S1–S3)	Exits pelvis through the greater sciatic foramen inferior to the piriformis, just medial to the sciatic nerve	NA	Skin on the posterior thigh	*posterior*, behind, + *femur*, the thigh, + *cutis*, the skin
Pudendal (S2–S4)	Exits the pelvis through the greater sciatic foramen inferior to the piriformis muscle, then travels through the lesser sciatic foramen to enter the perineum	Perineal muscles, external anal sphincter, and external urethral sphincter	External genitalia in both males and females	*pudendal*, that which is shameful
Sciatic (L4–S3)	Exits the pelvis through the greater sciatic foramen inferior to the piriformis muscle, then enters the posterior compartment of the thigh through a groove between the ischial tuberosity and the greater trochanter of the femur	Posterior compartment of the thigh (the tibial division of the sciatic is responsible for innervation of all posterior compartment muscles except for the short head of the biceps femoris)	NA	*sciaticus*, the hip joint
Superficial Fibular (L5–S2)	Arises from the common fibular nerve at the neck of the fibula and descends within the lateral compartment of the leg	Lateral compartment of the leg	Skin on the distal, lateral surface of the leg and the dorsal surface of the foot	*superficialis*, the surface, + *fibular*, relating to the fibula
Superior Gluteal (L4–S1)	Exits the pelvis through the greater sciatic foramen superior to the piriformis	Gluteus medius, gluteus minimus, and tensor fascia lata muscles	NA	*superus*, above, + *gloutos*, the buttock
Tibial (S) (L4–S3)	Begins at the bifurcation of the sciatic nerve proximal to the popliteal fossa, runs along the tibialis posterior muscle, and then branches into two plantar nerves at the ankle	Posterior compartment of the leg, plantar musculature of the foot	NA	*tibial*, relating to the tibia

(continued on next page)

(continued from previous page)

the rules. The general rules for the lumbar plexus are as follows:

Compartment	Nerve
Anterior thigh	Femoral nerve
Medial thigh	Obturator nerve

3. *Lumbar Plexus:* Identify the nerves and muscles listed in **figure 16.8** in the lower limb of a human cadaver or a model of the lower limb, using table 16.6 and the textbook as guides. Then label them in figure 16.8.
4. *Sacral plexus:* Identify the nerves and muscles listed in **figure 16.9*a*** in the lower limb of a human cadaver or a model of the lower limb, using table 16.7 and the textbook as guides. Then label them in figure 16.9*a*.
5. *Nerves of the Sacral Plexus Within the Gluteal Region:* The **gluteal nerves** (superior and inferior) are named for their exit location relative to the piriformis muscle of the deep buttock (figure 16.9*b*). The **superior gluteal nerve** arises superior to the piriformis, while the **inferior gluteal nerve** arises inferior to the piriformis. A small yet important nerve that also arises in this region is the **pudendal nerve** (*pudendal*, that which is shameful). The **sciatic nerve** arises inferior to the piriformis muscle and travels into the posterior compartment of the thigh, passing close to the ischial tuberosity. The sciatic nerve is the largest nerve in the body because it is actually two nerves, the **common fibular nerve** and the **tibial nerve**, bundled together in a common connective tissue sheath. Most commonly the two nerves separate from each other just proximal to the popliteal fossa of the knee. It is the **tibial division** of the sciatic nerve that is responsible for innervating the posterior compartment of the thigh. The general rules for the sacral plexus are as follows:

Compartment	Nerve
Posterior thigh	Sciatic nerve (tibial division)
Anterior leg	Deep fibular nerve
Lateral leg	Superficial fibular nerve
Posterior leg	Tibial nerve

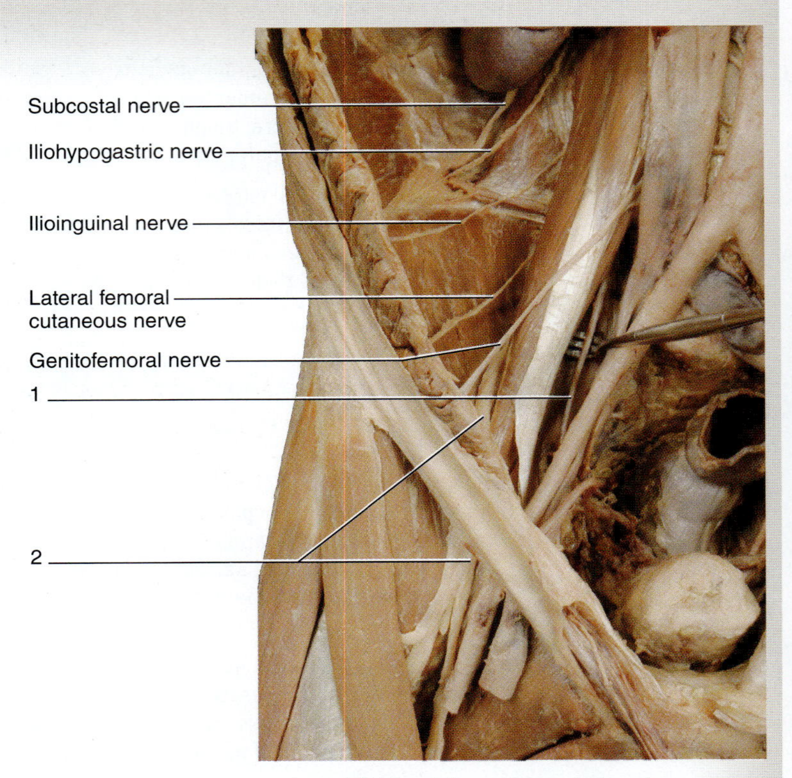

Figure 16.8 Nerves of the Lumbar Plexus. Deep dissection of the pelvis and anterior thigh. Use the terms to fill in the numbered labels in the figure.

- ☐ femoral nerve
- ☐ obturator nerve

6. Identify the nerves and muscles of the gluteal region listed in figure 16.9*b* on a human cadaver or a model of the gluteal region, using table 16.7 and the textbook as guides. Then label them in figure 16.9*b*.

INTEGRATE

LEARNING STRATEGY

Remember that the word *pudendal* means "that which is shameful." Then, it will be easy to remember the structures it innervates. It innervates all of our "shameful." structures—namely, the external genitalia. In addition, its fibers control the contraction of the external sphincters surrounding the anus and urethra. Wouldn't it be "shameful" if this nerve were damaged? A very minimal anesthetic given during childbirth is a **pudendal nerve block.** It blocks sensation from the birth canal (vagina), but the mother still receives sensations from the contracting uterus.

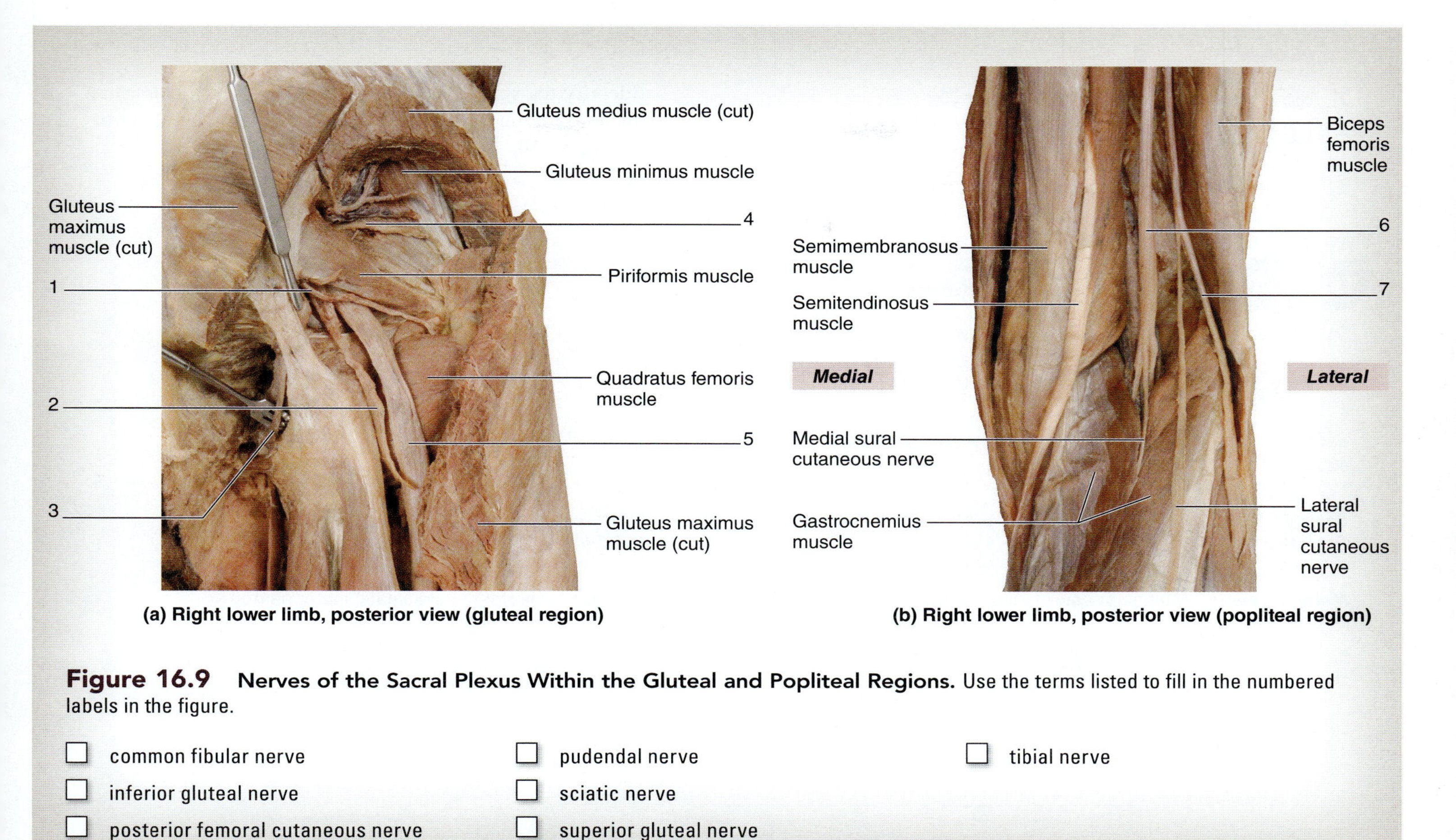

Figure 16.9 Nerves of the Sacral Plexus Within the Gluteal and Popliteal Regions. Use the terms listed to fill in the numbered labels in the figure.

- ☐ common fibular nerve
- ☐ inferior gluteal nerve
- ☐ posterior femoral cutaneous nerve
- ☐ pudendal nerve
- ☐ sciatic nerve
- ☐ superior gluteal nerve
- ☐ tibial nerve

Somatic Reflexes

Reflexes involve a rapid response to a stimulus that is automatic and involuntary. All reflexes involve the same key components: receptor, sensory neuron, control center, motor neuron, and effector. In somatic reflexes, the sensory receptors are commonly cutaneous receptors within the skin (e.g., lamellated corpuscles, Meissner corpuscles, or free nerve endings) and the motor output is always to skeletal muscle (remember: somatic is associated with *voluntary* movement). This is precisely the type of reflex that allows the rapid withdrawal of a hand from a hot or painful stimulus. The reflex is the automatic movement that is separate from the person's conscious awareness of the stimulus. The following exercise explores the components of a reflex. Excercises within the physiology section of this chapter allow students to elicit various somatic reflexes using standard reflex tests. Abnormal reflex responses may indicate certain diseases (e.g., syphilis, late-stage alcoholism, diabetic neuropathy), damage to peripheral nerves, or damage to the spinal cord.

INTEGRATE

CONCEPT CONNECTION

Spinal nerves, like cranial nerves, are part of the peripheral nervous system. All sensory information coming into the central nervous system "travels" within peripheral nerves. Although processing by the brain is necessary to provide this conscious awareness of sensory input, the brain is not required for "processing" *all* sensory and motor information. For example, spinal reflexes, such as the patellar reflex are neural circuits that require only the functioning of spinal nerves and the spinal cord. The overall process for the flow of information in reflexes and more complex sensory modalities is essentially the same: (1) peripheral nerves transmit sensory information to the central nervous system, (2) the central nervous system integrates this sensory information, and (3) peripheral nerves carry motor information away from the central nervous system to the appropriate effector organs.

EXERCISE 16.6

IDENTIFYING COMPONENTS OF A REFLEX ON A CLASSROOM MODEL

1. Observe a classroom model or a diagram of a reflex arc.
2. Identify the structures listed in **table 16.8** on a classroom model or a diagram of the reflex arc. Then label them in **figure 16.10.**

Table 16.8 Components of a Reflex

Component	Description and Function	Example(s)
Receptor	Sensory organ designed to detect a specific stimulus	Tactile corpuscle in skin
Sensory Neuron	Unipolar neuron that carries information from a receptor to the control center	Somatic sensory neuron
Control Center	Area within the brain or spinal cord that controls a particular reflex response	Superior colliculus in brain
Motor Neuron	Multipolar neuron that carries information from the control center to an effector	Somatic motor neuron
Effector Organ	Muscles or glands that carry out the response to a stimulus	Skeletal, smooth, or cardiac muscle; glands

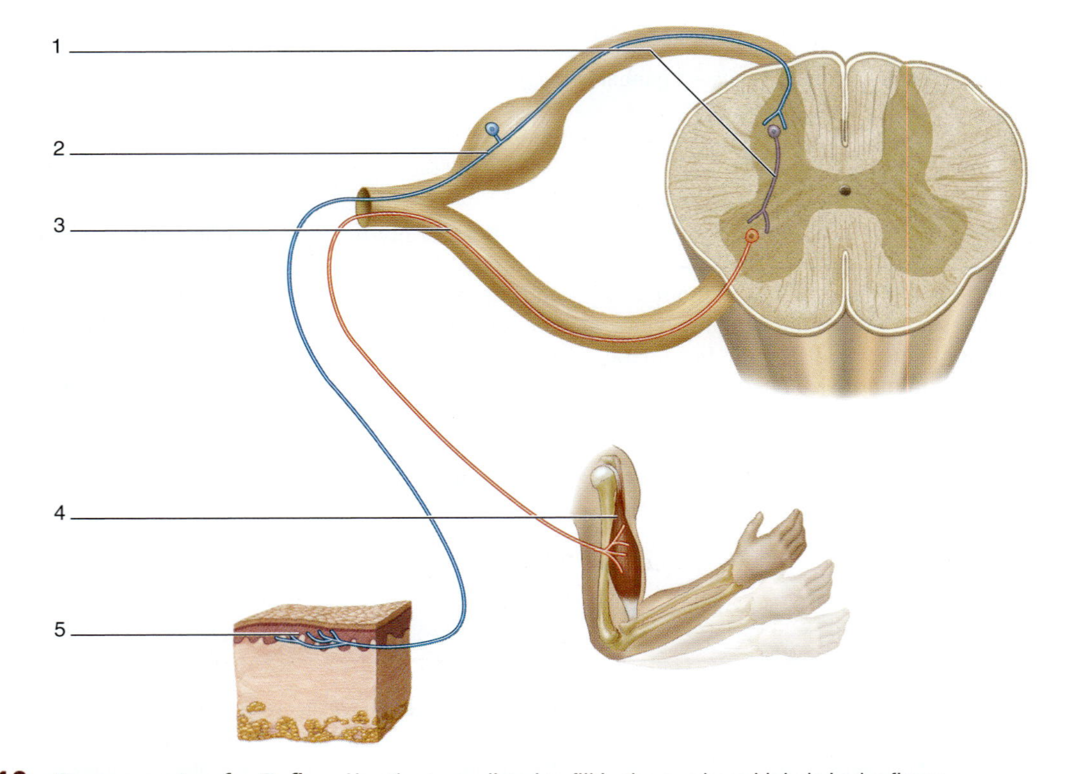

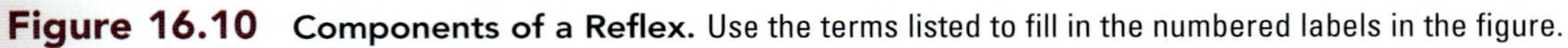
Figure 16.10 Components of a Reflex. Use the terms listed to fill in the numbered labels in the figure.

- ☐ control center/interneuron
- ☐ effector
- ☐ motor neuron
- ☐ receptor
- ☐ sensory neuron

PHYSIOLOGY

The following exercises explore common somatic spinal reflexes: the **patellar reflex**, the **withdrawal** and **crossed-extensor reflex**, and the **plantar reflex**. When performing each of the reflex tests, think about the anatomical structures that are involved. For example, when testing a reflex involving the triceps brachii recall that the nerve that innervates the triceps brachii is the radial nerve, which is a branch of the brachial plexus. The brachial plexus arises from spinal nerves C5–T1. Thus, testing this reflex is a way of testing the function of the spinal cord levels between C5 and T1. In an actual clinical situation, a physician is trained to identify spinal cord levels even more precisely than this. However, simply going through the thought process will allow a deeper understanding of how to assess the function of certain regions of the CNS using reflex tests.

Reflex Physiology

EXERCISE 16.7

PATELLAR REFLEX

1. Obtain a rubber percussion hammer and choose a student to be the subject.
2. Have the subject sit on the lab table or a chair with knees bent, legs completely relaxed, and feet not touching the floor (this is why a lab table is a better location than a chair).
3. Gently strike the subject's patellar ligament with the *blunt* side of the percussion hammer **(figure 16.11)** and note observations:

   ____

   ____

   ____

4. This exercise tests the function of the components of the stretch reflex, namely muscle spindles located within the muscles that are stretched when the patellar ligament is tapped. Recall from chapter 13 that the patellar ligament is an extension of the patellar tendon. The patellar tendon is associated with what group of muscles?

   ____

   What somatic nerve innervates this group of muscles?

   ____

   From what somatic nerve plexus does this nerve extend?

   ____

   What spinal nerves contribute to this plexus?

   ____

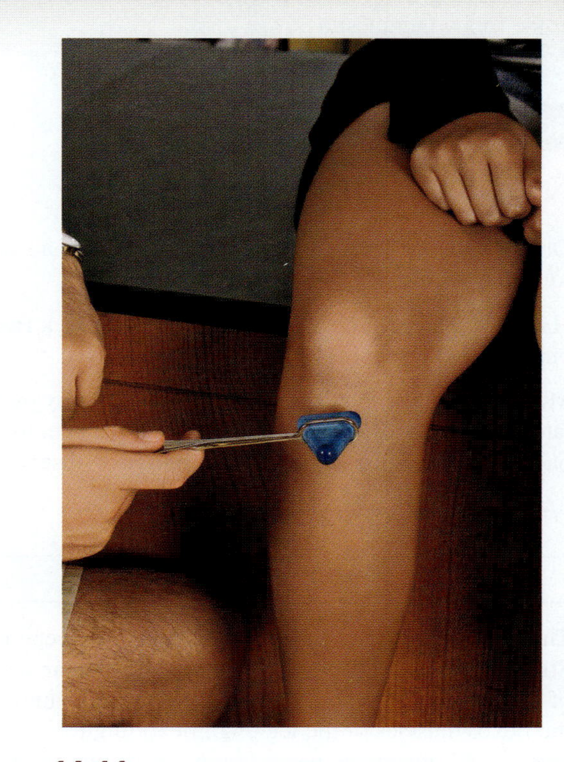

Figure 16.11 Testing the Patellar Reflex. The subject's leg should be completely relaxed and feet should not touch the floor. Using the blunt end of a percussion hammer, gently strike the patellar ligament of the subject and note what happens.

EXERCISE 16.8

WITHDRAWAL AND CROSSED-EXTENSOR REFLEX

The withdrawal reflex is difficult to reproduce because it involves withdrawing a limb away from a noxious or painful stimulus, such as a hot, cold, or sharp object. When nociceptors (*noci-*, pain) are stimulated, the subject withdraws quickly and unconsciously from the painful stimulus, exciting flexor muscles ipsilaterally (on the same side of the body). Simultaneously, extensor muscles are excited on the contralateral side (the opposite side of the body). To elicit a true withdrawal reflex, the subject must be unaware that the stimulus is coming, which makes it particularly difficult to reproduce in a laboratory setting.

Rather than eliciting a withdrawal and crossed-extensor reflex for this laboratory exercise, recall one of the following scenarios: touching a hot surface or stepping on a sharp object. Answer the following questions:

1. What happened to the limb (upper or lower) that experienced the painful stimulus?
___
___
2. What happened to the opposite limb?
___
___
3. Which happened first, withdrawal of the limb away from the stimulus or perception of the stimulus (circle one)? Based on an understanding of the anatomy of spinal reflexes, justify your answer.
___
___
___

EXERCISE 16.9

PLANTAR REFLEX

1. Obtain a rubber percussion hammer and choose a student to be the subject.
2. Have the subject remove his or her shoe and sock from one foot and lie supine on the lab table.
3. Move the metal end of the rubber hammer firmly over the lateral aspect of the sole of the foot from the heel to the base of the great toe **(figure 16.12)** and note observations:
___
___
___
4. This exercise tests the function of cutaneous receptors. Stimulation of cutaneous receptors on the plantar surface of the foot evokes a spinal reflex that, in turn, stimulates the flexor muscles of the leg (e.g., flexor digitorum longus). What somatic nerve innervates this group of muscles (Hint: it extends down the posterior leg)?
___
___

From what somatic nerve plexus does this nerve extend?
___

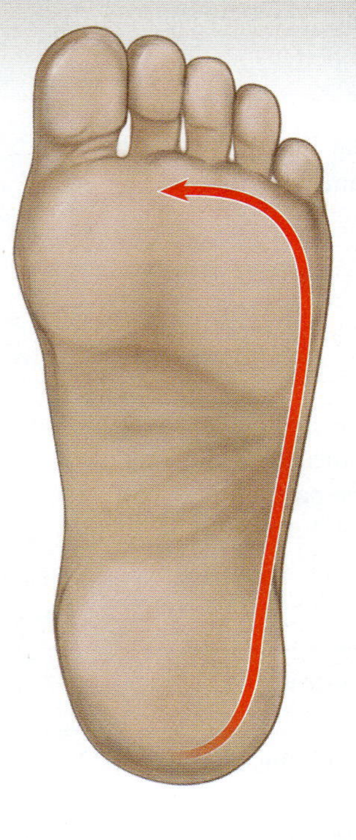

Figure 16.12 Testing the Plantar Reflex. Have the subject lie down on the lab table. Use the metal end of the percussion hammer so that you can press firmly on the foot. Press the bottom of the foot, following the arc denoted by the arrow in this figure. Note the reaction of the subject.

What spinal nerves contribute to this plexus? ___
___

INTEGRATE

CLINICAL VIEW
Babinski Reflex

In infants, the normal response to the plantar reflex test (figure 16.12) is opposite the normal response of adults because the axons within this nerve are not yet completely myelinated. That is, the infant response to the plantar reflex test involves the toes spreading apart (abduction) and the foot dorsiflexing. This reflex is called the *Babinski reflex* (toe abduction and foot dorsiflexion denotes a "positive" Babinski sign). If an adult responds to the plantar reflex test with a positive Babinski sign, it indicates an abnormal response. Specifically, this abnormality suggests damage to the corticospinal pathway (or pyramidal neurons) within the central nervous system, because this motor pathway is responsible for modulating the spinal reflex. The cause of such damage must be explored further by the physician.

Chapter 16: The Spinal Cord, Spinal Nerves, and Reflexes

Name: ______________________________

Date: ____________ **Section:** ___________

POST-LABORATORY WORKSHEET

The ❶ corresponds to the Learning Objective(s) listed in the chapter opener outline.

Do You Know the Basics?

Exercise 16.1: Histological Cross Sections of the Spinal Cord

1. The outer cortex of the spinal cord consists of ______________________ (gray/white) matter, whereas the inner medulla of the spinal cord consists of ______________________ (gray/white) matter. ❶

2. Match the type of neuron listed in column A with the corresponding part of the spinal cord where the soma of the neuron may be found, listed in column B. ❷

Column A	*Column B*
______ 1. interneuron	a. anterior gray horn
______ 2. somatic motor neuron	b. lateral gray horn
______ 3. somatic sensory neuron	c. posterior gray horn
______ 4. sympathetic motor neuron	d. posterior root ganglion

3. The ______________________ (cervical/lumbar) enlargement of the spinal cord is associated with the upper limbs, whereas the ______________________ (cervical/lumbar) enlargement is associated with the lower limbs. ❸

Exercise 16.2: Gross Anatomy of the Spinal Cord

4. White matter is located ______________________ (centrally/peripherally) within the brain, and ______________________ (centrally/peripherally) within the spinal cord. ❹

5. Which of the following is a structure that consists of posterior and anterior roots extending from the inferior end of the spinal cord into the lower vertebral canal and the sacral canal? (Circle one.) ❺
   a. cauda equina
   b. conus medullaris
   c. filum terminale
   d. posterior root ganglion

Exercise 16.3: The Cervical Plexus

6. The cervical plexus is formed from the anterior rami of these spinal nerves. (Circle one.) ❻
   a. C1–C3
   b. C1–C4
   c. C1–C5
   d. C1–C6
   e. C1–C7

Exercise 16.4: The Brachial Plexus

7. Rank the levels of organization of the brachial plexus in order, beginning with anterior rami and ending with branches. ❼
   ______ a. branches
   ______ b. cords
   ______ c. divisions
   ______ d. roots (anterior rami)
   ______ e. trunks

8. The ulnar nerve forms from the ______________________ (medial/lateral) cord of the brachial plexus. ❽

9. Using colored pencils, color in the segments of the brachial plexus in the illustration as per the colors in the key. Then label the nerves indicated by the leader lines. 9

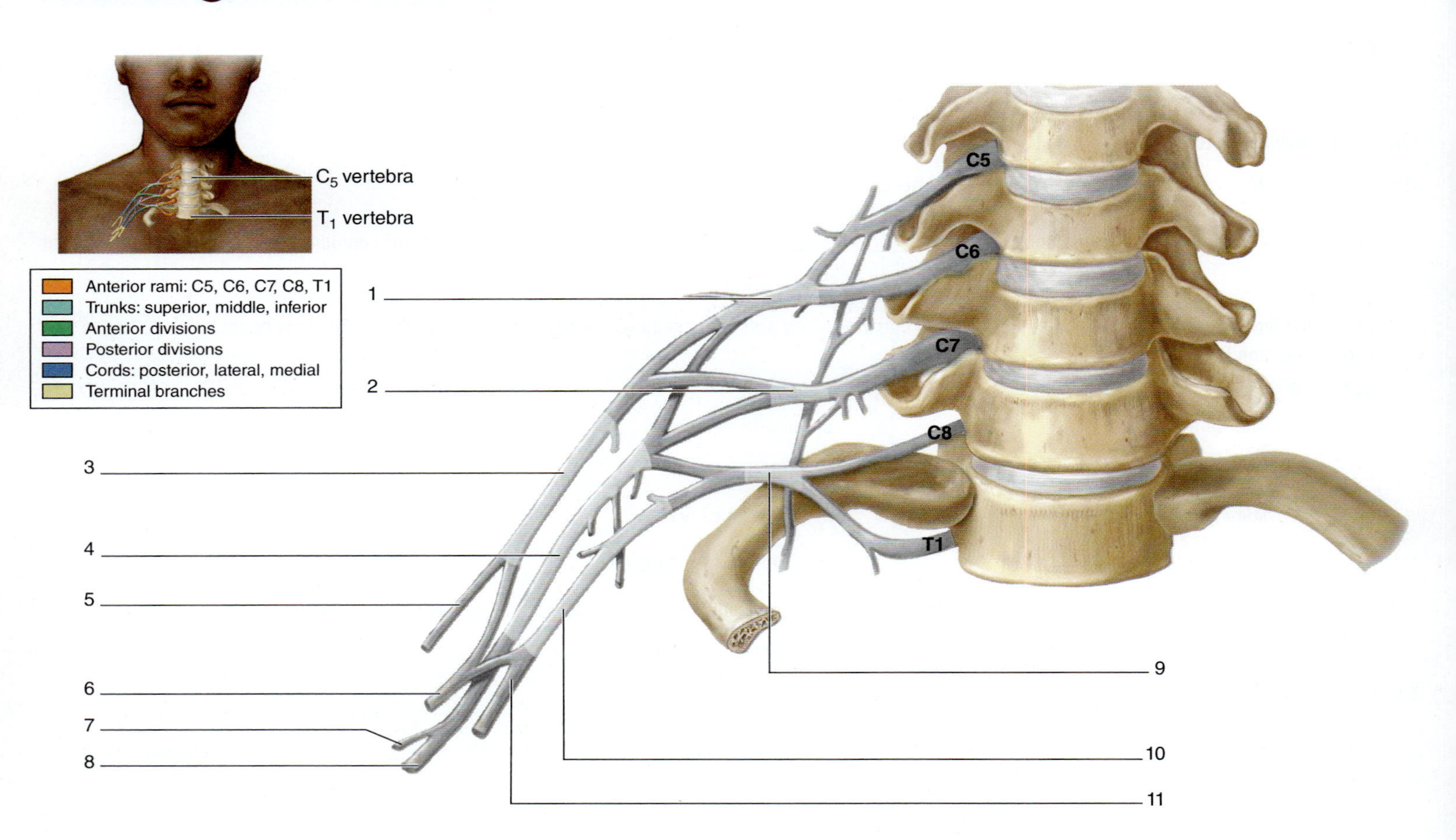

10. Match the region or compartment of the upper limb listed in column A with the nerve that innervates the majority of muscles in that region/compartment listed in column B. (Answer choices may be used more than once.) 9

Column A	***Column B***
_______ 1. anterior arm (e.g., biceps brachii)	a. axillary n.
_______ 2. anterior forearm (medial aspect and intrinsic muscles of the hand)	b. median n.
_______ 3. most of the anterior forearm (lateral aspect)	c. musculocutaneous n.
_______ 4. posterior arm (e.g., triceps brachii muscle)	d. radial n.
_______ 5. posterior forearm	e. ulnar n.
_______ 6. shoulder (e.g., deltoid muscle)	

Exercise 16.5: The Lumbar and Sacral Plexuses

11. Identify a nerve that arises from the lumbar plexus. ______________________ 10

12. The femoral nerve innervates the ______________________ (anterior/medial/posterior) thigh, whereas the obturator nerve innervates the ______________________ (anterior/medial/posterior) thigh. 11

13. Identify a nerve that arises from the sacral plexus. ______________________ 12

14. Match the region or compartment of the lower limb listed in column A with the nerve that innervates the majority of muscles in that region/compartment listed in column B. (13)

Column A	*Column B*
______ 1. anterior leg	a. deep fibular n.
______ 2. lateral leg	b. sciatic n.
______ 3. posterior leg	c. superficial fibular n.
______ 4. posterior thigh	d. tibial n.

Exercise 16.6: Identifying Components of a Reflex on a Classroom Model

15. Rank the components of a reflex arc in the correct order from stimulus to response. (14)

______ a. control center

______ b. effector organ

______ c. motor neuron

______ d. receptor

______ e. sensory neuron

Exercise 16.7: Patellar Reflex

16. The muscles that are rapidly stretched with a patellar ligament tap (testing the patellar reflex) compose the hamstrings muscle group.

______________________ (True/False) (15)

17. The patellar reflex tests spinal nerves in which of the following regions of the spinal cord? (Circle one.) (16)
    a. cervical
    b. lumbar
    c. sacral
    d. thoracic

Exercise 16.8: Withdrawal and Crossed-Extensor Reflex

18. Identify the type of sensory receptor that is stimulated when eliciting the withdrawal and crossed-extensor reflex. (Circle one.) (17)
    a. chemoreceptors
    b. mechanoreceptors
    c. nociceptors
    d. photoreceptors

19. When eliciting the withdrawal and crossed-extensor reflexes, flexion withdrawal occurs in the ______________________ (contralateral/ipsilateral) limb, whereas extension occurs in the ______________________ (contralateral/ipsilateral) limb. (18)

Exercise 16.9: Plantar Reflex

20. Which muscles tested in the plantar reflex compose the triceps surae group? (Check all that apply.) (19)

______ a. fibularis longus

______ b. gastrocnemius

______ c. plantaris

______ d. soleus

______ e. tibialis anterior

21. The plantar reflex tests spinal nerves in which of the following regions of the spinal cord? (Check all that apply.) (20)

______ a. cervical

______ b. lumbar

______ c. sacral

______ d. thoracic

Can You Apply What You've Learned?

22. Fill in the table below with features of the main nerve plexuses.

Plexus	Formed from Anterior Rami of These Spinal Nerves	Major Nerves Formed from the Plexus
Cervical		
Brachial		
Lumbosacral		

23. For each of the following, identify the nerve that innervates the compartment:

    Anterior compartment of the arm: ______________________

    Lateral compartment of the leg: ______________________

    Medial compartment of the thigh: ______________________

    Posterior compartment of the forearm: ______________________

    Posterior compartment of the thigh: ______________________

Can You Synthesize What You've Learned?

24. If a posterior *root* is severed, what loss of function will result?

    ____

    ____

    ____

25. As relates to spinal cord injuries, what is an advantage of having the phrenic nerve arise from the cervical plexus instead of arising from the thoracic part of the spinal cord (even though the throacic spinal cord is closer in physical location to the diaphragm)?

    ____

    ____

26. Susan, a 50-year-old female, is experiencing motor and sensory deficits in her limbs. Specifically, she exhibits hyperactive stretch reflexes and bilateral positive Babinski signs. Explain each of these results. Include in your explanation a description of how each can be tested in a clinical setting. Finally, discuss possible locations (i.e., spinal cord levels) of the injury.

    ____

    ____

    ____

    ____

    ____

    ____

    ____

CHAPTER 17

The Autonomic Nervous System

OUTLINE AND LEARNING OBJECTIVES

MODULE 7: NERVOUS SYSTEM

INTRODUCTION

Why is it that heart rate increases and sweat glands increase secretions during stressful or frightening conditions? How is it that we begin salivating at the sight, smell, or even thought of something tasty? The answer is the **autonomic nervous system (ANS)**. The human **nervous system** is divided into two major functional components: the somatic nervous system and the autonomic nervous system. The **somatic nervous system (SNS)** can be thought of as the "voluntary" division—it involves sensations that are usually conscious. This includes sensory input from the five special senses (touch, taste, hearing, vision, and smell) and proprioceptors within joints, tendons, and muscles. Motor output to skeletal (voluntary) muscle tissue is also considered part of the somatic nervous system. In contrast, the ANS can be thought of as the "involuntary" division—it involves sensory stimuli that we are usually unaware of (such as blood pressure or oxygen levels in the blood) and motor output to cardiac muscle, smooth muscle, and glands. The ANS is composed of two divisions: the parasympathetic division, which controls "rest-and-digest" activities, and the sympathetic division, which controls the "fight-or-flight" response.

This laboratory session begins by reviewing the histological and gross anatomical organization of the parasympathetic and sympathetic divisions of the ANS. Exercises that are designed to test the function of specific autonomic reflexes will follow. Use knowledge from the previous chapters on the nervous system when completing these relatively condensed but very important laboratory exercises.

Reference Table

Chapter 17: The Autonomic Nervous System

Name: ______________________

Date: __________ Section: __________

These Pre-Laboratory Worksheet questions may be assigned by instructors through their connect course.

PRE-LABORATORY WORKSHEET

1. The ________________ (parasympathetic/sympathetic) division of the autonomic nervous system (ANS) controls "rest-and-digest" activities, whereas the ________________ (parasympathetic/sympathetic) division of the ANS controls the "fight-or-flight" response.

2. The parasympathetic division of the ANS is always inhibitory to cardiac muscle, smooth muscle, and glands. ________________ (True/False)

3. Which of the following statement(s) is/are accurate about the sympathetic division of the ANS? (Check all that apply.)

   _____ a. the adrenal medulla acts as a modified ganglion

   _____ b. ganglia are located at or within target organs

   _____ c. preganglionic neuron cell bodies are located in the brainstem and sacral region of the spinal cord

   _____ d. preganglionic neuron cell bodies are located in the lateral horns of the thoracic and lumbar regions of the spinal cord

   _____ e. preganglionic axons are short, whereas postganglionic axons are long

4. The division of the ANS also known as the *craniosacral* division is the ________________ (parasympathetic/sympathetic) division. The division of the ANS also known as the *thoracolumbar* division is the ________________ (parasympathetic/sympathetic) division.

5. Which of the following statements describes an action performed by the parasympathetic division of the ANS? (Check all that apply.)

   _____ a. increased gastrointestinal (GI) motility

   _____ b. increased heart rate

   _____ c. increased salivation

   _____ d. pupillary constriction

6. Match the feature listed in column A with the corresponding division of the ANS listed in column B. (Answers will be used more than once.)

Column A	***Column B***
_____ 1. cell bodies of preganglionic neurons located in the lateral horns of the T1–L2 segments of the spinal cord	a. parasympathetic
_____ 2. extensive divergence of axons	b. sympathetic
_____ 3. long preganglionic axons	
_____ 4. paravertebral ganglia located on either side of vertebral column	
_____ 5. preganglionic neurons located in the brainstem and lateral gray matter of the S2–S4 segments of the spinal cord	
_____ 6. prevertebral ganglia located anterior to vertebral column and descending aorta	
_____ 7. terminal ganglia located close to target organs	

7. Activation of the sympathetic division of the ANS will cause the pupils of the eye to ________________ (constrict/dilate).

8. Identify the structure(s) that is/are innervated by autonomic motor neurons. (Check all that apply.)

   _____ a. cardiac muscle

   _____ b. glands

   _____ c. skeletal muscle

   _____ d. smooth muscle

GROSS ANATOMY

Autonomic Nervous System

The autonomic nervous system regulates cardiac muscle, smooth muscle, and glands. Because of the involuntary nature of the ANS, most visceral effectors are dually innervated by the ANS. This means that effector organs (e.g., heart, gastrointestinal tract) receive innervation from both divisions of the ANS. These two divisions are the **parasympathetic** (craniosacral) **division,** which functions to control "rest-and-digest" functions, and the **sympathetic** (thoracolumbar) **division,** which is activated in emergency situations and during the "fight-or-flight" response **(figure 17.1).**

The term "parasympathetic" refers to the fact that the nerve fibers of this division originate from the CNS adjacent to where the nerve fibers of the sympathetic division originate (*para*, alongside). Recall from chapter 16 that cell bodies of autonomic (sympathetic) motor neurons are located in the lateral horns of the thoracic and lumbar regions of the spinal cord. The cell bodies of the parasympathetic division arise from portions of the CNS that are superior and inferior to the regions that give rise to the sympathetic division. Specifically, cell bodies of neurons that compose the parasympathetic division arise from the brain (with axons that travel within cranial nerves III, VII, IX, and X) and also from lateral horns within the sacral region of the spinal cord. This is the origin of the alternate terms *craniosacral* for the parasympathetic division and *thoracolumbar* for the sympathetic division. In addition to the difference in the locations of preganglionic neuron cell bodies, there is a difference in the location of the ganglia for each of the divisions. Parasympathetic ganglia are located either close to the effector (these are called **terminal ganglia**), or within the wall of the target organs (these are called **intramural ganglia**) **(table 17.1).** Sympathetic ganglia are located in either the **sympathetic trunk ganglia** (also called paravertebral ganglia) or the **prevertebral ganglia**. The following exercise examines the gross anatomy of each division of the ANS, and involves noting locations of preganglionic and postganglionic neurons and tracing the pathway of motor output to target organs of the ANS throughout the body.

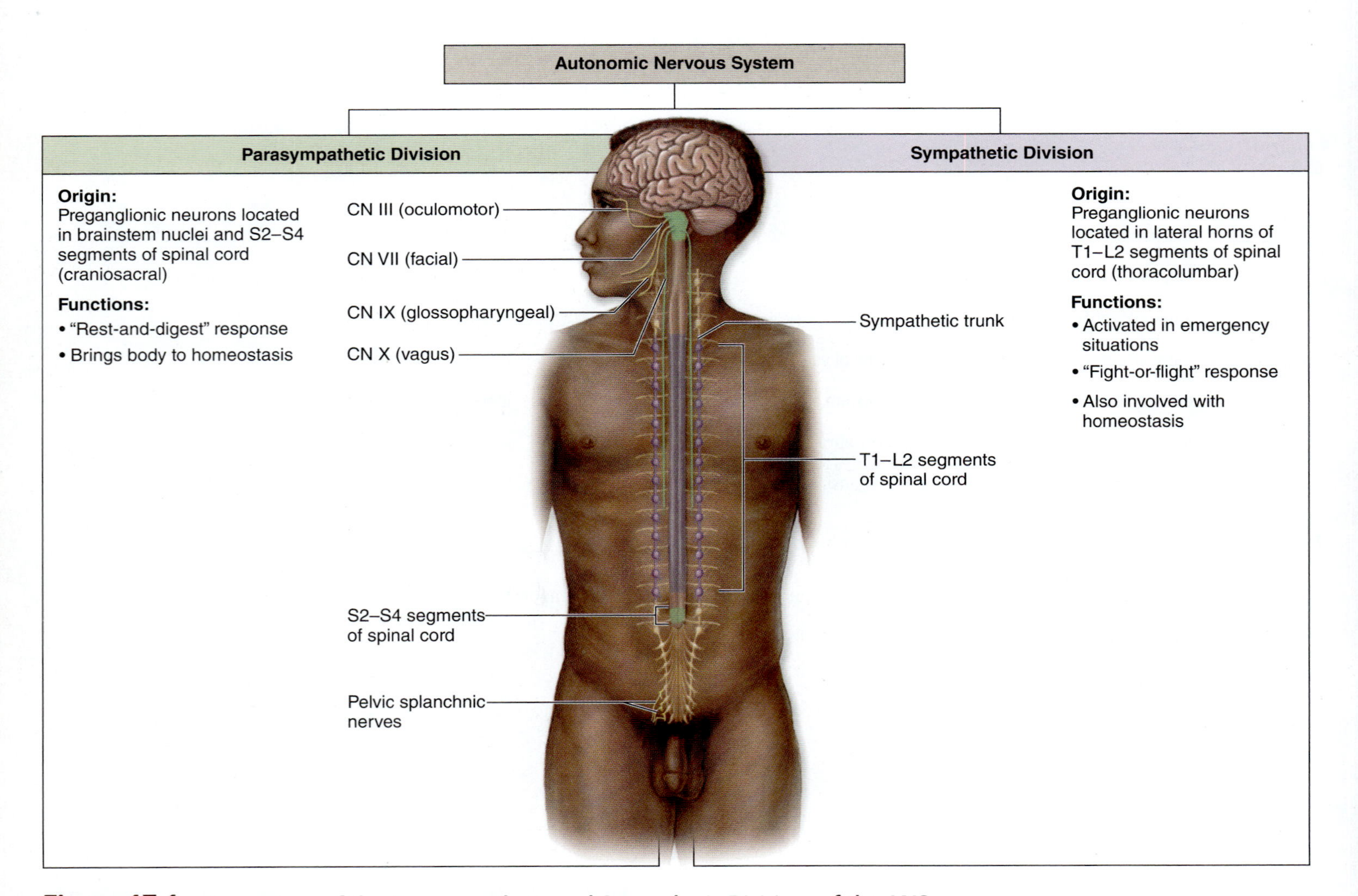

Figure 17.1 Comparison of the Parasympathetic and Sympathetic Divisions of the ANS.

INTEGRATE

LEARNING STRATEGY

When learning the two divisions of the ANS, remember "rest-and-digest" for the parasympathetic division and "fight-or-flight" for the sympathetic division. Based on these descriptions, many assume that the parasympathetic division is always inhibitory and that the sympathetic division is always excitatory. However, that is not the case. For example, while the parasympathetic division is inhibitory to pacemaker cells of the heart, causing heart rate to decrease, it is stimulatory to smooth muscle in the digestive tract, promoting digestion and motility of the GI tract. Likewise, while the sympathetic division is stimulatory to pacemaker cells of the heart, causing heart rate to increase, it is inhibitory to smooth muscle surrounding the bronchioles (small airways), which causes dilation of the airways. Rather than simply associating "inhibition" or "excitation" with each division, think about what makes sense in light of the type of response being elicted and remember that the actions of autonomic nerves can be inhibitory or excitatory for either division of the ANS.

Table 17.1 Comparison of Parasympathetic and Sympathetic Divisions

Feature	Parasympathetic Division	Sympathetic Division
Divergence of Axons	Few (1 axon innervates < 4 ganglionic cell bodies)	Extensive (1 axon innervates > 20 ganglionic cell bodies)
Function	Conserves energy and replenishes energy stores; maintains homeostasis; "rest-and-digest" division	Prepares body to cope with emergencies and intensive muscle activity; "fight-or-flight" division
Length of Postganglionic Axon	Short	Long
Length of Preganglionic Axon	Long	Short
Location of Ganglia	Terminal ganglia located close to the target organ; intramural ganglia located within wall of target organ	Sympathetic trunk (paravertebral) ganglia located on either side of vertebral column; prevertebral (collateral) ganglia located anterior to vertebral column and descending aorta
Location of Ganglionic Neuron Cell Bodies	Terminal or intramural ganglion	Sympathetic trunk ganglion (paravertebral) or prevertebral ganglion
Location of Preganglionic Neuron Cell Bodies	Brainstem and lateral gray matter in S2–S4 segments of spinal cord	Lateral horns in T1–L2 segments of spinal cord
Rami Communicantes	None	White rami attach to T1–L2 spinal nerves; gray rami attach to all spinal nerves

EXERCISE 17.1

PARASYMPATHETIC DIVISION

1. Observe **figure 17.2,** which is an overview of the parasympathetic pathways. Identify the structures listed in figure 17.2, using the textbook as a guide.
2. In the spaces provided, list the structures that are innervated by parasympathetic fibers.
   - **a.** Structures innervated by parasympathetic fibers within cranial nerves:

     CN III (oculomotor) ____________________

     CN VII (facial) ____________________

     CN IX (glossopharyngeal) ____________________

     CN IX (vagus) ____________________

     ____

     ____
   - **b.** Structures innervated by parasympathetic fibers originating in the sacral region of the spinal cord:

     ____

     ____

(continued on next page)

(continued from previous page)

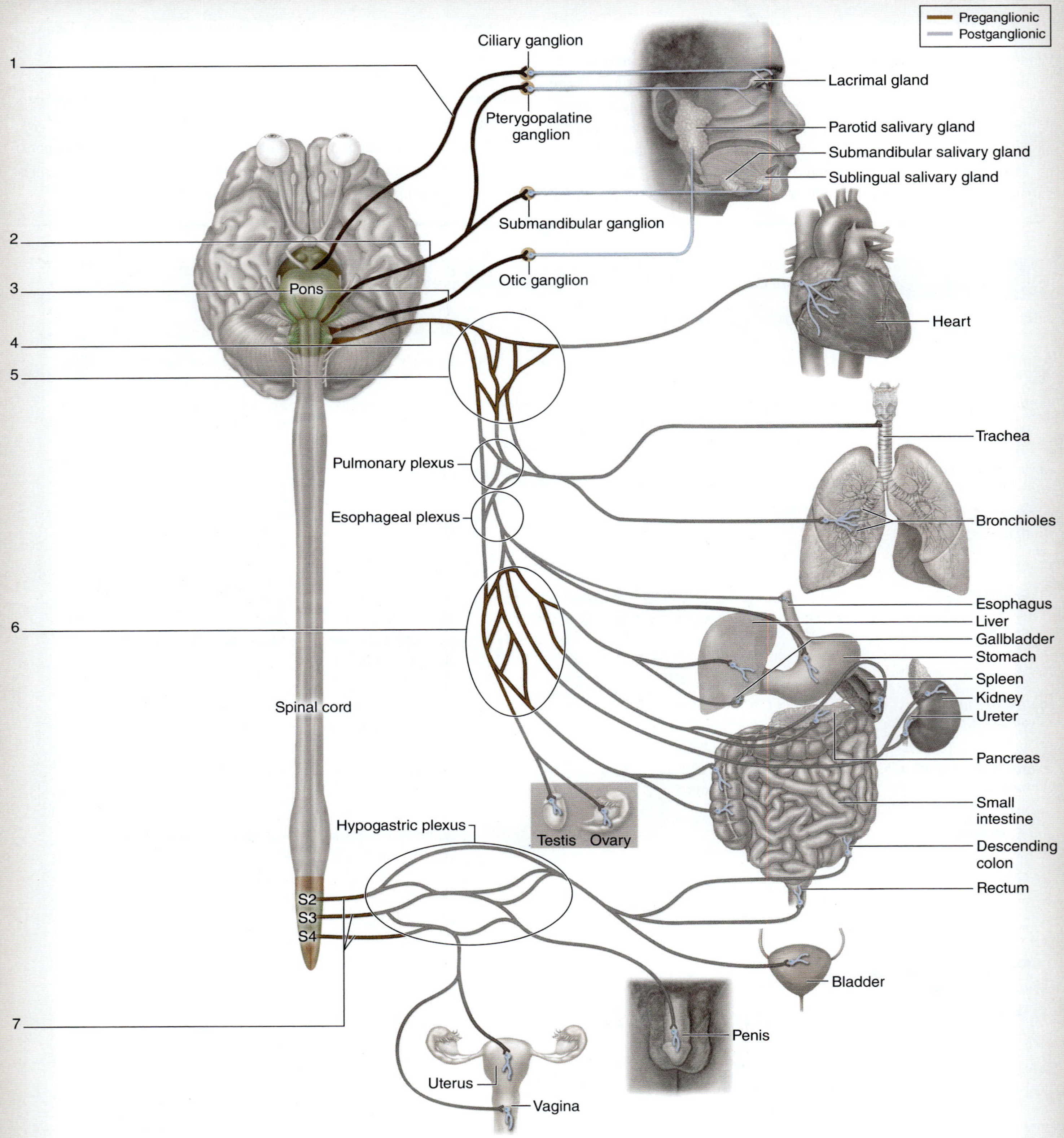

Figure 17.2 Overview of the Parasympathetic Division of the ANS. The parasympathetic division of the ANS controls functions involved with "resting and digesting." Use the terms listed to fill in the numbered labels in the figure.

- ☐ abdominal aortic plexus
- ☐ cardiac plexus
- ☐ facial nerve (CN VII)
- ☐ glossopharyngeal nerve (CN IX)
- ☐ oculomotor nerve (CN III)
- ☐ pelvic splanchnic nerves
- ☐ vagus nerve (CN X)

INTEGRATE

LEARNING STRATEGY

A "big picture" way of remembering where parasympathetic innervation of visceral organs arises is to learn a couple of basic rules:

1. Specialized structures within the head (e.g., salivary glands) receive parasympathetic innervation through cranial nerves III, VII, IX, and X.
2. Nearly all thoracic and abdominal viscera receive parasympathetic innervation via the vagus nerve (CN X). Recall from chapter 15 that vagus means "wanderer" (because it wanders throughout the body). The vagus nerve is the only cranial nerve to innervate structures below the head.
3. Viscera within the pelvis are innervated by fibers that arise from the sacral spinal cord (e.g., structures of both the reproductive and urinary systems).

Clinically, this means that someone who has been paralyzed from the neck down will still have parasympathetic innervation of the majority of the vital viscera (e.g., heart, small intestine). However, the individual will lose function of involuntary sphincters such as those that control the flow of urine from the urinary bladder.

EXERCISE 17.2

SYMPATHETIC DIVISION

The pathways for sympathetic fibers are more complex than those for parasympathetic fibers. One reason for this is that in order for the sympathetic response to have widespread and immediate effects, it must activate organs throughout the entire body in unison to initiate the "fight-or-flight" response. There are three separate pathways for preganglionic fibers to reach these ganglia from the spinal cord, with additional nerve fibers that extend to the **adrenal medulla.** This exercise involves tracing these pathways and considering the response of each organ that is innervated by each pathway.

1. Identify the structures listed in **figure 17.3,** using the textbook as a guide.
2. Use table 17.1 as a guide and think about the "fight-or-flight" response initiated by the sympathetic division. Then, for each of the following organs, first identify the pathway for nerve signals to reach the organ, then name the specific effect created by the nervous stimulation.

Heart (both rate and strength of contraction):

Pathway: ______________________________

Response: ______________________________

Small intestine:

Pathway: ______________________________

Response: ______________________________

Blood vessels within skeletal muscle:

Pathway: ______________________________

Response: ______________________________

Adrenal medulla:

Pathway: ______________________________

Response: ______________________________

(continued on next page)

(continued from previous page)

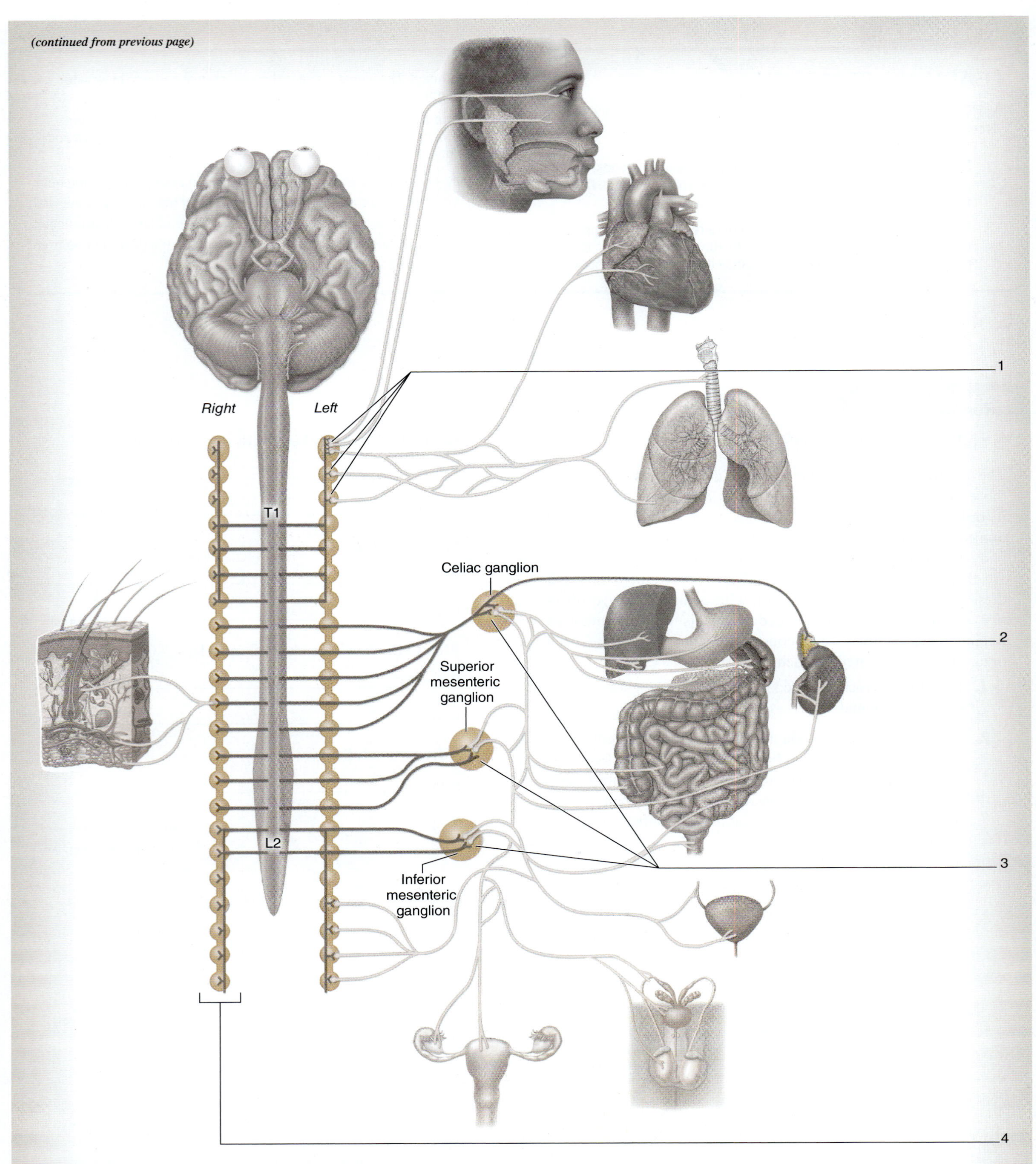

Figure 17.3 Overview of the Sympathetic Division of the ANS. The sympathetic division of the ANS controls the "fight-or-flight" response. Use the terms listed to fill in the numbered labels in the figure.

- ☐ adrenal medulla
- ☐ prevertebral ganglia
- ☐ sympathetic trunk
- ☐ sympathetic trunk ganglia (paravertebral)

INTEGRATE

CLINICAL VIEW
Pheochromocytoma

Pheochromocytoma (*phaios*, dark; *chroma*, color; *kytos*, cell; *oma*, tumor) is a tumor that develops in the adrenal medulla, specifically in chromaffin cells. Chromaffin cells are responsible for synthesizing and releasing epinephrine and norepinephrine into the systemic circulation. Recall that the adrenal medulla receives input from preganglionic neurons of the sympathetic division of the ANS. Therefore, the adrenal medulla acts as a modified sympathetic ganglion. As expected, pheochromocytomas result in increased blood levels of epinephrine and norepinephrine. Symptoms mimic those of overstimulation of the sympathetic division of the ANS. For example, patients may exhibit increases in heart rate, blood pressure, sweat production, and blood glucose levels. Patients also experience weight loss.

Pheochromocytomas are sometimes inherited, or may be the result of metastases from primary tumors that originated in other endocrine glands, such as the parathyroid or thyroid glands. Diagnosis is typically confirmed by noting elevated levels of epinephrine, norepinephrine, and their metabolites in blood and urine. In addition, magnetic resonance imaging (MRI) or computed tomography (CT) imaging may be done to confirm the presence of a tumor within the adrenal medulla. Surgical removal of the tumor is the typical treatment for patients suffering from pheochromocytoma.

INTEGRATE

CONCEPT CONNECTION

Recall that somatic motor neurons as well as ANS motor neurons (sympathetic division) are located in the gray matter of the spinal cord. Specifically, somatic motor neurons are located in the anterior (ventral) horn of the spinal cord gray matter, whereas ANS motor neurons are located in the lateral horn of the spinal cord gray matter. Whether projecting to voluntary muscles (somatic) or involuntary muscles and/or glands (autonomic), these motor neurons have axons that exit the spinal cord via the same route. Since both somatic and autonomic neurons are providing motor output, through which root (anterior or posterior) will the fibers travel out of the spinal cord? ______________________

PHYSIOLOGY

Autonomic Reflexes

Autonomic reflexes, like somatic reflexes, involve both sensory input and motor output. However, because these reflexes are *autonomic*, the motor output is to smooth muscle, cardiac muscle, or glands. Thus, the reflex occurs below the level of conscious awareness. Autonomic reflexes can involve either the spinal cord (spinal reflexes) or the brain (cranial reflexes) as the control center. The reflexes observed in this exercise are examples of cranial reflexes. **Cranial reflexes** involve transmission of nerve signals along cranial nerves. They are important in clinical tests to determine the function of specific cranial nerves or regions of the brain where the appropriate cranial nerves arise.

EXERCISE 17.3

PUPILLARY REFLEXES

The pupillary reflex is an example of a protective reflex. The **pupillary reflex** involves photoreceptors located within the retina, which detect light entering the eye. Sensory input (afferent signal) is relayed along the *optic nerve* (CN II) to nuclei within the brain (control center). Motor output (efferent signal) is then relayed from the brain along the *oculomotor nerve* (CN III) to the pupillary sphincter muscle within the iris of the eye (effector). When a bright light is shone into the eye, the normal pupillary response is constriction of the pupil. This is a protective function because too much light entering the eye can damage the retina. Testing for the pupillary reflex is important when brain trauma or other brain damage (e.g., a stroke) is suspected. For example, increased intracranial pressure can crush the cranial nerves that transmit either afferent or efferent signals. If this happens, the pupils will remain dilated when a light is shone into the eye. In addition, the size of the pupil is controlled

(continued on next page)

(continued from previous page)

by both divisions of the autonomic nervous system. Activation of the sympathetic division of the ANS results in dilation of the pupil, so as to increase visual acuity. Conversely, activation of the parasympathetic division of the ANS results in the constriction of the pupil.

1. Obtain a flashlight and a small ruler, and choose a study subject.
2. Measure the size of the subject's pupil using the ruler. DO NOT touch the ruler to the subject's eyes. Simply hold it up in front of each eye and get as close as possible to a precise measurement. Record the initial pupil diameters:

   Right pupil diameter: __________ mm   Left pupil diameter: __________ mm

3. Shielding one eye (by holding a hand vertically between the subject's eyes), shine a flashlight into the other eye **(figure 17.4).**
4. Measure the resulting pupil diameters:

   Right pupil diameter: __________ mm   Left pupil diameter: __________ mm

Figure 17.4 Testing the Pupillary Reflex. Have the subject shield one eye by holding his or her hand in front of the nose as shown. Shine a light in one eye and note the change in pupil diameter in *both* eyes.

EXERCISE 17.4

BIOPAC GALVANIC SKIN RESPONSE

The **galvanic skin response (GSR),** also known as the *electrodermal response*, is a response of the skin to arousing stimuli. That is, the skin is known to conduct electricity better when a person is having a "sympathetic response." This exercise involves observing the changes in the GSR during varying emotional states. The GSR is altered due to changes in sympathetic nervous system stimulation, particularly to sweat glands and cutaneous blood vessels. That is, increasing sympathetic output increases sweat production and dilates cutaneous blood vessels, resulting in increased blood flow to the skin. This is the underlying physiology behind a polygraph test, which is a test used to detect whether or not a subject is lying or telling the truth. Typically, lying increases sympathetic output, which alters the galvanic skin response. A true polygraph test also assesses respiratory and heart rates to determine whether or not a subject is telling the truth. The following BIOPAC exercise measures GSR, heart rate, and respiratory rate (breaths per minute, BMP) during a series of tasks.

Obtain the Following:

- **BIOPAC electrode lead set (SS2L)**
- **BIOPAC EDA Setup–EDA Lead (SS57L) and 2 gelled electrodes (EL507)**
- **BIOPAC respiratory transducer (SS5LB)**
- **Colored paper (9 colors: white, black, red, blue, green, yellow, orange, brown, purple)**
- **BIOPAC general purpose electrodes (EL503)**

Before beginning the experiment, become familiar with the following concept (use the textbook as a reference):

- **Ohm's law**

1. Prepare the BIOPAC equipment by turning the computer ON and the MP36/35 unit OFF. Plug in the following: respiration transducer (SS5LB) (CH 1), electrode lead set (SS2L) (CH 2), and EDA (SS3LA) (CH 3). Turn the MP36/35 unit ON.
2. Place the respiratory transducer around the subject's chest. The band should fit firmly around the chest, below the subject's armpits and above the subject's nipples. The band can be placed directly over the subject's skin or over thin clothing (e.g., a thin t-shirt) **(figure 17.5*a*).**
3. For best recording results, it is helpful if the subject's hands are slightly sweaty before beginning calibration or data collection. Fill the two electrode (EL507) cavities with electrode gel to enhance conduction between the skin and the electrodes. Connect the gelled electrodes to the subject's index and middle finger. Attach the leads (SS57L) to the electrodes and wrap the Velcro tape around each finger. The contact should be firm, but not so tight that it cuts off circulation. Wait at least five minutes prior to recording. Be sure that the transducer is placed on the

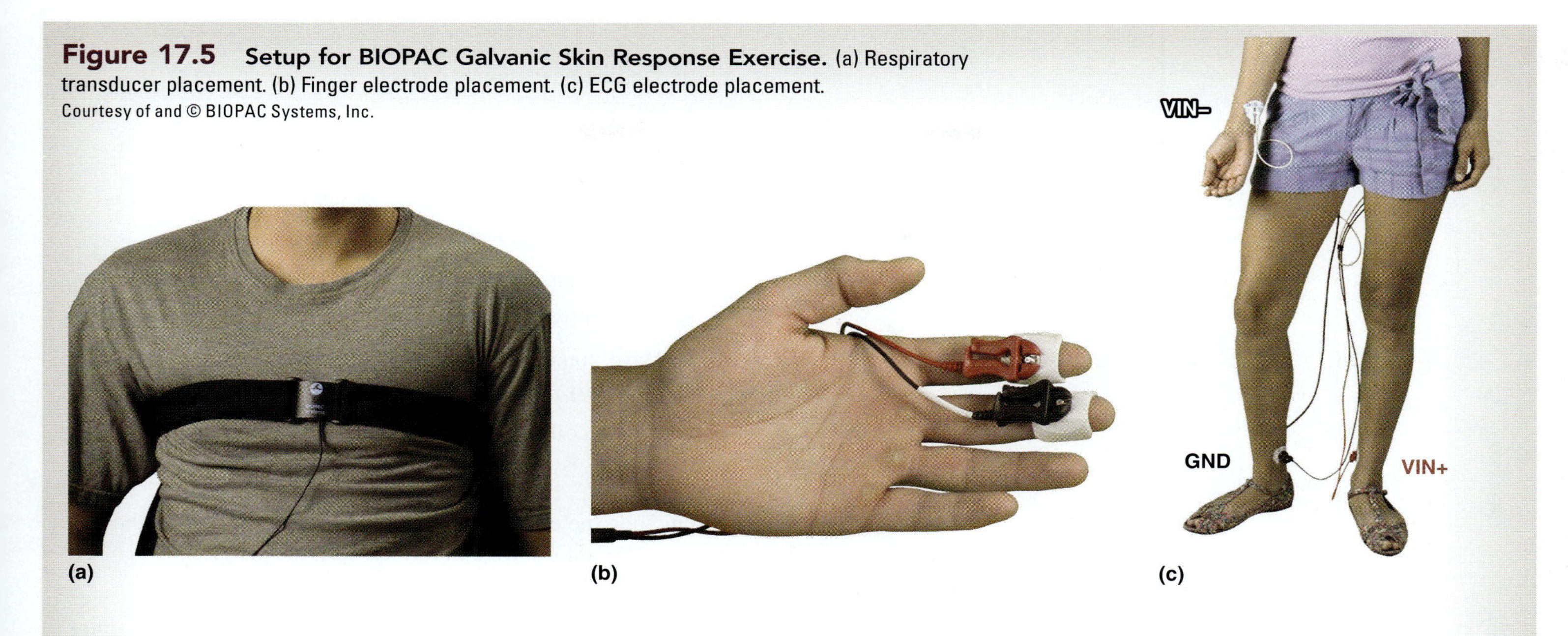

Figure 17.5 Setup for BIOPAC Galvanic Skin Response Exercise. (a) Respiratory transducer placement. (b) Finger electrode placement. (c) ECG electrode placement.
Courtesy of and © BIOPAC Systems, Inc.

skin beneath the fingertip rather than under the fingernail (figure 17.5*b*).

4. Prepare the subject's skin for electrode placement by abrading with an alcohol pad on the electrode sites (figure 17.5*c*). Apply a small quantity of electrode gel to the center of the surface ECG electrodes. Place the electrodes firmly on the subject's skin in the designated locations. Attach the leads (SS2L) to each electrode according to the color code: right anterior forearm above the wrist (white), medial surface of the right leg just above the ankle (black), and medial surface of the left leg just above the ankle (red).

5. Start the BIOPAC Student Lab Program. Select "L09 EDA & Polygraph" from the drop-down menu, enter the file name, and click "OK."

6. When beginning data collection, have the subject assume a relaxed position in a chair and make sure the subject cannot see the computer screen. This is important, because visual feedback may skew results. The room must also be quiet, and the subject should remain as quiet as possible during data collection. One lab member will serve as the **director,** asking questions and performing specific tasks, and one lab member will serve as the **recorder.**

7. Click "Calibrate" in the upper-left corner of the Setup window. Have the subject breathe deeply in and out one time, and then return to normal breathing. The calibration will stop automatically after 10 seconds of recording. Once the calibration has stopped, check the data **(figure 17.6).** The recording should show some fluctuations, which correspond to the deep breathing. If no fluctuations are present, click "Redo Calibration."

8. OBTAIN DATA: The director will instruct the subject to perform a series of tasks. Prior to each task, the recorder will mark the event on the trace. The director will ensure that the data trace returns to baseline conditions before having the subject initiate another task. Click "Record" and begin the following tasks 5 seconds after recording begins:

   a. Director: Ask the subject to say his/her name quietly; Recorder: press F2 and wait 5 seconds.

   b. Director: Ask the subject to count backward from 10; Recorder: Press F3 and wait 5 seconds.

   c. Director: Ask the subject to count backward from 30 subtracting increasing odd numbers;
   Recorder: Press F4 and wait 5 seconds.

   d. Director: Touch the side of the subject's face;
   Recorder: Press F5 and wait 5 seconds.

   Once data collection is complete, the recorder will click on "Suspend." At this time, review the data to ensure that the traces resemble **figure 17.7*a*** If not, click "Redo." If so, click, "Continue."

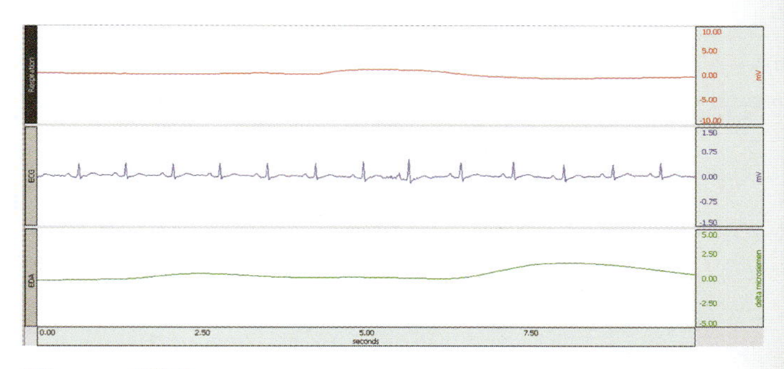

Figure 17.6 Sample Calibration Data for BIOPAC Galvanic Skin Response Exercise. Courtesy of and © BIOPAC Systems, Inc.

9. The director will then resume data collection for another task: showing the subject different colored papers. The recorder will mark the trace (F9) with event markers.

(continued on next page)

(continued from previous page)

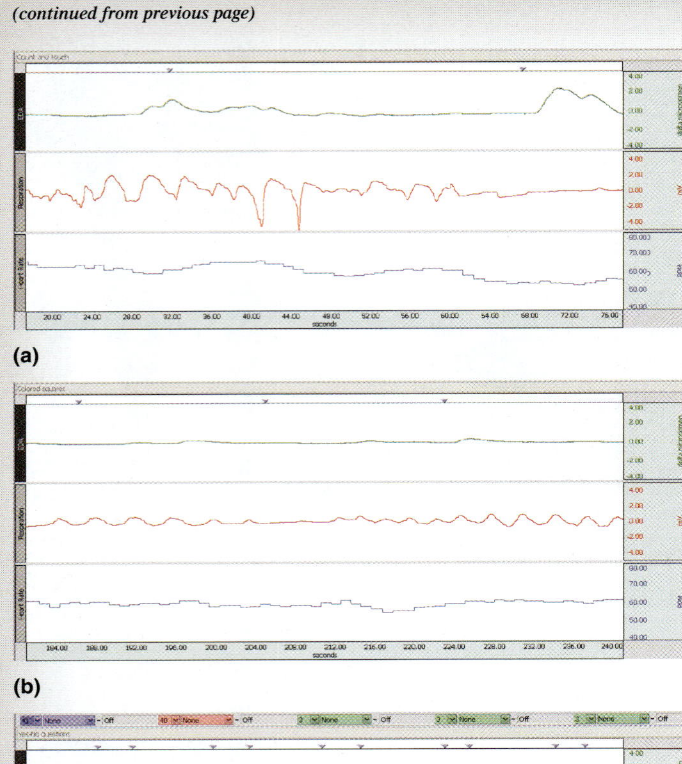

(a)

(b)

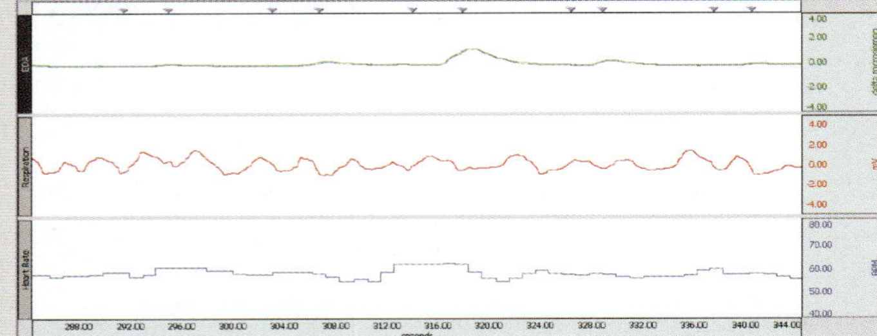

(c)

Figure 17.7 Sample Data for Galvanic Skin Response Exercise. (a) Count and touch. (b) Colored squares. (c) Yes-No.

Nine colors are shown to the subject, in this order: white, black, red, blue, green, yellow, orange, brown, and pink. Once data collection is complete, the recorder will click on "Suspend." Review the data to ensure that the traces resemble figure 17.7*b* If not, click "Redo." If so, click, "Continue."

10. The director will ask the subject a series of ten "yes" or "no" questions. The recorder will insert markers (F6 when question is asked; F7 if answered "Yes"; F8 if answered "No") at the end of each of the following questions:

a. Are you currently a student?

b. Are your eyes blue?

c. Do you have any brothers?

d. Did you earn an 'A' on the last physiology exam?

e. Do you drive a motorcycle?

f. Are you less than 25 years of age?

g. Have you ever traveled to another planet?

h. Have aliens from another planet visited you?

i. Do you like dogs?

j. Have you answered all of the preceding questions truthfully?

Once data collection is complete, the recorder will click on "Suspend." Review the data to ensure that traces resemble figure 17.7*c*. If not, click "Redo." If so, click "Done" to conclude the experiment. Be sure to save the data to the computer's hard drive before closing the program.

11. ANALYZE THE DATA: If analyzing data that was just collected, click on "Analyze Current Data." If opening data that was collected previously, click on "Review Saved Data." To prepare to analyze the data, do the following:

a. Use the zoom tool to select the first five seconds of data.

b. Click "Display" menu at the top of the screen and select "Autoscale Waveforms."

c. Set up the measurement boxes at the top of the screen using the drop-down menus. Set "CH 41" to "Heart Rate." Set "CH 40" to "BPM." Set "CH 3" to "EDA."

12. Locate the "I-beam" tool. Use the activated I-beam tool to select the baseline EDA value and corresponding heart rate (at two seconds). Record this value in **table 17.2**. Then, select the portion of the respiration trace that corresponds to one respiratory cycle, from one inhalation to the next inhalation. Record this baseline respiratory rate (BPM) in table 17.2.

13. Scroll through all of the segment 1 data, repeating step 12 for each subsequent task. For each task, there should be a corresponding heart rate, respiratory rate, and EDA. Record these values in table 17.2.

14. Repeat steps 12 and 13 for the second and third segments of data, which correspond to the colored paper task and the series of "yes" or "no" questions, respectively. Record values for heart rate, respiratory rate, and EDA for each condition in **tables 17.3** and **17.4.** Save or print the data and exit the program.

15. Make note of any pertinent observations here: ____________

____

____

____

Table 17.2 Segment 1 Data

Procedure	Heart Rate (CH 41 Value)	Respiratory Rate (CH 40 BPM)	EDA (CH 3 Value)
Resting (baseline)			
Quietly say name			
Count from 10			
Count from 30			
Face touched			

Table 17.3 Segment 2 Data

Color	Heart Rate (CH 41 Value)	Respiratory Rate (CH 40 BPM)	EDA (CH 3 Value)
White			
Black			
Red			
Blue			
Green			
Yellow			
Orange			
Brown			
Pink			

Table 17.4 Segment 3 Data

Question	Answer	Truth	Heart Rate (CH 41 Value)	Respiratory Rate (CH 40 BPM)	EDA (CH 3 Value)
Student?	Y N	Y N			
Blue eyes?	Y N	Y N			
Brothers?	Y N	Y N			
Earn "A"?	Y N	Y N			
Motorcycle?	Y N	Y N			
Less than 25?	Y N	Y N			
Another planet?	Y N	Y N			
Aliens visit?	Y N	Y N			
Like dogs?	Y N	Y N			
Truthful?	Y N	Y N			

Chapter 17: The Autonomic Nervous System

Name: ______________________

Date: __________ Section: __________

POST-LABORATORY WORKSHEET

The 1 corresponds to the Learning Objective(s) listed in the chapter opener outline.

Do You Know the Basics?

Exercise 17.1: Parasympathetic Division

1. The parasympathetic division of the ANS is also called the ______________ (craniosacral/thoracolumbar) division because of its anatomical location. 1

2. Identify the cranial nerves that carry parasympathetic information. (Check all that apply.) 1

   _____ a. facial (CN VII)

   _____ b. glossopharyngeal (CN IX)

   _____ c. oculomotor (CN III)

   _____ d. optic (CN II)

   _____ e. vagus (CN X)

Exercise 17.2: Sympathetic Division

3. The sympathetic division of the ANS is also called the ______________ (craniosacral/thoracolumbar) division because of its anatomical location. 2

4. The major sympathetic ganglia that lie on top of the unpaired abdominal blood vessels (celiac trunk, superior mesenteric artery, inferior mesenteric artery) are also known as ______________ (paravertebral/prevertebral) ganglia because of their anatomic location relative to the vertebral column. 2

Exercise 17.3: Pupillary Reflexes

5. The optic nerve (CN II) carries ______________ (afferent/efferent) input of the pupillary reflex, whereas the oculomotor nerve (CN III) carries ______________ (afferent/efferent) output of the pupillary reflex. 3

6. The function of the pupillary reflex is to protect the retina from UV radiation damage. ______________ (True/False) 3

Exercise 17.4: BIOPAC Galvanic Skin Response

7. When a subject is not telling the truth, the GSR, heart rate, and BPM should all ______________ (increase/decrease). 4

8. The division of the ANS responsible for these corresponding changes in GSR, heart rate, and BPM is the ______________ (parasympathetic/sympathetic) division. 5

Can You Apply What You've Learned?

9. Match each of the effectors listed in column A with the cranial nerve that innervates the effector listed in column B.

Column A	*Column B*
_____ 1. bronchioles	a. facial (CN VII)
_____ 2. ciliary muscles and iris of eye	b. glossopharyngeal (CN IX)
_____ 3. heart	c. oculomotor (CN III)
_____ 4. kidneys	d. vagus (CN X)
_____ 5. lacrimal gland	
_____ 6. salivary gland (parotid)	
_____ 7. salivary glands (under tongue)	
_____ 8. stomach and other digestive organs	

Can You Synthesize What You've Learned?

10. Patients experiencing hypertension (high blood pressure) may be prescribed drugs called beta-blockers. These drugs bind to receptors on cardiac muscle cells and block the action of the sympathetic nerves to the heart. Describe how administration of a beta-blocker will influence heart rate and strength of contraction.

11. A visit to the eye doctor often involves "dilation" of the pupils. Doctors will place drops of a drug called *tropicamide* in each eye and wait a period of time for the drops to take effect. The drug binds to receptors on the pupillary sphincter muscles to block action of the parasympathetic nerves to these muscles. Describe why application of these eye drops leads to increased sensitivity to light.

12. John, a 28-year-old male, enters the emergency room after experiencing a severe blow to the head while playing football. Doctors test pupillary reflexes in both eyes and they are abnormal. Describe these results, including how the pupillary reflex can be tested in a clinical setting. Finally, discuss possible locations of the injury.

13. When a person is accused of a crime, he/she may be subjected to a polygraph test. Describe the three physiological measurements used in a polygraph test, and describe how each measurement should change when a person is not telling the truth.

CHAPTER 18

General and Special Senses

OUTLINE AND LEARNING OBJECTIVES

HISTOLOGY 462

General Senses 462

EXERCISE 18.1: TACTILE (MEISSNER) CORPUSCLES 462

1. Identify dermal papillae and tactile corpuscles when viewed with a microscope
2. Describe the structure and function of tactile corpuscles

EXERCISE 18.2: LAMELLATED (PACINIAN) CORPUSCLES 463

3. Identify the reticular layer of the dermis and lamellated corpuscles when viewed with a microscope
4. Describe the structure and function of lamellated corpuscles

Special Senses 464

EXERCISE 18.3: GUSTATION (TASTE) 464

5. Identify different types of tongue papillae, and identify taste buds when viewed with the microscope
6. Describe the structure and function of taste buds

EXERCISE 18.4: OLFACTION (SMELL) 466

7. Identify olfactory epithelium when viewed with a microscope
8. Identify the general location and describe the function of basal cells, olfactory receptor cells, and supporting cells within olfactory epithelium

EXERCISE 18.5: VISION (THE RETINA) 467

9. Identify the retina when viewed with a microscope
10. Identify the rod and cone layer, bipolar cell layer, and ganglion cell layer in a histology slide of the retina
11. Describe the functional importance of the fovea centralis and the optic disc, and describe how the histology of the retina is modified in these areas to match those functions

EXERCISE 18.6: HEARING 470

12. Identify the cochlea when viewed with a microscope
13. Identify the scala vestibuli, scala media, and scala tympani of the cochlea when viewed with a microscope
14. Identify the histological structure of the spiral organ when viewed with a microscope
15. Describe how cells within the spiral organ function to transmit the special sense of hearing to the brain

GROSS ANATOMY 472

General Senses 472

EXERCISE 18.7: SENSORY RECEPTORS IN THE SKIN 473

16. Describe the general location and function of tactile discs, tactile corpuscles, lamellated corpuscles, and free nerve endings

Special Senses 474

EXERCISE 18.8: GROSS ANATOMY OF THE EYE 474

17. Identify external and internal structures of the eye on a classroom model of the eye

EXERCISE 18.9: EXTRINSIC EYE MUSCLES 478

18. Identify the six extrinsic eye muscles, and describe the actions performed by each muscle.

EXERCISE 18.10: COW EYE DISSECTION 479

19. Identify internal and external eye structures on a fresh cow eye
20. Describe the ways in which a cow eye is both similar to and different from a human eye

EXERCISE 18.11: GROSS ANATOMY OF THE EAR 481

21. Identify structures of the ear on a classroom model of the ear
22. Describe the gross structure, location, and functions of the external, middle, and inner ear

PHYSIOLOGY 485

General Senses 485

EXERCISE 18.12: TWO-POINT DISCRIMINATION 485

23. Describe the relationship between receptive field and the density of sensory receptors in the skin
24. Explain what areas of the body have the greatest density of sensory receptors in the skin

EXERCISE 18.13: TACTILE LOCALIZATION 486

25. Explain why it is more difficult to locate the exact source of sensory input when a stimulus is applied to the tips of the fingers as compared to the back of the forearm

EXERCISE 18.14: GENERAL SENSORY RECEPTOR TESTS: ADAPTATION 487

26. Explain the process of sensory adaptation

Special Senses 488

EXERCISE 18.15: GUSTATORY TESTS 488

27. List the five basic taste sensations and describe the type of substance that is responsible for causing each sensation

EXERCISE 18.16: OLFACTORY TESTS 489

28. Describe the effect that reduced olfactory sensation has on the ability to taste

EXERCISE 18.17: VISION TESTS 491

29. Describe how to test for visual acuity and how to report the results of the test
30. Explain the relationship between flexibility of the lens and near-point accommodation

31 Describe how to test for the blind spot and colorblindness, and explain how each test works

EXERCISE 18.18: HEARING TESTS 493

32 Explain the use of the Rinne and Weber tests, and what abnormal findings for each of the tests may indicate

EXERCISE 18.19: EQUILIBRIUM TESTS 494

33 Explain the purpose of the Romberg and Barany tests, and what abnormal findings for each of the tests may indicate

Anatomy & Physiology REVEALED® aprevealed.com **MODULE 7: NERVOUS SYSTEM**

INTRODUCTION

Reading this chapter and performing the described laboratory activities in this book are both activities that are dependent upon the proper functioning of sensory systems. The special sense of vision is used to read the words typed on the page. The special sense of hearing is used to respond to the instructions delivered by the instructor and to discuss the material with classmates. The special sense of equilibrium maintains the body's three-dimensional position in space so the body does not fall over when moving about the classroom. In addition, sensory receptors in the skin constantly perceive stimuli from the environment relaying touch, pressure, and temperature sensations, thus allowing the body to remain oriented and protected from harm. Indeed, reading this chapter introduction while enjoying a cup of coffee or other food or drink requires the special senses of olfaction (smell) and gustation (taste). In fact, enjoyment of things like music, warm baths, food, drink, and a good book are all completely dependent upon the variety of sensory receptors that are used to detect stimuli from the environment. Of course, it is the brain that actually *interprets* such sensory input. However, without the ability to detect the appropriate input, there would be nothing for the brain to interpret. The laboratory exercises within this chapter explore the intricate structure and function of many of the amazing, beautiful, and incredibly *functional* sensory receptors and organs found throughout our bodies.

The histology and gross anatomy exercises in this chapter explore the histology of a select few somatic sensory receptors and special sensory organs of the body and the gross structure of two special sensory organs (the eye and ear) through observations of classroom models and by dissection of a cow eye. The physiology exercises explore specific tests that are used to examine the function of both general and special senses.

List of Reference Tables

Chapter 18: General and Special Senses

Name: ______________________

Date: __________ Section: __________

These Pre-Laboratory Worksheet questions may be assigned by instructors through their connect course.

PRE-LABORATORY WORKSHEET

1. Which of the following sensory receptors is/are involved in general sensation? (Check all that apply.)

_____ a. cochlear hair cells _____ b. free nerve endings _____ c. olfactory receptors _____ d. photoreceptors _____ e. tactile corpuscles

2. Which of the five special senses have receptor organs located within the petrous part of the temporal bone? (Check all that apply.)

_____ a. equilibrium _____ b. gustation _____ c. hearing _____ d. olfaction _____ e. vision

3. Tactile corpuscles are located within the ______________ (papillary/reticular) layer of the dermis.

4. Lamellated corpuscles are located within the ______________ (papillary/reticular) layer of the dermis.

5. Match the special sense listed in column A with the corresponding cranial nerve that transmits the sensation to the brain, listed in column B. (Some answers may be used more than once because some special senses are transmitted by more than one cranial nerve.)

	Column A	***Column B***
_____	1. equilibrium	a. facial nerve (CN VII)
_____	2. gustation	b. glossopharyngeal nerve (CN IX)
_____	3. hearing	c. olfactory nerve (CN I)
_____	4. olfaction	d. optic nerve (CN II)
_____	5. vision	e. vestibulocochlear nerve (CN VIII)

6. Olfactory nerves extend through which part of the ethmoid bone? (Circle one.)

   a. crista galli
   b. cribriform plate
   c. middle nasal concha
   d. perpendicular plate
   e. superior nasal concha

7. Which of the following are internal structures of the eye? (Check all that apply.)

   _____ a. cornea
   _____ b. fovea centralis
   _____ c. lens
   _____ d. optic nerve
   _____ e. retina
   _____ f. sclera

8. The size of receptive fields and the density of sensory receptors varies in different regions of the body. ______________ (True/False)

9. The ______________ (Rinne/Weber) test is a test for conduction deafness.

10. The sense of olfaction does *not* influence the sense of gustation. ______________ (True/False)

HISTOLOGY

General Senses

Sensory receptors can be classified based on receptor distribution in the body. The two distribution types are the general senses and the special senses. Receptors of the **general senses** are simple structures composed of dendritic endings of sensory neurons that are located throughout the body in the skin and internal organs. They are organized into two categories based upon their location in the body. **Somatic sensory receptors** are housed within the skin for monitoring tactile sensations (e.g., pressure, vibrations, and pain) and within joints, muscle, and tendons for detection of stretch and pressure relative to position and movement of the skeleton and muscles. **Visceral sensory receptors** are located within the wall of the viscera (internal organs); they respond to temperature, chemicals, stretch, and pain. In comparison, receptors of the **special senses** are complex structures located only in the head, and are associated with taste, smell, vision, hearing, and equilibrium.

This section focuses on somatic sensory receptors in the skin **(table 18.1).**

Table 18.1 Sensory Receptors in Thick Skin

Receptor	Location	Structure	Senses	Word Origin
Unencapsulated Tactile Receptors				
Free Nerve Ending	Primarily the papillary layer of the dermis with some extending into the epidermis; associated with glands and hair follicles	An unmodified, unencapsulated nerve ending	Temperature, pain, and pressure	*free*, referring to the fact that there are no connective tissue coverings
Tactile (Meissner) Corpuscle	Within dermal papillae, especially in lips, palms of the hand, eyelids, nipples, and external genitalia	An oval structure consisting of modified neurolemmocytes and connective tissue encapsulating a nerve ending	Fine, discriminative touch to determine textures and shapes; light touch	*tactus*, to touch, + *corpus*, body
Encapsulated Tactile Receptors				
Lamellated (Pacinian) Corpuscle	Deep within the reticular layer of the dermis	Concentric layers of inner neurolemmocytes and outer connective tissue surrounding a nerve ending	Deep pressure and high-frequency vibration	*lamina*, plate, + *corpus*, body
Tactile Disc (Merkel Disc)	At the junction between the dermis and the epidermis	An association between a modified keratinocyte in the epidermis, called a tactile cell, and a specialized nerve ending in the dermis, called a tactile disc	Fine, light touch	*tactus*, to touch

EXERCISE 18.1

TACTILE (MEISSNER) CORPUSCLES

1. Obtain a histology slide of thick skin and place it on the microscope stage. Bring the tissue sample into focus using the scanning objective.

2. Observe the slide on low power and review the layers of the skin (**figures** 6.1 and **18.1**).

   Epidermis: stratum basale, stratum spinosum, stratum granulosum, stratum lucidum, and stratum corneum

   Dermis: papillary layer and reticular layer

3. Move the microscope stage so the junction between the dermis and epidermis is in the center of the field of view, and locate the **dermal papillae** (figure 18.1).

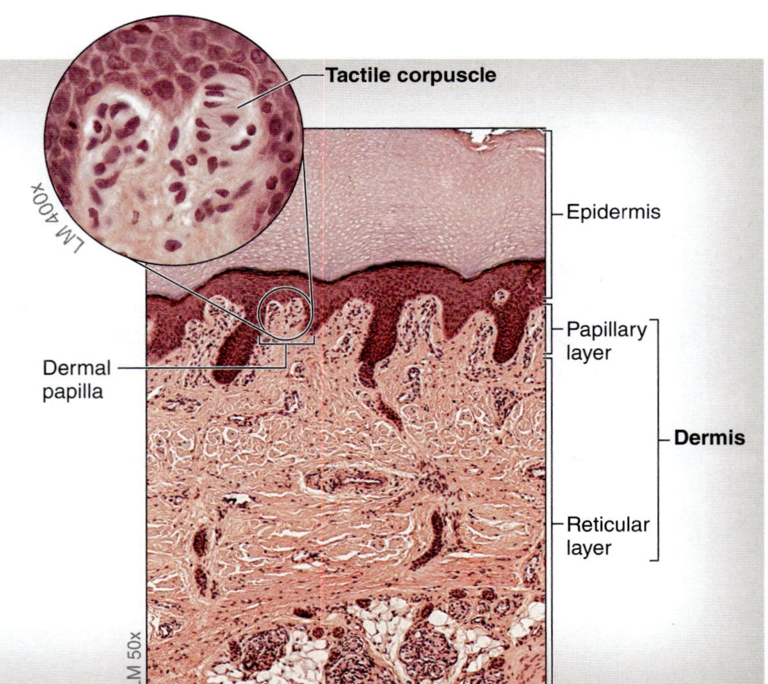

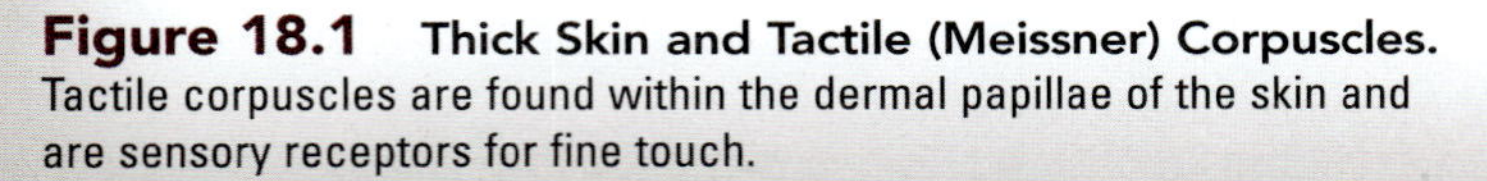
Figure 18.1 Thick Skin and Tactile (Meissner) Corpuscles. Tactile corpuscles are found within the dermal papillae of the skin and are sensory receptors for fine touch.

Next, move the stage so a single papilla is in the center of the field of view, then change to high power.

4. Sensory receptors called **tactile corpuscles** are located within the dermal papillae of thick skin. These receptors are oval in shape with a surrounding capsule of connective tissue (figure 18.1 and table 18.1), and they function in sensing light touch. Look for tactile corpuscles within the dermal papilla. If a tactile corpuscle is not visible, scan the slide for other papillae that may have tactile corpuscles within.
5. Sketch a tactile corpuscle within a dermal papilla as seen through the microscope in the space provided.

EXERCISE 18.2

LAMELLATED (PACINIAN) CORPUSCLES

1. Obtain a histology slide of thick skin and place it on the microscope stage.
2. Observe the slide on low power and identify the dermis and epidermis (figure 18.1).
3. Move the microscope stage so the deepest part of the reticular layer of the dermis is in the center of the field of view. Within this portion of the dermis there are cross sections through numerous sweat glands and blood vessels. Cross sections of **lamellated corpuscles** should also be visible. Lamellated corpuscles resemble onions in cross section because they are composed of concentric layers of connective tissues surrounding an inner core of neurolemmocytes that ensheath dendritic endings of sensory neurons (**figure 18.2** and table 18.1).
4. Locate a lamellated corpuscle and then move the microscope stage so the lamellated corpuscle is in the center of the field of view. Change to a higher power to view its structure in more detail. Lamellated corpuscles function in the sensation of deep pressure. When enough pressure is applied to the surface of the skin, the layers of connective tissue surrounding the central sensory receptor are compressed, and initiate nerve signals to the brain.
5. Sketch a lamellated corpuscle as seen through the microscope, making note of its location in the skin, in the space provided.

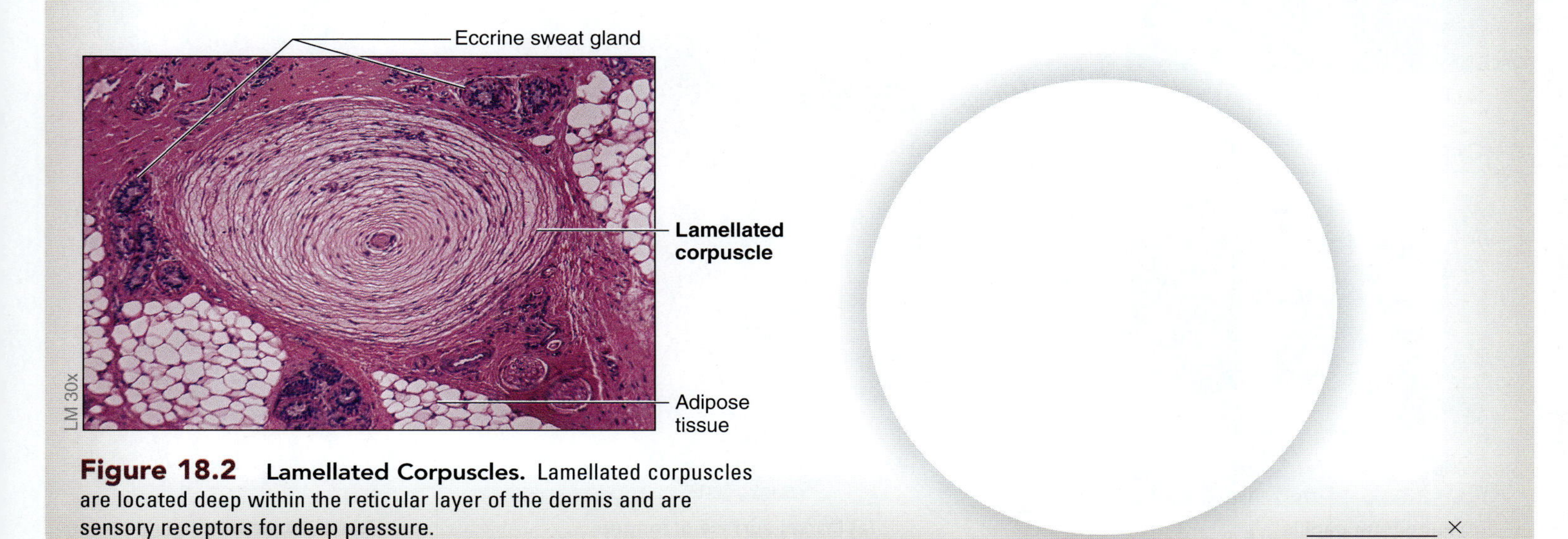

Figure 18.2 Lamellated Corpuscles. Lamellated corpuscles are located deep within the reticular layer of the dermis and are sensory receptors for deep pressure.

Special Senses

Special senses are specialized organs within the head that respond specifically to the modalities of olfaction, taste, vision, hearing, and equilibrium. The next set of laboratory exercises explores the structure and function of these special sensory organs.

EXERCISE 18.3

GUSTATION (TASTE)

Gustation (taste) is one of the most pleasurable sensations humans can experience. However, to have a complete sense of gustation, our olfactory sense must also be functioning. In fact, without the ability to smell, the ability to taste suffers tremendously. Perhaps one reason gustation is so pleasurable is that the gustatory and olfactory pathways relay sensory input to the limbic system, the emotional brain.

There are four types of papillae located on the tongue: filiform, fungiform, vallate (circumvallate), and foliate. The detailed structure, function, and locations of the tongue papillae are shown in **figure 18.3.** The large papillae are the **foliate** and **vallate (circumvallate)** papillae. The vallate papillae house more than half of our taste buds in the walls of the crypts that surround each papilla. The sensory receptor specialized for gustatory sensation is a **taste bud** (gustatory bud) (**table 18.2** and **figure 18.4**). Taste buds are particularly concentrated on the papillae of the tongue, but are also located throughout the oral cavity and pharynx. This exercise explores the types of tongue papillae and the location and function of taste buds associated with the papillae.

1. Obtain a histology slide of the tongue or a histology slide demonstrating mammalian *vallate papillae* (figure 18.3*d*). Scan the slide at low power and identify the papillae on the surface of the tongue. Move the microscope stage so one or two papillae are in the center of the field of view. Then increase the magnification, first to medium and then to high power. Locate a taste bud at the edge of the papilla (figure 18.4).
2. Identify the following structures, using figures 18.3 and 18.4, and table 18.2 as guides. (Note that all of the types of papillae may not be visible on a single slide.)

- ☐ **basal cells**
- ☐ **filiform papillae**
- ☐ **foliate papillae**
- ☐ **fungiform papillae**
- ☐ **gustatory cells**
- ☐ **supporting cells**
- ☐ **taste pores**
- ☐ **vallate (circumvallate) papillae**

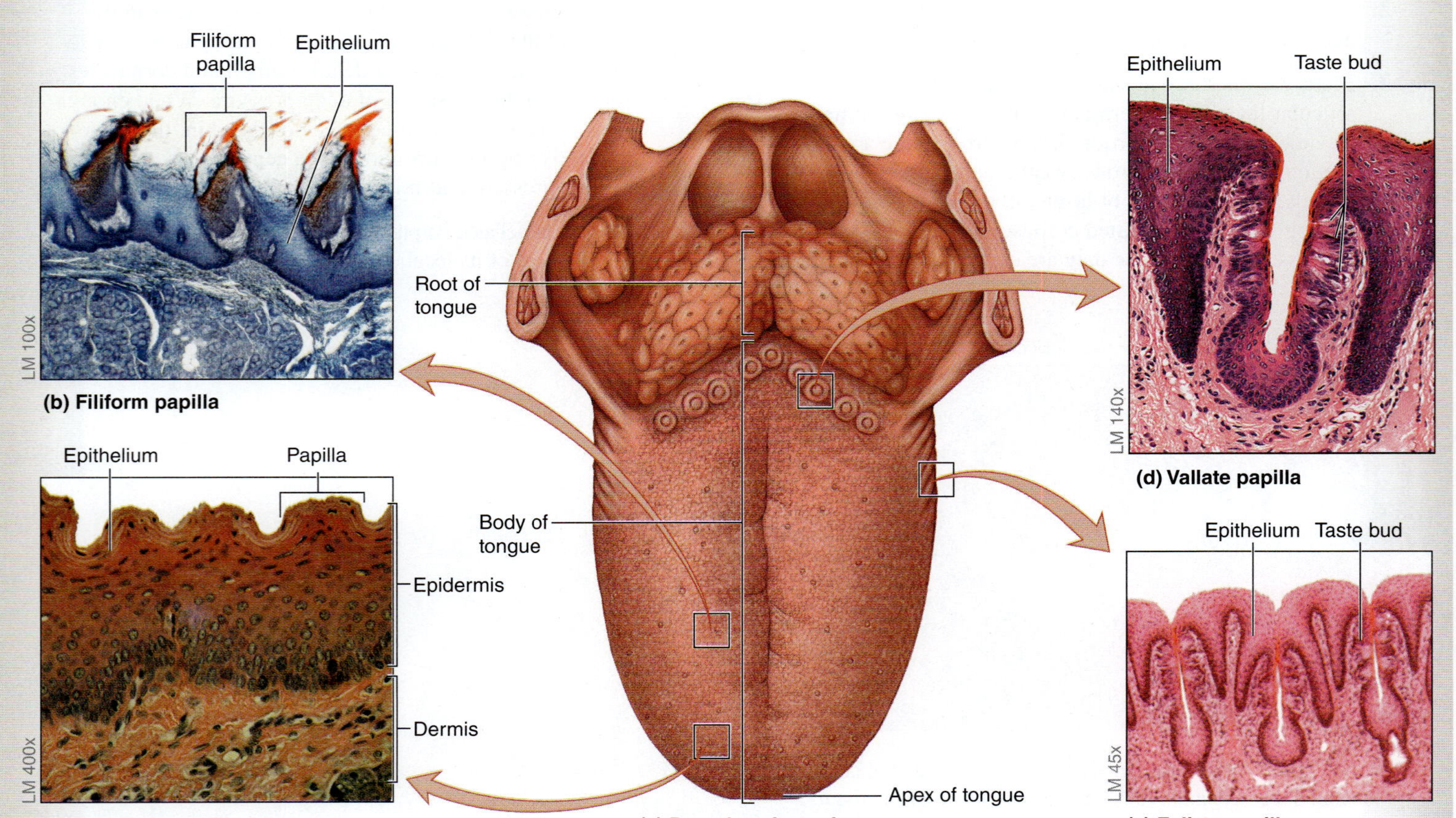

Figure 18.3 **Taste Buds.** Gustation (taste) requires taste buds, which are associated with tongue papillae. (a) Dorsal surface of tongue, (b) Filiform papilla, (c) Fungiform papilla, (d) Vallate papilla, (e) Foliate papilla.

3. Locate a taste bud at the edge of the papilla in the crevice (crypt) between two papillae (figure 18.4).
4. Sketch a vallate papilla as seen through the microscope in the space provided. Be sure to label the taste buds.

Table 18.2	Cells Associated with Taste Buds		
Structure	**Description**	**Function**	**Word Origin**
Basal Cells	Small stem cells found at the base of the taste bud	Precursor cells to the supporting cells and gustatory cells	*basalis*, situated near the base, + *cella*, a chamber
Gustatory Cells	Light-staining cells with round nuclei; contain modified microvilli (sometimes called *taste hairs*) on the apical surface that extend through taste pores to detect tastants	Detect chemicals dissolved in solution; transmit nerve signals for taste sensation to the CNS	*gustus*, a tasting, + *oriusi*, having to do with, + *cella*, a chamber
Supporting Cells (Sustentacular Cells)	Dark-staining cells with oval-shaped nuclei; located between gustatory cells	Support the gustatory cells by producing a glycoprotein; may also function in taste sensation	*supporto*, to carry, + *cella*, a chamber

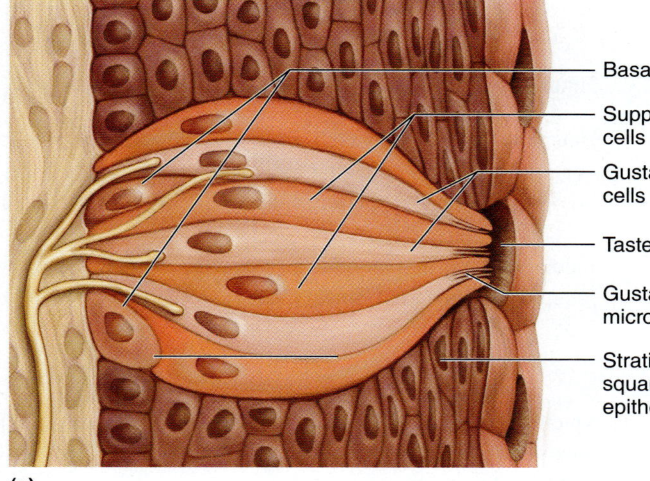

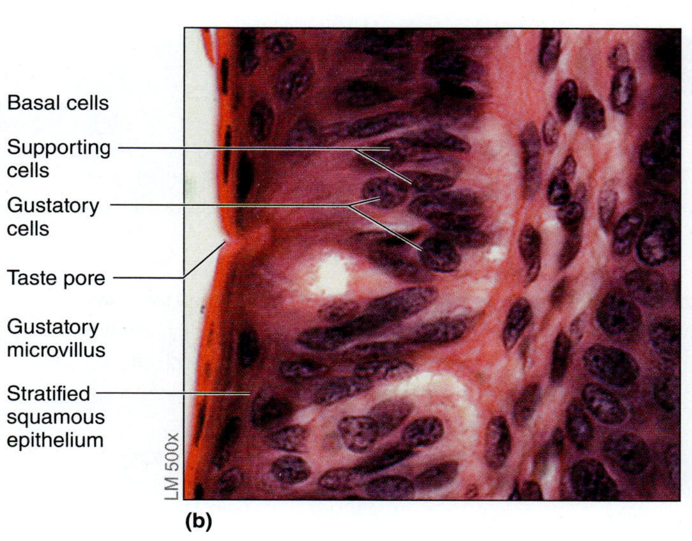

Figure 18.4 **Detailed Structure of a Taste Bud.** (a) Illustration. (b) Photomicrograph.

EXERCISE 18.4

OLFACTION (SMELL)

Olfaction is the sense of smell. It is an important sensory modality, not just for smell, but also for gustation (taste). Olfactory sensation is detected by special sensory cells found within the epithelium lining the roof of the nasal cavity—**olfactory epithelium** (**figure 18.5*a*** and **table 18.3**). The **olfactory receptor cells** are neurons, but they are unique because they are continuously replaced. The olfactory receptor cells compose the **olfactory nerves (CN I)**.

1. Obtain a histology slide of olfactory epithelium (figure 18.5*b*) and place it on the microscope stage.
2. Bring the tissue sample into focus on low power. Increase magnification to medium power, and then to high power so the cells composing the olfactory epithelium are clearly visible. It may be difficult to distinguish between the three major cell types of the olfactory epithelium: basal cells, olfactory receptor cells, and supporting cells. In general, the nuclei closest to the basement membrane of the epithelium are the nuclei of *basal cells* (the cells appear triangular in shape), the nuclei in the middle of the epithelium are the nuclei of *olfactory receptor cells*, and the nuclei closest to the apical surface of the epithelium are the nuclei of *supporting cells*.

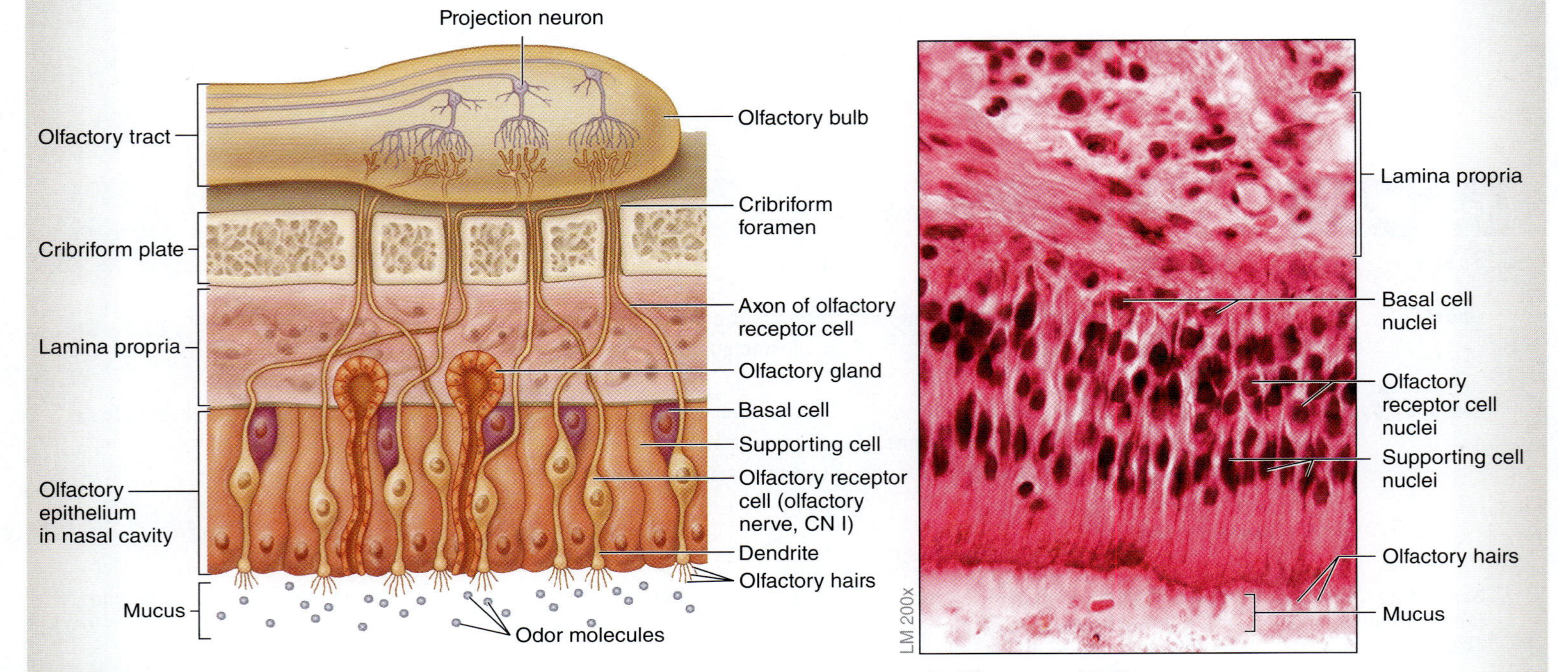

Figure 18.5 **Olfactory Epithelium.** (a) Cell types within the olfactory epithelium. (b) Histology of the olfactory epithelium.

Table 18.3 Olfactory Epithelium

Structure	Description	Function	Word Origin
Basal Cells	Cuboidal (triangular) cells whose nuclei are located near the basal surface of the olfactory epithelium	Function as neural stem cells to continually replace both olfactory receptor cells and supporting cells	*basalis*, situated near the base, + *cella*, a chamber
Olfactory Receptor Cells	Bipolar neurons that have undergone extensive differentiation; contain large, centrally located nuclei; have both a single dendrite with olfactory hairs (modified cilia) and an unmyelinated axon	Bipolar neurons that function as sensory receptors for olfaction (smell)	*olfactus*, to smell, + *recipio*, to receive, + *cella*, a chamber
Olfactory Hairs (Cilia)	Nonmotile cilia of apical ends of olfactory receptor cells	Apical ends are the site of interaction between dissolved odoriferous substances and the olfactory receptor cells	*olfactus*, to smell
Supporting (Sustentacular) Cells	Columnar epithelial cells with nuclei located near the apical surface of the olfactory epithelium	Surround and support the specialized olfactory receptor cells	*supporto*, to carry, + *cella*, a chamber

3. Identify the following structures on the histology slide of olfactory epithelium, using figure 18.5 and table 18.3 as guides:

- ☐ **basal cells**
- ☐ **olfactory receptor cells**
- ☐ **ofactory hairs (cilia)**
- ☐ **supporting cells**

4. Sketch the olfactory epithelium as seen through the microscope in the space provided. Label all of the structures listed in step 3.

__________ ×

INTEGRATE

CONCEPT CONNECTION

An anatomic relationship exists between the structures associated with olfaction and the ethmoid bone, which forms the roof and lateral walls of the nasal cavity. The superior and middle nasal conchae cause incoming air to swirl within the nasal cavity, which provides more time for odors to stimulate olfactory receptor cells within the olfactory epithelium. In addition, the superior portion of the ethmoid bone contains the **cribriform foramina,** tiny holes within the cribriform plate. Axons of the olfactory nerves (CN I) extend through the cribriform foramina to synapse with neurons within the **olfactory bulbs,** which rests on the superior aspect of the cribriform plate. Neurons within the olfactory bulbs subsequently transmit nerve signals to the brain via the **olfactory tract.** Damage to any of these structures, including damage to the ethmoid bone itself (as might occur with a fracture to the bone), may cause a reduced ability or inability to smell (anosmia).

EXERCISE 18.5

VISION (THE RETINA)

The **retina (figure 18.6)** of the eye is called the **neural tunic,** because this layer is composed of neural tissue. The retina develops as a direct outgrowth of the brain. Thus, the retina is the only part of the brain visible without surgical intervention (though an ophthalmoscope is required). Axons from neurons within the retina travel to the brain through the **optic nerve (CN II).** The retina is responsible for transducing light rays into electrical signals (action potentials) to the brain. This information is relayed from photoreceptor cells (rods and cones) to bipolar cells to the ganglion cells within the retina. Axons of ganglion cells extend from the back of the eye as the optic nerve to synapse with neurons within the **thalamus.** These neurons extend to the **occipital lobe** of the brain. Here, the visual information is processed and interpreted. The retina is a very complex yet beautifully organized structure. Rods function in dim light and cones function in high-intensity light and in color vision. This laboratory exercise explores the structure and function of the retina by observing the cells that are visible histologically. Laboratory exercises within the gross anatomy section of this chapter will place the retina in the context of other structures of the eye.

1. Obtain a histology slide of the retina and place it on the microscope stage.

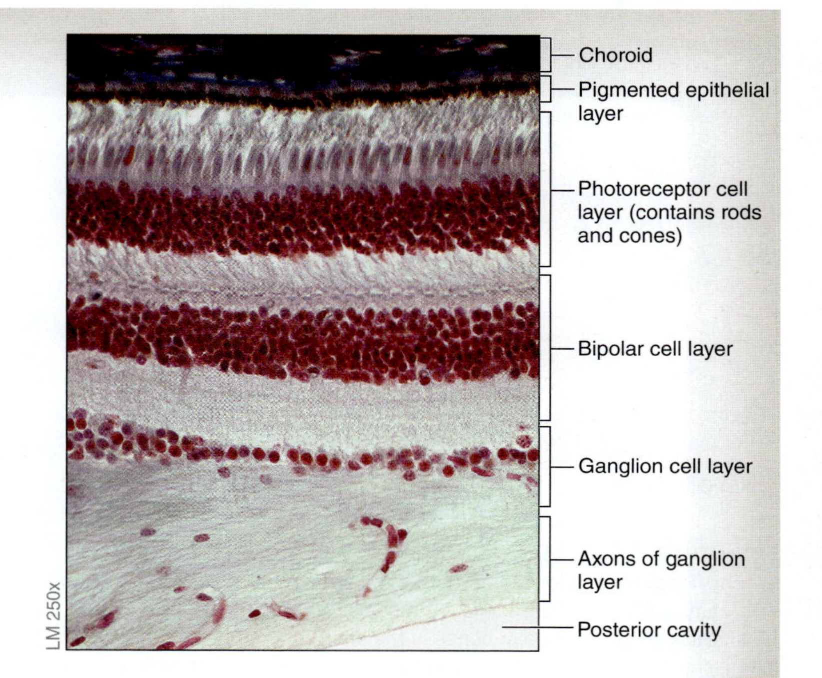

Figure 18.6 Histology of the Retina.

(continued on next page)

(continued from previous page)

2. Bring the tissue sample into focus on low power, then move the microscope stage so the retina (figure 18.6) is in the center of the field of view. Switch to medium power and bring the tissue sample into focus once again.
3. Identify the following structures on the slide of the retina, using **table 18.4** and figure 18.6 as guides. (High power may be required to view all of the structures or to see them in greater detail.)

- ☐ **bipolar cell layer**
- ☐ **choroid**
- ☐ **ganglion cell layer**
- ☐ **photoreceptor cell layer**
- ☐ **pigmented layer**
- ☐ **sclera**

4. With the medium-power objective in place, scan the slide and locate the **fovea centralis (figure 18.7*a*).** The fovea centralis is a thinner than normal area of the retina. The fovea centralis contains photoreceptor cells and is devoid of bipolar and ganglion cell layers. This area has the highest concentration of cones of the entire retina, providing the highest visual acuity. We turn our head so that we are using the fovea centralis to generate the sharpest image of the object of interest.
5. Scan the slide and locate the **optic disc,** the location where the **optic nerve** leaves the eye (figure 18.7*b*). The **optic disc** (blind spot) is easily identifiable because all retinal layers are absent at this location. Notice that the optic disc and optic nerve are approximately the same color as the cells in the ganglion cell layer of the retina. This is helpful in understanding that the ganglion cell layer, the optic disc, and the optic nerve all are composed of the axons of ganglion cells. The axons of the ganglion cell layer leave the eye at the optic disc and extend from the eye to the brain as the optic nerve. Axons of the optic nerve are the only portion of the ganglionic axons that are myelinated.
6. Sketch the retina as seen through the microscope in the space provided. Label the following layers: rod and cone layer, bipolar cell layer, and ganglion cell layer.

7. *Optional Activity:* **AP|R 7: Nervous System**—Watch the "Vision" animation to learn the sequence of events involved in vision and the functions of cells in the retina.

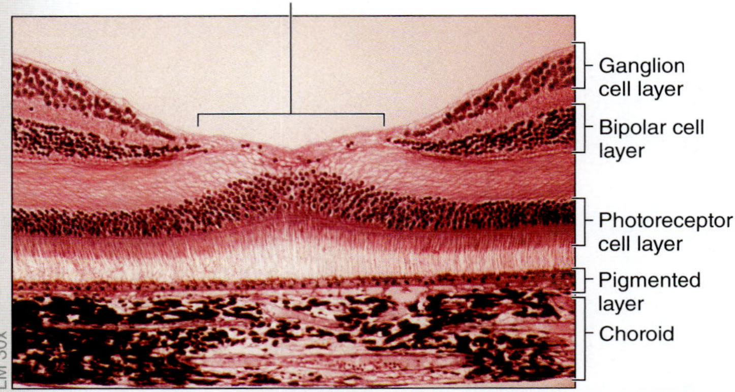

(a) Fovea centralis

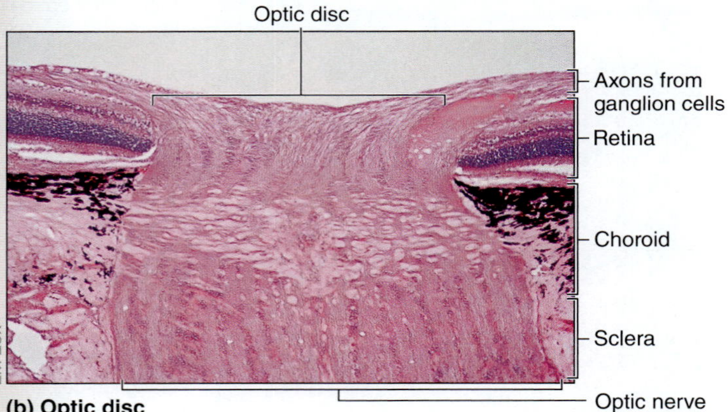

(b) Optic disc

Figure 18.7 Specialized Areas of the Neural Tunic of the Eye. (a) The fovea centralis is the area of the retina where visual acuity is the highest. Ganglion and bipolar cell layers are absent, and there is an abundance of cones in the photoreceptor layer. (b) The optic disc is the area of the retina where the axons of ganglion cells exit the retina to become the optic nerve. There are no rods or cones in this area, which is why the optic disc is also referred to as the "blind spot" of the retina.

Table 18.4 The Retina

Structure	Description	Function	Word Origin
Bipolar Cell Layer	Middle cell layer of the retina, composed of cells with intermediate-sized nuclei; as the name suggests, these neurons are bipolar neurons	Receives signals from rods and cones and transmits electrical signals to ganglion cells	*bipolar*, relating to bipolar neurons
Choroid*	The vascular and pigmented layer of the eye located between the retina (internal) and the sclera (external); recognized histologically by numerous blood vessels and by dark staining characteristics of the melanin	Blood vessels of the choroid supply nutrients to the tissues of the retina and sclera, and the pigment absorbs excess light waves	*choroideus*, like a membrane
Fovea Centralis	An area of the retina devoid of bipolar and ganglion cell layers; photoreceptor layer is composed exclusively of cones	Area of highest visual acuity in the eye. Focusing on an object requires moving the eyes so the light entering the eye is focused on the fovea.	*fovea*, a pit, + *centralis*, in the center
Ganglion Cell Layer	Innermost layer of the retina, composed of cells with very large nuclei	Receives information from bipolar cells and sends that information to the brain	*ganglion*, a swelling or knot
Pigmented Layer	Outermost portion of retina attached to choroid	Absorbs extraneous light; provides vitamin A for photoreceptor cells	NA
Retina	Referred to as the "neural tunic" of the eye; consists of numerous layers of neurons involved in phototransduction	Phototransduction: transduction of light waves that enter the eye into nerve signals (action potentials) that can be interpreted by the brain	*rete*, a net
Photoreceptor Cell Layer	Outermost layer of the retina (closest to the choroid and sclera), containing the light-transducing portions of photoreceptor cells (rods and cones); the layer immediately internal to this layer contains the nuclei of rods and cones, the smallest and most numerous nuclei of the retina	Layer of the retina where light waves are initially transduced into neuronal action potentials; the cells in this layer synapse with neurons in the bipolar cell layer	NA
Cones	Photoreceptor cells with a light-transducing portion located in the outermost layer of the retina and with nuclei located in the layer just internal to that; not possible to distinguish rods from cones using a light microscope	Photoreceptor cell specializing in color vision	*conus*, shaped like a cone
Rods	Photoreceptor cells with a light-transducing portion located in the outermost layer of the retina and with nuclei located in the layer just internal to that; not distinguishable from cones using a light microscope	Photoreceptor cells specializing in black-and-white vision; very sensitive, most useful when light is dim	*rod*, shaped like a rod
Sclera*	Dense irregular connective tissue that surrounds the entire eye except for its anterior aspect where the cornea is located; histologically, the most external layer, composed of collagen fibers and fibroblasts	Protects the eye, serves as an attachment point for extraocular eye muscles, and helps maintain the round shape of the eye	*skleros*, hard

*Note that the choroid and sclera are tunics of the eye, rather than structures of the retina. However, both tunics are visible on histological slides of the retina.

EXERCISE 18.6

HEARING

Hearing is a function of the **cochlea.** This special sensory organ is located within the petrous part of the temporal bone. The gross anatomy section of this chapter focuses on the location, gross structure, and function of this organ. This section focuses on the histological features of the highly specialized epithelium that lines the cochlea, the **spiral organ** (organ of Corti), which will provide an insight into how this organ performs its function: transformation of sound waves into nerve signals that can be interpreted by the brain.

The cochlea **(figure 18.8)** is the organ responsible for transducing fluid vibrations received at the oval window into electrical signals that are sent to the thalamus and then on to the temporal lobe of the brain, where they are interpreted. Within the cochlea, the spiral organ rests upon the **basilar membrane,** within the scala media (cochlear duct).

1. Obtain a slide of the cochlea **(figure 18.9)** and place it on the microscope stage. Bring the tissue sample into focus on low power and then increase the magnification. Move the microscope stage until a single cross section through the cochlea is in the center of the field of view.

2. Identify the three chambers within the cochlea and the membranes that separate the chambers from each other, using figures 18.8 and 18.9 and **table 18.5** as guides:

- ☐ **basilar membrane**
- ☐ **scala media (cochlear duct)**
- ☐ **scala tympani**
- ☐ **scala vestibuli**
- ☐ **vestibular membrane**

3. Once the *scala media* (cochlear duct) has been identified, move the microscope stage so the scala media is in the center of the field of view. Increase the magnification to high power and focus in on the *spiral organ* (figure 18.9).

4. Identify the following structures on the slide of the cochlea, using table 18.5 and figure 18.9 as guides:

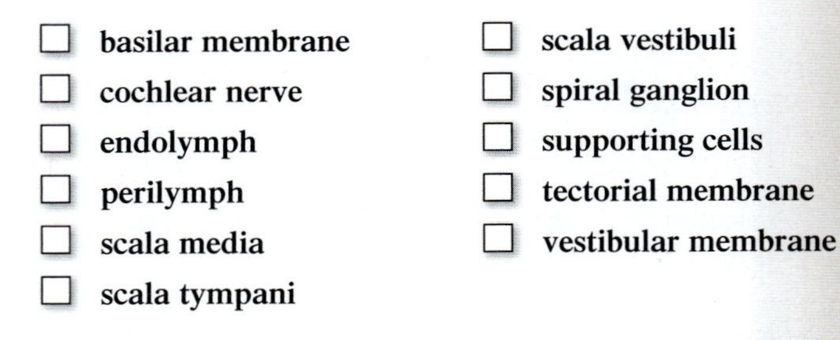

- ☐ **basilar membrane**
- ☐ **cochlear nerve**
- ☐ **endolymph**
- ☐ **perilymph**
- ☐ **scala media**
- ☐ **scala tympani**
- ☐ **scala vestibuli**
- ☐ **spiral ganglion**
- ☐ **supporting cells**
- ☐ **tectorial membrane**
- ☐ **vestibular membrane**

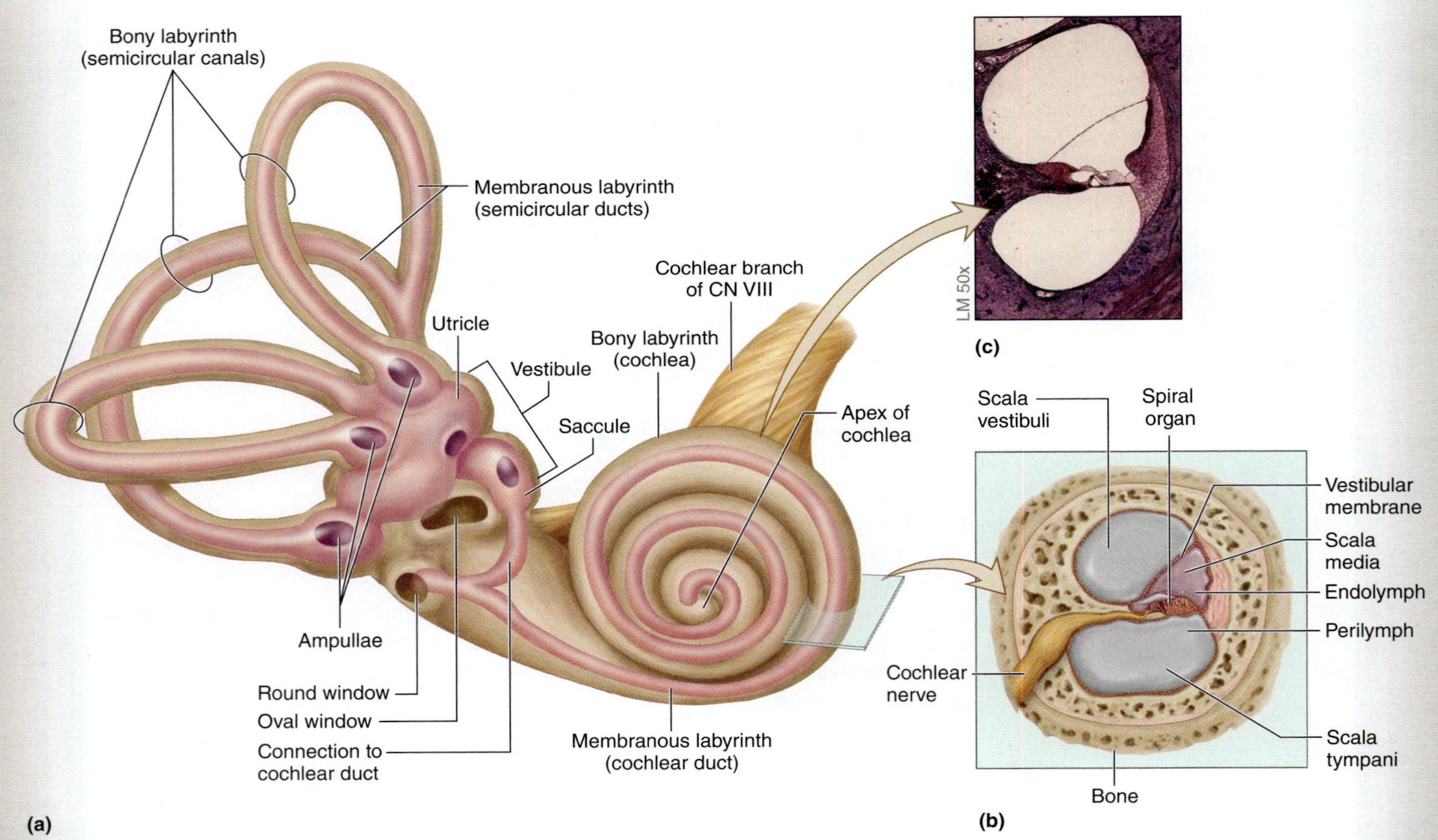

Figure 18.8 The Cochlea. (a) The semicircular canals and cochlea are part of the inner ear. (b) The cochlea houses the spiral organ, which contains specialized cells that translate sound waves into sensory impulses. (c) Light micrograph demonstrating a cross section through the cochlea.

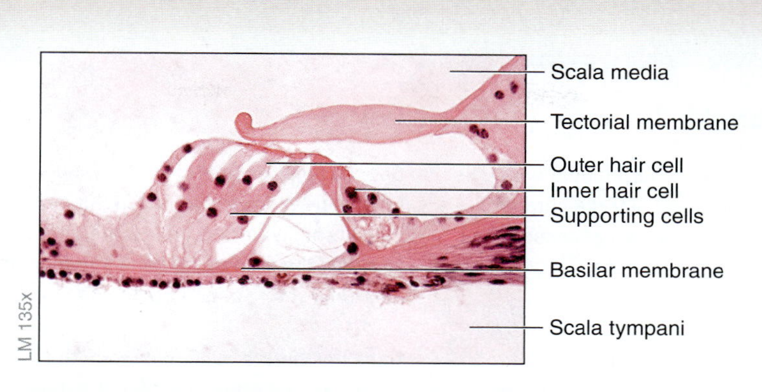

Figure 18.9 Histology of the Spiral Organ.

Table 18.5 The Cochlea

Structure	Description	Function	Word Origin
Basilar Membrane	Forms the floor of the scala media and supports the hair cells of the spiral organ	Vibrates in response to movement of the perilymph within the scala tympani	*basalis*, situated near the base, + *membrana*, a membrane
Endolymph	Fluid within the scala media (cochlear duct); composition is similar to that of intracellular fluid (high in potassium)	Nourishes the epithelial cells of the spiral organ	*endon*, within, + *lympha*, a clear fluid
Hair Cells (Sensory Cells)	Highly specialized epithelial cells of the spiral organ that contain stereocilia, which are embedded in the tectorial membrane	Bending of the stereocilia generates action potentials in the hair cells; axons of sensory cells project into the spiral ganglion and synapse with neurons that relay the signals to the brain	*sensus*, to sense
Helicotrema	A connection between the scala vestibuli and scala tympani at the end of the cochlea; shaped like a half moon	Allows perilymph from the scala tympani to communicate with perilymph from the scala vestibuli	*helix*, a spiral, + *trema*, a hole
Perilymph	Fluid contained within the scala vestibuli and scala tympani; composition is similar to that of extracellular fluid (high in sodium)	Transmits pressure waves through the scala tympani and scala vestibuli; vibrations of the stapes at the oval window cause the pressure waves, and they are dampened when they reach the round window	*peri-*, around, + *lympha*, a clear fluid
Scala Media (Cochlear Duct)	Middle chamber of the cochlea, containing the spiral organ and filled with endolymph	Contains the special sensory organ of sound, the spiral organ	*scala*, a stairway, + *medialis*, middle
Scala Tympani	Chamber inferior to the scala media; filled with perilymph	Transmits pressure waves of the perilymph from the helicotrema to the round window	*scala*, a stairway, + *tympani*, relating to the tympanic membrane
Scala Vestibuli	Chamber superior to the scala media; filled with perilymph	Transmits pressure waves of the perilymph from the oval window to the helicotrema	*scala*, a stairway, + *vestibulum*, entrance court
Spiral Ganglion	Group of neuron cell bodies located on the cochlear part of the cochlear nerve	Contains the cell bodies of bipolar neurons, which receive input from sensory cells of the spiral organ and then relay those signals to the brain	*spiralis*, a coil, + *ganglion*, a swelling or knot

(continued on next page)

(continued from previous page)

Table 18.5 The Cochlea *(continued)*

Structure	Description	Function	Word Origin
Spiral Organ (Organ of Corti)	Composed of specialized epithelium that rests on the basilar membrane; the epithelium is composed of sensory hair cells and supporting cells	Special inner and outer sensory hair cells have stereocilia embedded in the tectorial membrane; when the basilar membrane vibrates, the stereocilia bend and the sensory hair cells send action potentials to the brain	*spiralis*, a coil, + *organum*, a tool, instrument
Tectorial Membrane	Gelatinous membrane in which the cilia of hair cells of the spiral organ are embedded	Does not vibrate itself, but when the basilar membrane vibrates, the cilia of the hair cells embedded in the tectorial membrane bend, causing the hair cells to generate action potentials	*tectus*, to cover, + *membrana*, a membrane
Vestibular Membrane	Thin membrane between the scala vestibuli and the scala media	Forms a partition separating endolymph within the scala media from perilymph within the scala vestibuli	*vestibulum*, relating to the vestibule, + *membrana*, a membrane
Vestibulocochlear Nerve (CN VIII)	Cranial nerve arising from axons of sensory cells of the spiral organ and from the vestibular apparatus	Transmits nerve signals from the cochlea and the vestibular apparatus to the brain	*audio*, to hear

5. Sketch a cross section of the cochlea as seen through the microscope in the space provided. Label the locations of the structures listed in step 4.

_______ ×

GROSS ANATOMY

General Senses

The definitions of general senses and special senses are given at the beginning of the histology section of this chapter. If starting laboratory observations with the gross anatomy exercises, read the introduction to general senses on p. 462 before proceeding.

Most sensory receptors responsible for general sensation (things such as touch, pain, pressure, temperature) are located in the skin. Exercise 18.7 involves observing a classroom model of the skin, with special emphasis on sensory receptors located within the skin.

EXERCISE 18.7

SENSORY RECEPTORS IN THE SKIN

1. Observe a classroom model of thick skin **(figure 18.10).** Some somatic sensory receptors in the skin, such as tactile menisci and free nerve endings, are too small to view under the microscope. Therefore, these structures will be observed on classroom models.

2. Locate a tactile disc. **Tactile discs** (Merkel discs) are located at the dermal/epidermal junction of thick skin, although their location is not restricted to dermal papillae, unlike tactile corpuscles. The function of tactile discs is the sensation of light touch.

3. Find some free nerve endings on the skin model. **Free nerve endings** are just that—nerve endings with no specialized cells surrounding them. The ends of these neurons are located near the epidermal/dermal junction, with many of them extending into the epidermis. They also have endings in hair follicles and glands. Free nerve endings function in the sensation of sustained touch, temperature, itching, and pain. Recall a time when you had a blister that ripped open. A blister forms when the dermis separates from the epidermis, and fluid accumulates between the layers. When the epidermis is removed from a blister, it is very painful! Why? The free nerve endings in the dermis are now exposed to the environment, and this causes them to generate action potentials. This is why large superficial wounds ("scrapes") on the skin are much more painful than deep wounds. A deep wound is perceived as a more severe wound (as it typically is), but is often confusing because it causes less pain than a more superficial wound. The greater the surface area exposed, the greater the number of nerve endings stimulated. Hence, the common exclamation, "It's only a scrape, but it hurts like crazy!"

4. Identify the following on a model of the skin, using figure 18.10, table 18.1, and the textbook as guides:

- ☐ **dermal papillae**
- ☐ **free nerve endings**
- ☐ **lamellated corpuscle**
- ☐ **papillary dermis**
- ☐ **reticular dermis**
- ☐ **stratum basale**
- ☐ **stratum corneum**
- ☐ **stratum granulosum**
- ☐ **stratum lucidum**
- ☐ **stratum spinosum**
- ☐ **tactile corpuscle**
- ☐ **tactile disc**

5. Sketch thick skin and label the locations of the sensory receptors listed in table 18.1 in the space provided.

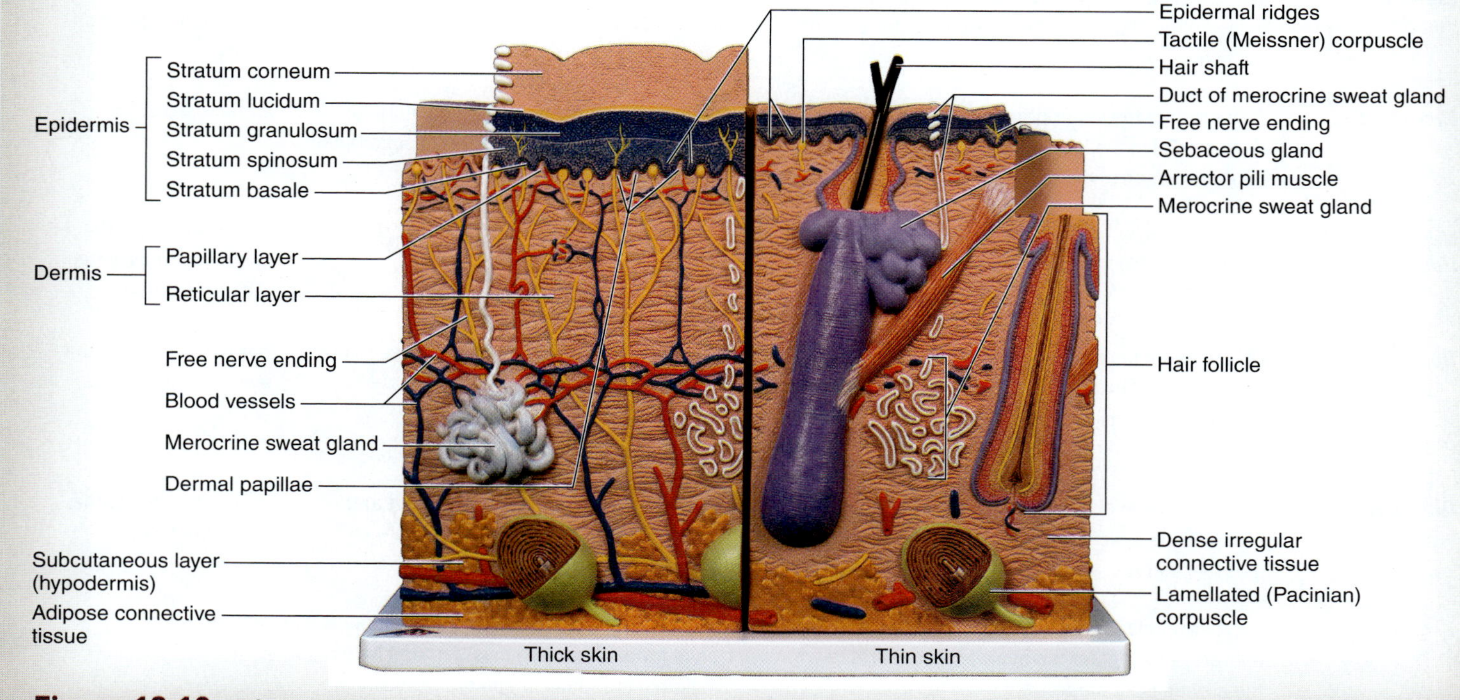

Figure 18.10 **Skin.** Model of the skin demonstrating sensory receptors, such as free nerve endings and lamellated corpuscles.

Special Senses

The definitions of general senses and special senses are given at the beginning of the histology section of this chapter. If starting laboratory observations with the gross anatomy exercises, read the introduction to special senses on p. 462 before proceeding. The gross anatomy exercises in this section focus on two special sensory organs: the eye and the ear.

EXERCISE 18.8

GROSS ANATOMY OF THE EYE

This laboratory exercise explores the accessory structures of the eye. Identify all structures listed in **table 18.6** on a classroom model or on yourself. Subsequent exercises will explore internal structures of the eye and structure and function of both external and internal structures of a cow eye.

Table 18.6 Accessory Structures of the Eye

Structure	Description	Function	Word Origin
Cornea	Transparent tissue on the anterior surface of the eye consisting of an external layer of stratified squamous epithelium, a middle layer of regularly arranged collagen fibers, and an inner layer of endothelium	The primary structure used to refract (bend) light waves in the eye	*corneus*, horny
Extrinsic Muscles	Muscles that originate from a common tendinous ring and insert on the sclera of the eye	Responsible for movement of eye within the orbit	*extra*, outside of, + *oculus*, the eye
Fibrous Tunic	Tough outer connective tissue covering of the eye, composed of the sclera and cornea	Included elsewhere in this table, for sclera and cornea	*fibro-*, fiber + *tunic*, a coat
Lacrimal Caruncle	Small, fleshy mound of tissue at the medial aspect of the eye containing ciliary glands (modified sweat glands) and tarsal glands	Ciliary glands that form secretions to lubricate eyelashes	*caruncula*, a small fleshy mass
Lacrimal Gland	Almond-shaped serous gland located in the superior and lateral aspect of the orbit	Secretes tears	*lacrima*, a tear
Lacrimal Puncta	Two small openings in the lacrimal caruncle; the tiny holes on the "bump" at the inferomedial aspect of the eye	Opening for the drainage of lacrimal fluid into the lacrimal sac	*lacrima*, a tear, + *punctum*, a prick or point
Lacrimal Sac	A swelling at the superior part of the nasolacrimal duct, medial to the lacrimal bone, lateral to the nasal bone, and deep to the maxilla	Receives lacrimal fluid from the lacrimal canals and transports it to the nasolacrimal duct	*lacrima*, a tear
Nasolacrimal Duct	A duct that runs from the lacrimal sac into the nasal cavity	Conducts lacrimal fluid from the lacrimal sac into the nasal cavity	*nasal*, relating to the nose, + *lacrima*, a tear, + *ductus*, to lead
Optic Nerve	CN II; a large nerve exiting the posteromedial region of the eye that exits the orbit through the optic foramen; consists of myelinated axons of ganglion cells	Transmits nerve signals from the eye to the brain	*optikos*, relating to the eye or vision
Orbital Fat Pad	A thick capsule of adipose connective tissue that fills in all the spaces between the eye, extrinsic eye muscles, nerves, and orbit	Cushions the eye and helps support and hold it in place	*orbital*, relating to the orbit of the eye
Sclera	Dense irregular connective tissue that surrounds the entire eye except for the eye's anterior surface where the cornea is located	Protects the eye, serves as an attachment point for extrinsic eye muscles, and helps maintain the round shape of the eye	*skleros*, hard

EXERCISE 18.8A Accessory Structures of the Eye

1. Observe a classroom model of the eye (**figure 18.11*a***).
2. Identify the following structures on a model of the eye, using figure 18.11*a*, table 18.6, and the textbook as guides:

- ☐ cornea
- ☐ iris
- ☐ lacrimal gland
- ☐ optic disc
- ☐ optic nerve
- ☐ sclera

3. Obtain a mirror and observe the externally visible structures of your eye. If you have no mirror, perform this observation on a lab partner. Using table 18.6 and the textbook as guides, identify all of the structures listed in figure 18.11 on your eye (or your lab partner's eye), and then label them in figure 18.11*b*.

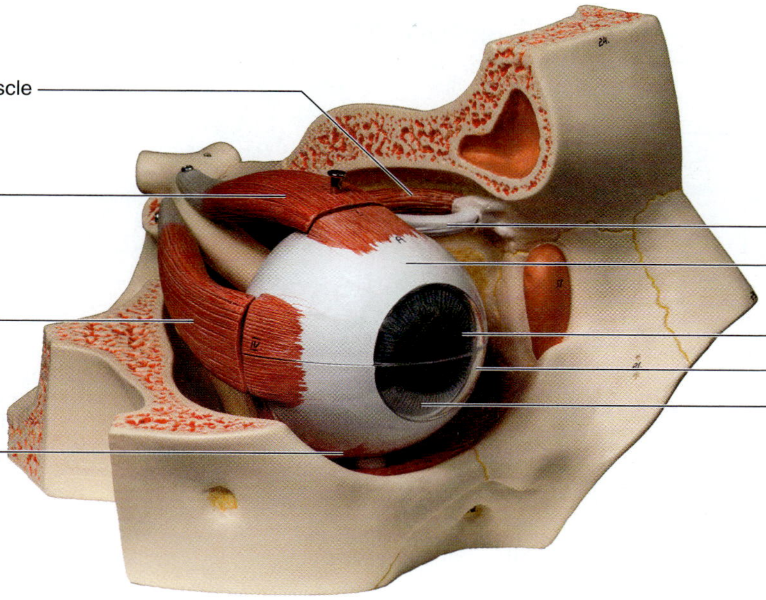

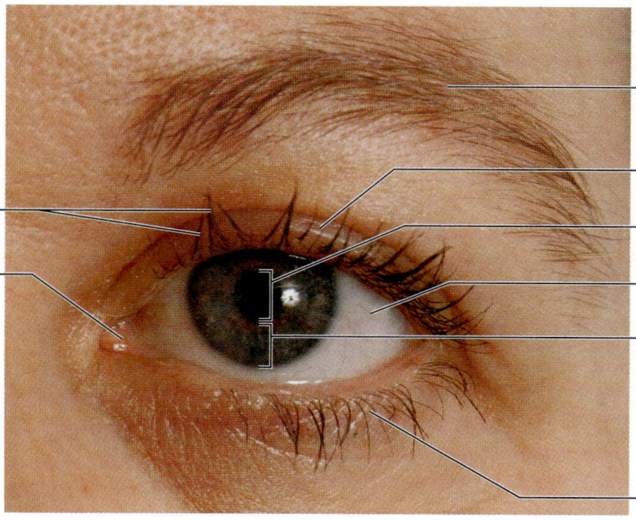

Figure 18.11 **Accessory Structures of the Eye.** (a) Classroom model. (b) Human eye.

- ☐ eyebrow
- ☐ eyelashes
- ☐ inferior eyelid
- ☐ iris
- ☐ medial canthus
- ☐ pupil
- ☐ sclera
- ☐ superior eyelid

(continued on next page)

(continued from previous page)

EXERCISE 18.8B Internal Structures of the Eye

The internal structures of the eye **(table 18.7)** are structures that function in the transmission of light, nourishment of the eye, and the processing of visual information.

1. Observe a classroom model of the eye where internal structures are visible **(figure 18.12).** Many classroom models of the eye contain both external and internal eye structures. Viewing internal eye structures may require disassembly of the eye model to access the structures.

2. Identify the following structures on the model of the eye, using figure 18.12, table 18.7, and the textbook as guides:

- ☐ **anterior cavity**
- ☐ **anterior chamber**
- ☐ **choroid**
- ☐ **ciliary body**
- ☐ **fovea centralis**
- ☐ **iris**
- ☐ **lens**

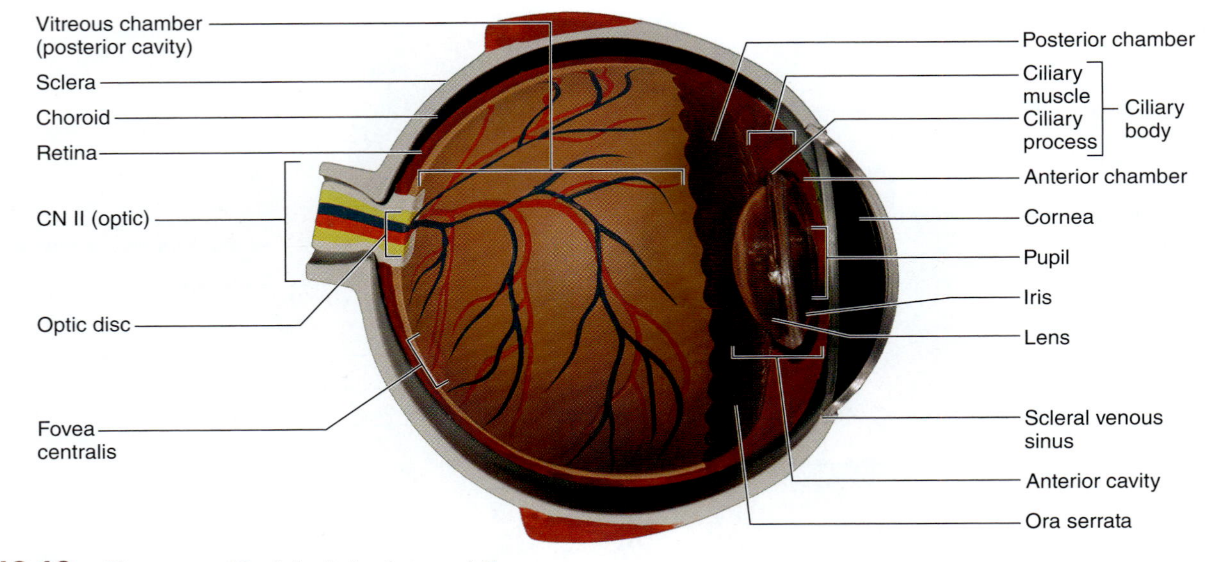

Figure 18.12 Classroom Model of the Internal Eye.

Table 18.7 Internal Structures of the Eye

Structure	Description	Function	Word Origin
Anterior Cavity	The space anterior to the lens and posterior to the cornea. It is subdivided by the iris into the anterior and posterior chambers.	Filled with aqueous humor	*anterior*, the front surface, + *cavus*, hollow
Anterior Chamber	The space between the cornea and the iris	Filled with aqueous humor, which is described elsewhere in this table	*anterior*, the front surface, + *camera*, an enclosed space
Posterior Chamber	The space between the lens and the iris	Filled with aqueous humor, which is described elsewhere in this table	*posterior*, the back surface, + *camera*, an enclosed space
Aqueous Humor	Watery fluid, similar in composition to cerebrospinal fluid. It is secreted by the ciliary processes and circulates within the anterior and posterior chambers of the eye.	Provides nourishment to the avascular lens and cornea	*aqueous*, watery, + *humor*, fluid
Choroid Layer	The pigmented, vascular layer located between the retina and the sclera	Blood vessels of the choroid supply nutrients to the tissues of the retina and sclera; pigment in the choroid absorbs light after it passes through the retina	*choroideus*, like a membrane
Ciliary Body	The thickened extension of the vascular tunic, located between the choroid and the iris; composed of both the ciliary process and the ciliary muscle	Produces aqueous humor; contraction of the ciliary muscle within the ciliary body alters the shape of the lens	*cilium*, eyelid, + *bodig*, a thing or substance

Table 18.7 Internal Structures of the Eye *(continued)*

Structure	Description	Function	Word Origin
Ciliary Muscle	Smooth muscle found within the ciliary body that is composed of both circular and radial fibers	Contraction of this muscle relaxes the suspensory ligaments that attach it to the lens, which increases the lens curvature to accommodate for near vision	*cilium*, eyelid
Fovea Centralis	The depression ("central pit") in the macula lutea that contains only cones and lacks blood vessels	The area of highest visual acuity in the eye	*fovea*, a pit, + *centralis*, in the center
Iris	The colored portion of the eye, which makes up the anterior portion of the vascular tunic; the dilator pupillae and sphincter pupillae muscles are located within the iris	Controls the amount of light entering the eye. Contraction of the radially arranged dilator pupillae muscle (under sympathetic stimulation) increases pupil diameter, whereas contraction of the circularly arranged sphincter pupillae muscle (under parasympathetic stimulation) decreases the diameter of the pupil.	*iris*, rainbow
Lens	A transparent, biconvex structure composed of a highly specialized, modified epithelium	Bends light waves (refraction) so that they hit the retina optimally for clear vision	*lens*, a lentil
Macula Lutea	A "yellow spot" on the retina located medial to the optic disc on the posterior wall of the eye, which contains the fovea centralis within it.	Contains the fovea centralis	*macula*, a spot, + *luteus*, yellow
Optic Disc ("Blind Spot")	An area of the retina where there is an absence of photoreceptors because it is where the axons of ganglion cells exit the eye to become the optic nerve	The location where axons of ganglion cells exit the eye	*optikos*, the eye, + *discus*, disc
Ora Serrata	Anteriormost portion of the retina, which appears serrated (hence the name)	Demarcates the division of the visual retina from the nonvisual retina	*ora*, an edge, + *serratus*, a saw
Posterior Cavity (Vitreous Chamber)	A space posterior to the lens and anterior to the retina	Occupied by the vitreous humor, which is described elsewhere in this table	*vitreus*, glassy, + *camera*, an enclosed space
Pupil	The space (opening) in the center of the iris	The size of the pupil (which is controlled by the smooth muscle within the iris) determines the amount of light entering the eye	*pupilla*, pupil
Retina	Also called the neural tunic of the eye; it is the inner layer of the eye composed of a pigmented layer, rods, cones, bipolar cells, and ganglion cells	Transduces light that enters the eye as light waves into nerve signals (action potentials) that are interpreted by the brain	*rete*, a net
Suspensory Ligaments	Ligaments that extend between the ciliary muscles and the lens	Attaches the lens to the ciliary muscles so that contraction and/or relaxation of ciliary muscles can alter the shape of the lens	*suspensio*, to hang up, + *ligamentum*, a bandage
Tapetum Lucidum	Metallic-appearing, opalescent inner layer of the sclera; present in many animals (e.g., the cow eye) but not in humans	Scatters light waves within the eye; allows for better vision in dim limited light (humans do not have this layer)	*tapeta*, a carpet, + *lucidus*, clear
Vascular Tunic	Middle layer of the wall of the eye; consists of the choroid, the ciliary body, and the iris	Provides nourishment to structures within the eye	*vasculum*, a small vessel
Vitreous Humor	Clear, gelatinous mass within the vitreous chamber (posterior cavity)	Helps maintain the round shape of the eye and is critical in keeping the retina against the wall of the eye	*vitreus*, glassy, + *humor*, fluid

EXERCISE 18.9

EXTRINSIC EYE MUSCLES

1. Observe a model of the eye with extrinsic eye muscles.
2. The **extrinsic,** or extraocular (*extra-*, outside of, + *oculus,* eye), muscles of the eye **(table 18.8)** allow us to move our eyes up, down, side to side, and at an angle. These muscles originate on bone and insert onto the sclera of the eye. They are named based on location and shape, so they are relatively easy to identify and remember. Recall from chapter 15 that cranial nerves innervate the extrinsic muscles of the eye. Therefore, observations of impairment in eye movements are helpful in assessing cranial nerve disorders.
3. Ask a laboratory partner to look in different directions and observe his or her eye movements. As his or her eyes move, name the muscles (in *both eyes* because they will be different!) used to cause the movement (use table 18.8 as a guide).
4. Identify the **extrinsic eye muscles** listed in **figure 18.13** and on the model of the eye, using table 18.8 and the textbook as guides. Then label them in figure 18.13.

Table 18.8 Extrinsic Eye Muscles

Muscle	Origin	Insertion	Action	Innervation	Word Origin
Inferior Oblique	Maxilla (anterior portion of orbit)	Sclera on the anterior, lateral surface of the eyeball, deep to the lateral rectus muscle	Elevates, abducts, and laterally rotates the eyeball	Oculomotor (CN III)	*inferior,* lower, + *obliquus,* slanting
Inferior Rectus	Sphenoid (tendinous ring around optic canal)	Sclera on the anterior, inferior surface of the eyeball	Depresses, adducts, and medially rotates the eyeball	Oculomotor (CN III)	*inferior,* lower, + *rectus,* straight
Lateral Rectus	Sphenoid (tendinous ring around optic canal)	Sclera on the anterior, lateral surface of the eyeball	Abducts the eyeball	Abducens (CN VI)	*lateralis,* lateral, + *rectus,* straight
Medial Rectus	Sphenoid (tendinous ring around optic canal)	Sclera on the anterior, medial surface of the eyeball	Adducts the eyeball	Oculomotor (CN III)	*medialis,* middle, + *rectus,* straight
Superior Oblique	Sphenoid (tendinous ring around optic canal)	Sclera on the posterior, superiolateral surface of the eyeball just deep to the belly of the superior rectus muscle	Depresses, abducts, and laterally rotates the eyeball	Trochlear (CN IV)	*superus,* above, + *obliquus,* slanting
Superior Rectus	Sphenoid (tendinous ring around optic canal)	Sclera on the anterior, superior surface of the eyeball	Elevates, adducts, and medially rotates the eyeball	Oculomotor (CN III)	*superus,* above, + *rectus,* straight
Levator Palpebrae Superioris*	Sphenoid (lesser wing anterior and superior to the optic canal)	Skin of the superior eyelid	Elevates the upper eyelid	Oculomotor (CN III)	*levatus,* to lift, + *palpebra,* eyelid, + *superus,* above

*This muscle, while associated with the eye, does not attach to, or move, the eyeball itself.

INTEGRATE

LEARNING STRATEGY

The following "chemical formula" can help you remember the eye muscle innervation:

$$[(SO_4)(LR_6)]_3$$

In words, the superior oblique **(SO)** is innervated by cranial nerve IV **(4)**, the lateral rectus **(LR)** is innervated by cranial nerve VI **(6)**, and the rest of the eye muscles are innervated by cranial nerve III **(3)**.

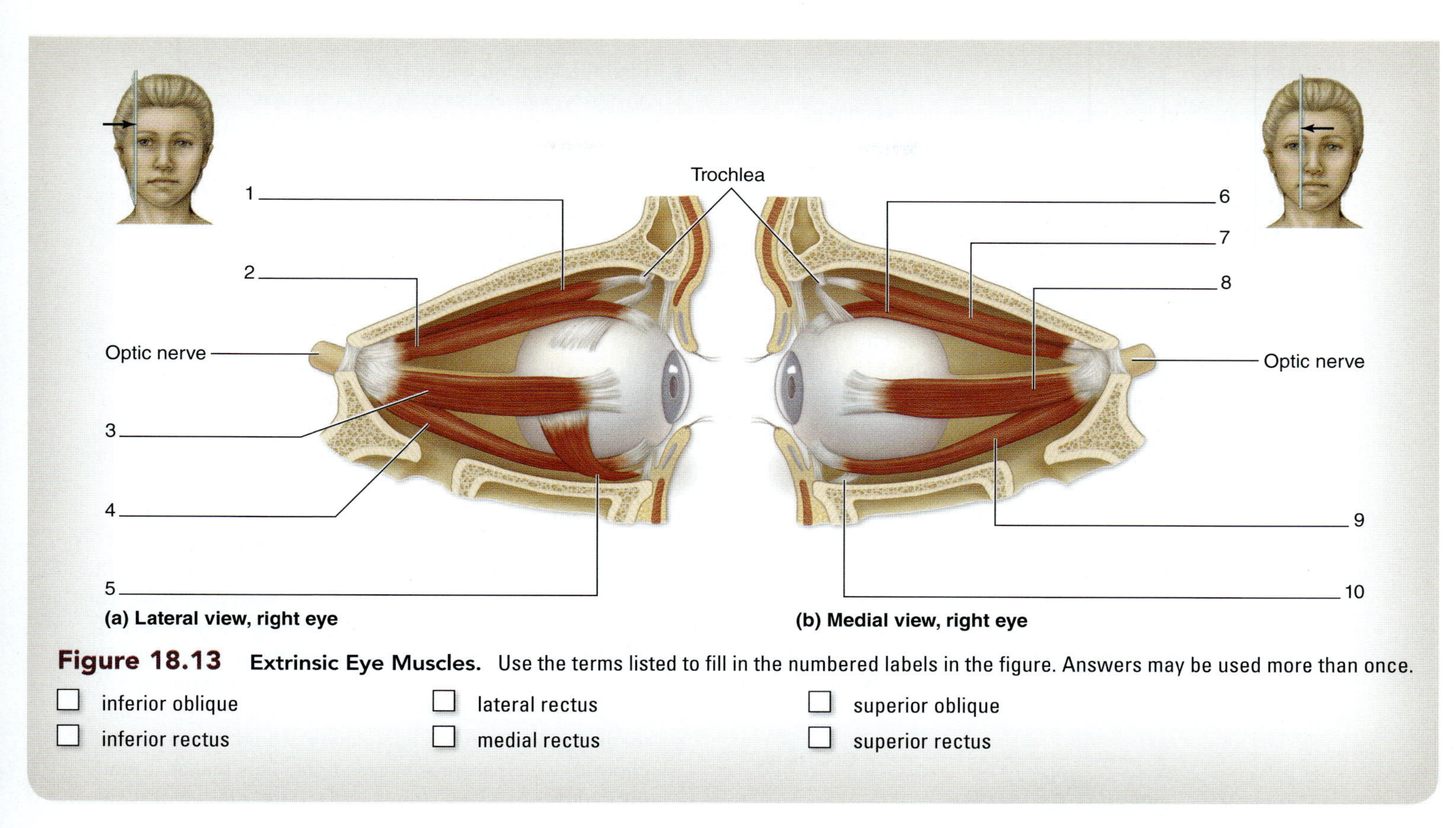

Figure 18.13 Extrinsic Eye Muscles. Use the terms listed to fill in the numbered labels in the figure. Answers may be used more than once.

- ☐ inferior oblique
- ☐ inferior rectus
- ☐ lateral rectus
- ☐ medial rectus
- ☐ superior oblique
- ☐ superior rectus

EXERCISE 18.10

COW EYE DISSECTION

In exercise 18.10 we will explore the anatomy of the eye, using a cow eye as a model organ. Although there is some variance in structure between a cow eye and a human eye, cow eyes are much easier to obtain and they are larger, which greatly facilitates making internal observations of the eye. This exercise is designed to be done using fresh cow eyes, although preserved cow or pig eyes may be substituted if necessary. If dissecting a preserved cow or pig eye, the tissues will be tougher and the cornea will be opaque instead of transparent.

1. Obtain a dissecting pan, dissecting tools, and a fresh cow eye. Observe the gross structure of the eye before making any cuts. Identify the following structures, using tables 18.7 and 18.8 and **figure 18.14** as guides:

   - ☐ **cornea**
   - ☐ **extrinsic eye muscles**
   - ☐ **optic nerve**
   - ☐ **orbital fat pad**
   - ☐ **sclera**

2.  Using scissors and forceps (the tissue will be slippery!), remove the orbital fat pad and extraocular muscles, leaving the optic nerve intact (figure 18.14*b*). Once these structures have been removed, the entire eye should be visible. Notice the toughness of the outer covering or **sclera** of the eye. This is the layer of tissue to cut through in order to see structures within the eye.

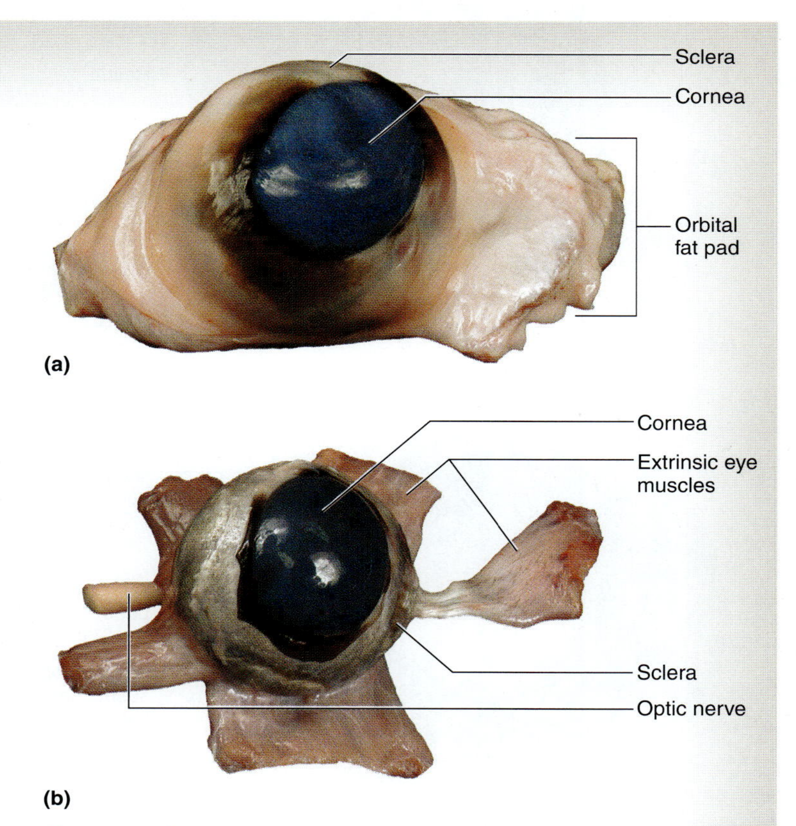

Figure 18.14 Fresh Cow Eye. (a) Before dissection with orbital fat pad intact. (b) After dissecting away the orbital fat pad to expose the extraocular muscles and optic nerve.

(continued on next page)

(continued from previous page)

3. Using scissors and forceps, cut the eye open by making a coronal incision through the sclera that completely encircles the eye approximately 1/4 of an inch posterior to the cornea. Once this has been done, notice that a jelly-like fluid oozes out of the posterior cavity of the eye. This fluid is the **vitreous humor,** which fills the posterior cavity **(figure 18.15*b*).** This fluid's functions include holding the **retina** against the posterolateral walls of the eye. In the cow eye, much of the **choroid,** which contains black pigment, becomes mixed into the vitreous humor once the eye is cut open. Thus, it has the potential to create a gooey, black mess almost immediately. Before things get too "mixed up," identify structures in the posterior half of the dissected eye. Look for a yellowish, thin membrane that is connected to the posterior wall of the eye at only one spot (figure 18.15*a*). This is the retina. It contains neurons responsible for detecting visual stimuli and initiating nerve signals to the brain (histology of the retina was covered in detail in exercise 18.5). The retina is very delicate and easily falls away from the posterior wall of the eye when the vitreous humor is not present to hold it in place.

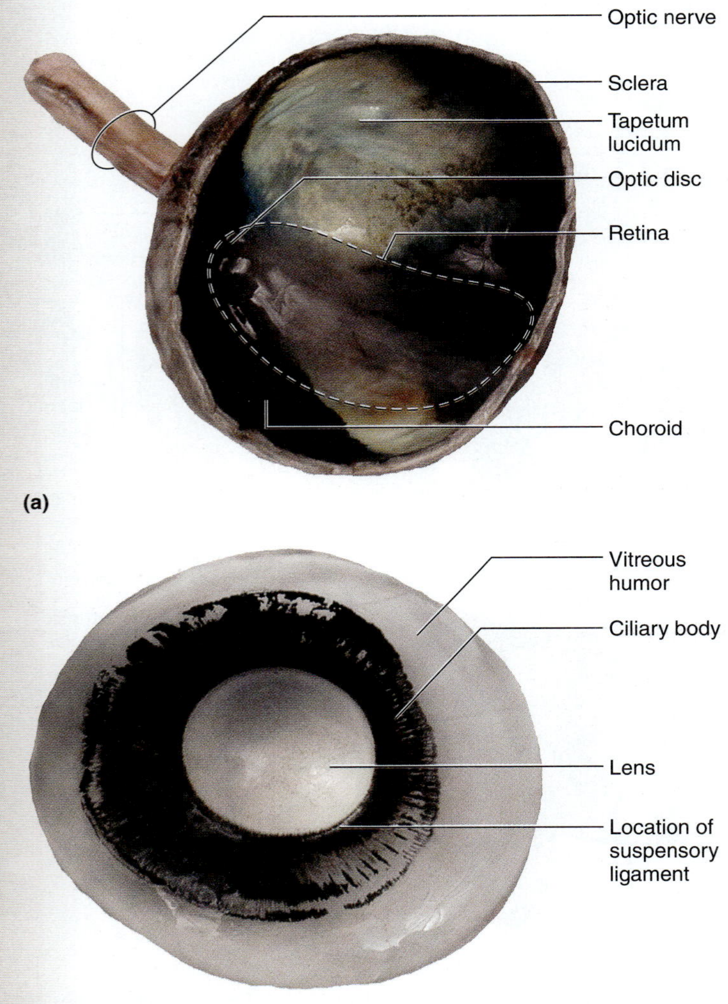

Figure 18.15 Coronal Sections of Cow Eye. (a) Posterior part demonstrating the retina, optic disc, and tapetum lucidum. (b) Anterior part demonstrating the choroid, lens, and iris.

4. Find the location where the retina attaches to the posterior wall of the eye (it will "pucker" in this area). If there is difficulty finding this spot, first find the optic nerve on the outside of the eye and then look inside the eye for the location where the optic nerve leaves the eye. This spot within the eye is the **optic disc** (blind spot). It is called a "blind" spot because it is devoid of photoreceptors. This is where axons from the ganglion cell layer of the retina (see exercise 18.5) leave the eye and extend to the brain as the optic nerve (CN II).

5. Observe the inner walls of the posterior half of the eye. Notice the very colorful, iridescent **tapetum lucidum** (figure 18.15*a*). This structure is not present in humans, but it is present in animals that must be able see well in dim light, such as cows. The tapetum lucidum reflects light. Thus, when it is dark outside and very little light is entering the eye, the tapetum lucidum causes light waves to bounce around within the eye and increases the frequency with which light rays stimulate the retina. This makes things more visible, but the image is not sharp. It is the reflection of light waves from the tapetum lucidum that causes a cow's eyes (or those of other animals) to "glow" when a light shines on them at night. In humans, the inside of the eye is completely coated with a black choroid, which absorbs excess light. This makes it more difficult for us to see things in the dark. On the plus side, the images seen by humans are sharper.

6. Now focus on the anterior portion of the eye (figure 18.15*b*). Again, it may be somewhat difficult to see many structures because the choroid covers everything, making the structures very dark. Notice the semitransparent **lens,** which is suspended in place by a ring of black-colored tissue. This tissue is the **ciliary body,** whose function is to suspend the lens. The cavity anterior to the lens and posterior to the cornea is the **anterior cavity** of the eye. The anterior cavity is further subdivided by the iris into an **anterior chamber** (between the cornea and the iris) and a **posterior chamber** (between the iris and the lens). In a living organism, the anterior cavity is filled with a clear, watery fluid called **aqueous humor.** Try to find the fine, delicate structures composing the **suspensory ligament,** which extends between the ciliary body and the lens.

7. Carefully remove the lens from the eye. Notice that it is somewhat, but not completely, transparent. Place it on a piece of paper containing text. As shown in **figure 18.16,** place the lens over a letter or two of text and make note of the change in appearance of the text, if any, as seen through the lens:

____

____

____

____

8. Identify the following structures on the interior of the dissected cow eye (use tables 18.6 and 18.7, and figures 18.14 through 18.16 as guides):

- ☐ **anterior cavity**
- ☐ **choroid**
- ☐ **ciliary body**
- ☐ **lens**
- ☐ **optic disc**
- ☐ **posterior cavity**
- ☐ **retina**
- ☐ **suspensory ligament**
- ☐ **tapetum lucidum**
- ☐ **vitreous humor**

9. When finished with the dissection, clean up the workspace: Dispose of the cow eye debris in the organic waste receptacle. Dispose of used scalpel blades in the sharps container. Dispose of used paper towels and other paper waste in the wastebasket. Rinse off the dissecting tray and dissection instruments, and lay them out to dry. Finally, wipe down the laboratory workstation with disinfectant so it is clean for the next person who comes into the laboratory.

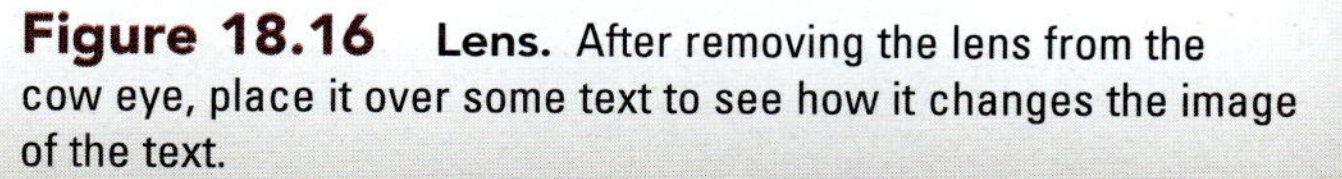

Figure 18.16 **Lens.** After removing the lens from the cow eye, place it over some text to see how it changes the image of the text.

EXERCISE 18.11

GROSS ANATOMY OF THE EAR

The ear is responsible for two important special sensory modalities: equilibrium (balance) and hearing. It houses the organs for both within the *petrous part of the temporal bone.* These tiny sensory organs are nearly impossible to identify on a cadaver; thus, exploration of the gross anatomy of the ear will be accomplished using models of the ear.

1. Obtain a model of the ear **(figure 18.17).** On the model, first distinguish between the external-ear, middle-ear, and inner-ear cavities, and the structures that link the cavities to each other **(table 18.9).** The **tympanic membrane** is the link between the external-ear and the middle-ear cavities, while the **oval window** is the link between the middle- and inner-ear cavities.

2. Identify the following structures on the model, using figure 18.17 and table 18.9 as guides:

- ☐ **auditory tube**
- ☐ **auricle**
- ☐ **cochlea**
- ☐ **endolymph**
- ☐ **external acoustic meatus**
- ☐ **incus**
- ☐ **malleus**
- ☐ **ossicles**
- ☐ **oval window**
- ☐ **perilymph**
- ☐ **round window**
- ☐ **saccule**
- ☐ **semicircular canals**
- ☐ **stapedius muscle**
- ☐ **stapes**
- ☐ **tensor tympani muscle**
- ☐ **tympanic membrane**
- ☐ **utricle**

3. Sketch the ear and label the locations of the structures listed in step 2 in the space provided.

(continued on next page)

(continued from previous page)

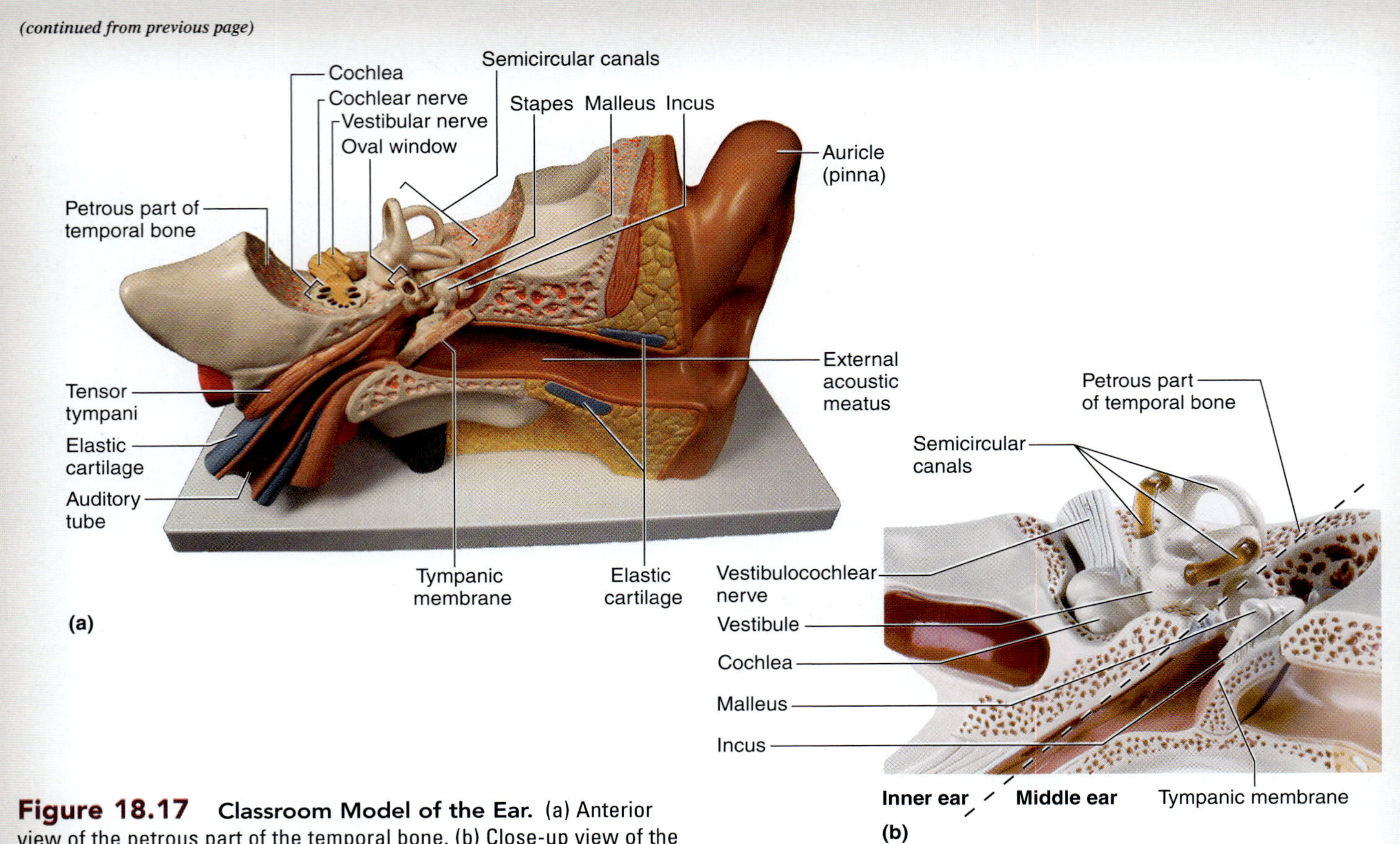

Figure 18.17 Classroom Model of the Ear. (a) Anterior view of the petrous part of the temporal bone. (b) Close-up view of the inner ear on the model.

Table 18.9 Structures of the External, Middle, and Internal Ear

Structure	Description	Function	Word Origin
External Ear			
Auricle (Pinna)	External ear, composed of an elastic cartilage skeleton that is covered with skin	Collects sound waves from the environment and funnels them into the external auditory meatus	*pinna*, a wing
External Acoustic Meatus	Canal leading from the auricle of the ear to the tympanic membrane	Transmits sound waves that arrive at the auricle to the tympanic membrane, where they cause it to vibrate	*externa*, outside, + *acoustic*, relating to sound, + *meatus*, a passage
Tympanic Membrane	Drum-like, tight, thin membrane that separates the external ear cavity from the middle-ear cavity	Vibrates in response to sound waves that strike it as they reach the end of the external acoustic meatus; these vibrations cause vibrations in the ossicles of the middle-ear cavity (malleus, incus, and stapes)	*tympanon*, drum, + *membrana*, a membrane
Middle Ear	Air-filled cavity between the external ear and the inner ear	Contains the ear ossicles	
Auditory Ossicles	Three tiny bones (malleus, incus, and stapes) found within the middle ear	Transmit movements caused by pressure vibrations from the tympanic membrane to the oval window of the cochlea, causing fluid pressure waves in the perilymph of the scala vestibuli	*ossiculum*, a bone
Incus	Tiny bone located within the middle-ear cavity that is shaped like an anvil	Transmits movements caused by pressure vibrations from the malleus to the stapes, thus participating in the transmission and amplification of the vibrations of the tympanic membrane	*incus*, an anvil

Table 18.9	Structures of the External, Middle, and Internal Ear *(continued)*		
Structure	**Description**	**Function**	**Word Origin**
Middle Ear *(continued)*			
Malleus	Tiny bone located within the middle-ear cavity that is shaped like a hammer	Transmits movements caused by pressure vibrations from the tympanic membrane to the incus, thus participating in the transmission and amplification of the vibrations that arrive as sound waves at the tympanic membrane	*malleus*, a hammer
Stapes	Tiny bone located within the middle-ear cavity that is shaped like a stirrup	Transmits movements caused by pressure vibrations from the incus to the oval window of the cochlea, causing fluid pressure waves in the perilymph of the scala vestibuli	*stapes*, a stirrup
Auditory Tube (Pharyngotympanic or Eustachian tube)	Tube lined with elastic cartilage that connects the middle-ear cavity to the nasopharynx	Opening of this channel allows air to enter or leave the middle-ear cavity such that the pressure in the middle ear equilibrates with the environmental pressure; this allows the tympanic membrane to vibrate freely	*audio*, to hear, + *tubus*, a canal
Oval Window	Membrane-covered opening into the scala vestibuli that is covered by the foot of the stapes	Vibrations of the stapes at the oval window cause fluid pressure waves in the perilymph of the scala vestibuli	*oval*, egg-shaped, + *window*, an opening
Stapedius Muscle	Small muscle connecting the neck of the stapes to the temporal bone	Contraction of this muscle acts to dampen vibrations of the stapes as a protective measure against excessive movement at the oval window from very loud noises	*stapedius*, relating to the stapes
Tensor Tympani Muscle	Small muscle connecting the handle of the malleus to the cartilage of the auditory tube	Contraction of this muscle pulls the malleus medially and tenses the tympanic membrane as a protective measure against excessive vibration from very loud noises	*tensus*, to stretch, + *tympani*, relating to the tympanic membrane
Inner Ear	Fluid-filled space located within the petrous part of the temporal bone that contains the cochlea, vestibule, and semicircular canals	Holds the organs responsible for the sensation of hearing (cochlea) and balance and equilibrium (vestibule and semicircular canals)	
Cochlea	Spiral-shaped organ found within the inner ear	Contains the spiral organ and associated structures that are involved in the special sense of hearing	*cochlea*, a snail shell
Saccule	Smallest membranous sac in the vestibule; connects with the cochlear duct	Contains receptors that sense linear vertical acceleration	*saccus*, a sac
Semicircular Canals	Three ring-like canals that are oriented at right angles to each other	Detect angular dynamic equilibrium	*semicircular*, shaped like a half circle, + canalis, a duct or channel
Spiral Organ (Organ of Corti)	Organ composed of specialized epithelium that is found within the scala media (cochlear duct) of the cochlea	Special sensory organ for hearing	*spiralis*, a coil, + *organon*, a tool or instrument
Utricle	The largest membranous sac in the vestibule	Contains receptors that sense linear horizontal acceleration	*uter*, leather bag
Vestibule	Located between the cochlea and the semicircular canals; contains the saccule and utricle	Detects both static equilibrium and linear, dynamic equilibrium	*vestibulum*, entrance court
Vestibulocochlear Nerve	CN VIII; travels through the internal acoustic meatus	Cranial nerve transmitting nerve signals associated with balance, equilibrium, and hearing to the brain	*vestibulo-*, referring to the vestibule, + *cochlea*, referring to the cochlea

(continued on next page)

(continued from previous page)

4. After identifying all of the gross structures of the ear, review the sequence of events required for the transmission of sound waves from the environment to the cochlea **(figure 18.18).** Name all of the structures involved in the sequence. The sequence is as follows:

   (1) Sound waves are "funneled" into the **external acoustic meatus** by the contours of the outer ear **(auricle)** and cause vibrations of the **tympanic membrane.** The **auditory tube** ensures that air pressure in the middle ear is the same as air pressure in the environment so the tympanic membrane can vibrate freely.

   (2) Vibrations of the tympanic membrane cause the **auditory ossicles** malleus, incus, and stapes to vibrate. Excessive vibrations (e.g., from a loud noise) cause a reflexive contraction of the **tensor tympani** and **stapedius** muscles to dampen the vibrations of the ear ossicles and help protect the delicate cells of the inner ear.

   (3) Vibration of the foot of the stapes against the **oval window** causes pressure waves of the **perilymph** within the scala vestibuli.

   (4) Vestibular membrane movements cause pressure waves in the endolymph within the scala media (cochlear duct). This displaces the basilar membrane (in different regions depending on the frequency). Hair cells of the spiral organ bend, initiating nerve signals that are transmitted along the cochlear division of the vestibulocochlear nerve (CN VIII) to the brain.

   (5) Pressure waves are absorbed by the round window.

5. *Optional Activity:* **AP|R 7: Nervous System**—Watch the "Hearing" animation to review the sequence of events involved in hearing.

INTEGRATE

CLINICAL VIEW

Pressure Changes in the Middle Ear

The auditory tube is lined with elastic cartilage and remains collapsed unless there is a large difference in pressure between the environment and the middle-ear cavity. When a difference in pressure exists, the auditory tube opens briefly and air moves to equalize the pressure in the middle ear with the pressure in the environment. The opening and closing of the auditory tube is what accounts for the "popping" sound made when pressure is equalized.

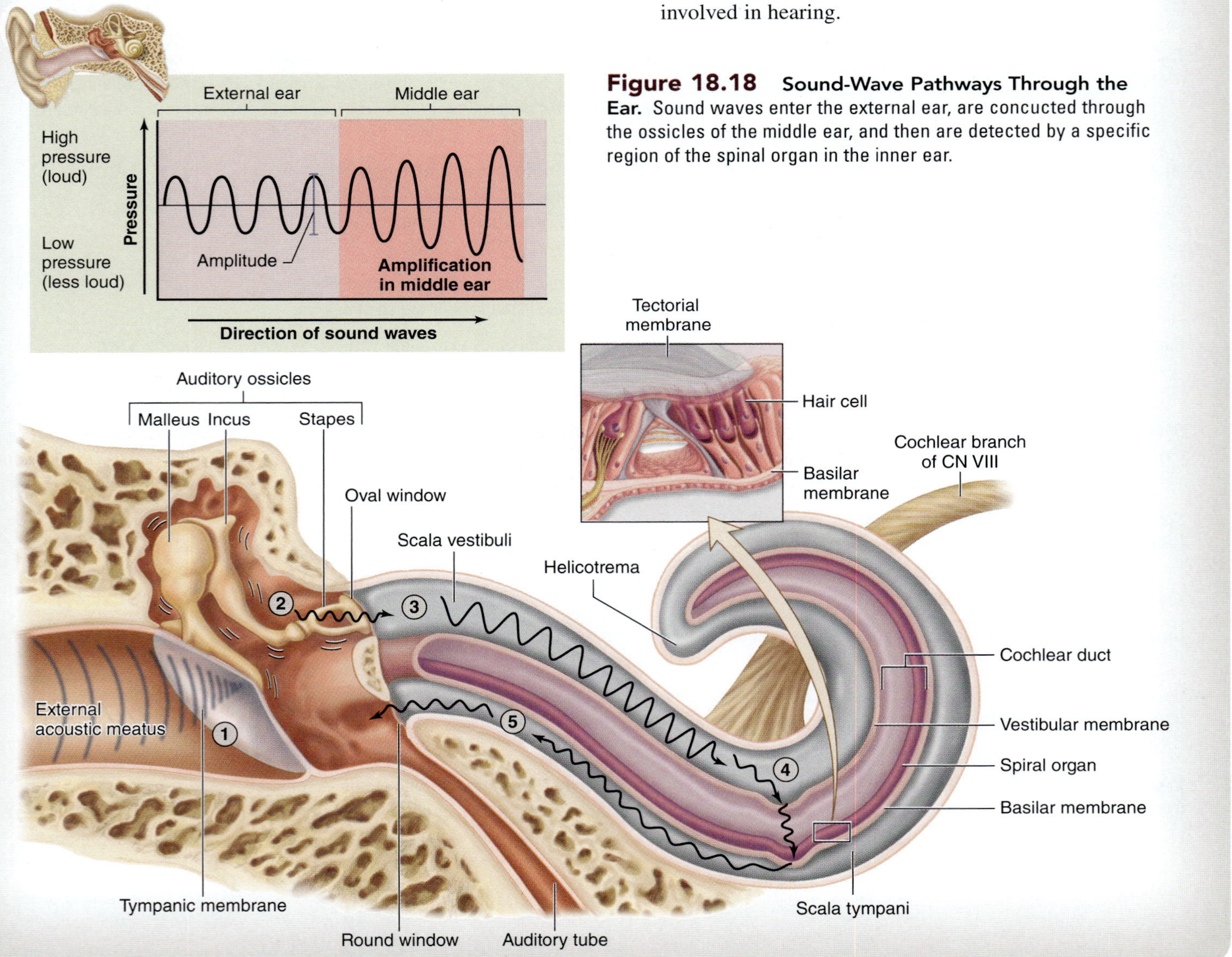

Figure 18.18 Sound-Wave Pathways Through the Ear. Sound waves enter the external ear, are concucted through the ossicles of the middle ear, and then are detected by a specific region of the spinal organ in the inner ear.

PHYSIOLOGY

General Senses

The following laboratory activities explore some of the functional properties of sensory receptors, such as the tactile (Meissner) and lamellated (Pacinian) corpuscles observed in the histology sections of this manual. General sensory perception can be complicated because several sensory pathways merge to share the same pathways as they extend to the brain. This convergence of pathways is the mechanism behind referred pain and "burning" cold. Due to time limitations in the laboratory, the general sensation physiology exercises in this chapter will be those that demonstrate the **size of receptive fields** in different areas of the body, the association between size of receptive fields and **density of sensory receptors**, and the process of sensory **adaptation.**

The remaining physiology exercises explore some of the functional properties of the sensory receptors involved with the special senses (gustation, olfaction, vision, hearing, and equilibrium).

EXERCISE 18.12

TWO-POINT DISCRIMINATION

This exercise tests the ability of an individual to discriminate between tactile sensations that are applied to two different locations on the skin at the same time (two-point discrimination). This involves pressing a semi-sharp object into the skin. Which of the general sensory receptors is responsible for responding to this mode of sensation (assume the test will not cause pain)?

____

____

The two-point discrimination test is an easy way to determine the size of the **receptive field,** which is associated with the relative density of sensory receptors in the skin. The test will be performed on skin covering different regions of the body. Then, the two-point discrimination results from these different regions will be compared to evaluate the difference in receptive field size between different regions of the body.

Before beginning, state a hypothesis regarding the difference (if any) in the density of sensory receptors on the back of the hand as compared to the palm of the hand:

____

____

____

____

1. Obtain a two-point discrimination tool, or a metric ruler and a set of calipers **(figure 18.19).**

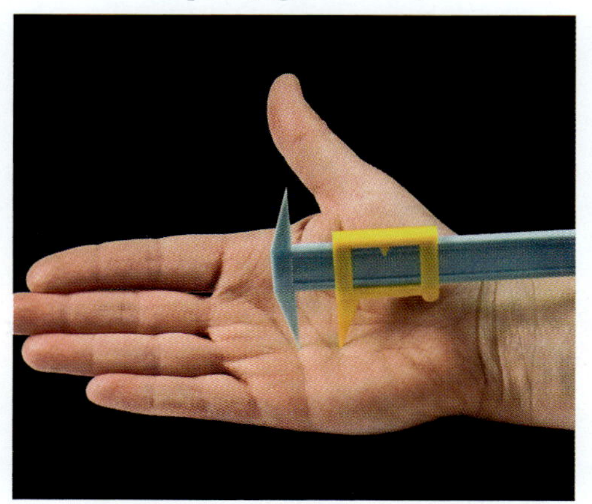

Figure 18.19 Two-Point Discrimination Test. Testing two-point discrimination on the palm of the hand using calipers.

2. Choose a lab partner to be the subject. First test for two-point discrimination on skin of the subject's cheek. Start with the calipers or two-point discrimination tool all the way closed. Next, gently touch the subject's cheek with the calipers. The subject will feel only one point being touched.
3. Remove the calipers, open them one segment, then touch the subject's cheek again. Continue this process until the subject reports feeling two distinct points being touched instead of one. Now remove the calipers while holding the calipers to ensure they don't move. Place the tip of the calipers on a metric ruler and measure the distance. When using a two-point discrimination tool, simply take the final measurement from the tool. Record the results in **table 18.10.**
4. Complete steps 2 and 3 for each of the regions listed in table 18.10. Then record the results in table 18.10.

Table 18.10 Results of the Two-Point Discrimination Test

Body Region	Two-Point Discrimination Distance (mm)
Face (cheek)	
Face (lips)	
Posterior Neck	
Forearm (anterior)	
Forearm (posterior)	
Hand (posterior/ back surface)	
Hand (anterior/ palmar surface)	
Fingertip	
Leg	

EXERCISE 18.13

TACTILE LOCALIZATION

This exercise tests the ability of the brain to detect the precise location on the body that has been touched by a stimulus. This is called **tactile localization.** Perception of the locale of a stimulus differs for areas of the body. This variation in stimulus locale is related to the size of the receptive field. When touching the subject during the tactile localization task, the subject will not necessarily know exactly what point was touched. Instead, the subject will perceive that a larger area surrounding the touch point was stimulated. This larger area is a *receptive field.*

Before beginning, state a hypothesis regarding the difference (if any) in the receptive field of sensory receptors within the skin on the back of the hand as compared to the skin on the palm of the hand:

____

____

____

____

1. Obtain two or three markers of different colors.
2. Choose a person to be the subject. Have the subject close his or her eyes.
3. Touch the palm of the subject's hand with a washable marker **(figure 18.20).**
4. Hand the subject a marker of a different color than the one previously used. Then ask the subject (still keeping his or her eyes closed) to try to touch the exact point previously touched, using his or her marker.
5. Measure the distance between the two points in millimeters (mm) and record it in **table 18.11**.
6. Repeat the process two more times. As the examiner, be sure to always touch the same point. However, the subject will not necessarily touch that same point. Calculate the average distance for the three trials and record it in table 18.11.
7. Complete steps 2–6 for each of the regions listed in table 18.11. Then record the results in table 18.11.

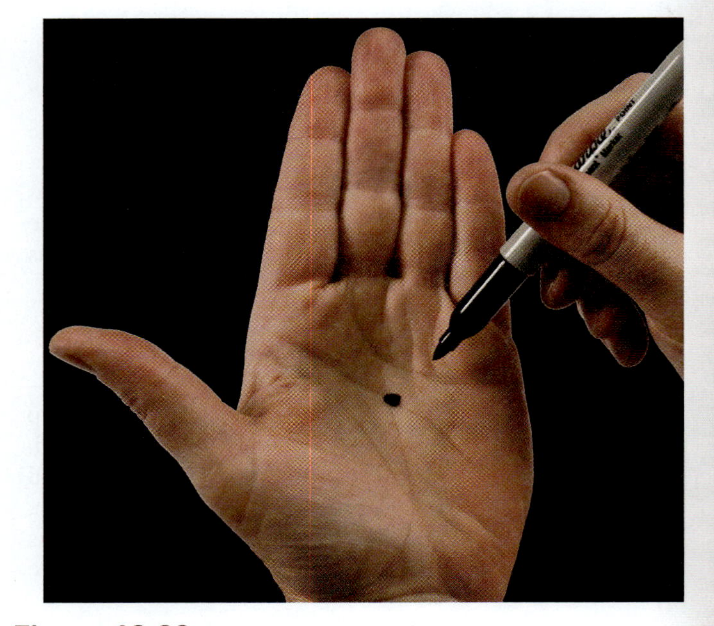

Figure 18.20 Tactile Localization Test. Testing tactile localization on the palm of the hand.

Table 18.11 Results of the Tactile Localization Test

Body Region	Test 1 Distance (mm)	Test 2 Distance (mm)	Test 3 Distance (mm)	Average Distance (mm)
Palm of Hand				
Back of Hand				
Fingertip				
Anterior Forearm				
Back of Neck				
Anterior Leg				

EXERCISE 18.14

GENERAL SENSORY RECEPTOR TESTS: ADAPTATION

This exercise tests the ability of the brain to detect a constantly applied stimulus over time. A general property of the nervous system is that the brain tends to place emphasis and focus on incoming stimuli that are *changing* as opposed to stimuli that remain constant. Thus, when initially putting on clothing, glasses, jewelry, or other items in the morning, you feel the clothing slide across your skin, the glasses rest upon your ears and nose, and your watch rest on your wrist. Soon afterwards, if these items don't move considerably, then you stop noticing them. Have you ever found yourself looking for your glasses only to discover they are right there on your head? This phenomenon is called **sensory adaptation.** Here, sensory receptors adapt so that the initial sensation (e.g., pressure) becomes a new "set point" for the brain. The brain subsequently ignores the constant signals coming in from that stimulus. This allows the brain to respond only when there is a *change* in the stimulus.

1. Obtain five pennies.
2. Choose a person to be the subject. Have the subject close his or her eyes.
3. Place a coin on the anterior surface of the subject's forearm approximately 2 cm proximal from the wrist **(figure 18.21*a*).**
4. Ask the subject to state when he or she can no longer feel the coin on his or her arm. Record the time elapsed in **table 18.12.**
5. Repeat the test, placing coins at the locations on the forearm listed in table 18.12. Record the time elapsed for each test in table 18.12.
6. Finally, perform the test one more time on the spot used in step 3, only this time perform the test using five pennies stacked on top of each other (figure 18.21*b*).

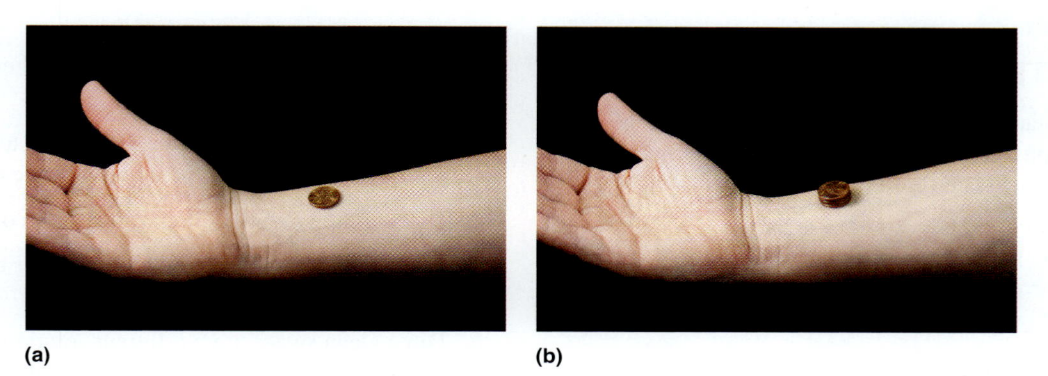

(a) (b)

Figure 18.21 **Sensory Adaptation Test.** Testing sensory adaptation on the anterior surface of the forearm with (a) one object and (b) multiple objects.

Table 18.12 Results of the Adaptation Test

Coin Placement on Anterior Forearm	Number of Coins	Time Elapsed When Subject No Longer Perceives the Coins (Seconds)
2 cm Proximal from Wrist	1	
Midway Between Wrist and Elbow	1	
1 cm Distal from Elbow	1	
2 cm Proximal from Wrist	5	
Midway Between Wrist and Elbow	5	
1 cm Distal from Elbow	5	

Special Senses

Gustation (Taste)

There are approximately 4000 gustatory receptors ("taste buds") in the body, many of which are associated with the papillae of the tongue observed in the histology section of this chapter. In addition to this location of receptors on the tongue, there are also taste receptors located on the soft palate, pharynx, and epiglottis. Each receptor is specialized to respond to one of five primary taste sensations: salty, sweet, sour, bitter, and umami ("meaty"). **Table 18.13** summarizes the types of substances that elicit each of the five basic taste sensations. This laboratory exercise tests the specificity and location of the different types of taste sensed by these taste receptors.

Table 18.13 The Five Basic Taste Sensations

Taste Sensation	Substances Stimulating the Sensation	Examples
Bitter	Alkaloids, which are plant-derived, nitrogen-containing basic compounds	Quinine, nicotine, caffeine, unsweetened cocoa
Salt	Metal ions	Table salt (NaCl)
Sour	Acids, which are substances in foods that release hydrogen ions (H^+)	Lemon juice, vinegar
Sweet	Natural sugars or artificial sweeteners	Corn syrup, Splenda®, granulated sugar
Umami	Amino acids such as glutamate and aspartate	Meats, cheeses, tomatoes, anchovies

EXERCISE 18.15

GUSTATORY TESTS

Before beginning, state a hypothesis regarding the region(s) of the oral cavity and pharynx that may have the largest number of receptors for each of the taste sensations.

____

____

____

1. Obtain seven cotton swabs and a small sample of each of the following:
   - ☐ monosodium glutamate solution (umami)
   - ☐ salt solution (salt)
   - ☐ sugar solution (sweet)
   - ☐ tonic water (bitter)
   - ☐ vinegar (sour)
   - ☐ water (for rinsing mouth)
2. One student will be the tester and one student will be the subject. As the tester, do not tell the subject which solutions are being tested. Dip a cotton swab in one of the solutions and apply it to the roof of the subject's mouth (the palate), the inside of his or her cheeks, and the surface of his or her tongue.
3. Place a check mark in **table 18.14** next to all of the regions where the subject is able to taste the solution.
4. Discard the swab in a biohazard waste container.
5. Have the subject drink some plain water to rinse out his or her mouth.
6. Dip a clean swab into a different solution and repeat step 2. Continue to do this until the subject has tasted all of the solutions.
7. Use two cotton swabs. Dip one in the salt solution and one in the monosodium glutamate solution. Swab the subject's tongue with both solutions simultaneously and ask the subject to describe the taste. Record the subject's description of the taste here:

____

____

____

Table 18.14 Results of the Gustatory Tests

Place a check in the boxes where the subject was able to detect each taste.

Taste	Palate	Cheeks	Tongue
Bitter			
Salty			
Sour			
Sweet			
Umami			

Olfaction (Smell)

As observed in exercise 18.4, the sense of olfaction involves sensory receptors that are embedded in a special epithelium that lines the roof of the nasal cavity, the **olfactory epithelium.** This epithelium also contains goblet cells, which produce mucin that when hydrated forms mucus. As odorants enter the nasal cavity, they dissolve in the mucus. The olfactory receptor cells bind the odorant molecules and are stimulated. Nerve signals are initiated in the brain that are interpreted as the sense of olfaction (smell). As likely experienced before, the sense of smell is highly associated with the sense of gustation (taste). Recall a time when you had a "stuffy nose." The blockage of the sense of smell may have greatly affected your appetite because of its effect on gustation. This exercise will test the effect of olfaction on the perception of gustatory (taste) sensation.

EXERCISE 18.16

OLFACTORY TESTS

EXERCISE 18.16A Effect of Olfaction on Taste I

1. Obtain six cotton swabs, a blindfold, and vials containing samples of each of the following:
   - ☐ almond extract
   - ☐ clove oil
   - ☐ lemon extract
   - ☐ peppermint oil
   - ☐ vanilla
   - ☐ wintergreen oil
2. One student will be the tester and one student will be the subject. As the tester, do not tell the subject which substances are being tested. Have the subject close his or her eyes (or use the blindfold if that is easier) and pinch his or her nose closed.
3. Dip a cotton swab in one of the solutions and apply it to the subject's tongue. Ask the subject if he or she can taste the substance. Place a check in the appropriate box in **table 18.15** if the subject was able to taste the odorant with the nose closed.
4. After a few seconds, have the subject release his or her nose and breathe in. Ask him or her to identify the odorant again. Place a check in the appropriate box in table 18.15 if the subject was able to taste the odorant with the nose open.
5. Briefly note the conclusions in the space below:

____

____

____

Table 18.15 Results of the Tests for the Effect of Olfaction on Taste I

Place a check in the boxes where the subject was able to detect each taste.

Odorant	Nose Closed	Nose Open
Almond		
Clove		
Lemon		
Peppermint		
Vanilla		
Wintergreen		

(continued on next page)

(continued from previous page)

EXERCISE 18.16B Effect of Olfaction on Taste II

1. Obtain the following:
   - ☐ **apple slices**
   - ☐ **potato slices**
   - ☐ **blindfold**
2. One student will be the tester and one student will be the subject. As the tester, do not tell the subject which substances are being tested. Have the subject close his or her eyes (or use the blindfold if that is easier). Then have the subject pinch his or her nose closed and stick out his or her tongue.
3. Apply either the apple or the potato slice to the subject's tongue. Ask the subject if the substance can be tasted. Place a check in the appropriate box in **table 18.16** if the subject was able to correctly identify the taste of the substance.
4. After a few seconds, have the subject once again close his or her eyes and stick out his or her tongue. Place the apple on the tongue and the potato under the nose (or vice versa). Place a check in the appropriate box in table 18.16 if the subject was able to correctly identify the substance.
5. Briefly note the conclusion in the space below:

____

____

____

Table 18.16 Results of Tests for the Effect of Olfaction on Taste Sensation

Subject pinches nose closed and tester places item on subject's tongue.

Correct Identification of Taste?	Yes	No
Apple		
Potato		

Tester places one item on subject's tongue and the other under subject's nose (subject's nose is open).

Item on Tongue	Item Under Nose	Correct Taste Perceived?
Apple	Potato	
Potato	Apple	

EXERCISE 18.16C Olfactory Adaptation

1. Obtain six cotton swabs, a blindfold, a timer or other mechanism for keeping time, and vials containing samples of each of the following:
   - ☐ **almond extract**
   - ☐ **peppermint oil**
   - ☐ **clove oil**
   - ☐ **vanilla**
   - ☐ **lemon extract**
   - ☐ **wintergreen oil**
2. One student will be the tester and one student will be the subject. As the tester, do not tell the subject which odors are being tested. Have the subject close his or her eyes (or use the blindfold if that is easier).
3. Dip a cotton swab in one of the solutions and hold the swab 1–2 inches away from the subject's nose. Start the timer as soon as the swab is placed next to the subject's nose. Ask the subject to identify the odor and state when he or she can no longer smell the odor. The odor will "go away" with time because of adaptation of the olfactory receptor cells to the odor. Record the time taken for adaptation to occur in **table 18.17.**
4. Repeat step 3 for all of the remaining odorants.
5. Briefly note the conclusion in the space below:

____

____

____

Table 18.17 Results of Olfactory Adaptation Tests

Record the time for adaptation for each of the following odorants.

Odorant	Time for Adaptation (Seconds)
Almond	
Clove	
Lemon	
Peppermint	
Vanilla	
Wintergreen	

Vision

Recall from exercise 18.5 that the special sense of vision involves stimulation of photoreceptors housed in the neural tunic of the eye, called the **retina.** Specifically, light changes the conformation of light-sensitive proteins in the photoreceptors of the retina: the rods and cones. **Rods** are photoreceptors that specialize in black-and-white vision, whereas **cones** are photoreceptors that specialize in color vision. When light stimulates these photoreceptors, nerve signals are transmitted along the optic nerve to the thalamus and then to the occipital lobe of the cerebrum, where the information is perceived as a visual image. Incidentally, there are no rods or cones where the optic nerve exits the eye, making that location a visual *blind spot.* Refer to the textbook as a guide when reviewing the specific pathways involved in processing visual information.

EXERCISE 18.17

VISION TESTS

The vast majority of us have first-hand knowledge of the special sense of vision. A trip to the eye doctor involves testing **visual acuity.** An ophthalmologist assesses **vision** with a Snellen eye chart **(figure 18.22),** and from there determines if corrective lenses are necessary. Also, sometimes age forces us to hold things we are reading farther away from the eyes in order to focus on an image or the words on a page. This **near-point accommodation** decreases dramatically with age due to the decreased elasticity in the lens of the eye. The result is an inability to change the shape of the lens, as is required when focusing on near objects. The following exercises test such aspects of vision as acuity, near-point accommodation, and the visual blind spot.

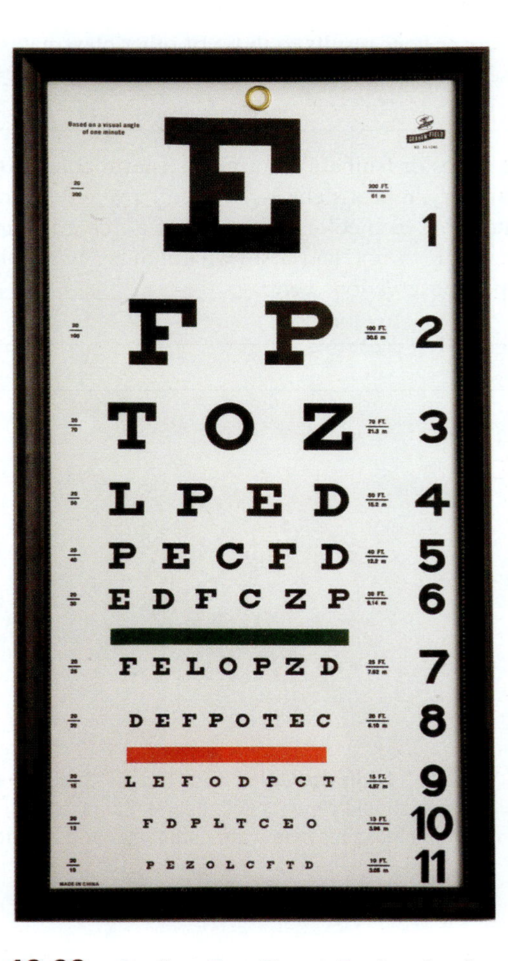

Figure 18.22 Snellen Eye Chart. Testing visual acuity using the Snellen eye chart.

EXERCISE 18.17A Visual Acuity

1. Obtain a Snellen eye chart to test visual acuity. Hang the eye chart at eye level on a wall. Be sure that the wall is well illuminated.
2. Have the subject stand 20 feet from the eye chart, while a partner remains near the eye chart to validate the subject's responses.
3. Have the subject cover his or her left eye and read the lowest possible line of letters with his or her right eye. Record the visual acuity in **table 18.18.**
4. Repeat step 3 with the subject's left eye. Note that if the subject wears corrective lenses, this test can be conducted with and/or without glasses. A ratio of 1 (e.g., 20/20) indicates normal vision. A ratio of greater than 1 (20/15) indicates greater visual acuity; a ratio of less than 1 (20/30) indicates lesser visual acuity.

Table 18.18 Results of Visual Acuity Tests

Record visual acuity for the right and left eye, with and without corrective lenses.

Eye	Visual Acuity
Right Eye	
Right Eye (with corrective lenses)	
Left Eye	
Left Eye (with corrective lenses)	

EXERCISE 18.17B Near-Point Accommodation

1. Obtain a pen or pencil and hold the object at arm's length in front of you. Cover the left eye and slowly move the object toward the right eye until the image of the object is no longer clear (i.e., is blurry or appears as two objects). Have a lab partner measure the distance

(continued on next page)

(continued from previous page)

from the right eye to the object. Record the near-point accommodation for the right eye in **table 18.19.**

2. Repeat step 1 by covering the right eye and moving the object toward the left eye. Record near-point accommodation for the left eye in table 18.19. Note that if the subject wears corrective lenses, this test can be conducted with and/or without glasses.

Table 18.19 Results of Near-Point Accommodation Tests

Record near-point accommodation for the right and left eye, with and/or without corrective lenses.

Eye	Near-Point Accommodation (cm)
Right Eye	
Right Eye (with corrective lenses)	
Left Eye	
Left Eye (with corrective lenses)	

EXERCISE 18.17C Blind Spot Determination

1. Hold **figure 18.23** an arm's length away from your face, approximately 46 cm (18 inches) away. Close the left eye and focus on the "X" with your right eye.
2. Move the figure toward your face while continuing to focus on the "X." Stop when the black dot no longer appears in your field of view. Have a lab partner measure the distance between the eye and the image. Record the distance between the image and the right eye in **table 18.20.**
3. Flip the lab book upside-down so the dot appears on the left of the image in figure 18.23. Repeat steps 1–2 by closing the right eye and focusing on the "X" with the left eye. Record the distance between the image and the left eye in table 18.20.

Figure 18.23 Blind Spot Determination.

Table 18.20 Results for Blind Spot Determination Test

Record the blind spot distance for the right and for the left eye.

Eye	Distance (cm)
Right Eye	
Left Eye	

EXERCISE 18.17D Color Blindness

1. Have a lab partner hold **figure 18.24** (Ishihara color plate number 7) approximately 30" away from your face directly in your line of vision.
2. Examine the Ishihara color plate number 7 (figure 18.24) for 3 seconds. Is a number visible on the plate? ________________ (yes/no). If yes, what number is visible? ______________
3. Have a lab partner repeat steps 1–2. Are the results the same? ____
4. Compare your results to those of other classmates. Note observations here: ________________________________

____
5. If a series of Ishihara color plates is available in the laboratory, repeat steps 1–4 with all available color plates. For each color plate, be sure to record if numbers are visible or not, and if so, what number is visible. Report observations here: _________________________

____

____

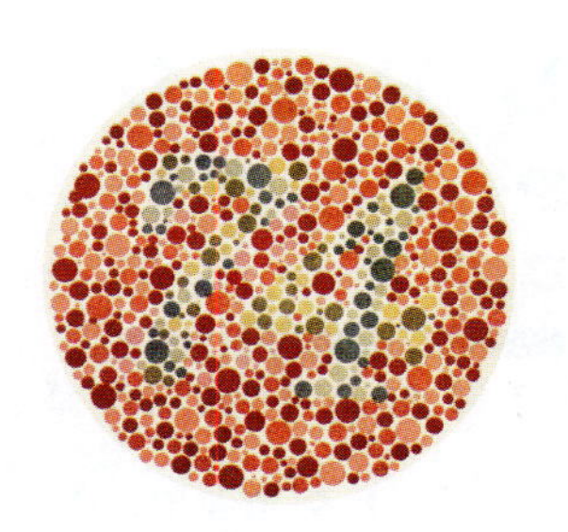

Figure 18.24 **Ishihara Color Test Plate.** Subjects with normal color vision should view the number *74.* Subjects with some degree of color blindness may view the number *21* or no number at all.

Hearing and Equilibrium

The inner ear is composed of the cochlea, vestibule, and semicircular canals. The **cochlea** is responsible for the sense of hearing (see p. 470).

The vestibule and semicircular canals are also contained within the inner ear. The **vestibule** monitors *static equilibrium*, when the body is not in motion, and *linear dynamic equilibrium*, when the body is accelerating or decelerating in a linear plane. The **semicircular canals** detect *angular dynamic equilibrium*. The semicircular canals are three fluid-filled canals that are arranged orthogonally (at right angles) to detect motion in three dimensions. For example, the vestibule detects acceleration or deceleration (i.e., traveling along a straight path), whereas semicircular canals detect angular motion (i.e., sharply turning a corner). Interestingly, hair cells, the same type of receptors responsible for detecting hearing in the cochlea, also detect static and dynamic equilibrium in the vestibule and semicircular canals.

Stimulation of the receptors within the vestibule and semicircular canals initiates nerve signals that are transmitted along the vestibular branch of the vestibulocochlear nerve (CN VIII) to the thalamus and cerebral cortex, as well as to the nuclei for the three cranial nerves that control the movement of the eye: oculomotor (CN III), trochlear (CN IV), and abducens (CN VI). Refer to your textbook as a guide for reviewing the specific neural pathways involved in hearing and equilibrium.

INTEGRATE

CLINICAL VIEW
Cochlear Implant

A cochlear implant is a medical device that attempts to mimic the vibrations experienced by hair cells of the cochlea, thereby performing the same amplification of sound performed typically by the cochlea. The implanted structure then transmits this signal by stimulating the vestibulocochlear nerve, which in turn allows the signal to reach the cerebral cortex, thereby allowing someone to perceive sound.

EXERCISE 18.18

HEARING TESTS

Health care providers may perform tests for proper hearing function if they suspect a problem. The following laboratory exercises involve performing tests of hearing. The sensory receptors for hearing are housed within the inner ear, and sensory information is transmitted along the vestibulocochlear nerve. There are tests for the different types of **deafness** (absence of sound perception) and for vestibular dysfunction.

Distinguishing between **nerve deafness** and **conduction deafness** requires important clinical tests, namely, the Rinne and Weber tests. The **Rinne test** evaluates a person's ability to perceive sound from vibrations in the air as compared to vibration applied directly to the temporal bone. The **Weber test** evaluates a person's ability to perceive sound in both ears equally, which is a specific test for nerve conduction deafness.

EXERCISE 18.18A Hearing Test: Rinne

1. Obtain a tuning fork and choose an individual to be the subject. When the tuning fork is tapped on a hard surface, it will begin to vibrate at a certain wavelength and produce sound.

2. Tap the tuning fork on a hard surface and hold it near the subject's ear (**figure 18.25*a***). Ask the subject to tell you about the sound and make a note of the subject's response here: ______________________________

   ____

3. Gently tap the tuning fork on a hard surface and place the vibrating instrument on the subject's mastoid process (figure 18.25*b*). Again, ask the subject about the sound and about any differences in perceived sound. Make a note of the subject's response here:

   ____

   ____

EXERCISE 18.18B Hearing Test: Weber

1. Obtain a tuning fork and choose an individual to be the subject.

2. Tap the tuning fork on a hard surface. Place the vibrating instrument on the center of the subject's forehead (**figure 18.26**). Ask the subject to compare his or her perception of sound in the right versus left ear, and make a note of the subject's response here:

   ____

   ____

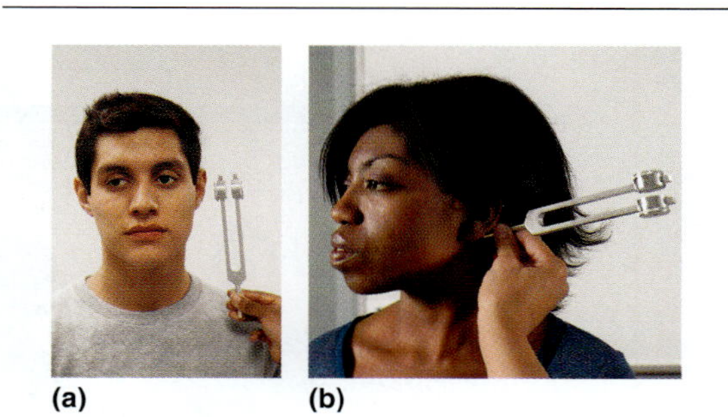

(a) (b)

Figure 18.25 **Rinne Hearing Test.** (a) Strike the tuning fork and place the device near the ear. (b) Place the vibrating tuning fork on the mastoid process.

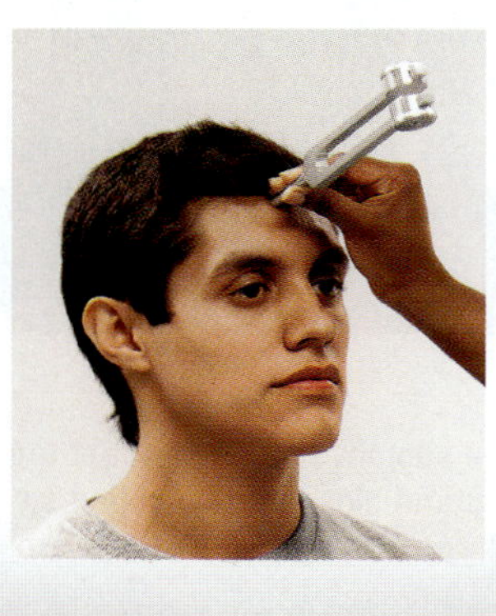

Figure 18.26 **Weber Hearing Test.** Place the vibrating tuning fork on the center of the forehead.

EXERCISE 18.19

EQUILIBRIUM TESTS

Health care providers may perform tests for proper vestibular function if they suspect a problem. Like those for hearing, sensory receptors for equilibrium are housed within the inner ear, and sensory information from these receptors is transmitted along the vestibulocochlear nerve. Equilibrium tests may be complicated to interpret because many anatomical structures, such as the vestibular apparatus, eyes, proprioceptors, cerebellum, and cranial nerves are involved. The following laboratory exercises explore the Romberg test and the Barany test. A **Romberg test** is administered during a neurological exam when a patient exhibits motor or sensory deficits. It is also performed when a person is suspected of drunk driving. There is an exercise that demonstrates the influence of vestibular input on eye movement. The **Barany test** demonstrates the reflex between the movement of the fluid in the semicircular canals and the extrinsic eye muscles.

EXERCISE 18.19A Equilibrium Test: Romberg

1. Choose an individual to be the subject. Ask the subject to stand for one minute with both feet together and both hands by his side **(figure 18.27*a*).** For ease, have the subject stand in front of a surface where the subject's shadow is visible or mark the subject's location (i.e., on a whiteboard). Note any exaggerated swaying movements to the left or right here:

   ____

   ____

   ____

2. Have the subject repeat step 1 with his or her eyes closed. Compare the magnitude of sway in the eyes open versus eyes closed condition. Make a note of it here:

   ____

   ____

   ____

3. Have the subject rotate 90 degrees, such that his or her shoulder is adjacent to the whiteboard (figure 18.27*b*). Be sure that the position is easily visible with a shadow or outline of the initial position. Look for any forward-backward swaying motion when the subject's eyes are open. Make a note of it here:

   ____

   ____

   ____

4. Have the subject close his or her eyes and repeat step 3. Look for any differences in forward-backward sway in the eyes open versus eyes closed condition. Make a note of any differences in sway here:

   ____

   ____

   ____

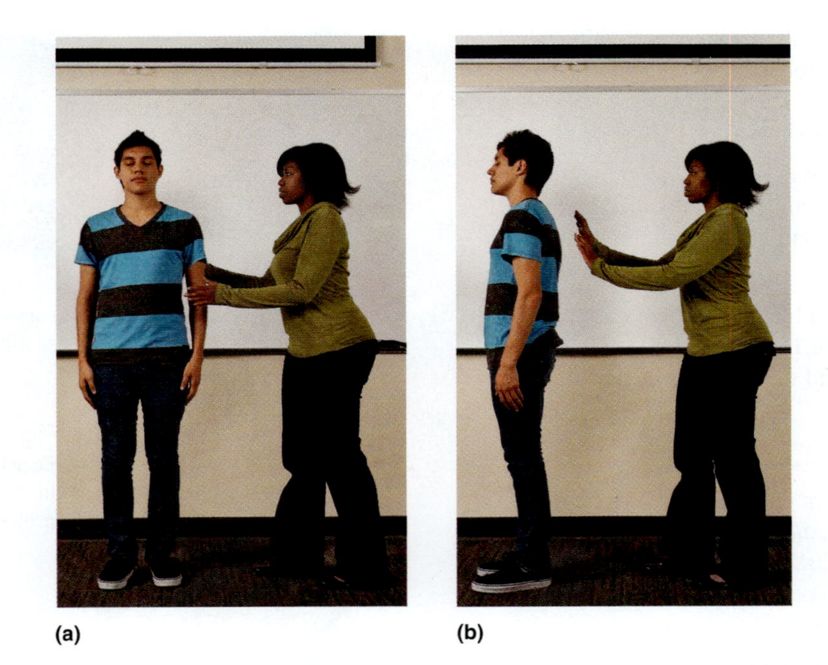

Figure 18.27 Romberg Test. The subject stands with feet together and hands by his side, with his (a) back toward the whiteboard and (b) shoulders adjacent to the whiteboard. Stand close to the subject so as to catch him if he loses his balance.

INTEGRATE

CONCEPT CONNECTION

A Romberg test does more than simply test the signals of the vestibular system. Rather, it evaluates the integration of vestibular information, visual information, and proprioception. All three signals are used to maintain an upright posture. When one or more of these senses is compromised, our ability to balance (particularly when impaired) is compromised. A Romberg test involves intentionally removing one signal, the visual input, and then observing any disruption in balance. Disruptions in balance are observed when there is an increase in sway. If a person exhibits a positive Romberg sign (increased sway with eyes closed), it indicates that there may be deficits with proprioception. Specifically, there is likely impairment of the dorsal column motor pathway, which transmits proprioceptive information from the sensory receptors to the central nervous system. The Romberg test demonstrates that vestibular input is not sufficient by itself to allow a person to maintain an upright posture.

EXERCISE 18.19B Equilibrium Test: Barany

CAUTION:

Any person subject to dizziness or nausea should not perform this test. Any rotation should cease if the subject reports dizziness or nausea. Testers should be prepared to hold or catch a subject if the test causes the subject to lose balance.

1. Choose an individual to be the subject. Have the subject sit in a swivel chair holding on to the arms of the chair firmly. Be sure to have several students standing nearby to catch the subject in the event that the subject loses his or her balance.

2. Have the subject sit with head tilted forward approximately 30 degrees. This position of the head is optimal for stimulation of the lateral semicircular canals **(figure 18.28*a*).** Spin the chair ten revolutions to the right while the subject keeps his or her eyes open.

3. Stop the chair and have the subject sit up and look forward. Note the movement, if any, of the subject's eyes here: ______________________

   Have the subject describe his or her perception of movement while sitting still. Make a note of it here:

   ____

   ____

4. Wait several minutes. Repeat steps 2 and 3 with the head tilted forward at 90 degrees (figure 18.28*b*). Note eye movements and the subject's perception of movement here: ______________________

   ____

   ____

(a)

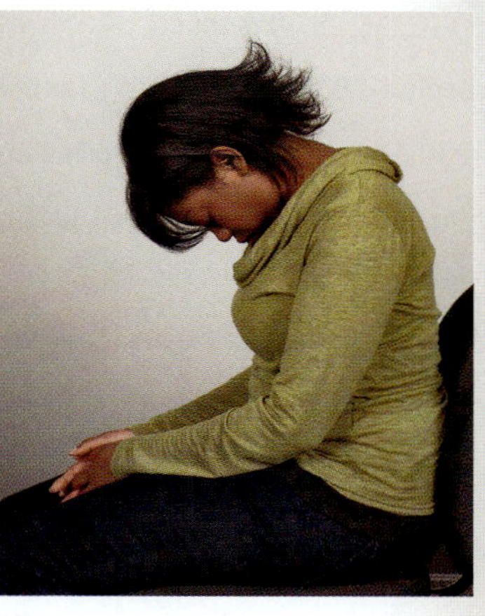

(b)

Figure 18.28 Barany Test. The subject sits in a swivel chair with the head tilted forward (a) 30 degrees or (b) 90 degrees.

INTEGRATE

CLINICAL VIEW
Vertigo

Vertigo is a condition in which a person experiences sensations of dizziness or unsteadiness even when standing still. That is, the individual is not experiencing any outward rotational movement. In essence, vertigo occurs when the individual's vestibular pathways are stimulated in the absence of outward movement. Some sensations of vertigo may occur immediately following rotational movement. For example, an individual may experience vertigo immediately after riding a roller coaster when experiencing the sensation of still moving even after the ride is over. People may also experience vertigo when they have a common cold or the flu. Ménierè's disease, which is characterized by excessive fluid in the inner ear, causes a severe sensation of vertigo. In Ménierè's disease the excess fluid in the inner ear causes excessive depolarization of hair cells within the vestibule and semicircular canals, which leads to the increased perception of spinning. Often vertigo will also impact hearing, because both hearing and equilibrium are transmitted to the central nervous system via the vestibulocochlear nerve (CN VIII). Sensations of vertigo may also be associated with extreme nausea and vomiting as well as *nystagmus*, a rapid, involuntary, horizontal movement of the eyes.

Chapter 18: General and Special Senses

Name: ______________________

Date: __________ Section: __________

POST-LABORATORY WORKSHEET

The (1) corresponds to the Learning Objective(s) listed in the chapter opener outline.

Do You Know the Basics?

Exercise 18.1: Tactile (Meissner) Corpuscles

1. Match the location listed in column A with its appropriate sensory receptor listed in column B. (1) (2) (3)

Column A	*Column B*
_____ 1. located at dermal/epidermal junction	a. free nerve ending
_____ 2. located deep in the reticular layer of the dermis	b. lamellated corpuscle
_____ 3. located within the dermal papillae	c. tactile corpuscle
_____ 4. located throughout the dermis	d. tactile disc

2. Which of the following corresponds to the mode of sensation detected by tactile (Meissner) corpuscles? (Check all that apply.) (2)

_____ a. deep pressure　　_____ c. light touch

_____ b. fine touch　　_____ d. vibration

Exercise 18.2: Lamellated (Pacinian) Corpuscles

3. Which of the following corresponds to the mode of sensation detected by lamellated (Pacinian) corpuscles? (Check all that apply.) (3) (4)

_____ a. deep pressure　　_____ c. light touch

_____ b. fine touch　　_____ d. vibration

Exercise 18.3: Gustation (Taste)

4. The foliate and vallate papillae are considered large. ________________ (True/False) (5)

5. Cells that detect chemicals dissolved in solution and transmit nerve signals for taste sensation to the CNS are known as ________________ (basal/gustatory) cells. (6)

Exercise 18.4: Olfaction (Smell)

6. The following table lists the cell types associated with olfactory epithelium. Next to each cell type, give a brief description of the location of the cell within the olfactory epithelium and the function of the cell. (7) (8)

Cell Type	Location	Function
Basal Cells		
Olfactory Receptor Cells		
Supporting Cells		

Exercise 18.5: Vision (The Retina)

7. Rank the following layers of the retina from the innermost to outermost layer. (9) (10)

_____ a. bipolar cell layer

_____ b. choroid

_____ c. ganglion cell layer

_____ d. pigmented layer

_____ e. photoreceptor cell layer

8. The area of highest visual acuity and largest density of cones in the retina is the ________________ (fovea centralis/optic disc). (11)

Exercise 18.6: Hearing

9. The cochlea is innervated by CN VIII. ____________________ (True/False)

10. The scala tympani and vestibuli both contain ____________________ (endolymph/perilymph), whereas the scala media contains ____________________ (endolymph/perilymph).

11. Which of the following structures in the cochlea corresponds to the location of the spiral organ? (Circle one.)
   a. scala media
   b. scala tympani
   c. scala vestibuli

12. Vibration of which membrane allows for sound transmission in the cochlea? (Circle one.)
   a. basilar membrane
   b. tectorial membrane
   c. vestibular membrane

Exercise 18.7: Sensory Receptors in the Skin

13. Match the location listed in column A with its appropriate sensory receptor listed in column B.

Column A	***Column B***
_______ 1. located at the dermal/epidermal junction	a. free nerve ending
_______ 2. located deep in the reticular layer of the dermis	b. lamellated corpuscle
_______ 3. located within the dermal papillae	c. tactile corpuscle
_______ 4. located throughout the dermis	d. tactile disc

Exercise 18.8: Gross Anatomy of the Eye

14. Match the description listed in column A with the part of the eye listed in column B.

Column A	***Column B***
_______ 1. anterior-most part of the retina, which appears serrated	a. aqueous humor
_______ 2. colored part of the eye	b. ciliary muscle
_______ 3. ligament extending between ciliary muscles and the lens	c. iris
_______ 4. metallic-appearing, opalescent inner layer of the sclera; it is present in many animals (e.g., the cow eye), but not the human eye	d. lens
_______ 5. neural tunic of the eye; composed of several layers of neurons involved with transducing light energy into nerve signals	e. ora serrata
_______ 6. smooth muscle within the ciliary body composed of both circular and radial muscle fibers	f. retina
_______ 7. transparent, biconvex structure composed of highly specialized, modified epithelium	g. suspensory ligament
_______ 8. watery fluid that circulates within the anterior and posterior chambers of the eye	h. tapetum lucidum

Exercise 18.9: Extrinsic Eye Muscles

15. Which of the following extrinsic eye muscles is/are innervated by CN III? (Check all that apply.) 

_______ a. inferior oblique

_______ b. inferior rectus

_______ c. lateral rectus

_______ d. medial rectus

_______ e. superior oblique

_______ f. superior rectus

Exercise 18.10: Cow Eye Dissection

16. A cow eye has a prominent, colorful structure on its posterior wall, which is not present in the human eye, called the tapedum lucidum.

    ____________________ (True/False) 19 20

Exercise 18.11: Gross Anatomy of the Ear

17. Match the description in column A with the part of the ear listed in column B. 21 22

Column A	*Column B*
______ 1. cavity between the external ear and inner ear; contains ossicles	a. auricle
______ 2. drum-like, tight, thin membrane that separates the external-ear cavity to the middle ear	b. inner ear
______ 3. external ear, composed of an elastic cartilage skeleton that is covered with skin	c. middle ear
______ 4. largest membranous sac in the vestibule; contains receptors for sensing horizontal acceleration	d. saccule
______ 5. portion of ear located within the petrous part of the temporal bone that includes the cochlea, vestibule, and semicircular canals	e. semicircular canals
______ 6. three ring-like canals that are oriented at right angles to each other and communicate with the vestibule	f. tympanic membrane

Exercise 18.12: Two-Point Discrimination

18. There is a(n) ____________________ (direct/inverse) relationship between the density of sensory receptors in the skin and the size of the receptive field in that area of the body. 23

19. The density of sensory receptors in the fingertip is ____________________ (larger/smaller) compared to the density of sensory receptors in the anterior forearm. 24

Exercise 18.13: Tactile Localization

20. When stimulating the tips of the fingers, it may be difficult to locate the exact source of sensory input because only one receptive field will likely be stimulated. ____________________ (True/False) 25

Exercise 18.14: General Sensory Receptor Tests: Adaptation

21. The nervous system adapts to a constant stimulus by becoming ____________________ (less/more) sensitive to that stimulus over time. 26

Exercise 18.15: Gustatory Tests

22. Match each of the substances listed in column A with an example of a taste sensation listed in column B. 27

Column A	*Column B*
______ 1. dark chocolate bar	a. bitter
______ 2. dill pickle	b. salty
______ 3. maple syrup	c. sour
______ 4. seasoning salt	d. sweet
______ 5. steak	e. umami

Exercise 18.16: Olfactory Tests

23. When tasting two substances that have similar consistencies (such as an apple and a potato), closing one's nose ____________________ (decreases/increases) the ability to distinguish one from the other. 28

Exercise 18.17: Vision Tests

24. An individual has the following results on a visual acuity test: 20/10. This individual's vision is ____________________ (better/worse) than "normal" vision. 29

25. As we age, the flexibility of the lens of the eye decreases. This causes the near-point of accommodation to move ____________________ (farther from/closer to) the eye. 30

26. When reflected light focuses on an individual's blind spot, he or she is unable to see the source of the light because there are no photoreceptors in that area. ____________________ (True/False) 31

27. A person that suffers from color blindness has a normal number and distribution of rods and cones. ____________________ (True/False) 31

Exercise 18.18: Hearing Tests

28. The ____________________ (Rinne/Weber) test is a test for neural deafness. 32

Exercise 19.19: Equilibrium Tests

29. The ____________________ (Barany/Romberg) test is used to determine if vertigo is caused by a disorder in the patient's inner ear or somewhere in the patient's brain. 33

Can You Apply What You've Learned?

30. A fracture of the cribiform plate of the ethmoid bone can result in a loss of the sense of smell. Given your knowledge of the location of olfactory receptor cells and the pathway taken by their axons to reach the brain, explain why this can happen.

____

____

____

____

31. The optic disc is referred to as the "blind spot" of the eye. Based on your histological observation of the optic disc, explain why this is the case.

____

____

____

____

Can You Synthesize What You've Learned?

32. Why do you think tactile (Meissner) corpuscles are located relatively close to the surface of the skin rather than deep within the dermis?

____

____

____

____

33. An individual who suffers a strong blow to the head may end up with a detached retina that causes visual problems. Why do you think the retina easily detaches from the posterior wall of the eye? What structure normally holds the retina in place?

____

____

____

____

34. Susan, a 67-year-old female, is experiencing ataxia, or poor muscle coordination. Doctors perform a neurological exam, and they find that Susan is not positive for the Romberg test. Describe what these findings suggest about the source of Susan's ataxia.

____

____

____

____

35. Anosmia is a condition characterized by an inability to smell. While some may be unable to perceive a particular odorant, most suffering from this condition are unable to perceive multiple odorants. Based on your knowledge of the olfactory sense, propose several scenarios that would lead to a patient's diagnosis of anosmia.

____

____

____

____

36. Patients suffering from hearing loss or impairment may be eligible for a cochlear implant, a device that mimics the action of the cochlea. Specifically, the cochlear implant contains external components that detect, process, and transmit sounds to an internal receiver. The internal components stimulate the vestibulocochlear nerve directly. Based on your knowledge of hearing, describe how the cochlear implant allows for the perception of sound.

____

____

____

____

CHAPTER 19

The Endocrine System

OUTLINE AND LEARNING OBJECTIVES

Anatomy & Physiology REVEALED® aprevealed.com **MODULE 8: ENDOCRINE SYSTEM**

INTRODUCTION

How long has it been since your last meal? Even if it has been several hours, blood glucose and blood calcium levels remain remarkably stable, fluctuating only minor amounts around the body's normal physiological level (unless, of course, there is an underlying condition such as diabetes). Maintenance of blood glucose and blood calcium levels are physiologic imperatives, for if their levels are too high or too low, severe impairment of nervous and muscular activity will occur. However, it is not often that blood glucose or calcium levels move drastically out of the normal range. This is because such variables are tightly regulated by the **endocrine system.** For example, the hormones insulin and glucagon, produced by the pancreas, regulate blood glucose levels, and the hormones calcitonin and parathyroid hormone, produced by the thyroid and parathyroid glands, respectively, regulate blood calcium levels.

This brief description of these hormones and the variables they regulate provides only a glimpse into the functioning of the endocrine system, a system of chemical messengers (called hormones) that are transported in the blood and act on distant target cells. The endocrine system consists of a number of "classical" endocrine organs, such as the pituitary gland, adrenal glands, pancreatic islets, and thyroid gland. However, many organs in the body also contain cells or tissues that produce and secrete hormones. For instance, cells in the walls of the stomach secrete hormones that regulate appetite, gastric motility, and acid secretion. Cells in the testes (males) and ovaries (females) are responsible for the secretion of hormones that regulate the maturation of sperm and eggs, respectively. While the thymus is considered an endocrine organ due to its secretion of the hormone *thymosin,* it plays a larger role in the immune system, as described in chapter 23.

The exercises in this chapter explore the structure and function of the classical endocrine organs. Many of the organs explored contain cells that secrete hormones (thus the organ has an endocrine role). However, the entire organ is not necessarily referred to as an endocrine *gland* because the organ has other functions as well. The few organs that are strictly endocrine in nature are generally quite small. Only the thyroid and adrenal glands can be viewed easily on a cadaver. Thus, the exploration of the endocrine system in the anatomy and physiology laboratory will be carried out predominantly at the microscopic level. However, there is also an exercise that involves locating the major (large) endocrine organs on a cadaver or on classroom models. When examining histology slides, make associations between the tissues viewed under the microscope and the gross anatomic location where the tissue is found. In addition, consider the names and function(s) of the hormone(s) secreted by each organ. This chapter also contains an activity that explores the physiology of metabolism, and a clinical case study that involves applying the principles of the endocrine system to a problem involving an imbalance in hormones. Subsequent chapters covering the cardiovascular, lymphatic, digestive, urinary, and reproductive systems contain exercises that further explore the structure and function of endocrine cells and tissues (and associated hormones) related to each specific system.

List of Reference Tables

Chapter 19: The Endocrine System

Name: ______________________

Date: __________ Section: __________

These Pre-Laboratory Worksheet questions may be assigned by instructors through their connect course.

PRE-LABORATORY WORKSHEET

1. Glands that produce and release chemical messengers (hormones) to be transported in the blood are ____________ (endocrine/exocrine) glands.

2. Glands that produce a product that is released into a duct, which transports the product directly to its target organ or tissue, are ____________ (endocrine/exocrine) glands.

3. Identify the gland where parafollicular cells are located. (Circle one.)
   a. adrenal
   b. pineal
   c. pituitary
   d. thymus
   e. thyroid

4. Match the description listed in column A with the appropriate endocrine gland listed in column B.

Column A		*Column B*
______	1. consists of a cortex and medulla, with each part having different embryological origins	a. adrenal gland
______	2. consists of follicles lined with a simple cuboidal epithelium	b. hypothalamus
______	3. four small endocrine glands that secrete a hormone that regulates blood calcium levels	c. parathyroid glands
______	4. secretes hormones to regulate hormone release by the anterior pituitary gland	d. pineal gland
______	5. secretes the hormone melatonin	e. thyroid gland

5. Which of the following is the major endocrine gland that secretes sex steroid hormones in the female? (Circle one.)
   a. adrenal
   b. ovaries
   c. pineal
   d. pituitary
   e. testes

6. Which of the following hormone(s) is/are secreted by the posterior pituitary gland? (Check all that apply.)
   ______ a. antidiuretic hormone
   ______ b. growth hormone
   ______ c. oxytocin
   ______ d. prolactin
   ______ e. thyroid stimulating hormone

7. The hormone released by the anterior pituitary gland that induces ovulation in females is ____________ (follicle stimulating hormone/luteinizing hormone).

8. The pancreas has both endocrine and exocrine functions. ____________ (True/False).

9. Which of the following occurs when thyroid hormone levels increase? (Check all that apply.)
   ______ a. decreased metabolic rate
   ______ b. increased glycogenesis
   ______ c. increased body temperature
   ______ d. increased oxygen consumption
   ______ e. increased lipolysis

10. A decrease in thyroid hormone ____________(inhibits/stimulates) the release of thyrotropin releasing hormone by the hypothalamus.

HISTOLOGY

Endocrine Glands

In the following exercises, the microscopic anatomy of the various endocrine glands is described in detail. The tables in these exercises list the glands, the hormones they produce, and the functions of each of the listed hormones. A major goal in completing the following exercises will be to learn to differentiate the various endocrine glands from each other when viewing them under the microscope.

EXERCISE 19.1

THE HYPOTHALAMUS AND PITUITARY GLAND

The **pituitary** gland, or hypophysis (*hypophysis*, an undergrowth), is a remarkable organ. This small organ plays a significant role in endocrine regulation of the body. The pituitary gland is only about the size and shape of a pea, but it secretes hormones that regulate the growth and development of nearly every other organ in the body. The secretion of pituitary hormones is tightly controlled by cells within the **hypothalamus,** a part of the brain whose structure and function was discussed in chapter 15. The textbook covers the structure of, function of, and relationships between the hypothalamus and pituitary gland in detail. Because it is not possible to visualize the details of these relationships in the laboratory, the focus here is on the structure and function of the pituitary gland alone. **Table 19.1** summarizes the cells and structures that compose the pituitary gland and lists the hormones secreted by each cell type, along with the releasing or inhibiting hormones secreted by the hypothalamus that influence the secretion of hormones by the pituitary gland.

1. Obtain a histology slide of the pituitary gland **(figure 19.1).** Before placing it on the microscope stage, observe the slide with the naked eye. Notice there is a distinctive difference in color between the two parts, or lobes, of the pituitary gland. The darker area is the **anterior pituitary** (also called the anterior lobe, or **adenohypophysis** [*adeno-*, a gland, + *hypophysis*, pituitary]), whereas the lighter area is the **posterior pituitary** (also called the posterior lobe, or **neurohypophysis** [*neuro-*, relating to nervous tissue, + *hypophysis*, pituitary]).
2. Place the slide on the microscope stage and bring the tissue sample into focus on low power. Once again, identify the anterior and posterior lobes (figure 19.1*a*). The anterior pituitary is derived embryologically from an outpocketing of the roof of the mouth and consists of epithelial tissue. Thus, the cells have an appearance that is characteristic of glandular epithelial tissue (figure 19.1*b*). In comparison, the posterior pituitary is derived embryologically from a downgrowth of the diencephalon of the brain and consists of nervous tissue. Thus, the cells have an appearance that is characteristic of nervous tissue (figure 19.1*b*).
3. *Anterior Pituitary*—Identify the anterior pituitary gland using figures 19.1 and 19.2 as guides. Then move the microscope stage so the anterior pituitary is in the center of the field of view.
4. Increase the magnification to observe the glandular nature of the cells **(figure 19.2).** There are three cell types within the anterior pituitary: acidophils, basophils, and chromophobes. The first two kinds of cells are named for their "love" (*-phil*, to love) of acidic or basic dyes. *Acidophils* attract acidic dyes, and appear red. *Basophils* attract basic dyes and appear blue. *Chromophobes* (*chroma*, color, + phobos, fear) are cells that attract neither acidic nor basic dyes, and are thought to be cells that have released their hormone(s). Identification of the various cell types within the anterior pituitary is not the goal of this exercise. However, the cell types and hormones produced by each cell form the basis for a couple of handy mnemonic devices that assist in recall of the names of the hormones produced by the anterior pituitary (see learning strategy on this page).
5. *Posterior Pituitary*—Identify the posterior pituitary using figure 19.1 as a guide. Then move the microscope stage so the posterior pituitary is in the center of the field of view.

INTEGRATE

LEARNING STRATEGY

Simple mnemonics for remembering which hormones are produced by which cells of the anterior pituitary are as follows:

Mnemonic for acidophils: **GPA** (as in Grade Point Average: If a student does not remember this information, it could be harmful to his or her GPA.)

G = growth hormone (GH)

P = prolactin (PRL)

A = acidophil

Mnemonic for basophils: ***B*-FLAT** (as in the musical note: If a student remembers this information, it will be beneficial to his or her GPA and he or she will be happily singing to the tune of B-flat!)

B = basophil

F = follicle-stimulating hormone (FSH)

L = luteinizing hormone (LH)

A = adrenocorticotropic hormone (ACTH)

T = thyroid-stimulating hormone (TSH)

Table 19.1 Histology of the Pituitary Gland

Pituitary Gland	Cells	Description	Hormones Produced	Action of Pituitary Hormone(s)	Hypothalamic Releasing or Inhibiting Hormone	Word Origin
Anterior Pituitary (Adenohypophysis)	Acidophils	Appear red in color due to their attraction for acidic stains	GH, PRL	NA	NA	*acidus*, sour (relating to acidic dyes), + *-phil*, to love
			Growth hormone (GH)	Stimulates the liver and other tissues to produce IGF-1 (insulin-like growth factor-1), which promotes bone and muscle growth	Growth-hormone-releasing hormone (GHRH) stimulates release, whereas growth-hormone-inhibiting hormone (GHIH, somatostatin) inhibits release	*grōthr*, growth, + *hormon*, to set in motion
			Prolactin (PRL)	Stimulates the mammary glands to develop and produce milk	Prolactin-releasing hormone (PRH) stimulates release, whereas prolactin-inhibiting hormone (PIH) inhibits release	*pro-*, before, + *lac*, milk
	Basophils	Appear blue in color due to their attraction for basic stains	FSH, LH, ACTH, TSH	See below	See below	*baso-*, basic (relating to basic dyes), + *-phil*, to love
			Follicle-stimulating hormone (FSH)	Stimulates the growth and maturation of ovarian follicles, which release estrogen (females); stimulates spermatogenesis (males)	Gonadotropin-releasing hormone (GnRH) stimulates release, whereas the hormone inhibin inhibits its release	*folliculus*, a small sac (referring to the ovarian follicles)
			Luteinizing hormone (LH)	Induces ovulation, stimulates the production of estrogen and progesterone by cells of the corpus luteum (females); stimulates interstitial cells to produce testosterone (males)	Gonadotropin-releasing hormone (GnRH) stimulates release	*luteus*, yellow (referring to the corpus luteum of the female ovary)
			Adrenocorticotropic hormone (ACTH)	Stimulates the growth, development, and secretion of steroid hormones by the adrenal cortex (e.g., cortisol)	Corticotropin-releasing hormone (CRH) stimulates release	*adrenocortico*, referring to the adrenal cortex, + *trophe*, nourishment
			Thyroid-stimulating hormone (TSH)	Stimulates the secretion of thyroid hormones by the thyroid gland	Thyrotropin-releasing hormone (TRH) stimulates release	*thyroid*, shaped like an oblong shield
	Chromophobes	Appear very light in color due to a lack of staining	Thought to be devoid of hormone, hence the lack of staining properties	NA	NA	*chroma*, color, + *phobos*, fear
Pituitary Gland	**Cells**	**Description**	**Hypothalamic Hormones Stored and Released**	**Action of Pituitary Hormone(s)**	**Hypothalamic Nucleus Containing Neuron Cell Bodies**	**Word Origin**
Posterior Pituitary (Neurohypophysis)	Axon terminals	Axon terminals store hormone that was produced in the cell bodies of the neurons within the hypothalamus	Oxytocin	Stimulates uterine contractions and milk ejection by mammary glands	Paraventricular nucleus and supraoptic nucleus	*axon*, axis, + *terminus*, the limit; *para-*, next to, + *ventricular*, relating to the third ventricle of the brain, + *nucleus*, a collection of neuron cell bodies; *okytckos*, swift birth
	Axon terminals	Axon terminals store hormone that was produced in the cell bodies of the neurons within the hypothalamus	Antidiuretic hormone (ADH, vasopressin)	Increases water retention by the kidneys (maintains blood volume and blood pressure); vasoconstriction	Paraventricular nucleus and supraoptic nucleus	*axon*, axis, + *terminus*, the limit; *supra-*, above, + *optic*, relating to the optic tract, + *nucleus*, a collection of neuron cell bodies; *anti*, against, + *diuresis*, excretion of urine
	Pituicytes	Derived from glial cells; have processes that surround axon terminals of the hormone-secreting neurons	NA	NA	NA	*pituita*, a phlegm (relating to the pituitary gland), + *-cyte*, cell

(continued on next page)

(continued from previous page)

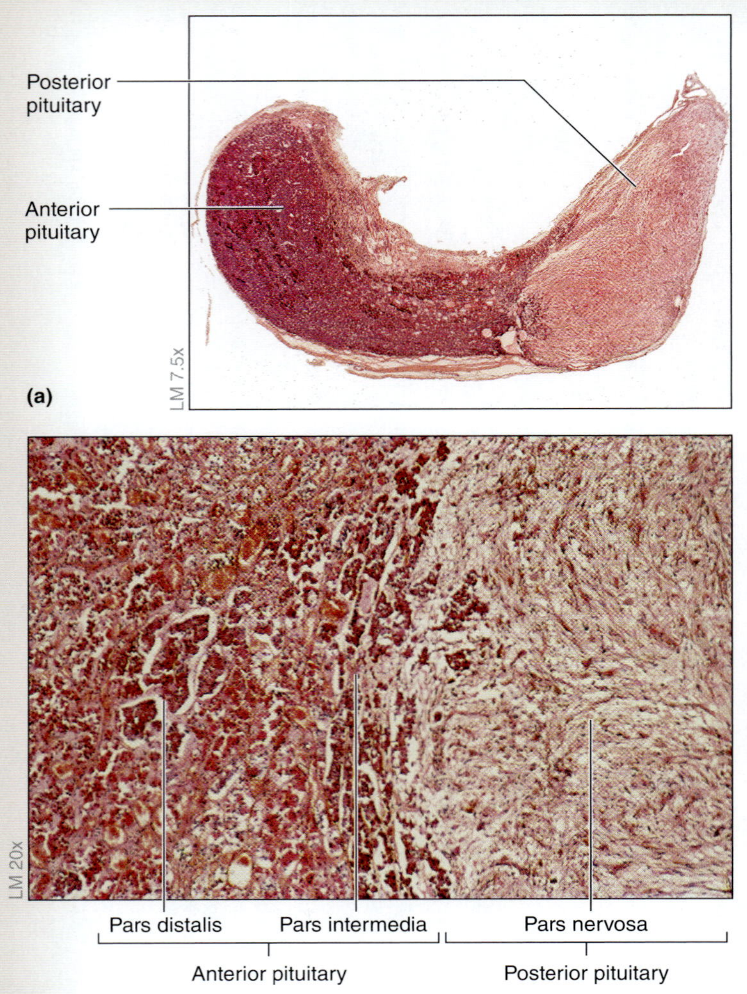

Figure 19.1 **Pituitary Gland.** (a) Low-magnification view of both parts of the pituitary. (b) Medium-magnification view of the anterior and posterior pituitary.

6. Increase the magnification so the cells are visible **(figure 19.3).** The majority of the nuclei visible within the slide are the nuclei of **pituicytes,** which are derived from glial cells. Pituicytes surround the axon terminals of neurons whose cell bodies are located in the paraventricular and supraoptic nuclei of the hypothalamus. These neurons secrete the hormones oxytocin and antidiuretic hormone (ADH, vasopressin).

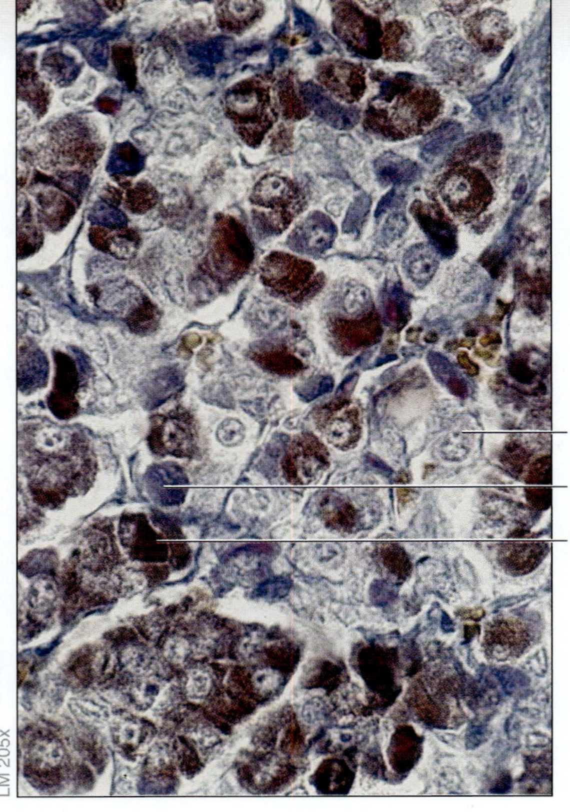

Figure 19.2 **Anterior Pituitary.** The three cell types in the anterior pituitary are basophils, which stain blue, acidophils, which stain red, and chromophobes, which don't take up biological stains.

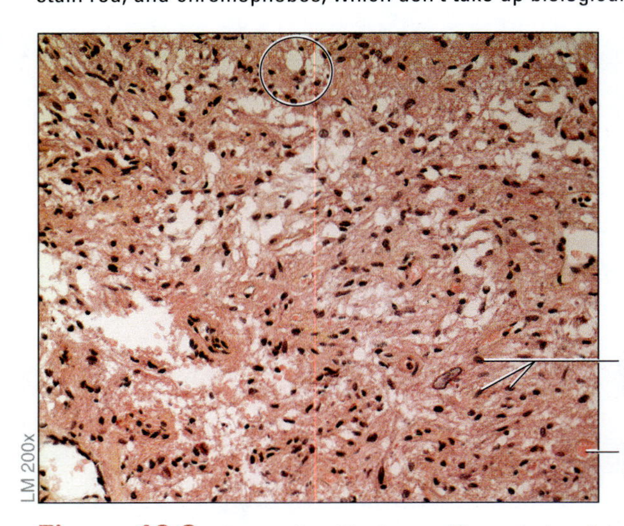

Figure 19.3 **Posterior Pituitary.** The majority of the nuclei seen in this micrograph are of pituicytes, which are glial cells.

INTEGRATE

CONCEPT CONNECTION

Hormones secreted by the pituitary gland circulate in the blood to target endocrine organs and glands all over the body. Pituitary hormones regulate functions such as growth, development, blood volume, and blood pressure. Several hormones secreted by the pituitary gland also influence the reproductive system in both males and females. The anterior pituitary gland releases follicle stimulating hormone (FSH), luteinizing hormone (LH), and prolactin, whereas the posterior pituitary gland releases oxytocin. Each of these hormones plays an important role in reproductive health. FSH regulates the development of ovarian follicles in females and the production of sperm by the testes in males. LH stimulates ovulation and the production of the sex steroid hormones estrogen and progesterone in females and the production of the sex steroid hormone testosterone in males. Prolactin promotes milk production in mammary glands of pregnant females. Oxytocin stimulates uterine contractions during childbirth, and milk ejection from mammary glands. More detailed explanations of the hormones related to the reproductive system are presented in chapter 27.

7. Identify the following structures on the slide of the pituitary gland, using figures 19.1 through 19.3 and table 19.1 as guides:

   - ☐ **acidophils**
   - ☐ **anterior pituitary**
   - ☐ **basophils**
   - ☐ **chromophobes**
   - ☐ **pituicytes**
   - ☐ **posterior pituitary**

8. Sketch the pituitary gland as seen through the microscope in the space provided, labeling the structures listed in step 7.

EXERCISE 19.2

THE PINEAL GLAND

The **pineal gland** (*pineus*, relating to a pine, shaped like a pinecone), also called the **pineal body,** is a small region of the epithalamus of the brain. Its primary cells, called **pinealocytes (figure 19.4),** secrete the hormone **melatonin.** Melatonin's effects in humans are unsubstantiated, but in other organisms melatonin is responsible for regulation of circadian rhythms. In humans, it may have a role in determining the onset of puberty. Pinealocytes are innervated by neurons from the sympathetic nervous system, and their secretion of hormone is affected by the amount of light received by the individual, which is relayed to the pineal gland through these neurons. Melatonin secretion increases when light levels are low (at night) and decreases when light levels are high (during the day). Pinealocytes appear in groups of cells within the pineal gland. They are surrounded by glial cells, whose function is similar to that of astrocytes in other parts of the brain. Clinically, one of the most important features of the pineal gland is the presence of calcium concretions, termed **"pineal sand"** (corpora arenacea). These concretions are easily visible in radiographs of the head, and provide radiologists with a landmark that is consistent and easy to identify. The number of concretions in the pineal gland increases with age.

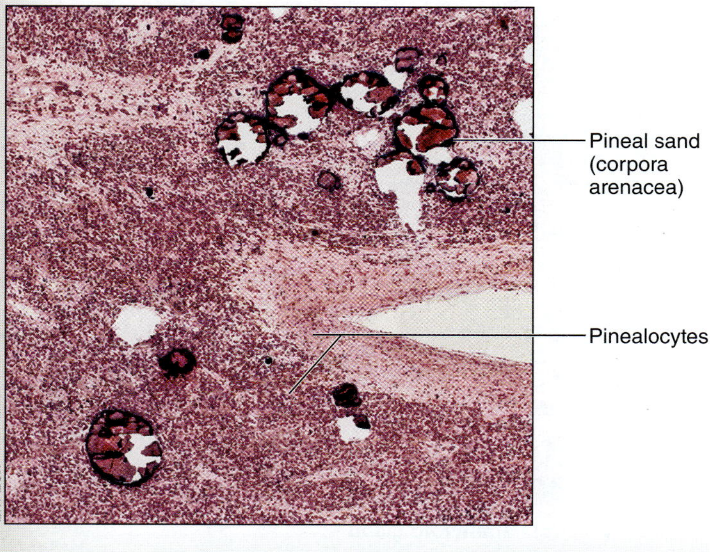

Figure 19.4 The Pineal Gland. Histology of the pineal gland.

1. Obtain a histology slide of the pineal gland. Place the slide on the microscope stage and bring the tissue sample into focus on low power.
2. Identify the following structures on the slide, using figure 19.4 as a guide:

   - ☐ **pinealocytes**
   - ☐ **pineal sand**

3. Sketch the pineal gland as seen through the microscope at medium or high magnification in the space provided, labeling pinealocytes and pineal sand.

______ ×

EXERCISE 19.3

THE THYROID AND PARATHYROID GLANDS

The **thyroid gland (figure 19.5; table 19.2)** is a butterfly-shaped gland located anterior to the trachea and inferior to the thyroid cartilage of the larynx. It consists of two main lobes connected to each other anteriorly by a narrow **isthmus** (*isthmus*, neck). The functional units of the thyroid gland are **thyroid follicles,** which are lined with a simple cuboidal epithelium. Inside each follicle is a mass of **colloid,** consisting largely of thyroglobulins. **Thyroglobulins** are the precursor molecules for the formation of the **thyroid hormones** (T3 and T4). The areas between the follicles contain another cell type, called **parafollicular cells.** These cells secrete the hormone **calcitonin.**

Embedded within the posterior aspect of the thyroid gland are a series of small glands (usually four) called **parathyroid glands.** These glands consist of two cell types: **chief (principal) cells,** which are smaller, more abundant cells with relatively clear cytoplasm that produce **parathyroid hormone** (PTH, parathormone), and **oxyphil cells,** which are larger, less abundant cells with granular pink cytoplasm, and whose function is unknown.

1. Obtain a histology slide of the thyroid and parathyroid glands and place it on the microscope stage. Bring the tissue sample into focus on low power and then change to high power.

2. Identify the following structures on the slide, using figure 19.5 and table 19.2 as guides:

Thyroid gland

- ☐ colloid
- ☐ follicular cells
- ☐ parafollicular cells
- ☐ thyroid follicles

Parathyroid gland

- ☐ chief (principal) cells
- ☐ oxyphil cells

3. Sketch the thyroid and parathyroid glands as seen through the microscope in the space provided, labeling all of the structures listed in step 2.

_______ ×

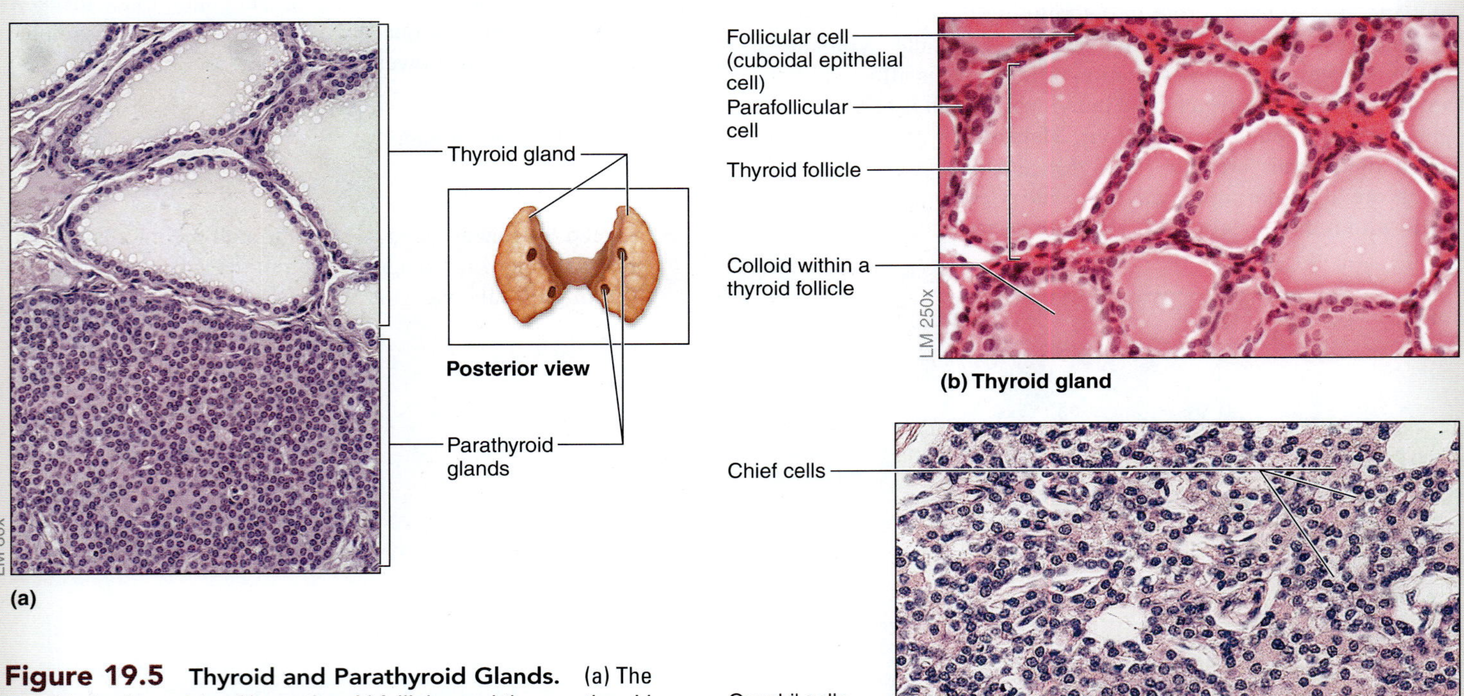

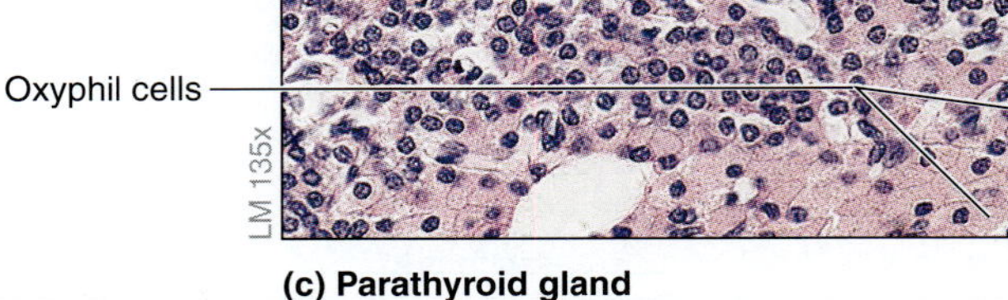

Figure 19.5 Thyroid and Parathyroid Glands. (a) The thyroid gland consists of large thyroid follicles and the parathyroid gland consists of tightly packed cells. (b) Micrograph showing both thyroid follicles, which produce the thyroid hormones, and parafollicular cells, which produce calcitonin. (c) High-magnification view of a parathyroid gland demonstrating chief cells and oxyphil cells.

Table 19.2 Histology of the Thyroid and Parathyroid Glands

Cell Types	Description	Hormones Produced	Action of Hormone	Mechanism of Action	Word Origin
Thyroid Gland	*A shield- or butterfly-shaped gland located anterior to the trachea and inferior to the thyroid cartilage. Consists of two lobes connected by a narrow isthmus anteriorly.*				*thyroid,* shaped like an oblong shield
Follicular Cells	Simple cuboidal epithelial cells that form the thyroid follicles; they have very dark nuclei	Thyroid hormones (T3 and T4/ Thyroxine)	Increase basal metabolic rate (BMR); important in early development of the central nervous system	Stimulates or inhibits transcription of certain genes in target cells	*folliculus,* a small sac
Parafollicular Cells	Lighter-staining cells found in the interstitial spaces between the thyroid follicles; larger than the follicular cells, with nuclei that have an appearance similar to a clock face	Calcitonin	Decrease blood calcium levels	Inhibits the action of osteoclasts, increases urinary excretion of calcium	*para,* next to, + *folliculus,* a small sac
Parathyroid Glands	*4–6 small glands located on the posterior surface of the thyroid gland*				*para,* next to, + *thyroid,* shaped like an oblong shield
Chief (Principal) Cells	Relatively small cells that contain a centrally located, round nucleus with one or more nucleoli	Parathyroid hormone (parathormone)	Increase blood calcium levels	Indirectly increases the action of osteoclasts, decreases urinary excretion of calcium, and stimulates synthesis of calcitriol, which increases dietary absorption of calcium	*principal,* the predominant cell type of a gland
Oxyphil Cells	Larger than chief cells and more reddish in color	Unknown	NA	NA	*oxys,* sour acid, + *-phil,* to love

EXERCISE 19.4

THE ADRENAL GLANDS

The adrenal (suprarenal) glands are located directly superior to each kidney (*ad,* to, + *ren,* kidney). They are similar to the pituitary gland in that they are composed of two regions **(figure 19.6),** each with a separate embryological origin. The outer region, the **adrenal cortex,** is derived from mesoderm and has the appearance of typical glandular epithelium. The inner region, the **adrenal medulla,** is derived from modified postganglionic sympathetic neurons (which are derived from neural crest cells) and has the appearance of nervous tissue. The cells of the adrenal cortex synthesize steroid hormones (specifically, **corticosteroids,** *cortico,* relating to the adrenal cortex, + *steroid,* steroid hormone such as cortisol), whereas the cells of the adrenal medulla synthesize **catecholamine** hormones (that is, hormones derived from the amino acid tyrosine and that contain a catechol ring; includes epinephrine and norepinephrine). The entire gland is surrounded by a dense irregular connective tissue **capsule,** which protects the gland and helps anchor it to the superior border of the kidney. The adrenal cortex has three recognizable zones, and each zone has cells that predominantly secrete one category of corticosteroid hormones. Characteristics of the zones, and descriptions of the hormones secreted by cells within each zone, are summarized in **table 19.3.**

1. Obtain a histology slide of the adrenal gland (figure 19.6) and place it on the microscope stage. Bring the tissue sample into focus on low power and identify the two major regions, the cortex and the medulla (figure 19.6*b*).
2. Move the stage so the *adrenal cortex* (figure 19.6*b*) is in the center of the field of view. Then change to high power.
3. Identify the following zones of the adrenal cortex from outermost to innermost: zona glomerulosa, zona fasciculata, and zona reticularis, using figure 19.6 and table 19.3 as guides.

(continued on next page)

(continued from previous page)

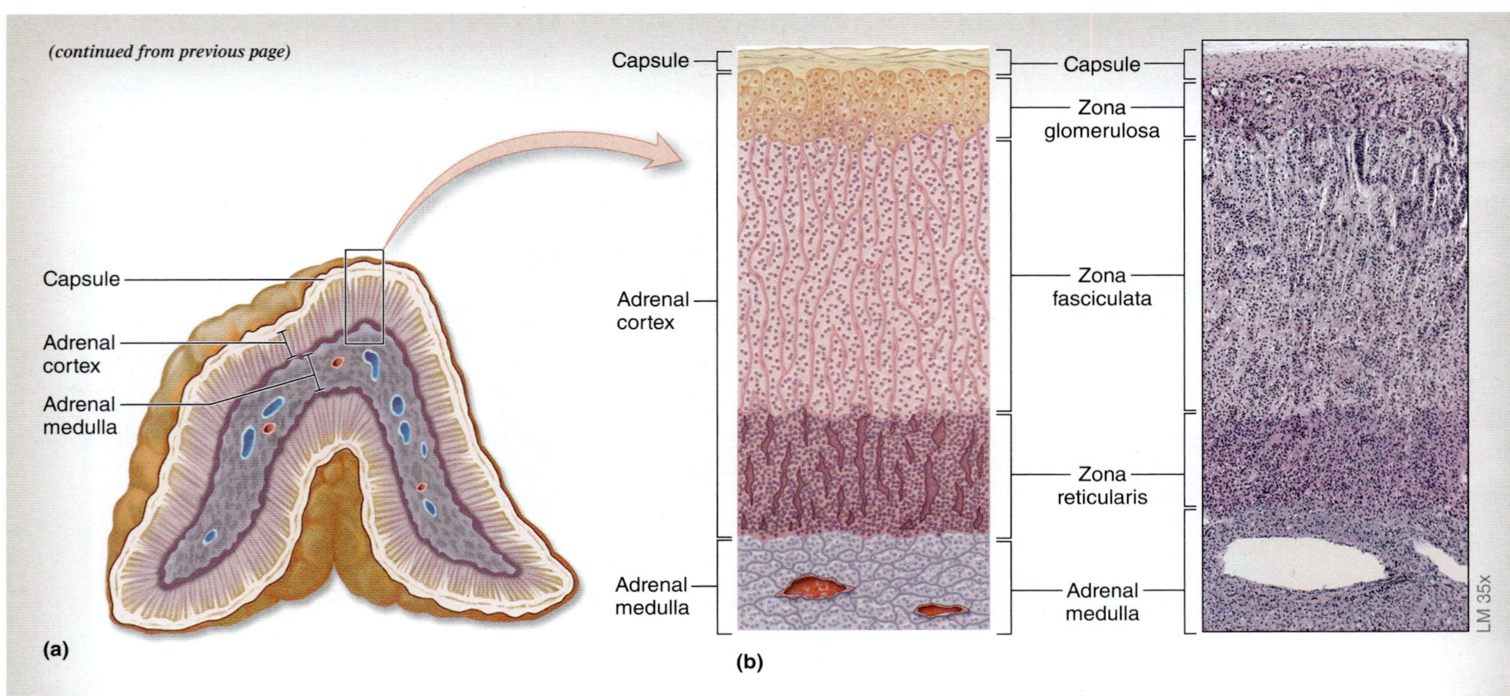

Figure 19.6 Adrenal Glands. (a) Cross section of the adrenal gland. (b) Micrograph demonstrating the three layers of the adrenal cortex and part of the adrenal medulla.

Table 19.3 Histology of the Adrenal Glands

Adrenal Gland Region	Zone and/or Cells	Description	Hormones Produced	Action of Hormone(s)	Word Origin
Adrenal Cortex	Zona glomerulosa	Outermost region of adrenal cortex; contains "balls" of cells located deep to the capsule of the adrenal gland	Mineralocorticoids: aldosterone	Increases sodium and water retention by the kidneys (thus increasing blood volume and blood pressure); vasoconstriction	*zona*, zone, + *glomus*, a ball of yarn
	Zona fasciculata	Middle and largest region of adrenal cortex; contains long cords, or "bundles," of cells	Glucocorticoids: cortisol	Increases glucose synthesis through gluconeogenesis (production of new glucose from amino acids), and lipolysis	*zona*, zone, + *fasciculus*, a bundle
	Zona reticularis	Innermost region of adrenal cortex; contains a "network" of cells, located between the zona fasciculata and cells of the adrenal medulla	Glucocorticoids and gonadocorticoids: androgens	Androgens are similar in structure and function to the male sex steroid hormone testosterone; secreted in very low amounts compared to testosterone secretion by the testes (males)	*zona*, zone, + *rete*, a net
Adrenal Medulla	Chromaffin cells	Large, spherical cells with yellowish-brown tint when stained due to their reaction with chrome salts	Catecholamines: epinephrine and norepinephrine	Epinephrine, secreted in the largest quantity (80–90% of all hormone secretion by the adrenal medulla), increases heart rate and contractility (force of heart muscle contraction); norepinephrine stimulates vasoconstriction	*chroma*, color, + *affinis*, affinity, attraction for

4. Change back to the low-power objective, move the microscope stage so the *adrenal medulla* is in the center of the field of view, and then change back to high power. The nuclei visible within the adrenal medulla are nuclei of *chromaffin cells*, which are modified post-ganglionic sympathetic neurons. What hormone(s) do these cells secrete? ______________________________

5. Identify the following structures on the slide of the adrenal glands, using figure 19.6 and table 19.3 as guides:

- ☐ **adrenal cortex**
- ☐ **adrenal medulla**
- ☐ **capsule**
- ☐ **chromaffin cells**
- ☐ **zona fasciculata**
- ☐ **zona glomerulosa**
- ☐ **zona reticularis**

6. Sketch the adrenal gland as seen through the microscope in the space provided, labeling all of the structures listed in step 5.

7. *Optional Activity:* **AP|R 8: Endocrine System—** Review the histology slides of the adrenal (suprarenal) gland, as well as the pituitary gland, thyroid gland, and endocrine pancreas.

INTEGRATE

CLINICAL VIEW
Corticosteroids

Excessive levels of cortisol may suppress the immune system by inhibiting inflammation and white blood cell activation and proliferation, and they impair repair of connective tissues. *Corticosteroids* are a class of drugs that are often prescribed for individuals who are suffering from widespread inflammation. For example, patients diagnosed with the autoimmune disease rheumatoid arthritis (RA) experience chronic inflammation in their synovial joints (see exercise 10.3). Since individuals suffering from this disease have abnormal activation and proliferation of immune cells, corticosteroids may be beneficial in limiting the progression of the disease. However, administration of the drug has some serious side effects, such as causing an increase in blood glucose levels and promoting the breakdown of tissue proteins over time. For these reasons, corticosteroids must be utilized with caution, particularly in elderly patients.

EXERCISE 19.5

THE ENDOCRINE PANCREAS—PANCREATIC ISLETS (OF LANGERHANS)

The **pancreas** is largely an exocrine gland. **Exocrine** glands produce substances that are secreted into ducts. The majority of the cells within the exocrine pancreas produce digestive enzymes, which are secreted into ducts that empty into the small intestine. The clusters of exocrine cells are **pancreatic acini** (singular: acinus). Interspersed among the acini are small islands of cells that have an endocrine function. **Endocrine** glands secrete hormones into the blood. The endocrine part of the pancreas consists of the **pancreatic islets** (islets of Langerhans). These islets contain hormone-secreting cells and a rich supply of blood capillaries. Four distinct cell types compose the islets, although they cannot be distinguished in a normal histological preparation. Thus, when observing the slide of the pancreas, the goal is to identify the pancreatic islets, not the specific cell types within. Nonetheless, know which cell type secretes each hormone. The

(continued on next page)

(continued from previous page)

cell types and hormones secreted by each cell type are listed in **table 19.4.**

1. Obtain a histology slide of the pancreas, place it on the microscope stage, and bring the tissue sample into focus on low power. The majority of the cells observed, which will look somewhat like cuboidal epithelial cells, are the exocrine cells of the pancreas **(figure 19.7).** Scan the slide to locate small clusters of cells, the **pancreatic islets,** which (typically) stain lighter in color than the exocrine cells.
2. Locate a pancreatic islet and move the microscope stage so the islet is in the center of the field of view.
3. Increase the magnification so the islet is visible in greater detail. Identify the following on the slide, using figure 19.7 and table 19.4 as guides:

- ☐ **pancreatic acini**
- ☐ **pancreatic islets**
- ☐ **secretory ducts**

Table 19.4 Histology of the Pancreatic Islets (of Langerhans)

Cell Types	Description	Hormones Produced	Action of Hormone	Mechanism of Action	Word Origin
Alpha Cells	Compose about 30% of islet cells; located on the periphery of the islet	Glucagon	Increases blood glucose levels	Stimulates glycogenolysis (breakdown of glycogen) and gluconeogenesis (formation of new glucose from amino acids) in the liver, and lipolysis in adipose connective tissue	*alpha,* the first letter of the Greek alphabet, + *glucose,* sugar, + *ago,* to lead
Beta Cells	Compose about 65% of islet cells; located in the center of the islet	Insulin	Decreases blood glucose levels	Stimulates glucose, amino acid, and potassium uptake by cells, stimulates glycogenesis (formation of glycogen from glucose) in the liver, and increases lipolysis in adipose connective tissue	*beta,* the second letter of the Greek alphabet, + *insula,* an island
Delta Cells	Compose about 4% of islet cells; located on the periphery of the islet	Somatostatin	Inhibits the release of glucagon and insulin by alpha and beta cells	Inhibits the release of glucagon and insulin when nutrient levels in the blood are high	*delta,* the 4th letter of the Greek alphabet, + *soma,* body, + *stasis,* standing still
F Cells	All other (rare) cell types in the pancreatic islets (~1%) are grouped together and given this name	Pancreatic polypeptide	Inhibits somatostatin release from delta cells	Regulates secretion of somatostatin from delta cells	NA

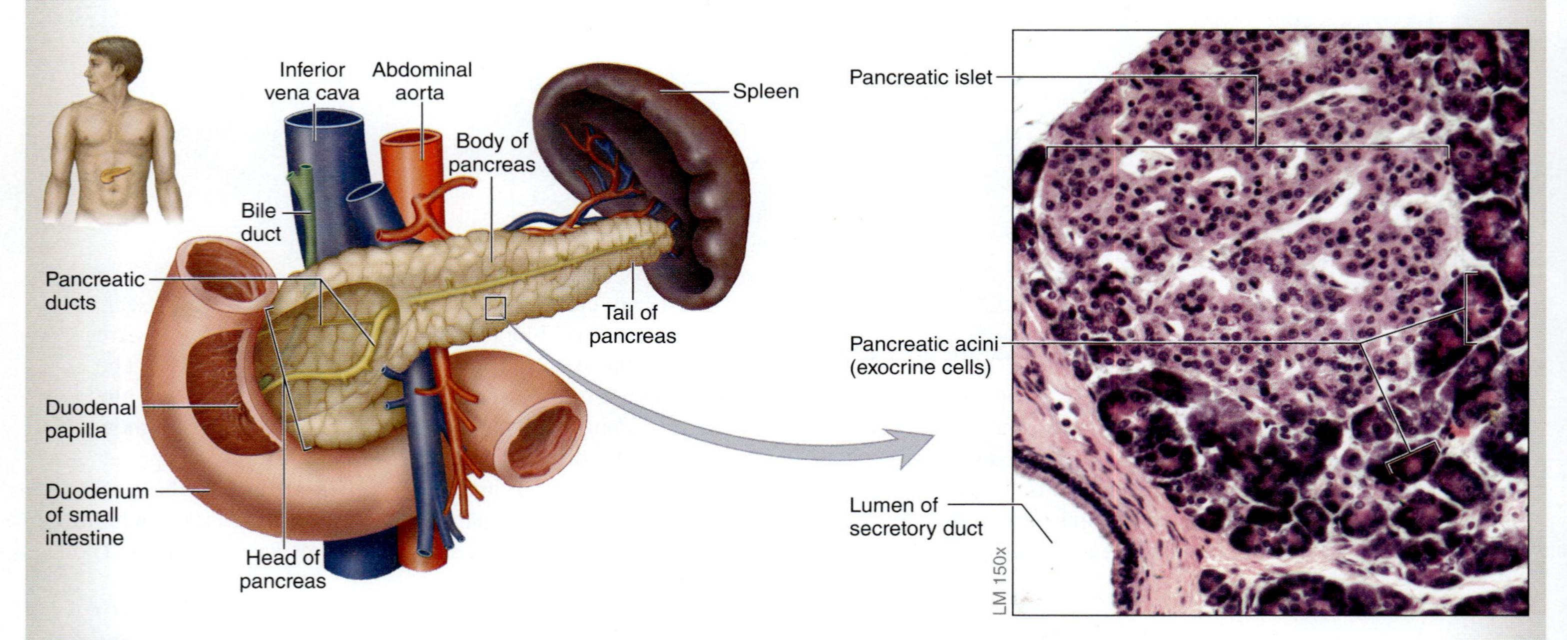

Figure 19.7 Endocrine Portion of Pancreas. The endocrine part of the pancreas consists of the pancreatic islets, which are the lighter-colored regions in this micrograph. The darker cells surrounding the islet are acinar cells, which compose the exocrine part of the pancreas.

4. Sketch a pancreatic islet and the exocrine cells that surround it as seen through the microscope in the space provided. Be sure to label the structures listed in step 3.

_______ ×

INTEGRATE

CONCEPT CONNECTION

Transport proteins are among the various proteins present in the plasma membrane of all cells. These proteins are responsible for movement of substances into and out of the cells. These proteins exhibit *specificity*, which means that they will assist only a specific type of substance across the plasma membrane. The movement of substances from an area of high concentration to an area of low concentration with the assistance of such transporters is called **facilitated diffusion** (see chapter 4). One substance that is moved into all cells by facilitated diffusion is glucose, which cells utilize as a substrate for cellular respiration.

GROSS ANATOMY

Some endocrine glands are difficult to identify on the cadaver. However, most classical glands and associated structures can be identified on the cadaver or on classroom models, so these will be the focus of this section.

Endocrine Organs

EXERCISE 19.6

GROSS ANATOMY OF ENDOCRINE ORGANS

1. Observe a human cadaver or classroom models of the brain, thorax, abdomen, and skull.
2. Identify the structures listed in **figure 19.8** on a human cadaver or on classroom models, using the textbook as a guide. Then label them in figure 19.8.
3. *Optional Activity:* AP|R **8: Endocrine System**—Watch the endocrine system animations to review the structure and function of the hypothalamus and pituitary gland, pancreas, thyroid and parathyroid glands, and adrenal (suprarenal) glands.

(continued on next page)

(continued from previous page)

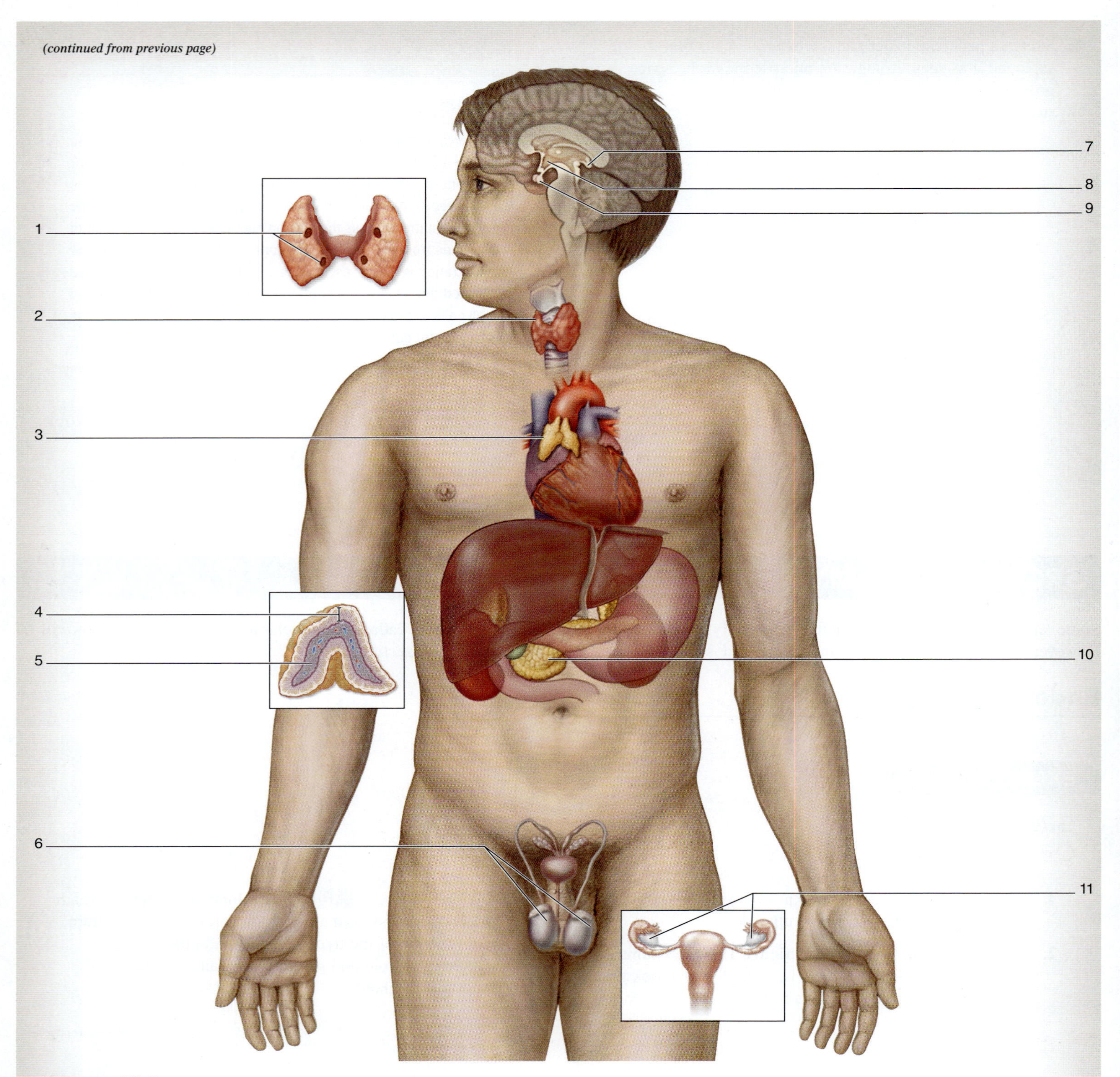

Figure 19.8 Labeling Major Endocrine Glands of the Body. (a) Major endocrine glands of the body. Use the terms listed to fill in the numbered labels in the figure. Some answers may be used more than once.

- ☐ adrenal cortex
- ☐ adrenal medulla
- ☐ anterior pituitary gland
- ☐ hypothalamus
- ☐ ovaries
- ☐ pancreas
- ☐ parathyroid glands
- ☐ pineal gland
- ☐ pituitary gland
- ☐ posterior pituitary gland
- ☐ testes
- ☐ thymus
- ☐ thyroid gland

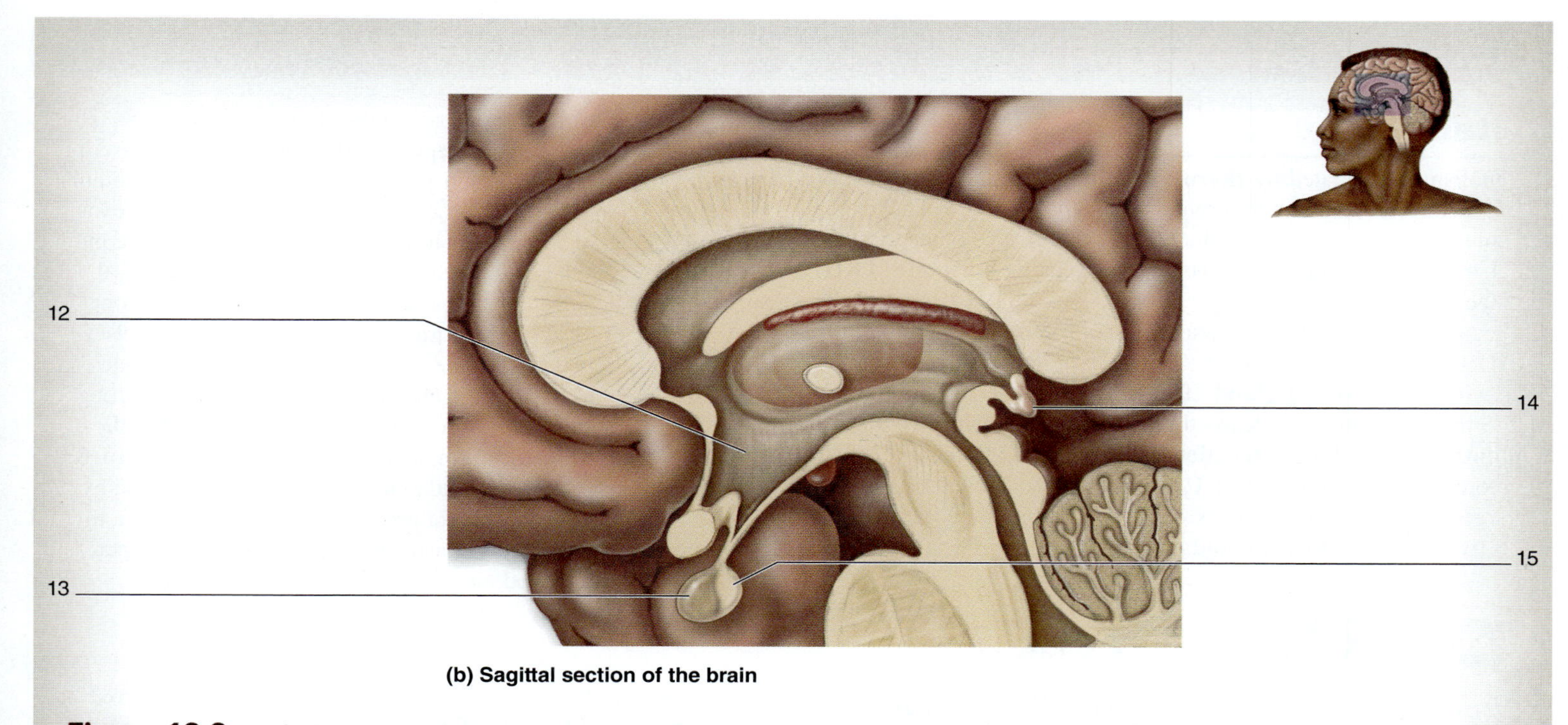

(b) Sagittal section of the brain

Figure 19.8 **Labeling Major Endocrine Glands of the Body *(continued)*.** (b) Illustration of sagittal section of the brain.

INTEGRATE

CLINICAL VIEW
Anabolic Steroids

Anabolic steroids are lipid-soluble hormones that are structurally and functionally similar to the endogenous hormone testosterone. When introduced exogenously (from outside the body) into the systemic circulation, either by injection, in pill form, or as a cream applied transdermally, anabolic steroids cross the plasma membrane of target cells and bind to androgen receptors. Overall, anabolic steroids promote protein synthesis, particularly in muscle cells. Other effects of anabolic steroids include increased appetite, increased bone growth, and increased erythropoiesis (red blood cell synthesis). Anabolic steroids also promote the development of secondary sex characteristics, including increased hair growth, increased sweat production by sweat glands, and elongation of vocal cords, which causes a deepening of the voice. Dangerous side effects of anabolic steroids include cardiovascular changes such as increased blood cholesterol levels and increased blood pressure.

Endogenous testosterone levels are regulated by negative feedback mechanisms. Therefore, exogenous administration of anabolic steroids acts as a stimulus to suppress the release of LH by the anterior pituitary gland, which in turn reduces testosterone production by the testes. In males, whose testosterone levels are typically much higher than in females, administration of anabolic steroids can reduce libido (sex drive), testicle size, and sperm production. They can also promote gynecomastia (enlargment of breast tissue). Anabolic steroids are sometimes used medically; however, they can also be abused, as when individuals take anabolic steroids to stimulate increases in body mass and strength (e.g., by bodybuilders). Individuals who use anabolic steroids recreationally may become dependent and may experience some moderate to severe psychological effects (e.g., "roid rage") during and following (i.e., withdrawal) use of the drugs.

PHYSIOLOGY

Metabolism

Metabolism is regulated by **thyroid hormone (TH).** The release of TH is controlled indirectly through the hypothalamus. The hypothalamus detects a decrease in blood levels of TH (and other stimuli), which triggers the release of **thyrotropin-releasing hormone (TRH)** into the hypothalamic-hypophyseal portal vein. TRH triggers cells in the anterior pituitary to release **thyroid-stimulating hormone (TSH).** TSH is transported in the general circulation to target cells within the **thyroid gland.** TSH stimulates the **follicular cells** of the thyroid gland to release thyroid hormone in its two forms: **triiodothyronine (T_3)** and **tetraiodothyronine (T_4, thyroxine).** T_3 is the more active hormone. Most T_4 is converted to T_3 when it reaches its target cells. For the purposes of this exercise, T_3 and T_4 will be collectively referred to as thyroid hormone (TH).

Thyroid hormone (a lipid-soluble hormone) exerts its effects by binding to intracellular receptors and stimulating the transcription of genes that code for certain proteins. For example, TH stimulates the transcription of genes that code for the Na^+/K^+ pump in excitable cells, which causes these cells to increase production of these pumps. To meet the demands of increased numbers of these pumps, the cell increases its rate of cellular respiration to synthesize the additional ATP that is required for this active transport process. TH acting elsewhere in the body stimulates the increase of heart and respiratory rates. This increases the transport of oxygenated blood to active cells throughout the body. The overall effect of TH is to **increase metabolic rate, increase oxygen consumption,** and **increase body temperature.** TH also decreases the level of circulating amino acids in the blood by promoting their uptake into cells for protein synthesis. In addition, thyroid hormone stimulates **lipolysis** and **glycogenolysis,** thereby increasing both fatty acid and blood glucose levels in the blood.

EXERCISE 19.7 Ph.I.L.S.

Ph.I.L.S. LESSON 19: THYROID GLAND AND METABOLIC RATE

The purpose of this laboratory exercise is to observe the effects of TH on body temperature and metabolic rate. Metabolic rates of two female white mice will be measured based on rates of oxygen consumption. This simulation will calculate the least square linear regression, which determines the *dependent* (y) and *independent* (x) variables, plots the data points, and determines the line of best fit for the data points. The equation that generates the line ($y = mx + b$) can then be used to predict the value of the dependent variable if the value of the independent variable is known.

The two mice in the simulation have eaten either (1) normal mouse "chow," or (2) mouse "chow" containing 0.15% propylthiouracil (PTU). The production of TH is significantly reduced by the compound PTU. Finally, the rate at which each mouse consumes oxygen will be measured at different temperatures in order to determine the effect of cooling on metabolic rate.

Prior to beginning the experiment, become familiar with the following concepts (use the textbook as a reference):

- negative feedback regulation
- the hypothalamus and its anatomical and physiological relationship with the anterior pituitary gland
- follicular tissue of the thyroid gland
- the cellular effects of thyroid hormone (TH)
- the relationship between oxygen consumption and metabolic rate

Prior to beginning the experiment, state a hypothesis regarding the influence of intake of mouse chow with TPU on the levels of TH, oxygen consumption, and metabolic rate in the mouse:

____

____

____

____

1. Open the Ph.I.L.S Lesson 19: Endocrine Function: Thyroid Gland and Metabolic Rate.
2. Read the objectives, introduction, and wet lab sections. Then take the pre-lab quiz. The laboratory exercise will open when the pre-lab quiz has been completed **(figure 19.9).**
3. Click the power switch to turn on the scale and then click the power switch to turn on the thermostat.
4. Click "Tare" to set the scale to zero. To weigh a mouse, click on one of the mice and drag it to the scale. After weighing the mouse, place it in the cage by clicking and dragging it to the cage in the lower left corner of the screen.
5. Remove the pipette from the beaker, which contains bubbles, and place the tip of the pipette to the calibration tube (at the 10).
6. Measure the initial position of the soap bubbles (if at the end of the tube it will be 10 mm) and record the position at 0:00 time in the table in the program.

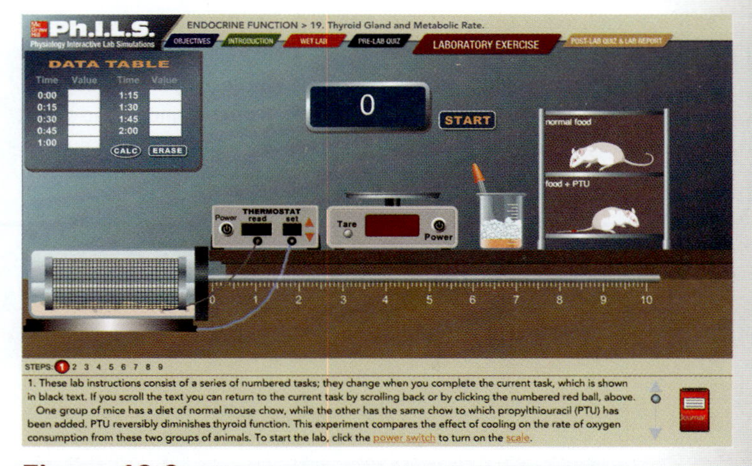

Figure 19.9 Opening Screen for the Laboratory Exercise on Endocrine Function: Thyroid Gland and Metabolic Rate.

7. After clicking the "Start" button, measure the position of the soap bubble within the calibration tube at 15-second intervals (a beep will be heard) and record it in the data table. (The timer will begin when you hit "Start.") It may be more manageable to click the "Pause" button after each 15-second interval, measure, record, and then click "Start" to resume the experiment. When finished, click "Pause."
8. Click "Calc" to see the graph (linear regression) in the program. The "Journal" will open and show a graph of the data collected up to this point in the laboratory exercise. The journal also includes a table. Notice that the table (with temperatures ranging from 8°C to 24°C) will have the first data point entered for the average oxygen consumed per minute (total amount of oxygen divided by 2 minutes). Enter this data in **table 19.5.** No points will appear on the graph until all data points are entered. Close the table and graph windows by clicking the "X" in the upper right-hand corner.
9. To complete the process again with the same animal, repeat steps 5–8 but decrease the temperature 2 degrees (by clicking on the down arrow on the thermostat) prior to beginning the steps until data is recorded for all temperatures. Be sure to record the data in table 19.5 every time one trial is completed.
10. After returning the first animal to the cage, repeat steps 4–9 for the second animal.
11. Construct a graph that plots the linear regression of oxygen consumption versus temperature in the space provided.
12. Open the post-lab quiz and lab report by clicking on it. Answer the post-lab questions that appear on the computer screen. Click "Print Lab" to print a hard copy of the report or click "Save PDF" to save an electronic copy of the report.
13. Make note of any pertinent observations here:

____

____

____

Table 19.5 Data for Ph.I.L.S. Lesson 19: Thyroid Gland and Metabolic Rate

Temp (°C)	Normal (24.2 g)	+PTU (g)
24		
22		
20		
18		
16		
14		
12		
10		
8		

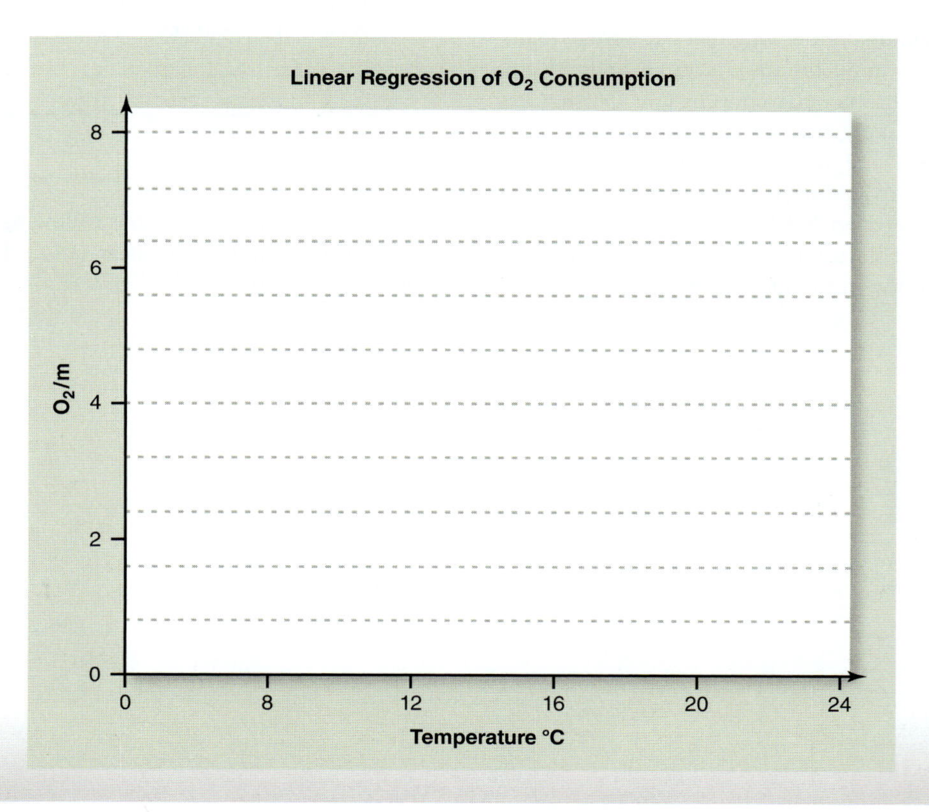

EXERCISE 19.8

A CLINICAL CASE IN ENDOCRINE PHYSIOLOGY

The purpose of this exercise is to apply concepts related to the endocrine system to a real scenario, or clinical case. This exercise requires information presented in this chapter, in the histology, gross anatomy, and physiology sections, as well as in previous chapters. Integrating multiple systems in a case study reinforces understanding of homeostatic processes.

Prior to beginning the case, ask a laboratory instructor if the exercise will be completed individually, in pairs, or in groups. Review the topics of the nervous system, as well as the endocrine system, including histology and gross anatomy of the pituitary gland, hormone production, and physiological effects of each hormone, as needed. Use the textbook and this lab manual as guides.

Case Study

Christopher, an 80-year-old male, was diagnosed with stage III non-small cell carcinoma of the lung. There was no evidence of metastasis on computed tomographic (CT) scans at the time of diagnosis. Even so, doctors treated his cancer aggressively with surgical removal of the tumor followed by radiation therapy. One year following his cancer diagnosis and treatment, Christopher complained of extreme fatigue, cold intolerance, headache, mental confusion, and visual disturbances. He experienced a severe fall, and was admitted to the hospital. Upon admission, blood tests revealed that Christopher's TSH and TH levels were low. In addition, he had elevated blood calcium levels as well as elevated PTH levels. He was prescribed *levothyroxine* (synthetic TH) and *prednisone* (a glucocorticoid). A magnetic resonance imaging (MRI) examination of his brain and spinal cord revealed slight abnormalities.

Doctors concluded that Christopher suffered from metastatic small cell carcinoma of the lung that had spread to his pituitary gland. The lesions in his pituitary protruded superiorly, which compressed nerves in his optic chiasm, thus causing his visual disturbances. It is likely that his altered consciousness and sudden onset of headache were caused by bleeding into the subarachnoid space surrounding his brain. To confirm the diagnosis, doctors performed an MRI of his head before and after administration of a contrast material. The MRI showed evidence of thickening of the infundibulum, enlargement of the pituitary gland, and enlargement of the third ventricle. There also was evidence of hypothalamic involvement. A biopsy was deemed too risky, and Christopher died just three weeks after his initial hospital admission.

Answer the following questions regarding Christopher's case:

1. Describe how a tumor in the hypothalamus and pituitary gland could lead to lowered levels of TSH and TH. Be sure to identify which cells are responsible for secreting each hormone, and describe feedback mechanisms that regulate their release.

____

____

____

2. Justify Christopher's cold intolerance and fatigue given lowered levels of TSH and TH. Be sure to state specifically how TH influences metabolism and temperature regulation.

____

____

____

3. Discuss the pros and cons of *prednisone* administration. Be sure to address specific targets of glucocorticoids, and positive and negative effects of the drug.

____

____

____

4. Are there any other hormones whose effects may be altered by a tumor in the hypothalamus and pituitary gland? If so, state which hormones may be altered, and discuss possible symptoms associated with altered levels of each hormone.

____

____

____

This case is adapted from the following: Case 35–2001. *The New England Journal Of Medicine,* 1483–1488, 2001 November 15, Vol. 345, Issue 20.

Chapter 19: The Endocrine System

Name: ______________________

Date: __________ Section: __________

POST-LABORATORY WORKSHEET

The (1) corresponds to the Learning Objective(s) listed in the chapter opener outline.

Do You Know the Basics?

Exercise 19.1: The Hypothalamus and Pituitary Gland

1. The anterior pituitary gland is composed of glandular tissue, which consists of acidophils, basophils, and chromophobes; thus it generally stains ________________(darker/lighter) than the posterior pituitary gland. (1)
2. Which of the following hormone(s) is/are produced by acidophils of the anterior pituitary gland? (Check all that apply.) (2)
   - ______ a. adrenocorticotropic hormone (ACTH)
   - ______ b. follicle stimulating hormone (FSH)
   - ______ c. growth hormone (GH)
   - ______ d. luteinizing hormone (LH)
   - ______ e. prolactin
3. The supraoptic and paraventricular nuclei contain cell bodies of neurosecretory cells whose axons extend along the infundibulum to axon terminals in the posterior pitutitary that release antidiuretic hormone (ADH) and oxytocin. ______________ (True/False) (3)

Exercise 19.2: The Pineal Gland

4. One defining histological feature of the pineal gland are the small islands of "crystal" looking structures that compose pineal sand (corpora arenacea). ____________________ (True/False) (4)
5. Melatonin, the hormone secreted by pinealocytes of the pineal gland, ______________(decreases/increases) when light levels are low, and ________________(decreases/increases) when light levels are high. (5)
6. Pineal sand is important to radiologists since it is an easily identifiable landmark. ______________ (True/False) (6)

Exercise 19.3: The Thyroid and Parathyroid Glands

7. The thyroid gland consists of follicles lined with simple columnar epithelial tissue. ______________ (True/False) (7)
8. Follicular cells of the thyroid gland produce _________________ (calcitonin/thyroid hormone), whereas parafollicular cells produce ________________ (calcitonin/thyroid hormone). (8)
9. The major action of parathyroid hormone, produced by the parathyroid gland, is to _________________ (decrease/increase) blood calcium levels by __________________ (decreasing/increasing) the action of osteoclasts and ____________________ (decreasing/increasing) urinary excretion of calcium. (9)

Exercise 19.4: The Adrenal Glands

10. Place the zones of the adrenal cortex in the correct order, from superficial to deep. (10)
    - ______ a. zona fasciculata
    - ______ b. zona glomerulosa
    - ______ c. zona reticularis
11. Match the region of the adrenal gland listed in column A with the corresponding hormones produced by that region, listed in column B. (11)

Column A	*Column B*
______ 1. medulla	a. androgens and glucocorticoids
______ 2. zona fasciculata	b. catecholamines
______ 3. zona glomerulosa	c. glucocorticoids
______ 4. zona reticularis	d. mineralcorticoids

12. Match the hormone description and/or action(s) listed in column A with the appropriate hormone, listed in column B. **12**

Column A	*Column B*
______ 1. increased heart rate, contractility, and vasoconstriction	a. aldosterone
______ 2. increased glucose synthesis through protein breakdown, gluconeogenesis, and lipolysis	b. androgens
______ 3. increased sodium and water retention by the kidneys	c. cortisol
______ 4. similar in structure and function to testosterone	d. epinephrine and norepinephrine

Exercise 19.5: The Endocrine Pancreas—Pancreatic Islets (of Langerhans)

13. The small islands of lighter-staining cells of the pancreas are known as pancreatic ______________ (acini/islets). **13**

14. The pancreatic islets compose the portion of the pancreas dedicated to ______________ (endocrine/exocrine) function, whereas the pancreatic acini compose the portion of the pancreas dedicated to ________________ (endocrine/exocrine) function. **14**

15. Insulin is released in response to a(n) ________________ (increase/decrease) in blood glucose levels, whereas glucagon is released in response to a(n) ________________ (increase/decrease) in blood glucose levels. **15**

16. Match the cell type of the pancreatic islets, listed in column A, with the hormones secreted by each cell, listed in column B. **16**

Column A	*Column B*
______ 1. alpha cells	a. glucagon
______ 2. beta cells	b. insulin
______ 3. delta cells	c. pancreatic polypeptide
______ 4. F cells	d. somatostatin

Exercise 19.6: Gross Anatomy of Endocrine Organs

17. The adrenal glands are located superior to each kidney. ______________ (True/False) **17**

Exercise 19.7: Ph.I.L.S. Lesson 19: Thyroid Gland and Metabolic Rate

18. Cooling the body ________________ (decreases/increases) thyroid hormone release, which in turn ________________ (decreases/increases) metabolic and oxygen consumption rate. **18** **19**

Exercise 19.8: A Clinical Case Study in Endocrine Physiology

19. Decreased levels of thyroid stimulating hormone and thyroid hormone could lead to which of the following symptoms? (Check all that apply.) **20**

______ a. cold intolerance

______ b. fatigue

______ c. weight gain

______ d. weight loss

Can You Apply What You've Learned?

20. Identify each of the following glands by writing in the name of the gland under the photo.

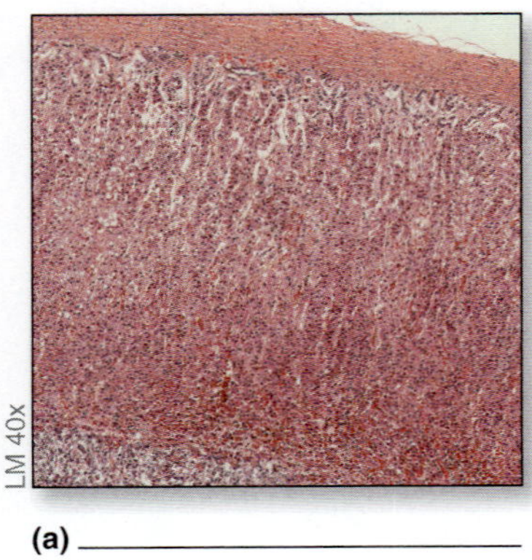

(a) ______________________

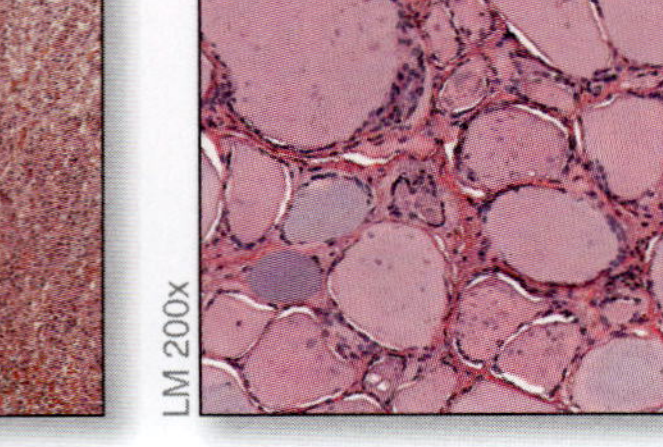
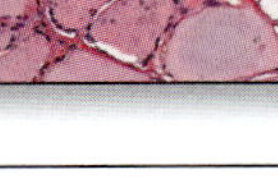

(b) ______________________

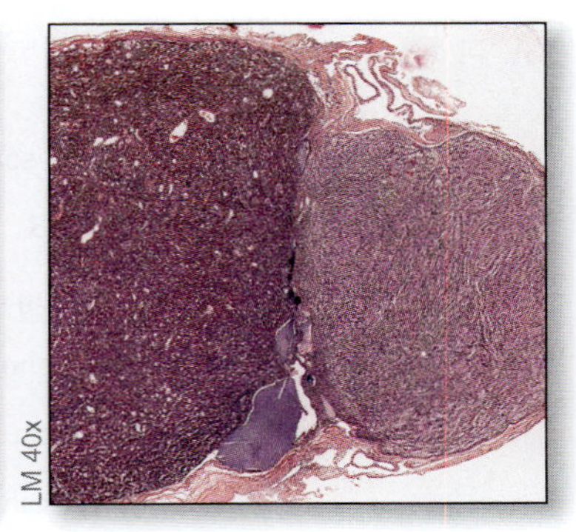

(c) ______________________

LM 400x

(d) ______________________

21. Based on your understanding of the location of the pituitary gland with respect to other brain structures, explain why individuals with pituitary gland tumors often also experience visual disorders.

22. Predict the oxygen consumption in an individual with hypothyroidism. Then compare the rate of oxygen consumption in an individual with hypothyroidism with that of an individual with hyperthyroidism.

23. Body builders have been known to inject insulin to increase muscle mass. Explain this reasoning and discuss any risks that may be posed by this practice.

24. Identify some health risks that may come from long-term use of corticosteroids.

Can You Synthesize What You've Learned?

25. Patients suffering from hyperparathyroidism are also at a greater risk for bone fracture. What do you predict blood calcium levels in a patient suffering from this condition would look like? Explain why these patients often experience a decrease in bone density.

26. Athletes may be interested in calculating their basal metabolic rate (BMR) to determine the number of calories their body utilizes at rest. This is done using a calorimeter, which measures oxygen consumption to measure BMR. The calorimeter calculates metabolic rate based on the rate of oxygen consumption. Describe how this machine likely accomplishes this task.

The Cardiovascular System: Blood

20

OUTLINE AND LEARNING OBJECTIVES

MODULE 9: CARDIOVASCULAR SYSTEM

INTRODUCTION

Blood is a bodily fluid that has fascinated humans for centuries. Its rich red color adds to both its beauty and its mystery. For many, the sight of blood is quite disturbing because it usually indicates that something is very wrong! It's no wonder that for centuries humans have been fascinated by this most colorful bodily fluid. Early Greek physicians believed that blood was one of the four bodily "humors" (*humor*, a bodily fluid). The other bodily fluids were phlegm, black bile, and yellow bile **(table 20.1)**. Each of these humors was associated with an element—air, water, earth, or fire—and had particular characteristics: hot, cold, wet, or dry. Before putting any confidence in this idea, know that modern science recognizes over ninety-two naturally occurring elements, and earth, air, fire, and water are not among them! Greek physicians believed that a balance of the four humors was required for optimal health of both the body and the mind, and that disease was the result of an imbalance between the bodily fluids. It was this concept of disease that made bloodletting a popular treatment for disease. By draining blood from a patient, the physician supposedly was putting the patient's humors back into balance. In those days, studying anatomy or medicine wasn't nearly as complicated as it is today. On the other hand, we can all be thankful that we didn't have to be patients back then. Ironically, to this day references to this early concept of medicine continue, even though most are largely unaware of their origins. An infant who cries all the time is referred to as being "colicky." Someone who is sad is said to be feeling "melancholy." For comparison, a patient living in the year 400 B.C. who was feeling depressed would have been diagnosed as having an excess of "black bile." When considering this approach to medicine, and the questionable results of treatment, it's easy to be thankful for twenty-first century medicine. If the idea of "black bile" isn't enough for a good laugh, consider what early Greeks thought to be the source of phlegm: the brain. That thought may be enough to transform one's mood from melancholy to "sanguine," or lively and optimistic. Are you feeling lively and optimistic yet? Then it is time to tackle the current "sanguine" topic: blood!

The exercises in this chapter explore the form and function of normal constituents of human blood. Knowledge of the normal appearance and abundance of the various types of blood cells provides a basis for recognizing when cells or amounts have become abnormal and for understanding what that means. **Tables 20.2** and **20.3** summarize the characteristics of the formed elements of blood, which are the structures that are the focus of this laboratory chapter. Exercises within this chapter involve identifying the formed elements of blood, exploring the functional relevance of each of the components, and learning how the different cell types play a critical role in both the cardiovascular and immune systems. Exercises also guide students in performing common blood tests such as determination of hemoglobin content of blood, determination of blood type, and measurement of blood cholesterol levels. Considerably more detail on the structure and function of blood is covered in the main textbook. Thus, it is important to read the section in the textbook that covers blood prior to performing these laboratory exercises.

List of Reference Tables

Table 20.1 The Four Greek Humors

Bodily Fluid (Humor)	Element	Characteristics	Source	Mood	Mood Characteristics	Word Origin
Black Bile	Earth	Dry and cold	Spleen	Melancholy	Sad, depressed	*melas*, black, + *chole*, bile
Blood	Air	Hot and wet	Heart	Sanguine	Lively and optimistic	*sanguis*, blood
Phlegm	Water	Cold and wet	Brain	Phlegmatic	Calm and unexcitable	*phlegma*, inflammation
Yellow Bile	Fire	Hot and dry	Liver	Choleric	Irritable	*chole*, bile

Chapter 20: The Cardiovascular System: Blood

Name: ____________________

Date: ____________ Section: ____________

PRE-LABORATORY WORKSHEET

These Pre-Laboratory Worksheet questions may be assigned by instructors through their connect course.

1. Number the following components of whole blood in order of abundance, from most abundant (1) to least abundant (3).

   _____ a. buffy coat

   _____ b. erythrocytes

   _____ c. plasma

2. Basophils, eosinophils, and neutrophils belong to the category of leukocytes known as ______________ (agranuloctyes/granulocytes).

3. Match the description of the leukocyte listed in column A with the type of leukocyte listed in column B.

Column A		***Column B***
_____	1. bilobed nucleus; blue cytoplasmic granules	a. basophil
_____	2. bilobed nucleus; red cytoplasmic granules	b. eosinophil
_____	3. large, horseshoe-shaped nucleus; much larger than an erythrocyte	c. lymphocyte
_____	4. multi-lobed nucleus; lavender cytoplasmic granules	d. monocyte
_____	5. round nucleus surrounded by pale blue cytoplasm; similar in size to an erythrocyte	e. neutrophil

4. Which of the following is the cell responsible for producing platelets? (Circle one.)

   a. basophils
   b. erythrocytes
   c. lymphocytes
   d. megakaryocytes
   e. neutrophils

5. Which of the following is the formed element responsible for transporting oxygen and carbon dioxide in the blood? (Circle one.)

   a. basophils
   b. erythrocytes
   c. lymphocytes
   d. megakaryocytes
   e. neutrophils

6. Which of the formed elements is responsible for initiating coagulation? (Circle one.)

   a. basophils
   b. erythrocytes
   c. megakaryocytes
   d. neutrophils
   e. platelets

7. A measure of the percentage of erythrocytes in whole blood is called erythropoiesis. ____________ (True/False)

HISTOLOGY

CAUTION:

The exercises in this laboratory may involve the use of human, animal, or artificial blood. Blood samples, whether human or animal, are considered biohazardous waste and should be handled with care. Be sure to take precautionary measures while handling blood samples by wearing nitrile gloves and safety goggles. Handle sharp objects with caution. Biohazardous wastes and sharps must be disposed of in the proper containers immediately following their use. Please review procedures for handling sharps and biohazardous wastes in chapter 1 of this manual and consult with the laboratory instructor for further instructions on laboratory policies and procedures for handling and disposing of these substances.

EXERCISE 20.1

IDENTIFICATION OF FORMED ELEMENTS ON A PREPARED BLOOD SMEAR

EXERCISE 20.1A Making a Human Blood Smear

Exercise 20.1 may involve use of a prepared slide of human blood, or a blood smear made from your own blood. To make a blood smear using your own blood, use the following procedure.

Obtain the Following:

- **four glass slides**
- **lancet**
- **cotton balls**
- **alcohol prep pads**
- **vial of Wright's stain**

1. Clean the slides with soap and water and let them dry. It takes two slides to make a blood smear, and may require several attempts to obtain a good preparation. For this reason, clean at least four slides.
2. Using the alcohol prep pad, clean the tip of the middle or ring finger on the hand you use the least (e.g., if you are right-handed, use the tip of the finger on your left hand).
3. Place the tip of the lancet on the clean finger, and pull the trigger so it pierces the skin. If using a lancet without a trigger **(figure 20.1),** poke the skin with a quick jab. Dispose of the lancet in a sharps container.
4. Squeeze the finger until a small drop of blood appears on the surface. Blot away the first drop of blood using an alcohol prep pad and then squeeze the finger again until another drop appears. Gently touch the drop of blood to the top surface of the microscope slide about 1.5–2 cm from the edge of the slide (**figure 20.2,** step 1). Preparing the blood smear does not require a lot of blood so try not to squeeze the finger too hard.
5. Orient the slide so the end with the drop of blood on it is closest to you. Then place the edge of a second microscope slide on the drop of blood and hold it at a 45-degree angle to the slide with the drop of blood on it (figure 20.2, step 2).
6. Quickly push the second slide over the first slide (push away from you) to smear the blood (figure 20.2, step 3). The goal is to obtain a very thin, transparent layer of blood on the slide. If the layer is too thick, it will be difficult to see individual cells because they will be too closely packed together. If the smear is not even or well spread out, start over with two new slides. It might take a couple of tries to create a good, even smear. Discard any slides that made contact with blood in a bowl containing a bleach solution or sharps container. Allow the slide with the blood smear on it to air dry.
7. Place a couple of drops of Wright's stain on the blood smear slide so there is enough stain on the slide to cover the smear, but not so much that it overflows the slide. Keep track of the number of drops used (number of drops = __________).

Figure 20.1 Lancing Finger for Blood Sample.

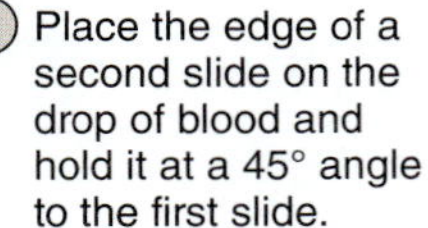

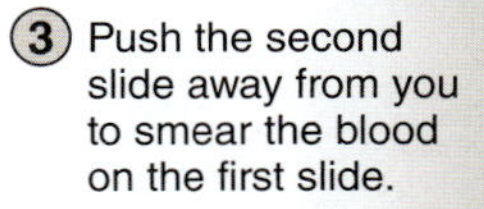

① Place a small drop of blood approximately 2 cm from the edge of the slide.
② Place the edge of a second slide on the drop of blood and hold it at a 45° angle to the first slide.
③ Push the second slide away from you to smear the blood on the first slide.

Figure 20.2 Preparation of a Blood Smear.

8. Let the slide stand for 3–4 minutes, until the smear begins to take on a blue/green color.
9. Add to the slide a volume of distilled water equal to the volume of Wright's stain used. Generally this means using the same number of drops of distilled water as Wright's stain. Gently move the slide back and forth and blow on it a little bit to mix the distilled water and stain on the slide.
10. Let the slide stand for 5–10 minutes. During this time, blow on the slide occasionally to keep the distilled water and stain mixed.
11. Rinse the slide with distilled water for 2–3 minutes, and then tilt the slide to allow the excess water to drip off of the slide. Allow the slide to air dry. Once the slide is dry, it will be ready to examine.

EXERCISE 20.1B Viewing Formed Elements on a Prepared Blood Smear

1. Obtain a prepared slide of human blood or use the slide prepared in exercise 20.1A. Place the slide on the microscope stage and observe at low power. Not much is visible at this magnification, but, as always, it is important to progress from low to high power to keep the sample in focus before progressing to the oil immersion objective.
2. Change from low to medium, then to high power, making sure to bring the slide into focus at each power. After the slide is in focus on high power, obtain a vial of immersion oil. Rotate the nosepiece (the part of the microscope that holds the objective lenses) so the high-power and oil immersion objectives lie on either side of the slide **(figure 20.3).** Place a drop of immersion oil over the center of the slide (where the light can be seen coming through the slide), and then carefully rotate the nosepiece to bring the immersion objective into the oil over the slide. Bring the slide into focus.
3. Identify the following formed elements on the prepared blood slide, using tables 20.2 and 20.3 and **figure 20.4** as guides. Note that some elements will be much easier to locate than others due to their relative abundance in whole blood.

- ☐ **basophils**
- ☐ **eosinophils**
- ☐ **erythrocytes**
- ☐ **lymphocytes**
- ☐ **monocytes**
- ☐ **neutrophils**
- ☐ **platelets**

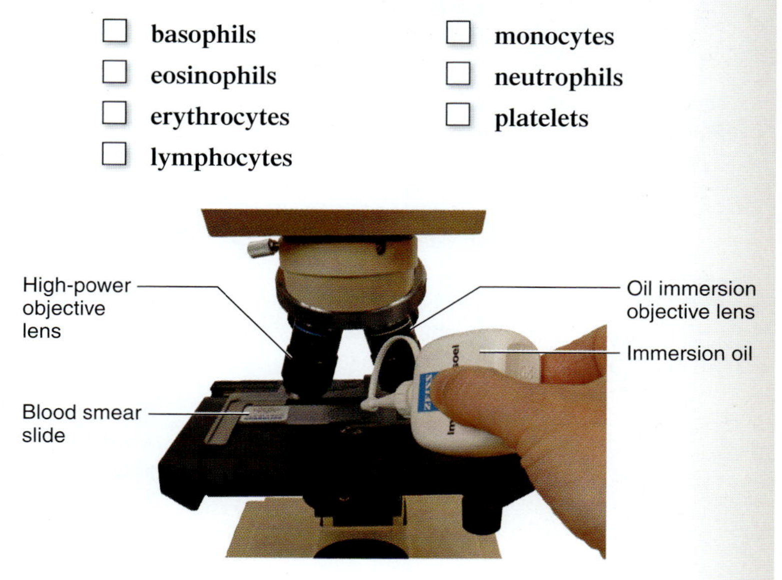

Figure 20.3 Orienting the Objective Lenses for Placement of a Drop of Immersion Oil on the Prepared Blood Slide.

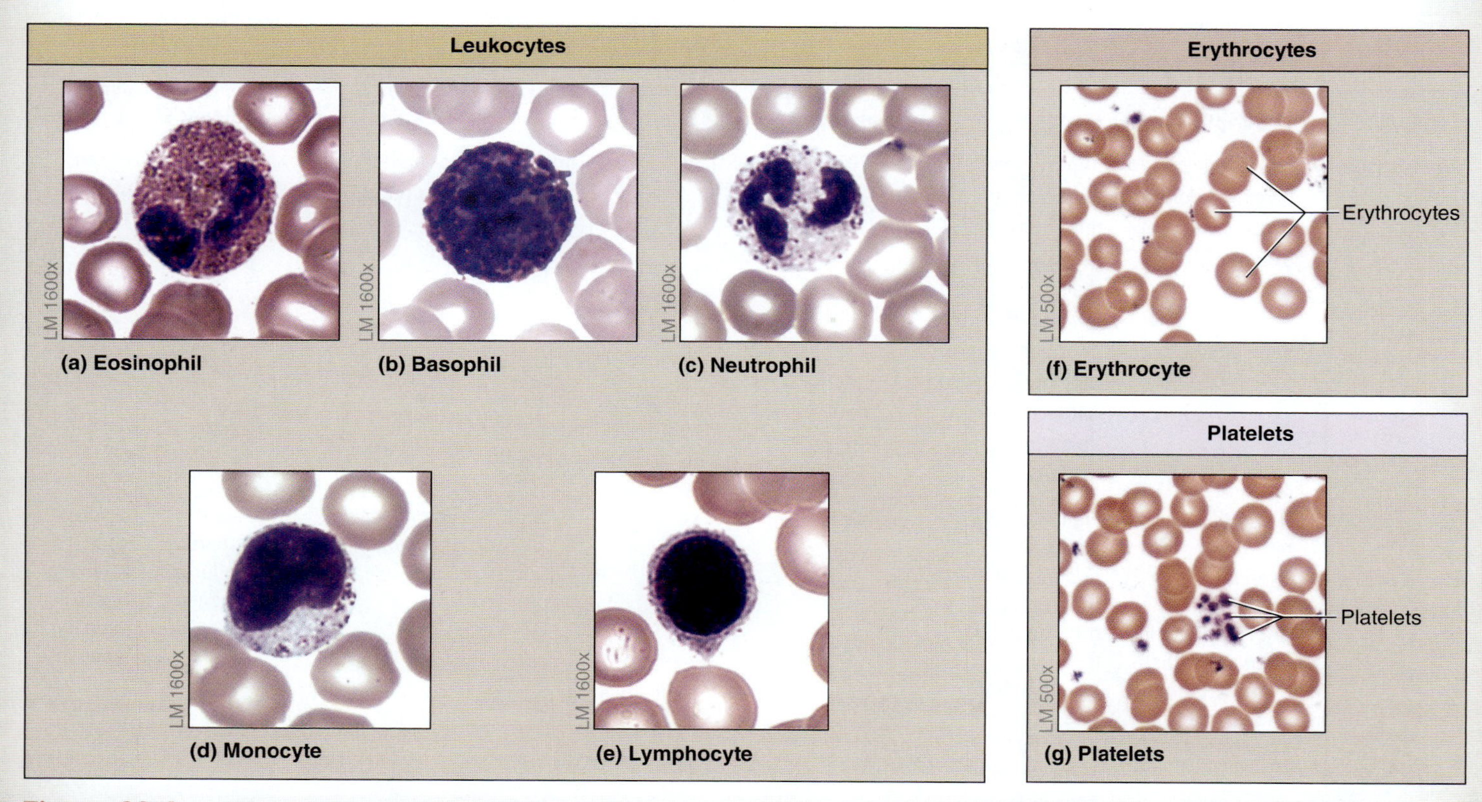

Figure 20.4 Formed Elements in the Blood.

(continued on next page)

(continued from previous page)

Table 20.2 Characteristics of the Formed Elements of Blood

Formed Element	Description	Function	Size (diameter)	Percentage of Formed Elements	Word Origin
Erythrocytes (Red Blood Cells)	Very uniform in size and shape. Shaped like biconcave discs, lack a nucleus, and are orange to red in color due to their attraction for eosinophilic stains.	Transport oxygen and carbon dioxide, and participate in regulation of the pH of blood	~7.5 μm	99%	*erythros*, red, + *kytos*, a hollow (cell)
Leukocytes (White Blood Cells)	Generally blue to purple in color due to the basophilic staining properties of their nuclei. A summary of the characteristics of the various leukocytes is provided in table 20.3.	Protect the body from pathogens, fight infections, and remove dead or damaged cells from the body	7–21 μm	<1%	*leukos*, white, + *kytos*, a hollow (cell)
Platelets	Small, purple-colored structures often mistaken for "junk" on the slide. Platelets are cell fragments of megakaryocytes and are small compared to erythrocytes and leukocytes.	Play a role in hemostasis (blood clotting). When activated, the surfaces become spiny instead of smooth and they become very "sticky," which helps to form the platelet plug that helps stop bleeding from the blood vessel wall.	~2 μm	1%	*platys*, flat

Table 20.3 Leukocyte Characteristics

Leukocyte	Description	Function	Percentage of Leukocytes	Diameter Relative to an Erythrocyte	Nuclear Shape	Color of Granules	Conditions Causing Increased Abundance	Word Origin
Neutrophils	Have very light, lavender-colored cytoplasmic granules; the multi-lobed nucleus can be seen easily.	Important in fighting bacterial infections. They migrate out of the blood toward the site of infection, where they phagocytize bacteria and damaged tissues.	50–70%	~1.3x	Multi-lobed (~5 lobes)	Lavender (pale blue/purple)	Bacterial infection	*neutro-*, neutral, + *philos*, to love
Eosinophils	Have orange to reddish cytoplasmic granules that are fairly light in color; the bluish-colored, bi-lobed nucleus can be seen easily.	Fight parasitic infections and mediate (neutralize) the effects of histamines. Phagocytize antigen–antibody complexes and allergens.	1–4%	~1.3x	Bi-lobed	Orange or red (eosinophilic)	Parasitic infections and allergic reactions	*acidus*, sour, referring to acidic dyes such as eosin, + *philos*, to love

Table 20.3 Leukocyte Characteristics *(continued)*

Leukocyte	Description	Function	Percentage of Leukocytes	Diameter Relative to an Erythrocyte	Nuclear Shape	Color of Granules	Conditions Causing Increased Abundance	Word Origin
Basophils	Very rare and therefore difficult to locate on a blood smear. Cytoplasmic granules stain very dark blue, so the nucleus is generally not visible, though it is usually bi-lobed.	Release histamine and heparin, and are involved in the inflammatory response	0.5–1%	~1.3x	Bi-lobed	Blue or purple (basophilic)	Tissue injury and allergic reactions	*baso-*, referring to basic dyes, + *philos*, to love
Lymphocytes	Nearly the same size as an erythrocyte. Nuclei are very large and blue/purple in color, and are surrounded by a halo of pale blue cytoplasm.	Responsible for the specific immune response to infection. Each type of lymphocyte has a specific function in fighting pathogens.	20–40%	~1–2x	Large and round, nearly fills the entire cell	NA	Viral infections, autoimmune diseases	*lympho-*, referring to the lymphatic system, + *kytos*, a hollow (cell)
Monocytes	Recognized by their very large size, at least twice the size of an erythrocyte. Have a large, blue/purple nucleus that often has a small indentation in it, making it "horseshoe-shaped," which helps distinguish it from a lymphocyte.	Monocytes are circulating cells. When they migrate out of the bloodstream, they become large, phagocytic cells called macrophages. They have little to no function in circulating blood.	2–8%	~2–3x	Large and horseshoe-shaped	NA	Bacterial infection	*monos*, single, + *kytos*, a hollow (cell)

4. After completing this activity, clean the microscope and slide thoroughly to remove all traces of oil from the instruments. To do this, rotate the nosepiece so the high-power and oil immersion lenses are on either side of the slide once again (figure 20.3). Use *lens paper* (do not use anything else or it may damage the lenses!) and lens cleaning solution to carefully and thoroughly clean the oil from both the oil immersion objective and the blood smear slide. Blood smear slides prepared using your own blood should be disposed of by placing them in a bleach solution for disinfection.

INTEGRATE

LEARNING STRATEGY

Although lymphocytes usually have circular nuclei and monocytes usually have indented, or "horseshoe-shaped," nuclei, sometimes monocytes can have nuclei that appear circular. In these cases, monocytes can appear very similar to lymphocytes. To keep from misidentifying a monocyte as a lymphocyte, always consider the size of the cell in addition to the shape of the nucleus. A monocyte is 2–3 times larger than an erythrocyte, whereas a lymphocyte is much closer to the same size as an erythrocyte.

EXERCISE 20.2

IDENTIFICATION OF MEGAKARYOCYTES ON A BONE MARROW SLIDE

All the formed elements of the blood are produced in the red bone marrow through the process of **hemopoiesis** (*hemo-*, blood, + *poiesis*, a making). Identifying the precursor cells of circulating blood cells is a complicated endeavor, which is generally undertaken only in upper-level histology courses. The goal of this laboratory exercise is to locate the precursor cells of platelets (*megakaryocytes*) and investigate their structure and function. However, while engaging in the process of identifying megakaryocytes, try to also obtain an appreciation for the general appearance of the precursor cells of erythrocytes and leukocytes that are also visible on the slide.

1. Obtain a prepared slide of red bone marrow and place it on the microscope stage. Bring the slide into focus on high power. Identify the following on the slide, using **figure 20.5** as a guide:

   ☐ **bone tissue** ☐ **platelets**
   ☐ **megakaryocytes** ☐ **red bone marrow**

2. **Megakaryocytes** (*megas*, big, + *karyo*, kernel [referring to the nucleus] + *kytos*, a hollow, cell) are extremely large cells with enormous nuclei. They are easily identifiable on prepared slides of bone marrow. Megakaryocytes remain within the bone marrow. However, their products, **platelets,** are continuously delivered to the bloodstream. Megakaryocytes produce long extensions called proplatelets. While still attached to the megakaryocyte, these proplatelets extend through the blood vessel wall (between the endothelial cells) in the bone marrow. The force from the blood flow "slices" these proplatelets into the fragments known as platelets.

3. Sketch a megakaryocyte as seen through the microscope in the space provided. Be sure to indicate the relative size of the megakaryocyte as compared to the size of the developing blood cells that are also visible on the bone marrow slide.

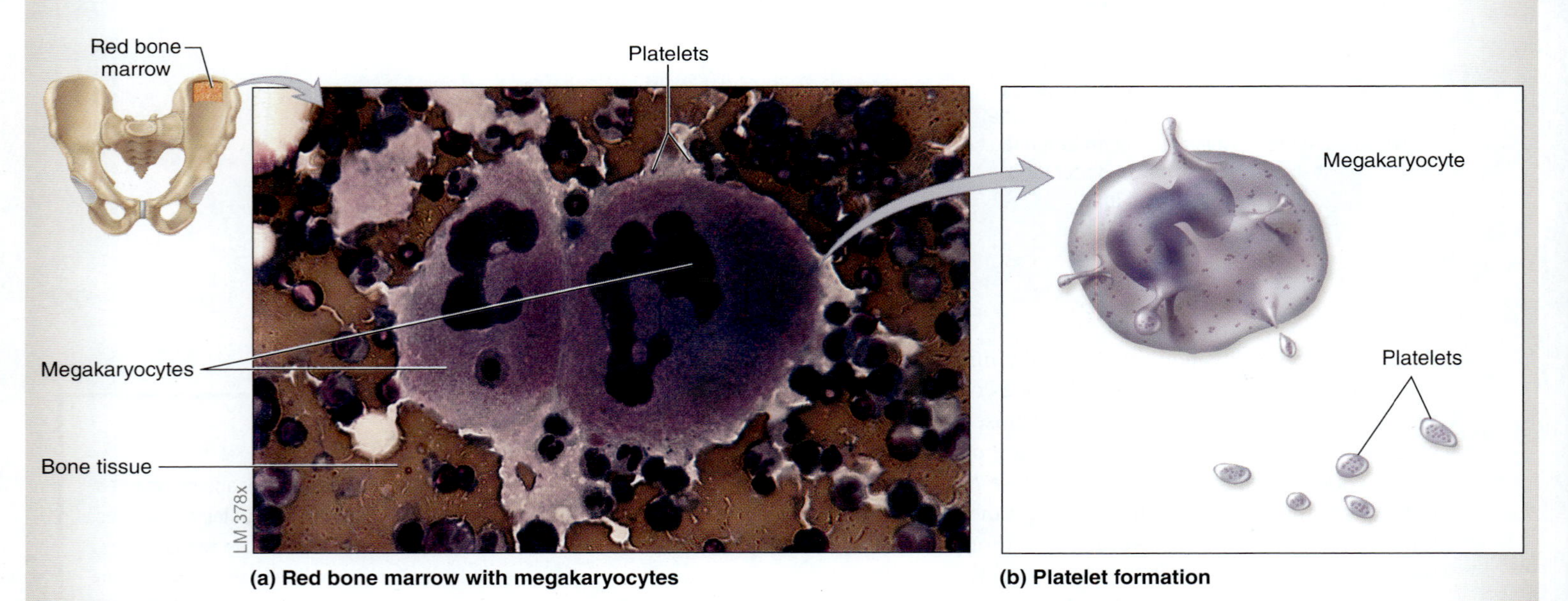

(a) Red bone marrow with megakaryocytes

(b) Platelet formation

Figure 20.5 **Bone Marrow Slide.** The megakaryocytes in red bone marrow give rise to platelets.

INTEGRATE

CLINICAL VIEW
Bone Marrow Transplants

Leukemia is a disease characterized by excessive proliferation of abnormal leukocytes. Patients who have experienced a progression of the disease may opt to undergo a bone marrow transplant. Leukocytes are produced in a process called leukopoiesis, which occurs in the bone marrow. A bone marrow transplant procedure involves first destroying the leukemia patient's existing bone marrow through radiation or chemotherapy, and then harvesting hemopoietic stem cells from a donor. The harvested cells are injected into the leukemia patient's blood. Recall that adult bone marrow still undergoing hemopoiesis is located in the proximal epiphyses of the humerus and femur, in the sternum, and in the iliac crest. Typically, bone marrow is harvested from the donor's iliac crest because it is relatively easy to access and causes the fewest complications for the donor.

After the donor's cells are transplanted into the recipient, they will normally migrate to the recipient's spleen and the process of leukopoiesis will begin anew. Care must be taken to ensure that the donor's bone marrow is a good match and will not elicit an immune response in the patient. While patients undergo this procedure, their leukocytes become depleted, making them very susceptible to infection. For this reason, patients must limit their exposure to pathogens until their leukocyte count has risen to safe levels once again.

GROSS ANATOMY

EXERCISE 20.3

IDENTIFICATION OF FORMED ELEMENTS OF THE BLOOD ON CLASSROOM MODELS OR CHARTS

1. Identify each of the following blood cell types on classroom models or charts demonstrating blood cells.

- ☐ **basophils**
- ☐ **eosinophils**
- ☐ **erythrocytes**
- ☐ **lymphocytes**
- ☐ **monocytes**
- ☐ **neutrophils**
- ☐ **platelets**

2. *Optional Activity:* APR **9: Cardiovascular System—** Watch the "Hemopoiesis" animation to review the formation and characteristics of each of the formed elements.

PHYSIOLOGY

Blood Diagnostic Tests

Blood is an amazing fluid with a beautiful red color due to the presence of hemoglobin within erythrocytes. **Hemoglobin** is a protein in erythrocytes that is essential for transporting oxygen throughout the body. However, blood transports much more than just oxygen. Blood also transports nutrients, waste products, carbon dioxide, hormones, and heat throughout the body. Proper circulation of blood to the tissues is essential for their survival. In medicine, a sample of a patient's blood can yield important clues as to what diseases may be affecting an individual **(table 20.4)**. A blood test involves first collecting a blood sample, typically performed by a *phlebotomist* (*phleps,* vein, + *tome-,* to cut), and placing the sample in a test tube (**figure 20.6,** step 1). The test tube is then placed in a centrifuge, which spins the sample at high speed to separate the component parts (figure 20.6, step 2). Because the formed elements like leukocytes and erythrocytes are heavy, they fall to the bottom of the test tube, while the blood plasma remains floating on the top (figure 20.6, step 3). Leukocytes and platelets form a thin layer called a *buffy* coat, which sits on top of the layer of erythrocytes. The iron present in the hemoglobin in erythrocytes makes erythrocytes heavier than leukocytes, which explains why erythrocytes lie in the layer below leukocytes in the test tube.

The following laboratory exercises explore purpose and procedure for performing several clinically relevant blood tests, including a differential leukocyte count, hematocrit, hemoglobin concentration, coagulation time, blood typing, cholesterol testing, and blood glucose testing.

Table 20.4	Normal Ranges for Laboratory Blood Tests
Blood Glucose (fasting)	70–100 mg/dl
Coagulation Time	5–8 min.
Hematocrit	Males: 39–49%, Females: 35–45%
Hemoglobin	Males: 13.5–17.5 g/dL, Females: 12.0–16.0 g/dL
Platelets	150,000–400,000 cells/mm^3
Erythrocytes	Males: ~5.4 million cells/mm^3; Females: ~4.8 million cells/mm^3
Leukocytes	4,500–11,000 cells/mm^3
Total Cholesterol = HDL + LDL (HDL = High density lipoprotein, LDL = Low density lipoprotein)	
LDL Cholesterol	Recommended: <130 mg/dL, Moderate risk: 130–159 mg/dL, High risk: >160 mg/dL
HDL Cholesterol	Males: >29 mg/dL, Females: >35 mg/dL

Reference: Goldman, L. and A.I. Schafer, *Goldman's Cecil Medicine 24e*, Elsevier: New York, 2012.

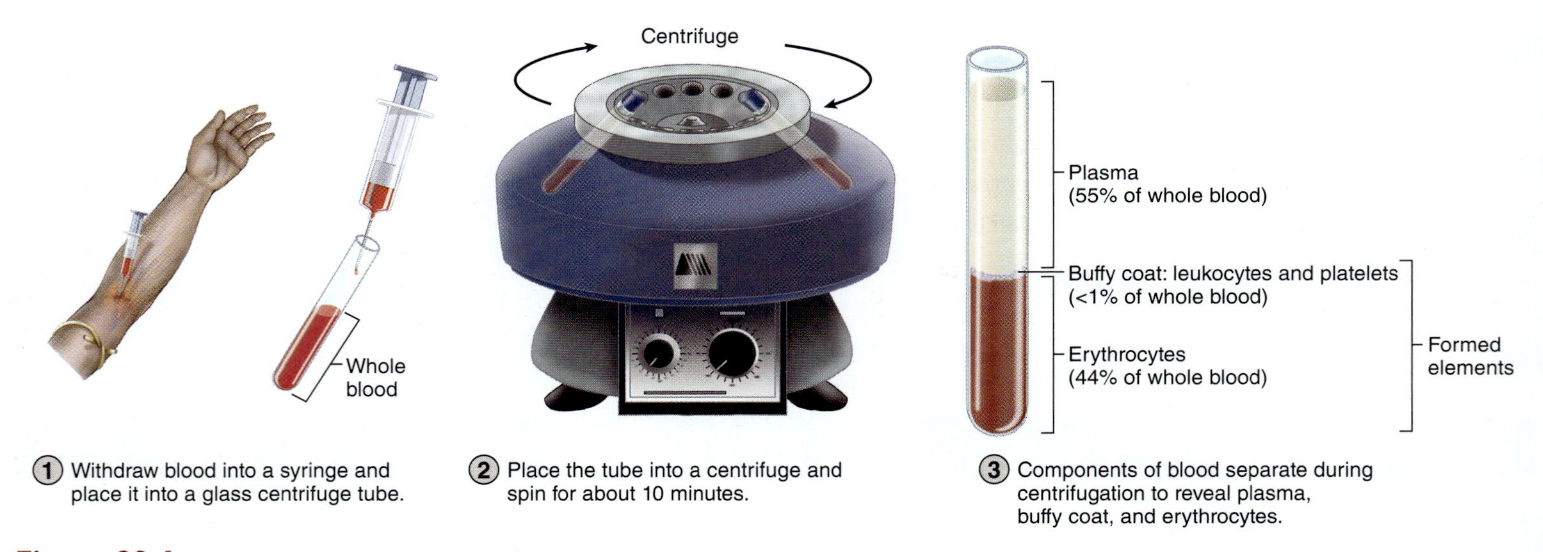

Figure 20.6 Separation of a Whole Blood Sample by Centrifugation. Using a centrifuge to separate whole blood into plasma (55%) and formed elements (45%) is the first step in determining the composition of whole blood.

INTEGRATE

CONCEPT CONNECTION

The formation of antibodies involves adaptive immunity (or acquired immunity), a process that exhibits the characteristics of *specificity* and *memory.* When a B- or T-lymphocyte recognizes an antigen on the surface of a cell, the lymphocyte becomes activated through a series of complex steps, which are outlined in the textbook. Once activated, lymphocytes proliferate, forming daughter cells (clones). B-lymphocyte proliferation yields plasma cells, which produce antibodies. Each subsequent exposure to a specific antigen causes a dramatic increase in the number of antibodies produced because of the increased number of clones and plasma cells circulating in the blood. Once antibodies are formed, they bind to antigens to form antigen–antibody complexes. While antibodies do not directly kill the invading pathogens, antigen–antibody complexes assist in the removal of pathogens from the body by making the pathogens become more susceptible to phagocytosis.

EXERCISE 20.4

DETERMINATION OF LEUKOCYTE COUNTS

A **leukocyte count** performed on a sample of the patient's blood can determine whether or not the patient has a bacterial infection. An excess of neutrophils will indicate that the patient has such an infection, whereas a normal abundance of neutrophils might be a clue that an infection is viral rather than bacterial in nature. An abnormally low number of leukocytes (leukopenia) may be indicative of a patient's inability to fight infection. In addition, certain pathological conditions may lead to the uncontrolled proliferation of abnormal leukocytes as in leukemia (see Clinical View: Bone Marrow Transplants, p. 531).

1. Obtain a slide of a human blood smear or prepare one using your own blood. See exercise 20.1A for preparing a human blood smear using your own blood.
2. Place the human blood smear slide on the microscope stage. Using the scanning objective (see exercise 3.1), move the slide on the stage so the highest density of blood cells appears in the field of view.
3. Rotate the coarse adjustment knob until leukocytes are visible and in focus within the field of view. Because of the stain, leukocytes will appear as small *purple* dots in a sea of erythrocytes.
4. To properly view blood cells, it is necessary to use the oil immersion (highest-power) objective of the microscope. See exercise 20.1B for instructions on how to view specimens using an oil immersion objective. Rotate the nosepiece so the highest-power or oil immersion objective is directed toward the blood smear slide.
5. Turn the fine-adjustment knob until the leukocytes come into focus.
6. Gently move the slide systematically in a back-and-forth motion making sure to span the entire width of the blood smear on each pass **(figure 20.7).** Count leukocytes as they cross the field of view. Be careful to count each cell only once. If blood counters are available in the laboratory, use these to make the process of counting leukocytes easier.

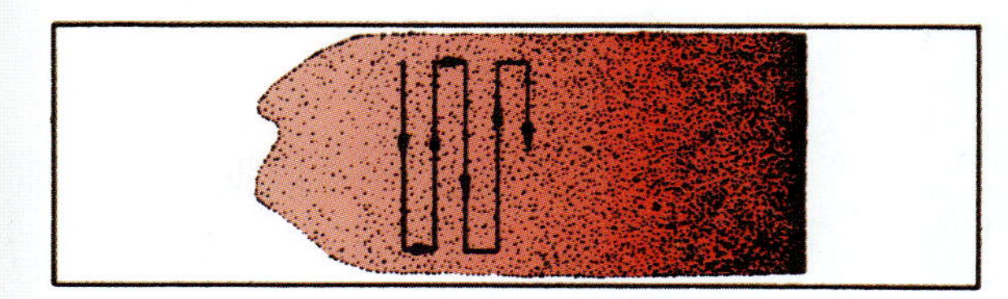

Figure 20.7 Method for Performing Differential Leukocyte Count. Scan the slide by moving the microscope stage so the view follows the pattern illustrated here.

Table 20.5 Differential Leukocyte Count Data Table

Type of Leukocyte	Number (#) Observed	Percent (%) Observed
Basophils		
Eosinophils		
Lymphocytes		
Monocytes		
Neutrophils		
Total number of leukocytes		

7. Record the number of each type of leukocyte observed in the appropriate row in **table 20.5**.
8. Calculate the total number of leukocytes observed by adding up the numbers of each type of leukocyte counted. Record this in table 20.5.
9. Calculate the percentage of each type of leukocyte and record it in table 20.5. For example, the following formula demonstrates how to calculate the percentage of neutrophils observed: Percent (%) = [# Neutrophils Observed / Total # Leukocytes Observed] × 100.
10. Compare the calculated percentages of each type of leukocyte with the normal values provided in table 20.3.
11. Make note of any pertinent observations here:

____

____

____

INTEGRATE

LEARNING STRATEGY

Mnemonic for remembering the relative abundance of leukocytes in the blood, from most abundant to least abundant: **N**ever **L**et **M**onkeys **E**at **B**ananas (**N**eutrophils, **L**ymphocytes, **M**onocytes, **E**osinophils, **B**asophils).

EXERCISE 20.5

DETERMINATION OF HEMATOCRIT

The number of erythrocytes in the blood (erythrocyte count) is clinically relevant. A low erythrocyte count is characterized by a reduced oxygen-carrying capacity (or anemia). The erythrocyte count can be approximated in a blood sample by measuring the **hematocrit,** or the packed cell volume (PCV). Because erythrocytes make up the majority of the formed elements in the blood, the PCV is positively correlated with the number of erythrocytes.

Obtain the Following:

- **heparinized capillary tube**
- **Seal-ease™**
- **alcohol swabs**
- **lancet**

1. Choose a subject. Before beginning, read and follow all instructions for safe handling of human blood (see box on p. 526). Use an alcohol swab to cleanse the tip of the subject's middle or ring finger and prick the finger with a lancet. Wipe away the first drop of blood with the alcohol swab. Dispose of the lancet in a sharps container.
2. Press the colored end of the heparinized capillary tube to the drop of blood. Be sure to hold the tube tilted slightly up, to allow the blood to fill the tube by capillary action. Be careful not to allow the capillary tube to drop below a horizontal plane, because this increases the potential for bubbles to form in the blood within the tube.
3. Fill the capillary tube with blood approximately three-fourths of the length of the tube **(figure 20.8*a*).**
4. Press the blood-containing end of the capillary tube into the Seal-ease™ to seal the tube (figure 20.8*a*).
5. Place the capillary tube in a groove of the microhematocrit centrifuge such that the sealed end of the tube is adjacent to the outer rubber gasket of the centrifuge. Be sure to balance the centrifuge by placing samples opposite the tube (figure 20.8*b*).
6. Place the inner centrifuge cover over the capillary tubes. Screw it down tight to ensure it does not come loose when the centrifuge is running. Close the outer centrifuge cover and run the centrifuge for 4 to 5 minutes.
7. Once the centrifuge has completed the cycle, remove the capillary tube(s).
8. The total volume of blood can be measured with either a microhematocrit reader or a metric ruler. If using a ruler, measure the length of total blood within the capillary tube in millimeters (figure 20.8*c*). This will include all three layers (plasma, buffy coat, and erythrocytes). Record the total blood length in **table 20.6.**
9. Measure the erythrocyte layer in millimeters. Record the length of the erythrocyte layer in table 20.6.
10. Compute the percentage of erythrocytes by dividing the length of the erythrocyte layer by the length of the total blood, and then multiplying by 100. Record the results in table 20.6.
11. Compare calculated hematocrits (percentage of erythrocytes in whole blood) with expected values from table 20.4 (males: 39–49%; females: 35–45%).
12. Make note of any pertinent observations here:

____

____

____

Table 20.6	Hematocrit Data Table
Total blood length (mm)	
Erythrocyte length (mm)	
Hematocrit (%)	

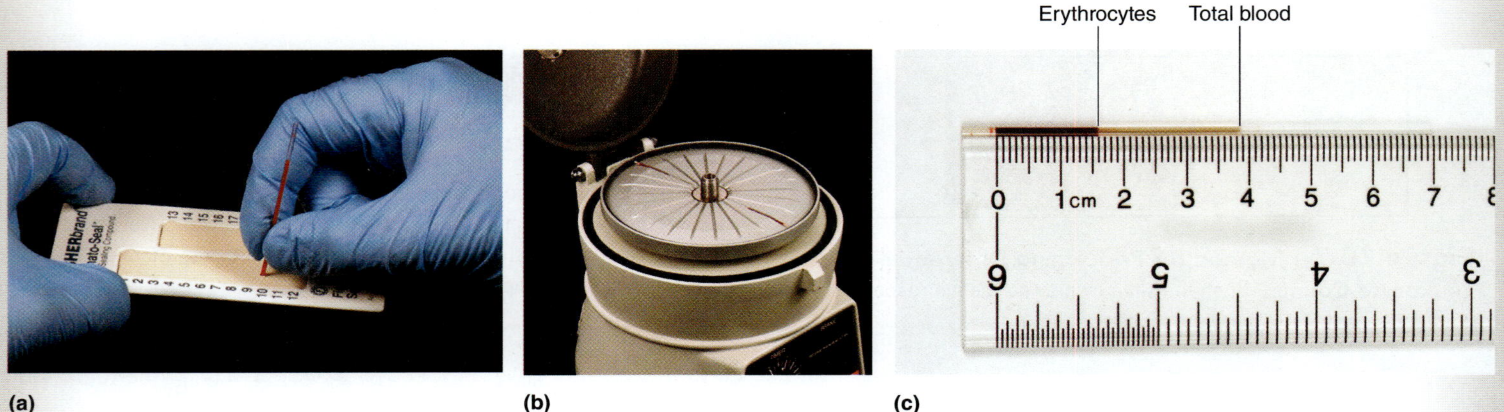

Figure 20.8 Determination of Hematocrit. (a) Fill the capillary tube with blood and plug one end with Seal-ease™. (b) Place the capillary tube in the microhematocrit centrifuge with the sealed end extending to the outside. (c) Measure the lengths of erythrocytes and total blood within the capillary tube.

EXERCISE 20.6

DETERMINATION OF HEMOGLOBIN CONTENT

Obtain the Following:

- **alcohol swab**
- **hemoglobinometer**
- **hemolysis applicator**
- **lancet**
- **lens paper**

1. Choose a subject. Before beginning, read and follow all instructions for safe handling of human blood (see box on p. 526).
2. Disassemble the hemoglobinometer by removing the blood chamber and separating the two plates of glass from the metal clips. Note that there is a larger piece of glass that contains a U-shaped area with surrounding depressions and a smaller piece of glass that is flat on both sides.
3. Use lens paper to clean the glass plates. Ensure that no streaks appear on the glass plates, as this can introduce artifacts when measuring blood hemoglobin levels.
4. Clean the tip of the middle finger or ring finger with an alcohol swab, place the tip of the lancet on the clean finger, and pull the trigger so it pierces the skin (figure 20.1). Wipe away the first drop of blood with the alcohol swab. Dispose of the lancet in a sharps container.
5. Place a drop of blood onto the U-shaped area of the larger piece of glass **(figure 20.9*a*).**
6. Use the hemolysis applicator to stir the blood such that the erythrocytes rupture (lyse) and the blood appears clear (~45 seconds) (figure 20.9*b*).
7. Place the smaller, flat piece of glass onto the U-shaped area as a cover slip.

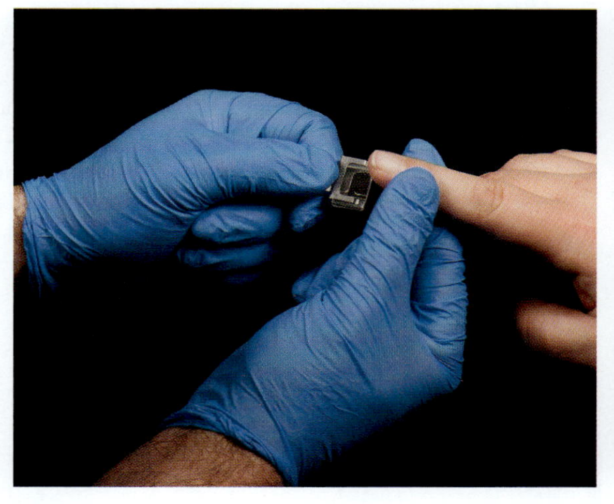

(a)

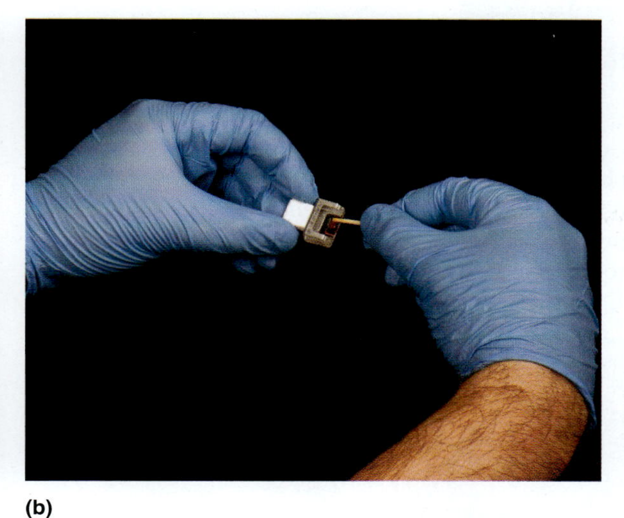

(b)

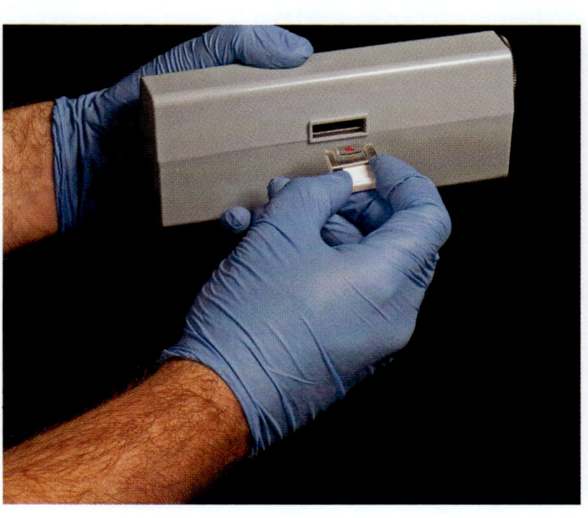

(c)

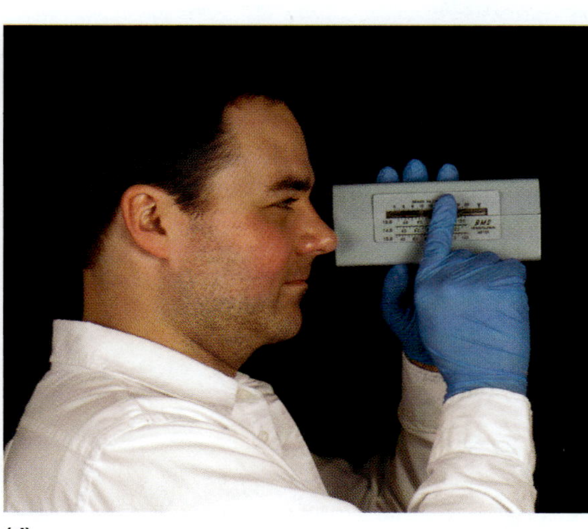

(d)

Figure 20.9 Determining Hemoglobin Content Using a Hemoglobinometer. (a) Gently place a drop of blood on the U-shaped part of the glass in the blood chamber. (b) Use a hemolysis applicator stick to stir the blood. (c) Slide the blood chamber into the slot on the side of the hemoglobinometer. (d) Use your thumb to turn on the light while using your index finger to move the slide until the colors seen through the window have matching intensity. Read the hemoglobin content from the side of the hemoglobinometer.

(continued on next page)

(continued from previous page)

8. Slide both pieces of glass into the metal clips of the blood chamber, and replace the blood chamber into the side slot of the hemoglobinometer (figure 20.9*c*). Be sure that the blood chamber is secure within the hemoglobinometer.
9. Locate the light switch on the underside of the hemoglobinometer. Using your left hand, hold the hemoglobinometer while resting your left thumb on the switch. Look into the eyepiece and turn the light on (figure 20.9*d*).
10. Notice there are two green areas in the field of view. Use your right hand to adjust the slide on the right side of the hemoglobinometer until the two halves of the green field match in color.
11. Record the resulting blood hemoglobin level: ____________________ g/dl
12. To clean, remove the blood chamber and place the glass plates and clip into a 10% bleach solution.
13. **Alternate Method:** The reasonably accurate and inexpensive *Tallquist method* may be used to estimate blood hemoglobin level. If performing this activity, use a **Tallquist test kit,** which contains absorbent paper and an associated color scale. Repeat step 5 and place the drop of blood on the absorbent paper. Allow the blood to dry (ensuring that the blood has not dried so much so that it appears brown in color) and match the blood to the color scale provided in the kit.

EXERCISE 20.7

DETERMINATION OF COAGULATION TIME

A test to determine the time required for blood to clot, or **coagulation time,** may be used to assess platelet number and function. When a blood vessel is damaged, platelets are the formed elements involved in **hemostasis** (*hemo-*, blood, + *stasis,* stability), the process that stops blood flow. Hemostasis involves three phases: vascular spasm, platelet plug formation, and coagulation. Both the formation of a platelet plug and the reactions required for coagulation involve platelets. **Platelet plug** formation occurs when a blood vessel is damaged and collagen fibers beneath the endothelium are exposed. Platelets stick to the exposed collagen, undergo morphologic changes, and secrete chemicals that attract additional platelets and promote both coagulation and vessel repair. **Coagulation** involves a cascade of reactions that lead to

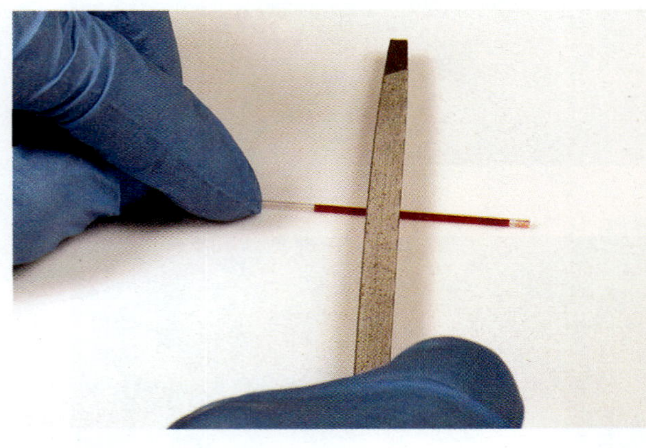

(a)

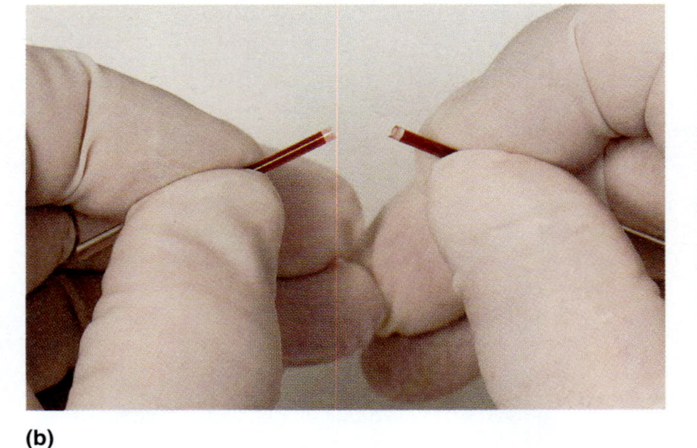

(b)

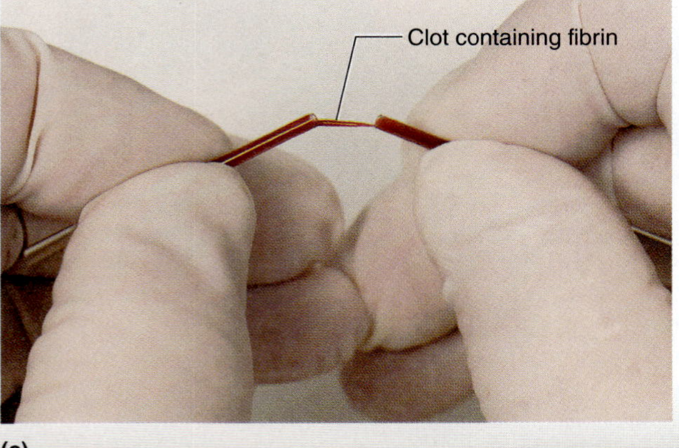

(c)

Figure 20.10 Determining Coagulation Time. (a) Use a triangular file to score the slide. (b) Break the slide to see if coagulation has occurred (no coagulation has yet occurred in this photo). (c) Appearance of fibrin within a clot after coagulation has occurred.

the formation of a blood clot, called a **thrombus.** In addition to platelets, these reactions require substances such as calcium, clotting factors, and vitamin K. The result is a meshwork of the insoluble protein, fibrin, which traps substances contained within the blood. When the inside of a vessel is damaged, the formation of a thrombus typically takes 3 to 6 minutes. Both the formation of the platelet plug and coagulation are regulated by positive feedback mechanisms in order to minimize blood loss.

Obtain the Following:

- **nonheparinized capillary tube**
- **alcohol swab**
- **lancet**
- **triangular file**

1. Choose a subject. Before beginning, read and follow all instructions for safe handling of human blood (see box on p. 526).
2. Use an alcohol swab to cleanse the tip of the subject's middle or ring finger and prick the finger with a lancet. Wipe away the first drop of blood with the alcohol swab. Dispose of the lancet in a sharps container.
3. Load the nonheparinized capillary tube with blood (exercise 20.5, step 2).
4. Fill the capillary tube with blood approximately three-fourths of the length of the tube.
5. Lay the capillary tube flat on a paper towel and record the initial time: ___________
6. After 1 minute, use the triangular file to weaken a small area of the capillary tube just adjacent to its end **(figure 20.10*a*).** Once the tube is filed slightly, use both hands to hold the tube on either side of the filed section. Break the tube away from you, ensuring that both ends remain close so that you can observe a fibrin clot (if any). If the clot has not yet formed, the tube will break cleanly (figure 20.10*b*). If the clot has formed, fibrin will extend from each end (figure 20.10*c*).
7. Repeat step 6 at 1-minute intervals until a fibrin clot is observed. Record the time required for coagulation: _______________ minutes

EXERCISE 20.8

DETERMINATION OF BLOOD TYPE

Another clinically relevant blood test that may be performed is determination of a person's blood type (A, B, AB, or O). **Blood type determination** requires some understanding of an immune response; when erythrocytes clump together (agglutination), it is indicative of an immune response called *agglutination*. Agglutination occurs when antibodies circulating in the blood bind to antigens on the surface of erythrocytes. As concerns blood types, people with blood type A have surface antigen A, and produce antibodies to type B blood. People with blood type B have surface antigen B, and produce antibodies to type A blood. People with blood type AB have both surface antigens A and B, and produce no antibodies. People with blood type O have no surface antigens and produce both type A and type B antibodies. Another common surface antigen on erythrocyte membranes determines the Rh blood type. The Rh blood type is determined by the presence or absence of the Rh surface antigen, often called either Rh factor, or surface antigen D. When the Rh factor is present, the individual is said to be Rh positive (Rh+). When the Rh factor is absent, the individual is said to be Rh negative (Rh–). Individuals with type O negative (O–) blood are typically referred to as *universal donors* because their blood can be donated to individuals of any blood type without causing a transfusion reaction in the recipient. Individuals with AB positive (AB+) blood are referred to as *universal recipients* because they produce no antibodies that would cause a transfusion reaction.

Obtain the Following:

- **blood sample (human, animal, or artificial)**
- **two clean microscope slides**
- **antibody sera (anti-A, anti-B, anti-Rh)**
- **droppers**
- **toothpicks**
- **wax pencil**
- **alcohol swabs and lancets (if using human blood)**

1. Before beginning, read and follow all instructions for safe handling of human blood (see box on p. 526). Draw a vertical line down the center of one microscope slide with a wax marking pencil, dividing the slide into equal halves. Label the left side "anti-A" and the right side "anti-B" **(figure 20.11*a*).**
2. Place one drop of blood on the left side of the dividing line and place one drop of blood on the right side of the dividing line of the first slide. If using human blood, use an alcohol swab to cleanse the finger and prick the finger with a lancet to obtain the drop of blood (figure 20.1). Dispose of the lancet in a sharps container.
3. Place one drop of blood in the center of the second slide.
4. Place one drop of anti-A serum on the drop of blood on the left side of the dividing line on the first slide. Place one drop of anti-B serum on the drop of blood on the right side of the first slide. Place one drop of anti-Rh serum (also called "Anti-D" serum) on the drop of blood on the second slide.
5. Mix each blood/anti-serum sample with a clean toothpick (figure 20.11*b*). Be sure to use a separate toothpick for each mixture. Transferring the toothpick to multiple mixtures will contaminate the samples. Note that an Rh typing box may be required if using human blood because thorough mixing of the blood and anti-Rh serum at a slightly elevated temperature is required.

(continued on next page)

(continued from previous page)

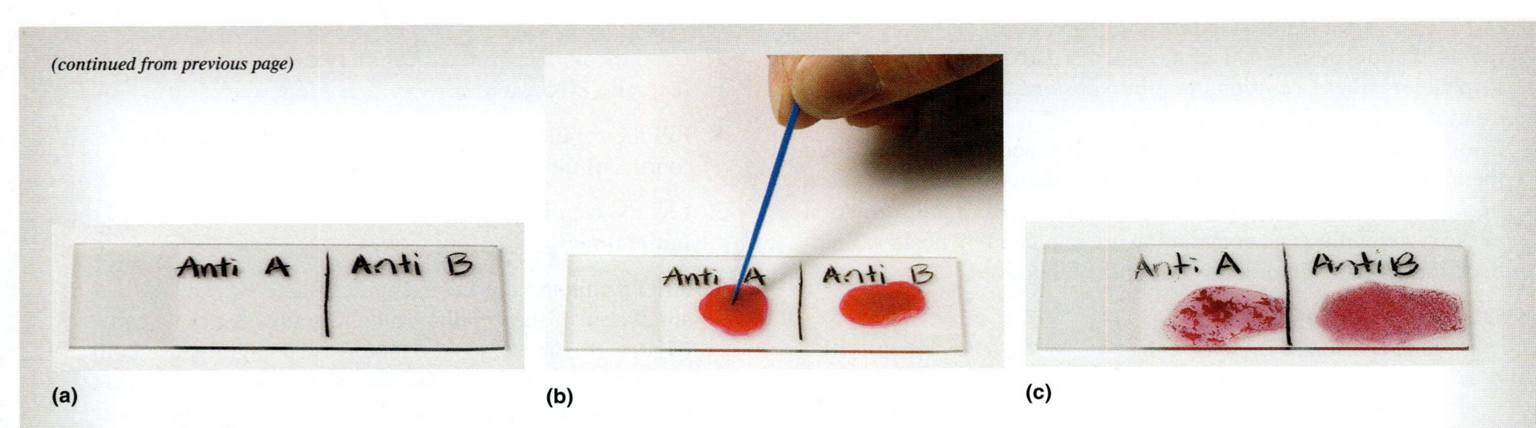

(a) (b) (c)

Figure 20.11 **Blood Typing.** (a) Prepare the slide as demonstrated here. (b) Mix anti-sera and blood sample with toothpick. (c) Observe the slides after 2–3 minutes, looking for agglutination reactions.

6. After 2 minutes, observe the blood samples. Agglutination in any of the samples indicates a positive result (figure 20.11*c*).
7. Place the microscope slides in a bowl containing a 10% bleach solution. Consult with the laboratory instructor for proper disposal of all contaminated materials.
8. If the entire class is reporting blood types to determine relative abundance of blood types in the the class (see Clinical View: Blood Type Abundance Within the U.S. Population on this page), record the blood type in the location indicated by the instructor. After all groups have reported their results, record the numbers and calculate abundances in the table on this page. Make note of any pertinent observations here:

___

___

___

INTEGRATE

CLINICAL VIEW

Blood Type Abundance Within the U.S. Population

In the United States, the most abundant blood type is O$^+$. However, there are slightly different relative abundances of blood types when broken down by ethnicity. Other antigens present in the blood may also cause agglutination reactions in some instances. In cases where the risk of such a reaction in a patient is particularly high, it may be important to transfuse blood from a donor that is not only matched by ABO blood type, but also comes from a donor of the same ethnic group as the patient.

Blood Type Abundance Within the U.S. Population

Blood Type	Class Results (%)	Caucasian (%)	African American (%)	Hispanic (%)	Asian (%)
O$^+$		37	47	53	39
O$^-$		8	4	4	1
A$^+$		33	24	29	27
A$^-$		7	2	2	0.50
B$^+$		9	18	9	25
B$^-$		2	1	1	0.40
AB$^+$		3	4	2	7
AB$^-$		1	0.30	0.20	0.10

Source: http://www.redcrossblood.org/learn-about-blood/blood-types

INTEGRATE

CLINICAL VIEW
Blood Transfusions

Blood transfusions involve the transfer of blood from one person, the **donor,** to another person, the **recipient.** It is critical that the blood from both donor and recipient are compatible, or the consequences of the transfusion may be lethal. For a recipient to safely receive a donor's blood, the donor blood must not be "recognized" as foreign. That is, the donor's erythrocytes must not contain antigens that the recipient's erythrocytes will bind to. If a recipient with A^+ blood receives A^- blood from a donor, the recipient will not reject the blood because both donor and recipient have surface antigen A on their erythrocytes and produce anti-B antibodies. On the other hand, if a recipient with A^- blood receives A^+ blood from a donor, the recipient will potentially reject the blood if the anti-Rh antibodies are present in the recipient's blood (A^-). If present, anti-D antibodies will bind to the Rh antigen that is on the surface of the donor's erythrocytes (A^+), thus identifying the donor blood as "foreign." Rejection involves an agglutination reaction, as observed in exercise 20.8.

EXERCISE 20.9

DETERMINATION OF BLOOD CHOLESTEROL

A test to determine **total blood cholesterol,** including the blood values for low density lipoproteins (LDLs) and high density lipoproteins (HDLs), may be important in evaluating an individual's risk for cardiovascular disease. Excess LDLs not used by peripheral tissues may be deposited in the walls of blood vessels, and are therefore considered to be "bad cholesterol." Conversely, HDLs *remove* lipids from the walls of blood vessels and transport the lipids to the liver for waste removal. Therefore, HDLs are considered to be "good cholesterol." Clinically, elevated levels of total blood cholesterol, increased numbers of LDLs, and decreased numbers of HDLs have all been implicated as risk factors for cardiovascular disease. Total blood cholesterol exceeding 200 mg/dl is considered "high." A ratio of HDL to LDL that exceeds 0.3 is considered healthy. A test to determine **blood glucose levels** is important to determine an individual's risk of developing diabetes (high blood sugar). A fasting blood glucose level of higher than 106 mg/dl is indicative of diabetes. Complications of long-term, unregulated diabetes include severe damage to blood vessels and nerves, which can lead to blindness and kidney failure, among other maladies.

Obtain the Following:

- **alcohol swab**
- **cholesterol test card**
- **color scale for cholesterol testing**
- **lancet**

1. Choose a subject. Before beginning, read and follow all instructions for safe handling of human blood (see box on p. 526). Use an alcohol swab to cleanse the tip of the subject's middle or ring finger and prick the finger with a lancet (figure 20.1). Wipe away the first drop of blood with the alcohol swab. Dispose of the lancet in a sharps container.
2. Place a drop of the blood onto the cholesterol test card and allow the blood to dry.
3. Wait 3 minutes and remove and dispose of the sample strip contained within the cholesterol test card.
4. Compare the remaining blood test spot with the color scale provided with the cholesterol test card. Record the blood cholesterol level: _______________ mg/dl.
   Note: Normally cholesterol tests are done after the subject has fasted for several hours. Therefore, a high blood cholesterol reading taken after a meal may not be accurate for diagnostic purposes.

EXERCISE 20.10

DETERMINATION OF BLOOD GLUCOSE

Obtain the Following:

- **alcohol swab**
- **glucose monitor**
- **glucose test strip**
- **lancet**

1. Choose a subject. Before beginning, read and follow all instructions for safe handling of human blood (see box on p. 526).
2. Obtain a glucose test strip and slide it into the appropriate spot on the glucose monitor. Use an alcohol swab to cleanse the tip of the subject's middle or ring finger and prick the finger with a lancet (figure 20.1). Wipe away the first drop of blood with the alcohol swab. Dispose of the lancet in a sharps container.
3. Have the subject gently touch the test strip with the drop of blood **(figure 20.12).** The test strip does not require a large volume of blood because it draws the blood into the slot by capillary action. Thus, take care not to get too much blood on the strip.
4. Wait approximately 3–4 seconds for the glucose meter to beep, indicating it is finished reading the strip. Record the blood glucose level: _______________ mg/dl.

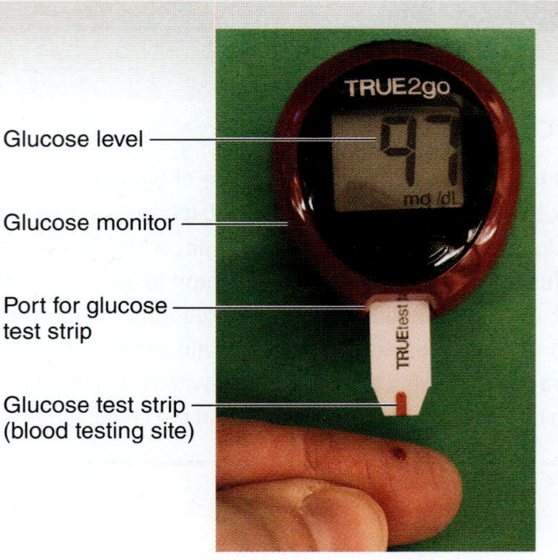

Figure 20.12 Blood Glucose Testing. A blood glucose monitor typically has a digital display and a slot for placing a test strip.

Note: Normally blood glucose tests are done after the subject has fasted for several hours. Therefore, a high blood glucose reading taken after a meal may not be accurate for diagnostic purposes.

INTEGRATE

CLINICAL VIEW

Hemoglobin A1c (Glycated Hemoglobin)

Knowing the lifespan of each of the different types of human blood cells is important for many diagnostic tests. While most leukocytes live only hours to days, most erythrocytes live 2–3 months. Therefore, erythrocytes are useful for monitoring long-term regulation of blood glucose levels in diabetics. The Hb A1c test is advantageous because it does not require the patient to have fasted for 12 hours, and it measures the average blood glucose levels for the patient over the previous 2–3 months. The Hb A1c test looks for glycated hemoglobin (i.e., hemoglobin with glucose "stuck" to it). A normal measure for Hb A1c is approximately 5% glycated glucose. Levels of Hb A1c greater than 7% indicate the patient's blood glucose levels have not been well controlled, that the patient likely has either type1 or type 2 diabetes, and that the patient must make lifestyle and other changes to ensure his or her long-term health.

Chapter 20: The Cardiovascular System: Blood

Name: ______________________

Date: ____________ Section: ____________

POST-LABORATORY WORKSHEET

The ❶ corresponds to the Learning Objective(s) listed in the chapter opener outline.

Do You Know the Basics?

Exercise 20.1: Identification of Formed Elements on a Prepared Blood Smear

1. Which of the following is the most abundant type of leukocyte? (Circle one.) ❶
   a. basophil
   b. eosinophil
   c. lymphocyte
   d. monocyte
   e. neutrophil

2. Which of the following is/are substances secreted by human basophils? (Check all that apply.) ❷
   ______ a. antihistamine
   ______ b. hemoglobin
   ______ c. heparin
   ______ d. histamine
   ______ e. platelets

Exercise 20.2: Identification of Megakaryocytes on a Bone Marrow Slide

3. Small cellular fragments that are involved in the process of hemostasis are platelets. ______________ (True/False) ❸

4. Platelets are derived from monocytes, which reside in the bone marrow. ________________ (True/False) ❹

Exercise 20.3: Identification of Formed Elements of the Blood on Classroom Models or Charts

5. A lymphocyte is a(n) ________________ (granulocyte/agranulocyte) with a large, spherical nucleus surrounded by pale blue cytoplasm. ❺

Exercise 20.4: Determination of Leukocyte Counts

6. Which formed element composes less than 1% of whole blood and is involved in defense against disease? (Circle one.) ❻
   a. basophil
   b. lymphocyte
   c. monocyte
   d. neutrophil
   e. platelet

7. A significant increase in the number of circulating neutrophils may indicate which of the following? (Check all that apply.) ❼
   ______ a. allergic reaction
   ______ b. bacterial infection
   ______ c. parasitic infection
   ______ d. viral infection

Exercise 20.5: Determination of Hematocrit

8. The hematocrit is the percentage of leukocytes in whole blood. ______________ (True/False) ❽

9. A(n) ________________ (increase/decrease) in hematocrit may indicate a person is anemic. ❾

Exercise 20.6: Determination of Hemoglobin Content

10. When tested with a hemoglobinometer, blood containing a decreased level of hemoglobin as compared to normal blood will appear ________________(darker/lighter) in color. (10)

Exercise 20.7: Determination of Coagulation Time

11. Which of the following are conditions that may cause coagulation time to increase? (Check all that apply.) (11)

______ a. decreased calcium

______ b. decreased megakaryocytes

______ c. decreased platelet count

______ d. decreased vitamin K

______ e. increased hematocrit

Exercise 20.8: Determination of Blood Type

12. If an agglutination reaction only occurs when Anti-B antibody is applied, then the blood type must be type _____ (A/B) ______ (+/–). (12)

Exercise 20.9: Determination of Blood Cholesterol

13. When considering components of total blood cholesterol, ____________ (HDLs/LDLs) are considered to be the "good" type. (13)

Exercise 20.10: Determination of Blood Glucose

14. A fasting blood glucose level of 150 mg/dl is considered normal. ______________ (True/False). (14)

Can You Apply What You've Learned?

15. In determining hematocrit, if the length of erythrocytes in the capillary tube is 45 millimeters and the length of total blood is 95, what is the hematocrit?

____

16. What are some symptoms you might expect an individual suffering from anemia to exhibit? ________________________

____

17. Since leukocytes and erythrocytes are produced from the same precursor cells, predict the change in leukocyte count that might occur in a patient who has an elevated erythrocyte count (polycythemia). ________________________

18. Predict the change in leukocyte count in a patient suffering from leukemia (cancer involving leukocytes). ________________

19. Erythropoietin (EPO), a hormone produced by the kidneys, is released into the blood when blood oxygen levels are low. EPO stimulates the production of erythrocytes by red bone marrow. Predict the resulting EPO level in the blood and erythrocyte count of a person who has spent considerable time living at high altitudes, where atmospheric oxygen levels are reduced. ________________________

____

20. A patient has been injured and is in need of a blood transfusion. When she arrives at the ER, doctors determine that she has type A^+ blood. What antigens are present on her erythrocytes? ________________________ What antibodies are present in her blood? ________________________ What blood type(s) can she receive in her blood transfusion?

____

Can You Synthesize What You've Learned?

21. Bonnie, a 50-year-old female, complains of severe fatigue, weakness, and dizziness. Her complete blood count (CBC) is shown below. Consult table 20.4 on page 532. Discuss which of her variables are out of the expected range and propose a diagnosis.

Leukocytes	1000 cells/mm^3
Erythrocytes	2 million cells/mm^3
Hemoglobin (Hb)	7.5 g/dl
Hematocrit	25%
Neutrophils	31%
Eosinophils	6%
Basophils	2%
Lymphocytes	53%
Monocytes	8%
Platelets	50,000 cells/mm^3

22. Elizabeth is 20 weeks pregnant with her second child. Elizabeth's blood type is AB negative (AB^-). Her physician is concerned about the possibility of a reaction between Elizabeth's blood and the baby's blood because of the blood type of the baby's father. The physician prescribes RhoGAM for Elizabeth. RhoGAM contains immunoglobulins that prevent Elizabeth from producing Rh antibodies (Anti-D antibodies). What does the physician fear may be a complication with this pregnancy in the event that RhoGAM is not administered?

CHAPTER

21

The Cardiovascular System: The Heart

OUTLINE AND LEARNING OBJECTIVES

Cardiac Cycle and Heart Sounds 572

EXERCISE 21.14: AUSCULTATION OF HEART SOUNDS 572

34 Describe the role of atrioventricular and semilunar valves in ensuring one-way blood flow through the heart

35 Determine the optimal placement of a stethoscope for auscultation of heart sounds

36 Determine the clinical relevance of abnormal heart sounds

EXERCISE 21.15: Ph.I.L.S. LESSON 26: THE MEANING OF HEART SOUNDS 573

37 Relate various heart sounds to the components of an ECG

Anatomy & Physiology REVEALED apreveiled.com **MODULE 9: CARDIOVASCULAR SYSTEM**

INTRODUCTION

The heart is an amazing organ that has been viewed with awe for ages. For many centuries physicians thought the heart, not the brain, was the control center and spiritual/emotional center of the body. Perhaps this is because the structural and functional relationships in the heart are relatively straightforward and easy to see, whereas such relationships are nearly impossible to discover through gross observation of the brain. Students of anatomy and physiology who have studied the brain know that our emotions come not "from the heart" but from the brain. Even though scientists and laypeople alike recognize that the brain, not the heart, controls the functioning of the rest of the body, they also recognize that the heart *is* essential for the survival of all organs and tissues in the body: If the heart fails to pump blood, and that failure results in a lack of flow of oxygenated blood to the tissues, the tissues will die.

Consider for a moment the incredible fact that the heart continues to beat, day and night, year after year, without stopping. It is an enormous job. Failure of this organ is often fatal. Perhaps it is no surprise that heart disease is the number one cause of death for Americans (it accounts for approximately 32% of all deaths, while cancer, at number two, accounts for 23% of all deaths). A thorough understanding of the structure and function of the heart is critical for everyone, whether an individual intends to go into a health science field or not. Nearly all health-care-related fields require practitioners to deal with individuals suffering from heart disease on a daily basis. Even if an individual's career does not involve health care, how an individual takes care of his or her heart now may very well determine the length and overall quality of that individual's life.

In this laboratory session, we will review the structure of cardiac muscle and identify heart structures on a preserved human heart or a model of the heart, and through dissection of a sheep heart. While working through the exercises, be aware that most textbook figures of the heart are drawn to make identification of the chambers and vessels very straightforward. When observing a real heart, identification of structures is more challenging because the chambers do not lie directly superior, inferior, or lateral to each other as they are often depicted in textbook drawings. In fact, most of the heart structures labeled "right" (such as the right atrium and right ventricle) lie not only on the right side of the heart, but also on the *anterior* surface of the heart. Likewise, most of the structures labeled "left" (such as the left atrium and left ventricle) lie not only on the left side of the heart, but also on the *posterior* surface of the heart.

Physiology exercises within this chapter will involve measuring electrical activity of the heart using electrocardiography (ECG), listening to heart sounds and relating these sounds to the events of the cardiac cycle, and exploring the length–tension relationship in cardiac muscle.

List of Reference Tables

Chapter 21: The Cardiovascular System: The Heart

Name: ______________________________

Date: ____________ **Section:** ___________

These Pre-Laboratory Worksheet questions may be assigned by instructors through their connect course.

PRE-LABORATORY WORKSHEET

1. The ______________________________ (pulmonary/systemic) circuit pumps blood to all organs of the body, and the ______________________________ (pulmonary/systemic) circuit pumps blood to the lungs.
2. The term *coronary* means "crown." ______________ (True/False)
3. A(n) ______________________________ (artery/vein) is a vessel that always carries blood away from the heart.
4. A(n) ______________________________ (artery/vein) is a vessel that always carries blood toward the heart.
5. The ______________________________ (pulmonary/systemic) circuit pumps blood at a higher pressure than the ______________________________ (pulmonary/systemic) circuit.
6. The muscles that attach to tendinous cords (chordae tendineae) are called:
   a. papillary muscles
   b. pectinate muscles
   c. pectoral muscles
   d. trabeculae carneae
7. The region where the great vessels such as the aorta and pulmonary trunk are attached to the heart is called the ______________________________ (apex/base) of the heart.
8. The wall of the ______________________________ (right/left) ventricle is much thicker than the wall of the ______________________________ (right/left) ventricle.
9. The layer of the pericardial sac that adheres to the heart is the ______________________________ (parietal/visceral) layer.
10. Most of the anterior surface of the heart receives oxygenated blood from a branch of the ______________________________ (left/right) coronary artery.
11. The valve between the right atrium and the right ventricle is the ______________________________ (bicuspid/tricuspid) valve.
12. The valve between the left atrium and the left ventricle is the ______________________________ (bicuspid/tricuspid) valve.
13. The left ventricle pumps blood into the ______________________________ (aorta/pulmonary trunk).
14. The superior and inferior vena cava drain blood into the ______________________________ (left/right) atrium.

HISTOLOGY

EXERCISE 21.1

CARDIAC MUSCLE

This activity is a review of observations of muscle tissue that were covered in chapter 11 of this laboratory manual. Depending on the time available in the laboratory, your instructor may want you to repeat your observations of cardiac muscle tissue, or may simply ask you to refer back to your notes on cardiac muscle tissue from chapter 11.

1. Obtain a slide of cardiac muscle and place it on the microscope stage. Bring the tissue sample into focus on low power, and then switch to high power.
2. Complete **table 21.1,** which compares features of cardiac muscle and skeletal muscle. Use table 11.1 and the textbook as guides.
3. Identify the structures listed in **figure 21.1** on the slide of cardiac muscle, using table 21.1 as a guide. Then label the structures in figure 21.1.

   What two types of cellular junctions are found in the intercalated discs?

   ____________________ and ____________________

   What are the functions of the cellular junctions in the intercalated discs?

   ____

   ____

4. Sketch cardiac muscle as seen through the microscope in the space provided. Be sure to include and label all of the structures listed in figure 21.1.

__________ ×

Table 21.1	Comparisons Between Cardiac and Skeletal Muscle Tissues	
Muscle Tissue	**Cardiac Muscle**	**Skeletal Muscle**
Nervous Control		
Appearance of Cells		
Number of Nuclei		
Location of Nuclei		

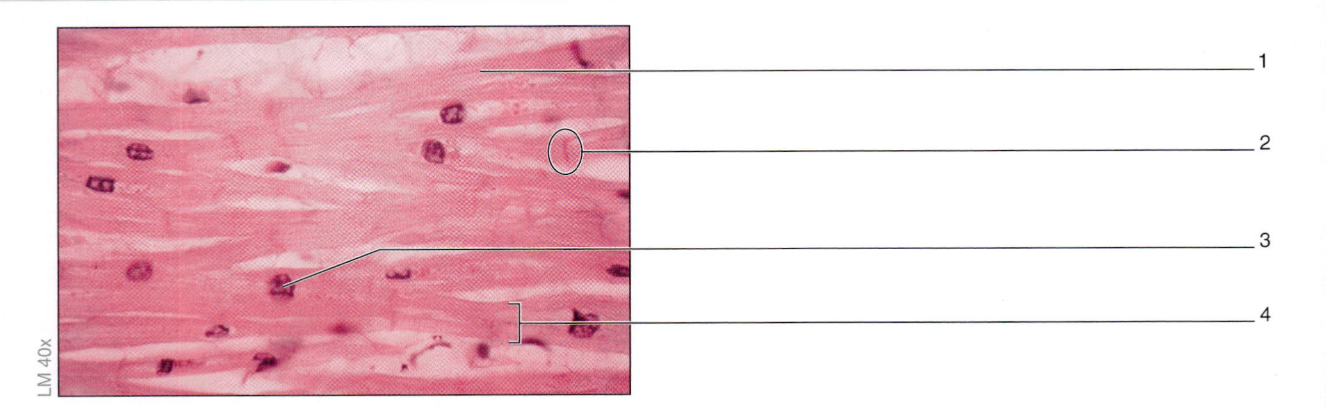

Figure 21.1 Cardiac Muscle Tissue. Cardiac muscle tissue is characterized by short, branching cells with single, centrally located nuclei. Intercalated discs are dark lines visible between the cells. A bands and I bands are also visible, indicating the presence of sarcomeres within the myofibers. Use the terms listed to fill in the numbered labels in the figure.

☐ cardiac muscle cell ☐ intercalated disc ☐ nucleus ☐ striations

INTEGRATE

CONCEPT CONNECTION

As discussed in exercise 21.1, there are important distinctions made between cardiac muscle tissue and skeletal muscle tissue. Cardiac muscle, like skeletal muscle, contains striations. The striations are due to the presence of sarcomeres, the fundamental contractile unit of cardiac and skeletal muscle. Sarcomeres are composed of overlapping thick filaments composed of myosin protein and thin filaments composed of actin, troponin, and tropomyosin.

The mechanism of muscle contraction in cardiac muscle is similar to that of skeletal muscle. Calcium binds to the regulatory protein troponin, causing a conformational change in troponin–tropomysin complex that exposes the myosin binding sites on actin. This allows myosin heads to bind to actin, forming crossbridges. The formation of crossbridges in both cardiac and skeletal muscle is what allows the muscle to produce force and generate movement.

EXERCISE 21.2

LAYERS OF THE HEART WALL

Both the atrial and ventricular heart wall are composed of three layers: the endocardium, myocardium, and epicardium. The layers of the heart wall are compared in **table 21.2.** Note that when referring to the outer layer of the heart wall as part of the pericardial sac, the appropriate term is *visceral layer of serous pericardium* (instead of epicardium).

1. Obtain a slide demonstrating the atria of the heart. Place it on the microscope stage and observe on low power.
2. Identify the following on the slide of the atrium, using **figure 21.2** and table 21.2 as guides. You may also see cross sections of the coronary vessels deep to the epicardium.

☐ **endocardium** ☐ **myocardium**
☐ **epicardium**

3. Sketch the layers of the heart wall as seen through the microscope in the space provided. Be sure to include and label all of the structures listed in step 2 in your drawing.

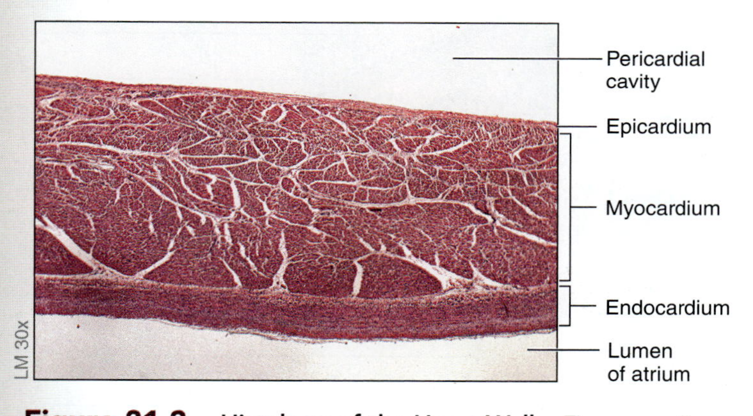

Figure 21.2 Histology of the Heart Wall. The myocardium is the thickest layer of the heart wall, and the endocardium is relatively thicker than the epicardium.

Table 21.2 Layers of the Heart Wall

Wall Layer	Tissue	Description	Word Origin
Endocardium	Simple squamous epithelium (called endothelium) plus a relatively thick underlying layer of areolar connective tissue	This layer is relatively thick in the atria, and thin in the ventricles	*endo*, within, + *kardia*, heart
Myocardium	Layers of cardiac muscle	This is the thickest layer of the heart wall; it is thicker in the ventricles than in the atria	*mys*, muscle, + *kardia*, heart
Epicardium	Simple squamous epithelium resting on a layer of areolar connective tissue.	This is the visceral pericardium. It is relatively thicker in the ventricles than in the atria.	*epi-*, upon, + *kardia*, heart

GROSS ANATOMY

EXERCISE 21.3

LOCATION OF THE HEART AND THE PERICARDIUM

The heart is located within the **thoracic cavity,** a cavity that also houses the lungs, trachea, esophagus, and a variety of nerves and blood vessels. Within the thoracic cavity, the heart is located in a space called the **mediastinum.** Within the mediastinum, the heart is enclosed in a double-layered **pericardial sac.** The outer portion of the pericardial sac is the **fibrous pericardium** composed of a tough, dense irregular tissue. This layer is attached inferiorly to the diaphragm and superiorly to the aorta and pulmonary trunk. The inner portion of the sac is the **parietal layer of serous pericardium.** Tightly adhered to the heart is the **visceral layer of serous pericardium**. Between the parietal and visceral layers is a fluid-filled space called the **pericardial cavity.**

1. Observe a model of a human thorax or the thoracic cavity of a human cadaver **(figures 21.3** and **21.4).**
2. Identify the structures listed in figure 21.3 on a classroom model of the thorax or on a human cadaver, using the textbook as a guide. Then label them in figure 21.3.
3. Observe the pericardium of a human cadaver (or simply view figure 21.4*a*). Remove the heart from the pericardial sac. The visceral layer of serous pericardium will remain attached to the heart. Note that both the parietal layer of serous pericardium and the visceral layer of serous pericardium are composed of simple squamous epithelium called **mesothelium.** These membranes produce a small amount of **pericardial fluid,** which lubricates the surface between the heart and the pericardial sac.

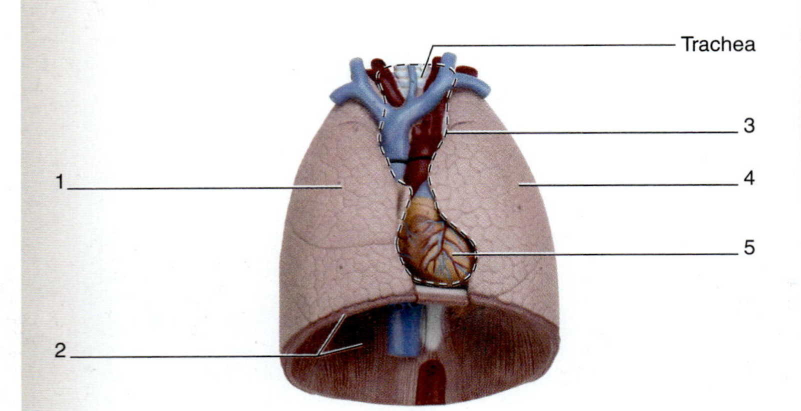

Figure 21.3 Location of the Heart Within the Thoracic Cavity. Within the thoracic cavity the heart is located in the mediastinum, the space between the two pleural cavities, which house the lungs. Use the terms listed to fill in the numbered labels in the figure.

- ☐ diaphragm
- ☐ heart
- ☐ left lung/pleural cavity
- ☐ mediastinum
- ☐ right lung/pleural cavity

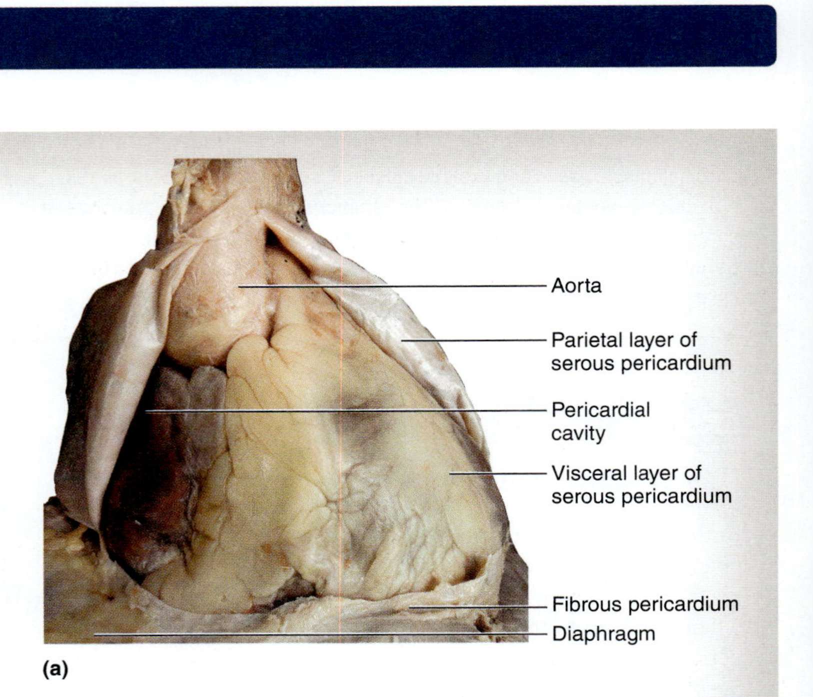

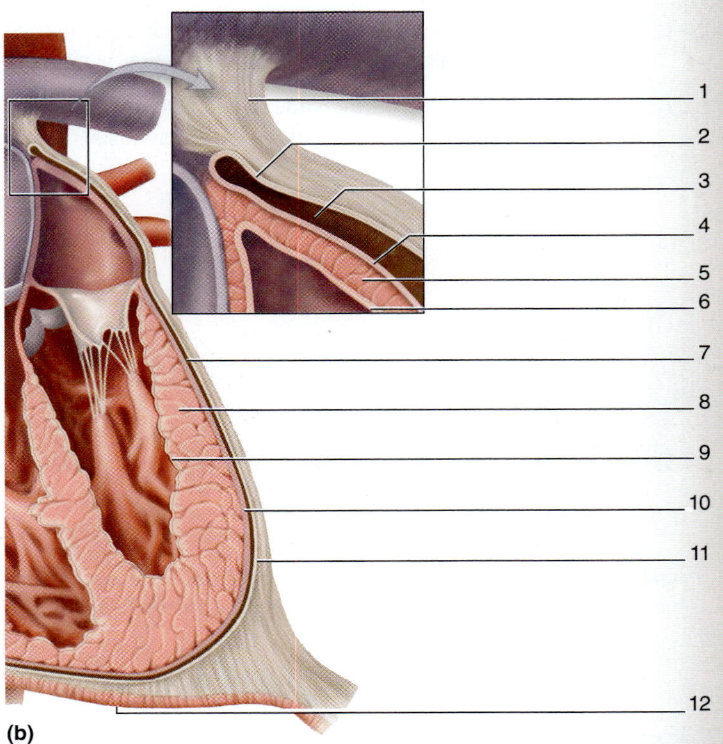

Figure 21.4 Pericardium. (a) Location of the heart within the pericardial sac of a cadaver. The parietal pericardium has been cut and reflected to reveal the heart, which is covered by visceral pericardium. (b) Layers and tissues composing the pericardial sac. Use the terms listed to fill in the numbered labels in the figure. Some terms may be used more than once.

- ☐ diaphragm
- ☐ endocardium
- ☐ fibrous pericardium
- ☐ myocardium
- ☐ parietal layer of serous pericardium
- ☐ pericardial cavity
- ☐ visceral layer of serous pericardium

4. Identify the structures listed in figure 21.4*b* on a model of the thorax or on a human cadaver, using figure 21.4*a* as a guide. Then label them in figure 21.4*b*.
5. *Optional Activity:* AP|R **9: Cardiovascular System—** Explore the "Thorax" dissections to view the heart in the thoracic cavity and the surrounding structures.

EXERCISE 21.4

GROSS ANATOMY OF THE HUMAN HEART

1. Obtain a preserved human heart from a cadaver or a classroom model of the heart. If using a preserved heart, place it in a dissecting pan and keep it moist while making observations. In addition, *use only a blunt probe* to point out structures on the heart so as not to damage the heart.
2. Note the size and shape of the heart. A normal heart is about the size and shape of a human fist. Based on this information, is the heart in the dissecting pan of normal size? ______________________ If the heart appears to be enlarged, make note of that observation, as it can be (though it is not necessarily) indicative of heart disease.
3. Notice that the heart may not look precisely like the drawings in the textbook. The heart is a twisted organ, so identification of chambers can be challenging at first. Identify the **apex** (the pointed, inferior portion of the heart) and the **base** (the superior point where the great vessels enter and leave). In most instances, the term *base* refers to the bottom of an organ or tissue. Is this generalization accurate for the heart? ______________________ The heart has a relatively flat inferior surface, the **diaphragmatic surface,** which is the region of the heart that lies on the superior part of the diaphragm.
4. Now place the heart in your right hand with the diaphragmatic surface in the palm of the hand, the apex directed toward the thumb and wrist, and the base directed toward the space between the tip of the thumb and the second digit **(figure 21.5).** This should place the heart in a position that very closely resembles the heart's orientation within the thorax. If you are unsure if the heart is in the proper orientation, check with the instructor before proceeding.

   Identify the four chambers of the heart: **right atrium, right ventricle, left atrium,** and **left ventricle** (figure 21.5).

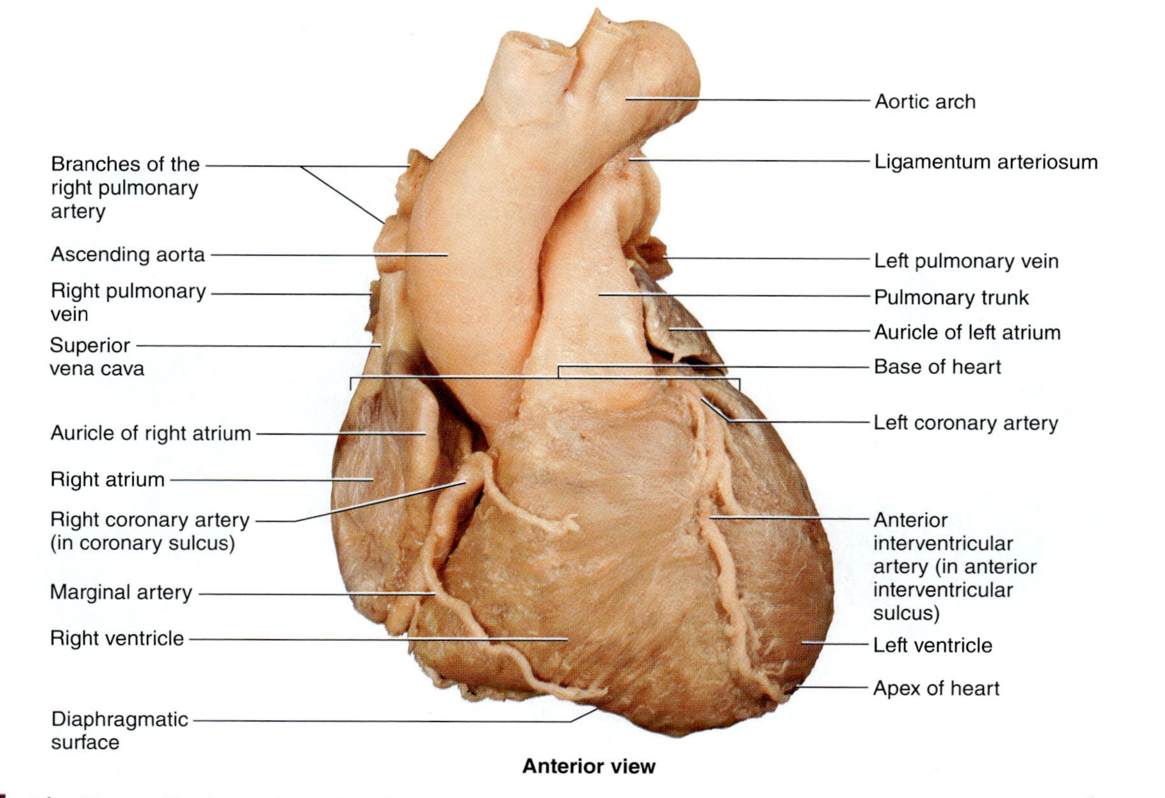

Figure 21.5 The Heart. The heart shown here is oriented with the anterior surface facing the observer, which is how it should look when it is in the observer's right hand with the apex pointed toward the thumb and wrist, and the great vessels directed toward the space between the tip of the thumb and the second digit.

(continued on next page)

(continued from previous page)

Keep in mind that the left atrium and left ventricle will be on *your* right, and the right atrium and ventricle will be on *your* left. The right ventricle is anterior and the left ventricle is posterior. Notice that the mass of the left ventricle fills up nearly the entire palm of your hand because it has a much thicker myocardium than the right ventricle. The **right atrium** is superior and lateral to the right ventricle. The left atrium is not visible in this view. You will need to rotate the heart to see the left atrium.

5. *Layers of the Heart Wall:* Identify the following components of the heart wall on a heart model or on a preserved human heart, using table 21.2 and figure 21.4 as guides:

☐ **endocardium** ☐ **myocardium**
☐ **epicardium**

INTEGRATE

LEARNING STRATEGY

One way to remember which of the atrioventricular (AV) valves is on the right side of the heart and which is on the left is to use the saying, "**Try** before you **Buy**." The **TRI**cuspid ("Try") valve, which is on the right side of the heart, comes before the **BI**cuspid ("Buy") valve, which is on the left side of the heart.

6. *The Right Atrium:* Holding the heart in your right hand once again, with the anterior surface directed toward you, identify the **superior vena cava** attached to the right atrium (figure 21.5). Within the thorax, the superior vena cava extends vertically and is attached to the upper portion of the right atrium. Now locate the inferior vena cava, which extends vertically and is attached to the lower portion of the right atrium. Next, look inside the right atrium. Notice the thin strands of **pectinate muscle** in the wall of the right atrium **(figure 21.6).** Pectinate muscle is found in the wall of the right atrium especially within its *auricles*, which are wrinkled, flat extensions of the atria. Locate the shallow depression covered with a thin membrane in the interatrial septum. This is the **fossa ovalis,** a remnant of a fetal shunt between the right and left atria called the **foramen ovale.** Is the opening completely closed off in the specimen? If not, what might some of the consequences be (this condition is called a *patent* foramen ovale [*pateo*, to lie open])?________________

Just inferior to the fossa ovalis, look for the small opening of the **coronary sinus,** a vein that drains nearly all deoxygenated blood from the heart wall. Finally, observe the **right atrioventricular (AV) valve** and count the cusps. How many cusps are there? ______________________
Based on that information, is the right AV valve a tricuspid or bicuspid valve? ___________________________

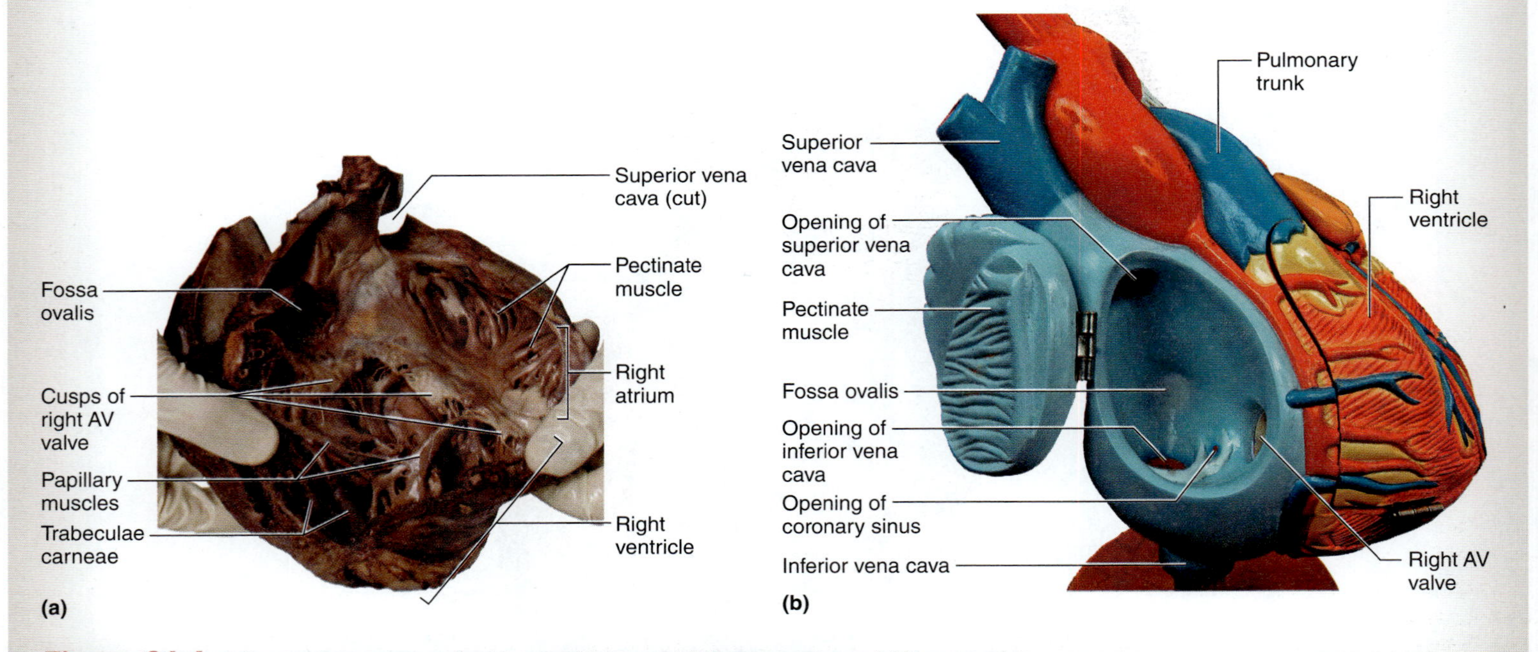

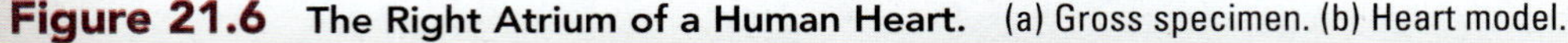

Figure 21.6 **The Right Atrium of a Human Heart.** (a) Gross specimen. (b) Heart model.

7. *The Right Ventricle:* Place a blunt probe in the right atrium, and direct the tip into the right ventricle **(figure 21.7).** If the right ventricle is not already cut open, ask the instructor to cut it open to identify the internal structures. The most prominent features in the walls of the right ventricles are the folds of cardiac muscle in the walls of the ventricle called **trabeculae carneae** (*trabs*, a beam, + *carneus*, fleshy), and the nipple-like **papillary muscles** (*papilla*, a nipple) that attach to the **cusps of the right AV valve** by string-like structures called **tendinous cords** (or *chordae tendineae*). When the ventricles contract, the papillary muscles also contract and pull down on the cusps of the AV valve. Because blood is being pushed up against the inferior portion of the valve cusps as the ventricles contract, the action of the papillary muscles pulling down on the valve cusps keeps the valve closed. This prevents blood from flowing back into the right atrium. Consequently, the blood is forced out through the **pulmonary trunk.**

8. Identify the following structures in the right ventricle of a human heart or heart model, using figure 21.7 as a guide:

   ☐ **cusps of the right AV valve** ☐ **tendinous cords**
   ☐ **papillary muscles** ☐ **trabeculae carneae**

9. Place the tip of a blunt probe in the right ventricle and pass it out through the **pulmonary trunk.** To enter the pulmonary trunk, the probe will have to pass through the **pulmonary semilunar valve.** If the vessel has been cut open, the cusps of the semilunar valve will be visible. Notice they are shaped like "half moons"—hence the name (*semi-*, half + *lunar*, moon). How many cusps are there?______________ Where is the blood in the pulmonary trunk transported?

   ____

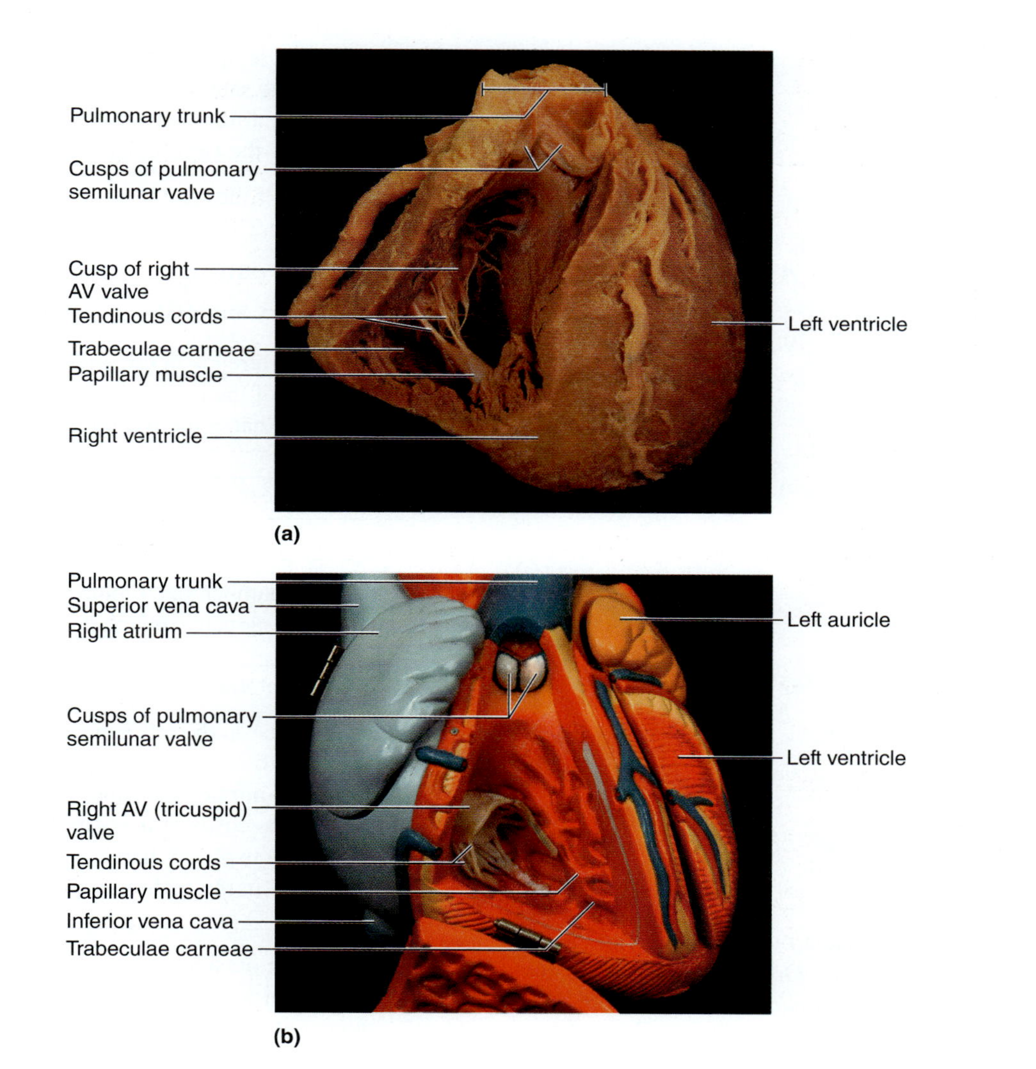

Figure 21.7 The Right Ventricle of a Human Heart. (a) Gross specimen. (b) Heart model.

(continued on next page)

(continued from previous page)

10. Sketch the right atrium and right ventricle in the space provided. Use arrows to indicate the flow of blood from the right atrium to the pulmonary trunk. Be sure to include and label all of the structures listed below in your drawing.

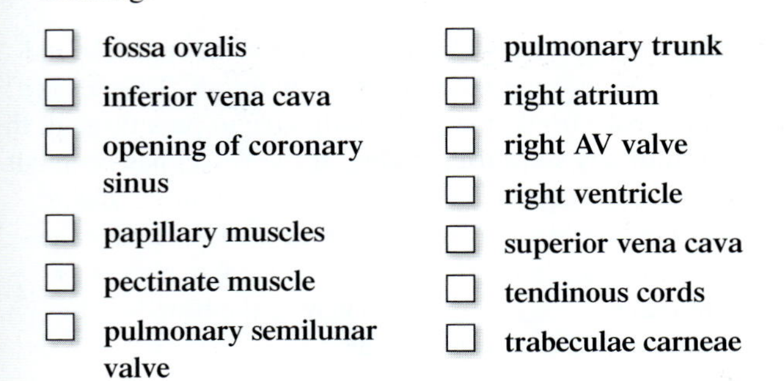

- ☐ **fossa ovalis**
- ☐ **inferior vena cava**
- ☐ **opening of coronary sinus**
- ☐ **papillary muscles**
- ☐ **pectinate muscle**
- ☐ **pulmonary semilunar valve**
- ☐ **pulmonary trunk**
- ☐ **right atrium**
- ☐ **right AV valve**
- ☐ **right ventricle**
- ☐ **superior vena cava**
- ☐ **tendinous cords**
- ☐ **trabeculae carneae**

11. *The Left Atrium:* Rotate the heart until its posterior surface is visible **(figure 21.8).** Open the left atrium and notice the four **pulmonary veins,** which collectively drain blood from the lungs into the right atrium. Look inside the left atrium. Notice that the wall of the left atrium is thin and smooth and does *not* have pectinate muscles (although the wall of the left *auricle* does). Recall that the right atrium has pectinate muscles in both its wall and its pectinate. This difference provides a means of distinguishing the right atrium from the left atrium when viewing the internal surface of the heart. The left atrium is little more than an expansion of the tissue where the four pulmonary veins come together. Thus, its walls are not always easy to identify. Now that both the right and left atria have been identified, place your index finger in the right atrium and your thumb in the left atrium to feel for the **fossa ovalis,** which lies in the **interatrial septum.** What is the name of the fetal structure of which the fossa ovalis is a remnant?____________________ Next, observe the **left atrioventricular (AV) valve** and count the cusps. How many cusps are there?____________________ Based on that information, is the left AV valve a tricuspid or bicuspid valve?____________________ What is another name for this valve?____________________________

12. *The Left Ventricle:* Place a blunt probe in the left atrium and pass it into the **left ventricle (figure 21.9).** If the left ventricle is not already cut open, ask the instructor to cut it open so you can identify the structures within. The left ventricle contains all the same structures as the right ventricle, and the **left atrioventricular (AV) valve** functions the same way the right AV valve functions. The biggest difference between the two chambers is the thickness of the ventricular walls.

13. Identify the following structures in the left ventricle, using figure 21.9 as a guide:

- ☐ **left AV valve**
- ☐ **papillary muscles**
- ☐ **tendinous cords**
- ☐ **trabeculae carneae**

14. Note the difference in thickness between the myocardium in the wall of the left ventricle and that of the right ventricle. What is the reason for this difference?

____

____

____

Does the chamber size (volume) of the left ventricle appear to be greater than that of the right ventricle? ______________ Do the two chambers pump the same volume of blood? ______________

15. Place a probe in the left ventricle and pass it out through the **aorta.** To enter the aorta, the probe will have to pass through the **aortic semilunar valve.** If the vessel has been cut open, the cusps of the semilunar valve will be visible. How many cusps are there?____________________ Where is the blood in the aorta transported? ____________________
____

16. Observe the outside of the arch of the aorta where it passes just superior to the pulmonary trunk. Look for a small ligament attaching the pulmonary trunk to the aorta in this location (see figure 21.5 and figure 21.9*b*). This is the **ligamentum arteriosum.** The ligamentum arteriosum is a remnant of what fetal structure?______________ This fetal structure shunts blood from the ______________ to the ______________, thereby allowing blood to bypass the ______________.

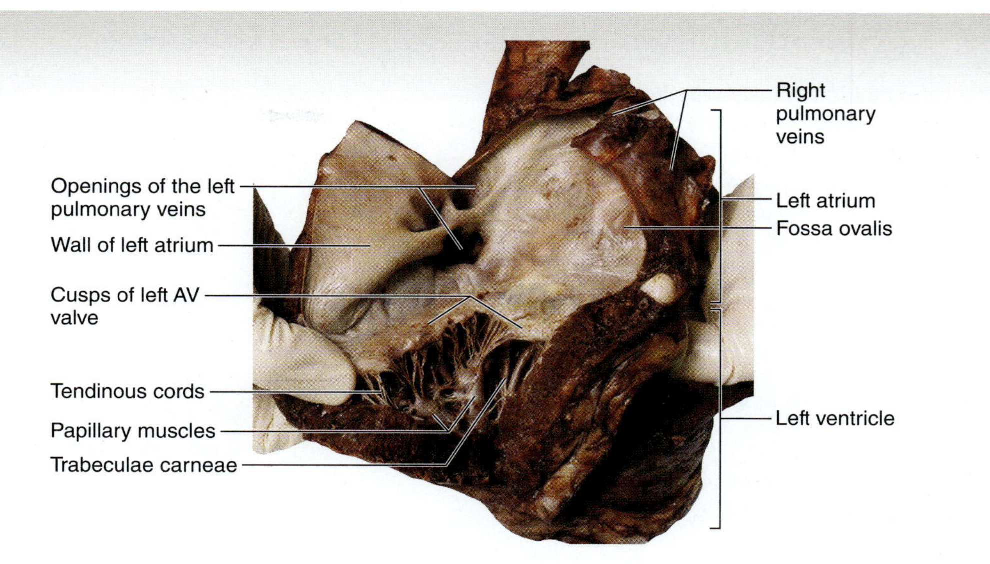

Figure 21.8 The Left Atrium of a Human Heart. Posterior view of a human heart with left atrium and left ventricle cut open.

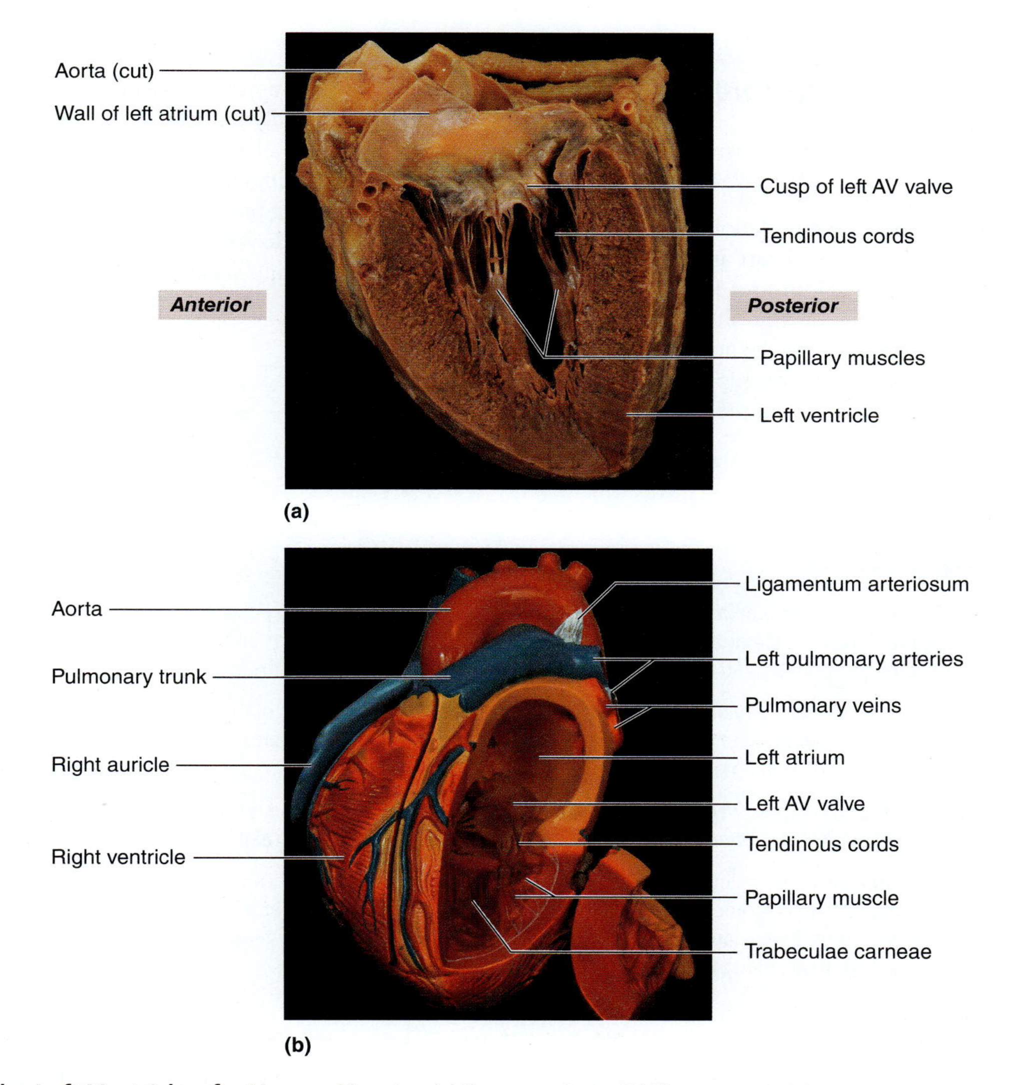

Figure 21.9 The Left Ventricle of a Human Heart. (a) Gross specimen. (b) Classroom model.

(continued on next page)

(continued from previous page)

17. Sketch the left atrium and left ventricle in the space provided. Use arrows to indicate the flow of blood from the lungs to the aorta. Be sure to include and label all of the structures listed below in your drawing.

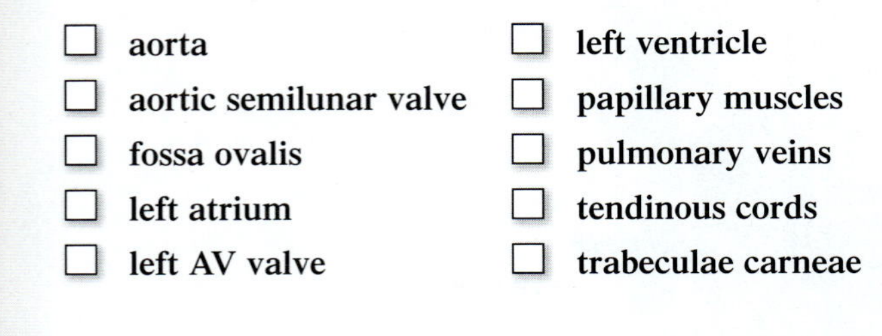

18. *Optional activity:* **AP|R 9: Cardiovascular System**—Watch the "Heart" animation to gain a 3D fly-through perspective of the internal heart.

EXERCISE 21.5

THE CORONARY CIRCULATION

1. Obtain a preserved human heart or a classroom model of the heart.

2. The **coronary circulation** is the circulation to the heart wall itself. The blood supply to the heart arises from two main coronary vessels, the **right and left coronary arteries (table 21.3).** The openings into these arteries arise immediately superior to the cusps of the aortic semilunar valve. **Figure 21.10** shows the relationship between the cusps of the aortic semilunar valve and the openings to the right and left coronary arteries. Observe the aortic semilunar valve and identify the openings for the right and left coronary arteries (figure 21.10). When the left ventricle contracts and pushes blood into the aorta, the cusps of the aortic semilunar valve cover the openings to the coronary arteries. What consequence does this have in terms of blood flow to the heart wall during ventricular contraction (systole)?

____

____

____

Table 21.3 Arterial Supply to the Heart

Vessel	Description	Areas Served	Word Origin
Left Coronary Artery	Located posterior to the pulmonary artery; branches into the anterior interventricular and circumflex arteries just after it emerges from behind the pulmonary artery	Branches into anterior interventricular and circumflex arteries	*corona*, a crown
Anterior Interventricular Artery	Branch of the left coronary artery; located in the anterior interventricular sulcus (groove). Physicians typically refer to this as the "LAD" (left anterior descending).	Anterior parts of the right and left ventricles	*inter*, between, + *ventricular*, referring to the ventricles of the heart (from *ventriculus*, belly)
Circumflex Artery	Branch of the left coronary artery; located in the coronary sulcus between the left atrium and left ventricle	Left atrium and left ventricle (lateral part)	*circum*, around, + *flexus*, to bend
Right Coronary Artery	Located in the coronary sulcus between the right atrium and the right ventricle. Branches include the marginal artery, posterior interventricular artery, and SA nodal artery, which supplies blood to the sinoatrial node within the right atrium.	Right atrium; branches into marginal and posterior interventricular arteries	*corona*, a crown
Marginal Artery	Branches off the right coronary artery at the right margin of the heart and is located on the lateral part of the right ventricle	Lateral part of the right ventricle	*margo*, border or edge
Posterior Interventricular Artery	Continuation of the right coronary artery located in the posterior interventricular sulcus	Posterior parts of the right and left ventricles	*inter*, between, + *ventricular*, referring to the ventricles of the heart (from *ventriculus*, belly)

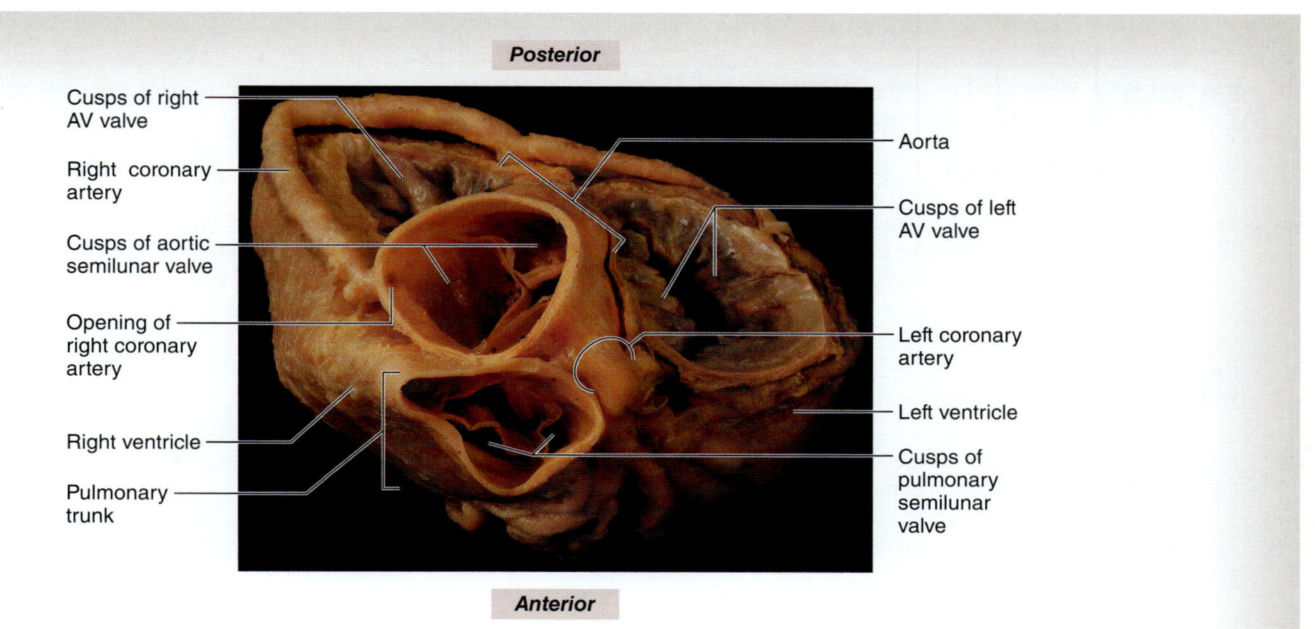

Figure 21.10 The Coronary Circulation. Superior view of the human heart with the atria removed. All four valves of the heart can be seen in this view. In this photo you can see where the right and left coronary arteries come off of the aorta, and you can see the cusps of both the pulmonary semilunar valve within the pulmonary trunk and the aortic semilunar valve within the aorta.

When the ventricles relax, the cusps of the aortic semilunar valves close as they fill with blood. Thus, they are no longer covering the openings to the coronary arteries. What consequence does this have in terms of blood flow to the heart wall during ventricular relaxation (diastole)? ______________________________

INTEGRATE

CLINICAL VIEW

Myocardial Infarction

Adequate blood flow through the coronary arteries is absolutely essential for the functioning of the heart. If any coronary vessel becomes blocked due to disease or other processes, the region of the heart served by the vessel may become **ischemic**, meaning it lacks blood flow (*ischio*, to keep back, + *chymos*, juice). Prolonged ischemia to heart muscle leads to **hypoxia** (*hypo-*, too little, + *oxia*, oxygen). When cardiac muscle lacks an oxygen supply for more than a few minutes, the tissue dies, or becomes **necrotic** (*nekrosis*, death). The area of dead tissue is referred to as a **myocardial infarction** (*myocardial*, referring to the myocardium, + *in-farcio*, to stuff into), otherwise known as a "heart attack" or "MI." If a myocardial infarction results in vast tissue destruction, the heart may no longer be effective as a pump, which can lead to the death of the individual. If the individual survives the attack, the body will replace the damaged tissue with scar tissue. Scar tissue is mainly composed of a dense collection of collagen fibers. **Figure 21.11** demonstrates a human heart with evidence of a healed myocardial infarction. If observing a human heart, does the specimen show any evidence of (scarring from) past myocardial infarctions (scar tissue is generally clear to whitish in appearance and is much tougher than muscle tissue)? ____________

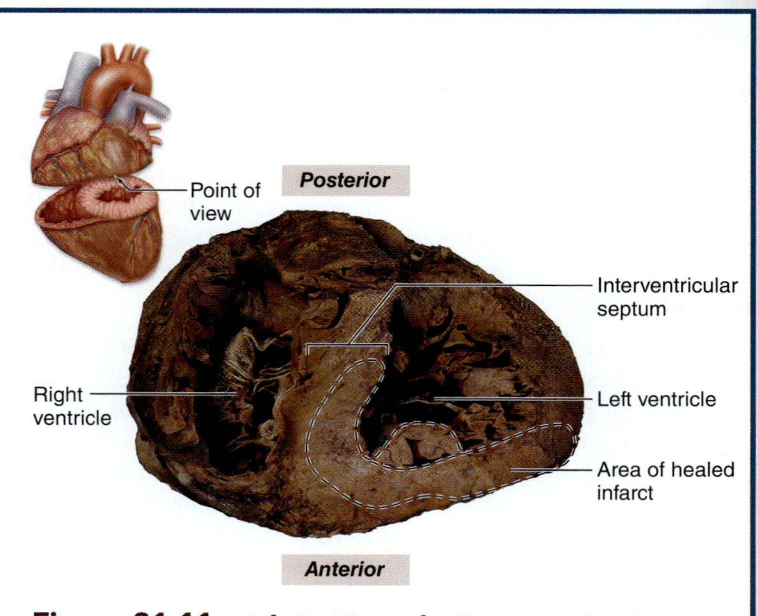

Figure 21.11 Inferior View of a Transverse Section Through a Human Heart. Scar tissue, which is evidence of a previous myocardial infarction, can be seen in the interventricular septum and posterior wall of the left ventricle.

(continued on next page)

(continued from previous page)

Table 21.4 Venous Drainage of the Heart

Vessel	Description	Areas Drained	Word Origin
Coronary Sinus	Located within the coronary sulcus on the posterior surface of the heart; it opens into the right atrium just inferior to the fossa ovalis	Entire heart; all veins of the heart drain into the coronary sinus, with the exception of a few small veins of the right ventricle, which drain directly into the right atrium	*corona*, a crown, + *sinus*, cavity
Great Cardiac Vein	Located in the anterior interventricular sulcus next to the anterior interventricular artery	Anterior parts of the right and left ventricles	*cardiacus*, heart, + *vena*, vein
Middle Cardiac Vein	Located in the posterior interventricular sulcus next to the posterior interventricular artery	Posterior parts of the right and left ventricles	*cardiacus*, heart, + *vena*, vein
Small Cardiac Vein	Located on the lateral part of the right ventricle, near the marginal artery	Lateral part of the right ventricle	*cardiacus*, heart, + *vena*, vein

3. *Cardiac Veins:* All venous blood draining the heart wall (with one exception; see **table 21.4**) eventually drains into one large vessel, the **coronary sinus.** The coronary sinus is located on the posterior surface of the heart within the coronary sulcus. What chamber does the coronary sinus empty into?

   ____

4. Identify the vessels shown in **figure 21.12** on the cadaver heart or on the classroom model of the heart, using tables 21.3 and 21.4 and the textbook as guides. Then label them in figure 21.12.
5. *Optional Activity:* **AP|R 9: Cardiovascular System—** Study the "Heart" dissections to review the vasculature and features of the heart, then use the Quiz feature to test yourself on these structures.

EXERCISE 21.6

SUPERFICIAL STRUCTURES OF THE SHEEP HEART

1. Obtain a dissecting pan, dissecting instruments, gloves, and a preserved sheep heart. Rinse the heart with water to remove any dried blood or other debris, and place it in the dissecting pan to begin observations of superficial structures.
2. **Figure 21.13** demonstrates superficial structures of the sheep heart from both anterior and posterior views. Begin by distinguishing the anterior surface from the posterior surface. One way to identify the anterior surface of the heart is the fairly distinctive, ruffled borders of both the right and left **auricles,** which are extensions of the right and left **atria.** Both auricles are visible in this view.
3. Observe the surface of the heart closely to locate the **visceral pericardium** (epicardium). Then observe the outer surfaces of the great vessels of the heart to identify remnants of the **parietal pericardium** where it was attached to these vessels. Note the large amount of fatty tissue deep to the epicardium. One of the functions of this fatty tissue, called **epicardial fat**, is to help cushion the heart within the pericardial cavity.
4. Identify the following superficial features on the sheep heart, using figure 21.13 as a guide:
   - ☐ **anterior interventricular sulcus**
   - ☐ **apex**
   - ☐ **coronary sulcus**
   - ☐ **left atrium**
   - ☐ **left auricle**
   - ☐ **left ventricle**
   - ☐ **posterior interventricular sulcus**
   - ☐ **right atrium**
   - ☐ **right auricle**
   - ☐ **right ventricle**
5. *Great vessels:* Identify the great vessels: the aorta, pulmonary trunk, pulmonary veins, superior vena cava, and inferior vena cava. These vessels are often cut very close to their attachments to the heart, which can make identification difficult. To make the task easier, carefully remove as much of the epicardial fat as possible from the superior aspect of the heart (leave the fat in place on the ventricles for now).

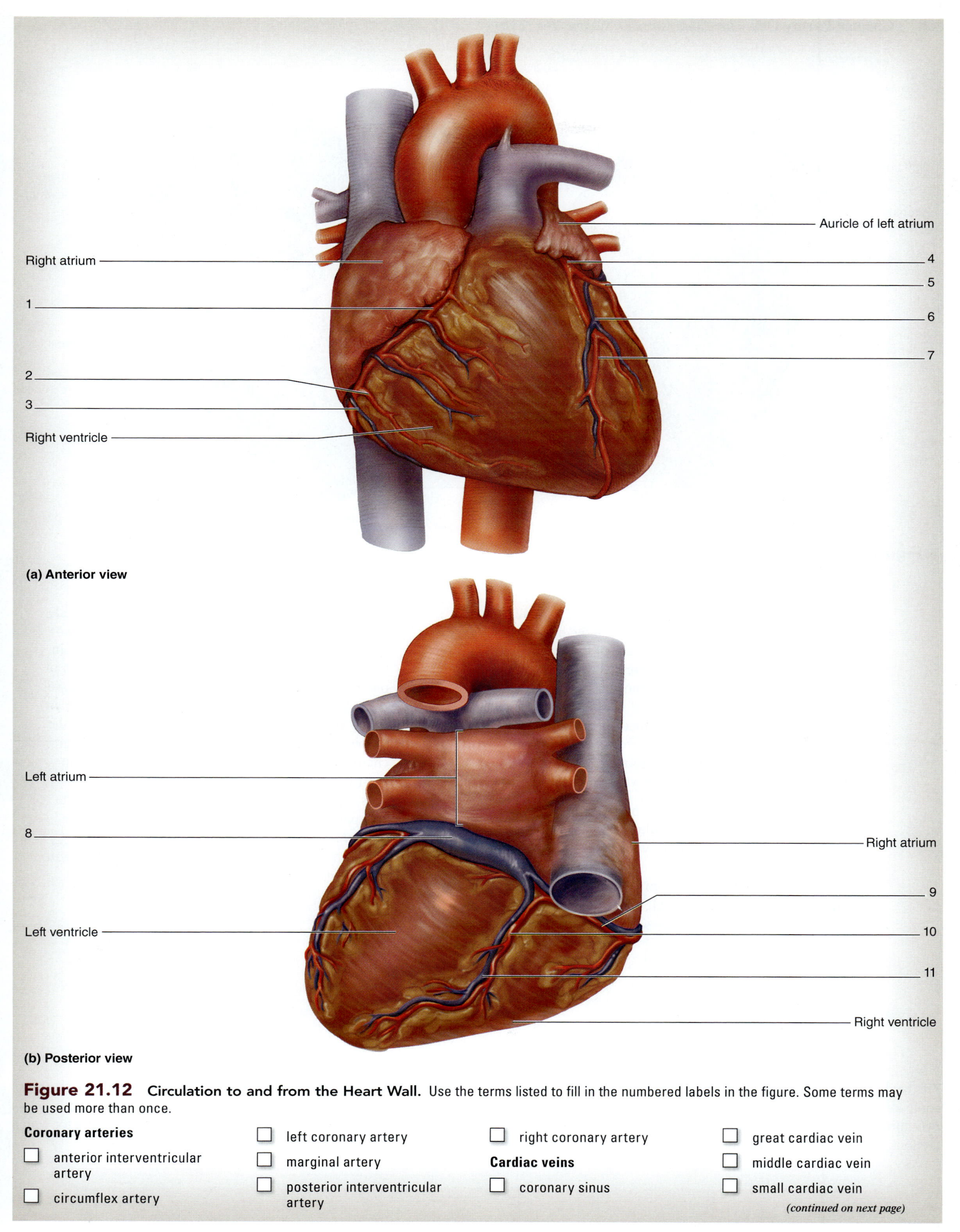

Figure 21.12 Circulation to and from the Heart Wall. Use the terms listed to fill in the numbered labels in the figure. Some terms may be used more than once.

Coronary arteries

- ☐ anterior interventricular artery
- ☐ circumflex artery
- ☐ left coronary artery
- ☐ marginal artery
- ☐ posterior interventricular artery
- ☐ right coronary artery

Cardiac veins

- ☐ coronary sinus
- ☐ great cardiac vein
- ☐ middle cardiac vein
- ☐ small cardiac vein

(continued on next page)

(continued from previous page)

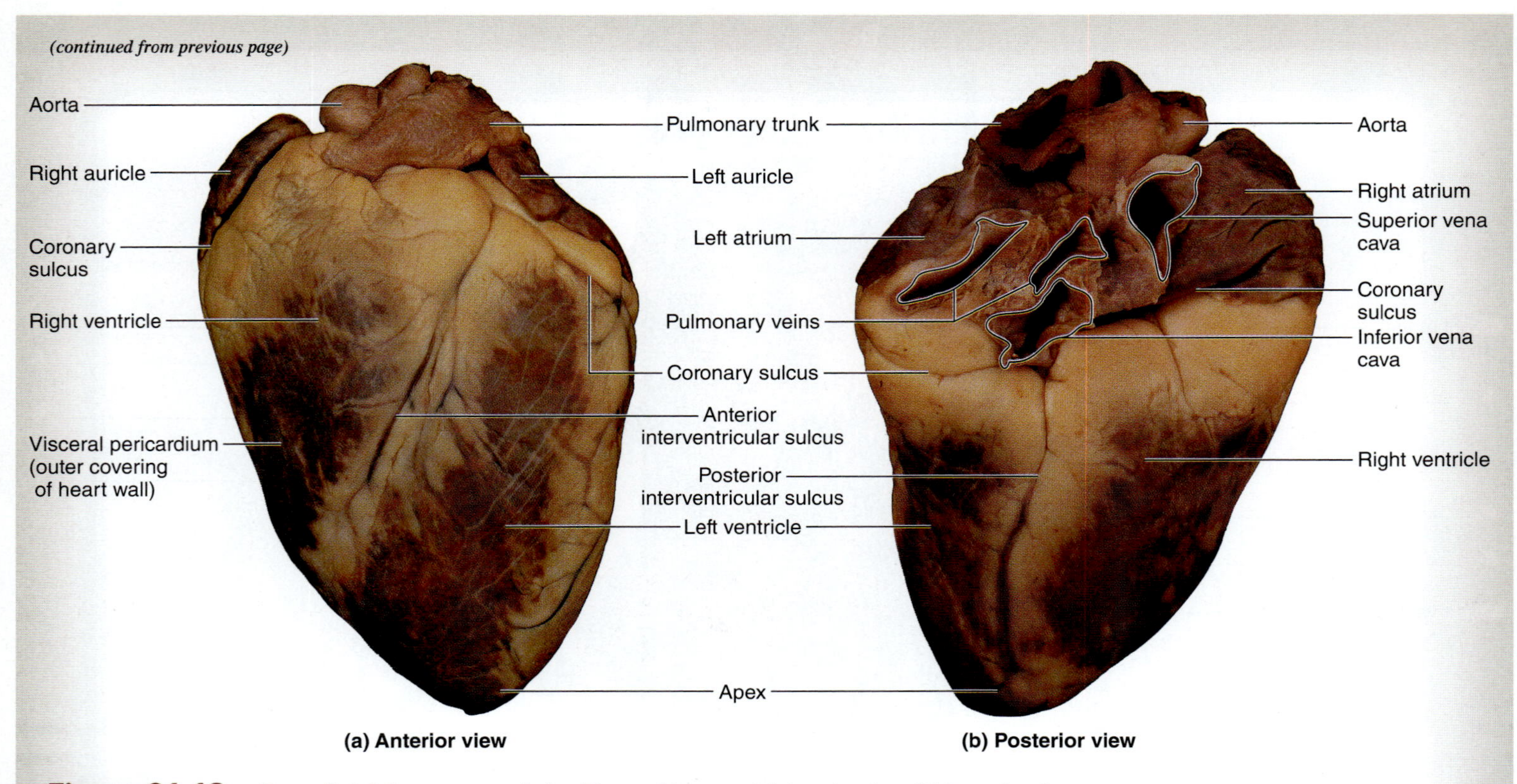

Figure 21.13 **Superficial Structures of the Sheep Heart.** (a) Anterior view. (b) Posterior view.

a. After cleaning away as much fat as possible, proceed with identification of the great vessels of the heart. Begin by viewing the anterior surface of the heart (figure 21.13*a*). The two most prominent vessels coming off the heart are the pulmonary trunk and the aorta. Both vessels have thick, tough walls, which helps with identification. The pulmonary trunk is located the most anteriorly and points to the right (from your point of view). The aorta is directly posterior to the pulmonary trunk and points to the left (from your point of view).

b. Turn the heart over to view the posterior surface (figure 21.13*b*). Locate the pulmonary veins, which will be more to the left (from your point of view), and the superior and inferior venae cavae, which will be more to the right (from your point of view). Before moving on, identify all of the great vessels:

- ☐ **aorta**
- ☐ **pulmonary veins**
- ☐ **inferior vena cava**
- ☐ **superior vena cava**
- ☐ **pulmonary trunk**

c. To verify that the great vessels have been identified correctly, use a blunt probe or your fingers (or both) to see where each vessel comes from, or leads to, in the heart. Place the tip of the probe into the lumen of one of the vessels and see where it goes. If the vessel is large enough, put your index finger into the lumen of the vessel and *feel* where it goes. Now answer the following questions by giving the name of the heart chamber the probe will pass into when placed into each vessel: A probe in the pulmonary trunk will pass into the _______________; a probe in the aorta will pass into the _______________; a probe in the pulmonary veins will pass into the _______________; and a probe in the superior or inferior vena cava will pass into the _______________.

The next two exercises involve cutting the entire sheep heart in half to compare wall thicknesses of the right and left ventricles. Approximately two-thirds of the dissection groups in the laboratory will make coronal sections of the heart, and the remaining third will make transverse sections. The different sections yield different views of the chambers and the structures within each chamber. All students will observe hearts that have been sectioned both ways. Ask the laboratory instructor which type of section to make before beginning the dissection.

EXERCISE 21.7

CORONAL SECTION OF THE SHEEP HEART

1. Obtain a scalpel or a knife with a 6-inch blade and a plastic ruler with millimeter increments. Either the scalpel or the knife will work for this next task, although the knife will make a cleaner cut. Turn the heart upside down so the base is on the dissecting pan, the apex is pointed toward you, and the anterior surface of the heart is facing your body. Make a coronal section through the entire heart to separate it into anterior and posterior portions **(figure 21.14).**

2. Identify the following structures within the ventricles, using figure 21.14 as a guide:

- ☐ **papillary muscles**
- ☐ **tendinous cords**
- ☐ **trabeculae carneae**

3. Using a small ruler, measure the thickness of the right ventricular wall approximately 1 cm below the valve ring (where the right atrioventricular valve is located) and record it here: ______________ cm. Next, measure the thickness of the left ventricular wall approximately 1 cm below the valve ring (where the left atrioventricular valve is located) and record it here: ______________ cm. Approximately how much thicker is the wall of the left ventricle than the wall of the right ventricle? ______________ What is the *functional* consequence of this difference? ___________________________________

____

____

4. Do the chambers of the right and left ventricles appear to differ in *volume* (that is, the amount of blood each could hold)? ______________ Explain the consequences, if any, of having each chamber pump a different volume of blood (Hint: Blood leaving the right side of the heart eventually enters the left side of the heart, and vice versa). ___________________________________

____

____

____

5. Sketch the coronal view of the ventricles in the space provided. Be sure to include and label all the structures listed below in the drawing.

☐ **aorta**	☐ **papillary muscle**
☐ **apex**	☐ **right atrium**
☐ **left atrium**	☐ **right AV valve**
☐ **left auricle**	☐ **tendinous cords**
☐ **left AV valve**	☐ **trabeculae carneae**

6. After completing the observations, discard the scalpel blade in the sharps container, clean the dissection instruments with soap and water and let them air dry, and put the sheep heart back into the container it came in—or dispose of it according to the instructor's directions.

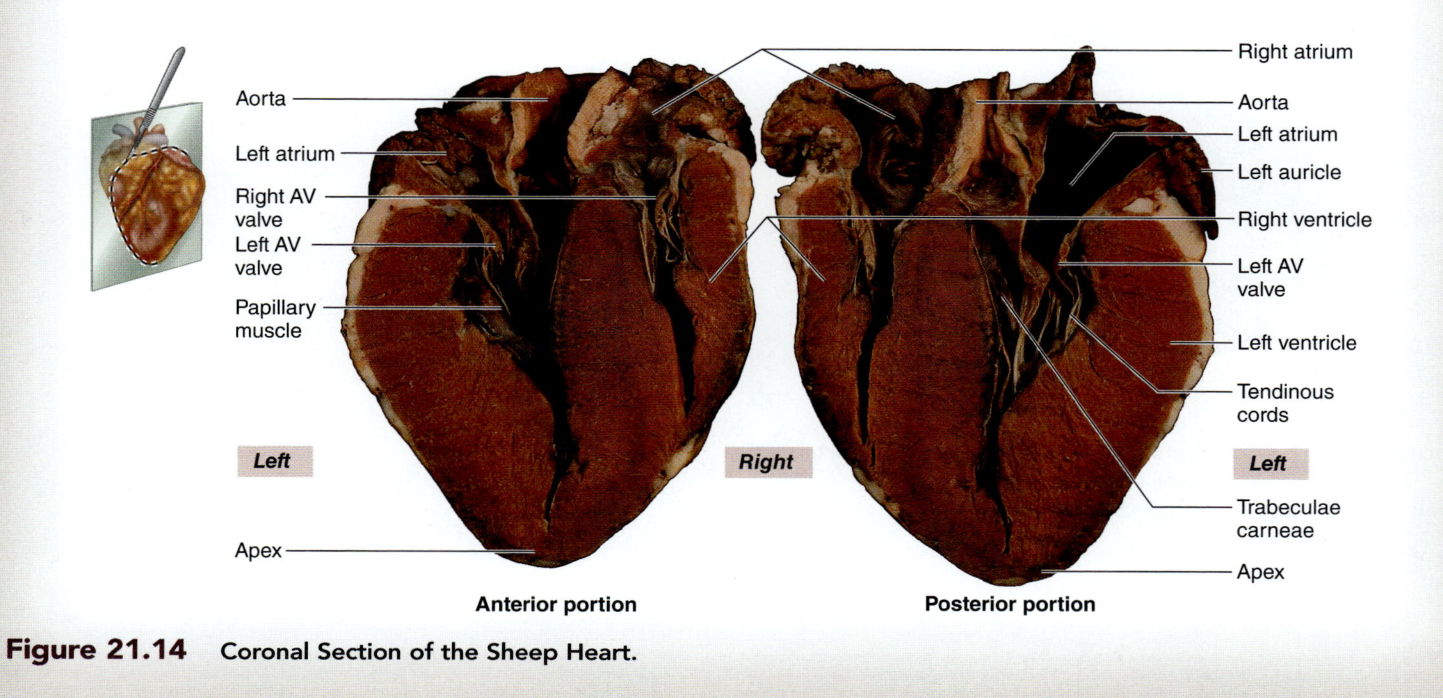

Figure 21.14 Coronal Section of the Sheep Heart.

EXERCISE 21.8

TRANSVERSE SECTION OF THE SHEEP HEART

1. Obtain an intact sheep heart, a scalpel or a knife, and a plastic ruler with millimeter increments. Either the scalpel or the knife will work for this next task, although the knife will make a cleaner cut. To make this cut, slice the heart transversely approximately 1 cm inferior to the coronary sulcus **(figure 21.15)** to separate the entire heart into superior and inferior portions. Observe the cut ends of the heart. Using a small ruler, measure the thickness of the right ventricular wall and record it here: ______________ cm. Next, measure the thickness of the left ventricular wall and record it here: ______________ cm. Approximately how much thicker is the wall of the left ventricle than the wall of the right ventricle? ______________ What is the *functional* consequence of this difference?

   ____

   ____

   ____

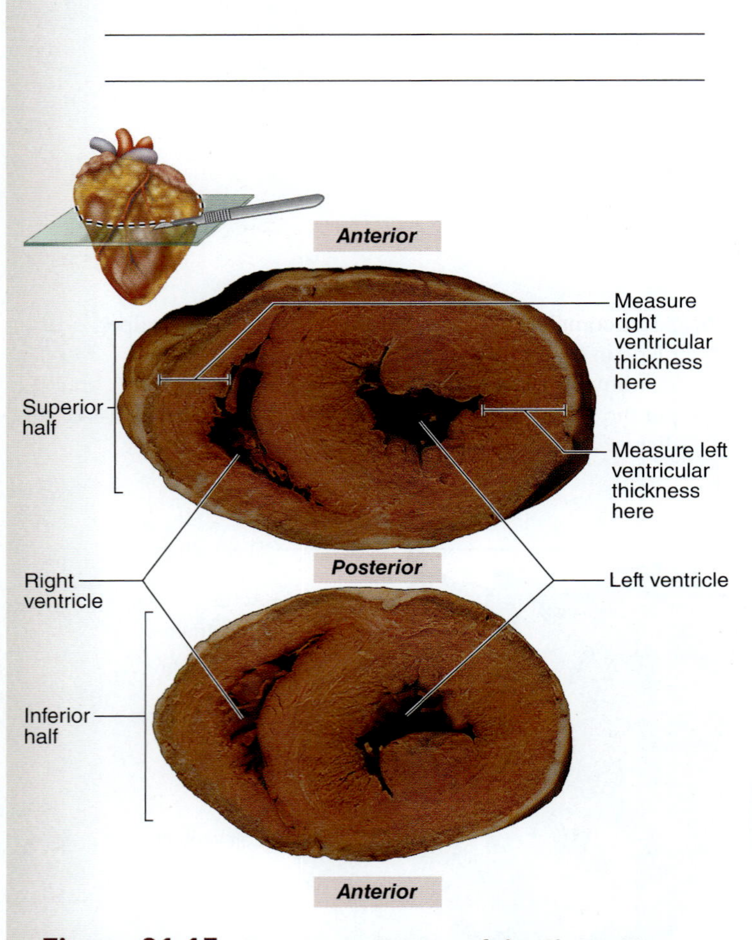

Figure 21.15 Transverse Section of the Sheep Heart. The locations for measuring the thickness of the right and left ventricles are shown in the upper portion of the figure.

2. Do the chambers of the right and left ventricles appear to differ in *volume* (that is, the amount of blood each could hold)? ____________________

   Explain the consequences, if any, of having each chamber pump a different volume of blood (Hint: Blood leaving the right side of the heart eventually enters the left side of the heart, and vice versa.).

   ____

   ____

   ____

3. Sketch the transverse view of the ventricles in the space provided. Be sure to include and label the right and left ventricles in your drawing.

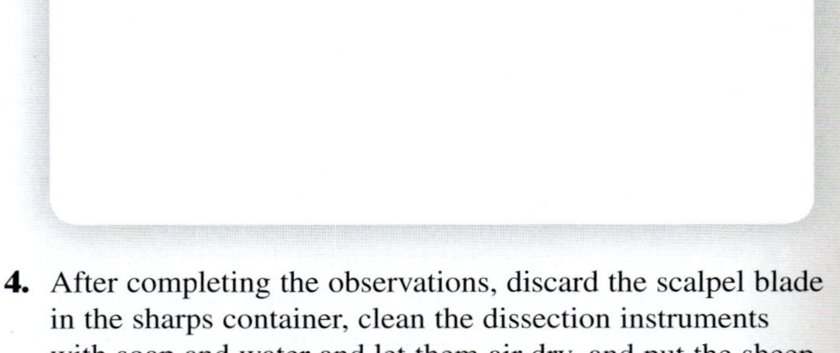

4. After completing the observations, discard the scalpel blade in the sharps container, clean the dissection instruments with soap and water and let them air dry, and put the sheep heart back into the container it came in—or dispose of it according to the instructor's directions.

PHYSIOLOGY

Electrical Conduction Within the Heart

Contraction of cardiac muscle results from a series of electrical potential changes (depolarization) that move rapidly through the heart. As with skeletal muscle cells, depolarization of cardiac muscle cells is what stimulates contraction. Cardiac muscle is unique in that the heart is stimulated to beat by *intrinsic* mechanisms. That is, cardiac muscle will continue to contract even when the nerves supplying the heart have been severed. The **intrinsic conduction system** ensures that cardiac muscle within the heart depolarizes in a sequential manner (atria to ventricles) so the heart can beat as a coordinated unit. The components of the conduction system of the heart are outlined in **table 21.5.**

The sinoatrial (SA) node, the atrioventricular (AV) node, the atrioventricular (AV) bundle, the right and left bundles, and the Purkinje fibers are all capable of stimulating the heart to contract. However, because the SA node has the fastest intrinsic rate, it controls the other parts of the conduction system. For this reason the SA node is called the heart's "pacemaker."

Stimulation of the heart is initiated at the SA node. As the pacemaker cells within the SA node spontaneously depolarize, an action potential is initiated and then transmitted through the atria via gap junctions until it reaches the AV node. The action potential is delayed briefly at the AV node and then transmitted sequentially to the **AV bundle** (*bundle of His*), the **right** and **left bundle** branches, the Purkinje fibers, and then via gap junctions through the walls of the ventricles **(figure 21.16).** A region of electrically isolated connective tissue separates the atria from the ventricles. Thus, the only mechanism for the propagation of depolarization within the atrial walls to reach the ventricles is via the AV node and AV bundle. Consequently, the atria are stimulated to contract prior to contraction of the ventricles. Note that it is depolarization that triggers the initiation of muscle contraction—first in the atria, and then in the ventricles.

Damage to any part of the conduction system of the heart may result in an irregular heart rate and other irregularities in cardiac function. In the event that the SA node is unable to set the pace of the heart, the AV node or even the Purkinje fibers may take over responsibility of setting the heart rate. While cells in other parts of the conduction system are able to depolarize spontaneously, the rhythm of each is much slower than the pace of the SA node; therefore, the heart rate decreases. Over time, if the AV node also becomes nonfunctional, the heart may be unable to deliver an adequate amount of blood to the body. An artificial pacemaker may be implanted to regulate a patient's heart rate.

The electrical changes associated with stimulation of the atria and ventricles can be detected on the body's surface using a diagnostic tool called an **electrocardiogram (ECG).** ECG electrodes (called "leads") are placed on the surface of the body in locations where it is easiest to pick up electrical signals from the heart. In a clinical setting, 12 leads

Table 21.5 Components of the Conduction System of the Heart

Conduction System Component	Location
Sinoatrial (SA) Node	Within the right atrium, inferior to the superior vena cava opening
Atrioventricular (AV) Node	Base of the right atrium
AV Bundle (Bundle of His)	Interventricular septum
Right and Left Bundles	Interventricular septum
Purkinje Fibers	Ventricular walls

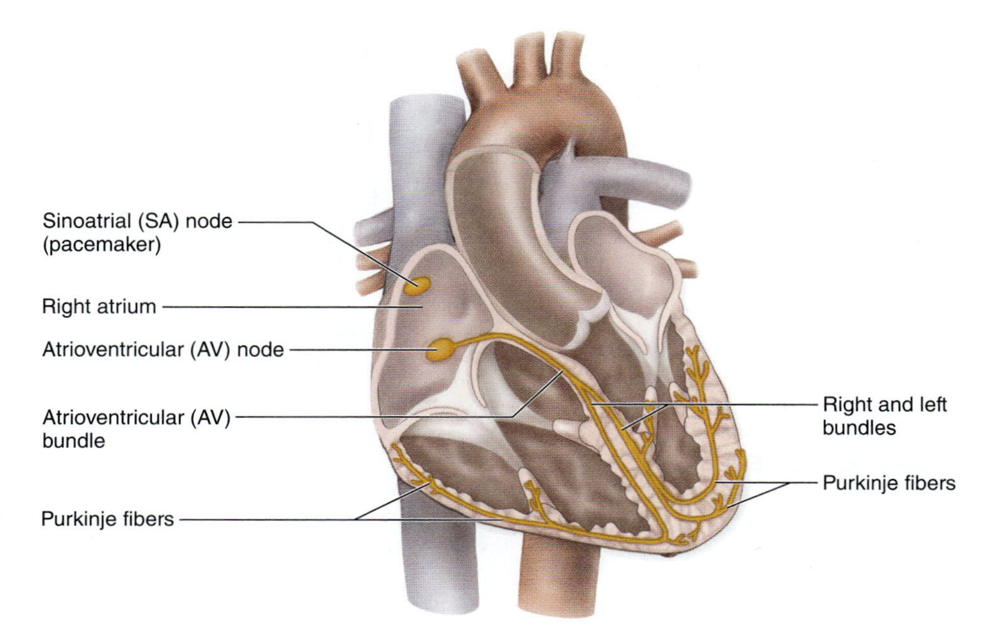

Figure 21.16 Conduction System of the Heart.

are used to measure an ECG (thus it is called a "12-lead ECG"). This laboratory exercise uses only three electrodes (left wrist, right wrist, and left ankle) to record the electrical activity associated with the heart. This arrangement roughly places the heart at the center of an equilateral triangle (*Einthoven's triangle*), where each side is the same length **(figure 21.17).** The electrodes on each wrist and the left leg represent the vertices of the triangle. Voltages are measured between the electrodes to record the ECG. Lead I corresponds to the voltage difference between the left arm and the right arm; lead II corresponds to the voltage difference between the right arm and the left leg; lead III corresponds to the voltage difference between the left arm and the left leg. **Einthoven's Law** states that the sum of two leads results in the magnitude of the third lead. Therefore, the third lead can be calculated as the sum of two measured leads.

A normal ECG measures an overall change in voltage across the atria and ventricles and consists of three deflection waves. Each deflection wave corresponds to a distinct electrical event within the heart **(figure 21.18).** The **P wave** corresponds to atrial depolarization, the **QRS complex** corresponds to ventricular depolarization, and the **T wave** corresponds to ventricular repolarization. Atrial repolarization does occur; however, the ECG deflection wave associated with it is masked by the large deflection caused by ventricular depolarization that occurs at approximately the same time. Figure 21.18 also depicts several other measures that reflect heart function: the P-Q segment, S-T segment, P-R interval, and Q-T interval. The **P-Q segment** corresponds to the atrial plateau that occurs on the sarcolemma of cardiac muscle cells when the muscle cells within the atria are contracting. The **S-T segment** corresponds to the ventricular plateau when the cardiac muscle cells within the ventricles are contracting. These segments can be measured directly and are physiologically relevant, as will be explored in the following exercises. The **P-R interval** is the time that it takes for the depolarization to spread from the SA node through the AV node, which typically ranges from 0.12 to 0.20 second. The **Q-T interval** is the time required for the ventricles to depolarize and repolarize, which typically ranges from 0.2 to 0.4 second. An ECG can be used to detect heart rates that are faster (*tachycardia*: >100 beats/min) or slower (*bradychardia*: <60 beats/min) than normal; the influence of drugs on cardiac function; the existence of ectopic pacemakers (ectopic = outside of the normal conduction system). Thus, the ECG is a valuable diagnostic tool.

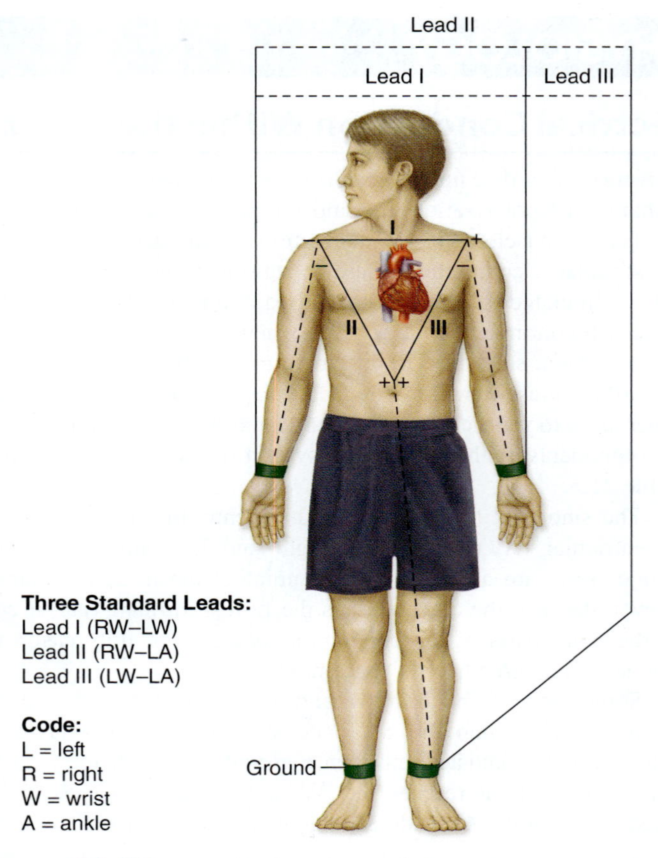

Figure 21.17 Standard Limb Leads for ECG. The standard limb leads for an ECG roughly place the heart at the center of Einthoven's triangle. Each lead (I–III) corresponds to a voltage difference between two measurement points.

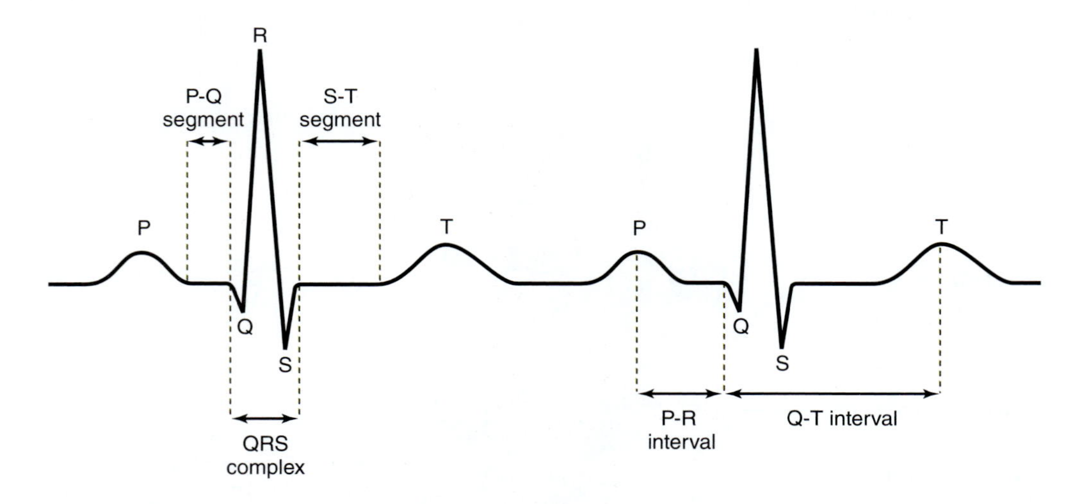

Figure 21.18 Normal Components of an ECG.

EXERCISE 21.9

ELECTROCARDIOGRAPHY USING STANDARD ECG APPARATUS

Obtain the Following:

- **alcohol swabs**
- **electrodes**
- **electrode gel**
- **lead selector switch**
- **electrocardiograph (or alternative ECG recording device)**

1. Choose a person to be the subject. Prepare to place the electrodes by cleaning the skin where the electrodes will be placed. To do this, rub the indicated area vigorously with an alcohol swab. First clean the skin on the anterior surface of the subject's forearms approximately 2–3 inches proximal to the left and right wrists. Next, clean the skin approximately 2–3 inches superior to the medial malleoli on both the left and right legs. Note that skin free of hair will make the best contact with the electrode.
2. If not using electrodes containing gel as a component of the electrode, place a small quantity of electrode gel on the center of each electrode. Place one electrode firmly on the skin in each of the following locations (that were cleaned in step 1): right arm, left arm, right leg, and left leg. The right leg will serve as the ground.
3. Observe the leads of the ECG machine. A lead is a wire that has a clip on the end that attaches to an electrode. The leads have labels that identify which limb each must be attached to. Attach the leads to the electrodes. RA = Right Arm, LA = Left Arm, RL = Right Leg, LL = Left Leg
4. Turn on the electrocardiograph. Set the paper speed to 25 mm/second, the standard for recording an ECG. At this speed, each millimeter of paper corresponds to a time interval of 1 second. Set the gain of the system to 10 mm/mV. Note: Depending on the machine in use, the gain may need to be set lower or higher. Ask the instructor if in doubt about where the gain should be set.
5. Have the subject lie supine or sit in a comfortable position.
6. Record the ECG for 1 minute. To begin recording, press the "record" button. If the machine is working properly, the ECG paper should start coming out of the machine and three ECG tracings (one for each lead: I–III) should be seen on the paper. When finished recording the ECG, write the subject's name and "Resting ECG" on the indicator paper.
7. Remove the leads from the electrodes (do *not* remove the electrodes at this time). Have the subject exercise vigorously for at least 3 minutes.
8. Have the subject return to the seated or supine position immediately post-exercise. Reconnect the leads to the electrodes, taking care to ensure the leads are connected to the proper locations. Record a post-exercise ECG for 3 minutes. When finished recording the ECG, write the subject's name and "Post-Exercise ECG" on the indicator paper.
9. Calculate the subject's heart rate at rest and post-exercise using the following procedure, which assumes the paper speed is set to 25 mm/second:
    a. Start as close as possible to the peak of one QRS complex and count out a 6-second interval on the ECG paper (**figure 21.19**). With a paper speed of 25 mm/second, one second corresponds to the length of five large squares (each large square = 0.2 second; each small square = 0.04 second).

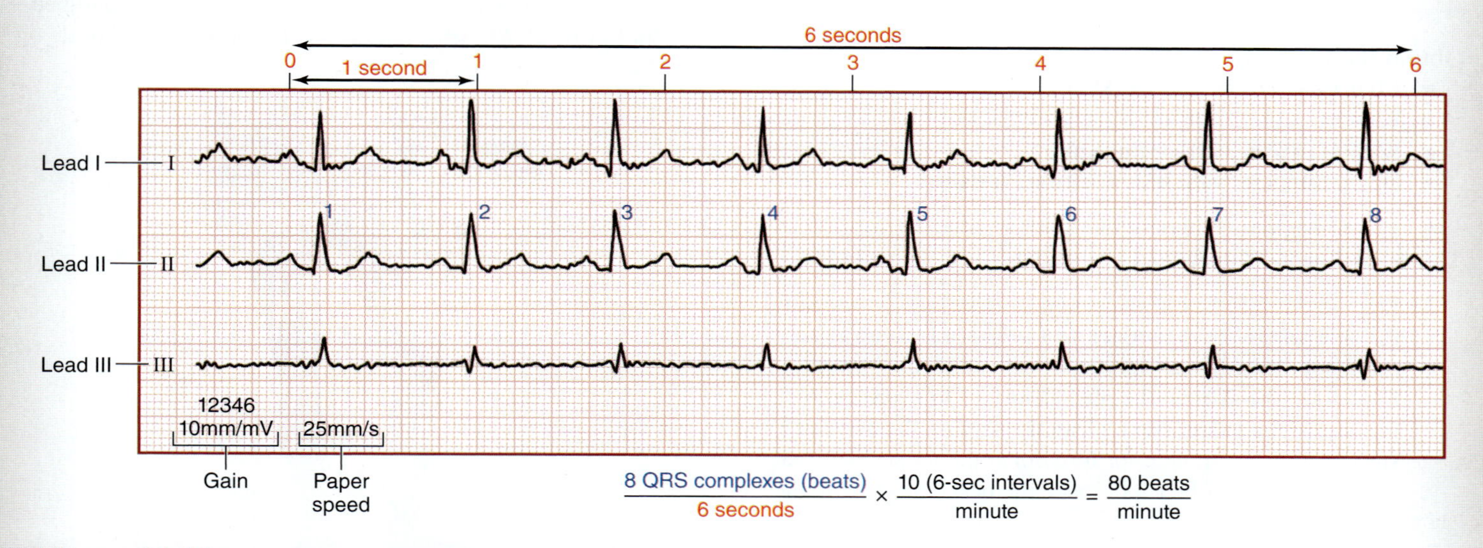

Figure 21.19 Interpreting an ECG Tracing. This ECG tracing was recorded at a paper speed of 25 mm/second. Red numbers indicate 1-second time intervals. Blue numbers indicate number of QRS complexes. The calculation demonstrates how to calculate heart rate in beats per minute using a 6-second time interval.

(continued on next page)

(continued from previous page)

b. Measure the number of QRS complexes for the Lead II tracing within the 6-second interval that was counted out in step a.

c. Multiply the number of QRS complexes counted in step b by 10 to get the heart rate (6 × 10 = 60 seconds = 1 minute)

d. Record the heart rate values in **table 21.6**.

10. Measure other important features of the ECG, including the durations of the QRS complex, Q-T interval, and P-T interval (see figure 21.18). Record these values in table 21.6.

11. Make note of pertinent observations comparing resting versus post-exercise ECG tracings here:

____

____

Table 21.6 ECG Component Data

ECG Component	Duration at Rest (sec)	Duration Post-Exercise (sec)
QRS Complex		
Q-T Interval		
P-T Interval		
Heart Rate		

EXERCISE 21.10

BIOPAC LESSON 5: ELECTROCARDIOGRAPHY I

This experiment introduces students to the procedure for measuring the electrical activity of the heart using a three-lead electrocardiogram (ECG). An ECG will be recorded for the subject under three conditions: lying down, sitting up, and after a bout of exercise. ECG parameters will be measured for each of the experimental conditions.

Obtain the Following:

- **BIOPAC electrode lead set (SS2L)**
- **BIOPAC general purpose electrodes (EL503)**
- **BIOPAC MP36/35 recording unit**
- **laptop computer with BSL 4 software installed**

1. Before beginning, choose one student to be the experimental subject and one student to be the recorder.

2. Prepare the subject for placement of electrodes by cleaning the skin where the electrodes will be placed. To do this, rub the indicated area vigorously with an alcohol swab. First clean the skin on the medial surface of each leg, just above the ankle. Next, clean the skin on the anterior surface of the right anterior forearm at the wrist. Note that skin that is free of hair will make the best contact with the electrodes.

3. Place one electrode (EL503) firmly on the skin in each of the locations that was cleaned in step 2. Wait approximately 5 minutes before attaching the leads (step 5) to give the electrodes a chance to affix firmly to the skin. Proceed with step 4 while waiting for the electrodes to affix firmly to the skin.

4. Turn the computer on. Plug the electrode lead set (SS2L) into channel 2 of the BIOPAC MP36/35 unit. Turn on the BIOPAC unit.

5. Have the subject lie down in a comfortable position. Attach the electrode leads (SS2L) to the electrodes that were placed in step 2 following the color scheme outlined in **figure 21.20.** Attach the WHITE lead to the electrode on

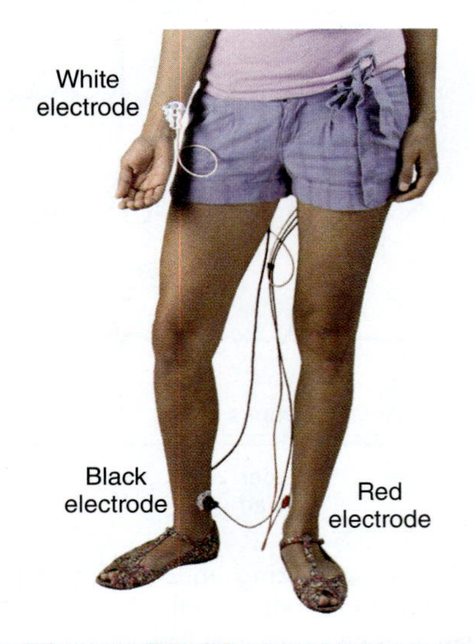

Figure 21.20 Lead Placement.
Courtesy of and © BIOPAC Systems, Inc.

the right forearm. Attach the BLACK lead to the electrode on the medial surface of the subject's right leg. Attach the RED lead to the electrode on the medial surface of the subject's left leg.

6. Clip the unit containing the lead cables to the subject's waistband. Be certain that the cables coming from the leads are not pulling on the electrodes and that clothing is not rubbing on the electrodes.
7. Start the Biopac Student Lab Program and choose "Lesson 5" (L05-ECG-I).
8. Type in a file name using the subject's first name and last initial. Click "OK."
9. Click on the "Calibrate" button in the upper left corner. Wait for the calibration procedure to finish. The calibration procedure stops automatically after about 8 seconds.
10. If the calibration readings look like an ECG **(figure 21.21)** and look stable, proceed to step 11. If the calibration readings do not look right, click "Redo Calibration" before proceeding to step 11.

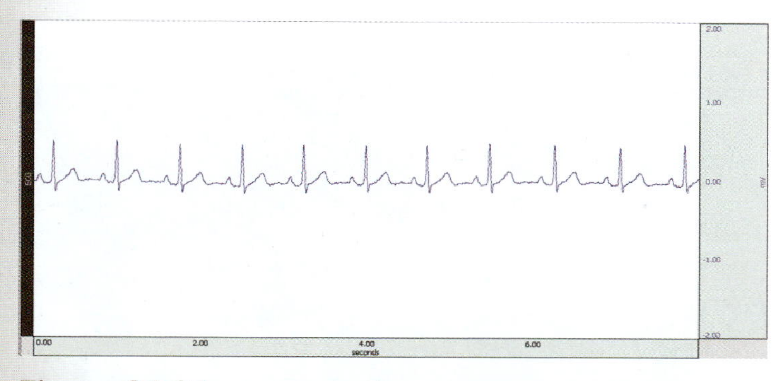

Figure 21.21 Correct Calibration Reading. The reading should look like a normal ECG. Courtesy of and © BIOPAC Systems, Inc.

11. **OBTAIN RESTING VALUES:** Ensure the subject is lying down and resting comfortably before starting to record data. Click on "Record" to start data collection. Record data for 20 seconds to obtain resting values. Next, click "Suspend" to stop the device from continuing to record the ECG. If the subject made any large movements, talked, or laughed during the recording (activities that interfere with collection of ECG data), click on "Redo" and record for another 20 seconds to obtain resting values. Otherwise, click "Done."
12. **OBTAIN VALUES FOR SITTING UP WITH NORMAL BREATHING:** Have the subject sit upright and breathe normally. When the subject is ready, click on "Record" to start data collection. Record data for 20 seconds to obtain values for the "sitting up" condition. Next, click "Suspend" to stop the device from continuing to record the ECG. If the subject made any large movements, talked, or laughed during the recording (activities that interfere with collection of ECG data), click on "Redo" and record for another 20 seconds. Otherwise, click "Done."
13. **OBTAIN POST-EXERCISE VALUES:** Unhook the leads from the ECG electrodes and unclip the unit containing the lead cables from the subject's waistband to "release" the subject. Have the subject exercise vigorously for at least 5 minutes (e.g., running up and down stairs, doing pushups or jumping jacks). When the subject has finished exercising, have the subject sit down next to the recording apparatus. Immediately reattach the leads to the electrodes, taking care to connect the leads to the proper locations (figure 21.20). Click on "Record" to start data collection. Record data for 60 seconds to obtain post-exercise values. Next, click "Suspend" to stop the device from continuing to record the ECG. If the subject made any large movements, talked, or laughed during the recording (activities that interfere with collection of ECG data), click on "Redo" and record for another 60 seconds.
14. When finished, click "Done." Remove the ECG leads and electrodes from the subject. The electrodes can be thrown away in the garbage and should not be reused. To record data for another subject, click on "Record data from another subject." When finished collecting data for all subjects, proceed to step 15: ANALYZE THE DATA.
15. **ANALYZE THE DATA:** If analyzing data that was just collected, click on "Analyze current data file." If opening data that was collected previously, click on "Review saved data." To prepare to analyze the data, do the following:
    - **a.** Use the zoom tool to select approximately 4 seconds worth of ECG waveforms.
    - **b.** Click on the "Display" menu at the top of the screen and select "Autoscale Waveforms."
    - **c.** Set up the measurement boxes at the top of the screen using the drop-down menus. Set "CH 40" to "Value." Set "CH 1" to "Delta T." Set "CH 1" to "P-P." Set "CH 1" to "BPM."
    - **d.** When using the I-beam tool, the indicator box for "Delta T" will show the time elapsed during the selected region in seconds; the indicator box for "P-P" will show the peak to peak measurement; the indicator box for "BPM" will show the heart rate in beats per minute (bpm). The computer calculates a heart rate using the time between two R-waves of adjacent ECG waveforms.
16. Scroll along the ECG tracing for the subject until the data from the first segment (20 seconds) of ECG recording is in the middle of the screen. Using the I-beam tool, measure the following parameters of the ECG and record them in **table 21.7:** Heart Rate; P wave; P-R Interval; P-R Segment; QRS Complex: S-T Segment; Q-T Interval; T Wave.
17. Scroll along the ECG tracing for the subject until the data from the second segment (20 seconds) of ECG recording is in the middle of the screen. Using the I-beam tool, measure the following parameters of the ECG and record them in table 21.7: Heart Rate; P wave; P-R Interval; P-R Segment; QRS Complex: S-T Segment; Q-T Interval; T Wave.

(continued on next page)

(continued from previous page)

18. Scroll along the ECG tracing for the subject until the data from the third segment (60 seconds) of ECG recording is in the middle of the screen. Using the I-beam tool, measure the following parameters of the ECG and record them in table 21.7: Heart Rate; P wave; P-R Interval; P-R Segment; QRS Complex: S-T Segment; Q-T Interval; T Wave.

19. Make a note of any pertinent observations here:

____

____

____

20. When data analysis is complete, click "Save" or "Print," then click "Quit" to close the program. Before leaving the workstation, throw away any garbage, clean the work area, and return the BIOPAC unit and computer to the condition in which they were found.

Table 21.7 Data for BIOPAC ECG Exercise

ECG Parameter	Resting ECG	Sitting Up ECG	Post-Exercise ECG	Expected Values (seconds)
Heart Rate				60–80 bpm resting; >100 bpm post-exercise
P Wave				0.06–0.11
P-R Interval				0.12–0.20
P-R Segment				0.08
QRS Complex				<0.12
S-T Segment				0.12
Q-T Interval				0.31–0.41
T Wave				0.16

EXERCISE 21.11 Ph.I.L.S.

Ph.I.L.S. LESSON 22: REFRACTORY PERIOD OF THE HEART

Cardiac muscle tissue must contract in a coordinated way to effectively pump the blood to the lungs and systemic tissues. Between each contraction, cardiac muscle must relax to allow time for the ventricles to fill with blood. Cardiac muscle, like skeletal muscle, has a period when it is insensitive to an electrical stimulus, known as the **refractory period.** The refractory period within cardiac muscle tissue is extended in comparison to skeletal muscle tissue. This extended refractory period results when cardiac muscle cells remain in the depolarized state because both voltage-gated K^+ channels and voltage-gated Ca^{2+} channels are open. This extended depolarized state is called the **plateau.** Because repolarization does not immediately follow depolarization, the mechanical events of cardiac muscle contraction and relaxation have time to occur prior to restimulation of cardiac muscle cells. (This prevents a sustained contraction [tetanus] from occuring.)

The purpose of this laboratory exercise is to measure the duration of the refractory period in cardiac muscle tissue. The exercise involves delivering brief electrical shocks to exposed frog cardiac muscle tissue and observing the contractions that are, or are not, evoked in the muscle as a result of stimulation. Shocks will be delivered at various times throughout the cardiac cycle to observe when evoked contractions are and are not possible.

Prior to beginning the experiment, familiarize yourself with the following concepts (use the textbook as a reference):

- Voltage-gated channels of cardiac cells
- Cardiac action potentials
- Refractory period

1. Open Ph.I.L.S Lesson 22: Refractory Period of the Heart.

2. Read the objectives, introduction, and wet lab sections. Then, take the pre-lab quiz. The laboratory exercise will open when the pre-lab quiz has been completed (**figure 21.22**).

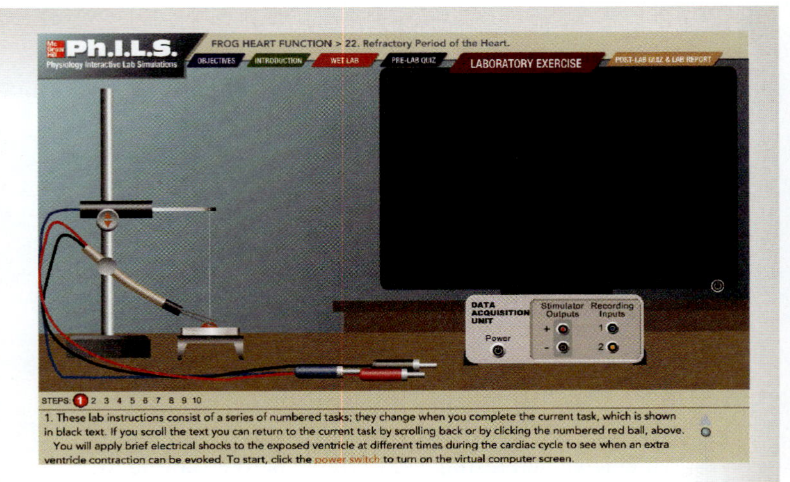

Figure 21.22 Opening Screen for the Laboratory Exercise on Refractory Period of the Heart.

3. Click the "power" button of the computer screen. Then, click the "power" button of the data acquisition unit (DAQ).

4. Click and drag the blue plug to recording input 1 on the DAQ.

5. Click and drag the red plug to the stimulator output "+" position on the DAQ. Click and drag the black plug to the stimulator output "–" position on the DAQ.

6. Click the "start" button on the control panel of the virtual monitor. A blue trace should appear, which represents ventricular contractions. The red trace indicates when a shock has been applied to the tissue. Click the orange "up" arrow on the clamp (left side of the screen) to increase the heart response on the screen.

7. Click the "shock" button to stimulate the heart approximately every two seconds. Click the "stop" button to cease stimulation. To further increase the amplitude of

the ventricular contractions, click the "up" arrow on the force transducer.

8. Once the recording has started, a normal contractile response will appear. Pay special attention after stimulating the heart because changes in the blue tracing will indicate the effect of the stimuli.

9. Open the post-lab quiz & lab report by clicking on it. Answer the post-lab questions that appear on the computer screen. Click "Print Lab" to print a hard copy of the report or click "Save PDF" to save an electronic copy of the report.

10. Make a note of any pertinent observations here:

____

____

____

INTEGRATE

CLINICAL VIEW
ECG

Clinicians use ECGs to diagnose abnormalities with the conduction system of the heart. While this laboratory exercise involves using a three-lead system, a more sophisticated 12-lead system is typically used in a clinic or hospital setting. Trained specialists evaluate the ECG trace, looking for the key components described in this exercise. To illustrate the possible abnormalities that can be detected, a series of abnormal ECG traces are depicted below **(figure 21.23).** The conditions illustrated include the following: atrial fibrillation (figure 21.23*a*), ventricular fibrillation (figure 21.23*b*), and premature ventricular contraction (figure 21.23*c*). **Fibrillation** involves chaotic depolarization of cardiac muscle fibers, which results in uncoordinated contraction of either the atria (atrial fibrillation) or the ventricles (ventricular fibrillation). A patient suffering from atrial fibrillation will exhibit many QRS complexes, but no clear P waves, because continuous and chaotic depolarization of the atria engages the conduction system of the heart. Observe the tracing in figure 21.23*a*. Are QRS complexes visible? (yes/no) Is each QRS complex followed by a T wave? (yes/no) Do all of the QRS complexes look the same? (yes/no) Look for regularly spaced P waves between the QRS complexes. Are they visible? (yes/no) In tracing (a) for example, the P waves are very irregular, indicating **atrial fibrillation.** Given this information, why do you think the QRS complexes also look a little bit irregular in spacing and amplitude?

____

Next, observe the tracing in figure 21.23*b*. Are QRS complexes visible? (yes/no) Is each QRS complex followed by a T wave? (yes/no) Next, look for P waves. Are any P waves visible? (yes/no)

Tracing (b) is a case of **ventricular fibrillation.** In ventricular fibrillation there are no clear P, QRS, or T waves due to aberrant electrical activity. While atrial fibrillation may lead to incomplete filling of the ventricles, ventricular fibrillation leads to inefficient ejection of blood into the systemic circulation. Without intervention, the latter results in certain death.

Observe the tracing in figure 21.23*c*. Are QRS complexes visible? (yes/no) Is each QRS complex followed by a T wave? (yes/no) Is each QRS complex preceded by a P wave? (yes/no)

Tracing (c) is a case of premature ventricular contraction (PVC). PVCs are characterized by QRS complexes that are not always preceded by a P wave. While not necessarily as detrimental as ventricular fibrillation, PVCs can also lead to a decreased cardiac output.

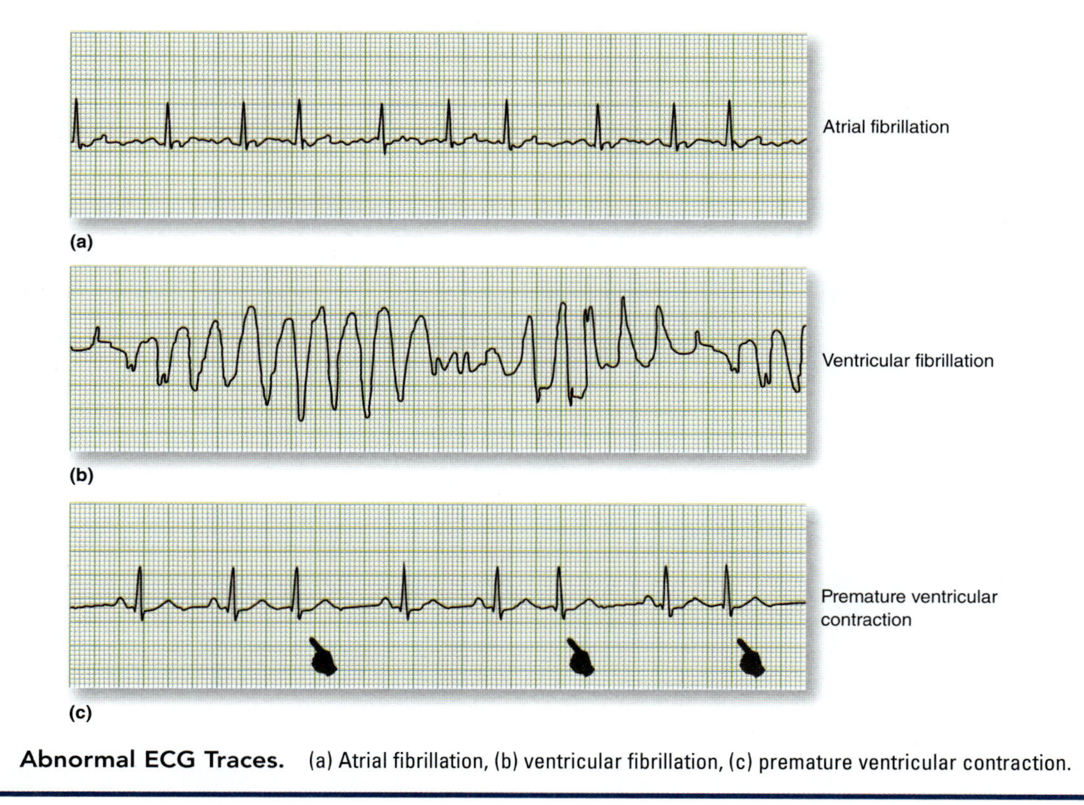

Figure 21.23 **Abnormal ECG Traces.** (a) Atrial fibrillation, (b) ventricular fibrillation, (c) premature ventricular contraction.

EXERCISE 21.12 Ph.I.L.S.

Ph.I.L.S. LESSON 29: ECG AND HEART BLOCK

The purpose of this laboratory exercise is to record one normal ECG and four abnormal ECGs in subjects with varying degrees of AV block (or "heart block"). The remainder of the exercise will explore the underlying physiological mechanisms that are responsible in each case of heart block.

Prior to beginning the experiment, familiarize yourself with the following concepts (use the main textbook as a reference):

- Electrical activity of the heart
- ECG
- Conduction block

1. Open Ph.I.L.S. Lesson 29: ECG and Heart Block.
2. Read the objectives, introduction, and wet lab sections. Then, take the pre-lab quiz. The laboratory exercise will open when the pre-lab quiz has been completed **(figure 21.24).**
3. Click the "power" button on the computer screen. Then, click the "power" button on the DAQ.

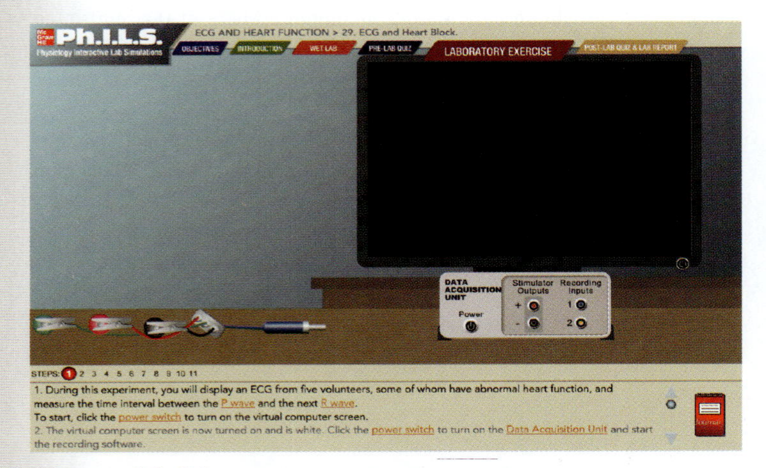

Figure 21.24 Opening Screen for the Laboratory Exercise on ECG and Heart Block.

4. Connect the ECG electrodes to the DAQ by clicking and dragging the blue plug to recording input 1.
5. Connect the ECG leads to the patient by clicking and dragging the black electrode to the left wrist, the red electrode to the right wrist, and the green electrode to the left ankle.
6. To record and measure the ECG, click the "start" button on the control panel.
7. Record at least eight complete cardiac cycles and then click the "stop" button on the control panel.
8. Click the "arrows" at the bottom of the control panel screen to scroll through the record. Center one QRS complex in the middle of the screen. Measure the P–R interval by moving one cursor to the beginning of a P wave. Move the other cursor to the peak of the R wave in the same cycle. The time interval between the P wave and the R wave is now displayed in the data window on the control panel. Enter the data in the journal by clicking the red "journal" button in the lower-right corner of the screen. Then enter the data in **table 21.8**.
9. Select a volunteer by moving the mouse over the patient list on the control panel. Click on one of the names. Repeat steps 5–8 for the four remaining patients and enter the data in table 21.8. Watch for abnormalities in the regular pattern. Measure each P-R interval and record them in table 21.8.
10. Open the post-lab quiz & lab report by clicking on it. Answer the post-lab questions that appear on the computer screen. Click "Print Lab" to print a hard copy of the report or click "Save PDF" to save an electronic copy of the report.
11. Make note of any pertinent observations here:

____

____

____

Table 21.8 P-R Interval Data

Patient	P-R Interval (normal P-R Interval = 0.12–20 seconds)		Patient	P-R Interval (normal P-R Interval = 0.12–20 seconds)	
Andrew	1.	5.	***David***	1.	5.
	2.	6.		2.	6.
	3.	7.		3.	7.
	4.	8.		4.	8.
Brianna	1.	5.	***Emily***	1.	5.
	2.	6.		2.	6.
	3.	7.		3.	7.
	4.	8.		4.	8.
Charlie	1.	5.			
	2.	6.			
	3.	7.			
	4.	8.			

EXERCISE 21.13 Ph.I.L.S.

Ph.I.L.S. LESSON 30: ABNORMAL ECGs

The purpose of this lab is to record normal (1) and abnormal (3) ECG traces in four virtual subjects. You will determine the physiological mechanism responsible for creating the ECG recording for each subject.

Prior to beginning the experiment, familiarize yourself with the following concepts (use the main textbook as a reference):

- Normal ECG recordings
- Causes of abnormal ECG recordings
- Heart block

State a hypothesis regarding the effect that damaged tissue has on the conduction system of the heart and the resulting contraction of cardiac muscle tissue. ______________________

____

____

____

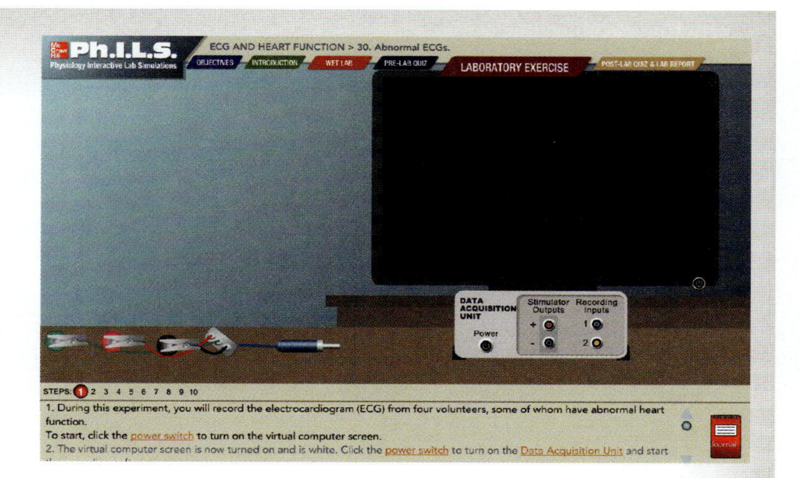

Figure 21.25 Opening Screen for the Laboratory Exercise on Abnormal ECGs.

1. Open Ph.I.L.S. Lesson 30: Abnormal ECGs.
2. Read the objectives, introduction, and wet lab sections. Then, take the pre-lab quiz. The laboratory exercise will open when the pre-lab quiz has been completed **(figure 21.25).**
3. Click the "power" button on the computer screen. Then click the "power" button on the data acquisition unit (DAQ).
4. Click and drag the blue plug to recording input 1.
5. Connect the ECG leads to the subject by dragging the black electrode to her left wrist, the red electrode to her right wrist, and the green electrode to her left ankle.
6. Click the "start" button on the control panel to begin recording the subject's ECG. Record a minimum of eight complete cycles.
7. Record observations for the first subject (Amy) in **table 21.9.**
8. Select another subject by clicking on the patient list in the control panel. Repeat steps 6 and 7 for each remaining subject.
9. Open the post-lab quiz & lab report by clicking on it. Answer the post-lab questions that appear on the computer screen. Click "Print Lab" to print a hard copy of the report or click "Save PDF" to save an electronic copy of the report.
10. Make note of any pertinent observations here:

____

____

____

Table 21.9 ECG Recordings for Four Virtual Subjects

Patient	ECG Observations	ECG Pattern (sketch the pattern in the spaces below)
Amy	P	
	QRS	
	T	
Brian	P	
	QRS	
	T	
Chris	P	
	QRS	
	T	
Deb	P	
	QRS	
	T	

Cardiac Cycle and Heart Sounds

The **cardiac cycle** refers to the events that occur in one heart beat. One complete cardiac cycle occurs when both atria and ventricles contract and then relax. The two main phases of the cardiac cycle are **diastole,** or relaxation, and **systole,** or contraction. These phases may be further subdivided as follows: (1) diastole = isovolumetric ventricular relaxation + ventricular filling, and (2) systole = isovolumetric ventricular contraction + ejection. A typical description of the cardiac cycle includes both atrial and ventricular events that occur on *the left side of the heart* only because the pressures on that side of the heart are much greater than those on the right. The volume of blood in the ventricles at the end of filling is the **end diastolic volume (EDV),** and the volume of blood remaining in the ventricles following contraction is the **end systolic volume (ESV).** The volume of blood ejected with each beat is the **stroke volume** (EDV – ESV). **Cardiac output (CO)** is simply the stroke volume (SV) multiplied by the heart rate (HR).

Blood is transported through the heart along pressure gradients, from areas of higher pressure to areas of lower pressure. **Atrioventricular (AV)** and **semilunar valves** ensure one-way flow of blood through the heart. Once pressure in the atria exceeds the pressure in the ventricles, the AV valves open and blood flows into the ventricles. Contraction of the right and left atria pushes any remaining blood into the ventricles. At this time the ventricles are relaxed and fill with blood. During ventricular systole, the pressure rises in the ventricles. When pressure in the ventricles exceeds pressure in the atria, the AV valves are forced closed. Once pressure in the ventricles exceeds pressure in the pulmonary trunk and aorta, blood pushes open the cusps of the pulmonary and aortic semilunar valves and blood is ejected from the heart into the pulmonary and systemic circuits.

The duration of a cardiac cycle varies among individuals, and also varies throughout one's lifetime. Typically, a cardiac cycle lasts 0.7–0.8 second, which corresponds to an average heart rate of 75–85 beats per minute. Heart sounds (*lub dupp*) can be associated with the cardiac cycle. Turbulent blood flow that occurs as the AV valves close creates the first heart sound (*lub*), whereas turbulent blood flow that occurs as the semilunar valves close creates the second heart sound (*dupp*). **Figure 21.26** shows the location of each of the heart valves with respect to thoracic wall surface anatomy. The yellow dots indicate the areas where each heart sound is best heard at the surface of the body. The semilunar valves are best heard in the second intercostal spaces on each side of the sternum, whereas the AV valves are best heard in the fifth intercostal spaces. Listen for the right AV valve in the fifth intercostal space, just to the right of the sternum. Listen to the left AV valve in the fifth intercostal space at the midclavicular line (see figure 12.9, p. 302). Auscultation is listening to the internal body sounds, such as the heart sounds.

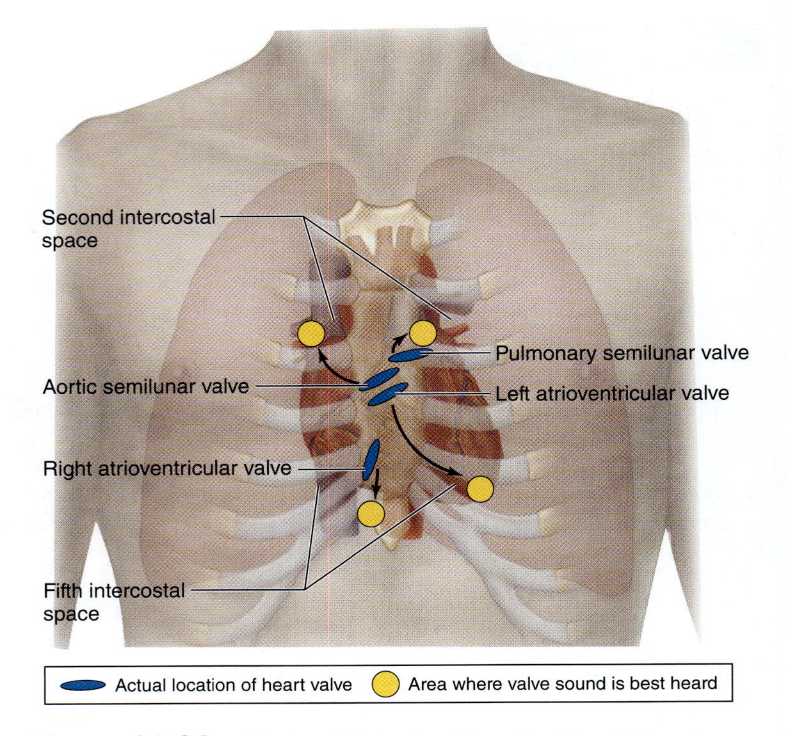

Figure 21.26 **Optimal Listening Sites for Heart Sounds.**

EXERCISE 21.14

AUSCULTATION OF HEART SOUNDS

1. Obtain alcohol swabs and a stethoscope.
2. Clean the earpieces and diaphragm of the stethoscope with the alcohol swabs and allow them to dry.
3. Place the earpieces of the stethoscope into the ears of the person doing the auscultating (external auditory canal), then press the diaphragm of the stethoscope to the subject's skin directly over the fifth intercostal space at the midclavicular line to the left of the subject's sternum (see figure 21.24). This space overlies the apex of the heart and is the optimal placement for auscultation of the first heart sound. Listen for a *lub* sound.
4. Next, press the diaphragm to the subject's skin directly over the second intercostal space to the left of the subject's sternum. (see figure 21.24). This is the optimal placement for auscultation of the second heart sound. Listen for a *dupp* sound.
5. The person doing the auscultating should inhale and exhale slowly and deeply while holding the diaphragm in each location, and should listen carefully for any differences in the heart sounds that occur as the subject breathes.
6. Make a note of any pertinent observations here:

____

____

____

EXERCISE 21.15 Ph.I.L.S.

Ph.I.L.S. LESSON 26: THE MEANING OF HEART SOUNDS

In this laboratory exercise, a virtual data acquisition unit (DAQ) will be used to record an ECG while listening to heart sounds in a virtual subject. The mechanical events of the heart, which result in audible sounds, will be correlated with electrical events as viewed by an ECG trace.

Prior to beginning the experiment, familiarize yourself with the following concepts (use the textbook as a reference):

- Cardiac cycle
- ECG
- Heart sounds

1. Open Ph.I.L.S. Lesson 26: The Meaning of Heart Sounds.
2. Read the objectives, introduction, and wet lab sections. View the videos that are indicated in red in the wet lab prior to beginning the experiment. Then take the pre-lab quiz. The laboratory exercise will open when the pre-lab quiz has been completed **(figure 21.27).**
3. Check that heart sounds generated by Ph.I.L.S. are audible on the computer by clicking the "sound test" button in the middle of the screen.

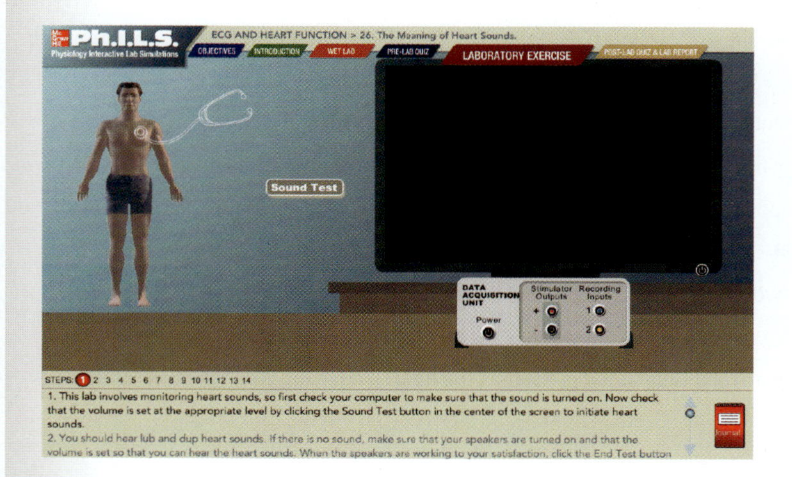

Figure 21.27 **Opening Screen for the Laboratory Exercise on ECG and Heart Function: The Meaning of Heart Sounds.**

4. Click the "power" button on the virtual computer screen. Then click the "power" button on the data acquisition unit (DAQ).
5. To begin the recording, connect the electrodes to the DAQ by inserting the orange plug into input 2 of the DAQ.
6. Click on the "practice" button in the middle of the screen to practice keeping time with the heart sounds. Once the heart sounds are audible, click on the red button to make a mark on the orange trace. Push the button when the first heart sound is audible (lub) and release the button when the second heart sound is audible (dupp). Click "end test" once the practice has been completed.
7. Insert the blue plug into input 1 of the DAQ. Attach electrodes to the subject (red electrode to the right wrist; black electrode to the left wrist; and green electrode to the left ankle).
8. Click "start" in the control panel. On the control panel the ECG is recorded (blue line) above the heart sounds (orange line).
9. As the line tracing scrolls across the control panel screen, click the mouse on the red recording button to mark the heart sounds. Record four to six cardiac cycles. Click "stop" in the control panel.
10. Use the arrows at the bottom of the monitor screen to scroll through the trace. Determine which heart sounds correspond to the QRS complex and the T wave. Print the graph of the ECG and heart sounds by clicking the "P" in the control panel.
11. Open the post-lab quiz & lab report by clicking on it. Answer the post-lab questions that appear on the computer screen. Click "Print Lab" to print a hard copy of the report or click "Save PDF" to save an electronic copy of the report.
12. Make a note of any pertinent observations here:

____

____

____

Chapter 21: The Cardiovascular System: The Heart

Name: ______________________

Date: __________ Section: __________

POST-LABORATORY WORKSHEET

The ❶ corresponds to the Learning Objective(s) listed in the chapter opener outline.

Do You Know the Basics?

Exercise 21.1: Cardiac Muscle

1. Cardiac muscle is ____________ (voluntary/involuntary), whereas skeletal muscle is ______________ (voluntary/involuntary). ❶

Exercise 21.2: Layers of the HeartWall

2. The endothelium, which lines the heart and blood vessels, is composed of simple squamous epithelium. ______________ (True/False). ❷

Exercise 21.3: The Pericardial Cavity

3. Match the following descriptions in column A with the corresponding structure in column B. ❸

Column A	***Column B***
_____ 1. inner portion of the pericardial sac	a. fibrous pericardium
_____ 2. layer of the pericardium that is tightly adhered to the heart	b. parietal layer of the serous pericardium
_____ 3. outer portion of the pericardial sac	c. visceral layer of the serous pericardium

Exercise 21.4: Gross Anatomy of the Human Heart

4. The atria and ventricles of the heart both contain three layers of the heart wall: endocardium, myocardium, and epicardium. ______________ (True/False). ❹
5. The chamber that contains pectinate muscles in its walls is the ____________ (atrium/ventricle). ❺
6. The structures that increase forcefulness of contraction within the right atrium are ________________ (papillary/pectinate) muscles, whereas the structures that prevent eversion of AV valves are ________________ (papillary/pectinate) muscles. ❻
7. Papillary muscles attach directly to __________ (atrioventricular/semilunar) valve cusps. ❼
8. The right atrioventricular valve is also known as the ______________ (bicuspid/tricuspid) valve, whereas the left atrioventricular valve is also known as the ______________ (bicuspid/tricuspid) valve. ❽
9. The right ventricle pumps a smaller volume of blood than the left ventricle. ________________ (True/False) ❾
10. Pulmonary and aortic semilunar valves each contain ____________ (two/three) cusps and prevent backflow of blood into the ____________ (atria/ventricles). ❿
11. Place the following structures in the order in which a drop of blood would travel through the right side of the heart, beginning with the superior and inferior venae cavae and ending with the lungs. ⓫

    _____ a. pulmonary arteries

    _____ b. pulmonary trunk

    _____ c. right atrioventricular valve

    _____ d. right atrium

    _____ e. pulmonary semilunar valve

    _____ f. right ventricle

12. Match the description listed in column A with the structure listed in column B. (Answers will be used more than once.) ⓬

Column A	***Column B***
_____ 1. located between the pulmonary trunk and aorta	a. fossa ovalis
_____ 2. located between the right and left atria	b. ligamentum arteriosum
_____ 3. remnant of the ductus arteriosis	
_____ 4. remnant of the foramen ovale	

Exercise 21.5: The Coronary Circulation

13. The anterior interventricular artery branches from the ______________ (right/left) coronary artery. (13)
14. The coronary sinus is located on the ____________________ (anterior/posterior) surface of the heart. (14)

Exercise 21.6: Superficial Structures of the Sheep Heart

15. The right ventricle is visible in an anterior view of the sheep heart. ______________ (True/False) (15)
16. The ______________ (parietal/visceral) layer of the serous pericardium is tightly adhered to the heart, whereas the ______________ (parietal/visceral) layer of the serous pericardium is tightly adhered to the fibrous pericardium. (16)
17. When viewing the anterior surface of the sheep heart, the ________________ (aorta/pulmonary trunk) is located most anteriorly. (17)
18. Oxygenated blood returns to the left atrium through the pulmonary ________________ (arteries/veins). (18)

Exercise 21.7: Coronal Section of the Sheep Heart

19. A coronal section of the sheep heart reveals that the left ventricular wall is significantly ______________ (thicker/thinner) than the right ventricular wall. (19)
20. The systemic circuit has ________________ (more/less) resistance than the pulmonary circuit. Therefore, the left ventricle contracts with a ______________ (greater/lesser) force than the right ventricle. (20)
21. Tendinous cords in the sheep heart tether the cusps of ____________________ (atrioventricular/semilunar) valves to prevent the backflow of blood. (21)

Exercise 21.8: Transverse Section of the Sheep Heart

22. When viewing a transverse section of the sheep heart, the right ventricular wall is ________________ (thicker/thinner) than the left ventricular wall. (22)
23. The right and left ventricles pump different volumes of blood due to the difference in ventricular wall thickness. ________________ (True/False) (23)

Exercise 21.9: Electrocardiography Using Standard ECG Apparatus

24. When placing electrodes on the skin to measure an ECG, the skin should be rubbed slightly with an abrasive pad to increase conductivity between the skin and surface electrode. ______________ (True/False) (24)
25. The QRS complex on a normal ECG trace represents ______________ (atrial/ventricular) depolarization. (25)

Exercise 21.10: BIOPAC Lesson 5: Electrocardiography I

26. The optimal limb lead and surface electrode placement for recording a three-lead ECG involves placing surface electrodes on the right arm, left arm, and left leg. ________________ (True/False) (26)
27. Heart rate should ______________ (increase/decrease) immediately following exercise. (27)

Exercise 21.11: Ph.I.L.S. Lesson 22: Refractory Period of the Heart

28. The plateau of a cardiac action potential extends the depolarization state of cardiac muscle. ______________ (True/False) (28)
29. Cardiac muscle has a ______________ (shorter/longer) refractory period than skeletal muscle, which prevents force summation in cardiac muscle. (29)

Exercise 21.12: Ph.I.L.S. Lesson 29: ECG and Heart Block

30. Which of the following accurately describes an ECG tracing for a patient who presents with an AV node block? (Circle one.) (30)
    a. absent P waves
    b. absent QRS complexes
    c. disconnected P waves and QRS complexes
    d. inverted T waves
31. A patient with a damaged SA node is likely to experience a(n) ______________ (decreased/increased) heart rate. (31)

Exercise 21.13: Ph.I.L.S. Lesson 30: Abnormal ECGs

32. Ectopic foci may result when damaged tissue is present in the nodal system of the heart. ______________ (True/False) 32

33. In ventricular fibrillation, all recognizable components of the ECG are absent. ______________ (True/False) 33

Exercise 21.14: Auscultation of Heart Sounds

34. The first heart sound corresponds to the closing of the ______________ (atrioventricular/semilunar) valves, whereas the second heart sound corresponds to the closing of ______________ (atrioventricular/semilunar) valves. 34

35. Optimal placement of a stethoscope for auscultation of the first heart sound is the ______ (second/fifth) intercostal space. 35

36. An incompetent semilunar valve may lead to an abnormal heart sound. ______________ (True/False) 36

Exercise 21.15: Ph.I.L.S. Lesson 26: The Meaning of Heart Sounds

37. The mechanical event of the heart associated with the P wave is __________ (atrial/ventricular) ______________ (contraction/relaxation). 37

Can You Apply What You've Learned?

38. A heart surgeon is about to perform a heart transplant. She has already cut through the sternum and entered the thoracic cavity and mediastinum. The heart, however, remains enclosed in the pericardial sac. The tissues the surgeon must cut through, from superficial to deep, to enter the pericardial cavity are these:

    a. ______________________ b. ______________________

39. Will the surgeon need to cut through the visceral layer of serous pericardium to remove the heart from the pericardial sac? Why or why not?

    ____

    ____

40. Describe the effect that scar tissue would have on the spread of depolarization through the ventricles. ______________________

    ____

41. If a patient suffers from heart block, would you expect the rate of ventricular contraction to be faster or slower than normal? Explain your answer.

    ____

    ____

42. When listening for heart sounds, only two heart sounds are typically audible. Propose a clinical scenario that might cause additional sounds to be heard.

    ____

    ____

Can You Synthesize What You've Learned?

43. Pulmonary edema (accumulation of fluid in the lungs) can result if there is a mismatch in the volumes of blood pumped by the right and left ventricles. Discuss how differences in these blood volumes may cause this condition. ______________________

    ____

    ____

44. External pacemakers are often used to correct problems related to electrical conduction from the atria to the ventricles. These pacemakers have two "leads," which are wires that are threaded into the heart. Where do you think the two leads are directed within the heart? Why would two leads be threaded into the heart rather than one? ______________________

    ____

    ____

45. A cardiac stress test is often used as a diagnostic tool to investigate underlying heart conditions. During a stress test, the doctor has the patient exercise on a treadmill at maximum intensity for a short duration while observing the patient's ECG for abnormalities. Describe how the stress test might influence components of the ECG, and how this may be used as a diagnostic tool. ______________________________

____

46. A patient complains of chest pains. Upon urging from his friends, he goes to the hospital. Doctors perform an ECG upon his arrival at the hospital. His ECG trace reveals that P waves and QRS complexes are occurring at different rates. More specifically, each QRS complex is not preceded by a P wave. Describe what is happening with the patient's electrical conduction system and propose a diagnosis.

____

____

The Cardiovascular System: Vessels and Circulation

OUTLINE AND LEARNING OBJECTIVES

Anatomy & Physiology REVEALED® aprevealed.com **MODULE 9: CARDIOVASCULAR SYSTEM**

INTRODUCTION

While proper functioning of the heart is necessary to be able to deliver adequate blood supply to working tissues, the vast system of blood vessels provides necessary conduits (channels) for the blood to flow to the body's tissues. **Arteries** are blood vessels that carry blood *away* from the heart, whereas **veins** are blood vessels that carry blood *toward* the heart. As arteries carry blood toward the tissues, they branch into smaller and smaller vessels, ultimately forming **arterioles** (*arteriole*, a small artery). Arterioles have the special function of controlling the flow of blood into **capillary beds.** Capillaries are the site of exchange for substances (e.g., gases, nutrients, wastes) between the blood and the tissues.

Blood flows out of capillaries into small veins called **venules,** which merge to form the larger veins that return blood to the heart. **Veins** transport blood back to the heart and serve as blood reservoirs (*reservoir,* a receptacle). Generally, a vein is positioned alongside each major artery and has the same name as the artery it accompanies. However, there are typically more veins draining a structure than there are arteries supplying it. In the limbs, most of the veins that do not accompany an artery are superficial veins, located just under the skin. For example, the brachium (arm) is supplied by the brachial artery, which has the brachial vein traveling next to it. The brachial artery and vein are located fairly deep within the arm, where they are protected by the musculature of the arm. In addition to the brachial vein, two superficial veins also drain blood from the arm. These are the cephalic and basilic veins. Note that there is considerably more variation among individuals in the branching patterns and locations of veins than there is with arteries. Such variation often has clinical significance. For example, when blood samples need to be collected from a patient, blood is commonly drawn from the median cubital vein. However, not all individuals have a median cubital vein. When tracing blood flow through the venous system, remember that veins *drain* blood from an area of the body. This means that descriptions of blood flow through veins start by naming the most distal veins first, and then name the veins blood travels through as it proceeds toward the heart.

The exercises in this chapter begin with an exploration of the histological characteristics of the different types of blood vessels (arteries, arterioles, capillaries, and veins). Subsequent exercises involve identification of the major arteries and veins of the body on a human cadaver or on classroom models of the cardiovascular system. Upon completion of the gross anatomy exercises in this chapter, a student should be able to describe the pathway a drop of blood takes as it is transported from the heart to a target organ and back to the heart once again. Finally, physiology exercises in this chapter explore the concept of **blood pressure,** which is the pressure that blood exerts against a blood vessel wall. The goal of these exercises is to investigate the relationships among blood pressure, blood flow, and blood vessel resistance.

List of Reference Tables

Chapter 22: The Cardiovascular System: Vessels and Circulation

Name: ______________________________

Date: ____________ Section: ____________

These Pre-Laboratory Worksheet questions may be assigned by instructors through their connect course.

PRE-LABORATORY WORKSHEET

1. The tunica ______________________________ (externa/intima/media) is the innermost layer of a blood vessel wall, whereas the tunica ______________________________ (externa/intima/media) is the outermost layer.

2. The wall of a capillary consists only of endothelium. ______________________________ (True/False)

3. The most permeable type of capillary is a ______________________________ (continuous/fenestrated/sinusoidal) capillary, whereas the least permeable type of capillary is a ______________________________ (continuous/fenestrated/sinusoidal) capillary.

4. Which of the following is an anatomic feature unique to veins?
   a. smooth muscle in the tunica media
   b. valves to prevent backflow of blood
   c. vasa vasorum in the tunica externa
   d. tunica media consists only of endothelium and a basement membrane

5. The hepatic portal system is a system of veins that drains blood into the ______________________________(inferior vena cava/liver).

6. Which of the following is *not* one of the unpaired branches of the abdominal aorta?
   a. celiac trunk
   b. inferior mesenteric artery
   c. renal artery
   d. superior mesenteric artery

7. The ligamentum arteriosum is a remnant of the fetal ______________________________ (ductus arteriosus/foramen ovale), whereas the fossa ovalis is a remnant of the fetal ______________________________ (ductus arteriosus/foramen ovale).

8. The major vein that drains blood from the inferior half of the body and empties into the right atrium of the heart is the ______________________________ (inferior/superior) vena cava.

9. Pulse pressure is a measure of the average pressure that blood exerts on a vessel during the course of a cardiac cycle. ______________________________ (True/False)

10. The first heart sound (lub) is created by turbulent blood flow that occurs when the ______________________________ (atrioventricular/semilunar) valves of the heart close.

HISTOLOGY

Blood Vessel Wall Structure

All blood vessels except capillaries have three tunics (layers) forming their walls. These layers, called the tunica intima, tunica media, and tunica externa, are analogous in both structure and function to the three layers of the heart wall (endocardium, myocardium, and epicardium). The differences in structure and function between the different types of blood vessels come mainly from modifications of these three wall layers. In particular, the type of tissue that composes the tunica media greatly affects the function of the vessel. **Table 22.1** describes the general composition of the three layers of a blood vessel wall, and **table 22.2** summarizes unique features in the different types of blood vessels.

Table 22.1 Layers of a Blood Vessel Wall

Wall Layer	Location	Components	Word Origin
Tunica Intima	Innermost layer; in contact with the lumen of the vessel	Endothelium (simple squamous epithelium) and a subendothelial layer composed of areolar connective tissue	*tunic*, a coat, + *intimus*, innermost
Tunica Media	Middle layer	Varied amounts of collagen fibers, elastic fibers, and smooth muscle cells	*tunic*, a coat, + *medius*, middle
Tunica Externa	Outermost layer	Areolar connective tissue that anchors the vessel to surrounding structures	*tunic*, a coat, + *externus*, on the outside

Table 22.2 Characteristics of Wall Layers in Specific Types of Blood Vessels

Type of Vessel	Tunica Intima	Tunica Media	Tunica Externa	Diameter	Characteristics and Special Functions
Elastic Artery	Endothelium and subendothelial layer; an internal elastic lamina is present but not easily distinguished from the elastic tissue of the tunica media	Contains numerous elastic and reticular fibers; also contains smooth muscle cells	Underdeveloped in contrast to other vessels; contains vasa vasorum, lymphatics, and nerves	2.5 cm–1 cm	Expansion and contraction of elastic tissues smooths out the flow of blood
Muscular Artery	Endothelium and subendothelial layer; contains a very prominent internal elastic lamina	Multiple layers of smooth muscle; numerous elastic fibers	Contains vasa vasorum, lymphatics, and nerves	1 cm–3 mm	Recoil of wall continues to propel blood through the arteries
Arteriole	Endothelium and subendothelial layer; an internal elastic lamina is present only in the largest arterioles	Contains less than six layers of smooth muscle, with no external elastic lamina	Very thin	3 mm–10 μm	Size of lumen is regulated to control the flow of blood into capillary beds
Capillary	Endothelium and a basement membrane only	NA	NA	8–10 μm	Thin wall allows for exchange between the blood and tissues
Postcapillary Venule	Endothelium and a thin subendothelial layer	Very thin with no smooth muscle cells	Very thin	10–50 μm	Drains blood from capillary beds. Site where leukocytes leave the circulation and enter the tissues via diapedesis.*
Venule	Endothelium and a thin subendothelial layer	Very thin with very few smooth muscle cells	Thickest layer of the wall	50–100 μm	Venules are simply small veins, and are the counterpart to arterioles
Vein	Endothelium and subendothelial layer; infoldings form valves, which prevent the backflow of blood. Not all veins have valves.	Very thin with a small amount of smooth muscle	Thickest layer of the wall; contains vasa vasorum	Greater than 100 μm	Low pressure conduits; valves aid in preventing backflow of blood

*diapedesis (*dia-*, through, + *pedesis*, a leaping): the passage of leukocytes through the walls of blood vessels

EXERCISE 22.1

BLOOD VESSEL WALL STRUCTURE

1. Obtain a slide showing an artery and a vein (they may both be on the same slide, or they may be on two different slides).
2. Place the slide on the microscope stage and bring the tissue sample into focus on low power. Scan the slide and look for the circular or oval cross section of a vessel. If there is more than one vessel on the slide, determine which vessel is an artery and which is a vein. In general, arteries have relatively thick walls and small lumens, whereas veins have relatively thin walls and large lumens **(figure 22.1).** In addition, the lumens of veins are often collapsed because of the relative thinness of the blood vessel wall.
3. After identifying an artery and a vein, move the microscope stage so the wall of the *artery* is in the center of the field of view. Increase the power on the microscope until all the layers of the artery wall are visible.
4. Identify the structures listed below using figure 22.1 and table 22.1 as guides. Keep in mind that the innermost layer of the vessel (the tunica intima) will be incredibly thin and difficult to identify except on high power. Most likely only the flattened nuclei of the endothelial cells, and very little of the rest of the cells, will be visible.

- ☐ **artery**
- ☐ **lumen**
- ☐ **tunica externa**
- ☐ **tunica intima**
- ☐ **tunica media**
- ☐ **vein**

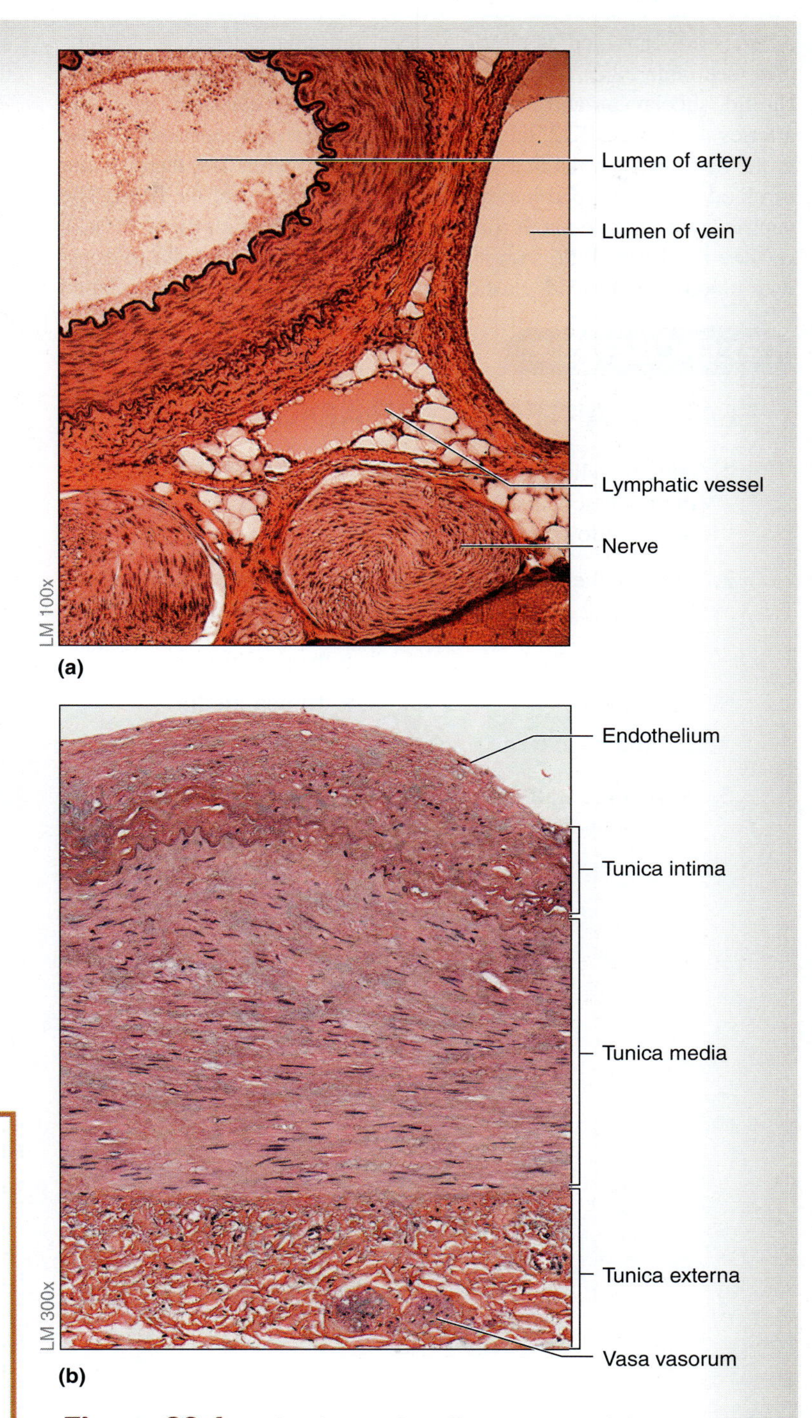

Figure 22.1 Blood Vessel Wall Structure. (a) Cross section through the center of a neurovascular bundle containing a nerve, artery, vein, and lymphatic vessel. (b) The three layers of the wall of a blood vessel: tunica intima, tunica media, and tunica externa. The tunica externa has its own blood supply, the vasa vasorum (literally, "the vessels of the vessels").

INTEGRATE

CONCEPT CONNECTION

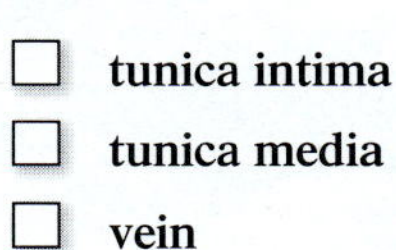

The tunica intima of a blood vessel consists of endothelium and a basement membrane. Recall from chapter 21 that the role of platelets is to form a platelet plug that stops bleeding from a damaged vessel. Normally, platelets are smooth, spherical structures that are not "sticky," which is a good thing. Sticky platelets would cause unwanted clots to form in vessels, which would severely impair blood flow. What, then, causes a platelet to become activated? Damage to the endothelium of a blood vessel wall exposes the underlying basement membrane. When platelets come in contact with the exposed collagen fibers of the basement membrane, the platelets become activated, thus making them "sticky." The activated platelets stick to each other, to the blood vessel wall, and to other formed elements within the blood creating a platelet plug. Chemicals released from the platelet plug will initiate the formation of fibrin to cause formation of a *thrombus*, or blood clot. Both a platelet plug and a thrombus are temporary structures that prevent bleeding from the vessel while the wall of the vessel is being repaired through other mechanisms. As the wall of the vessel is being repaired, both the platelet plug and thrombus are reabsorbed.

Elastic Arteries

Arteries are classified as elastic arteries, muscular arteries, and arterioles (see table 22.2). The aorta, pulmonary, brachiocephalic, common carotid, subclavian, and common iliac arteries are classified as elastic arteries (see figure 22.2, table 22.2). These arteries, located very close to the heart, have walls that are thick enough to withstand the pressure of blood that is pumped into them from the ventricles of the heart. The ventricles generate enough force to move blood through these vessels, so they need very little smooth muscle in their tunica media to assist with blood flow. Instead, elastic arteries have an abundance of collagen and elastic fibers in their tunica media, which makes them both tough (collagen fibers) and expandable (elastic fibers). The ability of the vessel wall to expand and recoil as it receives blood from the ventricles, and then recoil, greatly smooths out the flow of blood through the arteries.

Large vessels such as the aorta have tiny blood vessels called *vasa vasorum* (literally, "the vessels of the vessels," see figure 22.1*b*) in the tunica externa. The vasa vasorum are analogous in both structure and function to the coronary arteries in the outer layer of the heart wall (epicardium). That is, the vasa vasorum supplies blood to the walls of larger vessels just as the coronary arteries supply blood to cardiac muscle tissue.

EXERCISE 22.2

ELASTIC ARTERY—THE AORTA

1. Obtain a slide of the aorta **(figure 22.2)** and place it on the microscope stage. Bring the wall of the aorta into focus on low power.
2. Identify the following on the slide of the aorta, using figure 22.2 and tables 22.1 and 22.2 as guides:

   - ☐ **elastic fibers**
   - ☐ **lumen**
   - ☐ **tunica externa**
   - ☐ **tunica intima**
   - ☐ **tunica media**
   - ☐ **vasa vasorum**

3. Sketch the wall of the aorta as seen through the microscope in the space provided.

__________ ×

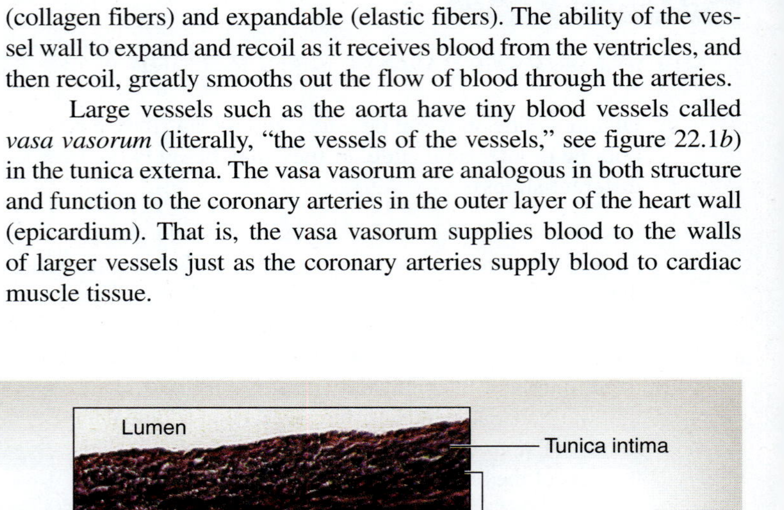

Figure 22.2 Elastic Artery. The wall of the aorta, an elastic artery, contains numerous elastic fibers (black) in the tunica media. No vasa vasorum are visible in the tunica externa in this micrograph.

Muscular Arteries

As blood moves through the elastic arteries and travels farther away from the heart, the pressure exerted by the heart is no longer great enough to keep the blood moving through the vessels. Thus, the amount of elastic tissue in the tunica media of the arteries starts to decrease and the amount of smooth muscle in the tunica media starts to increase. Contraction of the smooth muscle keeps blood moving through the arteries as the blood gets farther away from the heart. Most named vessels, including the brachial, anterior tibial, and inferior mesenteric arteries are muscular arteries. **Muscular arteries** are easily distinguished from elastic arteries by the presence of two prominent bands of elastic fibers: the **internal elastic lamina,** which is the outermost layer of the tunica intima, and the **external elastic lamina** (adjacent to the tunica externa). There are several layers of smooth muscle sandwiched in between the two prominent elastic laminae (see figure 22.3, table 22.2).

EXERCISE 22.3

MUSCULAR ARTERY

1. Obtain a slide of a small, muscular artery, and place it on the microscope stage. Bring the tissue sample into focus on low power, then locate the wall of the vessel **(figure 22.3)**.

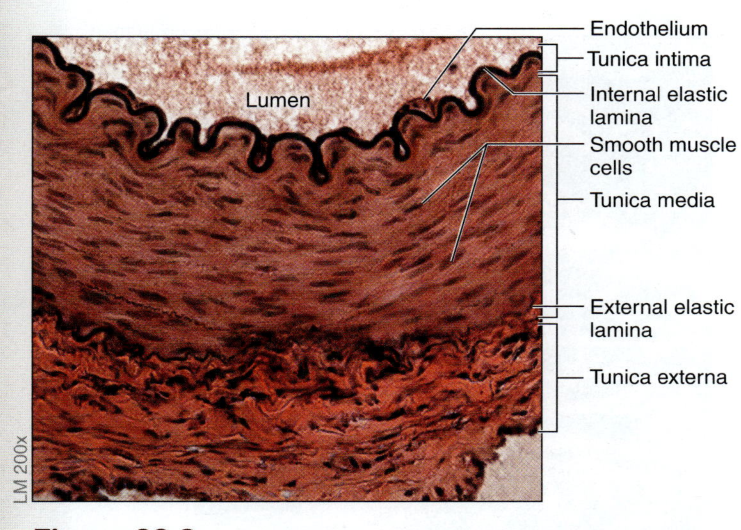

Figure 22.3 Muscular Artery. Muscular arteries contain distinct internal and external elastic laminae bordering the tunica media, which is predominantly smooth muscle.

2. Identify the following on the slide of the muscular artery, using figure 22.3 and table 22.2 as guides:

   - ☐ **external elastic lamina**
   - ☐ **tunica externa**
   - ☐ **internal elastic lamina**
   - ☐ **tunica intima**
   - ☐ **lumen**
   - ☐ **tunica media**
   - ☐ **smooth muscle cells**

3. Sketch the wall of a muscular artery as seen through the microscope in the space provided.

_________ ×

Arterioles

Arterioles have a unique function in the cardiovascular system. Arterioles control the flow of blood into capillary beds (see table 22.2). As such, the most prominent feature of the wall of an arteriole is layers of circular smooth muscle in the tunica media. The smooth muscle cells of these layers are regulated to contract and relax to change the diameter of the lumen of the vessel, thus regulating the flow of blood into the capillary beds.

EXERCISE 22.4

ARTERIOLE

1. Obtain a slide of an arteriole and place it on the microscope stage. Bring the tissue sample into focus on low power. Scan the slide to locate an arteriole in cross section **(figure 22.4).**

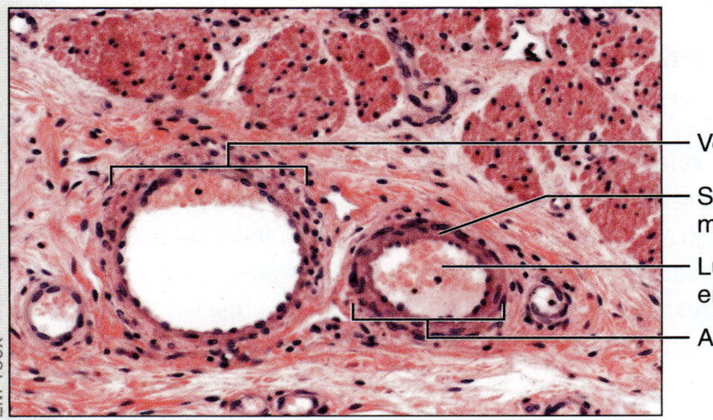

Figure 22.4 Arteriole. The tunica media of the arteriole contains circular smooth muscle that functions to alter the diameter of the vessel.

2. Identify the following on the slide of the arteriole, using figure 22.4 and tables 22.1 and 22.2 as guides:

   - ☐ **lumen**
   - ☐ **smooth muscle**

3. Sketch an arteriole as seen through the microscope in the space provided.

_________ ×

Veins

Veins are vessels that function to return blood to the heart at low pressure. They are characterized by having large lumens and thin walls (see figure 22.5, table 22.2). They may also contain valves, which are infoldings of the tunica intima. These valves prevent blood from flowing backward. Large veins, like large arteries, also contain vasa vasorum. **Venules** are small veins. The venules that come immediately after capillary beds, **postcapillary venules,** are the site where most white blood cells leave the circulation to enter the tissues (see table 22.2).

EXERCISE 22.5

VEIN

1. Place a slide of a large vein on the microscope stage and bring the tissue sample into focus on low power.
2. Identify the following on the slide of the large vein, using **figure 22.5** and table 22.2 as guides:
   - ☐ **elastic fiber**
   - ☐ **smooth muscle cells**
   - ☐ **tunica externa**
   - ☐ **tunica intima**
   - ☐ **tunica media**
   - ☐ **valve (may not be visible on the slide)**
   - ☐ **vasa vasorum**
3. Sketch a cross-section of a large vein as seen through the microscope in the space provided.

_____________ ×

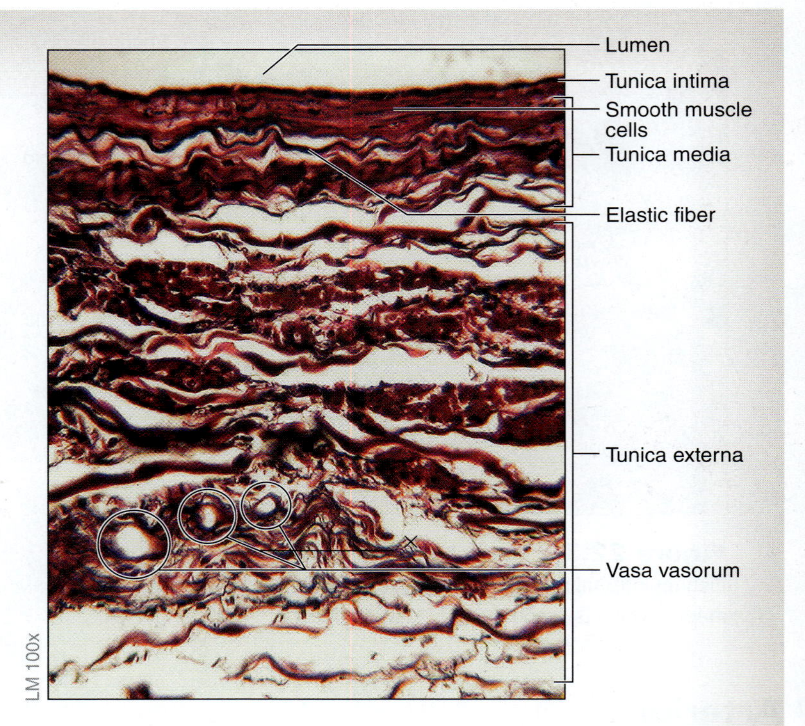

Figure 22.5 **Vein.** Cross section through the wall of a vein. The thickest layer of the vessel wall is the tunica externa, which contains vasa vasorum. (Spaces within the tunica externa are artifacts produced during slide preparation.)

INTEGRATE

CLINICAL VIEW

Great Saphenous Vein and Varicose Veins

Varicose veins are veins that become enlarged and engorged with blood as a result of incompetent valves. The most common site of varicose veins is in the great saphenous vein, the longest vein in the body. The great saphenous vein runs from the medial malleolus to the inguinal region to the femoral vein. As a superficial vein, the great saphenous vein has little external support to assist the movement of blood. Thus, venous valves are critically important to the proper functioning of the great saphenous vein. Eventually, the varicose veins become problematic because they remain engorged with blood and they form unsightly, tortuous masses beneath the skin of the medial leg. Treatment of varicose veins begins with application of compression stockings, which helps prevent the veins from expanding. Also, the veins can be functionally removed either by surgery or by injection of a substance that causes the vein to permanently collapse, no longer allowing blood to flow within it.

INTEGRATE

CLINICAL VIEW

Atherosclerosis in the Internal Carotid Artery

Atherosclerosis (*athere*, porridge + *sclerosis*, hardening) is a common condition affecting almost all individuals at some point in their lives, although genetics or behavioral factors (e.g., smoking) may accelerate the process. Because the internal carotid arteries operate at relatively high pressure, they tend to be affected by atherosclerosis to a significant degree. Atherosclerosis is a disease in which calcium and lipids are deposited into the wall of the arteries. This decreases the elasticity and diameter of the blood vessel and increases the resistance to blood flow through those arteries. Over time, this leads to a reduction in blood flow to the brain. The decreased blood flow is a significant cause of dementia in older adults.

Capillaries

Capillaries are unique blood vessels in that they consist *only* of a tunica intima (endothelium and basement membrane). This allows for the exchange of substances between the blood and the tissues. There are several different types of capillaries in the body, which vary in their degree of permeability. For instance, capillaries in the spleen and liver, organs where a lot of blood processing takes place, allow a great deal of exchange to occur. In contrast, capillaries in the brain, an organ that must be protected from harmful substances such as toxins and viruses that might be present in the blood, limit the exchange.

EXERCISE 22.6

OBSERVING ELECTRON MICROGRAPHS OF CAPILLARIES

1. Because capillaries are so small, the only way to truly appreciate their structure is to view them with an electron microscope. Obviously most anatomy and physiology laboratories do not have access to an electron microscope. Thus, this portion of the exercise will be performed by viewing the electron micrographs in **table 22.3.**

2. Observe the electron micrograph of a **continuous capillary** in table 22.3a. Continuous capillaries are the most common type of capillaries. They are found within muscle, the skin, the lungs, and central nervous system. They are composed of a continuous lining of simple squamous cells resting on a complete basement membrane. Identify the endothelial cells in the micrograph and then look for the areas where two endothelial cells come in contact with each other. Notice that the endothelial cells overlap each other in these areas. Where the cells come together to form a continuous lining around the lumen they do not form a complete "seal." The gaps between the endothelial cells are called **intercellular clefts.** Substances can move into and out of the blood either through endothelial cells by cellular transport processes (e.g., diffusion, pinocytosis) or between endothelial clefts by diffusion and bulk flow.

3. Observe the electron micrograph of a **fenestrated capillary** in table 22.3b. Identify endothelial cells in the micrograph. In contrast to the smooth surface of the endothelial cells of the continuous capillaries, the endothelial cells of the fenestrated capillaries appear wavy, particularly in the area adjacent to the nucleus of the endothelial cell. As with continuous capillaries, the endothelial cells of fenestrated capillaries form a continuous lining with intercellular clefts between the blood and the tissues. However, the endothelial cells themselves have **fenestrations,** which are regions where the endothelial cells are extremely thin. Fenestrations are approximately 10–100 nm in diameter, and allow any substance that is smaller than the size of the fenestrations to be exchanged easily between the blood and the tissues. Thus, fenestrated capillaries allow much greater exchange than continuous capillaries. Fenestrated capillaries are found in organs where a greater amount of exchange is necessary, such as within endocrine glands, in the small intestine for absorption of nutrients, and in the kidneys for filtering blood.

4. Observe the electron micrograph of a **sinusoid** in table 22.3c. Sinusoids are located within organs that do a lot of processing of the blood, such as the liver and spleen. Sinusoids are characterized by having *discontinuous* endothelial cells, which do not overlap each other. In fact, there are open spaces between endothelial cells. The endothelial cells themselves are also fenestrated. Thus, the open spaces and fenestrations together form a minimal barrier between the blood and the tissues, which allows for maximum exchange (including the exchange of formed elements and large plasma proteins). The micrograph in table 22.3c demonstrates a sinusoidal capillary within the liver. Inside the lumen of the capillary is a macrophage. The one large, prominent nucleus visible in the micrograph is that of a liver cell, or hepatocyte, which contains many small lipid droplets. There is another hepatocyte at the top of the micrograph. Observe the edges of the hepatocytes that face the lumen of the sinusoidal capillary. Observe the thin endothelial cells that line the capillary. Now observe the area where there should be endothelium between the macrophage and the hepatocyte containing the lipid droplets and nucleus. Notice that there is no endothelial cell in between the two. Also, in the location between the macrophage and the hepatocyte above it, toward the left side of the micrograph, fenestrations are visible when looking very closely. The magnification of this micrograph is not quite high enough to see the fenestrations clearly, but they are there.

(continued on next page)

(continued from previous page)

Table 22.3 Characteristics of the Three Types of Capillaries

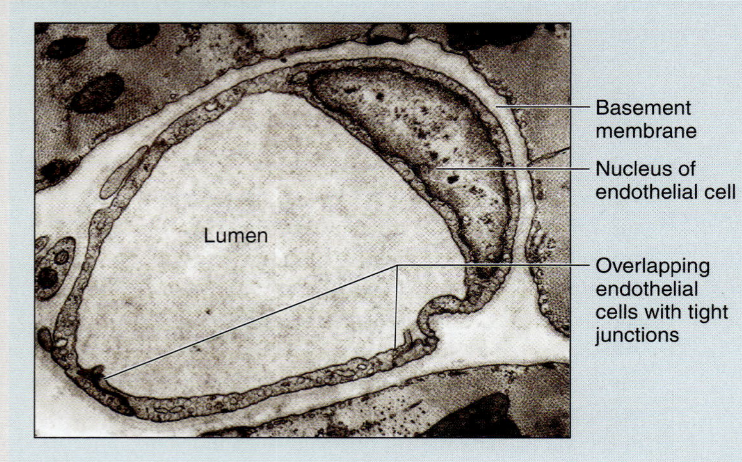

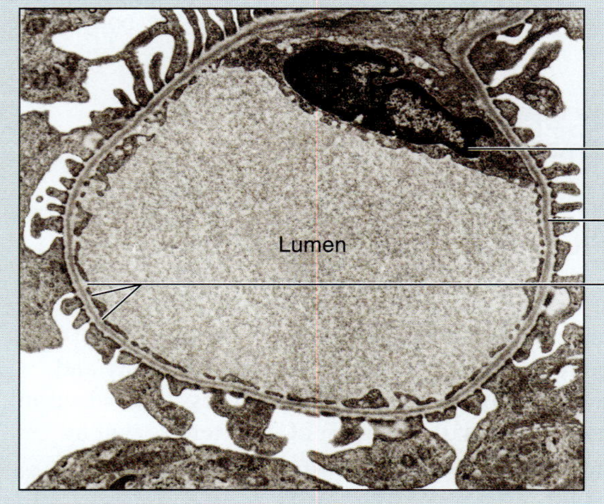

(a) Continuous Capillary		(b) Fenestrated Capillary	
Characteristics	Endothelial cells connected to each other by tight junctions	***Characteristics***	Same as continuous capillaries except also contain **fenestrations**
Endothelial Cells	Not fenestrated	***Endothelial Cells***	Fenestrated
Basement Membrane	Continuous	***Basement Membrane***	Continuous
Description and Function	Form a continuous lining between blood and tissues; exchange through endothelial cells and between endothelial cells via intercellular clefts	***Description and Function***	Allow for increased exchange between blood and tissues
Locations	Muscle tissue, the skin, connective tissues, exocrine glands, and central nervous system	***Locations***	Small intestine, kidneys, choroid plexus of brain, ciliary process of the eye
Word Origin	*continuus*, continued	***Word Origin***	*fenestra*, a window

Continuous

Fenestrated

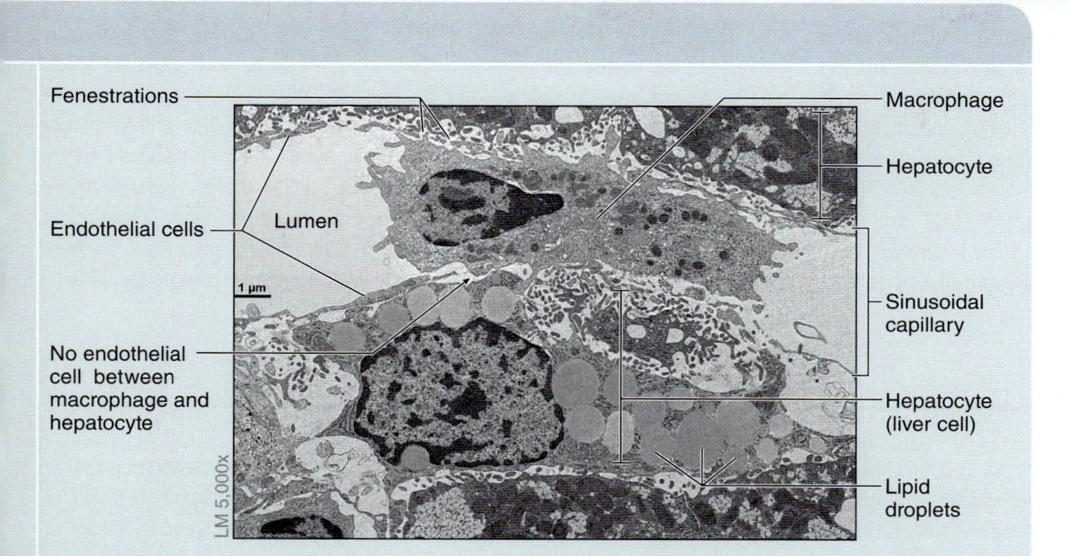

(c) Sinusoids

Characteristics	Endothelial cells are separated from each other by wide spaces. Endothelial cells are fenestrated without diaphragms.
Endothelial Cells	Fenestrated
Basement Membrane	Discontinuous
Description and Function	Allow blood and tissues to come into intimate contact with each other, which allows for maximum exchange between blood and tissues
Locations	Liver, spleen, bone marrow, and some endocrine glands
Word Origin	*sinus*, a channel, + *eidos*, resemblance

5. Sketch each of the three types of capillaries in the spaces provided. Be sure to label the following:

- ☐ **basement membrane**
- ☐ **endothelial cells**
- ☐ **fenestrations (if present)**
- ☐ **lumen of capillary**

Sinusoidal

GROSS ANATOMY

The following laboratory exercises guide the student in identification of the major arteries and veins of the body on a prosected human cadaver or on classroom models or charts. While identifying each vessel, consider the area of the body supplied by or drained by the vessel, and try to determine what tissue(s) or organ(s) would suffer damage if the vessel were blocked or cut.

Exercises then guide the student in tracing the flow of blood from the heart to an organ within each region, and back to the heart (see Integrate: Learning Strategy box on this page). Initial traces will be done using figures, and will include only vessels that are relatively close to the target organ. However, students will also be asked to write out the complete trace in words. This means completing the trace that was started in the figure by extending the list of blood vessels to include the vessels close to the heart that are not shown in the figures (for example, the aorta). When asked to do a *complete* trace, include blood flow through the heart and the pulmonary circulation, and name all of the heart valves blood travels through on its journey. A complete trace both begins and ends in the right atrium of the heart.

INTEGRATE

LEARNING STRATEGY

The task of tracing the flow of blood through the blood vessels of the body is challenging for many students. If you find yourself having difficulty with this task, use the analogy of driving a car to school. A drop of blood or a red blood cell represents the car, and the blood vessels represent the roads and highways. The organs represent the destination (e.g., school, the supermarket), and the heart represents your home. For example, to travel from your home to school, you must drive your car along a series of roadways that lead to the school. Each street has a name, which helps direct people driving their cars to their destinations. Similarly, as blood travels from the heart to a destination in the body (for example, the right hand), it flows along a given route. This route consists of several "streets," each with an identifying name. When tracing the flow of blood from the heart to the right hand, imagine driving a car inside the blood vessels, and write down the names of the arteries you "drive" through en route to your destination. Of course, once at your destination, you must eventually return "home" to the heart. Thus, you also need to visualize the trip through the veins that you would take to get from the hand back to the heart.

Pulmonary Circuit

The **pulmonary circuit** is the system of blood vessels that transports blood from the right ventricle of the heart to the lungs and back to the left atrium of the heart. Blood leaves the right ventricle relatively deoxygenated, and returns to the left atrium highly oxygenated, having picked up oxygen as it moves through the pulmonary capillaries.

EXERCISE 22.7

PULMONARY CIRCUIT

1. Observe the thoracic cavity of a human cadaver or observe classroom models or charts demonstrating blood vessels of the heart and lungs.
2. Identify the structures listed in **figure 22.6** on the cadaver, models, or charts, using the textbook as a guide. Then label them in figure 22.6.
3. *Optional Activity:* **AP|R 9: Cardiovascular System—** Watch the "Pulmonary and Systemic Circulation" animation to review the differences between these two circuits.
4. Trace the flow of blood from the right ventricle of the heart through the pulmonary circuit to the left atrium of the heart in words in the space provided. Be sure to include heart valves that are passed along the way.

____
____
____
____
____
____
____
____
____
____
____
____

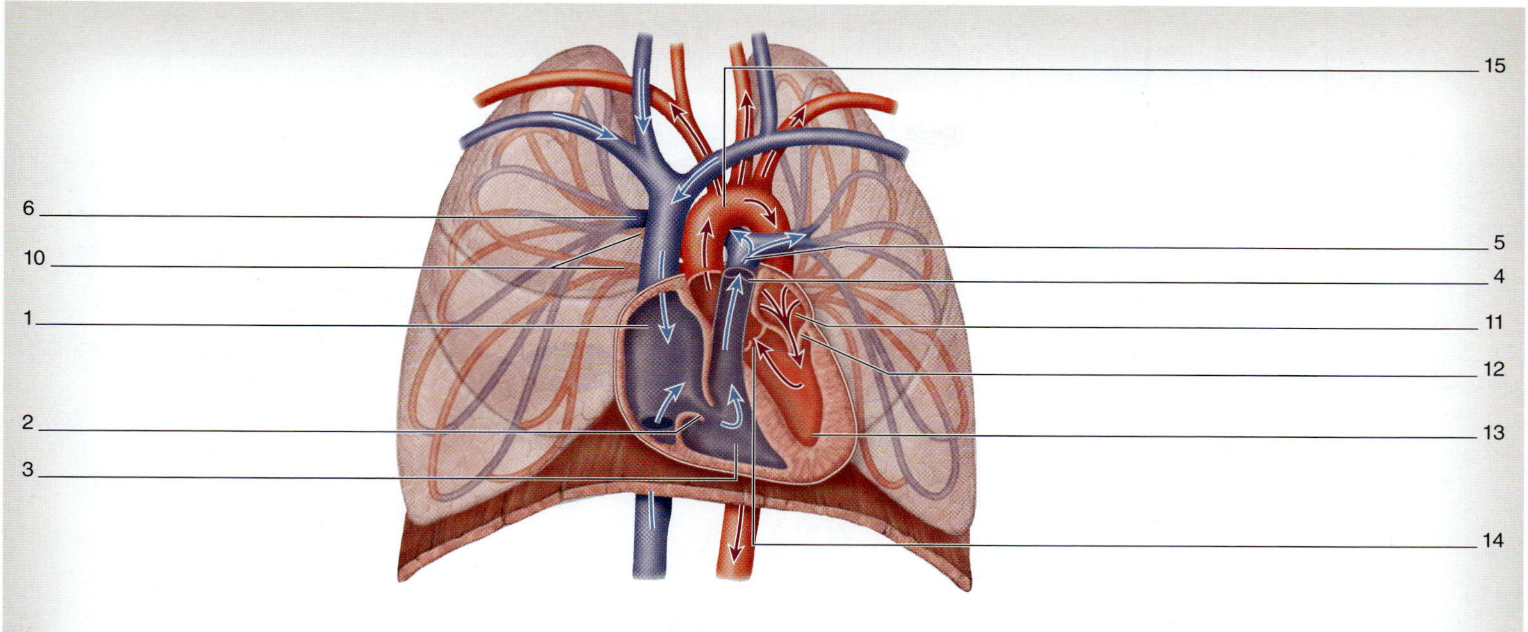

6
10
1
2
3
15
5
4
11
12
13
14

(a) Heart and lungs

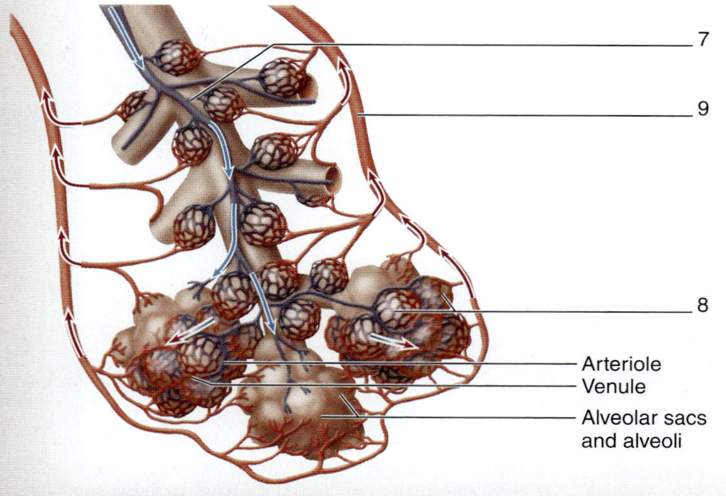

7
9
8
Arteriole
Venule
Alveolar sacs and alveoli

(b) Close-up of vessels within lung

Figure 22.6 Pulmonary Circuit. Blue arrows indicate the path of deoxygenated blood. Red arrows indicate the path of oxygenated blood. Use the terms listed to fill in the numbered labels in the figure.

- ☐ aorta
- ☐ aortic semilunar valve
- ☐ branch of pulmonary artery
- ☐ branch of pulmonary vein
- ☐ left atrium
- ☐ left AV valve
- ☐ left ventricle
- ☐ pulmonary arteries
- ☐ pulmonary capillaries
- ☐ pulmonary semilunar valve
- ☐ pulmonary trunk
- ☐ pulmonary veins
- ☐ right atrium
- ☐ right AV valve
- ☐ right ventricle

Systemic Circuit

The **systemic circuit** is the system of blood vessels that carry blood from the left ventricle of the heart to body organs and back to the right atrium of the heart. Blood leaves the left ventricle highly oxygenated, and returns to the right atrium relatively deoxygenated.

EXERCISE 22.8

CIRCULATION TO THE HEAD AND NECK

1. Observe the head and neck regions of a prosected human cadaver or observe classroom models or charts showing blood vessels of the head and neck.
2. The major arteries carrying blood to structures of the head and neck are the external and internal carotid arteries (**figure 22.7**). The **external carotid arteries** supply most superficial structures of the head and neck. The **external jugular veins** drain most superficial areas of the scalp, face, and neck.

(continued on next page)

(continued from previous page)

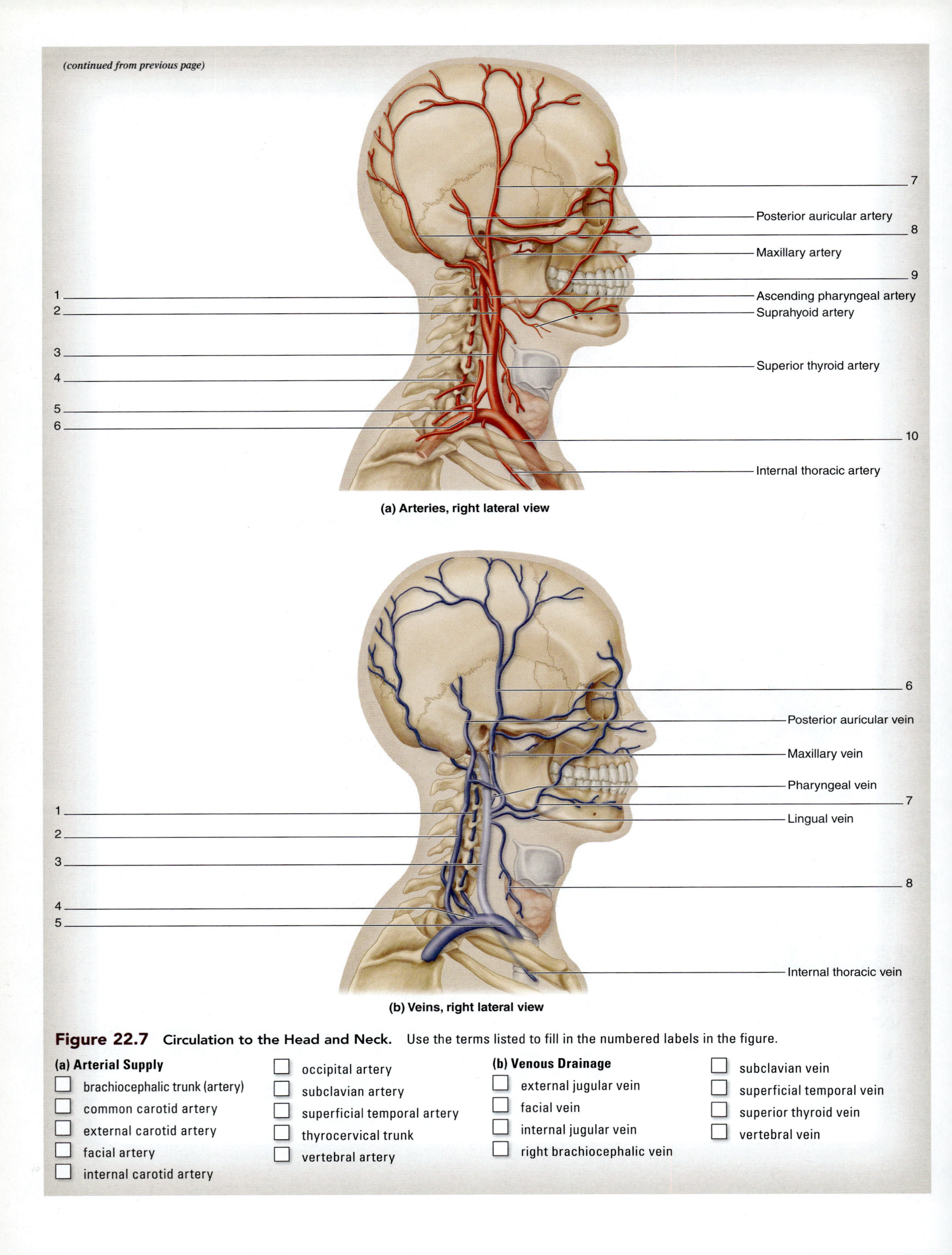

Figure 22.7 **Circulation to the Head and Neck.** Use the terms listed to fill in the numbered labels in the figure.

(a) Arterial Supply

- ☐ brachiocephalic trunk (artery)
- ☐ common carotid artery
- ☐ external carotid artery
- ☐ facial artery
- ☐ internal carotid artery
- ☐ occipital artery
- ☐ subclavian artery
- ☐ superficial temporal artery
- ☐ thyrocervical trunk
- ☐ vertebral artery

(b) Venous Drainage

- ☐ external jugular vein
- ☐ facial vein
- ☐ internal jugular vein
- ☐ right brachiocephalic vein
- ☐ subclavian vein
- ☐ superficial temporal vein
- ☐ superior thyroid vein
- ☐ vertebral vein

3. Identify the *arteries* listed in figure 22.7*a* that supply blood to the head and neck, using the textbook as a guide. Then label them in figure 22.7*a*.

4. Identify the *veins* listed in figure 22.7*b* that drain blood from the head and neck, using the textbook as a guide. Then label them in figure 22.7*b*.

5. The pathway a drop of blood takes to get from the aortic arch to the *skin overlying the anterior part of the right parietal bone of the skull* and back to the superior vena cava is shaded in **figure 22.8.** Trace this flow of blood by writing in the names of the vessels in the figure. Label the vessels in order, starting at number 1, so as to figuratively trace the pathway and name the vessels the blood flows through.

6. *Optional Activity:* **APR 9: Cardiovascular System—** *Anatomy & Physiology Revealed* includes numerous dissections showing vascular supply to all body regions; review these dissections and use the Quiz feature to review each region.

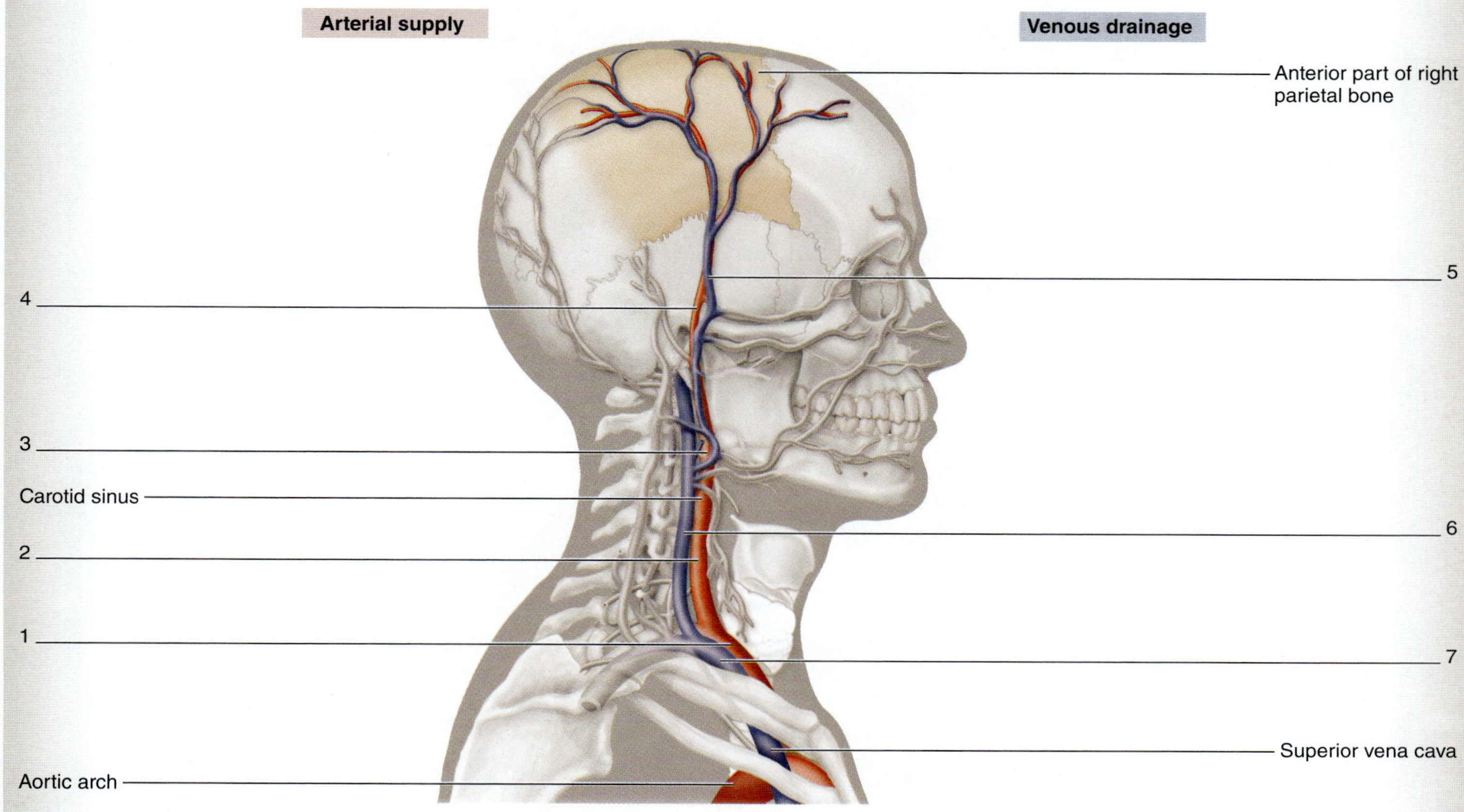

Arterial supply

Venous drainage

Anterior part of right parietal bone

5

4

3

Carotid sinus

6

2

1

7

Superior vena cava

Aortic arch

Figure 22.8 **Circulation from the Aortic Arch to the Anterior Part of the Right Parietal Bone and Back to the Superior Vena Cava.** Use the terms listed to fill in the numbered labels in the figure.

- ☐ brachiocephalic trunk (artery)
- ☐ right brachiocephalic vein
- ☐ right common carotid artery
- ☐ right external carotid artery
- ☐ right internal jugular vein
- ☐ right superficial temporal artery
- ☐ right superficial temporal vein

EXERCISE 22.9

CIRCULATION TO THE BRAIN

1. Observe the cranium and brain of a prosected human cadaver or observe classroom models or charts demonstrating blood vessels of the cranium and brain.

2. The major arteries carrying blood to structures of the brain are the internal carotid arteries and the vertebral arteries (see figure 22.7 and **figure 22.9**). The **internal carotid arteries** supply ~75% of the blood flow to the brain, while the **vertebral arteries** supply ~25% of the blood flow to the brain. Both pairs of vessels supply blood to the **cerebral arterial circle** (figure 22.9*a*). The major veins draining blood from the head, neck, and brain are the external and internal jugular veins. The **internal jugular veins** drain all blood from inside the cranial cavity plus some superficial areas of the face.

(continued on next page)

(continued from previous page)

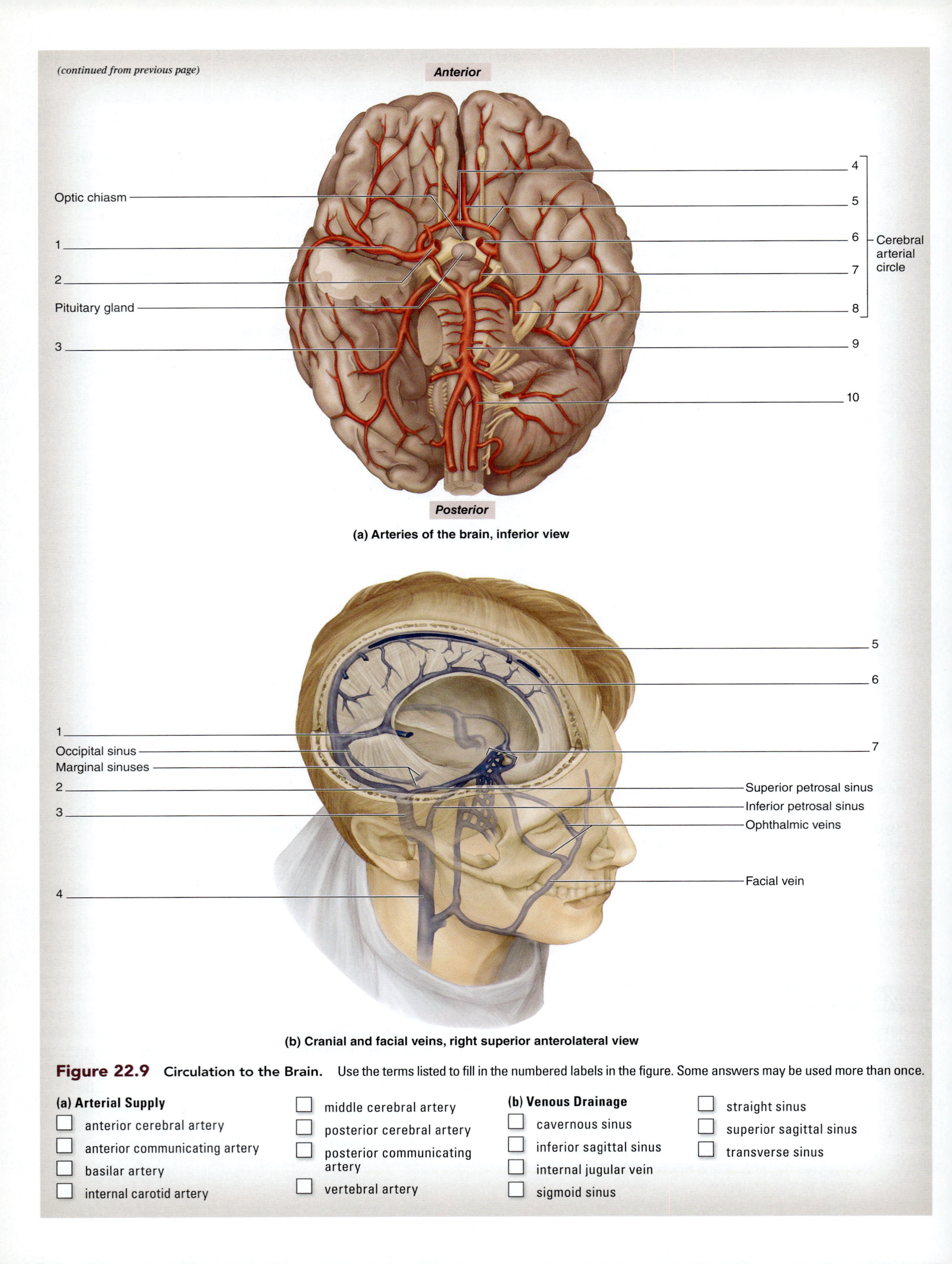

(a) Arteries of the brain, inferior view

(b) Cranial and facial veins, right superior anterolateral view

Figure 22.9 **Circulation to the Brain.** Use the terms listed to fill in the numbered labels in the figure. Some answers may be used more than once.

(a) Arterial Supply

- ☐ anterior cerebral artery
- ☐ anterior communicating artery
- ☐ basilar artery
- ☐ internal carotid artery
- ☐ middle cerebral artery
- ☐ posterior cerebral artery
- ☐ posterior communicating artery
- ☐ vertebral artery

(b) Venous Drainage

- ☐ cavernous sinus
- ☐ inferior sagittal sinus
- ☐ internal jugular vein
- ☐ sigmoid sinus
- ☐ straight sinus
- ☐ superior sagittal sinus
- ☐ transverse sinus

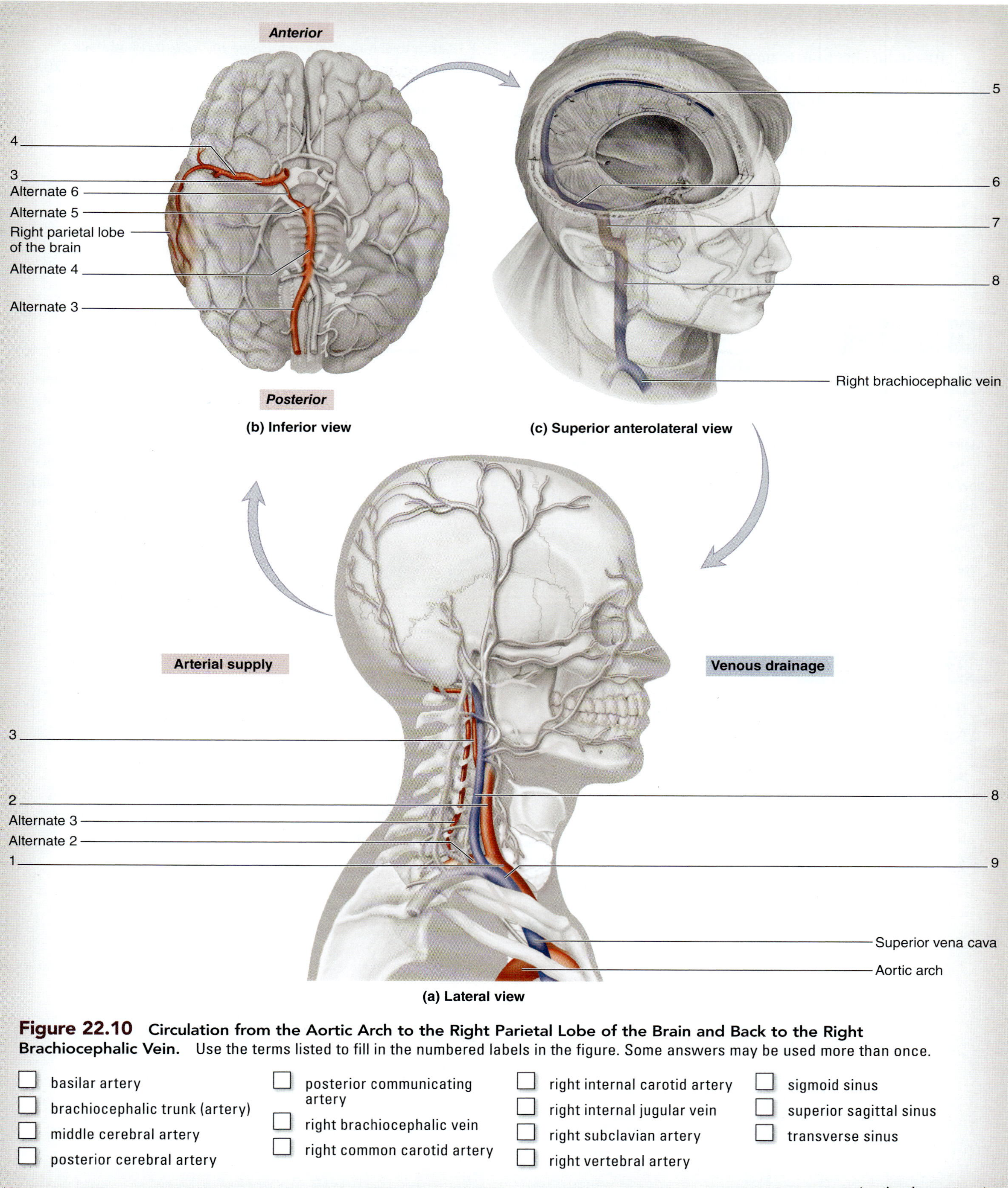

Figure 22.10 **Circulation from the Aortic Arch to the Right Parietal Lobe of the Brain and Back to the Right Brachiocephalic Vein.** Use the terms listed to fill in the numbered labels in the figure. Some answers may be used more than once.

- ☐ basilar artery
- ☐ brachiocephalic trunk (artery)
- ☐ middle cerebral artery
- ☐ posterior cerebral artery
- ☐ posterior communicating artery
- ☐ right brachiocephalic vein
- ☐ right common carotid artery
- ☐ right internal carotid artery
- ☐ right internal jugular vein
- ☐ right subclavian artery
- ☐ right vertebral artery
- ☐ sigmoid sinus
- ☐ superior sagittal sinus
- ☐ transverse sinus

(continued on next page)

(continued from previous page)

Blood draining from brain tissues first enters the **dural venous sinuses,** which collectively drain into the internal jugular vein (figure 22.9*b*).

3. Identify the *arteries* listed in figure 22.9*a* that supply blood to the brain, using the textbook as a guide. Then label them in figure 22.9*a*.
4. Identify the *veins* listed in figure 22.9*b* that drain blood from the brain, using the textbook as a guide. Then label them in figure 22.9*b*.
5. The pathway a drop of blood takes to get from the aortic arch to the *right parietal lobe of the brain* and back to the right brachiocephalic vein is shaded in **figure 22.10.** Trace this flow of blood by writing in the names of the vessels in the figure. Label the vessels in order, starting at number 1, so as to figuratively trace the pathway and name the vessels the blood flows through.

INTEGRATE

CLINICAL VIEW
Stroke

Maintaining adequate blood flow to organs ensures that oxygen and nutrients are delivered to the tissues. A stroke may occur when blood flow is restricted to the brain. This, in turn, restricts oxygen delivery to the tissues, a condition referred to as ischemia (*ischaemus,* stopping blood). Prolonged periods of ischemia may result in tissue death, or necrosis. As previously discussed, blood flow is influenced by changes in blood pressure and resistance. Common causes of ischemia include atherosclerotic plaques that dislodge and travel to the brain (embolism). When an embolus is present in a vessel, there is a decrease in blood vessel diameter, and an increase in resistance. Due to the inverse relationship between resistance and blood flow, this increased resistance results in a dramatic decrease in blood flow. Deficits observed following a stroke depend on the location of the ischemic event. For instance, a stroke in Broca's area may result in slurred speech, and an ischemic event in the vermis of the cerebellum may lead to motor deficits. Clinically, tests such as magnetic resonance imaging (MRI) or computed tomography (CT) are used to observe the area of infarct (tissue death).

EXERCISE 22.10

CIRCULATION TO THE THORACIC AND ABDOMINAL WALLS

1. Observe the thoracic cavity on a prosected human cadaver or observe classroom models or charts demonstrating blood vessels of the thoracic and abdominal cavities.
2. Identify the *arteries* listed in **figure 22.11*a*** that supply blood to the thoracic and abdominal walls, using the textbook as a guide. Then label them in figure 22.11*a*.
3. Identify the *veins* listed in figure 22.11*b* that drain blood from the thoracic and abdominal walls, using the textbook as a guide. Then label them in figure 22.11*b*.
4. Identify the paired and unpaired arteries that supply blood to the abdominal organs, pelvis, and perineum in **figure 22.12**, using the textbook as a guide. Then label them in figure 22.12.

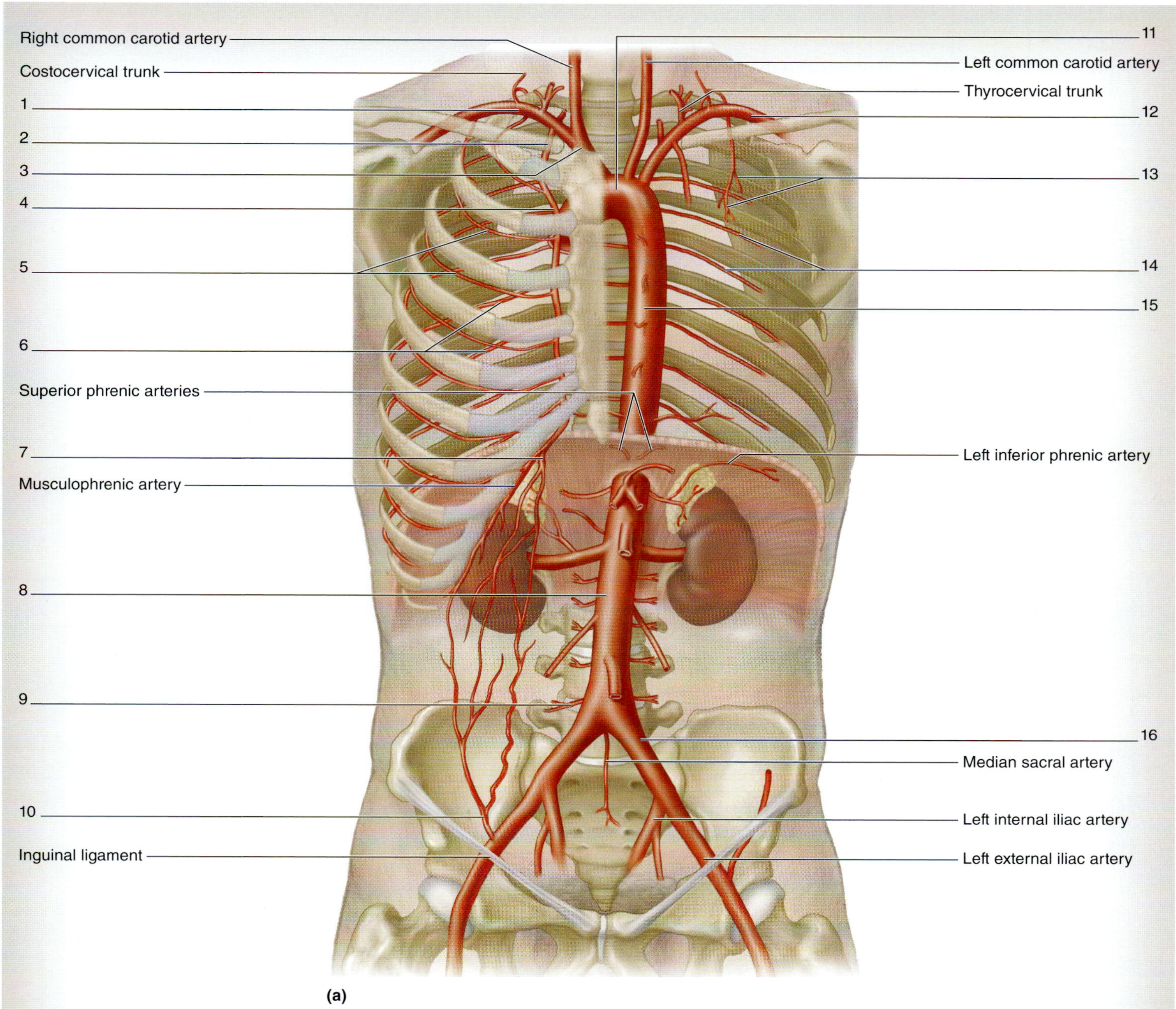

Figure 22.11 Circulation to the Thoracic and Abdominal Walls. Use the terms listed to fill in the numbered labels in the figure. Some answers may be used more than once.

(a) Arterial Supply

- ☐ anterior intercostal arteries
- ☐ aortic arch
- ☐ brachiocephalic trunk
- ☐ descending abdominal aorta
- ☐ descending thoracic aorta
- ☐ inferior epigastric artery
- ☐ internal thoracic artery
- ☐ left common iliac artery
- ☐ left subclavian artery
- ☐ posterior intercostal arteries (1–2)
- ☐ posterior intercostal arteries (3–11)
- ☐ right lumbar artery
- ☐ right subclavian artery
- ☐ superior epigastric artery

(continued on next page)

(continued from previous page)

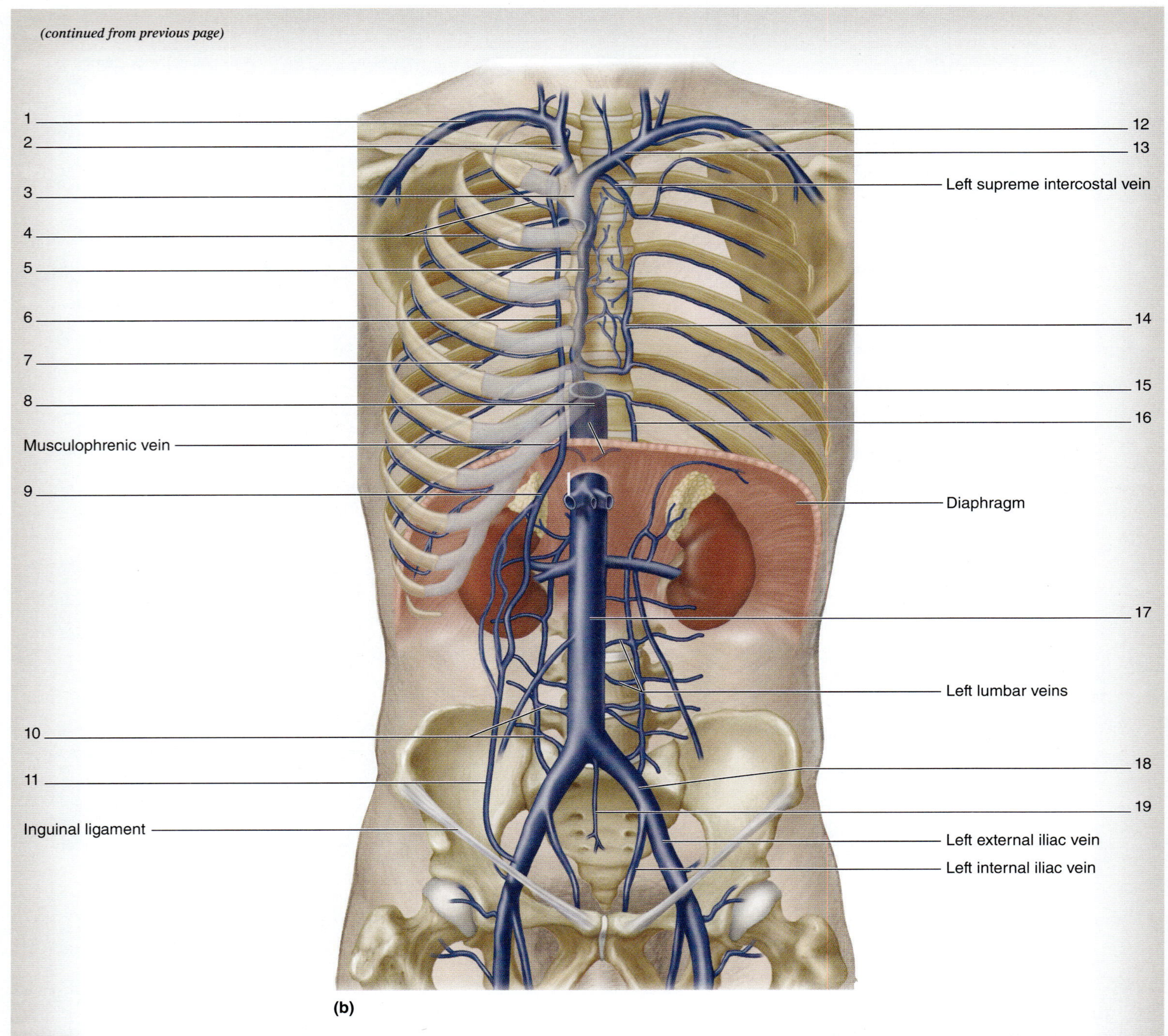

Figure 22.11 Circulation to the Thoracic and Abdominal Walls *(continued)*. Use the terms listed to fill in the numbered labels in the figure. Some answers may be used more than once.

(b) Venous Drainage

- ☐ accessory hemiazygos vein
- ☐ anterior intercostal veins
- ☐ azygos vein
- ☐ hemiazygos vein
- ☐ inferior vena cava
- ☐ internal thoracic vein
- ☐ left brachiocephalic vein
- ☐ left common iliac vein
- ☐ left posterior intercostal vein
- ☐ left subclavian vein
- ☐ median sacral vein
- ☐ right brachiocephalic vein
- ☐ right inferior epigastric vein
- ☐ right lumbar veins
- ☐ right posterior intercostal vein
- ☐ right subclavian vein
- ☐ right superior epigastric vein
- ☐ superior vena cava

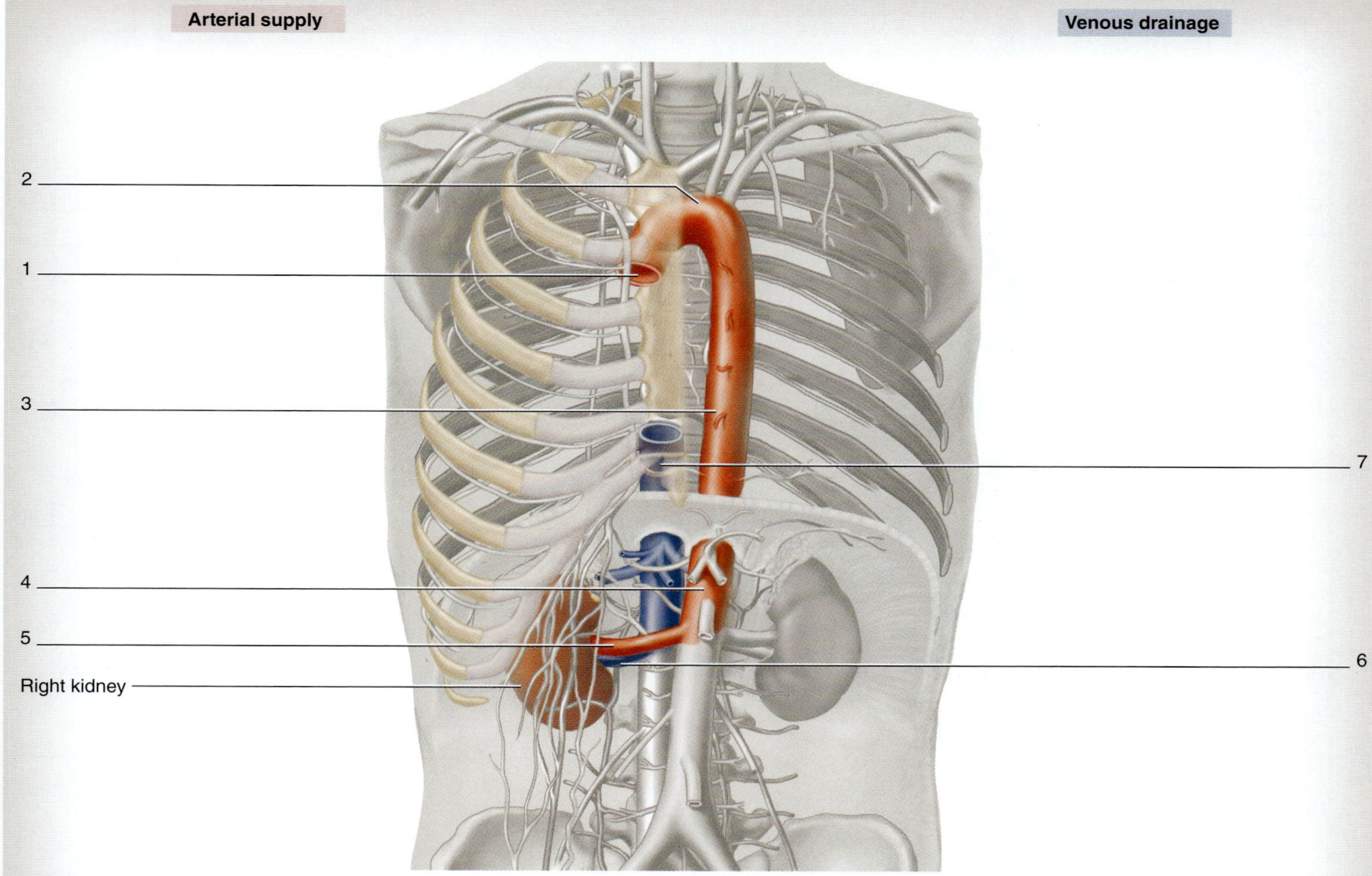

Arterial supply

Venous drainage

2
1
3
7
4
5
6
Right kidney

Figure 22.12 Circulation from the Left Ventricle of the Heart to the Right Kidney and Back to the Right Atrium of the Heart. Use the terms listed to fill in the numbered labels in the figure.

- ☐ aortic arch
- ☐ ascending aorta
- ☐ descending abdominal aorta
- ☐ descending thoracic aorta
- ☐ inferior vena cava
- ☐ right renal artery
- ☐ right renal vein

EXERCISE 22.11

CIRCULATION TO THE ABDOMINAL CAVITY

1. Observe the abdominal cavity on a prosected human cadaver and/or observe classroom models or charts demonstrating blood vessels of the abdominal cavity.
2. Identify the *arteries* listed in **figure 22.13** that supply blood to structures in the abdomen, using the textbook as a guide. Then label them in figure 22.13.
3. Venous drainage of abdominal organs is unique in that it is an example of a portal system, called the **hepatic portal system.** In this system there are two capillary beds—the first in an abdominal organ, and the second in the liver—connected to each other by a **portal vein.** An artery supplies blood to the first capillary bed, which is located in an abdominal organ such as the stomach, intestine, or spleen. Blood drains from abdominal organs into three veins: the **splenic, inferior mesenteric,** and **superior mesenteric** veins. These veins then drain into the **hepatic portal vein,** which carries blood to the second capillary bed in the liver. This blood is high in nutrient content, but also may be transporting toxins, bacteria, and other potentially dangerous substances. Because the blood flows from the abdominal organs directly to the liver, nutrients, drugs, and pathogens may be removed from the blood before the blood enters the general circulation. Thus, the liver is said to have "first pass" at the blood that drains from the abdominal organs. Finally, venous blood drains

(continued on next page)

(continued from previous page)

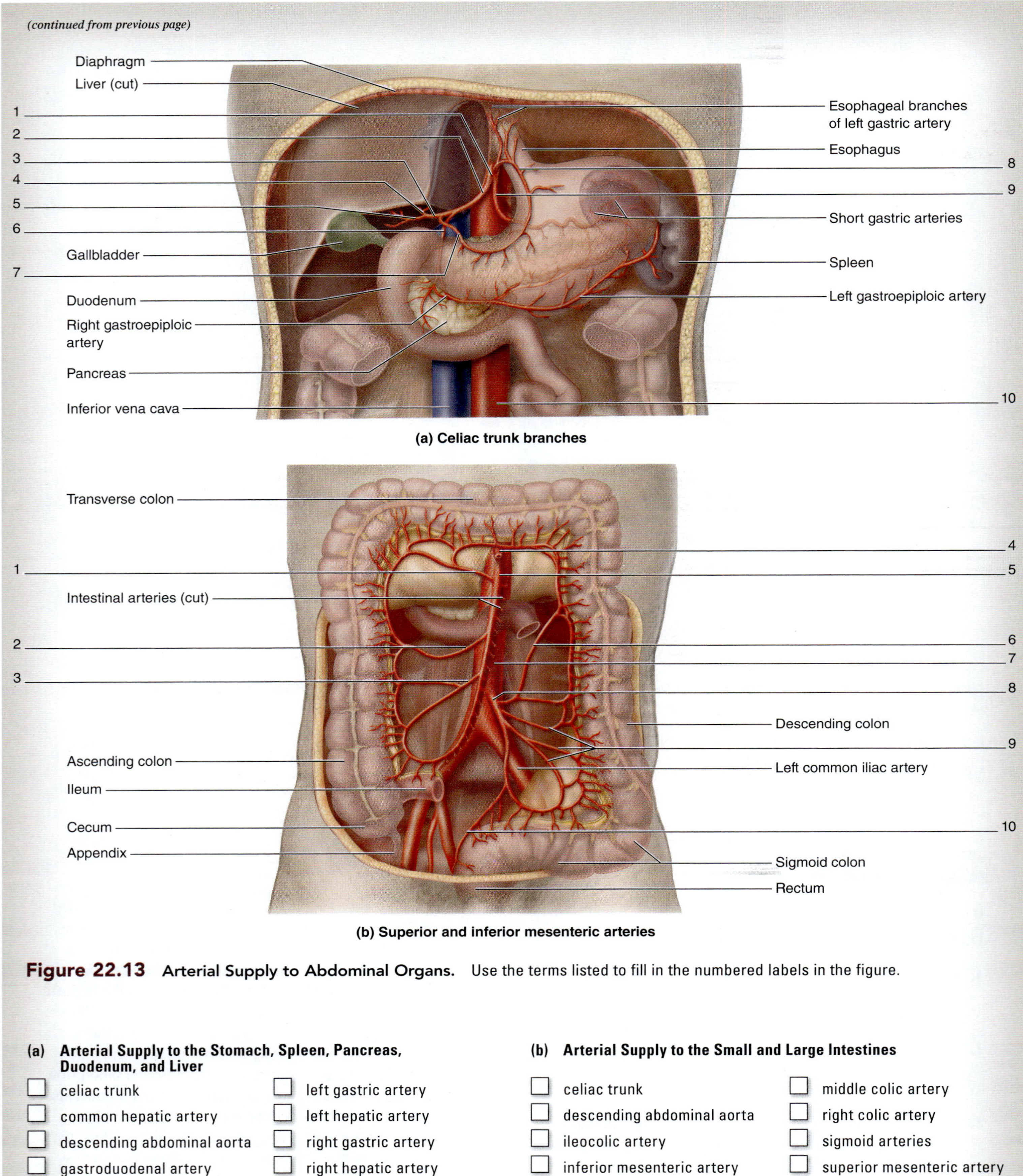

(a) Celiac trunk branches

(b) Superior and inferior mesenteric arteries

Figure 22.13 Arterial Supply to Abdominal Organs. Use the terms listed to fill in the numbered labels in the figure.

(a) Arterial Supply to the Stomach, Spleen, Pancreas, Duodenum, and Liver

- ☐ celiac trunk
- ☐ common hepatic artery
- ☐ descending abdominal aorta
- ☐ gastroduodenal artery
- ☐ hepatic artery proper
- ☐ left gastric artery
- ☐ left hepatic artery
- ☐ right gastric artery
- ☐ right hepatic artery
- ☐ splenic artery

(b) Arterial Supply to the Small and Large Intestines

- ☐ celiac trunk
- ☐ descending abdominal aorta
- ☐ ileocolic artery
- ☐ inferior mesenteric artery
- ☐ left colic artery
- ☐ middle colic artery
- ☐ right colic artery
- ☐ sigmoid arteries
- ☐ superior mesenteric artery
- ☐ superior rectal artery

from liver capillaries into **hepatic veins,** which carry it to the inferior vena cava and back to the heart.

4. Identify the *veins* listed in **figure 22.14** that compose the hepatic portal system, using the textbook as a guide. Then label them in figure 22.14.
5. The pathway a drop of blood takes to get from the abdominal aorta to the *spleen* and back to the right atrium of the heart is shaded in **figure 22.15.** Trace this flow of blood by writing in the names of the vessels in the figure. Label the vessels in order, starting at number 1, so as to figuratively trace the pathway and name the vessels the blood flows through.
6. The pathway a drop of blood takes to get from the abdominal aorta to the *duodenum* and back to the right atrium of the heart is shaded in **figure 22.16.** Trace this flow of blood by writing in the names of the vessels in the figure. Label the vessels in order, starting at number 1, so as to figuratively trace the pathway and name the vessels the blood flows through.
7. The pathway a drop of blood takes to get from the abdominal aorta to the *sigmoid colon* and back to the right atrium of the heart is shaded in **figure 22.17.** Trace this flow of blood by writing in the names of the vessels in the figure. Label the vessels in order, starting at number 1, so as to figuratively trace the pathway and name the vessels the blood flows through.

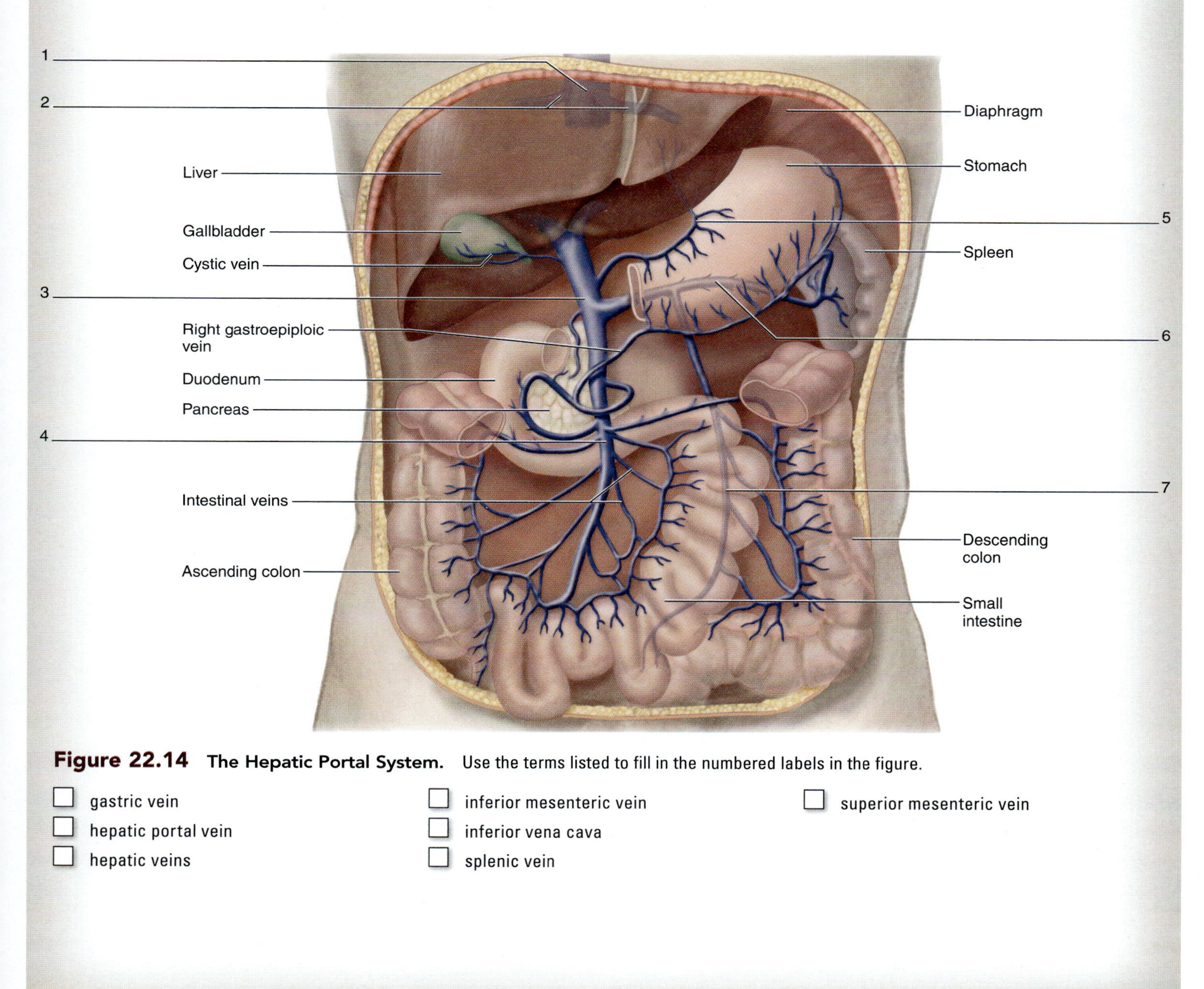

Figure 22.14 The Hepatic Portal System. Use the terms listed to fill in the numbered labels in the figure.

- ☐ gastric vein
- ☐ hepatic portal vein
- ☐ hepatic veins
- ☐ inferior mesenteric vein
- ☐ inferior vena cava
- ☐ splenic vein
- ☐ superior mesenteric vein

(continued on next page)

(continued from previous page)

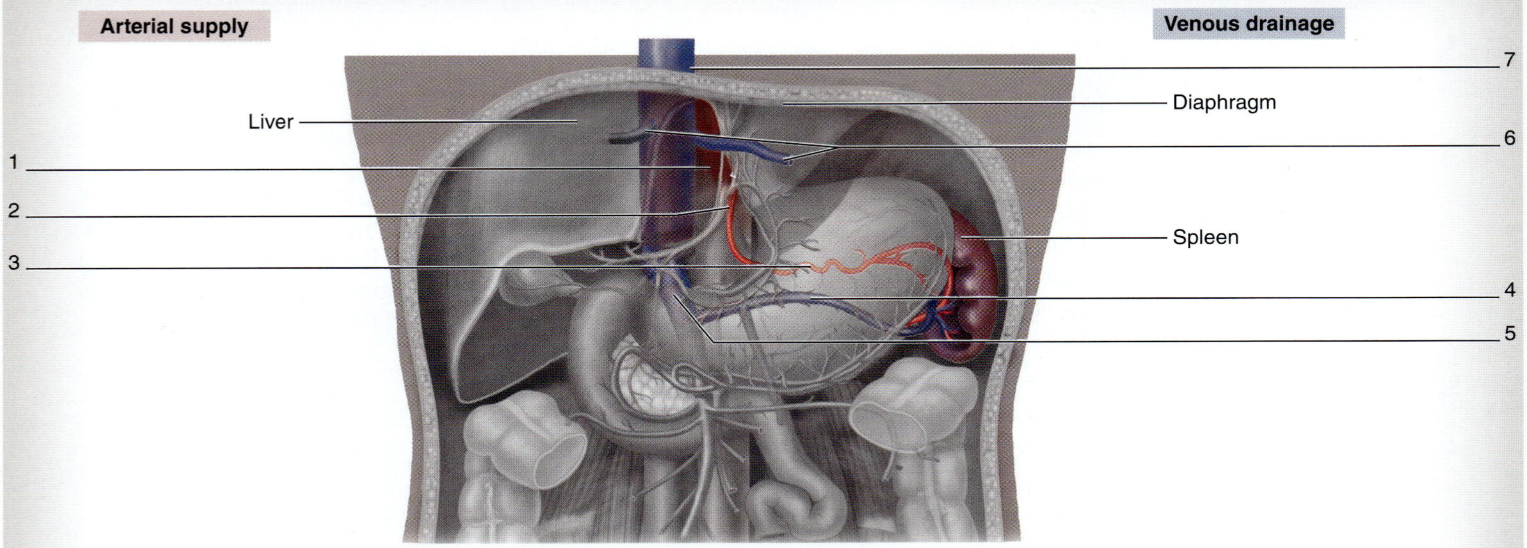

Figure 22.15 **Circulation from the Abdominal Aorta to the Spleen and Back to the Right Atrium of the Heart.** Use the terms listed to fill in the numbered labels in the figure.

- ☐ abdominal aorta
- ☐ celiac trunk
- ☐ hepatic portal vein
- ☐ hepatic veins
- ☐ inferior vena cava
- ☐ splenic artery
- ☐ splenic vein

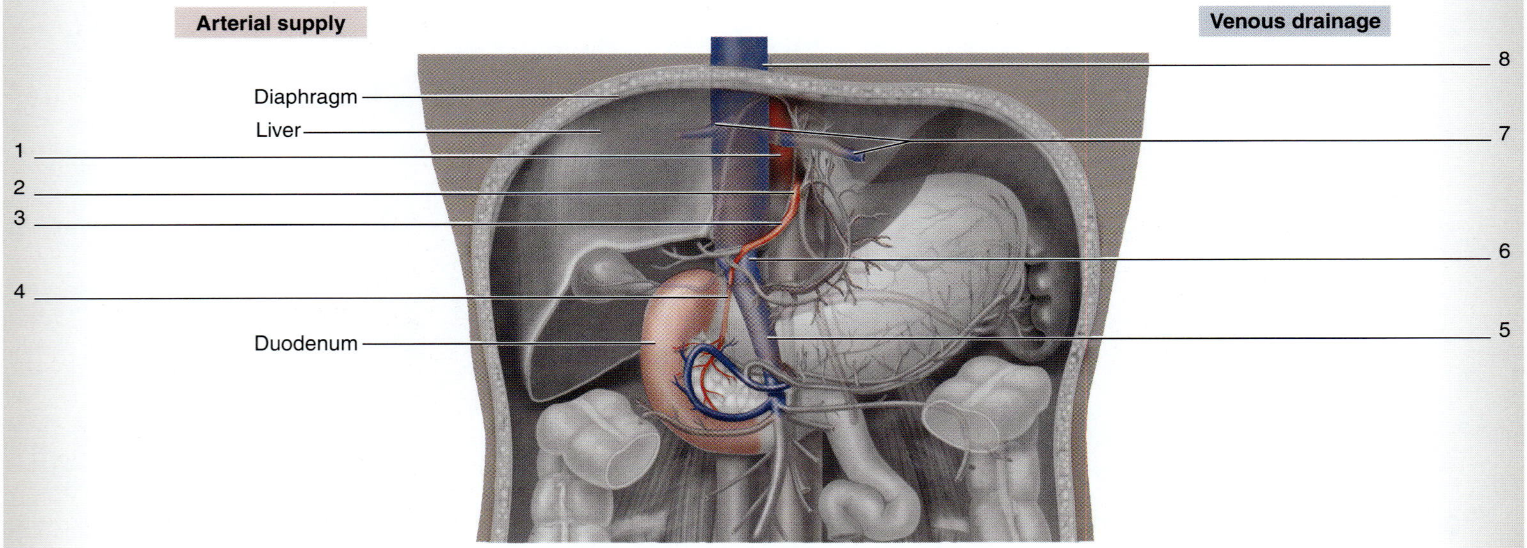

Figure 22.16 **Circulation from the Abdominal Aorta to the Duodenum and Back to the Right Atrium of the Heart.** Use the terms listed to fill in the numbered labels in the figure.

- ☐ abdominal aorta
- ☐ celiac trunk
- ☐ common hepatic artery
- ☐ gastroduodenal artery
- ☐ hepatic portal vein
- ☐ hepatic veins
- ☐ inferior vena cava
- ☐ superior mesenteric vein

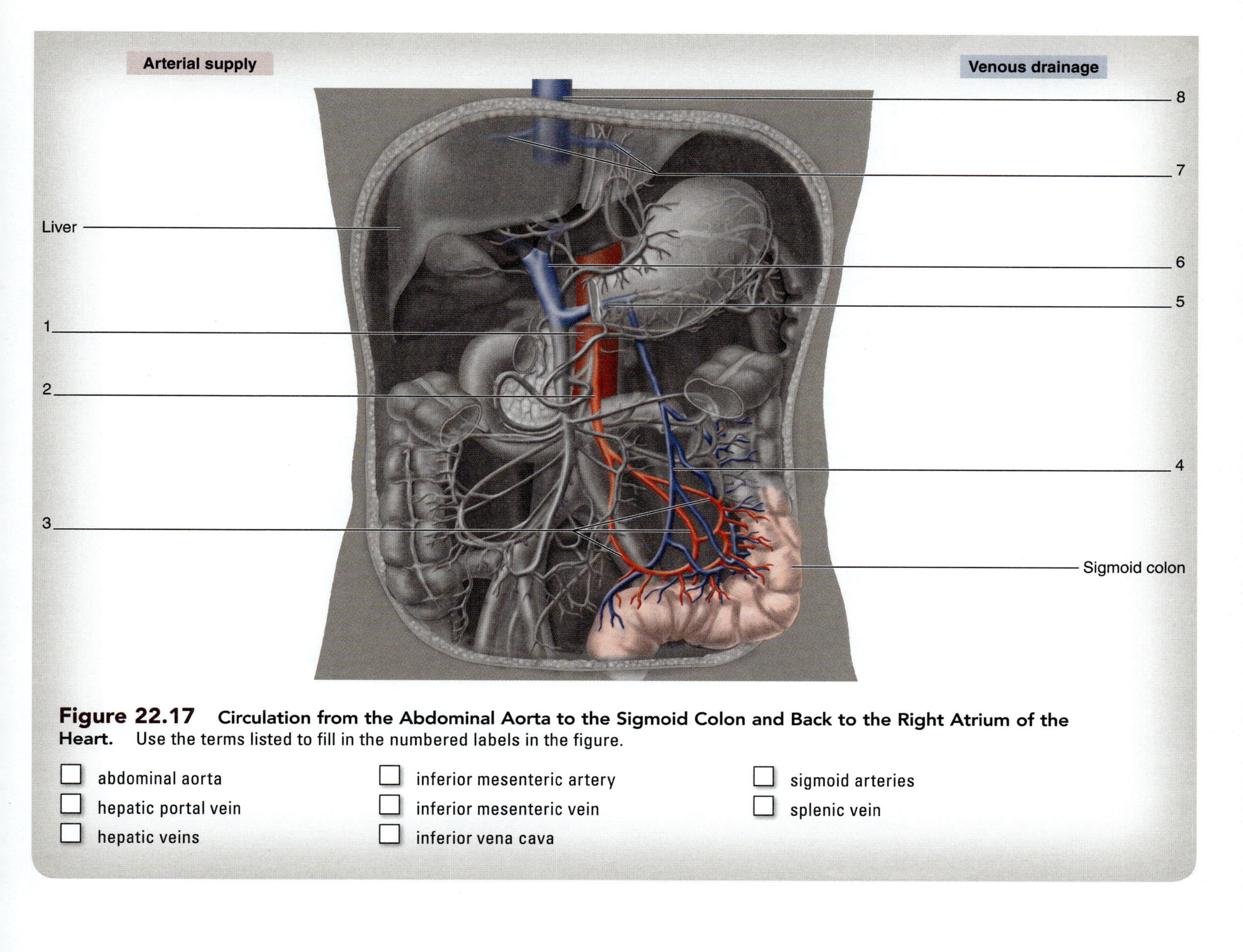

Figure 22.17 Circulation from the Abdominal Aorta to the Sigmoid Colon and Back to the Right Atrium of the Heart. Use the terms listed to fill in the numbered labels in the figure.

- ☐ abdominal aorta
- ☐ hepatic portal vein
- ☐ hepatic veins
- ☐ inferior mesenteric artery
- ☐ inferior mesenteric vein
- ☐ inferior vena cava
- ☐ sigmoid arteries
- ☐ splenic vein

INTEGRATE

CLINICAL VIEW
Portal-Systemic Anastomoses

Veins of the portal system are unique in that they have *no valves*. Thus, when blood backs up in this system, it can back up into systemic veins and will attempt to take an alternate route to the inferior vena cava through anastomoses (s. anastomosis). An **anastomosis** (*anastomo*, to furnish with a mouth) is a connection between two blood vessels. The major anastomoses of the portal circulation are found in veins of the esophagus and the rectum, and surrounding the umbilicus. These vessels are common sites of **varicosities** (*varix*, a dilated vein). Varicosities of the esophageal veins are extremely serious because rupture of varicose veins of the esophagus can result in an individual bleeding to death. Varicosities of the rectal veins cause **hemorrhoids** (*haimo*, blood, + *rhoia*, a flow), and varicosities of the umbilical veins form a *caput medusa*—the varicose vessels radiating out from the umbilicus resemble the snakes on Medusa's head (from Greek mythology). One cause of portal-systemic anastomoses comes from the long-term effects of alcoholism that result in liver damage leading to **cirrhosis** (*Kirrhos*, yellow, + *-osis*, condition). With cirrhosis, normal liver tissue is replaced over time with scar tissue, decreasing the size of the liver and the number of capillaries within. This creates resistance to blood flow through the liver, which causes blood to back up into the veins of the hepatic portal system.

EXERCISE 22.12

CIRCULATION TO THE UPPER LIMB

1. Observe the upper limb of a prosected human cadaver or observe classroom models or charts demonstrating blood vessels of the upper limb.
2. Identify the *arteries* listed in **figure 22.18*a*** that supply blood to the upper limb, using the textbook as a guide. Then label them in figure 22.18*a*.
3. Identify the *veins* listed in figure 22.18*b* that drain blood from the upper limb, using the textbook as a guide. Then label them in figure 22.18*b*.

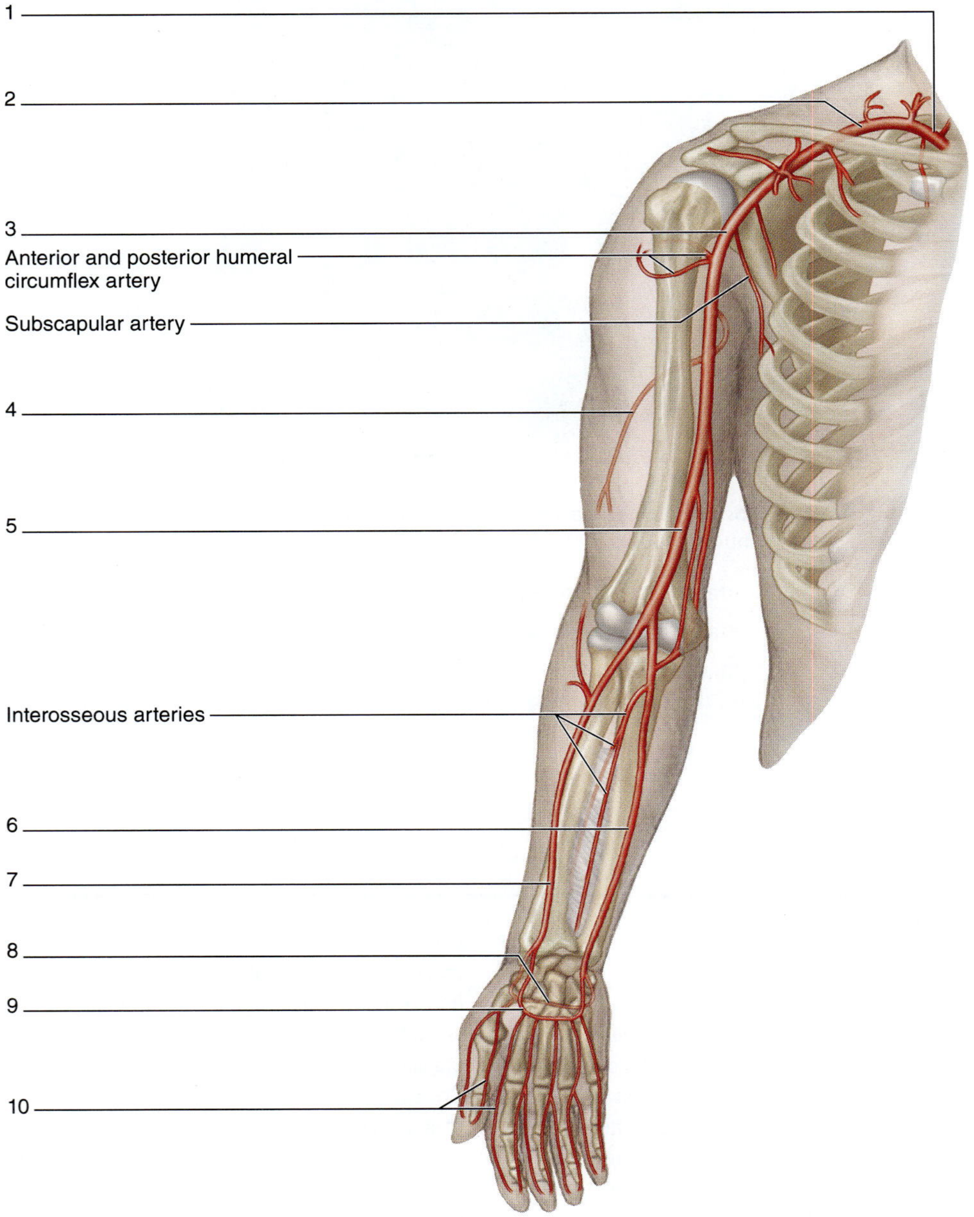

1 __________

2 __________

3 __________

Anterior and posterior humeral circumflex artery

Subscapular artery

4 __________

5 __________

Interosseous arteries

6 __________

7 __________

8 __________

9 __________

10 __________

(a) Arteries of right upper limb

Figure 22.18 **Circulation to the Upper Limb.** Use the terms listed to fill in the numbered labels in the figure.

(a) Arterial Supply

- ☐ axillary artery
- ☐ brachial artery
- ☐ brachiocephalic trunk (artery)
- ☐ deep brachial artery
- ☐ deep palmar arch
- ☐ digital arteries
- ☐ radial artery
- ☐ subclavian artery
- ☐ superficial palmar arch
- ☐ ulnar artery

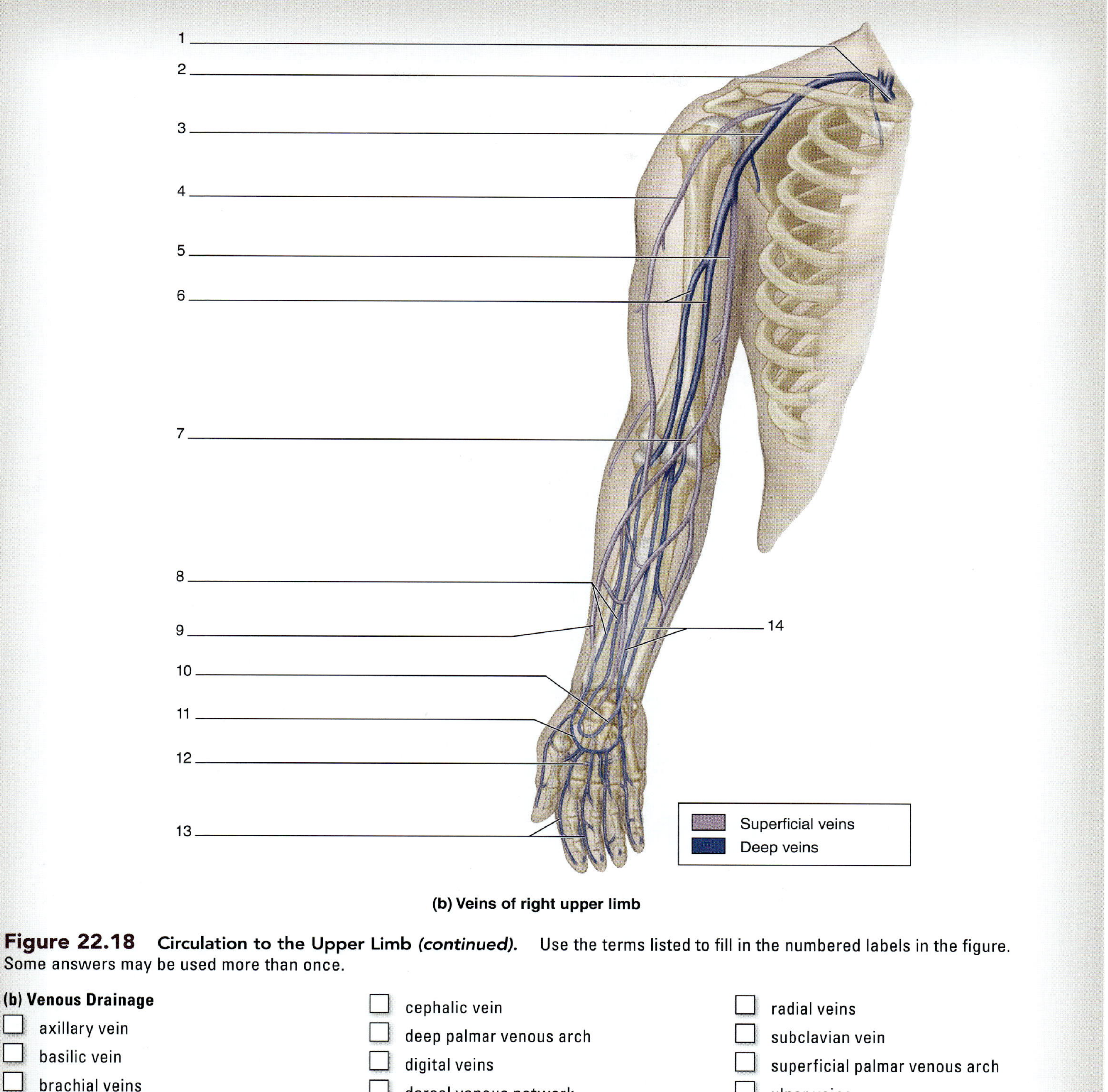

(b) Veins of right upper limb

Figure 22.18 Circulation to the Upper Limb *(continued)*. Use the terms listed to fill in the numbered labels in the figure. Some answers may be used more than once.

(b) Venous Drainage

- ☐ axillary vein
- ☐ basilic vein
- ☐ brachial veins
- ☐ brachiocephalic vein
- ☐ cephalic vein
- ☐ deep palmar venous arch
- ☐ digital veins
- ☐ dorsal venous network
- ☐ median cubital vein
- ☐ radial veins
- ☐ subclavian vein
- ☐ superficial palmar venous arch
- ☐ ulnar veins

(continued on next page)

(continued from previous page)

4. *Superficial Trace*: The pathway a drop of blood takes to get from the aortic arch to the *anterior surface of the index finger* and back along a superficial route to the superior vena cava is shaded in **figure 22.19.** Trace this flow of blood by writing in the names of the vessels in the figure. Label the vessels in order, starting at number 1, so as to figuratively trace the pathway and name the vessels the blood flows through.

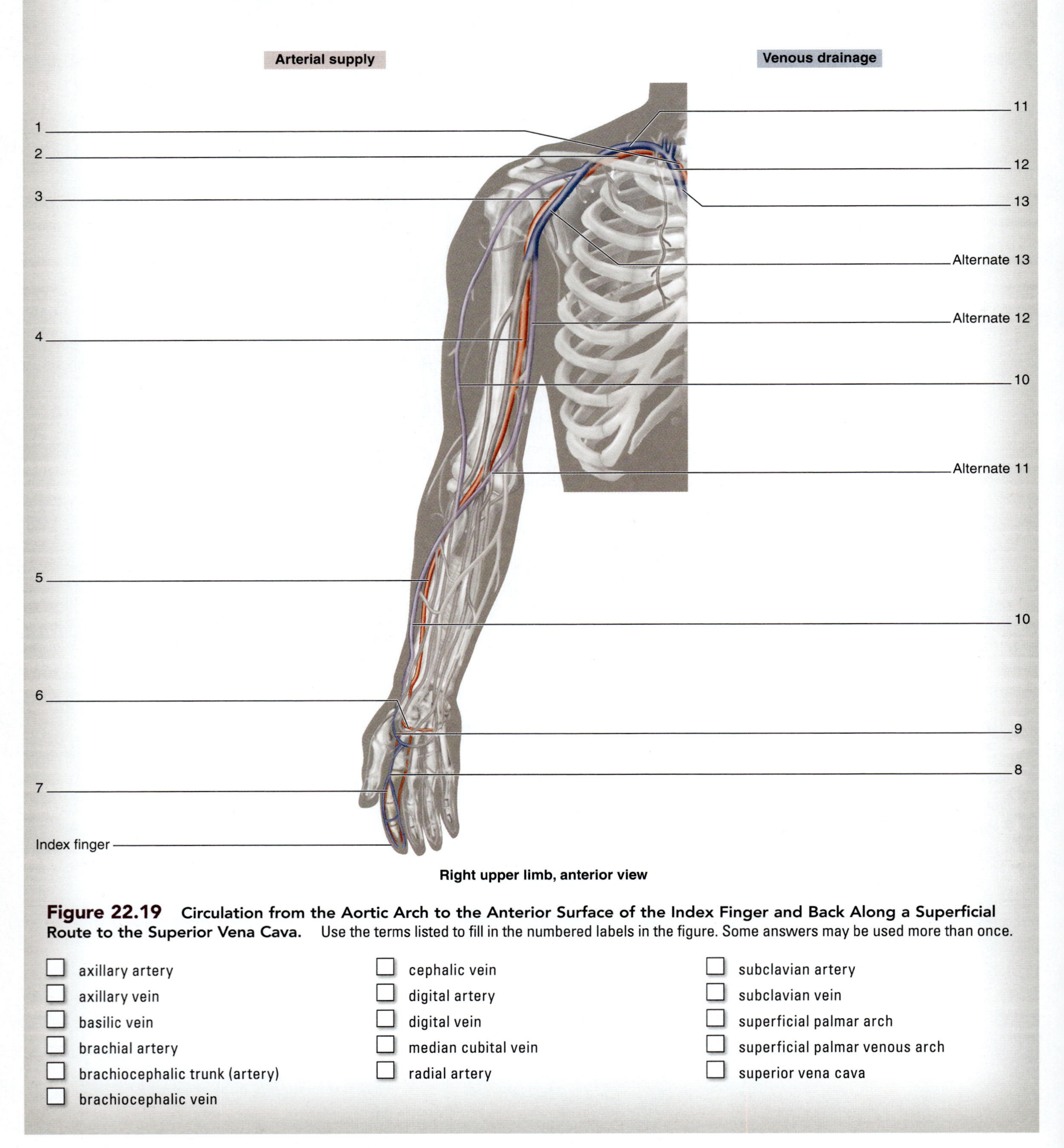

Figure 22.19 **Circulation from the Aortic Arch to the Anterior Surface of the Index Finger and Back Along a Superficial Route to the Superior Vena Cava.** Use the terms listed to fill in the numbered labels in the figure. Some answers may be used more than once.

- ☐ axillary artery
- ☐ axillary vein
- ☐ basilic vein
- ☐ brachial artery
- ☐ brachiocephalic trunk (artery)
- ☐ brachiocephalic vein
- ☐ cephalic vein
- ☐ digital artery
- ☐ digital vein
- ☐ median cubital vein
- ☐ radial artery
- ☐ subclavian artery
- ☐ subclavian vein
- ☐ superficial palmar arch
- ☐ superficial palmar venous arch
- ☐ superior vena cava

5. *Deep Trace:* The pathway a drop of blood takes to get from the aortic arch to the *capitate bone in the wrist* and back along a deep route to the superior vena cava is shaded in **figure 22.20.** Trace this flow of blood by writing in the names of the vessels in the figure. Label the vessels in order, starting at number 1, so as to figuratively trace the pathway and name the vessels the blood flows through.

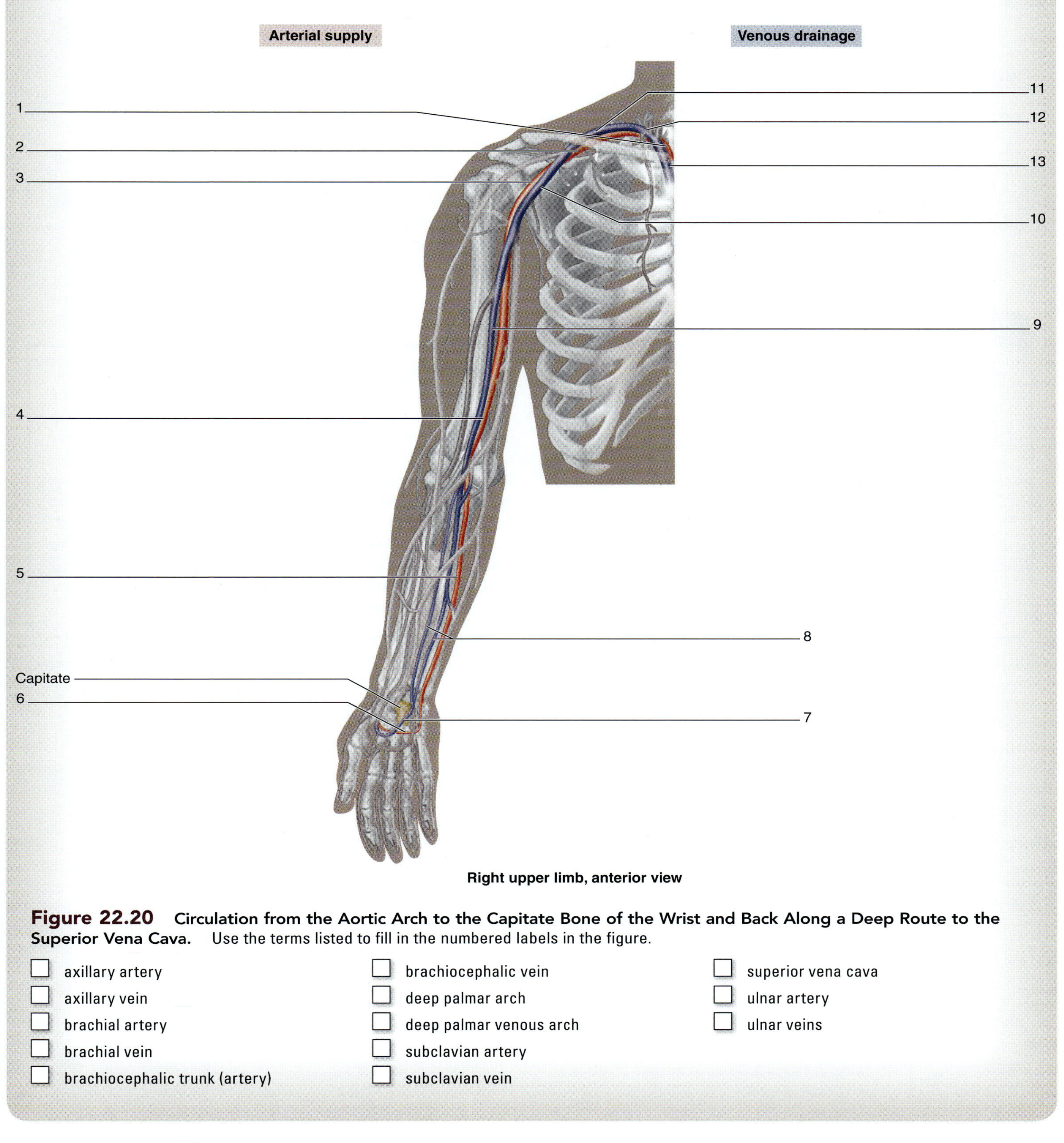

Figure 22.20 Circulation from the Aortic Arch to the Capitate Bone of the Wrist and Back Along a Deep Route to the Superior Vena Cava. Use the terms listed to fill in the numbered labels in the figure.

- ☐ axillary artery
- ☐ axillary vein
- ☐ brachial artery
- ☐ brachial vein
- ☐ brachiocephalic trunk (artery)
- ☐ brachiocephalic vein
- ☐ deep palmar arch
- ☐ deep palmar venous arch
- ☐ subclavian artery
- ☐ subclavian vein
- ☐ superior vena cava
- ☐ ulnar artery
- ☐ ulnar veins

EXERCISE 22.13

CIRCULATION TO THE LOWER LIMB

1. Observe the lower limb of a prosected human cadaver or observe classroom models or charts demonstrating blood vessels of the lower limb.
2. Identify the *arteries* listed in **figure 22.21*a*** that supply blood to the lower limb, using the textbook as a guide. Then label them in figure 22.21*a*.
3. Identify the *veins* listed in figure 22.21*b* that drain blood from the lower limb, using the textbook as a guide. Then label them in figure 22.21*b*.
4. *Superficial Trace:* The pathway a drop of blood takes to get from the abdominal aorta to the *dorsal surface of the big toe (hallux)* and back along a superficial route to the inferior vena cava is shaded in **figure 22.22.** Trace this flow of blood by writing in the names of the vessels in the figure. Label the vessels in order, starting at number 1, so as to figuratively trace the pathway and name the vessels the blood flows through.
5. *Deep Trace:* The pathway a drop of blood takes to get from the abdominal aorta to the *cuboid bone in the foot* and back along a deep route to the inferior vena cava is shaded in **figure 22.23.** Trace this flow of blood by writing in the names of the vessels in the figure. Label the vessels in order, starting at number 1, so as to figuratively trace the pathway and name the vessels the blood flows through.

INTEGRATE

CLINICAL VIEW

Cardiac Catheterization via the Femoral Artery

When a patient with a blockage in one of the coronary arteries requires balloon angioplasty and stenting to open up the blockage, the physician performing the procedure takes the route of least resistance to thread the catheter with the balloon through the circulatory system to get to the blocked artery. Ironically, this is not the shortest path. Instead, the catheter is introduced into the cardiovascular system via the femoral artery within the *femoral triangle* of the thigh (see chapter 13). The femoral artery is relatively easy to access because the only tissues that lie superficial to it are skin, connective tissue, and fat. Once a catheter is introduced into the femoral artery, it can then be threaded "backward" to the heart via the following pathway: femoral artery, external iliac artery, common iliac artery, abdominal aorta, thoracic aorta, aortic arch, right or left coronary artery. Although this seems like a long distance, the path is relatively straightforward, and the catheter travels through these relatively large vessels until it gets to the coronary artery itself. Once the catheter is within the target vessel, a balloon is inflated, which presses the blockage against the wall of the artery, thus opening up the blockage. After the angioplasty is complete, a stent may be placed to prevent the artery from closing off again. A stent is a piece of mesh that holds the artery open.

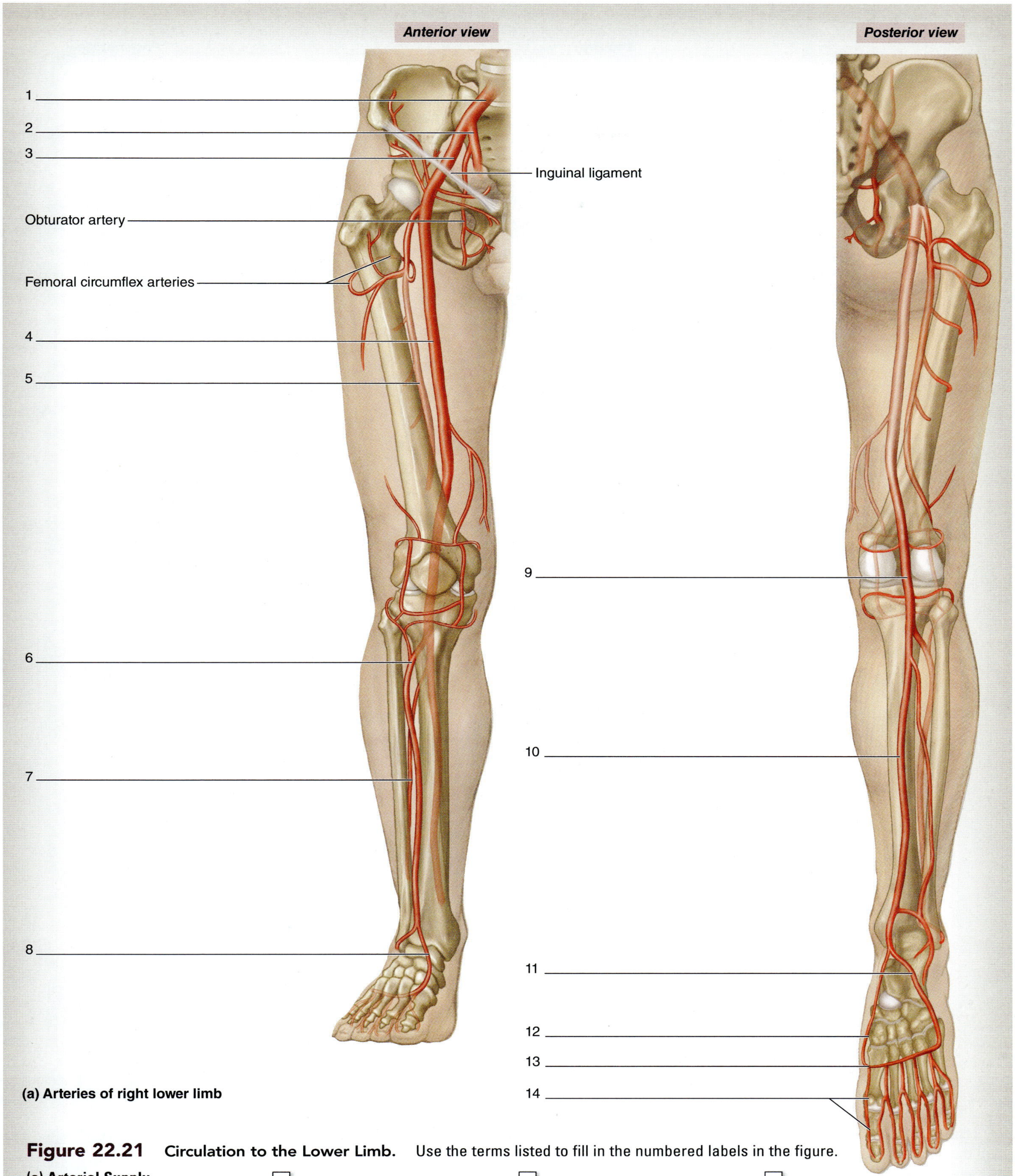

Anterior view

Posterior view

1
2
3
Inguinal ligament
Obturator artery
Femoral circumflex arteries
4
5
6
7
8
9
10
11
12
13
14

(a) Arteries of right lower limb

Figure 22.21 **Circulation to the Lower Limb.** Use the terms listed to fill in the numbered labels in the figure.

(a) Arterial Supply

- ☐ anterior tibial artery
- ☐ common iliac artery
- ☐ deep femoral artery
- ☐ digital arteries
- ☐ dorsalis pedis artery
- ☐ external iliac artery
- ☐ femoral artery
- ☐ fibular artery
- ☐ internal iliac artery
- ☐ lateral plantar artery
- ☐ medial plantar artery
- ☐ plantar arterial arch
- ☐ popliteal artery
- ☐ posterior tibial artery

(continued on next page)

(continued from previous page)

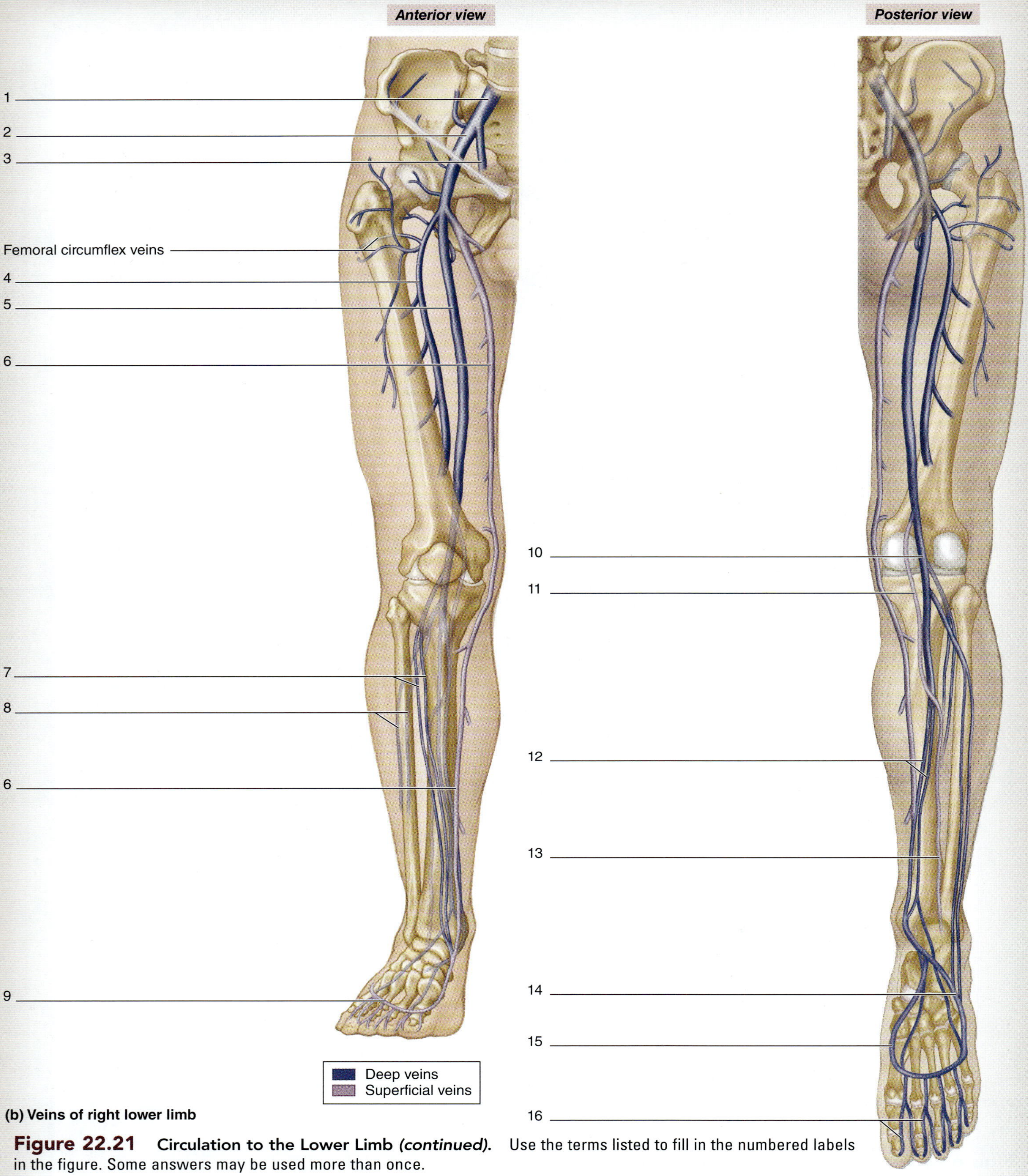

Anterior view

1

2

3

Femoral circumflex veins

4

5

6

7

8

6

9

Deep veins

Superficial veins

(b) Veins of right lower limb

Posterior view

10

11

12

13

14

15

16

Figure 22.21 **Circulation to the Lower Limb *(continued)*.** Use the terms listed to fill in the numbered labels in the figure. Some answers may be used more than once.

(b) Venous Drainage

- ☐ anterior tibial veins
- ☐ common iliac vein
- ☐ deep femoral vein
- ☐ digital veins
- ☐ dorsal venous arch
- ☐ external iliac vein
- ☐ femoral vein
- ☐ fibular veins
- ☐ great saphenous vein
- ☐ internal iliac vein
- ☐ lateral plantar vein
- ☐ medial plantar vein
- ☐ popliteal vein
- ☐ posterior tibial veins
- ☐ small saphenous vein

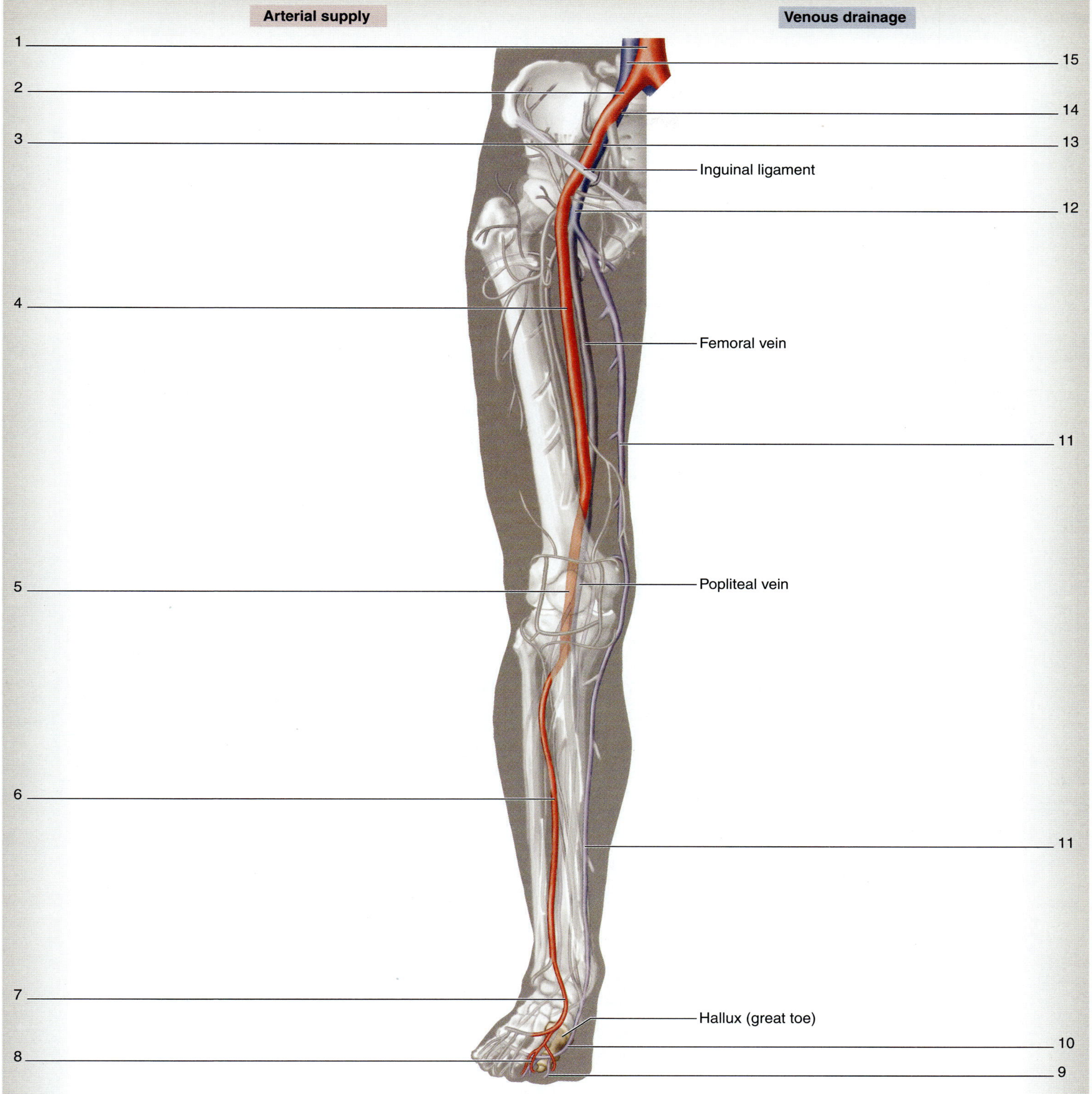

Arterial supply

Venous drainage

Inguinal ligament

Femoral vein

Popliteal vein

Hallux (great toe)

Right lower limb, anterior view

Figure 22.22 **Circulation from the Abdominal Aorta to the Dorsal Surface of the Big Toe (Hallux) and Back Along a Superficial Route to the Inferior Vena Cava.** Use the terms listed to fill in the numbered labels in the figure.

- ☐ abdominal aorta
- ☐ anterior tibial artery
- ☐ common iliac artery
- ☐ common iliac vein
- ☐ digital artery
- ☐ digital vein
- ☐ dorsal venous arch
- ☐ dorsalis pedis artery
- ☐ external iliac artery
- ☐ external iliac vein
- ☐ femoral artery
- ☐ femoral vein
- ☐ great saphenous vein
- ☐ inferior vena cava
- ☐ popliteal artery

(continued on next page)

(continued from previous page)

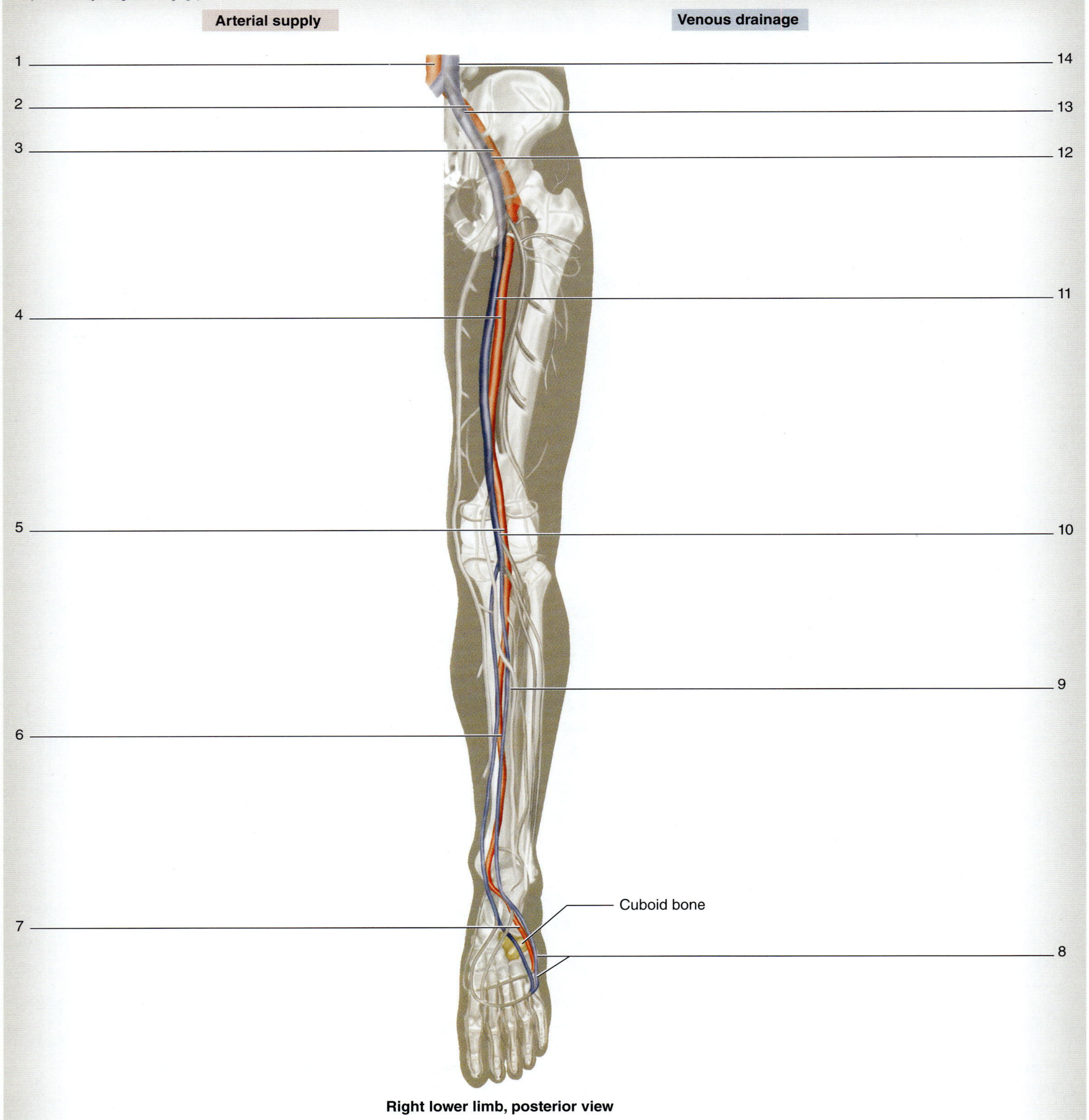

Arterial supply

Venous drainage

1 ____

2 ____

3 ____

4 ____

5 ____

6 ____

7 ____

8 ____

9 ____

10 ____

11 ____

12 ____

13 ____

14 ____

Cuboid bone

Right lower limb, posterior view

Figure 22.23 Circulation from the Abdominal Aorta to the Cuboid Bone in the Foot and Back Along a Deep Route to the Inferior Vena Cava. Use the terms listed to fill in the numbered labels in the figure.

- ☐ abdominal aorta
- ☐ common iliac artery
- ☐ common iliac vein
- ☐ external iliac artery
- ☐ external iliac vein
- ☐ femoral artery
- ☐ femoral vein
- ☐ inferior vena cava
- ☐ lateral plantar artery
- ☐ lateral plantar veins
- ☐ popliteal artery
- ☐ popliteal vein
- ☐ posterior tibial artery
- ☐ posterior tibial vein

Fetal Circulation

The lungs are nonfunctional in the fetus and need only a small amount of blood to support the developing lung tissue. This blood must be oxygenated blood coming from the fetal respiratory organ: the placenta. Once the fetus is born, the circulation must change as the lungs replace the placenta as the respiratory organs. In addition, the blood returning from the placenta via the umbilical vein bypasses the liver through the ductus venosus. Thus, there are a number of **shunts** present in the fetal circulation that direct blood away from the lungs, to and from the placenta, and away from the liver. These shunts must close at birth to establish the normal postnatal circulatory pathways. The following exercise involves identifying the unique cardiovascular structures of the fetal circulation, tracing the flow of blood through the fetal circulation, and identifying the postnatal structures that are remnants of the fetal circulation.

EXERCISE 22.14

FETAL CIRCULATION

1. Identify the fetal circulatory system structures listed in **figure 22.24,** using the textbook as a guide.
2. List the structures that the blood passes through as it is transported from the left ventricle of the fetal heart to the placenta and back to the right atrium of the fetal heart.

____
____
____
____
____
____
____
____
____
____
____
____

3. Write in the names of the postnatal structures that are remnants of the fetal circulation, and describe each structure's function in the fetus in **table 22.4.**

Table 22.4 Fetal Cardiovascular Structures and Associated Postnatal Structures

Fetal Cardiovascular Structure	Postnatal Structure	Function of Fetal Cardiovascular Structure
Ductus Arteriosus		
Ductus Venosus		
Foramen Ovale		
Umbilical Arteries		
Umbilical Vein		

(continued on next page)

(continued from previous page)

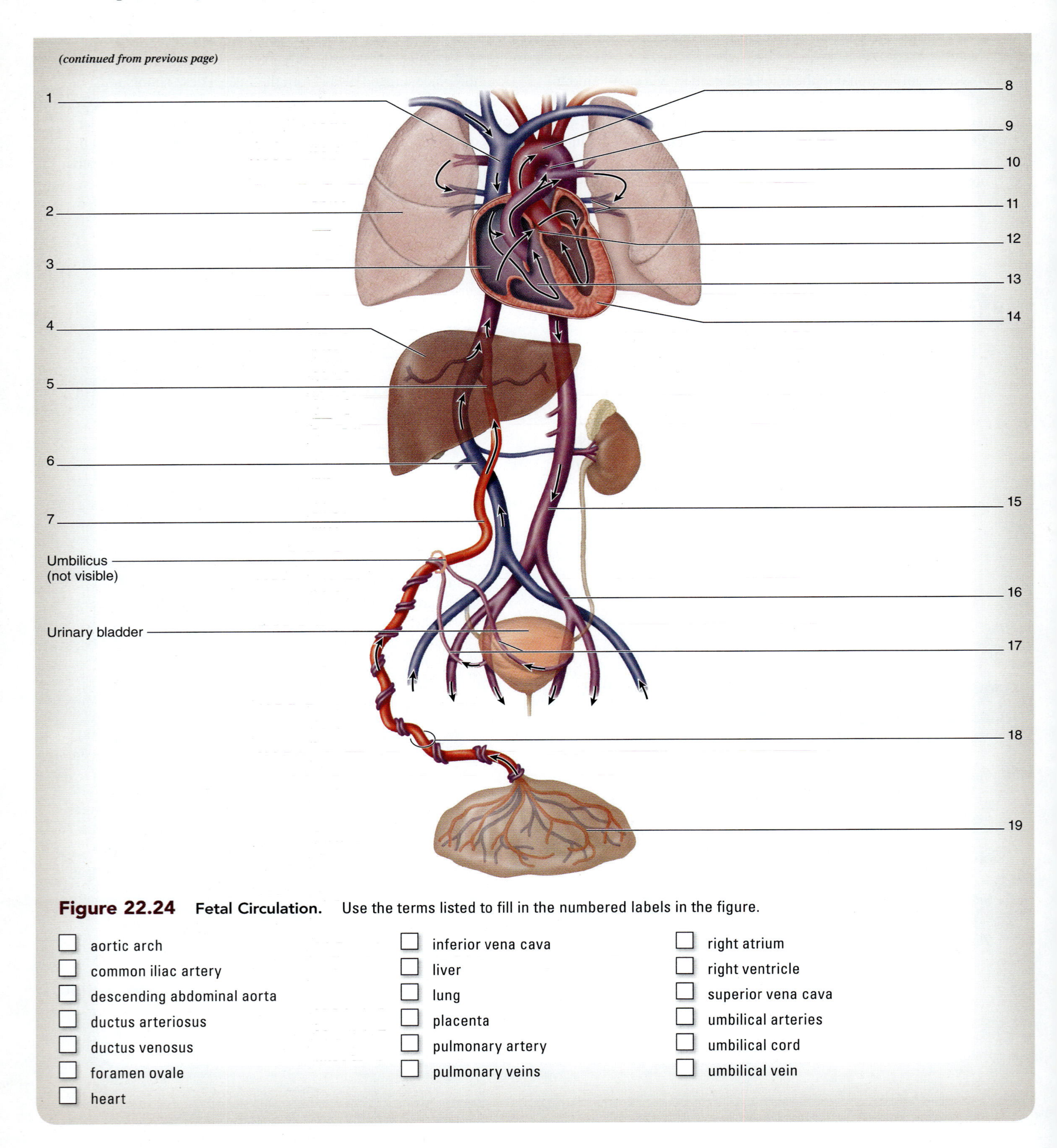

Figure 22.24 Fetal Circulation. Use the terms listed to fill in the numbered labels in the figure.

- ☐ aortic arch
- ☐ common iliac artery
- ☐ descending abdominal aorta
- ☐ ductus arteriosus
- ☐ ductus venosus
- ☐ foramen ovale
- ☐ heart
- ☐ inferior vena cava
- ☐ liver
- ☐ lung
- ☐ placenta
- ☐ pulmonary artery
- ☐ pulmonary veins
- ☐ right atrium
- ☐ right ventricle
- ☐ superior vena cava
- ☐ umbilical arteries
- ☐ umbilical cord
- ☐ umbilical vein

PHYSIOLOGY

Blood Pressure and Pulse

The blood vessels observed in the histology and gross anatomy exercises in this chapter are the conduits that provide the passageways for blood to flow through the body. As the blood flows through these vessels, it exerts force on the vessel walls. This force (per unit area) is termed **blood pressure.** Blood flow in both the systemic circulation and pulmonary circulation is driven by blood pressure. Specifically, there must be a pressure gradient to drive the flow from the heart through the vessels. Contraction of the ventricles of the heart generates this pressure gradient. It follows then, that the greater the blood pressure gradient, the greater the blood flow through the vessels. Significant drops in blood pressure may lead to insufficient oxygen and nutrient delivery to the tissues. Without sufficient oxygen and nutrients, tissue death (necrosis) may occur.

Blood pressure within blood vessels is pulsatile (not smooth); that is, blood pressure waxes and wanes with each heartbeat (cardiac cycle). Pressure is highest as the ventricles contract **(systole)** and lowest when the ventricles relax **(diastole).** In a clinical setting, we refer to these two pressures (maximum and minimum) as the **systolic** and **diastolic** pressures. A clinically "normal" or "safe" blood pressure is 110/70 (110 "over" 70), where 110 corresponds to the systolic pressure and 70 corresponds to the diastolic pressure. A clinically relevant pressure is the **pulse pressure,** or the difference between the systolic blood pressure and diastolic blood pressure. The expansion and recoil of an artery associated with the pulse pressure can be palpated in arteries that are close to the body's surface—locations called **pulse points.** This is referred to simply as "taking a pulse." Counting the number of "pulses" in an artery per minute is an indirect measure of heart rate. Another clinically significant measurement is the **mean arterial blood pressure (MAP),** which is the average pressure exerted on the arterial blood vessels over time. Because the ventricles spend significantly more time in relaxation than in contraction during a single cardiac cycle, the MAP is closer in magnitude to the diastolic blood pressure. Mathematically, MAP can be calculated as follows:

$$\text{MAP} = 1/3 \text{ pulse pressure} + \text{diastolic pressure}$$

Blood pressure can also be measured indirectly with the **auscultatory method** (*auscultatare*, to listen), which involves the use of a stethoscope (*stetho-*, chest + *skop,* to look at) and **sphygmomanometer** (*sphygmo-*, pulsation + *manometer*, sparse measure). The sounds heard are referred to as **Korotkoff sounds.** To measure blood pressure using the auscultatory method, an inflatable cuff is placed around the arm and inflated until the pressure in the cuff exceeds systolic blood pressure in the brachial artery. This causes arterial blood flow in the artery to cease. Air is then released from the cuff through a valve, causing the pressure in the cuff to drop. When pressure in the the cuff falls below systolic blood pressure, blood momentarily pushes open the artery and flows through. This creates a sharp tapping sound, the first Korotkoff sound. The pressure at which the first Korotkoff sound is heard is an approximate measure of systolic blood pressure. As pressure continues to drop in the cuff, blood pressure is able to keep the brachial artery open for longer periods of time with each subsequent cardiac cycle. Each time the artery closes (during diastole), the blood flow becomes turbulent (*turbulentus*, restless), and creates a noise that can be heard through the stethoscope. Sounds are audible until the pressure in the cuff drops below the diastolic blood pressure, at which time the blood flow is smooth (no detectable turbulence). The pressure at which there is a cessation (stopping) of the sounds is an approximate measure of diastolic blood pressure.

INTEGRATE

CONCEPT CONNECTION

Blood flow is directly proportional to *changes* in pressure and inversely proportional to the resistance of the vessels.* Thus, as blood pressure increases, blood flow increases and as blood pressure decreases, blood flow decreases. In contrast, as resistance increases, blood flow decreases and as resistance decreases, blood flow increases. The following equation shows this relationship: $F = \Delta P/R$, where F is flow, ΔP is a change in pressure, and R is the resistance. Resistance is a property of both the blood and the blood vessels and is influenced by three main factors: viscosity of the blood, length of the vessel, and diameter of the vessel. **Viscosity** is a measure of the "thickness" of the blood. Molasses, for instance, has a much higher viscosity than water. In blood, the number of formed elements, particularly red blood cells, is the major factor that affects blood viscosity. For instance, a patient with an elevated hematocrit would have an increase in blood viscosity, thereby increasing the resistance in the vessels. In addition, increasing the length of the vessels, as in the case of obesity, increases resistance.

While viscosity and length influence overall resistance, they are factors that do not change on a minute-to-minute basis. The most dramatic way to alter the resistance of a vessel is to alter the diameter of the lumen of the vessel. Specifically, decreasing the diameter of lumen of the vessel dramatically increases the resistance. Imagine water flowing through a garden hose (small diameter) versus a fire hose (large diameter). The flow of water out of the fire hose is much greater than that out of the garden hose. Based on the mathematical relationship relating flow to resistance, when there are dramatic changes in resistance, there will be dramatic changes in blood flow (when blood pressure is maintained). Therefore, it is no surprise that the fastest way to change blood flow to an organ is to change the diameter of the lumen of the vessel that supplies that organ with blood. This is controlled at the level of arterioles. Recall the smooth muscle in the tunica media of these vessels. When the smooth muscle contracts, it closes down the arteriole, dramatically increasing resistance and decreasing blood flow.

This relationship is also important when looking at the blood vasculature system as a whole. When systemic blood pressure drops (for example, in an accident victim who loses a great deal of blood), the sympathetic nervous system stimulates widespread constriction of blood vessels all over the body. This causes an increase in total peripheral resistance, which helps to increase blood pressure.

While the relationship between flow, pressure, and resistance may seem specific to the cardiovascular system, the same principles can be applied to any situation in which fluid flows through a tube. In chapter 24, air flow through the upper respiratory tract is described using the same mathematical formula. In the respiratory system, air flows due to changes in pressure (ΔP), and changes in resistance greatly influence air flow.

*In a direct relationship, when variable A increases, variable B also increases. In an inverse relationship, when variable A increases, variable B decreases.

EXERCISE 22.15 Ph.I.L.S.

Ph.I.L.S. LESSON 27: ECG AND FINGER PULSE

In this laboratory exercise, blood flow and pulse will be monitored in a virtual subject, and the subject's heart sounds and ECG will then be related to the events of the cardiac cycle.

Prior to beginning the experiment, familiarize yourself with the following concepts (use the main textbook as a reference):

- Cardiac cycle
- Systemic circuit of blood flow
- Pulsatile flow of blood
- Arterial distensibility

1. Open Ph.I.L.S. Lesson 27: ECG and Finger Pulse.
2. Read the objectives, introduction, and wet lab sections. Then take the pre-lab quiz. The laboratory exercise will open when the pre-lab quiz has been completed **(figure 22.25).**
3. Click to the "test sound" button in the middle of the window to check if the sound is working. Heart sounds should be audible; if not, contact the instructor or campus internet/computer support for assistance. Click "end test."
4. Click the "power" button on the computer screen. Then, click the "power" button on the data acquisition unit (DAQ) to begin the experiment.

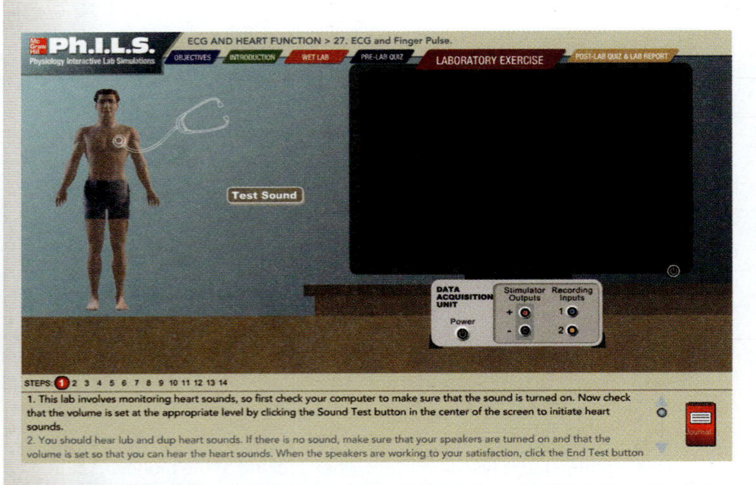

Figure 22.25 Opening Screen for Ph.I.L.S. Lesson 27: ECG and Finger Pulse.

5. Connect the finger pulse unit to the recording input by dragging the orange plug to input 2 on the DAQ.
6. Click and drag the finger pulse unit to the fingers of the subject's left hand.
7. Now, connect the ECG electrodes by clicking and dragging the blue plug to recording input 1 on the DAQ.
8. Connect the ECG recording electrodes to the subject by clicking and dragging the black electrode to the subject's left wrist, the red electrode to the subject's right wrist, and the green electrode to the subject's left ankle.
9. To record and measure ECG, pulse, and heart sounds, click the "start" button on the control panel to begin recording. On the recording, the ECG is shown at the top, the finger pulse appears in the middle, and the heart sounds are displayed on the bottom. Click the "stop" button on the control panel after at least six heart beat cycles.
10. Click the "arrows" at the bottom of the control panel screen to scroll through the record. Center one QRS complex in the middle of the screen. Measure the time interval between an R wave and the peak of the pulse recording. Move the cursor to the top of the peak of the R wave. Then, move the other cursor over the peak of the pulse wave. The time interval is displayed in the data window on the control panel. Enter the data in the journal by clicking the red "journal" button in the lower-right corner of the screen. Enter the data below:

    Time Interval: ______________________________

11. Open the post-lab quiz & lab report by clicking on it. Answer the post-lab questions that appear on the computer screen. Click "Print Lab" to print a hard copy of the report or click "Save PDF" to save an electronic copy of the report.
12. Make a note of any pertinent observations here:

    ____

    ____

    ____

EXERCISE 22.16

BLOOD PRESSURE AND PULSE USING A STANDARD BLOOD PRESSURE CUFF

CAUTION:

Cuff pressure should not exceed 160 mm Hg. Cuff pressures greater than 120 mm Hg should not be maintained for greater than 1 minute. Ensure that each subject is in good health and that he/she has refrained from activities or substances that may elevate heart rate and/or blood pressure (e.g., caffeine, exercise, smoking) at least 1 hour prior to testing.

Obtain the Following:

- **alcohol swabs**
- **sphygmomanometer**
- **stethoscope**

1. To clean the stethoscope wipe the earpieces and diaphragm with alcohol swabs.
2. Allow the subject's right arm to rest at heart level. The subject should remain as relaxed as possible throughout the exercise.
3. Locate the subject's radial artery (figure 22.18*a*) on the anterior surface of the right wrist.
4. Count the subject's pulse rate for one minute. Record this value in **table 22.5.**
5. Locate the right brachial artery, approximately 1.5–2 inches above the antecubital fossa (figure 22.18*a*). Wrap the cuff of the sphygmomanometer evenly and snugly around the subject's right arm such that the lower edge of the cuff is directly over the brachial artery, and attach Velcro® to hold in place. Ensure that the tubing and cables are not tangled or pinched.
6. Close the valve and inflate the cuff. Observe the sphygmomanometer while inflating the cuff and record the pressure at which the subject's pulse disappears. This corresponds to the systolic blood pressure. Deflate the cuff by opening the valve and letting air escape.
7. Firmly press the diaphragm of the stethoscope over the right brachial artery. Reinflate the cuff until reaching a pressure that is approximately 30 mm Hg greater than the systolic pressure observed in step 6.
8. Deflate the cuff at a rate of 2–3 mm Hg per second while listening for the first Korotkoff sound from the brachial artery. Make note of the pressure at which the first Korotkoff sound is heard. This is the systolic pressure. Continue to deflate the cuff while listening for the sounds to disappear. Make note of the pressure at which the sound can no longer be heard. This is the diastolic pressure. Record both the systolic blood pressure and the diastolic blood pressure in **table 22.6.**

Table 22.5 Pulse and Blood Pressure Readings While Sitting

	Sitting			
	Trial 1		Trial 2	
	Right Arm	Left Arm	Right Arm	Left Arm
Pulse Rate				
Systolic Blood Pressure				
Diastolic Blood Pressure				

Table 22.6 Pulse and Blood Pressure Readings While Lying Down and Standing, and After Vigorous Exercise

	Lying Down		Standing		Exercise	
	Right Arm	Left Arm	Right Arm	Left Arm	Right Arm	Left Arm
Pulse Rate						
Systolic Blood Pressure						
Diastolic Blood Pressure						

(continued on next page)

(continued from previous page)

9. Deflate the cuff completely.
10. Repeat this process in each arm (pulse rate: steps 4–5, blood pressure: steps 9–11). Allow the subject to rest for 2–3 minutes between recordings.
11. Repeat this process for the following three conditions: 1. after allowing the subject to lie down for 3–5 minutes; 2. after allowing the subject to stand for 3–5 minutes; 3. after having the subject perform vigorous exercise for 3–5 minutes. Record one trial using the left arm and the second using the right arm (or vice versa) (pulse rate: steps 4–5; blood pressure: steps 9–11). Record the results in table 22.6.
12. Make a note of any pertinent observations here:

____

____

____

EXERCISE 22.17

BIOPAC LESSON 16: BLOOD PRESSURE

CAUTION:

Cuff pressure should not exceed 160 mm Hg. Cuff pressures greater than 120 mm Hg should not be maintained for greater than 1 minute. Ensure that each subject is in good health and that he/she has refrained from activities or substances that may elevate heart rate and/or blood pressure (e.g., caffeine, exercise, smoking) at least 1 hour prior to testing.

Obtain the Following:

- **BIOPAC blood pressure cuff (SS19L) that fits securely around the subject's arm**
- **BIOPAC MP36/35 recording unit Laptop computer with BSL 4 software installed**
- **BIOPAC electrode lead set (SS2L)**
- **BIOPAC general purpose electrodes (EL503)**
- **BIOPAC stethoscope (SS30L)**

1. Ensure that all the air has been expelled from the blood pressure cuff by turning the release valve counterclockwise. Turn the valve clockwise to close.
2. Prepare the BIOPAC equipment by turning the computer ON and the MP36/35 unit OFF. Plug in the following: BP cuff (SS19L/LA)-CH 1, stethoscope (SS30L)-CH 2, and electrode lead set (SS2L)-CH 4. Turn the MP36/35 unit ON.
3. Clean the stethoscope earpieces and diaphragm using alcohol swabs. Prepare the subject for placement of electrodes by cleaning the skin where the electrodes will be placed. To do this, rub the indicated area vigorously with an alcohol swab. First, clean the skin on the medial surface of each leg, just above the ankle. Next, clean the skin on the anterior surface of the right anterior forearm at the wrist. Note that skin that is free of hair will make the best contact with the electrodes.
4. Have the subject sit in a relaxed position. Apply a small quantity of electrode gel to the center of the electrodes, and place the electrode firmly on the skin in the designated locations: right arm, right leg, and left leg. Wait approximately 5 minutes before attaching the leads (step 5) to give the electrodes a chance to affix firmly to the skin.

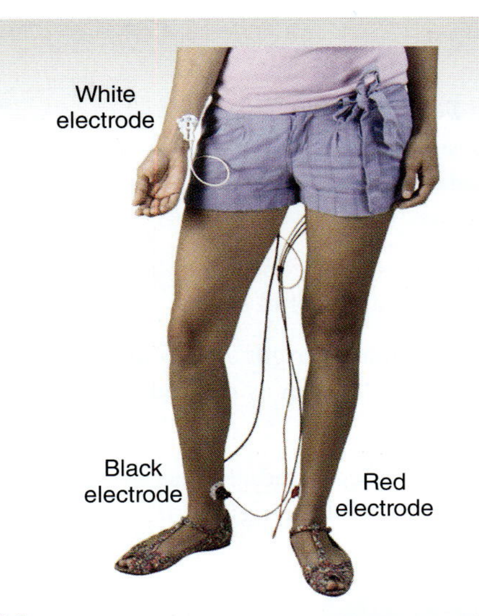

Figure 22.26 Lead Placement for ECG Recording.

Courtesy of and © BIOPAC Systems, Inc.

5. Attach the electrode leads (SS2L) to the electrodes, following the color scheme outlined in **figure 22.26.** Attach the WHITE lead (VIN-) to the electrode on the right forearm. Attach the BLACK (GND) lead to the electrode on the medial surface of the subject's right leg. Attach the RED (VIN+) lead to the electrode on the medial surface of the subject's left leg.
6. Start the BIOPAC Student Lab Program and choose "Lesson 16—Blood Pressure" and click "OK." Enter a file name.
7. To calibrate, ensure that the cuff is fully deflated. Click "calibrate" in the upper left corner of the Setup window. Inflate the cuff to 100 mm Hg and click "OK." Deflate the cuff to 40 mm Hg. Note that cuff pressure should be released at a rate of 2–3 mm Hg per second. In 20–30 seconds, the pressure should drop approximately 60 mm Hg. Click "OK." Once calibration data has started recording, tap the stethoscope diaphragm twice and wait for the calibration to stop. Check the calibration data to ensure that two sounds are recorded in the middle box, that the ECG is visible in the bottom box, and that the calibration data resembles **figure 22.27.** If so, continue with "Data Recording"; otherwise, click "Redo Calibration."

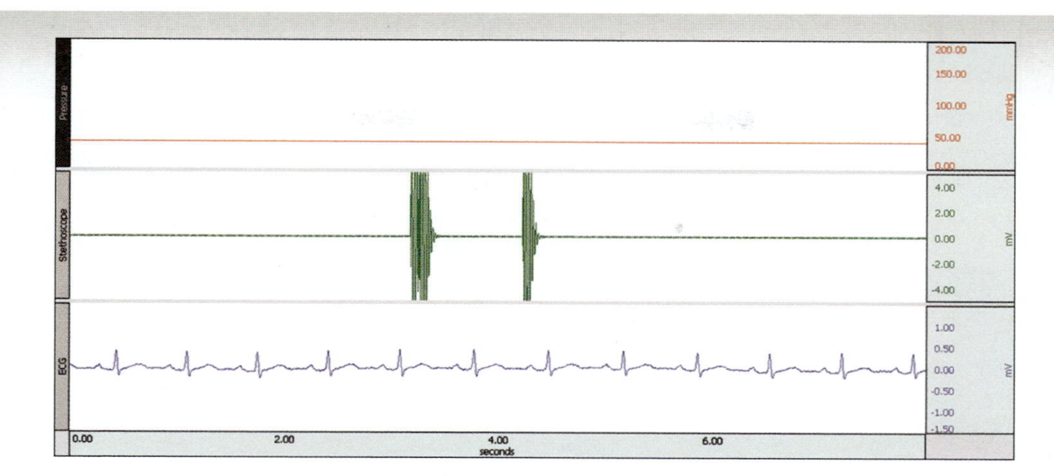

Figure 22.27 **Calibration Data for ECG Recording.**
Courtesy of and © BIOPAC Systems, Inc.

8. Allow the subject's left arm to rest at heart level and position the "artery" label over the subject's brachial artery, approximately 1.5–2 inches above the antecubital fossa. Wrap the cuff of the sphygmomanometer evenly and snugly around the subject's left arm such that the lower edge of the cuff is directly over the brachial artery, and attach Velcro® to hold in place. Ensure that the tubing and cables are not tangled or pinched. Position the pressure dial indicator such that it can be visualized easily. Firmly press the stethoscope diaphragm over the brachial artery.

9. OBTAIN VALUES FOR SITTING: Inflate the cuff to 160 mm Hg. Click "Record." Release pressure at a rate of 2–3 mm Hg/second and insert an event marker by clicking "F4" when the Korotkoff sound for systolic pressure is heard. Continue listening and note the time ("F5") when sounds are no longer heard. Click "Suspend." Fully deflate the cuff. If incorrect, click "Redo." Otherwise, click "Done."

10. Repeat step 9 for the left arm "sitting" condition. Remove the cuff and place on the subject's right arm. Repeat step 9 two times (Trials 1 and 2) for the right arm "sitting" condition.

11. OBTAIN VALUES FOR LYING DOWN: Have the subject lie down and relax. Repeat step 9 two times (Trials 1 and 2) for the right arm "lying down" condition.

12. OBTAIN POST-EXERCISE VALUES: Disconnect the electrode lead cables from the subject by releasing the electrodes at the metal clip. Have the subject perform an exercise to elevate her heart rate (e.g., running or jumping jacks). Reattach the electrode lead cables when exercise is complete. Click "resume."

13. ANALYZE THE DATA: If analyzing data that was just collected, click on "Analyze current data file." To perform the data analysis on data that was collected previously, click "review saved data." Note that CH 1 displays pressure (mm Hg), CH 2 displays the stethoscope readings (mV), and CH 3 displays ECG (mV).

14. To prepare to analyze the data, do the following:
   a. Use the zoom tool for optimal viewing of the first recording.
   b. Set up measurement boxes at the top of the screen using the drop-down menus. Set "CH 1" to "Value," set "CH 2" to "BPM," and set "CH 3" to "Delta T."

15. Locate the "I-beam" cursor. Select the point on the recording segment that corresponds to the first event marker, or systolic blood pressure. Obtain the amplitude in the "Value" measurement box, and record this value in **table 22.7.** Repeat for each event marker. Calculate the averages for each condition and record in table 22.7.

16. Use the "I-beam" tool to select an area from one R wave to the next R wave. View the "BPM" measurement box and record the value in **table 22.8.** Repeat for two additional R waves. Calculate the average BPM for three cycles within each trial, and calculate the average across trials for each condition. Record the answers in table 22.8. Use these values, and those obtained in step 14, to calculate the mean arterial pressures and pulse pressures. Record these answers in **table 22.9.**

17. Zoom in on one ECG trace between the systolic and diastolic pressure. Use the "I-beam" tool to select the area from the peak of the R wave to the beginning of the sound detected by the stethoscope. Record the Delta T. Zoom out, locate the next recording segment, and repeat the measurement. Repeat for each recording segment and record the values in **table 22.10.**

18. Make a note of any pertinent observations here:

(continued on next page)

(continued from previous page)

Table 22.7 Systolic and Diastolic Pressures

Condition	Trial	Systolic Pressure (mm Hg)		Diastolic Pressure (mm Hg)	
		Sound Detection	Sound Average (calculate)	Sound Detection	Sound Average (calculate)
Left Arm, Sitting	1				
	2				
Right Arm, Sitting	1				
	2				
Right Arm, Lying Down	1				
	2				
Right Arm, Exercise	1				
	2				

Table 22.8 BPM Measurements

Condition	Trial	BPM (R–R Measurements)			Average BPM (calculate)	
		Cycle 1	Cycle 2	Cycle 3	Across Cycles	Between Trials
Left Arm, Sitting	1					
	2					
Right Arm, Sitting	1					
	2					
Right Arm, Lying Down	1					
	2					
Right Arm, Exercise	1					
	2					

Table 22.9 MAP and Pulse Pressure Calculations

Condition	Systolic	Diastolic	BPM	Calculations	
	Sound Average (table 22.5)	Sound Average (table 22.5)	Average BPM Between Trials (table 22.6)	MAP	Pulse Pressure
Left Arm, Sitting					
Right Arm, Sitting					
Right Arm, Lying Down					
Right Arm, Exercise					

Table 22.10 Timing of Heart Sounds Relative to ECG Components

Condition	Trial	Timing of Sounds	
		Delta T	Mean (calculated)
Left Arm, Sitting	1		
	2		
Right Arm, Sitting	1		
	2		
Right Arm, Lying Down	1		
	2		
Right Arm, Exercise	1		
	2		

Courtesy of and © BIOPAC Systems, Inc. Data from BIOPAC Table 16.6.

Chapter 22: The Cardiovascular System: Vessels and Circulation

Name: ______________________

Date: ____________ **Section:** ____________

POST-LABORATORY WORKSHEET

The ❶ corresponds to the Learning Objective(s) listed in the chapter opener outline.

Do You Know the Basics?

Exercise 22.1: Blood Vessel Wall Structure

1. The tunic in blood vessels that is composed of simple squamous epithelium (or endothelium) and a subendothelial layer composed of areolar connective tissue is known as the tunica ______________________ (interna/externa). ❶ ❷

Exercise 22.2: Elastic Artery—The Aorta

2. Which of the following statements correctly describes an elastic artery? (Check all that apply.) ❸

_____ a. contains an internal elastic lamina

_____ b. contains multiple layers of smooth muscle

_____ c. tunica media contains numerous elastic fibers

_____ d. typical diameters are 1–2.5 cm in diameter

_____ e. very thin with no smooth muscle cells

3. Elastic fibers allow for turbulent flow in elastic arteries. ______________________ (True/False) ❹

4. The vasa vasorum refers to "the vessels of the vessels," and they function to deliver blood to larger vessels. ______________________ (True/False) ❺

Exercise 22.3: Muscular Artery

5. A characteristic feature of a muscular artery is the presence of an internal and external elastic lamina.______________________ (True/False) ❻

6. Elastic arteries are typically located closest to the heart where pressures are ______________________ (highest / lowest). ❼

Exercise 22.4: Arteriole

7. Constriction of arterioles ______________________ (decreases/increases) blood flow into capillary beds. ❽

Exercise 22.5: Vein

8. Veins are considered ______________________ (high/low) pressure vessels. ❾

9. A venous valve is formed by an infolding of the tunica intima. ______________________ (True/False) ❿

Exercise 22.6: Observing Electron Micrographs of Capillaries

10. Rank the following types of capillaries from most to least permeable. ⓫

_____ a. continuous

_____ b. fenestrated

_____ c. sinusoidal

Exercise 22.7: Pulmonary Circuit

11. Trace the flow of blood from the right ventricle of the heart through the pulmonary circulation and back to the left atrium of the heart. Be sure to identify heart valves through which the blood flows. ⓬

right ventricle ⟶ ______________ ⟶ ______________ ⟶ ______________ ⟶ ______________ ⟶ ______________ ⟶ left atrium

Exercises 22.8–22.13: Systemic Circuit

12. Label the diagram with the appropriate artery names. 13 – 25

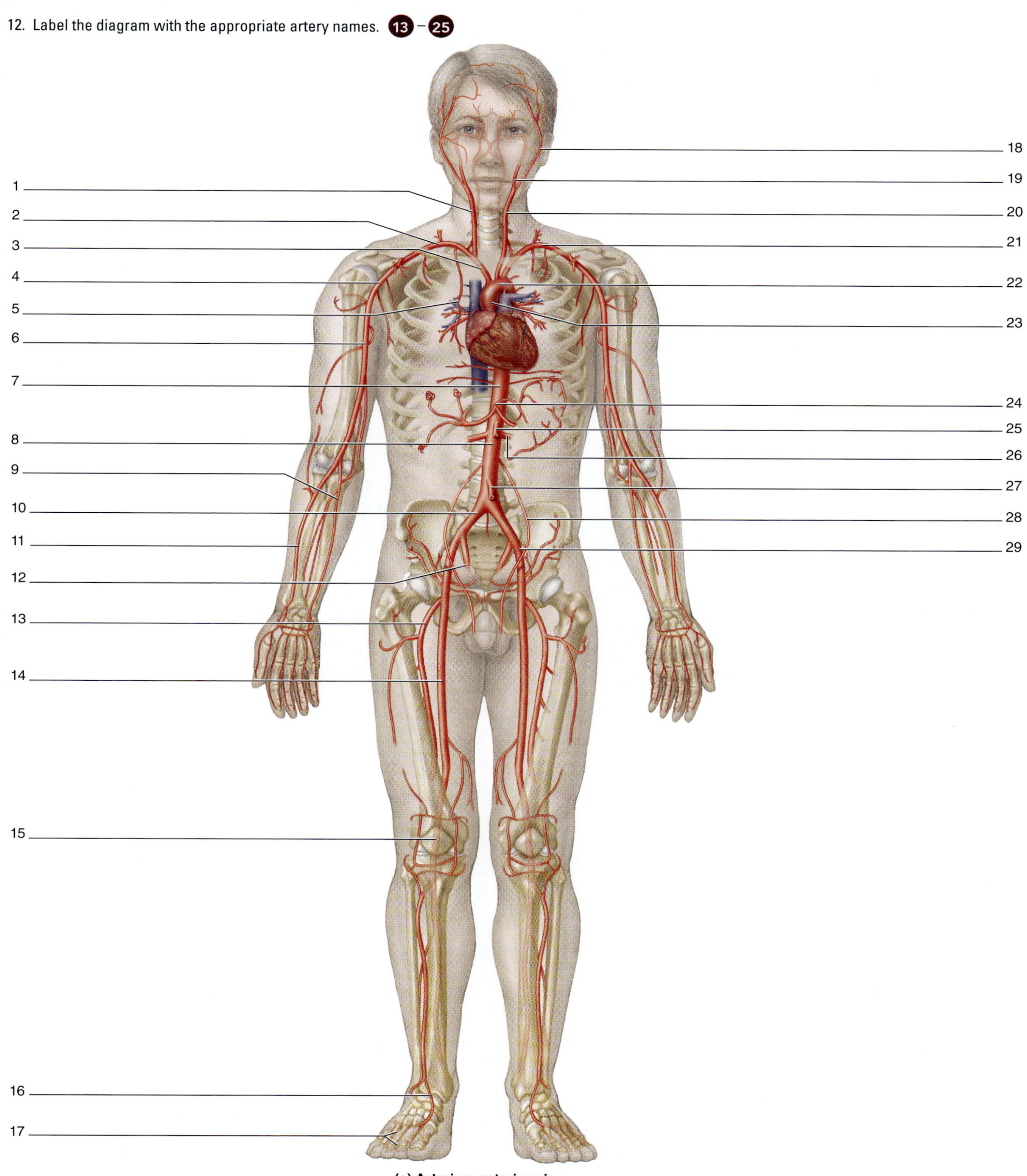

(a) Arteries, anterior view

13. Label the diagram with the appropriate vein names. 13 - 25

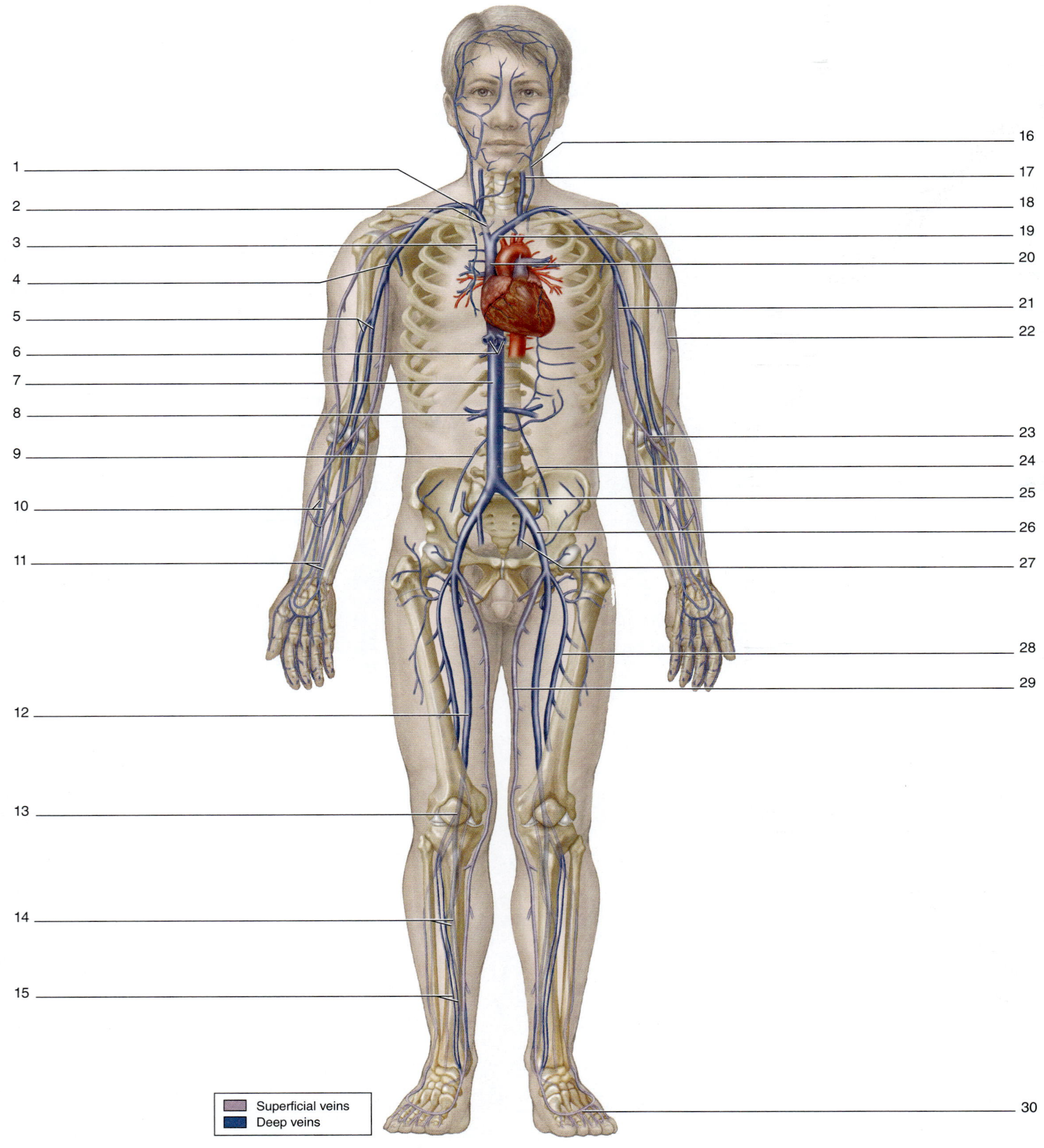

(b) Veins, anterior view

14. Which of the following are veins that compose the hepatic portal circulation? (Check all that apply.) 20

____ a. celiac trunk

____ b. inferior mesenteric vein

____ c. inferior vena cava

____ d. splenic vein

____ e. superior mesenteric vein

Exercise 22.14: Fetal Circulation

15. In the fetus, the umbilical vein carries ____________________ (oxygenated/deoxygenated) blood, whereas the umbilical arteries carry ____________________ (oxygenated/deoxygenated) blood. 26

16. In fetal circulation, oxygenated blood flows from the placenta, to the umbilical ____________________ (arteries/vein), to the right atrium of the heart. Blood returns to the placenta via the umbilical ____________________ (arteries/vein). 27

17. When referencing a structure present in the interatrial septum, the prenatal structure is called the ____________________ (foramen ovale/fossa ovalis), whereas the postnatal structure is called the ____________________ (foramen ovale/fossa ovalis). 28

Exercise 22.15: Ph.I.L.S. Lesson 27: ECG and Finger Pulse

18. Blood is propelled forward through vessels during ____________________ (diastole/systole). 29

19. An increase in heart rate should correspond to a(n) ____________________ (increase/decrease) in a R-R interval and a(n) ____________________ (increase/decrease) in pulse rate. 30

Exercise 22.16: Blood Pressure and Pulse Using a Standard Blood Pressure Cuff

20. The first Korotkoff sound corresponds to ____________________ (diastolic/systolic) pressure, whereas the second Korotkoff sound corresponds to ____________________ (diastolic/systolic) pressure. 31 33

21. When measuring blood pressure using a blood pressure cuff, the blood pressure changes that occur during the cardiac cycle are detected by measuring the pulse in the ____________________ artery. 32

Exercise 22.17: BIOPAC Lesson 16: Blood Pressure

22. Mean arterial pressure (MAP) is closer in value to a subject's ____________________ (diastolic/systolic) blood pressure. 34

23. Vigorous exercise ____________________ (decreases/increases) systemic blood pressure. 35

Can You Apply What You've Learned?

24. Explain why large blood vessels have their own blood vessels (vasa vasorum) in the tunica externa.

____

____

25. A physician wishes to place a balloon catheter into the left coronary artery of his patient. A catheter is placed into the femoral artery just inferior to the inguinal ligament, and is threaded backward through the arterial system until it reaches the ascending aorta. From there, the catheter will be threaded into the left coronary artery. List all of the arteries, in order, that the catheter will pass through as it travels from the femoral artery to the left coronary artery.

Femoral artery ⟶ ____________ ⟶ ____________ ⟶ ____________ ⟶ ____________ ⟶ ____________ ⟶ ____________ ⟶Left coronary artery

26. A physician wishes to place a central line (a catheter used to repeatedly administer drugs such as chemotherapy drugs into a patient's circulatory system) into the right atrium of the patient's heart. To do this, the physician will place the catheter into the basilic vein just superior to where it branches from the median cubital vein. The physician will then guide the catheter along the venous system until it reaches the right atrium. List, in order, the veins through which the central line passes as it travels from the basilic vein to the right atrium of the heart.

Basilic vein ⟶ ____________ ⟶ ____________ ⟶ ____________ ⟶ ____________ ⟶ Right atrium

27. After the central line was placed (question 26), it pinched off and failed to work when the patient held a heavy weight in her hand and the inferior movement of her clavicle compressed the central line within the subclavian vein. The physician decided to remove the central line from the basilic vein and place another central line into the internal jugular vein (IJV). The IJV provides a more direct route to the heart and avoids the problem of catheter pinch-off. List the veins, in order, through which the central line passes as it travels from the internal jugular vein to the right atrium of the heart.

IJV ⟶ ____________ ⟶ ____________ ⟶ ____________ ⟶ Right atrium

28. A patient suffers a stroke of the middle cerebral artery that cuts off the blood supply to the right parietal lobe of the brain. What neurological deficits would result?

29. Discuss why blood pressure might differ in the right and left arms.

30. As we age our arteries become stiff and less flexible. Discuss what might happen to resistance and blood flow through the arteries in aging adults.

Can You Synthesize What You've Learned?

31. Local factors are important for regulating blood flow to tissue capillary beds. For example, in skeletal muscle tissue that is actively contracting (as during exercise), levels of oxygen start to decrease as the oxygen is utilized, and levels of carbon dioxide increase as it is produced during cellular respiration. Would increasing levels of carbon dioxide in the tissues cause the smooth muscle forming the sphincters in arterioles to contract or relax? Justify your answer.

32. Specialized capillary beds are located throughout the body and are adapted for their specific physiological function. Recall from chapter 14 that the blood vessels that supply nervous tissues of the brain are surrounded by glial cells called astrocytes, which collectively form the blood-brain barrier. Similarly, blood vessels that form the choroid plexus within the brain ventricles are surrounded by glial cells called ependymal cells, which collectively form the blood-cerebrospinal fluid (CSF) barrier. What type of capillary (continuous, fenestrated, or sinusoids) would you expect to find in each of these areas of the brain, and why?

33. Given what you have just learned about the contents of the blood entering the liver from the hepatic portal system, discuss the type of capillaries located within the liver and state why these capillaries might be advantageous given their structure and function.

34. Preeclampsia is a condition that can occur in pregnant women that results in elevated blood pressure. While the origin of preeclampsia is unclear, there is evidence that substances that cause vasoconstriction may be released in the blood. Describe how these substances will increase blood pressure. Then discuss why this could be life threatening for both mother and child.

35. A patient enters a clinic for a routine physical examination. The nurse assigned to the patient begins to record the patient's blood pressure using a sphygmomanometer and stethoscope. As the nurse inflates the cuff to 160 mm Hg, there is an emergency down the hall that requires her attention. Rather than releasing the valve, she keeps the cuff inflated. Describe the risks to the patient that are involved when leaving the cuff inflated for greater than 1 minute.

36. Dan is diagnosed with atherosclerosis and coronary artery disease. He undergoes coronary bypass surgery to restore blood flow to his heart. Discuss why doctors might suggest that Dan limit vigorous exercise during his recovery.

CHAPTER 23

The Lymphatic System and Immunity

OUTLINE AND LEARNING OBJECTIVES

Anatomy & Physiology REVEALED® aprevealed.com **MODULE 10: LYMPHATIC SYSTEM**

INTRODUCTION

The **lymphatic system** functions closely with two other body systems: the **cardiovascular system** and the **immune system.** The lymphatic system aids the cardiovascular system by returning excess fluid to the blood to maintain fluid balance, blood volume, and blood pressure. The lymphatic system facilitates the immune system in defending against foreign substances.

The terms *lymph* and *lymphatic* come from the Latin word *lympha*, which means "pure spring water." Although lymph is not exactly comparable to spring water in its composition, the fluid is relatively clear and normally free of suspended material because it contains no red blood cells and only limited amounts of plasma proteins. Lymphatic vessels return approximately 1–3 L of fluid to the cardiovascular system each day. When lymphatic vessels become blocked or damaged so lymph cannot flow freely through them, severe **edema** (*oidema*, a swelling) typically develops in the body part distal to the location of the obstruction.

Because most lymphatic structures are anatomically small and difficult to see in a human cadaver or on isolated organ preparations, the majority of the exercises in this chapter involve histological observations of lymphatic tissues and organs. Histological observations will be complemented by observations of classroom models that show lymphatic organs such as lymph nodes, the spleen, and the thymus. If human cadavers are used in the laboratory, most lymphatic structures will be studied on the cadaver in conjunction with other organ systems, such as the respiratory and digestive systems, rather than within this laboratory exercise. Exercises within this chapter will also explore the functional relevance of lymphatic tissues and lymph, with a review of the immune system and the cell types involved in eliminating infectious agents.

List of Reference Tables

Chapter 23: The Lymphatic System and Immunity

Name: ____________________

Date: ____________ Section: ____________

These Pre-Laboratory Worksheet questions may be assigned by instructors through their connect course.

PRE-LABORATORY WORKSHEET

1. Which of the following is *not* a function of the lymphatic system?
   a. absorption of dietary fats
   b. production and proliferation of lymphocytes
   c. production of erythrocytes
   d. transport excess interstitial fluid back into the blood

2. Which of the following is *not* a named aggregation of MALT (Mucosa-Associated Lymph Tissue)?
   a. lymph node
   b. palatine tonsil
   c. Peyer patch
   d. pharyngeal tonsil
   e. vermiform appendix

3. The structure of lymphatic vessels most closely resembles ________________ (arteries/veins).

4. The right lymphatic duct drains lymph from the right side of the head and neck, the right lower limb, and the right side of the thoracic and abdominopelvic cavities. ________________________ (True/False)

5. Which of the following is a lymphatic organ that filters blood?.
   a. lymph node
   b. Peyer patch
   c. spleen
   d. thymus
   e. vermiform appendix

6. Lymphatic vessels contain valves, which prevent the backflow of lymph. ________________________ (True/False)

7. Which of the following is a lymphatic organ located in the thoracic cavity just deep to the sternum?
   a. Peyer patch   b. spleen   c. thymus   d. tonsil   e. vermiform appendix

8. The basic unit of lymphatic tissue is called a lymph node. ________________ (True/False)

9. Which of the following is a lymphatic organ that has both afferent and efferent lymphatic vessels?
   a. lymph node   b. spleen   c. thymus   d. tonsil   e. vermiform appendix

10. Which of the following is a lymphatic organ that contains crypts?
    a. lymph node   b. spleen   c. thymus   d. tonsil   e. vermiform appendix

HISTOLOGY

Lymphatic Vessels

Lymphatic vessels are very similar in structure to veins because they carry fluid at low pressure from the tissues back toward the heart. Lymphatic vessels are thin-walled and have valves, which prevent the backflow of fluid. Similar to veins, the contraction of skeletal muscle also assists the movement of fluid within lymphatic vessels toward the heart.

EXERCISE 23.1

LYMPHATIC VESSELS

1. Obtain a *dissecting* microscope and a slide demonstrating lymphatic vessels.
2. Place the slide on the microscope stage. Bring the tissue sample into focus on low power and locate a lymphatic vessel **(figure 23.1).** Then move the microscope stage to scan the length of the vessel until a valve is visible. If necessary, change to high power and focus in on the valve to see its structure more clearly.

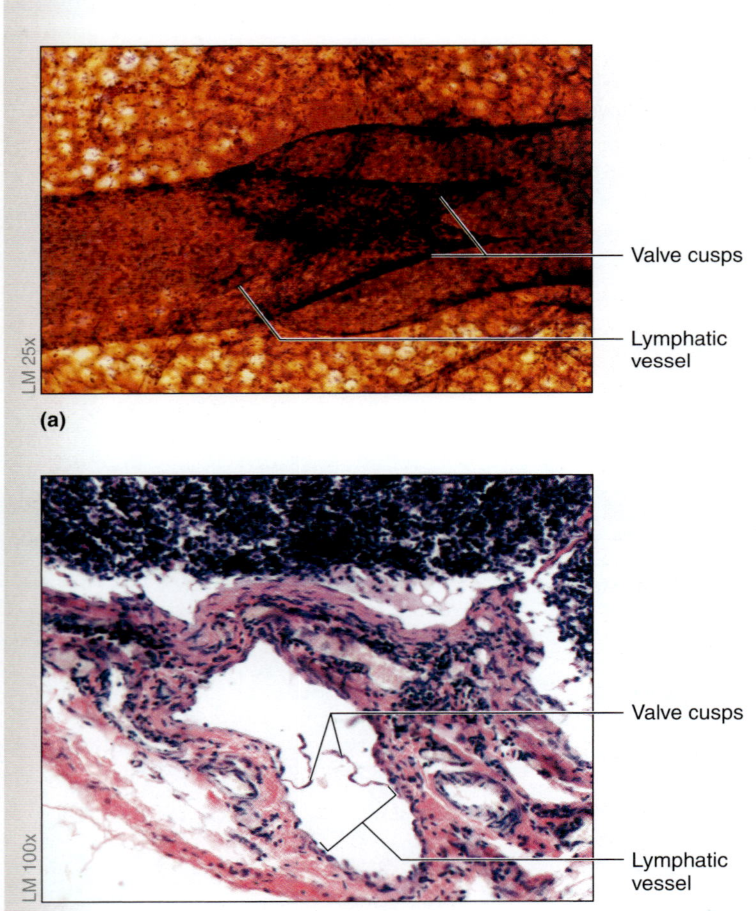

Figure 23.1 Lymphatic Vessels. (a) View of a lymphatic vessel and valve as seen through a dissecting microscope. (b) Cross section of a lymphatic vessel demonstrating a valve as seen through a compound microscope.

3. Identify the following structures, using figure 23.1*a* as a guide:

   ☐ **lymphatic vessel** ☐ **valve cusps**

4. Observe a slide of lymphatic vessels using a compound microscope. Place the slide on the microscope stage. Bring the tissue sample into focus on low power and locate a cross section through a lymphatic vessel (figure 23.1*b*). The walls of lymphatic vessels contain the same three tunics as the walls of blood vessels, but they are not as well defined. The valves are extensions of the tunica intima of the vessel (as they are in veins). Are there any red blood cells visible within the lumen of the vessel? (yes/no) Are red blood cells normally found within lymphatic vessels? (yes/no) Are there any white blood cells visible within the lumen of the vessel? (yes/no) Are white blood cells normally found within lymphatic vessels? (yes/no)

6. Identify the following structures, using figure 23.1*b* as a guide:

   ☐ **lymphatic vessel** ☐ **valve cusps**

7. Sketch a lymphatic vessel as viewed through both the dissecting and the compound microscopes in the spaces provided.

_______ × _______ ×

Mucosa-Associated Lymphatic Tissue (MALT)

The basic functional unit of lymphatic tissue is a **lymphatic nodule (follicle).** A lymphatic nodule is a cluster of cells that are predominantly B-lymphocytes, with some macrophages and T-lymphocytes on its outer borders. **Figure 23.2** demonstrates a lymphatic nodule as seen through a compound microscope on high power. Notice that the central region stains lighter than the surrounding area. The central region, called the **germinal center** (*germen*, sprout), is where B-lymphocytes are most actively proliferating. Lymphatic nodules are found in many locations throughout the body. They are commonly found just deep to epithelial tissues and are even more common in locations where the epithelium changes from one type to another. Lymphatic nodules are particularly abundant in the nasal and oral cavities, and in the walls of the digestive tract. Such tissue is collectively referred to as **MALT,** or **m**ucosa-**a**ssociated **l**ymphatic **t**issue. In several regions of the respiratory and digestive tracts, the aggregations of MALT are so consistent in structure and so regular in location that they are given names. Such named aggregations of MALT include **tonsils, Peyer patches,** and the **vermiform appendix. Table 23.1** summarizes the characteristics of MALT.

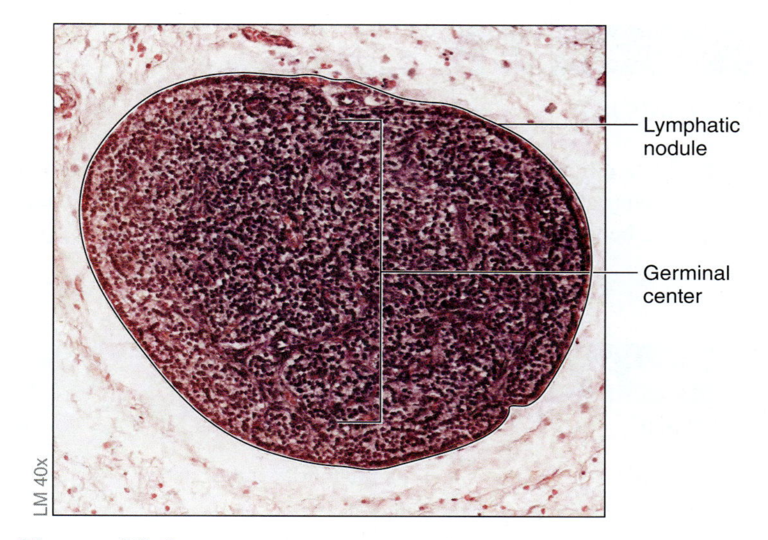

Figure 23.2 Lymphatic Nodule. The germinal center of the nodule consists of actively proliferating B-lymphocytes.

Table 23.1 Named Aggregations of MALT (Mucosa-Associated Lymphatic Tissue)

Named Aggregation of MALT	Description and Location	Function	Word Origin
Peyer Patches	Lymphatic nodules in the submucosa (deep to the epithelium) of the ileum of the small intestine	Protect against pathogens that enter the body through the intestinal mucosa	*Peyer*, Johann K. Peyer, a Swiss anatomist (1653–1712)
Tonsils	Lymphatic nodules deep to the epithelium lining the pharynx. See table 23.2 for descriptions and locations of specific tonsils.	Protect against pathogens that enter the body through the mucosa of the pharynx	*tonsilla*, a stake
Vermiform Appendix	A diverticulum (blind-ended pouch) extending from the cecum just past the ileocecal junction that contains lymphatic nodules in its walls	Protect against pathogens that enter the body through the mucosa of the cecum	*vermis*, worm-like, + *forma*, shape, + *appendix*, appendage

INTEGRATE

CLINICAL VIEW
Appendicitis

Appendicitis refers to a condition in which the vermiform appendix is inflamed (*-itis*, inflammation). Appendicitis, most commonly caused by a blockage of the opening of the appendix into the cecum of the large intestine, can happen to individuals of any age. Patients with appendicitis generally present with pain in the lower right quadrant of the abdomen near McBurney's point, a surface landmark used to approximate the location of the vermiform appendix. McBurney's point is located by drawing a line from the umbilicus to the anterior superior iliac spine and marking a point approximately one-third of the way along the line away from the umbilicus. If a patient has appendicitis, pressure placed on the abdomen near McBurney's point is likely to elicit sharp, intense pain and rebound tenderness (tenderness associated with the "rebound" of the anterior abdominal wall as pressure is taken away).

The appendix is covered with visceral peritoneum (peritoneum is the lining of the abdominal cavity). Irritation of visceral peritoneum is associated with dull, diffuse pain because visceral peritoneum is innervated by autonomic nerves. In contrast, irritation of the parietal peritoneum (which lines the inside of the anterior abdominal wall) is associated with sharp, localized pain because parietal peritoneum is innervated by somatic nerves. When contact is made between visceral and parietal peritoneum at a source of irritation or inflammation, the dull, diffuse autonomic pain is suddenly felt as sharp, intense somatic pain. It is for this reason that pressure placed around McBurney's point causes the sharp, intense pain of appendicitis. Presence of a fever and blood tests showing elevated leukocytes (particularly neutrophils) may further confirm appendicitis and help distinguish it from other potential sources of right-lower quadrant pain.

EXERCISE 23.2

TONSILS

Tonsils are lymphatic nodules located within the walls of both the oral cavity and pharynx. They consist of lymphatic nodules interspersed between deep **crypts** (*crypt*, a pit-like depression) that open to the surface of the tonsil. The tonsils are covered superficially by the same epithelium that lines the part of the body in which they are located. There are three sets of tonsils: **pharyngeal tonsils** (called *adenoids* when they are swollen), **palatine tonsils,** and **lingual tonsils. Table 23.2** lists characteristics of the three kinds of tonsils.

Table 23.2 Tonsils

Tonsil	Description and Location	Crypts	Epithelium	Word Origin
Lingual	Small, paired tonsils at the base of the tongue	1–2 short crypts	Stratified squamous nonkeratinized	*lingua*, tongue
Palatine	Paired tonsils located in the fauces (the space between the mouth and pharynx) just posterior to the soft palate	10–20 deep crypts	Stratified squamous nonkeratinized	*palatum*, palate
Pharyngeal (Adenoid)	Single tonsil projecting from the roof of the nasopharynx	None	Respiratory (ciliated pseudostratified columnar)	*pharynx*, the throat, *adenos*, a gland, + *eidos*, appearance

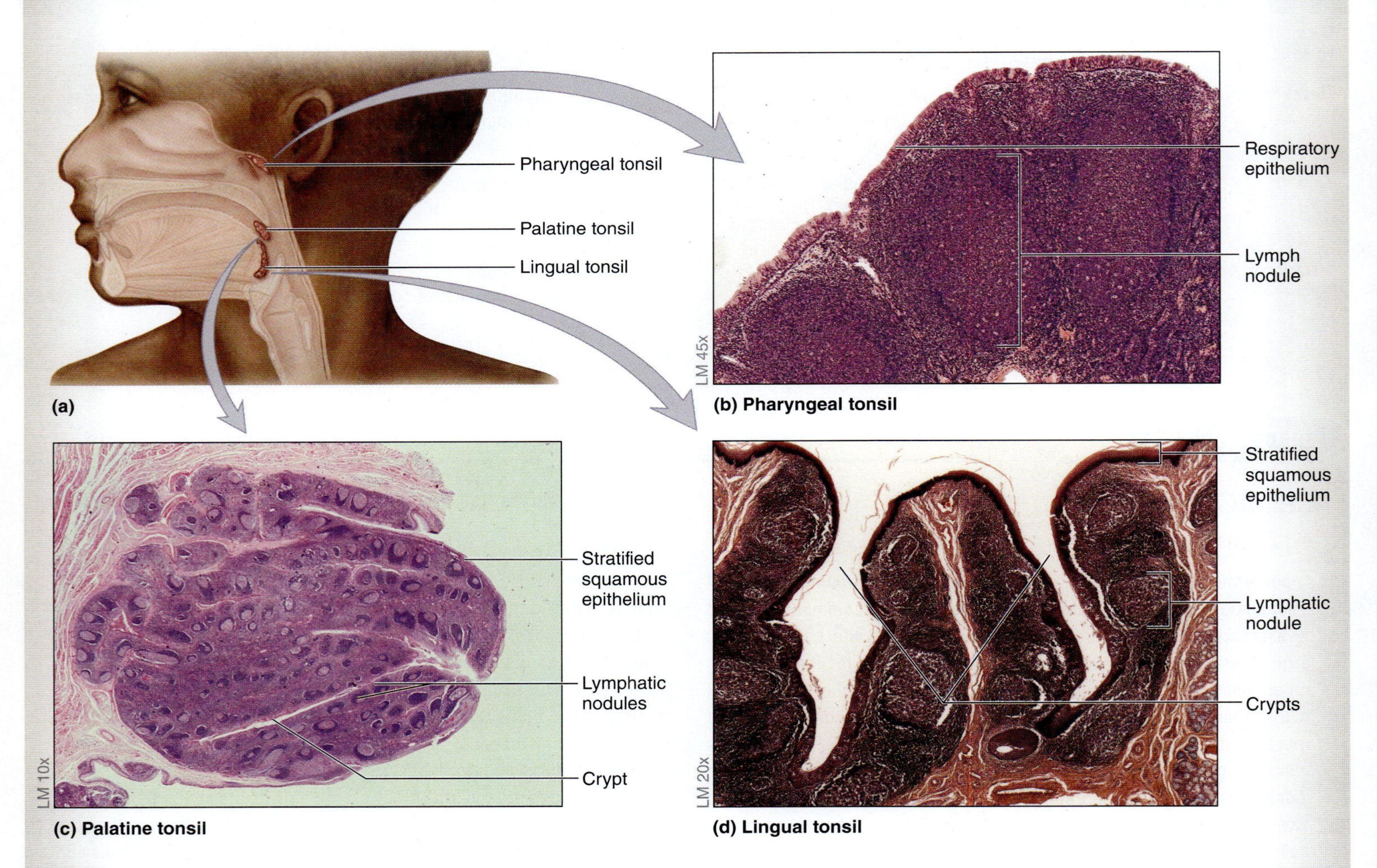

Figure 23.3 **Tonsils.** (a) Location of tonsils within the pharynx. (b) Pharyngeal tonsils are lined with respiratory epithelium (pseudostratified columnar with cilia and goblet cells). (c) Palatine tonsils contain numerous deep crypts and are lined with stratified squamous epithelium. (d) Lingual tonsils contain a few shallow crypts and are lined with stratified squamous epithelium.

1. Obtain a slide of the tonsils (lymphatic tissue) and place it on the microscope stage.
2. Observe the slide at low magnification. Locate the following structures, using **figure 23.3** and table 23.2 as guides:

   ☐ **crypt** ☐ **lymphatic nodule**
   ☐ **epithelium**

3. Determine what type of tonsil (pharyngeal, palatine, or lingual) is shown on the slide based on the type of epithelium covering the tonsil and the number and length of the crypts (table 23.2). List the type of tonsil observed:

   ____

4. Sketch the tonsil(s) observed through the microscope in the space provided.

EXERCISE 23.3

PEYER PATCHES

1. Obtain a slide of the ileum and place it on the microscope stage.
2. Bring the tissue sample into focus on low power and scan the slide to locate the epithelial lining of the ileum, which is a simple columnar epithelium. Deep to the epithelium, look for aggregations of cells that are stained purple **(figure 23.4).** These are the aggregations of lymphocytes that compose the **Peyer patches** (aggregated lymphatic nodules of the small intestine).

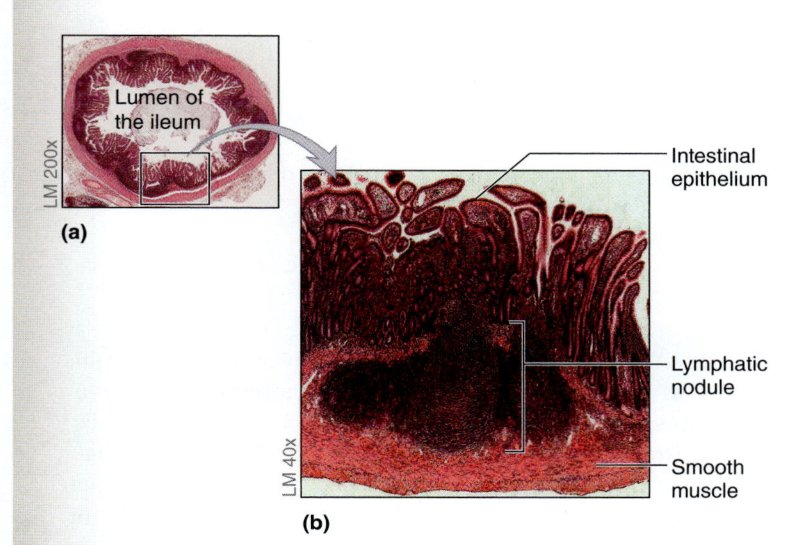

Figure 23.4 Peyer Patches. A cross section of the ileum of the small intestine is shown in (a). Deep to the intestinal epithelium and superficial to the smooth muscle that surrounds the ileum are the lymphatic nodules called Peyer patches. (b) Peyer patches shown in detail.

3. Move the microscope stage until a Peyer patch is at the center of the field of view. Switch to medium or high power to observe the cells within the Peyer patch.

   What kind of cells are found within the Peyer patches?______________________

4. Identify the following structures on the slide of the ileum, using table 23.2 and figure 23.4 as guides:

   ☐ **epithelium** ☐ **Peyer patch**
   ☐ **lymphatic nodule**

5. Sketch a Peyer patch as seen through the microscope in the space provided.

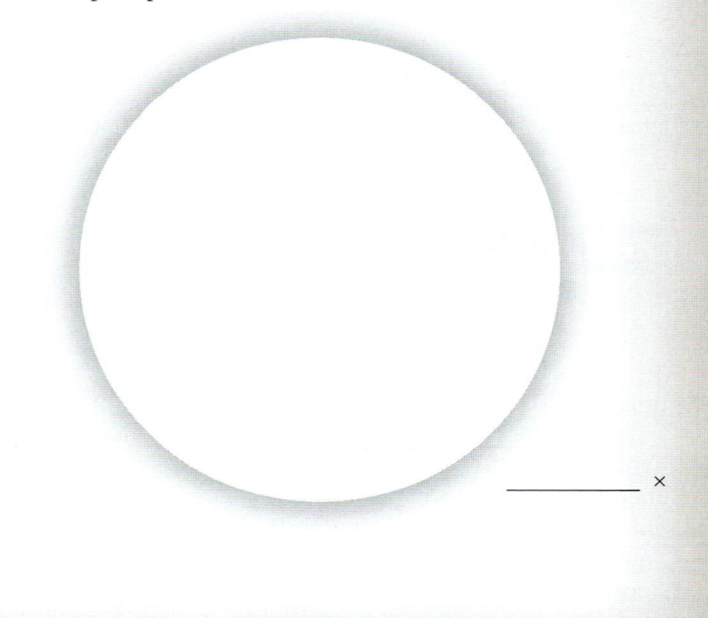

EXERCISE 23.4

THE VERMIFORM APPENDIX

The **vermiform appendix** (*vermiform*, shaped like a worm) projects from the inferior region of the cecum (first part of the large intestine). It is usually about 2–4 centimeters in length and about half a centimeter in diameter. It is an evagination of the cecum and, as such, has a lumen that opens into the lumen of the cecum. Its epithelium is also a continuation of the epithelium of the cecum, and it contains numerous lymphatic nodules in its walls.

1. Obtain a slide showing a cross section of the appendix and place it on the microscope stage.
2. Bring the tissue sample into focus on low power and scan the slide to locate the lumen of the appendix **(figure 23.5).** Then change to medium power and bring the tissue sample into focus once again. One identifying feature of the appendix is that its lumen contains cellular debris, bacteria, and other breakdown products of food that are left over after digestion in the small intestine. Look for this "junk" within the lumen of the appendix. What type of epithelium lines the inside of the appendix?

____

____

3. Observe the tissues located deep to the epithelial tissue to locate the lymphatic nodules, which are aggregates of purple-staining lymphocytes.
4. Identify the following structures on the slide of the appendix, using table 23.2 and figure 23.5 as guides:

☐ **epithelium**
☐ **lymphatic nodule**
☐ **lumen of the appendix**

5. Sketch a cross section of the appendix as seen through the microscope in the space provided.

_____________ ×

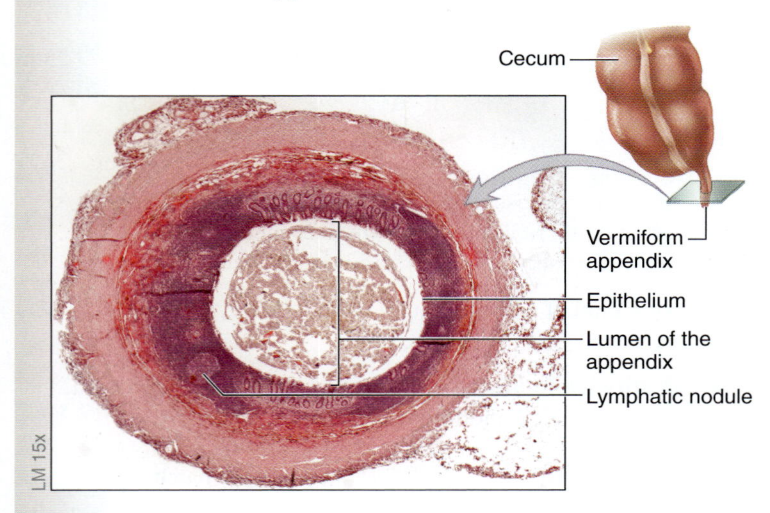

Figure 23.5 Vermiform Appendix. A cross section through the appendix, demonstrating epithelium, lymphatic nodules, and a lumen, which contains breakdown products of food.

Lymphatic Organs

Lymphatic organs differ from mucosa-associated lymphatic tissue (MALT) in that they are covered by a dense connective tissue capsule and are more highly organized. Lymphatic organs include the thymus, lymph nodes, and the spleen.

EXERCISE 23.5

THE THYMUS

The thymus is a gland located in the superior mediastinum **(figure 23.6).** It is the site of T-lymphocyte maturation, and it is also an endocrine organ, secreting the hormones **thymosin, thymulin**, and **thymopoietin**. The thymus is composed of two **lobes,** each separated into smaller **lobules** by connective tissue **septae** (singular: septa) **(figure 23.7).** Each lobule has a darker-staining outer cortex and a lighter-staining inner medulla. The majority of the cells within the thymus are T-lymphocytes, which are surrounded by cells of epithelial origin. Lymphocytes migrate to the thymus from the red bone marrow. Once in the thymus, T-lymphocytes move from the outer cortex to the inner medulla as they complete the process of **selection.** Selection is a process by which only those T-lymphocytes that are able to recognize the MHC protein and do not bind with self-antigens are permitted to leave the thymus. Selection occurs in the outer cortex. The majority of T-lymphocytes that migrate to the thymus from the bone marrow either fail to recognize the MHC protein or react to self-antigen. These cells are destroyed through a process called **apoptosis** (programmed cell death), which occurs within the thymic cortex. T-lymphocytes that survive the selection process leave the thymus through venules located in the medulla to migrate to secondary lymphatic structures (e.g., lymph nodes, spleen). After puberty the thymus begins a gradual decline in size with age. Epithelial-derived cells in the medulla come together and form small, onion-like masses of tissue called **thymic corpuscles.** A large number of thymic corpuscles in the thymic medulla indicates an aging thymus.

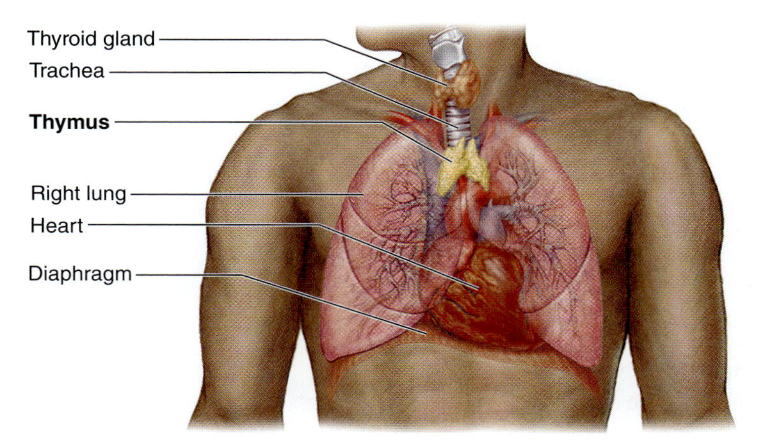

Figure 23.6 Location of the Thymus. The thymus is located in the superior mediastinum, deep to the sternum.

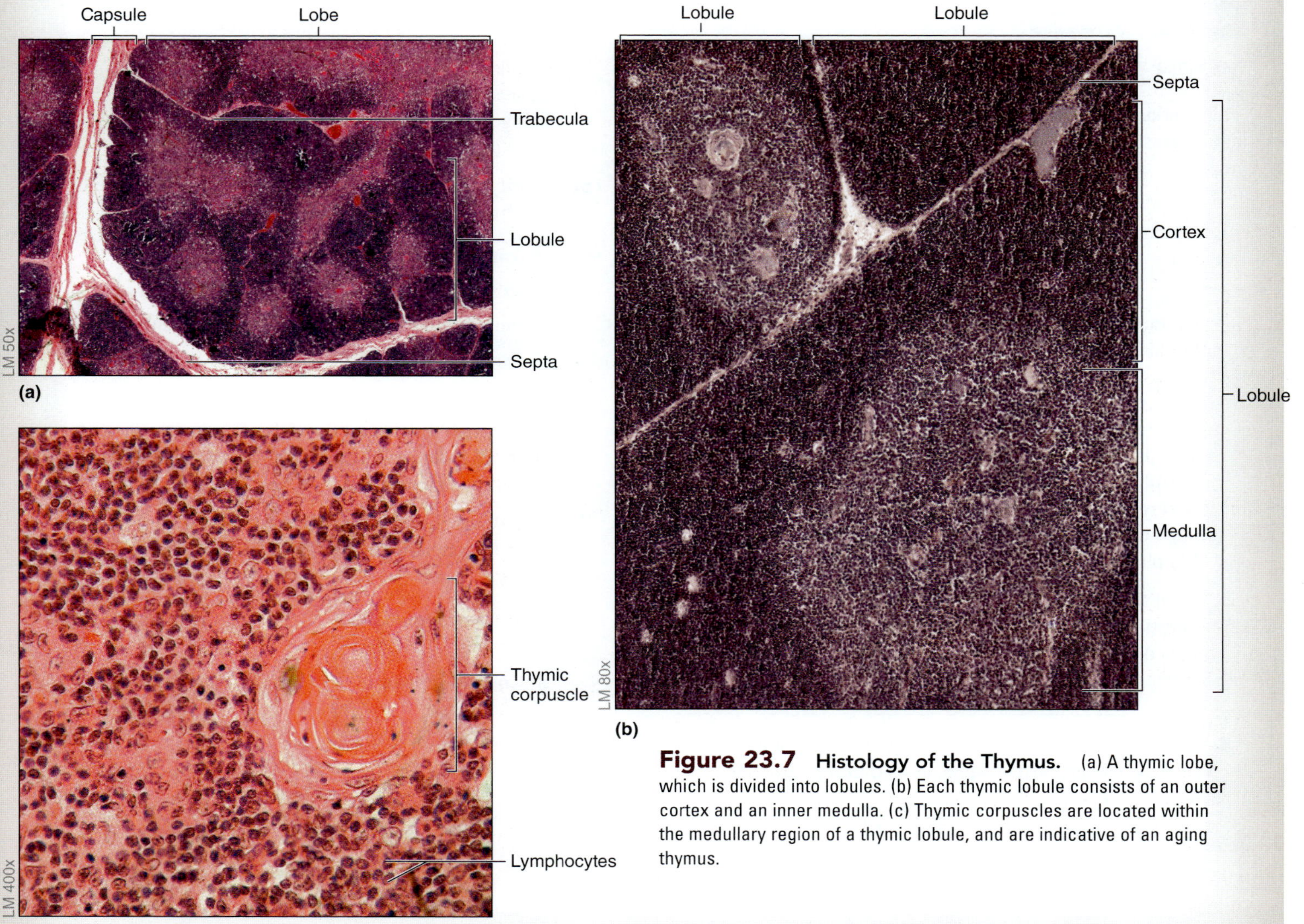

Figure 23.7 Histology of the Thymus. (a) A thymic lobe, which is divided into lobules. (b) Each thymic lobule consists of an outer cortex and an inner medulla. (c) Thymic corpuscles are located within the medullary region of a thymic lobule, and are indicative of an aging thymus.

(continued on next page)

(continued from previous page)

1. Obtain a slide of the thymus and place it on the microscope stage.
2. Bring the tissue sample into focus at low power, then scan the slide to locate the connective tissue trabeculae that separate the gland into lobules (figure 23.7).
3. Identify the following structures, using figure 23.7 as a guide:
   - ☐ **cortex**
   - ☐ **lobule**
   - ☐ **medulla**
   - ☐ **thymic corpuscle**
   - ☐ **trabeculae**
4. Sketch the thymus as seen through the microscope in the space provided. Be sure to label all the structures listed in step 3.

_______________ ×

EXERCISE 23.6

LYMPH NODES

Lymph nodes are small lymphatic organs located along lymphatic vessels. They filter lymph that flows through the lymphatic vessels to identify and attack lymph-borne foreign antigens.

1. Obtain a slide showing a lymph node **(figure 23.8)** and place it on the microscope stage.
2. Bring the tissue sample into focus on low power. Identify the dense irregular connective tissue **capsule,** and notice the kidney-bean shape of the lymph node (if an entire node is on the slide). A lymph node has two main regions: an outer **cortex** and an inner **medulla.** The indented region of the lymph node is called the **hilum.** This is where blood vessels enter and leave the node, and where the **efferent lymphatic vessels** drain lymph from the lymph node. **Afferent lymphatic vessels** bring lymph into the lymph node on the regions that lie opposite the hilum of the lymph node.
3. Note the dense irregular connective tissue **trabeculae** that partition the cortex into smaller regions. Each region of the cortex contains one or two **lymphatic nodules.** What types of cells are found within lymphatic nodules?

   ____

4. Scan the slide until the medulla is at the center of the field of view. Locate the dark-staining **medullary cords.** The medullary cords are strands of B-lymphocytes, T-lymphocytes, and macrophages. Can B- and T-lymphocytes be distinguished from each other using light microscopy?

   ____

5. Identify the following structures on the tissue slide of the lymph node, using figure 23.8 as a guide:
   - ☐ **capsule**
   - ☐ **cortex**
   - ☐ **hilum**
   - ☐ **lymphatic nodules**
   - ☐ **medulla**
   - ☐ **medullary cords**
   - ☐ **trabeculae**
6. Observe the numerous spaces, called **medullary sinuses,** where lymph travels as it flows through the lymph node. **Figure 23.9** demonstrates the pathway taken by lymph as it flows through a lymph node.
7. Identify the following spaces and vessels on the slide of the lymph node, using figure 23.8 and **table 23.3** as guides:
   - ☐ **afferent lymphatic vessels**
   - ☐ **cortical sinuses**
   - ☐ **medullary sinuses**

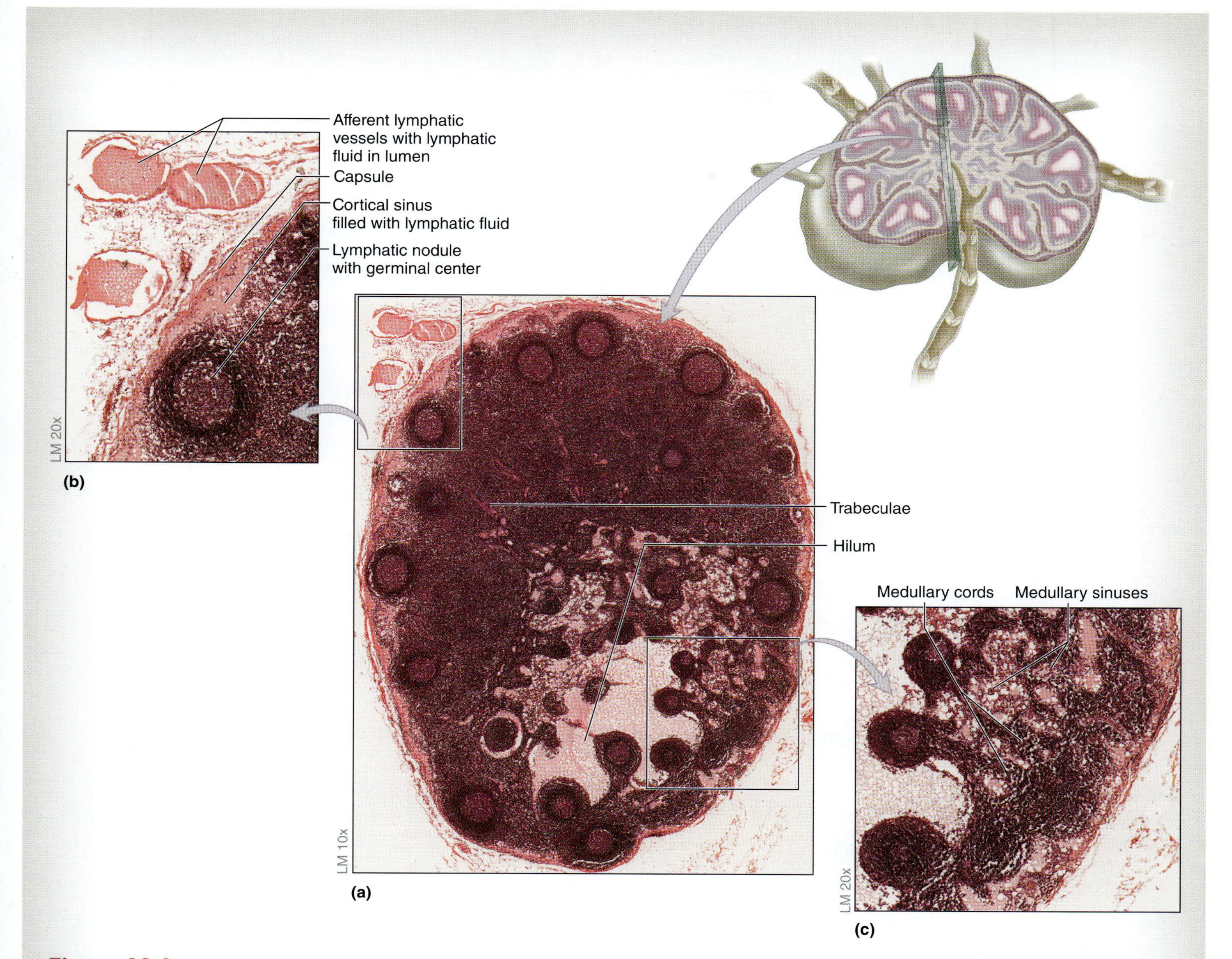

Figure 23.8 **Lymph Node** (a) Photomicrograph of a cross section of a lymph node that has been sectioned through the hilum. On the surface opposite the hilum, cross sections of three afferent lymphatic vessels are visible, as well as the capsule and a cortical sinus filled with lymphatic fluid (b). Medullary cords and medullary sinuses are visible in (c).

(continued on next page)

(continued from previous page)

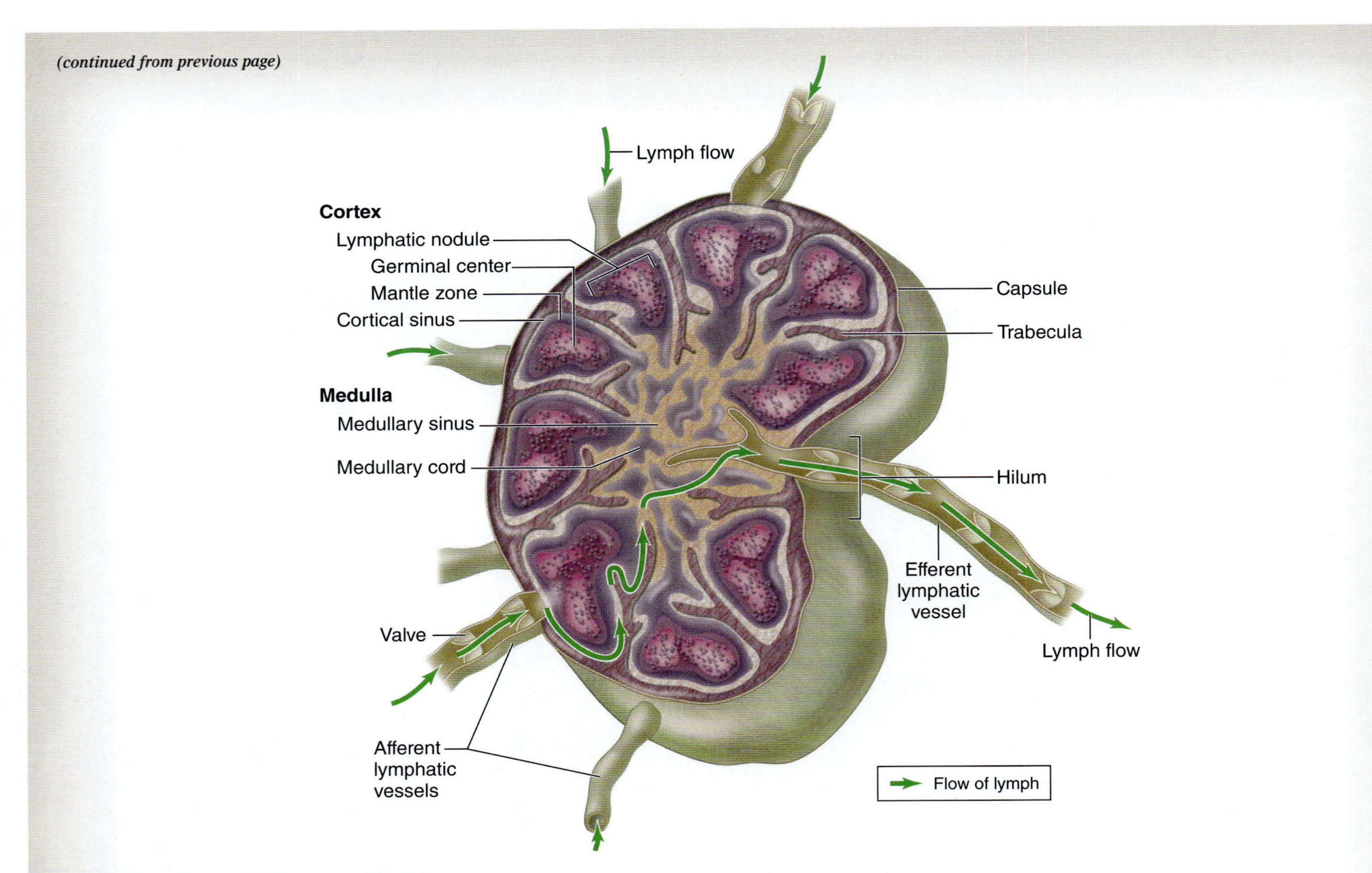

Figure 23.9 **Lymph Node.** Pathway that lymph takes as it flows through a lymph node.

Table 23.3 Parts of a Lymph Node

Structure	Description	Word Origin
Afferent Lymphatic Vessel	Vessel that transports lymph into a lymph node	*affero,* to bring to
Capsule	Dense irregular connective tissue that surrounds the entire lymph node to support and protect it	*capsula*, a box
Cortex	Outer portion of the lymph node that lies deep to the capsule and contains lymphatic nodules	*cortex*, bark
Cortical sinuses	Spaces located between lymphatic nodules and the capsule or the trabecula in a lymph node through which lymphatic fluid flows	*cortex*, bark + *sinus*, hollow curve
Efferent Lymphatic Vessel	Vessel that transports lymph away from the lymph node; located at the hilum	*effero*, to bring out
Hilum	Indented area where efferent lymphatic vessels (and blood vessels, nerves) enter	*hilum,* a small bit
Lymphatic Nodule	Aggregate of B-lymphocytes within a germinal center, surrounded by a mantle zone, which contains T-lymphocytes, macrophages, and dendritic cells	*lympha*, clear spring water, + *nodulus*, a small knot
Medulla	Inner region of the lymph node containing medullary cords of B-lymphocytes and macrophages	*medius*, middle
Medullary Cord	Strands of B-lymphocytes, T-lymphocytes, and macrophages supported by connective tissue that are located in the medulla of a lymph node	*medius*, middle, + *chorda,* a string
Medullary Sinuses	Spaces between the medullary cords in a lymph node through which lymphatic fluid flows	*medius*, middle + *sinus*, hollow curve
Trabeculae	Invaginations of the dense irregular connective tissue capsule of a lymph node that partition the cortex into smaller compartments	*trabs*, a beam

8. Sketch a lymph node as seen through the microscope in the space provided. Label all of the structures listed in steps 5 and 7.

9. Describe the pathway taken by lymph as it flows through a lymph node in the space provided.

______ ×

INTEGRATE

CONCEPT CONNECTION

As discussed in chapter 5, reticular tissue is a type of connective tissue found in lymphatic organs, including the spleen and lymph nodes. Recall that reticular tissue is composed of cells and reticular fibers. Reticular fibers are a fine type of collagen that provide a loose supporting framework for the cells and aid in trapping substances. The reticular fibers help to slow the flow of lymph, allowing the contents of the lymph to come in contact with immune cells within the organ. In addition, the structure of the lymph node is such that there are many more afferent lymphatic vessels than efferent lymphatic vessels. Lymph enters the organ and meanders through the sinuses, thus increasing the likelihood that a pathogen will encounter an immune cell. Once pathogens become trapped by the reticular fibers of the lymph nodes, B- and T-cells can launch an immune response. The details of the immune response are discussed in the physiology section of this chapter.

EXERCISE 23.7

THE SPLEEN

The spleen is similar to a lymph node in many ways, with one major exception: The spleen filters *blood*, whereas a lymph node filters *lymph*. However, both organs filter their fluid for the purpose of identifying and eliminating foreign antigens that may be present in the fluid. The spleen receives blood through the **splenic artery,** which is a branch of the celiac trunk. Blood leaves the spleen through the **splenic vein,** which drains into the hepatic portal vein (see figures 22.13 and 22.14, pp. 600–601).

When a fresh spleen is cut, two different types of tissue are observed: red pulp and white pulp. **Red pulp** consists of splenic sinusoids, and is red due to the large amount of blood contained within the sinusoids. Red pulp contains erythrocytes, platelets, macrophages, and B-lymphocytes. It composes the majority of the splenic tissue. **White pulp** consists of aggregations of T-lymphocytes, B-lymphocytes, and some macrophages, and is white due to the white blood cells (leukocytes) within each mass of tissue. Because the spleen tissue sample that will be viewed through the microscope has been stained, red and white pulp will not appear red and white in color. Instead, the red pulp will usually be reddish-pink in color, but the white pulp will be darker purple, with a lighter-colored central region.

1. Obtain a slide of the spleen and place it on the microscope stage. Bring the tissue sample into focus on low power and locate the dense irregular connective tissue capsule of the spleen, if present in the section on the slide **(figure 23.10).** Look for several connective tissue trabeculae, which are invaginations of the capsule that separate the spleen into distinct regions.
2. Notice the circular purple aggregations of cells. These aggregations constitute the white pulp of the spleen, which consists of B-lymphocytes, T-lymphocytes, and some macrophages.
3. Look for a **central artery** within each mass of white pulp. Central arteries receive blood from **trabecular arteries.** Blood coming into the spleen travels from the splenic artery to the trabecular arteries to the central arteries, and finally empties into the sinusoids or surrounding tissues of the spleen before draining into splenic veins.
4. Move the microscope stage until a mass of white pulp is in the center of the field of view and then change to high power. At this magnification individual small purple-staining cells, which are lymphocytes, will be visible.

(continued on next page)

(continued from previous page)

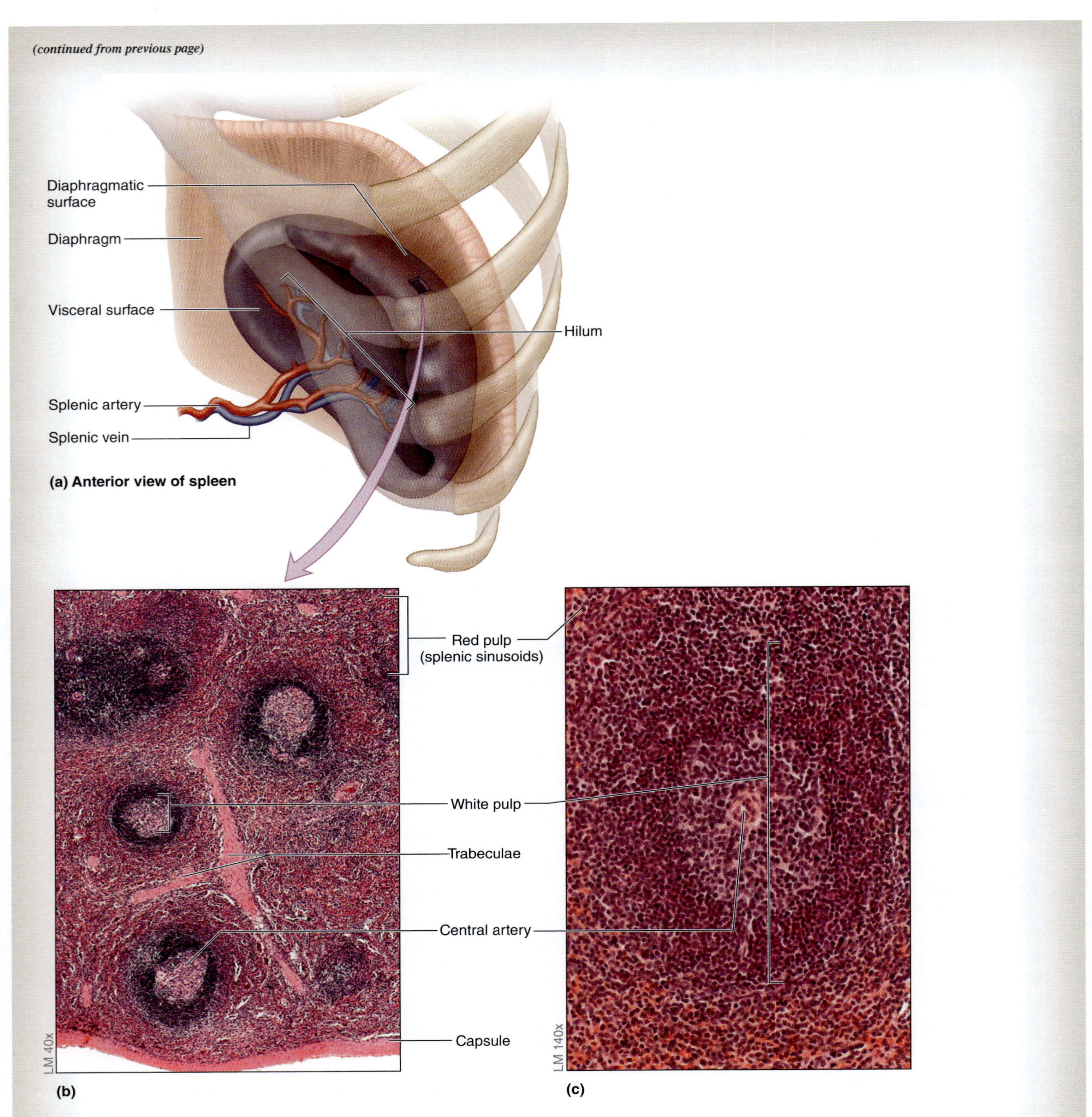

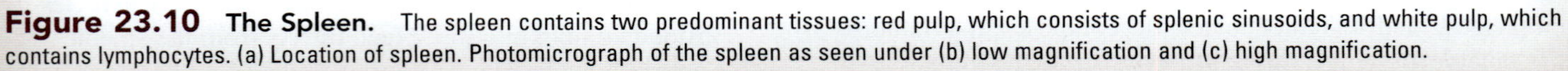

Figure 23.10 The Spleen. The spleen contains two predominant tissues: red pulp, which consists of splenic sinusoids, and white pulp, which contains lymphocytes. (a) Location of spleen. Photomicrograph of the spleen as seen under (b) low magnification and (c) high magnification.

5. Move the microscope stage so the red pulp of the spleen is in the center of the field of view. The red pulp of the spleen consists mainly of *splenic sinusoids* containing numerous erythrocytes. The blood-filled splenic sinusoids also contain numerous lymphocytes and macrophages. As blood travels slowly through the spleen, the lymphocytes and macrophages monitor the blood for **antigens,** which, if present, will be removed by these immune cells. In addition, the structure and function of the splenic sinusoids is designed to destroy old or abnormal erythrocytes. What is the life span of a typical erythrocyte? __________ days. (Hint: The answer is in chapter 20.)

6. Identify the following structures on the histology slide of the spleen, using **table 23.4** and figure 23.10 as guides:

- ☐ **capsule**
- ☐ **central artery**
- ☐ **red pulp**
- ☐ **sinusoids**
- ☐ **trabecula**
- ☐ **white pulp**

7. Sketch the spleen as seen through the microscope in the space provided. Be sure to label all the structures listed in step 6.

Table 23.4 Parts of the Spleen

Structure	Description	Word Origin
Capsule	Dense irregular connective tissue that supports and protects the outside of the spleen; surrounds the entire spleen	*capsula*, a box
Central Artery	Artery located within the white pulp of the spleen	*centrum*, center, + *arteria*, the windpipe
Hilum	The indented area of the spleen where the splenic artery enters and the splenic vein exits	*hilum*, a small bit
Red Pulp	Splenic tissue consisting of splenic sinusoids lined predominantly by T-lymphocytes; also contains erythrocytes, platelets, macrophages, and B-lymphocytes	*pulpa*, flesh
Splenic Sinusoids	Very permeable capillaries within the red pulp, which consist of fenestrated endothelial cells and a discontinuous basement membrane	*sinus*, cavity, + *eidos*, resemblance
Trabeculae	Invaginations of the connective tissue capsule of the spleen that partition the cortex into smaller compartments	*trabs*, a beam
White Pulp	Splenic tissue consisting of lymphatic nodules that contain T-lymphocytes, B-lymphocytes, and macrophages; each mass of white pulp contains a central artery	*pulpa*, flesh

GROSS ANATOMY

EXERCISE 23.8

GROSS ANATOMY OF LYMPHATIC STRUCTURES

1. *Major Lymph Vessels of the Body:* Obtain a model of the thorax and abdomen. Remove the heart and lungs from the thorax, and the stomach and intestines from the abdomen. Identify the structures listed in **figure 23.11** on the model, using **table 23.5** and the textbook as guides. Then label them in figure 23.11.

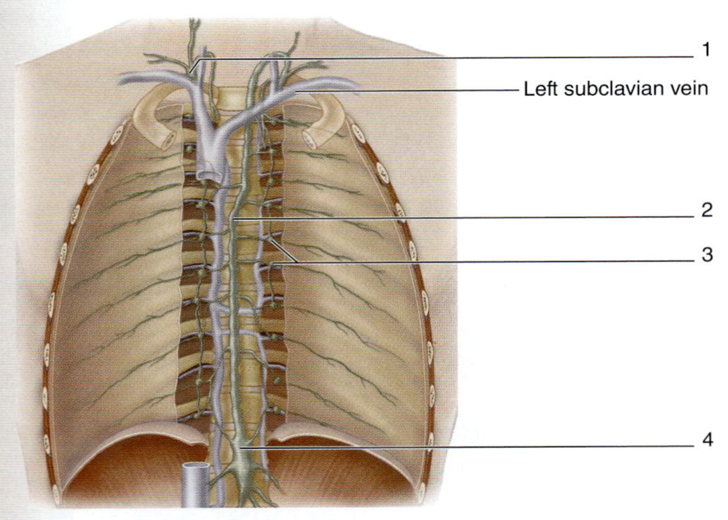

Figure 23.11 Major Lymph Vessels of the Body. Use the terms listed to fill in the numbered labels in the figure.

- ☐ cisterna chyli
- ☐ lymph nodes
- ☐ right lymphatic duct
- ☐ thoracic duct

INTEGRATE

LEARNING STRATEGY

Identification of the thoracic duct within the thoracic cavity of a cadaver can be challenging because of its relatively small diameter and lack of color (remember lymph does not contain erythrocytes). It travels between the esophagus and the azygos vein. Anatomists thus refer to it as the "duck between two gooses." The "duck" is the thoracic duct, and the "gooses" are the "azygoose" (azygos vein) and the "esophagoose" (esophagus).

2. *Mucosa-Associated Lymphatic Tissue (MALT):* Obtain a model demonstrating a midsagittal section of the head and neck, and a model of the abdomen. Identify the structures listed in **figure 23.12** on the model, using the textbook as a guide. Then label them in figure 23.12.

3. *Lymph Node:* Obtain a classroom model of a lymph node. Identify the structures listed in **figure 23.13** on the model, using the textbook as a guide. Then label them in figure 23.13.

4. *The Spleen:* Obtain a classroom model of the abdominal cavity, or observe the abdominal cavity of a human cadaver. Locate the spleen and identify the structures listed in **figure 23.14** on the model or the cadaver. Then label them in figure 23.14.

5. *Optional Activity:* **AP|R 10: Lymphatic System—** Watch the "Lymphatic System Overview" animation for a summary of the primary lymphatic structures.

Table 23.5 Major Lymphatic Vessels of the Body

Lymphatic Vessel	Location	Function	Word Origin
Cisterna Chyli	A"swelling" located at approximately vertebral level L1/L2. Lymph draining from the cisterna chyli flows into the thoracic duct.	Receives lymph from multiple smaller ducts that drain the lower limbs and abdominal cavity	*cista*, box + *chylos*, juice
Right Lymphatic Duct	A small duct that returns lymph to the cardiovascular system at the junction of the right internal jugular vein and the right subclavian vein.	Receives lymph from the right side of the head, neck, and thorax	*right*, the right side, + *lympha*, clear water + *ductus*, a leading
Thoracic Duct	Begins at the cisterna chyli, extends along the posterior thoracic wall between the azygous vein and the esophagus, and returns lymph to the cardiovascular system at the junction of the left internal jugular vein and the left subclavian vein	Receives lymph from the entire body below the diaphragm, and the left side of the head, neck, and thorax	*thoracic*, the thoracic cavity, + *ductus*, a leading

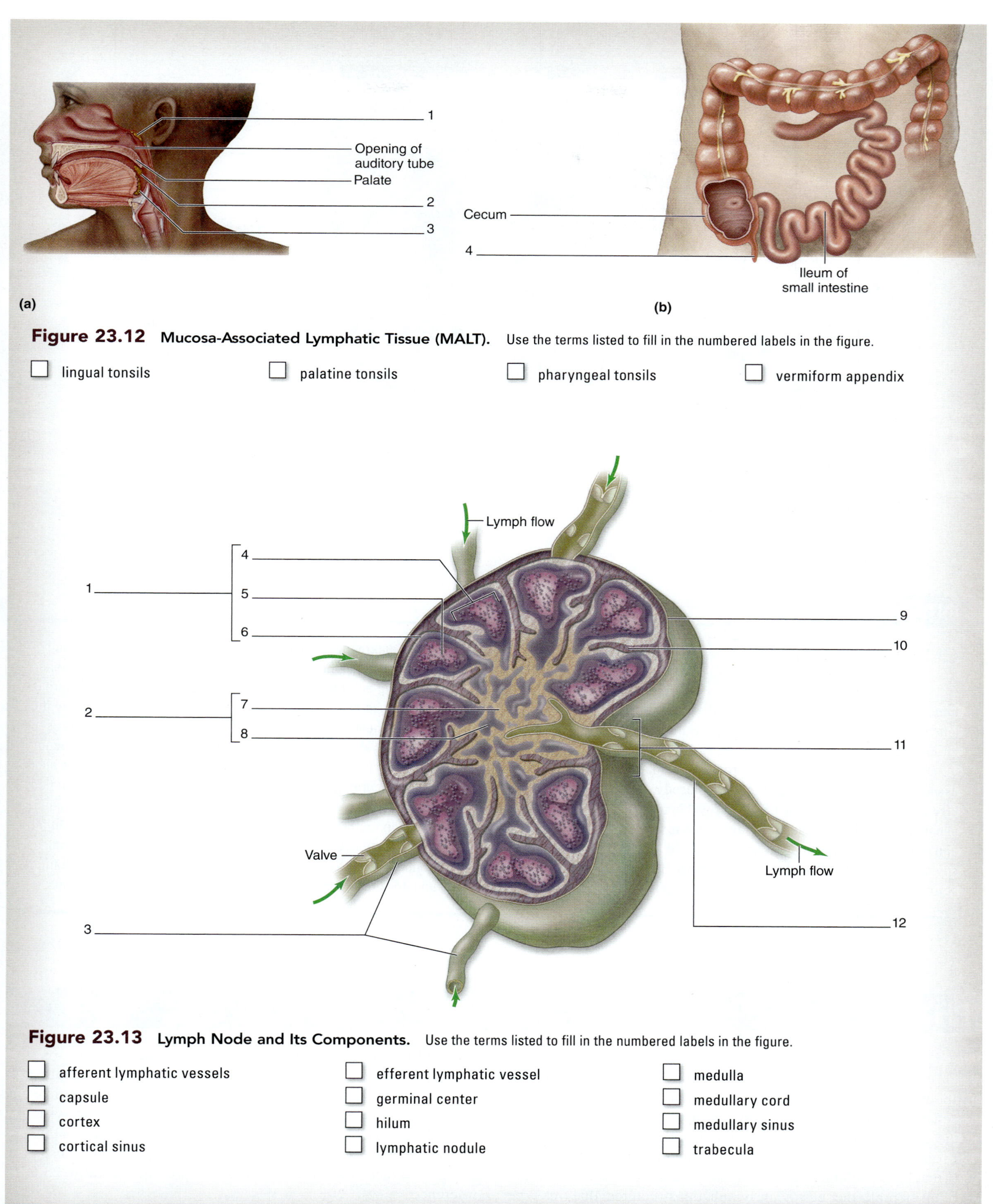

Figure 23.12 Mucosa-Associated Lymphatic Tissue (MALT). Use the terms listed to fill in the numbered labels in the figure.

- ☐ lingual tonsils
- ☐ palatine tonsils
- ☐ pharyngeal tonsils
- ☐ vermiform appendix

Figure 23.13 Lymph Node and Its Components. Use the terms listed to fill in the numbered labels in the figure.

- ☐ afferent lymphatic vessels
- ☐ capsule
- ☐ cortex
- ☐ cortical sinus
- ☐ efferent lymphatic vessel
- ☐ germinal center
- ☐ hilum
- ☐ lymphatic nodule
- ☐ medulla
- ☐ medullary cord
- ☐ medullary sinus
- ☐ trabecula

(continued on next page)

(continued from previous page)

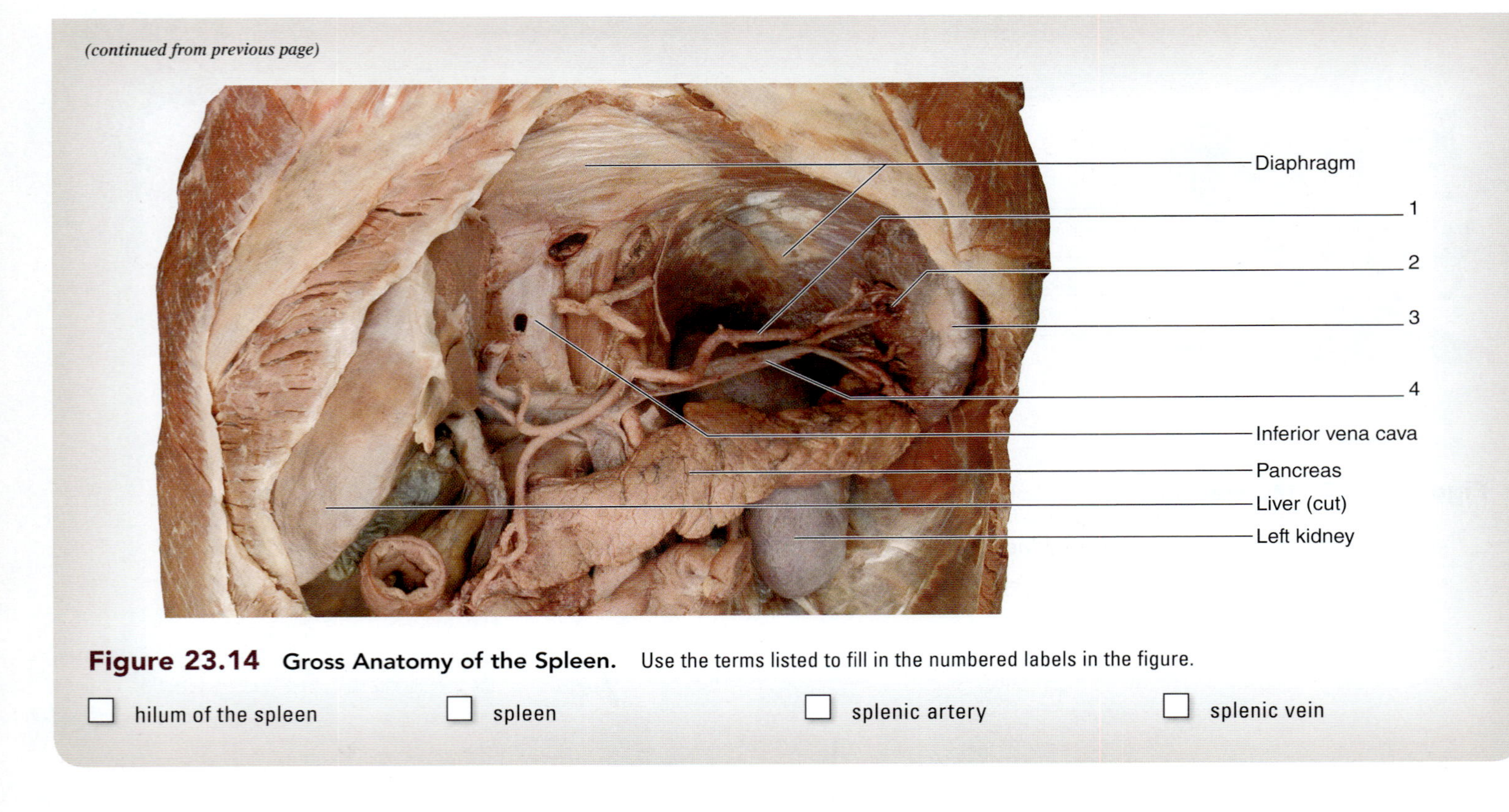

Figure 23.14 Gross Anatomy of the Spleen. Use the terms listed to fill in the numbered labels in the figure.

☐ hilum of the spleen ☐ spleen ☐ splenic artery ☐ splenic vein

INTEGRATE

CLINICAL VIEW
Mononucleosis

Infectious mononucleosis, or "mono" (i.e., the "kissing disease") is an infectious disease caused by the Epstein-Barr virus (EBV) and is spread through saliva. Typical symptoms include a mild fever, fatigue, severe sore throat, and lymphadenopathy (enlarged lymph nodes). The sore throat associated with mononucleosis is often confused with strep throat because EBV preferentially attacks lymphocytes within the palatine tonsils in the initial phase of the disease. This causes the tonsils to become enlarged and the crypts to fill with pus. The pus leads to the appearance of white spots on the tonsils, which are similar to those observed in a patient with strep throat. As the disease progresses, the spleens of infected individuals may become enlarged (splenomegaly) due to the proliferation of lymphocytes within, similar to the enlargement that occurs in lymph nodes when they are fighting an active infection. An enlarged spleen is a dangerous condition because the spleen becomes susceptible to rupture due to the stretching of its delicate outer capsule. A mononucleosis spot test is used to detect specific antibodies within the blood to confirm an active infection and distinguish it from strep throat. Unfortunately, the mononucleosis spot test may not accurately detect the disease during the incubation period (4–7 weeks) of the virus, or after active disease symptoms have subsided. For this reason, outbreaks of mononucleosis in college dorms or other places where people are in close contact with each other are common, because individuals may be unknowingly spreading the disease by sharing eating and drinking utensils before they present with active symptoms.

PHYSIOLOGY

The immune system is a complex system involved with bodily defenses against foreign or abnormal antigens (**antigen** = **anti**body **gen**erating substance). The majority of the exercises in this chapter focus on the anatomical structures that house the cells of the immune system (namely: lymphocytes and macrophages). The following laboratory exercise requires the student to consider how the body defends itself against a viral infection. In doing so, consider all the pertinent lymphatic structures that may be involved in housing components of the immune response (e.g., if a virus invades the nasopharynx, what lymphatic structures are located nearby that may house lymphocytes that can initiate the immune reponse?). A detailed discussion of the immune system is beyond the scope of this laboratory manual. However, a summary of the major cells responsible for initiating an immune response, along with their associated functions, is provided in **table 23.6** for reference.

Table 23.6 Cells of the Immune System

Cell Type	Location	Function	Word Origin
B-lymphocyte	Housed in secondary lymphatic structures including lymph nodes, spleen, tonsils, and MALT. Maturation is completed in the bone marrow.	When activated, B-lymphocytes form memory cells and plasma cells. Plasma cells actively secrete antibodies to help fight off the infection that initiated their formation.	*B,* matures in the bone marrow, + *lymphocyte,* cell of the lymphatic system
Cytotoxic T-lymphocyte (CD 8 cell)	Housed in secondary lymphatic structures including lymph nodes, spleen, tonsils, and MALT. Maturation is completed in the thymus.	Destroys abnormal or unwanted cells (e.g., cancerous or virus-infected cells) through perforin-mediated apoptosis	*T,* matures in the thymus, + *lymphocyte,* cell of the lymphatic system
Helper T-lymphocyte (CD 4 cell)	Housed in secondary lymphatic structures including lymph nodes, spleen, tonsils, and MALT. Maturation is completed in the thymus.	Secretes cytokines that activate and regulate other immune cells (e.g., B-lymphocytes and Cytotoxic T-lymphocytes)	*T,* matures in the thymus, + *lymphocyte,* cell of the lymphatic system
Macrophage	Housed in secondary lymphatic structures including lymph nodes, spleen, tonsils, and MALT, as well as within numerous body organs (e.g., alveolar macrophages, dendritic cells in the skin, microglia in the brain)	Engulfs antigen and presents antigen to T-lymphocytes	*macro,* huge + *phagos,* to eat

EXERCISE 23.9

A CLINICAL CASE STUDY IN IMMUNOLOGY

The purpose of this exercise is to apply concepts from the immune system to a real scenario, or clinical case. This exercise requires information presented in the histology, gross anatomy, and physiology sections of this chapter, as well as information from previous chapters. Integrating multiple systems in a case study reinforces concepts from several chapters.

Prior to beginning the case, consult with a laboratory instructor to see if this exercise will be completed individually, in pairs, or in groups. Review topics from the autonomic nervous system (chapter 17), endocrine system (chapter 19), and cardiovascular system (chapters 20–22), in addition to the histology and gross anatomy of the lymphatic system and the physiology of immunity. Use the textbook and this laboratory manual as guides.

Case Study

Greg, a 29-year-old male, had been in overall good health until recently when he began to develop a nonproductive cough, myalgia (muscle pain), sore throat, nasal congestion, headache, and fever (39.4 °C). In addition, he complained of mild thoracic pain while breathing. When Greg's pain spread to his lower back and scrotum, he sought medical attention. Upon presentation to the physician's office, Greg was found to have an elevated pulse (tachycardia) and fluid in his lungs. Greg told doctors that 5 days prior to his onset of symptoms, he had been exposed to a child who had an upper respiratory tract infection. A buccal swab test was negative for both influenza A and B. Suspecting that Greg was suffering from pneumonia and lymphadenopathy (enlarged lymph nodes), doctors prescribed a broad-spectrum antibiotic.

The antibiotic failed to treat Greg's symptoms. Instead, his symptoms worsened and he began to vomit blood. His pulse remained elevated and his respiratory rate also became elevated. In addition, the amount of fluid in his lungs increased, causing him to experience dyspnea (shortness of breath) and labored breathing. An electrocardiogram revealed sinus tachycardia, and a transthoracic echocardiogram revealed a decreased ejection fraction of his heart. In addition, Greg's lymphadenopathy worsened, and mediastinal and hilar (lung) lymph nodes became enlarged. Greg also suffered from splenomegaly (enlarged spleen). Greg's mean arterial pressure dropped suddenly, and he developed hypoxemia and renal failure. Doctors administered *epinephrine* and high doses of *methylprednisolone* (a glucocorticoid).

Still in search of the pathogen responsible for Greg's rapid decline, doctors placed a bronchoscope into his lungs to obtain a specimen. From the specimen, doctors determined that Greg tested positive for the H1N1 influenza virus (i.e., "swine flu"). Once the diagnosis was confirmed, doctors prescribed an antiviral drug. Unfortunately, Greg's condition continued to decline, and he died 9 days following his onset of symptoms. Greg was one of many victims of the 2009 H1N1 pandemic.

(continued on next page)

(continued from previous page)

Answer the following questions regarding Greg's case:

1. Describe innate and adaptive defenses that would be employed to prevent entry of a pathogen such as the H1N1 influenza virus into the respiratory tract.

2. Why did Greg suffer from lymphadenopathy, particularly in his mediastinal and hilar lymph nodes, given his diagnosis of H1N1 influenza? Be sure to describe the structure and function of lymph nodes, and describe why some lymph nodes may become enlarged but not others.

3. Why was Greg's spleen enlarged, given his diagnosis of H1N1 influenza? Describe the spleen's role in both cardiovascular and immune function.

4. Justify the administration of *methylprednisolone* given Greg's symptoms. Where are glucocorticoids typically synthesized, and what is their typical function?

Adapted from Case 40-2009, *The New England Journal of Medicine*, 361, no. 26 (2009): 2558–69.

Chapter 23: The Lymphatic System and Immunity

Name: ____________________

Date: __________ Section: __________

POST-LABORATORY WORKSHEET

The (1) corresponds to the Learning Objective(s) listed in the chapter opener outline.

Do You Know the Basics?

Exercise 23.1: Lymphatic Vessels

1. Lymphatic vessels have valves. ____________ (True/False) (1) (2)
2. Lymphatic vessels are most similar in structure to ______________ (arteries/veins). (3)

Exercise 23.2: Tonsils

3. Match the description listed in column A with the type of tonsil listed in column B. (Some answers may be used more than once.) (4) (5) (6)

Column A

_____ 1. covered with pseudostratified ciliated columnar epithelium

_____ 2. covered with stratified squamous (nonkeratinized) epithelium

_____ 3. located at the base of the tongue

_____ 4. located at the junction of the nasopharynx and oropharynx at the posterior aspect of the soft palate

_____ 5. projects from the roof of the nasopharynx

Column B

a. lingual

b. palatine

c. pharyngeal

Exercise 23.3: Peyer Patches

4. Peyer patches have a darker-staining germinal center and a lighter-staining outer region. ______________ (True/False) (7)
5. Peyer patches are located in the duodenum. ____________ (True/False) (8)

Exercise 23.4: The Vermiform Appendix

6. The vermiform appendix is lined with which type of epithelial tissue? (Circle one.) (9)
   a. pseudostratified ciliated columnar epithelium
   b. simple columnar epithelium
   c. simple cuboidal epithelium
   d. simple squamous epithelium
   e. stratified squamous epithelium
7. An individual suffering from an inflamed appendix will most commonly have pain in the ______________ (upper/lower) ______________ (left/right) abdominopelvic quadrant. (10)

Exercise 23.5: The Thymus

8. The thymus contains connective tissue septae that separate the gland into lobules. ______________ (True/False) (11)
9. The thymus is the site of ______________ (B-cell/T-cell) maturation. (12)
10. An increase in the number of thymic corpuscles indicates that the thymus is ______________(older/younger). (13)

Exercise 23.6: Lymph Nodes

11. Match the descriptions listed in column A with the structure listed in column B. (14) (15)

Column A

_____ 1. space through which lymph travels within a lymph node

_____ 2. structure in the medulla of the lymph node that contains T-lymphocytes and macrophages

_____ 3. vessel bringing lymph into the lymph node

_____ 4. vessel taking lymph away from the lymph node

Column B

a. afferent

b. efferent

c. medullary cord

d. sinus

12. Imagine a virus is circulating in the lymph. Place the following structures in the order in which the virus travels. (16)

_____ a. afferent lymphatic vessel

_____ b. cortical sinus

_____ c. efferent lymphatic vessel

_____ d. medullary sinus

Exercise 23.7: The Spleen

13. Splenic sinusoids filter __________ (blood/lymph). (17)

14. The __________ (white/red) pulp of the spleen contains lymphocytes and serves to fight pathogens, whereas the __________ (white/red) pulp of the spleen contains sinusoids and filters blood. (18)

Exercise 23.8: Gross Anatomy of Lymphatic Structures

15. Imagine a bacterium has entered a lymph capillary in the right arm. Place the following structures in order as the bacterium travels from a lymph node in the axilla to the right atrium of the heart. (19)

_____ a. efferent lymphatic vessel

_____ b. right brachiocephalic vein

_____ c. right lymphatic duct

_____ d. right subclavian vein

_____ e. superior vena cava

Exercise 23.9: A Clinical Case Study in Immunology

16. *Lymphadenopathy* is a clinical term for enlarged lymph nodes. __________ (True/False) (20)

17. Which of the following systems would be affected by splenomegaly (enlarged spleen)? (Check all that apply.) (21)

_____ a. cardiovascular

_____ b. digestive

_____ c. lymphatic

_____ d. respiratory

Can You Apply What You've Learned?

18. When pharyngeal tonsils become inflamed, they are referred to as adenoids. Based on the location of the adenoids (pharyngeal tonsils), what kinds of symptoms might a patient experience due to the presence of swollen adenoids?

____

____

____

19. When an individual suffers from a ruptured spleen, the spleen is surgically removed so the patient does not die from internal bleeding. Using knowledge of the circulation of the spleen, explain why a ruptured spleen bleeds so profusely. (Hint: What type of capillaries are found in the spleen?)

____

____

20. Discuss why lymphatic nodules are so abundant in the wall of the small intestine. __________

____

____

21. Lacteals are specialized lymphatic capillaries found in the small intestine that have the special function of absorbing dietary fats. Absorbed fats first enter lymphatic capillaries and are transported as a component of lymph within lymph vessels. Using knowledge of the flow of both lymph and blood in the body, describe the route a dietary lipid would take to get from its location of absorption in the small intestine to its entry into the right atrium of the heart.

Can You Synthesize What You've Learned?

22. Acute appendicitis (inflammation of the vermiform appendix) requires immediate surgical removal of the appendix. In the event that the appendix bursts, its contents are released into the abdominal cavity, which can lead to widespread infection, sepsis, and death. Discuss the events that might lead to an enlarged and inflamed vermiform appendix.

23. Following the 2009 H1N1 pandemic, annual influenza vaccines began to include strains for H1N1 influenza. Discuss the purpose of vaccines and state how vaccines may protect against disease outbreaks.

CHAPTER 24

The Respiratory System

OUTLINE AND LEARNING OBJECTIVES

INTRODUCTION

Breathe in deeply. What muscles were used to do this? Recall from chapter 11 that the **diaphragm** and the **external intercostal muscles** are the key muscles used for **inspiration.** Contraction of these muscles increases the volume of the thoracic cavity, causing the lungs to expand. As the lungs expand and the volume of the lungs increases, the pressure within the lungs becomes lower than atmospheric pressure and air flows into the lungs. This air brings a fresh supply of oxygen that is picked up by **erythrocytes** in the blood flowing through the lung capillaries. Freshly oxygenated blood will be transported back to the heart and pumped out to the body. Take another deep breath. This time focus on the process of **expiration** (breathing out). Quiet expiration is a passive process, which does not require muscular effort. As soon as the diaphragm and external intercostal muscles relax, the elastic tissues within the lungs and chest cavity wall recoil causing a decrease in the volume of the thoracic cavity (and the lungs), which increases intrapulmonary pressure. As intrapulmonary pressure becomes greater than atmospheric pressure, air flows out of the lungs. This air has a relatively high concentration of carbon dioxide, a waste product from cellular respiration, which must be removed from the body.

The exercises in this chapter explore the histological and gross structure of the respiratory tract and the lungs. They begin with an investigation into the general layering pattern of the walls of the respiratory tract to prepare the student to identify specific histological structures located within each layer. This will be followed by analysis of the histology of major respiratory tract structures, including the lungs. Histological observations will be followed by observations of the gross anatomy of the upper airways and the lungs, using cadaveric specimens, fresh sheep specimens, models, or some combination of these. Finally, the processes of ventilation, gas exchange, and gas transport will be explored through a series of physiological experiments.

List of Reference Tables

Chapter 24: The Respiratory System

Name: ____________________

Date: __________ Section: __________

These Pre-Laboratory Worksheet questions may be assigned by instructors through their connect course.

PRE-LABORATORY WORKSHEET

1. Respiratory epithelium, which lines the nasal cavity and upper respiratory tract, is classified as: (Circle one.)
   a. simple columnar with cilia
   b. simple cuboidal with microvilli
   c. simple columnar with cilia and goblet cells
   d. pseudostratified columnar with cilia and goblet cells
   e. pseudostratified columnar with microvilli
2. The left lung contains ______________ (two/three) lobes, whereas the right lung contains ______________ (two/three) lobes.
3. Which of the following bones does *not* form a border of the nasal cavity? (Circle one.)
   a. ethmoid b. mandible c. maxilla d. palatine e. vomer
4. The pulmonary arteries carry ______________ (oxygenated/deoxygenated) blood from the ______________ (right/left) ventricle of the heart to the lungs, and the pulmonary veins carry ______________ (oxygenated/deoxygenated) blood from the lungs to the ______________ (right/left) atrium of the heart. This circuit is referred to as the ______________ (pulmonary/systemic) circuit.
5. Contraction of the diaphragm decreases the volume of the thoracic cavity. ______________ (True/False)
6. Which of the following is *not* a component of the respiratory membrane? (Circle one.)
   a. capillary endothelial cell
   b. fused basement membranes
   c. type I cell
   d. type II cell
7. Which of the following are cells that produce surfactant? (Circle one.)
   a. alveolar macrophages
   b. capillary endothelial cell
   c. fused basement membranes
   d. type I cell
   e. type II cell
8. A ______________ (lobar/main/segmental) bronchus leads into each lung; a ______________ (lobar/main/segmental) bronchus leads into a lobe of a lung; a ______________ (lobar/main/segmental) bronchus leads into a bronchopulmonary segment.
9. The ______________ (parietal/visceral) pleura covers the outer surface of each lung, whereas the ______________ (parietal/visceral) pleura lines the inner surfaces of the thoracic cavity.
10. Structures that compose the ______________ (conducting/respiratory) division of the respiratory tree contain alveoli.

HISTOLOGY

Upper Respiratory Tract

The respiratory system is organized into two structural regions: an upper respiratory tract and a lower respiratory tract. The upper respiratory tract includes the nose, nasal cavity, paranasal sinuses, and pharynx. The lower respiratory tract includes the larynx, trachea, bronchi, bronchioles, alveolar ducts, and alveoli. The structures of the respiratory system are also categorized based on function. Passageways that serve to transport or conduct air compose the **conducting zone;** these structures include the passageways from the nasal cavity to the terminal bronchioles. Structures that participate in gas exchange with the blood—including the respiratory bronchioles, alveolar ducts, and alveoli—compose the **respiratory zone.**

The walls of the respiratory passageways are lined by a mucous membrane, called a mucosa**.** The **mucosa** (*mucosus*, mucus) consists of the epithelium resting on a basement membrane, and an underlying layer of areolar connective tissue called the **lamina propria** (*lamina*, layer, + *proprius*, one's own). Typically, there are more layers deep to the mucosa including both the **submucosa** (*sub-*, under, + mucosa), and the **adventitia** (*adventicius,* coming from abroad). The submucosa consists of areolar connective tissue, in addition to glands, blood vessels, and nerves. The adventitia is composed of loose connective tissue, blood vessels, and nerves. In some portions of the respiratory passageway, including the trachea and bronchi, there are partial rings or plates of hyaline cartilage, which provide support to the airway. The cartilage is ensheathed in dense irregular connective tissue, and is positioned between the submucosa and adventitia. **Figure 24.1** shows the layers of the tracheal wall. While viewing slides of respiratory system structures, it is helpful to first identify the three layers, particularly the mucosa and submucosa, This will make subsequent identification of specific cell types within each layer easier.

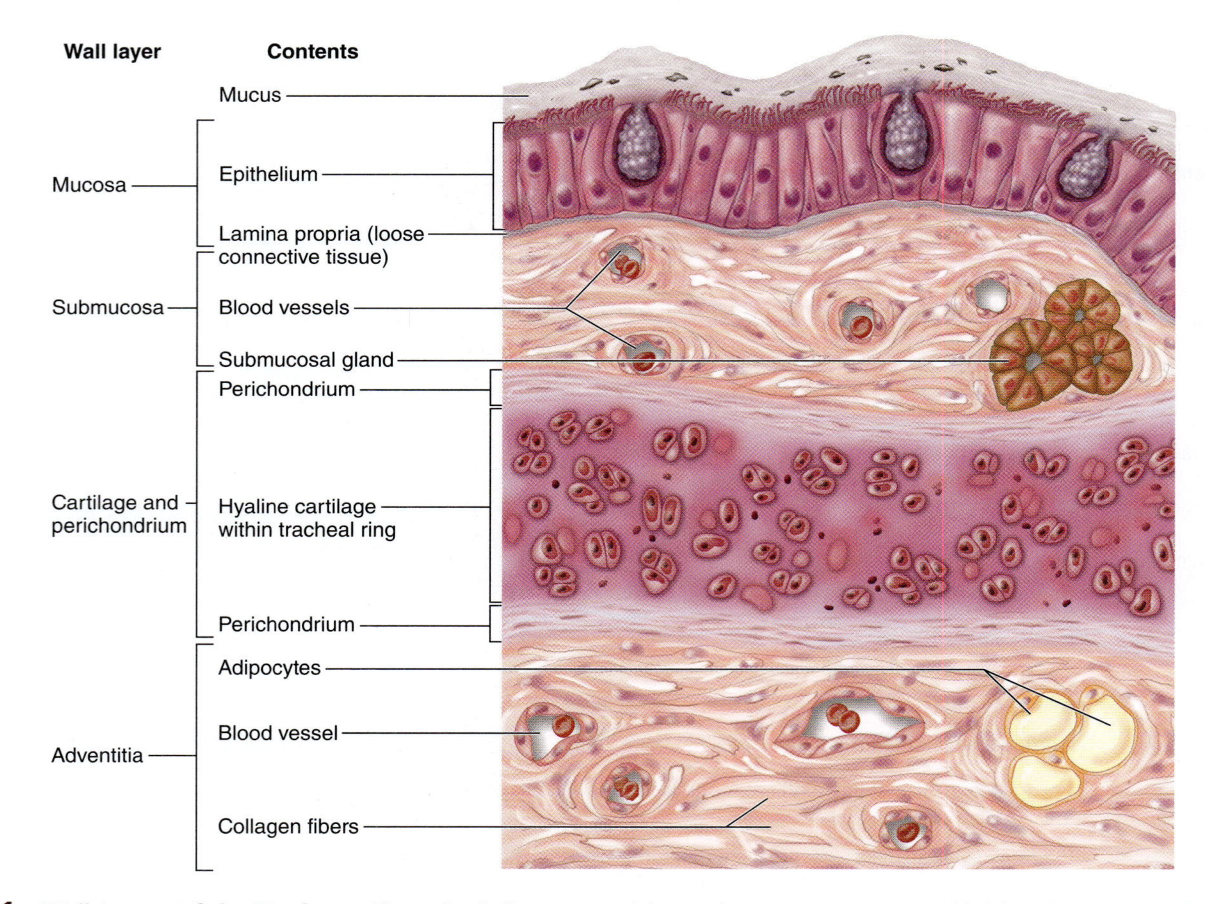

Figure 24.1 Wall Layers of the Trachea. The walls of all structures of the respiratory tract are composed of three layers: mucosa, submucosa, and adventitia/serosa. This drawing depicts the wall of the trachea, which has an outer covering of connective tissue referred to as an adventitia.

EXERCISE 24.1

OLFACTORY MUCOSA

The epithelium that lines the nasal cavity is highly specialized. It serves to warm, moisten, and filter the air as it enters the respiratory passageway. In addition, the epithelium located in the superior portions of the nasal cavity **(olfactory epithelium)** acts as a special sensory organ to detect odors **(figure 24.2)**. Table 18.4 (p. 469) describes the appearance and function of the various cell types and structures within the olfactory epithelium.

1. Obtain a compound microscope and a slide of olfactory epithelium. Place the slide on the microscope stage and bring the tissue sample into focus on low power.
2. Identify the **mucosa** and **epithelium.** Move the microscope stage so the epithelium is at the center of the field of view, then change to high power.
3. Identify the following structures on the slide, using table 18.4 and figure 24.2 as guides:

- ☐ **basal cells**
- ☐ **olfactory glands**
- ☐ **olfactory receptor cells**
- ☐ **supporting cells (ciliated)**

4. Sketch the olfactory epithelium as viewed through the microscope in the space provided. Be sure to label all the structures listed in step 3 in the drawing.

________ ×

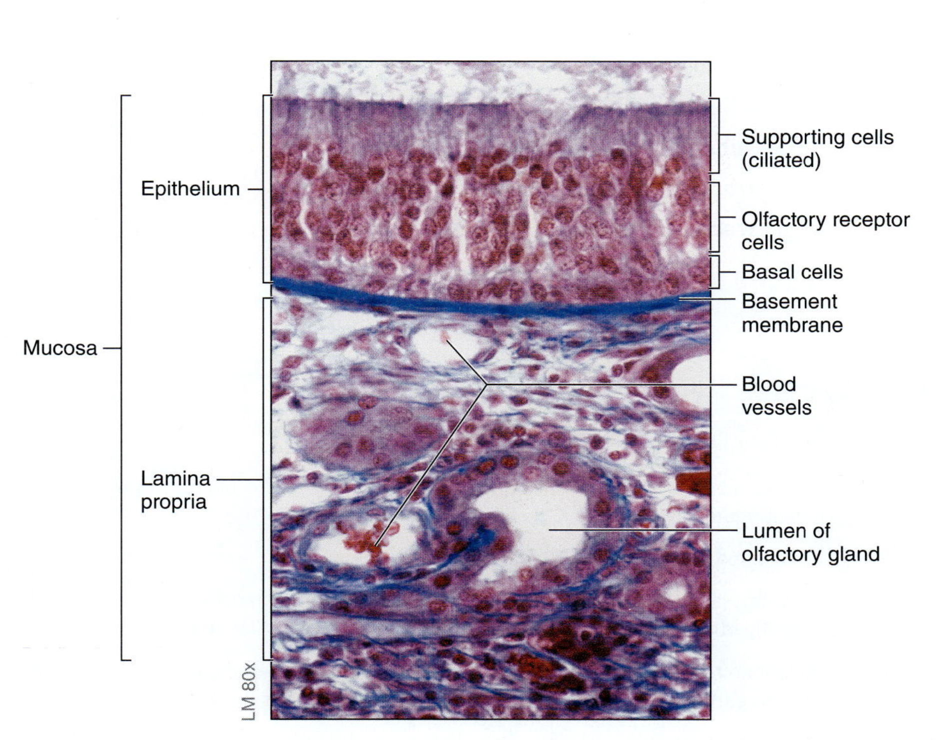

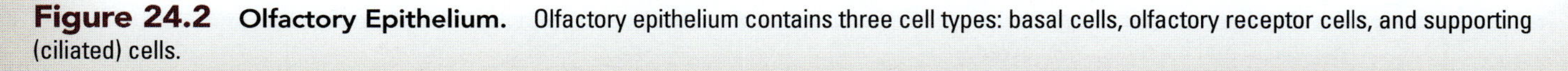
Figure 24.2 Olfactory Epithelium. Olfactory epithelium contains three cell types: basal cells, olfactory receptor cells, and supporting (ciliated) cells.

Lower Respiratory Tract

The lower respiratory tract includes the larynx, **trachea, bronchi,** bronchioles, alveolar ducts, and alveoli. As the airways extend progressively deeper into the lungs from the trachea to the bronchi and bronchioles, the amount of cartilage in the submucosa decreases and the relative amount of smooth muscle increases. In addition, the epithelium transitions from pseudostratified ciliated columnar epithelium of the trachea to the simple squamous epithelium of the alveoli within the lungs. Table 24.1 describes the wall structure of the trachea, bronchi, and bronchioles.

EXERCISE 24.2

THE TRACHEA

1. Obtain a slide of the trachea (*tracheia*, rough artery) and place it on the microscope stage. This slide will also contain the esophagus (*oisophagos*, gullet) because the two structures are adjacent to each other *in vivo*.
2. Bring the tissue sample into focus on low power and distinguish the trachea from the esophagus. Move the microscope stage so the lumen and epithelium of the trachea are in the center of the field of view. Then switch to high power.
3. While observing the trachea, first identify the layers of the wall of the trachea (mucosa, submucosa, and adventitia). Then identify the structures located within each layer. **Table 24.1** describes the wall structure of the trachea and the structures located within each layer, and **figure 24.3** demonstrates the histological appearance of the wall of the trachea.
4. Identify the following structures on the slide of the trachea, using table 24.1 and figure 24.3 as guides:

- ☐ **adventitia**
- ☐ **basal cells**
- ☐ **ciliated cells**
- ☐ **epithelium**
- ☐ **goblet cells**
- ☐ **lamina propria**
- ☐ **mucosa**
- ☐ **submucosa**
- ☐ **submucous glands**
- ☐ **tracheal cartilages**
- ☐ **trachealis muscle**

5. Sketch a cross section of the trachea as viewed through the microscope in the space provided. Be sure to label all of the structures listed in step 4 in the drawing.

_____________ ×

Table 24.1	Histology of the Trachea
Structure	**Description and Function**
Mucosa	Innermost layer that lines the tracheal wall; consists of respiratory epithelium resting on a basement membrane and underlying lamina propria
Epithelium	Pseudostratified ciliated columnar epithelium
Basal Cells	Small triangular cells that lie on the basal lamina and do not reach the apical surface of the epithelium. These cells are the precursor cells to the other cell types and are responsible for regeneration of the epithelium.
Ciliated Cells	The most prominent and abundant cells. These are pseudostratified ciliated columnar cells of the epithelium. They function to propel mucus and particulate matter via the "mucus escalator" along the epithelial sheet toward the pharynx.
Goblet Cells	Large round cells that appear white or clear. They secrete mucin onto the surface of the epithelium. The mucus traps particulate matter that enters the trachea while the cilia move the mucus superiorly (a "mucus escalator") toward the pharynx so that it may be swallowed.
Lamina Propria	A layer of areolar connective tissue that underlies the respiratory epithelium. In the trachea this layer can be seen in the area between the epithelial folds and the trachealis muscle or C-shaped cartilages.
Submucosa	The middle layer of the tracheal wall containing seromucous glands, blood vessels, and nerves
Submucous Glands	These produce a substance that is part watery (serous) and part viscous (mucus)

Table 24.1	Histology of the Trachea *(continued)*
Structure	**Description and Function**
Tracheal Cartilages	C-shaped sections of cartilage that serve to support the wall of the respiratory tract.
Trachealis Muscle	A layer of smooth muscle found on the posterior aspect of the trachea where the trachea lies against the esophagus. Laterally, the trachealis muscle is anchored to the ends of the cartilage C rings, and its contraction decreases the diameter of the trachea, which is important for coughing and sneezing. The decreased diameter of the trachea causes the air to exit more forcefully, which helps dislodge substances within the airways.
Adventitia	Loose connective tissue on the outermost surface of the tracheal wall

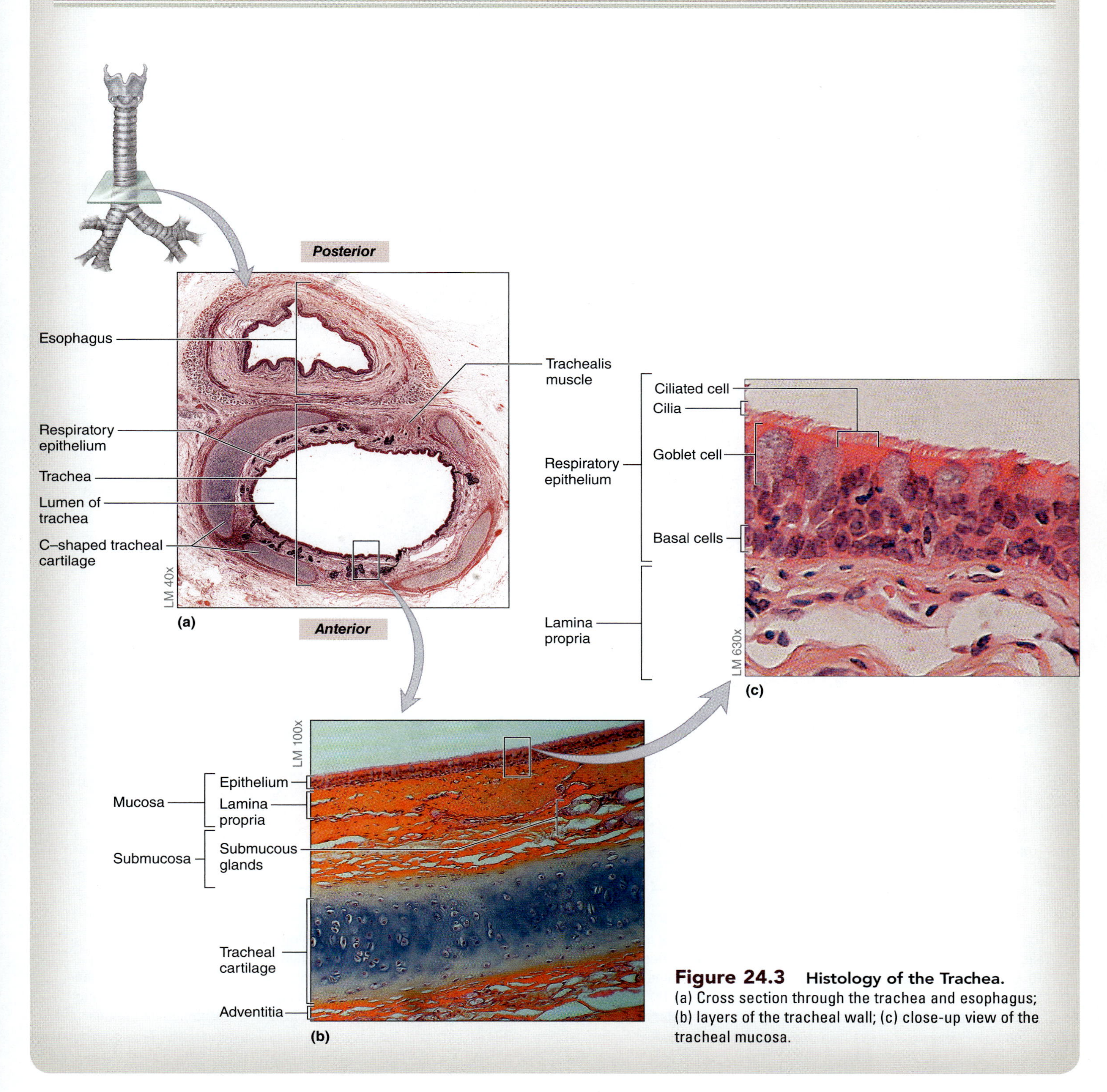

Figure 24.3 **Histology of the Trachea.** (a) Cross section through the trachea and esophagus; (b) layers of the tracheal wall; (c) close-up view of the tracheal mucosa.

INTEGRATE

LEARNING STRATEGY

The trachea conducts air from the larynx to the bronchi within the lungs. The structure of the trachea is ideal for this function in that C-shaped hyaline cartilages provide semirigid support to maintain a patent (open) airway. Posterior to the cartilages lies a band of smooth muscle, the trachealis muscle. Contraction of the trachealis muscle, such as occurs during a cough reflex, rapidly changes the diameter of the airway. The rapid decrease in the diameter of the airway causes a sudden increase in velocity of the air moving through the trachea. This increase in velocity is important for forcefully expelling foreign objects or mucus. In addition, the absence of cartilage on the posterior portion of the trachea provides flexibility to the airway. This flexibility allows the esophagus to expand anteriorly when food is being transported within.

EXERCISE 24.3

THE BRONCHI AND BRONCHIOLES

1. Obtain a slide of the lungs and place it on the microscope stage. Bring the tissue sample into focus on low power and scan the slide to locate cross sections of **large bronchi, small bronchi,** and **bronchioles (figure 24.4).**

2. The different characteristics of the conducting zone structures are listed in **table 24.2.** In general, as the airways travel deeper into the lung and become progressively smaller, three changes take place: (1) The epithelium transitions from pseudostratified ciliated columnar to simple ciliated columnar, and the number

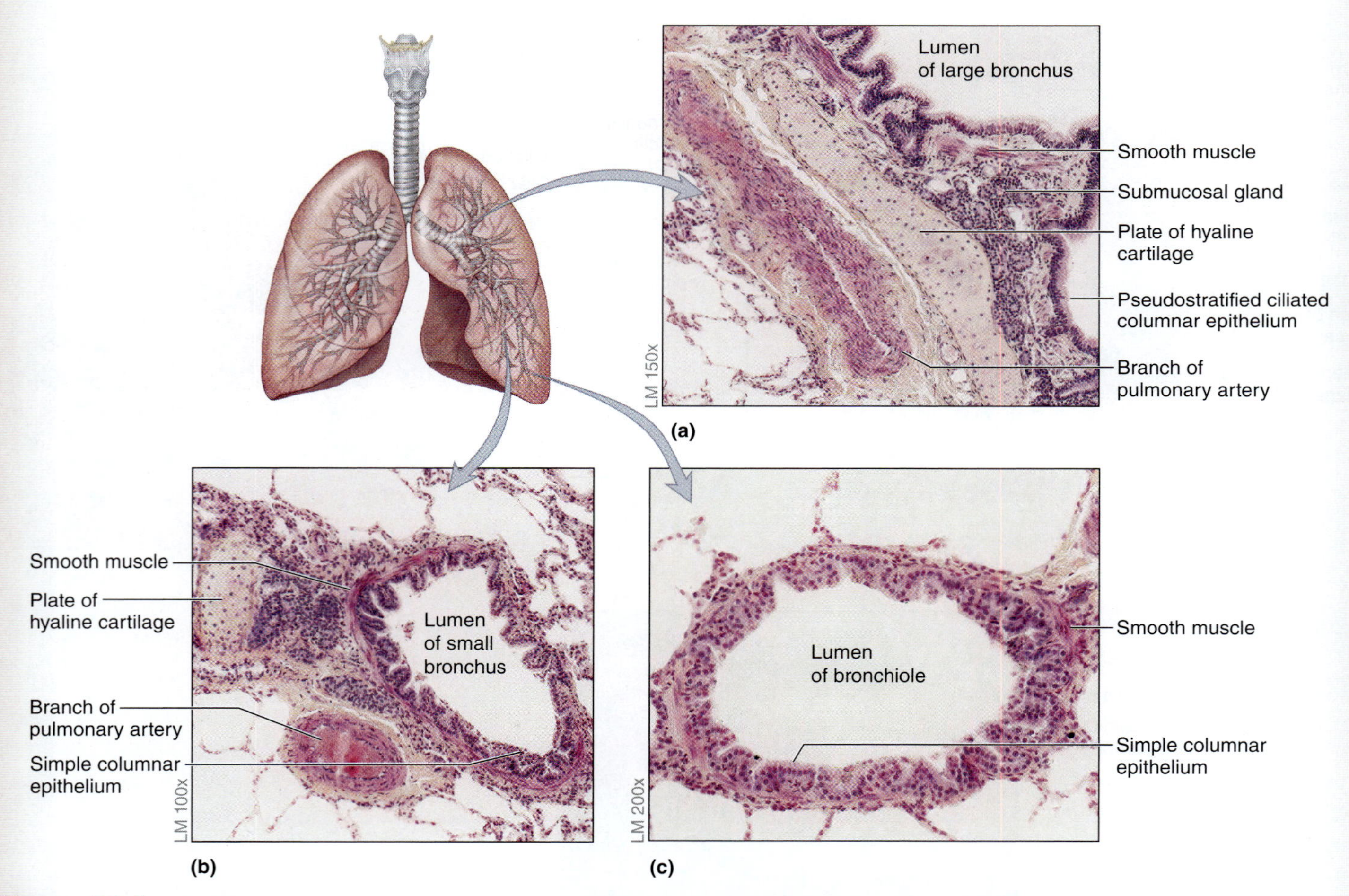

Figure 24.4 The Bronchial Tree. Cross sections of portions of the bronchial tree are visible in a histological image of a lung. (a) The walls of larger bronchi are lined with ciliated pseudostratified columnar epithelium, and include relatively large plates of hyaline cartilage along with smooth muscle. (b) The walls of smaller bronchi are lined with simple ciliated columnar epithelium, and contain smaller plates of cartilage and relatively more smooth muscle than in the larger bronchi. (c) Bronchioles feature simple ciliated columnar epithelium, no cartilage, and a thin ring of smooth muscle. Note: The lack of cartilage in bronchioles distinguishes these portions of the respiratory tract from bronchi, which do contain cartilage.

of cilia decrease; (2) large plates of hyaline cartilage in the walls give way to smaller and smaller pieces of cartilage, with no cartilage present in bronchioles; and (3) the relative amount of smooth muscle in the airways increases.

3. Identify the following structures on the slide of the lungs, using figure 24.4 and table 24.2 as guides:

- ☐ **branch of pulmonary artery**
- ☐ **bronchiole**
- ☐ **hyaline cartilage**
- ☐ **large bronchus**
- ☐ **small bronchus**
- ☐ **smooth muscle**

4. Sketch a cross section of a bronchiole as viewed through the microscope in the space provided.

_______________ ×

Table 24.2 Histology of the Bronchial Tree

Structure	Epithelium	Hyaline Cartilage	Smooth Muscle
Large Bronchi	Pseudostratified ciliated columnar	Large plates, which keep airway open	Encircles the lumen
Small Bronchi	Simple ciliated columnar	Small plates	Encircles the lumen
Bronchioles	Simple ciliated columnar	None	Encircles the lumen

Lungs

The lungs consist of functional units called **alveoli** (*alveus*, hollow sac), which are the sites of gas exchange between the air within the alveoli and the blood within the pulmonary capillaries. The alveoli are lined with a simple squamous epithelium, which provides the thinnest possible barrier to diffusion between the air within the alveoli and the blood within the pulmonary capillaries that surround them. In a slide of the lung, numerous alveoli and cross sections of some of the smaller airways (such as respiratory bronchioles) will be visible scattered throughout the slide. The transition from conducting zone structures to respiratory zone structures occurs deep within the lungs. The key feature in distinguishing respiratory zone structures from conducting zone structures is the presence of alveoli. Any structure that has at least one alveolus coming off of it (as may be the case with a respiratory bronchiole) participates in gas exchange and, thus, is part of the respiratory zone.

INTEGRATE

CLINICAL VIEW

Tuberculosis

Pulmonary tuberculosis, typically caused by *Mycobacterium tuberculosis*, is an infection that impacts the lungs. While the majority of individuals infected with *M. tuberculosis* lack symptoms, an active infection can cause severe complications, even death. Immune compromised individuals are particularly susceptible to developing an active tuberculosis infection. Typically, *M. tuberculosis* invades alveolar macrophages and begins replication, forming a tubercle (small nodule). Additional immune cells (i.e., B- and T-lymphocytes) surround the macrophages, forming a granuloma. Necrosis may occur at the center of the granuloma. A chest radiograph of a patient suffering from tuberculosis will typically show nodules near the apex of the lungs. Interestingly, presence of nodules may also indicate a latent infection. Signs of an active infection include persistent cough, fever, and fatigue. Tuberculosis may be spread easily to others through contact with bodily fluids. While pulmonary tuberculosis is the most common type of tuberculosis, the disease may impact other organs. *M. tuberculosis* may enter the blood and/or lymph vessels and easily spread from the lungs to other organs. Treatment of an active infection often involves the use of multiple antibiotics.

EXERCISE 24.4

THE LUNGS

Table 24.3 describes the structure and function of the airways that compose the respiratory portion of the bronchial tree.

Table 24.4 describes the structure and function of the cell types within the alveoli. For gas exchange to occur, gas molecules must pass between the alveoli and the capillaries that surround them. The structures lining the alveoli and the capillaries collectively form the **respiratory membrane**. The respiratory membrane is composed of (1) alveolar type I cells (2) the fused basement membrane of alveolar type I cells and endothelial cells of the capillaries, and (3) the endothelial cells lining the capillaries.

1. Obtain a slide of the lungs and place it on the microscope stage. Bring the tissue sample into focus on low power and scan the slide to locate cross sections of **bronchioles (figure 24.5)**.
2. Increase the magnification to observe the smaller airways and the air sacs (alveoli) in greater detail.
3. Identify the following structures on the slide of the lungs, using table 24.3, table 24.4, and figure 24.5 as guides:

- ☐ alveolar ducts
- ☐ alveolar macrophages
- ☐ alveolar sacs
- ☐ alveolar type I cell
- ☐ alveolar type II cell
- ☐ alveoli
- ☐ capillary
- ☐ respiratory bronchioles

4. Sketch the respiratory membrane in the space provided. Then label the three structures that make up the respiratory membrane.

5. *Optional Activity:* AP|R **11: Respiratory System**—Watch the "Diffusion Across Respiratory Membrane" animation to visualize the microscopic structure and function of lung tissue.

Table 24.3 Structures Composing the Respiratory Zone

Structure	Description	Epithelium
Respiratory Bronchioles	Thin-walled ducts with alveoli scattered along the passageway	Ciliated simple cuboidal
Alveolar Ducts	Ducts with alveoli lining the entire passageway	Simple squamous
Alveolar Sacs	Shaped like a bunch of grapes with several alveoli (the "grapes") bunched together	Simple squamous
Alveoli	A sac-like structure that is the site of gas exchange	Simple squamous

Table 24.4 Microscopic Structures Within the Lungs

Structure	Description
Alveolar Macrophage	Cell derived from a monocyte, which is housed within alveoli. Engulf particulate matter and microorganisms.
Alveolar Type I Cell	Simple squamous epithelial cells covering 97% of the alveolar surface area. They are connected to each other with tight junctions.
Alveolar Type II Cell	Compose about 3% of alveolar surface area. Joined to type I cells by desmosomes and tight junctions. Contain a large nucleus, foamy cytoplasm, and vesicles containing **pulmonary surfactant** (a mixture of phospholipids, glycosaminoglycans, and proteins) that functions to reduce surface tension within alveoli.

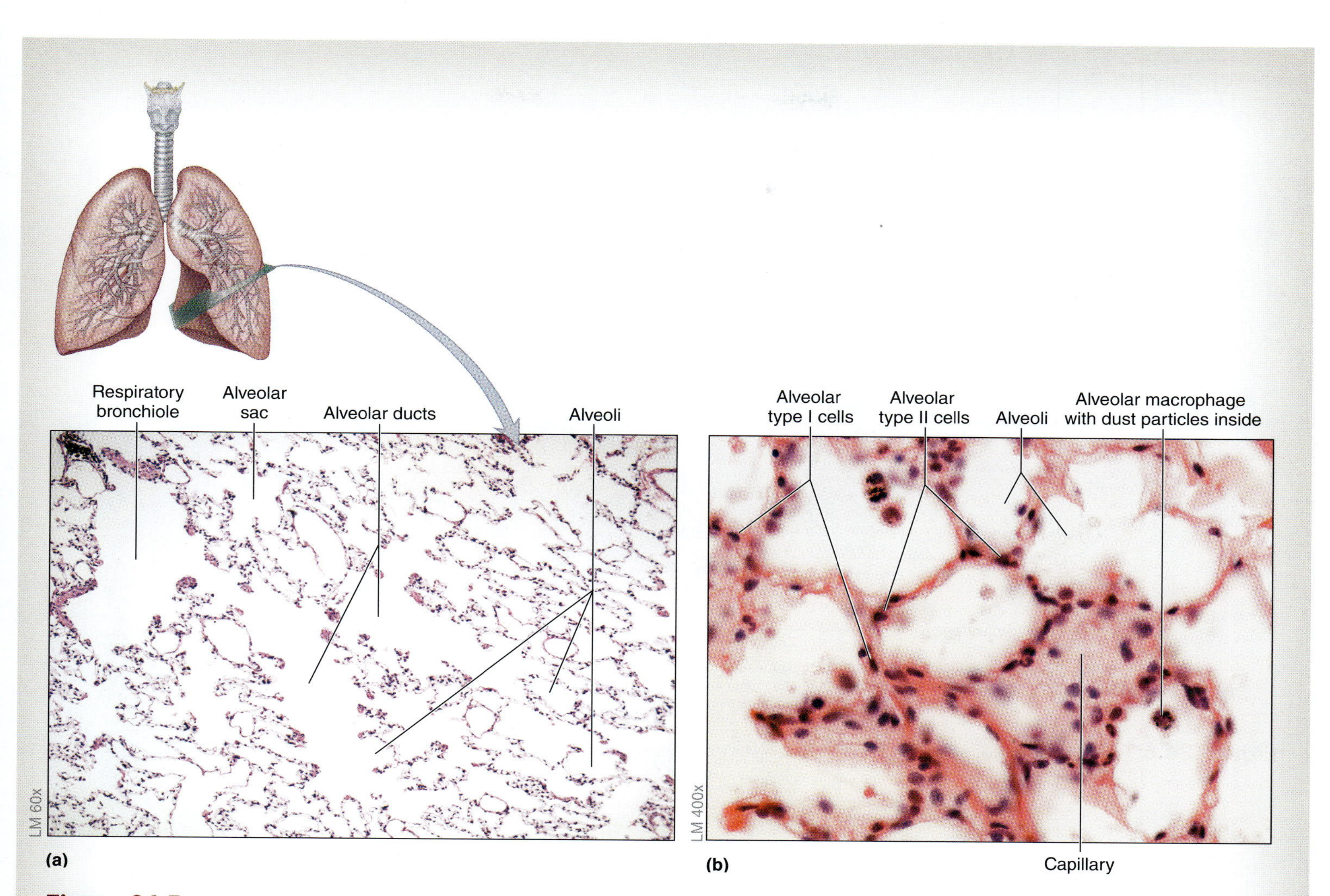

Figure 24.5 Histology of the Lungs. (a) Low-magnification view demonstrating respiratory zone structures. (b) High-magnification view showing alveolar type I cells (composed of simple squamous epithelium), alveolar type II cells (composed of simple cuboidal epithelium), and alveolar macrophages with particles of dust inside the cells.

GROSS ANATOMY

Upper Respiratory Tract

The following exercises explore the gross anatomy of the respiratory structures that convey air into the lungs, the cavities that house the lungs, and the lungs themselves. Recall, however, that the process of breathing requires muscular action. Details of the structure, location, and actions of the respiratory muscles are covered in chapter 12. This is a good point in time to go back to exercise 12.7 on p. 301 review the actions of the respiratory musculature, paying particular attention to the location and actions of the diaphragm.

Structures of the upper respiratory tract are best seen through gross observation of a midsagittal section of the head and neck. This provides the best view for observing the gross structures within the nasal cavity that warm, moisten, and filter the air before it enters the respiratory passageways.

EXERCISE 24.5

MIDSAGITTAL SECTION OF THE HEAD AND NECK

1. Obtain a classroom model of a midsagittal section of the head and neck (or a cadaveric specimen that has been sectioned along the sagittal plane). **Figure 24.6** shows respiratory system structures that are visible in a midsagittal section of the head and neck.
2. Identify the structures listed in **figure 24.7** on the classroom model or cadaveric specimen, using figure 24.6 and the textbook as guides. Then label figure 24.7.
3. *Optional Activity:* **AP|R 11: Respiratory System—** Watch the "Respiratory System Overview" to review the structures of the upper and lower respiratory tracts.

(continued on next page)

(continued from previous page)

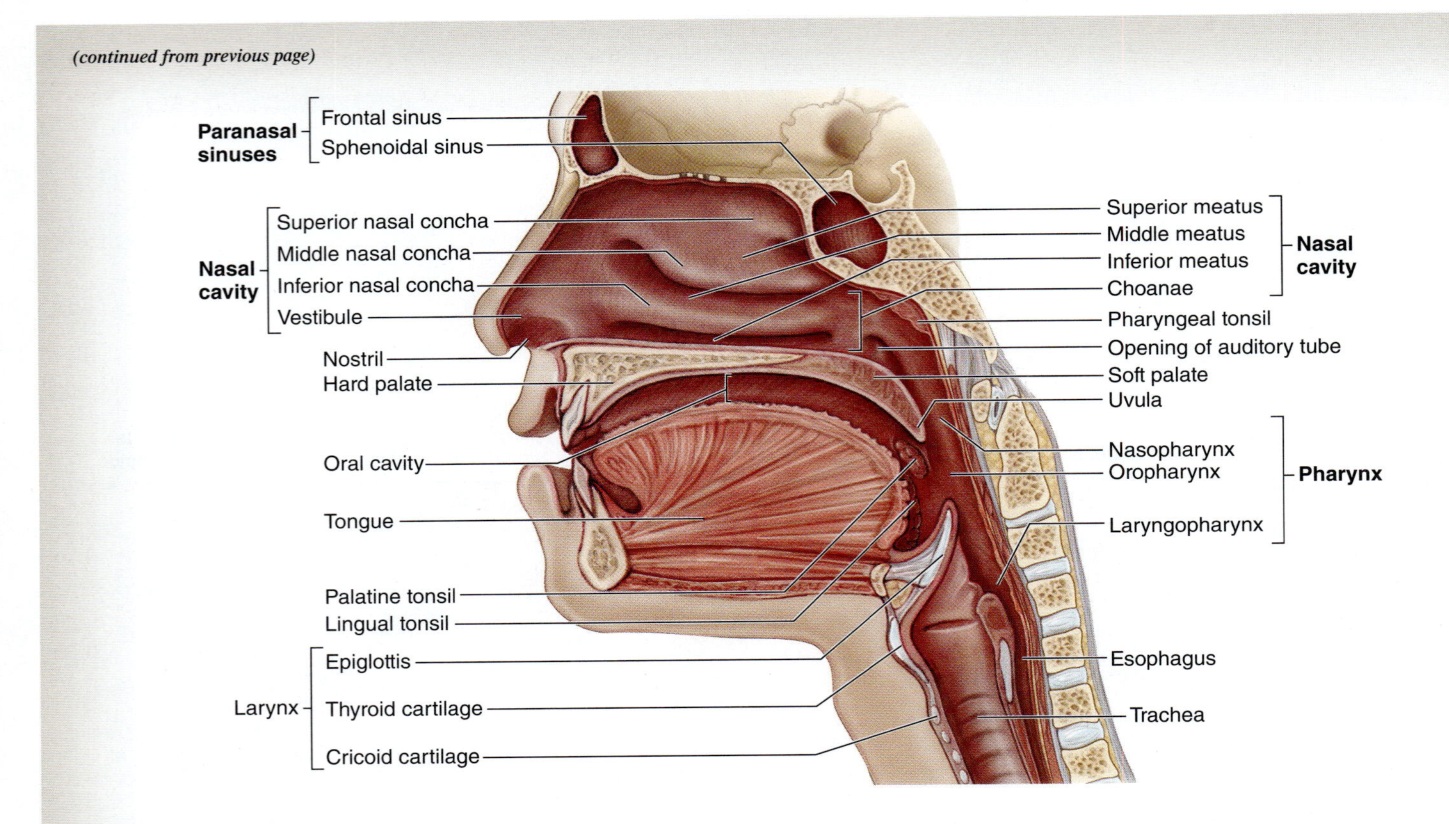

Figure 24.6 The Upper Respiratory Tract. Respiratory system structures visible in a midsagittal view of the head and neck.

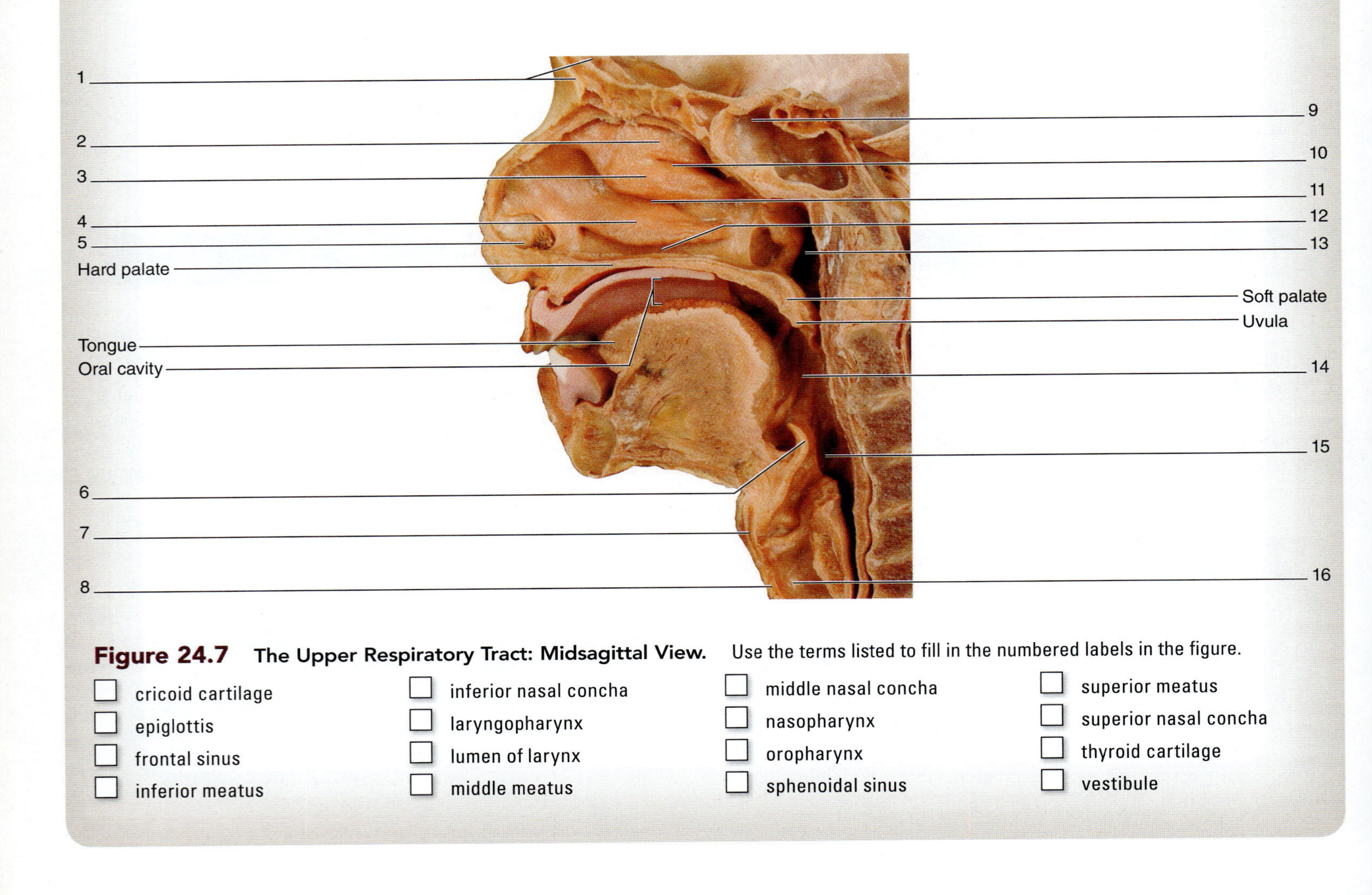

Figure 24.7 The Upper Respiratory Tract: Midsagittal View. Use the terms listed to fill in the numbered labels in the figure.

- ☐ cricoid cartilage
- ☐ epiglottis
- ☐ frontal sinus
- ☐ inferior meatus
- ☐ inferior nasal concha
- ☐ laryngopharynx
- ☐ lumen of larynx
- ☐ middle meatus
- ☐ middle nasal concha
- ☐ nasopharynx
- ☐ oropharynx
- ☐ sphenoidal sinus
- ☐ superior meatus
- ☐ superior nasal concha
- ☐ thyroid cartilage
- ☐ vestibule

Lower Respiratory Tract

The lower respiratory tract is composed of structures of both the conducting portion (larynx, trachea, bronchi, and bronchioles) and the respiratory portion (respiratory bronchioles, alveolar ducts, and alveoli). The **larynx** (*larynx*, organ of voice production) is much more than just an opening into the lower respiratory tract. It also houses the **vocal folds,** referred to as the "true vocal cords," which are responsible for phonation (sound production) and intrinsic muscles that control the length and tension of the vocal folds. The structural framework of the larynx is composed of several paired and unpaired cartilages, which act as attachment sites for the intrinsic musculature. The muscles of the larynx are all innervated by branches of the vagus nerve (CN X). Thus, lesions of the vagus nerve result in problems with phonation, such as hoarseness. **Table 24.5** describes the cartilaginous structures of the larynx, and **table 24.6** describes the noncartilaginous structures of the larynx.

Table 24.5 Cartilaginous Structures of the Larynx

Cartilage	Description	Word Origin
Arytenoid	Small paired pyramid-shaped cartilages found on the superior, posterior aspect of the cricoid cartilage. The vocal ligaments (true vocal cords) attach to them.	*arytania*, a ladle, + *eidos*, resemblance
Corniculate	Small paired cartilages found superior to the arytenoid cartilages. The vestibular folds ("false vocal cords") attach to them.	*cornicatus*, horned
Cricoid	The second largest laryngeal cartilage, this ring-shaped cartilage serves as an attachment point for muscles	*krikos*, a ring, + *eidos*, resemblance
Cuneiform	Small paired cartilages found within the aryepiglottic fold	*cuneus*, a wedge, + *forma*, form
Epiglottis	A plate of elastic cartilage at the superior aspect of the larynx that closes over the opening to the larynx during swallowing to prevent substances from entering the larynx. It is covered by stratified squamous epithelium on its superior aspect and by respiratory epithelium on its inferior aspect.	*epi-*, above, + *glottis*, the mouth of the windpipe
Thyroid	The largest of the laryngeal cartilages, located superior to the isthmus of the thyroid gland. The vocal ligaments (true vocal cords) attach to it.	*thyreos*, an oblong shield, + *eidos*, resemblance

Table 24.6 Noncartilaginous Structures of the Larynx

Structure	Description and Function	Word Origin
Glottis	Consists of the rima glottidis plus the vocal folds	*glottis*, the mouth of the windpipe
Rima Glottidis	The space between the vocal folds; also known as the true glottis	*rima*, a slit, + *glottis*, the mouth of the windpipe
Vestibular Folds	The vestibular ligaments plus the folds of mucous membrane that cover them; also called the "false vocal cords"	*vestibulum*, a small cavity at the entrance of a canal
Vestibular Ligaments	Ligaments that stretch between the angle of the thyroid cartilage and the corniculate cartilages	*vestibulum*, a small cavity at the entrance of a canal, + *ligamentum*, a bandage
Vocal Folds ("true vocal cords")	The vocal ligaments plus the mucosa covering them. Involved directly in voice production; also called "true vocal cords"	*vocalis*, pertaining to the voice, + *chorda*, cord
Vocal Ligaments	Ligaments that stretch between the thyroid and arytenoid cartilages	*vocalis*, pertaining to the voice, + *ligamentum*, a bandage

EXERCISE 24.6

THE LARYNX

1. Obtain a classroom model of the larynx **(figure 24.8).**
2. Identify the structures listed in figure 24.8 on the model of the larynx, using tables 24.6 and 24.7 and the textbook as guides. Then label figure 24.8.
3. *Optional Activity:* **APR 11: Respiratory System—** Take the "Lower Respiratory Tract" test in the Quiz section to review the larynx and other lower respiratory structures.

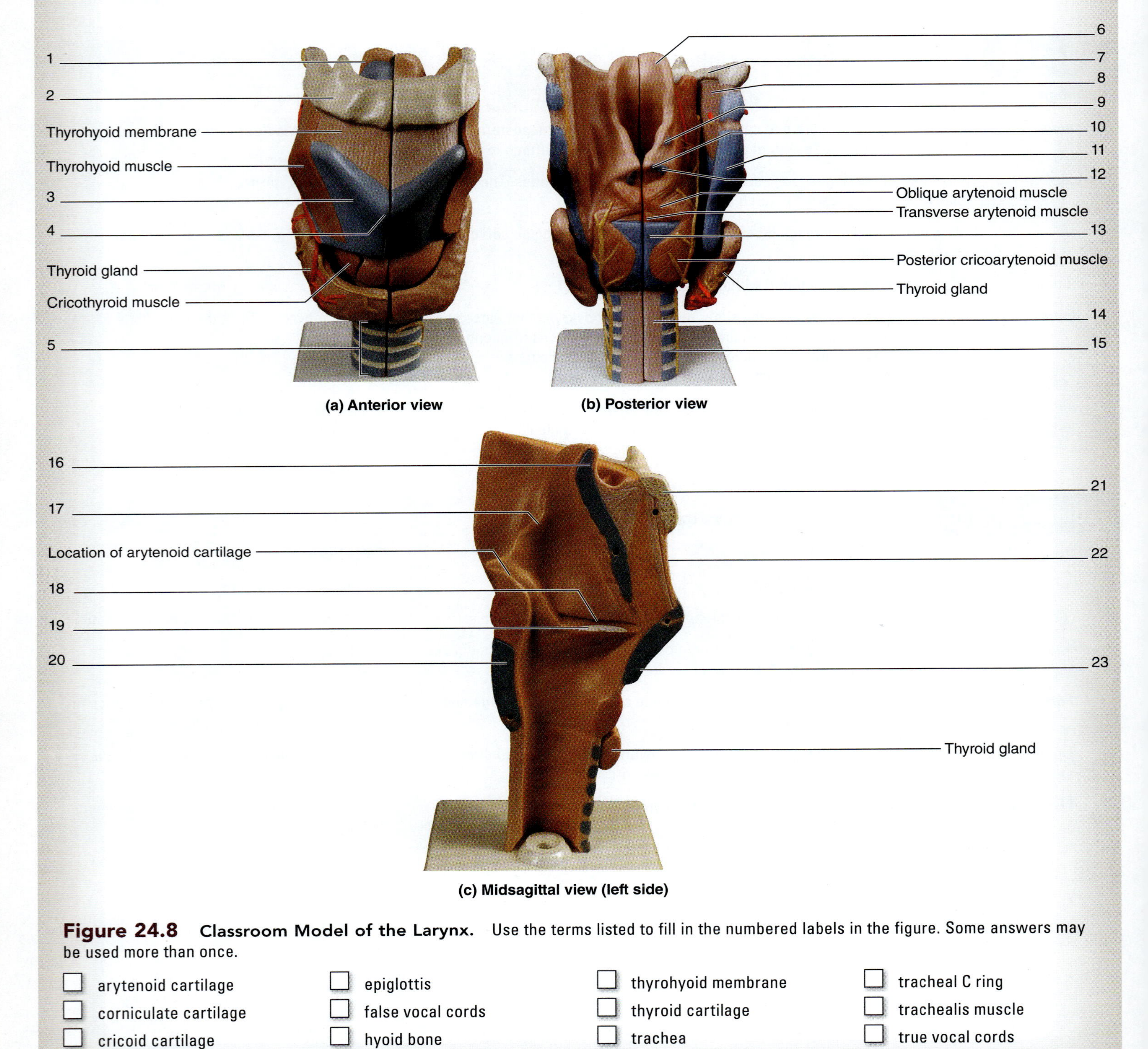

Figure 24.8 Classroom Model of the Larynx. Use the terms listed to fill in the numbered labels in the figure. Some answers may be used more than once.

- ☐ arytenoid cartilage
- ☐ corniculate cartilage
- ☐ cricoid cartilage
- ☐ cuneiform cartilage
- ☐ epiglottis
- ☐ false vocal cords
- ☐ hyoid bone
- ☐ laryngeal prominence
- ☐ thyrohyoid membrane
- ☐ thyroid cartilage
- ☐ trachea
- ☐ tracheal C ring
- ☐ trachealis muscle
- ☐ true vocal cords

The Pleural Cavities and the Lungs

Within the thoracic cavity, the lungs are located within separate **pleural cavities** (*pleura*, a rib). The space within the thoracic cavity between the two lungs is the **mediastinum** (*medius*, middle). If one of the lungs collapses, the structures within the mediastinum will shift toward the side of the collapsed lung. Tightly adhered to each lung is a serous membrane called the **visceral pleura.** Adhered to the inner wall of the thoracic cavity is a serous membrane called the **parietal pleura.**

On a human cadaver the parietal pleura can often be seen as a shiny tissue attached to the innermost part of the rib cage or the superior surface of the diaphragm. Between the two serous membranes is a fluid-filled space called the **pleural cavity.** Note that the lungs are contained within the pleurae, which contain serous fluid. The serous fluid reduces friction and increases surface tension of the pleurae. The latter allows the lungs to expand and contract as the thoracic cavity changes volume, because it keeps the visceral and parietal layers "stuck" together. Incidentally, if air is introduced into this pleural space, a condition called pneumothorax (*pneumo*-, air + thorax), the lungs will collapse, thereby making ventilation impossible.

EXERCISE 24.7

THE PLEURAL CAVITIES

1. Observe the thoracic cavity of a human cadaver or a model of the thorax **(figure 24.9).**
2. Identify the structures listed in figure 24.9 on a cadaver or model of the thorax, using the textbook as a guide. Then label figure 24.9.

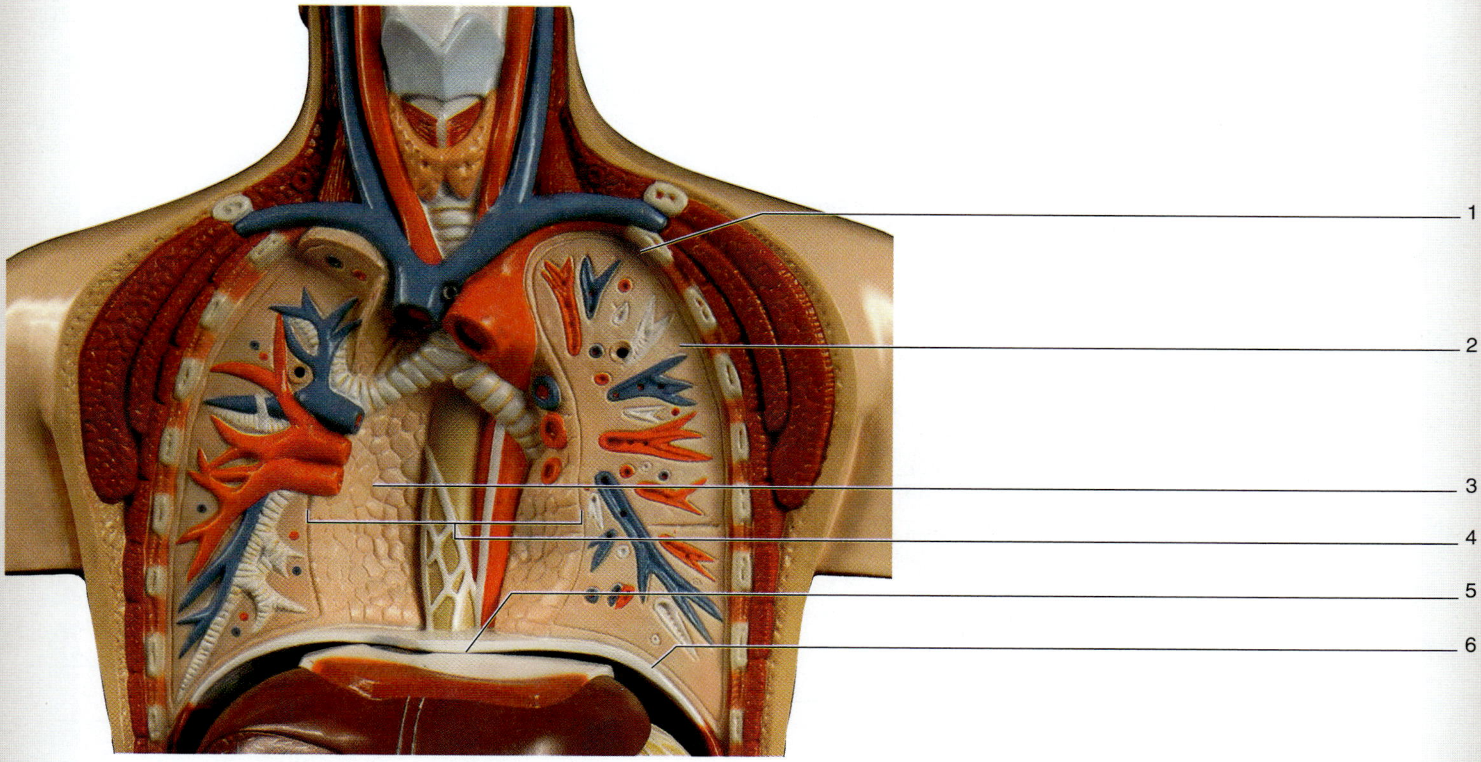

1 ____
2 ____
3 ____
4 ____
5 ____
6 ____

Figure 24.9 **The Pleural Cavities.** Use the terms listed to fill in the numbered labels in the figure.

- ☐ diaphragm
- ☐ left lung
- ☐ mediastinum
- ☐ parietal pleura
- ☐ right lung
- ☐ visceral pleura

EXERCISE 24.8

THE LUNGS

This exercise involves comparing and contrasting the structures of the right and left lungs, and observing the branching pattern of the respiratory tree. Although it is easy to distinguish the right and left lungs from each other based on the number of lobes (two for the left, three for the right), locating the structures that enter the hilum of the lung (pulmonary arteries, pulmonary veins, and bronchi) is more challenging. There are patterns for recognizing these structures that are described in this laboratory exercise. In addition, observations of several **impressions** made in the lungs by adjacent structures will be made. These impressions are visible in preserved human cadaver lungs and on classroom models of the lungs, but may not be visible in fresh lungs.

(continued on next page)

(continued from previous page)

This is because the process of fixing the lungs with preservative also fixes the impressions of adjacent organs. The impressions are not found in fresh lungs because they have not been fixed with preservative.

EXERCISE 24.8A The Right Lung

1. Observe the lungs of a human cadaver, a fresh or preserved sheep pluck (a *pluck* contains the heart, lungs, and trachea), or a classroom model of the lungs.
2. Begin by observing the right lung **(figure 24.10).** How many lobes does the right lung have? ______________
3. Turn the lung so the hilum is visible (medial view; figure 24.10*b*). The hilum of the right lung contains branches of the pulmonary arteries and pulmonary veins, bronchi, and small bronchial arteries and veins (which represent the systemic circulation to the lungs). Which of the vessels (pulmonary arteries or pulmonary veins) should have thicker walls? Explain your answer:

____

____

____

In general, the pulmonary arteries are located on the superior aspect of the hilum of the right lung, the pulmonary veins are located on the inferior and anterior aspect of the hilum of the right lung, and the bronchi are located on the superior and posterior aspect of the hilum of the right lung. If viewing cadaveric lungs, these structures will be more difficult to differentiate from each other because they are not color-coded. Thus, relying on the texture of the vessels and their locations will be necessary for proper identification.

4. Identify the structures listed in figure 24.10 on the right lung, using the textbook as a guide. Then label figure 24.10.

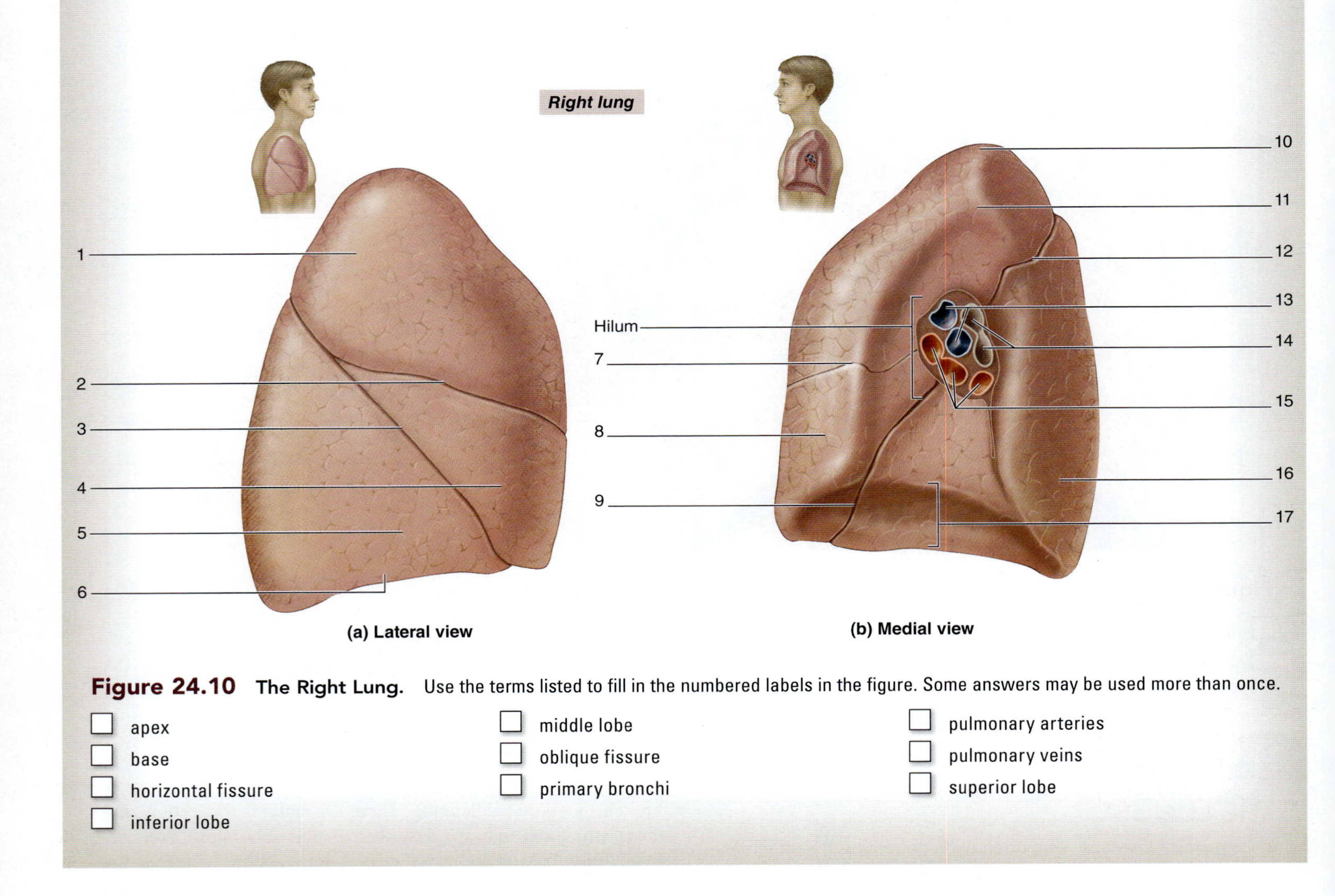

Figure 24.10 The Right Lung. Use the terms listed to fill in the numbered labels in the figure. Some answers may be used more than once.

- ☐ apex
- ☐ base
- ☐ horizontal fissure
- ☐ inferior lobe
- ☐ middle lobe
- ☐ oblique fissure
- ☐ primary bronchi
- ☐ pulmonary arteries
- ☐ pulmonary veins
- ☐ superior lobe

EXERCISE 24.8B The Left Lung

1. Observe the lungs of a human cadaver, a fresh or preserved sheep pluck, or a classroom model of the lungs.
2. Observe the left lung **(figure 24.11).** How many lobes does the left lung have? ______________ Is the left lung larger or smaller than the right lung? ________________ Explain the reason for this difference. ________________

   ____
3. Turn the lung so the hilum is visible (medial view; figure 24.11*b*). Identify the branches of the pulmonary arteries and veins, bronchi, and small bronchial arteries and veins (which represent the systemic circulation to the lungs).
4. In general, the pulmonary arteries are located on the superior aspect of the hilum of the left lung, the pulmonary veins are located on the inferior and anterior aspect of the hilum of the left lung, and the bronchi are located on the superior and posterior aspect of the hilum of the left lung. If viewing cadaveric lungs, these structures will be more difficult to differentiate from each other because they are not color-coded. Thus, relying on the texture of the vessels and their locations is necessary for proper identification. The medial view of the left lung allows visualization of several impressions that accommodate nearby adjacent structures in the living human. The most prominent of these impressions is the **cardiac impression** made by the heart.
5. What features distinguish the left lung from the right lung? ________________________________

   ____

   ____
6. Identify the structures listed in figure 24.11 on the left lung, using the textbook as a guide. Then label figure 24.11.

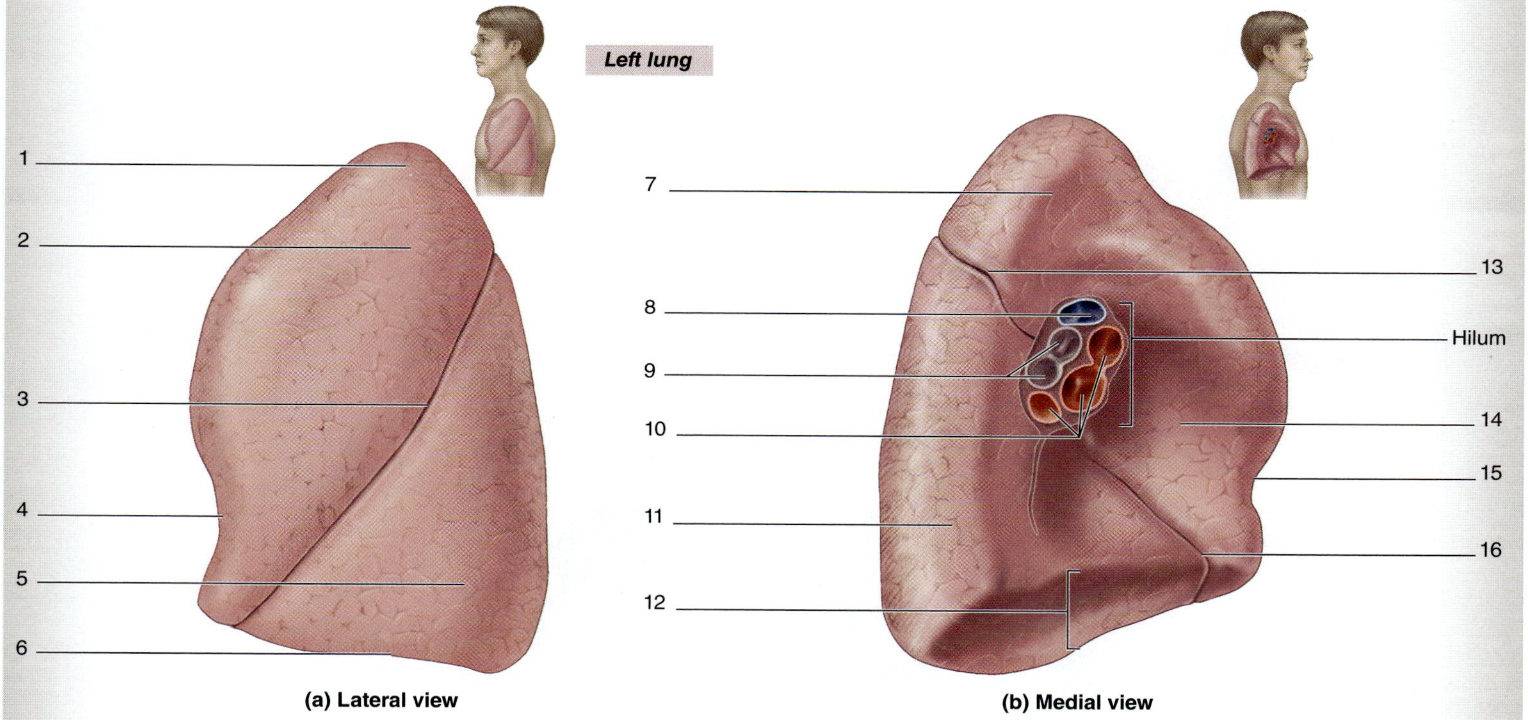

(a) Lateral view

(b) Medial view

Figure 24.11 The Left Lung. Use the terms listed to fill in the numbered labels in the figure. Some answers may be used more than once.

- ☐ apex
- ☐ base
- ☐ cardiac impression
- ☐ cardiac notch
- ☐ inferior lobe
- ☐ oblique fissure
- ☐ primary bronchi
- ☐ pulmonary artery
- ☐ pulmonary veins
- ☐ superior lobe

INTEGRATE

LEARNING STRATEGY

When learning the number of lobes in the right versus the left lung, think of heart anatomy as a guide. Recall the *tri*cuspid valve of the heart, which has *three* cusps, is located on the *right* side of the heart. Similarly, the *right* lung has *three* lobes. Recall the *bi*cuspid valve of the heart, which has *two* cusps, is located on the *left* side of the heart. Similarly, the *left* lung has two lobes.

EXERCISE 24.9

THE BRONCHIAL TREE

The lungs are subdivided, from larger to smaller units, into **lobes, bronchopulmonary segments,** and **lobules.** The branching pattern of the bronchial tree follows the segmentation of the lungs, **(figure 24.12).** For example, lobar (secondary) bronchi lead into lobes of the lung so there are three lobar bronchi leading into the three lobes of the right lung, and two lobar (secondary) bronchi leading into the two lobes of the left lung. **Table 24.7** describes the levels of the bronchial tree that serve each segment of the lungs. This segmental nature of the lungs and bronchial tree makes it relatively easy to remove a segment of the lung that contains a tumor (for example) without interfering with the other

Table 24.7 The Bronchial Tree

Portion of Bronchial Tree	Description
Main (primary) Bronchi	One to each lung. The right is more vertical than the left, so foreign objects are more likely to lodge in it.
Lobar (secondary) Bronchi	One to each lobe (2 on the left, 3 on the right)
Segmental (tertiary) Bronchi	One to each bronchopulmonary segment

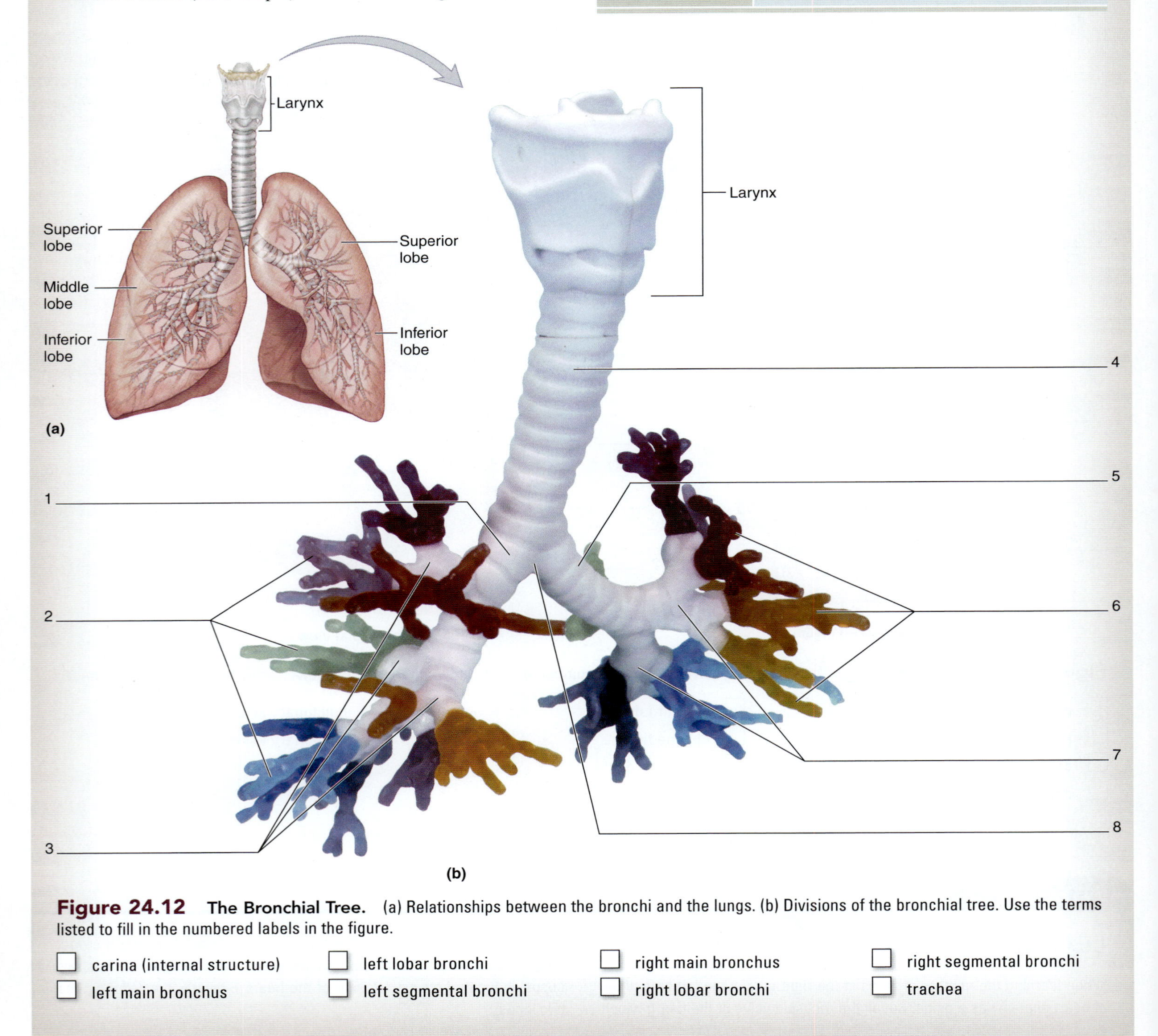

Figure 24.12 The Bronchial Tree. (a) Relationships between the bronchi and the lungs. (b) Divisions of the bronchial tree. Use the terms listed to fill in the numbered labels in the figure.

- ☐ carina (internal structure)
- ☐ left main bronchus
- ☐ left lobar bronchi
- ☐ left segmental bronchi
- ☐ right main bronchus
- ☐ right lobar bronchi
- ☐ right segmental bronchi
- ☐ trachea

parts of the lung. The **carina** (*carina,* keel of a ship) is the *internal* ridge between the most inferior tracheal cartilage and the start of the main bronchi. In the laboratory the location of the carina will be identified externally, where the main (primary) bronchi originate from the trachea. However, it is important to realize that the actual structure is internal. The carina is an important landmark for physicians performing bronchoscopy.

1. Observe the lungs of a human cadaver, a fresh or preserved sheep pluck, or a classroom model of the lungs.
2. Identify the structures listed in figure 24.12 on the cadaver lungs, sheep pluck, or classroom model of the lungs, using table 24.7 and the textbook as guides. Then label figure 24.12.

PHYSIOLOGY

Respiratory Physiology

Breathing is critical for life and seems like a simple process to most of us. However, the mechanisms underlying much of respiratory physiology are complex. Respiration is the movement of both oxygen and carbon dioxide between the atmosphere and the body tissues, and involves four major processes: (1) pulmonary ventilation, (2) alveolar gas exchange, (3) gas transport, and (4) systemic gas exchange. **Pulmonary ventilation** is the process of air moving between the atmosphere and the alveoli. The movement of air is dependent upon differences in pressure between the environment and the thoracic cavity/lungs. Air flow through airways is dependent upon pressure gradients. Changes in thoracic pressure are related to changes in thoracic volume and obey the law of physics called **Boyle's law** ($P_1V_1 = P_2V_2$). When pressure inside the lungs (intrapulmonary pressure) is less than atmospheric pressure, air flows into the lungs. In contrast, when intrapulmonary pressure is greater than atmospheric pressure, air flows out of the lungs.

During **inspiration,** the diaphragm and external intercostal muscles contract, the volume of the thoracic cavity increases, and intrapulmonary pressure decreases. Air then flows from the atmosphere into the alveoli from the area of higher pressure to the area of lower pressure. During **expiration,** the diaphragm and external intercostal muscles relax and the elastic tissue of the lungs and chest cavity wall recoil, causing a decrease in the volume of the thoracic cavity. This causes intrapulmonary pressure to increase. Air then moves from the alveoli to the atmosphere.

Clinically, a **spirometer** can be used to record measurements that assess ventilation, such as tidal volume (TV), inspiratory reserve volume (IRV), and expiratory reserve volume (ERV). These measured volumes can then be used to calculate lung capacities, including inspiratory capacity, functional residual capacity, vital capacity, and total lung capacity. Measured volumes and capacities can be compared to "normal" or "expected" values for an individual. **Table 24.8** is a quick reference guide of respiratory volumes and capacities. The table contains brief descriptions of each measurement/calculation, any pertinent formulas for calculating lung capacities, and reference values for males and females. **Figure 24.13** illustrates the four lung volumes and four lung capacities graphically.

Table 24.8 Respiratory Volumes and Capacities

Volumes

Volume	Definition	Normal Values (Male)	Normal Values (Female)
Expiratory Reserve Volume (ERV)	The amount of air expelled from the lungs during a forced expiration, following a quiet expiration; ERV is a measure of lung and chest wall elasticity	1200 mL	700 mL
Inspiratory Reserve Volume (IRV)	The amount of air taken into the lungs during a forced inspiration, following a quiet inspiration; IRV is a measure of lung compliance	3100 mL	1900 mL
Residual Volume (RV)	Amount of air left in lungs following a forced expiration	1200 mL	1100 mL
Tidal Volume (TV)	Volume of air taken into or expelled out of lungs during a quiet breath	500 mL	500 mL

Capacities

Capacity	Formula	Definition	Normal Values (Male)	Normal Values (Female)
Functional Residual Capacity	ERV + RV	Amount of air normally left (residual) in lungs after quiet expiration	2400 mL	1800 mL
Inspiratory Capacity	TV + IRV	Total ability to inspire	3600 mL	2400 mL
Total Lung Capacity	TV + IRV + ERV + RV	Total amount of air that can be in lungs	6000 mL	4200 mL
Vital Capacity	TV + IRV + ERV	Measure of the strength of respiration	4800 mL	3100 mL

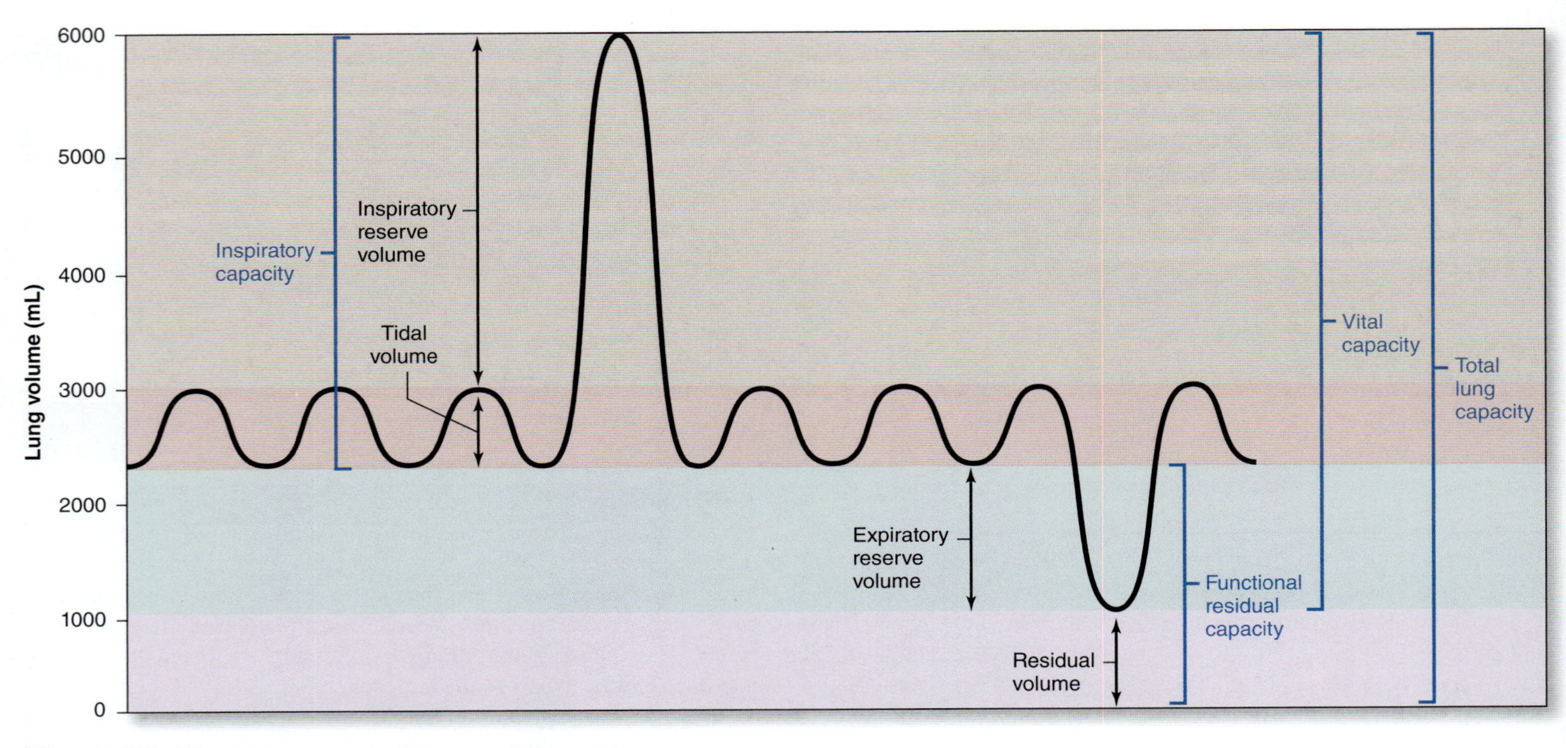

Figure 24.13 Respiratory Volumes and Capacities.

INTEGRATE

CLINICAL VIEW

Obstructive versus Restrictive Respiratory Diseases

Spirometry tests can be used clinically to distinguish between obstructive and restrictive pulmonary diseases. Obstructive diseases, such as emphysema, involve a physical barrier (obstruction) in the airway, whereas restrictive diseases, such as pulmonary fibrosis (scarring), involve an inability to physically expand the lungs (restriction). Both types of diseases decrease exchange of respiratory gases at the respiratory membrane, but for different reasons. If a patient presents with respiratory distress, doctors can measure respiratory volumes using a spirometer and calculate lung capacities to determine the disease classification. Two clinical measurements commonly used include **forced expiratory volume** in 1 second $(FEV)_1$, which corresponds to the amount of air forcefully exhaled in 1 second, and **forced vital capacity** (FVC), the amount of air forcefully exhaled after a maximal inhalation. The ratio of these two values, or $FEV_1\%$, is typically 75–80% in normal, healthy adults. Because patients with obstructive pulmonary disease have difficulty breathing out, their $FEV_1\%$ is typically reduced. Patients with an $FEV_1\%$ of less than 75% are diagnosed with an obstructive disease. In comparison, patients suffering from a restrictive disease have difficulty breathing in because of the limited ability to expand the lungs. These patients typically will have lower than normal inspiratory capacity and vital capacity. Once a patient's disease is classified as either obstructive or restrictive, further diagnostic testing can be performed and treatment plans may be devised to address the specific disease that caused the patient's respiratory distress.

INTEGRATE

CONCEPT CONNECTION

The respiratory system plays an important role in maintaining blood pH within a narrow physiological range (7.35 – 7.45). It does so by altering breathing patterns to change the levels of carbon dioxide in the blood. When tissue cells are more metabolically active, they produce more carbon dioxide, which has a tendency to decrease the pH of the blood. Increases in P_{CO_2} in the blood directly or indirectly stimulate respiratory control centers in the brain, which increase the breathing rate so as to "blow off" the excess carbon dioxide. Sensory receptors that detect carbon dioxide (and pH) levels are located strategically throughout the body including the aorta, the carotid arteries, and the medulla oblongata. These receptors detect small changes in carbon dioxide (and pH) levels, and adjust the respiratory rate accordingly. This adjustment involves transmitting increased frequency of nerve signals along motor neurons that innervate the muscles responsible for inspiration, including the diaphragm and external intercostal muscles. This activation will increase either the respiratory rate, the depth of breathing, or both.

When the respiratory rate decreases below physiological demand **(hypoventilation),** carbon dioxide accumulates in the blood **(hypercapnia),** leading to a drop in blood pH **(acidosis).** Conversely, when the respiratory rate increases above physiological demand **(hyperventilation),** carbon dioxide decreases in the blood **(hypocapnia),** leading to an increase in blood pH **(alkalosis).** Alteration of the respiratory rate, and consequently the blood pH, is a relatively rapid process to maintain acid-base balance. The concept of acid-base balance will appear again in the urinary system, as the kidneys also play a role in regulating blood pH by secreting hydrogen ions (or bicarbonate ion) into the filtrate for excretion in the urine. Overall, the respiratory system aids in short-term regulation of acid-base balance making adjustments within minutes, whereas the urinary system aids in long-term regulation of acid-base balance typically making adjustments over hours to days.

EXERCISE 24.10

MECHANICS OF VENTILATION

This exercise explores the mechanics of ventilation using a model that simulates the diaphragm, thoracic cavity, and lungs. Before begining, state the relationship between thoracic volume and intrapulmonary pressure, and how it relates to regulating the flow of air into and out of the lungs. ______________________

____

____

1. Obtain a model lung **(figure 24.14).** In this model, the balloons represent the lungs, the rubber sheeting represents the diaphragm, The Y tube represents the trachea and bronchi, and the bell jar represents the thoracic cavity.
2. Increase the volume of the thoracic cavity by pulling down on the "diaphragm" (rubber sheeting). Does the pressure inside the thoracic cavity increase or decrease?

   ____

   What happens to the balloons? ______________________

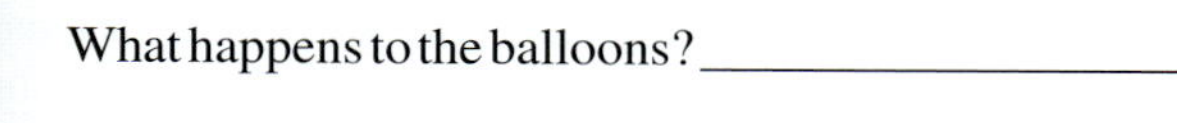

   ____

   Describe the direction of airflow. ______________________

   ____

3. Repeat step 2 by pushing up on the "diaphragm." Does the pressure inside the "thoracic cavity" increase or decrease? ______________________

   What happens to the balloons? ______________________

   ____

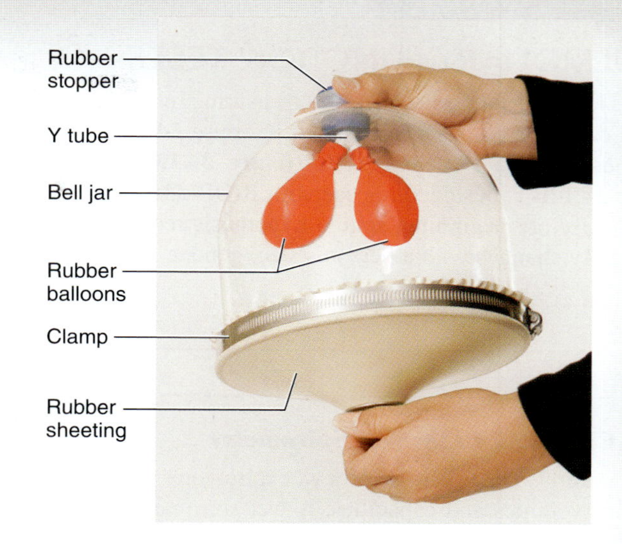

Figure 24.14 **Model Lung.**

Describe the direction of airflow. ______________________

____

4. Repeat step 2 and introduce air into the "thoracic cavity" by partially releasing the rubber stopper, thereby simulating a pneumothorax (air within pleural space). What happens to the balloons? ______________________

   ____

EXERCISE 24.11

AUSCULTATION OF RESPIRATORY SOUNDS

The sound of air flowing through respiratory structures can be detected by placing the diaphragm of a stethoscope on strategic locations of the chest and back, a process called **auscultation** (*auscultatio*, a listening). Physicians evaluating a patient with respiratory distress frequently perform this test. Physicians are listening for two main respiratory sounds: bronchial and vesicular. **Bronchial sounds,** caused by airflow through the bronchi, are often high-pitched and loud during both inspiration and expiration. **Vesicular sounds,** caused by air filling the alveolar sacs, are lower in pitch, quiet, and heard primarily during inspiration. When auscultation is performed between the first and second intercostal spaces, bronchovesicular sounds of intermediate pitch can be heard during inspiration and expiration. When inflammation or infection is present in the respiratory structures, these sounds take on different characteristics. Crackles, or popping noises, particularly during inspiration, may be indicative of fluid present in the lungs. Wheezing, or high-pitched sounds, particularly during expiration, may be indicative of increased resistance in the airway. The goal of this exercise is to auscultate these clinically relevant respiratory sounds.

Before beginning, describe the optimal placement of the stethoscope diaphragm that will allow for auscultation of respiratory sounds in the space provided.

____

____

1. Obtain a stethoscope and alcohol swabs, and choose a study subject. First, clean the stethoscope by wiping the earpieces and diaphragm with an alcohol swab. Allow the stethoscope to dry. Insert the earpieces into your ears.
2. Locate the subject's larynx. Press the stethoscope diaphragm on the skin just inferior to the subject's larynx. Auscultate bronchial sounds as the subject takes several deep breaths through the mouth. While the subject is breathing, slowly move the stethoscope diaphragm inferiorly, following the anatomical course of the larynx and trachea, until the sounds are no longer audible.
3. To auscultate vesicular sounds, press the stethoscope diaphragm firmly on the subject's skin at the following locations on the right side of the body: (1) inferior to the clavicle, (2) in the intercostal spaces, and (3) along the medial border of the scapula. Be sure to keep the stethoscope diaphragm in one location during a complete cycle of inspiration and expiration.
4. Repeat the process described in step 3 on the left side of the body.

EXERCISE 24.12

PULMONARY FUNCTION TESTS

This laboratory exercise involves measuring respiratory volumes and capacities using a spirometer. Exercise 24.12A is designed for use with a wet spirometer **(figure 24.15)**, whereas exercise 24.12B is designed for use with BIOPAC laboratory equipment. Before beginning, state a hypothesis regarding how vital capacity changes with a person's size, gender, and age.

___

___

___

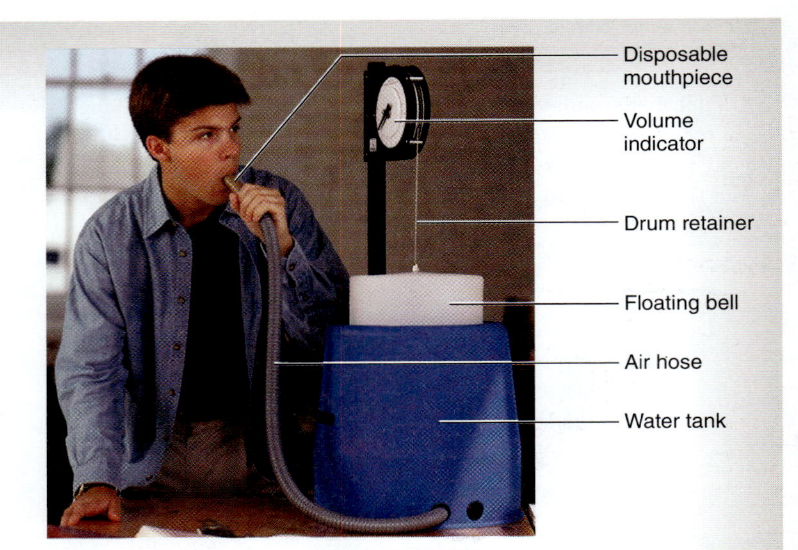

Figure 24.15 Wet Spirometer. A bell floats on top of a tank of water. As air enters the bell, it displaces the water. This displacement is measured using the volume indicator, which is used to suspend the bell over the water.

EXERCISE 24.12A Wet Spirometry

This exercise involves use of a wet spirometer to measure respiratory volumes and capacities. A wet spirometer gets its name from the fact that the apparatus involves floating a chamber (the "bell") on top of a tank of water and measuring the volume of air that flows into and out of the bell (figure 24.15). A "dry" spirometer (as is used in the BIOPAC exercise in this chapter) measures air volumes by *calculating the volume of air that flows through a measuring device per unit time and comparing that to calibration values.*

1. Obtain a wet spirometer, a disposable mouthpiece, alcohol swabs, and a nose clip. Clean the nose clip by wiping it down with the alcohol swab and allowing it to air dry.

   Use a new (or clean) mouthpiece for each subject to ensure that potentially infectious microorganisms are not transmitted between subjects. Used mouthpieces should either be disinfected in an alcohol bath prior to reuse, or thrown away. Ask the laboratory instructor for instructions on proper disposal of used mouthpieces.

2. Become familiar with the parts of the spirometer by comparing it to figure 24.15 and identifying the following parts:

   - ☐ **air hose**
   - ☐ **disposable mouthpiece**
   - ☐ **drum retainer**
   - ☐ **floating bell**
   - ☐ **volume indicator**
   - ☐ **water tank**

3. Prepare the wet spirometer by pushing the bell of the spirometer down as far as it will go. This will remove any excess air that lies inside the bell (and above the water). Note that air leaving the bell flows out of the air hose. When performing this exercise, as air flows into or out of the bell it will displace the bell above the water tank. Air volumes are recorded by noting the movement of the volume indicator that is attached to the bell by the drum retainer. Set the volume indicator to zero.

4. Place a disposable mouthpiece on the end of the air hose. Prepare the subject by placing a clean nose clip over the subject's nose. Have the subject sit quietly next to the wet spirometer. Instruct the subject to maintain an upright posture while performing the tests.

5. **Measure tidal volume (TV):** First, have the subject become relaxed with breathing into the mouthpiece by expiring several normal breaths through the mouthpiece. Be sure to push the bell of the spirometer down as far as it will go and reset the volume indicator to zero after each breath. Observe the movement of the volume indicator to determine tidal volume. When the subject is comfortable, have the subject expire a normal breath through the mouthpiece. Record the TV in **table 24.9.** Push the bell jar down as far as it will go and repeat for a total of five trials. Record each of the five TVs in table 24.9. Calculate the average TV from the five breaths and record it in table 24.9. Compare the results to expected values, which are listed in Table 24.8.

Table 24.9 Tidal Volume Measurements

Breath #	Tidal Volume (ml)
1	
2	
3	
4	
5	
Average TV	

6. Have the subject remove the mouthpiece and air hose from his or her mouth. Reset the spirometer by pushing the bell down to expel the air inside. Set the volume indicator to zero.

7. **Measure expiratory reserve volume (ERV):** Instruct the subject to start by taking five normal breaths through the mouthpiece. Then, have the subject perform a maximum exhalation following the fifth normal breath. Observe the movement of the volume indicator to determine the volume of air exhaled above and beyond the normal TV (= ERV). Record the ERV in **table 24.10**..Repeat this procedure for a total of five trials, recording the results from each trial in table 24.10. Calculate the average ERV from the five trails and record it as well. Compare the results to expected values listed in table 24.8. Repeat step 6 to reset the spirometer to zero.

Table 24.10 Expiratory Reserve Volume Measurements

Breath #	Expiratory Reserve Volume (ml)
1	
2	
3	
4	
5	
Average ERV	

8. **Measure vital capacity (VC):** Before beginning, explain to the subject that he/she must use as much effort as possible (including changing body position by bending over upon exhalation and leaning back upon inhalation) when performing a *maximum* exhalation/inhalation to obtain good results. To measure VC, instruct the subject to start by taking five normal breaths *without* the mouthpiece. Then, still without the mouthpiece in his/her mouth, have the subject perform a maximum exhalation, followed by a maximum inhalation following the fifth normal breath. Immediately after the maximum inhalation, have the study subject insert the mouthpiece into his/her mouth and exhale through the air hose as hard as possible to obtain a maximum exhalation.

   Observe the movement of the volume indicator to determine the volume of air exhaled during the maximal exhalation that followed maximum inhalation (= VC). Record the VC table 24.11. Reset both the bell jar down as far as it will go and the volume indicator to zero. Repeat this procedure for a total of five trials, recording the results from each trial in table 24.11. Calculate the average VC from the five trials and record it as well. Compare the results to expected values listed in table 24.8.

9. **Calculate inspiratory reserve volume (IRV).** IRV can be calculated using the formula: IRV = VC – (TV + ERV). Use the average values for ERV, TV, and VC measured in steps 6–8 to calculate IRV. Record the subject's IRV here: ____________________ ml. Compare the results to expected values listed in table 24.8.

Table 24.11 Vital Capacity Measurements

Trial #	Vital Capacity (ml)
1	
2	
3	
4	
5	
Average VC	

10. **Calculate remaining lung capacities.** Use the volumes measured in steps 6–9 to calculate the following lung capacities. Assume that residual volume (RV) is 1 liter.

    - ☐ **functional residual capacity (FRC) = RV + ERV**
    - ☐ **inspiratory capacity (IC) = TV + IRV**
    - ☐ **total lung capacity (TLC) = TV + IRV + ERV + RV**
    - ☐ **vital capacity (VC) = TV + IRV + ERV**

    Record the calculated lung capacities in **table 24.12.** Compare the results to expected values listed in table 24.8.

Table 24.12 Calculation of Lung Volumes and Capacities

Lung Volume/ Capacity	Calculated Value (ml)
FRC	
IC	
TLC	

11. Upon completion of the exercise, dispose of the disposable mouthpiece, reset the spirometer to zero, and clean the workstation.

EXERCISE 24.12B BIOPAC Lesson 12: Pulmonary Function Tests

Obtain the Following:

- **BIOPAC Airflow transducer (SS11LA)**
- **BIOPAC Bacteriological filter (AFT1)**
- **BIOPAC Disposable mouthpiece (AFT2)**
- **BIOPAC Noseclip (AFT3)**
- **BIOPAC Calibration Syringe (0.6-L or 2-L) (AFT6 or AFT6A)**
- **BIOPAC MP36/35 recording unit**
- **Laptop computer with BSL 4 software installed**

1. Prepare BIOPAC equipment by turning the computer ON and MP36/35 unit OFF. Plug the airflow transducer (SS11LA) into channel 1 (CH 1). Turn the MP36/35 unit ON.

(continued on next page)

(continued from previous page)

2. Start the BIOPAC Student Lab Program. Select "L12-Pulmonary Function I" from the dropdown menu, enter the filename, and click "OK."
3. To calibrate, hold the airflow transducer horizontally such that it remains upright during the procedure. Click "Calibrate." Ensure that data appears as a flat and centered line. If not, click "Redo." If so, click "Continue."
4. For stage 2 of the calibration, place a disposable filter (AFT1) on the end of the calibration syringe and insert the entire assembly into the airflow transducer on the side labeled "inlet" **(figure 24.16).** Hold the airflow transducer/calibration syringe assembly horizontally such that the assembly remains upright during the entire procedure. Pull the calibration syringe plunger out to its maximal position. Click "Calibrate." Pull the plunger in and out for five cycles, waiting 2 seconds between each cycle. Click "End Calibration." Compare the calibration data to **figure 24.17.** If there are differences, click "Redo Calibration." Otherwise, click "Continue."
5. OBTAIN THE DATA: Have the subject assume a relaxed, seated position. The subject should remain still and quiet, and the screen should not be visible to the subject during data collection. Prepare the airflow transducer by first removing the calibration syringe and then inserting a disposable mouthpiece on the side labeled "inlet." Have the subject put on a noseclip.
6. Click "Record" as the subject does the following: (1) takes five normal breaths, (2) inspires maximally once, (3) expires maximally once, (4) takes five normal breaths. Click "Stop." Review the data to ensure that positive spikes appear during inspiration and negative spikes appear during expiration. If the data do not show positive spikes during inspiration and negative spikes during expiration, click "Redo." Otherwise, click "Done."

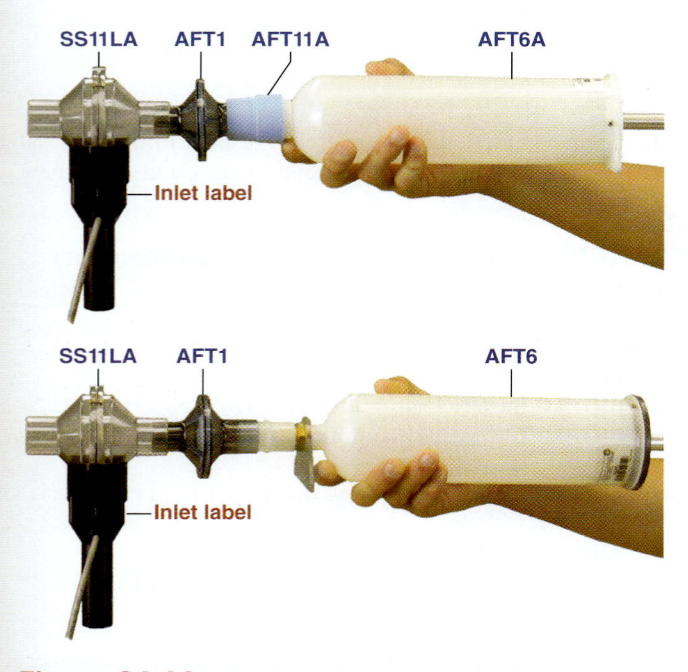

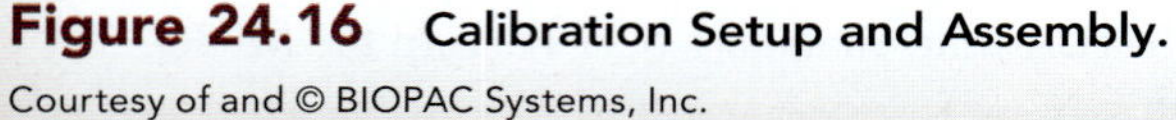
Figure 24.16 Calibration Setup and Assembly.

Courtesy of and © BIOPAC Systems, Inc.

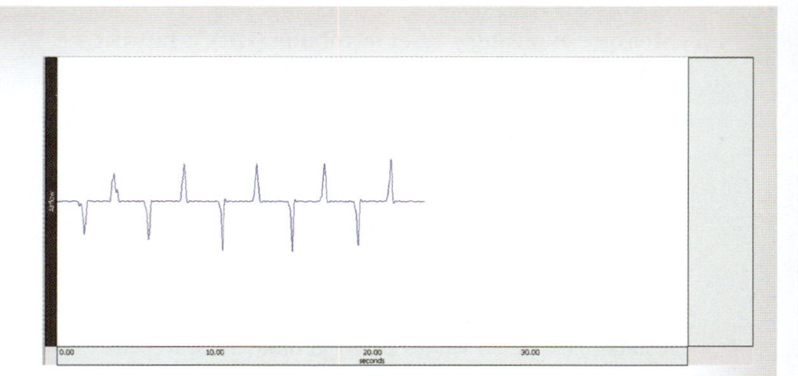

Figure 24.17 Sample Calibration Data.

Courtesy of and © BIOPAC Systems, Inc.

7. ANALYZE THE DATA: To perform data analysis on data that was collected previously, click "Review Saved Data." Note that CH 1 displays airflow and CH 2 displays volume. Review the airflow data and then turn this channel off by clicking on the channel number box and "Alt + click" (PC) or "Option + click" (Mac). Note that data analysis is easier when airflow data is not visible.
8. To prepare to analyze the data, do the following:
   - **a.** Use the zoom tool for optimal viewing of the recording.
   - **b.** Set up measurement boxes at the top of the screen using the drop-down menus. Set "CH 2" to each of the four boxes: "P-P" (max–min), "Max," "Min," "Delta" (last point–first point).
9. Locate the "I-beam" cursor. Select the point on the recording segment that corresponds to the vital capacity (VC; see figure 24.13) and record the P-P measurement in **table 24.13.**
10. To calculate the average tidal volume (TV), use the "I-beam" cursor to highlight the inspiration phase of cycle 3, which should correspond to the minimum and to the maximum of the third cycle. Record the P-P measurement as trial "1" in table 24.13. Repeat this procedure for the expiration phase of cycle 3, and record the P-P measurement as trial "2" in table 24.13. Calculate the average for these two trials, and record the calculation in table 24.13.
11. Measure the following respiratory volumes and capacities by using the "I-beam" cursor for the following values: IRV (Delta), ERV (Delta), RV (Min), IC (Delta), EC (Delta), TLC (Max). Consult figure 24.13 to review the portion of the recording that corresponds to each value. Record the data in table 24.13. Click "Exit."

12. Consult table 24.8 as a guide for the formulas to calculate each respiratory capacity. Calculate each capacity in table 24.13 and compare the measurements with expected values.

13. Make note of any pertinent observations here:

____

____

____

Table 24.13 Measured and Calculated Respiratory Volumes and Capacities

Abbreviation	Volume/Capacity	Measurement		Calculated	Reference
Volumes					
ERV	Expiratory Reserve Volume				700 – 1200 mL
IRV	Inspiratory Reserve Volume	Approximately 1 L			1900 – 3100 mL
RV	Residual Volume				1100 – 1200 mL
TV	Tidal Volume	(1)	(2)	AVG =	500 mL
Capacities					
IC	Inspiratory Capacity				2400 – 3600 mL
FRC	Functional Residual Capacity				1800 – 2400 mL
TLC	Total Lung Capacity				4200 – 6000 mL
VC	Vital Capacity				3100 – 4800 mL

Courtesy of and © BIOPAC Systems, Inc. Data from BIOPAC Table 12.2

EXERCISE 24.13 Ph.I.L.S.

Ph.I.L.S. LESSON 34: pH AND Hb-OXYGEN BINDING

Respiratory gases are exchanged between the air in the alveoli and the blood in the pulmonary capillaries during **alveolar gas exchange.** The epithelium of the alveoli, the endothelium of the alveolar capillaries, and the fused basement membranes between the two collectively form the **respiratory membrane.** For oxygen to enter the blood, or carbon dioxide to leave the blood, the gases must diffuse across this respiratory membrane. The rate at which each gas diffuses is governed by **Dalton's law of partial pressures** which states that each gas in a mixture of gases exerts an individual, or partial, pressure in relation to its abundance in the mixture of gases. The partial pressure of each gas is calculated by multiplying the total pressure of the gas by the percentage of the individual gas in the mixture. For example, if the total pressure is 760 mm Hg, and oxygen is 21% of the mixture, then the partial pressure of oxygen (P_{O_2}) = 760 mm Hg × 0.21 = 159 mm Hg. Gases will diffuse along their individual partial pressure gradients from an area of high partial pressure to an area of low partial pressure.

Other factors that influence alveolar gas exchange include the total number of alveoli (surface area for gas exchange), the thickness of the respiratory membrane, and the solubility coefficient for the particular gas. The solubility coefficient dictates how readily a gas will dissolve in a liquid, and is governed by **Henry's law.** Carbon dioxide has a solubility coefficient that is roughly twenty-four times greater than that of oxygen; this means that carbon dioxide dissolves much more easily in liquid than does oxygen. Functionally, this means that carbon dioxide diffuses across the respiratory membrane much more readily than oxygen, or requires much smaller partial pressure differences to diffuse. A larger partial pressure gradient, increased surface area, thinner respiratory membrane, and greater solubility coefficient will all increase the rate of diffusion across the respiratory membrane.

Gas exchange also occurs between systemic arterial blood and tissue cells. Here the same concepts for gas exchange apply. However, in the tissues, the membrane the gases must cross consists of the capillary endothelium (and its basement membrane) and the plasma membrane of the cell.

In addition to **exchange** of gases between the air and the blood, and the blood and the tissues, gases must also be transported within the blood. The process of moving the gases through the cardiovascular system is known as **gas transport.** Because of the low solubility of oxygen in water, 98% of the oxygen transported in the blood is bound to the heme group of hemoglobin, the major component of erythrocytes (see chapter 20). The majority of carbon dioxide (~70%) is transported as bicarbonate ion. The conversion of carbon dioxide to bicarbonate ion occurs within erythrocytes, which contain the enzyme *carbonic anhydrase.* Carbonic anhydrase catalyzes the reaction between carbon dioxide (CO_2) and water (H_2O) to form the weak acid carbonic acid (H_2CO_3). In solution, carbonic acid easily dissociates into hydrogen ions (H^+) and bicarbonate ions (HCO_3^-). This reaction is represented by the following equation:

$$CO_2 + H_2O \leftrightarrow H_2CO_3 \leftrightarrow H^+ + HCO_3^-$$

This reaction is important in understanding not only carbon dioxide transport as bicarbonate ion in the blood, but also the relationship between the carbon dioxide levels in the blood and the pH of the blood. As more carbon dioxide accumulates in the blood, more of it is converted into hydrogen ions. As the concentration of hydrogen ions in the blood increases, the pH decreases. Use the textbook as a guide while performing the following exercises. These exercises will guide exploration of the major respiratory processes and the relationship between pH and respiration.

Hemoglobin within red blood cells binds oxygen so oxygen can be transported to the tissues. Hemoglobin's affinity for oxygen increases when the partial pressure of oxygen (P_{O_2}) is high, as occurs in the lungs, and decreases when P_{O_2} is low, as occurs in the tissues. The affinity of hemoglobin for oxygen is also influenced by blood pH. Any shift in blood pH will dramatically affect the degree to which hemoglobin binds oxygen. In this exercise the partial pressure of oxygen in a blood sample will be altered using a vacuum pump to decrease the P_{O_2} of a blood sample. A spectrophotometer will then be used to evaluate the relationship between P_{O_2} of the blood sample and the percent saturation of hemoglobin. In addition, the percent saturation of hemoglobin at three different pH values (6.8, 7.4, and 8.0) will be analyzed to determine the effect of pH on hemoglobin saturation.

Before beginning, become familiar with the following concepts (use the textbook as a reference):

- Structure of hemoglobin
- Oxygen-hemoglobin dissociation curve
- Meaning of a right-shift or left-shift of the oxygen-hemoglobin dissociation curve

1. Open Ph.I.L.S. Lesson 34: pH and Hb-Oxygen Binding.
2. Read the objectives, introduction, and wet lab sections. Be sure to click "open" and view the videos that are indicated in red in the wet laboratory. Then, take the pre-lab quiz. The laboratory exercise will open when the pre-lab quiz has been completed **(figure 24.18)**.

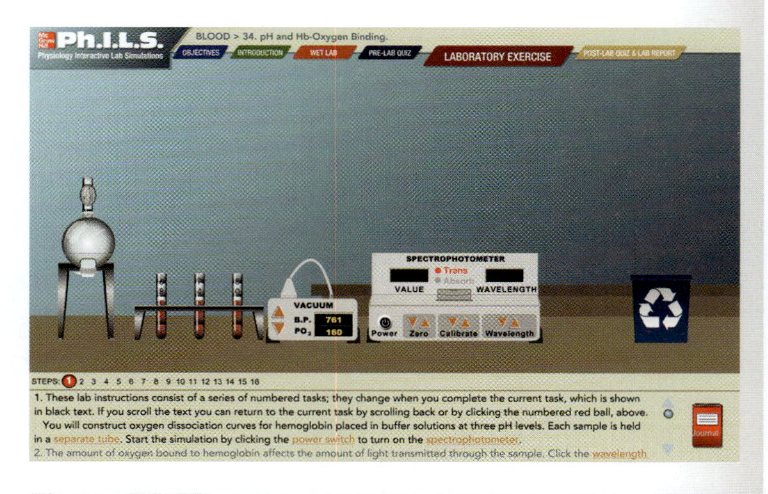

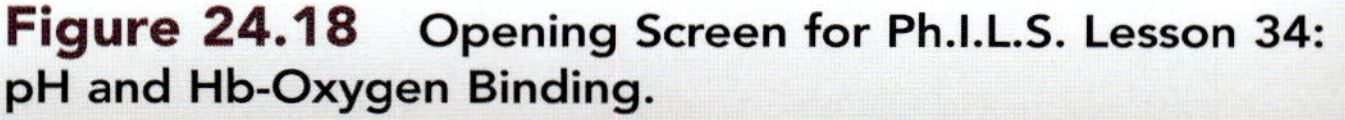
Figure 24.18 Opening Screen for Ph.I.L.S. Lesson 34: pH and Hb-Oxygen Binding.

3. Click the "power" switch to turn on the spectrophotometer.
4. To calibrate the spectrophotometer, set the wavelength at 620 nm (wavelength at which intact hemoglobin absorbs light) and set the transmittance value to 0 using the arrows.
5. Open the lid of the spectrophotometer holder (if unsure, click on the word "spectrophotometer holder" in the instructions at the bottom of the screen).
6. Click on the globe and drag it to the top of one of the tubes (it will snap in place) to form the tonometer.
7. Click and drag the tonometer into the spectrophotometer. The journal will open a table showing partial pressures of oxygen ranging from 160 mm Hg to 0 mm Hg for three different pH values—6.8, 7.4, and 8.0. Set the transmittance to 100% using the arrows at "Calibrate."
8. To record the transmittance through the sample, click "Journal" (red rectangle at the bottom right of screen).
9. Click on the tonometer to remove it. The tonometer will come up out of the spectrophotometer and the journal will close automatically. Then click and drag the tonometer back to the rack.
10. To change the partial pressure of oxygen in the sample, position the end of the vacuum tube on the top of the tonometer.
11. Click the down arrow on the vacuum once to reduce the barometric pressure (which simulates a decrease in partial pressure of oxygen). Click on the vacuum tube to return it to its original position.
12. Place the tonometer in the spectrophotometer. Click "Journal" to record.
13. Repeat steps 9–12 over the pressure range (160–0 mm Hg) for that sample until all values for the different partial pressures of oxygen have been recorded in the table on the computer.
14. When all values have been recorded in the table for a given sample, drag the tonometer to place the sample into the recycling can (three arrows in a circle in the bottom right of the screen) to dispose of the sample.
15. Repeat steps 5–14 with the remaining blood samples using different pH values.
16. Click "Journal" to see the complete table. To view the graph, click on "graph" in the lower right of the journal window. Complete **table 24.14** from the table created on the screen and graph the data in the space provided.
17. Open the post-lab quiz & lab report by clicking on it. Answer the post-lab questions that appear on the computer screen. Click "Print Lab" to print a hard copy of the report or click "Save PDF" to save an electronic copy of the report.

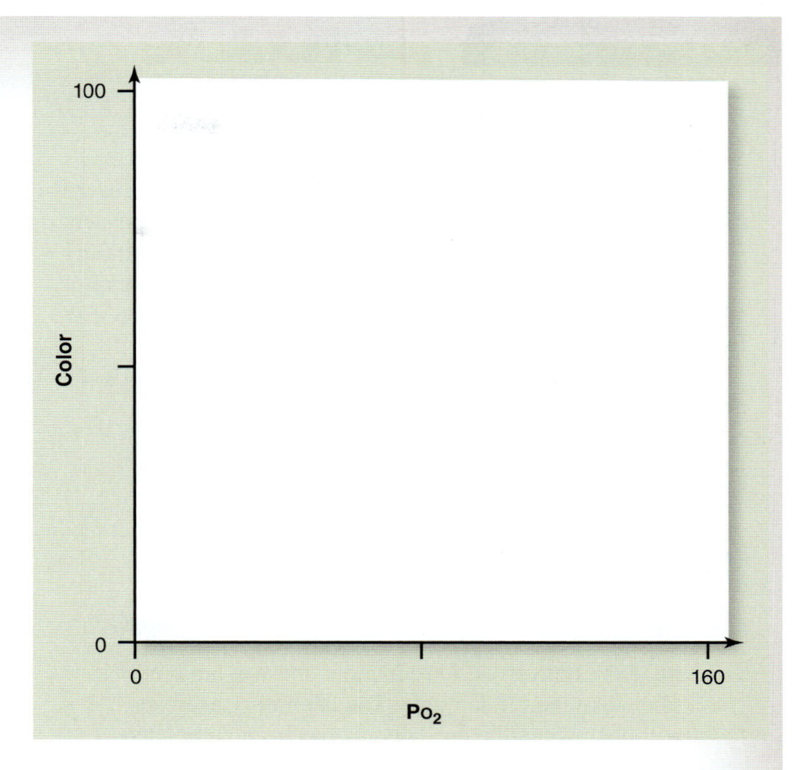

Table 24.14 Data Table for Partial Pressures of Oxygen and Blood pH

Po_2	pH 6.8	pH 7.4	pH 8.0
160			
140			
120			
100			
80			
60			
40			
20			
0			

18. Make note of any pertinent observations here:

____

____

____

EXERCISE 24.14 Ph.I.L.S.

Ph.I.L.S. LESSON 39: EXERCISE-INDUCED CHANGES

During exercise, the respiratory system must adapt to increased oxygen demands of working muscles. The respiratory system accomplishes this by increasing tidal volume and respiratory rate. The purpose of this laboratory exercise is to monitor changes in respiratory volumes and rates of a virtual subject before and after exercise.

Before beginning, become familiar with the following concepts (use the textbook as a reference):

- **inspiratory reserve volume**
- **expiratory reserve volume**
- **tidal volume**
- **vital capacity**
- **respiratory rate**

1. Open Ph.I.L.S. Lesson 39: Exercise-Induced Changes.
2. Read the objectives, introduction, and wet lab sections. Then take the pre-lab quiz. The laboratory exercise will open when the pre-lab quiz has been completed **(figure 24.19).**
3. Click the "power" button on the computer. Then click the "power" button on the data acquisition unit (DAQ) to begin the experiment.
4. Click and drag the blue plug to recording input 1 on the DAQ.
5. Click and drag the mouthpiece to the subject's mouth.
6. Click the button marked "resting" on the control panel of the virtual monitor.
7. To begin recording normal breathing patterns, click "start" on the control panel of the virtual monitor. A blue trace will appear representing the subject's breathing pattern.

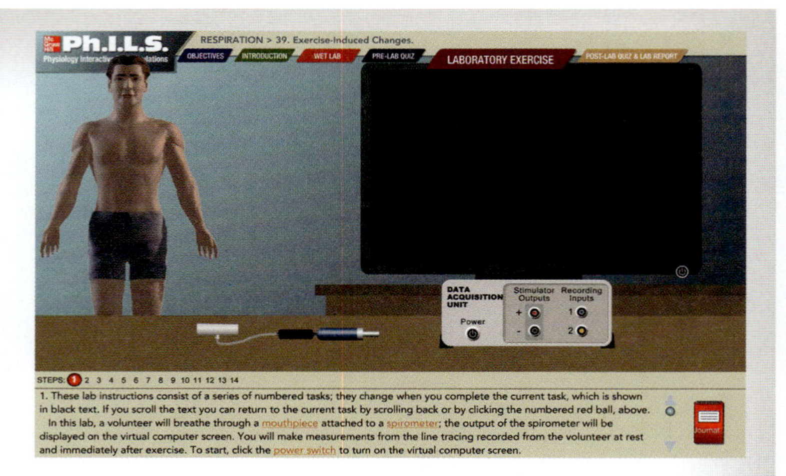

Figure 24.19 Opening Screen for Ph.I.L.S. Lesson 39: Exercise-Induced Changes.

8. The apparatus will measure three tidal volumes (TV), then the subject will maximally exhale and maximally inhale. At this point the recording will stop and measurements of respiratory volumes may be made.
9. Measure the TV for the first breath by dragging the cursor over the first peak. Then move the other cursor over the first trough. Data will appear in the window on the virtual monitor. Click the red "Journal" button in the lower right of the screen. Repeat this measurement for all three breaths. Enter the TV for each of the three breaths, the mean, and the calculated breaths per minute in **table 24.15.**
10. To measure the expiratory reserve volume (ERV), move the cursor to the third trough. Then move the other cursor to the lowest point on the tracing. The data appears in the data window on the virtual monitor. Click the red "Journal" button to enter it in the journal, and enter the data in **table 24.16.**

Table 24.15 Tidal Volumes Before and After Exercise

Trial	TV Resting (mL)	TV Post-Exercise (mL)	Time Interval at Rest (s)	Time Interval Post-Exercise(s)
#1				
#2				
#3				
Mean				
Breaths per Minute				

Table 24.16	Lung Volumes Before and After Exercise	
	Resting (mL)	**Post-Exercise (mL)**
IRV		
TV		
ERV		
VC		

11. To measure the inspiratory reserve volume (IRV), move the cursor to the third peak. Then move the other cursor to the highest part of the trace. Click the red "Journal" button to enter the data in the journal, and enter the data in table 24.16.

12. To record the lung volumes following exercise, click the button marked "After Exercise" on the control panel of the virtual monitor. Repeat steps 7–11 and record the data in tables 24.15 and 24.16.

13. Open the post-lab quiz & lab report by clicking on it. Answer the post-lab questions that appear on the computer screen. Click "Print Lab" to print a hard copy of the report or click "Save PDF" to save an electronic copy of the report.

14. Make note of any pertinent observations here:

____

____

____

Chapter 24: The Respiratory System

Name: ______________________________

Date: ____________ **Section:** ____________

POST-LABORATORY WORKSHEET

The ❶ corresponds to the Learning Objective(s) listed in the chapter opener outline.

Do You Know the Basics?

Exercise 24.1: Olfactory Mucosa

1. Place the following layers of the wall of the respiratory tract in order from innermost to outermost. ❶

   ______ a. adventitia

   ______ b. mucosa

   ______ c. submucosa

2. Stratified squamous epithelial tissue lines the nasal cavity. _________________________ (True/False) ❷

Exercise 24.2: The Trachea

3. The trachea is ______________ (anterior/posterior) to the esophagus. ❸

4. Which of the following cell surface modifications and/or specialized cells may be found within the epithelium that lines the trachea? (Check all that apply.) ❹

   ______ a. cilia

   ______ b. goblet cells

   ______ c. keratinization

   ______ d. microvilli

5. The C-shaped tracheal cartilages allow the trachea to _________________________ (close/open), whereas the trachealis muscle allows the trachea to _____________ (close/open). ❺

Exercise 24.3: The Bronchi and Bronchioles

6. Which of the following epithelia and/or cell surface modifications are present in large bronchi? (Check all that apply.) ❻ ❼

   ______ a. cilia

   ______ b. goblet cells

   ______ c. microvilli

   ______ d. pseudostratified ciliated columnar epithelium

   ______ e. stratified squamous epithelium

Exercise 24.4: The Lungs

7. Alveoli are sac-like structures that serve as the site of gas exchange. _________________________ (True/False) ❽

8. The respiratory membrane, through which a molecule of carbon dioxide must travel in order to diffuse from the blood into the alveoli, is composed of alveolar epithelium, a fused basement membrane, and a capillary endothelium. _________________________ (True/False) ❾

9. Alveolar type _________________________ (I/II) cells are responsible for producing surfactant. ❿

Exercise 24.5: Midsagittal Section of the Head and Neck

10. The nasal cavity contains the superior, middle, and inferior nasal _________________________, (conchae/meatuses) and the spaces between, known as the superior, middle, and inferior _________________________ (conchae/meatuses). ⓫

Exercise 24.6: The Larynx

11. The structures located more superior in the larynx are the __________________ (vestibular/vocal) folds. 12

12. Which of the following are unpaired cartilages of the larynx? (Check all that apply.) 13

    ______ a. arytenoid

    ______ b. cricoid

    ______ c. cuneiform

    ______ d. epiglottis

    ______ e. thyroid

13. The larynx houses vestibular folds, which are responsible for sound production. __________________ (True/False) 14

Exercise 24.7: The Pleural Cavities

14. The outermost layer of the pleura is the __________________ (parietal/visceral) pleura. 15

Exercise 24.8: The Lungs

15. The left lung is smaller than the right lung. __________________ (True/False) 16

16. The hilum of the right or left lung contains pulmonary arteries and veins. __________________ (True/False) 17

17. The right lung contains __________________ (one/two) fissures, whereas the left lung contains __________________ (one/two) fissures. 18

Exercise 24.9: The Bronchial Tree

18. Rank the following branches of the bronchial tree in the order a molecule of oxygen would encounter them as it moves from the trachea into the left lung. 19

    ______ a. left lobar bronchus

    ______ b. left main bronchus

    ______ c. left segmental bronchus

19. Two lobar bronchi lead into the right lung. __________________ (True/False) 20

Exercise 24.10: Mechanics of Ventilation

20. A(n) __________________ (increase/decrease) in thoracic volume leads to an increase in intrapulmonary pressure. 21

21. A decrease in intrapulmonary pressure allows air to flow __________________ (into/out of) the lungs. 22

Exercise 24.11: Auscultation of Respiratory Sounds

22. The optimal position for auscultation of bronchial sounds is over the xiphoid process of the sternum. ______________ (True/False) 23

Exercise 24.12: Pulmonary Function Tests

23. Match the appropriate definitions listed in column A with the terms listed in column B. 24 25

Column A	*Column B*
______ a. amount of air remaining in the lungs after quiet expiration	1. expiratory reserve volume (ERV)
______ b. measure of the strength of respiration	2. functional residual capacity (FRC)
______ c. total ability to inspire	3. inspiratory capacity (IC)
______ d. total amount of air that can be in the lungs	4. inspiratory reserve volume (IRV)
______ e. volume of additional inspired air, above and beyond the tidal volume	5. residual volume (RV)
______ f. volume of air left in the lungs following forced expiration	6. total lung capacity (TLC)
______ g. volume of air maximally expired above and beyond the tidal volume	7. tidal volume (TV)
______ h. volume of air that enters and leaves the lungs with each normal breath	8. vital capacity (VC)

24. Reduced forced expiratory volume in 1 second (FEV_1) typically indicates __________________ (obstructive/restrictive) pulmonary disease. 26

Exercise 24.13: Ph.I.L.S. Lesson 34: pH & Hb-Oxygen Binding

25. As the partial pressure of oxygen (P_{O_2}) increases, the percent transmittance of light ______________________ (increases/decreases). 27

26. There is a direct relationship between percent transmittance of light and percent saturation of hemoglobin. As the percent transmittance of light increases, the percent saturation of hemoglobin ______________________ (increases/decreases). Thus, as the partial pressure of oxygen increases, the percent saturation of hemoglobin ______________________ (increases/decreases). 28

Exercise 24:14: Ph.I.L.S. Lesson 39: Exercise-Induced Changes

27. Respiratory rate and tidal volume ______________________ (increase/decrease) with exercise, whereas expiratory reserve volume and inspiratory reserve volume ______________________ (increase/decrease) with exercise. 29

Can You Apply What You've Learned?

28. Trace the pathway a molecule of oxygen must take to travel from the nasal cavity to an alveolus in the inferior lobe of the right lung. Be sure to name all the conducting and respiratory portion structures the molecule will pass through along the way.

29. A toddler coughs when attempting to swallow a bite of hot dog, and the hot dog is directed into the respiratory tree instead of the esophagus. The piece of hot dog is most likely to become lodged in the airways leading to the ______________________ lung because the primary bronchus to this lung is more vertically oriented than the primary bronchus to the other lung. (Note: In 2010 the FDA listed hot dogs as a major choking hazard for children. It even suggested that hot dogs should have warning labels on them because of the large number of children who choke on them.)

30. When a person suffers from an asthma attack, there is often excessive inflammation of respiratory structures, which leads to increased mucus production and smooth muscle contraction. Assuming that asthma is an obstructive respiratory disease, predict whether a person suffering from asthma would have more difficulty inhaling or exhaling, and justify your answer. ______________________

31. Discuss abnormal respiratory sounds that might be heard in a patient suffering from asthma. ______________________

32. A patient is rushed to the hospital with a gunshot wound to the chest. Upon arrival, doctors discover that the patient is suffering from a collapsed right lung. Explain why the lung collapsed when the bullet penetrated the patient's thoracic cavity. ______________________

33. Opening a can of carbonated soda leads to the loss of "fizz" over time. Describe why this occurs, and relate this phenomenon to alveolar gas exchange.

34. When a person is hyperventilating, a home remedy is to have the person breathe deeply into and out of a paper bag. Describe how this "treatment" may be beneficial in regulating blood pH and respiratory rate. ______________________

35. Describe why a patient with a restrictive lung disease would have a decreased vital capacity. ______

Can You Synthesize What You've Learned?

36. The world's tallest free-standing mountain, Mount Kilimanjaro in Tanzania, Africa, stands at approximately 19,340 feet, where atmospheric pressure is roughly half that of pressure at sea level. Discuss how partial pressures of oxygen will change at the peak of the mountain as compared to sea level. Then describe how these partial pressures may affect respiratory rate and alveolar gas exchange. ______

37. Exposure to large quantities of carbon monoxide (CO) in a poorly ventilated space can be lethal. Unfortunately, this is a common occurrence during winter months because furnaces run the risk of emitting large quantities of CO. When inhaled, CO has a higher affinity for the heme group on hemoglobin than does oxygen. Describe how exposure to CO will influence percent oxygen saturation of hemoglobin and affect systemic gas exchange.

38. A hypersensitivity to peanuts can lead to severe inflammation of the tissues lining the respiratory tract. Often, doctors will prescribe *epinephrine* to be administered by auto injection (EpiPen®) in case of exposure. When *epinephrine* is injected intramuscularly, it binds to adrenergic receptors throughout the body (see chapter 17), including those located on the smooth muscle of respiratory structures. Based on your knowledge of the autonomic nervous and respiratory systems, predict what will happen to both resistance and airflow through the respiratory structures of a patient when *epinephrine* is administered.

CHAPTER

25

The Urinary System

OUTLINE AND LEARNING OBJECTIVES

Anatomy & Physiology REVEALED® aprevealed.com **MODULE 13: URINARY SYSTEM**

INTRODUCTION

The urinary system is responsible for maintaining blood volume and composition. It accomplishes this task through the process of filtration, which involves forcing fluid out of the blood across a membrane called the **filtration membrane.** The fluid thus formed is called **filtrate.** The filtrate flows through the **nephrons,** the structural and functional units of the kidney. Many substances, including over 90% of the water in the filtrate, are **reabsorbed** back into the blood so they are not lost from the body. Other substances, such as **urea** (a breakdown product of protein metabolism), are **secreted** into the filtrate so they can be removed from the body. Ultimately, a small (relative to the volume of blood that is filtered) amount of fluid leaves the kidney as **urine,** which will be transported via the **ureters** to the **urinary bladder** for storage. At a time that is convenient for the individual, the urine is emptied from the bladder and exits the body through the **urethra.**

The structural and functional unit of the kidney is the nephron. Each kidney contains more than 1.25 million nephrons. Remarkably, the kidneys can maintain their function even when 85–90% of their nephrons have been destroyed through disease. However, further losses will result in **kidney failure**. If an individual's kidneys fail, he or she must be placed on **dialysis** (*dialyo,* to separate). This involves filtering the blood using a dialysis machine, or artificial kidney. A patient on dialysis must undergo three to four sessions a week, each session lasting approximately 4 hours. Without dialysis, the individual cannot survive because the balance of fluid, electrolytes, and waste products in the blood cannot be maintained at appropriate levels. The consequences of kidney failure underscore the enormous role the kidneys play in maintaining health.

These laboratory exercises focus on the structures of the urinary system with a special emphasis on the kidney. Nearly all functions of the urinary system are functions of the kidney. The remaining structures of this system (ureter, urinary bladder, and urethra) transport or store the urine formed by the kidney. Because the structural and functional units of the kidney—the nephrons—are actively involved in altering the composition of the blood, it is critical to understand the pattern of blood flow through the kidney and how this pattern of blood flow parallels the parts of the nephron. This is achieved by observing the gross anatomy of the kidney and its blood supply, and correlating gross structures with histological observations of the parts of the nephron (renal corpuscles, proximal convoluted tubules, distal convoluted tubules, and nephron loops). Subsequent exercises guide the student in exploring the processes involved in urine formation, identifying normal and abnormal constituents of urine, and describing the role of the urinary system in regulating acid-base balance in the blood. These concepts will be integrated in a clinical case that applies concepts previously covered in multiple chapters. This exercise will facilitate further understanding of complex physiological systems and their integration to maintain overall homeostasis.

List of Reference Tables

Chapter 25: The Urinary System

Name: ____________________

Date: __________ Section: __________

PRE-LABORATORY WORKSHEET

These Pre-Laboratory Worksheet questions may be assigned by instructors through their connect course.

1. Which of the following is a function of the urinary system? (Check all that apply.)

   _______ a. acid-base balance of the blood

   _______ b. long-term blood pressure regulation

   _______ c. regulation of erythrocyte production

   _______ d. regulation of leukocyte production

   _______ e. excretion of wastes

2. Urine flows through the ____________________ (ureters/urethra) to exit the kidneys.

3. Which of the following is the structural and functional unit of the kidney? (Circle one.)

   a. collecting duct
   b. glomerulus
   c. minor calyx
   d. nephron
   e. renal pelvis

4. Identify the type of epithelial tissue found lining the urinary bladder. (Circle one.)

   a. pseudostratified ciliated columnar epithelium
   b. simple squamous epithelium
   c. stratified squamous epithelium
   d. transitional epithelium

5. Which of the following structures is lined with simple cuboidal epithelium with microvilli? (Check all that apply.)

   _______ a. collecting duct

   _______ b. distal convoluted tubule

   _______ c. parietal layer of the glomerular capsule

   _______ d. proximal convoluted tubule

6. The glomerulus of the kidney consists of ____________________ (continuous/fenestrated) capillaries.

7. The kidneys, ureters, and urinary bladder are all retroperitoneal structures. ____________________ (True/False)

8. Place the following vessels in the correct order in which a drop of blood flowing through the kidneys would encounter them.

   _______ a. afferent arteriole

   _______ b. arcuate vein

   _______ c. interlobar artery

   _______ d. peritubular capillaries

   _______ e. renal artery

9. Glucose reabsorption in the nephron occurs in the ____________________ (distal/proximal) convoluted tubule.

10. Which of the following is considered an abnormal constituent of urine? (Check all that apply.)

    _______ a. erythrocytes

    _______ b. glucose

    _______ c. ketone bodies

    _______ d. urea

    _______ e. water

HISTOLOGY

The Kidney

Each kidney is composed of two major regions: an outer **renal cortex** and an inner **renal medulla (figure 25.1).** The arrangement of nephrons along the **corticomedullary junction** means that some nephron components fall predominantly in the renal cortex and others in the renal medulla. Thus, the two regions exhibit distinct histological features. The renal cortex contains the renal corpuscles, proximal convoluted tubules (PCTs), distal convoluted tubules (DCTs), and peritubular capillaries. The renal medulla contains the nephron loops, collecting ducts (CDs), and vasa recta. Most structures within the kidney can be identified histologically by recognition of both the region (cortex or medulla) and the type of epithelium forming the structure. **Table 25.1** summarizes the type of epithelium that forms each of the structures and lists the major functions of each structure.

Table 25.1 Histological Features of the Kidney

Structure	Epithelium	Function
Renal Cortex		
Distal Convoluted Tubule (DCT)	Simple cuboidal with few, short microvilli. Nuclei are located near the apical surface of the cells.	Secretes H^+ and K^+. Reabsorbs Na^+ and water. Contains the macula densa of the juxtaglomerular apparatus, which is involved with the regulation of blood pressure.
Glomerulus	Fenestrated endothelium (simple squamous)	Filtration
Glomerular Capsule		
Visceral Layer	Simple squamous modified to form podocytes	Extensions of podocytes called pedicels contain actin filaments. The membrane-covered openings between the pedicels, called filtration slits, participate in the filtration process.
Parietal Layer	Simple squamous	Forms an outer, impermeable wall of the glomerular capsule
Proximal Convoluted Tubule (PCT)	Simple cuboidal with long, dense microvilli. Nuclei are located near the basal surface of the cells.	Reabsorbs glucose, amino acids, Ca^{2+}, PO_4, HCO_3^-, and 80% of the water and NaCl present in the filtrate. Secretes substances like penicillin and toxins after they have undergone modification by the liver. Also secretes organic acids and bases.
Renal Medulla		
Collecting Tubule and Duct (CD)	Simple cuboidal or simple columnar epithelium. Cells have very precise boundaries. Overall tube diameter is the same as the PCT, but the CD has a larger lumen and no microvilli. Cells are paler than those of the thick segment of the nephron loop.	Concentrates urine under the influence of antidiuretic hormone (ADH). Decreases urine volume through the action of aldosterone. Both hormones act on principal cells. Intercalated cells participate in regulating acid-base balance.
Nephron Loop		
Descending Limb—Thick Segment	Simple cuboidal	Epithelial cells are impermeable to sodium, but water is drawn out into the interstitial spaces by osmosis. Thus, the filtrate becomes more concentrated as it moves down the descending nephron loop.
Descending Limb—Thin Segment	Simple squamous	Epithelial cells are impermeable to sodium, but water is drawn out into the interstitial spaces by osmosis. Thus, the filtrate becomes more concentrated as it moves down the descending nephron loop.
Ascending Limb—Thin Segment	Simple squamous	Epithelial cells are impermeable to water, but sodium passively diffuses out. Thus, the filtrate becomes less concentrated as it moves up the ascending nephron loop.
Ascending Limb—Thick Segment	Simple cuboidal. Cells are darker than in the collecting duct.	The epithelial cells are impermeable to water, and they actively transport sodium out of the tubule. Thus, the filtrate becomes less concentrated as it moves up the ascending nephron loop.

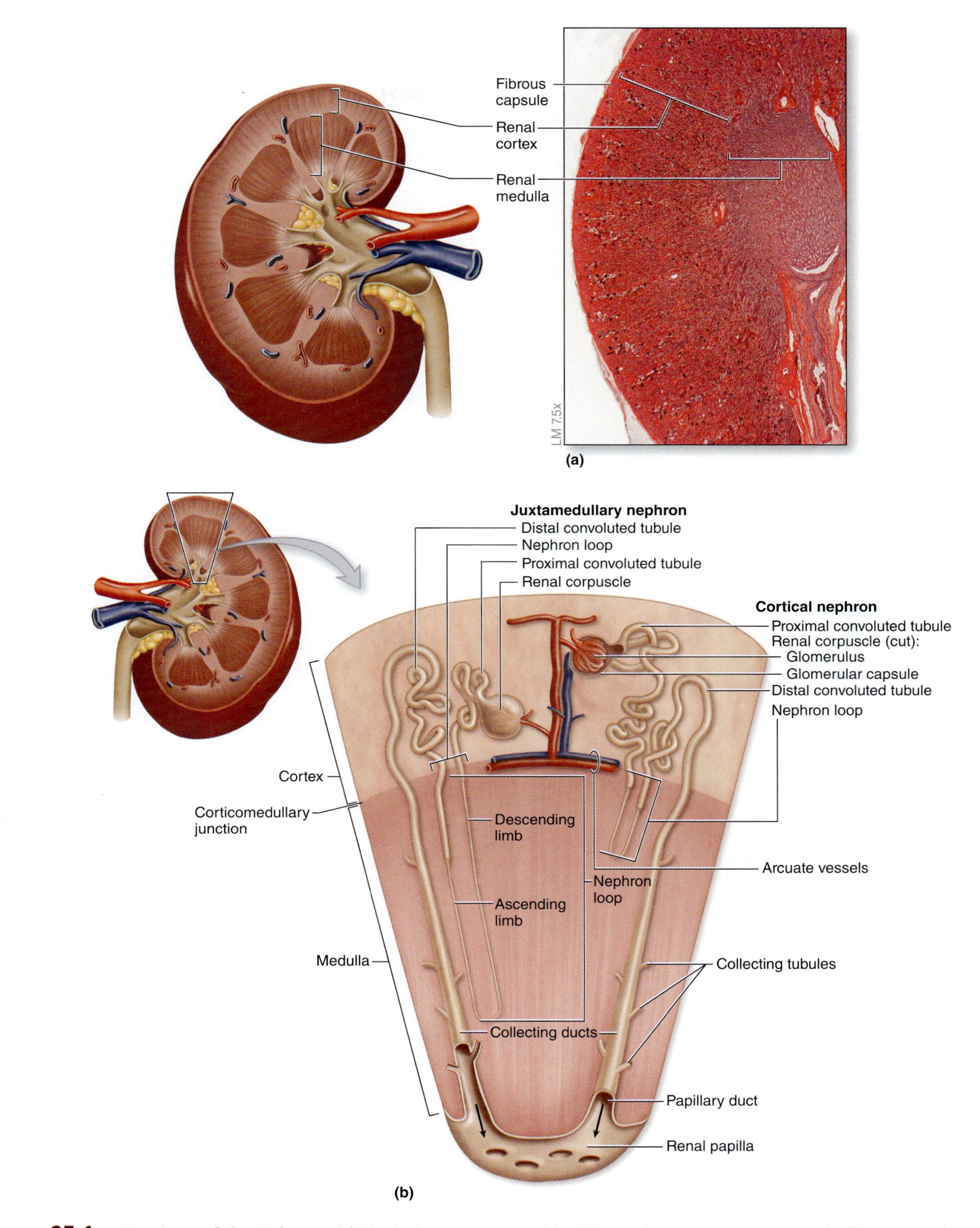

Figure 25.1 Histology of the Kidney. (a) Histological appearance of the kidney at low power demonstrates the fibrous capsule, outer cortex, and inner medulla. (b) Position of nephron structures within the renal cortex and renal medulla.

EXERCISE 25.1

HISTOLOGY OF THE RENAL CORTEX

The renal cortex contains **renal corpuscles,** which are the site of filtration. Each renal corpuscle is composed of a **glomerulus** (a tuft of capillaries), surrounded by a **glomerular capsule** (also known as Bowman's capsule). The glomerular capsule itself is composed of an inner **visceral layer,** which consists of modified simple squamous epithelial cells called **podocytes,** and an outer **parietal layer,** which consists of unmodified simple squamous epithelium. The renal cortex also contains **proximal convoluted tubules (PCTs)**, which are the site of most reabsorption and secretion, and **distal convoluted tubules (DCTs),** which are involved in both reabsorption and secretion and compose part of the **juxtaglomerular apparatus.**

EXERCISE 25.1A The Renal Corpuscle

1. Obtain a compound microscope and a histology slide of the kidney and place the slide on the microscope stage.
2. Bring the tissue sample into focus at low power and scan the slide to distinguish between the outer cortex and inner medulla (figure 25.1). Next, move the microscope stage so the renal cortex is in the center of the field of view.
3. Scan the slide in the region of the cortex to locate a circular **renal corpuscle (figure 25.2).** Bring the renal corpuscle into the center of the field of view and then change to high power. Although the visceral layer of the glomerular capsule is indistinguishable from the glomerular capillaries, identify the tissue that contains both the visceral layer of the glomerular capsule and the glomerular capillaries (figure 25.2). Next, identify the parietal layer of the glomerular capsule. What type of epithelium composes the parietal layer of the glomerular capsule? ______________________

   What is the name of the space located between the visceral and parietal layers of the glomerular capsule?

   ____

   The space just identified becomes continuous with the lumen of which of the following? (Circle one.)

   PCT **nephron loop** **DCT** **collecting duct**

4. Sketch the renal corpuscle as seen through the microscope in the space provided. Be sure to label the glomerulus (and visceral layer of the glomerular capsule), the parietal layer of the glomerular capsule, and the capsular space on the drawing.

5. *Optional Activity:* **AP|R 13: Urinary System—**Review the "Kidney—Microscopic Anatomy" animation to visualize the parts of the nephron and their placement in the renal cortex and medulla.

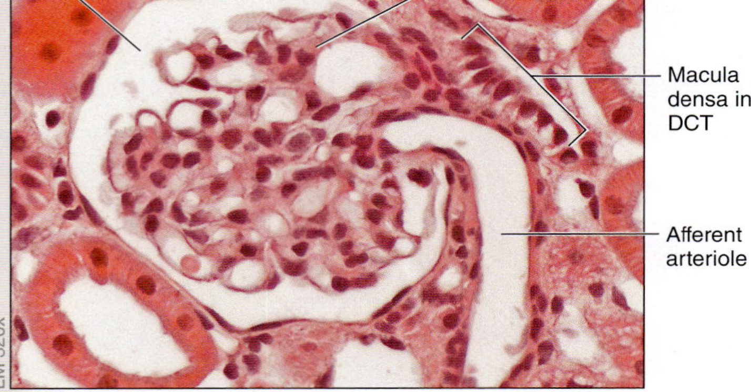

Figure 25.2 The Renal Corpuscle. The renal corpuscle consists of the glomerulus and the glomerular capsule.

EXERCISE 25.1B Proximal and Distal Convoluted Tubules

1. After completing exercise 25.1A, there should be a slide of the kidney on the microscope stage that has the outer cortex in the center of the field of view. If you did not perform exercise 25.1A, go through steps 1–2 of exercise 25.1A and then continue with this exercise.

2. Scan the slide in the region of the cortex to locate **proximal** and **distal convoluted tubules** (PCTs and DCTs) **(figure 25.3).** When viewing a histology slide of the cortex of the kidney, cross sections of PCTs and DCTs, with a few renal corpuscles scattered throughout, should be visible. Although both PCTs and DCTs are lined with simple cuboidal epithelium, the *proximal* convoluted tubules have long, dense microvilli, which make the lumens of the PCTs appear "fuzzy." The distal convoluted tubules have very few, short microvilli, which make the lumens appear to be clear. What does the presence of microvilli indicate about the function of an epithelium? ______________________________

____

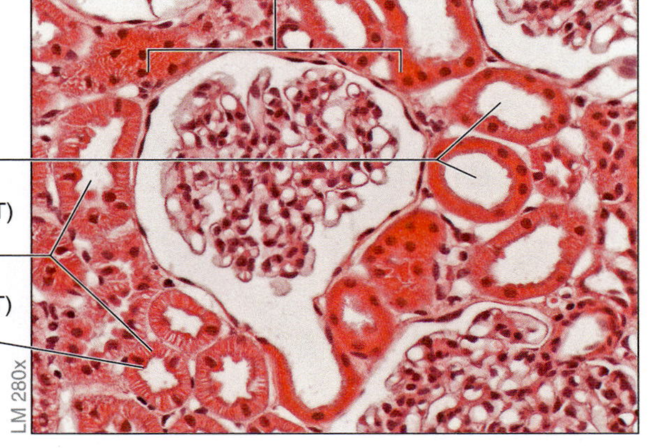

Figure 25.3 The Renal Cortex. The renal cortex contains renal corpuscles, proximal convoluted tubules (PCT), and distal convoluted tubules (DCT). PCTs have "fuzzy" lumens because of the presence of microvilli.

3. Sketch cross sections of proximal and distal convoluted tubules as seen through the microscope in the space provided.

__________ ×

4. Identify the following structures on the slide of the cortex of the kidney, using figures 25.1 to 25.3 and table 25.1 as guides:

- ☐ capsular space
- ☐ distal convoluted tubule
- ☐ microvilli
- ☐ parietal layer of glomerular capsule
- ☐ proximal convoluted tubule
- ☐ renal corpuscle
- ☐ visceral layer of glomerular capsule and glomerular capillaries

EXERCISE 25.2

HISTOLOGY OF THE RENAL MEDULLA

The renal medulla contains **nephron loops** (loops of Henle) and **collecting ducts,** with surrounding capillaries called the **vasa recta.** These structures are all elongated tubules that lie next to each other and function together to concentrate the urine formed by the nephrons.

1. Obtain a compound microscope and a histology slide of the kidney. Be sure to note the section type, whether a longitudinal or cross-section of the kidney. Place the slide of a longitudinal section on the microscope stage.

2. Bring the tissue sample into focus at low power and scan the slide to distinguish between the outer cortex and inner medulla. Next, move the microscope stage so the renal medulla is in the center of the field of view (figure 25.1*a*).

3. With the medulla in the center of the field of view, switch to high power and bring the tissue sample into focus once again. In this region of the kidney, thick and thin limbs of the nephron loops, collecting ducts, and possibly vasa recta should be visible (table 25.1, **figure 25.4**). Notice how, in longitudinal section, these structures appear as row after row of cells lined up next to each other. The main objective in viewing this part of the kidney is to gain an appreciation for the way these structures line up next to each other, which is critical to their function. An additional objective is to be able to distinguish thick limbs of the nephron loops from the collecting ducts based on the diameter of the lumens. Both structures are typically lined with simple cuboidal

(continued on next page)

(continued from previous page)

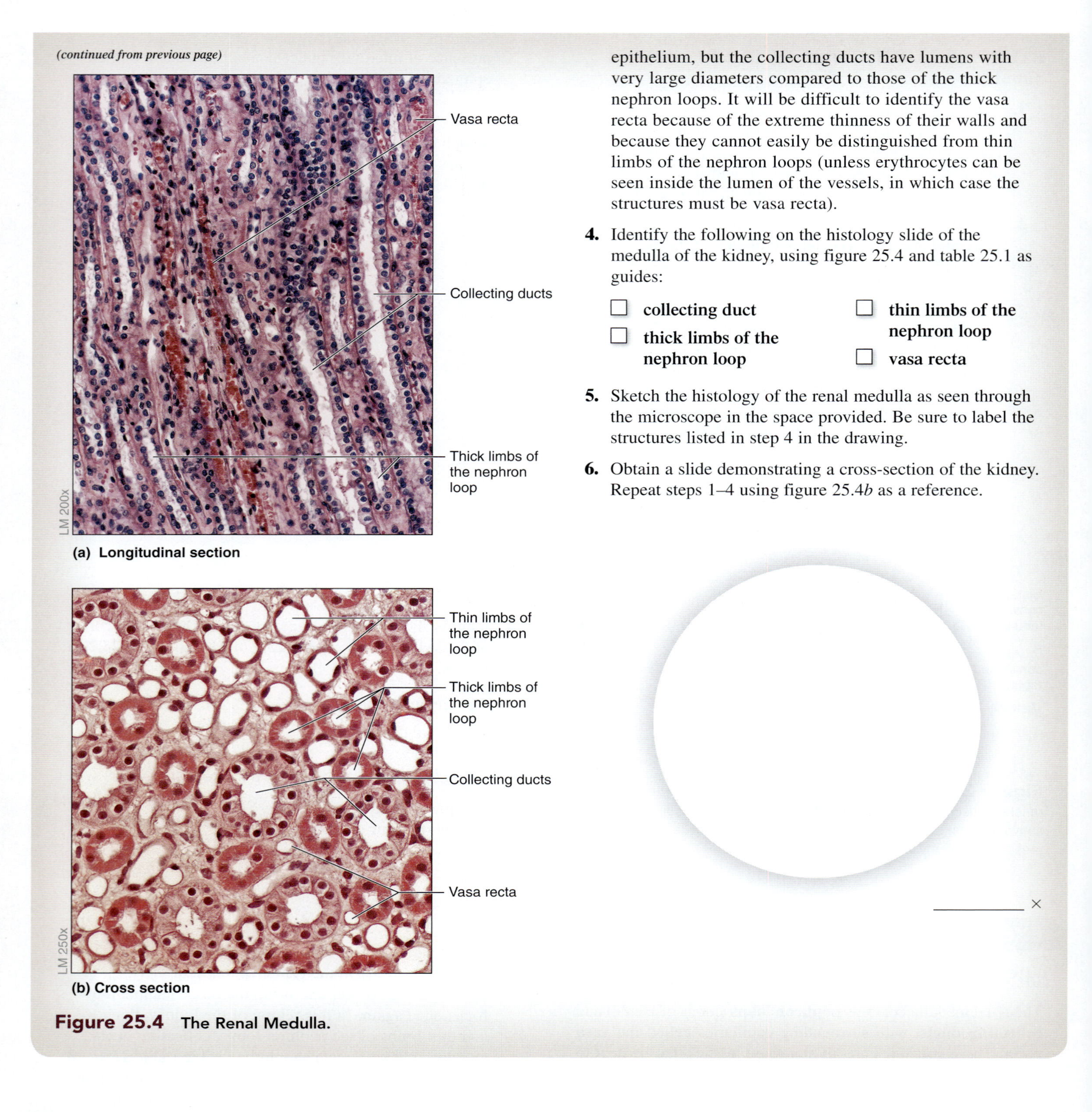

Figure 25.4 The Renal Medulla.

epithelium, but the collecting ducts have lumens with very large diameters compared to those of the thick nephron loops. It will be difficult to identify the vasa recta because of the extreme thinness of their walls and because they cannot easily be distinguished from thin limbs of the nephron loops (unless erythrocytes can be seen inside the lumen of the vessels, in which case the structures must be vasa recta).

4. Identify the following on the histology slide of the medulla of the kidney, using figure 25.4 and table 25.1 as guides:

   ☐ **collecting duct**
   ☐ **thick limbs of the nephron loop**
   ☐ **thin limbs of the nephron loop**
   ☐ **vasa recta**

5. Sketch the histology of the renal medulla as seen through the microscope in the space provided. Be sure to label the structures listed in step 4 in the drawing.

6. Obtain a slide demonstrating a cross-section of the kidney. Repeat steps 1–4 using figure 25.4*b* as a reference.

______ ×

The Urinary Tract

The **urinary tract** consists of structures that function to transport or store urine: the ureters, urinary bladder, and urethra. This section involves making observations of the histological structure of the walls of the ureters and urinary bladder. **Table 25.2** lists the type of epithelium that lines the calyces of the kidney (the areas of kidney that drain urine from the kidney into the urinary tract) and each of the structures of the urinary tract.

Table 25.2 Urine-Draining Structures

Structure	Epithelium	Number of Cell Layers	Word Origin
Calyces	Transitional	2–3	*calyx*, cup of a flower
Ureter	Transitional	4–5	*oureter*, urinary canal
Urinary Bladder	Transitional	> 6	*urinary*, relating to urine
Urethra	In males, the proximal portions are lined with transitional epithelium and the distal portions are lined with stratified squamous. In females, the epithelial tissue lining the entire urethra is primarily stratified squamous.	NA	*ourethra*, canal leading from the bladder

EXERCISE 25.3

HISTOLOGY OF THE URETERS

The **ureters** are long, epithelial-lined, fibromuscular tubes that transport urine from the hilum of the kidney to the urinary bladder. They are lined with **transitional epithelium,** which has the ability to stretch when urine is being transported through the ureters. Like other tubular structures in the body, the wall of the ureter is composed of multiple layers. The walls of the ureters, however, do not have a submucosa. There are three layers to the walls of the ureters: the mucosa, muscularis, and adventitia **(figure 25.5).** Contraction of smooth muscle in the muscularis layer of the ureters transports urine from the renal pelvis to the urinary bladder. The arrangement of smooth muscle layers in the muscularis of the ureters is just the opposite of that of the digestive tract organs. The inner layer is composed of **longitudinal** smooth muscle, whereas the outer layer is composed of **circular** smooth muscle. In addition, the outer layer is an **adventitia** (and not serosa) because the ureters are **retroperitoneal** (*retro-*, behind, + *peritoneal*, referring to the peritoneum) and are not covered by a visceral peritoneum.

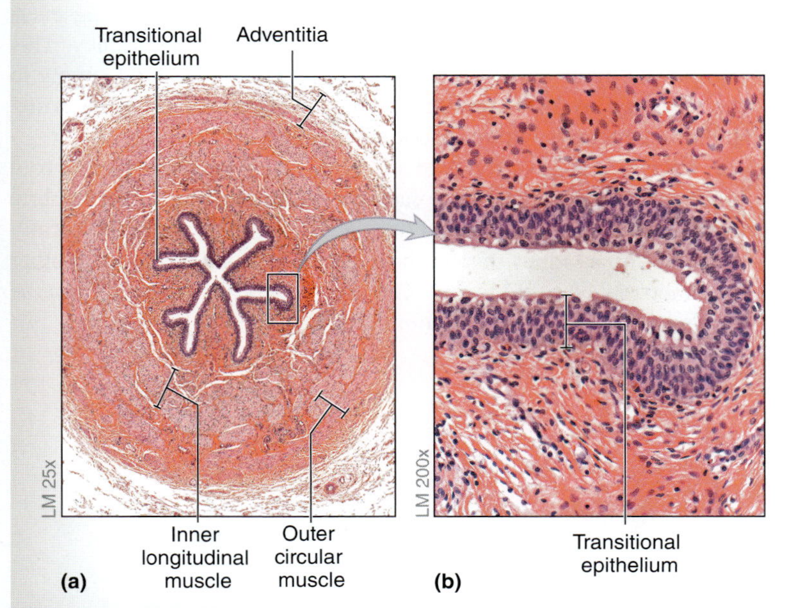

Figure 25.5 The Ureter. (a) Cross section through the entire ureter. (b) Close-up of the mucosa of the ureter.

1. Obtain a histology slide demonstrating a cross-sectional view of a ureter. Place it on the microscope stage and bring the tissue sample into focus on low power.
2. Identify the following structures on the slide of the ureter, using figure 25.5 as a guide:

   - ☐ **adventitia**
   - ☐ **inner longitudinal muscle**
   - ☐ **outer circular muscle**
   - ☐ **transitional epithelium**

3. Sketch a cross section of a ureter as seen through the microscope in the space provided. Be sure to label all the structures listed in step 2 in the drawing.

_______________ ×

EXERCISE 25.4

HISTOLOGY OF THE URINARY BLADDER

The wall of the urinary bladder has a more typical arrangement of layers, similar to that of other organs in the body: the mucosa, submucosa, muscularis, and adventitia (with serosa only on the superior surface of the urinary bladder) **(figure 25.6).** Similar to the ureters, the urinary bladder is lined with **transitional epithelium,** which allows it to stretch as it fills with urine, and its outermost layer is an adventitia because the bladder lies outside the peritoneal cavity. The muscularis layer of the bladder wall is composed of several individual layers of smooth muscle that are collectively referred to as the **detrusor muscle** (*detrudo,* to drive away).

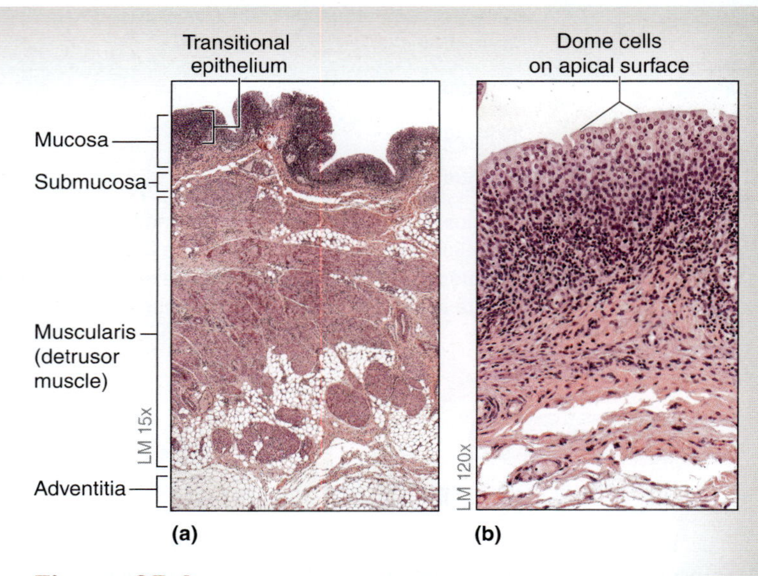

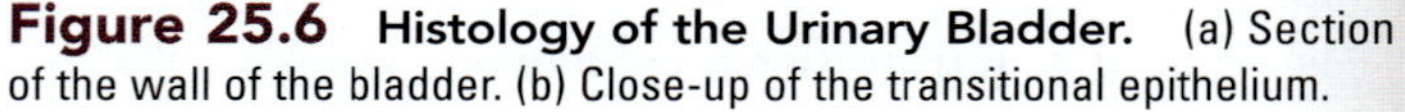
Figure 25.6 Histology of the Urinary Bladder. (a) Section of the wall of the bladder. (b) Close-up of the transitional epithelium.

1. Obtain a histology slide of the wall of the urinary bladder. Place it on the microscope stage and observe at low power. Identify the following structures, using figure 25.6 as a guide:

   - ☐ **adventitia**
   - ☐ **mucosa**
   - ☐ **muscularis (detrusor muscle)**
   - ☐ **serosa**
   - ☐ **submucosa**
   - ☐ **transitional epithelium**

2. Move the microscope stage so the transitional epithelium is at the center of the field of view, then change to high power. Look for rounded epithelial cells on the apical surface of the epithelium, which are often binucleate. These cells are sometimes referred to as **dome cells** because of their shape.

GROSS ANATOMY

This set of exercises involves observing the gross anatomy of the kidneys, ureters, urinary bladder, and urethra. Particular attention will be focused on observing the location of these structures within the abdominopelvic cavity and their relationships with other organs.

The Kidney

The kidneys are located along the posterior body wall from vertebral levels T_{12}–L_3, with the right kidney slightly lower than the left kidney because of the location of the liver. The kidneys, like the ureters and urinary bladder, are retroperitoneal structures. The following exercises involve first observing the gross anatomy of the kidney. This is followed by an exercise focused on observing the blood supply to the kidney. The final exercise focuses on observations of the urine-draining structures within the kidney.

Each kidney is surrounded and protected by several layers. From innermost to outermost these include the renal capsule, perinephric fat, renal fascia, and paranephric fat. Only the renal capsule typically remains with a kidney that has been removed from a cadaver or other specimen (e.g., a beef kidney). The other layers are removed from the kidney when it is extracted from the body.

EXERCISE 25.5

GROSS ANATOMY OF THE KIDNEY

1. Obtain a preserved animal kidney that has been sectioned along a coronal plane, or a classroom model of a coronal section of the kidney **(figure 25.7).** Whether viewing an actual kidney or a model of the kidney, the outermost structure that is visible is the **fibrous capsule,** which is composed of dense irregular connective tissue. This is because the other protective layers have been removed.

2. Identify the structures listed in figure 25.7 on the kidney or classroom model of the kidney. Then label them in figure 25.7.

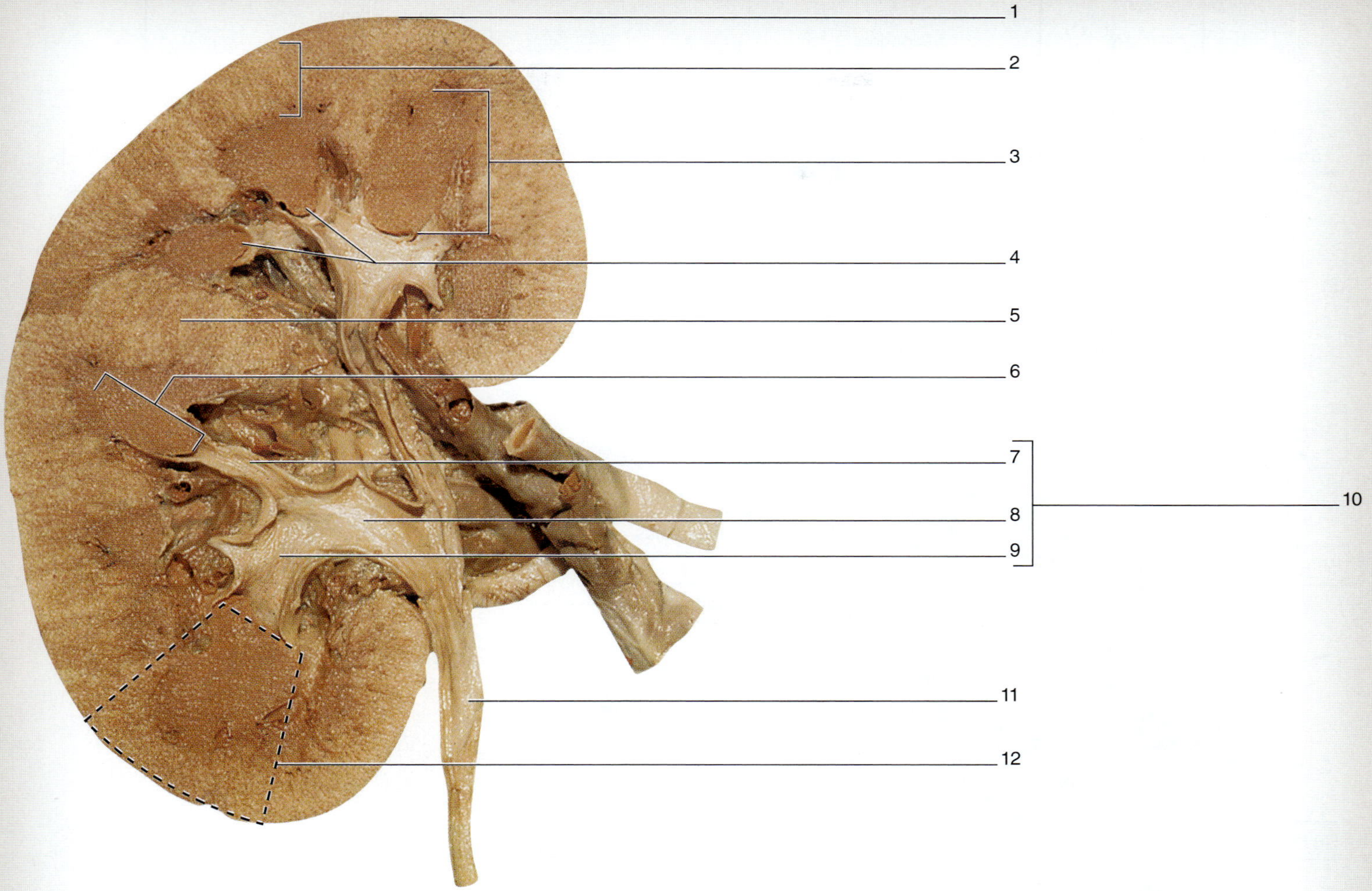

Figure 25.7 Coronal Section Through the Right Kidney. Use the terms listed to fill in the numbered labels in the figure.

- ☐ fibrous capsule
- ☐ major calyx
- ☐ minor calyx
- ☐ renal column
- ☐ renal cortex
- ☐ renal lobe
- ☐ renal medulla
- ☐ renal (medullary) pyramid
- ☐ renal papilla
- ☐ renal pelvis
- ☐ renal sinus
- ☐ ureter

EXERCISE 25.6

BLOOD SUPPLY TO THE KIDNEY

The kidneys receive approximately 20–25% of the blood pumped by the heart each minute. This is not because they have a huge demand for oxygen or nutrients, but instead because their major function is to filter the blood to alter its volume and composition. The kidney accomplishes this through the processes of filtration, reabsorption, and secretion, which all involve transport of fluids and other substances between the functional units of the kidney (nephrons) and the blood. Thus, an understanding of blood flow into and out of the kidney is critical to understanding how the kidney functions.

1. Observe a classroom model of the kidney that demonstrates blood vessels of the kidney.
2. **Figure 25.8** diagrams the flow of blood through the kidney. Blood enters the kidney through the renal artery (a branch off the abdominal aorta). Blood is transported into the kidney by progressively smaller arteries until reaching the capillaries. The arrangement of capillaries around the nephrons of the kidney is unique. There are three capillary beds associated with each nephron: the glomerulus (within the renal corpuscle), the peritubular capillaries, and the vasa recta. The **glomerulus,** the first capillary bed, is the site of *filtration,* which is the movement of fluid (filtered plasma) from the blood across the filtration membrane into the capsular space of the renal corpuscle (figure 25.8*a,b*). Blood enters the glomerulus through an afferent arteriole and exits through an efferent

(continued on next page)

(continued from previous page)

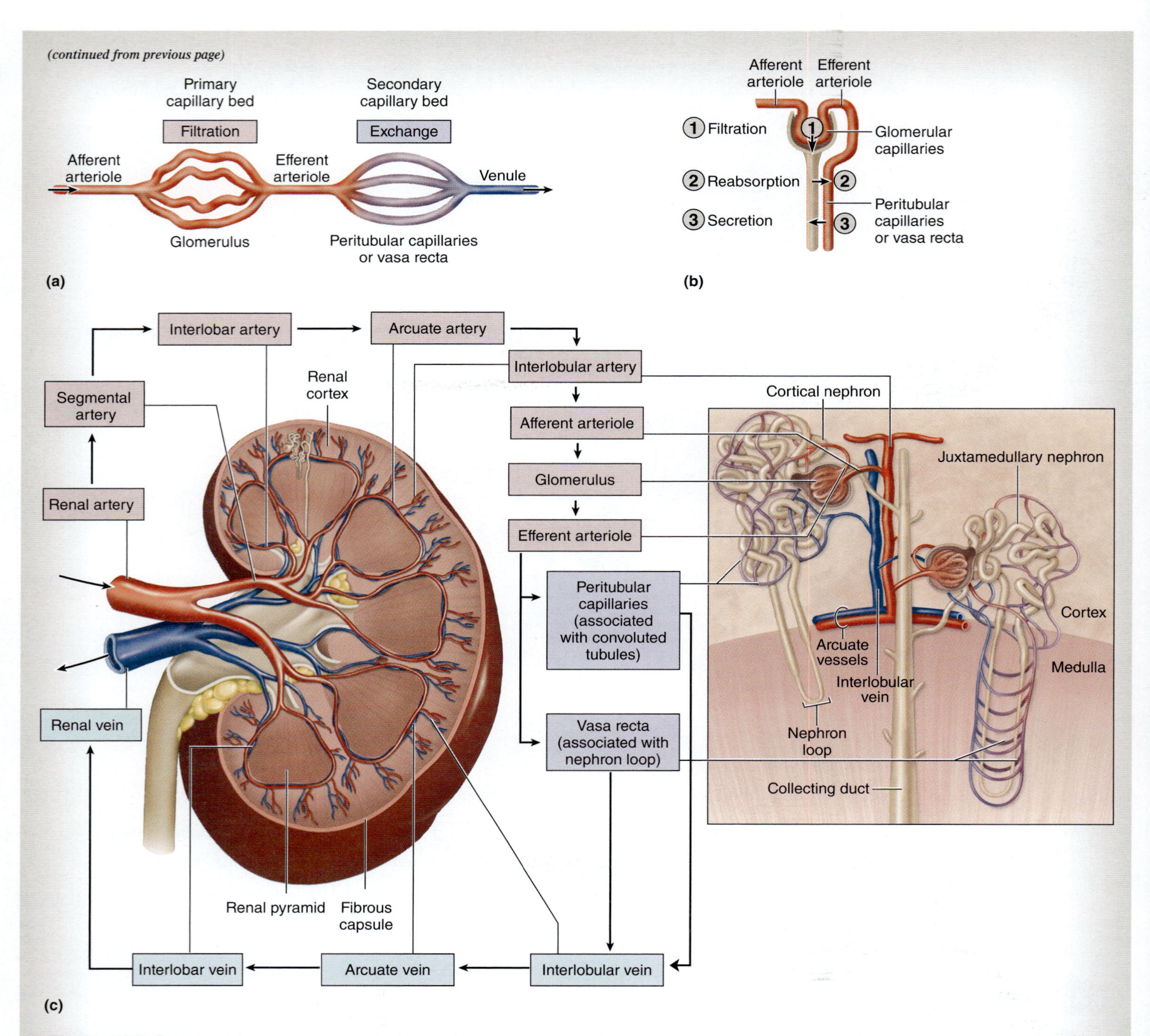

Figure 25.8 Blood Supply to the Kidney. (a) Blood flows through two capillary beds in the kidney. The first capillary bed, the glomerulus, is the site of filtration. The second capillary bed, peritubular capillaries in the cortex or vasa recta in the medulla, is the site of exchange (reabsorption or secretion). (b) The basic renal processes are filtration, reabsorption, and secretion. (c) Blood flow through the kidney.

arteriole. The efferent arteriole then leads into the second capillary bed, which is either the **peritubular capillaries** (within the cortex) or the **vasa recta** (within the medulla). The second capillary bed is the site of *exchange* of fluids, electrolytes, respiratory gases, and nutrients between the tubular portions of the nephron and the blood. Exchange in these capillaries involves both *reabsorption* (movement of substances from the tubular lumen into the blood) and *secretion* (movement of substances from the blood into the tubular lumen). Finally, blood leaving the peritubular capillaries and vasa recta drains into progressively larger veins that carry the blood out of the kidney to the inferior vena cava.

3. Identify the blood vessels listed in **figure 25.9** on a classroom model of the kidney, using the textbook as a guide. Then label the structures in figure 25.9. What parts of the nephron do the peritubular capillaries surround?

   ____

   What parts of the nephron do the vasa recta surround?

   ____

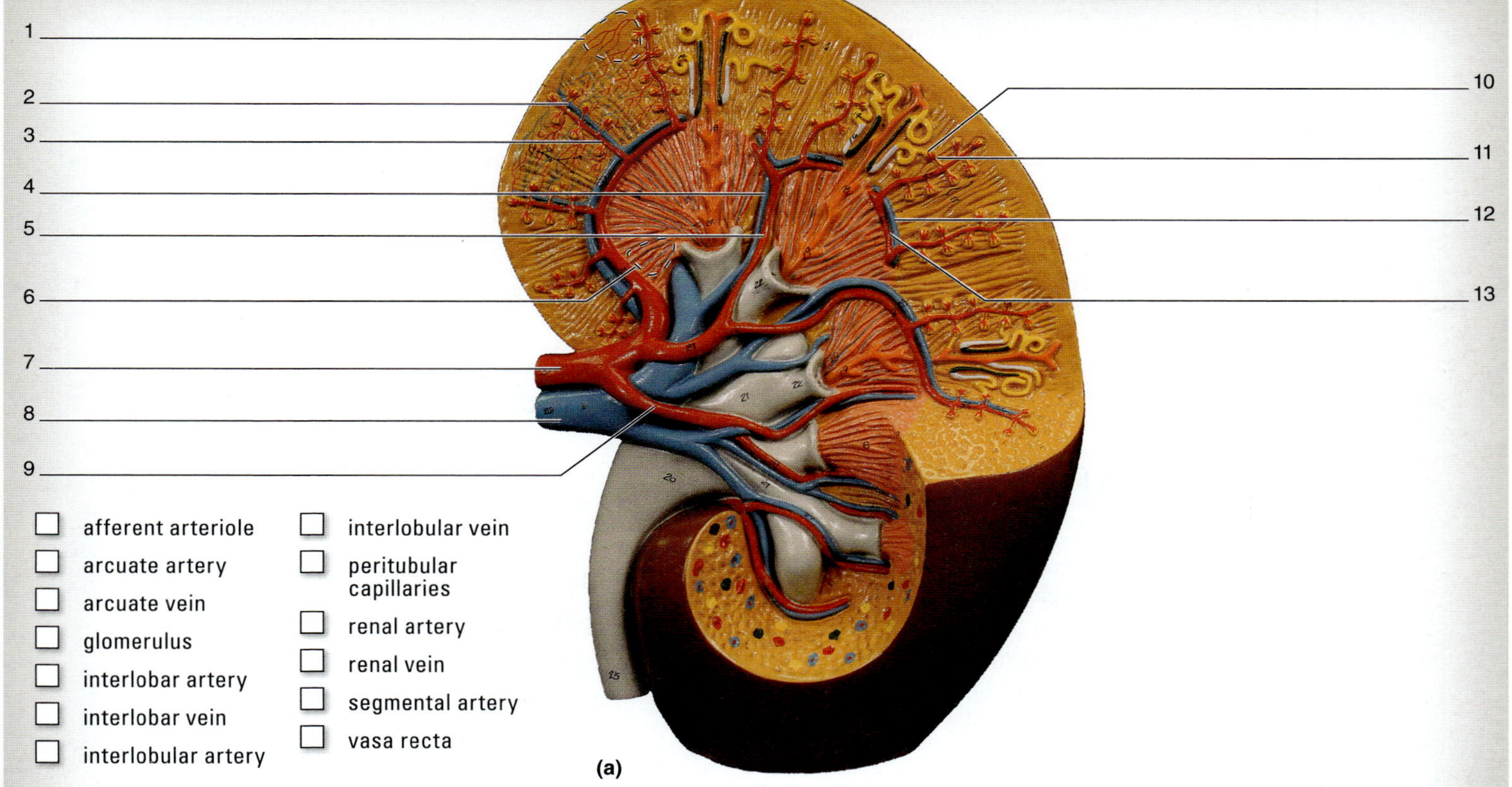

1
2
3
4
5
6
7
8
9
10
11
12
13

- afferent arteriole
- arcuate artery
- arcuate vein
- glomerulus
- interlobar artery
- interlobar vein
- interlobular artery
- interlobular vein
- peritubular capillaries
- renal artery
- renal vein
- segmental artery
- vasa recta

(a)

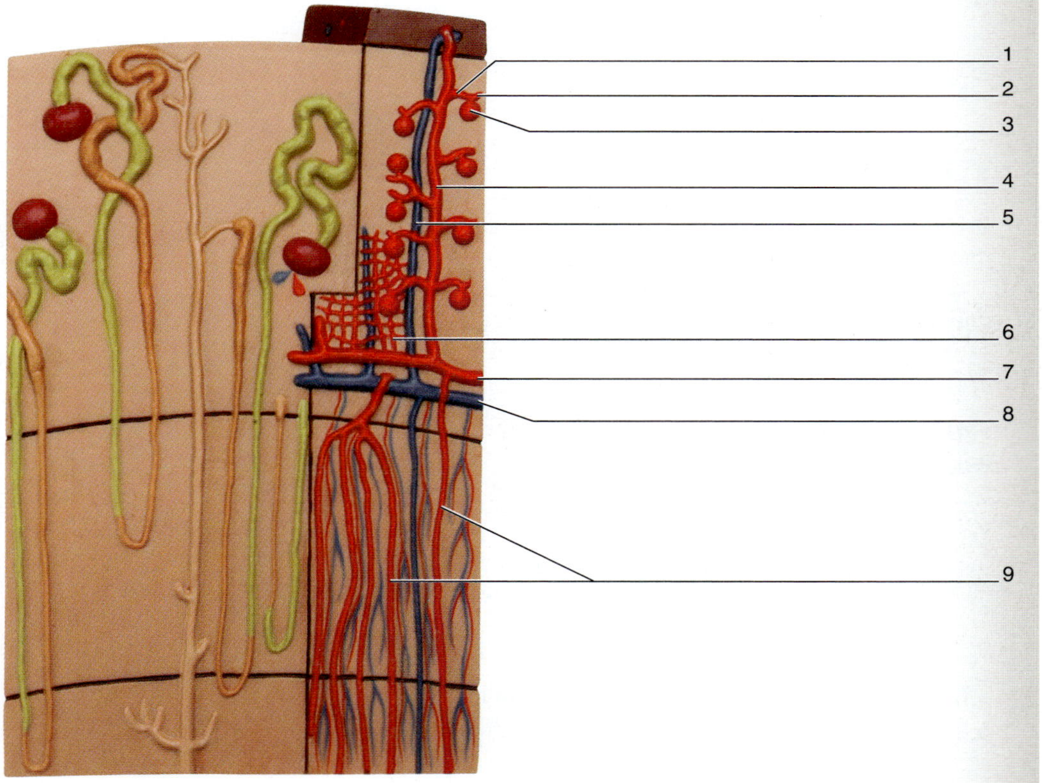

1
2
3
4
5
6
7
8
9

- afferent arteriole
- arcuate artery
- arcuate vein
- efferent arteriole
- glomerulus
- interlobular artery
- interlobular vein
- peritubular capillaries
- vasa recta

(b)

Figure 25.9 Models of the Kidney Demonstrating the Blood Supply to the Kidney. (a) Coronal section. (b) Close-up of the renal cortex and renal medulla. Use the terms listed to fill in the numbered labels in the figure.

(continued on next page)

(continued from previous page)

4. In the space below, list the vessels that blood moves through as it travels through the kidney. Start at the abdominal aorta and end at the inferior vena cava.

____

____

____

____

EXERCISE 25.7

URINE-DRAINING STRUCTURES WITHIN THE KIDNEY

Once filtrate has been formed by the nephrons, the fluid is processed as it passes through the nephron tubules (PCT, nephron loop, DCT), collecting tubules, and collecting ducts, and is called **tubular fluid.** Fluid leaving the collecting ducts and entering the papillary ducts is called **urine.** Urine will drain through the minor calyx, major calyx, and renal pelvis within the kidney before entering a ureter.

1. Observe a classroom model of a coronal section of a kidney, or a gross sample of a kidney that has been sectioned along a coronal plane.

2. Identify the following urine-draining structures of the kidney, using figures 25.7 and 25.9 and table 25.2 as guides:

- ☐ **collecting duct**
- ☐ **papillary duct**
- ☐ **major calyx**
- ☐ **minor calyx**
- ☐ **renal papilla**
- ☐ **renal pelvis**
- ☐ **renal pyramid**
- ☐ **ureter**

3. *Optional Activity:* **APR 13: Urinary System**—Watch the "Kidney—Gross Anatomy" and "Urine Formation" animations for an overview of kidney anatomy, vasculature, and function.

The Urinary Tract

The urinary tract consists of the ureters, urinary bladder, and urethra. Within the kidney, urine draining from the calyces begins to collect in the renal pelvis. Every 2 to 3 minutes, peristaltic contractions of the smooth muscle lining the ureters conveys the urine through the ureters, which transport it into the urinary bladder. Urine is then stored within the urinary bladder until a time when it is convenient to allow the urine to exit the body through the urethra. The process of voiding urine is called **micturition.** This section involves observing the gross anatomy of the structures composing the urinary tract.

EXERCISE 25.8

GROSS ANATOMY OF THE URETERS

The ureters are long, thin, epithelia-lined fibromuscular tubes that travel from the hilum of the kidney to the posterior, inferior surface of the urinary bladder **(figure 25.10).** Like the kidney, they are located retroperitoneally.

1. Observe a human cadaver or a classroom model of the abdomen.

2. Identify the structures listed in figure 25.10 on the cadaver or classroom model. Then label figure 25.10.

3. Follow the pathway of the ureter from its site of origin near the hilum of the kidney to its entry into the urinary bladder. What major muscle does the ureter cross over to get to the urinary bladder?

____

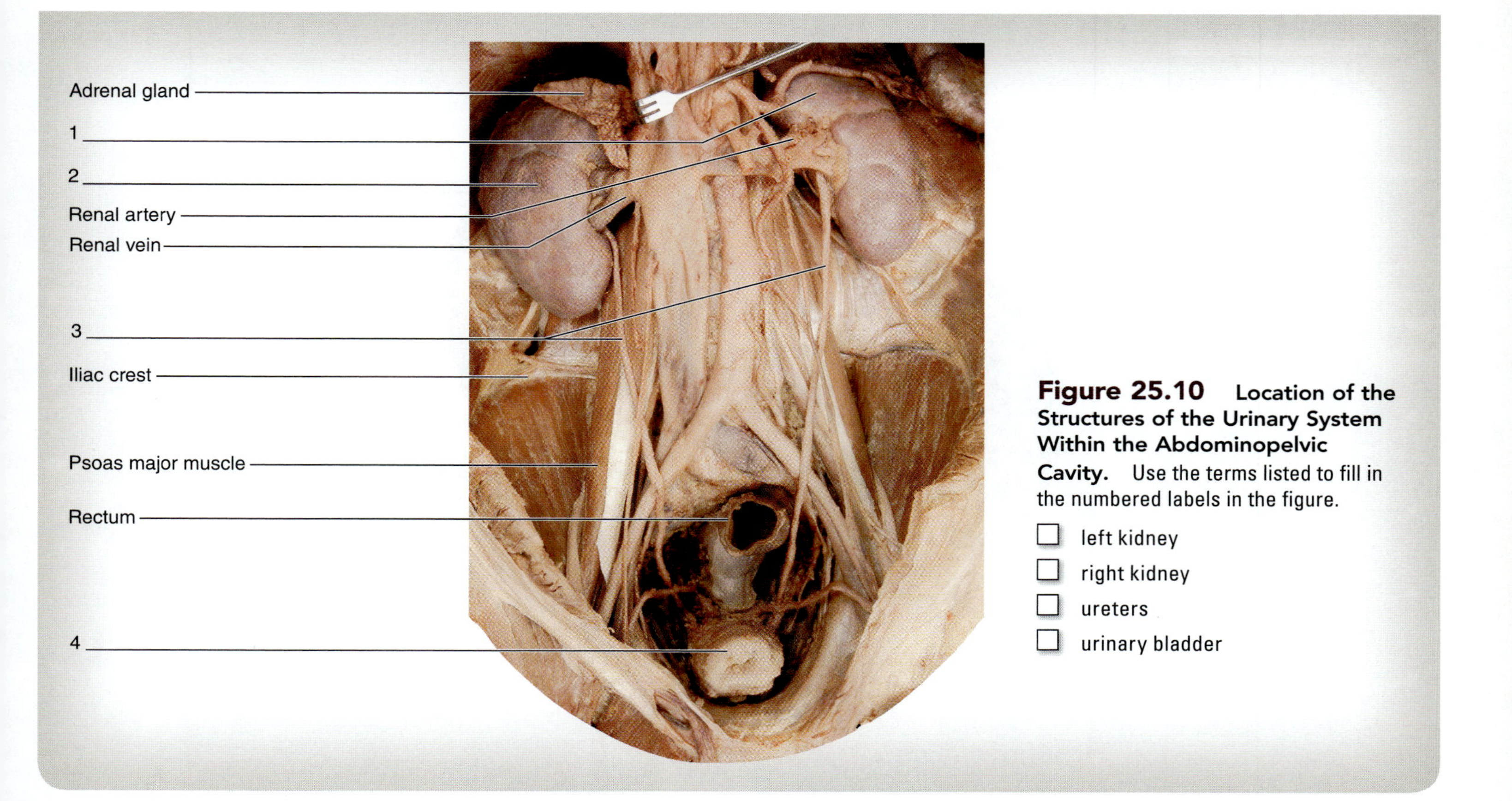

Figure 25.10 Location of the Structures of the Urinary System Within the Abdominopelvic Cavity. Use the terms listed to fill in the numbered labels in the figure.

- ☐ left kidney
- ☐ right kidney
- ☐ ureters
- ☐ urinary bladder

EXERCISE 25.9

GROSS ANATOMY OF THE URINARY BLADDER AND URETHRA

Like the kidneys and ureters, the **urinary bladder** is a retroperitoneal structure. The superior surface of the bladder is covered with peritoneum, but the rest is not. The urinary bladder lies in the pelvic cavity, just posterior to the pubic symphysis. The **urethra** is a passageway extending from the inferior surface of the urinary bladder to the external urethral orifice. It is much longer in males than in females because it extends through the length of the penis. (The urethra also serves as the passageway for sperm during ejaculation.) In both males and females there are both internal and external urethral sphincters. The **internal urethral sphincter** is composed of smooth muscle in the wall of the urinary bladder that encloses the entrance into the urethra, and is under involuntary control. The **external urethral sphincter** is composed of skeletal muscle of the **urogenital diaphragm,** and is under voluntary control.

1. Observe the abdominopelvic cavity of a male human cadaver or a classroom model of the abdominopelvic cavity of a male **(figure 25.11).** The **trigone** is the area in the floor of the urinary bladder (in both males and females) that is enclosed by imaginary lines that extend between the openings of the two ureters and the urethra. This area is clinically significant because urinary tract infections are common in this area. If the urinary bladder is cut open on the cadaver, or if there is a classroom model that demonstrates the interior of the urinary bladder, identify the urinary trigone in the floor of the urinary bladder.
2. Identify the structures listed in figure 25.11 on the cadaver or classroom model, using the textbook as a guide. Then label them in figure 25.11.
3. Sketch the structures forming the boundaries of the urinary trigone as seen from the interior of the urinary bladder in the space provided.

(continued on next page)

(continued from previous page)

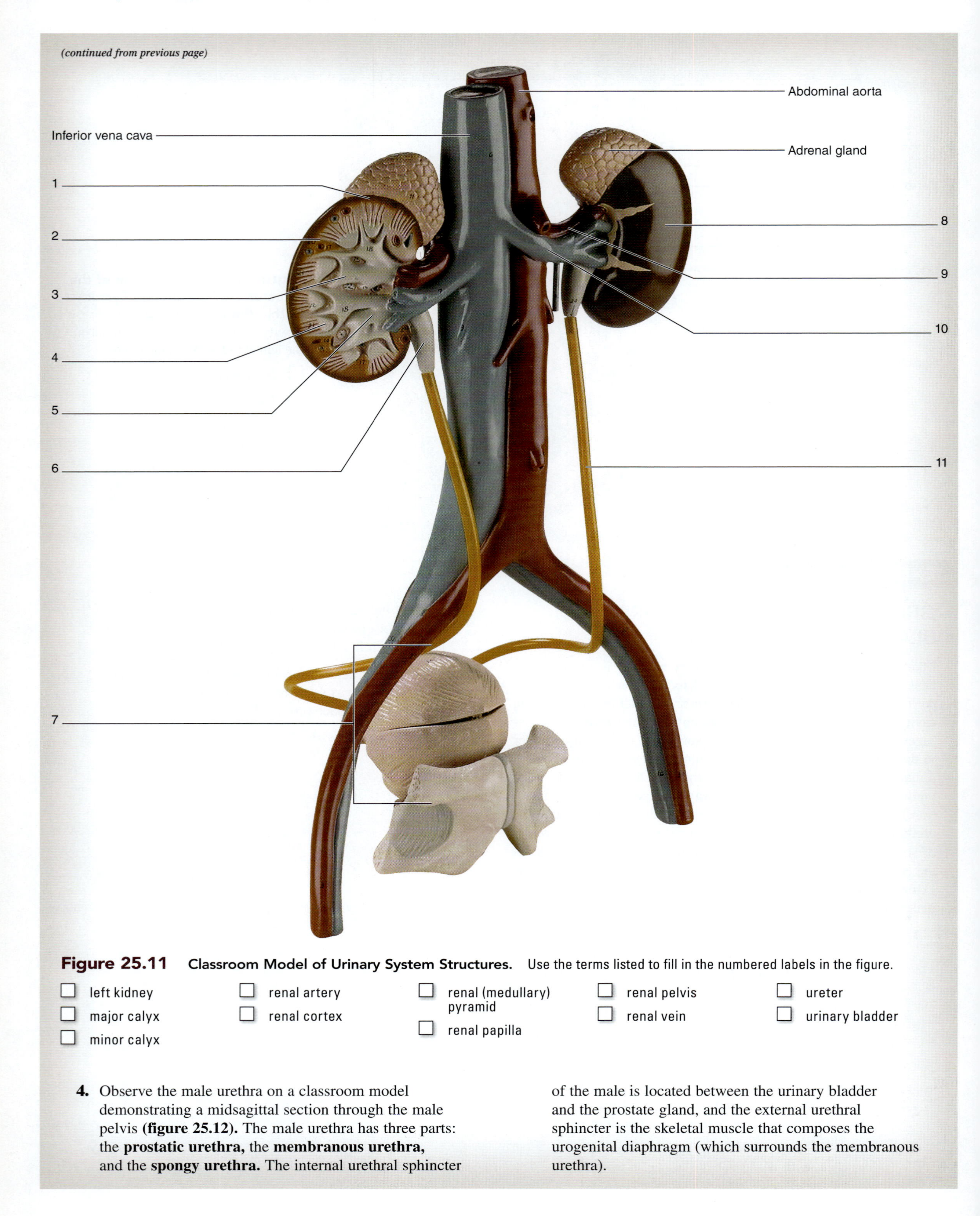

Figure 25.11 **Classroom Model of Urinary System Structures.** Use the terms listed to fill in the numbered labels in the figure.

- ☐ left kidney
- ☐ major calyx
- ☐ minor calyx
- ☐ renal artery
- ☐ renal cortex
- ☐ renal (medullary) pyramid
- ☐ renal papilla
- ☐ renal pelvis
- ☐ renal vein
- ☐ ureter
- ☐ urinary bladder

4. Observe the male urethra on a classroom model demonstrating a midsagittal section through the male pelvis **(figure 25.12).** The male urethra has three parts: the **prostatic urethra,** the **membranous urethra,** and the **spongy urethra.** The internal urethral sphincter of the male is located between the urinary bladder and the prostate gland, and the external urethral sphincter is the skeletal muscle that composes the urogenital diaphragm (which surrounds the membranous urethra).

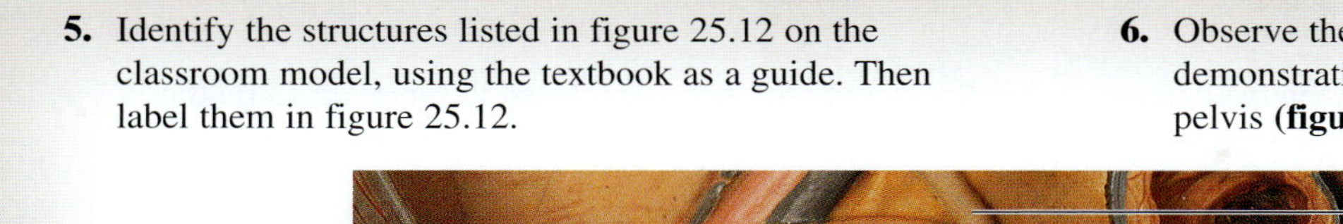

5. Identify the structures listed in figure 25.12 on the classroom model, using the textbook as a guide. Then label them in figure 25.12.

6. Observe the female urethra on a classroom model demonstrating a midsagittal section through the female pelvis **(figure 25.13).** The female urethra is short, and

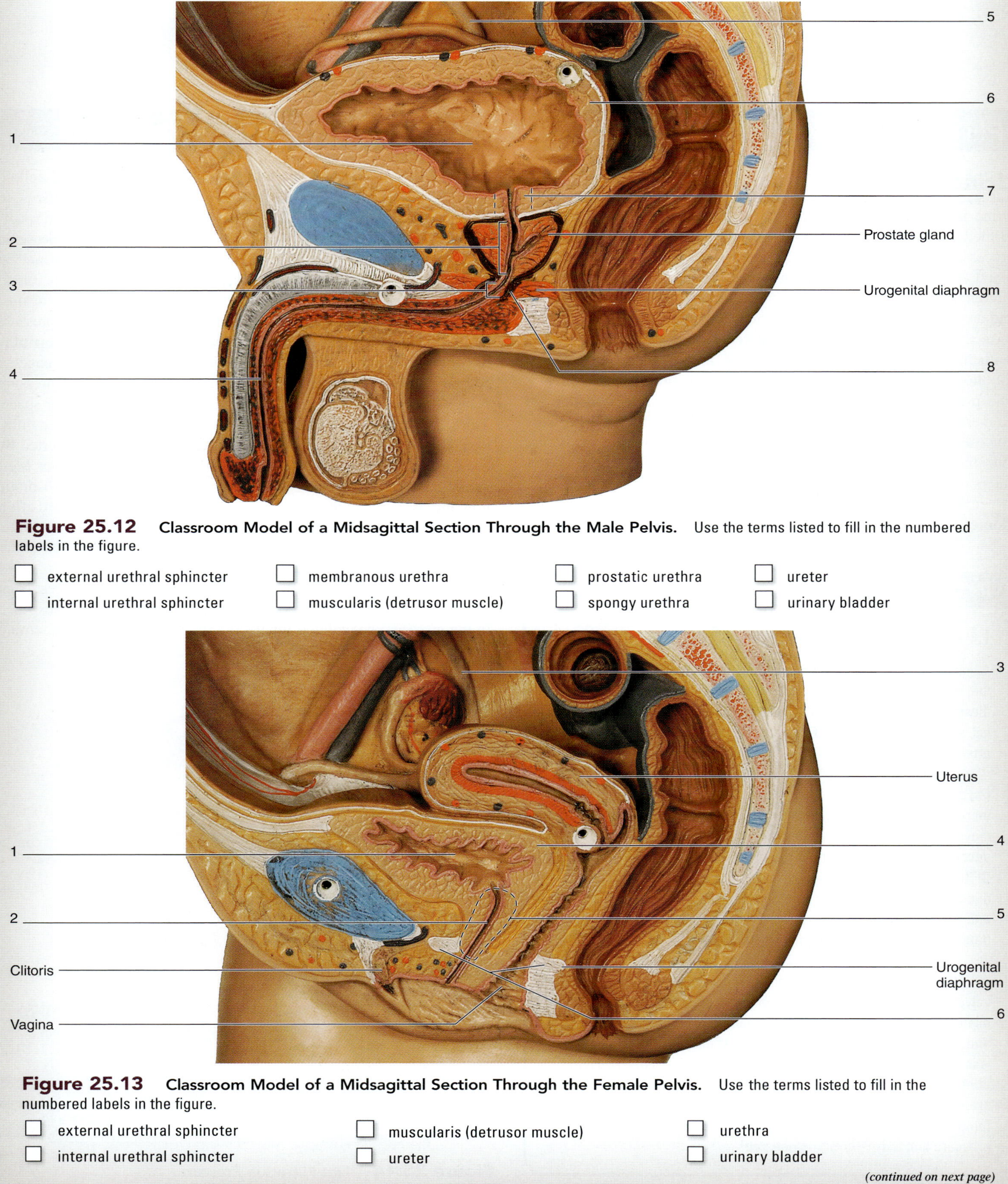

Figure 25.12 Classroom Model of a Midsagittal Section Through the Male Pelvis. Use the terms listed to fill in the numbered labels in the figure.

- ☐ external urethral sphincter
- ☐ internal urethral sphincter
- ☐ membranous urethra
- ☐ muscularis (detrusor muscle)
- ☐ prostatic urethra
- ☐ spongy urethra
- ☐ ureter
- ☐ urinary bladder

Figure 25.13 Classroom Model of a Midsagittal Section Through the Female Pelvis. Use the terms listed to fill in the numbered labels in the figure.

- ☐ external urethral sphincter
- ☐ internal urethral sphincter
- ☐ muscularis (detrusor muscle)
- ☐ ureter
- ☐ urethra
- ☐ urinary bladder

(continued on next page)

(continued from previous page)

the external urethral orifice is located between the clitoris and the vagina. The entire urethra is surrounded by the internal urethral sphincter. The external urethral sphincter is skeletal muscle of the urogenital diaphragm, just as it is in the male.

7. Identify the structures listed in figure 25.13 on the classroom model, using the textbook as a guide. Then label them in figure 25.13.

PHYSIOLOGY

Urine Formation

Urine is formed by three main physiologic processes that are performed by the nephron: filtration, reabsorption, and secretion. The first process, **filtration,** occurs in the renal corpuscle. Within the nephron, blood enters a specialized capillary bed called a **glomerulus** via the afferent arteriole. The glomerular capillaries are fenestrated capillaries, and are covered by specialized cells called **podocytes**. The podocytes form slits through which blood is filtered. Collectively, the fenestrated epithelial cells, their basement membrane, and the membrane-covered filtration slits formed by podocytes compose the **filtration membrane**. The process of filtration occurs when blood is transported into the glomerular capillaries. Small substances such as water, electrolytes, glucose, and amino acids pass freely across the filtration membrane into the capsular space (Bowman's space). Collectively the fluid that enters the surrounding capsular space is known as **filtrate.**

The rate at which filtration occurs in the glomerulus, the **glomerular filtration rate (GFR),** is determined by glomerular hydrostatic pressure, capsular hydrostatic pressure, and blood colloid osmotic pressure. **Hydrostatic pressure** *(hydro-,* water + *statikos-,* to make stand) is a force per unit area exerted by a fluid in a closed space. For instance, blood flowing through blood vessels exerts a hydrostatic pressure against the vessel wall. You measured "blood pressure" in chapter 22. The **glomerular hydrostatic pressure (HP_g)** is the blood pressure within the glomerulus, and is the pressure pushing against the glomerular capillary walls. Thus, glomerular hydrostatic pressure is the primary force that *promotes* filtration **(figure 25.14). Capsular hydrostatic pressure (HP_c)** is pressure caused by the presence of filtrate within the capsular space. Think of this as a fluid pressure pushing back against the the flow of fluid entering the capsular space. Thus, capsular hydrostatic pressure is a pressure that *opposes* filtration. **Blood colloid osmotic pressure (OP_g)** is a pressure exerted by proteins such as albumin within the blood plasma. These proteins exert osmotic pressure to "hold on" to water. Thus, blood colloid osmotic pressure is a force that also *opposes* filtration. The **net filtration pressure** is determined by subtracting both the capsular hydrostatic pressure and the blood colloid osmotic pressure from the glomerular hydrostatic pressure (see figure 25.14).

Tubular **reabsorption** involves transport of substances from the filtrate back into the blood. Substances are primarily reabsorbed in the proximal convoluted tubule and include water, glucose, amino

INTEGRATE

LEARNING STRATEGY

A unique feature of the glomerulus is that it is a specialized capillary bed that is both supplied *and* drained by arterioles. Constriction or dilation of these arterioles can greatly affect the fluid pressure within the glomerulus, and thus filtration. Specifically, constriction or dilation of the afferent arteriole directly influences GFR and filtration/urine output by the kidneys.

Diameter of the afferent arteriole is self-regulated by a myogenic (*myo,* muscle) response. As systemic blood pressure increases, it increases the stretch on the wall of the afferent arteriole (AA). Smooth muscle in the AA reflexively contracts, thereby constricting the AA and decreasing GFR. The result of the myogenic response is to maintain GFR over a wide range of systemic blood pressures so urinary output does not vary with changes in systemic blood pressure. The concepts and rules governing pressure, flow, and resistance, as described in detail in chapter 22, apply to this situation.

The afferent arteriole is not unlike an outdoor water spigot. When you open the spigot valve, water flows freely. When you close the valve, water merely trickles out. Similarly, when the afferent arteriole dilates, more blood flows into the glomerulus, which increases GFR and filtrate/urine production. When the afferent arteriole constricts, less blood flows into the glomerulus, which decreases GFR and filtrate/urine production. Remember that the blood flowing through the afferent arteriole is representative of the systemic blood pressure. Here is one fun quip to help you easily remember the relationship between blood pressure and GFR: "No BP, no pee pee!"

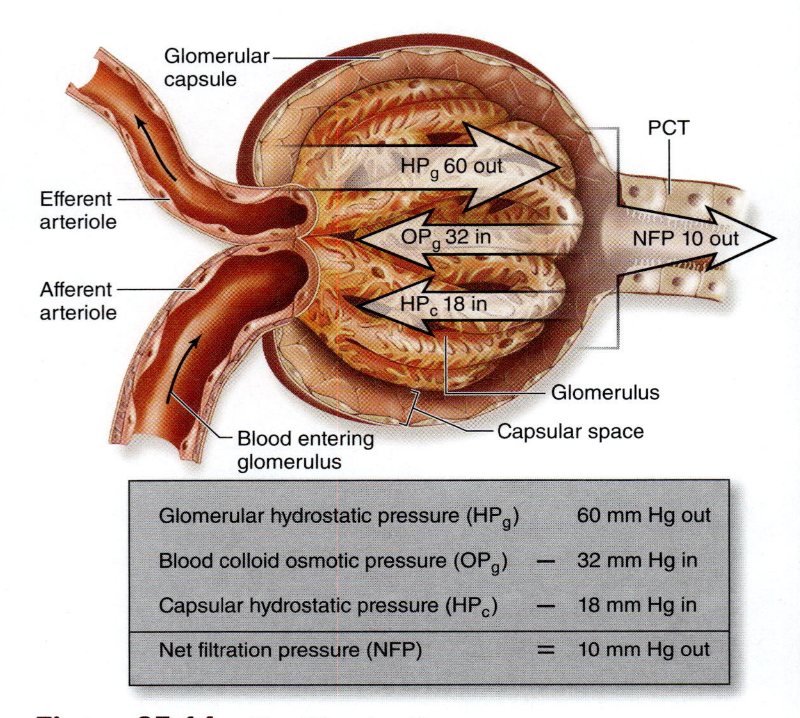

Glomerular hydrostatic pressure (HP_g)		60 mm Hg out
Blood colloid osmotic pressure (OP_g)	−	32 mm Hg in
Capsular hydrostatic pressure (HP_c)	−	18 mm Hg in
Net filtration pressure (NFP)	=	10 mm Hg out

Figure 25.14 Net Filtration Pressure.

INTEGRATE

CLINICAL VIEW

Glomerulonephritis

Glomerulonephritis involves inflammation of glomeruli. Inflammation increases the permeability of fenestrated capillaries that compose the glomerulus. This increases the likelihood that protein and erythrocytes pass through the filtration membrane and are excreted as an abnormal constituent of urine (see table 25.5). One significant, and often observable, sign of glomerulonephritis is edema (swelling). Edema occurs in patients suffering from glomerulonephritis because the loss of protein from the blood decreases the blood colloid osmotic pressure. The reduced blood colloid osmotic pressure allows more fluid to move from the blood into the interstitial spaces during capillary exchange. Glomerulonephritis may be acute or chronic, and may be primary (i.e., originating within the kidney) or secondary (i.e., resulting from infections or diseases such as lupus or diabetes).

INTEGRATE

CONCEPT CONNECTION

Net filtration pressure results from opposing pressures, primarily, glomerular, hydrostatic pressures and colloid osmotic pressure. This is the same mechanism that regulates movement of fluid (blood plasma) from systemic capillaries into the interstitial space on the arterial end of systemic capillaries, and from the interstitial space into post-capillary venules. This process is not 100% efficient, and results in a net loss of fluid from the blood into the interstitial spaces. Recall that this remaining fluid is generally picked up by lymphatic vessels and carried as lymph. Fluid movement from systemic capillaries into the interstitial spaces is driven by hydrostatic pressure within the capillaries. Therefore, hydrostatic pressure "pushes" the fluid into the interstitial space, thereby overcoming the colloid osmotic pressure within the capillaries that tends to want to "hold on" to the water within the blood plasma. The net effect is that fluid flows from the blood capillaries into the tissues. As fluid leaves the capillaries, larger solutes, including albumin (protein), remain in the vessel. On the venous side, hydrostatic pressure within the capillary is quite low because of the net loss of water to the interstitial spaces. However, the colloid osmotic pressure, mainly due to the presence of albumin in the capillaries, is quite high. Therefore, water is drawn back into the vessels via osmosis.

In the kidney, the balance of glomerular hydrostatic pressure and colloid osmotic pressures is impacted when there is significant damage to the filtration membrane of the nephron. That is, damage to the filtration membrane may allow larger molecules, including albumin, to enter the filtrate. Because there are no mechanisms for reabsorbing albumin within the nephron, any albumin that is filtered ends up excreted in the urine. In addition, albumin in the filtrate draws more water out of the glomerular capillaries, thus increasing water loss through the kidneys.

The loss of albumin from the blood causes a decrease in the colloid osmotic pressure of systemic capillaries as well. Without the normal ability to "hold on" to water, more water will leave systemic capillaries and enter the interstitial spaces. This results in fluid accumulation in the interstitial space surrounding systemic capillary beds, called **edema.**

acids, and electrolytes. **Table 25.3** provides a quick reference of the parts of the nephron, the substances reabsorbed in each part, and the mechanism of reabsorption for each substance.

In comparison, tubular **secretion** involves transport of substances from the blood into the filtrate. Numerous substances are secreted including hydrogen ions, drugs, toxins, and ammonia. **Table 25.4** provides a quick reference for the urinary structures, substances secreted, and the site of secretion for each substance.

Once the blood has been filtered, and "processed" through tubular reabsorption and tubular secretion, the remaining fluid (now called urine) is drained from collecting ducts into the cayces and renal pelvis of each kidney. Urine then drains through the ureters into the urinary bladder where it is stored until it is eliminated from the body. Typically, the composition of urine is slightly acidic, with a pH ranging from 4.5 to 8.0. Given the processes outlined previously, it makes sense that anything that may alter the blood pH might also alter urine pH. For instance, a diet high in protein and low in carbohydrates results in

Table 25.3 Reabsorption in the Nephron

Structure	Substance Reabsorbed	Mechanism of Reabsorption (Basolateral Membrane)
Proximal Convoluted Tubule	Amino Acids	Facilitated diffusion
	Glucose	
	K^+	Paracellular transport
	Na^+	Active transport
	Water	Osmosis
Descending Limb of Nephron Loop	Water	Osmosis
Ascending Limb of Nephron Loop	Cl^-	Facilitated diffusion
	K^+	
	Na^+	Active transport
Distal Convoluted Tubule	Cl^-	Facilitated diffusion
	K^+	
	Na^+	Active transport
	Water (regulated by ADH)	Osmosis
Collecting Duct	Water (regulated by ADH)	Osmosis + transport through proteins called aquaporins

Table 25.4 Secretion in the Nephron

Structure	Substance Secreted	Mechanism of Secretion
Proximal Convoluted Tubule	H^+	Antiport (with Na^+)
	Drugs and toxins	Active transport
	Ammonia	Simple diffusion
Distal Convoluted Tubule	K^+	Active transport
	H^+	Antiport (with Na^+)

INTEGRATE

CLINICAL VIEW

Diabetes

Most people associate *diabetes* (*diabetes*, siphon) with either type I or type II diabetes *mellitus* (*mellitus*, sweet). However, there are two other forms of diabetes as well: gestational diabetes and diabetes *insipidus*. The common link for all forms of diabetes is the production of copious amounts of diluted urine **(polyuria)**. While the etiology of each subtype of diabetes is different, the outcome is the same. These patients are often extremely thirsty with a high urine output. As previously described, diabetes mellitus, whether type I or type II, results from an inability to regulate blood glucose levels. When blood levels of glucose are high, urine production may also be high because water is attracted to the glucose (and sodium) in the filtrate, leading to greater volumes of urine.

During pregnancy, some women may experience high blood glucose levels, a condition called gestational diabetes. To test for gestational diabetes, it is standard practice to give a pregnant woman a glucose tolerance test between 24 and 28 weeks gestation. With this test, the woman drinks a solution with known (high) concentrations of glucose, and her urine is collected 1 hour later. The urine is then tested for the presence of glucose. Those women with elevated levels of glucose in their urine must return for a more prolonged glucose tolerance test that requires collection of both urine and blood samples.

Diabetes *insipidus* is a disease characterized by an inability to produce antidiuretic hormone (ADH), a hormone that targets kidney tubule cells to promote water reabsorption (through aquaporins—*aqua-*, water + *-poros*, passage) in the distal convoluted tubule and collecting ducts. Failure to release ADH means that water reabsorption is dramatically reduced. The excess water is excreted as urine.

utilization of fats and proteins for energy. The products of fat metabolism may result in the production of ketone bodies, which are acidic and therefore lower the pH of the blood. The increased concentration of hydrogen ions in the blood results in an increased rate of secretion of hydrogen ions into the urine, thereby decreasing the pH of the urine.

Specific gravity is a measure of the density of a substance (g/mL) when compared to the density of water 1.000 g/mL. The specific gravity of urine typically ranges from 1.003 to 1.035. Concentrated (darker) urine generally has a higher specific gravity than does dilute (lighter) urine. Substances that are normally found in the urine include salts (Na^+, Cl^-, K^+, Mg^{2+}, Ca^{2+}, SO_4^{2-}, $H_2PO_4^-$, and NH_4^+), nitrogenous wastes (urea), and toxins. However, substances such as glucose, albumin, red blood cells, white blood cells, or abnormally high levels of the above substances may be cause for concern. For instance, glucose, which is typically reabsorbed via facilitated diffusion in the PCT, may appear in the urine when the carrier molecules are completely saturated. This is often a sign of uncontrolled diabetes mellitus. When larger molecules, such as albumin, red blood cells, or white blood cells appear in the urine, this may be a sign of damage to the glomerular capillaries; such is often the case in chronic hypertension or traumatic injury. The presence of white blood cells may also be indicative of inflammation of the urinary tract. **Table 25.5** identifies substances that are not normally detected in urine, the clinical term associated with each pathology, and possible causes for each condition.

Table 25.5 Abnormal Urine Constituents

Abnormal Urine Constituent	Clinical Term	Possible Cause
Bile Pigment	Bilirubinuria	Hepatitis or cirrhosis
Erythrocytes (RBCs)	Hematuria	Increased glomerular permeability (hypertension, trauma, toxins)
Glucose	Glucosuria or Glycosuria	Diabetes mellitus
Hemoglobin (Hb)	Hemoglobinuria	Increased glomerular permeability (hypertension, trauma, toxins)
Ketone Bodies	Ketonuria	Diabetes mellitus; diet (low carbohydrate)
Leukocytes (WBCs)	Pyuria	Urinary tract infection
Nitrites	Nitrituria	Urinary tract infection
Protein	Proteinuria	Increased glomerular permeability (glomerulonephritis, hypertension, trauma, toxins)

EXERCISE 25.10

URINALYSIS

CAUTION:

The following laboratory exercises may involve the use of human urine samples. These are considered biohazard wastes and should be handled with care. Be sure to take precautionary measures while handling urine samples by wearing nitrile gloves and safety goggles. Careful disposal of biohazard wastes is required. Please consult with the laboratory instructor for further instructions on laboratory policies and procedures for handling and disposing of these substances.

These laboratory exercises involve analyzing urine samples. The following steps may be completed using your own urine and/or artificial urine. The urine will be analyzed for physical characteristics and tested for abnormal constituents. There is also an optional exercise (25.10B) that involves measuring specific gravity of urine using a urinometer.

EXERCISE 25.10A Urinalysis—Testing Characteristics and Constituents of Urine

1. Obtain a urine sample. If collecting your own urine, be sure to wash your hands prior to collecting the sample. Observe all laboratory safety procedures when handling

urine samples. Obtain a clean, transparent collection cup. Be sure to collect a midstream sample (at least 50 mL) to avoid possible contamination by bacteria contained within the urethra. Refrigerate any unused urine immediately.

2. Observe the **odor** of the urine sample. Note if the urine smells like acetone or ammonia, fruity or sweet, or particularly pungent. To do this, gently wave your hand over the cup toward your nose. Record your observations in **table 25.6.** Compare your observations with the description of normal odor presented in the table. Note that a normal odor may be aromatic but not unpleasant.
3. Observe the **color** of the urine sample. Record your observations in table 25.6, and compare your observations with the description of normal color presented in the table. Note that a normal color is yellow, ranging from light to dark. Abnormal colors may include yellow-brown, due to bile pigments, or red to brown, due to the presence of blood.
4. Observe the **transparency** of the urine. Typically, urine is clear. A slightly cloudy or very cloudy sample may indicate the presence of substances such as bacteria or mucus. Record your observations in table 25.6.
5. Test the following urine properties or constituents using individual test strips or combination test strips (such as Chemstrip® or Multistix®): pH, specific gravity, glucose, protein, ketones, bilirubin, leukocytes, and hemoglobin. Submerge a dipstick into the urine as directed on the strip container. Remove excess liquid on the strip by touching the strip to the inside rim of the urine container. Remove the strip and wait for the recommended time. Compare the color strip to the available color scale, as directed on the container. Record your observations in table 25.6.
6. Repeat steps 2–5 for an "unknown" sample provided by the instructor.
7. Make note of any pertinent observations here:

____

____

____

Table 25.6 Data for Urinalysis Using Test Strips

Characteristic	Urine Sample 1	Urine Sample 2	Normal
Odor (e.g., acetone, ammonia, fruity, sweet, or pungent)			Aromatic (not unpleasant)
Color			Yellow (light, medium, or dark)
Transparency			Clear
pH			4.5–8.0
Specific Gravity (Exercise 25.10A or 25.10B)			1.003–1.035
Glucose			Negative
Protein			Trace
Ketones			Negative
Bilirubin			Negative
Leukocytes			Negative
Hemoglobin			Negative

(continued on next page)

(continued from previous page)

EXERCISE 25.10B Urinalysis—Specific Gravity

1. Obtain a urine sample. If collecting your own urine, be sure to wash your hands prior to collecting the sample. Observe all laboratory safety procedures when handling urine samples. Obtain a clean, transparent collection cup. Be sure to collect a midstream sample (at least 50 mL) to avoid possible contamination by bacteria contained within the urethra. Refrigerate any unused urine immediately.
2. Pour the urine into a urinometer, filling it approximately two-thirds full.
3. Lower the specific gravity tube into the urinometer glass vial. Note that the specific gravity tube must float. If it does not float, continue to add urine until the urinometer glass vial floats.
4. Read the specific gravity by aligning the meniscus of the urine with the readings on the floating specific gravity tube **(figure 25.15).** Record the measurement in table 25.6 and compare this to normal values presented in the table.
5. Repeat steps 2–4 for an "unknown" urine sample provided by the instructor.
6. Make note of any pertinent observations here:

____

____

____

Figure 25.15 Measuring Specific Gravity of Urine Using a Urinometer. Fill the glass tube with enough urine to make the specific gravity tube float. Then, observe the meniscus of the urine with the marks on the specific gravity tube to determine the specific gravity.

Acid-Base Balance

Maintaining proper blood pH (7.35–7.45) is a critical homeostatic process that involves both short- and long-term regulation and employs multiple physiologic processes. Because blood pH constantly changes, the first means of helping to prevent pH changes are the buffer systems. The most robust of the chemical buffers is the **carbonic acid** buffering system within the blood, which is described in detail in chapter 24. Additional chemical buffers in the blood include proteins and phosphates, which act in a similar manner to rapidly buffer against pH changes.

The respiratory system functions as a physiologic buffering system. The respiratory system is tightly linked to the carbonic acid buffering system. As discussed in chapter 24, respiratory rates directly impact the amount of carbonic acid in the blood, thereby impacting blood pH. Accumulation of carbon dioxide, whether due to **hypoventilation** or resulting from reduced gas diffusion, leads to the increased formation of carbonic acid and the possible accumulation of hydrogen ions in the blood. This results in a drop in blood pH, or **respiratory acidosis**. Conversely, **hyperventilation** results in low levels of carbon dioxide (blowing off too much carbon dioxide), which may lead to **respiratory alkalosis**. When the source of acid does not involve changes in carbon dioxide levels due to alterations in the respiratory system, it is considered **metabolic**. Clinically, reviewing blood pH as well as carbon dioxide and bicarbonate levels reveals whether or not the patient is in acidosis (pH < 7.35) or in alkalosis (pH > 7.45), and if the source is respiratory (normal P_{CO_2} = 35–45 mm Hg) or metabolic (normal bicarbonate = 22–26 mEq/L blood). **Table 25.7** provides a quick reference guide for classifiying respiratory and metabolic acidosis and alkalosis.

The urinary system also functions as a physiologic buffering system. For example, excessive fat metabolism and the loss of bicarbonate due to excessive diarrhea both significantly lower the blood pH. As the blood becomes more acidic, the kidneys secrete excess hydrogen ions and reabsorb more bicarbonate ions in the distal convoluted tubules and collecting tubules, thereby lowering urine pH. Conversely, as the blood becomes more basic, the rate of hydrogen secretion and bicarbonate reabsorption in the kidneys decreases.

Table 25.7 Classification of Respiratory versus Metabolic Acidosis/Alkalosis

Condition		P_{CO_2}	Bicarbonate
Respiratory	Acidosis	> 45 mm Hg	22–26 mEq/L blood (normal range)
	Alkalosis	< 35 mm Hg	
Metabolic	Acidosis	35–45 mm Hg (normal range)	< 22 mEq/L blood
	Alkalosis		> 26 mEq/L blood

EXERCISE 25.11

A CLINICAL CASE STUDY IN ACID-BASE BALANCE

The purpose of this exercise is to apply the concept of acid-base balance to a real scenario, or clinical case. This exercise requires information presented in this chapter, as well as previous chapters. Integrating multiple systems in a case study reinforces a concept that is so important in physiology: homeostasis.

Prior to beginning, state how both the respiratory system and the urinary system function to maintain homeostatic pH levels in the blood.

____

____

____

____

Consult with the laboratory instructor as to whether this exercise will be completed individually, in pairs, or in groups. Review the topics of diffusion and osmosis and the cardiovascular, lymphatic, respiratory, digestive, and urinary systems in the textbook and in previous chapters in this laboratory manual for assistance with this case study.

Case Study

Robert, now 60 years old, was diagnosed with type I diabetes mellitus at age 10 and has managed the disease with few complications by administering daily insulin injections. Fifteen years ago, Robert complained of extreme fatigue and his health began to decline. Within a few years, Robert was diagnosed with a failing heart, chronic hypertension (high blood pressure), and peripheral artery disease (PAD), in which the peripheral vessels become narrowed or blocked.

Just two months ago, Robert exhibited signs of intermittent confusion followed by bouts of diarrhea and intense feelings of weakness. He was unable to regulate blood glucose levels, and he began experiencing respiratory symptoms. Eventually, his breathing became labored and difficult, particularly in the evenings as he would lie down to sleep. Sometimes, these symptoms would wake him up, sometimes after only 1 or 2 hours of sleep. Robert went to the hospital for evaluation.

Upon examination, doctors heard crackles (see chapter 24) bilaterally in his lungs. Doctors ordered a chest radiograph and a series of laboratory tests (see **table 25.8,** 1st Admission). The results of these tests indicated that Robert was suffering from pneumonitis, or inflammation of the lung tissue. Robert was given a diuretic, which prevented the reabsorption of sodium in kidney tubule cells, and a broad-spectrum antibiotic to fight the infection. He was released from the hospital just two days later.

Robert's condition worsened during the following weeks. The frequency of diarrhea increased, and he was lethargic (lacking energy). Robert had used up his supply of insulin injections when his sister found him unresponsive in his home. Emergency technicians arrived on the scene and transported him to the hospital. They recorded a blood pressure of 90/50 mm Hg (reference = 110/70 mm Hg), a pulse of 90 beats per minute, and oxygen saturation of 90%. Soon, Robert's systolic blood pressure dropped to 70 mm Hg. Robert exhibited poor inspiratory effort, and crackles were heard bilaterally.

Robert was transported to the coronary care unit for further evaluation. Doctors performed an echocardiogram, which revealed some insufficiencies in the atrioventricular and semilunar valves, narrowing of the aorta, and hypertrophy of the left ventricle. Computed tomography (CT) of the chest revealed possible infectious agents in the right upper lobe of the lungs. Doctors performed a series of laboratory tests (see table 25.8, 2nd Admission). Robert's respiratory distress and mental confusion persisted, and he experienced several more bouts of hypotension (low blood pressure). A lumbar puncture revealed elevated glucose and protein levels in the cerebrospinal fluid. Pneumonitis and hypotension persisted, and Robert was found unresponsive just days later. Efforts to resuscitate Robert were unsuccessful.

Answer the following questions regarding Robert's case.

1. Describe how Robert's narrowed aorta might have contributed to chronic respiratory problems, such as pulmonary edema. ______________________

   ____

   ____

   ____

2. Based on Robert's chest radiograph results, describe how gas exchange across the respiratory membrane might have been affected by his chronic respiratory problems.

   ____

   ____

   ____

   ____

3. Justify Robert's blood and urine glucose levels (1st admission) based on the details of his case. Based on this data, estimate the volume of Robert's urinary output at that time. ______________________

   ____

   ____

   ____

4. Discuss what Robert's glomerular filtration rate revealed about his systemic blood pressure. ______________________

   ____

   ____

   ____

5. Describe how continued diarrhea might have influenced Robert's blood pH. What compensatory mechanisms could have been initiated? ______________________

   ____

   ____

   ____

(continued on next page)

(continued from previous page)

6. Describe how chronic respiratory problems might have influenced Robert's blood pH. What compensatory mechanisms could have been initiated? ______________

____

____

____

7. Based on the data provided, did Robert suffer from respiratory or metabolic acidosis or alkalosis? Justify your answer. ______________

____

____

____

Table 25.8 Patient Data for Acid-Base Case Study

Test	Variable	Reference Range	1st Admission	2nd Admission
Arterial Blood Gases	P_{CO_2}	35–45 mm Hg	50	65
	pH	7.35–7.45	7.2	7.1
	P_{O_2}	80–100 mm Hg	120	160
Blood	Bicarbonate (mEq/L)	22–26	25	25
	Chloride (mmol/liter)	100–108	90	105
	Glomerular filtration rate (mL/min)	> 60	20	10
	Glucose (mg/dl)	70–110	680	170
	Hematocrit (%)	42.0–52.0 (men)	35	30
	Hemoglobin (g/dl)	13–18 (men)	10.5	8.5
	Leukocytes (per mm^3)	5000–10,000	9000	10,000
	Potassium (mmol/liter)	3.4–4.8	6	5
	Sodium (mmol/liter)	135–145	130	135
Urine	Albumin (mg/dl)	Negative	1+	2+
	Bacteria	Negative	Moderate	Many
	Color	Yellow	Yellow	Other
	Erythrocytes	0–2	10–25	10–25
	Glucose (mg/dl)	Negative	1000	Negative
	Ketones (mg/dl)	Negative	3+	1+
	Leukocytes	0–2	250	Packed
	pH	4.5–9.0	5	4.5
	Specific gravity	1.003–1.035	1.02	> 1.035

Chapter 25: The Urinary System

Name: ____________________

Date: __________ Section: __________

POST-LABORATORY WORKSHEET

The 1 corresponds to the Learning Objective(s) listed in the chapter opener outline.

Do You Know the Basics?

Exercise 25.1: Histology of the Renal Cortex

1. Which of the following are components of the nephron? (Check all that apply.) 1

    _____ a. collecting duct

    _____ b. glomerulus

    _____ c. distal convoluted tubule

    _____ d. nephron loop

    _____ e. proximal convoluted tubule

2. Fill in the table below with the type of epithelium lining each structure and a brief description of the function of each structure. 1 2

Structure	Epithelium	Function
Glomerulus		
Visceral Layer of Glomerular Capsule		
Parietal Layer of Glomerular Capsule		
PCT		
DCT		

3. The renal corpuscle consists of the glomerular capsule and glomerular capillaries. ______________ (True/False) 3

4. Filtration slits in the filtration membrane are formed by which of the following structures? (Circle one.) 4

    a. basement membrane of glomerular capillaries

    b. endothelium of glomerular capillaries

    c. podocytes of the glomerular capsule

5. Which of the following structures are located in the renal cortex? (Check all that apply.) 5

    _____ a. collecting ducts

    _____ b. distal convoluted tubule

    _____ c. nephron loop

    _____ d. proximal convoluted tubule

    _____ e. renal corpuscle

Exercise 25.2: Histology of the Renal Medulla

6. Which of the following structures are located in the renal medulla? (Check all that apply.) 6

    _____ a. collecting ducts

    _____ b. distal convoluted tubule

    _____ c. nephron loop

    _____ d. proximal convoluted tubule

    _____ e. renal corpuscle

7. The thick limbs of the nephron loop are composed of simple ______________ (cuboidal/squamous) epithelium, whereas the thin limbs of the nephron loop are composed of simple ______________ (cuboidal/squamous) epithelium. 7

Exercise 25.3: Histology of the Ureters

8. Place the layers of the wall of the ureter in order, from innermost to outermost. **8**

   ______ a. adventitia

   ______ b. mucosa

   ______ c. muscularis

9. The muscularis layer of the ureter consists of an inner ____________________ (circular/longitudinal) layer of smooth muscle and an outer ____________________ (circular/longitudinal) layer of smooth muscle. **9**

Exercise 25.4: Histology of the Urinary Bladder

10. The urinary bladder has ____________________ (three/four) layers in its wall. **10**

11. The urinary bladder is lined with which type of epithelial tissue? (Circle one.) **11**

    a. simple cuboidal epithelium
    b. simple squamous epithelium
    c. stratified squamous epithelium
    d. transitional epithelium

Exercise 25.5: Gross Anatomy of the Kidney

12. The layer of dense irregular connective tissue that surrounds the entire kidney is called the renal corpuscle. ____________________ (True/False) **12**

13. The ____________________ (right/left) kidney is located more superiorly than the other. **13**

14. Place the following structures in order from innermost to outermost. **14**

    ______ a. paranephric fat

    ______ b. perinephric rat

    ______ c. renal capsule

    ______ d. renal fascia

15. Retroperitoneal structures are located behind the peritoneum. ____________________ (True/False) **15**

Exercise 25.6: Blood Supply to the Kidney

16. The renal ____________________ (artery/vein) supplies oxygen-rich blood to each kidney, whereas the renal ____________________ (artery/vein) drains blood from each kidney to the inferior vena cava. **16**

17. Place the following vessels in the correct order, from the entry into the kidney via the renal artery to the glomerulus. **17**

    ______ a. afferent arteriole

    ______ b. arcuate artery

    ______ c. interlobar artery

    ______ d. interlobular artery

    ______ e. segmental artery

Exercise 25.7: Urine-Draining Structures Within the Kidney

18. Fluid that passes through nephron tubules, collecting tubules, and collecting ducts is called urine. ____________________ (True/False) **18**

19. Place the following structures in the correct order of urine flow. **19**

    ______ a. major calyx

    ______ b. minor calyx

    ______ c. renal papilla

    ______ d. renal pelvis

Exercise 25.8: Gross Anatomy of the Ureters

20. Place the following structures in the correct order of urine flow from the kidney to the exterior of the body. 20

_______ a. ureter

_______ b. urethra

_______ c. urinary bladder

21. The urinary bladder is located in the ____________________ (abdominal/pelvic) cavity. 21

Exercise 25.9: Gross Anatomy of the Urinary Bladder and Urethra

22. Identify the three structures that form the boundaries of the urinary trigone. (Check three.) 22

_______ a. right renal artery

_______ b. left renal artery

_______ c. right ureter

_______ d. left ureter

_______ e. urethra

23. Place the three sections of the male urethra in the correct order from proximal to distal. 23

_______ a. prostatic urethra

_______ b. membranous urethra

_______ c. spongy urethra

24. The internal urethral sphincter is composed of ____________________ (skeletal/smooth) muscle, whereas the external urethral sphincter is composed of ____________________ (skeletal/smooth) muscle. 24

Exercise 25.10: Urinalysis

25. Which of the following is the site of filtration in the nephron? (Circle one.) 25

a. distal convoluted tubule
b. glomerulus
c. nephron loop
d. proximal convoluted tubule

26. Which of the following is an abnormal constituent of urine? (Circle one.) 26

a. glucose
b. sodium ions
c. trace amounts of protein
d. urea
e. water

27. Urine should normally test negative for the presence of glucose. ____________________ (True/False) 27

28. A greater concentration of substances in urine (e.g., glucose or protein) would cause the urine to have a ____________________ (higher/lower) specific gravity than water. 28

Exercise 25.11: A Clinical Case Study in Acid-Base Balance

29. The urinary system aids in ____________________ (short-/long-) term regulation of acid-base balance. 29

30. Which of the following systems is/are employed to regulate acid-base balance in the blood? (Check all that apply.) 30

_______ a. cardiovascular system

_______ b. lymphatic system

_______ c. respiratory system

_______ d. urinary system

31. Blood pH of 7.1 with elevated partial pressure of carbon dioxide and normal bicarbonate levels is indicative of ____________________ (metabolic/respiratory) ____________________ (acidosis/alkalosis). 31

Can You Apply What You've Learned?

32. A surgeon has the job of removing a patient's kidney. His approach to the kidney will be to cut through skin and adipose tissue of the lower back, remove portions of the most inferior ribs and a portion of the quadratus lumborum muscle, and then enter the retroperitoneal space. Once the surgeon has entered the retroperitoneal space, what are the layers of tissue (in order) that the surgeon must cut through to reach the kidney?

33. Describe why muscular contractions are needed to transport urine through the ureters.

34. Patients with chronic hypertension are at risk of suffering damage to the kidneys. Describe the specific structures of the nephron that might be damaged in a patient with hypertension.

35. Doctors will often first advise patients with hypertension to eat a diet low in sodium. Discuss why this may be beneficial in reducing systemic blood pressure. (Hint: Remember that an increased blood volume will correspond to an increase in systemic blood pressure.)

36. Patients in kidney failure require dialysis to perform the processes that are normally accomplished by the nephron. A dialysis membrane does this by mimicking the filtration membrane in the nephron. Describe the essential function typically accomplished by the kidneys that is replaced by the dialysis machine.

37. Describe why a person suffering from kidney failure would experience edema, which would be alleviated by dialysis.

38. Among its many effects, alcohol inhibits ADH secretion from the posterior pituitary gland. Describe the typical action of this hormone. Then discuss how ingestion of alcohol will affect urine formation.

39. A patient with suspected internal bleeding is admitted to the hospital. His systemic blood pressure is 80/50 mm Hg. Predict the patient's GFR given his current systemic blood pressure. Justify your answer.

Can You Synthesize What You've Learned?

40. In severe cases of hypertension, an angiotensin-converting enzyme (ACE) inhibitor may be prescribed for a patient. The enzyme ACE converts angiotensin I to angiotensin II. One target of angiotensin II is the adrenal cortex, where angiotensin II stimulates the release of the hormone aldosterone (see chapter 19). Aldosterone increases sodium reabsorption by kidney tubule cells. Describe how an ACE inhibitor will influence blood volume and blood pressure (as it relates to aldosterone). ____________________

41. In addition to stimulating the release of aldosterone, angiotensin II is a widespread vasoconstrictor. This means that smooth muscle in the walls of blood vessels will be stimulated to contract when angiotensin II is formed.

    a. Describe what will happen to GFR when angiotensin II stimulates the afferent arterioles within the nephron to constrict. ____________________

    b. Describe what effect widespread vasoconstriction will have on systemic blood pressure. ____________________

42. Susan, a 25-year-old female, contracted food poisoning likely due to eating undercooked eggs. She has been vomiting violently and cannot keep food or liquids down. Susan sought medical attention. Doctors measured her blood pH at 7.5, P_{CO_2} at 40 mm Hg, P_{O_2} at 95 mm Hg, O_2 saturation at 97%, and bicarbonate at 32 mEq/L.

    a. Classify Susan's case as respiratory acidosis, respiratory alkalosis, metabolic acidosis, or metabolic alkalosis, and justify your answer. ____________________

    b. Describe compensatory mechanisms this may have been initiated to adjust Susan's blood pH back toward normal homeostatic levels. ____________________

CHAPTER 26

The Digestive System

OUTLINE AND LEARNING OBJECTIVES

PHYSIOLOGY 741

Digestive Physiology 741

MODULE 12: DIGESTIVE SYSTEM

INTRODUCTION

The digestive (*digero*, to force apart, dissolve) system is responsible for the breakdown and absorption of molecules that are needed by the body for energy, maintenance, and ultimately survival. The gastrointestinal (GI) tract, or digestive tract, consists of a long tube that is composed of the mouth, pharynx, esophagus, stomach, small intestine, and large intestine. In addition, accessory structures including the salivary glands, liver, gallbladder, and pancreas add secretions to the GI lumen, which are required for digestion of ingested food. The wall that composes each portion of the GI tract is composed of four layers that include the mucosa, submucosa, muscularis, and adventitia or serosa. Consider that the GI tract is open to the environment at both ends (the oral cavity and the anus). This means that the lining of the GI tract is open to the external environment. Thus, the interface between the lumen of the GI tract and the internal environment of the body presents a special problem. The GI tract must transport needed substances from the lumen of the GI tract into the blood capillaries and lymph capillaries within the GI wall while also preventing pathogens from entering. For this reason, the wall of the entire GI tract is densely populated with lymphatic tissues.

The laboratory exercises in this chapter explore how the wall layers of the various parts of the GI tract are modified to suit the particular needs of the organ. Further exercises explore the structure and function of organs such as the liver and pancreas. While observing the structures of the digestive system, various circulatory and lymphatic structures associated with these structures will also be reconsidered in the context of their specific roles within the digestive system.

The exercises in this chapter begin with an exploration of both the histological structure and the gross structure of the GI tract. The materials in the "Histology" and "Gross Anatomy" sections are organized so the order of structures studied begins at the mouth and ends at the anus. In other words, the structures are studied in the same order that a bolus of food encounters them as it moves through the GI tract. It is not imperative that the structures are observed in this order. The advantage of studying the structures in this order is that it allows one to reflect on how the different parts of the GI tract work together to accomplish the process of digestion. The Physiology section of this chapter explores the processes of chemical digestion by digestive enzymes, including the effect of temperature on enzyme activity.

List of Reference Tables

Chapter 26: The Digestive System

Name: ______________________

Date: __________ Section: __________

PRE-LABORATORY WORKSHEET

These Pre-Laboratory Worksheet questions may be assigned by instructors through their connect course.

1. Humans have ________________ (two/three) pairs of salivary glands.

2. The oral cavity opens into the ________________ (nasopharynx/laryngopharynx/oropharynx).

3. The esophagus is located ________________ (anterior/posterior) to the trachea.

4. The wall of the stomach contains ________________ (two/three) layers of smooth muscle.

5. Which of the following types of epithelial tissue lines the small and large intestine? (Circle one.)
   a. simple cuboidal with goblet cells
   b. simple columnar with goblet cells
   c. simple cuboidal with goblet cells and cilia
   d. simple columnar with goblet cells and microvilli
   e. simple columnar with goblet cells and cilia

6. Which of the following places the three segments of the small intestine in order? (Circle one.)
   a. jejunum, duodenum, ileum
   b. ileum, jejunum, duodenum
   c. duodenum, jejunum, ileum
   d. duodenum, ileum, jejunum
   e. jejunum, ileum, duodenum

7. The ducts from the liver, gallbladder, and pancreas empty in to the ________________ (pyloris/duodenum).

8. The liver consists of ________________ (two/four) lobes. The largest lobe of the liver is the ________________ (right/left) lobe.

9. The cecum is located at the beginning of the ________________ (ascending/descending) colon.

10. Which of the following is the monomer that makes up proteins? (Circle one.)
    a. monosaccharides
    b. nucleic acids
    c. amino acids
    d. disaccharides
    e. fatty acids

11. Glucose is a ________________ (monosaccharide/disaccharide), whereas sucrose is a ________________ (monosaccharide/disaccharide).

12. Which of the following is the structural component of triglycerides to which fatty acids are attached? (Circle one.)
    a. glucose
    b. glycerol
    c. fatty acids
    d. amino acids
    e. glycogen

HISTOLOGY

Salivary Glands

The **salivary glands (figure 26.1)** are accessory digestive glands composed of modified epithelial tissue that produce **saliva,** a watery secretion that contains a mixture of substances including the enzyme **salivary amylase,** which initiates the digestion of carbohydrates. There are three pairs of extrinsic salivary glands: parotid, submandibular, and sublingual. The glands empty secretions into the oral cavity through their associated ducts. Salivary glands contain two cell types: serous cells and mucous cells. **Serous cells** produce watery secretions containing glycoproteins, electrolytes, lysozyme, and the enzymes salivary amylase. **Mucous cells** produce mucin, which, when hydrated, becomes mucus. Both serous fluid and mucus function as lubricants to ease the passage of the wet mass of food called the **bolus** (*bolos*, lump) through the pharynx and esophagus. Definitions of structures related to salivary glands are described in **table 26.1.** Details of the structure and function of each of the salivary glands are listed in **table 26.2.**

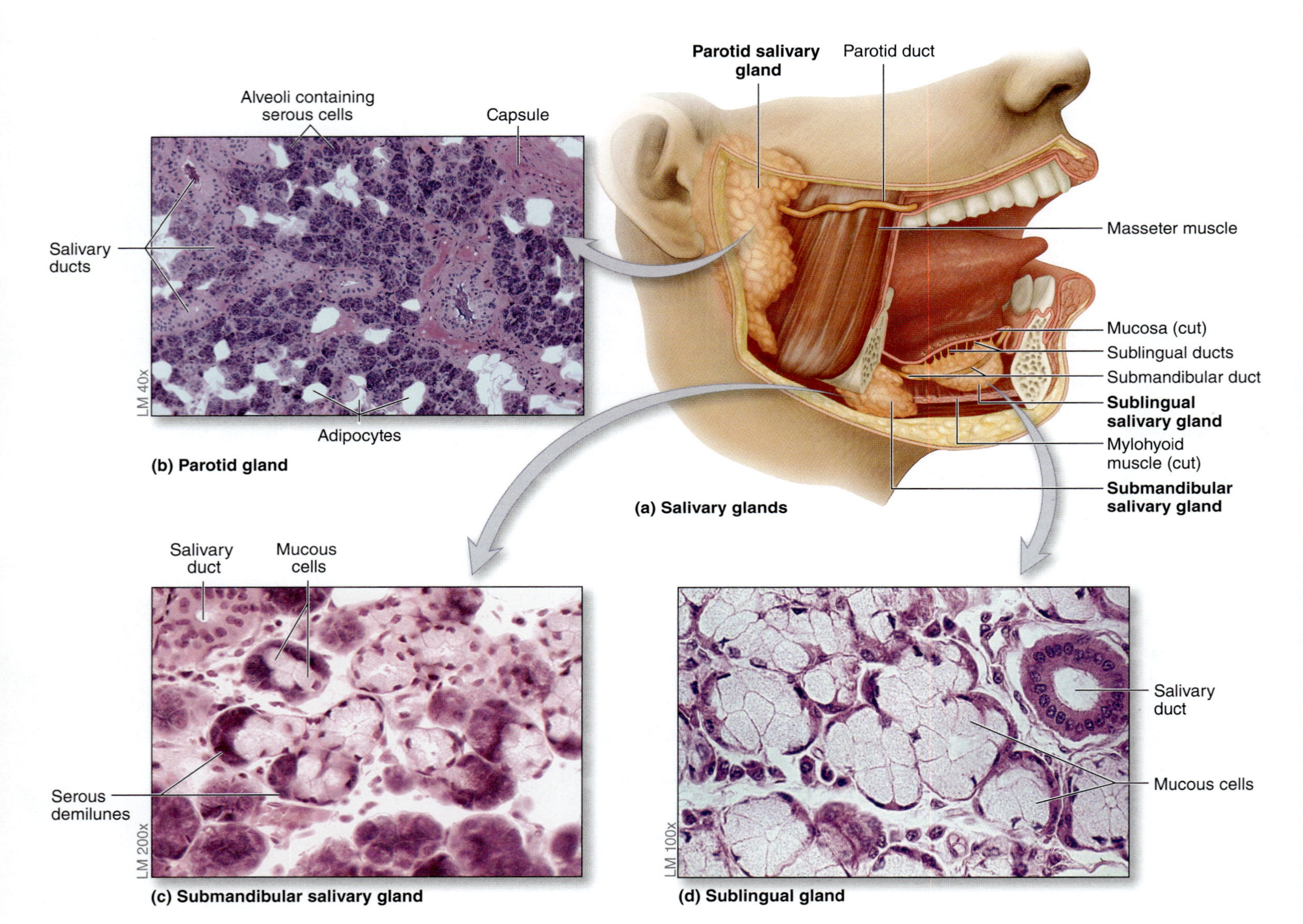

Figure 26.1 **Salivary Glands.** (a) Location of the salivary glands. Histological features of the (b) parotid, (c) submandibular, and (d) sublingual glands.

Table 26.1 Salivary Gland Structures

Structure	Description and Location	Function	Word Origin
Alveolus	The grape-shaped secretory portion of a gland	NA	*acinus*, grape
Mucous Cells	Cells have flattened nuclei that are located on the basal surface. Mainly located along the tubules of salivary glands.	Secrete mucin	*mucosus*, mucous
Myoepithelial Cells	Cells are not visible in light microscopy. Located around the alveoli and the long axes of the ducts.	Contraction of these cells expels the secretions from salivary glands	*mys*, muscle, + *epithelial*, relating to epithelial tissues
Serous Cells	Cells have round nuclei and contain numerous secretory granules. Located in the alveoli of the parotid gland and in the demilunes of the submandibular and sublingual glands.	Secrete glycoproteins, electrolytes, and salivary amylase	*serous*, having a watery consistency
Serous Demilunes	Crescent- or moon-shaped groups of serous cells located at the periphery of a mucous alveolus	Secrete glycoproteins, electrolytes, and salivary amylase	*serous*, having a watery consistency, + *demilune*, half-moon

Table 26.2 Histological Characteristics of Salivary Glands

Gland	Secretory Cells	% of Saliva	Opening	Word Origin
Parotid	**Serous:** serous cells occupy ~90% of the gland's volume (the rest is adipose connective tissue)	26–30%	Empties via the parotid duct opposite the second upper molar	*para*, beside, + *ous*, ear
Sublingual	**Mixed:** 60–70% serous, 30–40% mucus; serous cells are located in serous demilunes	3–5%	Empties via multiple ducts into either the submandibular duct or directly into the oral cavity	*sub-*, under, + *lingual*, the tongue
Submandibular	**Mixed:** 80% serous, 20% mucus, contains a few serous demilunes	60–70%	Empties via the submandibular ducts between the lingual frenulum and the mandible	*sub-*, under, + *mandible*, the mandible

EXERCISE 26.1

HISTOLOGY OF THE SALIVARY GLANDS

1. Obtain a compound microscope and histology slides of the parotid, submandibular, and sublingual salivary glands.
2. Place the slide of the **parotid gland** on the microscope stage. Bring the tissue sample into focus on low power and then switch to high power.
3. Identify the following structures on the slide of the parotid gland, using figure 26.1*b* and tables 26.1 and 26.2 as guides:

   ☐ **adipocytes** ☐ **serous cells**

4. Place the slide of the **submandibular gland** on the microscope stage. Bring the tissue sample into focus on low power and then switch to high power.
5. The serous cells in the submandibular glands are located surrounding the mucous cells and are shaped like half moons. Thus, they are more specifically referred to as *serous demilunes* (*demi*, half, + *luna*, moon). Scan the slide to locate serous demilunes (figure 26.1*c*).
6. Identify the following structures on the slide of the submandibular gland, using figure 26.1*c* and tables 26.1 and 26.2 as guides:

   ☐ **mucous cells** ☐ **serous demilunes**
   ☐ **salivary duct**

7. Place the slide of the **sublingual gland** on the microscope stage. Bring the tissue sample into focus on low power and then switch to high power.
8. The sublingual gland is similar to the submandibular gland in that it contains serous demilunes. Two characteristics help distinguish the sublingual gland from the submandibular gland. The sublingual gland contains fewer adipocytes and greater numbers of mucous cells than the submandibular gland (table 26.2 and figure 26.1*d*).
9. Identify the following structures on the slide of the sublingual gland, using figure 26.1*d* and tables 26.1 and 26.2 as guides:

   ☐ **mucous cells** ☐ **salivary duct**

The Stomach

The **stomach** is an organ that mixes the bolus with gastric juice, which is released from the stomach wall. The bolus is changed to a liquid puree called **chyme** (*chymos*, juice). In this section the histology of the stomach wall will be observed first, followed by observations of specific regional characteristics of the stomach epithelium.

EXERCISE 26.2

WALL LAYERS OF THE STOMACH

The walls of the GI tract are composed of four concentric tunics (layers): **mucosa, submucosa, muscularis,** and **serosa. Figure 26.2** shows the tunics that typically compose the GI tract. **Figure 26.3** shows a section of the stomach wall. Notice the three layers of smooth muscle composing the muscularis of the stomach wall. (Other regions of the GI tract contain only two layers—circular and longitudinal.) **Table 26.3** describes the types of tissues that are located in each of the layers of the stomach wall.

1. Obtain a histology slide of the stomach (figure 26.3) and place it on the microscope stage. Bring the tissue sample into focus on low power.
2. Identify the following layers of the stomach wall on the slide, using figure 26.3 and table 26.3 as guides:
   - ☐ **epithelium**
   - ☐ **lamina propria**
   - ☐ **mucosa**
   - ☐ **muscularis**
   - ☐ **muscularis: circular layer**
   - ☐ **muscularis: longitudinal layer**
   - ☐ **muscularis mucosae**
   - ☐ **muscularis: oblique layer**
   - ☐ **serosa**
   - ☐ **submucosa**

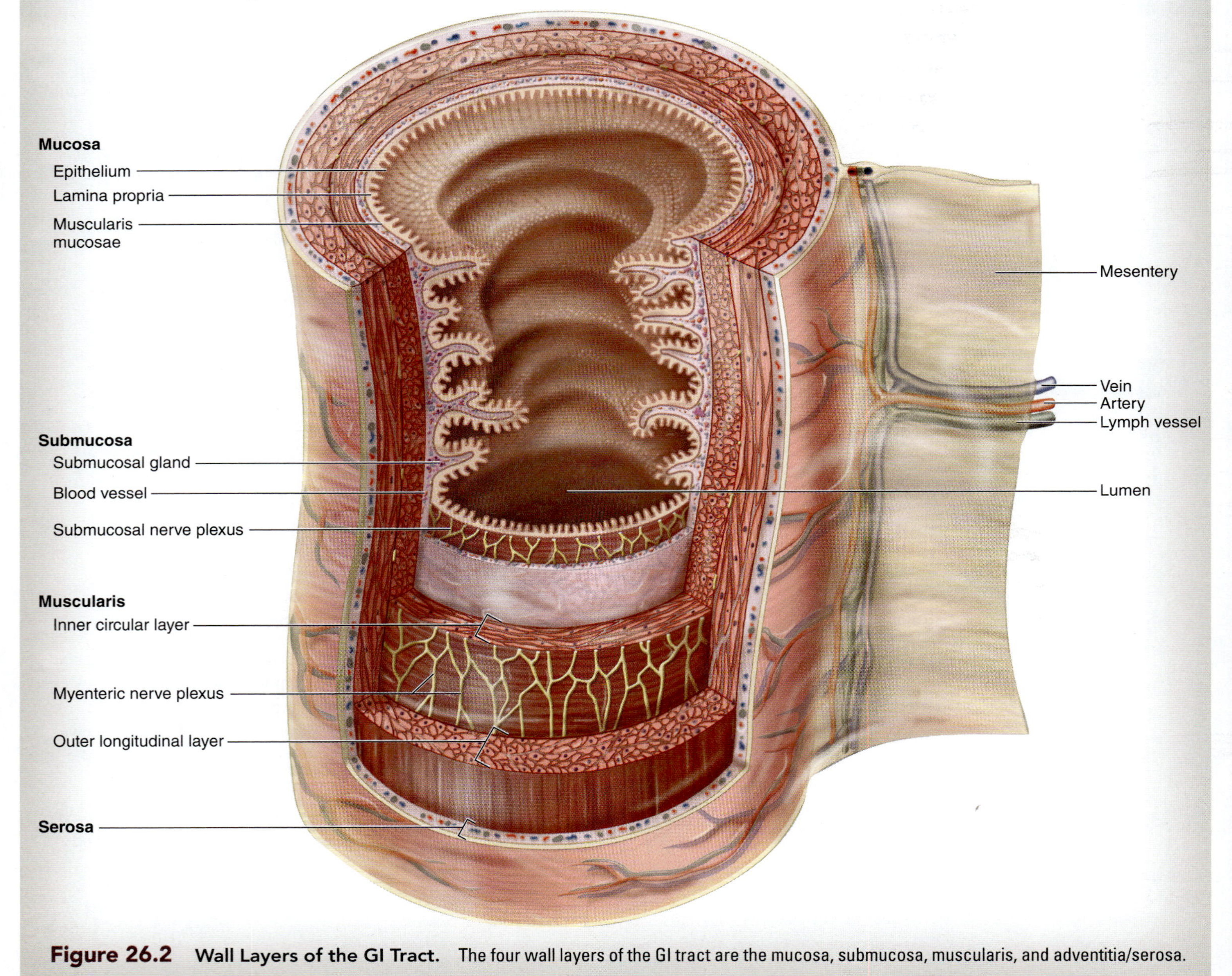

Figure 26.2 **Wall Layers of the GI Tract.** The four wall layers of the GI tract are the mucosa, submucosa, muscularis, and adventitia/serosa.

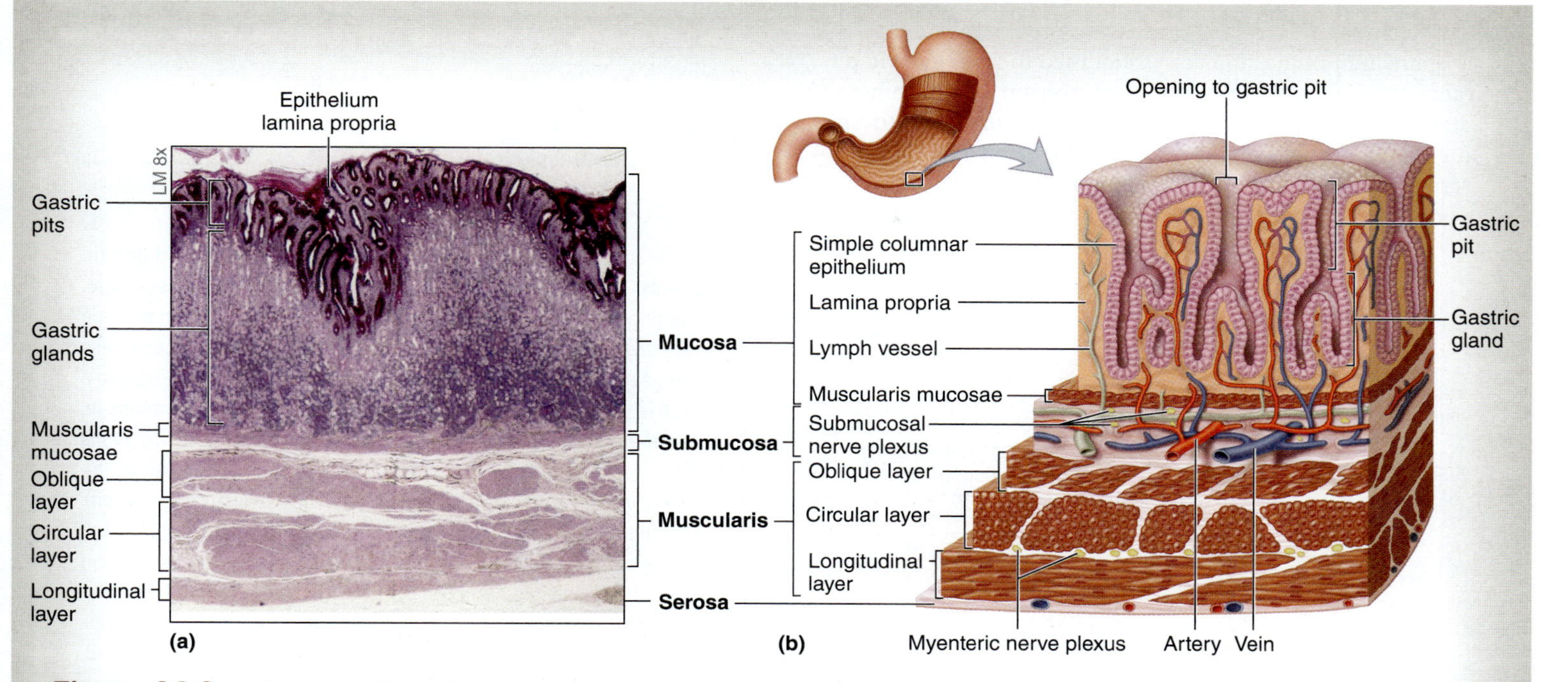

Figure 26.3 **The Stomach Wall.** (a) Histology of the stomach wall. (b) Layers of the stomach wall.

Table 26.3 Wall Layers of the Stomach

Layer	Sublayer	Characteristics
Mucosa	Epithelium	Simple columnar epithelium containing five distinct cell types (see table 26.5)
	Lamina propria	Areolar connective tissue that contains blood vessels and nerves. Lymphatic nodules are common.
	Muscularis mucosae	Thin layer of smooth muscle
Submucosa	NA	Areolar and dense irregular connective tissue that contains blood vessels, lymph vessels, and the submucosal nerve plexus
Muscularis	Three layers of smooth muscles	Myenteric plexus is located between the layers of smooth muscle
	Inner oblique layer	Muscle fibers responsible for causing the "twisting" action of the stomach
	Middle circular layer	Thickest layer of the three layers of muscle. Thickenings of this layer form the inferior esophageal (cardiac) and pyloric sphincters.
	Outer longi-tudinal layer	Located in the greater and lesser curvatures only
Serosa	NA	Areolar connective tissue with dispersed collagen and elastic fibers; covered by visceral peritoneum (a serous membrane)

3. Sketch the histology of the stomach wall as seen through the microscope in the space provided. Be sure to label the structures listed in step 2 in the drawing.

_____________ ×

EXERCISE 26.3

HISTOLOGY OF THE STOMACH

The epithelium of the stomach is modified to form **gastric pits,** invaginations of the epithelium that contain the openings of **gastric glands (figure 26.4).** The cell types in the gastric pits and glands vary in different regions of the stomach. **Table 26.4** describes the structure of the pits and glands in each of the major regions of the stomach (cardia, fundus/body, and pylorus).

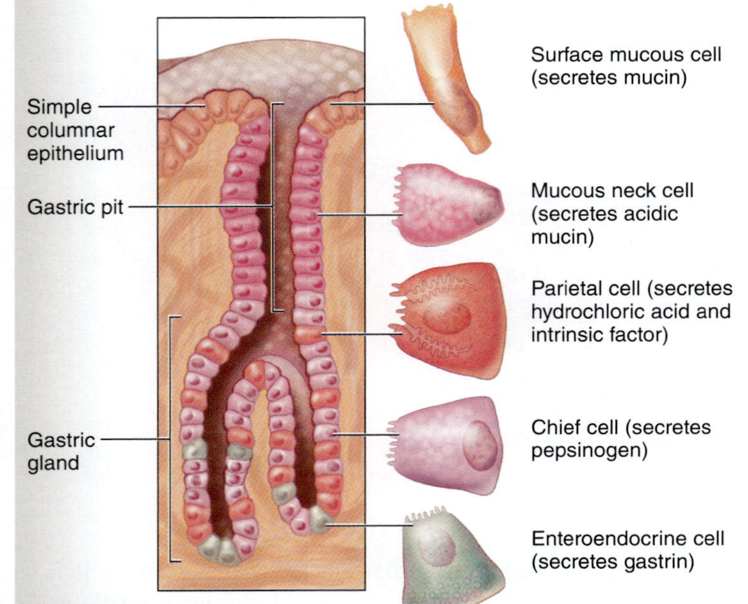

Table 26.5 describes the five cell types that compose the gastric pits and glands, and the types and functions of the secretions produced by each cell.

INTEGRATE

LEARNING STRATEGY

Understanding the relationship between **gastric pits** and **gastric glands** can be difficult. Let's imagine a hypothetical scenario: Imagine a person standing inside the stomach (mind you: this is a very tiny person!). The person is standing on surface mucous cells and sees a "hole" in the distance. This "hole" is the opening into a **gastric pit**. The person moves toward the gastric pit and jumps in. As the person falls into the pit, he/she is surrounded by the walls of the gastric pit, which are lined with mucous neck cells (thus, the person would probably be "slimed" as he/she moved through the pit). Eventually, the gastric pit narrows such that the person's movement will be stopped. At this point the person is standing at the junction between a gastric pit and the opening of a **gastric gland**. Looking down, the person sees an even smaller hole below, which is too small for him/her to move into. This hole is the lumen of a **gastric gland**. The gastric gland is lined with parietal cells and chief cells, which make HCl and pepsinogen, respectively. From the point of view of the person standing at the junction between the gastric pit and the gastric gland, a mixture of HCl and pepsinogen can be seen squirting upward into the lumen of the gastric pit. Luckily, the mucous neck cells protect the lumen of the gastric pit from the incoming acid. Unfortunately, the tiny person may not be so protected!

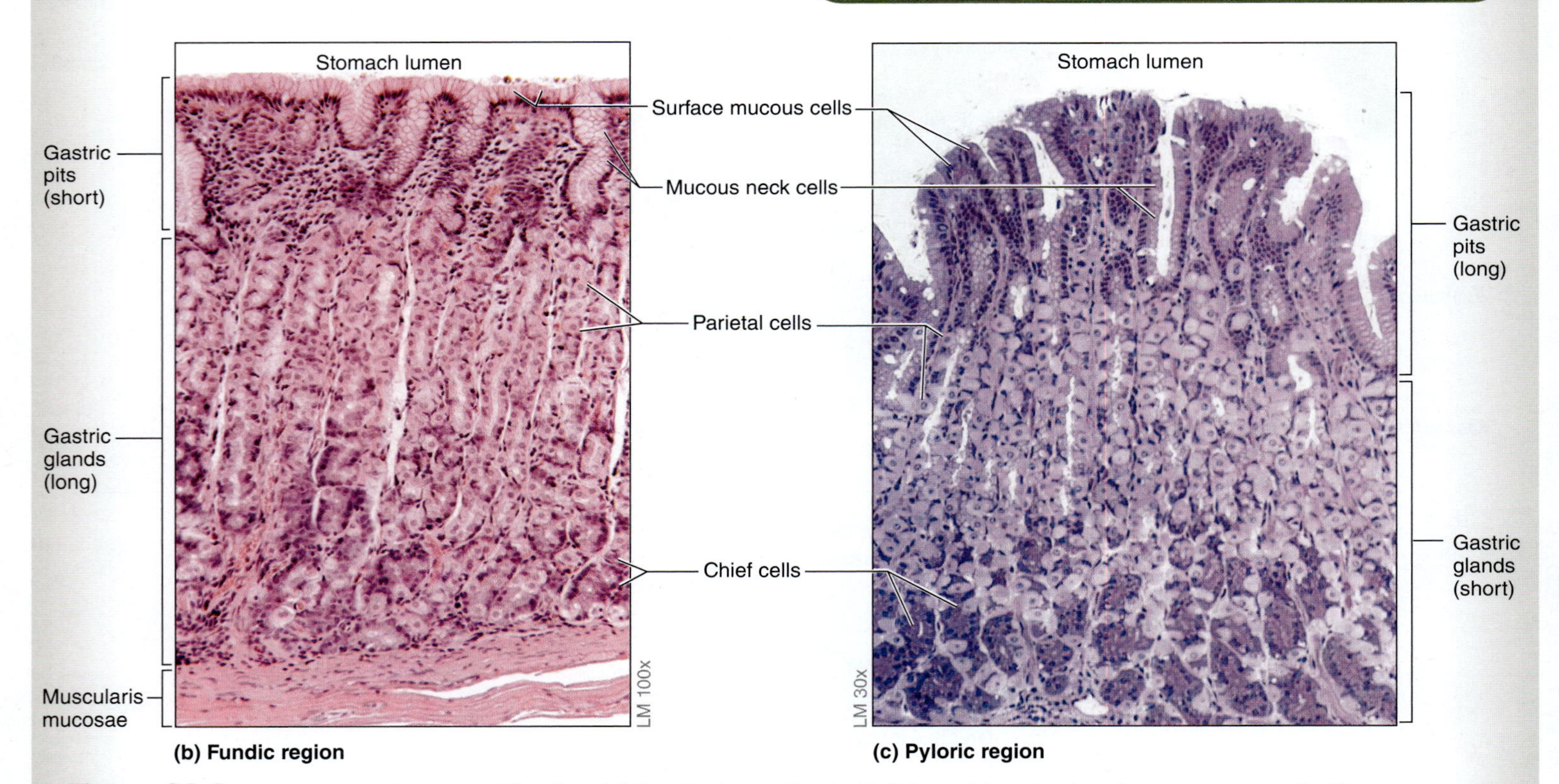

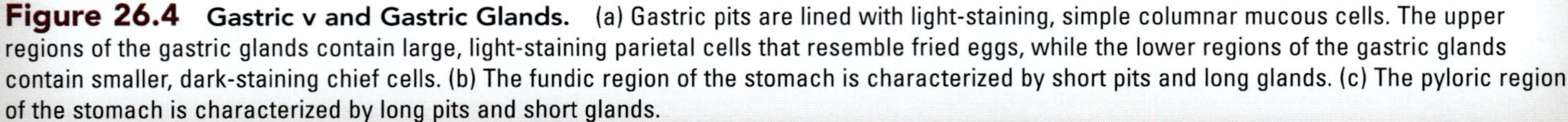

Figure 26.4 Gastric v and Gastric Glands. (a) Gastric pits are lined with light-staining, simple columnar mucous cells. The upper regions of the gastric glands contain large, light-staining parietal cells that resemble fried eggs, while the lower regions of the gastric glands contain smaller, dark-staining chief cells. (b) The fundic region of the stomach is characterized by short pits and long glands. (c) The pyloric region of the stomach is characterized by long pits and short glands.

Table 26.4	Regional Characteristics of the Gastric Pits and Glands of the Stomach		
Region	**Pit Structure**	**Main Cell Type(s) in Pits/Glands**	**Gland Structure**
Cardia	Short pits with long glands	All mucous	Simple branched, tubular glands
Fundus/Body	Short pits with long glands	Parietal and chief cells with mucous neck cells	Branched, tubular glands
Pylorus	Long pits with short, coiled glands	Mostly mucous	Branched, tubular glands

1. Obtain a histology slide of the stomach and place it on the microscope stage. Bring the tissue sample into focus on low power and then scan the slide to locate the epithelial lining of the stomach (figure 26.4). Once the epithelium of the stomach is at the center of the field of view, change to high power and bring the tissue sample into focus once again. Fully classify the epithelium that lines the stomach:

____

____

2. After identifying the epithelium lining the stomach wall, locate a gastric pit and gland and move the microscope stage so the pit/gland is in the center of the field of view.
3. Identify the following structures, using figure 26.4 and table 26.5 as guides:

- ☐ **chief cells**
- ☐ **gastric glands**
- ☐ **gastric pits**
- ☐ **mucous neck cells**
- ☐ **parietal cells**
- ☐ **surface mucous cells**

4. Sketch gastric glands and pits as seen through the microscope in the space provided. Be sure to label the locations of the specialized cell types in the drawing.

Which region of the stomach wall is being observed based on the structure of the gastric pits and glands?

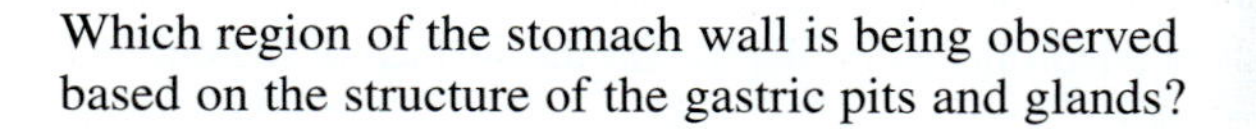

____

________ ×

5. *Optional Activity:* **AP|R 12: Digestive System**—Watch the "Stomach" animation to review the stomach wall layers and their histology.

Table 26.5	Cell Types in the Gastric Pits and Glands of the Stomach		
Cells	**Location**	**Secretions**	**Action of Secretion**
Chief Cells (zymogenic)	The most common secretory cell; located deep within the gastric glands. Contain numerous eosinophilic (red) granules.	Pepsinogen	A zymogen* that is converted to pepsin when it encounters the acidic environment of the stomach. Pepsin is a protease (it breaks down proteins).
		Gastric lipase	Enzyme with limited role in triglyceride digestion
G-Cells	Enteroendocrine cells scattered throughout the gastric glands	Gastrin	Hormone that stimulates chief cells and parietal cells to secrete their products, and stimulates the smooth muscle in the stomach walls to contract
Mucous Neck Cells	Lining the base of the gastric pits	Mucin	Glycoprotein that protects the mucosa from HCl
Parietal Cells	Superficial within the gastric glands in the fundus and body of the stomach. A few are found in the pylorus. None are found in the cardia.	HCl (hydrochloric acid)	Decreases the pH of the stomach to about 2 (very acidic).
		Intrinsic factor	Necessary for vitamin B_{12} absorption in the small intestine.
Surface Mucous Cells	Cells that line both the superior portion of gastric pits and area surrounding the gastric pits	Mucin	Glycoprotein that protects the mucosa from HCl

*Zymogen is a general term for an inactive protein. Generally, the names of zymogens end in -ogen, as in pepsinogen.

The Small Intestine

The liquid puree called chyme (*chymos*, juice) originates in the stomach, flows through the pyloric sphincter, and enters the small intestine. It is in the small intestine where the greatest amount of chemical digestion occurs and almost all absorption of nutrients into the blood or lymph takes place. To facilitate absorption of nutrients, the walls of the small intestine contain several modifications that greatly increase the inner surface area: circular folds, villi, and microvilli. The mucosal and submucosal layers of the small intestine are thrown into folds called **circular folds** (also called plicae circularis). On the surface of the circular folds, there are folds of mucosa called **villi** (s., *villus*). Each villus is lined with cells forming a simple columnar epithelium that contain **microvilli** on their apical surface. Intestinal glands (also known as intestinal crypts) are invaginations of the mucosa and are located between the villi. These glands resemble the structure and function of gastric glands. Goblet cells within the epithelium are found in these intestinal glands and produce mucin, which helps protect the epithelial lining and lubricates the passage of chyme through the small intestine. **Figure 26.5** shows the three segments of the small intestine (duodenum, jejunum, and ileum).

INTEGRATE

LEARNING STRATEGY

The three parts of the small intestine can be distinguished from each other by characteristic features present or absent in the submucosal layer. The **duodenum** is characterized by the presence of **duodenal glands** in the submucosa. Duodenal glands produce an alkaline secretion that protects the duodenum from acidic chyme. The **jejunum** is characterized by its **lack of specialized structures** in the submucosa. The term *jejunum* comes from the Latin word *jejunus*, which means "empty." Remember that the submucosa of the jejunum is "empty" because it lacks specialized structures in the submucosa. The **ileum** is characterized by the presence of **Peyer patches** (aggregates of lymphatic tissue) scattered throughout the submucosa.

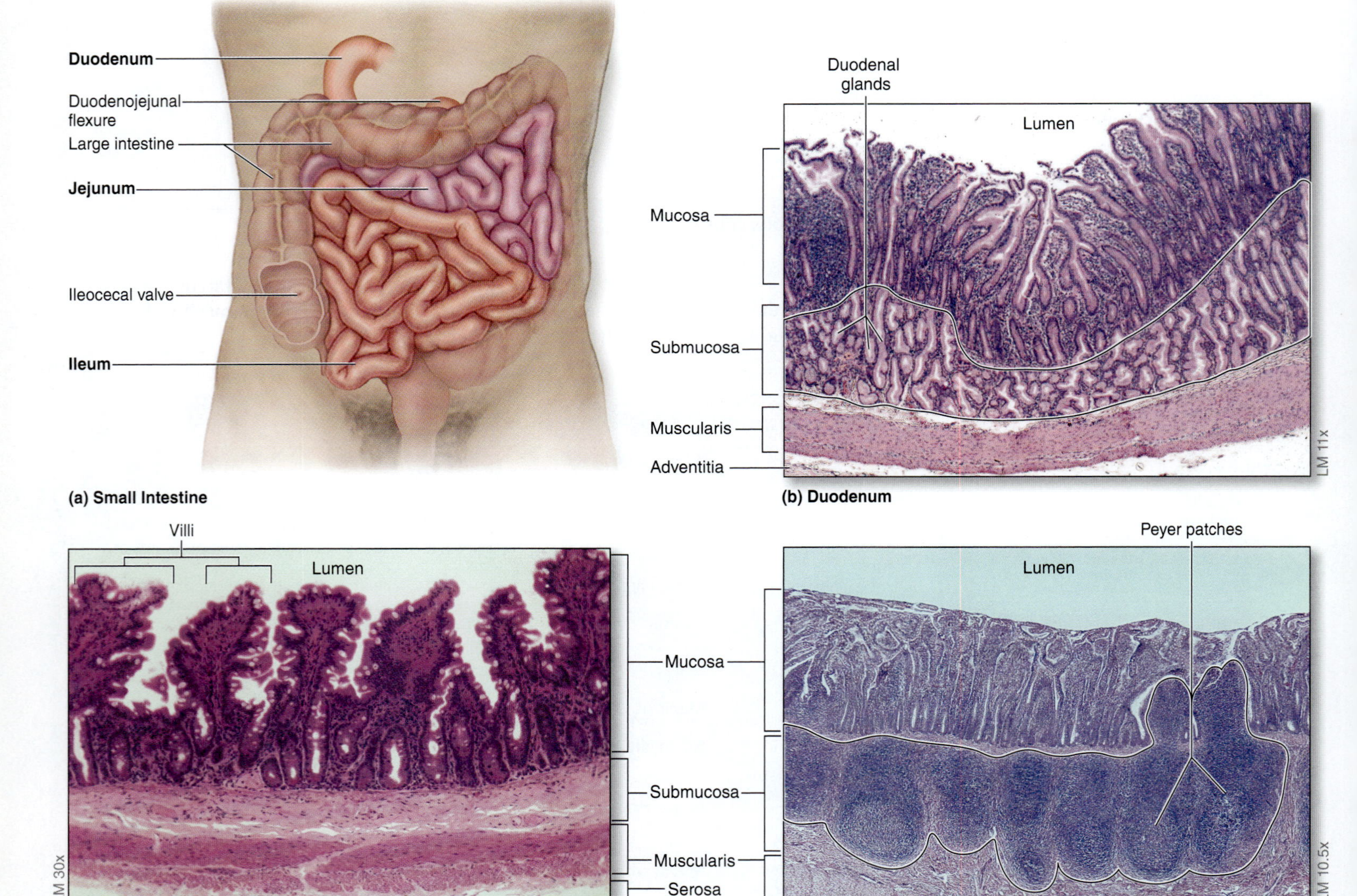

Figure 26.5 The Small Intestine. (a) Segments of the small intestine. (b) Histology of the duodenum, demonstrating duodenal glands in the submucosa. The outermost layer of the wall of the duodenum is an adventitia instead of serosa because the duodenum is retroperitoneal, and therefore not covered by peritoneum. (c) Histology of the jejunum, which lacks duodenal glands and Peyer patches. (d) Histology of the ileum, demonstrating Peyer patches in the submucosa.

EXERCISE 26.4

HISTOLOGY OF THE SMALL INTESTINE

1. Obtain a histology slide of the small intestine and place it on the microscope stage. If the laboratory is equipped with slides of each section of the small intestine (duodenum, jejunum, and ileum), be sure to view all three. If not, use figure 26.5 to decide which part of the small intestine is on the slide.
2. Bring the tissue sample into focus at low power and move the microscope stage until the epithelium is at the center of the field of view. Then change to high power. What type of epithelium lines the small intestine?

   ____

   ____

   What surface modifications are present? ______________

   ____

   What is the purpose of these surface modifications?

   ____

3. Identify the following structures on the slide(s), using figure 26.5 as a guide:

   - ☐ **duodenal glands**
   - ☐ **epithelium**
   - ☐ **lamina propria**
   - ☐ **mucosa**
   - ☐ **muscularis**
   - ☐ **muscularis mucosae**
   - ☐ **Peyer patches**
   - ☐ **serosa**
   - ☐ **submucosa**
   - ☐ **villi**

4. Sketch the histology of the duodenum, jejunum, and ileum of the small intestine in the space provided. Identify the histological features that differentiate these three portions of the small intestine from each other (refer to figure 26.5 for reference).

Duodenum

__________ ×

Jejunum

__________ ×

Ileum

__________ ×

The Large Intestine

Chyme leaving the small intestine enters the **large intestine** at the iliocecal junction. A valve, the **iliocecal valve,** controls the passage of chyme from the small intestine to the large intestine. Recall that the vast majority of nutrients are absorbed in the small intestine. The function of the large intestine is mainly to absorb water (and some electrolytes) from the chyme that remains, and compact the waste products as **feces** for elimination from the body. The epithelium of the large intestine contains many goblet cells, which help lubricate the epithelium to ease the passage of feces through the large intestine.

EXERCISE 26.5

HISTOLOGY OF THE LARGE INTESTINE

1. Obtain a histology slide of the large intestine (colon) and place it on the microscope stage.
2. Bring the tissue sample into focus on low power and identify the epithelial layer **(figure 26.6).** Move the microscope stage so the epithelium is at the center of the field of view and then change to high power.
3. Identify the following structures in the slide of the large intestine, using figure 26.6 and the textbook as guides:

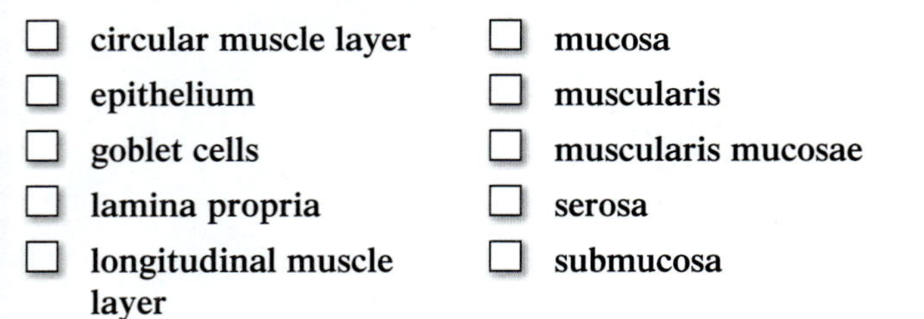

- ☐ **circular muscle layer**
- ☐ **epithelium**
- ☐ **goblet cells**
- ☐ **lamina propria**
- ☐ **longitudinal muscle layer**
- ☐ **mucosa**
- ☐ **muscularis**
- ☐ **muscularis mucosae**
- ☐ **serosa**
- ☐ **submucosa**

4. Sketch the histology of the large intestine as seen through the microscope in the space provided. Label the four major layers of the wall of the large intestine in the drawing.

_______ ×

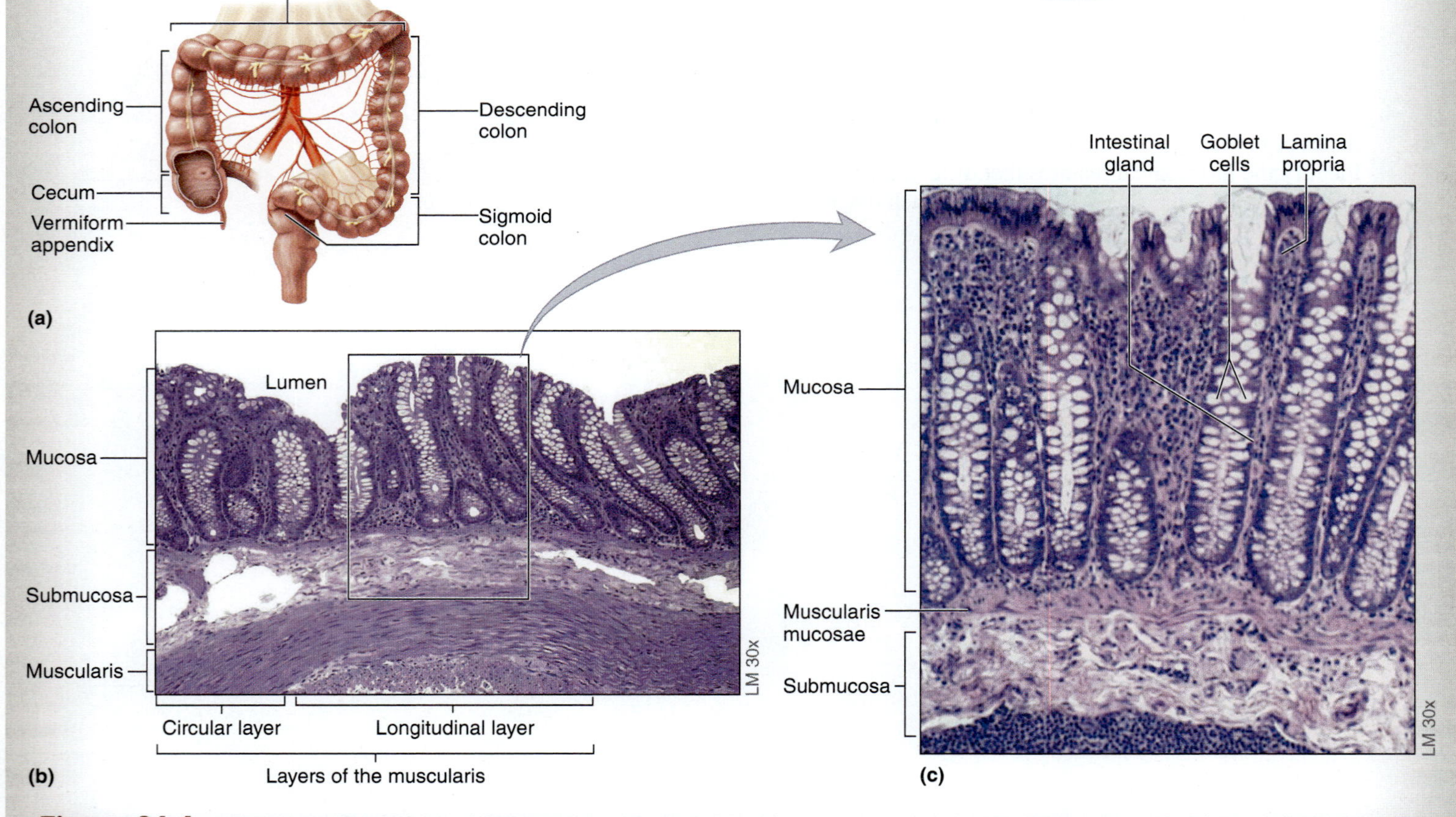

Figure 26.6 The Large Intestine. (a) Parts of the large intestine. (b) Histology of the wall of the large intestine. (c) Close-up of the epithelium lining the large intestine; numerous goblet cells are present.

The Liver

The **liver** is the largest accessory organ in the digestive system. Indeed, it is the largest internal organ within the human body. The liver performs numerous vital functions that include detoxifying the blood, storing nutrients, and producing plasma proteins. However, its primary function in digestion is the production of bile. The structural and functional unit of the liver is a **hepatic lobule,** which is a hexagonally-shaped structure consisting of strands of **hepatocytes** (the strands of hepatocytes are **hepatic cords**) radiating from a **central vein** in the middle **(figure 26.7).** In the areas where the outer edges of the hepatic lobules come together there are **portal triads,** which consist of a branch of the hepatic artery, a branch of the hepatic portal vein, and a bile duct. Within the hepatic lobules, in between the hepatic cords, are **hepatic sinusoids:** capillaries that carry blood from the branches of the hepatic artery and hepatic portal vein to the central veins. Along the sinusoids are several macrophage-like cells, **reticuloendothelial** (Kupffer) cells. These cells engulf microorganisms that enter the liver from the portal circulation.

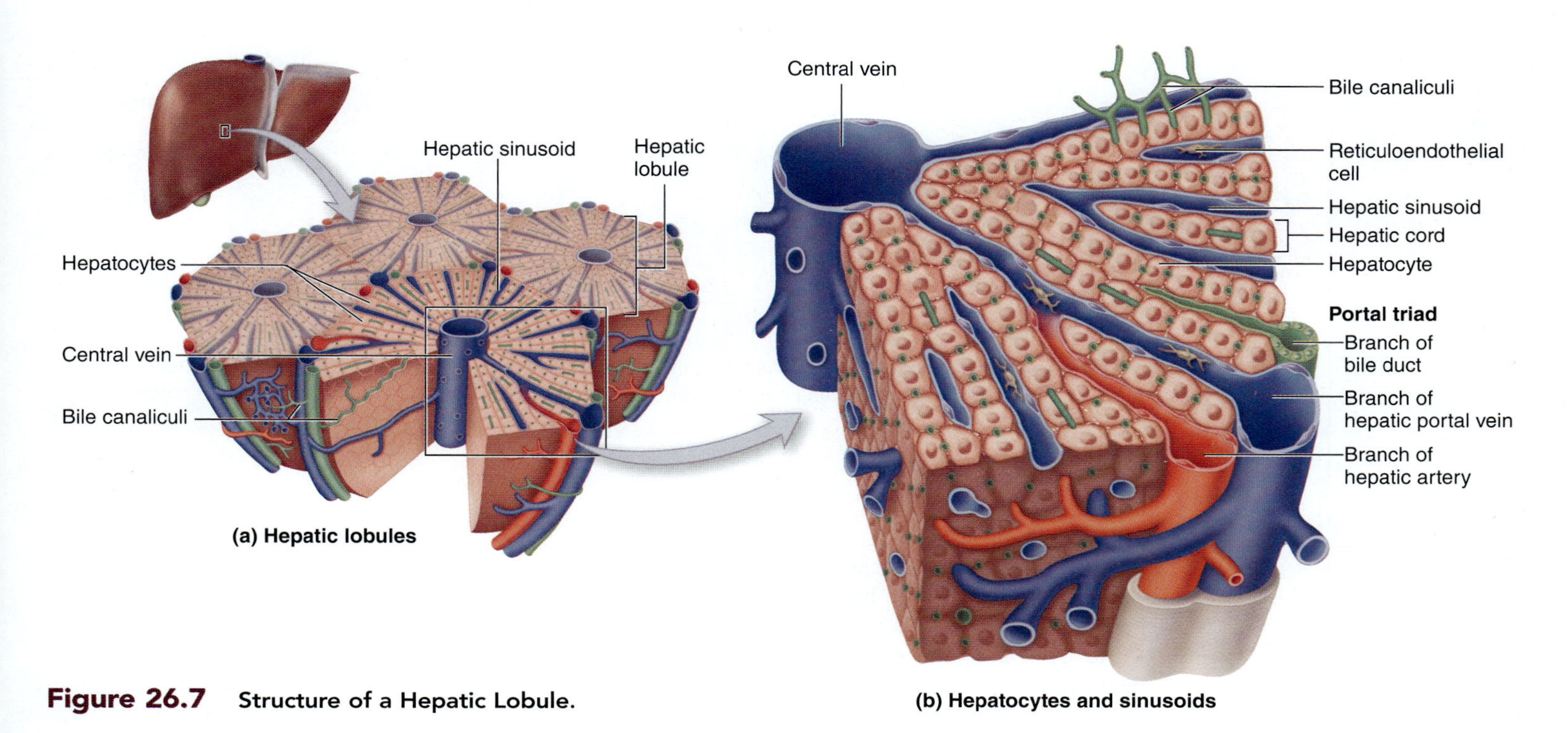

Figure 26.7 Structure of a Hepatic Lobule.

EXERCISE 26.6

HISTOLOGY OF THE LIVER

1. Obtain a histology slide of the liver and place it on the microscope stage. Bring the tissue sample into focus on low power and locate a hepatic lobule **(figure 26.8*a*).**
2. Move the microscope stage until the hepatic lobule is at the center of the field of view, and then change to high power (figure 26.8*b*). Identify the following structures on the slide of the liver, using figures 26.7 and 26.8 as guides:

- ☐ **branch of bile duct**
- ☐ **branch of hepatic artery**
- ☐ **branch of hepatic portal vein**
- ☐ **central vein**
- ☐ **hepatic cords**
- ☐ **hepatic lobule**
- ☐ **hepatic sinusoids**
- ☐ **hepatocyte**
- ☐ **portal triad**
- ☐ **reticuloendothelial cells**

3. Sketch the liver as seen through the microscope in the space provided. Be sure to label all of the structures listed in step 2 in the drawing.

_____________ ×

4. *Optional Activity:* **AP|R 12: Digestive System**—Watch the "Liver" animation to visualize the organization and structure of a liver lobule.

(continued on next page)

(continued from previous page)

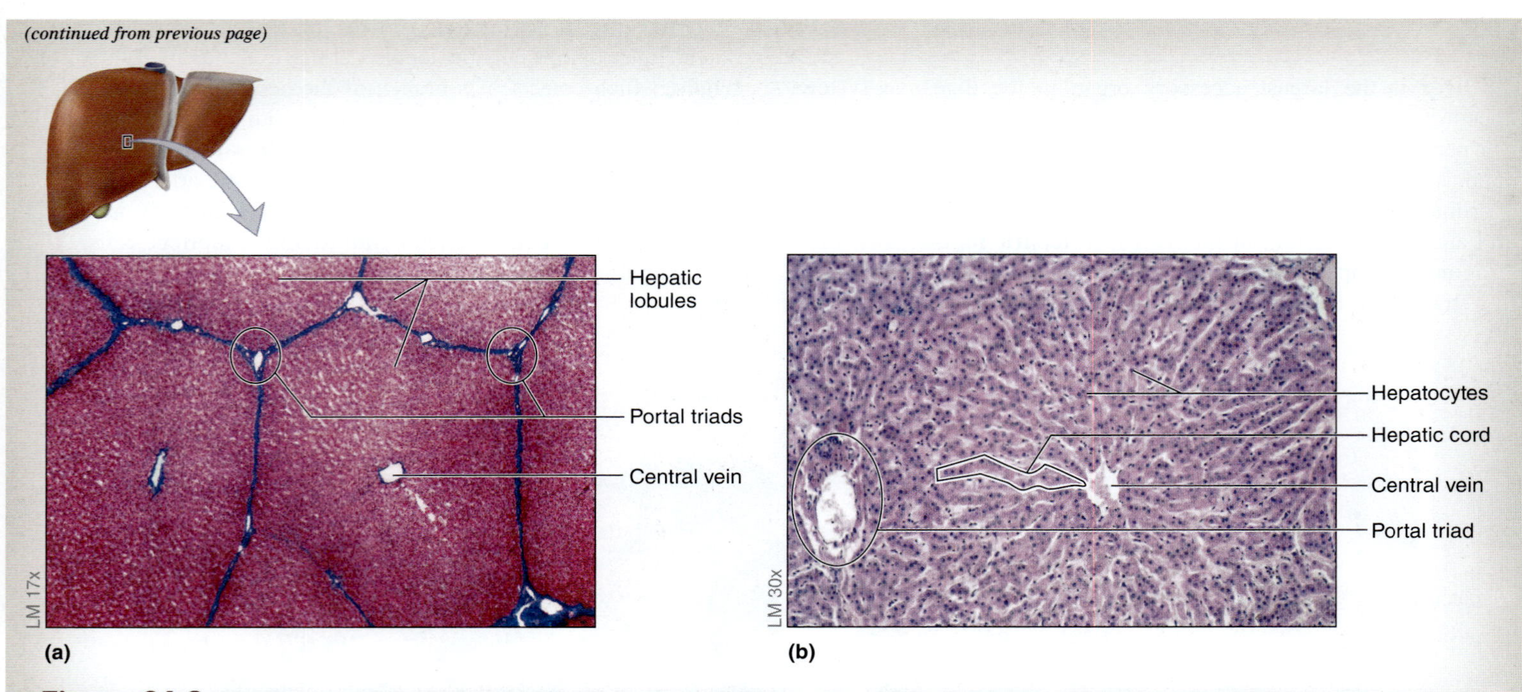

Figure 26.8 The Liver. (a) Low-magnification histology slide demonstrating multiple hepatic lobules with portal triads in the spaces between lobules. (b) Medium-magnification histology slide demonstrating a central vein, hepatocytes arranged into hepatic cords, and a portal triad.

The Pancreas

The **pancreas** is the second-largest accessory organ in the digestive system. It is both an endocrine and an exocrine gland. The histology of the endocrine part of the pancreas (the pancreatic islets) was covered in chapter 19 ("The Endocrine System"). The **exocrine** portion of the pancreas consists of grape-like bunches of cells called **acini** (s., *acinus*), which are similar in many ways to the cells that compose the salivary glands. The **acinar cells** composing acini produce many substances important for digestion, including digestive enzymes (e.g., pancreatic amylase). Cells of pancreatic ducts produce bicarbonate ion (HCO_3^-). Collectively, these secretions are called **pancreatic juice.** Pancreatic juice is transported from the acinar cells into small ducts that become larger ducts, and eventually dump the secretions into the duodenum via the **main pancreatic duct** (or accessory pancreatic duct) **(figure 26.9).**

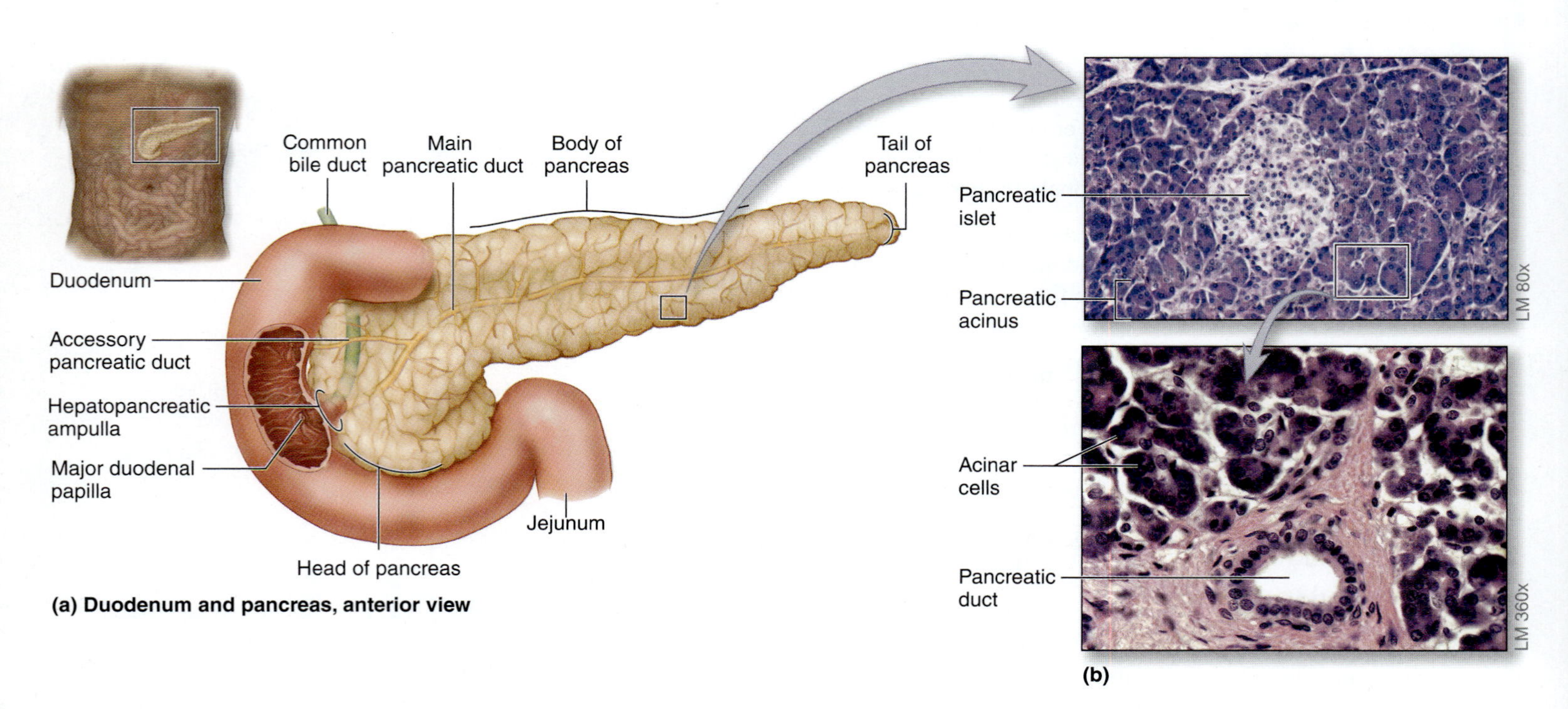

Figure 26.9 The Pancreas. (a) Location of the pancreas. (b) Histology of the pancreas demonstrating acinar cells, which secrete digestive enzymes, and pancreatic islets, which contain the hormone-secreting cells of the pancreas.

EXERCISE 26.7

HISTOLOGY OF THE PANCREAS

1. Obtain a histology slide of the **pancreas** and place it on the microscope stage. Bring the tissue sample into focus on low power. Scan the slide until a pancreatic islet surrounded by acinar cells is at the center of the field of view (figure 26.9*a*).
2. Move the microscope stage so the acinar cells that surround the pancreatic islet are in the center of the field of view, and then change to high power. Identify the following structures on the slide, using figure 26.9 as a guide:

- ☐ **acinar cells**
- ☐ **pancreatic duct**
- ☐ **pancreatic acinus**
- ☐ **pancreatic islet**

3. Sketch the pancreas as seen through the microscope in the space provided. Be sure to label all of the structures listed in step 2 in the drawing.

_____________ ×

INTEGRATE

LEARNING STRATEGY

After observing the histology of the pancreas, go back and review the histology of salivary glands. Note that the two tissues are extremely similar in structure. This is because they are also very similar in function. Acinar cells of the pancreas and serous cells of salivary glands both produce the enzyme *amylase*, which breaks down carbohydrates. When in doubt about which organ is under the microscope, look for pancreatic islets. The *presence* of pancreatic islets is the best way to confirm the organ being viewed is the pancreas. If they are *absent*, the organ is most likely a salivary gland. In addition, the presence of mucous cells also indicates the organ is a salivary gland.

GROSS ANATOMY

The Gastrointestinal (GI) Tract

EXERCISE 26.8

OVERVIEW OF THE GI TRACT

Before covering the details of individual organs that compose the digestive system, it is useful to do a short review of the structures encountered in the GI tract as one moves from mouth to anus.

1. Identify the structures listed in **figure 26.10** on a torso model, using the textbook as a guide.
2. Label the organs of the GI tract in figure 26.10. When labeling each organ, identify at least one digestive function of that organ.

(continued on next page)

(continued from previous page)

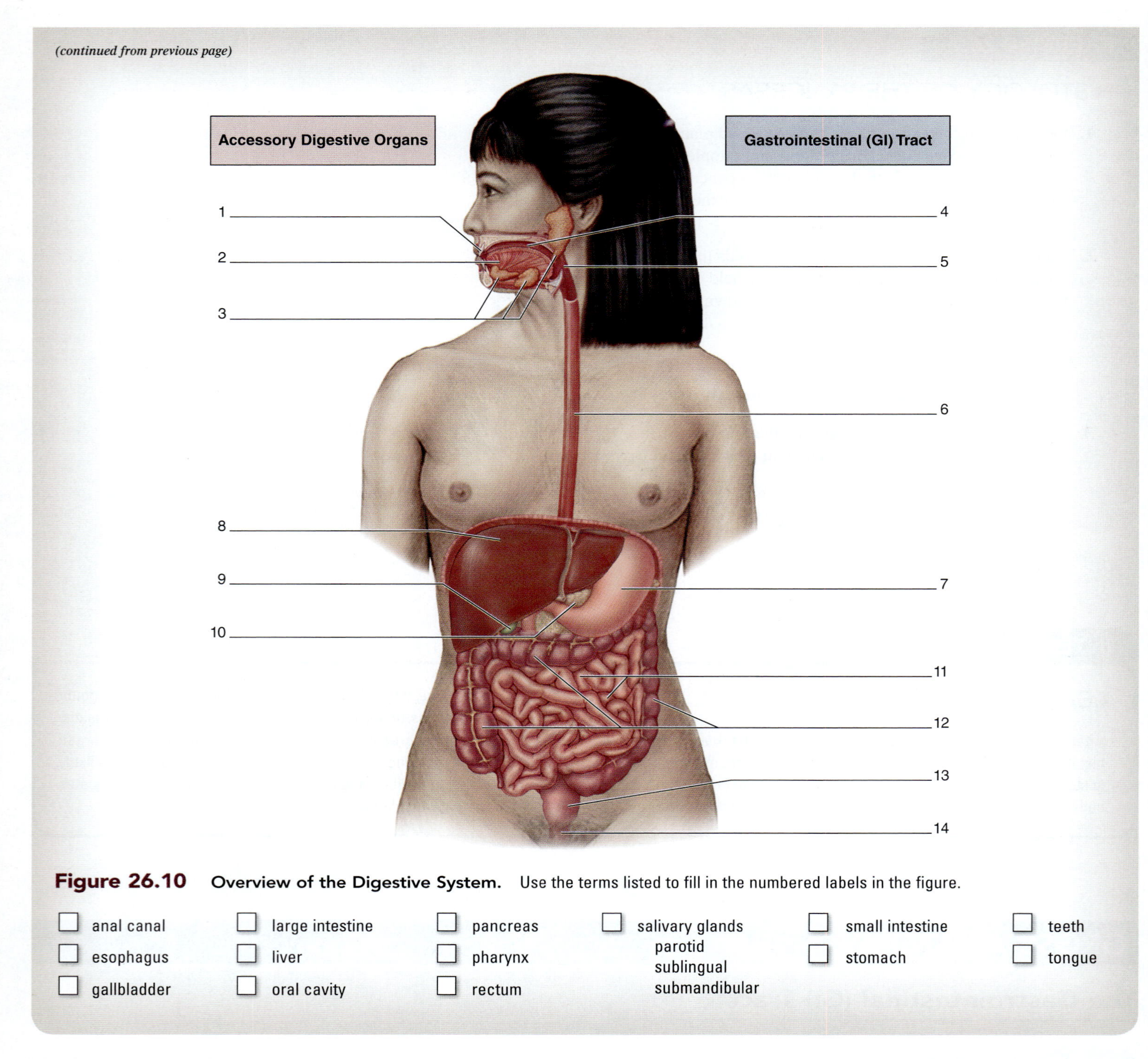

Figure 26.10 Overview of the Digestive System. Use the terms listed to fill in the numbered labels in the figure.

- ☐ anal canal
- ☐ esophagus
- ☐ gallbladder
- ☐ large intestine
- ☐ liver
- ☐ oral cavity
- ☐ pancreas
- ☐ pharynx
- ☐ rectum
- ☐ salivary glands
  - parotid
  - sublingual
  - submandibular
- ☐ small intestine
- ☐ stomach
- ☐ teeth
- ☐ tongue

The Oral Cavity, Pharynx, and Esophagus

The oral cavity contains a number of digestive system structures, including the teeth, salivary glands, lips, and tongue. These structures are important for wetting and manipulating food as it enters the GI tract. The resulting **bolus** (*bolos*, lump; bolus = a wet mass of food) leaves the oral cavity, travels through the oropharynx (*oris*, mouth, + *pharynx*, throat), and enters the esophagus (*oisophagos*, gullet), which transports the bolus to the stomach. The esophagus meets up with the stomach just after it passes through the esophageal hiatus in the diaphragm.

EXERCISE 26.9

GROSS ANATOMY OF THE ORAL CAVITY, PHARYNX, AND ESOPHAGUS

1. Observe a cadaver or a classroom model demonstrating the head and neck.
2. Using the textbook as a guide, identify the structures listed in **figure 26.11** on the cadaver or classroom model. Then label them in figure 26.11.

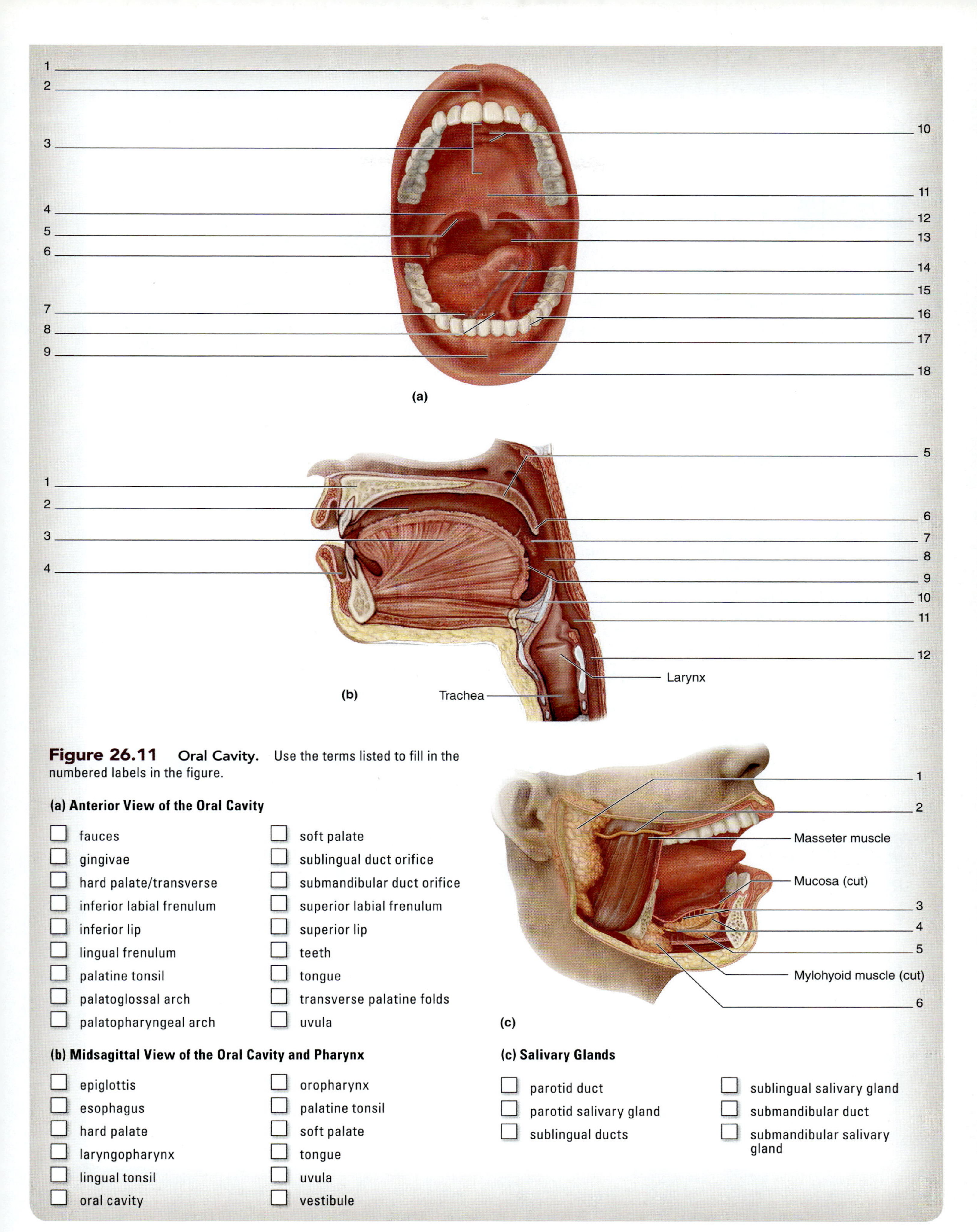

Figure 26.11 **Oral Cavity.** Use the terms listed to fill in the numbered labels in the figure.

(a) Anterior View of the Oral Cavity

- ☐ fauces
- ☐ gingivae
- ☐ hard palate/transverse
- ☐ inferior labial frenulum
- ☐ inferior lip
- ☐ lingual frenulum
- ☐ palatine tonsil
- ☐ palatoglossal arch
- ☐ palatopharyngeal arch
- ☐ soft palate
- ☐ sublingual duct orifice
- ☐ submandibular duct orifice
- ☐ superior labial frenulum
- ☐ superior lip
- ☐ teeth
- ☐ tongue
- ☐ transverse palatine folds
- ☐ uvula

(b) Midsagittal View of the Oral Cavity and Pharynx

- ☐ epiglottis
- ☐ esophagus
- ☐ hard palate
- ☐ laryngopharynx
- ☐ lingual tonsil
- ☐ oral cavity
- ☐ oropharynx
- ☐ palatine tonsil
- ☐ soft palate
- ☐ tongue
- ☐ uvula
- ☐ vestibule

(c) Salivary Glands

- ☐ parotid duct
- ☐ parotid salivary gland
- ☐ sublingual ducts
- ☐ sublingual salivary gland
- ☐ submandibular duct
- ☐ submandibular salivary gland

The Stomach

The stomach is a large, sac-like organ where both mechanical and chemical digestion continues on the bolus. The stomach is located in the epigastric abdominopelvic region, superficial to the pancreas, and deep to the anterior abdominal wall. **Table 26.6** lists the major features of the stomach and describes the function of each.

Table 26.6 Gross Anatomic Regions and Features Associated with the Stomach

	Description	Word Origin
Regions		
Body	Main part of the stomach located between the fundus and the pylorus	*body*, the principal mass of a structure
Cardia	Small, narrow, superior portion of the stomach where esophagus enters	*kardia*, heart; relating to the part of the stomach nearest the heart
Fundus	The dome-shaped part of the stomach that lies superior to the cardiac notch	*fundus*, bottom
Pylorus	The region of the stomach that opens into the duodenum	*pyloros*, a gatekeeper
Features		
Gastric Folds (rugae)	Folds of the mucosal lining of the stomach	*ruga*, a wrinkle
Greater Curvature	The large, inferior convex portion of the stomach. It is one attachment point for the greater omentum.	*greater*, larger
Greater Omentum	A fold of four layers of peritoneum that attaches to the greater curvature of the stomach, drapes over the abdominal contents, and folds back upon itself to attach to the transverse colon	*omentum*, the membrane that encloses the bowels
Inferior Esophageal (cardiac) Sphincter	A physiologic sphincter composed of the part of the diaphragm that surrounds the esophagus. When the diaphragm contracts, it closes off this opening, preventing reflux of stomach contents back into the esophagus. Some circular smooth muscle in the wall of the esophagus also contributes to this sphincter, but its contribution is weak.	*kardia*, heart; relating to the sphincter of the stomach nearest the heart
Lesser Curvature	The small, superior concave portion of the stomach. It is one attachment point of the lesser omentum.	*lesser*, smaller
Lesser Omentum	A fold of four layers of peritoneum that extends between the liver and the lesser curvature of the stomach	*omentum*, the membrane that encloses the bowels
Pyloric Sphincter	An anatomic sphincter composed of smooth muscle within the wall of the pylorus. It controls passage of chyme from the stomach to the duodenum.	*pyloros*, a gatekeeper, + *sphinkter*, a band

EXERCISE 26.10

GROSS ANATOMY OF THE STOMACH

1. Observe a cadaver or a classroom model demonstrating the stomach.
2. Identify the structures listed in **figure 26.12** on the cadaver or classroom model, using the textbook as a guide. Then label them in figure 26.12.
3. Sketch the stomach in the space provided. Be sure to label all of the structures listed in figure 26.12 in the drawing.

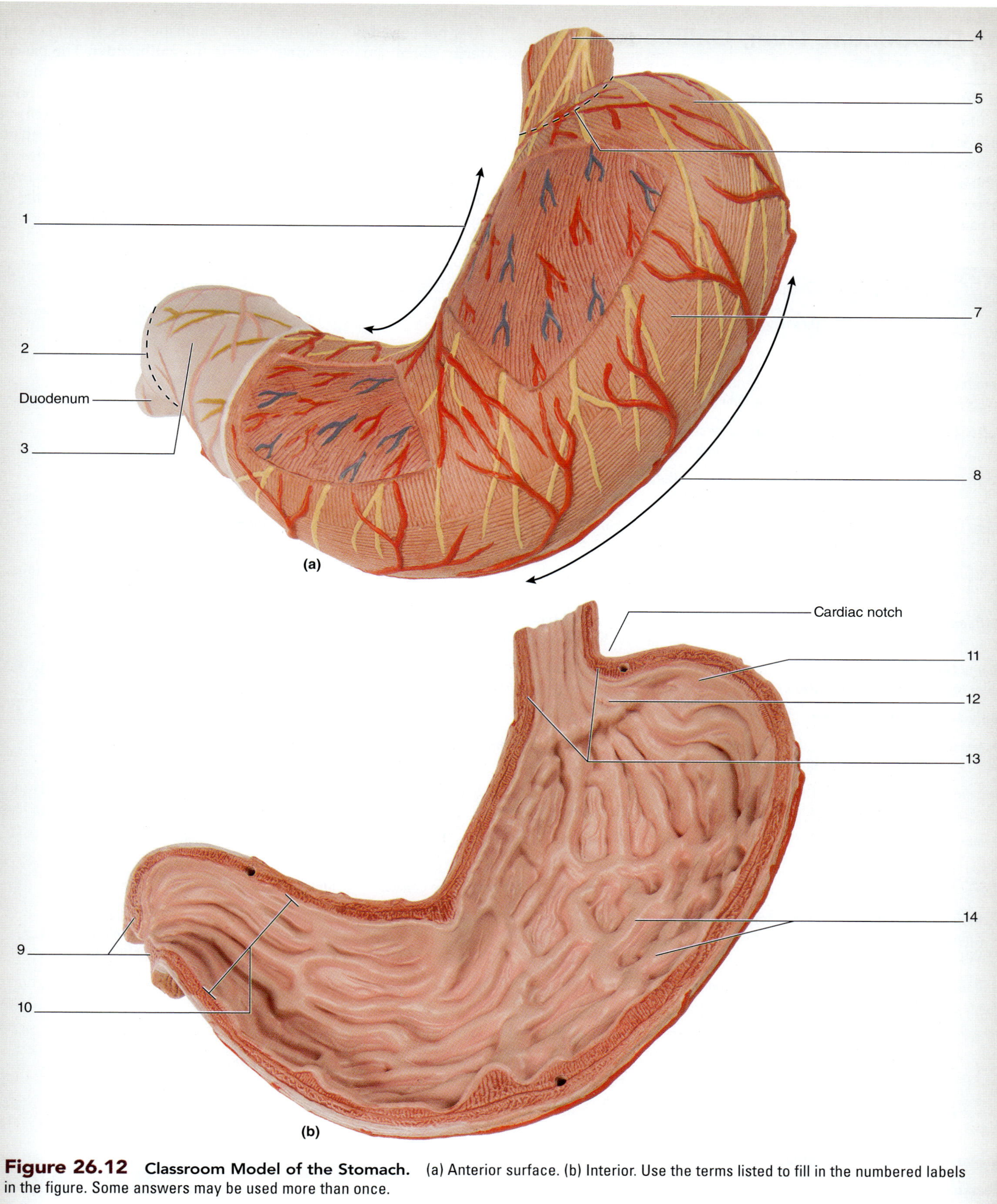

Figure 26.12 Classroom Model of the Stomach. (a) Anterior surface. (b) Interior. Use the terms listed to fill in the numbered labels in the figure. Some answers may be used more than once.

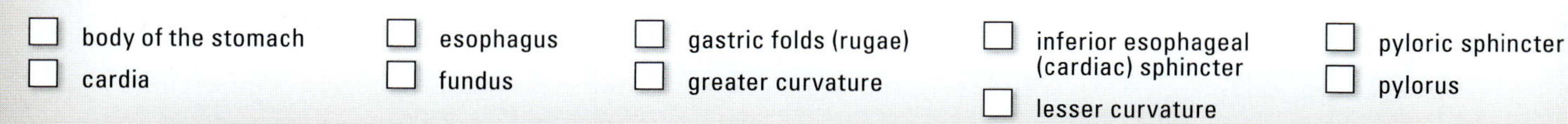

The Duodenum, Liver, Gallbladder, and Pancreas

The **duodenum** (*duodeno-*, breadth of twelve fingers) is the first part of the small intestine. It is C-shaped, and mostly retroperitoneal, which allows it to be anchored to the posterior abdominal wall. This is advantageous because a number of ducts coming from the liver, gallbladder, and pancreas empty their contents into the duodenum. The relationships between the duodenum and the liver, gallbladder, and pancreas are critically important for the process of digestion. The liver produces **bile,** a substance that emulsifies fats, which is temporarily stored and concentrated within the **gallbladder.** The pancreas produces **pancreatic juice**, which contains digestive enzymes and bicarbonate ion. When the liver, gallbladder, and pancreas release their secretions into the duodenum, the acidity of the chyme (*chymos*, juice) that has entered the duodenum from the stomach is neutralized and the digestion of proteins and carbohydrates continues. The digestion of fats and nucleic acids also begins in the duodenum. **Table 26.7** lists the major features of the liver, pancreas, and the duct system, and describes the functions of each component.

Table 26.7 Gross Anatomic Features of the Liver, Gallbladder, Pancreas, and Their Associated Ducts

	Description	Word Origin
Liver		
Falciform Ligament	A fold of peritoneum that extends from the diaphragm and anterior abdominal wall to the liver. Its free inferior border contains the round ligament of the liver.	*falx*, sickle, + *forma*, form
Left Lobe	The second largest lobe of the liver. Extends from the falciform ligament toward the midline of the body.	*lobos*, lobe
Porta Hepatis	A depression on the inferomedial part of the liver that contains the hepatic artery, hepatic portal vein, and common bile duct	*porta*, gate, + *hepatikos*, liver
Right Lobe	The largest lobe of the liver, it is on the right side of the abdomen and composes over half of the mass of the liver	*lobos*, lobe
Caudate Lobe	A small lobe of the liver located between the right and left lobes and on the posterior, inferior part of the liver	*caudate*, possessing a tail, + *lobos*, lobe
Quadrate Lobe	A small lobe of the liver located between the right and left lobes and on the anterior, inferior part of the liver between the gallbladder and the round ligament	*quadratus*, square, + *lobos*, lobe
Round Ligament of the Liver (ligamentum teres)	A remnant of the fetal umbilical vein, which connects to the umbilicus. Located within the free edge of the falciform ligament on the anterior abdominal wall.	*ligamentum*, a bandage, + *teres*, round
Gallbladder and Duct System		
Accessory Pancreatic Duct	Excretory duct located in the head of the pancreas. Empties into duodenum at the minor duodenal papilla.	*pankreas*, the sweetbread
Common Bile Duct	The bile duct formed from the union of the common hepatic duct and the cystic duct. Empties into the hepatopancreatic ampulla.	*bilis*, a yellow/green fluid produced by the liver
Common Hepatic Duct	Formed by the joining of the right and left hepatic ducts. Drains bile into the common bile duct.	*hepatikos*, liver
Cystic Duct	The bile duct that transports bile from the gallbladder to the common bile duct	*cystic*, relating to the gallbladder
Hepatopancreatic Ampulla	A duct formed by the joining of the common bile duct and the main pancreatic duct	*hepatikos*, relating to the liver, + *pancreatic*, relating to the pancreas, + *ampulla*, a two-handled bottle
Gallbladder	A sac-like appendage of the liver that stores and concentrates bile	*gealla*, bile, + *blaedre*, a distensible organ
Main Pancreatic Duct	The main excretory duct of the pancreas. Runs longitudinally in the center of the gland and empties into the duodenum at the major duodenal papilla.	*pankreas*, the sweetbread
Major Duodenal Papilla	A raised "nipple-like" bump located on the posterior wall of the descending part of the duodenum. The hepatopancreatic ampulla empties its contents here.	*major*, great, + *papilla*, a nipple
Minor Duodenal Papilla	A small raised "nipple-like" bump located superior to the major duodenal papilla. Contains the opening of the accessory pancreatic duct.	*minor*, smaller, + *papilla*, a nipple
Right/Left Hepatic Duct	Drain bile from the right lobe and left lobe of the liver, respectively	*hepatikos*, liver
Pancreas		
Body	The main portion of the pancreas extending between the head and the tail	*pankreas*, the sweetbread
Head	The portion of the pancreas that sits in the depression formed by the curvature of the duodenum	*pankreas*, the sweetbread
Tail	The tapered, right end of the pancreas located near the hilum of the spleen	*pankreas*, the sweetbread

EXERCISE 26.11

GROSS ANATOMY OF THE DUODENUM, LIVER, GALLBLADDER, AND PANCREAS

1. Obtain a classroom model demonstrating the relationships among the duodenum, liver, gallbladder, and pancreas **(figure 26.13),** or view these structures in the superior abdominal cavity of a prosected human cadaver.
2. Identify the structures listed in figure 26.13 on the classroom model or human cadaver, using table 26.7 and the textbook as guides. Then label them in figure 26.13.
3. Sketch the ducts coming from the liver, gallbladder, and pancreas in the space provided. Use table 26.7 and the textbook as guides. Show how the ducts merge to eventually empty their contents into the duodenum. Label each duct and organ in the drawing.
4. Obtain a model of the liver **(figure 26.14),** or view the liver from a human cadaver.
5. Identify the structures listed in figure 26.14 on the model of the liver or human cadaver, using table 26.7 and the textbook as guides. Then label them in figure 26.14.

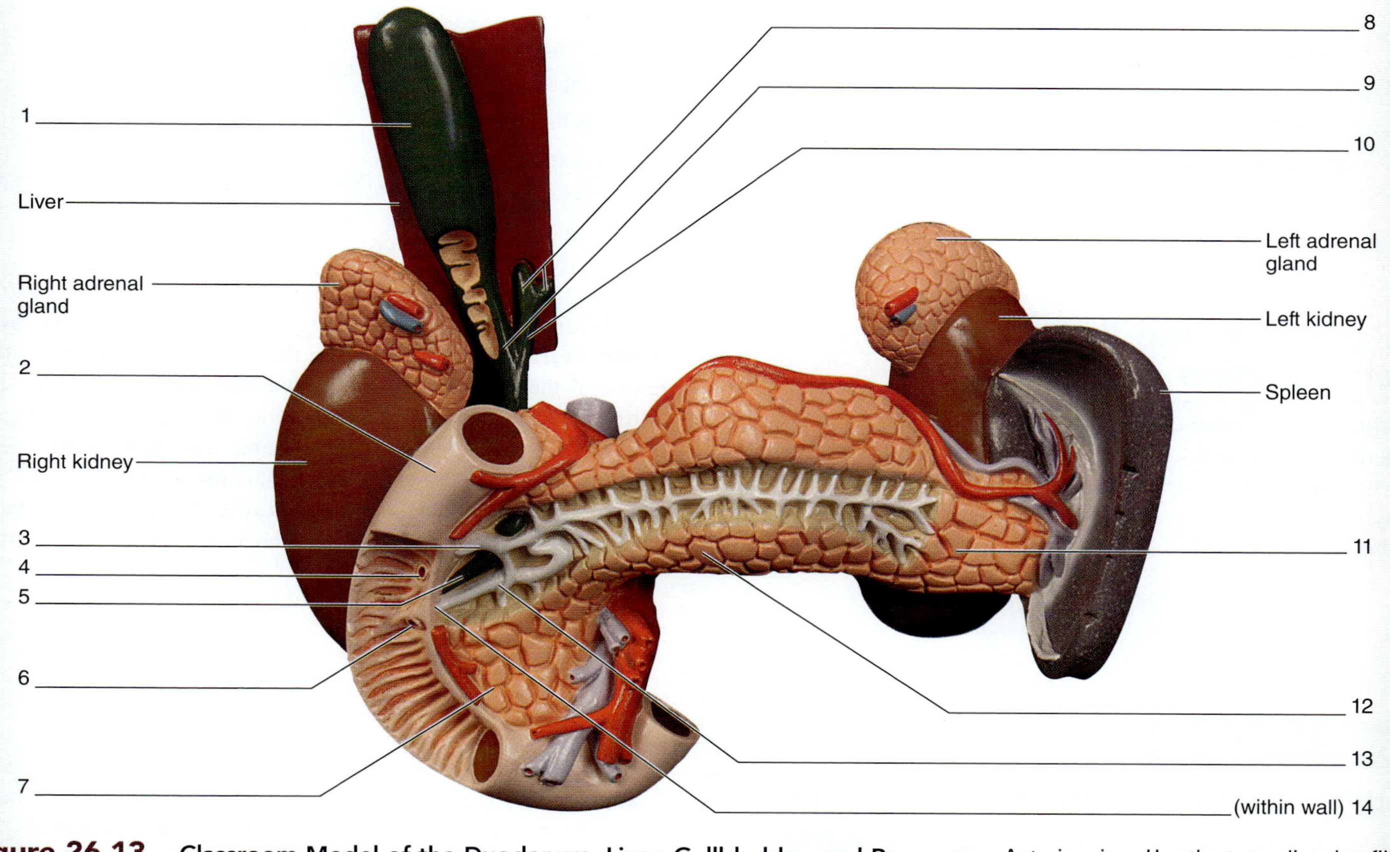

Figure 26.13 Classroom Model of the Duodenum, Liver, Gallbladder, and Pancreas. Anterior view. Use the terms listed to fill in the numbered labels in the figure.

- ☐ accessory pancreatic duct
- ☐ body of pancreas
- ☐ common bile duct
- ☐ common hepatic duct
- ☐ cystic duct
- ☐ duodenum
- ☐ gallbladder
- ☐ head of pancreas
- ☐ hepatopancreatic ampulla
- ☐ left and right hepatic ducts
- ☐ main pancreatic duct
- ☐ major duodenal papilla
- ☐ minor duodenal papilla
- ☐ tail of pancreas

(continued on next page)

(continued from previous page)

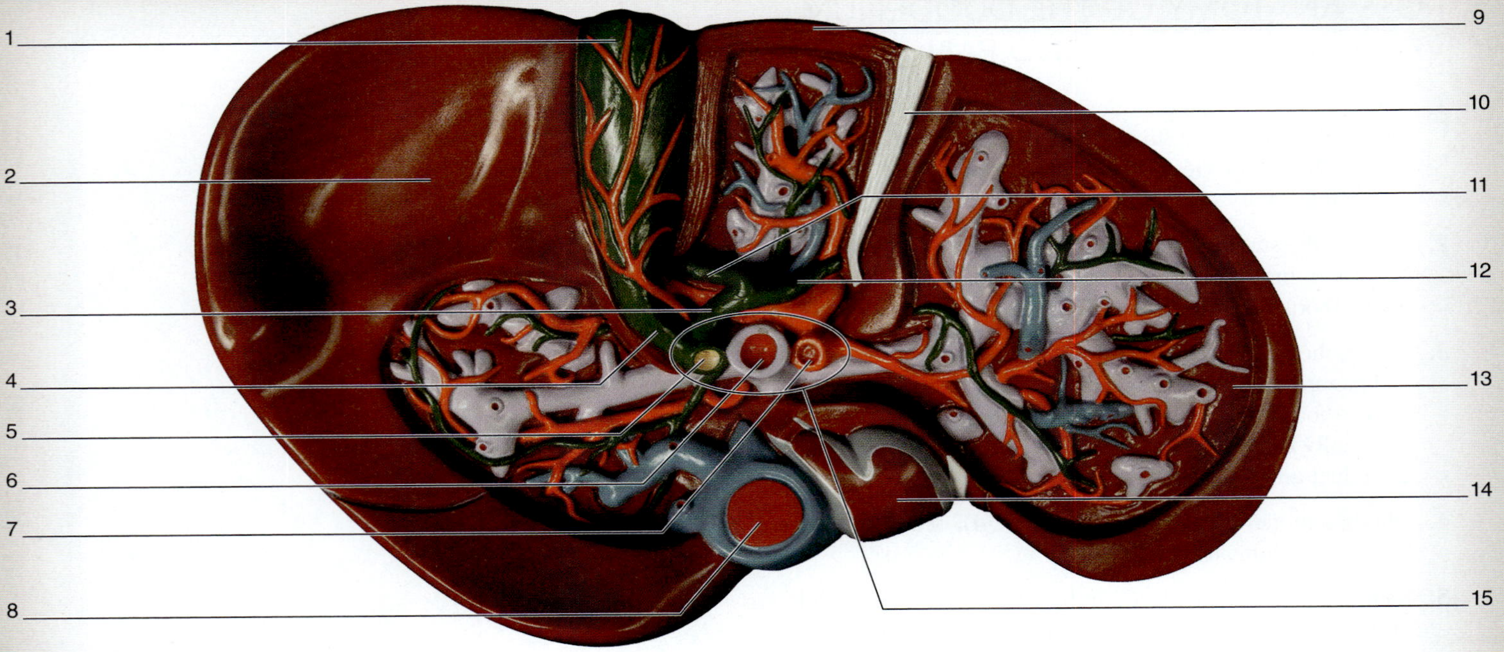

Figure 26.14 Classroom Model of the Liver and Gallbladder. Anteroinferior view. In this view, the anterior surface of the liver has been rotated superiorly so the structures on the inferior surface of the liver are visible.

- ☐ caudate lobe of liver
- ☐ common bile duct
- ☐ common hepatic duct
- ☐ cystic duct
- ☐ gallbladder
- ☐ hepatic artery proper
- ☐ hepatic portal vein
- ☐ inferior vena cava
- ☐ left hepatic duct
- ☐ left lobe of liver
- ☐ porta hepatis
- ☐ quadrate lobe of liver
- ☐ right hepatic duct
- ☐ right lobe of liver
- ☐ round ligament of liver

The Jejunum and Ileum of the Small Intestine

The **jejunum** and **ileum** compose the majority of the small intestine. In general the jejunum is located in the upper left part of the abdominal cavity and the ileum is located in the lower right part of the abdominal cavity. The two segments can be distinguished from each other both histologically and grossly. Distinguishing anatomic features include: circular folds, encroaching fat, arterial arcades, and vasa recta **(figure 26.15). Circular folds** are the mucosal folds found within the lumen of the small intestine. **Encroaching fat** is mesenteric fat that "rides up" upon the wall of the intestine. **Arterial arcades** are arching branches of the mesenteric arteries, and **vasa recta** (*vasa*, vessel, + *rectus*, straight) are straight vessels that come off of the arterial arcades and enter the small intestine proper. **Table 26.8** summarizes the gross anatomical features that distinguish the jejunum from the ileum.

EXERCISE 26.12

GROSS ANATOMY OF THE JEJUNUM AND ILEUM OF THE SMALL INTESTINE

1. Observe a classroom model of the abdominal cavity or the abdominal cavity of a prosected human cadaver in which the small intestine is intact.
2. Identify the gross structures listed in figure 26.15 on the classroom model of the abdomen or in the abdominal cavity of the human cadaver. Use table 26.8 and the textbook as guides.

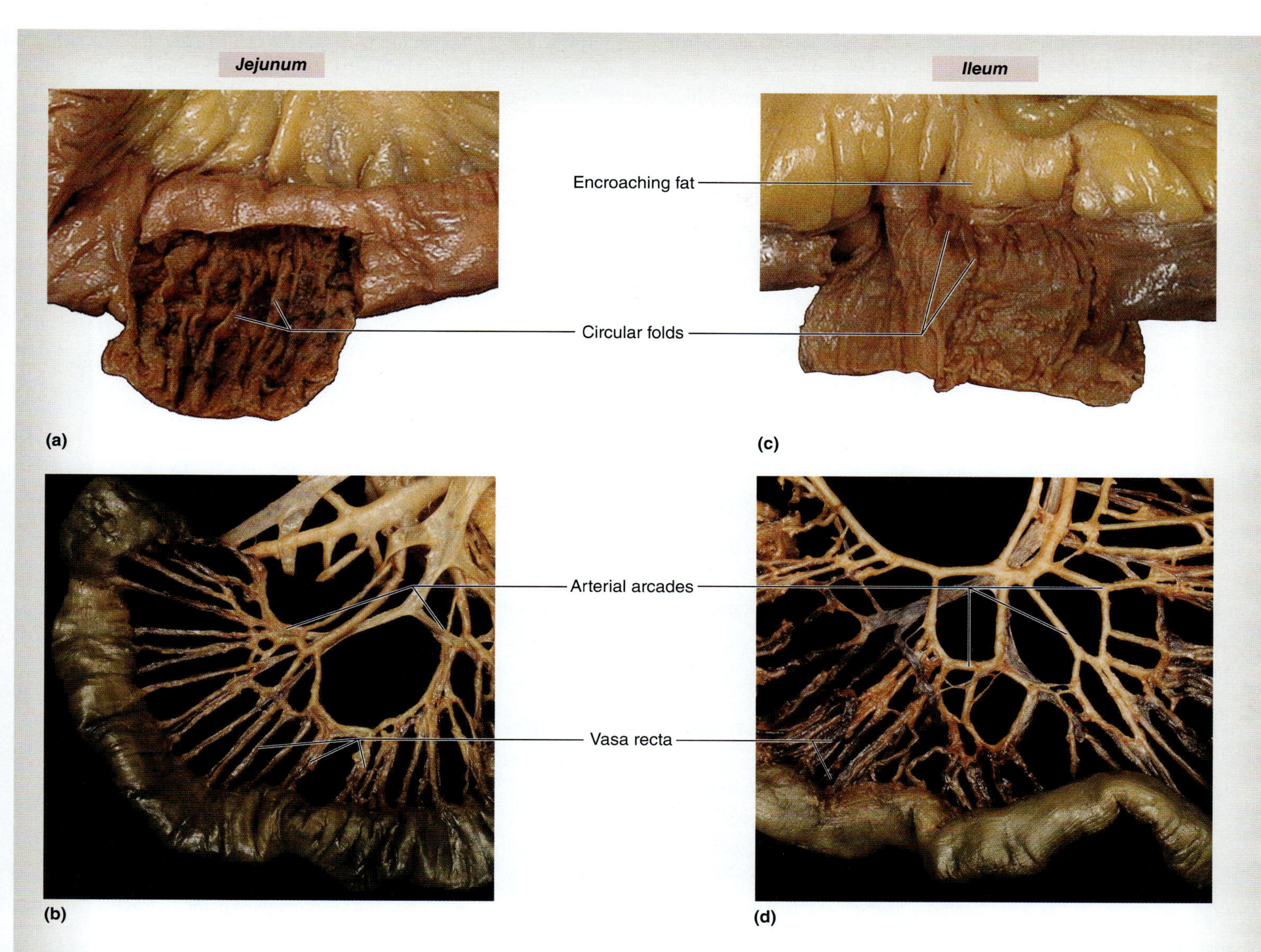

Figure 26.15 The Jejunum and Ileum. (a) The jejunum contains many deep circular folds and has no encroaching fat. (b) The blood vessels serving the jejunum consist of short arterial arcades and long vasa recta. (c) The ileum contains few shallow circular folds and has encroaching fat. (d) The blood vessels serving the ileum consist of large arterial arcades and short vasa recta.

Table 26.8 Gross Anatomic Differences Between the Jejunum and the Ileum

Part of the Small Intestine	Circular Folds	Encroaching Fat	Arterial Arcades	Vasa Recta
Jejunum	Deep, many	No	Fewer, larger	Longer
Ileum	Shallow, few	Yes	More, smaller, stacked upon each other	Shorter

The Large Intestine

The large intestine begins as a large sac called the **cecum,** which is located in the right inferior abdominopelvic quadrant. Extending from the cecum, the **colon** extends along the borders of the abdominal cavity as the ascending colon, transverse colon, and descending colon before entering the pelvic cavity via the sigmoid colon. The sigmoid colon empties into the **rectum,** which is located within the pelvic cavity. The rectum ends at the **anus. Table 26.9** summarizes these structures.

Table 26.9 Regions of the Large Intestine

Structure	Description	Word Origin
Cecum	Blind-ended sac located at the junction between the ileum of the small intestine and the ascending colon	*caecus*, blind
Ileocecal Valve	Smooth muscle sphincter located where the ileum opens into the cecum	*ileo-*, ileum, + *cecal*, cecum
Colon	The part of the large intestine that extends from the cecum to the rectum	*kolon*, large intestine
Ascending Colon	The part of the colon that extends from the cecum to the liver	*ascendere*, to climb up
Right Colic (hepatic) Flexure	Curve of the colon medial to the liver, where the transverse colon begins	*hepatikos*, relating to the liver, + *flexura*, a bend
Transverse Colon	The part of the colon that extends between the liver and the spleen	*transversus*, crosswise
Left Colic (splenic) Flexure	A curve of the colon medial to the spleen, where the descending colon begins	*splenic*, relating to the spleen, + *flexura*, a bend
Descending Colon	The part of the colon that extends from the left colic (splenic) flexure to the sigmoid colon	*descendere*, come down
Sigmoid Colon	The S-shaped part of the colon that extends from the descending colon to the rectum	*sigma*, the letter *S*, + *eidos*, resemblance
Rectum	Final portion of the GI tract, located within the pelvic cavity and extending from the sigmoid colon to the anus	*rectus*, straight
Anus	Inferior opening of the GI tract	*anus*, the lower opening of the GI tract
Associated Structures		
Haustra	Pouches of the colon formed when the taenia coli (longitudinal smooth muscle) contract	*haustus*, to draw up
Omental (epiploic) Appendices	Small, fatty appendages that hang off of the colon	*omentum*, the membrane that encloses the bowels; *epiploic*, related to the omentum
Taenia Coli	Three small bands of longitudinal smooth muscle of the muscularis externa of the colon; contraction of this muscle creates pouches (haustra) in the colon	*tainia*, a band, + *coli*, colon

EXERCISE 26.13

GROSS ANATOMY OF THE LARGE INTESTINE

1. Observe either a classroom model of the abdominal cavity that includes the large intestine, or the large intestine of a prosected human cadaver **(figure 26.16).**
2. Identify the gross structures listed in figures 26.16 and **26.17** on the classroom model of the abdomen or in the abdominal cavity of the human cadaver, using table 26.9 and the textbook as guides. Then label figures 26.16 and 26.17.
3. *Optional Activity:* APR **12: Digestive System**—Watch the "Digestive System Overview" animation to review the locations and functions of the major organs of the digestive system.

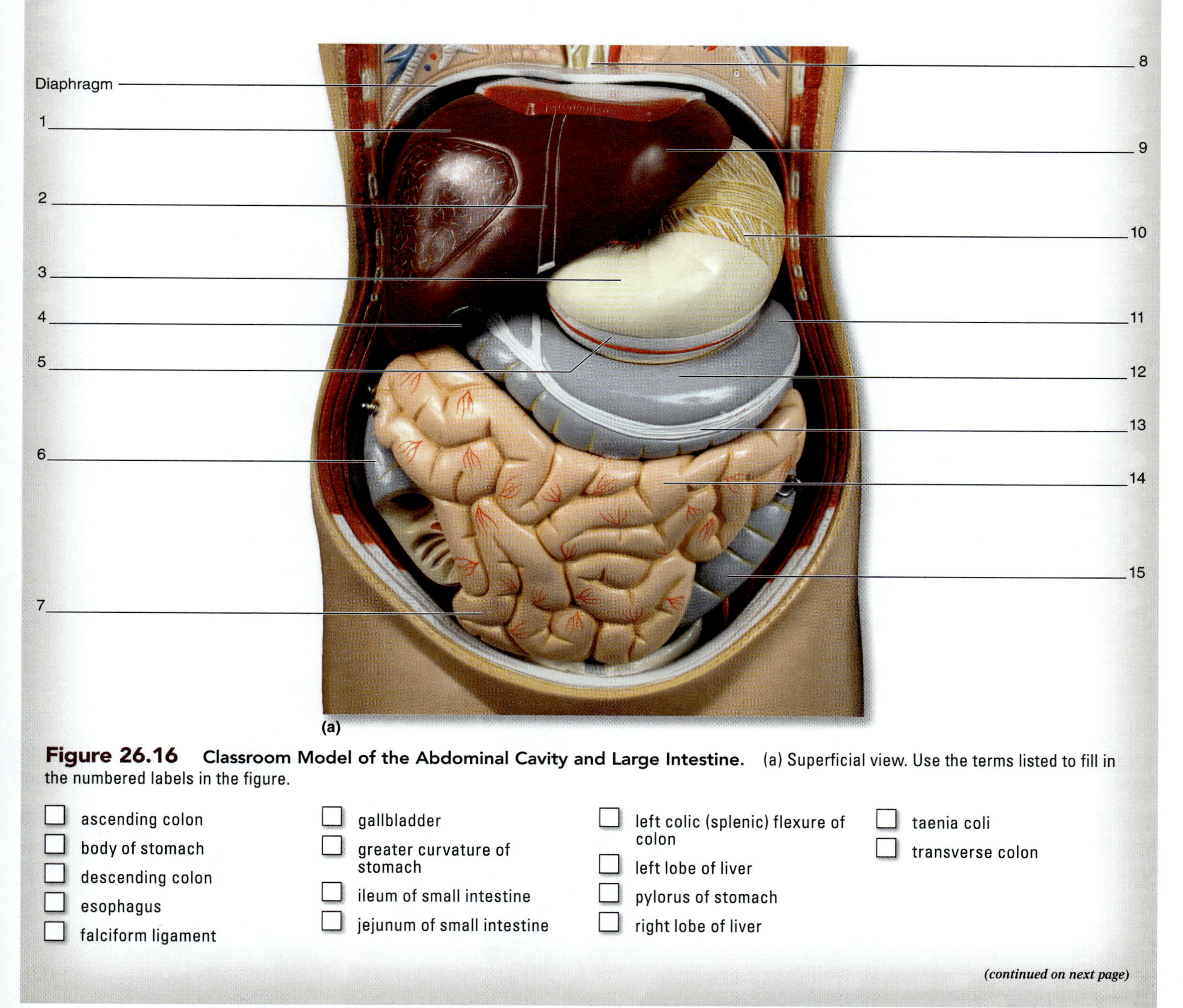

Figure 26.16 **Classroom Model of the Abdominal Cavity and Large Intestine.** (a) Superficial view. Use the terms listed to fill in the numbered labels in the figure.

- ☐ ascending colon
- ☐ body of stomach
- ☐ descending colon
- ☐ esophagus
- ☐ falciform ligament
- ☐ gallbladder
- ☐ greater curvature of stomach
- ☐ ileum of small intestine
- ☐ jejunum of small intestine
- ☐ left colic (splenic) flexure of colon
- ☐ left lobe of liver
- ☐ pylorus of stomach
- ☐ right lobe of liver
- ☐ taenia coli
- ☐ transverse colon

(continued on next page)

(continued from previous page)

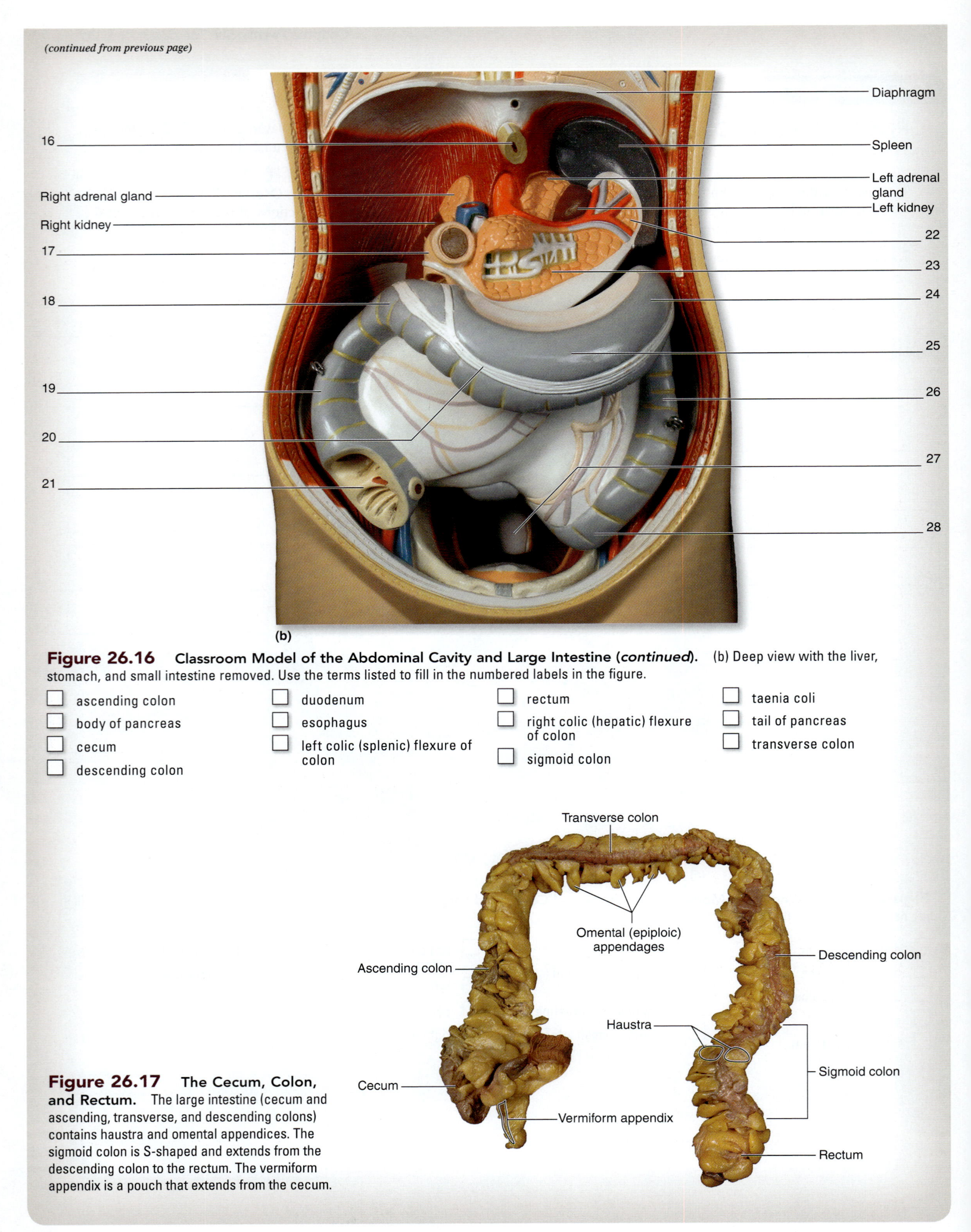

Figure 26.16 **Classroom Model of the Abdominal Cavity and Large Intestine (*continued*).** (b) Deep view with the liver, stomach, and small intestine removed. Use the terms listed to fill in the numbered labels in the figure.

- ☐ ascending colon
- ☐ body of pancreas
- ☐ cecum
- ☐ descending colon
- ☐ duodenum
- ☐ esophagus
- ☐ left colic (splenic) flexure of colon
- ☐ rectum
- ☐ right colic (hepatic) flexure of colon
- ☐ sigmoid colon
- ☐ taenia coli
- ☐ tail of pancreas
- ☐ transverse colon

Figure 26.17 **The Cecum, Colon, and Rectum.** The large intestine (cecum and ascending, transverse, and descending colons) contains haustra and omental appendices. The sigmoid colon is S-shaped and extends from the descending colon to the rectum. The vermiform appendix is a pouch that extends from the cecum.

PHYSIOLOGY

Digestive Physiology

Previous exercises in this chapter have detailed the anatomy of each of the digestive and accessory organs of the GI tract. The following laboratory exercises will explore some physiologic functions of these organs. The overall function of the digestive system is to supply the body with nutrients, electrolytes, and water that are obtained from food and drink. Overall, the digestive system provides the means to **ingest** food, mechanically and chemically **digest** macromolecules within the ingested food **(table 26.10)**, **propel** food along the GI tract, **secrete** products that aid in digestion, **absorb** nutrients, and **eliminate** wastes. Specialized functions of each of the digestive and accessory organs allow these overall processes to take place. The following paragraph summarizes the fate of a food item from the moment it enters the mouth to the moment it exits the anus.

As food enters the mouth, mechanical digestion of the entire food and chemical digestion of its carbohydrate components begins. Mechanical digestion (mastication) breaks down the food item into smaller pieces, which will allow enzymes to act on the macromolecules in the food and assist in their degradation (chemical digestion). The act of swallowing moves a bolus of moistened, partially digested food into the esophagus. This bolus is then transported through the esophagus by coordinated contractions of the longitudinal and circular layers of muscle that line the esophageal walls (peristalsis). From the esophagus, the bolus enters the stomach, where mechanical digestion (churning) continues and chemical digestion of proteins and fat begins. In addition, the highly acidic environment of the stomach promotes digestion of proteins. Digestion within the stomach converts the bolus that entered into a substance called chyme. From the

INTEGRATE

CLINICAL VIEW
Gallstones

Gallstones are hard stones that form from cholesterol and bile deposits in the gallbladder **(figure 26.18).** The condition of having gallstones is called *cholelithiasis* (*chole*, bile, + *lithos*, stone, + *-iasis*, condition). One of the most serious complications resulting from a gallstone occurs when the stone passes into the cystic duct and makes its way toward the duodenum, but becomes lodged somewhere en route to the duodenum. This blocks the flow of bile from the liver and gallbladder to the duodenum. If it lodges farther down, near the hepatopancreatic ampulla, it can also block the flow of pancreatic juice from the pancreas. Normally some of the pancreatic enzymes (proteolytic enzymes) are not activated until they enter the duodenum. However, when these enzymes build up within the pancreas because of a blockage of flow of pancreatic juice from the pancreas, these enzymes become activated and begin to digest the pancreas itself. This causes *pancreatitis* (inflammation of the pancreas), which can be life threatening.

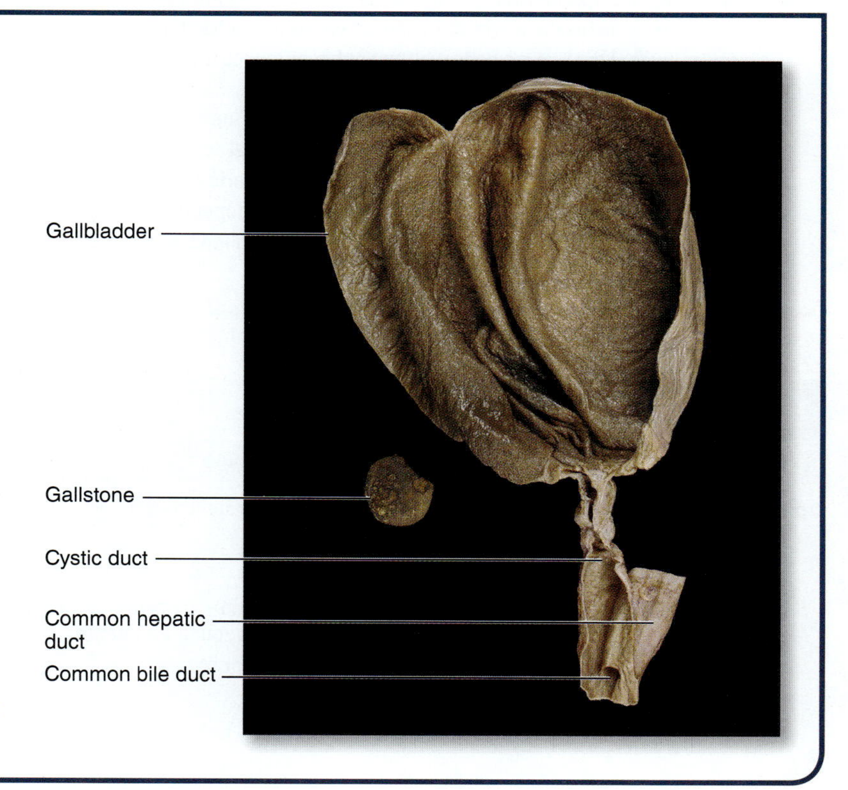

Figure 26.18 **Gallbladder and Gallstone.** The gallstone is a mass of bile salts that have crystallized and condensed. Note the small crystals that have formed on the larger stone. A stone of this size might not cause problems for the patient if it were too large to enter into the cystic duct.

Table 26.10 Digestion and Absorption of Nutrients

Substance	Subunits	Site of Digestion	Site of Absorption	Mechanism of Absorption	Word Origin
Carbohydrate	Monosaccharides (glucose, galactose, fructose)	Mouth, small intestine	Small intestine	Cotransport with sodium (glucose and galactose); facilitated diffusion (fructose); transport to hepatic portal vein	*carbo-*, carbon + *-hydro*, water
Lipid	Fatty acids, monoglycerides, glycerol	Stomach, small intestine	Small intestine	Diffuse into lacteals via chylomicrons; transported via lymph	*lipos*, fat
Nucleic Acid	Pentose sugars, nitrogenous bases, phosphates	Small intestine	Small intestine	Active transport; transport to hepatic portal vein	*nucleus*, kernel
Protein	Amino acids	Stomach, small intestine	Small intestine	Cotransport with sodium; transport to hepatic portal vein	*prote-*, primary + *-in*, in

stomach the acidic, liquid chyme is squeezed (3 mL at a time) into the duodenum of the small intestine. In the duodenum, secretions from accessory digestive organs—the liver, gallbladder, and pancreas—mix with the acidic chyme from the stomach to continue the chemical digestion of proteins, carbohydrates, fats, and nucleic acids. Pancreatic secretions (pancreatic "juice") contain enzymes that are required for protein, carbohydrate, fat, and nucleic acid digestion. Pancreatic secretions also contain bicarbonate ion, which is necessary to neutralize acids. The liver and gallbladder secrete bile into the duodenum. Bile is required for emulsification of fats. While all of this chemical digestion begins to take place, the small intestine also continues to mechanically digest the chyme using peristaltic contractions of its smooth muscle.

Following the digestion of carbohydrates, proteins, lipids, and nucleic acids into their absorbable form (monosaccharides, amino acids, fatty acids and monoglycerides, and the individual components of a nucleotide, respectively), each is transported from the GI tract into the blood or lymph. The surface area of the small intestine that is available for absorption is immense due to the presence of villi and microvilli. All nutrients (except lipids) are absorbed into the blood and then immediately enter the hepatic portal circulation (see chapter 20). In comparison, triglycerides (and other lipids) are packaged into specialized structures (chylomicrons) and absorbed into specialized lymph vessels called **lacteals** (see chapter 23). Water is reabsorbed along the length of the large intestine through the process of osmosis.

Any substances not absorbed remain within the lumen of the large intestine, are moved through the segments of the large intestine, and are eliminated as feces. **Table 26.11** summarizes the digestion of macromolecules, including the location and substances secreted that aid in the digestion of carbohydrates, proteins, lipids, and nucleic acids. Use the textbook as a guide while further exploring the processes of digestion, motility, and absorption in the following laboratory exercise.

INTEGRATE

CONCEPT CONNECTION

Motility in the GI tract is made possible by both an outer longitudinal and an inner circular layer of smooth muscle within the muscularis of the GI tract wall (see figure 26.2). Alternating waves of contraction of these two layers of muscle along the length of the tube moves substances along in a process termed peristalsis (*peri-*, around + *stalsis,* compression). Contraction of the outer, longitudinal layer decreases the length of the tube and increases its diameter; contraction of the inner, circular layer increases the length of the tube and decreases the diameter. This coordinated action is the same type of action that an earthworm uses to move about in the dirt. Contraction of the longitudinal layer of the worm's body wall musculature makes the worm short and fat, whereas contraction of the circular layer of the worm's body wall musculature makes the worm long and skinny (much like squeezing a tube of toothpaste).

These muscular contractions are coordinated by a network of nerves that lie adjacent to each muscle layer: the submucosal and myenteric plexuses. Collectively, these plexuses comprise the **enteric nervous system,** or "gut brain." While these nerves can operate independently, much like pacemaker cells in the heart, their activity is modulated by the autonomic nervous system.

Table 26.11 Digestion of Macromolecules

Location	Carbohydrates	Proteins	Lipids	Nucleic Acids
Oral Cavity (saliva)	Starch—*salivary amylase*⟶ partially digested starch	No protein digestion	*Lingual lipase* added but activated in low pH of stomach	No nucleic acid digestion
Stomach (gastric juice)	No additional enzymes added	Protein—*pepsin*⟶ polypeptide and peptide fragments	Triglyceride—*lingual lipase*⟶monoglyceride and fatty acids (limited amounts) Triglyceride—*gastric lipase*⟶monoglyceride and fatty acids (limited amounts)	No nucleic acid digestion
Small Intestine (pancreatic juice secreted into duodenum)	Partially digested starch—*pancreatic amylase*⟶ oligosaccharides, maltose, and glucose	Protein—*trypsin*⟶ polypeptide and peptide fragments Protein—*chymotrypsin*⟶ polypeptide and peptide fragments Protein—*carboxypeptidase*⟶ amino acids from carboxy-end of peptides	Triglyceride—*pancreatic lipase*⟶monoglyceride and fatty acids (within micelles)	DNA—*pancreatic deoxyribonuclease*⟶ deoxyribonucleotides RNA—*pancreatic ribonuclease*⟶ ribonucleotides
Small Intestine (brush border enzymes)	Oligosaccharides—*dextrinase* and *glucoamylase*⟶ maltose, glucose Maltose—*maltase*⟶glucose Lactose—*lactase*⟶glucose, galactose Sucrose—*sucrase*⟶glucose, fructose	Dipeptides—*dipeptidase*⟶ amino acids Peptides—*aminopeptidase*⟶amino acids from amino-end of peptides	No lipid digestion completed by brush border enzymes	Nucleotides—*phosphatase*⟶ nucleosides and phosphate Nucleosides—*nucleosidase*⟶ nitrogenous base and sugar (ribose or deoxyribose)

INTEGRATE

CLINICAL VIEW
Cholera

Diarrhea (*dia-*, through + *-rrhoea,* to flow) is a relatively common condition in which water is either not absorbed or is "pulled" by osmosis into the lumen of the GI tract. When substances such as ions or undigested materials remain within the lumen of the GI tract, water may be attracted to them, thereby preventing the absorption of water. This is what happens when some pathogens invade the intestinal wall and cause diseases such as cholera, a disease that occurs in areas where sanitary conditions are poor. **Cholera** is caused by a bacterium, *Vibrio cholerae*, which embeds itself within the mucosa of the small intestine. Once in place, the bacterium causes large quantities of ions, including sodium, potassium, and chloride, to be pumped into the lumen of the GI tract. The ions attract water via osmosis. The quantity of water within the GI tract becomes so great that the stretch receptors in the intestinal wall are stimulated, which increases gut motility. The result is production of massive amounts of watery diarrhea. Because dehydration is of great concern in a patient suffering from cholera, the "treatment" is administration of oral rehydration therapy (ORT). ORT promotes rehydration by taking advantage of the mechanisms of water absorption in the large intestine. That is, the solution contains sodium, glucose, and water. Sodium and glucose move across the epithelium of the intestinal wall via cotransport. Sodium is the body's greatest source of osmotic pressure. Thus, as sodium is reabsorbed from the lumen of the GI tract into the blood, glucose and, subsequently, water, will follow. Without treatment, a person suffering from cholera may lose up to 20 L of water per day, which can rapidly lead to death.

INTEGRATE

CONCEPT CONNECTION

Acidic and basic solutions play an important role in digestion. The stomach produces acidic chyme due to HCl produced by parietal cells. Acinar cells in the pancreas produce bicarbonate ion to neutralize the acidic chyme once it enters the duodenum of the small intestine. Both processes (production of acidic and basic secretions) involve the same reaction, the *bicarbonate reaction,* which was introduced in chapter 24: $CO_2 + H_2O \leftrightarrow H_2CO_3 \leftrightarrow H^+ + HCO_3^-$. In the stomach, hydrogen ions move into the lumen of the stomach; in the pancreas, bicarbonateions moves into the pancreatic duct to be dumped into the small intestine. While this reaction is important for digestion, it is also involved in regulating blood pH. As hydrogen ions are pumped to the lumen of the stomach, the bicarbonate ions move into the blood. Similarly, as bicarbonate ions move to the pancreatic duct, the hydrogen ions move into the blood. When a person vomits excessively, that person's parietal cells must compensate for the loss of acidic chyme by moving more H^+ into the lumen of the stomach and releasing more HCO_3^- into the blood. The excess HCO_3^- in the blood may raise the blood pH, which can lead to **metabolic alkalosis.** Likewise, when a person suffers from excessive diarrhea, pancreatic acinar cells must compensate to replace the loss of bicarbonate ion by moving more HCO_3^- into the lumen of the small intestine and moving more H^+ into the blood. The excess H^+ in the blood may decrease blood pH, which can lead to ***metabolic acidosis.*** Recall from chapter 24, that changes in breathing rates can lead to *respiratory* acidosis or *respiratory* alkalosis.

EXERCISE 26.14

DIGESTIVE ENZYMES

The purpose of this laboratory exercise is to examine the properties of the enzyme amylase, which is involved in the chemical digestion of starch, and to observe the effect that temperature has on amylase function.

Before beginning, state a hypothesis regarding the effect of temperature on the rate of chemical digestion.

____

____

____

Obtain the Following:

- **0.5% amylase solution**
- **0.5% starch solution**
- **37°C water bath**
- **6 test tubes**
- **Benedict's solution (or glucose test strips)**
- **hot plate**
- **ice**
- **Lugol's solution (iodine-potassium-iodide)**
- **porcelain spot plate**
- **test-tube clamp**
- **wax pencil**

1. Place three test tubes in a rack and label each with a wax pencil (#1, #2, and #3). Label the test tubes near the top of the test tube. The test tubes will be placed in boiling water. Thus, if the labels are below the water line they may come off. Fill test tube 1 with 6 mL of 0.5% amylase solution. Fill test tube 2 with 6 mL of 0.5% starch solution. Fill test tube 3 with 5 mL of 0.5% starch solution + 1 mL of 0.5% amylase solution. Shake each tube until the contents are mixed well.
2. To heat the tubes, place the three test tubes in a warm water bath, 37°C (98.6°F), for approximately 10 minutes. Remove the test tubes from the warm water bath.
3. To test solution #1 for the presence of starch, remove 1 mL of the solution from test tube #1 and place it in one depression of a spot plate. Add 1 drop of Lugol's solution to the test solution and observe the color. A blue-black color indicates the presence of starch. Record the color observations and conclusions regarding the presence (+) or absence (–) of starch in **table 26.12.**
4. Repeat step 3 for the solutions in test tubes #2 and #3.
5. **Test each solution for the presence of sugars (monosaccharides and disaccharides):** to do this,

(continued on next page)

(continued from previous page)

remove 1 mL of the solution from test tube #1 and place it in a clean test tube. Add 1 mL of Benedict's solution to the test tube, place the tube in boiling water for 2 minutes, and observe the color. Remember to always use a test-tube clamp when lowering the test tube into boiling water. The color may vary depending on the amount of sugar present in the solution. The range in color follows that of the visible light spectrum (red, orange, yellow, green, and blue), where red indicates most and green indicates least sugar present. A solution that is blue in color indicates no sugar present. Record the color observations and conclusions regarding the presence (+) or absence (–) of sugar in table 26.12. If using glucose test strips to test for glucose, do not add Benedict's solution and do not boil the test tube. Instead, simply dip the glucose test strip into the test tube containing solution #1 and pull it back out. Shake the test strip gently to remove excess solution. Then, compare the color of the test strip to the chart provided with the glucose test strips to determine the presence or absence of glucose in the solution.

6. Repeat step 5 for the solutions in test tubes #2 and #3.
7. **Test the effect of temperature on enzyme activity:** to do this, place 1 mL of 0.5% amylase solution in each of three clean test tubes. Label test tube 1 "cold," test tube 2 "warm," and test tube 3 "hot." Vary the temperature conditions for the solutions by placing test tube 1 in a beaker of ice water (~0°C), test tube 2 in a beaker with warm water (37°C), and test tube 3 in a beaker of boiling water (100°C). Use a test tube clamp to lower test tube 3 into the boiling water.
8. Remove test tube 1 from the water bath. Add 5 mL of 0.5% starch solution to the test tube and mix thoroughly. Note that 0.5% starch solution added to test tube 1 should be chilled to approximately 0°C prior to being added to the amylase solution. Return the test tube to the respective water bath for 10 minutes.
9. Remove the test tube from the water bath and repeat steps 3 and 5 to test for the presence of starch and sugar, respectively. Record the color of each solution and conclusions regarding the presence (+) or absence (–) of starch and sugar in table 26.12.
10. Repeat steps 8 and 9 for the solutions in test tubes 2 and 3. Be careful when handling and mixing the contents in test tube 3 because the solution is very hot.
11. Describe the effect of increasing temperature on enzymatic activity. ______________________________

____

____

____

Table 26.12 Data for Chemical Digestion Test

Test Tube #	Solution	Lugol's (Color)	Starch Present (+/–)	Benedict's (Color)	Sugar Present (+/–)
1	0.5% Amylase solution				
2	0.5% Starch solution				
3	0.5% Amylase & 0.5% Starch solution				
4	0.5% Amylase at 0°C				
5	0.5% Amylase at 37°C				
6	0.5% Amylase at 100°C				

EXERCISE 26.15

A CLINICAL CASE STUDY IN DIGESTIVE PHYSIOLOGY

The purpose of this exercise is to apply concepts related to digestive physiology to a real scenario, or clinical case study. This exercise requires information presented in this chapter as well as in previous chapters. Integrating multiple systems in a case study reinforces understanding of homeostatic processes.

Prior to beginning the exercise, ask the laboratory instructor if this exercise is to be completed individually, in pairs, or in groups. Use the textbook and this laboratory manual as guides to review the following topics that relate to the case study: digestive system, membrane transport, autonomic nervous system, immune system, cardiovascular system, respiratory system, and acid-base balance.

Case Study

Beth, an 18-month-old female, was developing normally until she started having episodes of vomiting and diarrhea. Her symptoms worsened over the six months since the time they began. At times, she suffered from ten to fifteen bouts of diarrhea per day. A physical exam revealed that Beth's abdomen was soft and nontender, and had no palpable masses. Fecal examinations were negative for parasites, and blood tests were normal. However, Beth's weight steadily dropped. Her pediatrician noted that she had decreased from the fiftieth percentile in weight at her 12-month checkup to the tenth percentile at her 18-month checkup. Her parents shared with the physician that Beth was becoming increasingly irritable, and at times could not be consoled.

Beth was referred to a gastroenterologist to determine the cause of the gastrointestinal symptoms and weight loss. The gastroenterologist performed an upper gastrointestinal (GI) endoscopy and a colonoscopy to visualize the mucosal layers of the GI tract. The examination revealed normal mucosa in Beth's esophagus, stomach, and duodenum. Multiple nodules were observed in the descending colon, sigmoid colon, and rectum. However the nodules were determined to be nonpathologic. These findings led the gastroenterologist to suspect that Beth was suffering from lactose intolerance, because milk had been introduced into Beth's diet around the same time that she began exhibiting gastrointestinal symptoms. However, blood tests revealed normal levels of lactase, the brush border enzyme required to digest lactose (the sugar found in dairy products). Gluten sensitivity (i.e., Celiac disease) was also ruled out, because Beth did not exhibit an abnormal immune response to gluten.

After ruling out the most common sources of diarrhea, the gastroenterologist next considered a more serious and rare diagnosis: presence of a hormonally active tumor, which could alter membrane transport across the intestinal epithelium. A computed tomography (CT) scan of Beth's abdomen revealed a 4.6 cm mass in the right paraspinal region, adjacent to the right kidney. A biopsy of the tumor demonstrated that the mass was a ganglioneuroma, a benign tumor that secretes the hormone vasoactive intestinal peptide (VIP). VIP is a hormone that resembles secretin, which is normally released by the small intestine. Secretin increases electrolyte secretion into the lumen of the GI tract, and increases gut motility. The mass was removed surgically, and Beth was thriving just 1 month later. Her prognosis for a full recovery was excellent.

Answer the following questions concerning the case study just presented:

1. Failure to gain weight may be due to a low intake of nutrients, increased metabolic needs, or malabsorption. Which of these possibilities was the cause of Beth's weight loss? Justify the answer. ______________________

____

____

2. The gastroenterologist initially suspected either lactose intolerance or gluten sensitivity. Describe how each condition would affect the processes of digestion and absorption.

____

____

____

3. Increased electrolyte secretion into the lumen of the gastrointestinal tract leads to secretory diarrhea. Describe how increasing the secretion of electrolytes into the lumen of the gastrointestinal tract could lead to excessive, watery stool and increased motility.

____

____

____

4. Chronic diarrhea is a common cause of acid-base and fluid balance disturbances in the blood. Based on Beth's symptoms, predict her blood pH (acidic/basic), respiratory rate (high/low), heart rate (high/low), and urine pH prior to treatment. Justify these predictions.

____

____

____

5. Which of the four conditions was she at most risk of: (a) metabolic acidosis, (b) metabolic alkalosis, (c) respiratory acidosis, or (d) respiratory alkalosis? Explain your answer. ______________________

____

____

6. Based on the name (ganglioneuroma) and location of the tumor, from what cells did the tumor likely originate?

____

Chapter 26: The Digestive System

Name: ______________________________

Date: ____________ **Section:** ____________

POST-LABORATORY WORKSHEET

The (1) corresponds to the Learning Objective(s) listed in the chapter opener outline.

Do You Know the Basics?

Exercise 26.1: Histology of the Salivary Glands

1. The cells within salivary glands that produce salivary amylase are ____________________ (mucous/serous) cells. (1)
2. Mary likes sour foods, and she decided to eat a slice of lemon. As she bit down on the lemon, she felt an uncomfortable squeezing-type sensation in her cheek as one of her salivary glands emptied its secretions into her mouth. Which of the three pairs of salivary glands did she feel? (Circle one.) (2)
   a. parotid
   b. sublingual
   c. submandibular

Exercise 26.2: Wall Layers of the Stomach

3. Place the layers of the stomach wall in the correct order, from innermost to outermost. (3)
   ______ a. mucosa
   ______ b. muscularis
   ______ c. serosa
   ______ d. submucosa
4. The stomach contains ____________________ (two/three) layers of smooth muscle in its muscularis layer. (4)
5. The mucosa of the stomach is lined with simple ____________________ (columnar/cuboidal) epithelial tissue. (5)

Exercise 26.3: Histology of the Stomach

6. In the cardia of the stomach, gastric pits are ____________________ (short/long), whereas gastric glands are ____________________ (short/long). (6)
7. Mucous neck cells are found in gastric ____________________ (glands/pits). (7)

Exercise 26.4: Histology of the Small Intestine

8. Which of the following structure(s) increase(s) the total surface area of the small intestine? (Check all that apply.) (8)
   ______ a. cilia
   ______ b. circular folds
   ______ c. microvilli
   ______ d. villi
9. ____________________ (Goblet/Transitional) cells in the small intestine produce mucous, which protects and lubricates the epithelium of the small intestine. (9)
10. The presence of Peyer patches is unique to the ileum. ____________________ (True/False) (10)

Exercise 26.5: Histology of the Large Intestine

11. Which of the following is the epithelial modification that is particularly abundant in the large intestine? (Circle one.) (11) (12)
    a. circular folds
    b. goblet cells
    c. microvilli
    d. submucosal glands

12. Place the wall layers of the large intestine in order from innermost to outermost. 13

_______ a. mucosa

_______ b. muscularis

_______ c. muscularis mucosa

_______ d. serosa

_______ e. submucosa

Exercise 26.6: Histology of the Liver

13. Goblet cells are abundant in the epithelium of the large intestine in order to provide lubrication for the passage of feces. ____________________ (True/False) 14

14. Hepatic lobules contain ____________________ (continuous, sinusoidal) capillaries, which receive both oxygen-rich blood from branches of the ____________________ (hepatic artery/hepatic portal vein) and nutrient-rich blood from branches of the ____________________ (hepatic artery/hepatic portal vein). This blood is transported to the central vein. 15 16

Exercise 26.7: Histology of the Pancreas

15. The two substances that are produced by the pancreas and released into the duodenum are ____________________ (bicarbonate/digestive enzymes), which is/are produced by acinar cells, and ____________________ (bicarbonate/digestive enzymes), which is/are produced by cells that line the pancreatic ducts. 17

16. An exocrine substance produced by ____________________ (acinar/islet) cells in the pancreas includes ____________________ (amylase/bicarbonate), which digests carbohydrates. 18

17. The endocrine part of the pancreas consists of ____________________ (acinar/islet) cells that produce ____________________ (hormones/pancreatic juice). 19

Exercise 26.8: Overview of the GI Tract

18. Place the organs of the GI tract in the correct order, from superior to inferior. 20

_______ a. anus

_______ b. esophagus

_______ c. large intestine

_______ d. small intestine

_______ e. rectum

_______ f. stomach

Exercise 26.9: Gross Anatomy of the Oral Cavity, Pharynx, and Esophagus

19. The esophagus passes through the esophageal ____________________ (hiatus/sphincter) in the diaphragm prior to entering the stomach. 21

Exercise 26.10: Gross Anatomy of the Stomach

20. A patient is suffering from gastroesophageal reflux disease (GERD). When she lies down, she feels a burning sensation in her esophagus caused by the reflux of stomach acids into the esophagus. This occurs because one of the sphincters in her stomach is not working properly. The sphincter that is not functioning properly is the ____________________ (cardiac/pyloric) sphincter. This sphincter is considered a(n) ____________________ (anatomic/physiologic) sphincter. The type of muscle tissue that composes this sphincter is ____________________ (skeletal/smooth) muscle. 22

Exercise 26.11: Gross Anatomy of the Duodenum, Liver, Gallbladder, and Pancreas

21. The ducts leading from the gallbladder and liver empty into the ____________________ (duodenum/pancreas). 23

22. Which of the following is/are a lobe of the liver? (Check all that apply.) 24

_______ a. caudate

_______ b. left

_______ c. posterior

_______ d. quadrate

_______ e. right

23. Place the following terms in the correct order in which bile is transported from its site of production in the liver to its entry into the duodenum. Assume that bile does not enter the gallbladder. 25

_______ a. common bile duct

_______ b. common hepatic duct

_______ c. left hepatic duct

_______ d. bile canaliculus

Exercise 26.12: Gross Anatomy of the Jejunum and Ileum of the Small Intestine

24. Loops of arteries that supply blood to the small intestine are called _______________ (arterial arcades/vasa recta), whereas straight vessels that supply blood to the small intestine are called _______________ (arterial arcades/vasa recta). 26

25. The ileum of the small intestine has _______________ (few/many) arterial arcades and _______________ (short/long) vasa recta, whereas the jejunum of the small intestine has _______________ (few/many) arterial arcades and _______________ (short/long) vasa recta. 27

Exercise 26.13: Gross Anatomy of the Large Intestine

26. Which of the following places the four parts of the colon in order? (Circle one.) 28

a. descending colon, transverse colon, ascending colon, sigmoid colon

b. ascending colon, transverse colon, descending colon, sigmoid colon

c. transverse colon, sigmoid colon, ascending colon, descending colon

d. ascending colon, sigmoid colon, transverse colon, descending colon

27. The _______________ (hepatic/splenic) flexure is located on the right side of the abdomen, whereas the _______________ (hepatic/splenic) flexure is located on the left side of the abdomen. 29

Exercise 26.14: Digestive Enzymes

28. Bile is produced by the _______________ (liver/pancreas), pepsin is produced by the _______________ (pancreas/stomach), and amylase is produced by both the _______________ (salivary glands/stomach) and the _______________ (pancreas/liver). 30

29. Within normal physiologic range, as temperature increases, enzyme activity _______________ (increases/decreases) and as temperature decreases, enzyme activity _______________ (increases/decreases). 31

30. Initial digestion of carbohydrates occurs in the oral cavity. _______________ (True/False) 32

Exercise 26.15: A Clinical Case Study in Digestive Physiology

31. Increased sodium secretion into the GI tract is likely to _______________ (increase/decrease) motility and _______________ (increase/decrease) absorption. 33

Can You Apply What You've Learned?

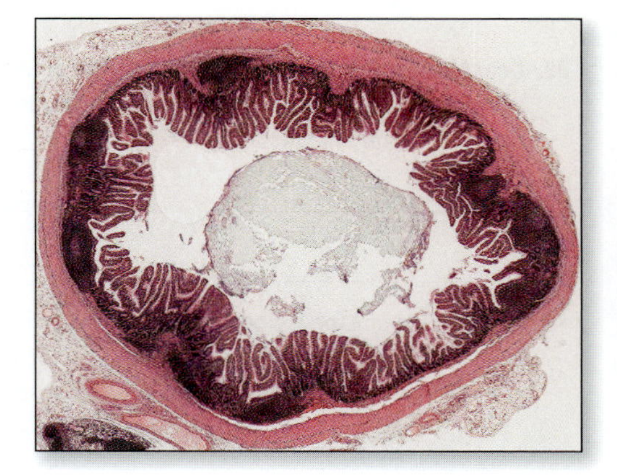

32. The image to the right is a cross section through part of the small intestine.
    a. What part of the small intestine is it (duodenum, jejunum, or ileum)?
    ___
    b. What characteristic(s) were used in determining which part of the small intestine this sample was taken from?
    ___
    ___
33. Using knowledge of the ducts draining bile from the liver and gallbladder and the pancreatic ducts, propose a location where a lodged gallstone might cause pancreatitis.___
    ___
    ___
34. A patient presented to his physician with pain in the left lower quadrant of his abdomen. The source of the pain was an adhesion* between the visceral peritoneum covering part of the patient's colon and the parietal peritoneum lining his anterolateral abdominopelvic wall in that region. What part of the colon was most likely adhered to the abdominopelvic wall?
    ___
    ___
35. Describe why a person suffering from massive bouts of diarrhea is also at risk for electrolyte imbalance.___
    ___
    ___
36. Acidic chyme can denature and inactivate enzymes in the small intestine. Describe how an inability to produce bicarbonate in the pancreas can lead to incomplete digestion and issues with absorption. ___
    ___
    ___
37. Lactose intolerance results from an inability to digest the disaccharide lactose because the enzyme lactase is lacking from the brush border of the small intestine. Lactase is responsible for breaking down lactose into its monomer units, glucose and galactose. Describe why a person suffering from lactose intolerance will commonly suffer from bouts of diarrhea. ___
    ___
    ___

Can You Synthesize What You've Learned?

38. Weight-loss drugs often work by reducing the amount of fat absorbed from the GI tract into the blood. For instance, Orlistat (Alli®) reduces the activity of lipases in the small intestine. Describe how Orlistat will influence the chemical digestion and absorption of fats, and discuss possible side effects when consuming a high-fat meal. ___
    ___
    ___

*An adhesion (*adhaereo,* to stick to) in the abdominopelvic cavity is an area where two layers of peritoneum are stuck to each other with connective tissue. It usually is the result of some sort of injury or inflammation.

39. Celiac disease is an autoimmune disorder whereby chronic inflammation leads to the destruction of the epithelium of the small intestine. More specifically, the villi in the small intestine begin to atrophy (*a-*, without + *-trophy*, nourishment). Describe how celiac disease will affect chemical digestion and absorption of ingested food. ________________

40. *Escherichia coli* (or *E. coli*) is a bacterium that frequently causes severe gastrointestinal symptoms. Ingesting food that is contaminated with *E. coli* often leads to bouts of diarrhea. If a person becomes infected with *E. coli*, describe what might happen to the blood pH as a result of diarrhea. Explain why the change in pH occurs, and predict how the respiratory rate may change to compensate for the difference. ________________

41. Metformin is a drug frequently prescribed to treat patients who are unable to keep their blood glucose levels under control, such as patients with type II diabetes. Among its actions, metformin works to decrease glucose absorption in the GI tract. Describe some potential gastrointestinal side effects that may occur when taking this drug. ________________

The Reproductive System and Early Development

OUTLINE AND LEARNING OBJECTIVES

MODULE 14: REPRODUCTIVE SYSTEM

INTRODUCTION

Of all the systems of the human body, the **reproductive system** (*re-*, again, + *productus*, to lead forth) is the only system that is not required for the survival of the human body. Instead, its function is to ensure the survival of the species by allowing for the formation of another human being. Reproduction requires two separate **sexes:** male and female. Each sex produces **gametes** (sperm in males; ova in females), which are haploid cells that combine to form a **zygote,** the first stage in the development of a new human. Male and female gametes are both formed by **meiosis,** which occurs within the **gonads** (*gone*, seed), which are the testes in the male and ovaries in the female. The human body invests a great deal of energy in the formation of gametes and in the maintenance of characteristics that are meant to attract members of the opposite sex. This much energy expenditure may seem counterintuitive when one thinks about homeostasis as the primary driving force for most organ system functions. However, when failure to reproduce could result in extinction of the species, the amount of energy invested makes more sense.

Although the reproductive structures in males and females are different, all human embryos develop from the same overall plan. Certain embryonic structures degrade in males but not in females. On the other hand, many embryonic structures persist in both males and females—forming apparently very different adult structures. Such structures are called **homologues** (*homos*, the same, + *logos*, relation). Ovaries and testes are examples of homologous structures. While exploring the male and female reproductive systems, pay particular attention to homologous structures. This will not only simplify the task of learning reproductive system structures, it will highlight the similarities between the male and female reproductive systems.

The exercises in this chapter explore the histological and gross structure of the male and female reproductive systems, with a focus on the histology of the primary reproductive organs of both females and males, the *ovaries* and *testes*. Observing the histology of the gonads provides background needed to understand the stages of gametogenesis (*gameto-*, referring to gametes + *genesis*, birth), which are covered in detail in the physiology section of this chapter. The focus of gross anatomic observations of the reproductive systems is on regional relationships between gross reproductive structures and structures of other body systems. Additional observations focus on homologous relationships between male and female reproductive structures. The study of anatomical structures will be related to the process of meiosis and hormonal regulation of gametogenesis for both males and females. A clinical case exercise will involve infertility and concepts of the reproductive system applied to a real-life scenario. Finally, two exercises will explore the processes involved in early development, from fertilization to embryonic development.

List of Reference Tables

Chapter 27: The Reproductive System and Early Development

Name: ____________________

Date: __________ Section: __________

These Pre-Laboratory Worksheet questions may be assigned by instructors through their connect course.

PRE-LABORATORY WORKSHEET

1. Which of the following is the female gonad? (Circle one.)
   a. clitoris
   b. oocyte
   c. ovary
   d. uterus
   e. vagina

2. Which of the following is the male gonad? (Circle one.)
   a. epididymis
   b. penis
   c. prostate
   d. seminiferous tubules
   e. testis

3. Human somatic cells contain twenty-two pairs of autosomal chromosomes and one pair of sex chromosomes. ____________ (True/False)

4. Spermatogenesis is the ____________ (maturation/production) of sperm, whereas spermiogenesis is the ____________ (maturation/production) of sperm.

5. Vesicular (Graafian) follicles contain a ____________ (primary/secondary) oocyte.

6. A secondary (maturing) follicle contains a ____________ (primary/secondary) oocyte.

7. Which of the following wall layers of the uterus is shed during menstruation? (Circle one.)
   a. endometrium
   b. myometrium
   c. perimetrium

8. Which of the following are male accessory reproductive structures? (Check all that apply.)
   ______ a. bulbourethral glands
   ______ b. epididymis
   ______ c. penis
   ______ d. prostate gland
   ______ e. seminal vesicles

9. Which of the following ligaments anchors the ovary to the body wall of the uterus? (Circle one.)
   a. broad ligament
   b. ovarian ligament
   c. round ligament of the uterus
   d. suspensory ligament of the ovary
   e. uterosacral ligament

10. A surge in ____________ (FSH/LH) causes ovulation in females.

11. In meiosis, there is/are ____________ (one/two) nuclear division(s), whereas in mitosis there is/are ____________ (one/two) nuclear division(s).

HISTOLOGY

Female Reproductive System

The female reproductive system has two important functions: produce the female gametes (ova) and support, protect, and nourish the developing embryo/fetus that is formed when an ovum becomes fertilized by a sperm. Components of the female reproductive system include the ovaries, uterine tubes, uterus, vagina, and mammary glands. Exercises 27.1 to 27.4 explore the anatomic details of the ovary, uterine tube, uterus, and vagina. Mammary glands are covered in Exercise 27.12.

EXERCISE 27.1

HISTOLOGY OF THE OVARY

The **ovary** is the primary reproductive organ in the female and functions in the production of the ova (eggs) and the female sex steroid hormones **estrogen** and **progesterone.** The ovaries contain several **ovarian follicles,** all at various stages of development. Each follicle contains a developing **oocyte (table 27.1).** The structure and function of the ovary is best viewed at the histological level because characteristics of ovarian follicles are evident at each stage of development.

1. Obtain a histology slide of an ovary and place it on the microscope stage.
2. Bring the tissue sample into focus on low power and identify the outer **cortex** and inner **medulla** of the ovary **(figure 27.1).** Next, move the microscope stage so the ovarian cortex is at the center of the field of view and then change to high power.
3. Observe the outermost region of the ovarian cortex. Locate the two layers of tissue that compose the outer coverings of the ovary: the germinal epithelium and the tunica albuginea **(figure 27.2).** The **germinal epithelium** (*germen*, a sprout) is a simple cuboidal epithelium that forms the outermost covering of the ovary. The name *germinal* refers to the fact that scientists once thought that the germ cells (oocytes) were formed from this epithelial layer. Most ovarian cancers arise in the germinal epithelium. Deep to the germinal epithelium is the **tunica albuginea** (*tunica*, a coat, + *albugineus*, white spot), which is composed of dense irregular connective tissue.
4. Focus on the cortex and identify follicles in each stage of follicular development listed in **table 27.2.**
5. Identify the following structures on the slide of the ovary, using tables 27.1 and 27.2 and figures 27.1 and 27.2 as guides:

- ☐ **corpus albicans**
- ☐ **corpus luteum**
- ☐ **germinal epithelium**
- ☐ **primary follicle**
- ☐ **primary oocyte**
- ☐ **primordial follicle**
- ☐ **secondary follicle**
- ☐ **secondary oocyte**
- ☐ **tunica albuginea**
- ☐ **vesicular (Graafian) follicle**

Table 27.1 Developmental Stages of an Oocyte

Cell Name/ Oocyte Stage	Description and Function	Stage of Mitosis/ Meiosis	Ploidy	Word Origin
Oogonia	Primitive germ cells that undergo mitosis in the second to fifth months of embryonic life form oogonia that will subsequently develop into primary oocytes. Approximately 70% of the primary oocytes will degenerate in a process called atresia.	Formed by mitosis; undergo mitosis to form primary oocytes	Diploid (2n)	*oon*, egg, + *gonia*, generation
Primary Oocyte	Oogonia has replicated DNA and has begun the process of meiosis and arrest in prophase of meiosis I. Primary oocytes will not undergo subsequent meiotic divisions unless the follicle is stimulated to mature by follicle-stimulating hormone (FSH) and luteinizing hormone (LH).	Arrested in Prophase I of meiosis	Diploid (2n)	*primary*, first, + *oon*, egg, + *kytos*, cell
Secondary Oocyte	Under the influence of FSH and LH (after puberty), primary oocytes complete meiosis I producing two cells of unequal sizes. The larger cell becomes a secondary oocyte and the smaller one becomes a polar body, which degenerates. The secondary oocyte will not undergo subsequent meiotic divisions unless fertilization occurs	Arrested in Metaphase II of meiosis	Haploid (1n)	*secondary*, second, + *oon*, egg, + *kytos*, cell
Definitive Oocyte	Upon fertilization, the secondary oocyte undergoes the second meiotic division to become an ovum	Formed after secondary oocyte is fertilized and completes meiosis	Haploid (1n)	*oon*, egg, + *kytos*, cell

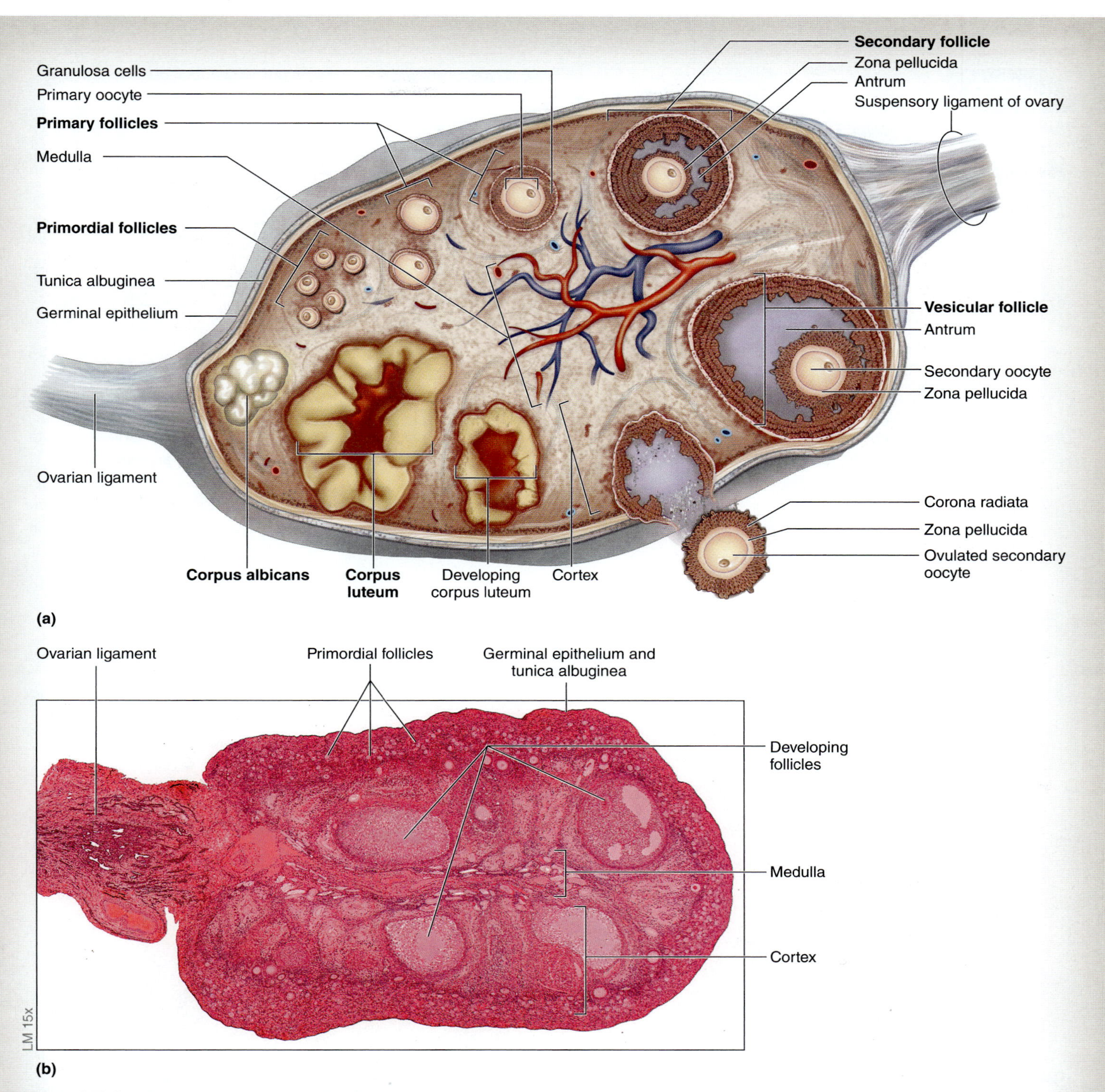

Figure 27.1 The Ovary. (a) The ovary is the primary reproductive organ of the female and produces the female gametes, the ova. (b) Histological structure of the ovary demonstrating the outer cortex containing follicles in various stages of development, and the inner medulla, which consists mainly of blood vessels and nerves.

6. Next move the microscope stage so a vesicular follicle is in the center of the field of view. A **vesicular (Graafian) follicle (figure 27.3)** is a mature follicle that is ready to be released during ovulation, after stimulation by a surge in luteinizing hormone (LH), which is secreted by the anterior pituitary gland. **Table 27.3** lists the histological features of a vesicular follicle. Identify the following parts of the vesicular follicle, using table 27.3 and figure 27.3 as guides:

- ☐ **antrum**
- ☐ **corona radiata**
- ☐ **cumulus oophorus**
- ☐ **granulosa cells**
- ☐ **secondary oocyte**
- ☐ **theca externa**
- ☐ **theca interna**
- ☐ **zona pellucida**

7. Identify a corpus luteum and corpus albicans on the slide, using table 27.2 as a guide.

(continued on next page)

(continued from previous page)

Table 27.2 Developmental Stages of Ovarian Follicles

Follicle Stage	**Primordial Follicle**	**Primary Follicle**	**Secondary (Maturing) Follicle**
Photograph	LM 500x	LM 500x	LM 50x
Description and Function	Contains a primary oocyte surrounded by a single layer of flattened follicular cells	Contains a primary oocyte surrounded by one or more layers of cuboidal follicular cells	Contains a full-size primary oocyte surrounded by an extracellular glycoprotein coat (the zona pellucida), which separates it from the corona radiata. Contains one or more fluid-filled antra. Thecal cells develop into internal and external layers.
Oocyte Stage	Primary oocyte	Primary oocyte	Primary oocyte
Word Origins	*primus*, first, + *ordior*, to begin, + *folliculus*, a small sac	*primus*, first, + *ordior*, to begin, + *folliculus*, a small sac	*secunda*, second, + *ordior*, to begin, + *folliculus*, a small sac

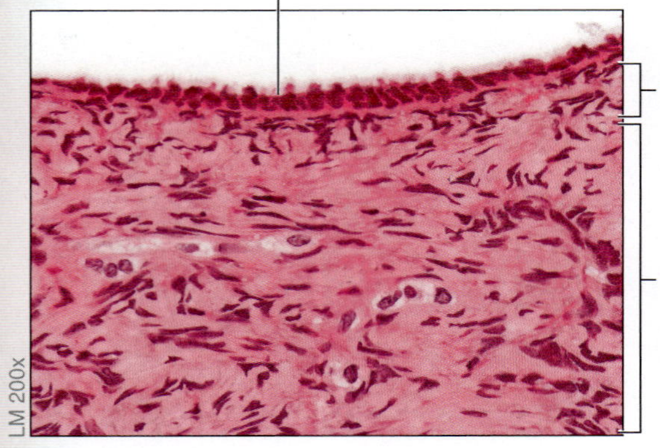

Figure 27.2 Germinal Epithelium and Tunica Albuginea of the Ovary. The germinal epithelium is the most common origin site for ovarian cancers. The tunica albuginea is a "white coat" of dense irregular connective tissue that surrounds the entire ovary internal to the germinal epithelium.

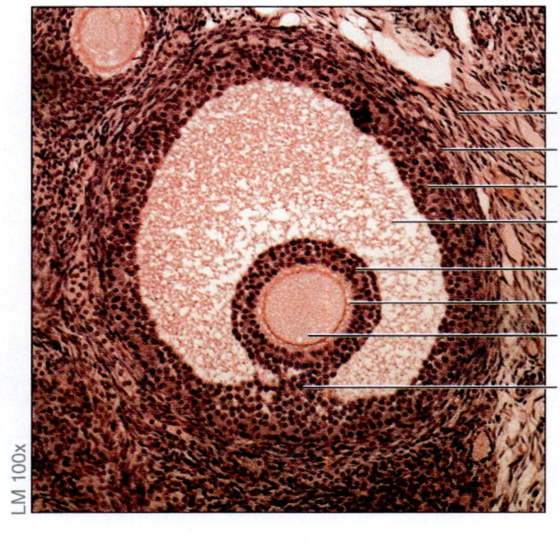

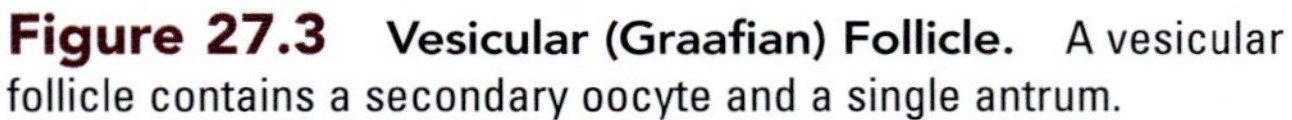

Figure 27.3 Vesicular (Graafian) Follicle. A vesicular follicle contains a secondary oocyte and a single antrum.

INTEGRATE

LEARNING STRATEGY

There are several distinguishing features used to identify follicles in various developmental stages. First, look for the number of cell layers and the shape of the cells that surround the oocyte. Primordial follicles contain a single layer of flattened cells, whereas primary follicles contain one or more layers of cuboidal cells. Next, look for the presence of fluid-filled antra. Secondary follicles contain one or more fluid-filled antra, whereas vesicular follicles are characterized by the presence of a single large antrum.

Vesicular (Graafian) Follicle	Corpus Luteum	Corpus Albicans
LM 100x	LM 25x	LM 80x
Has nearly the same structure and function as a secondary follicle, but contains a secondary oocyte and only a single large antrum	A "yellow body" that gets its color from the steroid hormones it secretes, which are lipids. After ovulation, the theca interna cells enlarge and continue to secrete the steroid hormones estrogen and progesterone. In the center of the corpus luteum is a large blood clot.	If fertilization does not occur, the corpus luteum stops secreting hormones after two weeks and becomes a smaller, *inactive* "white body" consisting mainly of scar tissue
Secondary oocyte	NA	NA
vesicular, a blister, + *folliculus*, a small sac	*corpus*, body, + *luteus*, yellow	*corpus*, body, + *albus*, white

Table 27.3 Components of a Vesicular Follicle

Structure	Description and Function	Word Origin
Antrum	The fluid-filled space in the center of the follicle	*antron*, a cave
Corona Radiata	A single layer of columnar cells derived from the cumulus oophorus that attach to the zona pellucida of the oocyte	*corona*, a crown, + *radiatus*, to shine
Cumulus Oophorus	A "mound" of granulosa cells that supports and surrounds the secondary oocyte within the follicle	*cumulus*, a heap, + *oophoron*, ovary
Granulosa Cells	Epithelial cells lining the follicle that will become the luteal cells of the corpus luteum after ovulation. They secrete the liquor folliculi, which is the fluid that fills the antrum.	*granulum*, a small grain
Theca Externa	The external fibrous layer of a well-developed vesicular follicle. The cells and fibers are arranged in concentric layers.	*theca*, a box, + *externus*, external
Theca Interna	The inner cellular layer of the vesicular follicle; these cells secrete androgen that is converted to estrogen by granulosa cells	*theca*, a box, + *internus*, internal
Zona Pellucida	A thick coat of glycoproteins that surrounds the oocyte	*zona*, zone, + *pellucidus*, clear

EXERCISE 27.2

HISTOLOGY OF THE UTERINE TUBES

The **uterine tubes** (fallopian tubes, or oviducts) are epithelia-lined fibromuscular tubes that extend from an ovary to the uterus. These tubes transport the ovum (or if fertilization occurs, the developing zygote) from the ovary to the uterus. The wall of the uterine tube is composed of three layers: mucosa, muscularis, and serosa.

1. Obtain a histology slide demonstrating a cross section of a uterine tube.
2. Place the slide on the microscope stage and bring the tissue sample into focus on low power. Note the highly folded **mucosa** of the tube **(figure 27.4).**
3. Move the microscope stage until an area of folded mucosa appears at the center of the field of view. Then switch to high power. The epithelium of the uterine tube contains two cell types, **ciliated epithelial cells** and **secretory cells. Table 27.4** describes the histological components of the uterine tubes. Focus on the epithelial cells that line the uterine tube, and distinguish ciliated cells from secretory cells (figure 27.4*b*).
4. Identify the following structures on the slide of the uterine tube, using table 27.4 and figure 27.4 as guides:
   - ☐ **ciliated cells**
   - ☐ **secretory cells**
   - ☐ **mucosal folds**
   - ☐ **serosa**
   - ☐ **muscularis**
5. Sketch a cross section of the uterine tube, as seen through the microscope in the space provided. Be sure to label the structures listed in step 4 in the drawing.

_______________ ×

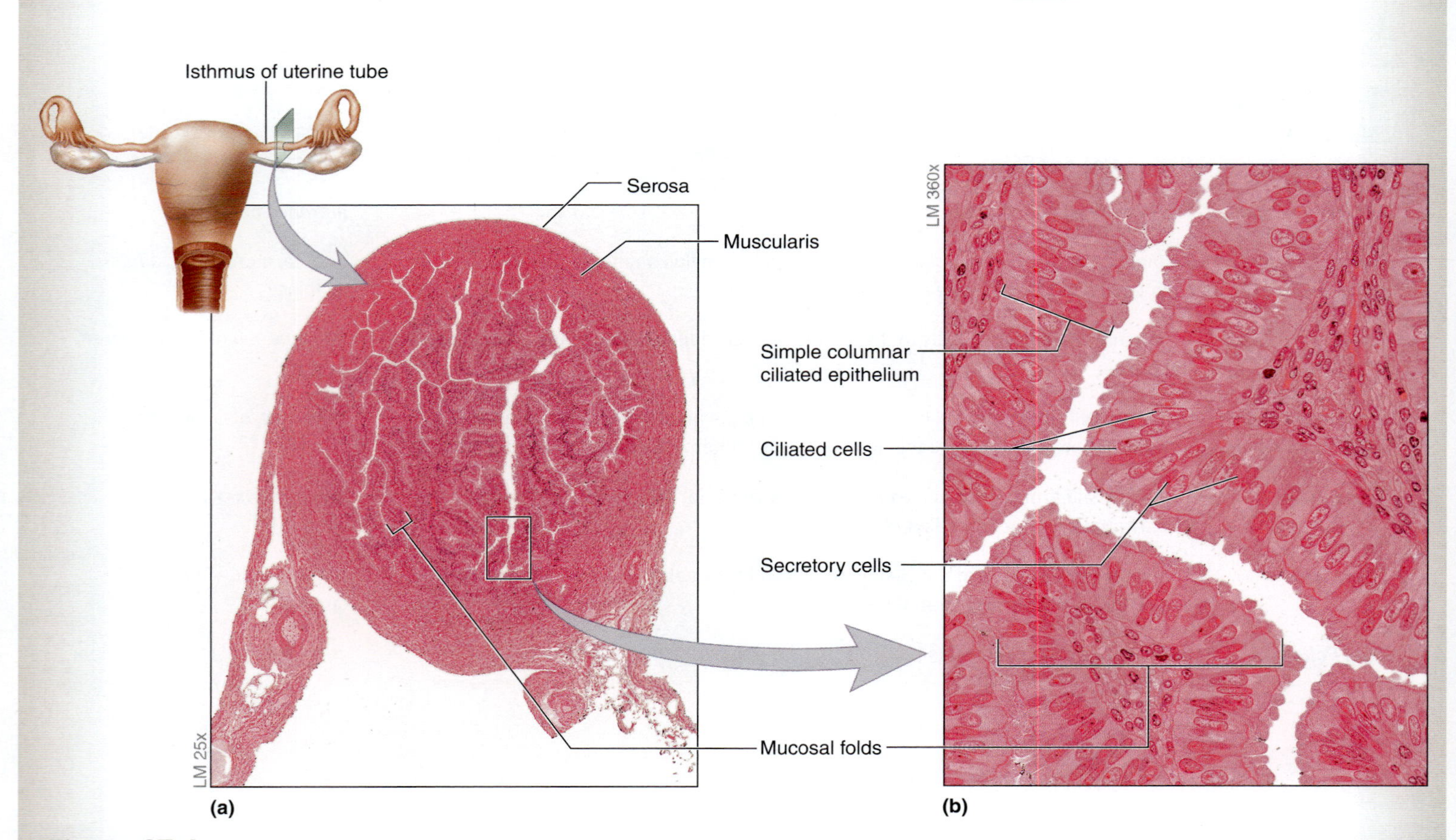

Figure 27.4 Histology of the Uterine Tubes. (a) Cross section through the isthmus of the uterine tubes demonstrating the many folds of the mucosa. (b) The epithelial cells lining the uterine tubes contain two cell types: ciliated cells and secretory cells. Ciliated cells have light nuclei, whereas secretory cells have dark nuclei.

Table 27.4 Components of the Uterine Tube

Uterine Tube Structure	Description and Function	Word Origin
Regions		
Infundibulum	The funnel-like expansion of the ovarian end of the uterine tube. Finger-like extensions of the infundibulum are fimbriae (*fimbria*, fringe).	*infundibulum*, a funnel
Ampulla	The wide part of the uterine tube located medial to the infundibulum. Its mucosa is highly folded and lined with simple ciliated columnar epithelium and secretory cells. Fertilization most commonly occurs in the ampulla.	*ampulla*, a two-handled bottle
Isthmus	The narrow part of the uterine tube located right next to the uterus	*isthmos*, a constriction
Uterine Part (Interstitial Segment)	The part of the uterine tube that penetrates the wall of the uterus	*intra–*, within, + *muralis*, wall (*inter*, between, + *sisto*, to stand)
Layers		
Serosa	The outer layer of the uterine tube. It consists of simple squamous epithelium (a fold of peritoneum).	*serosus*, serous
Muscularis	The middle layer of the wall of the uterine tube. It consists of smooth muscle whose contraction assists in transporting the ovum toward the uterus.	*musculus*, mouse (referring to muscle)
Mucosa	The inner layer of the wall of the uterine tube. It is highly folded in the infundibulum and ampulla of the uterine tube. Contains ciliated columnar epithelium and a layer of areolar connective tissue.	*mucosus*, mucous
Cells		
Ciliated Cells	Simple ciliated columnar epithelial cells; cilia beat toward the uterus to transport the ovum to the uterus. Nuclei stain lighter than those of the secretory cells.	*cilium*, eyelash
Secretory Cells	Nonciliated columnar epithelial cells that promote the activation of spermatozoa (capacitation) and provide nourishment for the ovum. Nuclei stain darker than those of the ciliated cells.	*secretus*, to separate

EXERCISE 27.3

HISTOLOGY OF THE UTERINE WALL

The **uterus** is a hollow, pear-shaped, muscular organ whose primary function is to support, protect, and nourish the developing embryo/fetus.

The uterine wall is composed of three layers: endometrium, myometrium, and perimetrium. The inner lining of the uterus, the **endometrium,** goes through its own cycle of growth during the ovarian cycle. Each time a woman's ovary undergoes a single ovarian cycle, the endometrium becomes prepared for the possibility that a fertilized egg will become implanted. If no implantation occurs, the innermost layer of the endometrium, the **functional layer,** sloughs off in the process of **menstruation** (*menstruus*, monthly). **Table 27.5** describes the layers of the uterine wall and the components that make up each layer. **Table 27.6** summarizes three phases of the menstrual cycle and demonstrates the appearance of the endometrium during each phase.

1. Obtain a histology slide that includes a portion of the uterine wall. Place it on the microscope stage and bring the tissue sample into focus on low power.
2. Identify the following structures on the slide of the uterus, using tables 27.5 and 27.6 as guides:
   - ☐ **endometrium (basal layer)**
   - ☐ **myometrium**
   - ☐ **endometrium (functional layer)**
   - ☐ **perimetrium**
3. Move the microscope stage so the functional layer of the endometrium is in the center of the field of view. Switch to high power and identify **uterine glands.**

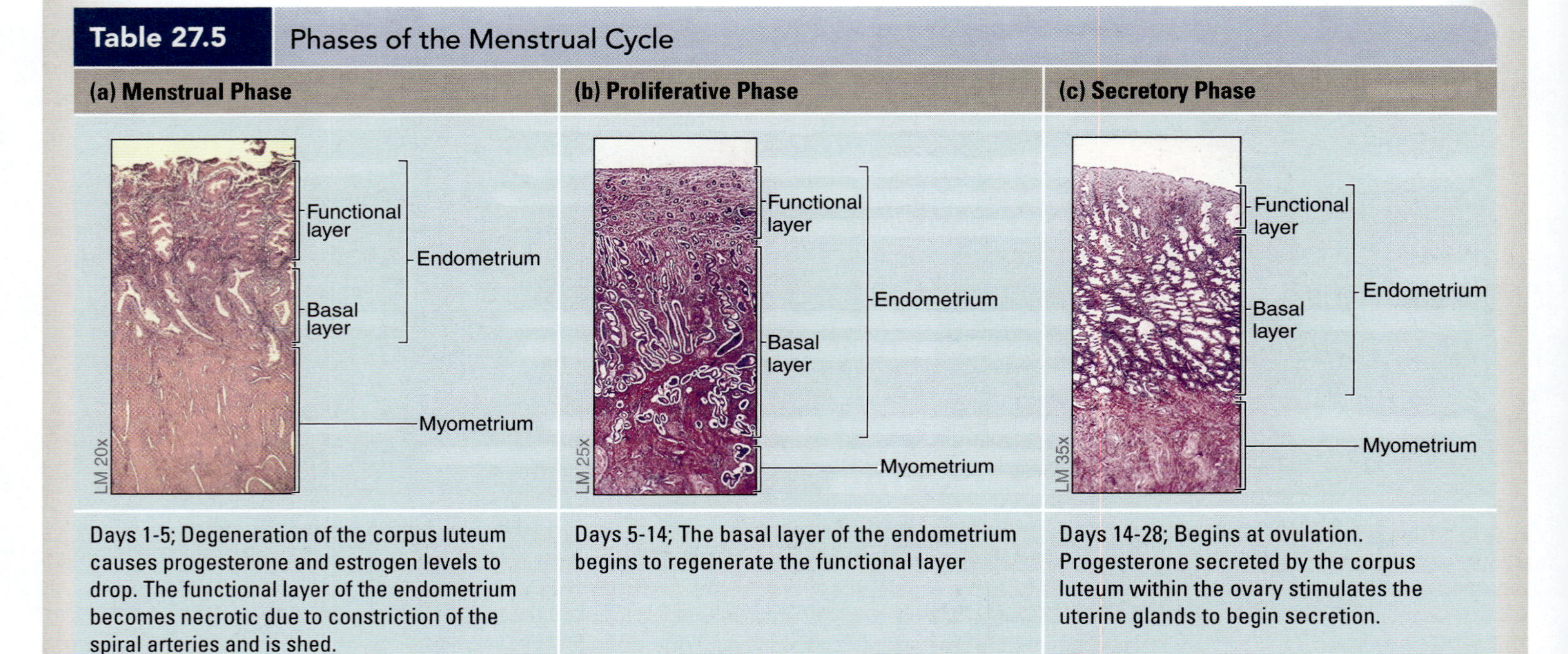

Table 27.5 Phases of the Menstrual Cycle

(a) Menstrual Phase	(b) Proliferative Phase	(c) Secretory Phase
Days 1-5; Degeneration of the corpus luteum causes progesterone and estrogen levels to drop. The functional layer of the endometrium becomes necrotic due to constriction of the spiral arteries and is shed.	Days 5-14; The basal layer of the endometrium begins to regenerate the functional layer	Days 14-28; Begins at ovulation. Progesterone secreted by the corpus luteum within the ovary stimulates the uterine glands to begin secretion.

Table 27.6 Wall Layers of the Uterus

Wall Layer	Description and Function	Word Origin
Endometrium	The mucous membrane composing the inner layer of the uterine wall. Consists of simple columnar epithelium and a lamina propria with simple tubular uterine glands. The structure, thickness, and state of the endometrium undergoes marked changes during the menstrual cycle.	*endon*, within, + *metra*, uterus
Functional Layer (Stratum Functionalis)	The apical layer of the endometrium. Most of this layer is shed during menstruation.	*stratum*, a layer, + *functus*, to perform
Basal Layer (Stratum Basalis)	The basal layer of the endometrium. It undergoes minimal changes during the menstrual cycle and serves as the basis for regrowth of the more apical stratum functionalis.	*stratum*, a layer, + *basalis*, basal
Myometrium	The muscular wall of the uterus composed of three layers of smooth muscle	*mys*, muscle, + *metra*, uterus
Perimetrium	The outermost covering of the uterus; a serous membrane formed from peritoneum	*peri-*, around, + *metra*, uterus

EXERCISE 27.4

HISTOLOGY OF THE VAGINAL WALL

The **vagina** (*vagina*, sheath) is a muscular tube that functions as the organ of copulation, serves as the passageway for menses, and serves as the birth canal during parturition (*parturio*, to be in labor).

1. Obtain a histology slide demonstrating a portion of the vaginal wall. Place it on the microscope stage and bring the tissue sample into focus on low power. The walls of the vagina are composed of a mucosa and muscularis; the mucosa is lined with a nonkeratinized stratified squamous epithelium **(figure 27.5).**
2. Identify the following structures on the slide of the vagina, using figure 27.5 as a guide:

- ☐ **lamina propria**
- ☐ **mucosa**
- ☐ **muscularis**
- ☐ **nonkeratinized stratified squamous epithelium**

3. *Optional Activity:* APR **14: Reproductive System—** Watch the "Female Reproductive System Overview" animation to review the female reproductive organs and their histological features.

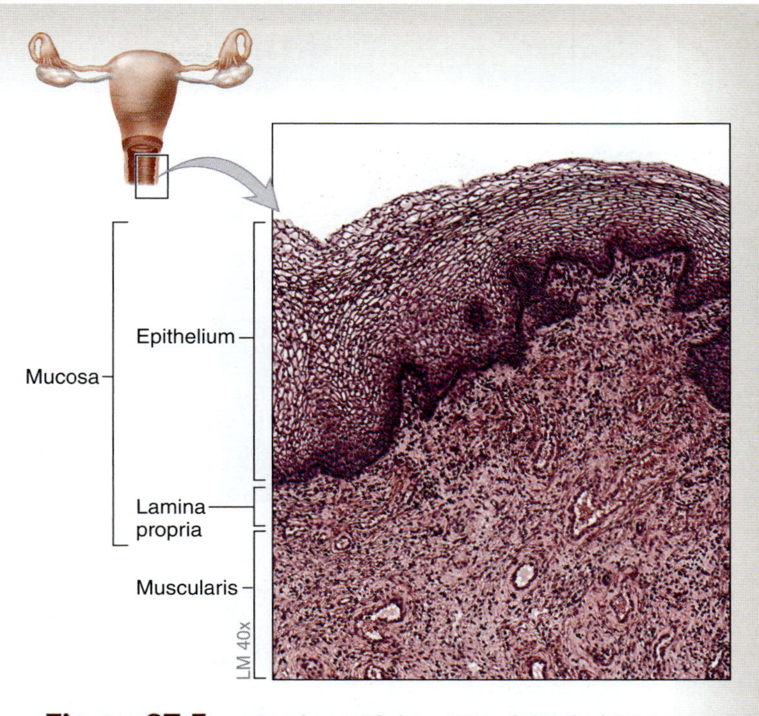

Figure 27.5 Histology of the Vaginal Epithelium. The epithelium is nonkeratinized stratified squamous.

Male Reproductive System

The male reproductive system consists of the testes, epididymis, ductus deferens, seminal vesicles, prostate gland, bulbourethral glands, and penis. The primary functions of the male reproductive system are to produce the male gamete (sperm), produce testosterone, and provide nourishment for the sperm and a mechanism for the sperm to be delivered to the female reproductive tract. Testes also produce testosterone hormone.

EXERCISE 27.5

HISTOLOGY OF THE SEMINIFEROUS TUBULES

The **testes** are the primary reproductive organ in the male, and function in the production of sperm and the male sex steroid hormone **testosterone.** Within each testis are several hundred, coiled **seminiferous tubules (figure 27.6),** which are the site of sperm production, or **spermatogenesis** (*sperma*, seed, + *genesis*, origin). Within each tubule, **spermatogonia** undergo successive meiotic divisions as they move from the basal to the apical surface of the tubule epithelium. **Table 27.7** describes the developmental stages of the sperm, and **table 27.8** describes the structure and function of the accessory cell types located within the testes.

1. Obtain a slide of the testes and place it on the microscope stage.
2. Observe the slide on low power and identify several cross sections of the seminiferous tubules. Move the microscope stage so that one tubule is at the center of the field of view, and change to high power.

(continued on next page)

(continued from previous page)

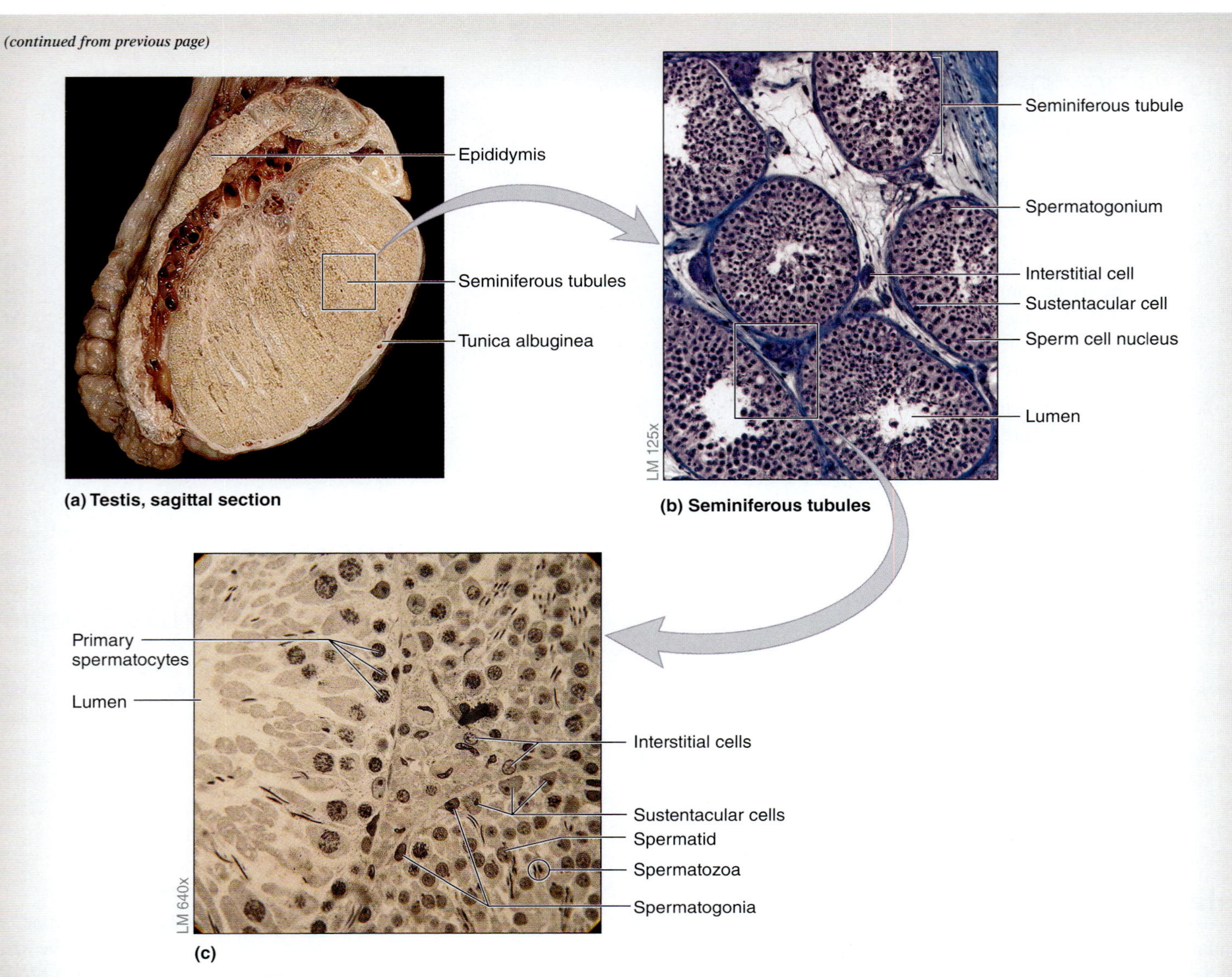

Figure 27.6 Testes and Seminiferous Tubules. (a) Testis and epididymis showing seminiferous tubules within the testes and tunica albuginea surrounding the testis. (b) Medium-power histological appearance of seminiferous tubules. (c) High-power histological view of seminiferous tubules demonstrating stages of spermatogenesis and sustentacular cells.

Table 27.7 Developmental Stages of Sperm

Cell Name	Description and Function	Stage of Mitosis/Meiosis	Ploidy	Word Origin
Spermatogonia	Primitive male gametes that are formed by mitosis from male germline stem cells	Formed by mitosis; undergo mitosis to form primary spermatocytes	Diploid (2n)	*sperma*, seed, + *gonia*, generation
Primary Spermatocyte	Male gamete that has replicated its DNA in preparation for meiosis. It will subsequently undergo meiosis I to form two secondary spermatocytes.	DNA is replicated in prophase of meiosis I	Diploid (2n)	*primary*, first, + *sperma*, seed, + *kytos*, cell
Secondary Spermatocyte	Male gamete that has completed meiosis I and will subsequently undergo meiosis II to form two spermatids	Formed after the first meiotic division	Haploid (1n)	*secondary*, second, + *sperma*, seed, + *kytos*, cell
Spermatid	Male gamete that has completed meiosis, but has not undergone spermiogenesis, the process of becoming a mature spermatozoon	Formed after the second meiotic division	Haploid (1n)	*sperma*, seed, + *-id*, a young specimen
Spermatozoon (Sperm)	A mature male gamete that has undergone spermiogenesis	Formed by maturation of spermatids	Haploid (1n)	sperma, seed, + *zoon*, animal

Table 27.8	Accessory Cells of the Testis	
Cell Name	**Description and Function**	**Word Origin**
Interstitial (Leydig) Cells	Produce the steroid hormone testosterone, which is required for proper sperm development and for the development of male secondary sex characteristics	*inter-*, between, + *sisto*, to stand
Sustentacular (Sertoli) Cells	Surround multiple developing spermatocytes to provide them with support and nourishment. Phagocytize excess cytoplasm from developing spermatocytes. Form the *blood-testis barrier*, which protects the developing spermatocytes from antigens that circulate in the blood. Produce androgen-binding protein (ABP), which concentrates testosterone around the developing spermatocytes.	*sustento*, to hold upright

3. Identify the following structures, using tables 27.7 and 27.8 and figure 27.6 as guides:

   - ☐ **interstitial cell**
   - ☐ **primary spermatocyte**
   - ☐ **seminiferous tubule**
   - ☐ **spermatids**
   - ☐ **spermatogonia**
   - ☐ **sustentacular cell**

4. Sketch a seminiferous tubule as seen through the microscope in the space provided. Be sure to label the structures listed in step 3 in the drawing.

5. *Optional Activity:* **APR 14: Reproductive System—** Watch the "Spermatogenesis" animation to visualize the formation of sperm in the seminiferous tubules.

_______ ×

EXERCISE 27.6

HISTOLOGY OF THE EPIDIDYMIS

After spermatozoa are formed within the seminiferous tubules, they are transported through the straight tubules, rete testis, and efferent ductules to enter the long, coiled tube of the **epididymis** (*epi-*, upon, + *didymos*, a twin [related to *didymoi*, testes]). Spermatozoa undergo the process of maturation and are stored in the epididymis until ejaculation takes place. One of the most characteristic features of the epididymis is the presence of thousands of sperm in the lumen of the tube **(figure 27.7). Table 27.9** describes the structure and function of the male accessory reproductive structures, including the epididymis.

1. Obtain a slide of the epididymis and place it on the microscope stage. Bring the tissue sample into focus on low power and identify several cross sections of the epididymis. Move the microscope stage so that one part of the lumen of the epididymis is at the center of the field of view, and then change to high power. Notice the sperm inside the lumen of the tubule, and how they do not come right up against the apical surface of the columnar epithelial cells. This is because of the stereocilia on the apical surface of the epithelial cells. **Stereocilia** (*stereo*, solid, + *cilium*, eyelid) are single, long microvilli (ironically, they are not cilia at all!) that increase the surface area of the epithelial cells for the purpose of secreting substances that nourish the sperm and absorbing substances from the sperm as they undergo maturation.

2. Identify the following structures on the slide of the epididymis, using table 27.9 and figure 27.7 as guides:

   - ☐ **pseudostratified ciliated columnar epithelial cells**
   - ☐ **sperm cells**
   - ☐ **stereocilia**

3. Sketch the epididymis as seen through the microscope in the space provided. Be sure to label the structures listed in step 2 in the drawing.

_______ ×

(continued on next page)

(continued from previous page)

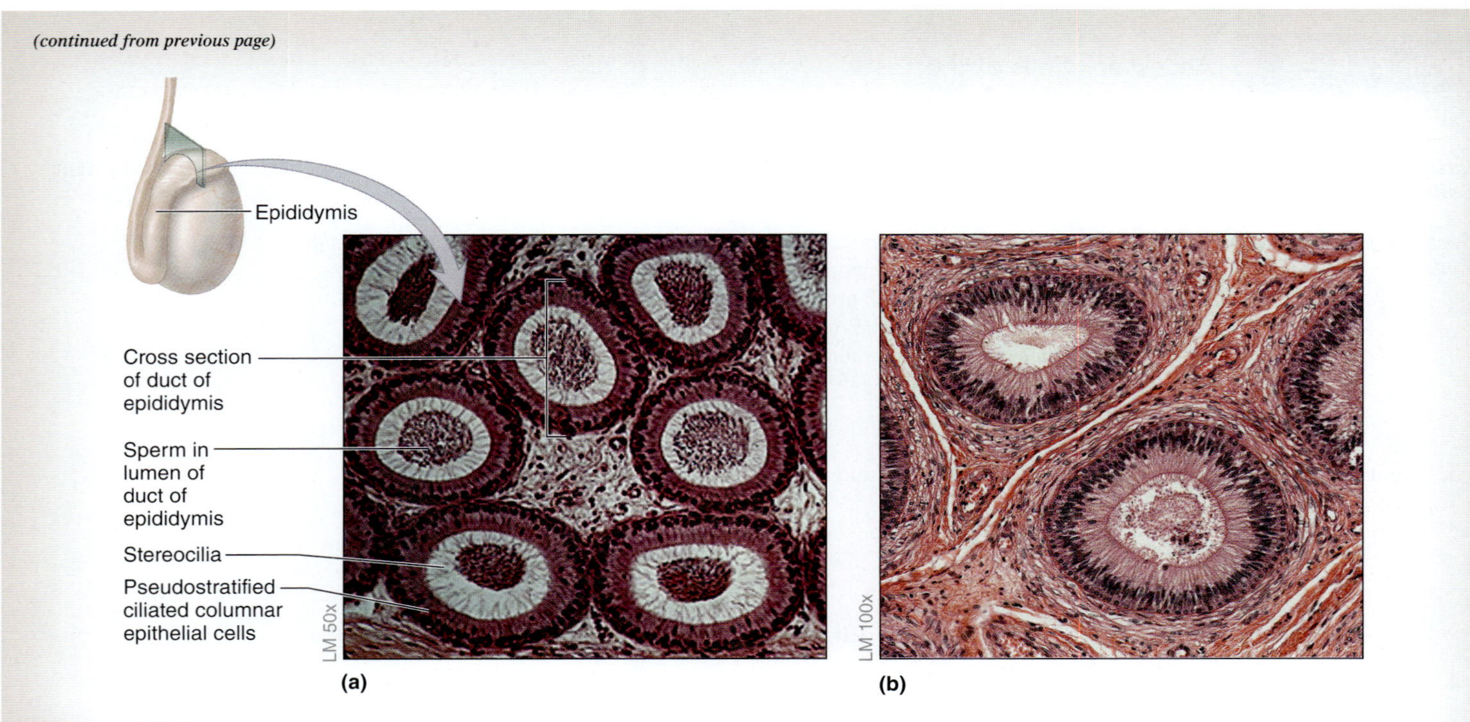

Figure 27.7 Cross Section Through the Epididymis. Several cross sections through the seminiferous tubules are visible. (a) Note the numerous sperm within the lumen. (b) A higher magnification view demonstrates pseudostratified columnar epithelium with stereocilia.

Table 27.9 Male Accessory Reproductive Structures

Structure	Epithelium	Function	Word Origin
Epididymis	Pseudostratified columnar epithelium with cilia	Site of storage and maturation of sperm. Walls contain some smooth muscle that will propel sperm into the ductus deferens during ejaculation.	*epi-*, upon, + *didymos*, a twin (related to *didymoi*, testes)
Ductus Deferens (Vas Deferens)	Pseudostratified columnar epithelium with stereocilia	A thick, muscular tube whose walls undergo peristaltic contractions during ejaculation to propel sperm into the urethra	*ductus*, to lead, + *defero*, to carry away
Prostate Gland	Glandular epithelium resembling pseudostratified or simple columnar	Accessory reproductive gland that contributes approximately 25% of the volume of semen. The fluid produced is rich in Vitamin C (citric acid) and enzymes (such as PSA, prostate-specific antigen) that are important for proper sperm function.	*prostates*, one standing before
Seminal Vesicles	Pseudostratified columnar	Accessory reproductive gland that contributes approximately 60% of the volume of semen. The fluid produced is rich in fructose, which is a source of energy for the sperm. They also produce prostaglandins, which are important in promoting sperm motility and may stimulate uterine contractions.	*seminal*, relating to sperm, + *vesicula*, a blister
Bulbourethral Glands	Glandular epithelium consisting of simple and pseudostratified columnar	Accessory reproductive gland that produces a mucus-like lubricating substance during sexual arousal (prior to ejaculation) that lubricates the urethra and neutralizes the acidity of the urine, thus preparing the way for the spermatozoa to pass	*bulbus*, bulb, + *urethral*, relating to the urethra

EXERCISE 27.7

HISTOLOGY OF THE DUCTUS DEFERENS

The **ductus deferens** (vas deferens) is a continuation of the epididymis and is the route by which sperm travel from the epididymis to the urethra during ejaculation (*ejaculo*, to shoot out). It is a highly muscular tube lined with pseudostratified columnar epithelium with cilia **(figure 27.8).**

1. Obtain a slide demonstrating the ductus deferens and place it on the microscope stage. Bring the tissue sample into focus on low power and identify the cross section of the ductus deferens (figure 27.8). Move the microscope stage so the lumen of the ductus deferens is at the center of the field of view, and then change to high power.
2. Identify the following structures on the slide of the ductus deferens, using table 27.9 and figure 27.8 as guides:
   - ☐ **cilia**
   - ☐ **lumen**
   - ☐ **mucosa**
   - ☐ **pseudostratified ciliated columnar epithelial cells**
   - ☐ **smooth muscle**
3. Sketch a cross section of the ductus deferens as seen through the microscope in the space provided. Be sure to label the structures listed in step 2 in the drawing.

_______ ×

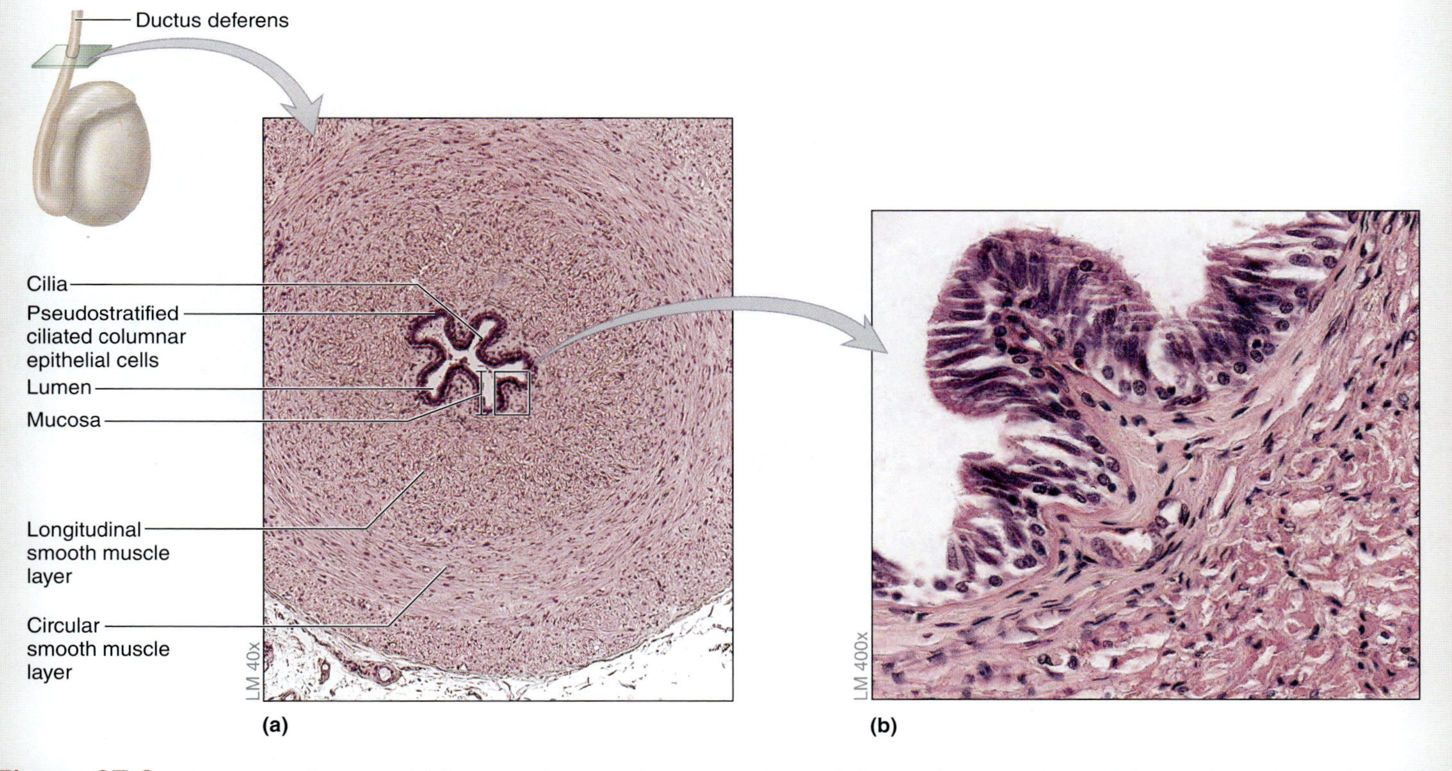

Figure 27.8 Ductus Deferens. (a) Cross section through the entire ductus deferens, demonstrating thick layers of smooth muscle surrounding the lumen. (b) Close-up view of the pseudostratified ciliated columnar epithelium.

EXERCISE 27.8

HISTOLOGY OF THE SEMINAL VESICLES

Near the posterior wall of the bladder, each ductus deferens comes together with the duct from a **seminal vesicle (figure 27.9).** These accessory reproductive glands are lined with pseudostratified columnar epithelium and produce substances (including fructose, which is a source of energy for the sperm) that are an important component of **semen.** Semen is the fluid expelled from the penis during ejaculation (table 27.9).

1. Obtain a slide of the seminal vesicles and place it on the microscope stage. Bring the tissue sample into focus on low power. Identify cross sections of the seminal vesicles. Move the microscope stage so the lumen of a portion of the seminal vesicle is at the center of the field of view, and then change to high power.

2. Identify the following structures on the slide of the seminal vesicles, using table 27.9 and figure 27.9 as guides:

   ☐ **lumen**
   ☐ **mucosal folds**
   ☐ **muscular wall**

3. Sketch the seminal vesicles as seen through the microscope in the space provided. Be sure to label the structures listed in step 2 in the drawing.

_____________ ×

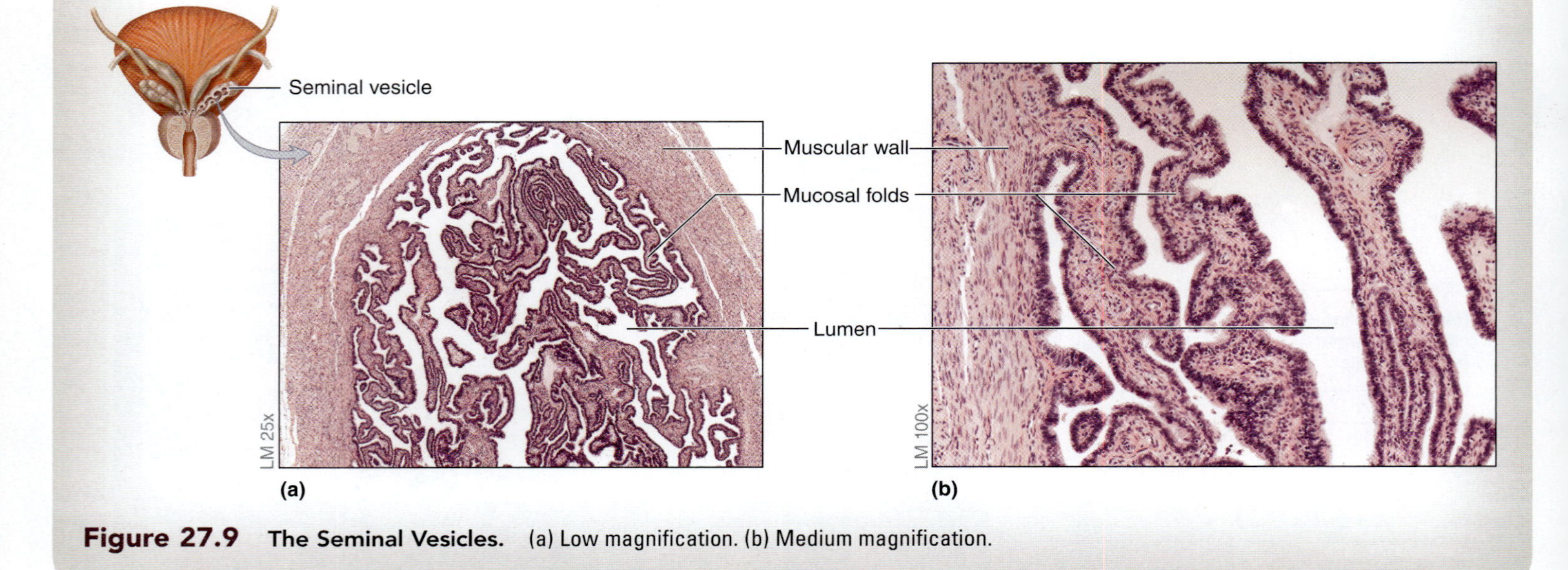

Figure 27.9 The Seminal Vesicles. (a) Low magnification. (b) Medium magnification.

EXERCISE 27.9

HISTOLOGY OF THE PROSTATE GLAND

The **prostate** gland is an accessory reproductive gland located immediately inferior to the urinary bladder. Like the seminal vesicles, the prostate produces substances that will become part of semen and that are important for proper sperm function.

1. Obtain a slide demonstrating the prostate gland and place it on the microscope stage. Bring the tissue sample into focus on low power and identify the prostatic urethra and the prostate gland itself **(figure 27.10).** Move the microscope stage so a part of the gland relatively far away from the urethra is at the center of the field of view, and then change to a higher power.
2. Identify the following structure on the slide of the prostate, using table 27.9 and figure 27.10 as guides:

   ☐ **tubuloalveolar glands**
3. As men age, calcifications often form in the prostate. Such calcifications are called **prostatic calculi** (corpora aranacea) (figure 27.10). Scan the slide and see if any of these are present in the specimen.

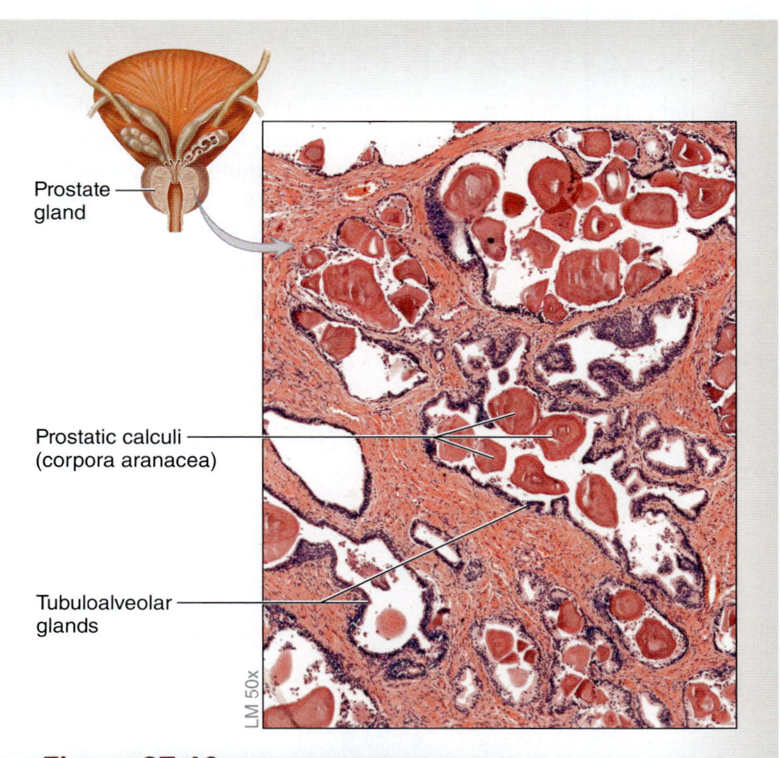

Figure 27.10 The Prostate Gland. Note the many prostatic calculi (corpora aranacea).

EXERCISE 27.10

HISTOLOGY OF THE PENIS

The **penis** is the male copulatory organ and is homologous to the female clitoris. It is composed of three erectile bodies, the paired **corpora cavernosa** and the single **corpus spongiosum,** which contains the male urethra **(figure 27.11).** In the center of each corpus cavernosum is a central artery that supplies blood to the penis. During an erection, the veins draining the penis become constricted and venous blood fills up the erectile tissues.

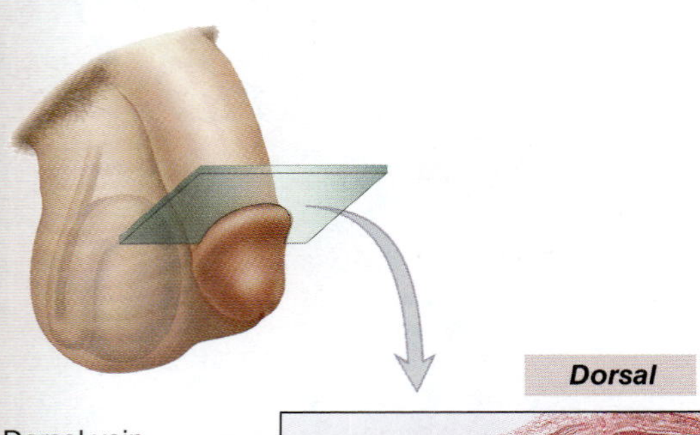

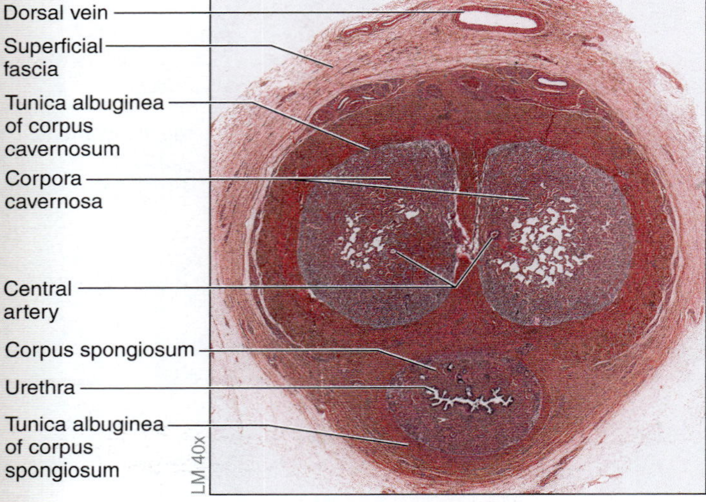

Figure 27.11 Cross Section Through the Shaft of the Penis. The penis contains two paired corpora cavernosa dorsally, and a single corpus spongiosum ventrally. Note the male urethra in the center of the corpus spongiosum.

1. Obtain a slide demonstrating a cross section of the penis and place it on the microscope stage. Observe at low power.
2. Identify the following structures on the slide of the penis, using figure 27.11 as a guide:
   - ☐ **central artery**
   - ☐ **corpus cavernosum**
   - ☐ **corpus spongiosum**
   - ☐ **dorsal surface**
   - ☐ **dorsal vein**
   - ☐ **superficial fascia (connective tissue)**
   - ☐ **urethra**
   - ☐ **ventral surface**
3. Sketch a cross section of the penis as seen through the microscope in the space provided. Be sure to label the structures listed in step 2 in the drawing.

4. *Optional Activity:* **AP|R 14: Reproductive System—** Watch the "Male Reproductive System Overview" animation to review the male reproductive structures and their relationships.

INTEGRATE

CLINICAL VIEW
Erectile Dysfunction

Erectile dysfunction (ED) occurs when a male is unable to achieve or maintain an erection. Stimulation of the parasympathetic nervous system causes central arteries within the corpora cavernosa to dilate. This dilation increases local blood flow to the erectile tissues. Concomitant constriction of veins at the base of the penis prevents the drainage of blood from the erectile tissues. The resulting engorging of the erectile tissues within the penis with blood causes an erection. The penis remains erect until ejaculation, a process stimulated by the sympathetic nervous system. Erectile dysfunction involves the interruption of the arousal response, and is most commonly caused by neurological, cardiovascular, or hormonal disturbances. Interestingly, some treatments for those suffering from erectile dysfunction, including sildenafil (i.e., Viagra), were first introduced as antihypertensive drugs. Sildenafil reduces systemic blood pressure by dilating blood vessels and decreasing total peripheral resistance. As concerns an erection, sildenafil dilates the central arteries within the penis, which facilitates the process of achieving and maintaining an erection.

GROSS ANATOMY

Female Reproductive System

The internal detail, wall structures, and functions of most of the female reproductive organs were covered in the histology section of this chapter. This section involves observing the gross structure of these organs to gain an appreciation of their location within the female pelvic cavity, and their relationships with each other. Note that there are numerous supporting ligaments that anchor the ovaries, uterine tubes, and uterus in place within the pelvic cavity. Primary anatomic features of the female breast will then be observed.

EXERCISE 27.11

GROSS ANATOMY OF THE OVARY, UTERINE TUBES, UTERUS, AND SUPPORTING LIGAMENTS

The ovary, uterine tubes, and uterus are all contained within the pelvic cavity of the female and are held in place by a number of supporting ligaments, many of which form from folds of the peritoneum as it drapes over these structures. **Table 27.10** summarizes the structure, function, and location of the supporting ligaments.

1. Observe a female human cadaver or classroom models of the female pelvis and female reproductive organs. Also observe a classroom model demonstrating a sagittal section through a female pelvis.
2. Identify the structures listed in **figures 27.12** and **27.13** on the cadaver or on classroom models, using table 27.10 and the textbook as guides. Then label them in figures 27.12 and 27.13.
3. *Optional Activity:* **AP|R 14: Reproductive System**—In the Quiz section, select the "Structures: Female" option to test yourself on gross anatomy of female structures.

Table 27.10 Ligaments Supporting the Ovary, Uterine Tubes, and Uterus

Supporting Ligament	Description and Function	Word Origin
Broad Ligament	Fold of peritoneum draped over the superior surface of the uterus. Portions of the broad ligament form the mesosalpinx and the mesovarium.	*broad*, wide, + *ligamentum*, a bandage
Mesosalpinx	The mesentery of the uterine tube, which is formed as a fold of the most superior part of the broad ligament	*meso-*, a mesentery-like structure, + *salpinx*, trumpet (tube)
Mesovarium	The mesentery of the ovary, which is formed as a posterior extension of the broad ligament	*meso-*, a mesentery-like structure, + *ovarium*, ovary
Ovarian Ligament	Ligament contained within folds of the broad ligament. It extends from the medial part of the ovary to the superolateral surface of the body of the uterus.	*ovarium*, ovary, + *ligamentum*, a bandage
Round Ligament of the Uterus	Ligament attached to the superolateral surface of the body of the uterus. It extends laterally to the deep inguinal ring, passes through the inguinal canal, and attaches to the skin of the labia majora.	*ligamentum*, a bandage, + *uterus*, uterus
Suspensory Ligament of the Ovary	A fold of peritoneum draping over the ovarian artery and vein superolateral to the ovary. Anchors the ovary to the lateral body wall.	*suspensio-*, to hang up, + *ligamentum*, a bandage
Transverse Cervical (Cardinal) Ligament	Ligament extending laterally from the cervix and vagina, connecting them to the pelvic wall	*transverse-*, across, + *cervix*, neck, + *ligamentum*, a bandage
Uterosacral Ligaments	Ligament connecting the inferior part of the uterus to the sacrum posteriorly	*utero-*, the uterus, + *sacral*, the sacrum

(continued on next page)

(continued from previous page)

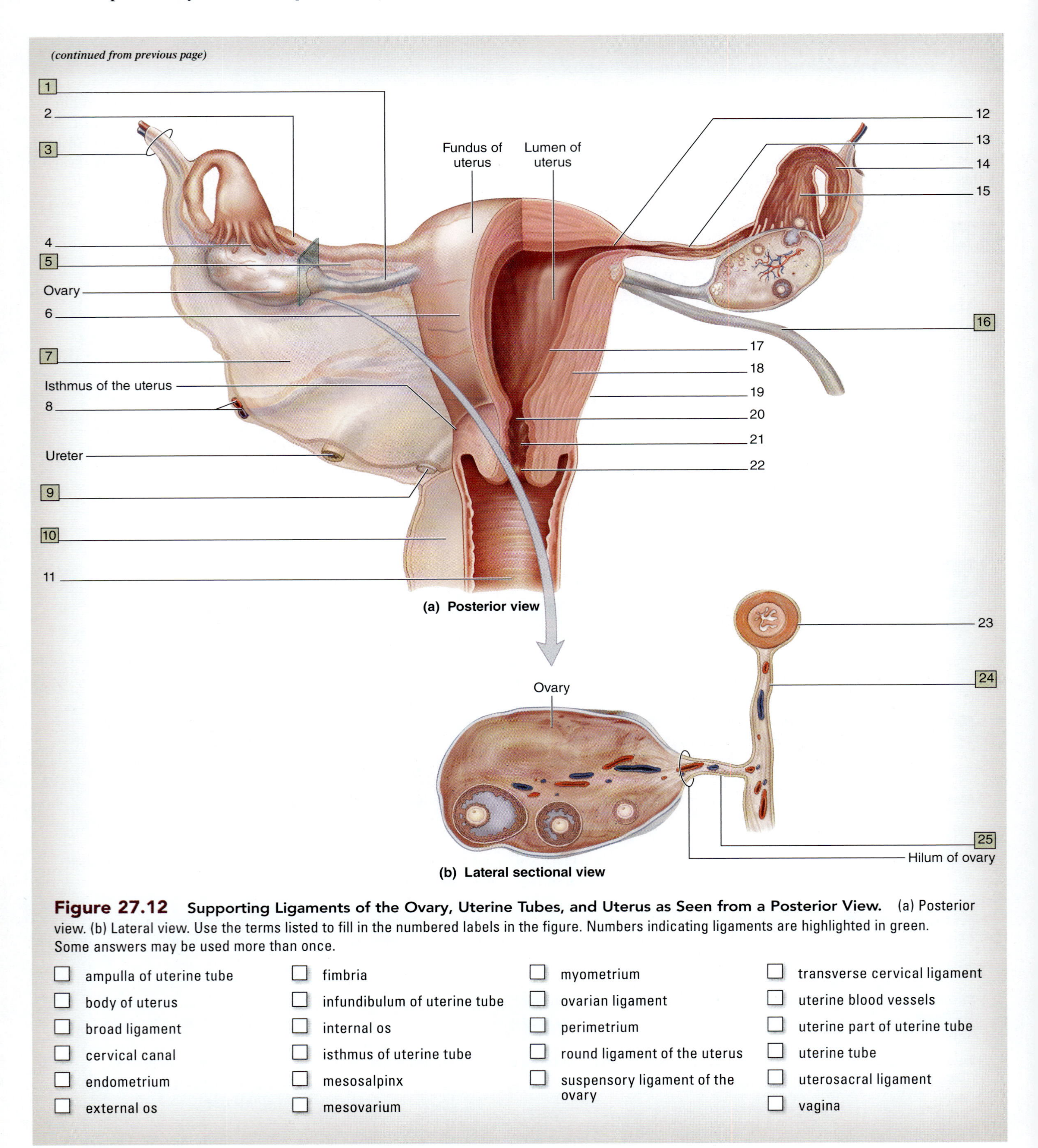

Figure 27.12 Supporting Ligaments of the Ovary, Uterine Tubes, and Uterus as Seen from a Posterior View. (a) Posterior view. (b) Lateral view. Use the terms listed to fill in the numbered labels in the figure. Numbers indicating ligaments are highlighted in green. Some answers may be used more than once.

- ☐ ampulla of uterine tube
- ☐ body of uterus
- ☐ broad ligament
- ☐ cervical canal
- ☐ endometrium
- ☐ external os
- ☐ fimbria
- ☐ infundibulum of uterine tube
- ☐ internal os
- ☐ isthmus of uterine tube
- ☐ mesosalpinx
- ☐ mesovarium
- ☐ myometrium
- ☐ ovarian ligament
- ☐ perimetrium
- ☐ round ligament of the uterus
- ☐ suspensory ligament of the ovary
- ☐ transverse cervical ligament
- ☐ uterine blood vessels
- ☐ uterine part of uterine tube
- ☐ uterine tube
- ☐ uterosacral ligament
- ☐ vagina

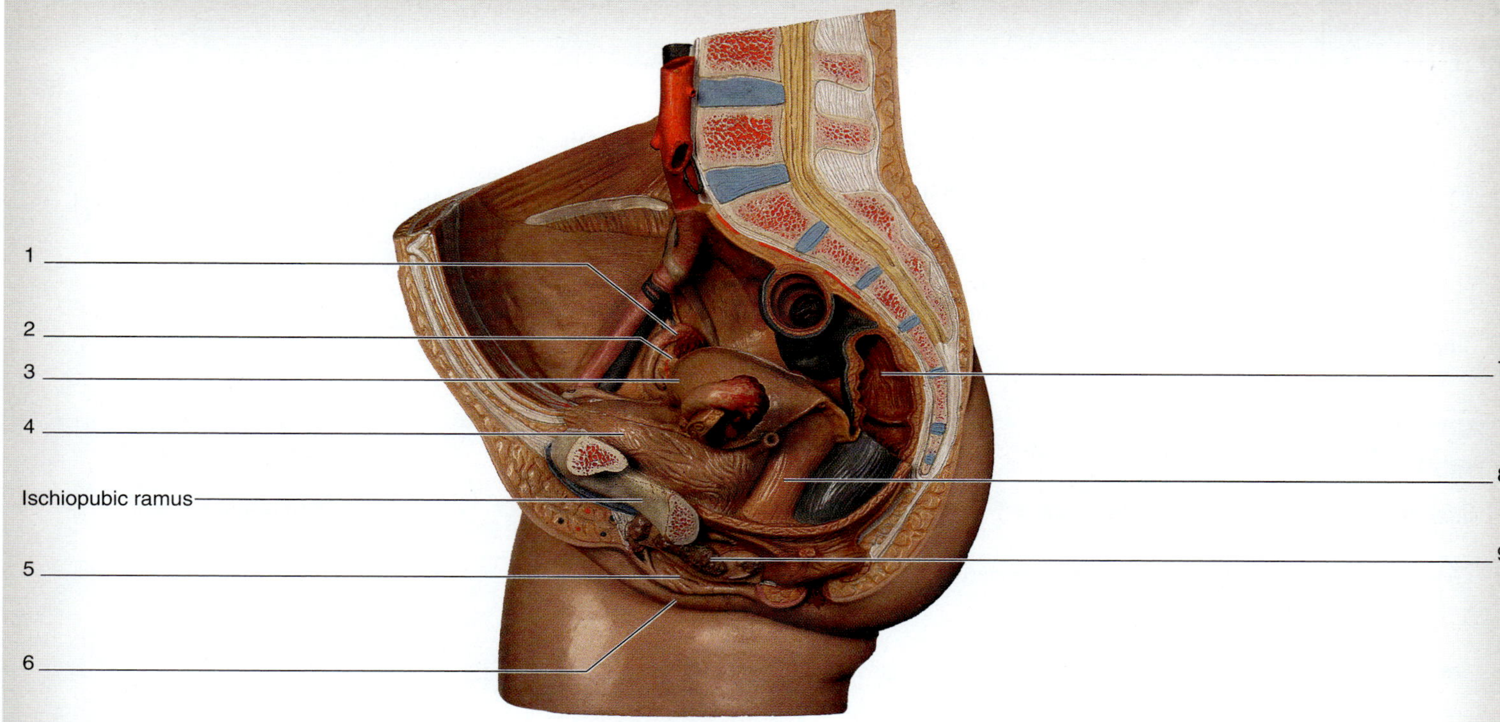

1
2
3
4
Ischiopubic ramus
5
6
7
8
9

(a) Sagittal view

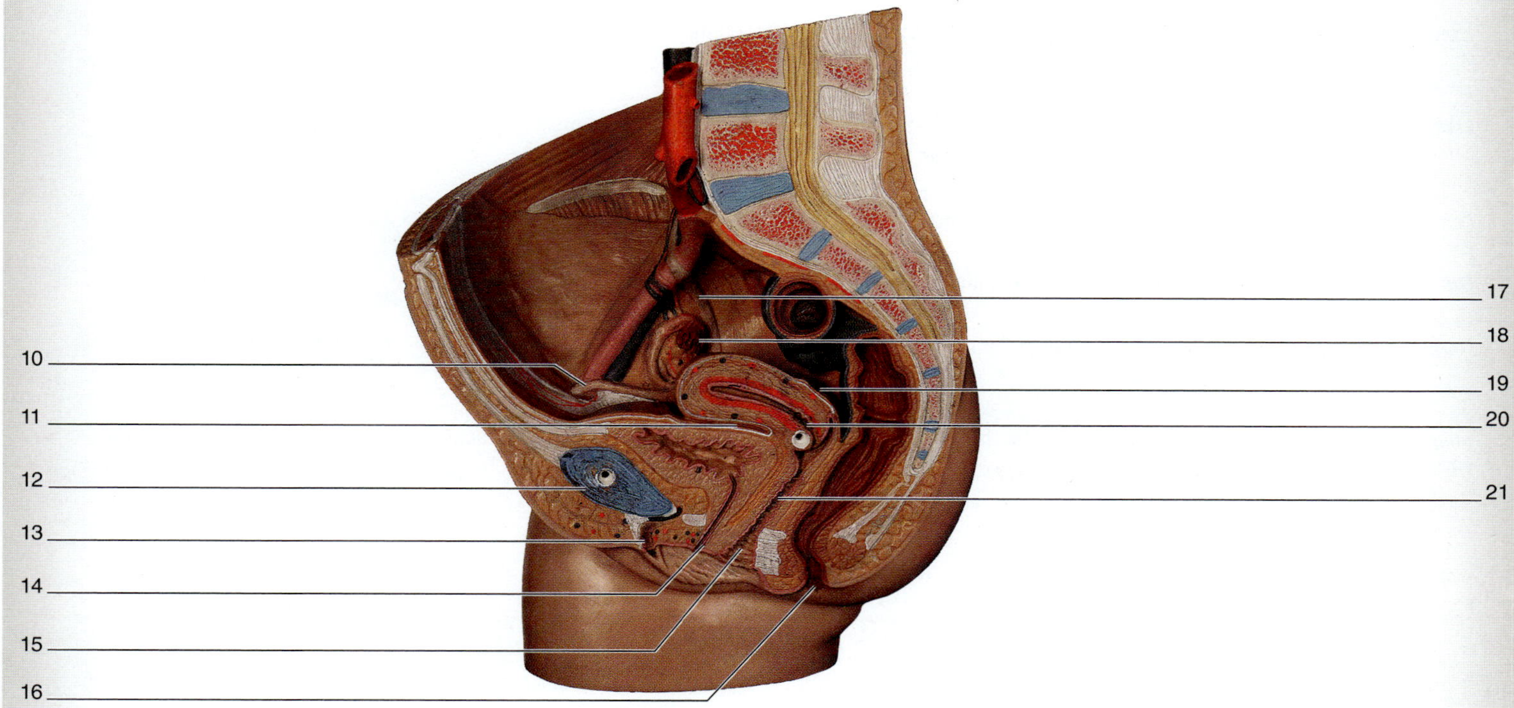

17
18
10
19
11
20
12
21
13
14
15
16

(b) Midsagittal view

Figure 27.13 **Classroom Model of the Female Pelvic Cavity.** (a) Sagittal view with pelvic structures intact. (b) Midsagittal view. Use the terms listed to fill in the numbered labels in the figure. Some answers may be used more than once.

- ☐ anus
- ☐ bulb of the vestibule
- ☐ cervix of uterus
- ☐ clitoris
- ☐ external urethral orifice
- ☐ fimbria of uterine tube
- ☐ labia majora
- ☐ labia minora
- ☐ ovary
- ☐ pubic symphysis
- ☐ rectouterine pouch
- ☐ rectum
- ☐ round ligament of the uterus
- ☐ ureter
- ☐ urinary bladder
- ☐ uterine tube
- ☐ uterus
- ☐ vagina
- ☐ vaginal orifice
- ☐ vesicouterine pouch

EXERCISE 27.12

GROSS ANATOMY OF THE FEMALE BREAST

The female breast consists largely of fatty tissue and suspensory ligaments. Imbedded within are numerous modified sweat glands, the **mammary glands** (*mamma*, breast), which are compound tubuloalveolar exocrine glands. These glands enlarge greatly during pregnancy, enabling them to produce milk to nourish the new baby. **Table 27.11** describes the structures that compose the female breast.

1. Observe the breast of a prosected female human cadaver or a classroom model of the female breast.
2. Identify the structures listed in **figure 27.14** on the cadaver or the classroom model of the female breast, using table 27.11 and the textbook as guides. Then label them in figure 27.14.

Table 27. 11 The Female Breast

Structure	Description and Function	Word Origin
Alveoli	Secretory units of the mammary glands, which produce milk	*alveolus*, a concave vessel, a bowl
Areola	The pigmented area of skin surrounding the nipple	*areola*, area
Areolar Glands	Sebaceous glands deep to the skin of the areola; produce sebum, which keeps the skin of the areola moist, particularly during lactation	*areolar*, relating to the areola of the breast
Lactiferous Ducts	Ducts that form from the confluence of small ducts draining milk from the alveoli and lobules	*lacto-*, milk + *ductus*, to lead
Lactiferous Sinuses	10–20 large channels that form from the confluence of several lactiferous ducts; the spaces where milk is stored prior to release from the nipple	*lacto-*, milk + *sinus*, cavity
Lobes	Large subdivisions of the mammary glands	*lobos*, lobe
Lobules	Smaller subdivisions of the mammary glands, which contain the alveoli	*lobulus*, a small lobe
Nipple	A cylindrical projection in the center of the breast that contains the openings of the lactiferous ducts	*neb*, beak or nose
Suspensory Ligaments	Bands of connective tissue that anchor the breast skin and tissue to the deep fascia overlying the pectoralis major muscle	*suspensio*, to hang up, + *ligamentum*, a band

INTEGRATE

CONCEPT CONNECTION

Lactation, the process of milk production and release from the mammary glands, involves both positive and negative feedback mechanisms. Lactation is regulated by hormones that are present during pregnancy and after childbirth. During pregnancy, estrogen and progesterone levels are elevated, which stimulates proliferation of the acini and branching of the lactiferous ducts within mammary glands. Following birth, estrogen and progesterone levels drop, however prolactin levels are elevated, which stimulates milk production. The baby suckling on the mother's breast stimulates the release of oxytocin from the posterior pituitary gland, which further increases milk production. Oxytocin also stimulates myoepithelial cells within mammary glands to contract, thereby releasing milk from the breast. Milk moves through the lactiferous ducts and sinuses, and exits the breast through openings in the nipple. Positive feedback mechanisms stimulate the production and ejection of milk. Negative feedback mechanisms may inhibit GnRH release by the hypothalamus, thereby inhibiting FSH and LH release from the anterior pituitary gland and preventing ovulation. Even so, women should use alternative methods of birth control while breast feeding because some women continue to ovulate during this time.

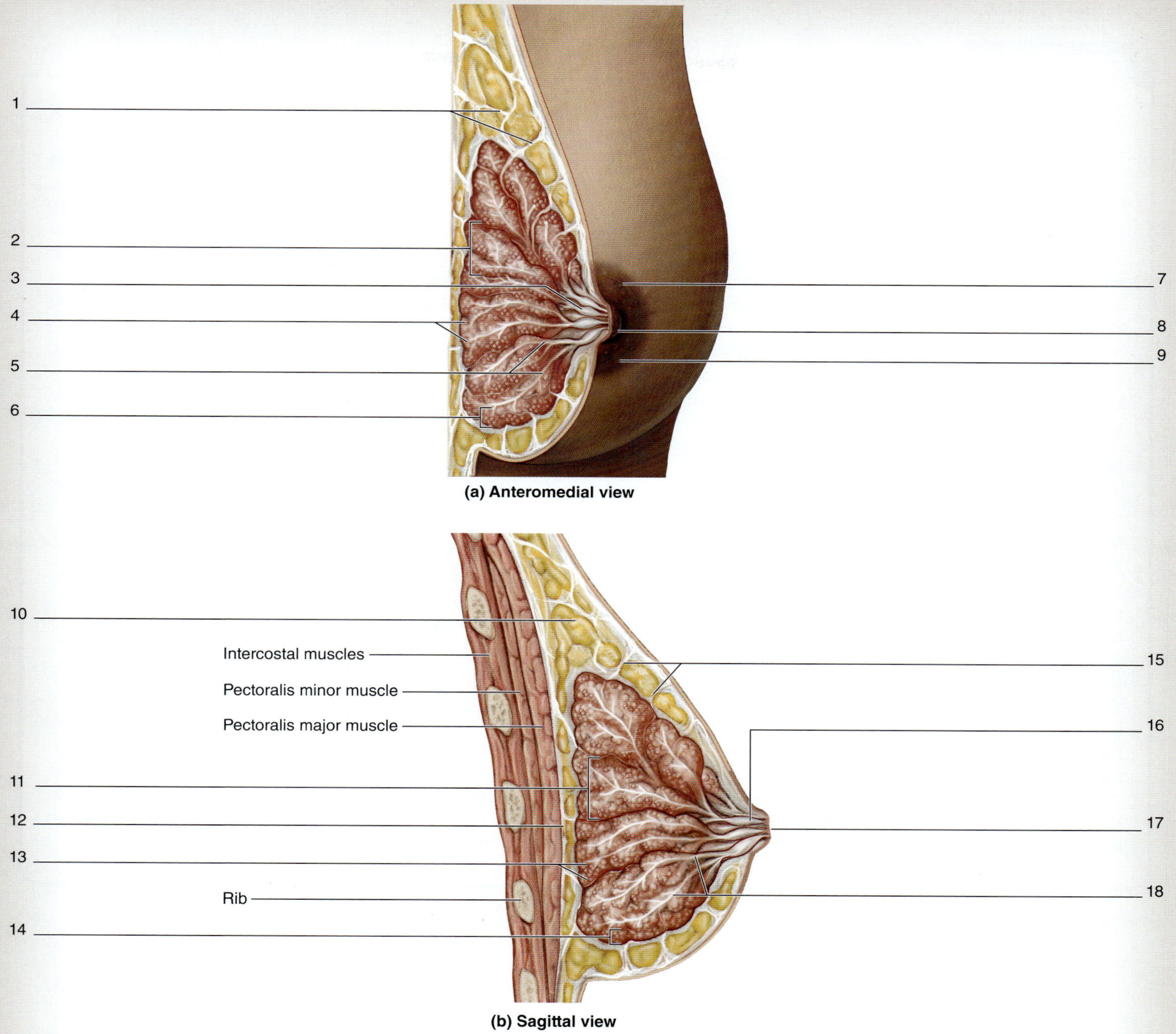

1
2
3
4
5
6
7
8
9

(a) Anteromedial view

10
Intercostal muscles
Pectoralis minor muscle
Pectoralis major muscle
11
12
13
Rib
14
15
16
17
18

(b) Sagittal view

Figure 27.14 The Female Breast. (a) Anterior view. (b) Sagittal view. Use the terms listed to fill in the numbered labels in the figure. Some answers may be used more than once.

- ☐ adipose tissue
- ☐ alveoli
- ☐ areola
- ☐ areolar gland
- ☐ deep fascia
- ☐ lactiferous ducts
- ☐ lactiferous sinus
- ☐ lobe
- ☐ lobule
- ☐ nipple
- ☐ suspensory ligaments

(continued on next page)

(continued from previous page)

INTEGRATE

CLINICAL VIEW

Breast Cancer Screening and Breast Self-Examination (BSE)

Clinical Breast Exam (CBE)

All women of reproductive age are encouraged to make annual visits to their gynecologist for a clinical breast exam (CBE) and a pap smear to test for early cervical changes that are risk factors for cervical cancer. The physician performing the CBE is palpating for abnormal lumps in the breast. However, the physician can also help teach a woman how to perform breast self-exam (BSE) so she can monitor the condition of her breasts monthly instead of just during yearly visits. The vast majority of breast cancers are discovered by the patient herself.

Mammography

A mammogram is an x-ray of the breast that is used to look for areas of extra density or calcifications, which can be early signs of breast cancer. Most women are advised to start having mammograms at age 40, but those with a family history of the disease are often advised to have their first mammogram at age 35. Mammography is good at locating very small tumors (especially DCIS), which may not be palpable on self-examinations or clinical examinations.

Breast Self-Exam (BSE)

Women should perform a breast self-exam at least once a month to look for any unusual bumps, lumps, or thickening of breast tissue. Upon palpation, a typical breast feels a bit lumpy, with some firmer areas and some softer areas. The firmer/harder areas can be thickenings of the suspensory ligaments, or fibrocystic changes to the breast. The softer areas are typically just adipose tissue. There are a number of benign (noncancerous) changes to the breast that can make breasts feel "lumpy." Specifically, fibrocystic changes in the breast consist of fluid-filled cysts and are often surrounded by dense fibrous tissue. This tissue, though often large and dense, is typically mobile (moves around easily), and often mirrors itself on the opposite breast. That is, if you find such changes in the lower-right quadrant of the right breast, you might also find it on the lower-right quadrant of the left breast. On the contrary, breast cancers tend to feel much firmer, like a kernel of unpopped popcorn, and they are immobile (they do not move when you touch them). They are generally painless and sometimes cause dimpling of the skin overlying the tumor.

The following are three procedures for performing a breast self-exam:

1. In the shower: Use soapy hands to palpate the breast while raising the arm on the same side as the breast you are palpating. Starting at the periphery of the breast, use two or three fingers to make small circles that progressively move toward the nipple. Squeeze the nipple to look for discharge. Be sure to palpate all the way up into the axilla, as the majority of tumors arise in the outer/upper quadrant toward the axillary region of the breast.
2. Lying down: Use the same procedure as in the shower. It is good to do an exam lying down because it makes the breasts lie flat, which may make it easier to feel certain lumps.
3. Standing in front of a mirror: Place your hands on your hips, tilt your elbows forward (anterior), and look for any indentations in the skin of the breast or "orange peel"–looking skin, which can be indicative of an underlying tumor.

If you discover anything that concerns you, make an appointment with your gynecologist as soon as possible to discuss your findings so the physician can help determine if what you felt was something benign or something that needs further exploration. Always remember that breast cancer is *highly curable* when caught early.

Male Reproductive System

The **testes,** the primary male reproductive organs, function in the production of sperm and testosterone. Sperm produced in the testes are stored in the **epididymis** and transported via the **ductus deferens** to the male urethra during ejaculation. The **seminal vesicles** and **prostate gland** are accessory reproductive glands that produce substances that nourish and protect the sperm and also compose the vast majority of semen. The **bulbourethral glands** are small glands located within the urogenital diaphragm that produce a mucus-like substance during sexual arousal that neutralizes the acidity of the male urethra and provides lubrication to ease the passage of semen during ejaculation.

Exercises 27.5 to 27.10 involved examining the histological appearance of many of these structures. The following exercise involves observing the male reproductive structures at the gross level.

EXERCISE 27.13

GROSS ANATOMY OF THE SCROTUM, TESTIS, SPERMATIC CORD, AND PENIS

1. Observe a prosected male human cadaver or classroom models of the male reproductive organs.

2. Identify the structures listed in **figures 27.15** and **27.16** on the cadaver or on classroom models, using the textbook as a guide. Then label them in figures 27.15 and 27.16.

3. *Optional Activity:* **AP|R 14: Reproductive System**—In the Quiz section, select the "Structures: Male" option to test your knowledge of male reproductive gross anatomy.

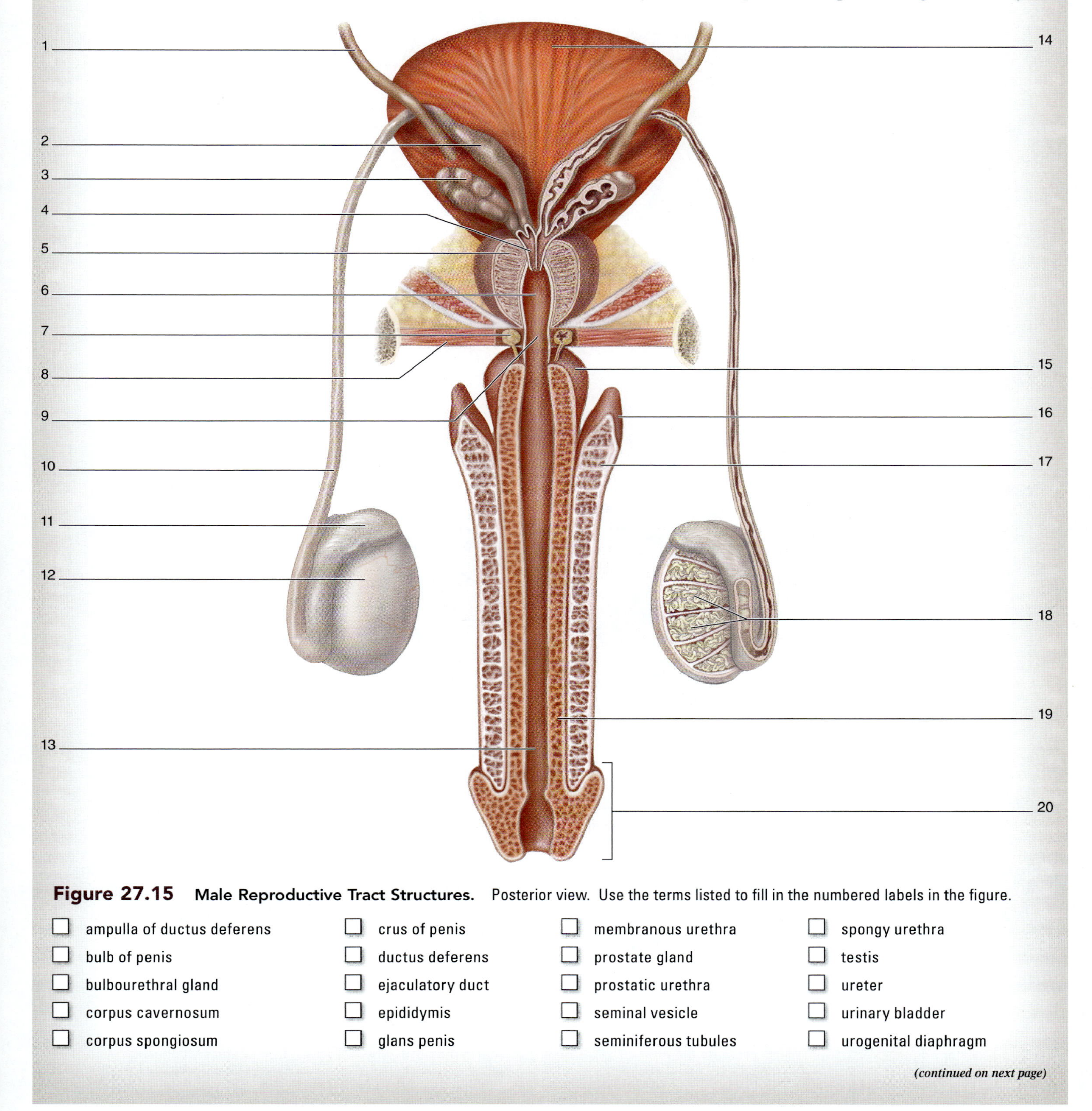

Figure 27.15 **Male Reproductive Tract Structures.** Posterior view. Use the terms listed to fill in the numbered labels in the figure.

- ☐ ampulla of ductus deferens
- ☐ bulb of penis
- ☐ bulbourethral gland
- ☐ corpus cavernosum
- ☐ corpus spongiosum
- ☐ crus of penis
- ☐ ductus deferens
- ☐ ejaculatory duct
- ☐ epididymis
- ☐ glans penis
- ☐ membranous urethra
- ☐ prostate gland
- ☐ prostatic urethra
- ☐ seminal vesicle
- ☐ seminiferous tubules
- ☐ spongy urethra
- ☐ testis
- ☐ ureter
- ☐ urinary bladder
- ☐ urogenital diaphragm

(continued on next page)

(continued from previous page)

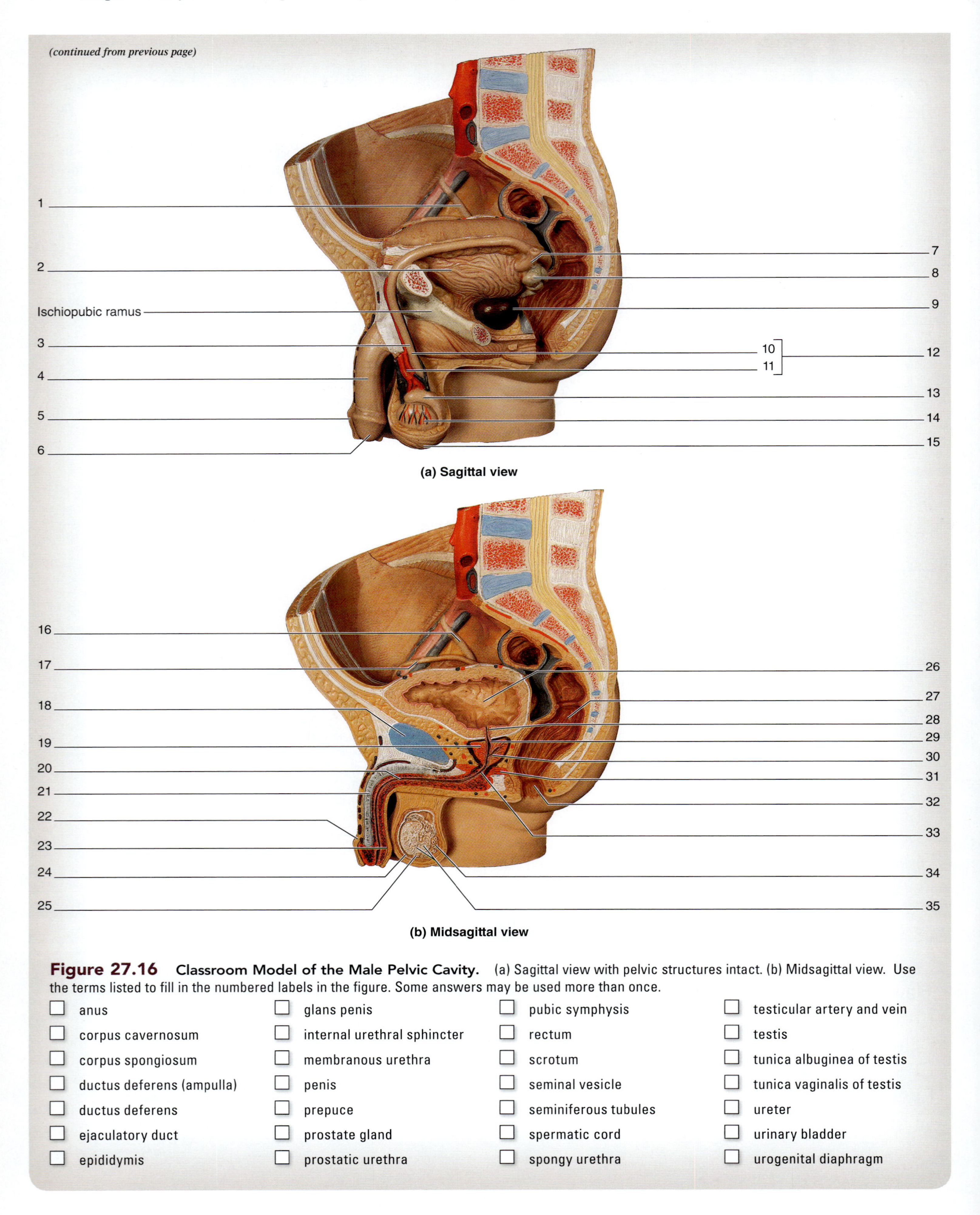

Figure 27.16 Classroom Model of the Male Pelvic Cavity. (a) Sagittal view with pelvic structures intact. (b) Midsagittal view. Use the terms listed to fill in the numbered labels in the figure. Some answers may be used more than once.

- ☐ anus
- ☐ corpus cavernosum
- ☐ corpus spongiosum
- ☐ ductus deferens (ampulla)
- ☐ ductus deferens
- ☐ ejaculatory duct
- ☐ epididymis
- ☐ glans penis
- ☐ internal urethral sphincter
- ☐ membranous urethra
- ☐ penis
- ☐ prepuce
- ☐ prostate gland
- ☐ prostatic urethra
- ☐ pubic symphysis
- ☐ rectum
- ☐ scrotum
- ☐ seminal vesicle
- ☐ seminiferous tubules
- ☐ spermatic cord
- ☐ spongy urethra
- ☐ testicular artery and vein
- ☐ testis
- ☐ tunica albuginea of testis
- ☐ tunica vaginalis of testis
- ☐ ureter
- ☐ urinary bladder
- ☐ urogenital diaphragm

PHYSIOLOGY

Reproductive Physiology

Gametogenesis, the production of the **gametes** (sperm and secondary oocytes) is accomplished by meiosis. **Meiosis** is similar in many ways to mitosis, but there are two nuclear divisions (I and II), which result in a reduction of chromosome number from a diploid number (2n, or 46) to a haploid number (n, or 23). Another difference in meiosis is the process of **synapsis** (crossing over) that occurs when homologous chromosomes are adjacent to each other. Synapsis is a process that introduces variability in the genetic code, which ensures that daughter cells are genetically different from parent cells. **Table 27.12** depicts and describes each phase of meiosis and the defining features of each phase.

Human somatic cells contain 23 pairs of chromosomes, or **homologous pairs.** These include 22 pairs of autosomal chromosomes, and one pair of sex chromosomes. Human gametes (egg and sperm) contain 22 autosomal chromosomes and one sex chromosome, an X chromosome for oocytes and an X or Y chromosome for sperm. Meiosis yields four genetically different, haploid daughter cells. The process of **fertilization** brings haploid sperm together with a haploid secondary oocyte. When a sperm cell has fertilized an oocyte, the resulting cell (called a zygote) is once again diploid.

Table 27.12 Phases of Meiosis

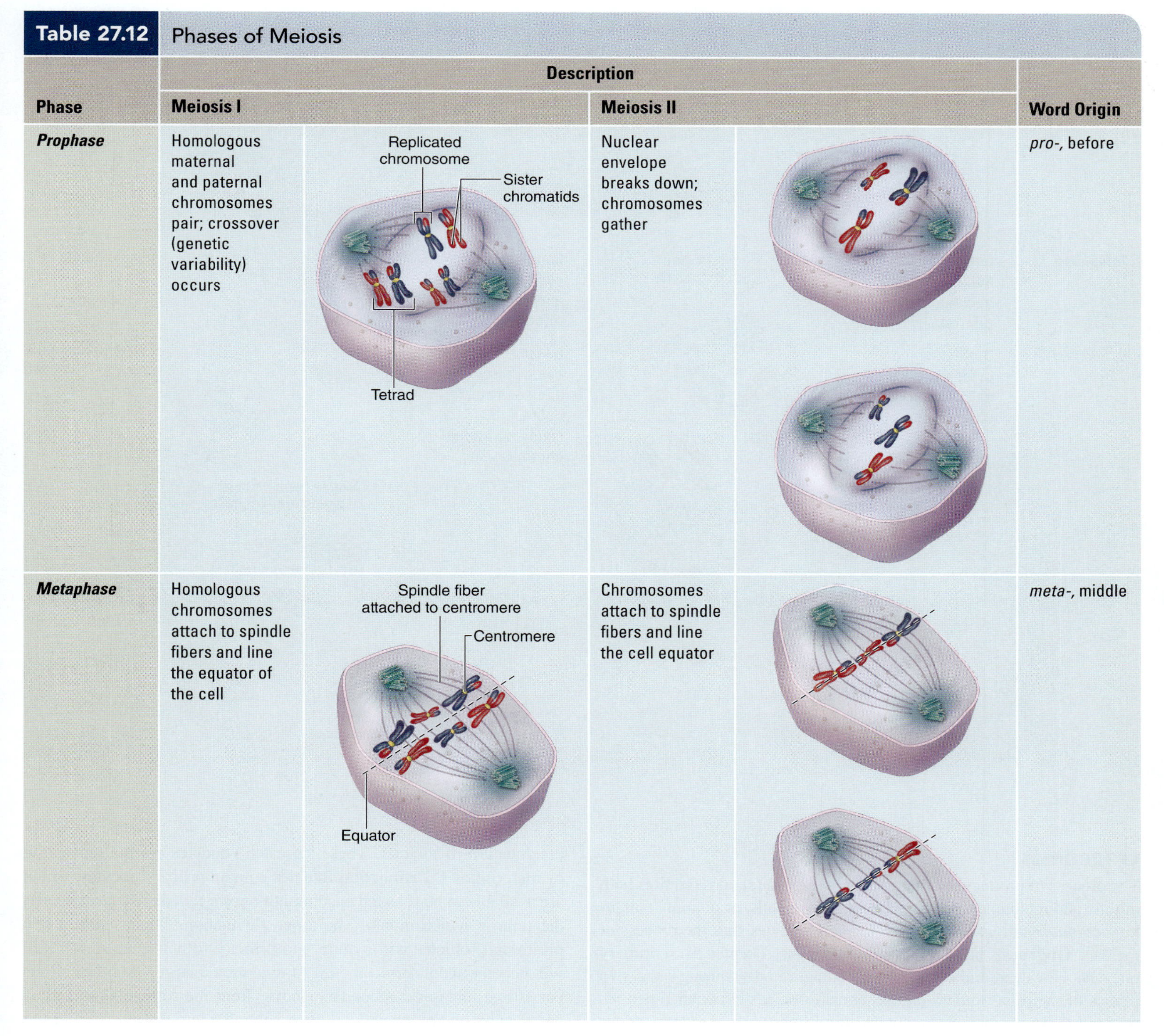

Phase	Description: Meiosis I	Description: Meiosis II	Word Origin
Prophase	Homologous maternal and paternal chromosomes pair; crossover (genetic variability) occurs	Nuclear envelope breaks down; chromosomes gather	*pro-*, before
Metaphase	Homologous chromosomes attach to spindle fibers and line the equator of the cell	Chromosomes attach to spindle fibers and line the cell equator	*meta-*, middle

(continued on next page)

Table 27.12 Phases of Meiosis *(continued)*

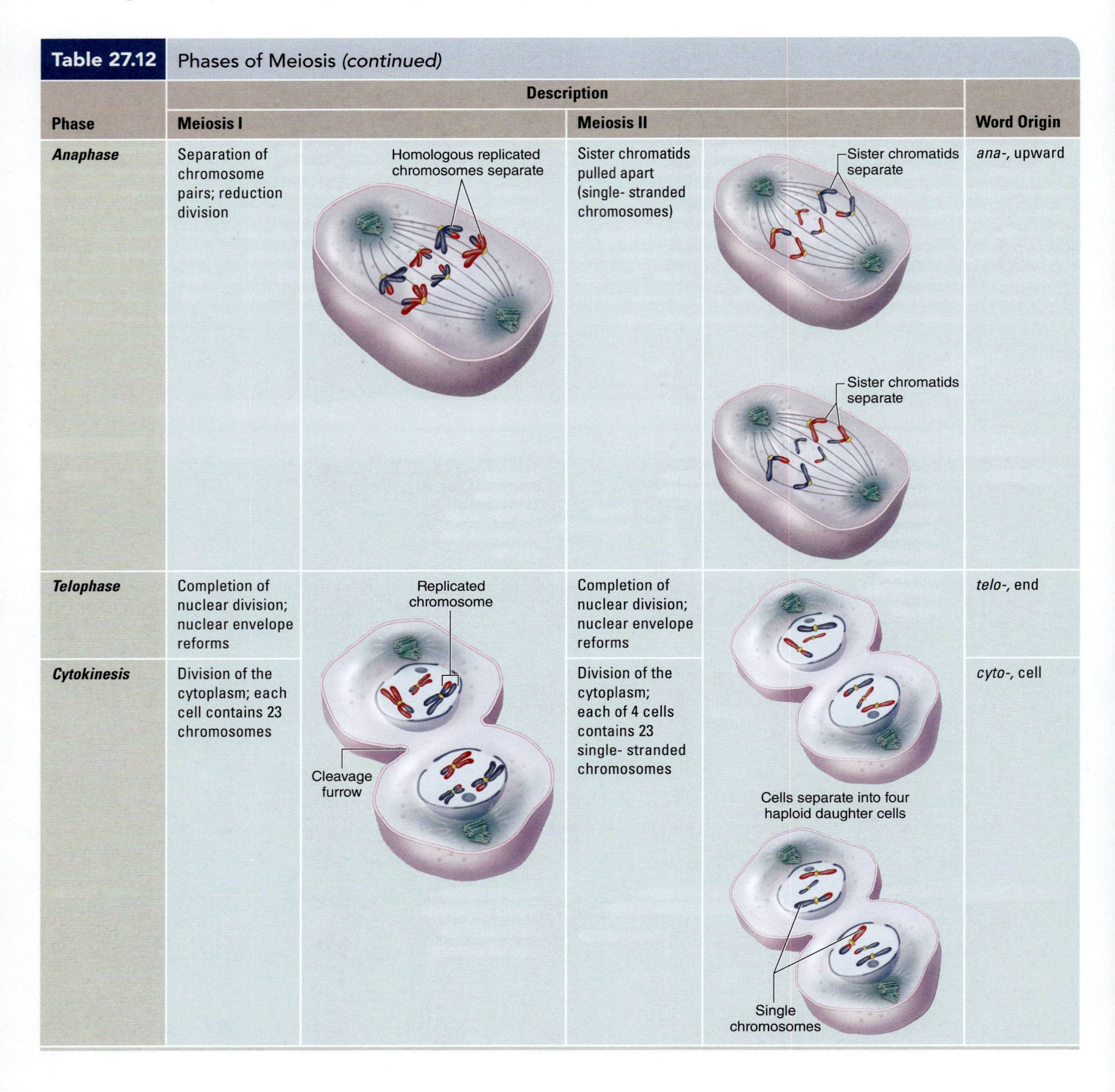

Phase	Description: Meiosis I	Description: Meiosis II	Word Origin
Anaphase	Separation of chromosome pairs; reduction division	Sister chromatids pulled apart (single- stranded chromosomes)	*ana-*, upward
Telophase	Completion of nuclear division; nuclear envelope reforms	Completion of nuclear division; nuclear envelope reforms	*telo-*, end
Cytokinesis	Division of the cytoplasm; each cell contains 23 chromosomes	Division of the cytoplasm; each of 4 cells contains 23 single- stranded chromosomes	*cyto-*, cell

Oogenesis

The process of producing a mature egg (ovum) in females requires both mitotic and meiotic division. Primordial germ cells, or oogonia (singular, *oogonium*), divide by mitosis. Primary oocytes start the process of meiosis. **Oogenesis** involves forming secondary oocytes from primary oocytes. However, rather than proceeding rapidly through the two phases of meiosis, primary and secondary oocytes experience periods of "arrest" where meiosis is stopped until hormones or events trigger continuation of meiosis. These "arrested" oocytes are housed within ovarian follicles. **Primordial follicles** contain primary oocytes, which are available to be selected each month once a female reaches puberty, the time at which monthly hormone fluctuations begin. Only a few primordial follicles will mature into primary follicles, and fewer still will become secondary follicles. At ovulation a large vesicular follicle bursts and releases a secondary oocyte from the follicle's fluid-filled sac. The secondary oocyte will then be drawn into the uterine tubes to potentially be fertilized by sperm. The final phases of meiosis occur after fertilization (see table 27.12).

The result of fertilization is formation of a **zygote,** which will subsequently undergo mitosis to form a mass of cells that may implant in the uterine wall. Development *in utero* continues, as described in exercise 27.16. **Figure 27.17** depicts the overall process of oogenesis, including identification of the various phases and arrests of meiosis, corresponding stages of the ovarian follicles, and histological images of each stage.

Hormones are responsible for regulating the monthly changes that occur both in the ovaries (ovarian cycle) and in the uterus (uterine cycle). The overall purpose is to produce viable eggs (ova) for fertilization and to provide an attractive place for implantation of the fertilized egg. The release of gonadotropin-releasing hormone (GnRH) from the hypothalamus stimulates the release of luteinizing hormone (LH) and follicle-stimulating hormone (FSH) from the anterior pituitary gland. It is LH and FSH that stimulate the development of the follicles during the **follicular phase** of the ovarian cycle. As the follicles develop, the granulosa cells surrounding the follicle secrete estrogen. Estrogen stimulates the cells in the uterine endometrium to proliferate during the **proliferative phase** of the uterine cycle.

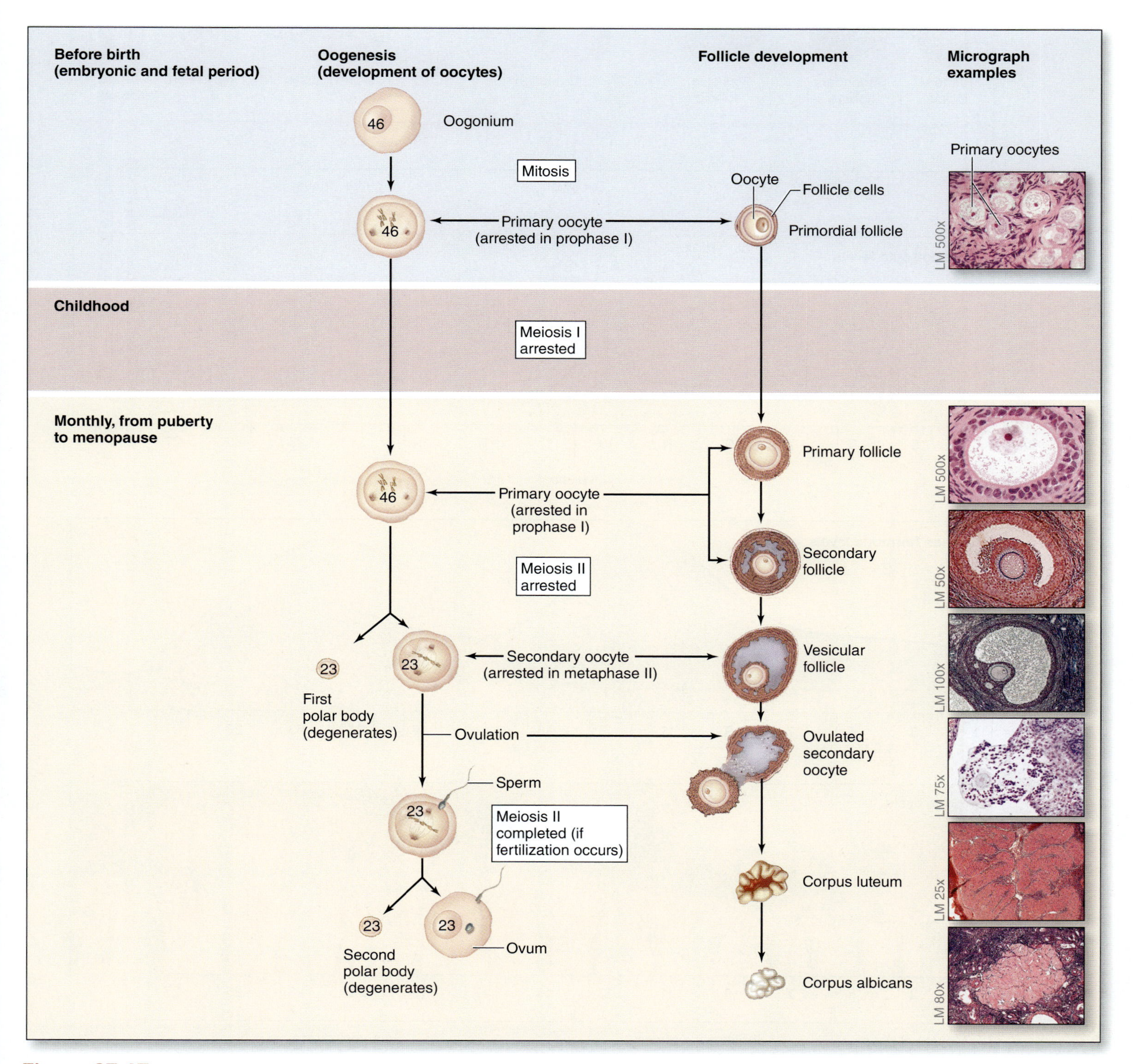

Figure 27.17 Oogenesis.

Once ovulation occurs, the ruptured vesicular follicle becomes the corpus luteum. This structure begins producing progesterone and estrogen throughout the **luteal phase** of the ovarian cycle. Increasing levels of progesterone result in thickening and increased vascularization of the endometrium. In addition, the corpus luteum maintains elevated progesterone and estrogen levels in the event fertilization occurs, but prior to the formation of the placenta, the structure responsible for maintaining a viable pregnancy. Human chorionic gonadotropin (hCG), the hormone secreted by the developing embryo, prevents the degradation of the corpus luteum in the event that fertilization and implantation occur. If fertilization and implantation do not occur, the corpus luteum degenerates into the corpus albicans. This degeneration also decreases levels of progesterone and estrogen. As the progesterone levels drop, blood vessels (spiral

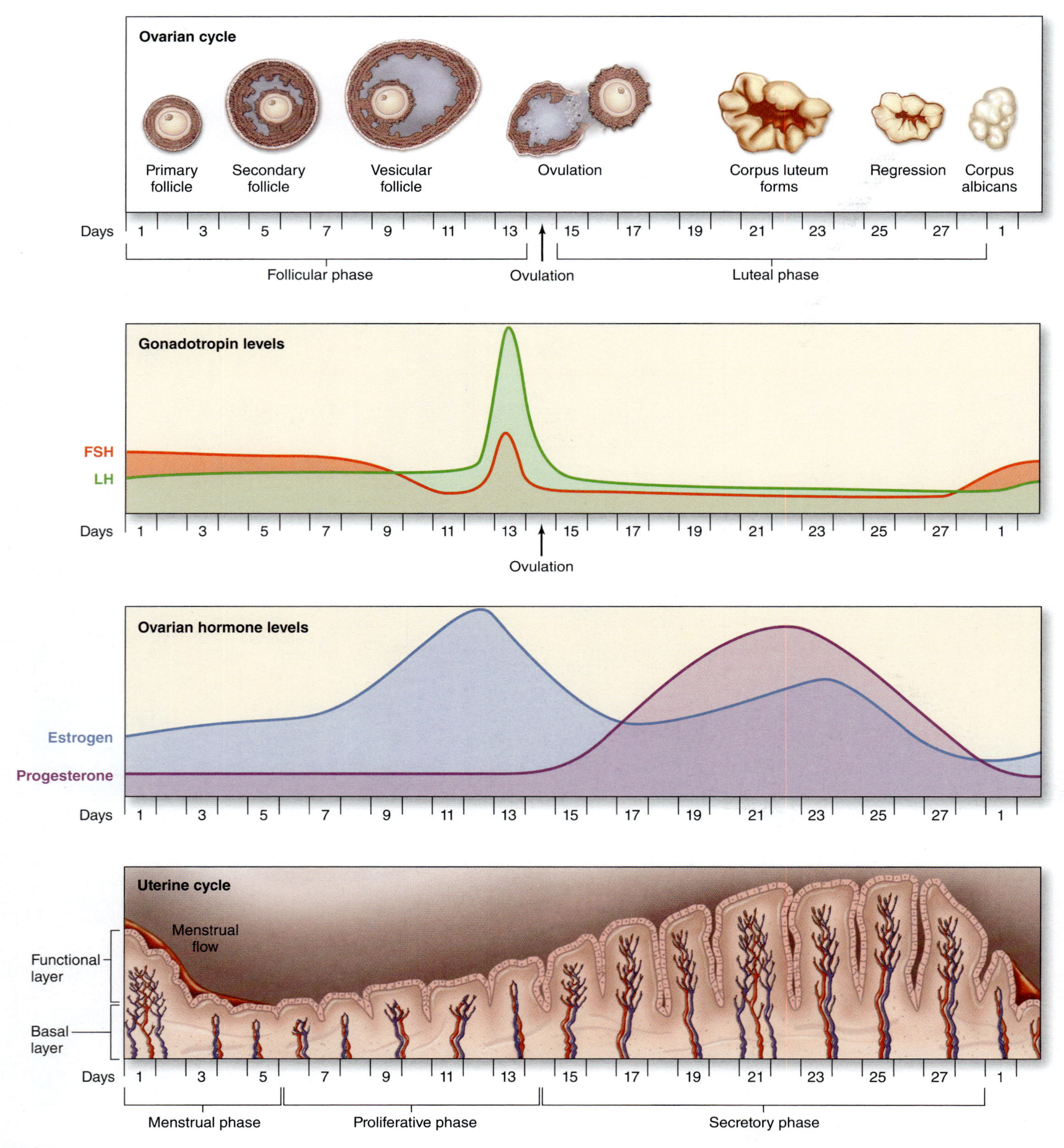

Figure 27.18 Ovarian and Uterine Cycles.

arterioles) in the uterine wall constrict, and the endometrial layer is shed for menses. Typically, these changes occur in 28-day cycles, as depicted in **figure 27.18.** However, cycles can vary in length. Research suggests that the duration of the luteal phase remains more consistent at roughly 14 days, whereas the duration of the foliicular phase may vary widely among women.

INTEGRATE

CONCEPT CONNECTION

Estrogen released by follicular cells acts directly on the hypothalamus and the anterior pituitary gland to regulate the amount of gonadotropin-releasing hormone (GnRH) released from the hypothalamus, and of luteinizing hormone (LH) and follicle-stimulating hormone (FSH) released from the anterior pituitary gland. Prior to ovulation, the increasing estrogen is stimulatory (positive feedback) for the release of GnRH, LH, and FSH. However, once ovulation occurs, estrogen is inhibitory (negative feedback) to the release of these hormones. It is a peak in LH that allows ovulation to occur; therefore, this switch from positive to negative feedback limits the number of eggs released each month. Incidentally, fertility drugs will often override these feedback mechanisms, thereby releasing multiple secondary oocytes with the chance for multiples (twins, triplets, or more). The release of these hormones was first introduced in the chapter on the endocrine system, chapter 19. The concepts of negative and positive feedback appear throughout the book, and are critical to understanding homeostasis.

Spermatogenesis

The production of sperm, **spermatogenesis,** in males is similar to oogenesis in females. It is interesting that the overall process is regulated by the same hormones (GnRH, LH, and FSH) that regulate oogenesis in females, with the exception of males secreting testosterone in place of estrogen. GnRH from the hypothalamus stimulates the release of LH and FSH from the anterior pituitary gland. LH stimulates interstitial cells in the seminiferous tubules of the testes to release testosterone, and FSH stimulates sustentacular cells to release androgen binding protein (ABP) and inhibin. The elevated levels of testosterone are stimulatory for sperm production, yet are inhibitory for GnRH production by the hypothalamus. The process begins as spermatogonia, or primordial germ cells, actively undergo mitosis and remain at the basement membrane of the seminiferous tubules. These spermatogonia divide to form primary spermatocytes, which are diploid cells. Primary spermatocytes then undergo meiosis. Unlike females, males have the ability to continuously form new primary spermatocytes for gamete formation. After meosis I, the spermatocyte is considered secondary and haploid, with only 23 chromosomes. Once the secondary spermatocyte undergoes meiosis II, it is considered a spermatid and is located closer to the lumen of the seminferous tubule. Each of the four spermatids produced then undergoes the maturation process, or **spermiogenesis.** This process involves the development of a head, or chromosome-containing acrosome, a midpiece containing abundant mitochondria, and a tail (flagellum) of the functional spermatozoon (plural, *spermatozoa*). The spermatozoon continues the maturation process in the epididymis before becoming a viable sperm cell.

EXERCISE 27.14

A CLINICAL CASE IN REPRODUCTIVE PHYSIOLOGY

The purpose of this exercise is to apply the concept of reproductive physiology to a real scenario, or clinical case study. This exercise requires information presented in this chapter, as well as in previous chapters. Integrating multiple systems in a case study reinforces a concept that is very important in physiology: homeostasis.

Before beginning, state how the ovarian cycle may be influenced by increasing levels of testosterone.

____

____

____

Consult with the laboratory instructor as to whether this exercise will be completed individually, in pairs, or in groups. Review the topics of the cardiovascular, lymphatic, respiratory, digestive, urinary, and reproductive systems in the textbook and in previous chapters in this laboratory manual as needed for assistance with the case study.

Case Study

Elizabeth, a 32-year-old newlywed, was eager to start a family. She was career focused, as was her husband, Frank, and both thrived under high-stress conditions. Their jobs required long hours, with little time to cook or clean at home. For that reason, they were constantly eating on the go, often getting takeout meals that were, admittedly, not the healthiest. Despite this hectic schedule, the couple was ready to have a child of their own. So, they began trying to conceive. Both Elizabeth and Frank were optimistic that they would be able to conceive despite having many friends that struggled with infertility. No women in Elizabeth's family had struggled before, so she had no real concerns.

Frank and Elizabeth's attempts to conceive were met with disappointment. They "tried" for 1 year with no luck, and ultimately sought medical advice, because Elizabeth was concerned that her age would soon make conception more difficult. Both Frank and Elizabeth visited a fertility clinic to investigate their case further. Doctors first tested Frank's sperm count and sperm viability, because this is the simplest test to conduct. There was no evidence of inadequate numbers of sperm or their inability to fertilize an egg. Frank and Elizabeth were relieved, yet still frustrated that they were unsure as to the cause of their infertility.

Doctors then made note of Elizabeth's physical health. At 5 feet, 0 inches, Elizabeth weighed 175 pounds. In addition to being overweight, she also suffered from chronic hypertension and an elevated heart rate. Doctors suggested a change in lifestyle to reduce stress, which they hoped would help Elizabeth lose weight and improve her cardiovascular health. Doctors also inquired about Elizabeth's reproductive health. Elizabeth admitted that she had been prescribed birth control pills

(continued on next page)

(continued from previous page)

at 15 years old to help regulate her irregular (unpredictable) periods. Since stopping the pill just 1 year ago, her periods were again irregular, with cycle durations ranging from 31 to 51 days. Menses would last 7 days with no remarkable pain or discomfort. She also noticed that she had an increased incidence of facial hair and acne, something that was quite unusual for her. Based on these symptoms, doctors ordered blood samples **(table 27.13)** to test for a condition called polycystic ovarian syndrome (PCOS).

PCOS is an endocrine disorder, characterized by elevated levels of testosterone, that is a major cause of infertility in females. Patients are often diabetic (type II), and exhibit signs of insulin resistance. Patients tend to be overweight and experience irregular periods due to a disruption in the positive and negative feedback mechanisms that regulate the ovarian and uterine cycles. Often, patients with PCOS do not ovulate, thereby preventing fertilization and implantation. Instead of ovulating, partially developed follicles may persist as "cysts" in the ovary. Clinically, ultrasound is used to observe ovarian cysts, and a diagnosis of PCOS requires that one ovary has more than twelve follicles ranging 2–9 mm in diameter. It is proposed that regulation of blood glucose levels and weight loss may allow ovulation to resume, although the exact mechanisms are unclear. In the event that dietary changes are unsuccessful, drugs can be administered to regulate blood glucose levels as well as stimulate ovulation.

Table 27.13 Elizabeth's Blood Test Results

Blood Test	Result	Reference Range
Cholesterol (mg/dl)	218	< 200
FSH (mLU/L)	6	3–20
Glucose (mg/dl)	141	70–110
Insulin (μU/mL)	167	0–20
LH (mLU/L)	18	3–20
Testosterone (pg/mL)	15	1.1–6.3
Triglycerides (mg/dl)	291	40–150

Answer the following questions regarding Elizabeth's case:

a. Doctors often prescribe metformin, a drug that decreases glucose transport in the GI tract, for patients suffering from PCOS. Describe how this would be beneficial, given the symptoms of the disease.

____

____

____

b. Elizabeth's blood samples were taken on day 3 of the menstrual cycle. Describe the relative levels of estrogen, progesterone, LH, and FSH during this phase of the cycle.

____

____

____

c. How might these hormone levels be different if samples were collected on day 14 (assume a 28-day cycle)?

____

____

____

d. Describe the potential consequences of elevated LH when compared to FSH levels.

____

____

____

e. Discuss why elevated testosterone may lead to anovulation. (Hint: Testosterone exhibits similar feedback mechanisms in males and females).

____

____

____

f. Do you think that Elizabeth is suffering from PCOS?

____

Fertilization and Development

The time period between **fertilization** and birth is considered the **prenatal period.** Fertilization yields a diploid cell, or **zygote,** that undergoes multiple mitotic divisions, known as **cleavage** events. Two cells become four cells, and then eight cells, until there is a mass of cells, called a **blastocyst,** which is ready for implantation in the uterine wall. This cleavage process takes approximately 2 weeks and constitutes the **pre-embryonic period.** Implantation marks the official start of gestation, a period that lasts 38 weeks in all. Weeks 3 through 8 constitute the **embryonic period,** a time in which the organism is called an **embryo**. During this time, major organs and organ systems develop. In the ninth week of development, the organism is classified as a **fetus.** The **fetal period** lasts for the remaining 30 weeks of gestation. During the fetal period, organs and systems continue to grow and develop in preparation for birth and to sustain life separate from the mother's womb. The following laboratory exercises explore the events of the pre-embryonic and embryonic periods.

EXERCISE 27.15

EARLY DEVELOPMENT: FERTILIZATION AND ZYGOTE FORMATION

The following laboratory exercise focuses on the changes that occur during the pre-embryonic period. Complete **table 27.14,** depicting the chronology of events in pre-embryonic development, using the textbook as a guide.

Table 27.14 Pre-Embryonic Period

Appearance	Developmental Stage	Location	Events
Ovum pronucleus; Sperm pronucleus; 120 µm			
Nucleus; 120 µm			
120 µm 4-cell stage → 120 µm 8-cell stage			
120 µm Morula			

(continued on next page)

(continued from previous page)

Table 27.14	**Pre-Embryonic Period *(continued)***		
Appearance	**Developmental Stage**	**Location**	**Events**
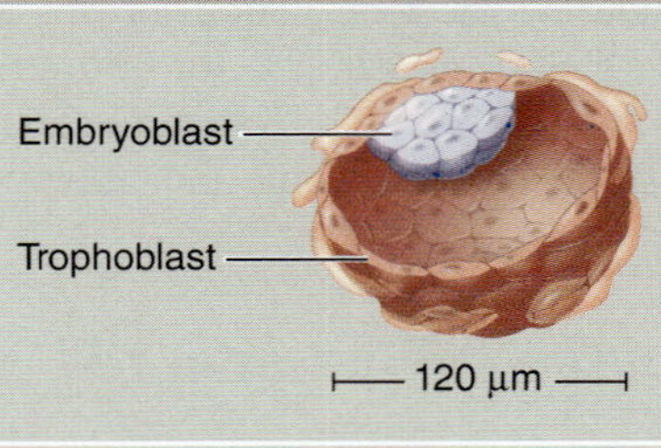 Embryoblast Trophoblast 120 μm			
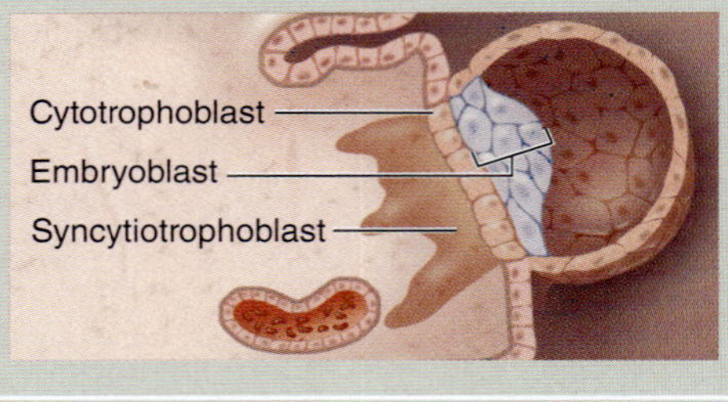 Cytotrophoblast Embryoblast Syncytiotrophoblast			

EXERCISE 27.16

EARLY DEVELOPMENT: EMBRYONIC DEVELOPMENT

This laboratory exercise focuses on the changes that occur in an organism during the embryonic period. Complete the following using the textbook as a guide.

1. Label the structures of the 3-week-old embryo identified in **figure 27.19.**
2. Complete **table 27.15** depicting the chronology of events in embryonic development.
3. Label the structures of the 4-week-old embryo identified in **figure 27.20.**

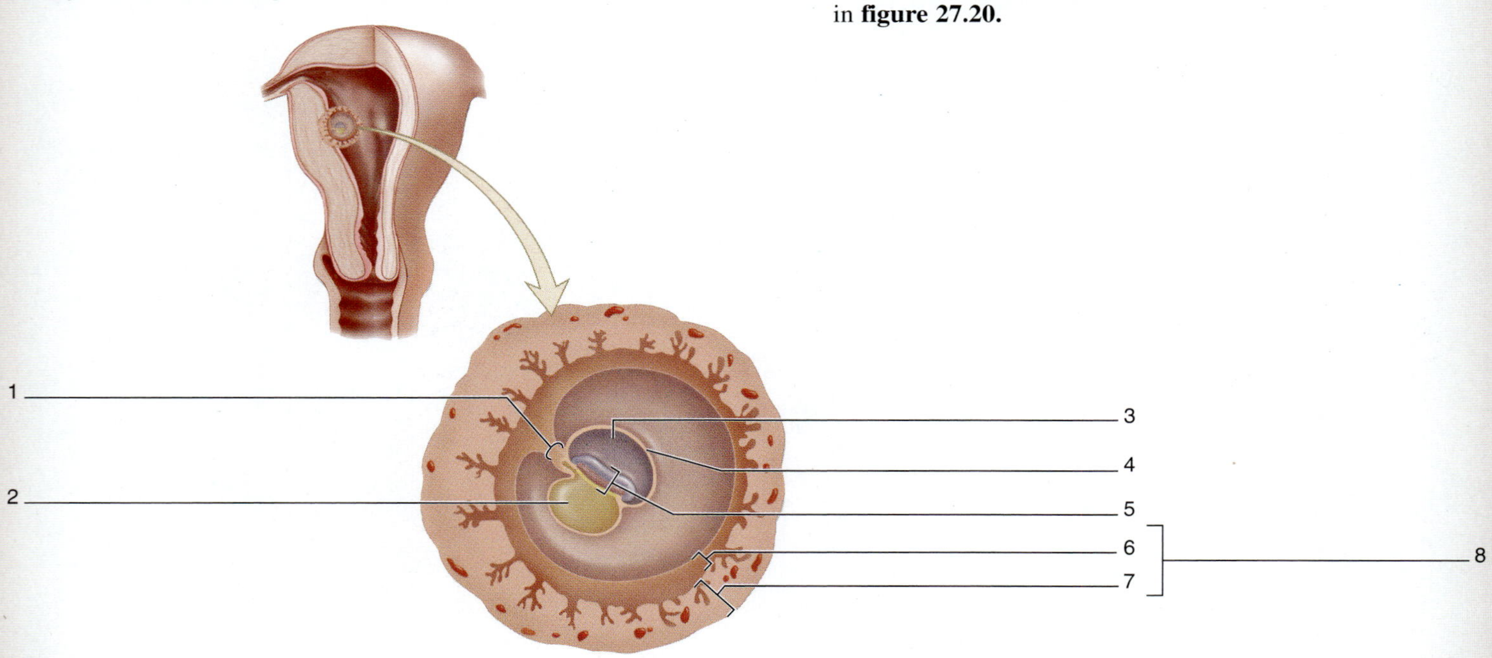

1
2
3
4
5
6
7
8

Figure 27.19 Structures of the 3-week-old Embryo. Use the terms listed to fill in the numbered labels in the figure.

- ☐ amnion
- ☐ amniotic cavity
- ☐ chorion
- ☐ connecting stalk
- ☐ embryo
- ☐ functional layer of uterus
- ☐ placenta
- ☐ yolk sac

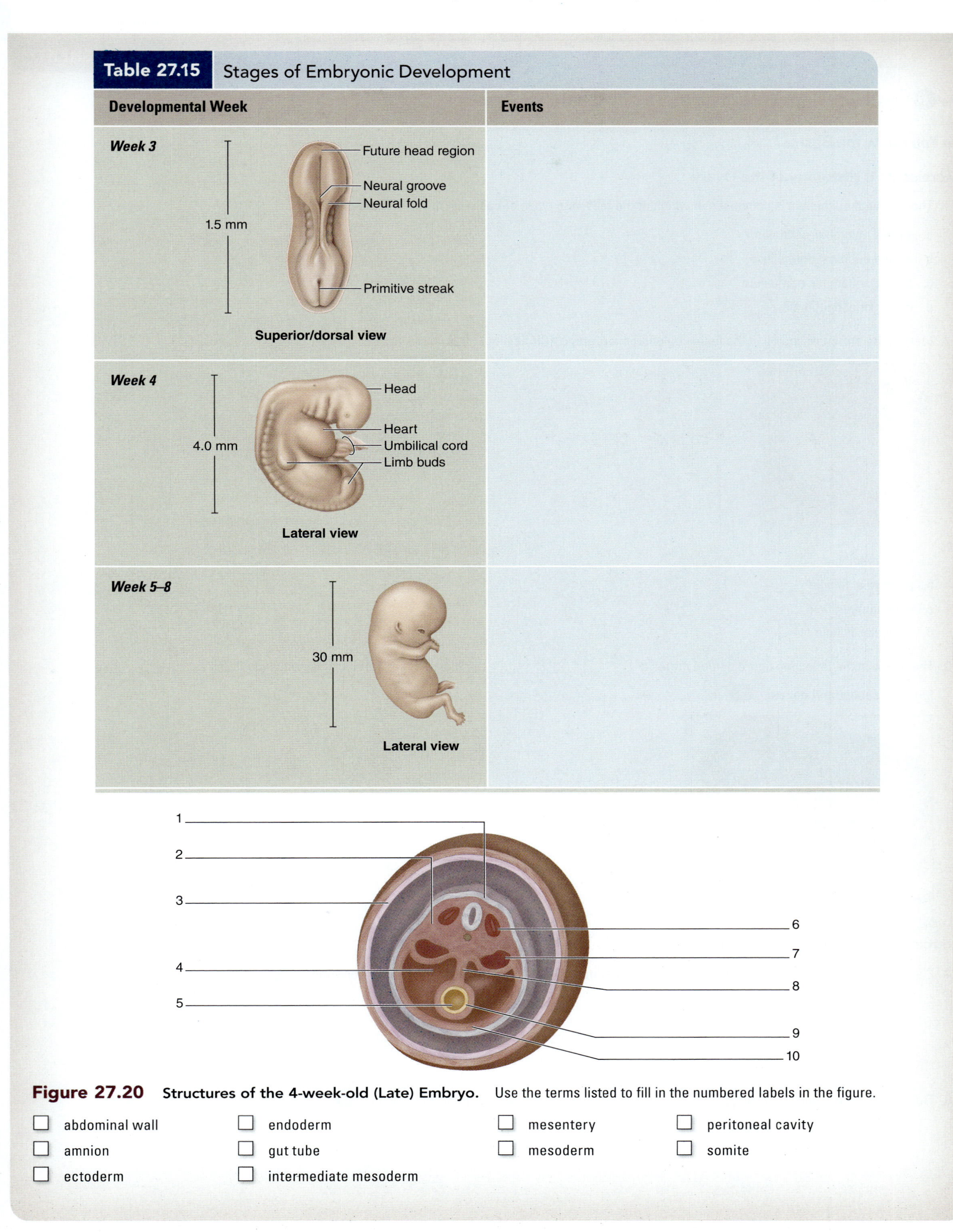

Table 27.15 Stages of Embryonic Development

Developmental Week	Events
Week 3	
Week 4	
Week 5–8	

Figure 27.20 **Structures of the 4-week-old (Late) Embryo.** Use the terms listed to fill in the numbered labels in the figure.

- ☐ abdominal wall
- ☐ amnion
- ☐ ectoderm
- ☐ endoderm
- ☐ gut tube
- ☐ intermediate mesoderm
- ☐ mesentery
- ☐ mesoderm
- ☐ peritoneal cavity
- ☐ somite

Chapter 27: The Reproductive System and Early Development

Name: ____________________

Date: __________ **Section:** __________

POST-LABORATORY WORKSHEET

The ❶ corresponds to the Learning Objective(s) listed in the chapter opener outline.

Do You Know the Basics?

Exercise 27.1: Histology of the Ovary

1. The germinal epithelium is composed of which of the following types of epithelial tissue? (Circle one.) ❶
   a. simple columnar epithelium
   b. simple cuboidal epithelium
   c. simple squamous epithelium
   d. transitional epithelium

2. Identify the structure shown in the following photomicrograph. (Circle one.) ❷

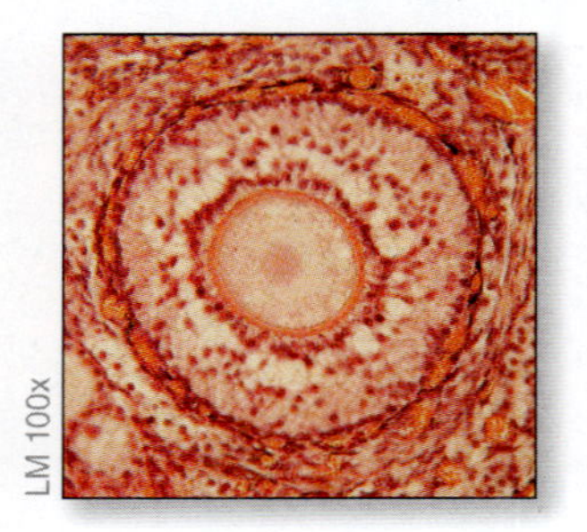

   a. primary follicle
   b. primordial follicle
   c. secondary follicle
   d. vesicular follicle

3. The arrow in the following figure demonstrates a __________________ (primary / secondary) follicle, which contains a __________________ (primary/secondary) oocyte. ❷

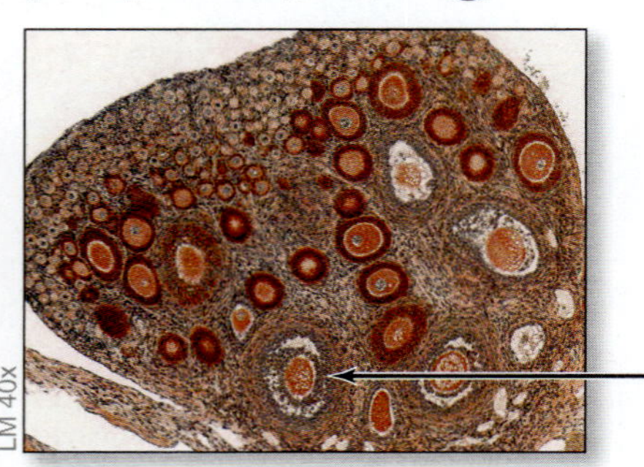

Exercise 27.2: Histology of the Uterine Tubes

4. Place the following layers that compose the wall of the uterine tube in order, from outermost to innermost. ❸ ❹
   ______ a. mucosa
   ______ b. muscularis
   ______ c. serosa

5. The uterine tube is lined with simple __________________ (cuboidal/columnar) epithelium. ❺

Exercise 27.3: Histology of the Uterine Wall

6. The two layers of the uterine endometrium are the functional layer and the basal layer. The ____________________ (functional/basal) layer is shed during menstruation. 6

7. The myometrium is composed of two layers of smooth muscle. ____________________ (True/False) 7

Exercise 27.4: Histology of the Vaginal Wall

8. The wall of the vagina is composed of which of the following layers? (Check all that apply.) 8

   ______ a. adventitia

   ______ b. mucosa

   ______ c. muscularis

   ______ d. submucosa

9. The innermost lining of the vagina is lined with which of the following type of epithelial tissue? (Circle one.) 9

   a. keratinized stratified squamous epithelium
   b. nonkeratinized stratified squamous epithelium
   c. simple columnar epithelium
   d. simple cuboidal epithelium
   e. simple squamous epithelium

Exercise 27.5: Histology of the Seminiferous Tubules

10. A cross section of a seminiferous tubule demonstrates a circular structure that contains layers of developing sperm. ____________________ (True/False) 10

11. Spermatogonia are located near the ____________________ (basement membrane/lumen) of the seminiferous tubules. 11

12. Sustentacular cells, located within seminiferous tubules, produce ____________________ (androgen-binding protein/testosterone), whereas interstitial (Leydig) cells, located in between seminiferous tubules, produce ____________________ (androgen-binding protein/testosterone). 12

Exercise 27.6: Histology of the Epididymis

13. Upon viewing the epididymis through a microscope, thousands of spermatazoa are visible in the lumen of the tube. ____________________ (True/False) 13

14. The epididymis is lined with ____________________ (simple/pseudostratified) columnar epithelium with stereocilia. 14

15. Which of the following is a function of stereocilia within the epidymis? (Check all that apply.) 15

    ______ a. enhance sperm maturation

    ______ b. increase surface area

    ______ c. propel sperm

    ______ d. secretion and absorption

    ______ e. sperm production

Exercise 27.7: Histology of the Ductus Deferens

16. Identify the structure shown in the following photomicrograph. (Circle one.) 16

    a. ductus deferens
    b. epididymis
    c. penis
    d. prostate gland

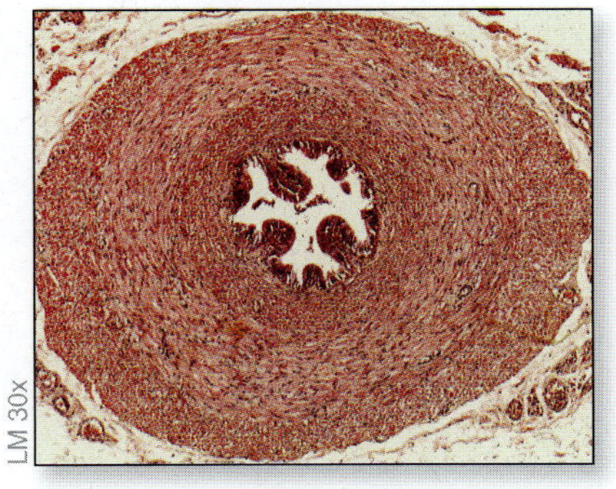

LM 30x

17. The ductus deferens contains a mucosa and layers of smooth muscle. The innermost smooth muscle layer is ____________________(circular/ longitudinal) muscle, whereas the outermost smooth muscle layer is ____________________(circular/longitudinal) muscle. 17

18. The ductus deferens is lined with pseudostratified ciliated columnar epithelium. ____________________ (True/False) 18

Exercise 27.8: Histology of the Seminal Vesicles

19. Seminal vesicles release substances, including fructose, which nourish sperm. ____________________ (True/False) 19

20. Seminal vesicles are lined with stratified cuboidal epithelium. ____________________ (True/False) 20

Exercise 27.9: Histology of the Prostate Gland

21. Which of the following are distinguishing features of the prostate gland of an aging male when viewed through a microscope? (Check all that apply.) 21 22

    ______ a. cilia

    ______ b. mucosal folds

    ______ c. muscular wall

    ______ d. prostatic calculi

    ______ e. tubuloalveolar glands

Exercise 27.10: Histology of the Penis

22. The urethra is surrounded by which structure in the penis? (Circle one.) 23

    a. central artery
    b. corpora cavernosa
    c. corpus spongiosum
    d. dorsal vein

23. When viewing the cross section of the penis through a microscope, there are three erectile bodies, paired corpora cavernosa, and a single corpus spongiosum. ____________________ (True/False) 24

Exercise 27.11: Gross Anatomy of the Ovary, Uterine Tubes, Uterus, and Supporting Ligaments

24. Place the following terms in the correct order to describe the pathway that an ovum takes as it travels from the ovary to the uterus. 25

    ______ a. ampulla of uterine tube

    ______ b. infundibulum of uterine tube

    ______ c. isthmus of uterine tube

    ______ d. ovarian cortex

    ______ e. uterine part of uterine tube

    ______ f. uterus

25. Which of the following structures travels through the inguinal canal in females? (Circle one.) 26

    a. broad ligament
    b. ovarian ligament
    c. round ligament of the uterus
    d. suspensory ligament of the ovary
    e. uterosacral ligament

26. The peritoneal cavity is considered an ____________________ (open/closed) cavity in the female because the uterine tube is ____________________ (open/closed) at its distal end (the infundibulum). 27

Exercise 27.12: Gross Anatomy of the Female Breast

27. Which of the following are the secretory units of the mammary glands, which produce milk? (Circle one.) 28

    a. alveoli
    b. areolar glands
    c. lactiferous sinuses
    d. lobules
    e. nipples

Exercise 27.13: Gross Anatomy of the Scrotum, Testis, Spermatic Cord, and Penis

28. Match the description listed in column A with the associated structure listed in column B. 29 32

Column A		Column B
______	1. located inferior to the urinary bladder	a. bulbourethral gland
______	2. produces fructose, an important component of semen	b. epididymis
______	3. produces a mucous-like substance that neutralizes acidity of the urethra	c. prostate gland
______	4. site of sperm maturation	d. seminal vesicles
______	5. site of sperm production	e. testes

29. Which of the following is the portion of the male urethra that passes through erectile tissue? (Circle one.) 30

    a. membranous urethra
    b. prostatic urethra
    c. spongy urethra

30. Which of the following are considered accessory reproductive structures that contribute substances to semen (other than sperm)? (Check all that apply.) 31 32

    ______ a. bulbourethral gland
    ______ b. epididymis
    ______ c. prostate gland
    ______ d. seminal vesicles
    ______ e. testes

31. The spermatic cord is composed of the testicular artery and nerve, the ductus deferens, lymphatic vessels, and the pampiniform plexus of veins.

    ____________________ (True/False) 33

32. Place the following structures in the order in which a sperm travels through the male reproductive tract during ejaculation. 34

    ______ a. ductus deferens
    ______ b. epididymis
    ______ c. membranous urethra
    ______ d. prostatic urethra
    ______ e. spongy urethra

Exercise 27.14: A Clinical Case in Reproductive Physiology

33. The product of mitosis is two ____________________ (diploid/haploid) daughter cells, whereas the product of meiosis is four ____________________ (diploid/haploid) daughter cells. 35

34. Rising levels of ____________________ (estrogen/progesterone) correspond to the ____________________(proliferative/secretory) phase of the uterine cycle. 36

35. Elevated levels of testosterone in females may inhibit ovulation. ____________________ (True/False) 37

Exercise 27.15: Early Development: Fertilization and Zygote Formation

36. The diploid cell that forms as a result of fertilization is called a ____________________(blastocyst/zygote), whereas the structure that implants in the uterine wall is called a ____________________ (blastocyst/zygote). 38

Exercise 27.16: Early Development: Embryonic Development

37. Embryonic development occurs during gestation weeks 3 through 8. ____________________ (True/False) 39

38. Which of the following are germ layers present in an embryo? (Check all that apply.) 40

    ______ a. amnion

    ______ b. ectoderm

    ______ c. endoderm

    ______ d. mesoderm

    ______ e. yolk sac

Can You Apply What You've Learned?

39. Unlike in the epididymis, it is highly unlikely that sperm cells are visible inside the lumen of the ductus deferens when viewed histologically. Why do you think this is? ____

40. The homologous structure to the male penis is the female clitoris. The clitoris consists of two paired erectile tissues, the corpora cavernosa, but it does not contain a corpus spongiosum. What structure is the clitoris "missing" as compared to the penis? ____________________

41. A woman that is trying to conceive may purchase ovulation kits in order to test for the release of the secondary oocyte. An ovulation kit requires placing a dipstick in a collected urine sample. A positive test indicates that ovulation will happen typically within 48 hours. For which hormone is this test sensitive?

42. One cause of miscarriage, or loss of pregnancy, is low levels of progesterone in the initial stages of development.

    a. Describe what mechanisms are in place to ensure adequate levels of progesterone in the event that fertilization occurs. ____________________

    b. What long-term mechanisms are in place to provide a continuous supply of progesterone (to maintain pregnancy)? ____________________

Can You Synthesize What You've Learned?

43. Viagra®, a drug known for its ability to enhance erection in males, is a drug that was first prescribed to treat hypertension. Using the concepts of pressure, flow, and resistance, describe why a drug used to counteract high blood pressure may also lead to an erection in males. ____________________

44. Birth control pills work to override the hormone cycles in order to prevent a secondary oocyte from being released. Based on your knowledge of hormonal regulation of the ovarian cycle, design a drug that could be used to prevent ovulation. Be specific about what hormones would be present in the female and why. Would you expect the levels of these hormones to change throughout the 28-day cycle? ____________________

CHAPTER 28

Cat Dissection Exercises

OUTLINE AND LEARNING OBJECTIVES

GROSS ANATOMY 794

Overview 794

EXERCISE 28.1: DIRECTIONAL TERMS AND SURFACE ANATOMY 794

1. Describe the proper handling and care of a preserved dissection specimen
2. Identify important surface anatomy landmarks in the cat
3. Note similarities and differences in directional terms as they concern the human and the cat
4. Determine the sex of a cat
5. Note modifications of the integument of a cat as compared to that of a human

Back and Limbs 796

EXERCISE 28.2: SKELETAL SYSTEM 796

6. Identify the major bones of the cat skeleton
7. Compare and contrast the skeletal system of the cat with that of the human

EXERCISE 28.3: SKINNING THE CAT 798

8. Observe the cutaneous maximus and platysma muscles in the cat
9. Note cutaneous nerves that run between muscles and skin

EXERCISE 28.4: MUSCLES OF THE HEAD AND NECK 800

10. Identify the major head and neck muscles of the cat and compare them to those of a human

EXERCISE 28.5: MUSCLES OF THE THORAX 802

11. Identify the major thoracic muscles of the cat and compare them to those of a human

EXERCISE 28.6: MUSCLES OF THE ABDOMINAL WALL 807

12. Identify the major abdominal muscles of the cat and compare them to those of a human
13. Observe the fiber orientation of the abdominal muscles and note the similarities with human abdominal muscles

EXERCISE 28.7: MUSCLES OF THE BACK AND SHOULDER 807

14. Identify the major muscles located in the back and shoulder regions of the cat and compare them to those of a human

EXERCISE 28.8: MUSCLES OF THE FORELIMB 809

15. Identify the major forelimb muscles of the cat and compare them to those of a human

EXERCISE 28.9: MUSCLES OF THE HINDLIMB 813

16. Identify the major hindlimb muscles of the cat and compare them to those of a human
17. List the three muscles that compose the triceps surae muscle in both cats and humans

Thoracic Cavity 819

EXERCISE 28.10: OPENING THE THORACIC AND ABDOMINAL CAVITIES 819

18. Prepare the thoracic and abdominal cavities for observation

EXERCISE 28.11: THE HEART, LUNGS, AND MEDIASTINUM 821

19. Identify the major structures of the cat heart and compare them to those of a human
20. Identify lung structures of the cat and compare them to those of a human
21. Identify the major thoracic blood vessels of the cat and compare them to those of a human
22. Identify the thoracic duct, esophagus, and trachea/bronchi in the cat and compare them to those of a human

Abdomen and Pelvis 825

EXERCISE 28.12: THE ABDOMINAL CAVITY 825

23. Identify digestive system structures of the cat and compare them to those of a human
24. Identify the major blood vessels of the abdominal cavity of the cat and compare them to those of a human

EXERCISE 28.13: THE DORSAL ABDOMINAL WALL 829

25. Identify structures of the dorsal abdominal wall (kidneys, adrenal glands, ureters) of the cat and compare them to those of a human

EXERCISE 28.14: UROGENITAL SYSTEMS 833

26. Identify the urinary bladder and urethra on male and female cats and compare them to those of a human
27. Identify external genitalia on both male and female cats and compare them to those of a human
28. Identify accessory reproductive organs in both male and female cats and compare them to those of a human

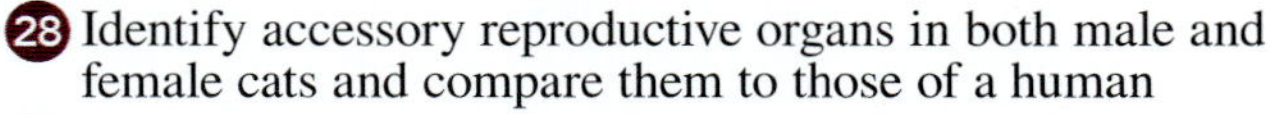

29. Identify the primary reproductive organs (testes and ovaries) on male and female cats, respectively, and compare them to those of a human

Head and Neck 838

EXERCISE 28.15: THE HEAD, NECK, AND ORAL CAVITY 838

30. Identify the major head and neck muscles of the cat and compare them to those of a human
31. Identify structures of the oral cavity of the cat and compare them to those of a human

INTRODUCTION

The exercises in this chapter cover the dissection of a vertebrate mammal that has anatomy similar to human anatomy: the domestic cat (*Felis domesticus*). The process of dissection provides an opportunity to feel the texture of tissues, to see the real size of structures, and to investigate relationships among anatomic structures that, until now, were studied separately. That is, the exercises in the beginning of this manual are all organized by system, whereas the cat dissection exercises are organized by region. A regional approach allows the dissector to focus on the relationships among organs and the blood and nerve supplies to those organs, at the same time. One beauty of this comparative vertebrate anatomy is that it allows one to realize just how much all vertebrates have in common. For example, although a cat's posture is not the same as a human's, the structure and function of the cat's skeletal system demonstrates remarkable similarities to those of a human. Although the process of dissection may cause apprehension at first, this feeling generally dissipates with experience. Dissection provides an unparalleled experience of discovery of the beauty of the human/animal form.

List of Reference Tables

Chapter 28: Cat Dissection Exercises

Name: ____________________

Date: __________ **Section:** __________

These Pre-Laboratory Worksheet questions may be assigned by instructors through their connect course.

PRE-LABORATORY WORKSHEET

1 The dorsal surface of a cat is the same as the ____________________ (posterior/superior) surface of a human, whereas the caudal surface of a cat is the same as the ____________________ (dorsal/posterior) surface of a human.

2. The anterior surface of a cat is the same as the ____________________ (cranial/ventral) surface of a human, whereas the inferior surface of a cat is the same as the ____________________ (caudal/ventral) surface of a human.

3. In a cat, a transverse plane separates ____________________ (anterior/superior) portions from ____________________ (inferior/posterior) portions.

4. In a cat, a frontal plane separates ____________________ (anterior/dorsal) portions from ____________________ (posterior/ventral) portions.

5. Superficially, a male cat can be identified by the presence of a ____________________ (scrotum/vestibule), whereas the female cat is identified by the presence of a ____________________ (scrotum/vestibule).

6. When skinning the pudendal region of the male cat, which structures may be at risk for damage? (Check all that apply.)

   ______ a. epididymis

   ______ b. penis

   ______ c. spermatic cord

   ______ d. testis

   ______ e. vestibule

7. Which of the following dissection techniques is most likely to allow for skinning the cat while protecting underlying structures? (Circle one.)

   a. blunt dissection
   b. dissection with scissors
   c. dissection with a scalpel
   d. open scissors technique

8. When handling preserved organisms, one should wear gloves and safety glasses. ____________________ (True/False)

GROSS ANATOMY

Overview

INTEGRATE

CLINICAL VIEW

Proper Handling and Care of Preserved Cats

The dissection specimens under study in the laboratory are embalmed, or preserved, to enable safe dissection without risk of exposure to necrotic (dead) tissue or harmful microorganisms. Before beginning the dissection, review table 1.6 in chapter 1 (p. 13), which describes the composition of embalming fluids, safe handling of embalming fluids, and safety procedures for dissection. Some specimens may have been preserved with "biosafe" solutions, which generally do not contain the fixative formalin, a substance that can be harmful if handled incorrectly. The dissection specimens used in the lab may or may not have been preserved using formalin. If you are uncertain what preservative was used, it is safest to use the same precautions used for formalin-preserved specimens. That is, always wear gloves before handling the specimen, wear safety glasses when there is a risk of fluid splashing from the specimen, and never touch your face or eyes with your hands if they may have preservative fluid on them. If you get the solution in your eyes, rinse your eyes out thoroughly at the eyewash station. Thoroughly clean your work area and be sure to wash your hands and forearms thoroughly with soap and warm water before leaving the laboratory for the day.

INTEGRATE

LEARNING STRATEGY

Figures in this chapter are color-coded so that common structures can be easily identified. Refer back to this color key often while studying the anatomic structures of the cat.

Color Key for Common Structures

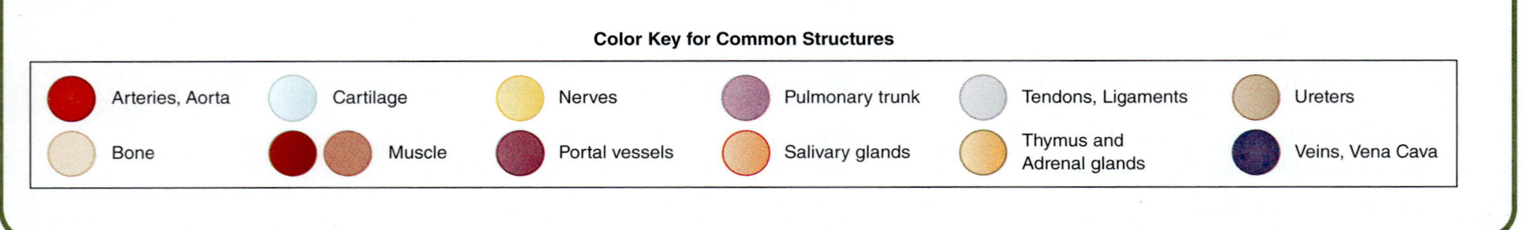

EXERCISE 28.1

DIRECTIONAL TERMS AND SURFACE ANATOMY

Before beginning dissection of the cat, be sure to read the Clinical View on the proper handling and care of preserved cats.

Directional Terms

One of the major differences encountered when comparing the anatomy of a cat to that of a human is the directional terminology used. This is because the normal anatomic position for the cat is with all four limbs on the ground **(figure 28.1*a*).** Humans, on the other hand, have a normal anatomic position with only the lower limbs on the ground (figure 28.1*b*).

Figure 28.1 Directional Terms. Note how directional terms in the cat (a) differ from the same terms in a human (b) because the cat stands on four legs whereas the human stands on only two.

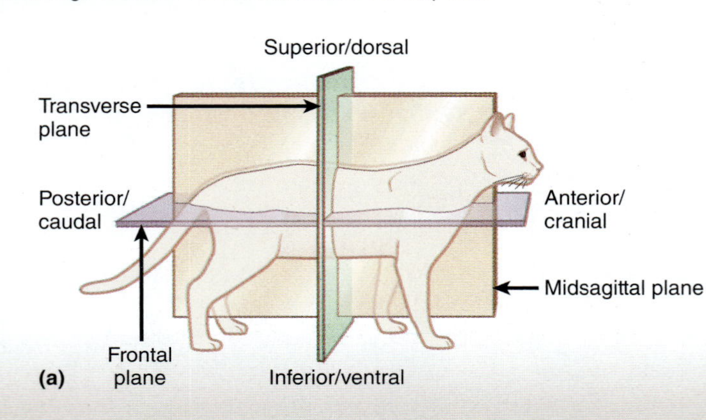

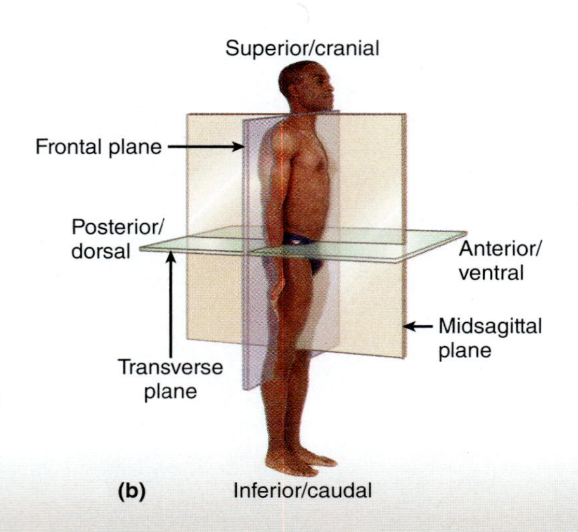

Obtain the Following:

- **dissecting tools**
- **dissecting tray**
- **gloves**
- **preserved cat**
- **storage bag**
- **paper towels**
- **wetting solution (for keeping the specimen moist)**

1. Lay the cat on the dissecting tray with its dorsal side up and start by observing the surface anatomy. On the head, make note of the **vibrissae** (whiskers), the **pinnae** (external ears), and the **nictitating membrane** of each eye **(figure 28.2).** The nictitating membrane is a membrane at the medial aspect of each eye that can brush across the anterior surface of the eye to keep it moist and protect it from debris when the eyes are open. Note that the fur superior to each eye is also elongated a bit to form "eyelashes." These, like human eyelashes, also help keep debris out of the cat's eyes.

2. Turn the cat over so it is ventral side up. Observe the neck and inguinal regions for any sign of incisions. Most commonly there will be an incision in the neck region. This is typically where injections of embalming fluid and any other substances have been introduced into the circulatory system. If there is an incision in the abdomen, the hepatic portal system may also have been injected.

3. Observe the skin adjacent to the midline of the ventral thorax and abdomen and look for the **mammary ridge**, which contains a series of **mammary glands** and **mammary papillae** (nipples). In a pregnant female cat, these ridges become raised because of the growth and development of mammary glands in the subcutaneous tissues. These stuctures come off with removal of the skin.

4. Determine the sex of the cat. This can be done by both visual inspection and palpation (*palpare*, to touch) of the inguinal and perineal regions. The external genitalia may not be easy to visualize because the cat is covered with fur. If the specimen is a male cat, a small, firm structure at the midline, which is the **penis,** can be palpated. Adjacent to the penis, two roundish structures can be palpated. These are the **testicles,** which are located within the **scrotum.** If the specimen is a female cat, there will be a small invagination anterior/cranial to the root of the tail, which is the **vestibule.** Note: If the specimen is a male, take extra care when removing the skin in this region so as not to destroy the contents of the **spermatic cord,** which extend from the posterior abdominal wall into the scrotum.

Figure 28.2 Surface Anatomy of the Cat. This photo shows several terms related to the surface anatomy of a cat's head.

(continued on next page)

(continued from previous page)

5. Record the sex of the specimen and make note of any pertinent incisions and/or surface anatomy observations that may be important in the space provided. If the cat has a unique identification number, be sure to record that number as well.

6. Observe a specimen of the opposite sex so that you will be able to identify external genitalia of both a male and a female cat.

Sex of Cat: ______________________________

Cat ID: ______________________________

Back and Limbs

Skeletal System

The cat skeleton has almost all the same bones as the human skeleton, although they often vary in both the size and shape, as well as their orientation within the skeleton compared with those in a human. Some of these differences reflect the cat's four-legged posture and locomotor capabilities.

EXERCISE 28.2

SKELETAL SYSTEM

1. Observe a cat skeleton **(figure 28.3).** Be sure to carefully observe all of the bones of the skeleton so that you are able to identify them later (e.g., on a laboratory exam).

2. Record in **table 28.1** the number of vertebrae in each region of the vertebral column in a human. Next, count and record the number of vertebrae in each region of the

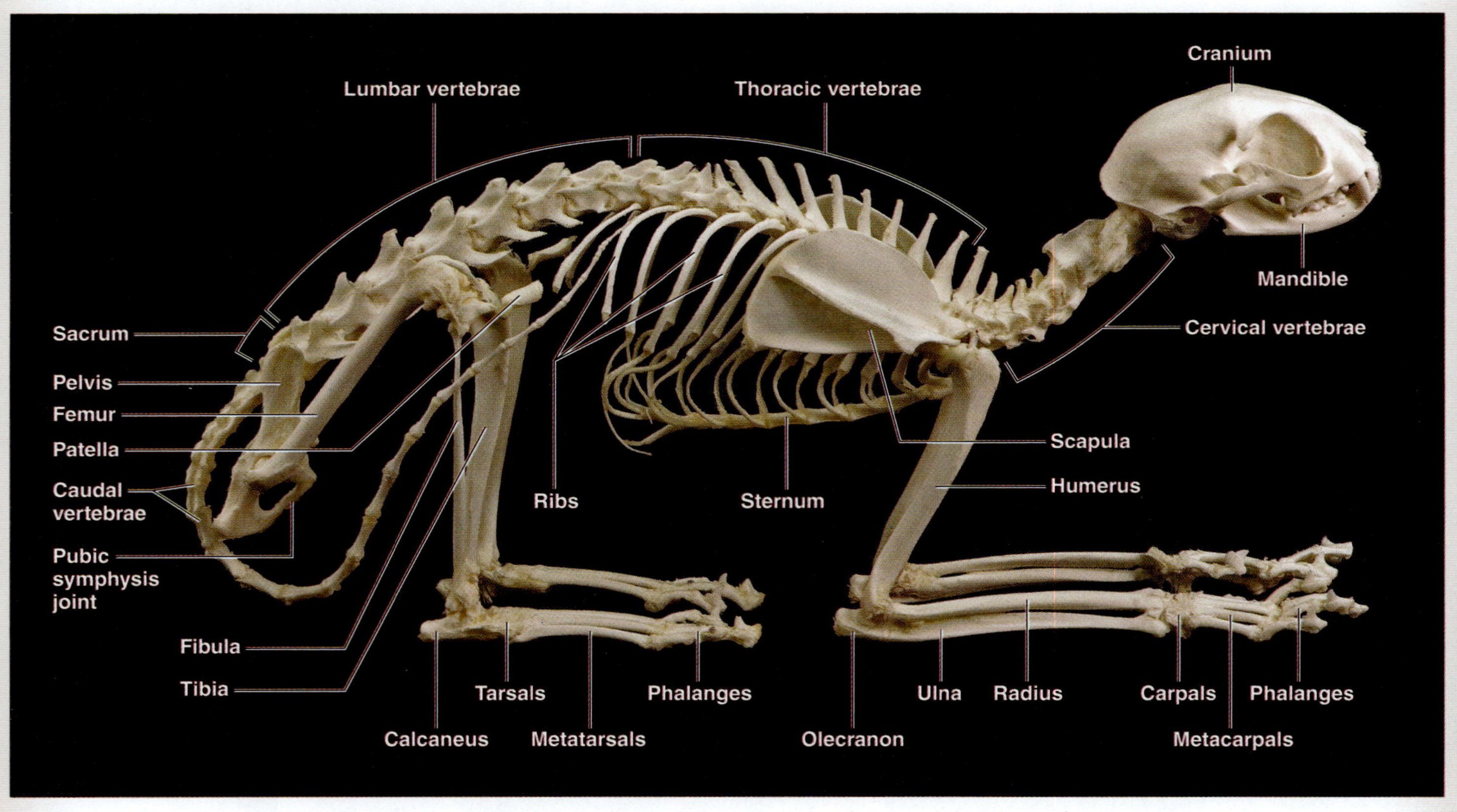

(a)

Figure 28.3 The Cat Skeleton. The cat skeleton is similar to the human skeleton in several ways. Several notable differences are the elongation of the pelvis, metacarpals, and metatarsals; the extra vertebrae; and a clavicle that does not articulate with any other bone. (a) Articulated cat skeleton in anatomic position.

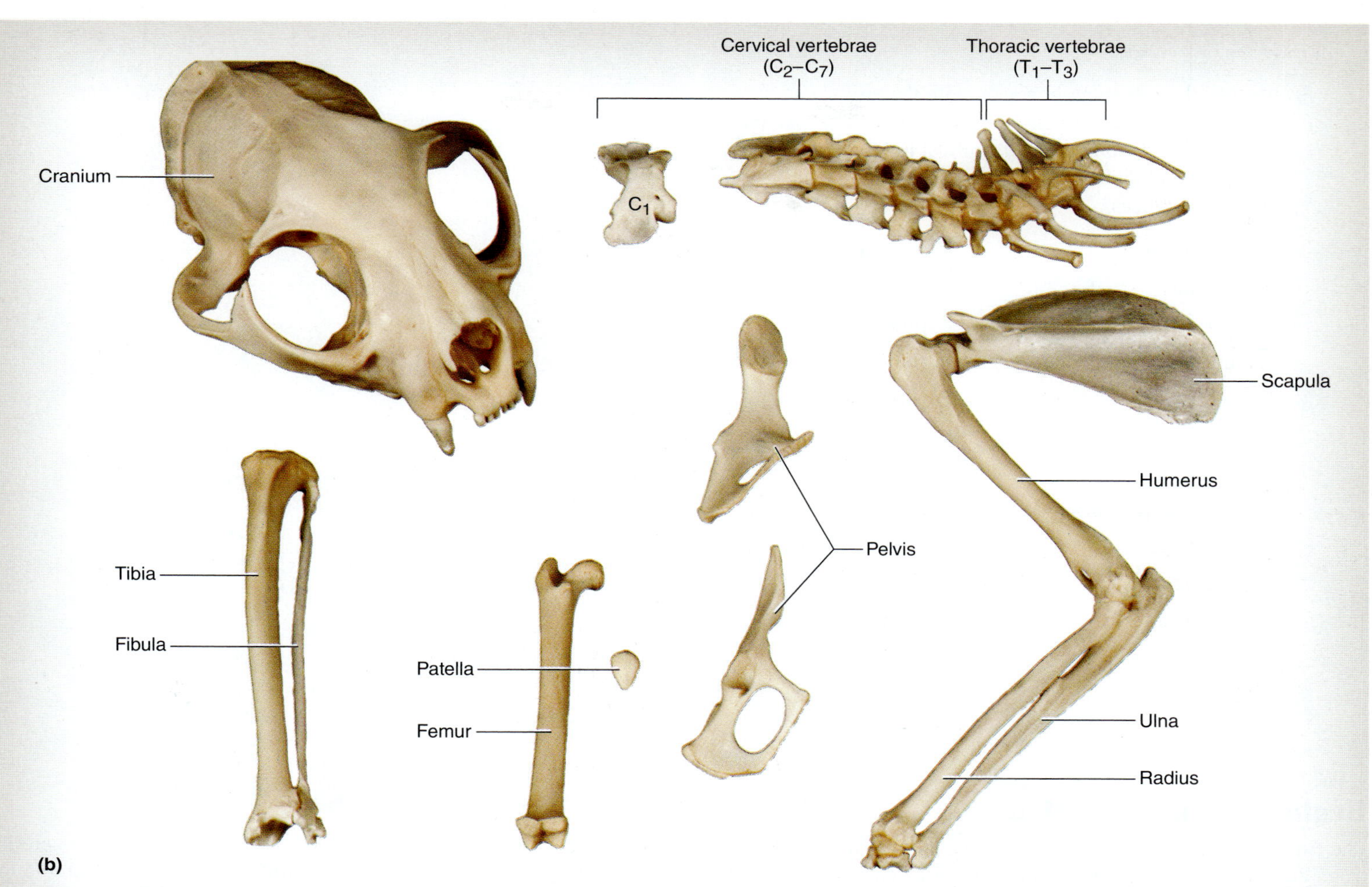

Figure 28.3 **The Cat Skeleton *(continued).*** (b) Disarticulated (partial) cat skeleton demonstrating features of individual bones.

vertebral column of the cat. Is the number of vertebrae in each region of the cat skeleton the same as the number of vertebrae in a human skeleton? Explain:

____

____

____

Table 28.1 Comparison of Human and Cat Vertebrae

Number of Vertebrae	Human	Cat
Cervical		
Thoracic		
Lumbar		
Sacral (fused)		
Coccygeal / Caudal		

INTEGRATE

CONCEPT CONNECTION

When observing the cat skeleton, pay particular attention to the bones of the distal portions of the limbs. The bones that compose hands (metacarpals and phalanges) and feet (metatarsals and phalanges) of humans are very elongated in cats. Cats, in essence, walk on the tips of their phalanges, and their claws are homologous to human fingernails. When a human is standing, the metatarsals lie flat on the ground. When a cat is standing, the metatarsals are lifted off the ground. It is only when the cat lies down that the metacarpals and metatarsals are able to lie flat on the ground (as in the photo of the cat skeleton, figure 28.3*a*).

(continued on next page)

(continued from previous page)

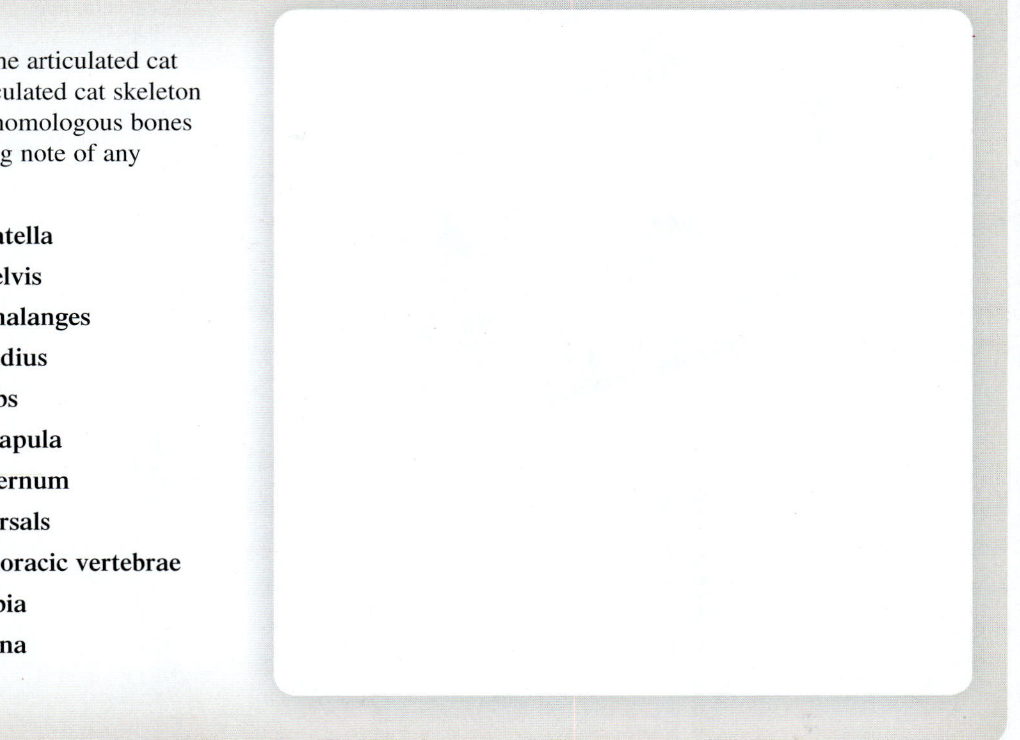

3. Identify the following bones on both the articulated cat skeleton (figure 28.3*a*) and the disarticulated cat skeleton (figure 28.3*b*). Compare and contrast homologous bones of the cat and human skeletons, making note of any major differences.

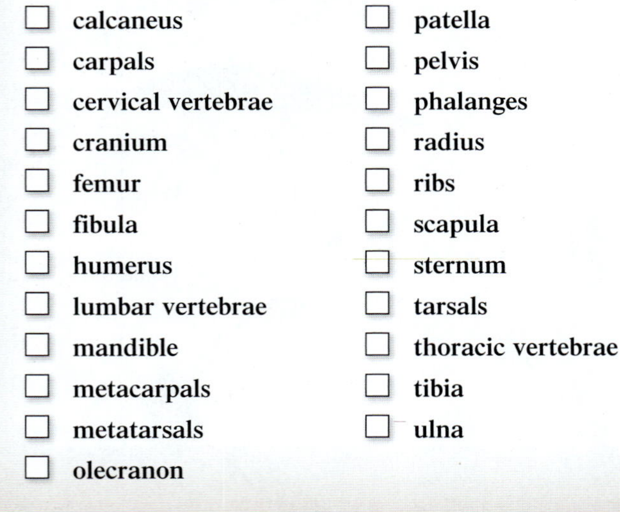

- ☐ **calcaneus**
- ☐ **carpals**
- ☐ **cervical vertebrae**
- ☐ **cranium**
- ☐ **femur**
- ☐ **fibula**
- ☐ **humerus**
- ☐ **lumbar vertebrae**
- ☐ **mandible**
- ☐ **metacarpals**
- ☐ **metatarsals**
- ☐ **olecranon**
- ☐ **patella**
- ☐ **pelvis**
- ☐ **phalanges**
- ☐ **radius**
- ☐ **ribs**
- ☐ **scapula**
- ☐ **sternum**
- ☐ **tarsals**
- ☐ **thoracic vertebrae**
- ☐ **tibia**
- ☐ **ulna**

Beginning the Dissection

The first step before dissection of the muscles is to remove the skin from the cat. Before beginning this exercise, check with the lab instructor to determine how he or she wants the cat to be skinned, because it may be different from the instructions provided here. When skinning the cat you will begin on one side, either dorsal or ventral, and then proceed to remove the skin from the trunk, the neck, and all four limbs down to the paws. The description given here is for skinning a cat starting on the ventral surface. It is preferable to remove the skin as one large piece in order to wrap the cat up again before placing it back in the storage bag. Keeping the cat wrapped in the skin during storage helps to keep the cat from drying out between laboratory sessions.

EXERCISE 28.3

SKINNING THE CAT

1. **Midline incision:** Place the cat supine (ventral side up) on the dissecting tray **(figure 28.4).** If the cat has an embalming/injection incision, begin the incision there, because it will be easier. If there is no embalming/injection incision, then use forceps to gently lift some of the skin away from the superior part of the sternum. Using scissors or a scalpel, make a small, vertical incision, taking care not to cut the underlying muscles (figure 28.4, step 1 on the diagram and figure 1.11, p. 19). Continue cranially with this incision, taking care to keep tension on the skin at all times. This stretches the underlying fascia and pulls it away from the underlying muscles. The target of your dissection instrument is the skin and underlying fascia, not the muscles. After using a scalpel to make the initial incision, put the scalpel away and change to blunt dissection techniques (see figure 1.16, p. 22). Otherwise, use the open scissors technique (see figure 1.15, p. 21) to disengage the skin from the underlying muscles without risking damage to the muscles. If you choose to continue using the scalpel, review the proper use of a scalpel in removing skin on page 19 (exercise 1.7) before beginning.

2. **Perineal incision:** Extend the cut along the midline from the chin to the perineum (figure 28.4, step 2). When the cut approaches the perineum, take care not to damage or cut the vestibule (female cats) or scrotum/penis (male cats). In male cats, also take care not to cut the spermatic cord while dissecting around the scrotum.

3. **Hindlimb incisions:** Begin by making an incision from the midline of the perineum to the ankle of the hindlimb

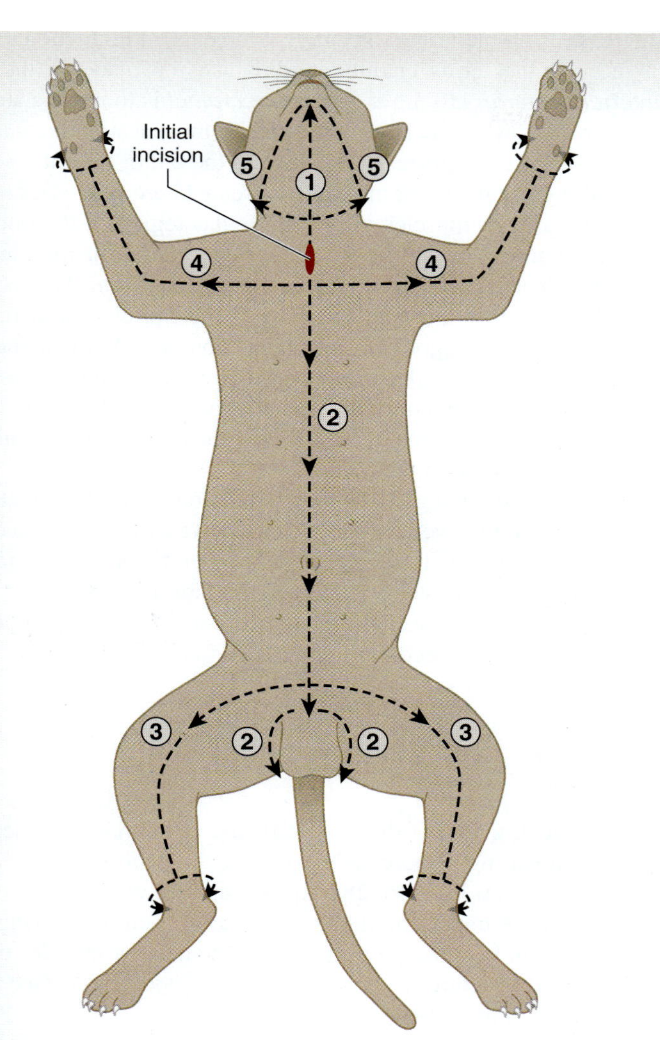

Figure 28.4 Ventral Surface Incisions of the Cat. If the cat has an incision in the neck from embalming/injecting, start your incision there and progress in the numbered sequence as shown.

(figure 28.4, step 3). Notice that the skin is both thinner and more tightly adhered to the underlying tissues near the distal part of the hindlimb. To prevent damaging underlying structures, start by pushing a probe under the skin (see figure 1.12, p. 20). Then make the incision superficial to the probe. When the incision reaches the ankle, *carefully* extend the cut in a circle around the ankle to free the skin from the limb. Beware: It is very easy to cut too deep in this area and damage important tendons, nerves, and blood vessels. Remove the skin from both hindlimbs.

4. **Forelimb incisions:** Begin an incision from the midline of the thorax to the wrist of the forelimb (figure 28.4, step 4). Similar to the hindlimbs, the skin is both thinner and more tightly adhered to the underlying tissues in the forelimb. To prevent damaging underlying structures, start by pushing a probe under the skin (see figure 1.12, p. 20). Then make the incision superficial to the probe. When the incision reaches the wrist, *carefully* extend the cut in a circle around the wrist to free it from the limb. Beware: It is very easy to cut too deep in this area and damage important tendons, nerves, and blood vessels. Remove the skin from both forelimbs.
5. **Neck incisions:** Start with the midline incision below the chin. Extend the incision along the inferior border of the mandible toward the pinnae of the ears (figure 28.4, step 5). Here, too, the skin is very tightly adhered to the underlying tissues, so use a probe when necessary to prevent cutting into underlying structures. Continue the incision dorsally until it encircles the back of the neck. The skin over the face and back of the head will remain intact.
6. **Removing the skin:** Once the neck incision is completed, it should be possible to remove the skin in one piece. As the skin is removed and the fascia between the skin and muscles becomes visible, look for muscle fibers within the skin of the back and anterior neck. These are cutaneous muscles (*cutis*, skin), which have no bony attachments. When these muscles contract, they cause the skin to move. The largest of the two muscles is the **cutaneous maximus,** which is located within the skin of the back. When this muscle contracts, it causes the skin of the back to "twitch" as when a cat tries to shake an insect off its back. Humans do not have a cutaneous maximus muscle. The smaller of the two cutaneous muscles is the **platysma,** a muscle that humans also have. When this muscle contracts, it causes the skin of the neck to wrinkle and fold. Some fibers of these two muscles may be adhered to underlying muscles. If so, gently cut the connections as close to the skin as possible, so as not to damage the underlying muscles.
7. **Cutanous nerves and vessels:** While pulling the skin away from the trunk, you may also notice several small, white tight bands that run between the body wall and the skin. There may even be tiny blood vessels that run next to them if the cat has been injected with colored latex. The whitish structures are **cutaneous nerves,** which are branches off underlying mixed spinal nerves. These nerves carry sensory information from receptors in the skin to the central nervous system and will be covered later in this chapter in more detail.
8. Once the skin is completely free from the body, place it in the storage bag so it is out of the way. As noted previously, at the end of the laboratory period, rewrap the skin around the cat to prevent the cat from drying out between laboratory sessions.

Muscular System

Just as with the skeletal system, the muscular system of the cat bears a remarkable resemblance to the human muscular system. Most of the differences we see in muscles are related to the adaption of the cat's muscular system for four-legged locomotion and the human muscular system for two-legged locomotion.

Before beginning the dissection, review the overall organization of the human muscular system presented in chapter 11, paying particular attention to major muscle groups. Similar muscle groups exit in the cat. Thus, taking a similar approach to learning the muscles of the cat will greatly assist the learning process. Information on the name, origin, insertion, and action of the muscles of the cat is contained in tables 28.2–28.4 and 28.6–28.7. These tables also make note of muscles that are different in the cat as compared with the human. Such muscles are different either because they are not found in the human, or because they have a slightly different name in the cat.

Recall that one of the best ways to remember the action of a muscle is to learn its attachment points (do not worry about distinguishing which attachment point is origin and which is insertion). Then, consider the action that is performed when the muscle shortens, using the terms for joint movements and directional terms. In the cat, the terms **cranial/caudal** take the place of the terms *anterior/posterior*. To move a structure *cranial* is to move a structure anterior and toward the head of the cat. To move a structure *caudal* is to move a structure posterior and toward the tail of the cat.

The musculature of the cat is presented here as a series of exercises that look at muscles in one particular region of the body at a time. Each time a new muscle is identified, refer to the muscle tables to review that muscle's origin, insertion, and action. Also consider how the size, location, and function of the muscle compares with those of the homologous muscle(s) in a human. Most of the time, these will be very similar. If you have already started to master the anatomy of the human muscular system, then it will be most efficient to give extra attention to only muscles of the cat that are considerably different from those in the human.

In the following exercises, dissection and identification of muscles start on the ventral surface and move from cranial to caudal. Subsequent exercises follow a progression of dissection and identification of muscles of the forelimb and the hindlimb from both lateral and dorsal views. In each case, dissections begin with superficial observations and dissections and proceed to deep dissections.

EXERCISE 28.4

MUSCLES OF THE HEAD AND NECK

1. **Preparation:** Place the cat ventral side up on the dissecting tray. Begin by observing the muscles of the anterior neck **(figure 28.5)**. On one half of the neck you will do a superficial dissection, and on the other, you will do a deep dissection to observe underlying muscles. Some of the superficial muscles may have been damaged by the embalming/injection incision. If that is the case, use that side for the deep dissection and the unaffected side for the superficial dissection. A large vein may be visible on both sides of the neck, lying superficial to the muscles. This is the **external jugular vein.** Next to this vein there may be several small, kidney-bean-size (and -shaped) structures. These are **lymph nodes.** While dissecting and cleaning the muscles, take care not to cut the internal jugular veins or remove the lymph nodes, because these will be observed and discussed further in the context of the cardiovascular and lymphatic systems.

2. **Superficial dissection:** Although the word "dissect" literally means "to cut into two pieces" (*di,* two + *sectio,* to cut), in reality very little dissecting of muscles happens in these exercises. The process is more about cleaning muscles and separating them from each other so as to best view them and understand their location and functions. Often the best tools for this job are forceps and a pair of scissors. Use the forceps to pull on the fascia (connective tissue) overlying the muscles; then use the scissors to cut the fascia away. When cutting, always cut in the direction of the muscle fibers, which prevents them from tearing. Before and during the dissection, preview the muscles in figure 28.5 to have an idea of what to look for. More important, this will also assist you in knowing where to be most careful, so as not to destroy muscles or related structures such as nerves and blood vessels (see the Learning Strategy on this page).

INTEGRATE

LEARNING STRATEGY

This learning strategy is more of a "dissection strategy." While dissecting the muscles, it is important not to destroy blood vessels and nerves that lie in the spaces around and between the muscles. **Nerves** appear white and somewhat stringy. **Arteries** will be red, if injected with latex, or white, like the nerves. Often the only way to distinguish a nerve from an artery is by palpation (touch). Nerves feel "cord-like," whereas arteries have a lumen and feel "tube-like." Veins will be blue, if injected with latex, or dark brownish-purple, if not. The dark brownish-purple color of veins comes from the dried blood that is still inside the vessels, which is visible due to the thin-walled nature of the veins.

Try to preserve (and clean) the arteries and nerves while dissecting. They are more likely to be kept intact if the dissection is performed from proximal to distal (rather than from distal to proximal). This is because the dissection begins where nerves and arteries are larger and follows their paths as they decrease in size.

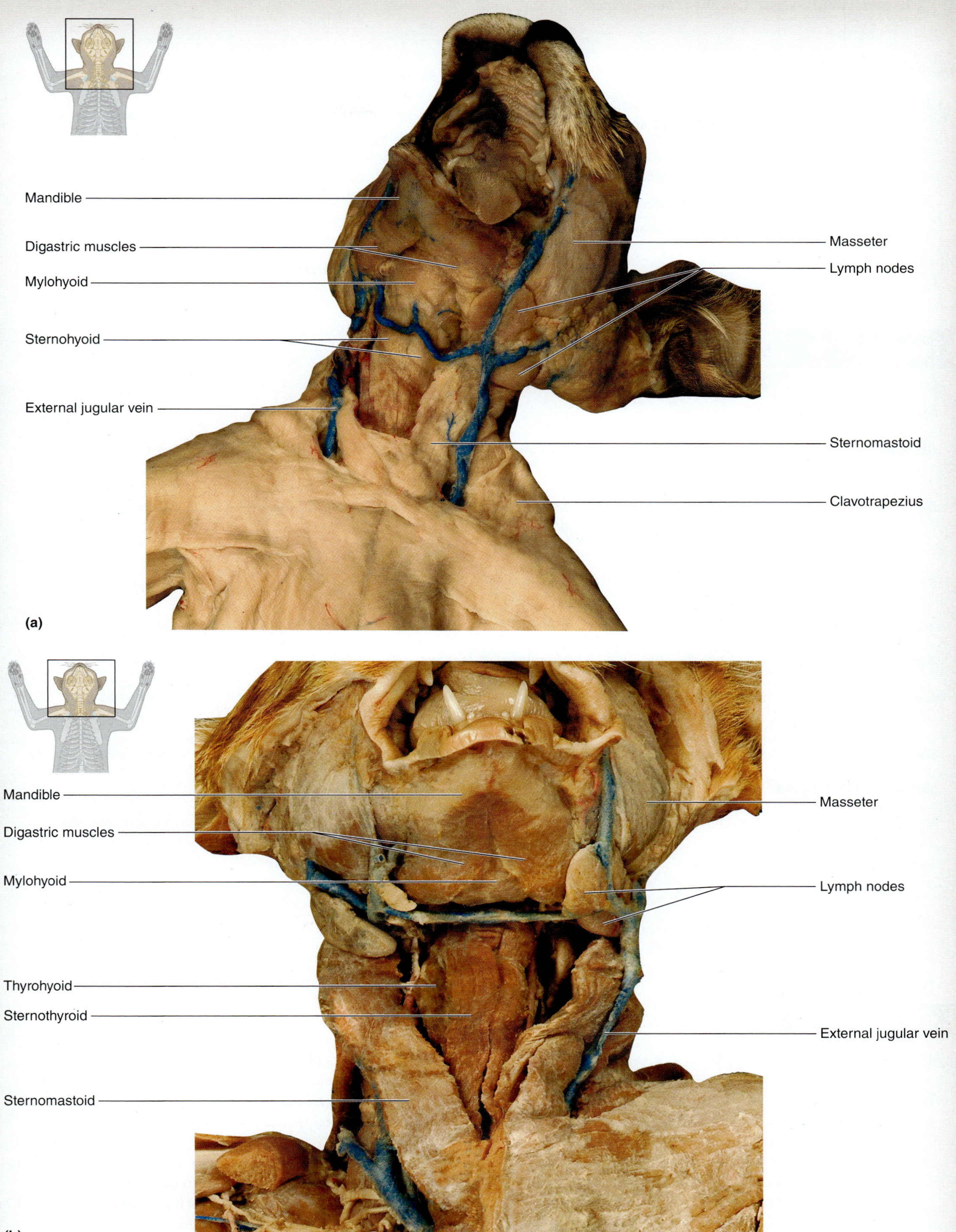

Figure 28.5 Muscles of the Head, Neck, and Thorax (Anterior View). Superficial muscles on the right side of the cat's neck are somewhat mangled from the embalming process, but the muscles can be seen clearly on the left side of the cat's neck (right side of the photo). (a) Superficial muscles. (b) Deep muscles.

(continued on next page)

(continued from previous page)

3. **Superficial muscles:** Clean, identify, and isolate the following muscles, using **table 28.2** and figure 28.5 as guides:

- ☐ clavotrapezius
- ☐ digastric
- ☐ masseter
- ☐ mylohyoid
- ☐ sternohyoid
- ☐ sternomastoid
- ☐ sternothyroid

4. **Deep muscles:** On the side of the neck that was previously determined to be most appropriate for a deep dissection, proceed to transect the sternomastoid and sternohyoid muscles. Clean, identify, and isolate the following two muscles, using table 28.2 and figure 28.5 as guides. Note that both of these muscles attach to the thyroid cartilage. The sternothyroid extends from the thyroid inferiorly to the sternum and the thyrohyoid extends from the thyroid superiorly to the hyoid bone.

- ☐ sternothyroid
- ☐ thyrohyoid

Table 28.2 Muscles of the Head and Neck

Cat Muscle	Human Muscle	Origin	Insertion	Action
Muscles of the Neck (Anterior)				
Cleidomastoid	Sternocleidomastoid-clavicular head	Clavicle	Mastoid process	Rotate head
Digastric	Digastric	Mastoid and jugular processes of occipital bone	Mandible	Depress mandible
Geniohyoid	Geniohyoid	Mandible	Hyoid	Move hyoid cranially
Mylohyoid	Mylohyoid	Mandible	Median raphe	Elevate floor of oral cavity
Sternohyoid	Sternohyoid	Manubrium of sternum	Hyoid	Move hyoid caudally
Sternomastoid	Sternocleidomastoid-sternal head	Manubrium of sternum	Lambdoidal ridge and mastoid process	Rotate head
Sternothyroid	Sternothyroid	Manubrium of sternum	Thyroid cartilage	Move larynx caudally
Muscles of Mastication				
Masseter	Masseter	Zygomatic arch	Mandible	Elevate mandible
Temporalis	Temporalis	Temporal bone	Coronoid process of mandible	Elevate mandible

EXERCISE 28.5

MUSCLES OF THE THORAX

1. **Superficial muscles:** With the cat still ventral side up on the dissecting tray, prepare to dissect the muscles of the anterior thorax. There are four main muscles in this group. These muscles collectively mimic one action of the pectoralis major and minor muscles in humans: adduction of the forelimb. While cleaning the superficial fascia around these muscles, note the direction of the fibers of the **pectoantebrachialis** (transverse), **pectoralis major** (slightly oblique), **pectoralis minor,** and **xiphihumeralis** (both very oblique). Note also that the pectoantebrachialis and xiphihumeralis are unique to the cat.

   While dissecting around the insertion of these muscles on the humerus, take care not to damage the latissimus dorsi and serratus ventralis muscles (**figures 28.6** and **28.7**), which will be studied with muscles of the back.

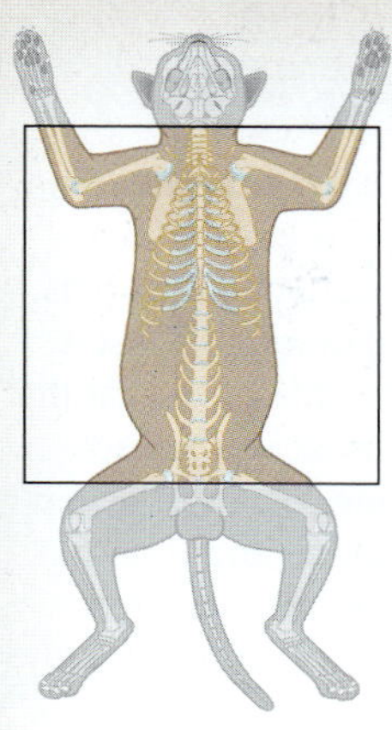

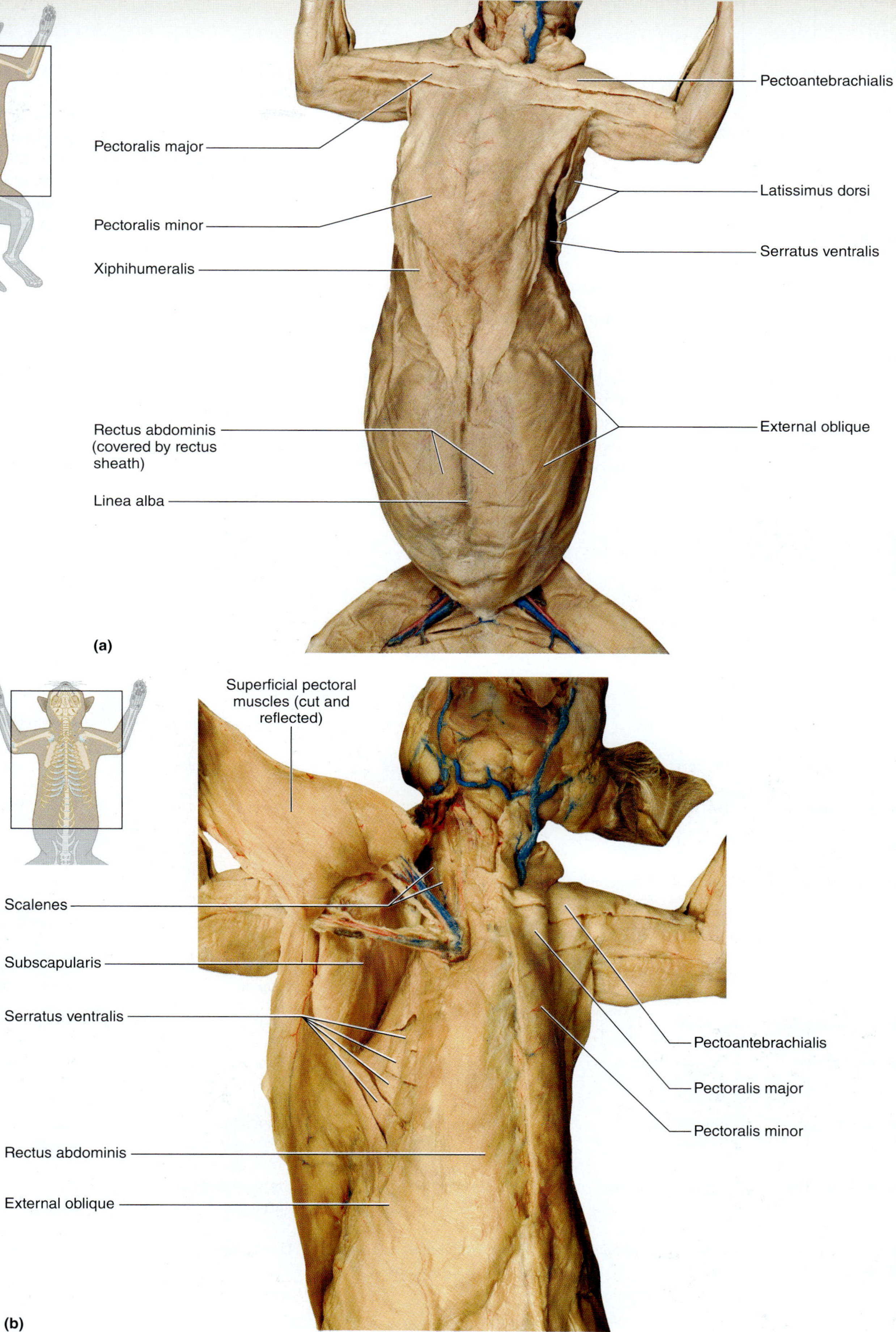

Figure 28.6 Muscles of the Thorax and Abdomen. (a) Superficial muscles of the thorax and abdomen. (b) Deep muscles of the thorax and abdomen. The superficial pectoral muscles (pectoantebrachialis, pectoralis major/minor, and xiphihumeralis) and the sternomastoid have been removed to show the deeper muscles.

(continued on next page)

(continued on previous page)

For what activity or activities would the pectoral group of muscles be most important for in the cat?

____

____

2. Clean, identify, and isolate the following muscles, using **table 28.3** and figure 28.6 as guides:
   - ☐ **pectoantebrachialis**
   - ☐ **pectoralis minor**
   - ☐ **pectoralis major**
   - ☐ **xiphihumeralis**

3. **Deep muscles:** Use blunt dissection to gently separate the pectoral muscles from the underlying muscles. It may be helpful to work a probe deep to the pectoral muscles, taking care not to tear the deep muscles of the thorax. The pectoral muscles should be bisected close to the midline to prevent damaging the nerves of the brachial plexus, which are located in the axillary region. Keeping the probe deep to the pectoral muscles, use scissors to make a vertical incision through the pectoantebrachialis, pectoralis major and minor, and xiphihumeralis muscles close to the midline of the body on the side of the cat that was chosen for the deep dissection. Bisect and reflect these muscles to expose the deeper muscles. In this view the scalene muscle group will be visible in addition to muscles that connect the scapula to the axial skeleton (serratus ventralis) and one of the muscles composing the rotator cuff: the subscapularis.

4. Clean, identify, and isolate the following muscles, using **table 28.4** and figure 28.6 as guides:
   - ☐ **scalenes**
   - ☐ **serratus ventralis**
   - ☐ **subscapularis**

Table 28.3 Muscles of the Thorax and Abdomen

Cat Muscle	Human Muscle	Origin	Insertion	Action
Superficial Muscles				
Pectoantebrachialis	NA	Manubrium	Fascia of forelimb near elbow	Adduct the forelimb
Pectoralis Major	Pectoralis major	Cranial sternebrae	Proximal humerus	Adduct the forelimb
Pectoralis Minor	Pectoralis minor	Caudal sternebrae	Proximal humerus	Adduct the forelimb
Xiphihumeralis	NA	Xiphoid process	Proximal humerus	Adduct the forelimb
Deep Muscles				
External Intercostals	External intercostals	Superior rib	Inferior rib	Draw ribs cranially
Internal Intercostals	Internal intercostals	Inferior rib	Superior rib	Draw ribs caudally
Serratus Ventralis	Serratus anterior	Ribs 1–10	Vertebral border of scapula	Depress scapula and move scapula cranially
Muscles of the Abdominal Wall				
External Oblique	External oblique	Lumbodorsal fascia and ribs	Linea alba	Compress abdominal viscera
Internal Oblique	Internal oblique	Lumbodorsal fascia and pelvis	Linea alba	Compress abdominal viscera
Rectus Abdominis	Rectus abdominis	Pubic symphysis	Sternum and costal cartilages	Flex trunk and compress abdominal viscera
Transversus Abdominis	Transversus abdominis	Ilium, lumbar vertebrae, and posterior ribs	Linea alba	Compress abdominal viscera

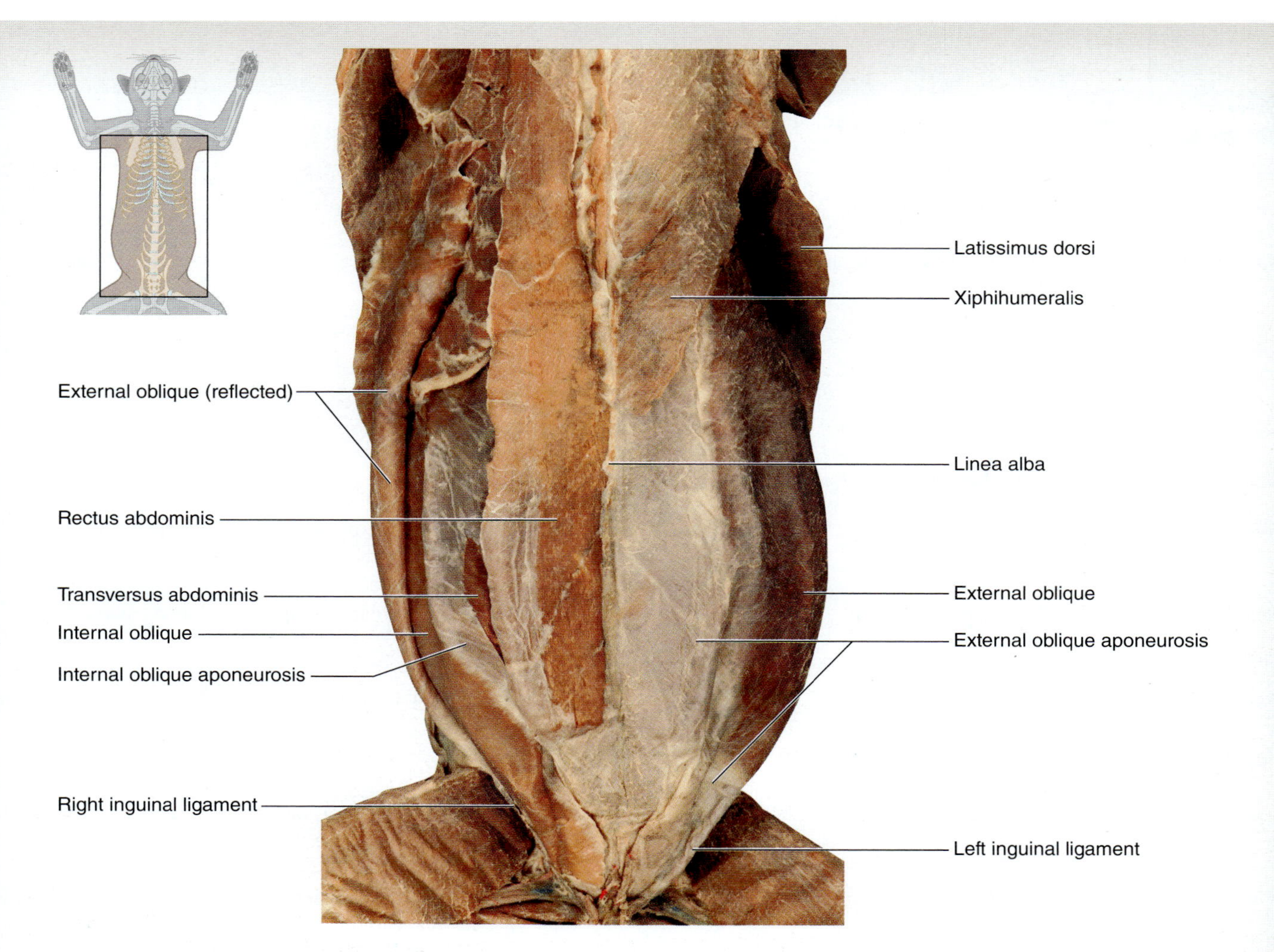

Figure 28.7 Deep Muscles of the Abdomen. In this photo the external oblique on the left side of the figure has been cut and reflected to show the rectus abdominis and internal oblique. The aponeurosis of the internal oblique has been removed to show the fibers of the deepest muscle, the transversus abdominis.

Table 28.4 Muscles of the Back and Shoulder

Cat Muscle	Human Muscle	Origin	Insertion	Action
Muscles of the Back and Shoulder (superficial)				
Acromiodeltoid	Deltoid—anterior portion	Acromium of scapula	Humerus	Flex humerus
Acromiotrapezius	Trapezius—superior portion	Spines of vertebra C_1–T_1	Acromium process and spine of scapula	Adduct and elevate scapula
Clavobrachialis (clavodeltoid)	Deltoid—middle portion	Clavicle	Proximal ulna	Abduct and rotate humerus
Clavotrapezius	Trapezius—middle portion	Nuchal line of skull and nuchal ligament	Clavicle	Move humerus cranially
Latissimus Dorsi	Latissimus dorsi	Lumbodorsal fascia and spines of lower thoracic and lumbar vertebrae	Medial humerus	Elevate and move humerus caudally
Levator Scapulae Ventralis	NA	Occipital bone, transverse process of atlas	Acromial process	Move scapula cranially

(continued on next page)

(continued from previous page)

Table 28.4 Muscles of the Back and Shoulder *(continued)*

Cat Muscle	Human Muscle	Origin	Insertion	Action
Muscles of the Back and Shoulder (superficial) *(continued)*				
Spinodeltoid	Deltoid—posterior portion	Spine of scapula	Humerus	Elevate and rotate humerus
Spinotrapezius	Trapezius—inferior portion	Spines of thoracic vertebrae	Scapula	Elevate and move scapula caudally
Teres Major	Teres major	Axillary border of scapula	Medial humerus	Adducts humerus
Muscles of the Back and Shoulder (deep)				
Infraspinatus	Infraspinatus	Infraspinous fossa	Greater tuberosity of humerus	Rotate humerus
Rhomboid	Rhomboid major and minor	Spines of thoracic vertebrae	Ventral border of scapula	Adduct scapula
Rhomboid Capitis	Levator scapulae	Superior nuchal line	Angle of scapula	Move scapula cranially
Scalenes	Scalenes	NA	NA	NA
Anterior	Anterior	Ribs 2–3	Transverse processes of cervical vertebrae	Flex neck and move ribs cranially
Medius	Middle	Ribs 6–9	Transverse processes of cervical vertebrae	Flex neck and move ribs cranially
Posterior	Posterior	Rib 3	Transverse processes of cervical vertebrae	Flex neck and move ribs cranially
Serratus Ventralis	Serratus anterior	Ribs 1–10	Vertebral border of scapula	Depress scapula and move scapula cranially
Splenius	Splenius capitis and cervicis	Lamboidal ridge of occipital bone	Nuchal line	Rotate and elevate/extend the head
Subscapularis	Subscapularis	Subscapular fossa	Lesser tuberosity of humerus	Adduct humerus
Supraspinatus	Supraspinatus	Supraspinous fossa	Greater tuberosity of humerus	Extend humerus
Teres Minor	Teres minor	Axillary border of scapula	Greater tuberosity of humerus	Rotate humerus
Transverse Costarum	NA	Sternum	First rib	Move sternum cranially

EXERCISE 28.6

MUSCLES OF THE ABDOMINAL WALL

1. **Superficial muscles:** With the cat still ventral side up on the dissecting tray, prepare to dissect the muscles of the abdominal wall. There are four main muscles in this group, just as there are in humans. Contraction of these muscles compresses the abdominal viscera, raising intra-abdominal pressure. Contraction of the rectus abdominis, as in humans, also flexes the vertebral column. These muscles have broad, flat **aponeuroses** for tendons and insert on the **linea alba,** just as in humans.
2. Clean, identify, and isolate the external oblique muscle and identify the following, using table 28.3 and figure 28.7 as guides:
   - ☐ **external oblique muscle**
   - ☐ **external oblique aponeurosis**
   - ☐ **linea alba**
3. **Deep muscles:** Using scissors and forceps, cut a small, vertical slit (a sagittal section) in the external oblique muscle close to its insertion on the linea alba. Carefully work a probe underneath the muscle so that it separates the external oblique from the underlying **rectus abdominis** and **internal oblique** muscle. Keeping the probe in place to protect the underlying muscles, continue the vertical incision of the external oblique from the inguinal ligament (figure 28.7) to the inferior border of the xiphihumeralis muscle. Reflect the external oblique to reveal the underlying rectus abdominis and internal oblique muscles.
4. Make an incision in the aponeurosis of the internal oblique muscle to reveal fibers of the **transversus abdominis** muscle. Either remove the aponeurosis of the internal oblique entirely or just cut a "window" through it to reveal the transversus abdominis (as shown in the left side of figure 28.7). When performing all of these dissections, take care not to cut too deep, which might result in exposing the abdominal cavity and possibly cutting into abdominal contents.
5. Clean, identify, and isolate the following muscles, using table 28.3 and figure 28.7 as guides:
   - ☐ **internal oblique**
   - ☐ **rectus abdominis**
   - ☐ **transverse abdominis**
6. Record the direction of fibers of the four abdominal muscles in **table 28.5.**

Table 28.5 Fiber Orientation of Cat Abdominal Muscles

Muscle	Fiber Orientation
External Oblique	
Internal Oblique	
Rectus Abdominis	
Transverse Abdominis	

EXERCISE 28.7

MUSCLES OF THE BACK AND SHOULDER

1. **Preparation:** Turn the cat over so that its ventral surface is on the dissecting tray and the dorsal surface is up. Begin at the head and progress toward the tail as was done in previous observations/dissections.
2. **Superficial muscles:** Recall that the most superficial "back" muscles in humans consist of the trapezius, deltoid, and latissimus dorsi muscles. These are not true "back" muscles because they do not move the vertebral column. Instead, their functions are to attach the scapula and upper limb to the axial skeleton, and to produce movements of the upper limb and scapula. Cats also have these muscles. The cat **deltoid** is divided into three separate muscles, each homologous to the three regions of the human deltoid muscle (table 28.4). The cat **trapezius** is similarly divided into three separate muscles, each homologous to the three regions of the human trapezius muscle. The cat also has a unique muscle, the **levator scapulae ventralis,** which has no human counterpart (table 28.4). The **latissimus dorsi** is similar in the cat and the human. While identifying and cleaning these muscles, think about the actions each performs and relate each muscle in the cat to its human homologue.
3. **Note about the clavicle:** In the cat, the clavicle is embedded within the two bellies of what appears to be a single muscle (at least superficially). These two muscle "bellies" are two different muscles: the **clavotrapezius** and the **clavodeltoid (figure 28.8).** While identifying these two muscles, do not be confused if the names appear to relate to two parts of what appears to be a single muscle. Palpate the midpoint of the two muscles to feel the clavicle. Once the clavotrapezius and clavodeltoid have been reflected, the clavicle will be clearly visible.

(continued on next page)

(continued from previous page)

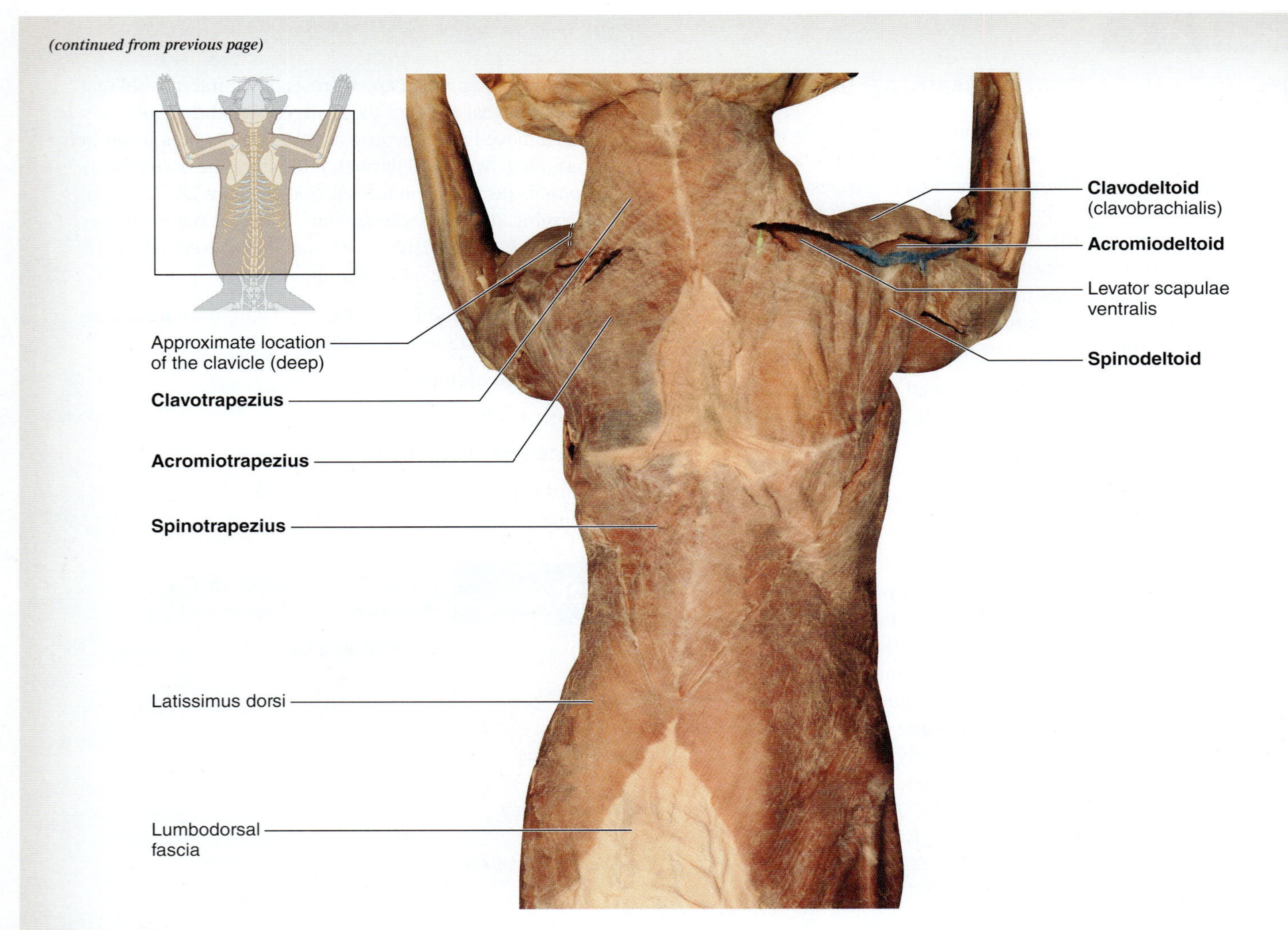

Figure 28.8 Superficial Back Muscles. The three parts of the trapezius muscle are labeled on the left side of the figure. The three parts of the deltoid muscle are labeled on the right side of the figure.

4. Clean, identify, and isolate the following muscles, using table 28.4 and figure 28.8 as guides:

- ☐ **acromiodeltoid**
- ☐ **acromiotrapezius**
- ☐ **clavobrachialis (clavodeltoid)**
- ☐ **clavotrapezius**
- ☐ **latissimus dorsi**
- ☐ **levator scapulae ventralis**
- ☐ **spinodeltoid**
- ☐ **spinotrapezius**

5. **Deep muscles:** Using scissors and forceps, cut a small, vertical slit (a sagittal section) in the **acromiotrapezius** and **spinotrapezius** muscles on both sides. To do this, slip a probe deep to the muscles to protect the underlying rhomboid muscles. Then cut superficial to the probe. When the acromiotrapezius and spinotrapezius muscles are cut, the scapula will pull away from the midline and the rhomboids will be visible. The **rhomboid** in the cat is homologous to the rhomboid major and minor in humans, and the **rhomboid capitis** muscle is homologous to the levator scapulae in humans. A third muscle that connects the scapula to the axial skeleton in the cat is the **serratus ventralis,** which is equivalent to the serratus anterior in humans. In the cat, however, this muscle is larger, with more attachments to the ribs, than the serratus anterior of humans.

6. Observe the muscles of the scapula that form part of the **rotator cuff** (see figure 13.1 on p. 316). Recall that these muscles form the major muscular support of the shoulder joint, and consist of the supraspinatus, infraspinatus, subscapularis, and teres minor. The **supraspinatus** and **subscapularis** are more easily visualized from a posterior view.

7. Clean, identify, and isolate the following muscles, using table 28.4 and **figure 28.9** as guides:

- ☐ **rhomboid**
- ☐ **rhomboid capitis**
- ☐ **splenius**
- ☐ **subscapularis**
- ☐ **supraspinatus**

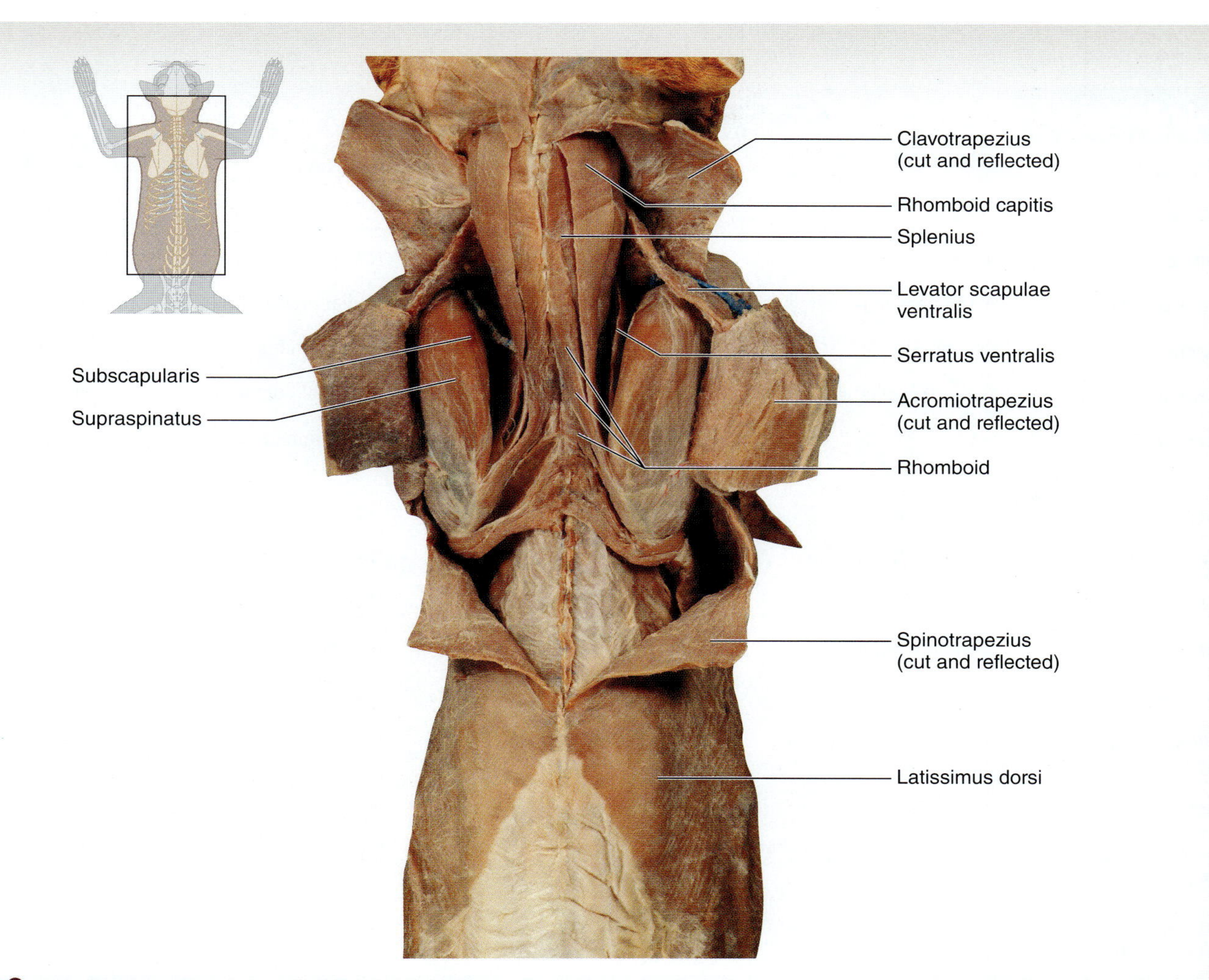

Figure 28.9 Deep Back Muscles. A better view of the deep muscles of the shoulder, back, and neck is possible after bisecting and reflecting the clavotrapezius, acromiotrapezius, and spinotrapezius muscles as shown here.

EXERCISE 28.8

MUSCLES OF THE FORELIMB

1. **Superficial muscles:** Recall that the human upper limb muscles are organized into four compartments: anterior and posterior *arm,* and anterior and posterior *forearm.* The muscles of the cat forelimb are similarly arranged. Thus, while dissecting out muscles in each region of the forelimb, always consider the common action of the muscles in that group before considering the specifics of each muscle. This will greatly facilitate the learning of these muscles. **Table 28.6** summarizes the human homologue, origin, insertion, and action of the forelimb muscles of the cat.
2. Observe the superficial muscles of the shoulder. The muscles of the deltoid group should already have been cleaned and identified. The dissection will now proceed distally along the forelimb, starting with the **triceps brachii (figure 28.10).** The triceps brachii in the cat has the same three heads and insertion as the triceps brachii of the human. While cleaning the anterior border of the lateral head of the triceps brachii near the elbow, note the origin of several other muscles: the **brachialis,** the **brachioradialis,** and several extensors of the forearm.
3. This dissection exercise involves cleaning, identifying, and isolating all of the arm/forearm muscles as a group. The dissection also requires working in a small area. Therefore it is most efficient to pay attention to all muscles in that area at the same time. To effectively complete the dissection, reposition the cat occasionally to make it easier to get at the muscles on the lateral and medial sides of the forelimb. While dissecting the forelimb muscles take particular care when cleaning

(continued on next page)

(continued from previous page)

Table 28.6 Muscles of the Forelimb

Cat Muscle	Human Muscle	Origin	Insertion	Action
Muscles of the Arm				
Aconeus	Aconeus	NA	NA	NA
Biceps Brachii	Biceps brachii—long head	Supraglenoid tubercle	Radial tuberosity	Flex forearm
Brachialis	Brachialis	Lateral humerus	Proximal ulna	Flex forearm
Coracobrachialis	Coracobrachialis	Coracoid process of scapula	Proximal humerus	Adduct humerus
Epitrochlearis	NA	Lateral surface of latissimus dorsi	Olecranon of ulna	Rotate ulna
Triceps Brachii	Triceps brachii	Humerus	Olecranon	
Lateral head	Lateral head	Deltoid ridge of humerus	Olecranon of ulna	Extend forearm
Long head	Long head	Axillary border of scapula	Olecranon of ulna	Extend forearm
Medial head	Medial head	Humerus	Olecranon of ulna	Extend forearm
Muscles of the Forearm (dorsal)				
Abductor Pollicis Longus	Abductor pollicis longus	Lateral radius and ulna	First metacarpal	Abduct the forelimb
Brachioradialis	Brachioradialis	Midhumerus	Distal radius	Supinate the manus
Extensor Carpi Radialis Brevis	Extensor carpi radialis brevis	Distal humerus	Third metacarpal	Extend the manus
Extensor Carpi Radialis Longus	Extensor carpi radialis longus	Distal humerus	Second metacarpal	Extend the manus
Extensor Digitorum Communis	Extensor digitorum	Distal humerus	Digits II–V	Extend the manus
Extensor Digitorum Lateralis	Extensor digiti minimi	Distal humerus	Digits II–V	Extend the digits
Extensor Carpi Ulnaris	Extensor carpi ulnaris	Lateral epicondyle of humerus	Fifth metacarpal	Extend the manus
Extensor Pollicis	Extensor pollicis brevis	Proximal ulna	Middle phalanx of first digit	Extend first digit
Supinator	Supinator	Lateral epicondyle of humerus	Radius	Supinate the manus
Muscles of the Forearm (ventral)				
Flexor Carpi Radialis	Flexor carpi radialis	Medial epicondyle of humerus	Digits II and III	Flex the manus
Flexor Carpi Ulnaris	Flexor carpi ulnaris	Medial epicondyle of humerus and ulna	Pisiform bone	Flex the manus
Flexor Digitorum Profundus	Flexor digitorum profundus	Ulna, humerus, and radius	Digits I–V	Flex the manus
Flexor Digitorum Superficialis	Flexor digitorum superficialis	Medial epicondyle of humerus	Digits II–V	Flex the manus
Pronator Teres	Pronator teres	Medial epicondyle of humerus	Medial border of radius	Pronate the manus

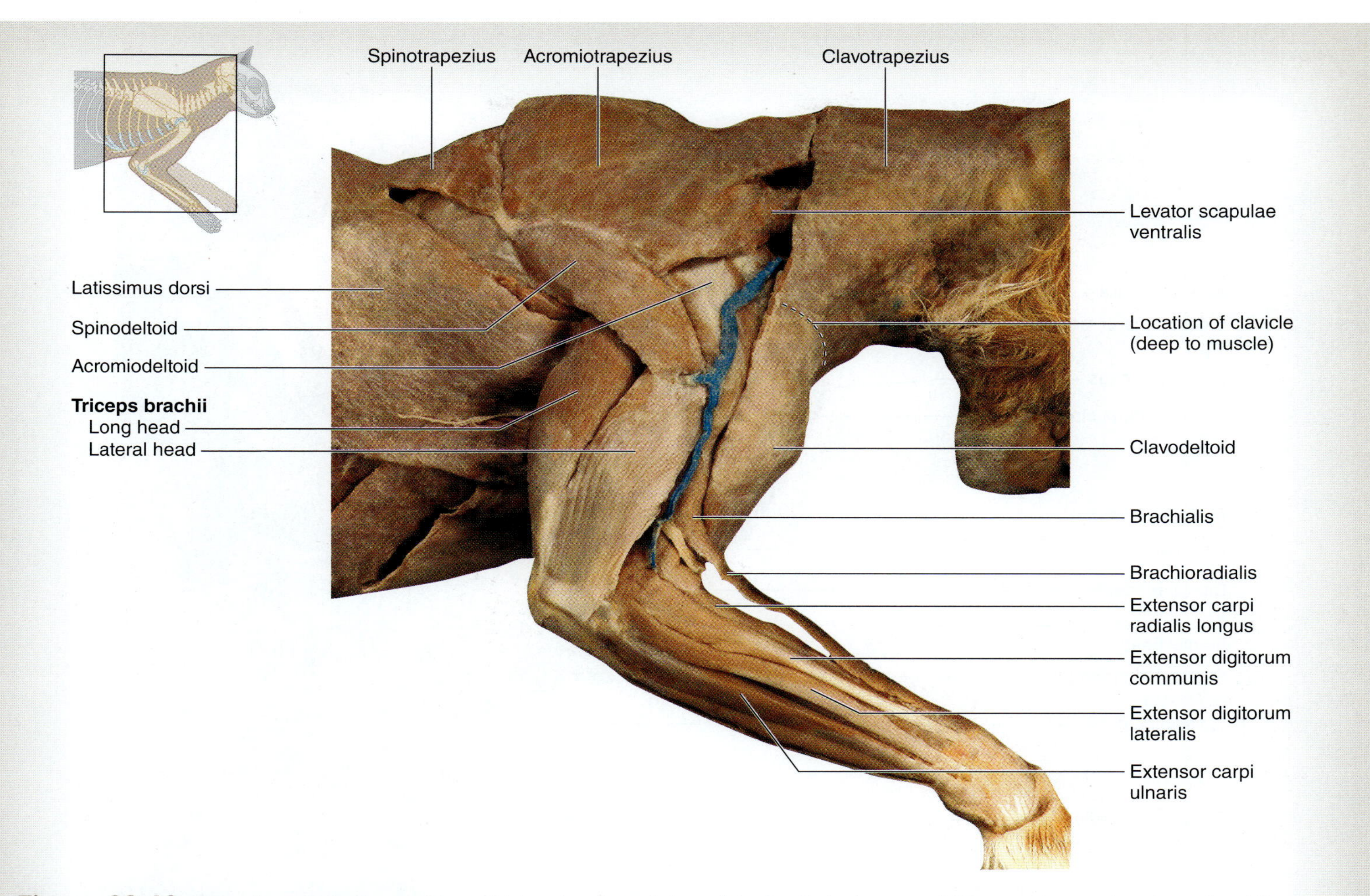

Figure 28.10 Muscles of the Lateral Forelimb. In a lateral view of the forelimb, the most visible muscles are the triceps brachii and the extensor muscles of the forearm.

the proximal attachments of the muscles that originate on the medial epicondyle of the humerus **(flexor carpi ulnaris, pronator teres, palmaris longus, flexor carpi radialis).** On the proximal end of these muscles, the fascia becomes embedded in the muscle, making it difficult to remove. Thus, do not remove the fascia when approaching that region. Instead, simply cut the fascia where it starts to embed the muscles so as not to tear the muscles. While dissecting these muscles, think about the homologous muscles in the human and note both similarities and differences.

4. The majority of muscles in the forelimb of the cat have counterparts in the human. The one major exception is the **epitrochlearis** muscle in the cat, which has no human counterpart. See more about this muscle in step 6.

5. Clean, identify, and isolate the following muscles of the arm and forearm, using table 28.6 and **figures** 28.10 and **28.11** as guides:

- ☐ **brachialis**
- ☐ **brachioradialis**
- ☐ **epitrochlearis**
- ☐ **extensor carpi radialis longus**
- ☐ **extensor carpi ulnaris**
- ☐ **extensor digitorum communis**

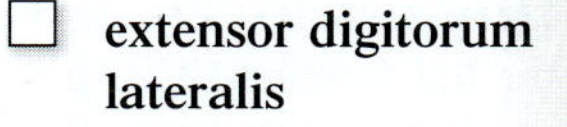

- ☐ **extensor digitorum lateralis**
- ☐ **flexor carpi radialis**
- ☐ **flexor carpi ulnaris**
- ☐ **triceps brachii—lateral head**
- ☐ **triceps brachii—long head**

(continued on next page)

(continued from previous page)

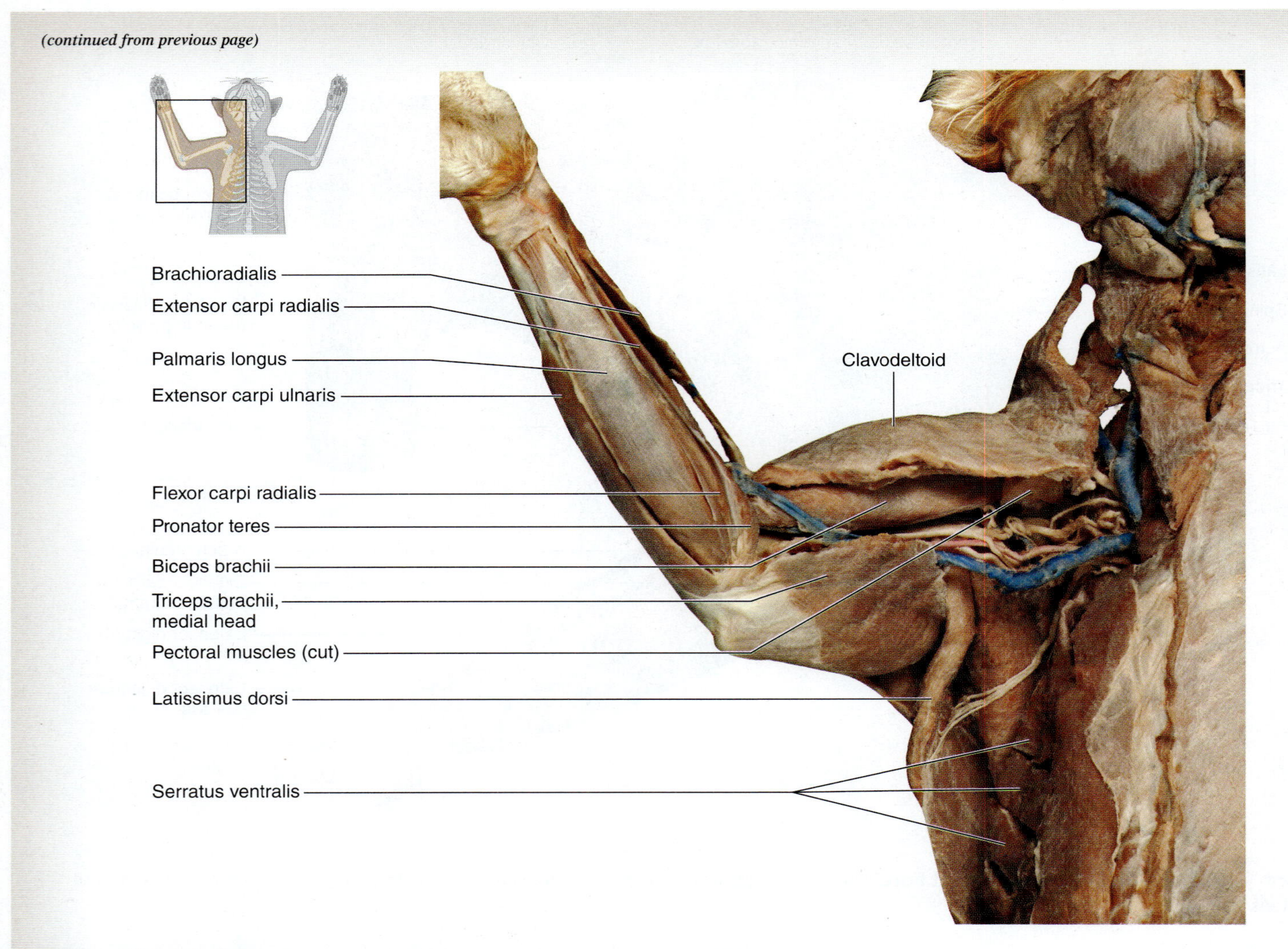

Figure 28.11 Muscles of the Medial Forelimb. In a medial view of the forelimb, the biceps brachii and most of the flexors of the forearm can be viewed.

6. **Deep muscles:** Locate the **epitrochlearis** muscle in the medial forearm. This muscle originates near the insertion of the **latissimus dorsi.** While cleaning the muscle near that location, take care not to damage any of the nerves of the brachial plexus that run in the axilla. Once the epitrochlearis has been identified and cleaned, push a blunt probe deep to the muscle to isolate it from the deeper muscles. Next, use scissors to bisect and reflect the epitrochlearis. This will allow some of the deeper muscles of the medial forearm to be visualized.

7. Clean, identify, and isolate the following muscles of the forelimb, using table 28.6 and figure 28.11 as guides:

- ☐ **biceps brachii**
- ☐ **pronator teres**
- ☐ **triceps brachii—medial head**

EXERCISE 28.9

MUSCLES OF THE HINDLIMB

EXERCISE 28.9A Muscles of the Hindlimb: Thigh

1. **Superficial muscles:** Begin the dissection of hindlimb with muscles of the posterior and lateral hip and thigh. Place the cat ventral side down on the dissection tray. Before starting the dissection, observe the hindlimb muscles in **figure 28.12.** Notice that while the cat has nearly the same gluteal and posterior thigh muscles as the human, the sizes of the muscles are quite different. These size differences have to do with the cat's four-legged posture as compared to the human's two-legged posture. That is, the gluteal muscles in the cat are comparatively small, and the posterior thigh, or "hamstring" muscles are comparatively large. This situation is largely reversed in humans, who tend to have larger gluteal muscles and smaller hamstring muscles (see Learning Strategy on p. 818). Using figure 28.12 as a guide, begin cleaning the fascia off the gluteal muscles and proceed toward the distal thigh. Take special care around the insertion of the **gluteus maximus** and **caudofemoralis** muscles. The insertions of these muscles are on the **fascia lata** of the thigh, which is really just a thickening of the deep fascia that envelops the thigh muscles. The goal in this part of the dissection is to preserve the band of connective tissue that composes the fascia lata, or **iliotibial (IT) band,** while removing the remaining fascia.

2. Clean, identify, and isolate the following muscles of the hindlimb, using table 28.6 and figure 28.12 as guides:

- ☐ **biceps femoris**
- ☐ **caudofemoralis**
- ☐ **gastrocnemius**
- ☐ **gluteus maximus**
- ☐ **gluteus medius**
- ☐ **sartorius**
- ☐ **tensor fasciae latae**

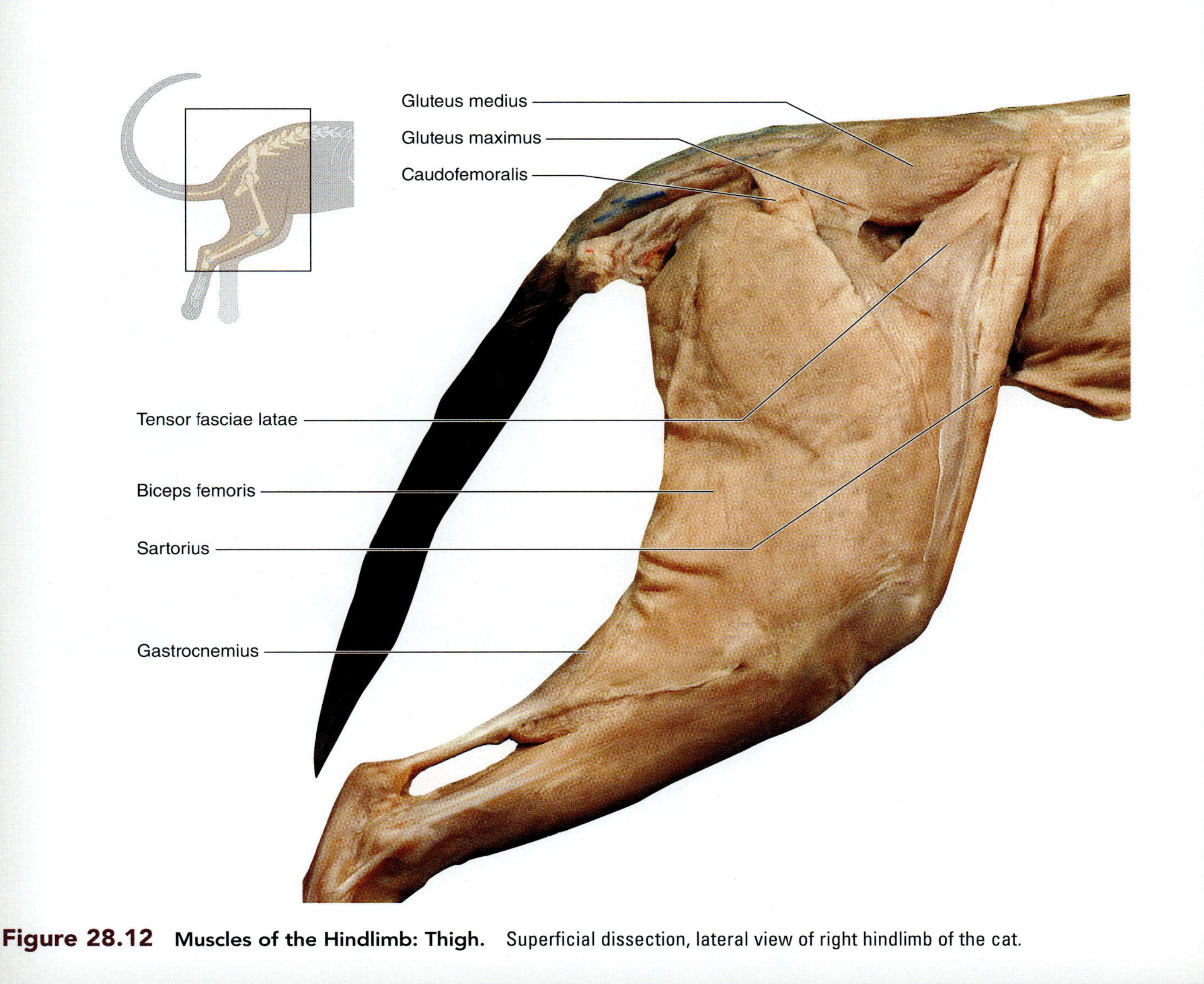

Figure 28.12 Muscles of the Hindlimb: Thigh. Superficial dissection, lateral view of right hindlimb of the cat.

(continued on next page)

(continued from previous page)

3. Turn the cat over so that the dorsal surface is down and the medial and anterior thigh muscles are visible **(figure 28.13*a*).** While cleaning muscles on the anterior thigh, note the origin of the **sartorius** muscle. Observe the small, triangular region that lies medial to the origin of the sartorius. This is the **femoral triangle,** a region that contains the femoral artery, nerve, and vein. It may be necessary to clean some fat superficial to the triangle to observe the neurovascular structures. The **inguinal ligament,** the **sartorius,** and the **adductors** form the borders of the femoral triangle. While cleaning the muscles, take care not to destroy the nerves and vessels in this region. Carefully trace the sartorius muscle from origin to insertion **(table 28.7),** clean the muscle, and separate it from the underlying muscles. Note that the location and function of the sartorius is similar in cats and humans, with one major exception: The human sartorius *flexes* the leg (weakly), whereas the cat sartorius *extends* the leg.

4. **Deep muscles:** Once the sartorius is isolated from the surrounding muscles, slip a blunt probe deep to the muscle to protect the deeper muscles. Then *transect* the **sartorius** across the middle of the muscle belly and *reflect* the two ends to expose underlying muscles. Next, isolate the **gracilis** from the deeper muscles. Using a probe to protect the underlying muscles, *transect* the gracilis, and *reflect* the two ends to expose the deeper muscles. Take care not to damage the femoral vessels and nerve while reflecting these two muscles.

5. **Anterior thigh:** Observe the **quadriceps femoris** group of muscles. These are homologous to the quadriceps femoris muscle group in humans. As in humans, all four muscles of the group come together to form a common tendon that encases the patella: the **patellar tendon.** This band of connective tissue extends distal to the patella to connect to the tibial tuberosity as the **patellar ligament.** Recall that the major action of the quadriceps femoris group of muscles is to extend the leg.

6. Clean, isolate, and identify the following structures, using figure 28.13*b* and table 28.7 as guides:

- ☐ **patella**
- ☐ **patellar ligament**
- ☐ **patellar tendon**
- ☐ **rectus femoris**
- ☐ **vastus intermedius**
- ☐ **vastus lateralis**
- ☐ **vastus medialis**

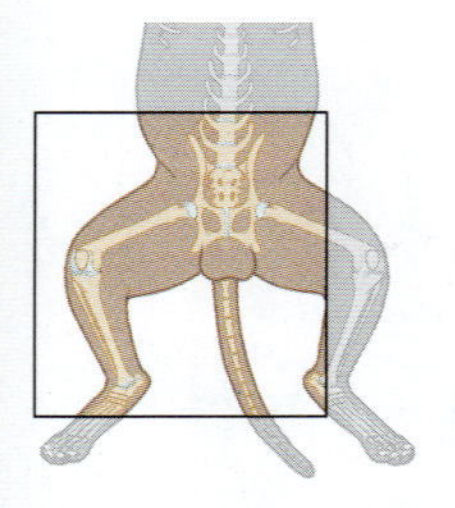

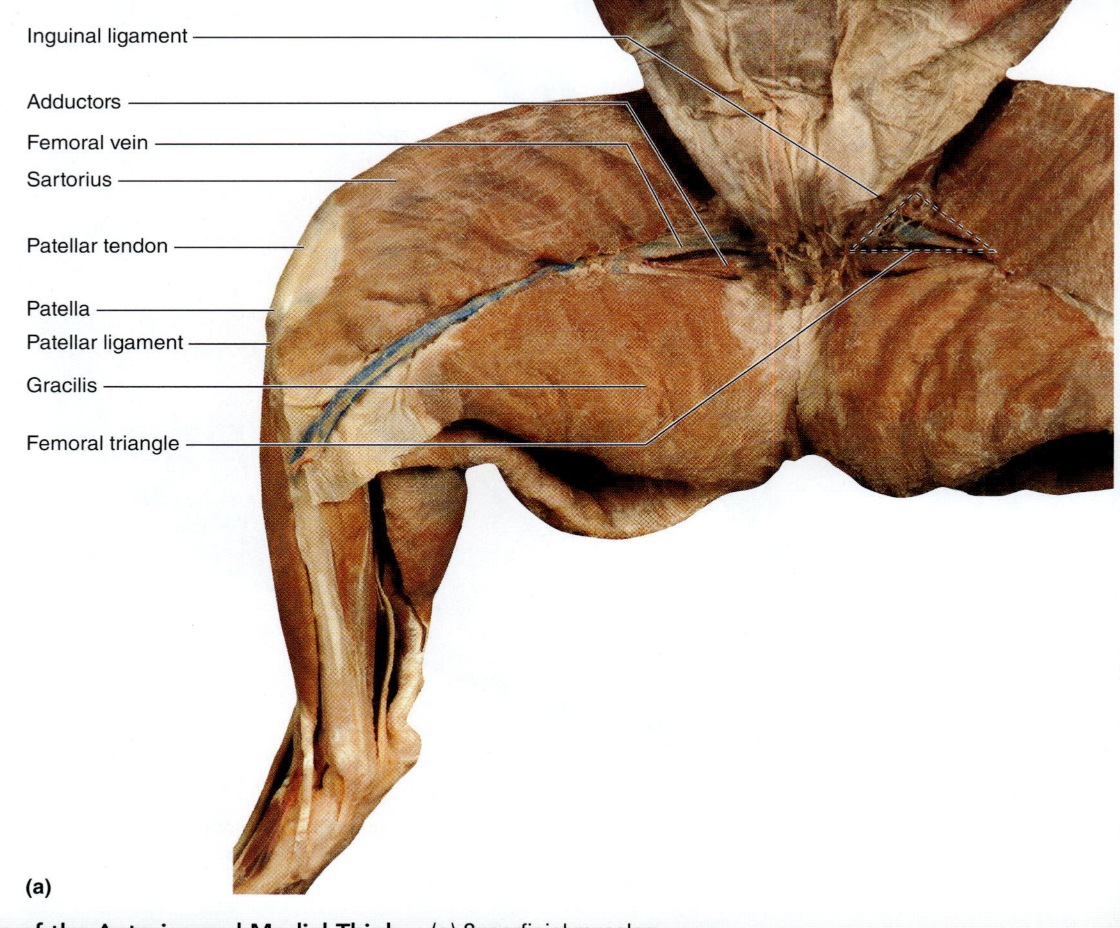

Figure 28.13 **Muscles of the Anterior and Medial Thigh.** (a) Superficial muscles.

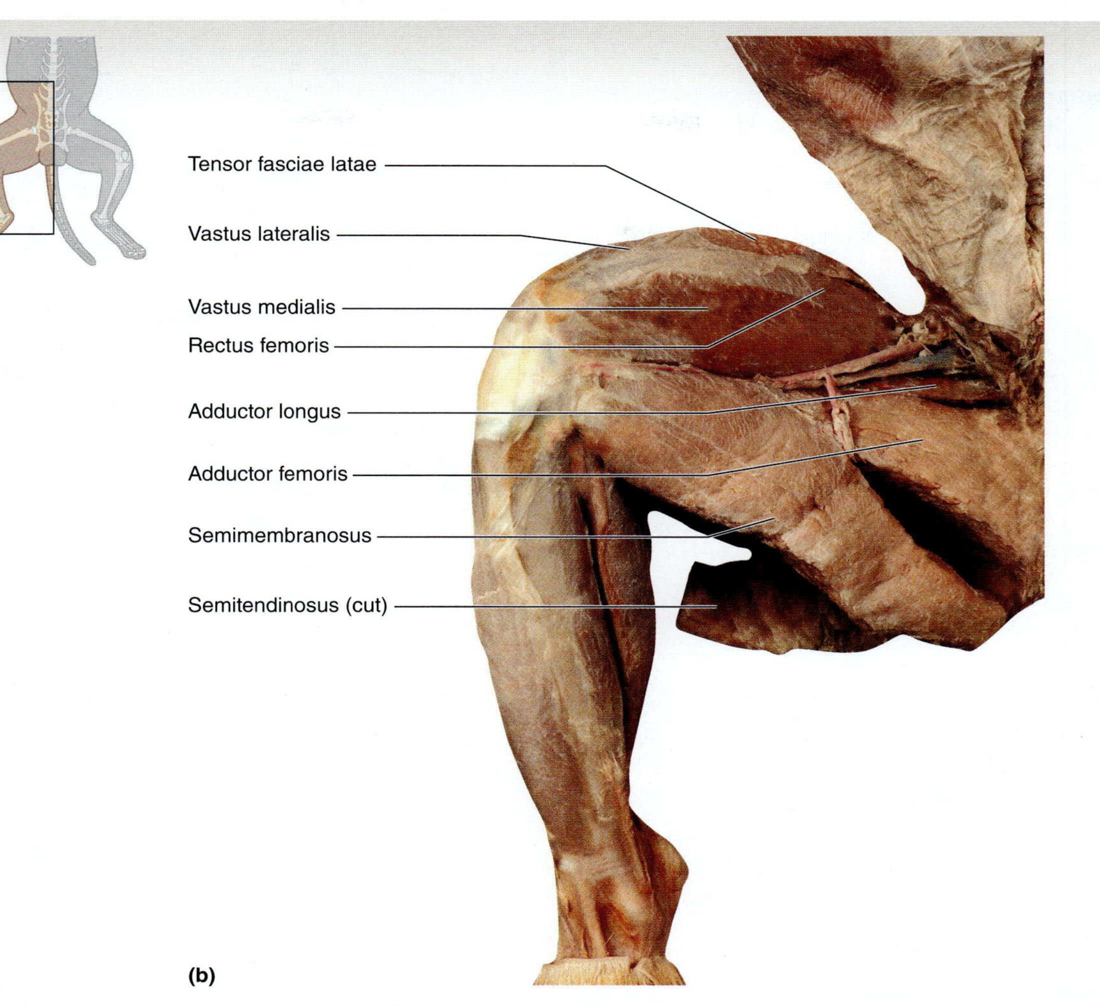

Figure 28.13 Muscles of the Anterior and Medial Thigh *(continued).* (b) Deep muscles. Note the location of the femoral artery and vein. Muscles lateral to the femoral vessels belong to the anterior compartment. Muscles medial to the femoral vessels belong to the medial compartment. Sartorius and gracilis have been cut and reflected, and thus are not visible in this photo. Vastus intermedius is deep to vastus lateralis and vastus medialis and thus is not visible in this photo.

7. **Medial thigh:** Observe the muscles medial to the femoral vessels (figure 28.13*b*). The medial compartment muscles of the cat, the **adductor** muscles, are again very similar to the homologous muscles in the human. The major exceptions are that the **gracilis** is much larger in the cat, and the cat has one large adductor, the **adductor femoris,** in place of the adductor brevis and adductor magnus muscles of the human. In the medial view of the thigh, parts of the large **semimembranosus** and **semitendinosus** muscle are also visible. These muscles are part of the posterior thigh, or "hamstring," muscles.

8. Clean, isolate, and identify the following muscles of the medial and posterior thigh, using figure 28.13*b* and table 28.7 as guides.

☐ **adductor femoris**
☐ **adductor longus**
☐ **semimembranosus**
☐ **semitendinosus**

Table 28.7 Muscles of the Hindlimb

Cat Muscle	Human Muscle	Origin	Insertion	Action
Muscles of the Hip				
Caudofemoralis	NA	Proximal caudal vertebra	Patella	Abduct thigh
Gluteus Maximus	Gluteus maximus	Last sacral and first caudal vertebra	Fascia lata	Abduct thigh
Gluteus Medius	Gluteus medius	Iliac crest, sacral vertebrae, and first caudal vertebra	Greater trochanter of femur	Abduct thigh

(continued on next page)

(continued from previous page)

Table 28.7 Muscles of the Hindlimb *(continued)*

Cat Muscle	Human Muscle	Origin	Insertion	Action
Gluteus Profundus/ Minimus	Gluteus minimus	NA	NA	NA
Gracilis	Gracilis	Pubic symphisis	Fascia of distal thigh	Adduct thigh
Ilipsoas	Ilipsoas	Lumbar vertebrae and ilium	Lesser trochanter of femur	Flex and rotate thigh
Tensor Fasciae Latae	Tensor fasciae latae	Ilium	Fascia lata	Extend fascia lata
Tenuissimus	Tenuissimus	Second caudal vertebra	Proximal tibia	Flex the shank
Muscles of the Thigh (anterior and medial)				
Adductor Femoris	Adductor brevis and magnus	Pubic symphysis	Ventral femur	Adduct thigh
Adductor Longus	Adductor longus	Pubic symphysis	Femur	Adduct thigh
Pectineus	Pectineus	Pubis	Shaft of femur	Adduct thigh
Rectus Femoris	Rectus femoris	Ilium	Patella	Extend leg
Sartorius	Sartorius	Iliac crest	Patella and proximal tibia	Adduct and laterally rotate thigh, extend the leg
Vastus Intermedius	Vastus intermedius	Femur	Patella	Extend leg
Vastus Lateralis	Vastus lateralis	Femur	Patella	Extend leg
Vastus Medialis	Vastus medialis	Femur	Patella	Extend leg
Muscles of the Thigh (posterior)				
Biceps Femoris	Biceps femoris	Ischial tuberosity	Proximal tibia	Flex shank, adduct thigh
Semimembranosus	Semimembranosus	Ischium	Medial epicondyle of femur, proximal tibia	Extend thigh
Semitendinosus	Semitendinosus	Ischium	Tibia	Flex leg
Muscles of the Leg (anterior and lateral)				
Extensor Digitorum Longus	Extensor digitorum longus	Lateral epicondyle of femur	Digits II–V	Extend the digits
Fibularis Brevis	Fibularis brevis	Fibula	Metatarsals	NA
Fibularis Longus	Fibularis longus	Fibula	First metatarsal	NA
Tibialis Anterior	Tibialis anterior	Proximal tibia and fibula	First metatarsal	Dorsiflex the foot
Muscles of the Leg (posterior)				
Flexor Digitorum Longus	Flexor digitorum longus	Tibia and head of fibula	Digits I–V	Flex the digits
Flexor Hallucis Longus	Flexor hallucis longus	Tibia and fibula	Tendon of flexor digitorum longus	Flex the digits

Table 28.7	Muscles of the Hindlimb *(continued)*			
Cat Muscle	**Human Muscle**	**Origin**	**Insertion**	**Action**
Muscles of the Leg (posterior) *(continued)*				
Gastrocnemius	Gastrocnemius	Distal, posterior femur, tendon of plantaris	Calcaneus	Extend the foot
Plantaris	Plantaris	Patella and femur	Calcaneus	Flex the digits
Soleus	Soleus	Proximal fibula	Calcaneus	Extend the foot
Tibialis Posterior	Tibialis posterior	Tibia and fibula	Tarsals	Extend the foot

EXERCISE 28.9B Muscles of the Hindlimb: Leg and Foot

1. **Superficial muscles:** Continue dissecting the hindlimb distally toward the leg. The muscles of the leg in cats are organized into compartments that are very similar to those of the human leg. These compartments are anterior, posterior, and lateral. Anterior compartment muscles dorsiflex the foot and extend the digits, posterior compartment muscles plantar flex the foot and flex the digits, and lateral compartment muscles evert the foot. The posterior compartment is the largest compartment of the leg, containing the triceps surae muscle group. The **triceps surae** consists of the gastrocnemius, soleus, and plantaris muscles, all of which insert on the calcaneus via the calcaneal tendon. As in the human, the triceps surae group is the most important group of muscles acting about the ankle. They are the primary muscles used for plantar flexion of the foot, an action that is necessary for standing, jumping, walking, and running.
2. Clean, identify, and isolate the following muscles, using **figure 28.14** and table 28.7 as guides:

- ☐ **extensor digitorum longus**
- ☐ **fibularis brevis**
- ☐ **fibularis longus**
- ☐ **gastrocnemius**
- ☐ **plantaris**
- ☐ **soleus**
- ☐ **tibialis anterior**

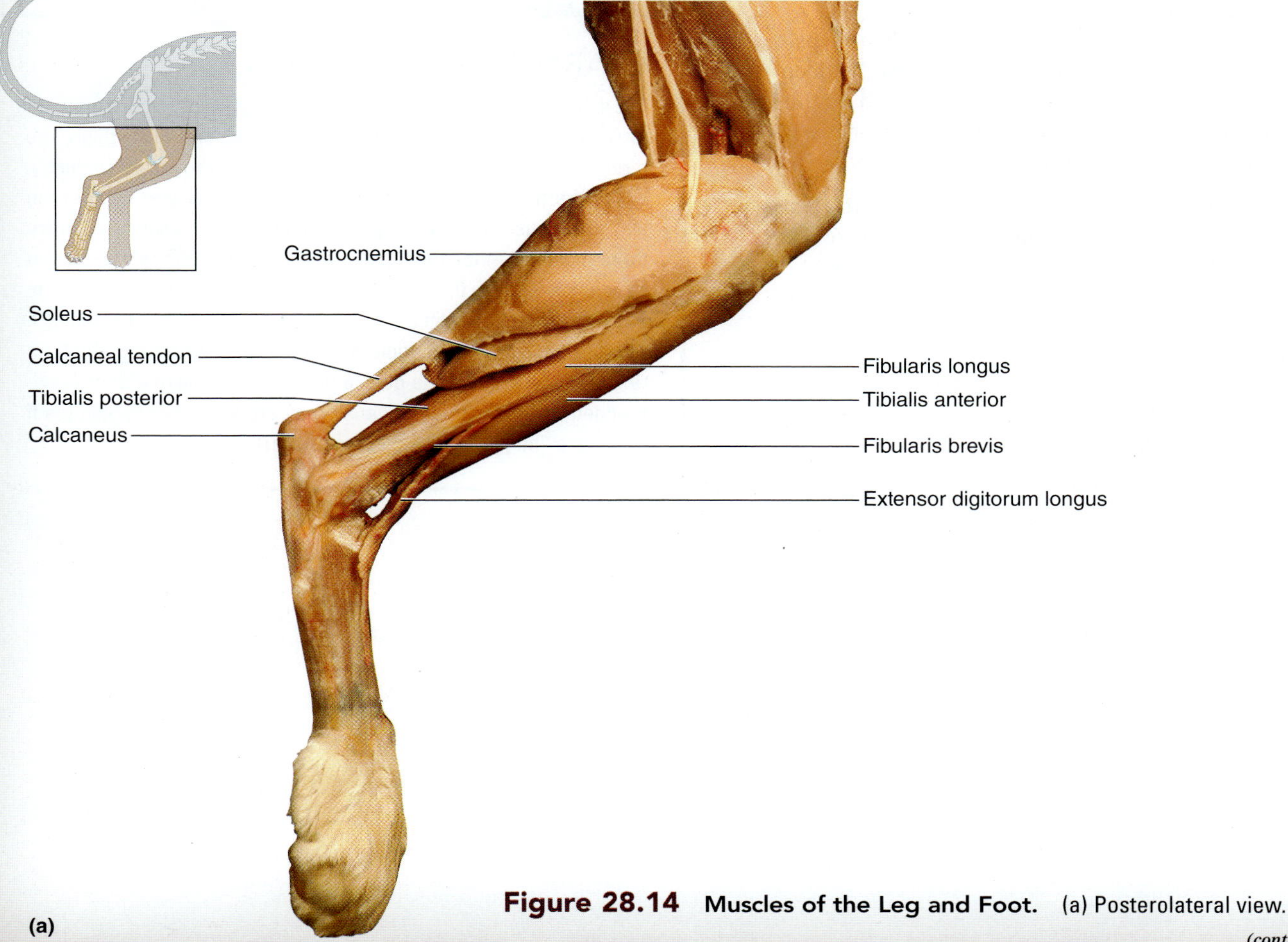

Figure 28.14 Muscles of the Leg and Foot. (a) Posterolateral view.

(continued on next page)

(continued from previous page)

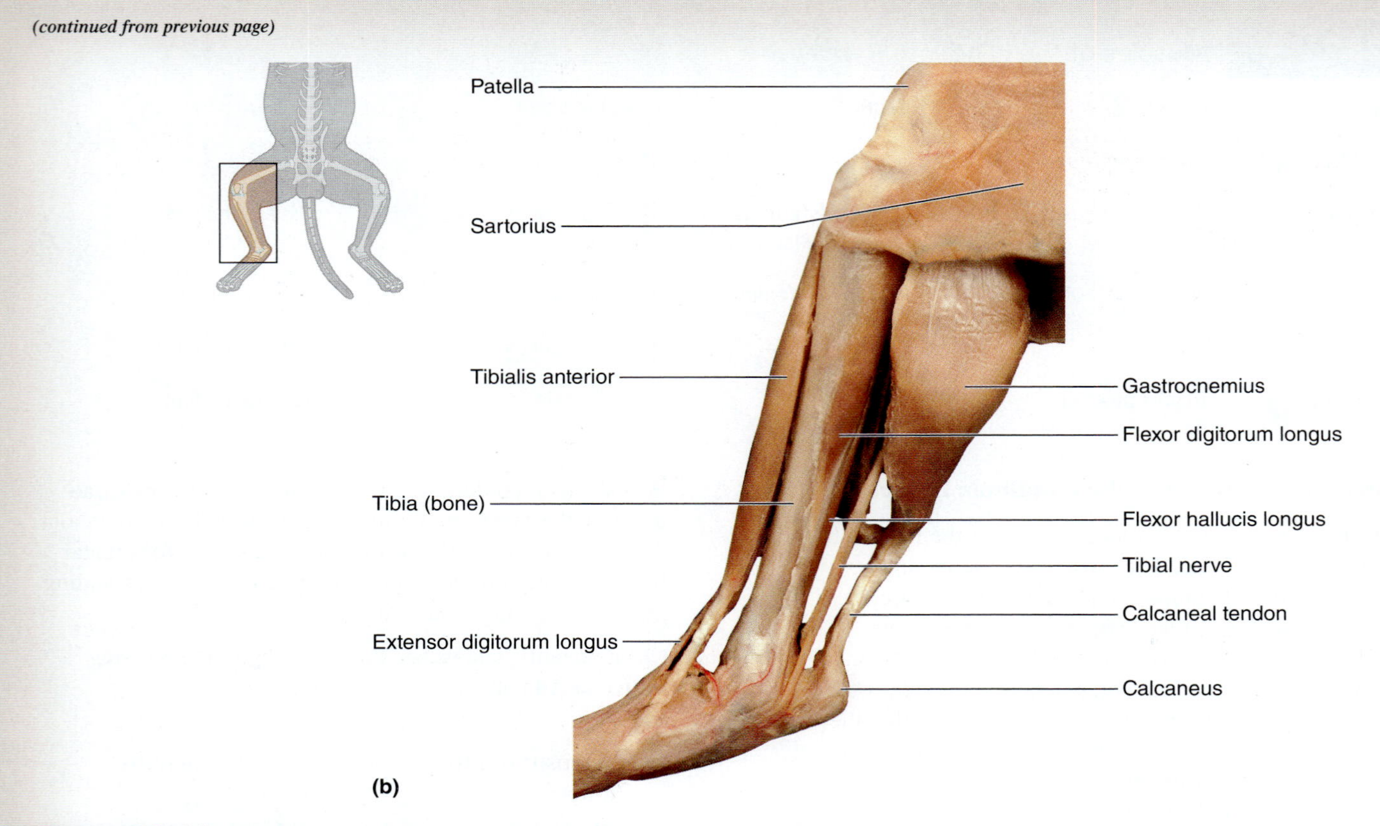

Figure 28.14 Muscles of the Leg and Foot *(continued).* (b) Medial view.

3. **Deep muscles:** Viewed earlier, the largest, most superficial muscles of the posterior leg are the triceps surae muscles. One must reflect these to view the deep muscles of the posterior leg. The easiest way to do this is to detach the calcaneal tendon from the calcaneus, and then reflect the triceps surae muscles cranially.
4. Once the superficial muscles have been reflected, clean, identify, and isolate the following deep muscles, which are quite small. Use figure 28.14 and table 28.7 as guides. While dissecting these muscles, consider the homologous muscles in the human leg and note any similarities and differences.
   - ☐ **flexor digitorum longus**
   - ☐ **flexor hallucis longus**
   - ☐ **tibialis posterior**

INTEGRATE

LEARNING STRATEGY

The semitendinosus, semimembranosus, and biceps femoris muscles are commonly referred to as the **hamstring muscles** of the thigh. The term *hamstring* comes from the animal meat processing industry. When the hindlimbs of animals, particularly pigs, are hung in the slaughterhouse, they are typically hung by the tendons of these three muscles—thus the term *ham* (the back of a hog's knee) and *string.* So, literally, the pig is strung up by these muscles. The term *hamstrung* has, over time, come to take on several other meanings. Most commonly when people are said to be *hamstrung,* the term refers to the fact that the people are unable to make a decision or proceed with something because they are stuck or crippled (as if they were "hung up" by their hamstrings!).

Thoracic Cavity

EXERCISE 28.10

OPENING THE THORACIC AND ABDOMINAL CAVITIES

1. The thoracic cavity is opened with a series of steps (see figure 28.15). The process begins with opening the thoracic cavity of the cat beginning with a large, vertical, midline incision in the anterior wall of the abdominal and thoracic cavities. Next, make an incision alongside the inferior border of the thoracic cage and diaphragm to free the abdominal muscles from the thoracic cage. Finally, open the thoracic cavity by reflecting the ribs laterally. **Figure 28.15** demonstrates the cuts required to open up the thoracic and abdominal cavities.

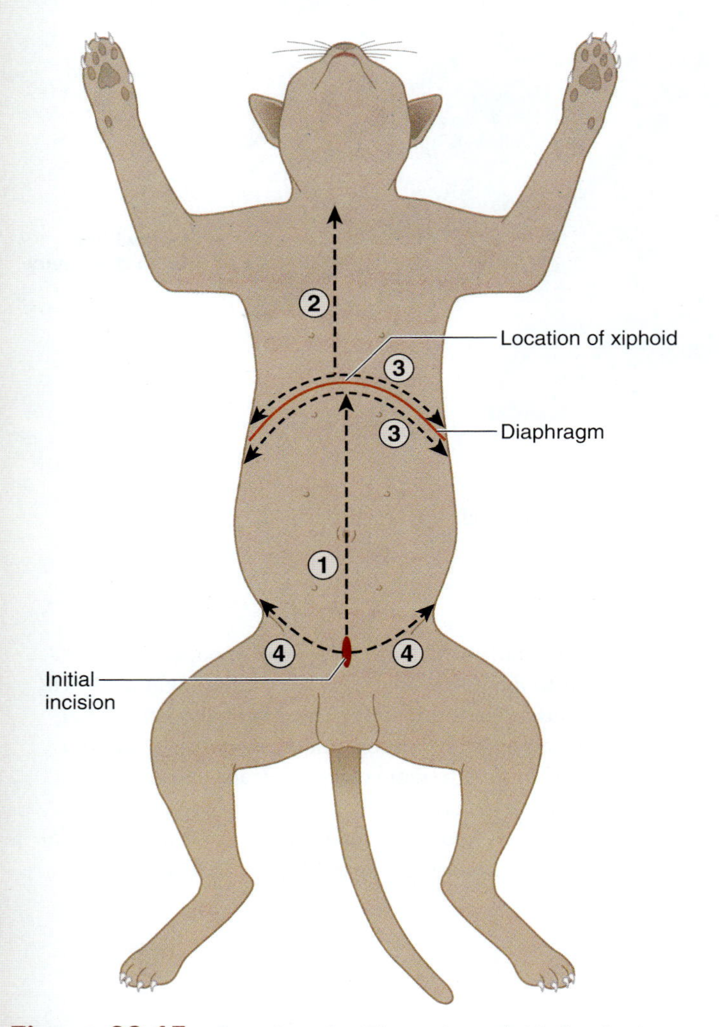

Figure 28.15 Opening the Thoracic and Abdominal Cavities. The numbered steps in this figure denote the order of cuts to be made.

2. Place the cat on the dissecting tray with the ventral side up. Obtain a pair of forceps, a blunt probe, and scissors or small bone cutters. Using figure 28.15 as a guide:

a. Make a vertical incision in the most inferior part of the linea alba of the abdominal wall. If dissecting a male cat, take care not to damage the spermatic cord or external genitalia. Using forceps, lift the tissue composing the abdominal wall away from the abdominal viscera. Then, using scissors, continue the vertical incision all the way to the **xiphoid process** (figure 28.15, step 1). While making the cut, use the scissors to continue to lift the abdominal wall away from the contents of the abdominal cavity to avoid accidentally cutting into abdominal organs.

b. When the incision reaches the xiphoid, its course needs to change slightly so that it does not continue directly on the midline. Instead, direct the incision slightly to the left of the sternum (figure 28.15, step 2). Here it is necessary to cut through the costal cartilages. Although this may be accomplished using scissors, if the cartilage is too thick or tough, bone cutters may be used in place of scissors. As with the incision in the anterior abdominal wall, take care not to damage deeper structures while making the incision. Continue the incision until the attachment of the most superior rib with the sternum has been cut.

c. Observe the inferior border of the ribs and locate the **diaphragm.** To preserve the diaphragm and its relations with the thoracic and abdominal cavities, separate its attachments to the body wall both above *and* below the muscle. To do this, make one lateral incision cranial to the diaphragm and another caudal to the diaphragm (figure 28.15, step 3).

d. Finally, make an oblique incision in the caudal, anterior abdominal body wall by cutting *just cranial* to the inguinal ligament (figure 28.5, step 4). If dissecting a male cat, make sure to locate the spermatic cord as it exits the abdominal cavity to enter the scrotum before making this cut. It is imperative that the spermatic cord and its relationship to the body wall are preserved for later study. Thus, take extra care to make this cut cranial to the spermatic cord.

(continued on next page)

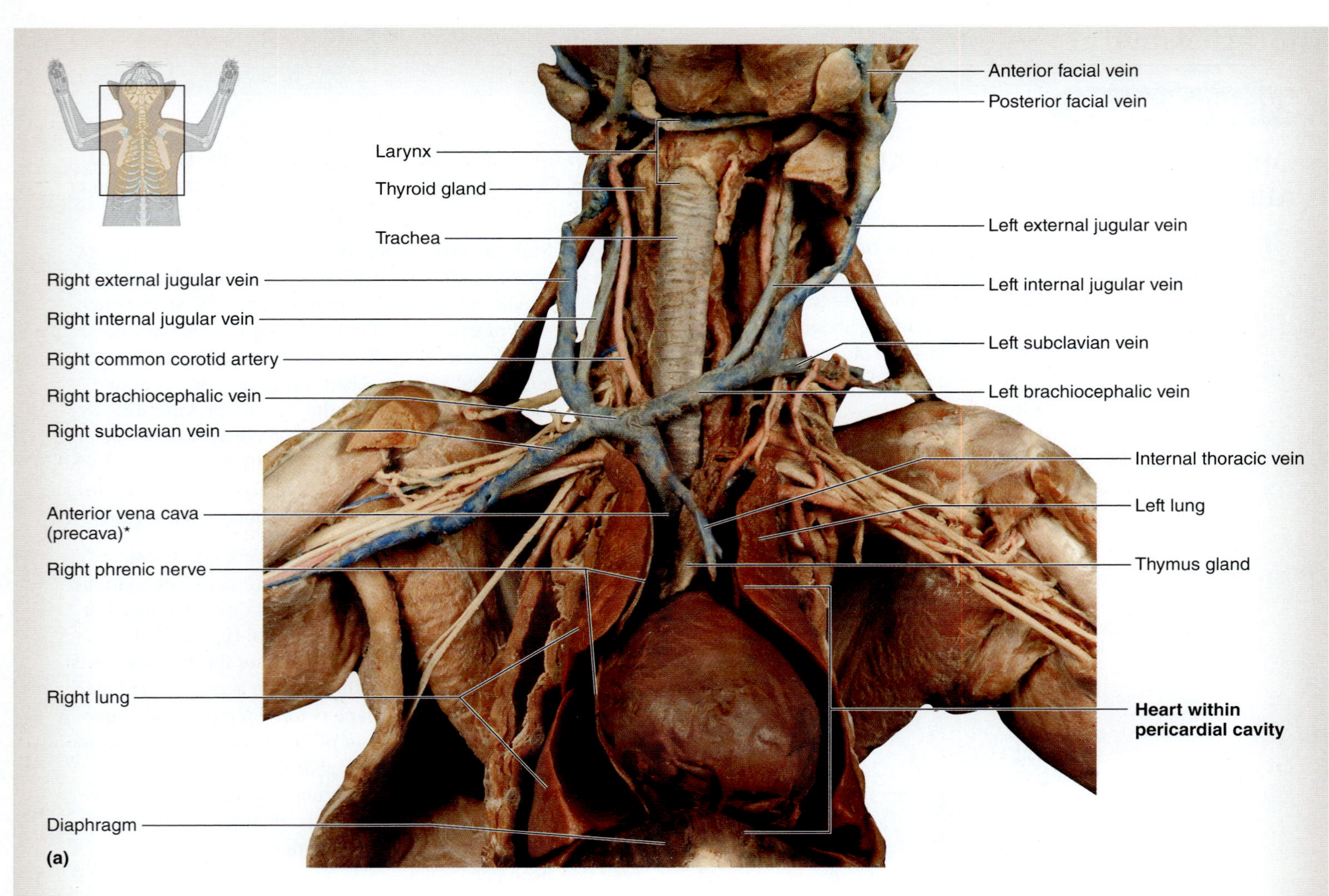

*The anterior vena cava (precava) in the cat is the same as the superior vena cava in the human.

Figure 28.16 **The Thoracic Cavity.** (a) Superficial dissection demonstrating the pleural and pericardial cavities.

3. After opening the thoracic and abdominal cavities, take the cat to the sink and rinse out the thoracic and abdominal cavities with water to rid the cavities of debris from embalming and any dried blood that may remain. Be sure to ask the lab instructor which sink is appropriate for this task before going ahead with the rinsing. Using a sink without proper disposal methods for dried blood and other debris can result in clogged drains. When finished, dry the cat with paper towels and return it to the dissecting pan. The abdominal cavity will not be observed at this time. Thus, keep the abdominal wall closed over the contents of the abdominal cavity and focus on the thoracic cavity. Recall that inside the thoracic cavity there are three cavities: two pleural cavities, which house the lungs, and a central pericardial cavity, which houses the heart **(figure 28.16).**

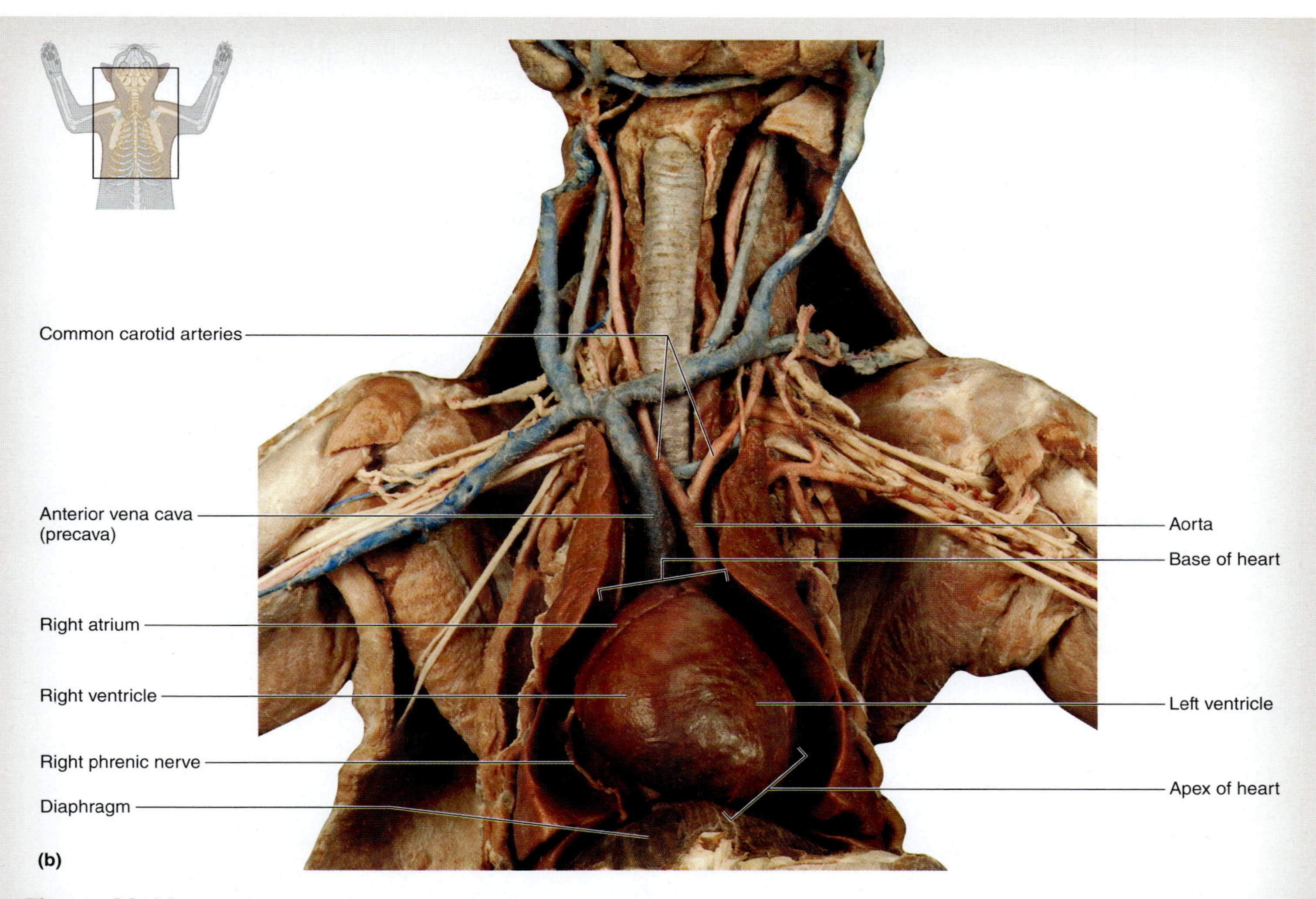

Figure 28.16 The Thoracic Cavity *(continued).* (b) Deep dissection demonstrating the heart and lungs and structures of the superior mediastinum (trachea, esophagus, blood vessels, and nerves).

EXERCISE 28.11

THE HEART, LUNGS, AND MEDIASTINUM

1. **Superficial dissection:** When removing the sternum, the anterior part of the **parietal pericardium** (outer covering of the pericardial sac) may be adhered to the bone. The goal is to keep the pericardium with the heart, not the body wall. Similarly, the **parietal pleura** (outer covering of the **pleural cavities**) is adhered to the inside of the thoracic cage. The pleura is more difficult to separate from the thoracic cage than the pericardium. Therefore make sure to observe the inside of the thoracic cage and make note of the parietal pleura. Where is the visceral pleura located? ______________________

2. Begin to clean the structures in the superior mediastinum (e.g., trachea, carotid arteries, esophagus), taking care to pull on any "stringy" structures very gently so as not to destroy them. While cleaning the region superficial and lateral to the pericardium, gently separate the **parietal pericardium** from the **parietal pleura** on both sides and look for the right and left **phrenic nerves,** which lie in the space between these two layers. What structure do the phrenic nerves innervate? ______________________

3. Observe the area just cranial to the pericardial cavity, where the great vessels enter and leave the heart. Depending on the age of the cat, the **thymus** may or may not be visible in this location. In a young cat, the thymus often looks like a small, extra lobe of the lung, although it is not part of the lung. In an older cat, the thymus may not be visible because it largely has been replaced with adipose connective tissue. Recall the function of the thymus from chapter 23. The thymus in humans, just as with cats, is also large in adolescents, but generally absent in older individuals. In the space provided, explain why the thymus is not usually visible in an older cat or human.

____

____

(continued on next page)

(continued from previous page)

4. Using scissors, carefully open the **pericardial sac** to reveal the heart within. Review the layers of the pericardium (exercise 21.4, p. 550) while dissecting away the sac. At the base of the heart (remember, the base of the heart is the most cranial part), the pericardial sac is adhered to the great vessels. Of note, the major veins entering the cat heart are the *anterior* vena cava (precava) and *posterior* vena cava (postcava). These are the same as the *superior* vena cava and *inferior* vena cava in the human, respectively. As you detach the sac from the vessels, identify each of the great vessels (**figures** 28.16 and **28.17**).

   - ☐ **anterior vena cava (precava)**
   - ☐ **aorta**
   - ☐ **pulmonary trunk**

5. Identify the four chambers of the heart, using figure 28.17 as a guide.

   - ☐ **left atrium**
   - ☐ **right atrium**
   - ☐ **left ventricle**
   - ☐ **right ventricle**

6. **Deep dissection:** After identifying the great vessels and heart chambers, the next step is to remove the heart and lungs from the thoacic cavity. To do this, follow these steps:

   a. Transect the **pulmonary arteries** and **veins** at the hilum of each lung. When doing this, also transect the **primary bronchi,** which enter the lungs adjacent to the pulmonary vessels.

   b. Transect the **anterior vena cava (precava)** and **aorta** just superior to the heart. This means cutting through the *ascending* part of the aorta.

   c. Bisect the **posterior vena cava (postcava),** which lies between the inferior surface of the heart and the diaphragm.

7. **The lungs:** Because all of the pulmonary vessels have been transected between the heart and the lungs, the lungs can now be removed from the thoracic cavity (figure 28.17). Remove both the heart and the lungs. Place them next to the cat in the dissecting tray or in another receptacle, and observe their features. How many lobes does the right lung have? _____________ How many lobes does the left lung have? _______________ Are these the same or different from the number of lobes in the human lungs?

____

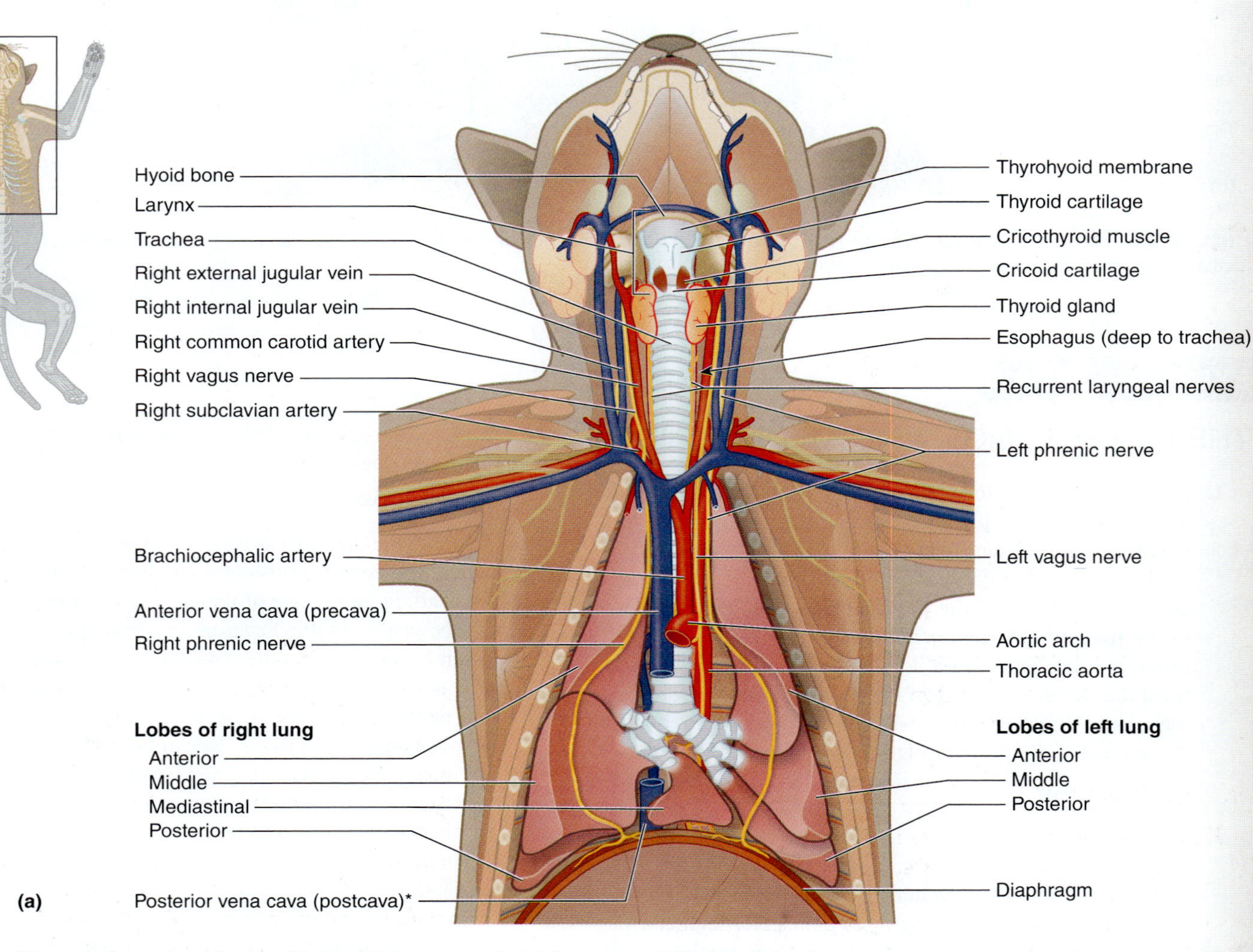

*The posterior vena cava (postcava) in the cat is the same as the inferior vena cava in the human.

Figure 28.17 The Respiratory System. (a) Illustration.

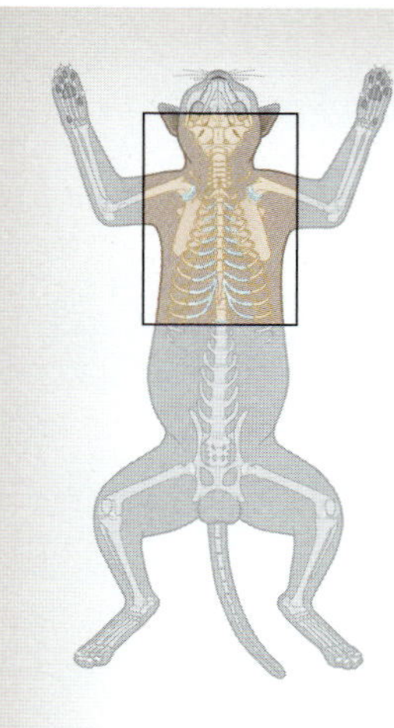

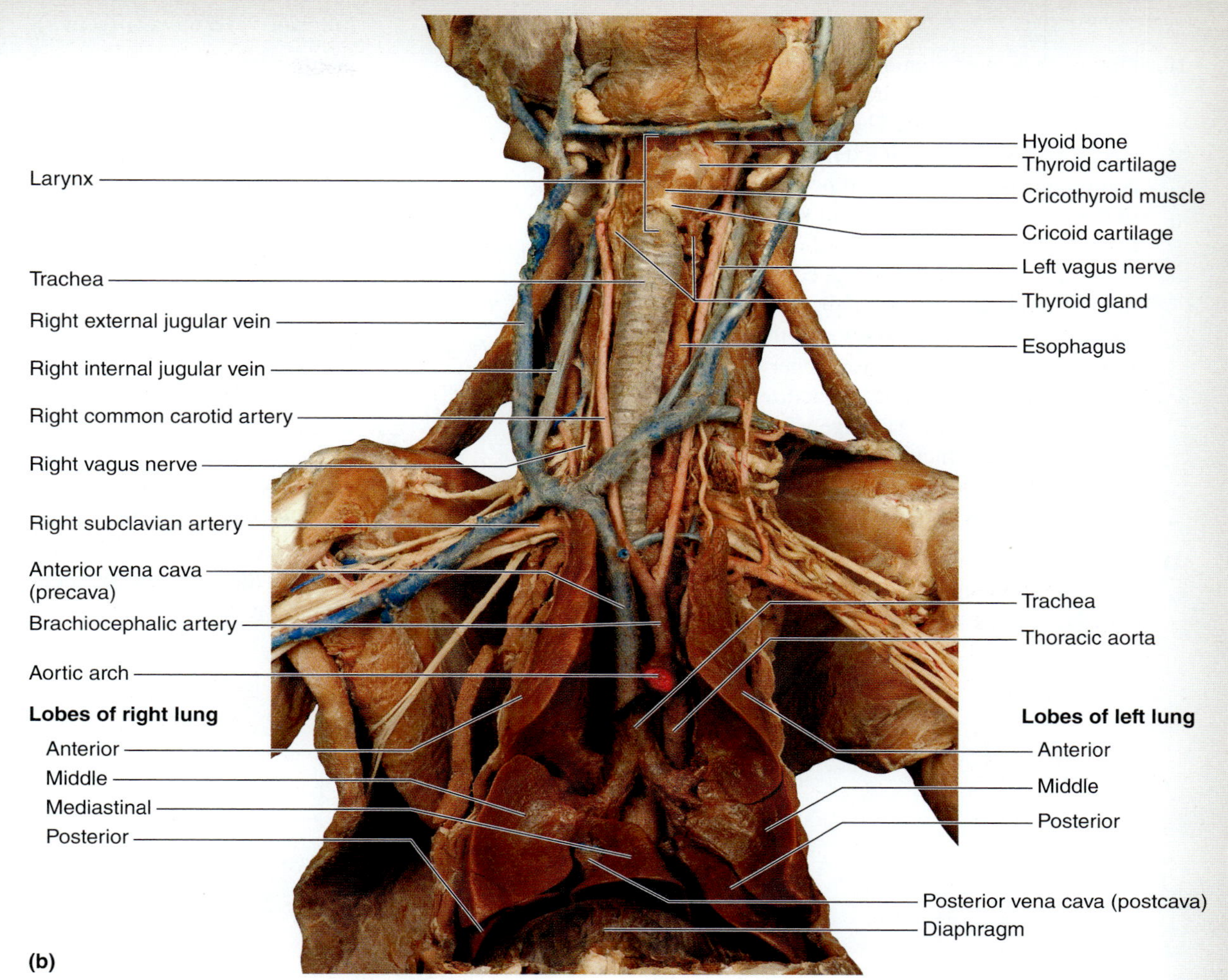

Figure 28.17 **The Respiratory System *(continued).*** (b) Photo.

8. **The heart:** Observe the heart and identify the four chambers of the heart. Notice the relative size of the different chambers. Optional step: To observe internal structures of the heart, use a scalpel to make a coronal section of the heart, starting from the apex and ending at the base. Identify all of the valves of the heart and note the relative size of the chambers. Note any similarities or differences between the cat heart and human heart in the space provided.

___

___

___

9. **Upper respiratory and digestive tracts:** This next step involves careful cleaning of structures that lie in the superior mediastinum: the **trachea,** the **esophagus,** the **thoracic duct,** and several blood vessels and nerves. Begin by locating the esophagus and the **azygos vein** on the posterior thoracic wall next to the vertebral column. The azygos vein is a vein that extends from caudal to cranial just to the right of the vertebral column. The azygos vein and thoracic duct are not visible in figure 28.17. Next, look for a thin, brown or colorless vessel located between the azygous vein and the esophagus, which is the **thoracic duct** (see Learning Strategy on this page). Do not worry if the thoracic duct cannot be located, as it is very small. The thoracic duct is the main lymphatic duct of the body The thoracic duct begins as the **cisterna chyli** in the abdominal cavity near vertebral level L_1, and runs along the dorsal abdominal and thoracic walls before it turns and enters the left subclavian vein.

INTEGRATE

LEARNING STRATEGY

The thoracic duct is a very small vessel, particularly when compared to other structures within the mediastinum. One way of remembering how to find the thoracic duct is to look for what anatomists call the "duck between two gooses." This is a play on words resulting from pronouncing the names of the three structures slightly different than normal. The "duck" is the thoracic **duc**t. The "two gooses" are the esopha**goose** (esophagus) and the azy**goose** (azygos vein). Thus, to find the thoracic duct, look for a small vessel that runs between the esophagus and the azygos vein.

(continued on next page)

(continued from previous page)

10. Observe the most cranial part of the mediastinum. Clean the trachea as it enters the neck. Carefully clean the **larynx** and the **thyroid gland.** The thyroid gland wraps around the larynx and portions of it extend between the larynx and the common carotid arteries. To perform most of this cleaning of connective tissue, use small scissors and forceps, taking care not to pull too hard on any structure so as not to tear it. While following the common carotid arteries into the neck, look for a small nerve that runs between the **common carotid artery** and the **internal jugular vein.** This is the vagus nerve (CN X), which was studied in chapter 15. Structures that are located cranial to the thyroid cartilage will be explored further in exercise 28.14, which covers structures of the head and neck. At this time, keep the rest of the head covered and hydrated, so it will be easy to dissect later.

11. Clean, identify, and isolate the following structures in the mediastinum and anterior neck, using **figures** 28.17 and **28.18** as guides:

- ☐ **anterior vena cava (precava)**
- ☐ **aortic arch**
- ☐ **common carotid arteries**
- ☐ **esophagus**
- ☐ **external jugular veins**
- ☐ **larynx**
- ☐ **primary bronchi**
- ☐ **subclavian arteries**
- ☐ **thoracic aorta**
- ☐ **thyroid gland**
- ☐ **trachea**

12. When finished with the dissection of the thoracic cavity, place the heart and lungs back into the thoracic cavity, close the thoracic cage, sprinkle the cat with wetting solution, and then wrap the cat in its skin and place it in the storage bag.

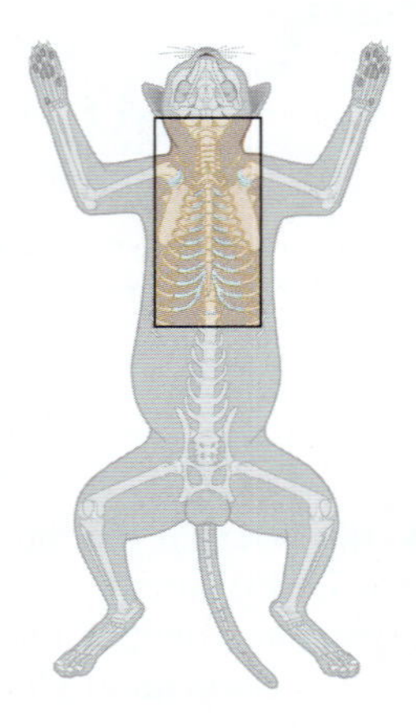

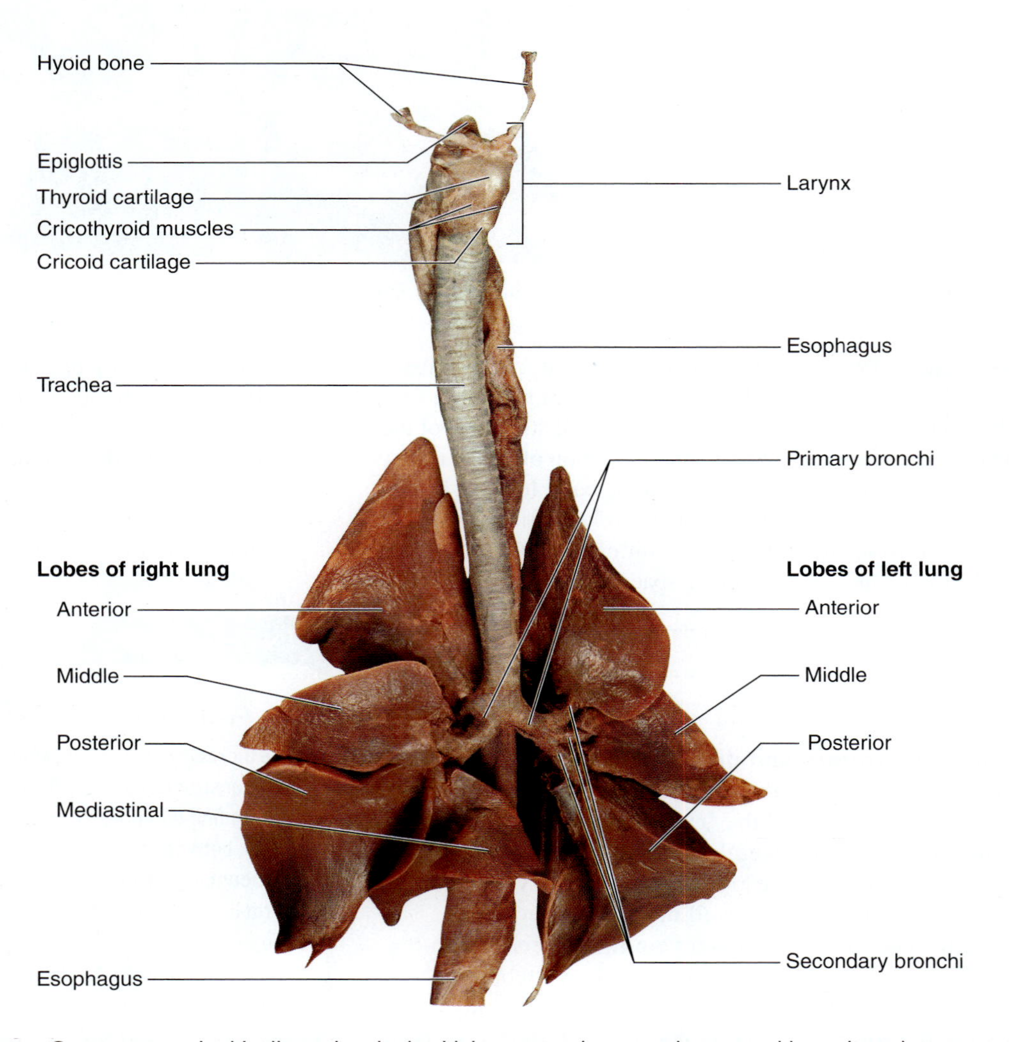

Figure 28.18 Respiratory System Structures. In this dissection the hyoid, larynx, trachea, esophagus, and lungs have been removed from the neck and thoracic cavity to make respiratory system structures more visible.

Abdomen and Pelvis

EXERCISE 28.12

THE ABDOMINAL CAVITY

The directions for making incisions to open both the thoracic and abdominal cavities are described in exercise 28.10, figure 28.16. While performing dissections on the abdominal cavity, make sure to keep the thoracic cavity closed so the heart and lungs don't fall out, and also to keep them hydrated so they do not dry out.

1. Observe the transverse cuts that were previously made below the diaphragm, and the vertical cut along the linea alba. Carefully reflect the anterior abdominal wall laterally along these lines.
2. **Superficial dissection:** One of the most prominent structures that becomes visible as soon as the abdominal cavity is opened is a huge "drape" of fat that overlies most of the abdominal contents. This is the **greater omentum (figures 28.19** and **28.20),** which is a quadruple fold of peritoneum and contains relatively large amounts of fat in many cats. Recall from chapter 25 that the greater omentum functions to help cushion and insulate abdominal organs. In addition, if an infection arises in the abdominal cavity, the greater omentum can move around until it "walls off" that area to prevent the spread of infection. What are the two structures to which the greater omentum is attached (see chapter 25 for review)?

   ______________________________________ and

   ____

   Gently lift the greater omentum from the abdominal cavity using forceps. Then use scissors to detach it from its connections to the abdominal organs.
3. **Deep dissection:** With the greater omentum removed, the contents of the abdominal cavity can be explored more easily. This process requires much less work than dissection of the muscles of the body because there is not a lot of connective tissue that needs to be removed. Instead, the majority of the organs move around freely because of the peritoneum that covers them. Recall that the **peritoneum** is to the abdominal cavity what the pericardium is to the pericardial cavity. That is, the **parietal peritoneum** is the inner lining of the abdominal

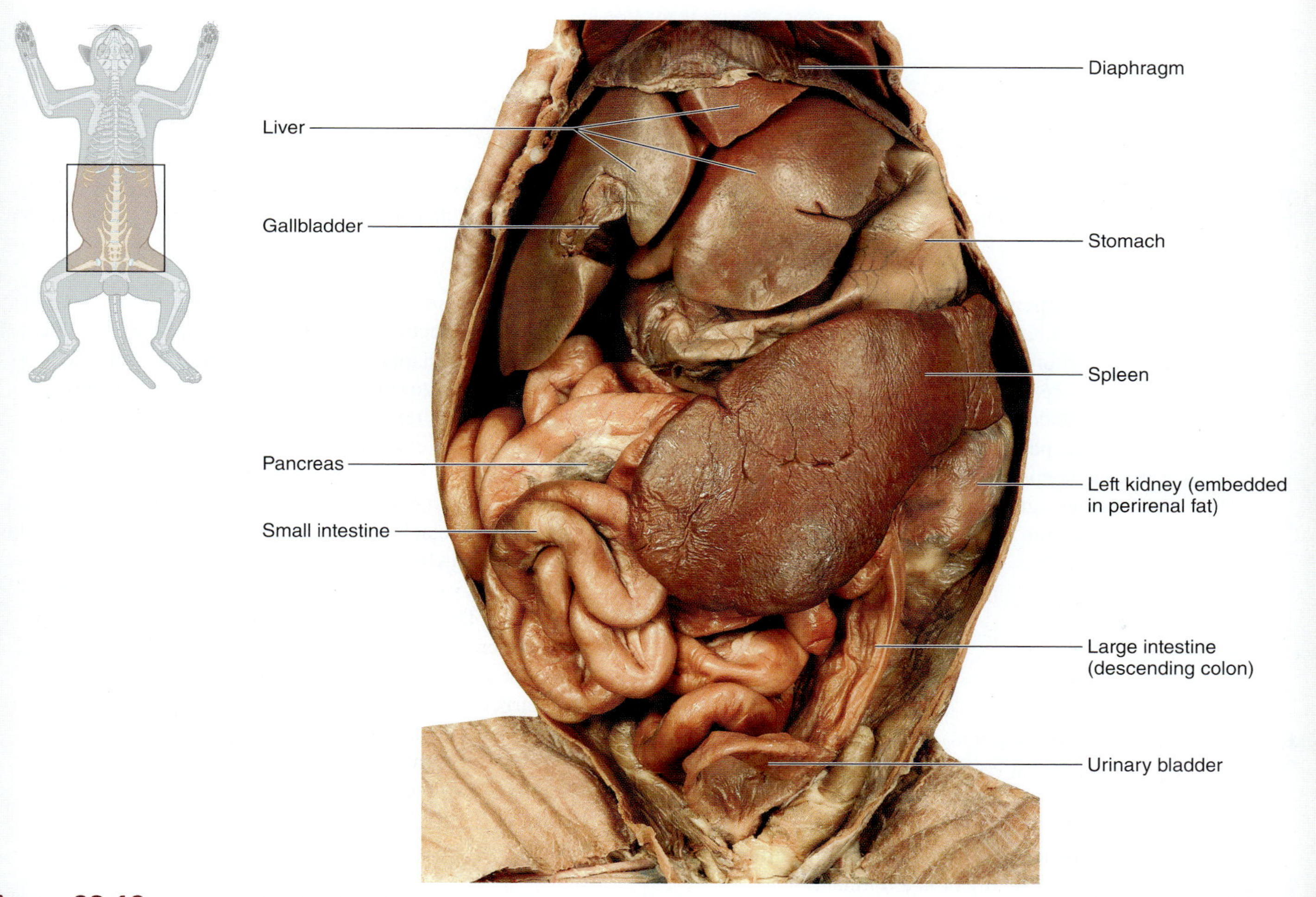

Figure 28.19 The Abdominal Cavity. The anterior abdominal wall and greater omentum has been removed to show a superficial view of the abdomen.

(continued on next page)

(continued from previous page)

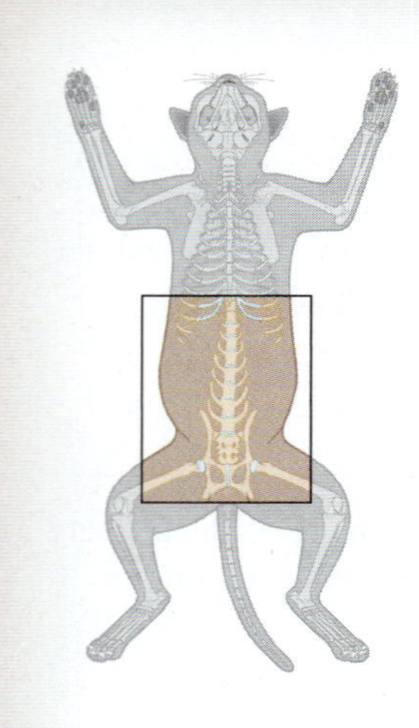

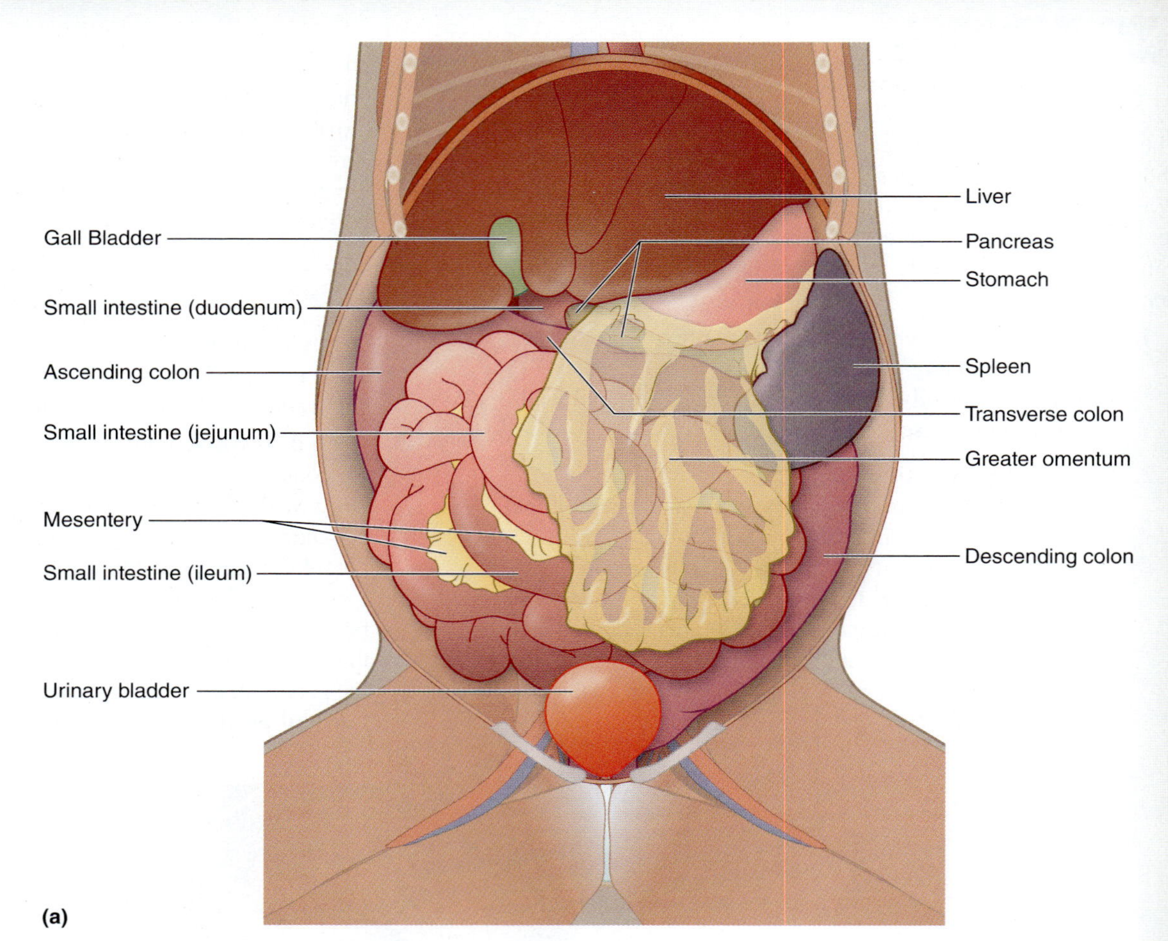

Figure 28.20 The Spleen, Jejunum, Ileum, and Large Intestine (Colon). Trace the tail of the pancreas to see where it enters the hilum of the spleen. Then observe both the small and large intestines and the folds of mesentery that anchor each to the body wall. (a) Illustration.

wall, and the **visceral peritoneum** is the shiny tissue covering most abdominal organs, such as the intestines. It is the visceral peritoneum that forms both omenta and mesenteries (more later on mesenteries). The "dissection" of the abdominal cavity largely consists of gently separating the peritoneum that anchors organs to each other to view deeper structures and to appreciate the relationships among organs. Identify the following organs or parts of organs within the abdominal cavity, using figure 28.20 as a guide:

- ☐ **large intestine**
- ☐ **liver**
- ☐ **small intestine**
- ☐ **spleen**
- ☐ **stomach**
- ☐ **urinary bladder**

4. **The liver and gallbladder:** Observe the liver. Note a band of peritoneum that connects the liver ventrally to the body wall. This is the **falciform ligament**. The posterior border of the falciform ligament contains another band of connective tissue, the **round ligament of the liver**. The round ligament is a remnant of the umbilical vein, which was present during fetal development. Observe the **lobes of the liver.** How many are there? ____________________ Note also the connection between the cranial surface of the liver and the diaphragm. Here, the liver is connected to the diaphragm via a ligament called the **coronary ligament.** The term "coronary" means "crown" and applies in this case because the ligament forms a circle or "crown" of sorts atop the liver. Feel around for the coronary ligament; then cut the coronary ligament with scissors to detach the cranial surface of the liver from the diaphragm. After completing this step, look carefully on the cranial surface of the liver and note the **bare spot** of the liver. The bare spot is the only part of the liver that is not covered with peritoneum; it is characteristically dull in appearance as compared with the shiny appearance of parts of the liver that are covered with peritoneum. Note also the **gallbladder,** which has a characteristic green color. What substance is stored in the gallbladder, which is also the substance that gives it a green color? ______________ What is the function of this substance? ______________________________

____

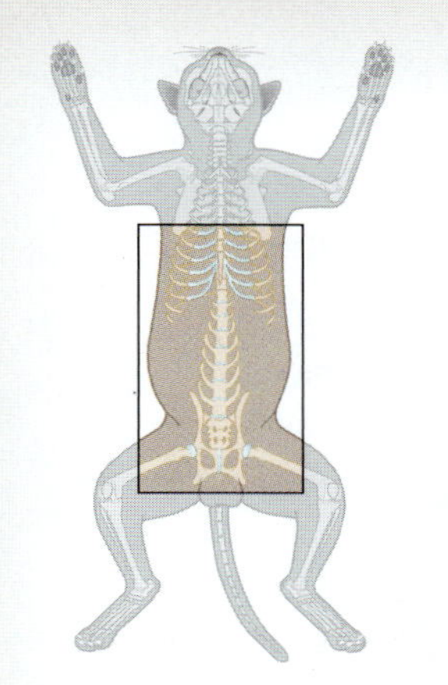

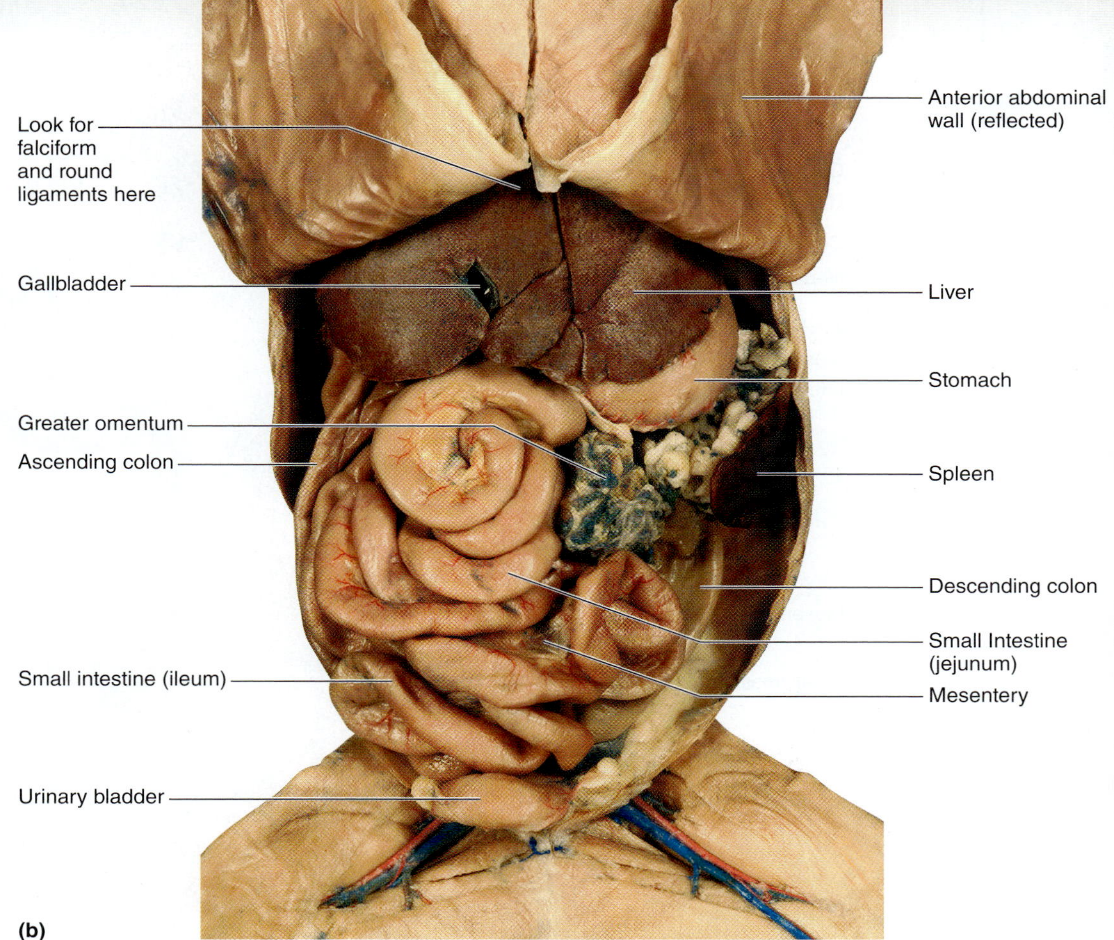

Figure 28.20 **The Spleen, Jejunum, Ileum, and Large Intestine (Colon)** ***(continued).*** (b) Dissection.

5. **The stomach, pancreas, spleen, and duodenum:** The next step requires lifting the liver cranially to view the structures that lie deep to it. If moving it is too difficult, use a scalpel to carefully remove the left lateral lobe of the liver. However, be sure to check with the lab instructor before doing this. Deep to the caudal part of the liver is the **stomach.** Deep to the stomach is the **pancreas,** and near the *head* of the pancreas is the first part of the small intestine: the **duodenum (figure 28.21).** Follow the *body* of the pancreas laterally to find the *tail* of the pancreas, which ends at the hilum of the **spleen.** Observe the area deep to the head of the pancreas and near the hilum of the liver to locate the following structures, using figure 28.21 as a guide:

☐ **common bile duct**	☐ **head of pancreas**
☐ **duodenum**	☐ **liver**
☐ **gallbladder**	☐ **stomach**

6. **The jejunum, ileum, and large intestine:** Lay the stomach and pancreas back into place and observe the spleen and the remainder of the intestinal tract. To fully appreciate the length of the small intestine, pick it up at the end of the duodenum and "walk" your fingers along the length of the small intestine until they reach the *ileocecal valve.* Note that it takes some time to reach that location! While doing this, make note of the **mesenteries,** which anchor the small intestine to the dorsal body wall (figure 28.20). It may be difficult to distinguish the **jejunum** (middle part of the small intestine) from the **ileum** (last part of the small intestine) in the cat, but attempt to use the criteria listed in figure 26.15 on p. 737 to distinguish the two from each other. A fun exercise is to have one dissection group remove the small intestine in its entirety from the abdominal cavity, dissect away its mesentery so it can be extended fully, and measure its length. If this is done, record the length of the entire small intestine here: ____________________ cm

(continued on next page)

(continued from previous page)

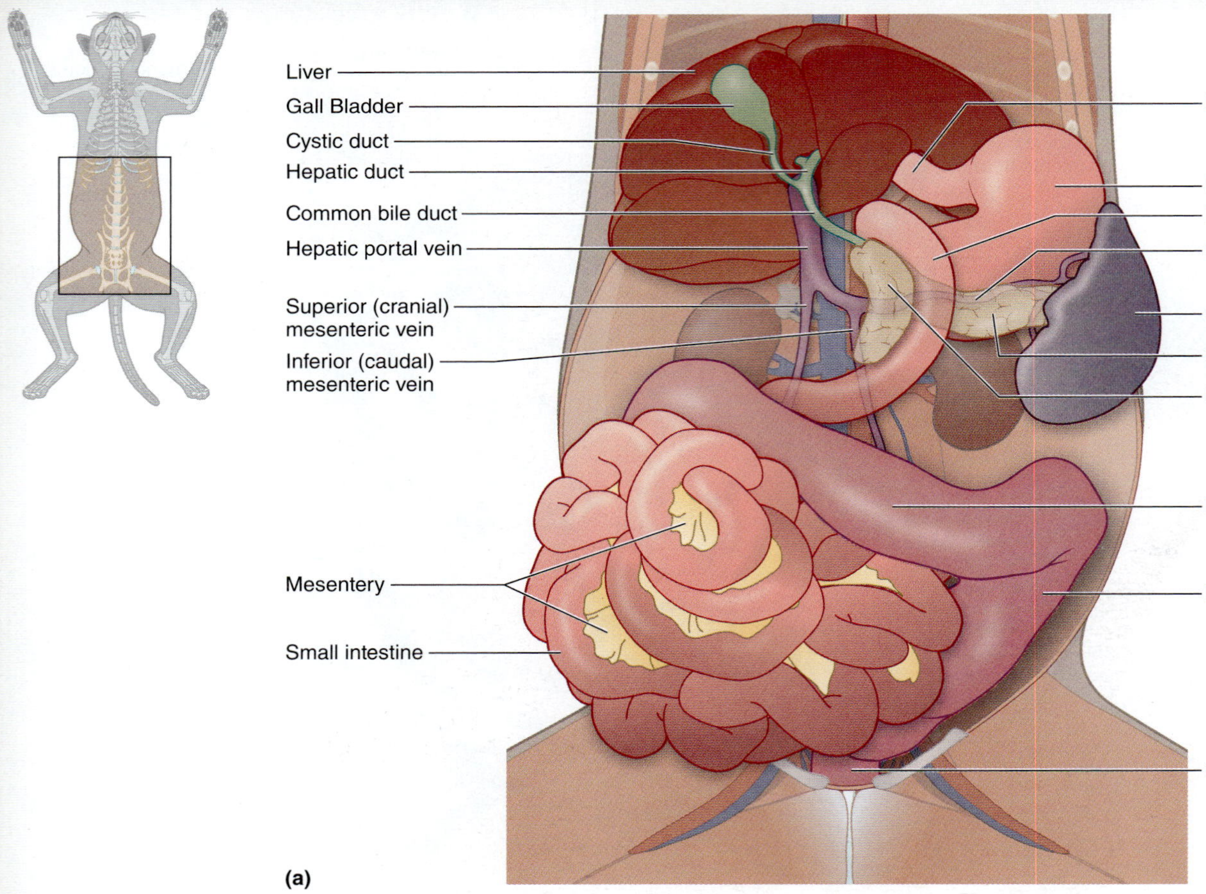

Liver
Gall Bladder
Cystic duct
Hepatic duct
Common bile duct
Hepatic portal vein
Superior (cranial) mesenteric vein
Inferior (caudal) mesenteric vein
Esophagus
Stomach
Duodenum
Gastrosplenic vein
Spleen
Body of pancreas
Head of pancreas
Transverse colon
Mesentery
Descending colon
Small intestine
Rectum

(a)

Figure 28.21 The Hepatic Portal System. The hepatic portal system consists of the gastrosplenic vein, inferior (caudal) mesenteric vein, superior (cranial) mesenteric vein, and the hepatic portal vein. (a) Illustration.

7. **The hepatic portal system:** Now that the organs of the gastrointestinal system within the abdominal cavity have been identified, focus on the blood supply to these organs. Before beginning, review the hepatic portal circulation of the human from chapter 22 on p. 601, figure 22.14. What are the three major veins that converge in the *human* to become the hepatic portal vein?

   a. ______________________________

   b. ______________________________

   c. ______________________________

   In the cat, there are two major veins that come together to form the hepatic portal vein, the **gastrosplenic vein** and the **anterior (cranial) mesenteric vein** (figure 28.21). The gastrosplenic vein receives blood largely from the spleen, stomach, and descending colon, thus making it homologous with both the splenic and inferior mesenteric veins in the human. The anterior mesenteric vein in the cat drains a region of the small and large intestines similar to the region drained by the superior mesenteric vein in the human. To locate these veins, use blunt dissection to gently tease away the mesentery of the small intestine. Both the arteries that supply arterial blood to the intestines (the **anterior (cranial) mesenteric artery** and the **posterior (caudal) mesenteric artery**), as well as the veins that drain blood from the intestine, extend between the two layers of peritoneum that compose the mesenteries. Identify the following blood vessels in the cat, using figure 28.21 as a guide:

   - ☐ **abdominal aorta**
   - ☐ **anterior (cranial) mesenteric artery**
   - ☐ **anterior (cranial) mesenteric vein**
   - ☐ **celiac trunk**
   - ☐ **gastrosplenic vein**
   - ☐ **hepatic portal vein**
   - ☐ **posterior (caudal) mesenteric artery**

8. After completing the dissection of the abdominal cavity, place the organs gently back into the abdominal cavity, close the abdominal wall, sprinkle the cat with wetting solution, and then wrap the cat in its skin and place it in the storage bag.

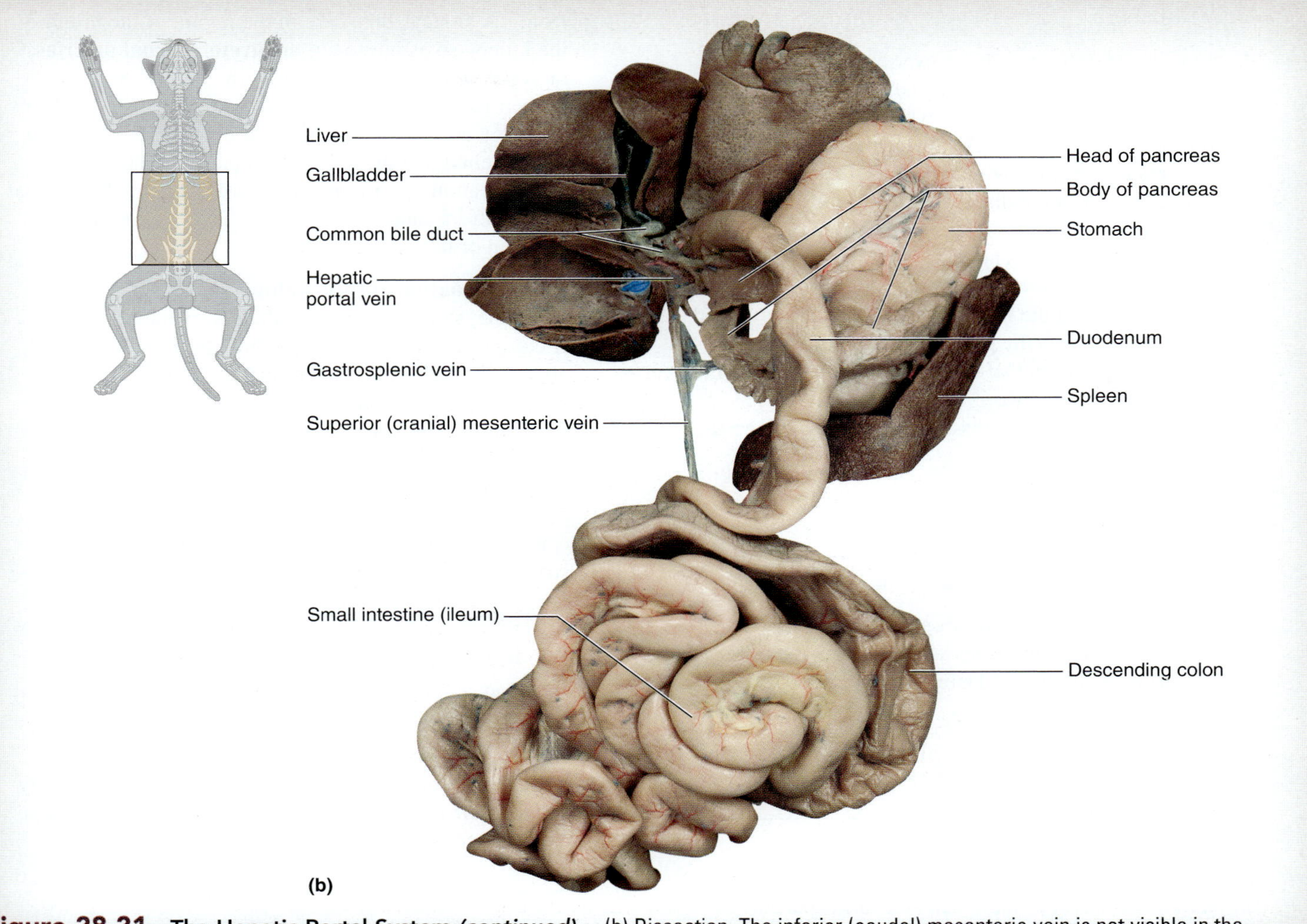

Figure 28.21 The Hepatic Portal System *(continued).* (b) Dissection. The inferior (caudal) mesenteric vein is not visible in the dissection photo.

EXERCISE 28.13

THE DORSAL ABDOMINAL WALL

This exercise begins with removal of the gastrointestinal organs from the abdominal cavity. This involves detaching them from the vascular and mesenteric connections that anchor them to the dorsal abdominal wall. They will then be removed as a unit so they can be replaced within the abdominal cavity when the dissection of the dorsal abdominal wall is completed. Structures of the dorsal abdominal wall, most of which belong to the urinary and endocrine systems, will be observed in addition to the main blood vessels that either arise from the abdominal aorta or drain into the posterior vena cava (postcava).

1. Begin with the cat supine and with the abdominal wall open. Removing the GI organs will first require separating them from their vasculature using the following steps:

a. Gently lift the pre-dissected (exercise 28.11) organs of the gastrointestinal system (liver, stomach, spleen, small and large intestines) away from the dorsal body wall. Observe the midline region next to the hilum of the liver to locate the blood vessels that enter the liver here.

b. Using scissors, detach the vessels where they enter the hilum of the liver while keeping a few millimeters of each attached to its vessel of origin for later identification.

c. Next, follow the abdominal aorta caudally until the origins of the **anterior** and **posterior mesenteric arteries** can be identified. Again, use scissors to detach these vessels from the aorta while keeping a few millimeters of each vessel attached to the aorta.

(continued on next page)

(continued from previous page)

d. Finally, carefully detach the liver from the **posterior vena cava** (postcava). This is the most difficult step, because the postcava is somewhat embedded in the dorsal wall of the liver. First place the scissors in the area between the cranial border of the liver and the caudal border of the diaphragm to cut the connection between the postcava and the right atrium of the heart (if this step was not already completed when the heart was removed). Next, gently lift the liver caudally and cut the **hepatic veins,** which drain blood from the liver into the postcava. This process may result in a postcava that ends up a little mangled. This is not something to worry about as it is somewhat unavoidable unless the dissector is very experienced.

e. With the vascular connections now severed, all of the GI organs can be removed from the abdominal cavity. While gently pulling them out, use blunt dissection and/or scissors (only when necessary) to detach any mesenteric connections that remain.

2. With the GI organs removed, observe the dorsal abdominal wall. The structures visible here are all **retroperitoneal** structures. Define retroperitoneal:

____

____

____

Notice that the **kidneys** are embedded in a large amount of fat. The majority of this fat is the **perirenal fat** (see figure 28.19, p. 825). This fat forms part of the protective layers of the kidney. Remove the fat to fully view the kidney. Before removing this fat, look carefully within the fat that lies cranial to each kidney. Embedded in the fat are the **adrenal glands** (chapter 19, figure 19.6, p. 510). To dig the adrenals out of the fat, the best technique to use is the open scissors technique (figure 1.15, p. 21) because it allows the dissector to loosen up the fat without damaging underlying structures—as long as the dissector uses a *gentle* hand. The adrenal glands have a similar consistency to that of the pancreas and a very thin capsule. Thus, they can be destroyed easily if one is not careful.

3. After isolating and cleaning the adrenal glands **(figure 28.22),** remove the remainder of the perirenal fat from the kidney. Take care in the region of the hilum of the kidney so as not to cut the **ureters, renal arteries,** or **renal veins.** While cleaning, trace these structures to their respective connections with the urinary bladder (ureters), aorta (renal arteries), and posterior vena cava (renal veins). On the left side, in particular, take care when cleaning the **renal vein.** Look for a small vein that comes off its caudal surface. This is the left **gonadal** (ovarian or testicular) **vein.**

4. Internal anatomy of the kidney: Choose one of the kidneys to bisect to view the internal anatomy. If completely removing one kidney from the dorsal abdominal wall, remove the right kidney. To do this, simply bisect the renal artery and vein midway between the hilum of the kidney and the aorta and posterior vena cava, respectively. Then bisect the ureter distal to the renal pelvis. Use a scalpel with a large blade or a dissecting knife to slice the kidney along the cranial/caudal plane (figure 28.22*b*). Identify the following structures, using figure 28.22 as a guide:

- ☐ **cortex**
- ☐ **fibrous capsule**
- ☐ **major calyx**
- ☐ **medulla**
- ☐ **minor calyx**
- ☐ **renal artery**
- ☐ **renal pelvis**
- ☐ **renal vein**

Are there observable differences between the cat kidney and the human kidney? _________ If so, describe them in the space provided:

____

____

____

5. While tracing the ureters caudally toward the urinary bladder, identify the **psoas major muscle,** which lies adjacent to the vertebral column. What is the function of this muscle?

____

____

____

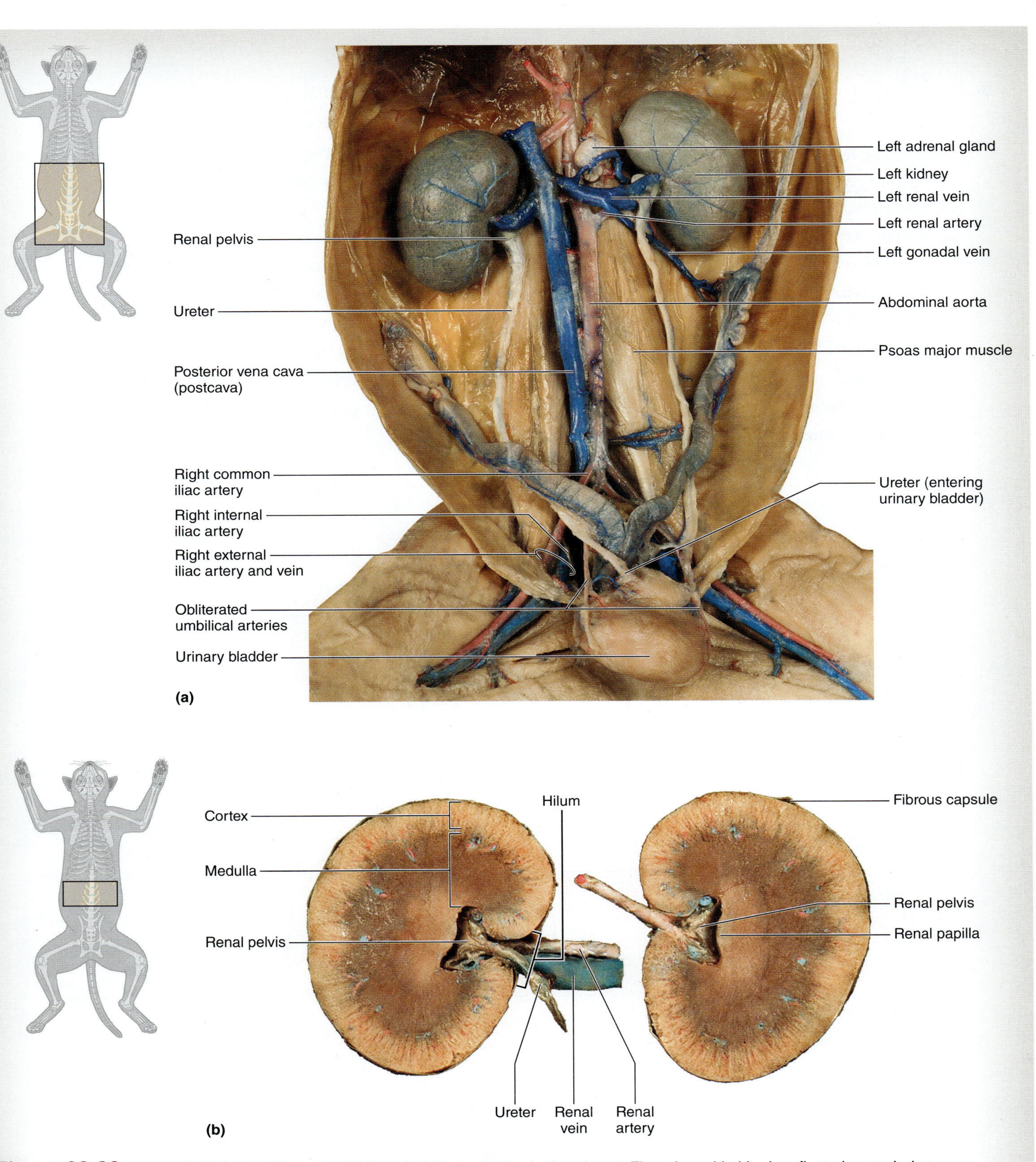

Figure 28.22 Dorsal Abdominal Wall. (a) Dorsal abdominal wall of a female cat. The urinary bladder is reflected posteriorly to demonstrate the course of the ureters superfical to the umbilical arteries. The ureters enter the dorsal, inferior surface of the urinary bladder. The umbilical arteries run alongside the urinary bladder and attach to the ventral abdominal wall internal to the umbilicus. (b) Bisected kidney.

(continued on next page)

(continued from previous page)

6. Identification of thoracic and abdominal vasculature: Before beginning, remove the heart and lungs from the thoracic cavity and the GI viscera from the abdominal cavity. Using **figure 28.23** as a guide, identify the following vessels in the abdominal and thoracic cavities:

- ☐ **abdominal aorta**
- ☐ **adrenal veins**
- ☐ **anterior vena cava (precava)**
- ☐ **aortic arch**
- ☐ **azygos vein**
- ☐ **brachiocephalic veins**
- ☐ **celiac trunk**
- ☐ **common iliac vein**
- ☐ **descending thoracic aorta**
- ☐ **external iliac artery**
- ☐ **external iliac vein**
- ☐ **hepatic veins**
- ☐ **iliolumbar artery**
- ☐ **iliolumbar vein**
- ☐ **internal iliac artery**
- ☐ **internal iliac vein**
- ☐ **left gonadal vein**
- ☐ **posterior vena cava (postcava)**
- ☐ **renal arteries**
- ☐ **renal veins**
- ☐ **right gonadal vein**

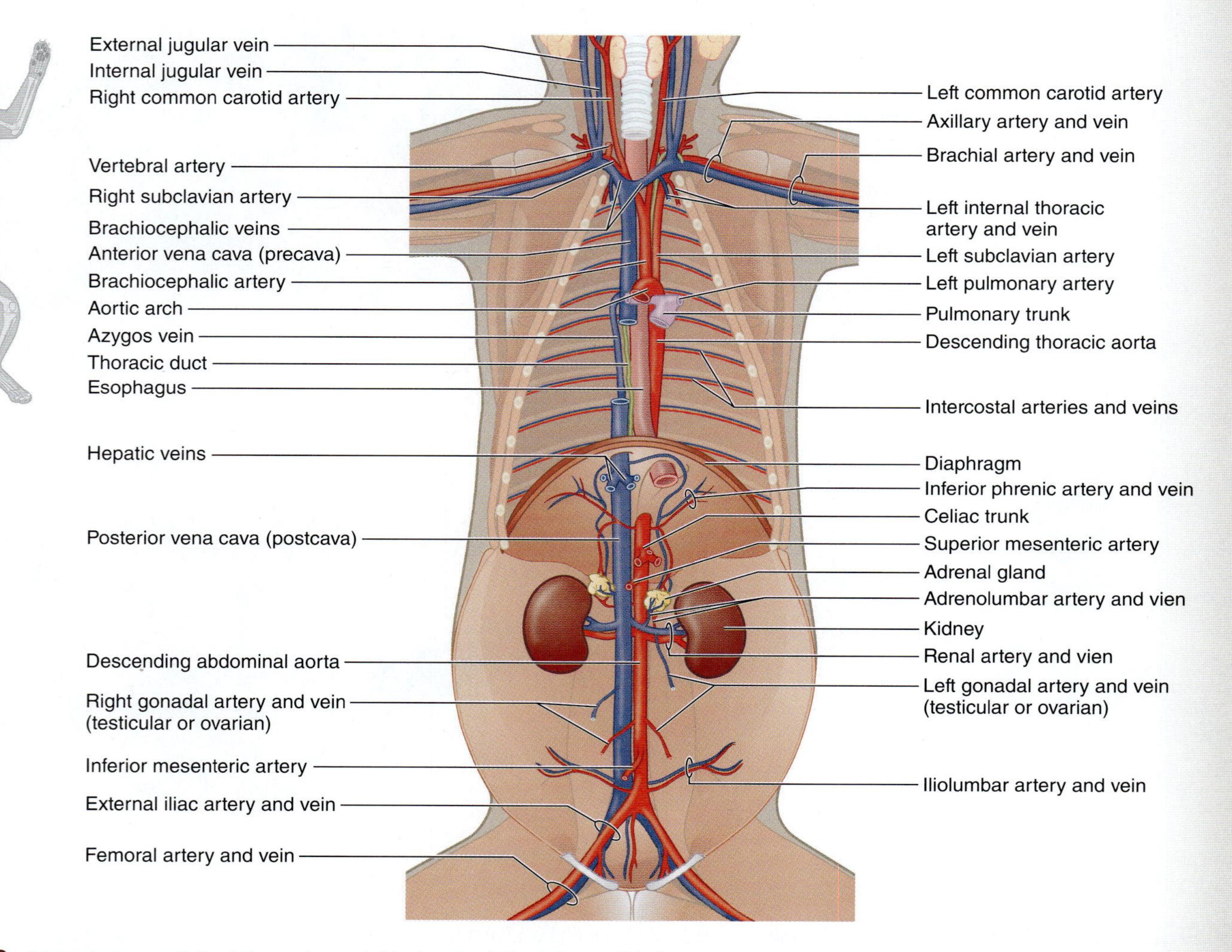

Figure 28.23 Vasculature of the Thoracic and Abdominal Cavities. This illustration demonstrates the arteries and veins of the thoracic and abdominal cavities (minus the hepatic portal system vessels).

EXERCISE 28.14

UROGENITAL SYSTEMS

EXERCISE 28.14A Male Urogenital System

1. Follow the instructions here to dissect a male cat. If dissecting a female cat, be sure to identify all of the structures listed in this section that are pertinent to the male cat, because knowledge of both male and female reproductive structures is required.

2. **External genitalia:** Lay the cat ventral side up on the dissecting tray and observe the pelvic and perineal regions. When the muscles of the ventral hindlimb and abdominal wall were dissected in a previous exercise, the spermatic cord and external genitalia were avoided. At this time attention will be focused on revealing these structures. Begin by observing the illustration and dissection photo in **figure 28.24,** which demonstrate the general location of male reproductive structures (figure 28.24*a*) and their appearance once the skin around them is removed. Next, refocus attention on the dissection specimen. Palpate the skin in the perineal region to feel for the penis and testes, then look for the **external urethral orifice** to definitively identify where the penis is located. The dissection in this area is somewhat difficult because the external genitalia are completely covered in thick, furry skin. Thus, before doing any cutting, be confident of what structures lie beneath the skin so they are not destroyed in the process of dissection. Also be sure to identify the spermatic cord extending from the inguinal region into the perineal region. The **spermatic cord** contains the testicular vessels, nerves, and the ductus deferens. Follow the path of the spermatic cord deep to the skin while dissecting caudally toward the testes.

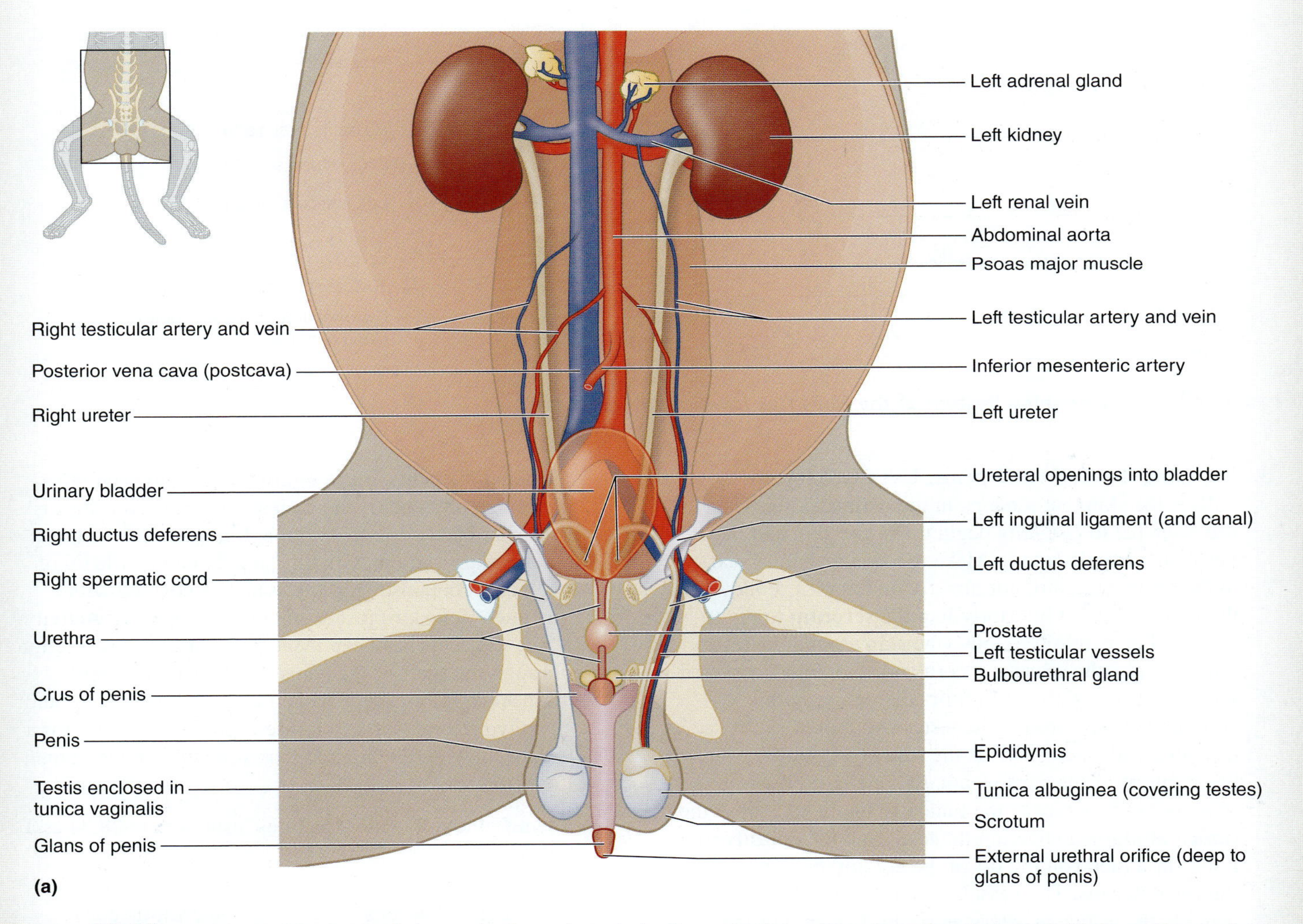

Figure 28.24 Urogenital System of the Male Cat. In both the illustration (a) and the dissection photograph (b), the pubic symphysis has been cut away to demonstrate the deep structures of the pelvis. Note in the dissection photo (b) that the hindlimb muscles have been removed for a better view of the male reproductive structures.

(continued on next page)

(continued from previous page)

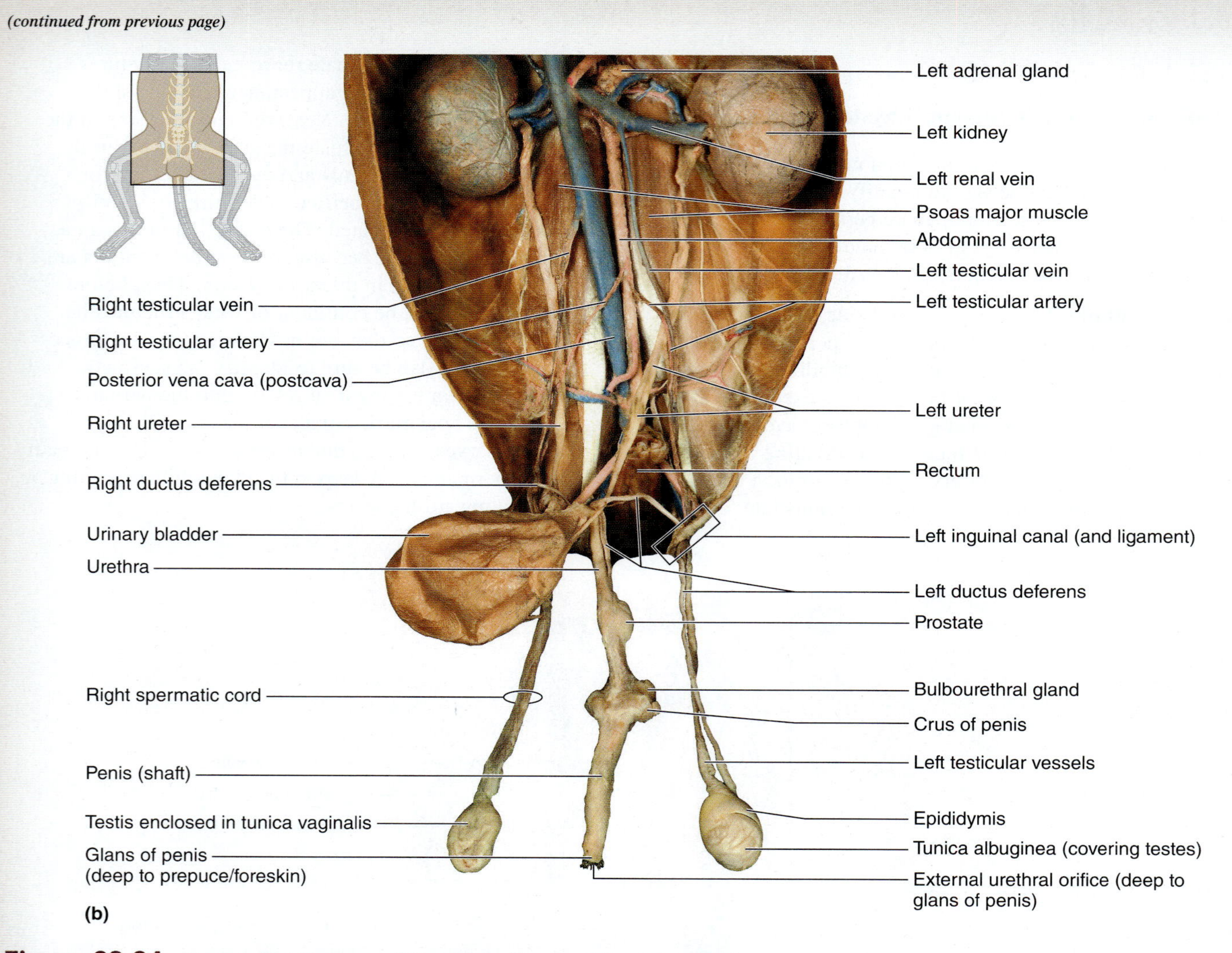

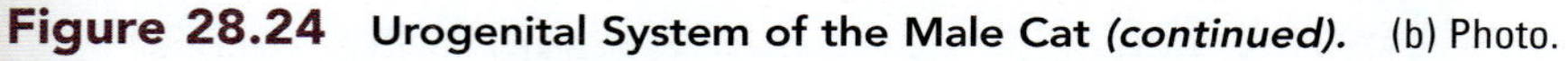

Figure 28.24 **Urogenital System of the Male Cat *(continued).*** (b) Photo.

3. **Exposing the testes and penis:** Use tissue forceps to pull on the skin that remains in the perineal area. Then use a scalpel to *carefully* begin to cut the skin and the fascia that lies deep to it. Make the skin incisions next to the spermatic cord, not directly on top of it. Follow the spermatic cord inferiorly into the **scrotum** to the **testes.** Then carefully remove the skin around the testes. Superficially, the spermatic cord and scrotum are covered with connective tissue. Carefully dissect away this connective tissue around the testes and look for the "sac" that surrounds them. This is the **tunica vaginalis,** which is a remnant of the peritoneum. On one side only, use scissors to dissect away the tunica vaginalis to expose the **tunica albuginea** covering the testis and the **epididymis.** Look for a tube coming off of the epididymis. This is the **ductus deferens.** The vessels visible within the spermatic cord are the testicular artery and vein. In step 5 of the dissection we trace these structures back into the pelvic cavity, which also facilitates identification of each of these structures. Medial to the testes and slightly cranial to them is the **penis.** To dissect the penis, carefully carve the skin away from the erectile tissues, because the skin and the penis are so tightly adhered to each other. Although a scalpel is the easiest tool to use for this task, it is sharp and easily cuts through important structures, so use extra caution with this step. Note that the tip of the penis contains the **glans,** which is covered with a **prepuce** (foreskin) just as in humans. Optional Activity: Once the penis has been identified, make a cross section through the shaft of the penis to identify the erectile tissues (corpora cavernosa and corpus spongiosum) and spongy urethra. Refer to figure 27.11 on page 768 for a figure showing the homologous structures in the human for reference.

4. Identify the following structures, using figure 28.24 as a guide:

- ☐ **ductus deferens**
- ☐ **epididymis**
- ☐ **glans of penis**
- ☐ **penis**
- ☐ **prepuce**
- ☐ **spermatic cord**
- ☐ **testes**
- ☐ **tunica albuginea**
- ☐ **tunica vaginalis**
- ☐ **urethra**

5. **Internal urogenital structures:** Lay the cat ventral side up on the dissecting tray. Before opening the abdominopelvic cavity, identify the **spermatic cord** once again. Carefully tease apart the structures that compose the spermatic cord and locate the **ductus deferens.** Although it can be difficult to distinguish the testicular blood vessels from the ductus deferens, palpating the structures will demonstrate that the texture of the ductus deferens is hard compared to that of the more elastic vessels. Once the ductus deferens is identified, trace it as it extends toward the ventral abdominal wall and try to locate where it enters the *superficial inguinal ring,* extends through the *inguinal canal,* and emerges within the abdominopelvic cavity. Continue to trace the ductus deferens within the abdominopelvic cavity as it continues superficial to the **ureters** and dorsal to the **urinary bladder,** and then enters the dorsal, posterior surface of the urinary bladder. Next, locate the testicular artery and testicular vein as they arise from their respective origins on the **abdominal aorta** (testicular arteries), **posterior vena cava/postcava** (right testicular vein), and **left renal vein** (left testicular vein). Trace these vessels along the dorsal abdominal wall, through the inguinal canal, and within the spermatic cord toward the testes. It may seem odd that these vessels are so long. However, the testes begin their development in the embryo adjacent to the kidneys and then subsequently descend along the dorsal abdominal wall, through the inguinal canal, and out into the scrotum. Thus, tracing the path of the testicular vessels is similar to tracing the path the testes took during their embryonic/fetal descent.

6. Next, refocus attention on the abdominopelvic cavity and lift abdominal structures such as the small intestine out of the way to obtain a better view of the pelvis and dorsal abdominal wall (or just remove them altogether if this was done with a previous dissection). Otherwise, pins or string may be used to hold the intestines out of the way while performing this part of the dissection. Locate the **kidneys.** Next, identify the hilum of the kidney and trace the ureters from there along the dorsal body wall to the dorsal aspect of the urinary bladder. To do this some of the **peritoneum** that covers the anterior surface of the urinary bladder and the ventral surface of the ureters may need to be removed (recall that urinary structures are all retroperitoneal). This is best accomplished by gently pulling on the peritoneum with tissue forceps. Once the peritoneum has been removed, clean the tissue around the urinary bladder and abdominopelvic wall. Using scissors, make a longitudinal incision in the ventral wall of the urinary bladder to view inside this organ. Within the urinary bladder, note the many folds of the walls, called **mucosal folds** (or *rugae*; similar to the folds within the stomach). Then, locate the openings of the two ureters, the opening of the urethra, and the triangular space delineated by the three openings: the **urinary trigone.** Does the wall of the urinary bladder look different in the region of the trigone?

____

If so, describe its appearance. ____________________

____

____

7. Locate the **urethra** as it emerges posterior to the urinary bladder (figure 28.24*b*). Next, look for a small, round gland surrounding the urethra just caudal to the urinary bladder: the **prostate gland.** Then trace the urethra caudally until it enters the corpus spongiosum of the penis. Adjacent to this location, look for the **crus of the penis.** The crus ("legs") are extensions of the corpora cavernosa of the penis into the perineal region. They are covered with muscle, but palpating them will demonstrate that they feel a little "spongy" to the touch because of the erectile tissues within. Adjacent to the crus of the penis, locate the two small, paired **bulbourethral glands.** These can be difficult to locate, so do not worry if they cannot be located.

INTEGRATE

CLINICAL VIEW
Urinary Blockage in Male Cats

Note the relative length of the urethra in the male cat as compared to the urethra of a human male. In particular, note the relative length of urethra extending between the bladder and the prostate gland in the cat, which is different from that in a human. One problem some male cats develop over time is a blockage that prevents urine from passing through the urethra when uric acid crystals form. (Uric acid crystals are similar to kidney stones that form in humans.) The crystals must travel through a very long and relatively narrow urethra, which predisposes them to clogging the urethra and blocking the flow of urine if the crystals become large and numerous. Treatment for this condition involves catheterization to clear the blockage; a change in diet to prevent the formation of uric acid crystals; and in very severe cases (particularly when reblockage occurs), removal of the penis to shorten the overall length of the urethra so as to prevent recurrence of the problem.

INTEGRATE

LEARNING STRATEGY

As you trace the ductus deferens within the pelvic cavity, you will see that it travels over the ureters en route to its entry into the posterior, inferior surface of the urinary bladder. When you are observing both the ureters and the ductus deferens, you can remember which structure is which by thinking of the following: "Water flows under the bridge." That is, the tubes containing the "water" (the ureters) travel inferior to the "bridge" (the ductus deferens).

(continued on next page)

(continued from previous page)

8. Identify the following structures, using figure 28.24 as a guide:

- ☐ bulbourethral glands
- ☐ crus of penis
- ☐ ductus deferens
- ☐ glans of penis
- ☐ prostate gland
- ☐ shaft of penis
- ☐ testicular arteries
- ☐ testicular veins
- ☐ ureters
- ☐ urethra
- ☐ urinary bladder

EXERCISE 28.14B Female Urogenital System

1. Follow the instructions here to dissect a female cat. If dissecting a male cat, be sure to identify all of the structures listed in this section that are pertinent to the female cat, because knowledge of both male and female reproductive systems is required.

2. **External genitalia:** Lay the cat ventral side up on the dissecting tray and observe the perineal region of the female **(figure 28.25*a*).** Just inferior to the tail is the *vulva,* which surrounds the opening to the urogenital sinus. The **urogenital sinus** is unique in the cat as compared to the human vulva, in that both the urethra and the vagina open into the urogenital sinus in the cat.

3. **Internal urogenital structures:** To reach the internal urogenital structures of the female cat, first dissect through the **pubic symphysis.** Observe figure 28.3 (cat skeleton) before beginning the dissection to have an idea of how the pubic symphysis of the cat relates to the muscles of the perineal region. Obtain a pair of bone cutters, tissue forceps, and a scalpel. Using the tissue forceps and scalpel, cut through the muscles of the medial thigh ~0.5 cm to either side of the midline (where the pubic symphysis lies) to remove the muscular attachments adjacent to the pubic symphysis. Next, taking care not to go too deep and cut the urogenital sinus, use the bone cutters to cut through the pubic symphysis. Parts of the pubic bone and pubic symphysis will likely end up being removed in several small pieces. The main goal is to open up a wide enough "channel" through the mid-pubic region to view the urogenital sinus.

4. Once the urogenital sinus has been exposed, use scissors to carefully cut open the last centimeter of the urogenital

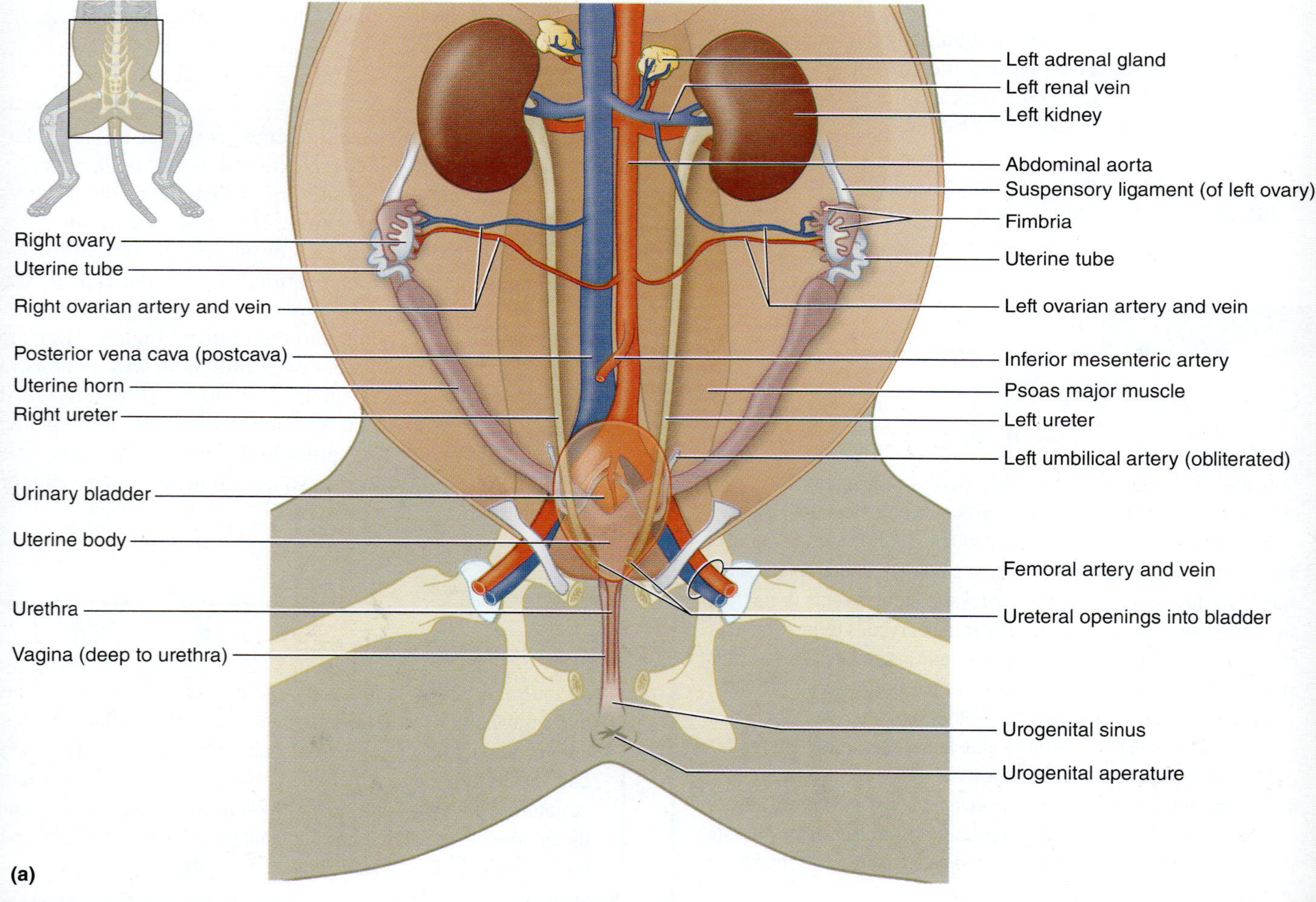

Figure 28.25 **Urogenital System of the Female Cat.** (a) Illustration.

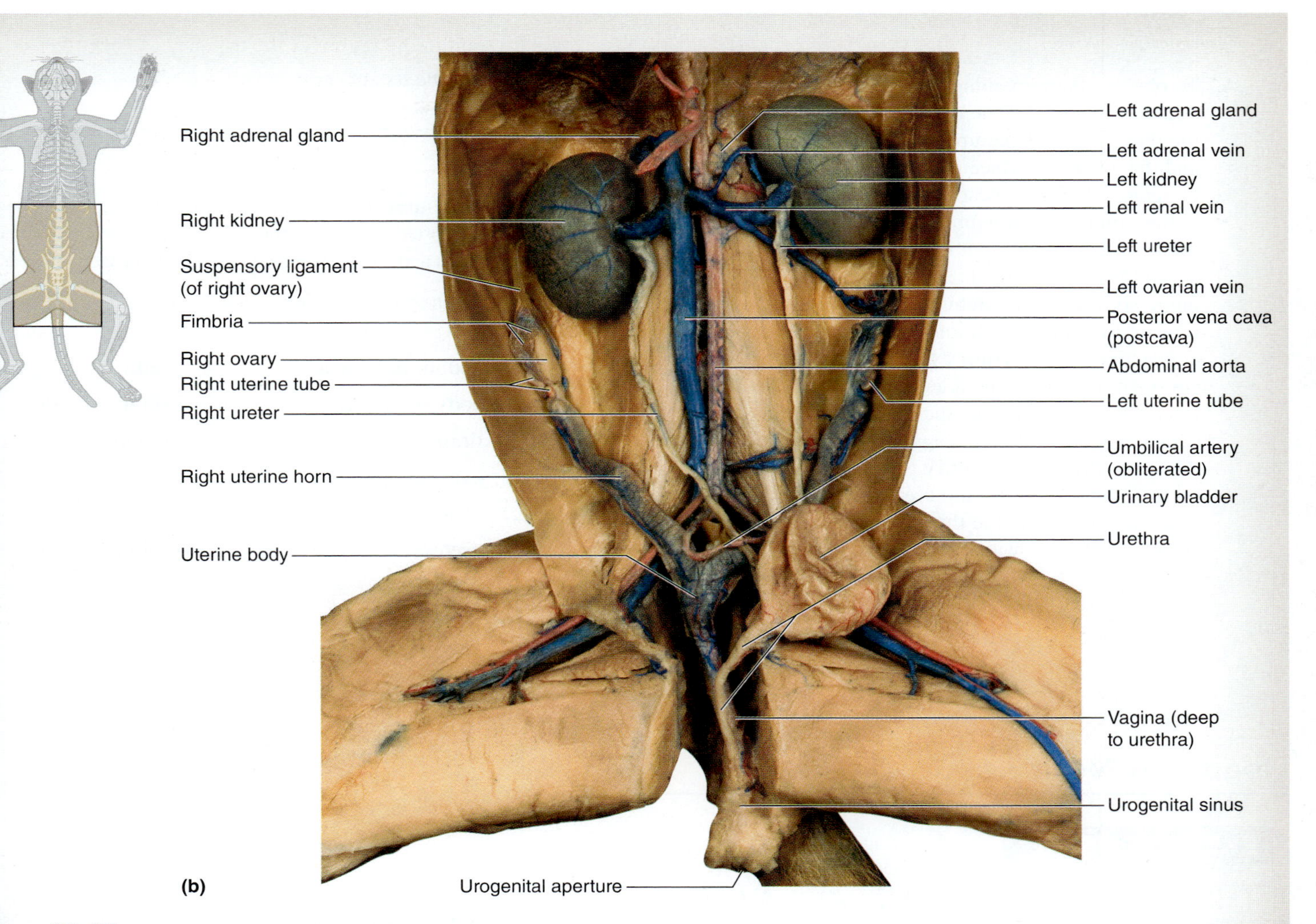

Figure 28.25 Urogenital System of the Female Cat *(continued).* (b) Photo. In this dissection, the pubic symphysis has been cut away to demonstrate the deep structures of the pelvis. The urogenital sinus has not been cut open in this specimen. Thus, the urethral orifice is not visible.

sinus. Start at the **urogenital aperture** (figure 28.25*a*) and cut toward the **vagina.** Once opened, note the following structures:

- ☐ **urethra**
- ☐ **urethral orifice**
- ☐ **urogenital aperture**
- ☐ **urogenital sinus**
- ☐ **vagina**

5. Next, refocus attention on the abdominopelvic cavity and lift abdominal structures such as the small intestine out of the way to obtain a better view of the pelvis and dorsal abdominal wall (or just remove them altogether if this was done in a previous dissection). Otherwise, pins or string may be used to hold the intestines out of the way while performing this part of the dissection. Locate the **kidneys.** Next, beginning at the hilum of the kidney trace the ureters from the renal pelvis and along the posterior body wall until they pass deep to the **uterine horns.** Unlike the human uterus, which has a single **fundus** superior to the **body,** the cat uterus has two long uterine horns, which converge at the midline to form the body of the uterus. These horns are where fertilized ova implant and where kittens develop during a pregnancy. The presence of these horns is what allows a female cat to produce a litter of several kittens, whereas the human uterus is structured to most effectively support only a single developing fetus. Note the fold of peritoneum that extends from the ovary to the dorsolateral abdominal wall. This is the **suspensory ligament** (of the ovary), which contains the ovarian vessels. The **broad ligament** is a fold of the peritoneum that helps support the uterus, uterine tubes, and ovaries within the abdominopelvic cavity. Trace the uterine horns laterally until you see them narrow where they converge with the **uterine tubes.** Continue to follow the uterine tubes until the openings into the tubes are visible. These are surrounded by small, delicate, finger-like projections called **fimbria(e).** Medial to the fimbriae is a small, hard, oval structure, the **ovary.** Locate the *ovarian ligament,* which attaches the ovary to the uterine horns.

(continued on next page)

(continued from previous page)

6. To observe the remaining urogenital structures of the female cat, some of the peritoneum that covers both the reproductive structures (such as the broad ligament) and the anterior surface of the urinary bladder must be removed. This is best accomplished by gently pulling on the peritoneum with tissue forceps. Once the peritoneum has been removed, clean the tissue around the urinary bladder and pelvic wall. While doing so, follow the ureters once again as they emerge inferior to the uterine horns and continue toward their connection to the **urinary bladder.** Use scissors to cut open the urinary bladder to look inside. Within the urinary bladder, note the many folds of the walls, called **mucosal folds** (or *rugae*; similar to those within the stomach). Then, identify the openings of the two ureters, the opening of the urethra, and the triangular space delineated by the three openings: the **urinary trigone.** Does the wall of the urinary bladder look different in the region of the trigone? ___________ If so, describe its appearance.

____

____

7. Locate the **urethra** as it emerges posterior to the urinary bladder. Posterior to the urethra, identify the vagina. Note how the urethra and vagina come together in the urogenital sinus.

8. Identify the following structures, using figure 28.25 as a guide:

- ☐ **abdominal aorta**
- ☐ **body of uterus**
- ☐ **fimbriae**
- ☐ **ovarian ligament**
- ☐ **ovary**
- ☐ **ureters**
- ☐ **urethra**
- ☐ **urinary bladder**
- ☐ **urogenital aperture**
- ☐ **urogenital sinus**
- ☐ **uterine horns**
- ☐ **vagina**

Head and Neck

EXERCISE 28.15

THE HEAD, NECK, AND ORAL CAVITY

1. **Superficial dissection:** Place the cat ventral side up on the dissecting tray. Identify the muscles that were first identified in exercise 28.4 (head and neck muscles) as well as the **lymph nodes** of the neck and the **external jugular veins.** Rotate the cat to obtain a clear view of the lateral aspect of the head. Identify the following structures, using **figures 28.26** and **28.27** as guides:

- ☐ **canine teeth**
- ☐ **external jugular veins**
- ☐ **incisor teeth**
- ☐ **lymph nodes**
- ☐ **masseter muscle**
- ☐ **molar teeth**
- ☐ **parotid duct**
- ☐ **parotid gland**
- ☐ **premolar teeth**
- ☐ **submandibular gland**
- ☐ **tongue**

2. **Deep dissection:** To expose the deeper respiratory and digestive structures of the neck, first remove the mandible. Check with the lab instructor before proceeding in case the lab instructor prefers some groups to keep this region undissected. Obtain tissue forceps, a scalpel, and bone cutters. Use the scalpel to cut through the masseter muscle just inferior to the zygomatic arch. It is easier to cut through the bone if the soft tissues attached to it (i.e., the masseter muscle in this case) are removed first. Obtain the bone cutters and place them in the corner of the mouth (the *angle* of the mouth) so they are positioned in the groove that was cut through the masseter muscle (figure 28.27*b*). Cut the mandible and any remaining associated tissues to free the **temperomandibular joint.** Do the same on the other side of the mouth. Now that the mandible is free, open the mouth as wide as it will go. Identify the following structures, using figure 28.27 as a guide:

- ☐ **auditory canal**
- ☐ **hard palate**
- ☐ **nasopharynx**
- ☐ **oropharynx**
- ☐ **rugae**
- ☐ **soft palate**
- ☐ **tongue**

3. Using a scalpel, make an incision in the groove around the tongue where it connects to the mandible, to free the mandible from any remaining soft tissues. Remove the mandible. Dissect away the connective tissue within the oropharynx and laryngopharynx to clean and identify the following structures. Use figures 28.17 and 28.27 as guides.

- ☐ **cricoid cartilage**
- ☐ **epiglottis**
- ☐ **esophagus**
- ☐ **internal carotid artery**
- ☐ **internal jugular vein**
- ☐ **internal nares**
- ☐ **larynx**
- ☐ **thyroid cartilage**
- ☐ **thyroid gland**
- ☐ **trachea**
- ☐ **vagus nerve**

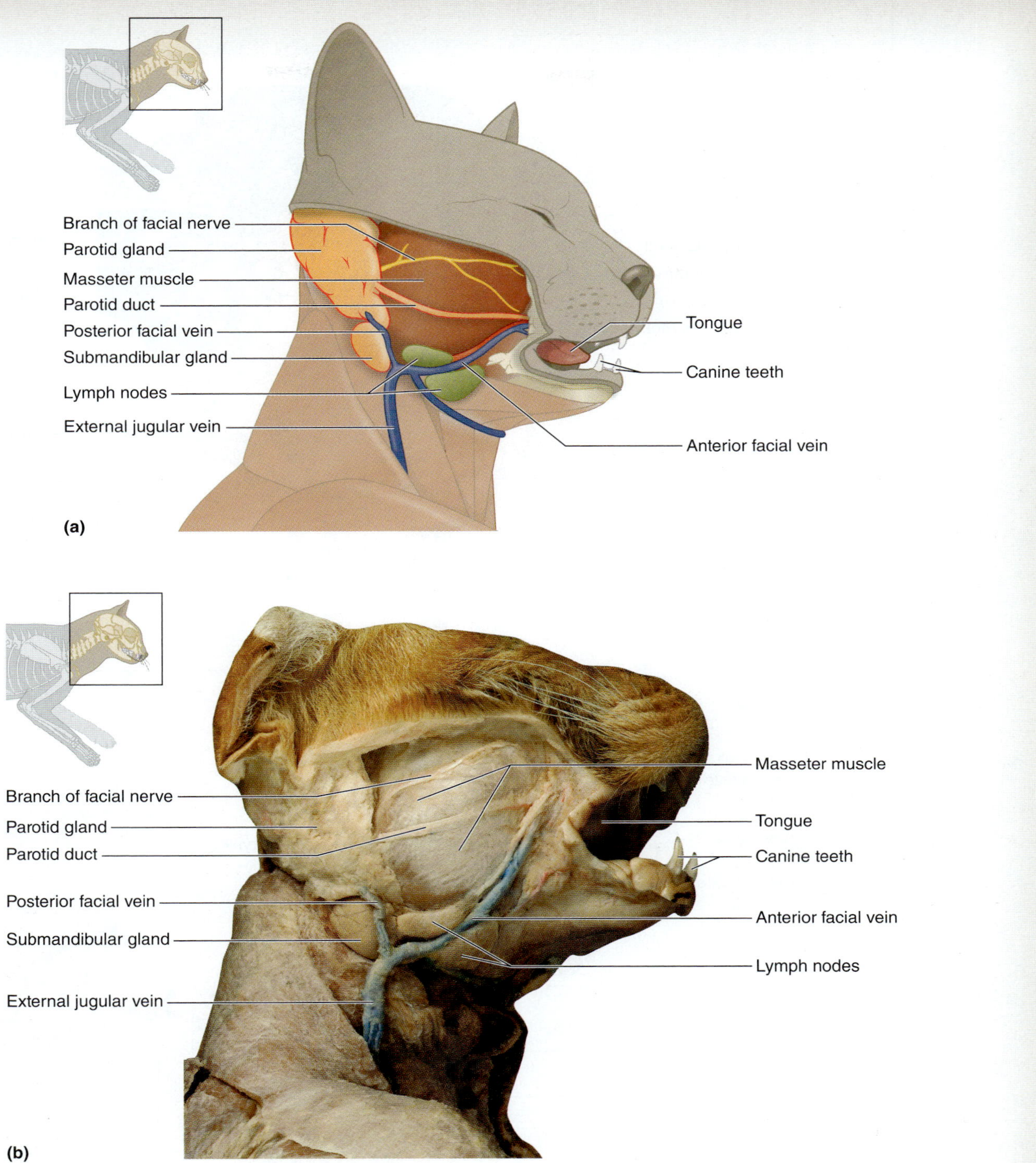

Figure 28.26 Head, Neck, and Oral Cavity. (a) Illustration showing structures of the head and neck, including the teeth. (b) Photo showing most of the structures in (a). All teeth are not visible.

(continued on next page)

(continued from previous page)

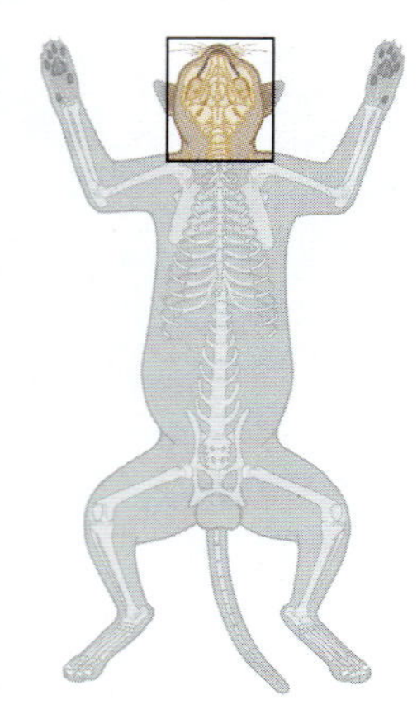

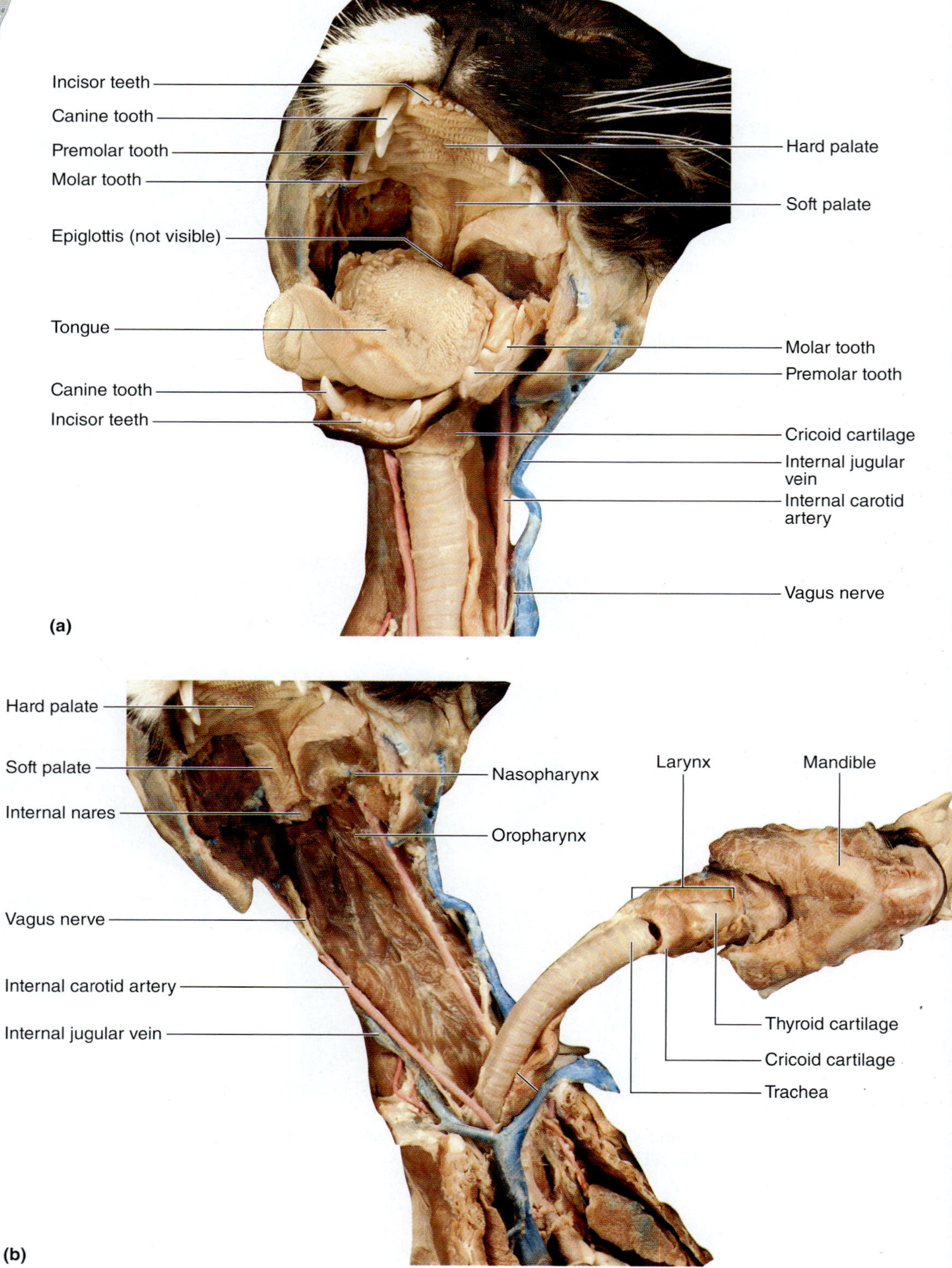

Figure 28.27 Deep Dissection of the Oral Cavity. Photos showing deep structures of the oral cavity of the cat. The thyroid gland has been removed. (a) Lower jaw in normal anatomic position. (b) Lower jaw, larynx, trachea, and esophagus displaced to the side for better view of deep structures.

Appendicular Neurovasculature

EXERCISE 28.16

NERVOUS SYSTEM: PERIPHERAL NERVES

1. **Brachial plexus:** Place the cat ventral side up on the dissecting tray. Briefly review the muscles of the limbs. While doing so, note the whitish, cord-like structures within the limbs. These are the **peripheral nerves.** This dissection focuses on the axillary region and the major components of the brachial plexus. Start by reflecting the pectoral muscles laterally (they may have already been removed, depending on the order of dissection exercises). Note the mass of connective tissue and "stringy stuff" in the axillary region. To distinguish nerves from blood vessels, use the open scissors blunt dissection technique to gently loosen the connective tissue in this region. Using this technique, connective tissue is pulled apart, leaving blood vessels and nerves intact. However, be gentle with the technique so as not to tear blood vessels and nerves.

2. Try to locate the **subclavian artery** as it emerges under the clavicle and enters the axilla as the **axillary artery.** Recall that the cords of the brachial plexus are named for their location relative to the axillary artery. After identifying the axillary artery, gently trace the nerves that surround it both toward and away from their origin in the spinal cord. Identify the following components of the brachial plexus, using **figure 28.28** as a guide:

   - ☐ **axillary artery (reference)**
   - ☐ **axillary nerve**
   - ☐ **lateral cord**
   - ☐ **medial cord**
   - ☐ **median nerve**
   - ☐ **musculocutaneous nerve**
   - ☐ **posterior cord**
   - ☐ **radial nerve**
   - ☐ **ulnar nerve**

3. **Lumbosacral plexus:** This dissection focuses on the pelvic and inguinal region, which is the location of the lumbosacral plexus. Start by observing the internal anatomy of the lower posterior abdominal wall. To view the nerves of the lumbosacral plexus, first remove the urinary bladder, the ureters, the kidney, and any reproductive structures that lie ventral to the posterior abdominal wall. Identify the **psoas major muscle,** a major landmark in the pelvis **(figure 28.29).** Trace the psoas major posteriorly toward its insertion on the femur. As it crosses the inguinal region, two nerves emerge adjacent to the muscle: the **femoral nerve** and the **obturator nerve.** Reflect one of the psoas major muscles superiorly to view the lumbar plexus from which these two nerves arise.

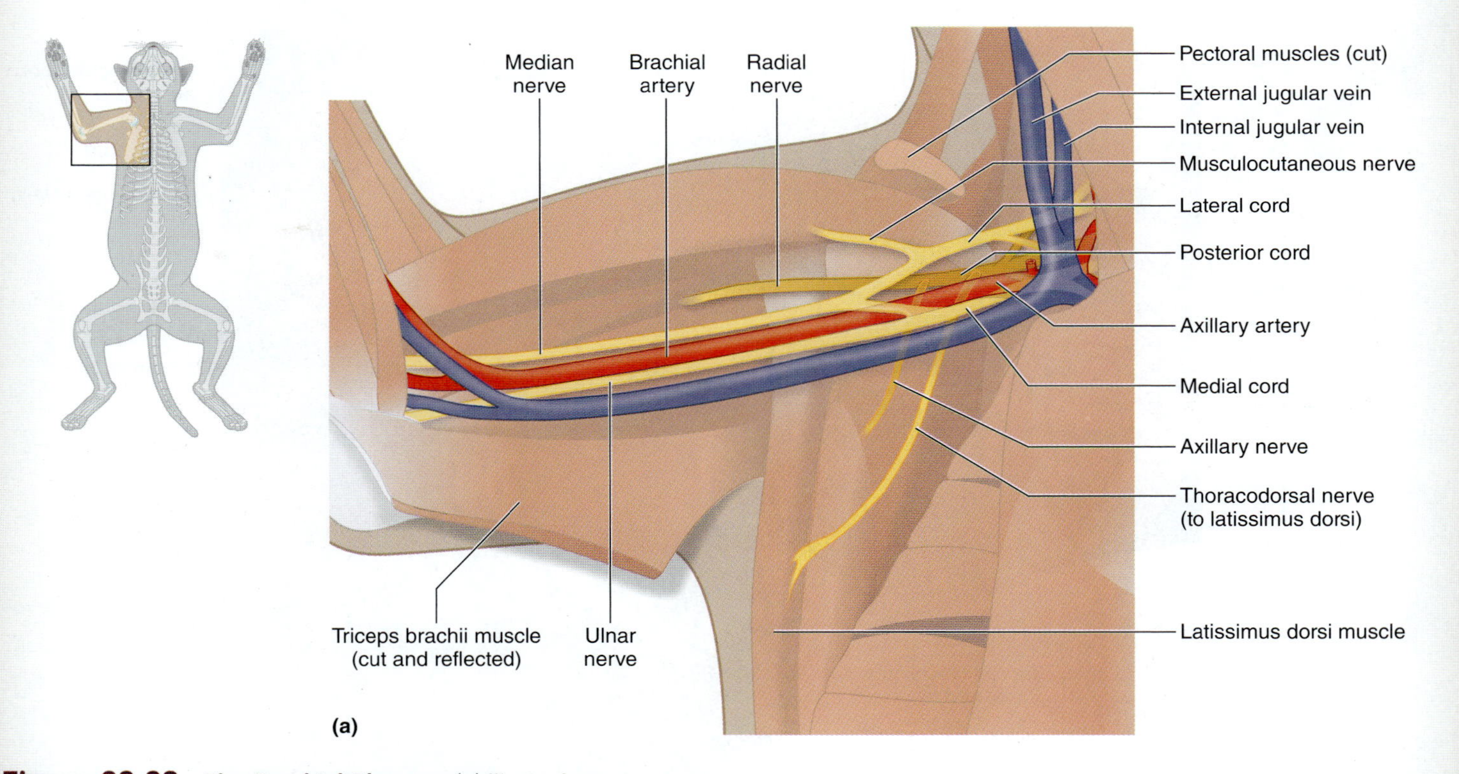

Figure 28.28 The Brachial Plexus. (a) Illustration.

(continued on next page)

(continued from previous page)

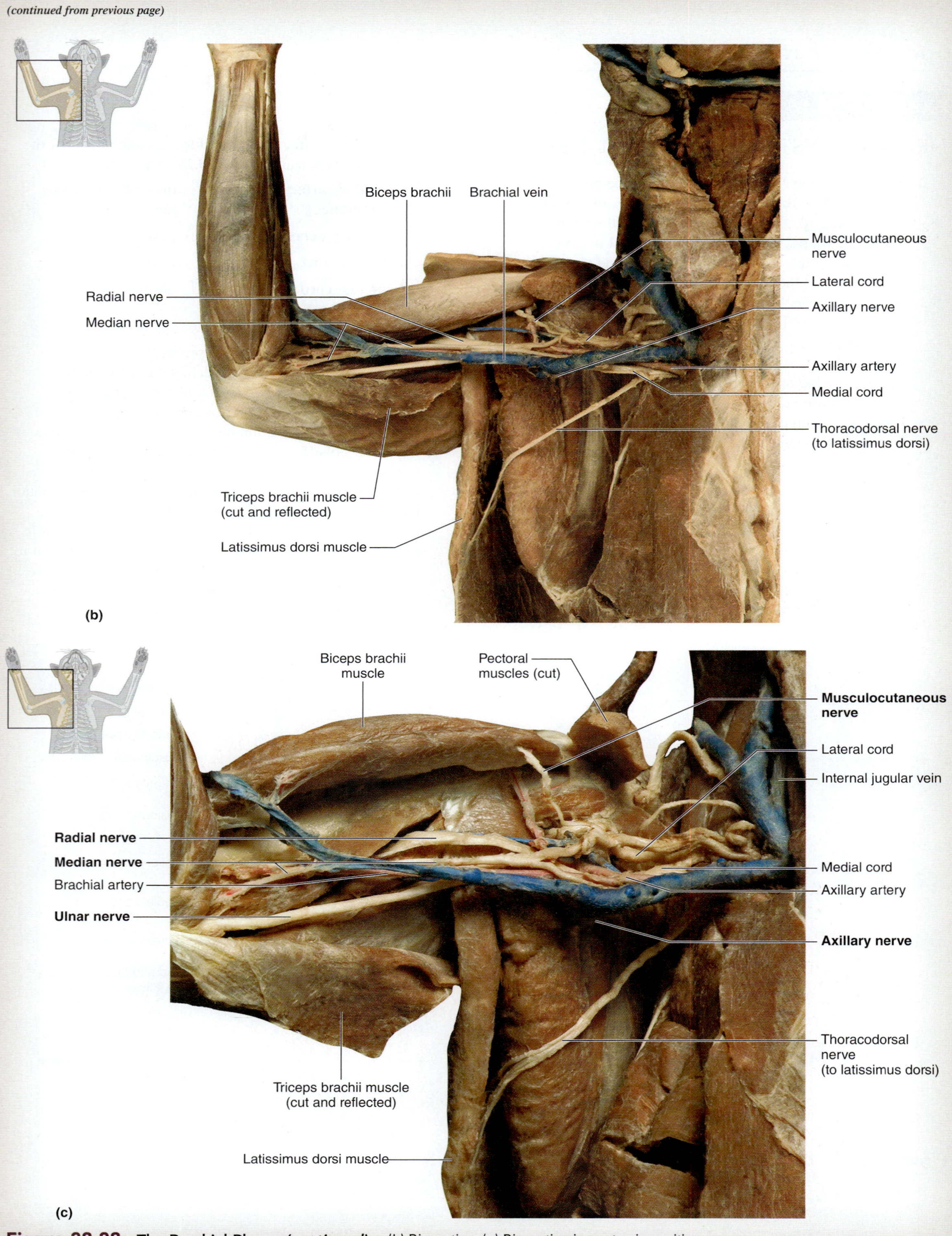

Figure 28.28 The Brachial Plexus *(continued)*. (b) Dissection. (c) Dissection in anatomic position.

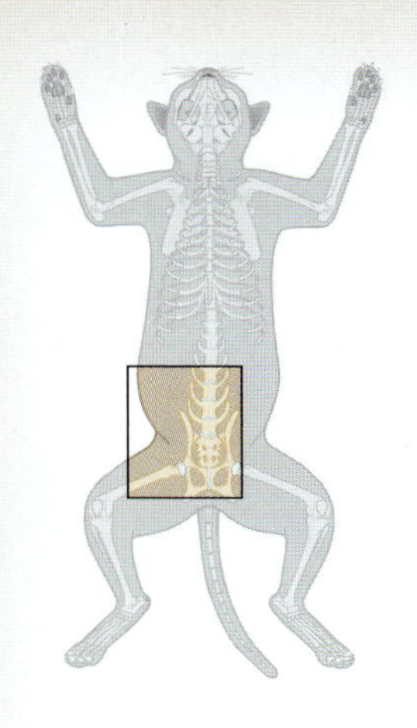

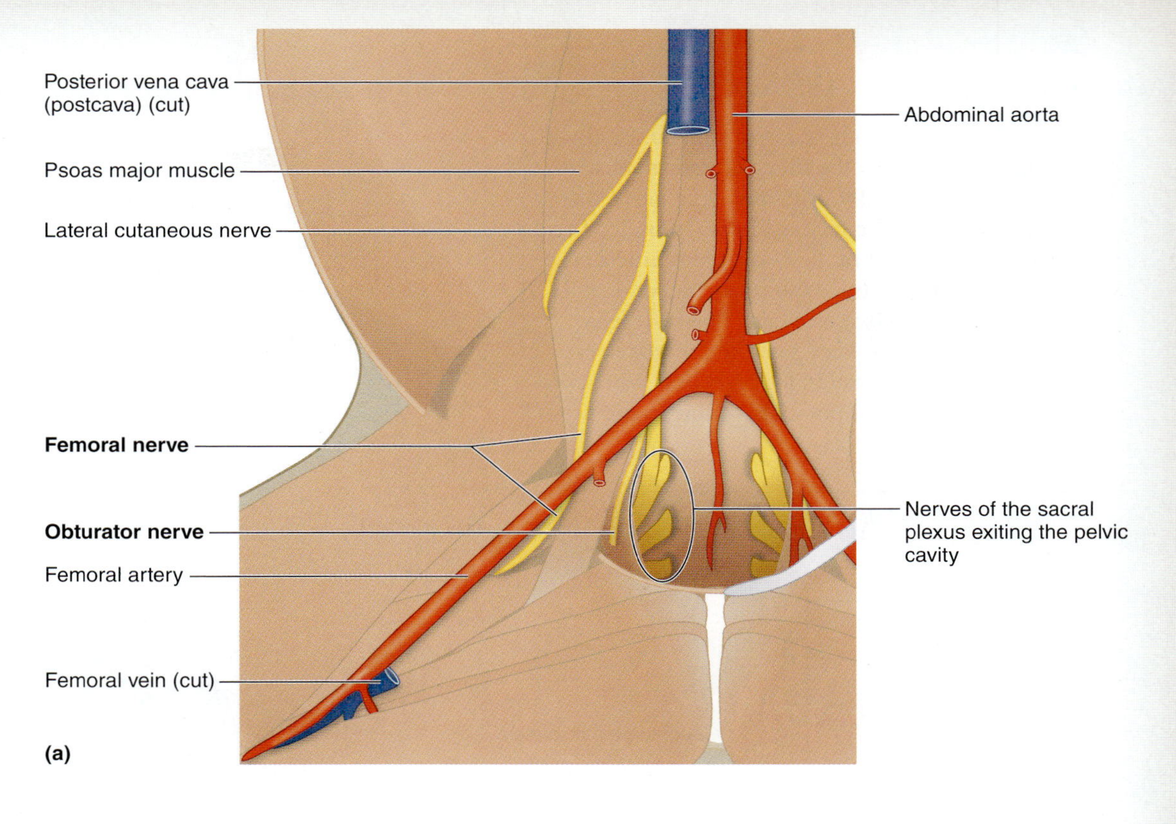

(a)

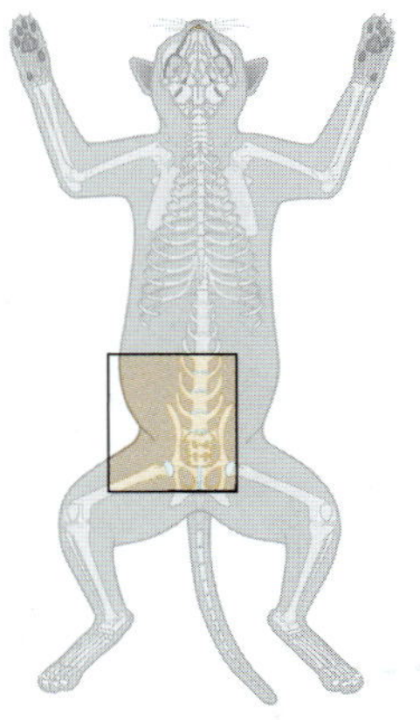

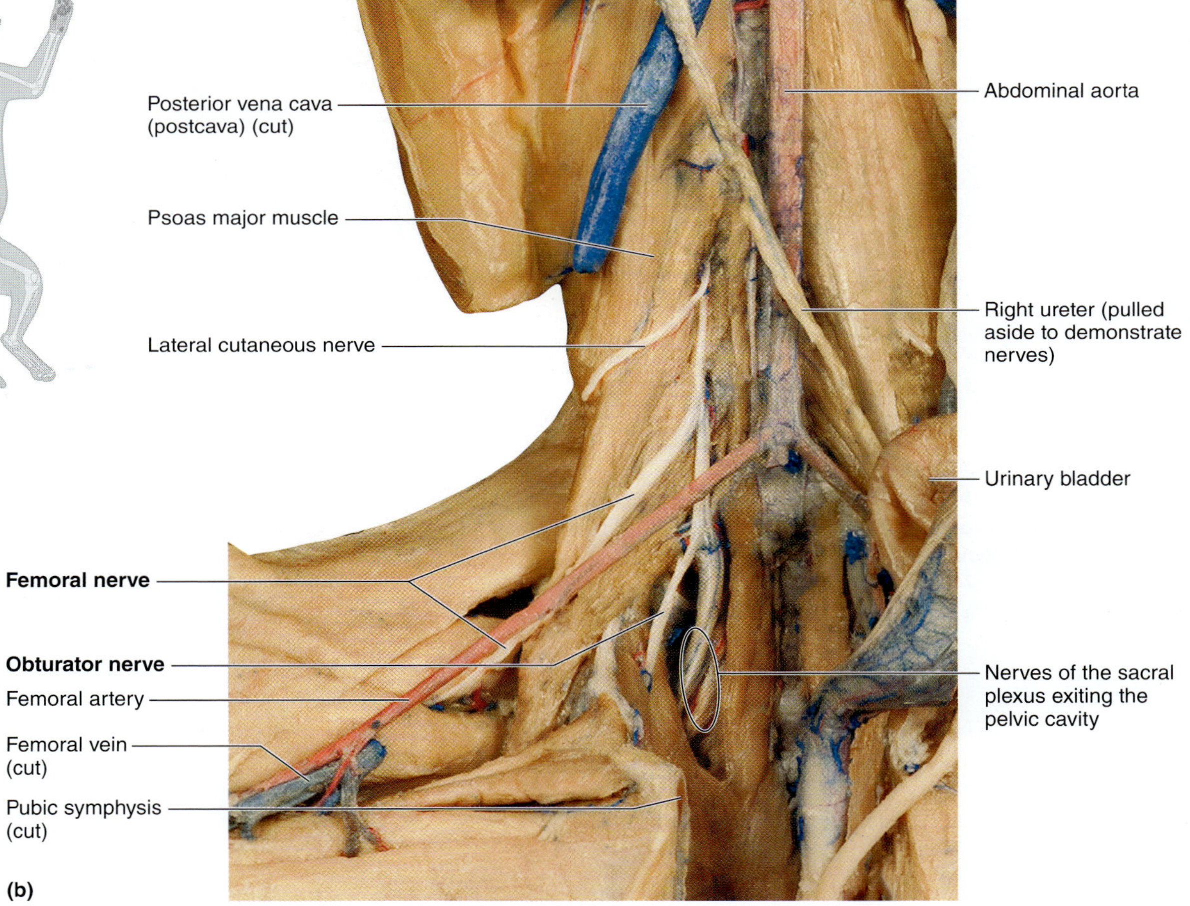

(b)

Figure 28.29 Lumbosacral Plexus: Lumbar Plexus. From an anterior view, the nerves of the lumbar portion of the lumbosacral plexus are best observed. (a) Illustration. (b) Dissection. Note the nerves of the sacral plexus that emerge within the pelvic cavity and run posterior to eventually exit within the gluteal region (figure 28.30).

(continued on next page)

(continued from previous page)

4. Turn the cat over so it is dorsal side up. Using a scalpel, cut through the **biceps femoris** muscle and reflect it laterally. Note the large nerve that emerges deep to the biceps femoris muscle. This is the largest nerve in the body, the **sciatic nerve (figure 28.30).** To trace the sciatic nerve distally the **biceps femoris** muscle must be bisecteds. Just superior to the knee joint, the sciatic nerve bifurcates into its two major branches. What two nerves are the major branches of the sciatic nerve?

_______________________________ and

____

5. Identify the following nerves of the lumbosacral plexus, using figures 28.29 and 28.30 as guides:

- ☐ **common fibular nerve**
- ☐ **femoral nerve**
- ☐ **obturator nerve**
- ☐ **sciatic nerve**
- ☐ **tibial nerve**

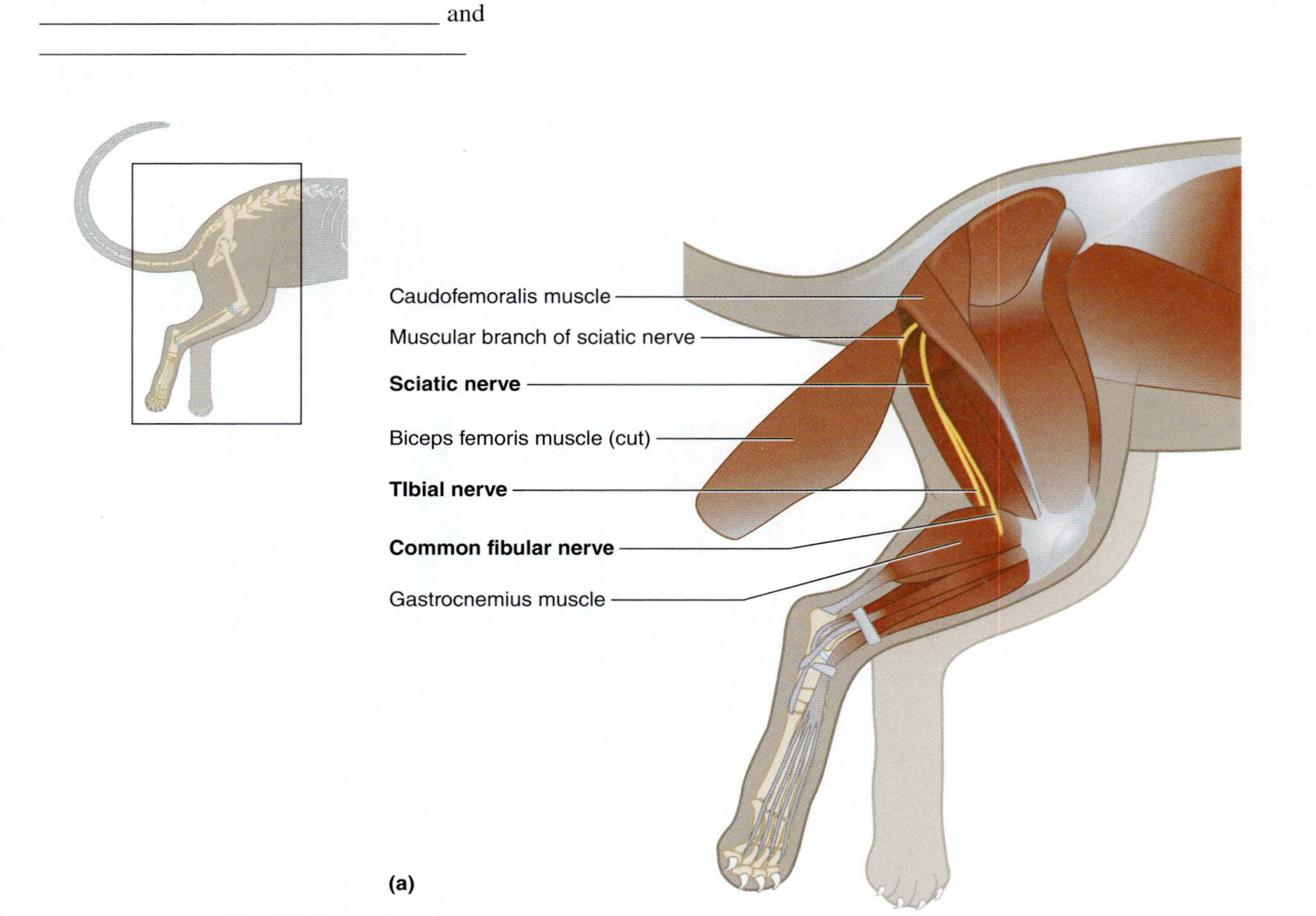

Figure 28.30 **Lumbosacral Plexus: Sacral Plexus.** From a posterior view, the nerves of the sacral portion of the lumbosacral plexus are best observed. (a) Illustration.

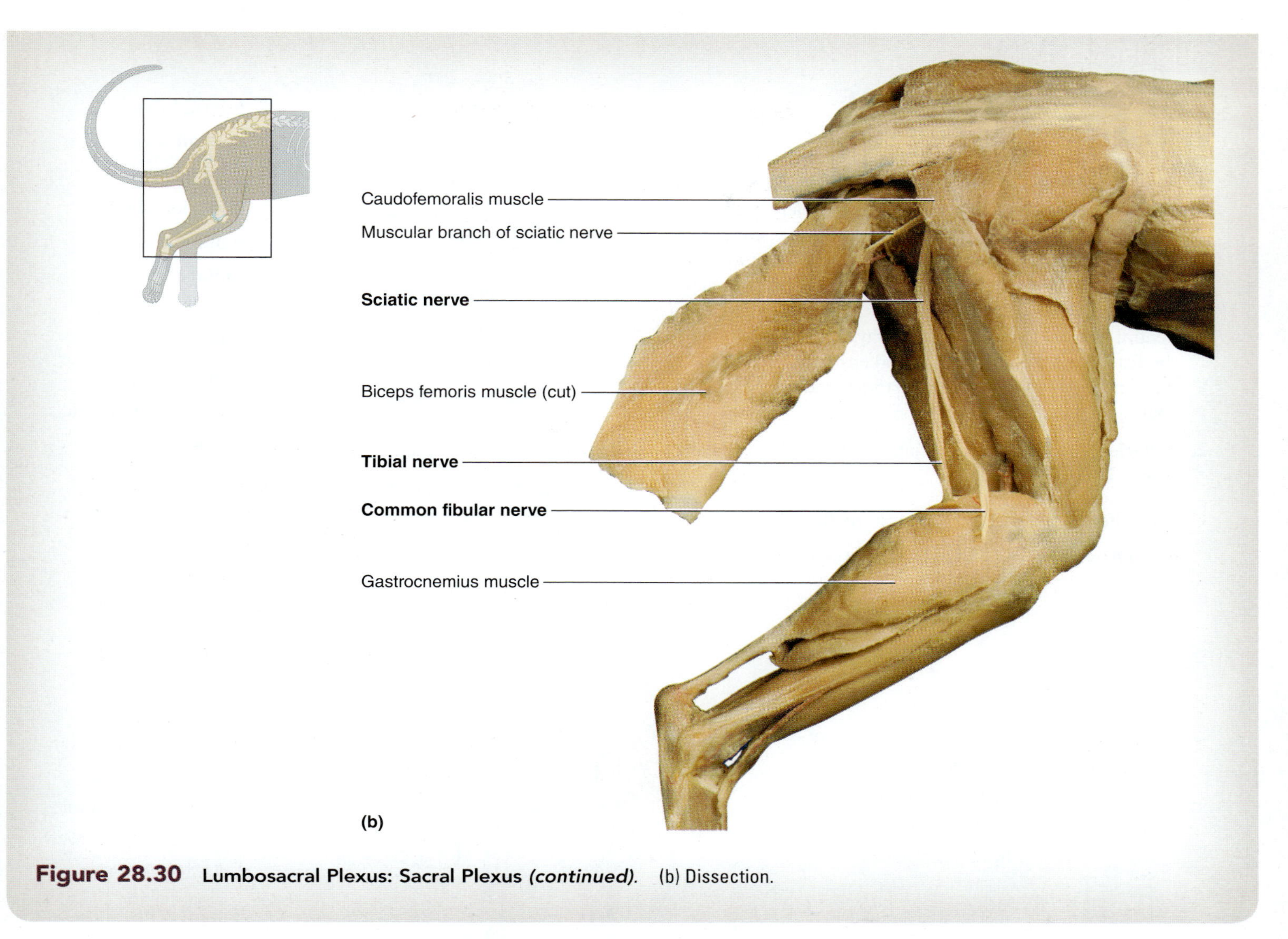

Figure 28.30 **Lumbosacral Plexus: Sacral Plexus *(continued).*** (b) Dissection.

EXERCISE 28.17

CARDIOVASCULAR SYSTEM: PERIPHERAL BLOOD VESSELS

1. This exercise reviews all of the major blood vessels of the cat. Many of these vessels have already been identified in the context of the structures they supply. Completion of this exercise will also ensure that all of the major blood vessels of the cat have been identified, and that the circulatory pathways are delineated. Place the cat ventral side up in the dissecting tray. While identifying the listed vessels, scissors and forceps may be used to clean any remaining connective tissue and fully trace the routes of each vessel.
2. **Arteries:** Identify the following arteries of the cat, using **figure 28.31** as a guide:

- ☐ **anterior tibial**
- ☐ **aortic arch**
- ☐ **axillary**
- ☐ **brachial**
- ☐ **brachiocephalic**
- ☐ **common carotid**
- ☐ **descending abdominal aorta**
- ☐ **descending thoracic aorta**
- ☐ **external carotid**
- ☐ **external iliac**
- ☐ **femoral**
- ☐ **gonadal (testicular or ova)**
- ☐ **inferior mesenteric**
- ☐ **internal carotid**
- ☐ **internal iliac**
- ☐ **median sacral**
- ☐ **popliteal**
- ☐ **posterior tibial**
- ☐ **pulmonary**
- ☐ **radial**
- ☐ **renal**
- ☐ **splenic**
- ☐ **subclavian**
- ☐ **superior mesenteric**
- ☐ **ulnar**

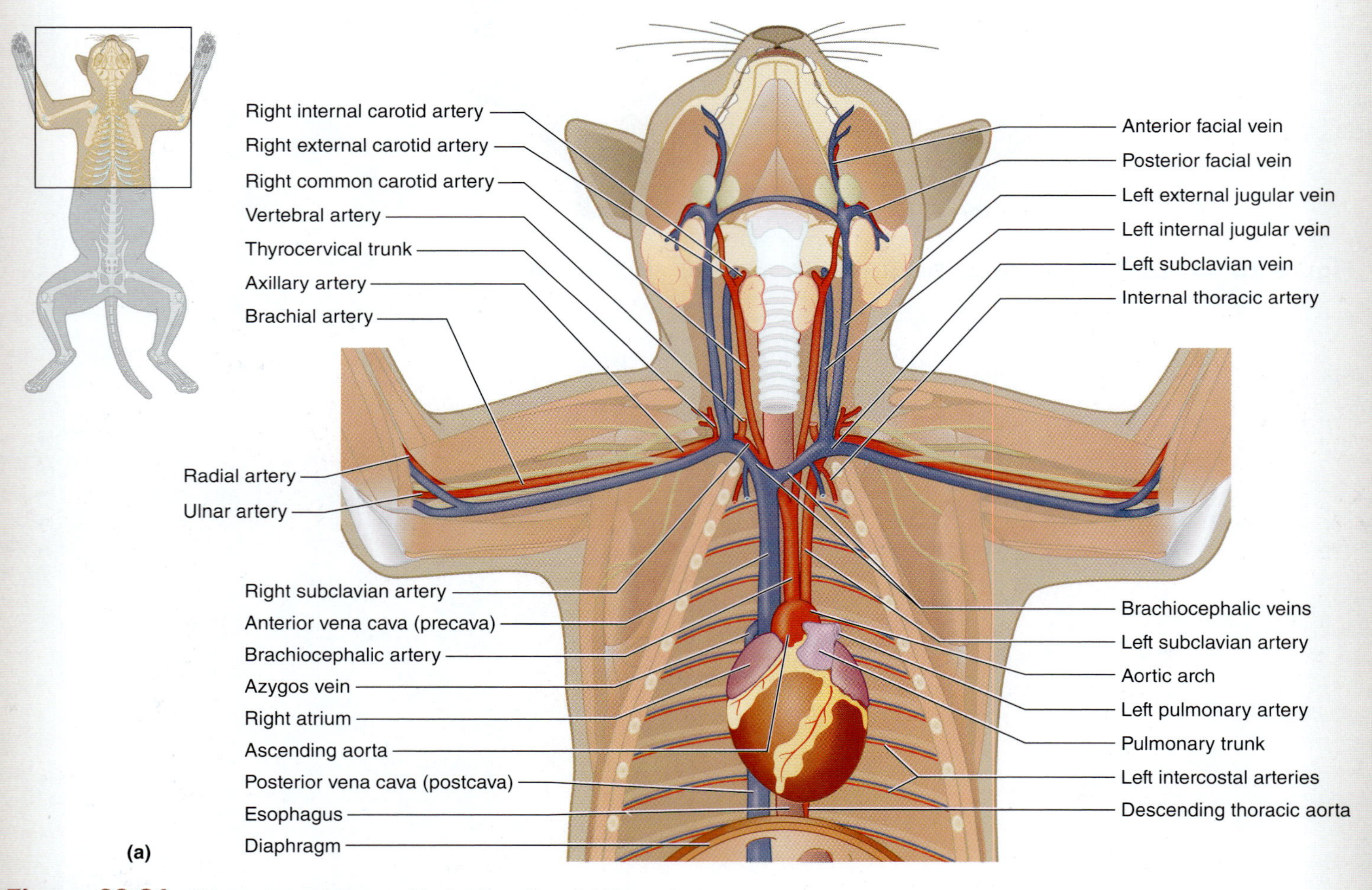

Figure 28.31 **Thoracic and Upper Limb Vessels.** (a) Illustration.

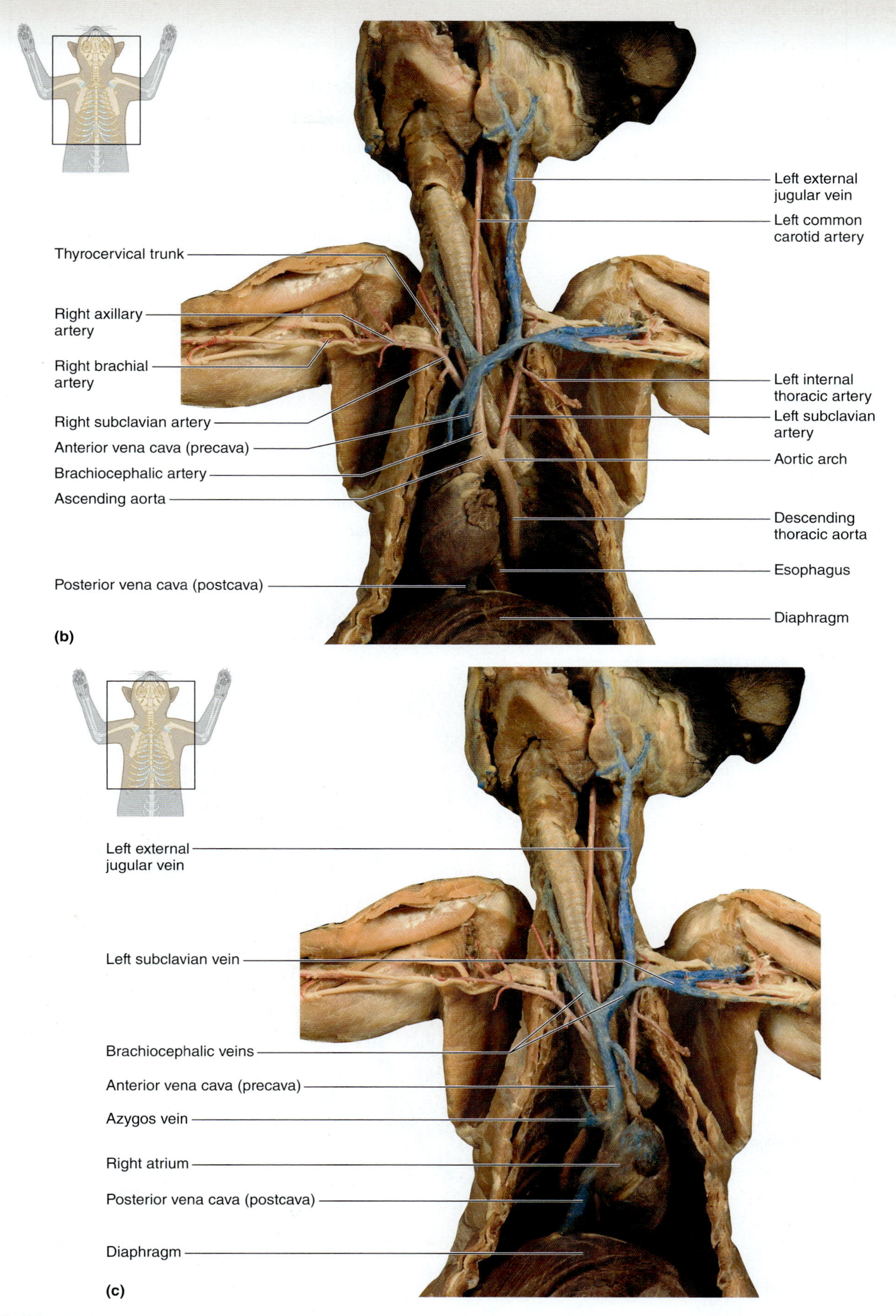

Figure 28.31 Thoracic and Upper Limb Vessels *(continued)*. (b) Dissection demonstrating the major arteries and veins of the thorax and upper limb of the cat in one figure. (c) Heart rotated out of anatomic position to better demonstrate the postcava, azygos vein, and right atrium of the heart. Observe the other dissection photos in this chapter to see individual vessels within specific regions of the body (e.g., upper limb, thorax, abdomen, etc.).

(continued on next page)

(continued from previous page)

3. **Veins:** Identify the following veins of the cat, using **figure 28.32** as a guide:

- ☐ anterior facial
- ☐ anterior tibial
- ☐ anterior vena cava (precava)
- ☐ axillary
- ☐ azygos
- ☐ brachial
- ☐ brachiocephalic
- ☐ cephalic
- ☐ common iliac
- ☐ external iliac
- ☐ external jugular
- ☐ femoral
- ☐ gastrosplenic
- ☐ gonadal (testicular or ovarian)
- ☐ great saphenous
- ☐ hepatic portal
- ☐ iliolumbar
- ☐ inferior (caudal) mesenteric
- ☐ internal iliac
- ☐ internal jugular
- ☐ median cubital
- ☐ popliteal
- ☐ posterior tibial
- ☐ posterior vena cava (postcava)
- ☐ radial
- ☐ renal
- ☐ subclavian
- ☐ superior (cranial) mesenteric
- ☐ ulnar

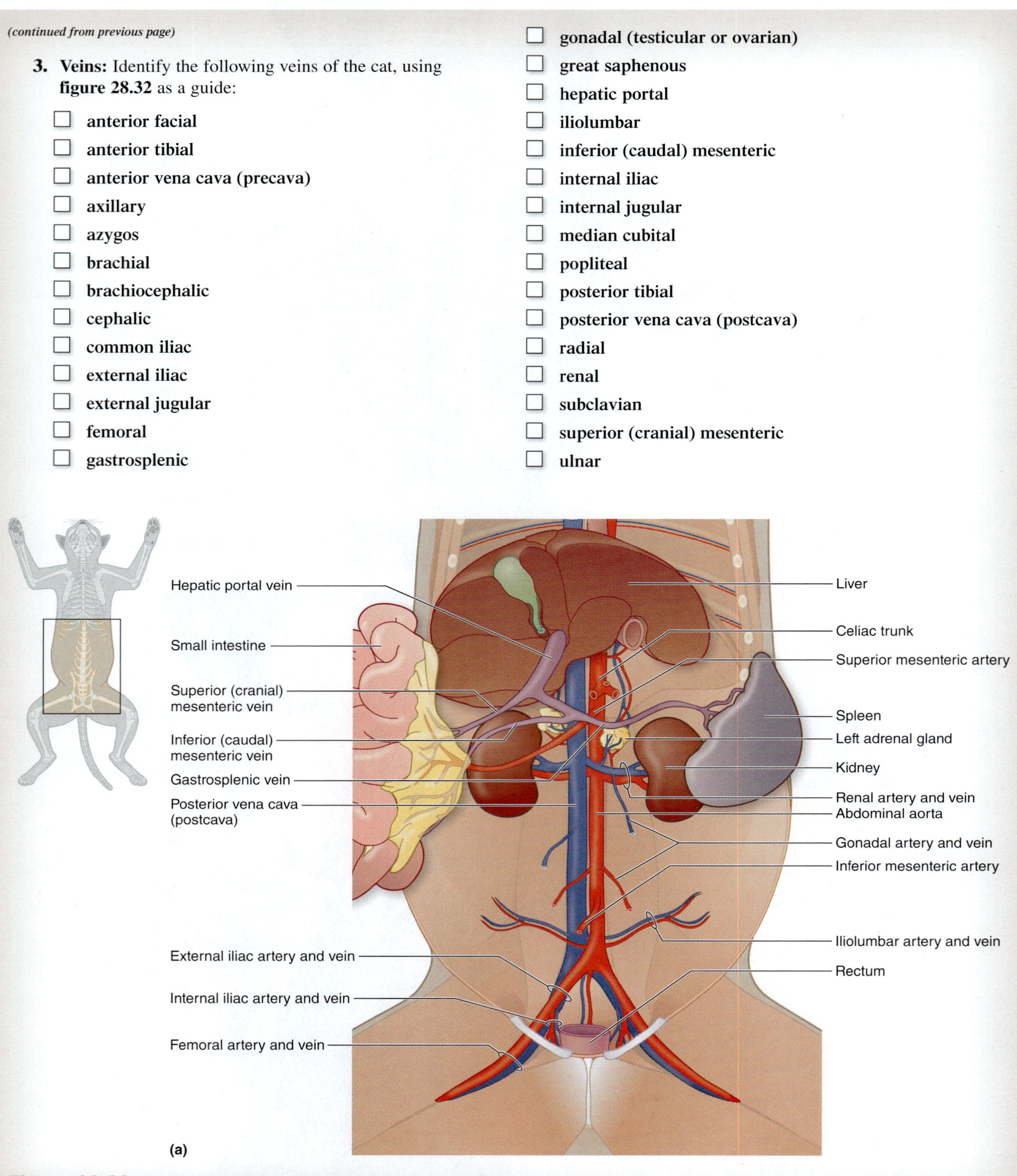

Figure 28.32 **Abdominal and Lower Limb Vessels.** (a) Illustration.

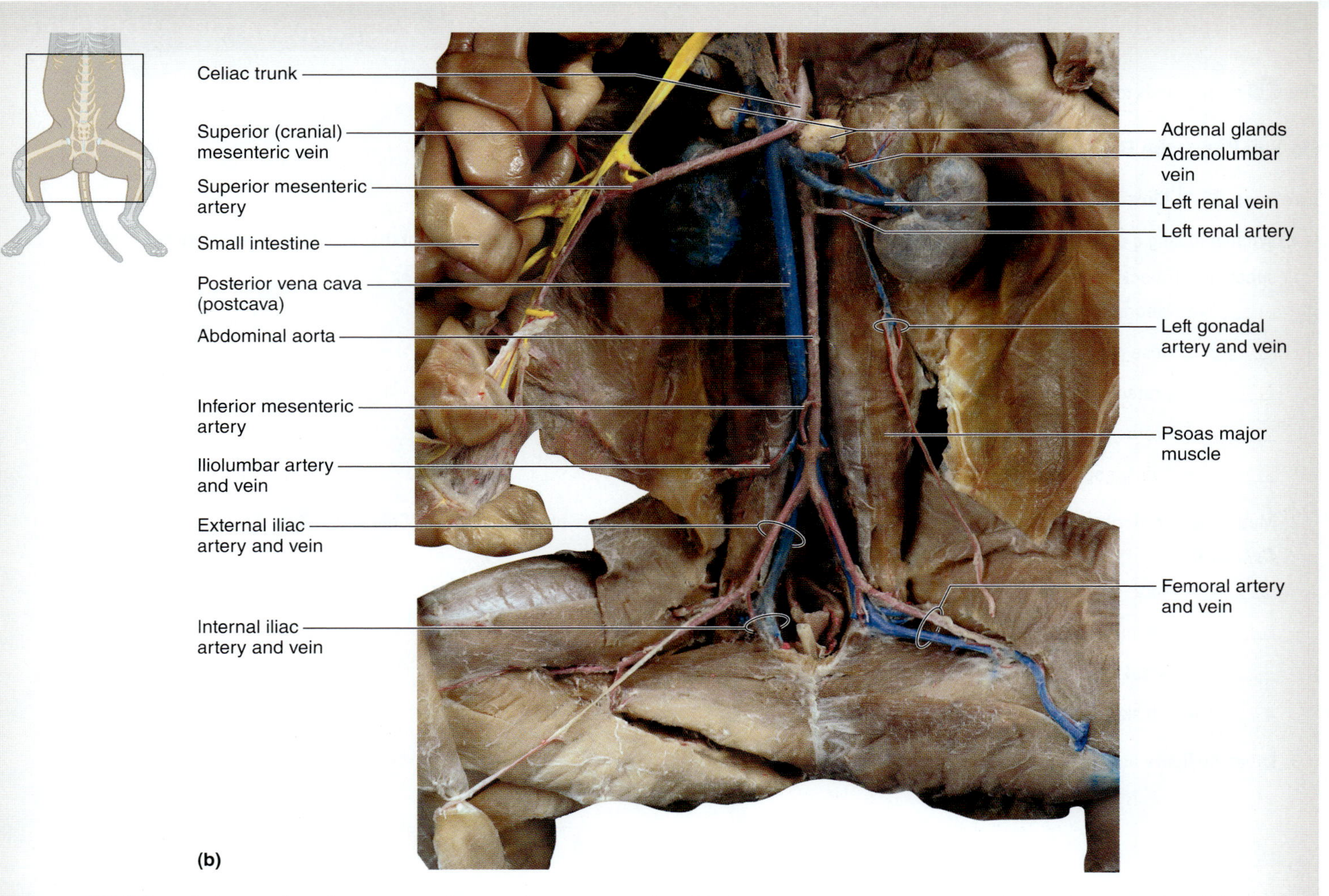

Figure 28.32 Abdominal and Lower Limb Vessels *(continued)*. (b) Dissection demonstrating the major arteries and veins of the abdomen and lower limb of the cat in one figure. Observe the other dissection photos in this chapter to see individual vessels within specific regions of the body (e.g., upper limb, thorax, abdomen, etc.).

Chapter 28: Cat Dissection Exercises

Name: ____________________

Date: __________ Section: __________

POST-LABORATORY WORKSHEET

The **1** corresponds to the Learning Objective(s) listed in the chapter opener outline.

Do You Know the Basics?

Exercise 28.1: Directional Terms and Surface Anatomy

1. When handling a preserved specimen, which of the following is an *essential* piece of safety equipment that should be used every time you handle a specimen? (Check all that apply.) **1**

   ______ a. laboratory coat

   ______ b. nose plugs

   ______ c. protective gloves

   ______ d. protective mask

   ______ e. safety goggles

2. Match the appropriate description in Column A with the surface anatomy structure of the cat listed in Column B. **2**

Column A	*Column B*
______ 1. ears	a. external nares
______ 2. eye membrane	b. nictitating membrane
______ 3. nasal openings	c. pinna
______ 4. whiskers	d. vibrissae

3. Label the figure using appropriate directional terms. **3**

1 ____________________

2 ____________________

3 ____________________

4 ____________________

5 ____________________

6 ____________________

7 ____________________

4. Which of the following is the *surface anatomy* landmark that can be used to distinguish a female from a male cat? (Circle one.) **4**

   a. mammary glands b. mammary ridge c. vagina d. vestibule

5. The integumentary structure that is unique to a cat as compared to a human is hair. ____________ (True/False) **5**

Exercise 28.2: Skeletal System

6. The cat has seven cervical vertebrae. ____________ (True/False) **6**

7. The clavicle is absent in the cat skeleton. ____________ (True/False) **7**

Exercise 28.3: Skinning the Cat

8. The platysma muscle of the cat is located on the skin of the dorsal surface of the body. ____________ (True/False) **8**

9. Structures that appear as tight bands that run between the body wall and the skin and carry sensory information from receptors in the skin to the central nervous system are known as cutaneous nerves. ____________ (True/False) **9**

Exercise 28.4: Muscles of the Head and Neck

10. Label the following photo of the head and neck muscles of the cat. **10**

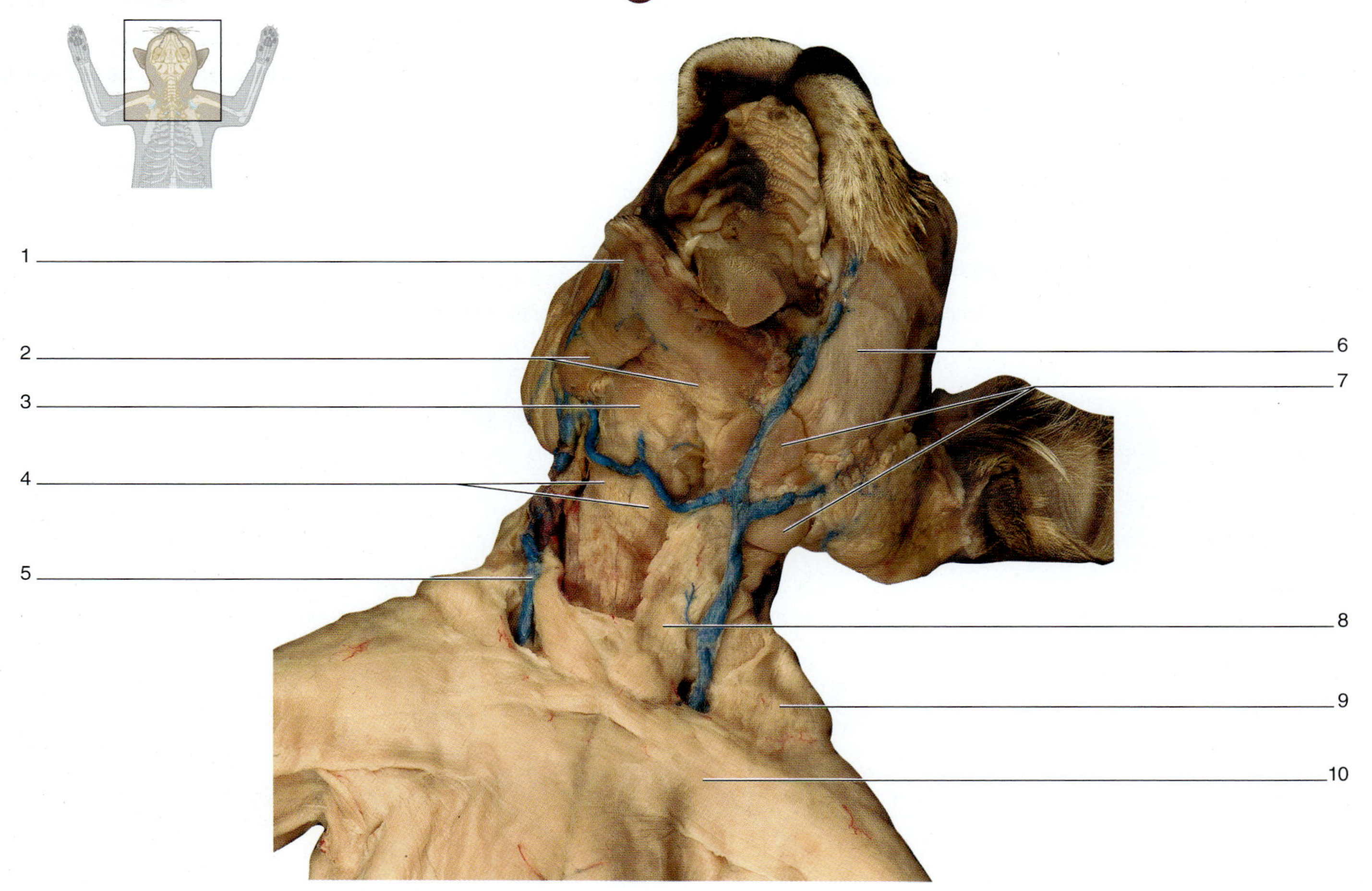

Exercise 28.5: Muscles of the Thorax

11. Label the following photo of the superficial thoracic muscles of the cat. **11**

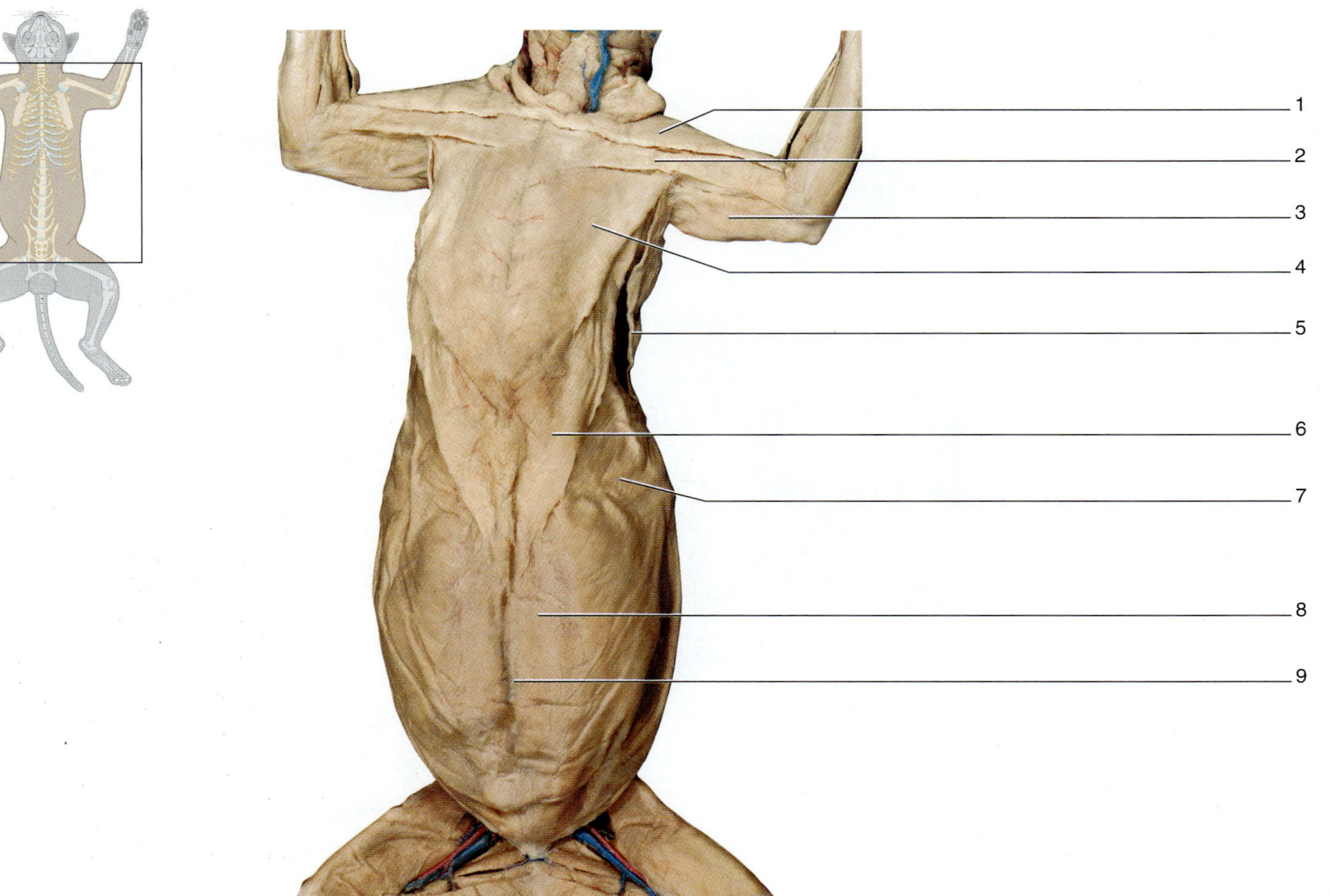

Exercise 28.6: Muscles of the Abdominal Wall

12. The cat abdominal wall muscles are identical to those of the human abdominal wall muscles with one exception; there is no rectus abdominis muscle. ____________________ (True/False) **12**

13. The external oblique muscle in the cat has fibers that are directed ____________________ (medial/lateral) and ____________________ (cranial/caudal). **13**

Exercise 28.7: Muscles of the Back and Shoulder

14. Complete the following table of the superficial muscles of the back and shoulder regions of the cat. For each muscle, identify the human analogue and the action. **14**

Cat Muscle	Human Muscle	Action
Acromiodeltoid		
Acromiotrapezius		
Clavobrachialis		
Clavotrapezius		
Latissimus Dorsi		
Levator Scapulae Ventralis		
Spinodeltoid		
Spinotrapezius		
Teres Major		

Exercise 28.8: Muscles of the Forelimb

15. Label the following photo of the lateral forelimb muscles of the cat. **15**

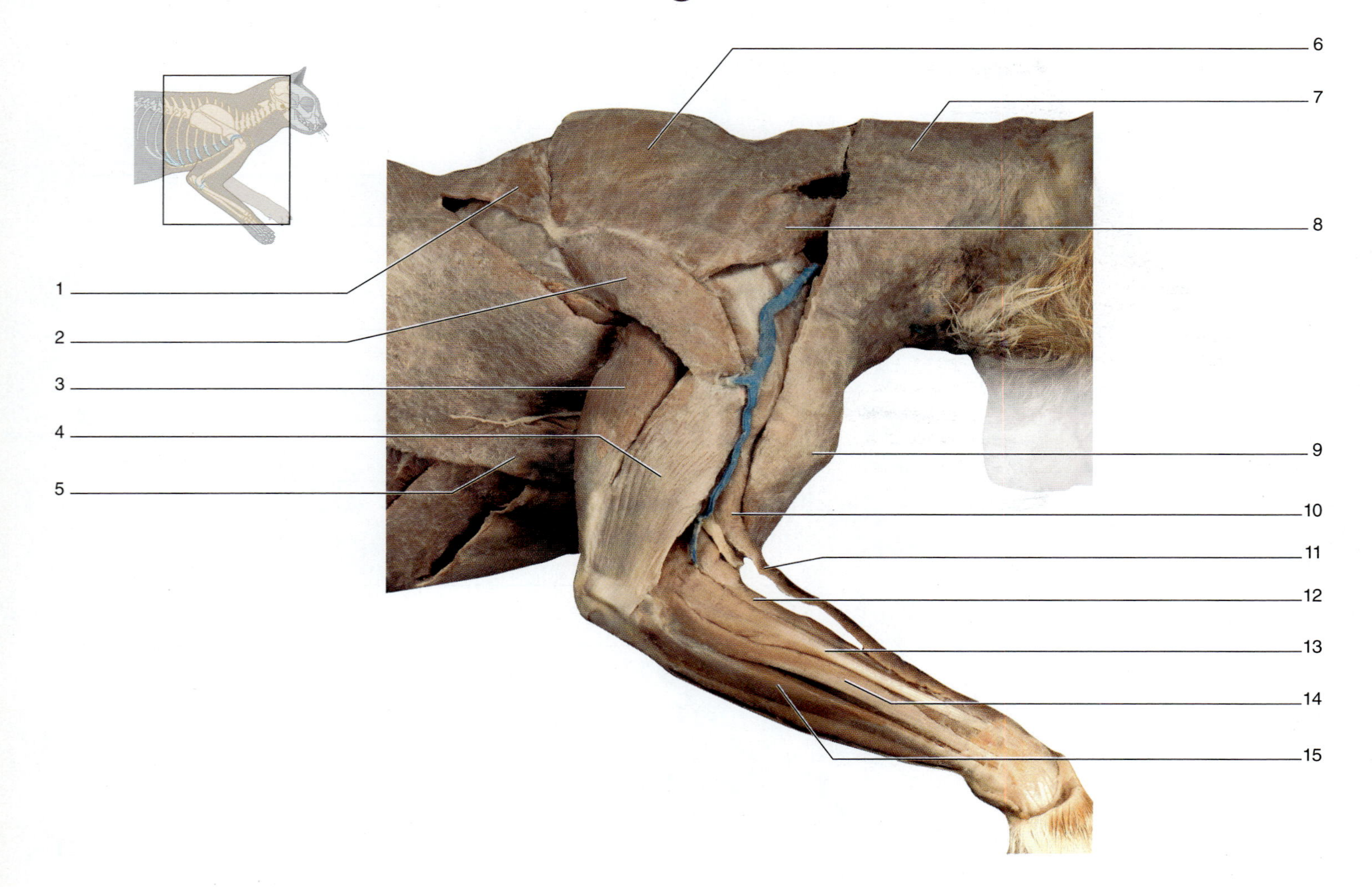

Exercise 28.9: Muscles of the Hindlimb

16. Label the following photo of the hindlimb muscles of the cat. 16

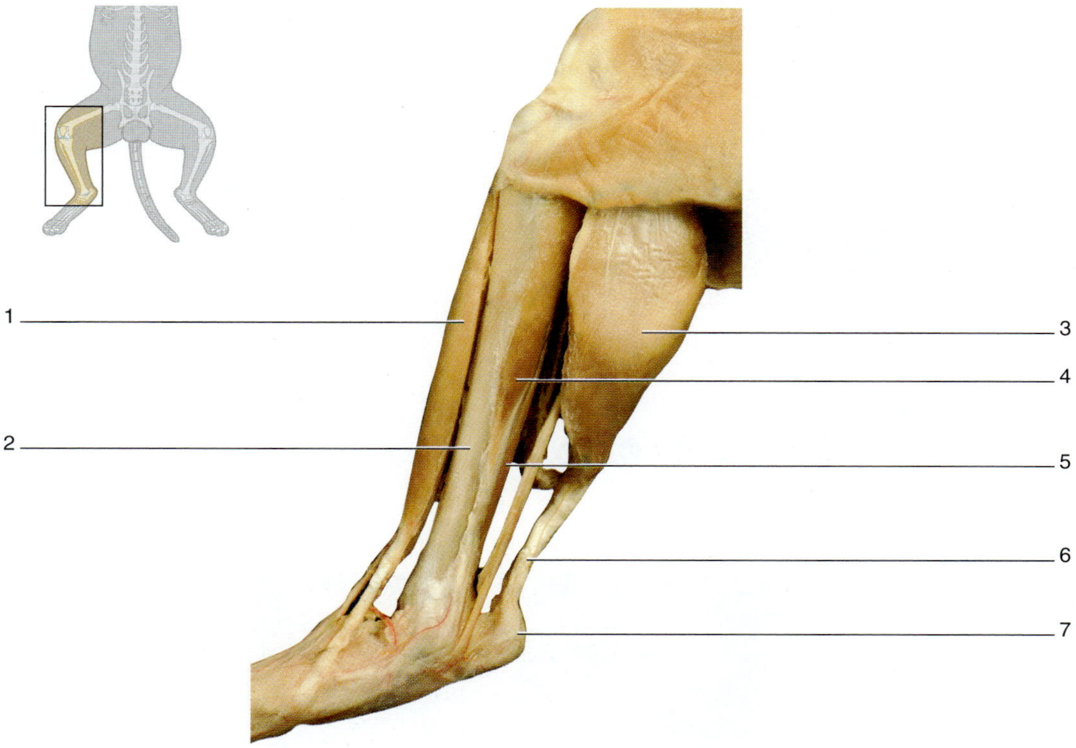

1 ____________________

2 ____________________

____________________ 3

____________________ 4

____________________ 5

____________________ 6

____________________ 7

17. Which of the following muscles is *not* a muscle that composes the triceps surae in both cats and humans? (Circle one.) 17

   a. gastrocnemius
   b. plantaris
   c. popliteus
   d. soleus

Exercise 28.10: Opening the Thoracic and Abdominal Cavities

18. Care must be taken when opening the thoracic cavity so as not to damage external genitalia. ____________________ (True/False) 18

Exercise 28.11: The Heart, Lungs, and Mediastinum

19. The cat heart consists of ____________________ (two/three) chambers. 19

20. The right lung in the cat has ____________________ (two/three) lobes and the left lung in the cat has ____________________ (two/three) lobes. This is compared to humans, where the right lung has ____________________ (two/three) lobes and the left lung has ____________________ (two/three) lobes. 20

21. Which of the following is the equivalent vessel to the superior vena cava in the human? (Circle one.) 21

   a. aorta
   b. postcava
   c. precava
   d. pulmonary trunk

22. Unlike humans, the cat does not have a thoracic duct. ____________________ (True/False) 22

Exercise 28.12: The Abdominal Cavity

23. Place the following digestive structures in sequential order beginning with the mouth. 23

______ a. duodenum
______ b. esophagus
______ c. ileocecal valve
______ d. ileum
______ e. jejunum
______ f. large intestine
______ g. oral cavity
______ h. rectum
______ i. small intestine
______ j. stomach

24. The two major veins that form the hepatic portal vein in the cat are the ______________________ and the ______________________. 24

Exercise 28.13: The Dorsal Abdominal Wall

25. These structures extend from the hilum of the kidney: 25

a. ________________ b. ________________ c. ________________

Exercise 28.14: Urogenital Systems

26. Label the photo of the male urogenital system of the cat. 26 27 28 29

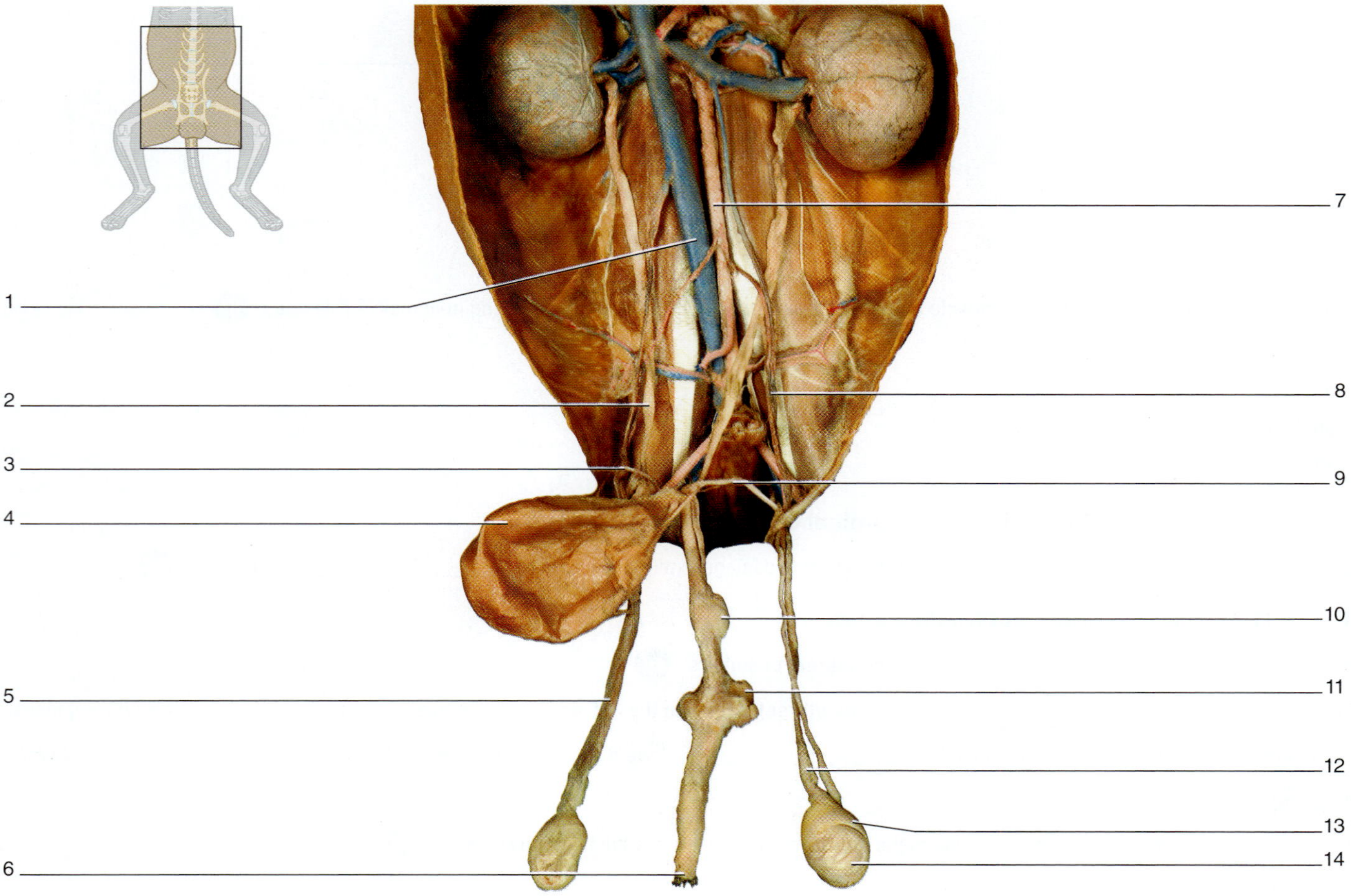

1 ________________
2 ________________
3 ________________
4 ________________
5 ________________
6 ________________
7 ________________
8 ________________
9 ________________
10 ________________
11 ________________
12 ________________
13 ________________
14 ________________

27. Label the photo of the female urogenital system of the cat. 26 27 28 29

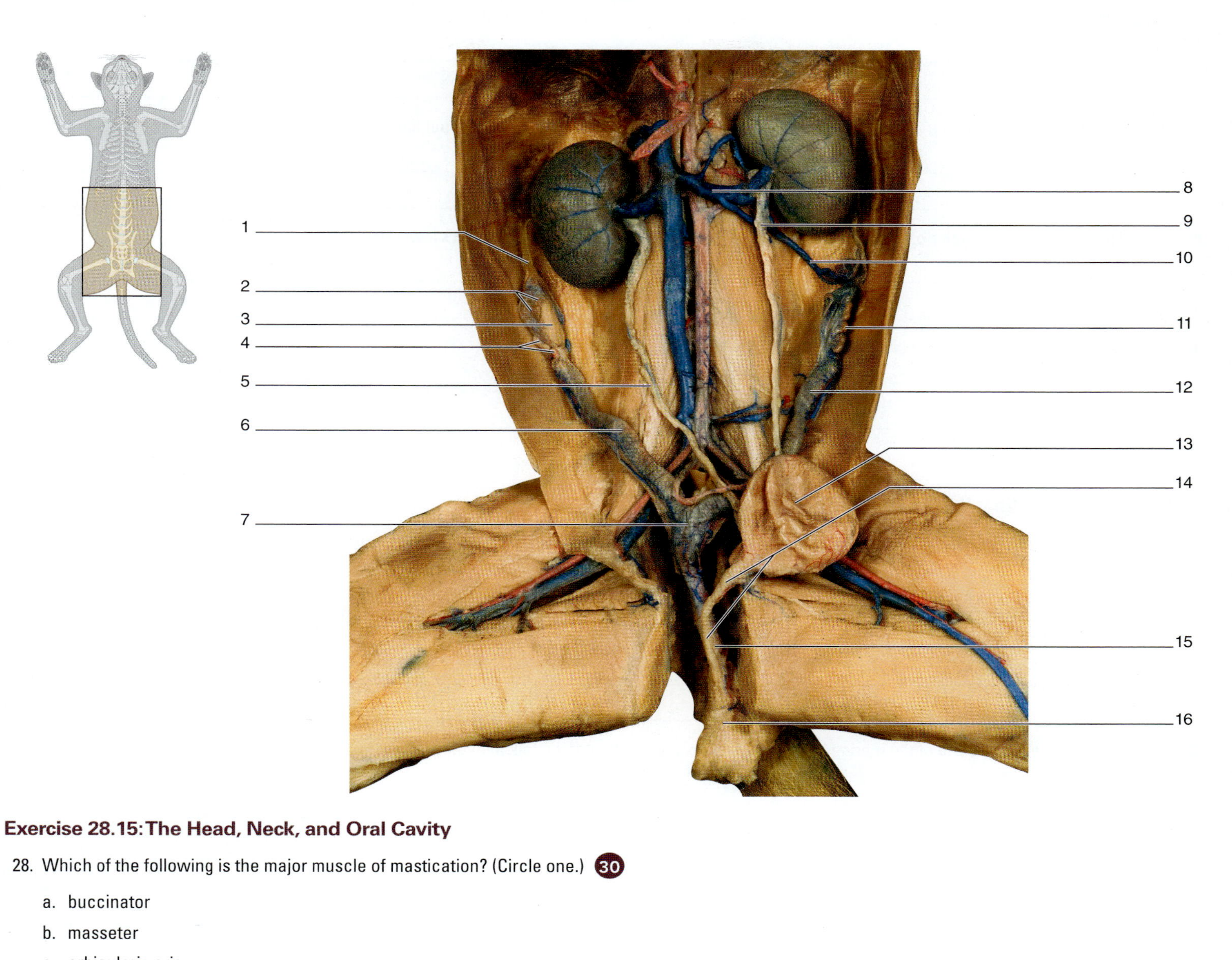

Exercise 28.15: The Head, Neck, and Oral Cavity

28. Which of the following is the major muscle of mastication? (Circle one.) 30

    a. buccinator

    b. masseter

    c. orbicularis oris

    d. zygomatic major

29. Which of the following covers the opening to the larynx in both humans and cats? (Circle one.) 31

    a. epiglottis

    b. glottis

    c. hyoid bone

    d. thyroid cartilage

Exercise 28.16: Nervous System: Peripheral Nerves

30. The group of nerves supplying the upper limb, which is found in the axillary region of the cat, is known as the ____________________. 32

31. The group of nerves supplying the lower limb, which is found in the pelvic region of the cat, is known as the ____________________. 33

Exercise 28.17: Cardiovascular System: Peripheral Blood Vessels

32. In humans, the left gonadal vein drains into the left renal vein. In cats, the right gonadal vein drains into the right renal vein.

    ____________________ (True/False) 34

Can You Apply What You've Learned?

33. Discuss two adaptations to the cat skeletal and muscular systems that allow for quadrupedal rather than bipedal locomotion.

34. Identify the nerves that branch from the brachial plexus to innervate the muscles of the cat's upper limb.

35. Identify the nerves that branch from the lumbosacral plexus to innervate the muscles of the cat's lower limb.

36. Compare hepatic portal circulation in humans and cats.

Can You Synthesize What You've Learned?

37. Trace the flow of urine from the ureters to the urethra in the male cat and in the female cat, identifying all structures along the path. Compare this path to that in humans.

38. Compare the size and structure of the uterine horn in the female cat with that of the fallopian tube in humans. Discuss why there is a dramatic difference between cats and humans.

appendix

ANSWER KEYS

This appendix includes answers to labeling exercises, table completion activities, and calculation questions found in each chapter. Answers to in-exercise questions prompting students to record their personal observations and drawings are not provided due to their variability.

Chapter 1

Exercise 1.3, #1

Practice:
- a 0.5 cm
- b 0.74 ounces
- c 0.5 L
- d 12.7 cm
- e 9.9 ft
- f 11.4 L
- g 3000 mL

Figure 1.2 Identification of Common Dissection Instruments

1. forceps
2. scalpel blades
3. scalpel (disposable)
4. scalpel blade handle (#3)
5. scalpel blade handle (#4)
6. scissors (curved)
7. scissors (pointed)
8. blunt probe
9. dissecting pins
10. dissecting needle
11. blunt probe
12. hemostat

Exercise 1.5

1. A
2. B
3. A
4. A
5. B
6. B
7. A
8. C
9. A

Chapter 2

Figure 2.2 Sections Through a Human Brain

1. coronal
2. midsagittal
3. transverse
4. sagittal

Figure 2.4 Posterior View of an Individual with Three Reference Locations Marked

1. thoracic posterior, superior surface of the thorax
2. antebrachial; posterior forearm distal to the elbow
3. femoral posterior thigh, distal to the hip

Figure 2.5 Regional Terms

1. oral
2. cervical
3. axillary
4. brachial
5. antebrachial
6. carpal
7. digital
8. femoral
9. crural
10. frontal
11. orbital
12. buccal
13. mental
14. mammary
15. pelvic
16. inguinal
17. tarsal
18. auricular
19. vertebral
20. sacral
21. sural
22. calcaneal
23. occipital
24. lumbar
25. perineal
26. popliteal

Figure 2.6 Body Cavities

1. posterior aspect
2. ventral cavity
3. cranial cavity
4. vertebral canal
5. thoracic cavity
6. abdominopelvic cavity
7. diaphragm
8. abdominal cavity
9. pelvic cavity
10. thoracic cavity
11. abdominopelvic cavity
12. mediastinum
13. pleural cavity
14. pericardial cavity
15. abdominal cavity
16. pelvic cavity

Exercise 2.5, #2

Organ	Quadrant(s)	Region(s)
Left kidney	Left upper	Left hypochondriac, left lumbar
Liver	Right upper, left upper	Right hypochondriac, right lumbar, epigastric
Pancreas	Right and left upper	Epigastric, left hypochondriac
Small intestine	Right and left lower	Umbilical, hypogastric, right and left lumbar and right and left iliac
Spleen	Left upper	Left hypochondriac
Stomach	Right and left upper	Epigastric, left hypochondriac, umbilical
Urinary bladder	Right and left lower	Hypogastric

Chapter 3

No Activities

Chapter 4

Concept Connection (p. 73)

Molecular weight of $KMnO_4$ = 158 g/mol
Molecular weight of $C_{16}H_{18}ClN_3S$ = 319 g/mol

Chapters 5–6

No Activities

Chapter 7

Figure 7.6 Identifying Bones Based on Structure

1. flat
2. short
3. long
4. irregular
5. long
6. irregular
7. irregular

Figure 7.10 The Human Skeleton

(a) Anterior View

1. skull
2. mandible
3. clavicle
4. scapula
5. sternum
6. humerus
7. rib
8. vertebra
9. ilium
10. sacrum
11. coccyx
12. ischium
13. radius
14. ulna
15. pubis
16. carpals
17. metacarpals
18. phalanges (of the hand)
19. femur
20. patella
21. fibula
22. tibia
23. tarsals
24. metatarsals
25. phalanges (of the foot)

(b) Posterior View

1. skull
2. mandible
3. scapula
4. humerus
5. rib
6. vertebra
7. ilium
8. sacrum
9. coccyx
10. ischium
11. radius
12. ulna
13. pubis
14. carpals
15. metacarpals
16. phalanges (of the hand)
17. femur
18. fibula
19. tibia
20. tarsals
21. metatarsals
22. phalanges (of the foot)

Chapter 8

Figure 8.1 Anterior View of the Skull

1. frontal bone
2. parietal bone
3. glabella
4. supraorbital foramen (notch)
5. superior orbital fissure
6. lacrimal bone
7. nasal bone
8. zygomatic bone
9. infraorbital foramen
10. maxilla
11. superciliary arch
12. supraorbital margin
13. sphenoid bone
14. temporal bone
15. optic canal
16. inferior orbital fissure
17. perpendicular plate of ethmoid
18. vomer
19. inferior nasal concha
20. alveolar processes
21. mandible
22. mental foramen
23. mental protuberance

Figure 8.2 The Orbit

1. nasal bone
2. supraorbital foramen (notch)
3. ethmoid bone
4. frontal bone
5. zygomatic process of frontal bone
6. optic foramen
7. superior orbital fissure
8. lacrimal bone
9. inferior orbital fissure
10. zygomatic bone
11. infraorbital foramen
12. maxilla

Figure 8.3 The Nasal Cavity

1. crista galli (of ethmoid bone)
2. cribriform plate of ethmoid bone
3. sella turcica
4. sphenoid sinus
5. palatine bone
6. sphenoid bone
7. frontal sinus
8. nasal bone
9. superior nasal concha
10. middle nasal concha
11. lacrimal bone
12. inferior nasal concha
13. maxilla
14. perpendicular plate of ethmoid bone
15. vomer

Figure 8.4 The Mandible

1. head of mandible
2. coronoid process
3. mandibular foramen
4. condylar process
5. mandibular notch
6. mylohyoid line
7. ramus
8. alveolar process
9. mental foramen
10. angle
11. body
12. mental protuberance

Figure 8.5 Lateral View of the Skull

1. parietal eminence
2. parietal bone
3. inferior temporal line
4. squamous suture
5. lambdoid suture
6. squamous part of temporal bone
7. occipital bone
8. external acoustic meatus
9. mastoid process
10. styloid process
11. head of mandible
12. zygomatic process of temporal bone
13. temporal process of zygomatic bone
14. body of mandible
15. coronal suture
16. frontal bone
17. superior temporal line
18. pterion
19. greater wing of sphenoid bone
20. nasal bone
21. lacrimal bone
22. ethmoid bone
23. lacrimal groove
24. zygomatic bone
25. maxilla
26. mental foramen
27. mental protuberance

Figure 8.6 Posterior View of the Skull

1. sagittal suture
2. parietal foramina
3. parietal bone
4. parietal eminence
5. lambdoid suture
6. sutural (Wormian) bone
7. occipital bone
8. temporal bone
9. external occipital protuberance
10. mastoid process

Figure 8.7 Superior View of the Skull

1. frontal bone
2. coronal suture
3. sagittal suture
4. parietal bone
5. parietal foramina
6. lambdoid suture
7. sutural (Wormian) bone
8. occipital bone

Figure 8.8 Inferior View of the Skull

1. maxilla
2. palatine bone
3. vomer
4. sphenoid bone
5. foramen ovale
6. foramen spinosum
7. foramen lacerum
8. jugular foramen
9. carotid canal
10. inferior nuchal line
11. occipital bone
12. incisive foramen
13. temporal process of zygomatic bone
14. zygomatic process of temporal bone
15. mandibular fossa
16. styloid process
17. temporal bone
18. mastoid process
19. occipital condyle
20. hypoglossal canal
21. basilar region of occipital bone
22. foramen magnum
23. external occipital crest
24. superior nuchal line
25. external occipital protuberance

Figure 8.9 Cranial Fossae

1. anterior cranial fossa
2. middle cranial fossa
3. posterior cranial fossa

Figure 8.11 Superior View of the Cranial Floor

1. frontal sinus
2. frontal bone
3. optic canal
4. lesser wing of sphenoid bone
5. foramen rotundum
6. greater wing of sphenoid bone
7. temporal bone
8. petrous part of temporal bone
9. parietal bone
10. internal acoustic meatus
11. foramen magnum
12. occipital bone
13. internal occipital protuberance
14. frontal crest
15. crista galli
16. cribriform plate of the ethmoid bone
17. sella turcica
18. foramen lacerum
19. foramen ovale
20. foramen spinosum
21. jugular foramen
22. hypoglossal canal
23. groove for sigmoid sinus
24. basilar part of occipital bone
25. groove for transverse sinus
26. internal occipital crest
27. anterior clinoid process
28. hypophyseal fossa
29. posterior clinoid process

Figure 8.12 The Hyoid Bone

1. greater cornu
2. lesser cornu
3. body

Figure 8.13 The Fetal Skull

1. frontal bone
2. parietal bone
3. sphenoidal fontanel
4. sphenoid bone
5. mandible
6. temporal bone
7. mastoid fontanel
8. occipital bone
9. frontal bone
10. anterior fontanel
11. parietal bone

Figure 8.14 Lateral View of the Vertebral Column

1. cervical vertebrae
   Number of vertebrae: 7
2. thoracic vertebrae
   Number of vertebrae: 12
3. lumbar vertebrae
   Number of vertebrae: 5
4. sacrum
   Number of vertebrae: 5
5. coccygeal vertebrae
   Number of vertebrae: 4
6. cervical curvature
7. thoracic curvature
8. lumbar curvature
9. sacral curvature

Figure 8.15 A Typical Vertebra

1. lamina
2. transverse process
3. superior articular process
4. body
5. spinous process
6. vertebral arch
7. transverse process
8. pedicle
9. vertebral (spinal) foramen
10. costal facet
11. superior articular process
12. body
13. inferior articular process

Figure 8.16 Cervical Vertebra

1. body
2. transverse process
3. superior articular process (and facet)
4. transverse foramen
5. pedicle
6. vertebral (spinal) foramen
7. lamina
8. spinous process
9. spinous process
10. superior articular process (and facet)
11. transverse foramen
12. body
13. inferior articular process (and facet)

Figure 8.17 The Atlas (C1)
1 anterior tubercle
2 anterior arch
3 superior articular facet
4 transverse process
5 transverse foramen
6 posterior tubercle
7 posterior arch
8 articular facet for dens
9 vertebral foramen

Figure 8.18 The Axis (C2)
1 dens (odontoid process)
2 superior articular process
3 transverse foramen
4 transverse process
5 pedicle
6 vertebral (spinal) foramen
7 lamina
8 spinous process

Figure 8.19 Thoracic Vertebra
1 body
2 vertebral (spinal) foramen
3 superior articular process
4 spinous process
5 costal facet
6 pedicle
7 costal facet
8 transverse process
9 lamina
10 costal facet
11 spinous process
12 inferior articular process
13 superior articular process
14 costal facet
15 body
16 costal demifacet

Figure 8.20 Lumbar Vertebra
1 body
2 vertebral (spinal) foramen
3 superior articular process (and facet)
4 spinous process
5 pedicle
6 transverse process
7 lamina
8 superior articular process (and facet)
9 transverse process
10 spinous process
11 inferior articular process (and facet)
12 body

Figure 8.21 Sacrum and Coccyx
1 superior articular process
2 ala
3 sacral promontory
4 anterior sacral foramina
5 transverse ridges
6 coccygeal cornu
7 coccyx
8 sacral canal
9 superior articular facet
10 median sacral crest
11 auricular surface
12 posterior sacral foramina
13 sacral hiatus
14 coccygeal cornu
15 coccyx

Figure 8.22 The Sternum
1 suprasternal notch
2 manubrium
3 sternal angle
4 body
5 second rib
6 xiphoid process

Figure 8.23 A Typical Rib
1 superior articular facet
2 inferior articular facet
3 shaft
4 intervertebral foramen
5 head
6 articular facet for transverse process
7 angle
8 body
9 head
10 neck
11 tubercle
12 articular facet for transverse process
13 angle
14 costal groove

Chapter 9

Figure 9.1 The Clavicle
1 acromial end (lateral)
2 sternal end (medial)
3 acromial end (lateral)
4 conoid tubercle
5 sternal end (medial)
6 costal tuberosity

Figure 9.2 The Right Scapula
1 acromion
2 coracoid process
3 glenoid cavity
4 lateral (axillary) border
5 inferior angle
6 suprascapular notch
7 spine
8 superior border
9 superior angle
10 subscapular fossa
11 medial (vertebral) border
12 acromion
13 spine
14 glenoid cavity
15 infraglenoid tubercle
16 lateral (axillary) border
17 superior angle
18 supraglenoid tubercle
19 coracoid process
20 subscapular fossa
21 inferior angle
22 coracoid process
23 suprascapular notch
24 superior border
25 supraspinous fossa
26 spine
27 infraspinous fossa
28 medial (vertebral) border
29 acromion
30 glenoid cavity
31 lateral (axillary) border
32 inferior angle

Figure 9.3 The Right Humerus
1 anatomical neck
2 greater tubercle
3 lesser tubercle
4 intertubercular sulcus
5 surgical neck
6 deltoid tuberosity
7 shaft
8 coronoid fossa
9 radial fossa
10 lateral epicondyle
11 capitulum
12 trochlea
13 head
14 medial epicondyle
15 head
16 greater tubercle
17 anatomical neck
18 surgical neck
19 deltoid tuberosity
20 medial epicondyle
21 olecranon fossa
22 lateral epicondyle
23 trochlea
24 lateral epicondyle
25 medial epicondyle
26 capitulum
27 trochlea
28 ulna
29 radius
30 humerus
31 medial epicondyle
32 lateral epicondyle
33 radius
34 ulna

Figure 9.4 The Radius
1 head
2 neck
3 radial tuberosity
4 shaft
5 styloid process of radius
6 ulnar notch

Figure 9.5 The Ulna
1 trochlear notch
2 coronoid process
3 radial notch
4 tuberosity of ulna
5 shaft of ulna
6 styloid process of ulna
7 olecranon
8 coronoid process

Exercise 9.2D, #3

Carpal Bone	Word Origin	Bone Shape/Appearance
Scaphoid	*skaphe*, boat	Shaped like a boat
Lunate	*luna*, moon	Moon-shaped
Triquetrum	*triquetrus*, three-cornered	Triangular
Pisiform	*pisum*, pea	Pea-shaped
Trapezium	*trapezion*, a table	Table-shaped
Trapezoid	*trapezion*, a table	Table-shaped
Capitate	*caput*, head	Head-shaped
Hamate	*hamus*, a hook	Hook-shaped

Figure 9.7 The Metacarpals and Phalanges
1 phalanges
2 metacarpals
3 carpals
4 distal phalanx
5 middle phalanx
6 proximal phalanx
7 distal phalanx of pollex
8 proximal phalanx of pollex
9 metacarpal V
10 metacarpal IV
11 metacarpal III
12 metacarpal I
13 metacarpal II

Figure 9.8 Surface Anatomy of the Upper Limb
1 clavicle
2 acromion of scapula
3 sternoclavicular joint
4 deltoid tuberosity
5 medial epicondyle of humerus
6 styloid process of ulna
7 olecranon
8 styloid process of radius
9 spine of scapula

Figure 9.9 The Right Os Coxae
1 ala
2 anterior gluteal line
3 posterior gluteal line
4 posterior superior iliac spine
5 posterior inferior iliac spine
6 greater sciatic notch
7 body of ischium
8 ischial spine
9 lesser sciatic notch
10 ischial tuberosity
11 iliac crest
12 anterior superior iliac spine
13 inferior gluteal line
14 anterior inferior iliac spine
15 lunate surface
16 acetabulum
17 superior pubic ramus
18 pubic crest

19 pubic tubercle
20 inferior pubic ramus
21 obturator foramen
22 ramus of ischium
23 iliac fossa
24 anterior superior iliac spine
25 anterior inferior iliac spine
26 arcuate line
27 pectineal line
28 superior pubic ramus
29 pubic tubercle
30 symphysial surface of pubic bone
31 obturator foramen
32 inferior pubic ramus
33 iliac crest
34 posterior superior iliac spine
35 auricular surface
36 posterior inferior iliac spine
37 greater sciatic notch
38 ischial spine
39 lesser sciatic notch
40 body of ischium
41 ischial tuberosity
42 ramus of ischium

Figure 9.11 The Right Femur

1 head
2 greater trochanter
3 fovea
4 neck
5 intertrochanteric crest
6 lesser trochanter
7 shaft
8 head
9 shaft
10 patellar surface
11 intercondylar fossa
12 lateral condyle
13 medial condyle
14 head
15 greater trochanter
16 neck
17 intertrochanteric line
18 lesser trochanter
19 shaft
20 adductor tubercle
21 lateral epicondyle
22 medial epicondyle
23 patellar surface
24 lateral condyle
25 medial condyle
26 head
27 fovea
28 greater trochanter
29 neck
30 intertrochanteric crest
31 lesser trochanter
32 pectineal line
33 gluteal tuberosity
34 linea aspera
35 medial supracondylar line
36 lateral supracondylar line
37 lateral epicondyle
38 popliteal surface
39 adductor tubercle
40 medial epicondyle
41 medial condyle
42 intercondylar fossa
43 lateral condyle

Figure 9.12 The Tibia

1 lateral condyle
2 intercondylar eminence
3 medial condyle
4 tibial tuberosity
5 anterior border
6 medial malleolus
7 medial condyle
8 intercondylar eminence
9 lateral condyle
10 fibular articular facet
11 medial malleolus

Figure 9.13 The Right Fibula

1 head
2 neck
3 shaft
4 lateral malleolus

Figure 9.14 The Tarsals

1 phalanges
2 metatarsals
3 medial cuneiform
4 navicular
5 intermediate cuneiform
6 lateral cuneiform
7 cuboid
8 talus
9 calcaneus

Exercise 9.5D, #5

Tarsal Bone	Bone Shape/Appearance	Word Origin
Talus	Convex, triangular	*talus*, ankle
Calcaneus	Elongated	*calcaneous*, heel
Navicular	Shaped like a boat	*navis*, ship
Medial Cuneiform	Wedge-shaped	*cuneus*, wedge
Intermediate Cuneiform	Wedge-shaped	*cuneus*, wedge
Lateral Cuneiform	Wedge-shaped	*cuneus*, wedge
Cuboid	Cube-shaped	*kybos*, cube

Figure 9.15 The Metatarsals and Phalanges

1 metatarsal II
2 metatarsal III
3 metatarsal IV
4 metatarsal V
5 lateral cuneiform
6 cuboid
7 distal phalanx
8 middle phalanx
9 proximal phalanx
10 metatarsal I
11 medial cuneiform
12 intermediate cuneiform
13 navicular
14 talus
15 calcaneus

Figure 9.16 Surface Anatomy of the Lower Limb

1 iliac crest
2 anterior superior iliac spine
3 sacrum
4 coccyx
5 greater trochanter of femur
6 ischial tuberosity
7 lateral epicondyle of femur
8 head of fibula
9 lateral malleolus
10 medial epicondyle of femur
11 patella
12 tibial tuberosity
13 shaft of tibia
14 medial malleolus
15 calcaneus
16 metatarsals
17 phalanges

Chapter 10

Figure 10.1 Fibrous Joints

1 syndesmosis
2 gomphosis
3 suture

Figure 10.2 Cartilaginous Joints

1 synchondrosis
2 symphysis
3 synchondrosis
4 symphysis

Figure 10.3 Diagram of a Representative Synovial Joint

1 periosteum
2 yellow bone marrow
3 fibrous layer of articular capsule
4 synovial membrane
5 synovial (joint) cavity
6 articular capsule
7 articular cartilage
8 ligament

Figure 10.4 Classifications of Synovial Joints

1 saddle
2 hinge
3 plane
4 ball-and-socket
5 condylar
6 pivot

Figure 10.5 A Representative Synovial Joint: Right Knee Joint

1 posterior cruciate ligament
2 lateral condyle of femur
3 lateral meniscus
4 fibular collateral ligament
5 anterior cruciate ligament
6 fibula
7 medial condyle of femur
8 medial meniscus
9 tibial collateral ligament
10 tibia
11 anterior cruciate ligament
12 medial condyle of femur
13 medial meniscus
14 posterior cruciate ligament
15 tibial collateral ligament
16 tibia
17 femur
18 lateral condyle of femur
19 fibular collateral ligament
20 lateral meniscus
21 fibula
22 femur
23 quadriceps femoris tendon
24 suprapatellar bursa
25 patella
26 prepatellar bursa
27 articular cartilage
28 patellar ligament
29 infrapatellar bursa
30 tibia

Chapter 11

Exercise 11.7 (Figure 11.7) Muscles of the Human Body

Muscle Number	Muscle Name	Architecture
1	Orbicularis oculi	Circular
2	Deltoid	Multipennate
3	Pectoralis major	Convergent
4	Sartorius	Parallel
5	Rectus femoris	Bipennate
6	Trapezius	Convergent
7	Triceps brachii	Bipennate
8	Extensors of wrist/hand	Parallel
9	Gastrocnemius	Bipennate (individual heads) Multipennate (entire muscle)

Figure 11.12 The Crossbridge Cycle

1 Calcium binds to troponin, causing a conformational change in troponin and tropomyosin. Myosin binding sites on actin are exposed.
2 Myosin head attaches to the binding site on actin.
3 ADP and P_i are released, resulting in a power stroke.
4 ATP binds to the ATP binding site on myosin, resulting in the detachment of the myosin head from actin.
5 The myosin ATPase hydrolyzes ATP, forming ADP and P_i. The myosin head remains in a "reset" position.

Chapter 12

Figure 12.1 Muscles of Facial Expression

1 epicranius (occipitofrontalis)
2 epicranial aponeurosis (connective tissue)
3 frontal belly of occipitofrontalis
4 procerus
5 orbicularis oculi
6 levator labii superioris
7 zygomaticus minor
8 zygomaticus major
9 depressor anguli oris
10 depressor labii inferioris
11 platysma
12 corrugator supercilii
13 nasalis
14 levator anguli oris
15 risorius
16 orbicularis oris
17 mentalis
18 epicranial aponeurosis (connective tissue)
19 frontal belly of occipitofrontalis
20 epicranius (occipitofrontalis)
21 buccinator
22 orbicularis oculi
23 levator labii superioris
24 zygomaticus minor
25 zygomaticus major
26 orbicularis oris
27 levator anguli oris
28 depressor labii inferioris
29 depressor anguli oris
30 platysma

Figure 12.2 Muscles of Mastication

1 temporalis
2 masseter
3 temporalis
4 lateral pterygoid
5 medial pterygoid

Figure 12.3 Muscles That Move the Tongue

1 palatoglossus
2 styloglossus
3 hyoglossus
4 genioglossus

Figure 12.4 Muscles of the Pharynx

1 tensor veli palatini
2 levator veli palatini
3 superior constrictor
4 stylopharyngeus
5 middle constrictor
6 inferior constrictor

Concept Connection (p. 292)

The right sternocleidomastoid rotates the neck to the left. The right splenius capitis rotates the neck to the right. In summary: To rotate your neck to the right, you use the sternal head of the sternocleidomastoid on the left side of the neck, and the splenius capitis muscle on the right side of the neck.

Figure 12.5 Muscles of the Head and Neck

1 mylohyoid
2 stylohyoid
3 digastric (anterior belly)
4 digastric (posterior belly)
5 omohyoid (superior belly)
6 sternohyoid
7 sternocleidomastoid
8 thyrohyoid
9 sternothyroid
10 scalenes
11 splenius capitis
12 sternocleidomastoid
13 scalenes
14 omohyoid (inferior belly)
15 digastric (posterior belly)
16 mylohyoid
17 digastric (anterior belly)
18 sternothyroid
19 omohyoid (superior belly)
20 sternohyoid
21 semispinalis capitis
22 sternocleidomastoid
23 splenius capitis

Figure 12.6 Muscles of the Vertebral Column

1 splenius capitis
2 splenius cervicis
3 iliocostalis
4 longissimus
5 spinalis
6 semispinalis capitis
7 semispinalis cervicis
8 semispinalis thoracis
9 multifidus
10 quadratus lumborum

Figure 12.7 Muscles of Respiration

1 external intercostals
2 internal intercostals
3 transversus thoracis
4 diaphragm
5 internal intercostals
6 external intercostals
7 diaphragm

Figure 12.8 Diaphragm

1 caval opening (for inferior vena cava)
2 aortic opening (hiatus)
3 esophageal opening (hiatus)

Figure 12.9 Muscles of the Abdominal Wall

1 tendinous intersections
2 rectus abdominis
3 transversus abdominis
4 internal oblique
5 external oblique

Chapter 13

Figure 13.1 Muscles That Move the Pectoral Girdle and Glenohumeral Joint

1 trapezius
2 deltoid
3 pectoralis major
4 biceps brachii (long head)
5 biceps brachii (short head)
6 subscapularis
7 coracobrachialis
8 pectoralis minor
9 serratus anterior
10 trapezius
11 rhomboid minor
12 rhomboid major
13 deltoid
14 rhomboid major
15 latissimus dorsi
16 levator scapulae
17 supraspinatus
18 infraspinatus
19 teres minor
20 teres major
21 serratus anterior

Figure 13.2 Anterior (Flexor) Compartment of the Arm

1 coracobrachialis
2 biceps brachii (short head)
3 biceps brachii (long head)
4 tendon of the long head of biceps brachii
5 coracobrachialis
6 brachialis

Figure 13.3 Posterior (Extensor) Compartment of the Arm

1 lateral head of triceps brachii
2 long head of triceps brachii

Figure 13.4 Anterior (Flexor) Compartment of the Forearm

1 pronator teres
2 brachioradialis
3 flexor retinaculum
4 flexor carpi radialis
5 palmaris longus
6 flexor carpi ulnaris
7 flexor digitorum superficialis
8 palmar aponeurosis
9 flexor pollicis longus
10 pronator quadratus
11 flexor digitorum profundus

Figure 13.6 Posterior (Extensor) Compartment of the Forearm

1. anconeus
2. extensor carpi ulnaris
3. extensor digiti minimi
4. extensor retinaculum
5. extensor digitorum tendons
6. extensor carpi radialis longus
7. extensor carpi radialis brevis
8. extensor digitorum
9. abductor pollicis longus
10. extensor pollicis brevis
11. extensor pollicis longus
12. extensor indicis
13. supinator
14. abductor pollicis longus
15. extensor pollicis brevis

Figure 13.7 Intrinsic Muscles of the Hand

1. lateral lumbricals
2. flexor digiti minimi brevis
3. abductor digiti minimi
4. first dorsal interosseous
5. medial lumbrical
6. adductor pollicis
7. flexor pollicis brevis
8. abductor pollicis brevis
9. thenar group
10. hypothenar group

Figure 13.9 Muscles That Act About the Hip Joint/Thigh

1. gluteus maximus
2. piriformis
3. superior gemellus
4. obturator internus
5. inferior gemellus
6. gluteus medius
7. gluteus minimus
8. gluteus medius
9. quadratus femoris
10. iliacus
11. psoas
12. iliopsoas
13. tensor fasciae latae
14. iliotibial tract

Figure 13.10 Medial Compartment of the Thigh

1. pectineus
2. adductor brevis
3. adductor longus
4. gracilis

Figure 13.11 Anterior Compartment of the Thigh

1. inguinal ligament
2. tensor fasciae latae
3. iliotibial tract
4. rectus femoris
5. vastus lateralis
6. pectineus
7. adductor longus
8. gracilis
9. sartorius
10. vastus medialis
11. quadriceps tendon

Figure 13.12 Muscles That Flex the Knee Joint/Leg

1. semimembranosus
2. semitendinosus
3. biceps femoris (long head)
4. biceps femoris (short head)

Figure 13.13 Anterior Compartment of the Leg

1. extensor digitorum longus
2. extensor hallucis longus
3. fibularis tertius tendon
4. tibialis anterior

Figure 13.14 Posterior Compartment of the Leg

1. plantaris
2. popliteus
3. tibialis posterior
4. flexor digitorum longus
5. flexor hallucis longus
6. calcaneal tendon

Figure 13.15 Lateral View of the Leg

1. gastrocnemius
2. soleus
3. fibularis longus
4. fibularis brevis
5. fibularis tertius
6. extensor digitorum brevis
7. fibularis tertius tendon
8. tibialis anterior
9. extensor digitorum longus
10. extensor hallucis longus
11. extensor hallucis brevis
12. extensor hallucis longus tendon
13. extensor digitorum longus tendons

Figure 13.16 Intrinsic Muscles of the Foot

1. flexor digitorum brevis
2. abductor hallucis
3. abductor digiti minimi
4. lumbricals
5. tendon of flexor hallucis longus
6. tendons of flexor digitorum longus
7. quadratus plantae
8. adductor hallucis
9. flexor hallucis brevis
10. flexor digiti minimi brevis
11. quadratus plantae
12. abductor digiti minimi
13. abductor hallucis
14. plantar interossei
15. dorsal interossei

Chapter 14

No Activities

Chapter 15

Figure 15.2 Meningeal Structures

(a) Coronal Section

1. superior sagittal sinus
2. falx cerebri
3. pia mater
4. subarachnoid space
5. arachnoid mater
6. falx cerebri
7. dura mater (periosteal layer)
8. dura mater
9. superior sagittal sinus
10. arachnoid villi
11. dura mater (meningeal layer)

(b) Midsagittal Section

12. falx cerebri
13. diaphragma sellae
14. superior sagittal sinus
15. inferior sagittal sinus
16. tentorium cerebelli
17. straight sinus
18. confluence of sinuses
19. falx cerebelli

Figure 15.4 Cast of the Ventricles of the Brain

1. lateral ventricles
2. interventricular foramen
3. third ventricle
4. cerebral aqueduct
5. fourth ventricle

Figure 15.5 Cerebrospinal Fluid (CSF) Production and Circulation

1. periosteal dura
2. arachnoid villus
3. superior sagittal sinus
4. meningeal dura
5. arachnoid mater
6. subarachnoid space
7. pia mater
8. brain/cerebral hemisphere
9. subarachnoid space
10. arachnoid villi
11. dura mater
12. superior sagittal sinus
13. pia mater
14. choroid plexus in third ventricle
15. choroid plexus in lateral ventricle
16. interventricular foramen
17. cerebral aqueduct
18. fourth ventricle
19. choroid plexus in fourth ventricle
20. median or lateral apertures
21. subarachnoid space
22. central canal of spinal cord

Exercise 15.3, #2

1. lateral ventricles
2. interventricular foramen
3. third ventricle
4. cerebral aqueduct
5. fourth ventricle
6. median and lateral apertures
7. subarachnoid space
8. arachnoid villi

Figure 15.6 Superior View of the Brain

1. frontal lobe
2. precentral gyrus
3. central sulcus
4. postcentral gyrus
5. longitudinal fissure
6. parietal lobe
7. occipital lobe

Figure 15.7 Lateral View of the Brain

1. frontal lobe
2. parietal lobe
3. central sulcus
4. precentral gyrus
5. postcentral gyrus
6. lateral sulcus
7. temporal lobe
8. pons
9. medulla oblongata
10. occipital lobe
11. transverse fissure
12. cerebellum

Figure 15.8 Inferior View of the Brain

1. frontal lobe
2. infundibulum
3. mammillary bodies
4. temporal lobe
5. pons
6. occipital lobe
7. olfactory bulb
8. olfactory tract
9. optic chiasm
10. optic nerve
11. optic tract
12. midbrain
13. cerebellum
14. medulla oblongata

Figure 15.9 Midsagittal View of the Brain

1. frontal lobe
2. cingulate gyrus
3. corpus callosum
4. septum pellucidum
5. interthalamic adhesion
6. third ventricle
7. hypothalamus
8. midbrain
9. infundibulum
10. cerebral peduncle
11. temporal lobe
12. pons
13. medulla oblongata
14. central sulcus
15. parietal lobe
16. thalamus
17. parieto-occipital sulcus
18. occipital lobe
19. pineal body (gland)
20. tectal plate (corpora quadrigemina)
21. mammillary body
22. cerebral aqueduct
23. fourth ventricle
24. cerebellum

Exercise 15.8, #4

Point of Exit	Cranial Nerve
Midbrain	III, IV
Pons	V, VI, VII, part of VIII
Medulla oblongata	part of VIII, IX, X, XI, XII

Chapter 16

Figure 16.2 Regional Gross Anatomy of the Spinal Cord

1 cervical plexus
2 cervical enlargement
3 brachial plexus
4 lumbosacral enlargement
5 L1 vertebra
6 conus medullaris
7 lumbar plexus
8 sacral plexus
9 filum terminale
10 cauda equina
11 brain (cerebellum)
12 posterior rootlets
13 posterior median sulcus
14 posterior roots
15 anterior roots
16 posterior median sulcus
17 pia mater
18 denticulate ligament
19 dura mater
20 conus medullaris
21 dura mater
22 cauda equina
23 filum terminale
24 posterior root ganglion

Figure 16.3 Model of a Spinal Cord in Cross Section

1 posterior funiculus
2 posterior horn
3 lateral horn
4 posterior rootlets
5 posterior root ganglion
6 spinal nerve
7 anterior root
8 anterior horn
9 posterior median sulcus
10 anterior rootlets
11 anterior median fissure

Figure 16.5 The Phrenic Nerves in the Thoracic Cavity

1 right phrenic nerve
2 diaphragm
3 left phrenic nerve

Figure 16.7 Major Nerves of the Brachial Plexus

1 lateral cord
2 posterior cord
3 medial cord
4 musculocutaneous nerve
5 axillary nerve
6 radial nerve
7 median nerve
8 ulnar nerve
9 superior trunk
10 middle trunk
11 inferior trunk

Figure 16.8 Nerves of the Lumbar Plexus

1 obturator nerve
2 femoral nerve

Figure 16.9 Nerves of the Sacral Plexus Within the Gluteal and Popliteal Regions

1 inferior gluteal nerve
2 posterior femoral cutaneous nerve
3 pudendal nerve
4 superior gluteal nerve
5 sciatic nerve
6 tibial nerve
7 common fibular nerve

Figure 16.10 Components of a Reflex

1 control center
2 sensory neuron
3 motor neuron
4 effector
5 receptor

Chapter 17

Figure 17.2 Overview of the Parasympathetic Division of the ANS

1 oculomotor nerve (CN III)
2 facial nerve (CN VII)
3 glossopharyngeal nerve (CN IX)
4 vagus nerve (CN X)
5 cardiac plexus
6 abdominal aortic plexus
7 pelvic splanchnic nerves

Figure 17.3 Overview of the Sympathetic Division of the ANS

1 sympathetic trunk ganglia (paravertebral)
2 adrenal medulla
3 prevertebral ganglia
4 sympathetic trunk

Chapter 18

Figure 18.11*b* Accessory Structures of the Eye

1 eyelashes
2 medial canthus
3 eyebrow
4 superior eyelid
5 pupil
6 sclera
7 iris
8 inferior eyelid

Figure 18.13 Extrinsic Eye Muscles

1 superior oblique
2 superior rectus
3 lateral rectus
4 inferior rectus
5 inferior oblique
6 superior rectus
7 superior oblique
8 medial rectus
9 inferior rectus
10 inferior oblique

Chapter 19

Figure 19.8 Labeling Major Endocrine Glands of the Body

1 parathyroid glands
2 thyroid gland
3 thymus
4 adrenal cortex
5 adrenal medulla
6 testes
7 pineal gland
8 hypothalamus
9 pituitary gland
10 pancreas
11 ovaries
12 hypothalamus
13 anterior pituitary gland
14 pineal gland
15 posterior pituitary gland

Chapter 20

No Activities

Chapter 21

Table 21.1 Comparisons Between Cardiac and Skeletal Muscle Tissues

Muscle Tissue	Cardiac Muscle	Skeletal Muscle
Nervous Control	Autonomic	Somatic
Appearance of Cells	Short, branched, striated	Long, cylindrical, striated
Number of Nuclei	Uninucleate	Multinucleate
Location of Nuclei	Central	Peripheral

Figure 21.1 Cardiac Muscle Tissue

1 striations
2 intercalated disc
3 nucleus
4 cardiac muscle cell

Figure 21.3 Location of the Heart Within the Thoracic Cavity

1 right lung/pleural cavity
2 diaphragm
3 mediastinum
4 left lung/pleural cavity
5 heart

Figure 21.4 Pericardium

1 fibrous pericardium
2 parietal layer of serous pericardium
3 pericardial cavity
4 visceral layer of serous pericardium
5 myocardium
6 endocardium
7 pericardial cavity
8 myocardium
9 endocardium
10 visceral layer of serous pericardium
11 fibrous pericardium
12 diaphragm

Figure 21.12 Circulation to and from the Heart Wall

1 right coronary artery
2 marginal artery
3 small cardiac vein
4 left coronary artery
5 circumflex artery
6 great cardiac vein
7 anterior interventricular artery
8 coronary sinus
9 right coronary artery
10 posterior interventricular artery
11 middle cardiac vein

Chapter 22

Exercise 22.7, #4

right ventricle → pulmonary semilunar valve → pulmonary trunk → pulmonary arteries → lungs → pulmonary veins → left atrium

Figure 22.6 Pulmonary Circuit

1 right atrium
2 right AV valve
3 right ventricle
4 pulmonary semilunar valve
5 pulmonary trunk
6 pulmonary arteries
7 branch of pulmonary artery
8 pulmonary capillaries
9 branch of pulmonary vein
10 pulmonary veins
11 left atrium
12 left AV valve
13 left ventricle
14 aortic semilunar valve
15 aorta

Figure 22.7 Circulation to the Head and Neck

(a) Arterial Supply

1 internal carotid artery
2 external carotid artery
3 common carotid artery
4 vertebral artery
5 thyrocervical trunk
6 subclavian artery
7 superficial temporal artery
8 occipital artery
9 facial artery
10 brachiocephalic trunk (artery)

(b) Venous Drainage

1 vertebral vein
2 external jugular vein
3 internal jugular vein
4 subclavian vein
5 right brachiocephalic vein
6 superficial temporal vein
7 facial vein
8 superior thyroid vein

Figure 22.8 Circulation from the Aortic Arch to the Anterior Part of the Right Parietal Bone and Back to the Superior Vena Cava

1 brachiocephalic trunk (artery)
2 right common carotid artery
3 right external carotid artery
4 right superficial temporal artery
5 right superficial temporal vein
6 right internal jugular vein
7 right brachiocephalic vein

Figure 22.9 Circulation to the Brain

(a) Arterial Supply

1 middle cerebral artery
2 internal carotid artery
3 posterior cerebral artery
4 anterior communicating artery
5 anterior cerebral artery
6 internal carotid artery
7 posterior communicating artery
8 posterior cerebral artery
9 basilar artery
10 vertebral artery

(b) Venous Drainage

1 straight sinus
2 transverse sinus
3 sigmoid sinus
4 internal jugular vein
5 superior sagittal sinus
6 inferior sagittal sinus
7 cavernous sinus

Figure 22.10 Circulation from the Aortic Arch to the Right Parietal Lobe of the Brain and Back to the Right Brachiocephalic Vein

1 brachiocephalic trunk (artery)
2 right common carotid artery
a. Alternate 2: right subclavian artery
3 right internal carotid artery
b. Alternate 3: right vertebral artery
4 middle cerebral artery
c. Alternate 4: basilar artery
d. Alternate 5: posterior cerebral artery
e. Alternate 6: posterior communicating artery
5 superior sagittal sinus
6 transverse sinus
7 sigmoid sinus
8 right internal jugular vein
9 right brachiocephalic vein

Figure 22.11 Circulation to the Thoracic and Abdominal Walls

(a) Arterial Supply

1 right subclavian artery
2 internal thoracic artery
3 brachiocephalic trunk
4 internal thoracic artery
5 anterior intercostal arteries
6 posterior intercostal arteries
7 superior epigastric artery
8 descending abdominal aorta
9 right lumbar artery
10 inferior epigastric artery
11 aortic arch
12 posterior intercostal arteries (1–2)
13 left subclavian artery
14 posterior intercostal arteries (3–11)
15 descending thoracic aorta
16 left common iliac artery

(b) Venous Drainage

1 right subclavian vein
2 right brachiocephalic vein
3 superior vena cava
4 anterior intercostal veins
5 azygos vein
6 internal thoracic vein
7 right posterior intercostal vein
8 inferior vena cava
9 right superior epigastric vein
10 right lumbar veins
11 right inferior epigastric vein
12 left subclavian vein
13 left brachiocephalic vein
14 accessory hemiazygos vein
15 left posterior intercostal vein
16 hemiazygos vein
17 inferior vena cava
18 left common iliac vein
19 median sacral vein

Figure 22.12 Circulation from the Left Ventricle of the Heart to the Right Kidney and Back to the Right Atrium of the Heart

1 ascending aorta
2 aortic arch
3 descending thoracic aorta
4 descending abdominal aorta
5 right renal artery
6 right renal vein
7 inferior vena cava

Figure 22.13 Arterial Supply to Abdominal Organs

(a) Arterial Supply to the Stomach, Spleen, Pancreas, Duodenum, and Liver

1 celiac trunk
2 common hepatic artery
3 hepatic artery proper
4 left hepatic artery
5 right hepatic artery
6 gastroduodenal artery
7 right gastric artery
8 left gastric artery
9 splenic artery
10 descending abdominal aorta

(b) Arterial Supply to the Small and Large Intestines

1 middle colic artery
2 right colic artery
3 ileocolic artery
4 celiac trunk
5 superior mesenteric artery
6 left colic artery
7 descending abdominal aorta
8 inferior mesenteric artery
9 sigmoid arteries
10 superior rectal artery

Figure 22.14 The Hepatic Portal System

1 inferior vena cava
2 hepatic veins
3 hepatic portal vein
4 superior mesenteric vein
5 gastric veins
6 splenic vein
7 inferior mesenteric vein

Figure 22.15 Circulation from the Abdominal Aorta to the Spleen and Back to the Right Atrium of the Heart

1 abdominal aorta
2 celiac trunk
3 splenic artery
4 splenic vein
5 hepatic portal vein
6 hepatic veins
7 inferior vena cava

Figure 22.16 Circulation from the Abdominal Aorta to the Duodenum and Back to the Right Atrium of the Heart

1 abdominal aorta
2 celiac trunk
3 common hepatic artery
4 gastroduodenal artery
5 superior mesenteric vein
6 hepatic portal vein
7 hepatic veins
8 inferior vena cava

Figure 22.17 Circulation from the Abdominal Aorta to the Sigmoid Colon and Back to the Right Atrium of the Heart

1 abdominal aorta
2 inferior mesenteric artery
3 sigmoid arteries
4 inferior mesenteric vein
5 splenic vein
6 hepatic portal vein
7 hepatic veins
8 inferior vena cava

Figure 22.18 Circulation to the Upper Limb

(a) Arterial Supply

1 brachiocephalic trunk (artery)
2 subclavian artery
3 axillary artery
4 deep brachial artery
5 brachial artery
6 ulnar artery
7 radial artery
8 deep palmar arch
9 superficial palmar arch
10 digital arteries

(b) Venous Drainage

1 brachiocephalic vein
2 subclavian vein
3 axillary vein
4 cephalic vein
5 basilic vein
6 brachial veins
7 median cubital vein
8 radial veins
9 cephalic vein
10 deep palmar venous arch
11 superficial palmar venous arch
12 dorsal venous network
13 digital veins
14 ulnar veins

Figure 22.19 Circulation from the Aortic Arch to the Anterior Surface of the Index Finger and Back Along a Superficial Route to the Superior Vena Cava

1 brachiocephalic trunk (artery)
2 subclavian artery
3 axillary artery
4 brachial artery
5 radial artery
6 superficial palmar arch
7 digital artery
8 digital vein
9 superficial palmar venous arch
10 cephalic vein
a. Alternate 11: median cubital vein
b. Alternate 12: basilic vein
c. Alternate 13: axillary vein
11 subclavian vein
12 brachiocephalic vein
13 superior vena cava

Figure 22.20 Circulation from the Aortic Arch to the Capitate Bone of the Wrist and Back Along a Deep Route to the Superior Vena Cava

1 brachiocephalic trunk (artery)
2 subclavian artery
3 axillary artery
4 brachial artery
5 ulnar artery
6 deep palmar arch
7 deep palmar venous arch
8 ulnar veins
9 brachial vein
10 axillary vein
11 subclavian vein
12 brachiocephalic vein
13 superior vena cava

Figure 22.21 Circulation to the Lower Limb

(a) Arterial Supply

1 common iliac artery
2 internal iliac artery
3 external iliac artery
4 femoral artery
5 deep femoral artery
6 anterior tibial artery
7 fibular artery
8 dorsalis pedis artery
9 popliteal artery
10 posterior tibial artery
11 lateral plantar artery
12 medial plantar artery
13 plantar arterial arch
14 digital arteries

(b) Venous Drainage

1 common iliac vein
2 external iliac vein
3 internal iliac vein
4 deep femoral vein
5 femoral vein
6 great saphenous vein
7 anterior tibial veins
8 fibular veins
9 dorsal venous arch
10 popliteal vein
11 small saphenous vein
12 posterior tibial veins
13 small saphenous vein
14 lateral plantar vein
15 medial plantar vein
16 digital veins

Figure 22.22 Circulation from the Abdominal Aorta to the Dorsal Surface of the Big Toe (Hallux) and Back Along a Superficial Route to the Inferior Vena Cava

1 abdominal aorta
2 common iliac artery
3 external iliac artery
4 femoral artery
5 popliteal artery
6 anterior tibial artery
7 dorsalis pedis artery
8 digital artery
9 digital vein
10 dorsal venous arch
11 great saphenous vein
12 femoral vein
13 external iliac vein
14 common iliac vein
15 inferior vena cava

Figure 22.23 Circulation from the Abdominal Aorta to the Cuboid Bone of the Foot and Back Along a Deep Route to the Inferior Vena Cava

1 abdominal aorta
2 common iliac artery
3 external iliac artery
4 femoral artery
5 popliteal artery
6 posterior tibial artery
7 lateral plantar artery
8 lateral plantar veins
9 posterior tibial vein
10 popliteal vein
11 femoral vein
12 external iliac vein
13 common iliac vein
14 inferior vena cava

Exercise 22.14, #2

left ventricle → aorta → common iliac arteries → internal iliac arteries → umbilical arteries → placenta → umbilical vein → ductus venosus → inferior vena cava → right atrium

Table 22.4 Fetal Cardiovascular Structures and Associated Postnatal Structures

Fetal Cardiovascular Structure	Postnatal Structure	Function of Fetal Cardiovascular Structure
Ductus Arteriosus	Ligamentum arteriosum	Shunt blood from the pulmonary trunk to the aorta, thereby bypassing the pulmonary circuit and nonfunctional lungs
Ductus Venosus	Ligamentum venosum	Carry oxygenated blood from umbilical vein to the inferior vena cava
Foramen Ovale	Fossa ovalis	Shunt blood from the right atrium to the left atrium, thereby bypassing the nonfunctional lungs
Umbilical Arteries	Medial umbilical ligaments	Carry deoxygenated blood from the fetus to the placenta so it can obtain oxygen from the mother's blood
Umbilical Vein	Round ligament of the liver (ligamentum teres)	Carry oxygenated blood from the placenta to the fetal circulation

Figure 22.24 The Fetal Circulation

1 superior vena cava
2 lung
3 right atrium
4 liver
5 ductus venosus
6 inferior vena cava
7 umbilical vein
8 aortic arch
9 ductus arteriosus
10 pulmonary artery
11 pulmonary veins
12 foramen ovale
13 right ventricle
14 heart
15 descending abdominal aorta
16 common iliac artery
17 umbilical arteries
18 umbilical cord
19 placenta

Chapter 23

Figure 23.11 Major Lymph Vessels of the Body

1 right lymphatic duct
2 thoracic duct
3 lymph nodes
4 cisterna chyli

Figure 23.12 Mucosa-Associated Lymphatic Tissue (MALT)

1 pharyngeal tonsils
2 palatine tonsils
3 lingual tonsils
4 vermiform appendix

Figure 23.13 Lymph Node and Its Components

1 cortex
2 medulla
3 afferent lymphatic vessels
4 lymphatic nodule
5 germinal center
6 cortical sinus
7 medullary sinus
8 medullary cord
9 capsule
10 trabecula
11 hilum
12 efferent lymphatic vessel

Figure 23.14 Gross Anatomy of the Spleen

1 splenic artery
2 hilum of the spleen
3 spleen
4 splenic vein

Chapter 24

Figure 24.7 The Upper Respiratory Tract: Midsagittal View

1 frontal sinus
2 superior nasal concha
3 middle nasal concha
4 inferior nasal concha
5 vestibule
6 epiglottis
7 thyroid cartilage
8 cricoid cartilage
9 sphenoidal sinus
10 superior meatus
11 middle meatus
12 inferior meatus
13 nasopharynx
14 oropharynx
15 laryngopharynx
16 lumen of larynx

Figure 24.8 Classroom Model of the Larynx

1 epiglottis
2 hyoid bone
3 thyroid cartilage
4 laryngeal prominence
5 tracheal C ring
6 epiglottis
7 hyoid bone
8 thyrohyoid membrane
9 cuneiform cartilage
10 corniculate cartilage
11 thyroid cartilage
12 arytenoid cartilage
13 cricoid cartilage
14 trachealis muscle
15 trachea
16 epiglottis
17 corniculate cartilage
18 false vocal cords

19. true vocal cords
20. cricoid cartilage
21. hyoid bone
22. thyrohyoid membrane
23. thyroid cartilage

Figure 24.9 The Pleural Cavities

1. visceral pleura
2. left lung
3. right lung
4. mediastinum
5. diaphragm
6. parietal pleura

Figure 24.10 The Right Lung

1. superior lobe
2. horizontal fissure
3. oblique fissure
4. middle lobe
5. inferior lobe
6. base
7. horizontal fissure
8. middle lobe
9. oblique fissure
10. apex
11. superior lobe
12. oblique fissure
13. pulmonary arteries
14. primary bronchi
15. pulmonary veins
16. inferior lobe
17. base

Figure 24.11 The Left Lung

1. apex
2. superior lobe
3. oblique fissure
4. cardiac notch
5. inferior lobe
6. base
7. superior lobe
8. pulmonary artery
9. primary bronchi
10. pulmonary veins
11. inferior lobe
12. base
13. oblique fissure
14. cardiac impression
15. cardiac notch
16. oblique fissure

Figure 24.12 The Bronchial Tree

1. right main bronchus
2. right segmental bronchi
3. right lobar bronchi
4. trachea
5. left main bronchus
6. left segmental bronchi
7. left lobar bronchi
8. carina (internal structure)

Chapter 25

Figure 25.7 Coronal Section Through the Right Kidney

1. minor calyx
2. renal pelvis
3. major calyx
4. renal (medullary) pyramid
5. renal column
6. fibrous capsule
7. renal cortex
8. renal medulla
9. renal papilla
10. renal sinus
11. ureter
12. renal lobe

Figure 25.9 Models of the Kidney Demonstrating the Blood Supply to the Kidney

(a) Coronal Section

1. peritubular capillaries
2. interlobular vein
3. interlobular artery
4. interlobar vein
5. interlobar artery
6. vasa recta
7. renal artery
8. renal vein
9. segmental artery
10. glomerulus
11. afferent arteriole
12. arcuate vein
13. arcuate artery

(b) Close-up of the Renal Cortex and Renal Medulla

1. peritubular capillaries
2. efferent arteriole
3. afferent arteriole
4. glomerulus
5. interlobular artery
6. interlobular vein
7. arcuate vein
8. vasa recta
9. arcuate artery
10. interlobar vein
11. interlobar artery

Exercise 25.6, #4

abdominal aorta → renal artery → segmental artery → interlobar artery → arcuate artery → interlobular artery → afferent arteriole → glomerulus → efferent arteriole → peritubular capillaries or vasa recta → interlobular vein → arcuate vein → interlobar vein → renal vein → inferior vena cava

Figure 25.10 Location of the Structures of the Urinary System Within the Abdominopelvic Cavity

1. left kidney
2. right kidney
3. ureters
4. urinary bladder

Figure 25.11 Classroom Model of Urinary System Structures

1. renal cortex
2. renal (medullary) pyramid
3. minor calyx
4. renal papilla
5. major calyx
6. renal pelvis
7. urinary bladder
8. left kidney
9. renal artery
10. renal vein
11. ureter

Figure 25.12 Classroom Model of a Midsagittal Section Through the Male Pelvis

1. urinary bladder
2. prostatic urethra
3. membranous urethra
4. spongy urethra
5. ureter
6. muscularis (detrusor muscle)
7. internal urethral sphincter
8. external urethral sphincter

Figure 25.13 Classroom Model of a Midsagittal Section Through the Female Pelvis

1. urinary bladder
2. urethra
3. ureter
4. muscularis (detrusor muscle)
5. internal urethral sphincter
6. external urethral sphincter

Chapter 26

Figure 26.11 Oral Cavity

(a) Anterior View of the Oral Cavity

1. superior lip
2. superior labial frenulum
3. hard palate
4. palatoglossal arch
5. palatopharyngeal arch
6. palatine tonsil
7. sublingual duct orifice
8. submandibular duct orifice
9. inferior labial frenulum
10. transverse palatine folds
11. soft palate
12. uvula
13. fauces
14. tongue
15. lingual frenulum
16. teeth
17. gingivae
18. inferior lip

(b) Midsagittal View of the Oral Cavity and Pharynx

1. hard palate
2. oral cavity
3. tongue
4. vestibule
5. soft palate
6. uvula
7. palatine tonsil
8. oropharynx
9. lingual tonsil
10. epiglottis
11. laryngopharynx
12. esophagus

(c) Salivary Glands

1. parotid salivary gland
2. parotid duct
3. sublingual ducts
4. submandibular duct
5. sublingual salivary gland
6. submandibular salivary gland

Figure 26.12 Classroom Model of the Stomach

1. lesser curvature
2. pyloric sphincter
3. pylorus
4. esophagus
5. fundus
6. inferior esophageal (cardiac) sphincter
7. body of the stomach
8. greater curvature
9. pyloric sphincter
10. pylorus
11. fundus
12. cardia
13. inferior esophageal (cardiac) sphincter
14. gastric folds (rugae)

Figure 26.13 Classroom Model of the Duodenum, Liver, Gallbladder, and Pancreas

1. gallbladder
2. duodenum
3. accessory pancreatic duct
4. minor duodenal papilla
5. common bile duct
6. major duodenal papilla
7. head of pancreas
8. left and right hepatic ducts
9. cystic duct
10. common hepatic duct
11. tail of pancreas
12. body of pancreas
13. main pancreatic duct
14. hepatopancreatic ampulla

Figure 26.14 Classroom Model of the Liver

1. gallbladder
2. right lobe of liver
3. common hepatic duct
4. cystic duct
5. common bile duct
6. hepatic portal vein
7. hepatic artery proper
8. inferior vena cava
9. quadrate lobe of liver
10. round ligament of the liver
11. right hepatic duct
12. left hepatic duct
13. left lobe of liver
14. caudate lobe of liver
15. porta hepatis

Figure 26.16 Classroom Model of the Abdominal Cavity and Large Intestine

1 right lobe of liver
2 falciform ligament
3 pylorus of stomach
4 gallbladder
5 greater curvature of stomach
6 ascending colon
7 ileum of small intestine
8 esophagus
9 left lobe of liver
10 body of stomach
11 left colic (splenic) flexure
12 transverse colon
13 taenia coli
14 jejunum of small intestine
15 descending colon
16 esophagus
17 duodenum
18 right colic (hepatic) flexure of colon
19 ascending colon
20 taenia coli
21 cecum
22 tail of pancreas
23 body of pancreas
24 left colic (splenic) flexure of colon
25 transverse colon
26 descending colon
27 rectum
28 sigmoid colon

Chapter 27

Figure 27.12 Supporting Ligaments of the Ovary, Uterine Tubes, and Uterus as Seen from a Posterior View

1 ovarian ligament
2 uterine tube
3 suspensory ligament of the ovary
4 fimbria
5 mesosalpinx
6 body of uterus
7 broad ligament
8 uterine blood vessels
9 uterosacral ligament
10 transverse cervical ligament
11 vagina
12 uterine part of uterine tube
13 isthmus of uterine tube
14 ampulla of uterine tube
15 infundibulum of uterine tube
16 round ligament of the uterus
17 endometrium
18 myometrium
19 perimetrium
20 internal os
21 cervical canal
22 external os
23 uterine tube
24 mesosalpinx
25 mesovarium

Figure 27.13 Classroom Model of the Female Pelvic Cavity

1 uterine tube
2 ovary
3 uterus
4 urinary bladder
5 labia minora
6 labia majora
7 rectum
8 vagina
9 bulb of the vestibule
10 round ligament of the uterus
11 vesicouterine pouch
12 pubic symphysis
13 clitoris
14 external urethral orifice
15 vaginal orifice
16 anus
17 ureter
18 fimbria of uterine tube
19 rectouterine pouch
20 cervix of uterus
21 vagina

Figure 27.14 The Female Breast

1 suspensory ligaments
2 lobe
3 lactiferous sinus
4 alveoli
5 lactiferous ducts
6 lobule
7 areolar gland
8 nipple
9 areola
10 adipose tissue
11 lobe
12 deep fascia
13 alveoli
14 lobule
15 suspensory ligaments
16 lactiferous sinus
17 nipple
18 lactiferous ducts

Figure 27.15 Male Reproductive Tract Structures

1 ureter
2 ampulla of ductus deferens
3 seminal vesicle
4 ejaculatory duct
5 prostate gland
6 prostatic urethra
7 bulbourethral gland
8 urogenital diaphragm
9 membranous urethra
10 ductus deferens
11 epididymis
12 testis
13 spongy urethra
14 urinary bladder
15 bulb of penis
16 crus of penis
17 corpus cavernosum
18 seminiferous tubules
19 corpus spongiosum
20 glans penis

Figure 27.16 Classroom Model of the Male Pelvic Cavity

1 ureter
2 urinary bladder
3 ductus deferens
4 penis
5 prepuce
6 glans penis
7 ductus deferens (ampulla)
8 seminal vesicle
9 prostate gland
10 ductus deferens
11 testicular artery and vein
12 spermatic cord
13 epididymis
14 testis
15 scrotum
16 ureter
17 ductus deferens
18 pubic symphysis
19 prostate gland
20 spongy urethra
21 corpus cavernosum
22 prepuce
23 corpus spongiosum
24 tunica albuginea of testis
25 tunica vaginalis of testis
26 urinary bladder
27 rectum
28 internal urethral sphincter
29 prostatic urethra
30 ejaculatory duct
31 urogenital diaphragm
32 anus
33 membranous urethra
34 epididymis
35 seminiferous tubules

Figure 27.19 Structures of the 3-week-old Embryo

1 connecting stalk
2 yolk sac
3 amniotic cavity
4 amnion
5 embryo
6 chorion
7 functional layer of uterus
8 placenta

Figure 27.20 Structures of the 4-week-old (Late) Embryo

1 ectoderm
2 mesoderm
3 amnion
4 peritoneal cavity
5 gut tube
6 somite
7 intermediate mesoderm
8 mesentery
9 endoderm
10 abdominal wall

Table 27.14 Pre-Embryonic Period

Developmental Stage	Location	Events
Fertilization	Ampulla of uterine tube	Sperm penetrates secondary oocyte; secondary oocyte completes meiosis and becomes an ovum
Zygote	Ampulla of uterine tube	Ovum and sperm pronuclei fuse to produce diploid cell
Cleavage	Uterine tube	Zygote undergoes cell division
Morula	Uterine tube	Structure formed resembles a solid ball of cells (16 or more cells present)
Blastocyst	Uterus	Hollow ball of cells; outer ring formed by trophoblast; embryoblast is an inner cell cluster
Implantation	Functional layer of endometrium of uterus	Blastocyst adheres to functional layer of uterus; placenta forms (trophoblast cells and functional layer)

Table 27.15 Stages of Embryonic Development

Developmental Week	Events
Week 3	Three primary germ layers form; notochord develops
Week 4	Derivatives of three germ layers form; limb buds appear
Weeks 5–8	Head enlarges; eyes, ears, and nose appear; major organs form

Chapter 28

Table 28.1 Comparison of Human and Cat Vertebrae

Number of Vertebrae	Human	Cat
Cervical	7	7
Thoracic	12	13
Lumbar	5	7
Sacral (fused)	5	3
Coccygeal/caudal	1–2	~20

Table 28.5 Fiber Orientation of Cat Abdominal Muscles

Muscle	Fiber Orientation
External Oblique	Medial and caudal
Internal Oblique	Medial and cranial
Rectus Abdominis	Straight along the sagittal plane
Transverse Abdominis	Straight along the transverse plane

credits

PHOTOGRAPHS

Front Matter

Page iv (top): © Christine Eckel; p. iv (middle): © Kyla Ross; p. iv (bottom) © Jay Bidle.

Chapter 1

All Photos: © Christine Eckel.

Chapter 2

Figure 2.1a: © McGraw-Hill Education/Eric Wise, photographer; 2.2 (1-2): © McGraw-Hill Education/ Photo and Dissection by Christine Eckel; 2.2 (3-4), 2.3a-c: © Christine Eckel.

Chapter 3

Figures 3.1-3.3: © Christine Eckel; 3.5a-c: © McGraw-Hill Education/Al Telser, photographer; 3.7 & p. 55: © Christine Eckel.

Chapter 4

Figure 4.1a-e: © Christine Eckel; 4.2: © Ed Reschke/ Getty Images; Table 4.2 (1, 5): © Michael Abbey/ Science Source; Table 4.2 (2-4): © Carolina Biological Supply Company/Phototake; 4.3: © The Science Source; 4.6, 4.8a-c, 4.10-4.13, 4.16: © Christine Eckel.

Chapter 5

Table 5.1 (1-3): © McGraw-Hill Education/Al Telser, photographer; Table 5.2 (1-2): © Victor P. Eroschenko; Table 5.2 (3-4): © McGraw-Hill Education/Al Telser, photographer; 5.1: © Christine Eckel; Table 5.3 (1): © McGraw-Hill Education/Dennis Strete, photographer; Table 5.3 (2): © Visuals Unlimited/Corbis; Table 5.3 (3): © Ed Reschke/Getty Images; Table 5.3 (4): © Victor P. Eroschenko; 5.3: © McGraw-Hill Education/Al Telser, photographer; 5.4: © Ed Reschke/ Getty Images; 5.5: © Victor P. Eroschenko; 5.6: © McGraw-Hill Education/Al Telser, photographer; 5.7a: © Visuals Unlimited/Corbis; 5.7b: © Victor P. Eroschenko; 5.7c: © Ed Reschke/Getty Images; 5.8: © McGraw-Hill Education/Al Telser, photographer; 5.9: © Dr. Frederick Skvara/Visuals Unlimited/Corbis; Table 5.4 (1-2): © McGraw-Hill Education/Al Telser, photographer; Table 5.4 (3): © Ed Reschke/Getty Images; 5.10: © Victor P. Eroschenko; 5.12-5.14: © McGraw-Hill Education/Al Telser, photographer; 5.15: © Ed Reschke/Getty Images; 5.16, 5.17, 5.19: © McGraw-Hill Education/Christine Eckel, photographer; 5.20: © Ed Reschke/Getty Images; 5.21: © McGraw-Hill Education/Al Telser, photographer; 5.22: © Ed Reschke/Getty Images; 5.23, 5.25-5.27a: © McGraw-Hill Education/Al Telser, photographer; 5.27b: © Christine Eckel; 5.28a: © Ed Reschke/Getty Images; 5.28b: © Biophoto Associates/Science Source; p. 121a, d, f-h: © McGraw-Hill Education/Christine Eckel, photographer; p. 121b, c: © Christine Eckel; p. 121e: © McGraw-Hill Education/Dennis Strete, photographer; p. 121i: © McGraw-Hill Education/Al Telser, photographer.

Chapter 6

Figure 6.1: © Astrid & Hanns-Frieder Michler/Science Source; 6.2: © Tom Caceci; 6.3a: © James Stevenson/ Science Source; 6.3b: © Christine Eckel; 6.4: © Ed Reschke; 6.5-6.6: © Ed Reschke/Getty Images; 6.7-6.8: © Biophoto Associates/Science Source; 6.9: © Fred Hossler, Ph.D./CustomMedical/Newscom; 6.10: © Christine Eckel; 6.11: Model # J16 [1000294] © 3B Scientific GmbH, Germany, 2013 www.3bscientific.com/Photo by Christine Eckel, Ph.D.; p. 141: © Christine Eckel.

Chapter 7

Figure 7.1: © Ed Reschke/Getty Images; 7.2: © McGraw-Hill Education/Al Telser, photographer; 7.3a: © Ed Reschke/Getty Images; 7.3b: © Michael Klein/Getty Images; 7.4: © Biophoto Associates/ Science Source; 7.6: © Christine Eckel; 7.7a: © McGraw-Hill Education/Christine Eckel, photographer; 7.7b: © Ralph Hutchings/Visuals Unlimited/Corbis; 7.8-7.10 & p. 159: © Christine Eckel.

Chapter 8

Figure 8.1: © McGraw-Hill Education/Christine Eckel, photographer; 8.2, 8.3b: © Christine Eckel; 8.4-8.8: © McGraw-Hill Education/Christine Eckel, photographer; 8.10: © Andy Crawford/Dorling Kindersley Media Library; 8.11a-b: © Christine Eckel; 8.12: © Alinari Archives/The Image Works; 8.13-8.22: © McGraw-Hill Education/Christine Eckel, photographer; 8.23a: © Christine Eckel; 8.23b & p. 193 (left-right): © McGraw-Hill Education/Christine Eckel, photographer.

Chapter 9

Figure 9.1a-b: © Christine Eckel; 9.2-9.3: © McGraw-Hill Education/Christine Eckel, photographer; 9.4-9.6: © Christine Eckel; 9.7-9.9: © McGraw-Hill Education/ Christine Eckel, photographer; 9.10 (top): © David Hunt/Smithsonian Institution; 9.10 (bottom): © VideoSurgery/Science Source; 9.11a-d: © McGraw-Hill Education/Christine Eckel, photographer; 9.12-9.13: © Christine Eckel; 9.14-9.15: © McGraw-Hill Education/Christine Eckel, photographer; 9.16a-c: © McGraw-Hill Education/JW Ramsey, photographer; p. 225: © McGraw-Hill Education/Christine Eckel, photographer.

Chapter 10

Figure 10.5a-c: © Christine Eckel; 10.6a-b: © McGraw-Hill Education/Christine Eckel, photographer.

Chapter 11

Figure 11.1e: © James Dennis/Phototake; 11.2: © Dr. David Phillips/Visuals Unlimited/Corbis; 11.3: © Rick Ash; 11.4c: © Dr. Thomas Caceci, Virginia-Maryland Regional College of Veterinary Medicine; 11.5: © Ed Reschke/Getty Images; 11.6a-b: © McGraw-Hill Education/Al Telser, photographer; 11.13: © Christine Eckel; 11.18-11.19: Courtesy of and © BIOPAC Systems, Inc. www.biopac.com.

Chapter 12

Figures 12.1a-b, 12.5a, c, 12.6, 12.7b, 12.10: © McGraw-Hill Education/Christine Eckel, photographer; 12.12: © Christine Eckel.

Chapter 13

Figures 13.1-13.4a: © McGraw-Hill Education/Christine Eckel, photographer; 13.4b: © Christine Eckel; 13.5: © McGraw-Hill Education/Christine Eckel, photographer; p. 322 (middle right): © McGraw-Hill Education/ JW Ramsey, photography; 13.6a: © McGraw-Hill Education/Christine Eckel, photographer; 13.7: © Christine Eckel; 13.9a: © McGraw-Hill Education/ Christine Eckel, photographer; 13.9b, 13.10: © Christine Eckel; 13.11-13.15: © McGraw-Hill Education/Christine Eckel, photographer.

Chapter 14

Figure 14.1: © Ed Reschke/Getty Images; 14.2-14.9: © Rick Ash.

Chapter 15

Figure 15.2b: © McGraw-Hill Education/Christine Eckel, photographer; 15.4, 15.6a: Models # VH410 [1001262] & C22 [1000228] © 3B Scientific GmbH, Germany, 2013 www.3bscientific.com/Photos by Christine Eckel, Ph.D.; 15.6b-15.9: © McGraw-Hill Education/Christine Eckel, photographer; 15.10-15.18: © Christine Eckel; 15.32-15.34: Courtesy of and © BIOPAC Systems, Inc. www.biopac.com; p. 413 (top, bottom): © Christine Eckel.

Chapter 16

Figure 16.1a, c: © Ed Reschke; 16.1b: © Lester V. Bergman/Corbis; 16.1d: © McGraw-Hill Education/Al Telser, photographer; 16.2 (top, bottom): © McGraw-Hill Education/Photo and Dissection by Christine Eckel; 16.2 (middle): © Christine Eckel; 16.3: © Copyright by Denoyer-Geppert; 16.5, 16.7-16.9: © McGraw-Hill Education/Photo and Dissection by Christine Eckel; 16.11: © Jill Braaten.

Chapter 17

Figure 17.4: © Jill Braaten; 17.5-17.7: Courtesy of and © BIOPAC Systems, Inc. www.biopac.com.

Chapter 18

Figure 18.1: © McGraw-Hill Education/Al Telser, photographer; 18.2: © Carolina Biological Supply Company/Phototake; 18.3b: © CNRI/Science Photo Library/Corbis RF; 18.3c: © Christine Eckel; 18.3d: © McGraw-Hill Education/Al Telser, photographer; 18.3e: © Dr. John D. Cunningham/Visuals Unlimited/ Corbis; 18.4b: © McGraw-Hill Education/Al Telser, photographer; 18.5b: © Visuals Unlimited/Corbis; 18.6: © McGraw-Hill Education/Al Telser, photographer; 18.7a: © Gene Cox/Science Source; 18.7b: © Victor P. Eroschenko; 18.8c: © Biophoto Associates/Science Source; 18.9: © Dr. John D. Cunningham/Visuals Unlimited/Corbis; 18.10: Model # J16 [1000294] © 3B Scientific GmbH, Germany, 2013 www.3bscientific.com/Photo by Christine Eckel, Ph.D.; 18.11a: Model # F13 [1000258] © 3B Scientific GmbH, Germany, 2015 www.3bscientific.com/Photo by Christine Eckel, Ph.D.; 18.11b: © McGraw-Hill Education/JW Ramsey, photographer; 18.12: Model # F15 [1000259] © 3B Scientific GmbH, Germany, 2015 www.3bscientific.com/Photo by Christine Eckel, Ph.D.; 18.14-18.16: © Christine Eckel; 18.17a: Model # E10 [1000250] © 3B Scientific GmbH, Germany, 2015 www.3bscientific.com/Photo by Christine Eckel, Ph.D.; 18.17b: © Copyright by Denoyer-Geppert. Photo by Christine Eckel; 18.19-18.21: © Christine Eckel; 18.22: © McGraw-Hill Education/Rick Brady, photographer; 18.24: © Steve Allen/Getty Images RF; 18.25-18.28: © McGraw-Hill Education/Jill Braaten, photographer.

Chapter 19

Figure 19.1a: © McGraw-Hill Education/Al Telser, photographer; 19.1b: © Astrid & Hanns-Frieder Michler/Science Source; 19.2: © Victor P. Eroschenko; 19.3: © Christine Eckel; 19.4: © McGraw-Hill Education/Al Telser, photographer; 19.5a, c: © Victor P. Eroschenko; 19.5b, 19.6b (right), 19.7: © McGraw-Hill Education/Al Telser, photographer; p. 520 (all): © McGraw-Hill Education/Christine Eckel, photographer.

Chapter 20

Figures 20.1, 20.3: © Christine Eckel; 20.4a-g: © McGraw-Hill Education/Al Telser, photographer; 20.5a: © Dr. Dorothea Zucker-Franklin/Phototake; 20.8-20.10a: © Christine Eckel; 20.10b-c: © J & J Photography; 20.11-20.12: © Christine Eckel.

Chapter 21

Figure 21.1: © Visuals Unlimited/Corbis; 21.2: © Rick Ash; 21.3: Model # G15 [1000270] © 3B Scientific GmbH, Germany, 2015 www.3bscientific.com; 21.4a: © Christine Eckel; 21.5: © McGraw-Hill Education/Photo and Dissection by Christine Eckel; 21.6a: © Christine Eckel; 21.6b: © Copyright by Denoyer-Geppert. Photo by Christine Eckel; 21.7a: © Christine Eckel; 21.7b: © Copyright by Denoyer-Geppert. Photo by Christine Eckel; 21.8, 21.9a: © Christine Eckel; 21.9b: © Copyright by Denoyer-Geppert. Photo by Christine Eckel; 21.10, 21.11, 21.13-21.15: © Christine Eckel; 21.20, 21.21: Courtesy of and © BIOPAC Systems, Inc. www.biopac.com; 21.23a-c: © doc-stock GmbH/Phototake.

Chapter 22

Figures 22.1-22.3: © Christine Eckel; 22.4: © McGraw-Hill Education/Al Telser, photographer; 22.5: © Christine Eckel; Table 22.3a-b: © Visuals Unlimited/Corbis; Table 22.3c: Image and © Dr. H. Jastrow from Dr. Jastrow's electron microscopic atlas http://www.drjastrow.de; 22.26-22.27: Courtesy of and © BIOPAC Systems, Inc. www.biopac.com.

Chapter 23

Figure 23.1a: © Ed Reschke; 23.1b: © McGraw-Hill Education/Al Telser, photographer; 23.2: © Christine Eckel; 23.3b, d: © McGraw-Hill Education/Al Telser, photographer; 23.3c: © Biophoto Associates/Science Source; 23.4a-b: © McGraw-Hill Education/Christine Eckel, photographer; 23.5, 23.7a-b: © McGraw-Hill Education/Al Telser, photographer; 23.7c: © Christine Eckel; 23.8a-c: © McGraw-Hill Education/Christine Eckel, photographer; 23.10b-c: © McGraw-Hill Education/Al Telser, photographer; 23.14: © McGraw-Hill Education/Photo and Dissection by Christine Eckel.

Chapter 24

Figure 24.2: © Victor P. Eroschenko; 24.3a-c: © Christine Eckel; 24.4a-c: © McGraw-Hill Education/Christine Eckel, photographer; 24.5a-b: © McGraw-Hill Education/Al Telser, photographer; 24.7: © McGraw-Hill Education/Photo and Dissection by Christine Eckel; 24.8a-c: Model # G21 [1000272] © 3B Scientific GmbH, Germany, 2015 www.3bscientific.com/Photos by Christine Eckel, Ph.D.; 24.9: © Copyright by Denoyer-Geppert. Photo by Christine Eckel; 24.12b: Model # G23 [1000274] © 3B Scientific GmbH, Germany, 2015 www.3bscientific.com; 24.14: © J & J Photography; 24.15: © Phipps & Bird, Inc., Richmond, VA. Used with permission; 24.16, 24.17: Courtesy of and © BIOPAC Systems, Inc. www.biopac.com.

Chapter 25

Figure 25.1a, 25.2-25.6: © McGraw-Hill Education/Al Telser, photographer; 25.7: © McGraw-Hill Education/Rebecca Gray, photographer/Don Kincaid, dissections; 25.9a: © Copyright by Denoyer-Geppert. Photo by Christine Eckel; 25.9b: Model # K11 [1000299] © 3B Scientific GmbH, Germany, 2015 www.3bscientific.com; 25.10: © Christine Eckel; 25.11: © Copyright by Denoyer-Geppert. Photo by Christine Eckel; 25.12-25.13: Models # H10 [1000281] & [1000282] © 3B Scientific GmbH, Germany, 2015 www.3bscientific.com/Photos by Christine Eckel, Ph.D.; 25.15: © Christine Eckel.

Chapter 26

Figure 26.1b: © Carolina Biological Supply Company/Phototake; 26.1c: © McGraw-Hill Education/Al Telser, photographer; 26.1d, 26.3a: © Victor P. Eroschenko; 26.4b: © Visuals Unlimited/Corbis; 26.4c, 26.5b: © Victor P. Eroschenko; 26.5c-d: © Carolina Biological Supply Company/Phototake; 26.6b-c, 26.8a-b, 26.9b (top): © Victor P. Eroschenko; 26.9b (bottom): © Alvin Telser, PhD/Corbis; 26.12a-b: Model # K15 [1000302] © 3B Scientific GmbH, Germany, 2015 www.3bscientific.com; 26.13-26.14: © Copyright by Denoyer-Geppert. Photos by Christine Eckel; 26.15a, c: © Christine Eckel; 26.15b, d: Courtesy of David A. Morton and Chris Steadman, University of Utah School of Medicine; 26.16a-b: © Copyright by Denoyer-Geppert. Photos by Christine Eckel; 26.17-26.18: © Christine Eckel; p. 749: © McGraw-Hill Education/Christine Eckel, photographer.

Chapter 27

Figures 27.1b, 27.2: © McGraw-Hill Education/Al Telser, photographer; 27.3: © Christine Eckel; Table 27.2 (Primordial Follicle, Primary Follicle, Corpus Luteum, Corpus Albicans): © McGraw-Hill Education/Al Telser, photographer; Table 27.2 (Secondary [Maturing] Follicle): © Ed Reschke/Getty Images; Table 27.2 (Vesicular [Graafian] Follicle): © Ed Reschke; 27.4a-b: © McGraw-Hill Education/Al Telser, photographer; Table 27.5a: © Educational Images LTD/CustomMedical/Newscom; Table 27.5b-c: © Biophoto Associates/Science Source; 27.5: © McGraw-Hill Education/Al Telser, photographer; 27.6a: From *Anatomy & Physiology Revealed*, © McGraw-Hill Education/The University of Toledo, photography and dissection; 27.6b: © McGraw-Hill Education/Al Telser, photographer; 27.6c: © Christine Eckel; 27.7a: © Ed Reschke/Getty Images; 27.7b: © McGraw-Hill Education/Christine Eckel, photographer; 27.8a: © McGraw-Hill Education/Al Telser, photographer; 27.8b: © McGraw-Hill Education/Christine Eckel, photographer; 27.9-27.10: © McGraw-Hill Education/Al Telser, photographer; 27.11: © McGraw-Hill Education/Christine Eckel, photographer; 27.13a-b, 27.16a-b: Models # H10 [1000281] & [1000282] © 3B Scientific GmbH, Germany, 2015 www.3bscientific.com/Photos by Christine Eckel, Ph.D.; 27.17 (1, 2, 6, 7): © McGraw-Hill Education/Al Telser, photographer; 27.17 (3): © Ed Reschke/Getty Images; 27.17 (4): © Ed Reschke; 27.17 (5): © Dr. Francisco Gaytan; p. 786 (top): © Christine Eckel; p. 786 (bottom): © McGraw-Hill Education/Christine Eckel, photographer; p. 787: © Christine Eckel.

Chapter 28

Figure 28.1b: © McGraw-Hill Education/Eric Wise, photographer; All other photos: © McGraw-Hill Education/Photos and Dissections by Christine Eckel.

index

A

F

G

N

O

S

U